THE HUTCHISON SERIES IN MATHEMATICS

Elementary and Intermediate ALGEBRA

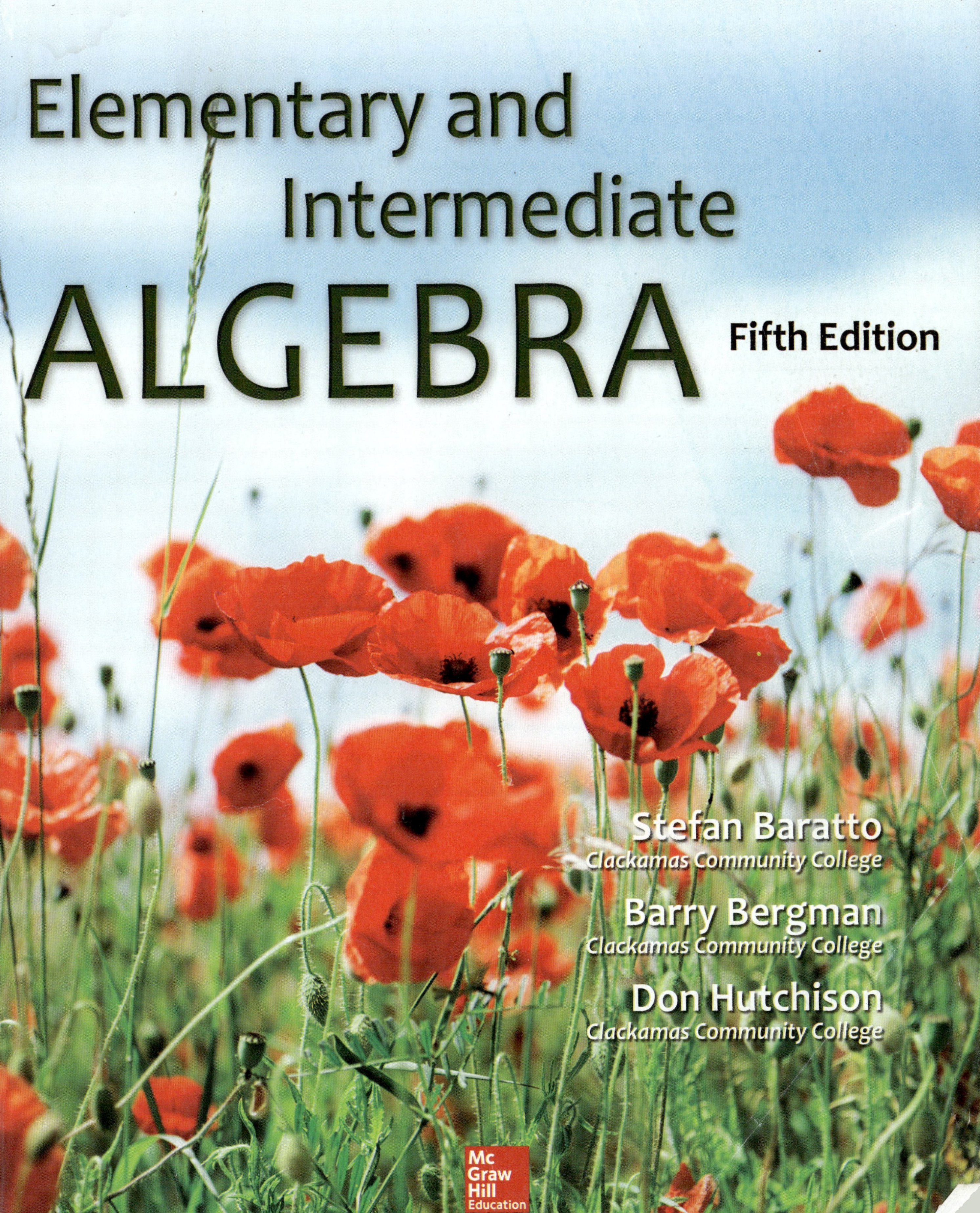
THE HUTCHISON SERIES IN MATHEMATICS
Elementary and Intermediate
ALGEBRA
Fifth Edition
Stefan Baratto
Clackamas Community College
Barry Bergman
Clackamas Community College
Don Hutchison
Clackamas Community College
Mc Graw Hill Education

ELEMENTARY AND INTERMEDIATE ALGEBRA, FIFTH EDITION

Published by McGraw-Hill Education, 2 Penn Plaza, New York, NY 10121.

This book is printed on acid-free paper.

1 2 3 4 5 6 7 8 9 0 QVS/QVS 1 0 9 8 7 6 5 4 3

ISBN 978–0–07–338446–7
MHID 0–07–338446–1

ISBN 978–0–07–757438–3 (Annotated Instructor's Edition)
MHID 0–07–757438–9

Senior Vice President, Products & Markets: *Kurt L. Strand*
Vice President, General Manager, Products & Markets: *Marty Lange*
Vice President, Content Production & Technology Services: *Kimberly Meriwether David*
Managing Director: *Ryan Blankenship*
Brand Manager: *Mary Ellen Rahn*
Director of Development: *Rose Koos*
Director of Digital Content: *Nicole Lloyd*
Senior Project Manager: *Vicki Krug*
Senior Buyer: *Sandy Ludovissy*
Senior Media Project Manager: *Sandra M. Schnee*
Design: *Tara McDermott*
Cover Photo Image: © *Peter Dazeley/Getty Images*
Senior Content Licensing Specialist: *Lori Hancock*
Compositor: *MPS Limited*
Typeface: *10/12 Times New Roman*
Printer: *Quad/Graphics*

About the Cover Photo
A flower symbolizes transformation and growth—a change from the ordinary to the spectacular! The Hutchison Series helps students in an arithmetic/basic math course grow their math skills ***from the ground up*** with a proven approach that motivates them to become stronger math students and to use their mathematical knowledge in everyday life.

Library of Congress Cataloging-in-Publication Data

Baratto, Stefan.
Elementary and intermediate algebra / Stefan Baratto, Clackamas Community College, Barry Bergman, Clackamas Community College, Donald Hutchison, Clackamas Community College. —Fifth edition.
pages cm
Includes index.
ISBN 978–0–07–338446–7 — ISBN 0–07–338446–1 (hardcopy : alk. paper) 1. Algebra–Textbooks. I. Bergman, Barry. II. Hutchison, Donald, 1948– III. Title.
QA152.3.H874 2013
372.7'1–dc23 2013009003

www.mhhe.com

ABOUT THE AUTHORS & DEDICATION

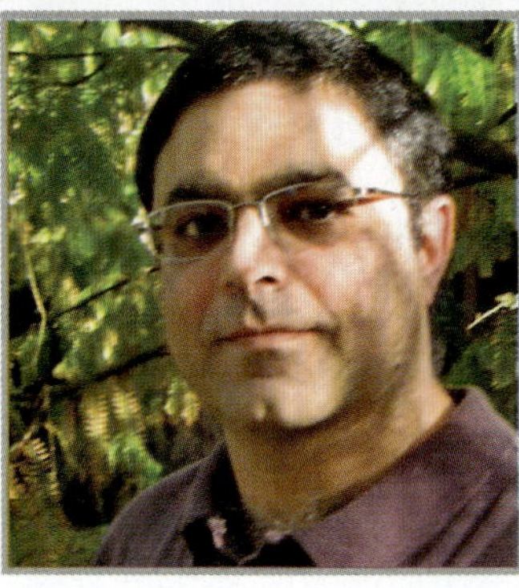

Stefan Baratto

Stefan began teaching math and science in New York City middle schools. He also taught math at the University of Oregon, Southeast Missouri State University, and York County Technical College. Currently, Stefan is a member of the mathematics faculty at Clackamas Community College where he has found a niche, delighting in the CCC faculty, staff, and students. Stefan's own education includes the University of Michigan (BGS, 1988), Brooklyn College (CUNY), and the University of Oregon (MS, 1996).

Stefan is currently serving on the AMATYC Executive Board as the organization's Northwest Vice President. He has been involved with ORMATYC, NEMATYC, NCTM, NADE, and the State of Oregon Math Chairs group, as well as other local organizations. He has applied his knowledge of math to various fields, using statistics, technology, and Web design. More personally, Stefan and his wife, Peggy, try to spend time enjoying the wonders of Oregon and the Pacific Northwest. Their activities include scuba diving and hiking.

Barry Bergman

Barry has enjoyed teaching mathematics to a wide variety of students over the years. He began in the field of adult basic education and moved into the teaching of high school mathematics in 1977. He taught high school math for 11 years, at which point he served as a K–12 mathematics specialist for his county. This work allowed him the opportunity to help promote the emerging NCTM standards in his region.

In 1990, Barry began the next portion of his career, having been hired to teach at Clackamas Community College. He maintains a strong interest in the appropriate use of technology and visual models in the learning of mathematics.

Throughout the past 35 years, Barry has played an active role in professional organizations. As a member of OCTM, he contributed several articles and activities to the group's journal. He has presented at AMATYC, OCTM, NCTM, ORMATYC, and ICTCM conferences. Barry also served 4 years as an officer of ORMATYC and participated on an AMATYC committee to provide feedback to revisions of NCTM's standards.

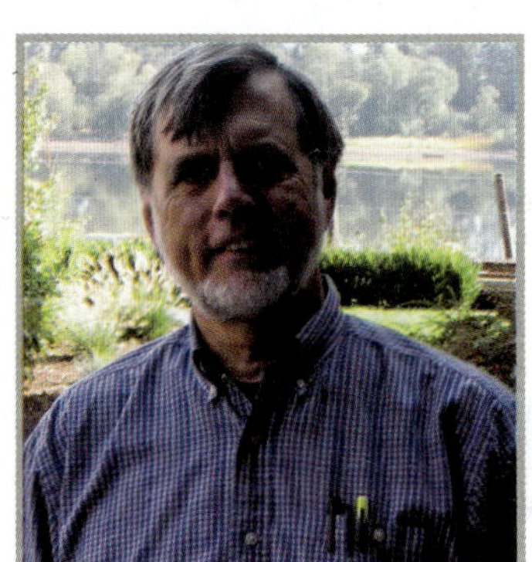

Don Hutchison

Don began teaching in a preschool while he was an undergraduate. He subsequently taught children with disabilities, adults with disabilities, high school mathematics, and college mathematics. Although each position offered different challenges, it was always breaking a challenging lesson into teachable components that he most enjoyed.

It was at Clackamas Community College that he found his professional niche. The community college allowed him to focus on teaching within a department that constantly challenged faculty and students to expect more. Under the guidance of Jim Streeter, Don learned to present his approach to teaching in the form of a textbook. Don has also been an active member of many professional organizations. He has been president of ORMATYC, AMATYC committee chair, and ACM curriculum committee member. He has presented at AMATYC, ORMATYC, AACC, MAA, ICTCM, and a variety of other conferences.

Above all, he encourages you to be involved, whether as a teacher or as a learner. Whether discussing curricula at a professional meeting or homework in a cafeteria, it is the process of communicating an idea that helps one to clarify it.

Dedication

We dedicate this text to the thousands of students who have helped us become better teachers, better communicators, better writers, and even better people. We read and respond to every suggestion we get—every one is invaluable. If you have any thoughts or suggestions, please contact us at

Stefan Baratto: sbaratto@clackamas.edu
Barry Bergman: bfbergman@gmail.com
Don Hutchison: donh@collegemathtext.com

Thank you all.

CONTENTS

Chapter 6

Factoring

Chapter 7

Radicals and Exponents

Chapter 8

Quadratic Functions

Chapter 9

Rational Expressions

Chapter 10

Exponential and Logarithmic Functions

Appendices

PREFACE

Welcome Students

Learning math can be an exciting adventure! As you embark on this journey, we are here with you. Whether this material is new to you or you are trying to master topics that previously eluded you, this text is designed to make it easier to learn the essential mathematics that you will need.

Learning math, learning to be a student (again), and growing as a person are all intertwined as part of your experience. We hope that you continue to grow as a student and as a person while you ***grow your math skills from the ground up***.

We have seen many students succeed in our math classes and we offer you some guidance that may help you to be one of the successful students. Through the first half of this text, we offer a series of *Tips for Student Success* features. These cover many of those skills and actions that successful college students exhibit. You can find a complete listing of the *Tips for Student Success* in the *Index*. We would like to highlight two of the more important items in these Tips.

Learning math takes time. Students are expected to study 2 to 3 hours per week, outside of class, for every credit hour in a math course. In order to learn the math necessary to succeed in your course, you will need to make this time commitment. Expect to spend an hour or more every day outside of class learning math. Because of the size of this commitment, you need to schedule your day and week to include these hours.

Second, while your instructor and this text are your primary resources, many other resources are available to you. Your college may offer study skills or new student experience courses. Take such a course if it's been a while since you've been a student. Your college may also offer tutoring or other helpful services.

You may also want to take advantage of the *Student's Solutions Manual* for this text as well as the online resources available to you. McGraw-Hill Higher Education offers the online course management system Connect Math Hosted by ALEKS Corp. and access to the ALEKS platform. These online resources offer the opportunity to practice as much as you need to and provide immediate feedback on your progress.

Practice is one key to learning to do math. Reading mathematics in a text teaches you to read math in a text. Watching instructors do mathematics teaches you to watch instructors do math. In order to learn to do mathematics, you must do math!

In our text, every example is followed by a *Check Yourself* exercise. It is imperative that you complete these exercises. Please do not move on to the next example until you can complete the *Check Yourself* exercise successfully. By being an active learner, you will learn to do math!

There are numerous other features in this text, each designed to help you succeed. Learning to think critically about the math and learning to read and communicate technical information are essential to becoming effective problem solvers. Our many applications and examples will help you achieve this by teaching you to apply the math you are learning to real-life situations.

At the end of each chapter, you will find materials such as a *Summary*, *Chapter Test*, and *Cumulative Review* that will help you coalesce your learning and maintain your knowledge base. They will help you succeed in ways you may have never thought possible.

We wish you fun and success as you continue your journey as a student and as a person.

To Our Colleagues

Over three careers, the author team has learned much about teaching and learning mathematics. Perhaps the biggest item we have learned is how little we know. For example, we can spot the successful students on day one, except when we can't. Just as often, we are pleasantly surprised by the success of a student who struggled at the beginning of the term.

There are some things we feel we do know. We are certain that one important key to learning mathematics is active participation. In our *Welcome* to the students, we write

> Practice is one key to learning to do math. Reading mathematics in a text teaches you to read math in a text. Watching instructors do mathematics teaches you to watch instructors do math. In order to learn to do mathematics, you must do math!

We feel that this may be the most important aspect of becoming a successful math student. We implore you to advise your students that being successful requires that they spend the appropriate amount of time engaged in the process of doing and learning math.

Often, students are tempted to utilize classroom time to complete their homework or engage in other activities that are best completed outside of class. We feel it is important to reinforce the idea that students achieve success by spending their time doing math outside of class. We stress this in our own classrooms and encourage you to do the same.

While we actively encourage students to imitate the habits of successful students, this text is a passive resource. You are their most important resource. We strive to provide you with assistance reiterating those things in print that we all say in the classroom. Between us, your students can ***grow their math skills from the ground up***.

Another key for us is helping students to see the relevance of mathematics to their daily experience. Our students are like your students. They don't always see how their math classes help them in their lives. With your help and ours, they have a world of growth ahead of them.

We offer many real-world applications for you and your students. We've included application examples in the exposition to assist newer instructors, who can demonstrate how to approach and solve these problems while in the classroom.

We feel that it is very important that students work with these applications. In addition to helping them see the relevance of what they learn, they gain invaluable training in critical thinking and problem solving.

Even more important, we see these applications as providing the student with true transferable skills. Being able to read, comprehend, and communicate technical information encompasses a set of skills that students can use throughout their college careers and their lives. This may be the most important thing they gain from their math classes.

Many people helped us to revise these texts. Our own classroom experiences, our students, and our colleagues at Clackamas Community College were our first line of resources, obviously. But, numerous instructors, users, and reviewers from around the country contributed their thoughts and ideas. Their input provided the impetus for many of the changes you will find. We believe that this helped us to write our strongest text yet and hope you agree. Please don't hesitate to contact the author team if you would like to provide your own input.

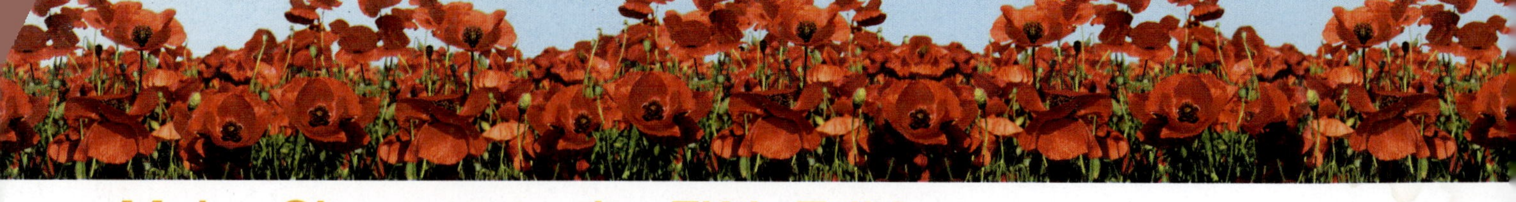

Major Changes to the Fifth Edition

Our revisions are based on our own experiences, as well as the comments and feedback that we receive from the many students, instructors, reviewers, and editorial personnel who have engaged with these materials.

Global Changes

Writing, instructions, and other materials edited for clarity

- Revised student learning objectives
- Improved and expanded instruction of reading, interpreting, and solving word problems and applications
- Improved calculator instruction throughout
- Increased emphasis placed on checking answers for accuracy and reasonableness
- Added and improved thought-provoking exercises

Applications integrated throughout

- Added more application examples to the exposition
- Updated data, current events, and applications

Reorganized and reformatted the exercise sets

- Over 725 new exercises; over 300 more exercises than the fourth edition

Revised additional learning materials

- Revised Tips for Student Success features to better meet the needs of today's students
- Reorganized the Chapter Tests to better reflect actual exams
- Revised and improved Chapter Activities

Chapter 0

Strong improvements to the introduction to real numbers (0.2)

Improved presentation and examples of order of operations (0.5)

Chapter 1

Moved formulas and applications to Chapter 2 (from 1.7)

Improved presentation and examples of evaluating expressions (1.2)

Improved presentation and examples of simplifying algebraic expressions (1.3)

Improved presentation and examples of solving equations (1.4 to 1.6)

Greater emphasis placed on checking solutions (1.4 to 1.6)

New material assists students to construct inequalities to model applications (1.7)

Chapter 2

Integrated applications material throughout chapter

Moved and revised percent instruction (2.2)

Greater emphasis placed on checking solutions (2.3)

Greater emphasis placed on the creation of linear inequality models (2.5)

Chapter 3

Expanded and improved graphing calculator exercises (3.1)

Greater emphasis placed on critical-thinking skills (e.g., interpreting slope and rate of change in the context of an application) (3.2 and 3.4)

Improved instruction and additional examples teaching rate of change and regression analysis (3.4)

Chapter 4

New examples and applications of graphing systems of equations (4.1)

Chapter 5

Improved graphing calculator instruction (5.1 and 5.2)

Improved presentation and examples on simplifying exponential terms (5.2)

Improved presentation and examples on adding and subtracting polynomials; wider set of notations used to help students become more familiar with notation (5.4)

> "This series is simple in its explanations, very readable, has very good organization, and contains very appropriate problem sets."
>
> —Sandi Tannen, *Camden County College*

Chapter 6

Major revisions improving the presentation, examples, and instruction introducing students to factoring (6.1)

Improved presentation and examples of factoring special products (6.2)

Nearly completely new presentation, examples, and exercises for trial-and-error factoring ($a = 1$) (6.3)

Greater emphasis placed on checking a factorization through multiplication (6.3 and 6.4)

More clearly separated factoring quadratics based on leading coefficient (6.3 and 6.4)

Major revisions improving the presentation, examples, and instruction for students learning to factor polynomials in which the leading coefficient is not one (6.4)

Improved presentation of factoring strategies and prime polynomials (6.5)

Using factoring to solve equations and problem solving combined into a single section (6.6)

Chapter 7

Provided more detail of steps in examples throughout chapter

Added content covering approximation (trapping) of roots (7.1)

Improved presentation of simplification of radical expressions (7.2)

Improved presentation and examples of rationalizing denominators (7.2)

Chapter 8

Emphasis placed on checking solutions to quadratic equations

Applications added throughout chapter

Expanded using the discriminant through examples and exercises (8.2)

Chapter 9

More progressive buildup of techniques to simplify rational expressions; better accounting of difficulty students have with fractions (9.1)

Improved presentation and examples for simplifying complex rational expressions (9.4)

Improved presentation and examples for graphing rational functions; improved graphing calculator instruction (9.5)

Added content covering proportions (9.6)

Many new applications of rational expressions (9.6)

Improved presentation and examples of work problems (9.6)

Chapter 10

Improved presentation and examples of algebra of functions (10.1)

Improved presentation and examples of function composition (10.2)

Emphasis placed on using composition to check that two functions are inverses (10.3)

Improved graphing calculator instructions for exponential and logarithmic functions (10.4 to 10.7)

> "An excellent textbook for students who have never taken higher levels of math in high school or have been out of school for a number of years. It would be a great reference book for topics covered in algebra and as a review of basic algebraic and arithmetic concepts needed in all areas of math and science. The text is student-friendly with many worked examples and great summary notes at the end of the chapters."
>
> —Linda Faraone, *Georgia Military College*

GROW YOUR MATHEMATICAL SKILLS

"Make the Connection"*—Chapter-opening vignettes* provide interesting, relevant scenarios that engage students in the upcoming material. Exercises and *Activities* related to the vignette revisit its themes to more effectively drive mathematical comprehension (marked with a 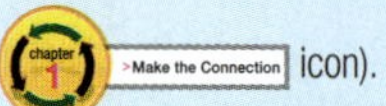 icon).

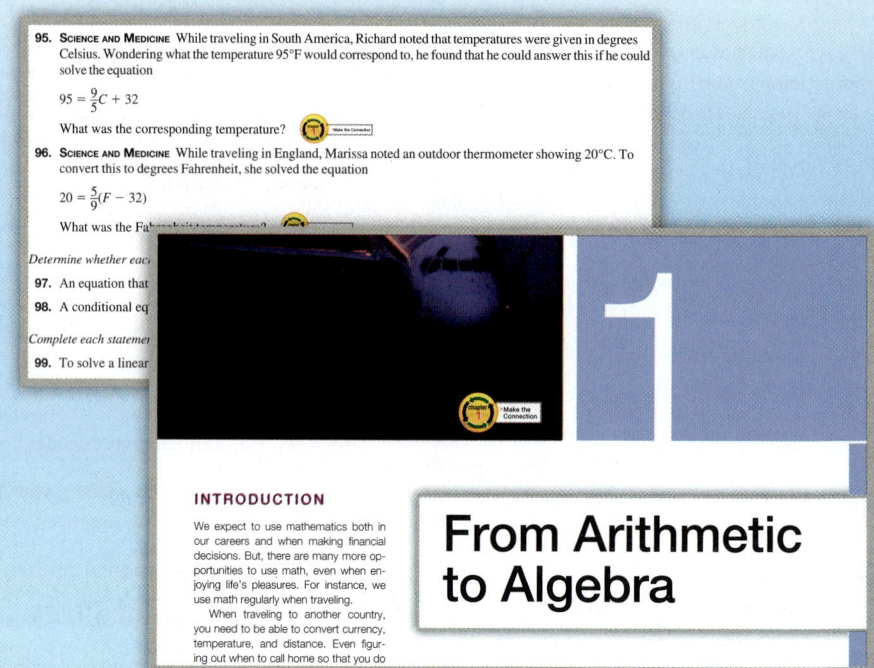

95. SCIENCE AND MEDICINE While traveling in South America, Richard noted that temperatures were given in degrees Celsius. Wondering what the temperature 95°F would correspond to, he found that he could answer this if he could solve the equation

$$95 = \frac{9}{5}C + 32$$

What was the corresponding temperature?

96. SCIENCE AND MEDICINE While traveling in England, Marissa noted an outdoor thermometer showing 20°C. To convert this to degrees Fahrenheit, she solved the equation

$$20 = \frac{5}{9}(F - 32)$$

What was the Fa...

Determine whether eac...

97. An equation that...

98. A conditional eq...

Complete each stateme...

99. To solve a linear...

1

INTRODUCTION

We expect to use mathematics both in our careers and when making financial decisions. But, there are many more opportunities to use math, even when enjoying life's pleasures. For instance, we use math regularly when traveling.

When traveling to another country, you need to be able to convert currency, temperature, and distance. Even figuring out when to call home so that you do

From Arithmetic to Algebra

Activities promote active learning by requiring students to find, interpret, and manipulate real-world data. The activities tie the chapter together with questions that sharpen mathematical and conceptual understanding of the chapter material. Students can complete the activities on their own or in small groups.

Activity 4 ::

Agricultural Technology

Nutrients and Fertilizers

When growing crops, it is not enough just to till the soil and plant seeds. The soil must be properly prepared before planting. Each crop takes nutrients out of the soil that must be replenished. Some of this is done with crop rotations (each crop takes some nutrients out of the soil while replenishing other nutrients), but maintaining proper nutrient levels often requires that some additional nutrients be added. This may be accomplished with fertilizers.

The three most vital nutrients are nitrogen, phosphorus, and potassium. Three different fertilizer mixes are available:

Urea: Contains 46% nitrogen

Growth: Contains 16% nitrogen, 48% phosphorus, and 12% potassium

Restorer: Contains 21% phosphorus and 62% potassium

A soil test shows that a field requires 115 lb of nitrogen, 78 lb of phosphorus, and

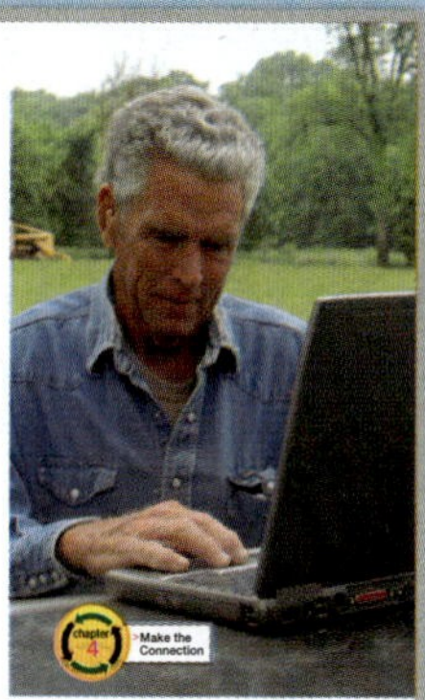

Reading Your Text offers a brief set of exercises at the end of each section to assess students' knowledge of key vocabulary terms, encourage careful reading, and reinforce understanding of core mathematical concepts. Answers are provided at the end of the book.

Reading Your Text

These fill-in-the-blank exercises will help you understand some of the key vocabulary used in this section. The answers to these exercises are in the Answers Appendix in the back of the text.

(a) An equation in two variables is an equation for which every ________ is a pair of values.

(b) Given an equation such as $x + y = 5$, there is an ________ number of solutions.

(c) To simplify writing the pairs that satisfy an equation, we use ________ notation.

(d) When an equation in two variables is solved for y, we say that y is the ________ variable.

THROUGH MORE CAREFUL PRACTICE

Chapter Tests let students check their progress and review important concepts so they can prepare for exams with confidence and proper guidance. Answers are given at the end of the book, with section references provided to help students review important material.

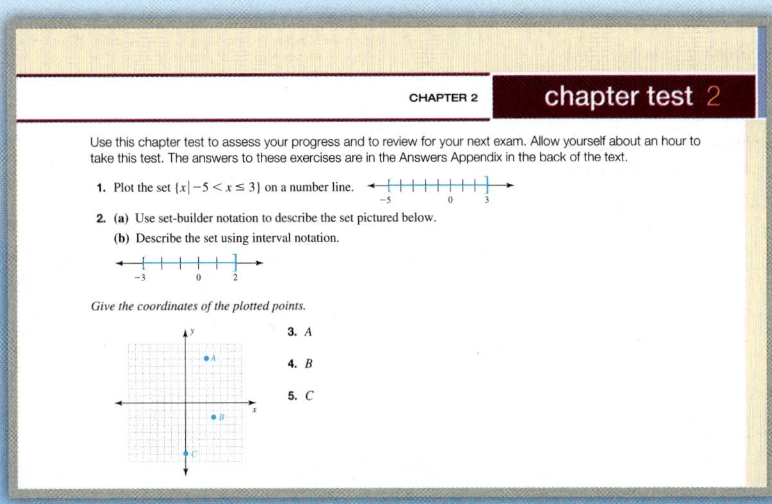

CHAPTER 2 **chapter test 2**

Use this chapter test to assess your progress and to review for your next exam. Allow yourself about an hour to take this test. The answers to these exercises are in the Answers Appendix in the back of the text.

1. Plot the set $\{x \mid -5 < x \le 3\}$ on a number line.

2. (a) Use set-builder notation to describe the set pictured below.
(b) Describe the set using interval notation.

Give the coordinates of the plotted points.

3. A

4. B

5. C

Cumulative Reviews, included starting with Chapter 2, follow the *Chapter Test* and reinforce previously covered material to help students retain knowledge throughout the course and identify skills necessary for them to review when preparing for exams. Answers are provided at the end of the book, along with section references.

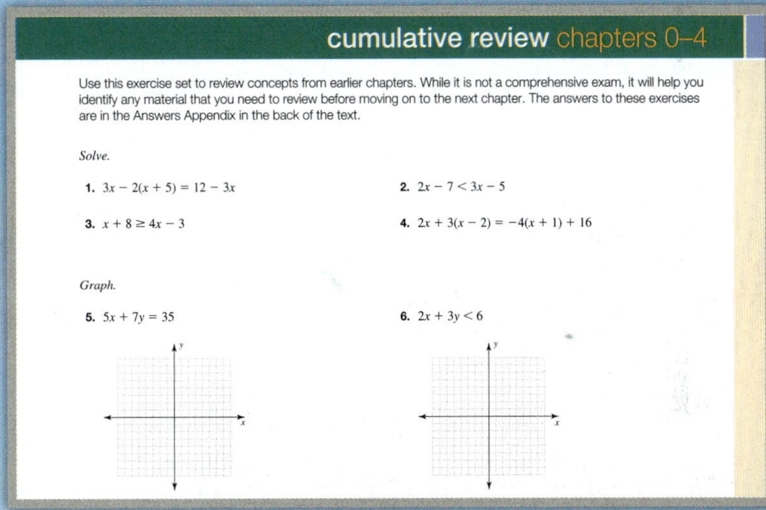

cumulative review chapters 0–4

Use this exercise set to review concepts from earlier chapters. While it is not a comprehensive exam, it will help you identify any material that you need to review before moving on to the next chapter. The answers to these exercises are in the Answers Appendix in the back of the text.

Solve.

1. $3x - 2(x + 5) = 12 - 3x$

2. $2x - 7 < 3x - 5$

3. $x + 8 \ge 4x - 3$

4. $2x + 3(x - 2) = -4(x + 1) + 16$

Graph.

5. $5x + 7y = 35$

6. $2x + 3y < 6$

"Check Yourself" Exercises actively involve students in the learning process. Every worked example in the book is followed by an exercise encouraging students to solve a problem similar to the one just presented and check, through practice, what they have just learned. Answers are provided at the end of the section for immediate feedback.

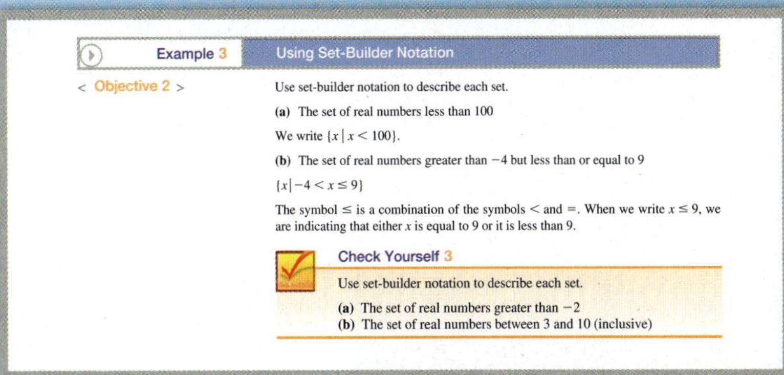

Example 3 Using Set-Builder Notation

< Objective 2 >

Use set-builder notation to describe each set.

(a) The set of real numbers less than 100

We write $\{x \mid x < 100\}$.

(b) The set of real numbers greater than -4 but less than or equal to 9

$\{x \mid -4 < x \le 9\}$

The symbol $\le$ is a combination of the symbols $<$ and $=$. When we write $x \le 9$, we are indicating that either x is equal to 9 or it is less than 9.

Check Yourself 3

Use set-builder notation to describe each set.

(a) The set of real numbers greater than -2
(b) The set of real numbers between 3 and 10 (inclusive)

GROW YOUR MATHEMATICAL SKILLS

End-of-Section Exercises help students evaluate their conceptual mastery of the section through practice. These comprehensive exercise sets are structured to highlight the progression in level, and organized by category and section learning objective to make it easier for instructors to plan assignments. Answers to odd-numbered exercises are provided at the end of the exercise set.

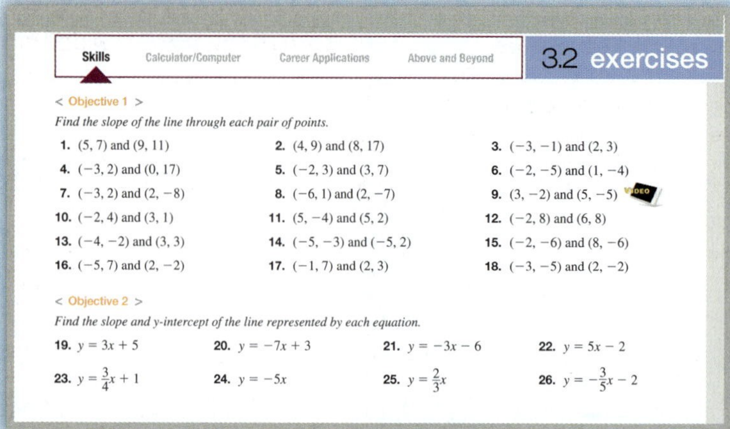

Skills | Calculator/Computer | Career Applications | Above and Beyond

3.2 exercises

< Objective 1 >

Find the slope of the line through each pair of points.

1. (5, 7) and (9, 11) **2.** (4, 9) and (8, 17) **3.** (−3, −1) and (2, 3)
4. (−3, 2) and (0, 17) **5.** (−2, 3) and (3, 7) **6.** (−2, −5) and (1, −4)
7. (−3, 2) and (2, −8) **8.** (−6, 1) and (2, −7) **9.** (3, −2) and (5, −5)
10. (−2, 4) and (3, 1) **11.** (5, −4) and (5, 2) **12.** (−2, 8) and (6, 8)
13. (−4, −2) and (3, 3) **14.** (−5, −3) and (−5, 2) **15.** (−2, −6) and (8, −6)
16. (−5, 7) and (2, −2) **17.** (−1, 7) and (2, 3) **18.** (−3, −5) and (2, −2)

< Objective 2 >

Find the slope and y-intercept of the line represented by each equation.

19. $y = 3x + 5$ **20.** $y = -7x + 3$ **21.** $y = -3x - 6$ **22.** $y = 5x - 2$
23. $y = \frac{3}{4}x + 1$ **24.** $y = -5x$ **25.** $y = \frac{2}{3}x$ **26.** $y = -\frac{3}{5}x - 2$

Summary and Summary Exercises—*Summaries* at the end of each chapter show students key concepts from the chapter that they need to review, and provide page references to where each concept is introduced. *Summary Exercises* give students practice on these important concepts, with section references showing where they can go back to review relevant worked examples. Answers to odd-numbered *Summary Exercises* are provided at the end of the book.

summary :: chapter 2

Definition/Procedure	Example	Reference
Formulas and Problem Solving		Section 2.1
Formula or **Literal Equation** An equation that expresses a relationship between two or more variables.	$a = \frac{2b + c}{3}$ is a formula or literal equation.	p. 153
Solving Formulas **Step 1** Remove any grouping symbols by applying the distributive or multiplication property. **Step 2** Multiply both sides of the equation … the equation of fractions or decim… **Step 3** Combine any like terms that appea… equation. **Step 4** Use the addition property of equalit… equation with the term containing t… one side of the equation and all oth… **Step 5** Use the multiplication property to … equation with the desired variable … its coefficient equal to 1.	Solve for *b*: $a = \frac{2b + c}{3}$	p. 155

summary exercises :: chapter 2

This summary exercise set will help ensure that you have mastered each of the objectives of this chapter. The exercises are grouped by section. You should reread the material associated with any exercises that you find difficult. The answers to the odd-numbered exercises are in the Answers Appendix in the back of the text.

2.1 *Solve for the indicated variable.*

1. $V = LWH$ (for W) **2.** $P = 2L + 2W$ (for L)
3. $ax + by = c$ (for y) **4.** $A = \frac{1}{2}bh$ (for h)
5. $A = P + Prt$ (for t) **6.** $m = \frac{n - p}{q}$ (for p)

Solve each application.

7. Number Problem The sum of 3 times a number and 7 is 25. What is the number?
8. Number Problem 5 times a number, decreased by 8, is 32. Find the number.
9. Number Problem If the sum of two consecutive integers is 85, find the two integers.
10. Problem Solving Larry is 2 years older than Susan, while Nathan is twice as old as Susan. If the sum of their ages is 30 years, find each of their ages.
11. Science and Medicine Lisa left Friday morning, driving on the freeway to visit friends for the weekend. Her trip took

Video Exercises offer guided video solutions to selected exercises in each section marked with a VIDEO icon.

Available through *Connect Math Hosted by ALEKS,* these videos feature a presenter working through the exercises just like an instructor would, following the solution methodology from the text. The videos are available closed-captioned for the hearing-impaired or subtitled in Spanish, and meet the Americans with Disabilities Act Standards for Accessible Design.

< Objective 2 >

Find the slope and y-intercept of the line represented by each equation.

19. $y = 3x + 5$ **20.** $y = -7x + 3$ **21.** $y = -3x - 6$ **22.** $y = 5x - 2$
23. $y = \frac{3}{4}x + 1$ **24.** $y = -5x$ **25.** $y = \frac{2}{3}x$ **26.** $y = -\frac{3}{5}x - 2$

Write each equation in function form. Give the slope and y-int…

27. $4x + 3y = 12$ **28.** $5x + 2y = 10$
30. $2x - 3y = 6$ **31.** $3x - 2y = 8$

< Objective 3 >

Write an equation of the line with given slope and y-intercept.

33. Slope 3; y-intercept: (0, 5) **34.** Slope −2; y-in…

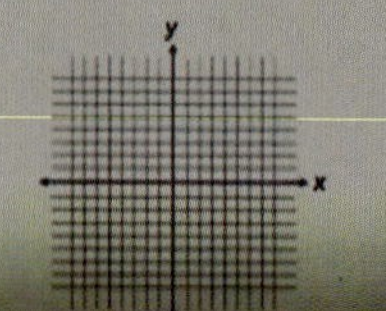

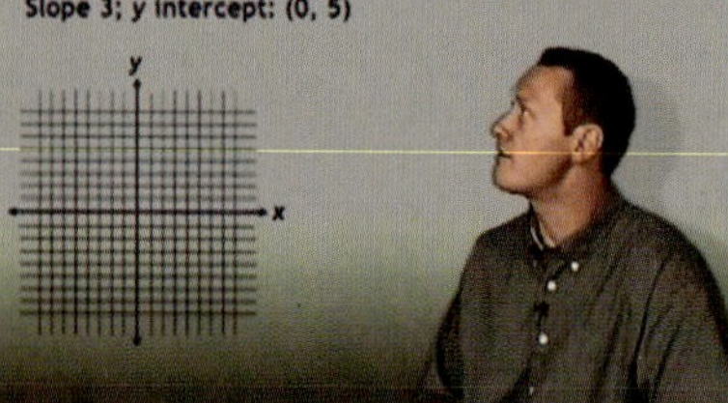
Write the equation of the line with given slope and y intercept. Then graph the line, using the slope and y intercept.
Slope 3; y intercept: (0, 5)
y
x

THROUGH BETTER ACTIVE LEARNING TOOLS

Tips for Student Success boxes offer valuable advice and resources to help students new to collegiate mathematics develop the study skills that will help them succeed. These class-tested suggestions provide students with extra direction on preparing for class, studying for exams, and familiarizing themselves with additional resources available outside of class.

1.4 Solving Equations with the Addition Property

< 1.4 Objectives >

1 > Determine whether a number is a solution to an equation
2 > Use the addition property to solve equations
3 > Use equations to solve applications

Tips for Student Success

Don't Procrastinate!

1. Complete your math homework while you are still fresh. Late at night, your mind is tired, making it difficult to understand new concepts.
2. Complete your homework the day it is assigned. The more recent the explanation, the easier it is to recall.
3. When you finish your homework, try reading the next section in the text. This will give you a sense of direction the next time you have class.

Remember, in a typical math class, you are expected to do 2 or 3 hours of homework for each hour of class time. This means 2 to 3 hours per day. Schedule the time and stay on schedule.

Notes and Recalls accompany the worked examples, providing just-in-time reminders that reinforce previously learned material and help students focus on information critical to their success.

NOTE

FOIL gives you an easy way of remembering the steps: *F*irst, *O*uter, *I*nner, and *L*ast.

RECALL

When you see the statement $-2 \le x \le 4$, think "all real numbers between -2 and 4, including -2 and 4."

Cautions are integrated throughout the textbook to alert students to common mistakes and how to avoid them.

> **CAUTION**

When a squared variable is replaced by a negative number, square the negative.

$(-5)^2 = (-5)(-5) = 25$

The exponent applies to -5!

(b) $7c^2$

$7c^2 = 7(-5$

(c) $b^2 - 4a$

$b^2 - 4ac =$

$=$

$=$

Graphing Calculators*—Graphing Calculator Option* boxes introduce students to key features of a graphing calculator, while the accompanying *Graphing Calculator Check* exercises help them master calculator techniques with practice. Throughout the text, calculator keystrokes and screenshots are provided for context-sensitive help, and worked examples that utilize a calculator are marked with a

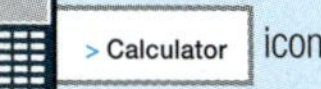

> Calculator icon.

Graphing Calculator Option

Using the Memory Feature to Evaluate Expressions

The memory features of a graphing calculator are a great aid when you need to evaluate several expressions, using the same variables and the same values for those variables.

Your graphing calculator can store variable values for many different variables in different memory spaces. Using these memory spaces saves a great deal of time when evaluating expressions.

Note: We use the TI-84 Plus model graphing calcula your instructor or the instruction manual. If you do manufacturer's website.

Evaluate each expression if $a = 4.6$, $b = -\frac{2}{3}$, an

(a) $a + \frac{b}{ac}$ (b) $b - b^2 + 3(a - c)$

Begin by entering each variable's value into a ca that has the same name as the variable you are saving

Step 1 Type the value associated with one variabl

Step 2 Press the store key, STO▸, the green alph

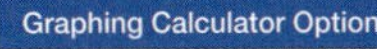

Graphing Calculator Check

Evaluate each expression if $x = -8.3$, $y = \frac{5}{4}$, and $z = -6$. Round your results to the nearest hundredth.

(a) $\frac{xy}{z} - xz$ (b) $5(z - y) + \frac{x}{x - z}$ (c) $x^2y^2z - (x + y)^2$ (d) $\frac{-2(x + z)^2}{y^2z}$

ANSWERS

(a) -48.07 (b) -32.64 (c) $-1{,}311.12$ (d) 34.9

Note: Throughout this text, we will provide additional graphing-calculator material. This material is optional. Th authors will not assume that students have learned this, but we feel that students using a graphing calculator wi benefit from these materials.

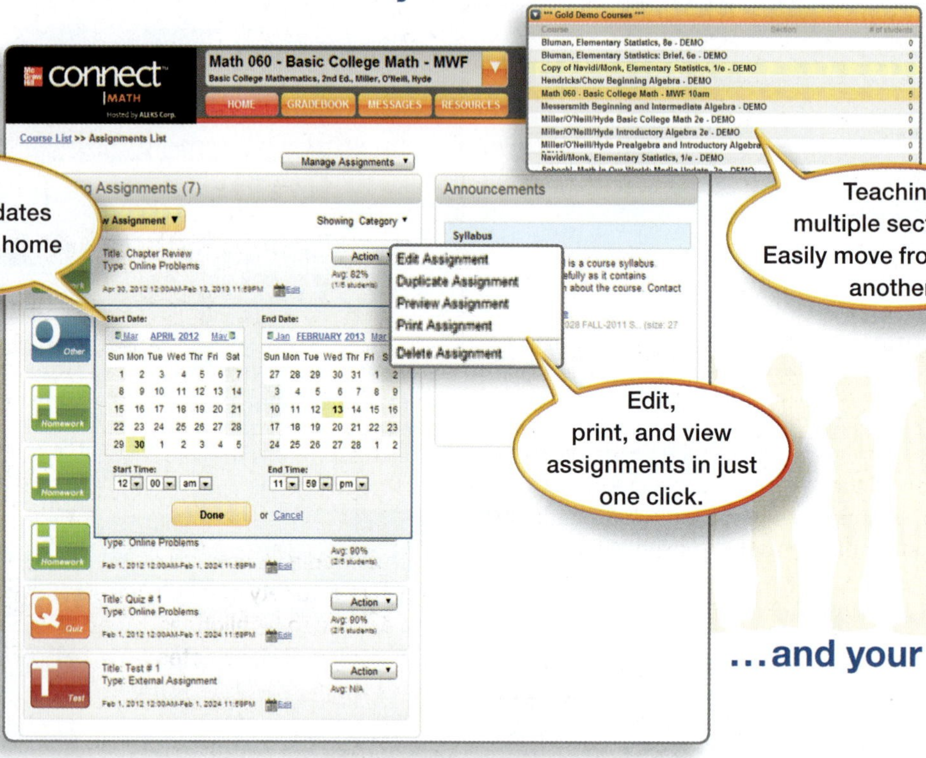

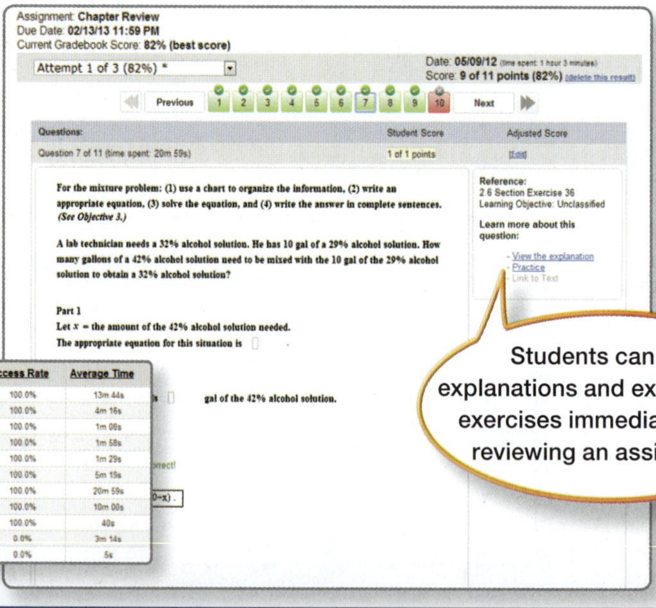

Quality Content For Today's Online Learners

Online Exercises were carefully selected and developed to provide a seamless transition from textbook to technology.

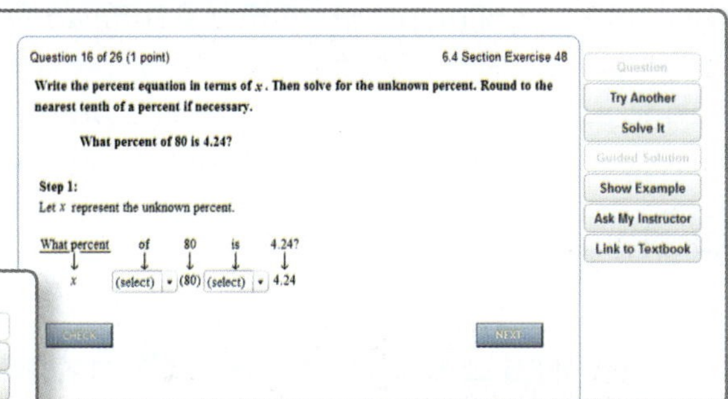

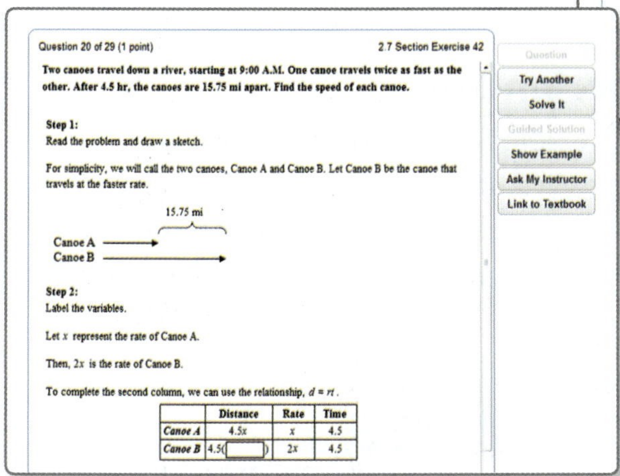

For consistency, the guided solutions match the style and voice of the original text as though the author is guiding the students through the problems.

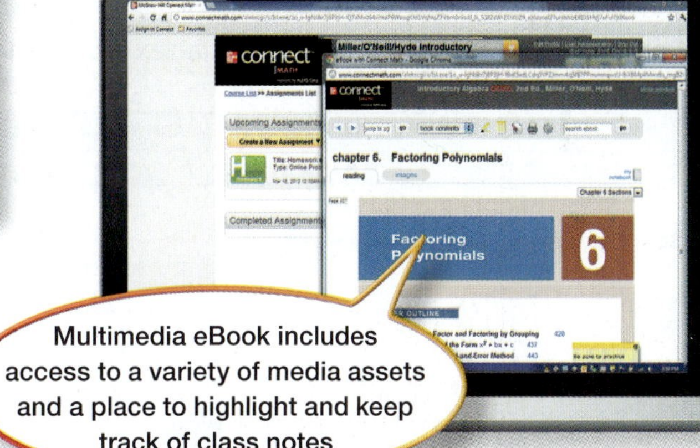

Multimedia eBook includes access to a variety of media assets and a place to highlight and keep track of class notes

Gradebook - ALEKS Initial Assessment #1 - Goulet, Robert

Assignment: **ALEKS Initial Assessment #1**
Completion Date: **04/18/11**
Gradebook Score: **100%**

ALEKS Assessment Report for Goulet, Robert

Course Mastery (209 of 300 Topics)

Student Readiness by Topic

This ALEKS Assessment report shows the percentage of students that have mastered the following topics

Ch.4-Linear Equations in Two Variables

ALEKS Corporation's experience with algorithm development ensures a commitment to accuracy and a meaningful experience for students to demonstrate their understanding with a focus towards online learning.

The ALEKS® Initial Assessment is an artificially intelligent (AI), diagnostic assessment that identifies precisely what a student knows. Instructors can then use this information to make more informed decisions on what topics to cover in more detail with the class.

www.successinmath.com

ALEKS is a unique, online program that significantly raises student proficiency and success rates in mathematics, while reducing faculty workload and office-hour lines. ALEKS uses artificial intelligence and adaptive questioning to assess precisely a student's knowledge, and deliver individualized learning tailored to the student's needs. With a comprehensive library of math courses, ALEKS delivers an unparalleled adaptive learning system that has helped millions of students achieve math success.

ALEKS Delivers a Unique Math Experience:

- **Research-Based, Artificial Intelligence** precisely measures each student's knowledge
- **Individualized Learning** presents the exact topics each student is most **ready to learn**
- **Adaptive, Open-Response Environment** includes comprehensive tutorials and resources
- **Detailed, Automated Reports** track student and class progress toward course mastery
- **Course Management Tools** include textbook integration, custom features, and more

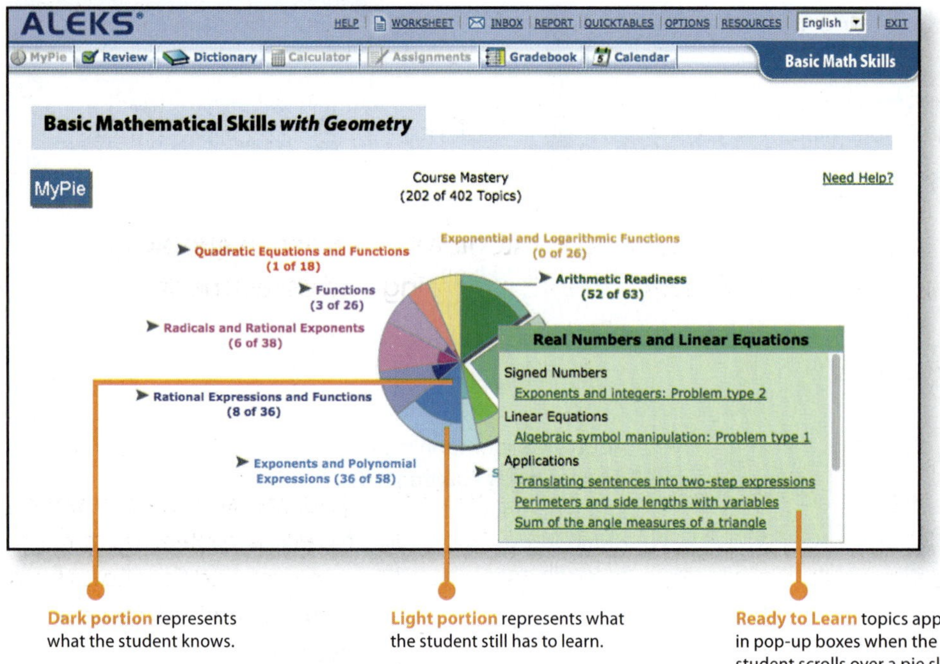

The ALEKS Pie summarizes a student's current knowledge, then delivers an individualized learning path with the exact topics the student is most ready to learn.

"My experience with ALEKS has been effective, efficient, and eloquent. **Our students' pass rates improved from 49 percent to 82 percent with ALEKS.** We also saw student retention rates increase by 12% in the next course. Students feel empowered as they guide their own learning through ALEKS."

—**Professor Eden Donahou, *Seminole State College of Florida***

To learn more about ALEKS, please visit: **www.aleks.com/highered/math**

ALEKS® Prep Products

ALEKS Prep products focus on prerequisite and introductory material, and can be used during the first six weeks of the term to ensure student success in math courses ranging from Beginning Algebra through Calculus. ALEKS Prep quickly fills gaps in prerequisite knowledge by assessing precisely each student's preparedness and delivering individualized instruction on the exact topics students are most ready to learn. As a result, instructors can focus on core course concepts and see improved student performance with fewer drops.

> "ALEKS is wonderful. It is a professional product that takes very little time as an instructor to administer. Many of our students have taken Calculus in high school, but they have forgotten important algebra skills. ALEKS gives our students an opportunity to review these important skills."
>
> —**Professor Edward E. Allen,** ***Wake Forest University***

A Total Course Solution

A cost-effective total course solution: fully integrated, interactive eBook combined with the power of ALEKS adaptive learning and assessment.

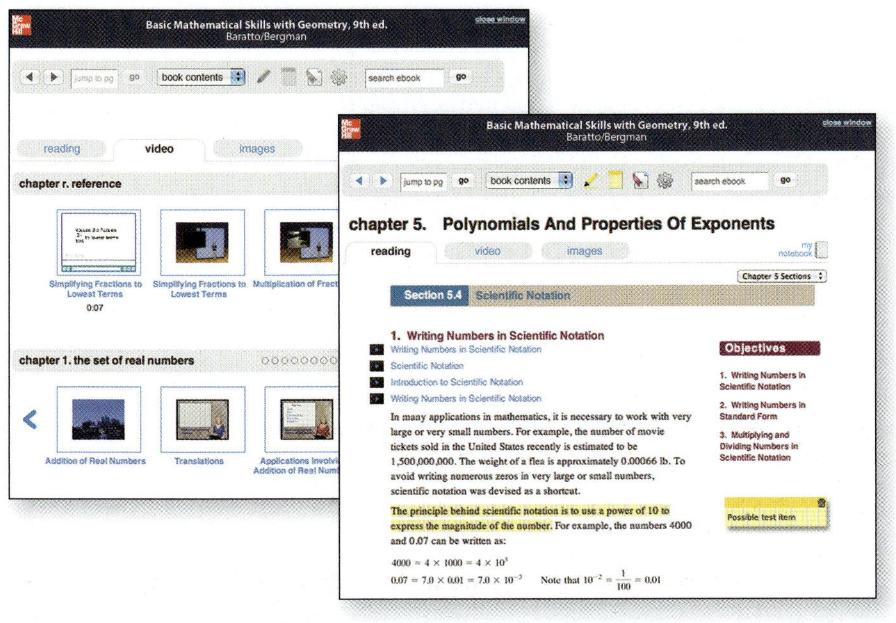

Students can easily access the full eBook content, multimedia resources, and their notes from within their ALEKS Student Accounts.

To learn more about ALEKS, please visit: **www.aleks.com/highered/math**

Our Commitment to Market Development and Accuracy

McGraw-Hill's Development Process is an ongoing, never-ending, market-oriented approach to building accurate and innovative print and digital products. We begin developing a series by partnering with authors that desire to make an impact within their discipline to help students succeed. Next, we share these ideas and manuscript with instructors for review for feedback and to ensure that the authors' ideas represent the needs within that discipline. Throughout multiple drafts, we help our authors adapt to incorporate ideas and suggestions from reviewers to ensure that the series carries the same pulse as today's classrooms. With any new series, we commit to accuracy across the series and its supplements. In addition to involving instructors as we develop our content, we also utilize accuracy checks through our various stages of development and production. The following is a summary of our commitment to market development and accuracy:

1. 3 drafts of author manuscript
2. 5 rounds of manuscript review
3. 2 focus groups
4. 1 consultative, expert review
5. 3 accuracy checks
6. 3 rounds of proofreading and copyediting
7. Toward the final stages of production, we are able to incorporate additional rounds of quality assurance from instructors as they help contribute toward our digital content and print supplements

This process then will start again immediately upon publication in anticipation of the next edition. With our commitment to this process, we are confident that our series has the most developed content the industry has to offer, thus pushing our desire for quality and accurate content that meets the needs of today's students and instructors.

Acknowledgments and Reviewers

The development of this textbook series would never have been possible without the creative ideas and feedback offered by many reviewers. We are especially thankful to the following instructors for their careful review of the manuscript.

Manuscript Review Panels

Over 150 teachers and academics from across the country reviewed the various drafts of the manuscript to give feedback on content, design, pedagogy, and organization. This feedback was summarized by the book team and used to guide the direction of the text.

Reviewers

Darla Aguilar, *Pima Community College*
Paul Ahad, *Antelope Valley College*
Carla Ainsworth, *Salt Lake Community College*
Muhammad Akhtar, *El Paso Community College*
Robin Anderson, *Southwestern Illinois College*
Nieves Angulo, *Hostos Community College*
Arlene Atchison, *South Seattle Community College*
Haimd Attarzadeh, *Kentucky Jefferson Community and Technical College*
Jody Balzer, *Milwaukee Area Technical College*
Rebecca Baranowski, *Estrella Mountain Community College*
Wayne Barber, *Chemeketa Community College*
Bob Barmack, *Baruch College*
Chad Bemis, *Riverside Community College*
Monika Bender, *Central Texas College*
Chris Bendixen, *Lake Michigan College*
Norma Bisluca, *University of Maine–Augusta*
Karen Blount, *Hood College*
Donna Boccio, *Queensborough Community College*
Steve Boettcher, *Estrella Mountain Community College*
Elena Bogardus, *Camden County College*
Karen Bond, *Pearl River Community College, Poplarville*
Laurie Braga Jordan, *Loyola University–Chicago*
Kelly Brooks, *Pierce College*
Dorothy Brown, *Camden County College*
Michael Brozinsky, *Queensborough Community College*
Debra Bryant, *Tennessee Technological University*
Susan Caldiero, *Consumnes River College*
Marc Campbell, *Daytona State College*
Amy Canavan, *Century Community and Technical College*
Faye Childress, *Central Piedmont Community College*
Pauline Chow, *Harrisburg Area Community College*
Kathleen Ciszewski, *University of Akron*
Bill Clarke, *Pikes Peak Community College*
Camille Cochrane, *Shelton State Community College*
William Coe, *Montgomery College*
Lois Colpo, *Harrisburg Area Community College*
Pat Cook, *Weatherford College*
Christine Copple, *Northwest State Community College*
Jonathan Cornick, *Queensborough Community College*
Julane Crabtree, *Johnson County Community College*
Carol Curtis, *Fresno City College*
Sima Dabir, *Western Iowa Tech Community College*

Reza Dai, *Oakton Community College*
Karen Day, *Elizabethtown Technical and Community College*
Mary Deas, *Johnson County Community College*
Anthony DePass, *St. Petersburg College*
Shreyas Desai, *Atlanta Metropolitan College*
Nancy Desilet, *Carroll Community College*
Deborah Detrick, *Lincoln College*
Robert Diaz, *Fullerton College*
Michaelle Downey, *Ivy Tech Community College*
Ginger Eaves, *Bossier Parish Community College*
Azzam El Shihabi, *Long Beach City College*
Joe Edwards, *Bevill State Community College–Hamilton Campus*
Kristy Erickson, *Cecil College*
Steven Fairgrieve, *Allegany College of Maryland*
Linda Faraone, *Georgia Military College*
Nerissa Felder, *Polk Community College*
Jacqui Fields, *Wake Technical Community College*
Bonnie Filer-Tubaugh, *University of Akron*
Carol Flakus, *Lower Columbia College*
Rhoderick Fleming, *Wake Tech Community College*
Matt Foss, *North Hennepin Community College*
Catherine Frank, *Polk Community College*
Ellen Freedman, *Camden County College*
Robert Frye, *Polk Community College*
Heather Gallacher, *Cleveland State University*
Matt Gardner, *North Hennepin Community College*
Lauryn Geritz, *Arizona Western College*
Jeremiah Gilbert, *San Bernardino Valley College*
Judy Godwin, *Collin County Community College–Plano*
Lori Grady, *University of Wisconsin–Whitewater*
Nancy Graham, *Rose State College*
Brad Griffith, *Colby Community College*
Jane Gringauz, *Minneapolis Community and Technical College*
Robert Grondahl, *Johnson County Community College*
Alberto Guerra, *St. Philip's College*
Shelly Hansen, *Mesa State College*
Kristen Hathcock, *Barton County Community College*
Mary Beth Headlee, *Manatee Community College*
Bill Heider, *Hibbing Community College*
Celeste Hernandez, *Richland College*
Tammy Higson, *Hillsborough Community College*
Kristy Hill, *Hinds Community College*
Mark Hills, *Johnson County Community College*
Lori Holdren, *Manatee Community College*
Sherrie Holland, *Piedmont Technical College*
Diane Hollister, *Reading Area Community College*
Steven Howard, *Rose State College*
Paul Hrabovsky, *Indiana University of Pennsylvania*
Janice Hubbard, *Marshalltown Community College*
Mathew Hudock, *St. Philip's College*
Nicholas Huerta, *Fullerton College*
Victor Hughes III, *Shepard University*
Denise Hum, *Canada College*
Byron D. Hunter, *College of Lake County*
Kelly Jackson, *Camden County College*
Patricia R. Jaquith, *Landmark College*
John D. Jarvis, *Utah Valley State College*
Nancy Johnson, *Manatee Community College–Bradenton*
Rashunda Johnson, *Pulaski Technical College*
Judith Jones, *Valencia Community College*
Joe Jordan, *John Tyler Community College–Chester*
Lisa Juliano, *El Paso Community College*
Judy Kasabian, *El Camino College*
Sandra Ketcham, *Berkshire Community College*
Lynette King, *Gadsden State Community College*
Kelly Kohlmetz, *University of Wisconsin–Milwaukee*
Chris Kolaczewski-Ferris, *University of Akron*
Jeff Koleno, *Lorain County Community College*
Randa Kress, *Idaho State University–Pocatello*
Donna Krichiver, *Johnson County Community College*
Nancy Krueger, *Central Community College*
Indra B. Kshattry, *Colorado Northwestern Community College*
Patricia Labonne, *Cumberland County College*
Ted Lai, *Hudson County Community College*
Debra Laraway, *Polk Community College*
Krynn Larsen, *Central Community College*
Pat Lazzarino, *Northern Virginia Community College*
Paul Wayne Lee, *St. Philip's College*
Richard Leedy, *Polk Community College*
Nancy Lehmann, *Austin Community College*
Jeanine Lewis, *Aims Community College–Main Campus*
Pam Lipka, *University of Wisconsin–Whitewater*
Michelle Christina Mages, *Johnson County Community College*
Jean-Marie Magnier, *Springfield Technical Community College*
Igor Marder, *Antelope Valley College*
Donna Martin, *Florida Community College–North Campus*
Amina Mathias, *Cecil College*
Robert Maxell, *Long Beach City College*
Robery Maynard, *Tidewater Community College*
Jean McArthur, *Joliet Junior College*
Carlea (Carol) McAvoy, *South Puget Sound Community College*
Tim McBride, *Spartanburg Community College*
Mikal McDowell, *Cedar Valley College*
Joan McNeil, *Quinebaug Valley Community College*
Sonya McQueen, *Hinds Community College*
Maria Luisa Mendez, *Laredo Community College*
Philip Meurer, *Palo Alto College*
Pam Miller, *Phoenix College*
Derek Milton, *Santa Barbara City College*
Madhu Motha, *Butler County Community College*
Shauna Mullins, *Murray State University*
Julie Muniz, *Southwestern Illinois College*
Kathy Nabours, *Riverside Community College*
Michael Neill, *Carl Sandburg College*
Nicole Newman, *Kalamazoo Valley Community College*
Said Ngobi, *Victor Valley College*
Denise Nunley, *Glendale Community College*
Deanna Oles, *Stark State College of Technology*
Staci Osborn, *Cuyahoga Community College–Eastern Campus*
Linda Padilla, *Joliet Junior College*
Karen D. Pain, *Palm Beach Community College*
Peg Pankowski, *Community College of Allegheny County South*
George Pate, *Robeson Community College*
Renee Patterson, *Cumberland County College*
Margaret Payerle, *Cleveland State University–Ohio*
Jim Pierce, *Lincoln Land Community College*

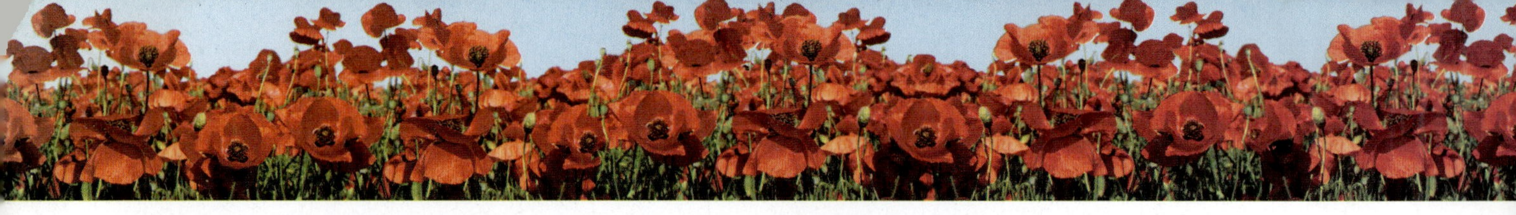

Andrew Pitcher, *University of the Pacific*
Linda Reist, *Macomb County Community College Center*
Tian Ren, *Queensborough Community College*
Nancy Ressler, *Oakton Community College*
Bob Rhea, *J. Sargeant Reynolds Community College*
Mary Richardson, *Bevill State Community College*
Minnie M. Riley, *Hinds Community College*
Matthew Robinson, *Tallahassee Community College*
Mary Romans, *Kent State University*
Melissa Rossi, *Southwestern Illinois College*
Anna Roth, *Gloucester County College*
Patricia Rowe, *Columbus State Community College*
Liz Russell, *Glendale Community College*
Alan Saleski, *Loyola University–Chicago*
Carol Saltsgaver, *University of Illinois at Springfield*
Kelly Sanchez, *Columbus State Community College*
Sheri Sanchez, *Great Basin College*
La Vache Scanlan, *Kapiolani Community College*
Sally Sestini, *Cerritos College*
Lisa Sheppard, *Lorain County Community College*
Mark A. Shore, *Allegany College of Maryland*
Mark Sigfrids, *Kalamazoo Valley Community College*
Amber Smith, *Johnson County Community College*
Leonora Smook, *Suffolk County Community College–Brentwood*
Linda Spears, *Rock Valley College*
Jane St. Peter, *Mount Mary College*
Renee Starr, *Arcadia University*
Daryl Stephens, *East Tennessee State University*
Bryan Stewart, *Tarrant County Community College*
Larry Stoneburner, *Indiana Institute of Technology*
Jennifer Strehler, *Oakton Community College*
Emily Sullivan, *Bates Technical College*
Renee Sundrud, *Harrisburg Area Community College*
Abolhassan Taghavy, *Richard J. Daley College*
Sandi Tannen, *Camden County College*
Patricia Taylor, *Thomas Nelson Community College*
Sharon Testone, *Onondaga Community College*
Harriet Thompson, *Albany State University*
Janet Thompson, *University of Akron*
John Thoo, *Yuba College*
Tanya Townsend, *Pennsylvania Valley Community College*
Fred Toxopeus, *Kalamazoo Valley Community College*
Dr. Joseph Tripp, *Ferris State University*
Linda Tucker, *Rose State College*
Sara Van Asten, *North Hennepin Community College*
Felix Van Leeuwen, *Johnson County Community College*
Josefino Villanueva, *Florida Memorial University*
Howard Wachtel, *Community College of Philadelphia*
Dottie Walton, *Cuyahoga Community College Eastern Campus*
Charles Wang, *California National University*
Walter Wang, *Baruch College*
Brock Wenciker, *Johnson County Community College*
Kevin Wheeler, *Three Rivers Community College*
Marjorie Whitmore, *Northwest Arkansas Community College*
Diane Williams, *Northern Kentucky University*
Latrica Williams, *St. Petersburg College*
Alma Wlazlinski, *McLennan Community College*
Rebecca Wong, *West Valley College*
Paul Wozniak, *El Camino College*
Christopher Yarrish, *Harrisburg Area Community College*
Kevin Yokoyama, *College of the Redwoods*
Steve Zuro, *Joliet Junior College*

Finally, we are truly grateful to the many people at McGraw-Hill Higher Education who have made our text a reality. The exceptional professionals who contributed to this text made it possible for us to reach page 2. We express our heartfelt thanks to the editorial personnel, project managers, proofreaders, accuracy checkers, art professionals, and everyone else who worked with us to better enable our students to succeed.

Supplements for the Student

Student's Solutions Manual (ISBN: 978-0-07-757440-6)

The *Student's Solutions Manual* provides comprehensive, worked-out solutions to the odd-numbered exercises in the Section Exercises, Summary Exercises, Chapter Test, and the Cumulative Reviews. The steps shown in the solutions match the style of solved examples in the textbook.

Supplements for the Instructor

Instructor's Solutions Manual

The *Instructor's Solutions Manual,* available online to adopting instructors, provides comprehensive, worked-out solutions to all exercises in the Section Exercises, Summary Exercises, Chapter Test, and the Cumulative Reviews. The methods used to solve the problems in the manual are the same as those used to solve the examples in the textbook.

Annotated Instructor's Edition

In the *Annotated Instructor's Edition (AIE),* answers to exercises and tests appear adjacent to each exercise set, in a color used *only* for annotations. Complete answers to all Reading Your Text, Summary Exercises, Chapter Tests, and Cumulative Reviews are also found at the back of the book.

Instructor's Testing and Resource Online

This computerized test bank, available online to adopting instructors, utilizes TestGen® cross-platform test generation software to quickly and easily create customized exams. Using hundreds of test items taken directly from the text, TestGen allows rapid test creation and the flexibility for instructors to create their own questions from scratch with the ability to randomize number values. Powerful search and sort functions help quickly locate questions and arrange them in any order, and built-in mathematical templates let instructors insert stylized text, symbols, graphics, and equations directly into questions without need for a separate equation editor. A separate CD version is available upon request.

> "The text is well suited to our curriculum and is an easy text to use. There are plenty of practice exercises for the student and the check yourself exercises and summary exercises are very good additional practice for the student."
>
> —Bonnie Filer-Tubaugh, *University of Akron*

applications index

GEOGRAPHY

GEOMETRY

HEALTH AND MEDICINE

INFORMATION TECHNOLOGY

MOTION AND TRANSPORTATION

POLITICS AND PUBLIC POLICY

SCIENCE AND ENGINEERING

SOCIAL SCIENCES AND DEMOGRAPHICS

SPORTS

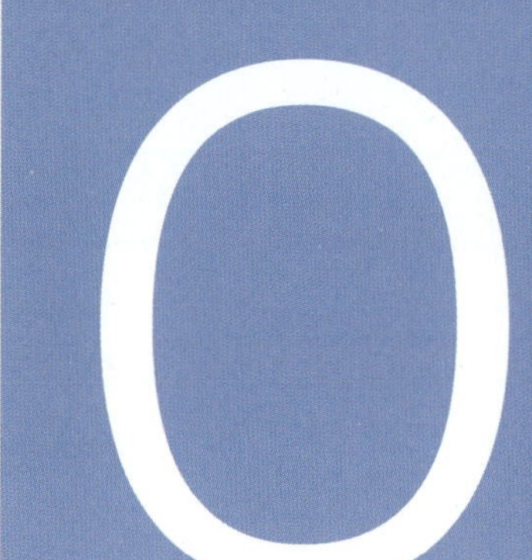

CHAPTER 0

Prealgebra Review

INTRODUCTION

Anthropologists and archeologists investigate both modern societies and cultures that existed so long ago that their characteristics must be inferred from objects found buried in lost cities or villages. With methods such as carbon dating, it has been established that large, organized cultures existed around 3000 B.C. in Egypt, 2800 B.C. in India, no later than 1500 B.C. in China, and around 1000 B.C. in the Americas.

Which is older, an object from 3000 B.C. or an object from 500 A.D.? An object from 500 A.D. is about 2,000 − 500 years old, or about 1,500 years old. But an object from 3000 B.C. is about 2,000 + 3,000 years old, or about 5,000 years old. Why subtract in the first case but add in the other? Because of the way years are counted before the common era (B.C.) and after the birth of Christ (A.D.), the B.C. dates must be considered as *negative* numbers.

Very early on, the Chinese accepted the idea that a number could be negative; they used red calculating rods for positive numbers and black for negative numbers. Hindu mathematicians in India worked out the arithmetic of negative numbers as long ago as 400 A.D., but western mathematicians did not recognize this idea until the sixteenth century. It would be difficult today to think of measuring things such as temperature, altitude, or money without using negative numbers.

CHAPTER 0 OUTLINE

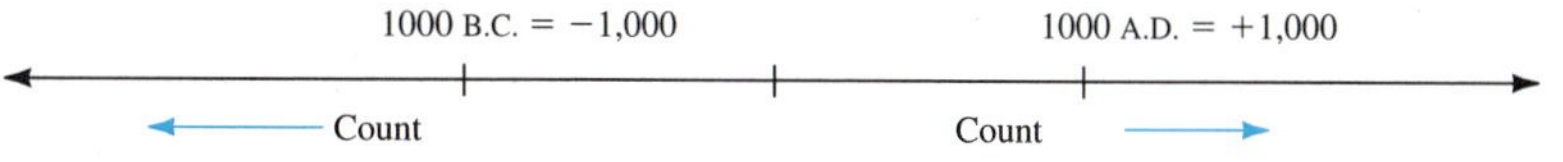

0.1 A Review of Fractions

< 0.1 Objectives >

1 > Simplify fractions
2 > Multiply and divide fractions
3 > Add and subtract fractions

Tips for Student Success

Throughout this text, we present you with a series of class-tested techniques designed to improve your performance in math classes.

Become Familiar with Your Textbook

Perform each task.

1. Use the Table of Contents to find the title of Section 5.1.
2. Use the Index to find the earliest reference to the term *factor*.
3. Find the answer to the first Check Yourself exercise in Section 0.1.
4. Find the answers to the odd-numbered end-of-section exercises in Section 0.1.
5. In the Margin Notes for Section 0.1, determine how to find the reciprocal of a fraction.
6. Find the Summary for Chapter 1.
7. Find the answers to the Chapter Test for Chapter 2.

Now you know where to find some of the more important features of the text. When you feel uncertain, think about using one of these features to clear up your confusion.

You may also be interested in the *Student's Solutions Manual* for your text. Many students find it helpful.

We begin with some assumptions about your mathematical training. We assume you are reasonably comfortable using the basic operations of addition, subtraction, multiplication, and division with whole numbers.

We also assume you are familiar with and able to perform these operations on the most common type of fractions, *decimal fractions* or *decimals*.

Finally, we assume that you have worked with fractions and negative numbers in the past.

In this chapter, we review basic operations and applications of fractions, signed numbers, exponents, and the order of operations. This is meant to be a brief review of these topics. If you would like a more in-depth discussion of this content, you should consider a course covering prealgebra material, a review of the text *Basic Mathematical Skills with Geometry* by Baratto and Bergman in this same Hutchison Series in Mathematics, or an online review using Connect Math or ALEKS.

The numbers used for counting are called the **natural numbers.** We write them as 1, 2, 3, 4, ... The three dots are called an ellipsis and indicate that the pattern continues in the obvious way.

> **NOTE**
>
> **Common fractions** can be written as $\frac{a}{b}$ in which a and b are integers and $b \neq 0$.

If we include zero in this set of numbers, we call them the **whole numbers.**

We review the **integers** in Section 0.2. This set includes the natural numbers, their negatives, and zero.

The **rational numbers** include the integers as well as every number that can be written as a *common fraction*. Common fractions include proper fractions such as $\frac{1}{2}$ and $\frac{2}{3}$ and improper fractions such as $\frac{5}{1}$ and $\frac{7}{4}$.

Interpreting a fraction as a division statement allows you to avoid some common careless errors. Simply recall that the fraction bar represents division.

$$\frac{5}{8} = 5 \div 8$$

We use this fact to explain some fraction basics.

$\frac{1}{6}$ is one-sixth of a whole, whereas

$\frac{6}{1}$ represents six "wholes" because this is $6 \div 1 = 6$.

> **NOTE**
>
> $\frac{0}{3}$ means a whole is divided into three parts and you have none of them.
>
> $\frac{3}{0}$ represents division by 0, which does not exist.

Similarly, division by 0 is not defined, but you can have no parts of a whole. $\frac{0}{3}$ means you have no thirds: $\frac{0}{3} = 0$.

On the other hand,

$\frac{3}{0} = 3 \div 0$ which does not exist. This expression has no meaning for us.

The number 1 has many different fraction forms. Any fraction in which the numerator and denominator are the same (and not zero) is another name for the number 1.

$$1 = \frac{2}{2} \qquad 1 = \frac{12}{12} \qquad 1 = \frac{257}{257}$$

To determine whether two fractions are equal or to find **equivalent fractions,** we use the **fundamental principle of fractions.** The fundamental principle of fractions states that multiplying the numerator and denominator of a fraction by the same number is the same as multiplying the fraction by 1. We express the principle symbolically.

Property

The Fundamental Principle of Fractions

$$\frac{a}{b} = \frac{a \times c}{b \times c} \quad \text{or} \quad \frac{a \times c}{b \times c} = \frac{a}{b} \qquad c \neq 0$$

Example 1 Rewriting Fractions

> **NOTE**
>
> Each representation is a name for the number. Each number has many names.

Use the fundamental principle to write three fractions equivalent to each number.

(a) $\frac{2}{3}$

Multiplying the numerator and denominator by the same number is the same as multiplying by 1.

$\frac{2}{3} = \frac{2 \times 2}{3 \times 2} = \frac{4}{6}$ Multiply the numerator and denominator by 2.

$\frac{2}{3} = \frac{2 \times 3}{3 \times 3} = \frac{6}{9}$ Multiply the numerator and denominator by 3.

$\frac{2}{3} = \frac{2 \times 10}{3 \times 10} = \frac{20}{30}$

> **RECALL**
>
> We can write any whole number as a fraction by putting it over 1. Thus,
>
> $5 = \frac{5}{1}$

(b) 5

$$5 = \frac{5 \times 2}{1 \times 2} = \frac{10}{2}$$

$$5 = \frac{5 \times 3}{1 \times 3} = \frac{15}{3}$$

$$5 = \frac{5 \times 100}{1 \times 100} = \frac{500}{100}$$

Check Yourself 1

Use the fundamental principle to write three fractions equivalent to each number.

(a) $\frac{5}{8}$ **(b)** $\frac{4}{3}$ **(c)** 3

The fundamental principle can also be used to find the simplest fraction equivalent to a number. Fractions written in this form are said to be **simplified.**

Example 2 Simplifying Fractions

< Objective 1 >

RECALL

A prime number is any whole number greater than 1 that has only itself and 1 as factors.

Use the fundamental principle to simplify each fraction.

(a) $\frac{22}{55}$ **(b)** $\frac{35}{45}$ **(c)** $\frac{24}{36}$

In each case, we first write the numerator and denominator as products of prime numbers.

(a) $\frac{22}{55} = \frac{2 \times 11}{5 \times 11}$

We then use the fundamental principle of fractions to "remove" the common factor of 11.

$$\frac{22}{55} = \frac{2 \times 11}{5 \times 11} = \frac{2}{5}$$

(b) $\frac{35}{45} = \frac{5 \times 7}{3 \times 3 \times 5}$

Removing the common factor of 5 yields

$$\frac{35}{45} = \frac{7}{3 \times 3} = \frac{7}{9}$$

(c) $\frac{24}{36} = \frac{2 \times 2 \times 2 \times 3}{2 \times 2 \times 3 \times 3}$

Removing the common factor $2 \times 2 \times 3$ yields

$$\frac{2}{3}$$

NOTE

Often, we use the convention of "canceling" a factor that appears in both the numerator and denominator to prevent careless errors. In part (b),

$$\frac{5 \times 7}{3 \times 3 \times 5} = \frac{\cancel{5} \times 7}{3 \times 3 \times \cancel{5}} = \frac{7}{3 \times 3} = \frac{7}{9}$$

Check Yourself 2

Use the fundamental principle to simplify each fraction.

(a) $\frac{21}{33}$ **(b)** $\frac{15}{30}$ **(c)** $\frac{12}{54}$

Fractions are often used in everyday situations. When solving an *application,* read the problem through carefully. Read the problem again and decide what you need to find and what you need to do. Then write out the problem completely and carefully. After completing the math work, be sure to answer the problem with a sentence.

Throughout this text, we use variations of this five-step process when working with applications. We will update this procedure after we introduce you to *algebra.*

Step by Step

Solving Applications

Step 1 Read the problem carefully to determine what you are being asked to find and what information is given in the application.

Step 2 Decide what you will do to solve the problem.

Step 3 Write down the complete (mathematical) statement necessary to solve the problem.

Step 4 Perform any calculations or other mathematics needed to solve the problem.

Step 5 Answer the question. Be sure to include units with your answer, when appropriate. Check to make certain that your answer is reasonable.

Example 3 Using Fractions in an Application

Jo, an executive vice president of information technology, already supervises 10 people and hires 2 more to fill out her staff. What fraction of her staff is new? Be sure to simplify your answer.

Step 1 We are asked to find the fraction of Jo's staff that is new. We know that her staff consisted of 10 people and 2 new people were hired.

Step 2 First, we figure out the size of her total staff. Then, we figure out the fraction comparing the new people to the total staff.

Step 3 Total staff: 10 original people and 2 new people

$10 + 2$

We construct the ratio,

$$\frac{\text{New people}}{\text{Total staff}} = \frac{2}{10 + 2}$$

Step 4

$$\frac{2}{10 + 2} = \frac{2}{12}$$

$$= \frac{1}{6}$$

Step 5 One-sixth of her staff is new.

This answer seems reasonable.

RECALL

We **cannot** simplify or "cancel" the twos in the sum.

$$\frac{2}{10 + 2} \neq \frac{\cancel{2}}{10 + \cancel{2}}$$

This is **incorrect**.

Check Yourself 3

There are 36 packaging machines in one division of Early Enterprises. At any given time, 4 of these machines are shut down for scheduled maintenance and service. What fraction of machines is operating at one time? Be sure your answer is simplified.

To simplify a fraction, use the fundamental principle of fractions, in reverse. In Example 3, we simplified the fraction in step 4 by factoring 2 from both the numerator and denominator.

$$\frac{2}{12} = \frac{2 \times 1}{2 \times 2 \times 3}$$ Prime factorization

$$= \frac{2}{2} \times \frac{1}{2 \times 3}$$ The fundamental principle of fractions

$$= 1 \times \frac{1}{6}$$ $\frac{2}{2} = 1$

$$= \frac{1}{6}$$

Usually, we combine these steps.

$$\frac{2}{12} = \frac{\cancel{2}^{1}}{\cancel{2}_{1} \times 6} = \frac{1}{6} \text{ or even } \frac{2}{12} = \frac{\cancel{2}^{1}}{\cancel{12}_{6}} = \frac{1}{6}$$

When multiplying fractions, we use the property

$$\frac{a}{b} \times \frac{c}{d} = \frac{a \times c}{b \times d}$$

We then write the numerator and denominator in factored form and simplify before multiplying.

Example 4 Multiplying Fractions

< Objective 2 >

RECALL

A **product** is the result of multiplication.

Find the product.

$$\frac{9}{2} \times \frac{4}{3}$$

$$\frac{9}{2} \times \frac{4}{3} = \frac{9 \times 4}{2 \times 3}$$

$$= \frac{3 \times 3 \times 2 \times 2}{2 \times 3} = \frac{3 \times 2}{1}$$

$$= \frac{6}{1}$$ The denominator of 1 is not necessary.

$$= 6$$

Check Yourself 4

Multiply and simplify.

(a) $\frac{3}{5} \times \frac{10}{7}$ **(b)** $\frac{12}{5} \times \frac{10}{6}$

The process describing fraction multiplication gives us insight into a number of fraction operations and properties.

RECALL

Multiplying or dividing a number by 1 leaves the number unchanged.

For instance, the fundamental principle of fractions is easily explained with the multiplication property. When applying the fundamental principle of fractions, all we are really doing is multiplying or dividing a given fraction by 1.

$$\frac{2}{3} = \frac{2}{3} \times 1$$

$$= \frac{2}{3} \times \frac{2}{2}$$ $1 = \frac{2}{2}$

$$= \frac{2 \times 2}{3 \times 2}$$ This is fraction multiplication.

$$= \frac{4}{6}$$

Another property that arises from fraction multiplication allows us to rewrite a fraction as a product using both the numerator and the denominator. For example,

$$\frac{3}{4} = \frac{3 \times 1}{1 \times 4} = \frac{3}{1} \times \frac{1}{4} = 3 \times \frac{1}{4} \quad \text{and} \quad \frac{3}{4} = \frac{1 \times 3}{4 \times 1} = \frac{1}{4} \times \frac{3}{1} = \frac{1}{4} \times 3$$

RECALL

We find a reciprocal by inverting the fraction.

To divide two fractions, the divisor is replaced with its **reciprocal;** then the fractions are multiplied.

$$\frac{a}{b} \div \frac{c}{d} = \frac{a}{b} \times \frac{d}{c} = \frac{a \times d}{b \times c}$$

Example 5 Dividing Fractions

NOTES

The divisor $\frac{5}{6}$ is inverted and becomes $\frac{6}{5}$.

The common factor 3 is removed from the numerator and denominator. This is the same as dividing by $\frac{3}{3}$ or 1.

Find the quotient.

$$\frac{7}{3} \div \frac{5}{6}$$

$$\frac{7}{3} \div \frac{5}{6} = \frac{7}{3} \times \frac{6}{5} = \frac{7 \times 6}{3 \times 5}$$

$$= \frac{7 \times 2 \times 3}{3 \times 5} = \frac{7 \times 2}{5} = \frac{14}{5}$$

Check Yourself 5

Find the quotient.

$$\frac{9}{2} \div \frac{3}{5}$$

NOTE

In algebra, improper fractions are preferred to mixed numbers. However, we prefer mixed numbers when answering many application exercises.

When adding two fractions, we need to find the **least common denominator (LCD)** first. The least common denominator is the smallest number that divides both denominators evenly. The process of finding the LCD is outlined here.

Step by Step

To Find the Least Common Denominator

Step 1 Write the prime factorization for each of the denominators.

Step 2 Find all the prime factors that appear in any one of the prime factorizations.

Step 3 Form the product of those prime factors, using each factor the greatest number of times it occurs in any one factorization.

Example 6 Finding the Least Common Denominator (LCD)

Find the LCD of fractions with denominators 6 and 8.

Our first step in adding fractions with denominators 6 and 8 is to determine the least common denominator. We begin by factoring 6 and 8.

$6 = 2 \times 3$

$8 = 2 \times 2 \times 2$ Because 2 appears 3 times as a factor of 8, it is used 3 times in writing the LCD.

The LCD is $2 \times 2 \times 2 \times 3$, or 24.

Check Yourself 6

Find the LCD of fractions with denominators 9 and 12.

The process is similar if there are more than two denominators.

Example 7 Finding the Least Common Denominator

Find the LCD of fractions with denominators 6, 9, and 15.

To add fractions with denominators 6, 9, and 15, we need to find the LCD. Factor the three numbers.

$6 = 2 \times 3$ 2 and 5 appear only once in any one factorization.

$9 = 3 \times 3$ 3 appears twice as a factor of 9.

$15 = 3 \times 5$

The LCD is $2 \times 3 \times 3 \times 5$, or 90.

Check Yourself 7

Find the LCD of fractions with denominators 5, 8, and 20.

To add two fractions, we use the property

$$\frac{a}{b} + \frac{c}{b} = \frac{a + c}{b}$$

Example 8 Adding Fractions

< Objective 3 >

RECALL

A sum is the result of addition.

RECALL

We use the LCD to write equivalent fractions.

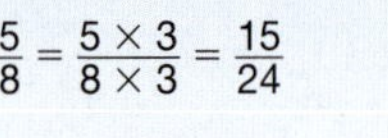

$$\frac{5}{8} = \frac{5 \times 3}{8 \times 3} = \frac{15}{24}$$

Find the sum.

$$\frac{5}{8} + \frac{7}{12}$$

The LCD of 8 and 12 is 24. Each fraction should be rewritten as a fraction with that denominator.

$\frac{5}{8} = \frac{15}{24}$ Multiply the numerator and denominator by 3.

$\frac{7}{12} = \frac{14}{24}$ Multiply the numerator and denominator by 2.

$\frac{5}{8} + \frac{7}{12} = \frac{15}{24} + \frac{14}{24} = \frac{15 + 14}{24} = \frac{29}{24}$ This fraction is simplified.

Check Yourself 8

Find the sum.

(a) $\frac{4}{5} + \frac{7}{9}$ **(b)** $\frac{5}{6} + \frac{4}{15}$

To subtract two fractions, use the rule

$$\frac{a}{b} - \frac{c}{b} = \frac{a - c}{b}$$

Subtracting fractions is treated exactly like adding them, except the numerator becomes the difference of the two numerators.

Example 9 Subtracting Fractions

Find the difference.

$$\frac{7}{9} - \frac{1}{6}$$

RECALL

A **difference** is the result of subtraction.

The LCD is 18. We rewrite the fractions with that denominator.

$$\frac{7}{9} = \frac{14}{18}$$

$$\frac{1}{6} = \frac{3}{18}$$

$$\frac{7}{9} - \frac{1}{6} = \frac{14}{18} - \frac{3}{18} = \frac{14-3}{18} = \frac{11}{18}$$ This fraction is simplified.

Check Yourself 9

Find the difference $\frac{11}{12} - \frac{5}{8}$.

We present a final application of fraction arithmetic before concluding this section.

Example 10 A Crafts Application

A potter uses $\frac{2}{3}$ pound (lb) of clay when making a bowl. How many bowls can be made from 15 lb of clay?

Step 1 The question asks for the number of $\frac{2}{3}$-lb bowls that the potter can make from a 15-lb batch of clay.

Step 2 This is a division problem. We divide to see how many full times $\frac{2}{3}$ goes into 15.

Step 3 $15 \div \frac{2}{3}$

Step 4 $15 \div \frac{2}{3} = 15 \times \frac{3}{2}$ Use the division property.

$= \frac{15 \times 3}{1 \times 2}$ Now multiply, $15 = \frac{15}{1}$.

$= \frac{45}{2}$ or $22\frac{1}{2}$ Complete the computation.

RECALL

$\frac{45}{2} = 45 \div 2$

$= 22\frac{1}{2}$

Step 5 The potter can complete 22 (whole) bowls from a 15-lb batch.

Reasonableness

Because each bowl uses less than a pound of clay, we would expect to get more than 15 bowls.

Because each bowl uses more than a half-pound of clay, we would expect to get fewer than $15 \times 2 = 30$ bowls.

22 bowls is a reasonable answer.

Check Yourself 10

A student survey at a community college found that $\frac{3}{4}$ of the students held jobs while going to school. Of those who have jobs, $\frac{5}{6}$ reported working more than 20 hours per week. What fraction of those surveyed worked more than 20 hours per week?

Check Yourself ANSWERS

1. Answers will vary. **2.** (a) $\frac{7}{11}$; (b) $\frac{1}{2}$; (c) $\frac{2}{9}$ **3.** $\frac{8}{9}$ **4.** (a) $\frac{6}{7}$; (b) 4 **5.** $\frac{15}{2}$ **6.** 36

7. 40 **8.** (a) $\frac{71}{45}$; (b) $\frac{11}{10}$ **9.** $\frac{7}{24}$ **10.** $\frac{5}{8}$

Reading Your Text

These fill-in-the-blank exercises will help you understand some of the key vocabulary used in this section. The answers to these exercises are in the Answers Appendix in the back of the text.

(a) The numbers used for counting are called the ____________ numbers.

(b) Multiplying the numerator and denominator of a fraction by the same nonzero number is the same as multiplying the fraction by ____________.

(c) A ____________ is the result of multiplication.

(d) Subtracting a pair of fractions is similar to adding them, except the numerator of the result is the ____________ of the two numerators.

0.1 exercises

Skills | Calculator/Computer | Career Applications | Above and Beyond

Use the fundamental principle of fractions to write three fractions equivalent to each number.

1. $\frac{3}{7}$ **2.** $\frac{2}{5}$ **3.** $\frac{4}{9}$ **4.** $\frac{7}{8}$

5. $\frac{5}{6}$ **6.** $\frac{11}{13}$ **7.** $\frac{10}{17}$ **8.** $\frac{2}{7}$

9. $\frac{9}{16}$ **10.** $\frac{6}{11}$ **11.** $\frac{7}{9}$ **12.** $\frac{15}{16}$

< Objective 1 >

Use the fundamental principle of fractions to simplify each fraction.

13. $\frac{10}{15}$ **14.** $\frac{12}{15}$ **15.** $\frac{10}{14}$ **16.** $\frac{18}{60}$

17. $\frac{12}{18}$ **18.** $\frac{28}{35}$ **19.** $\frac{35}{40}$ **20.** $\frac{28}{32}$

21. $\frac{11}{44}$ **22.** $\frac{10}{25}$ **23.** $\frac{11}{33}$ **24.** $\frac{18}{48}$

25. $\frac{24}{27}$ **26.** $\frac{27}{45}$ **27.** $\frac{32}{40}$ **28.** $\frac{17}{51}$

29. $\frac{75}{105}$ **30.** $\frac{62}{93}$ **31.** $\frac{24}{30}$ **32.** $\frac{48}{66}$

33. $\frac{105}{135}$ VIDEO **34.** $\frac{39}{91}$

< Objective 2 >

Multiply. Simplify each product.

35. $\frac{3}{7} \times \frac{4}{5}$ **36.** $\frac{2}{7} \times \frac{5}{9}$ **37.** $\frac{3}{4} \times \frac{7}{5}$ **38.** $\frac{3}{5} \times \frac{2}{7}$

39. $\frac{3}{5} \times \frac{5}{7}$ **40.** $\frac{6}{11} \times \frac{8}{6}$ **41.** $\frac{6}{13} \times \frac{4}{9}$ VIDEO **42.** $\frac{5}{9} \times \frac{6}{11}$

43. $\frac{3}{11} \times \frac{7}{9}$ **44.** $\frac{7}{9} \times \frac{3}{5}$ **45.** $\frac{4}{21} \times \frac{7}{12}$ **46.** $\frac{5}{21} \times \frac{14}{25}$

Divide. Write each result in simplest form.

47. $\frac{1}{7} \div \frac{3}{5}$ **48.** $\frac{2}{5} \div \frac{1}{3}$ **49.** $\frac{2}{5} \div \frac{3}{4}$ **50.** $\frac{5}{8} \div \frac{3}{4}$

51. $\frac{8}{9} \div \frac{4}{3}$ **52.** $\frac{4}{7} \div \frac{6}{11}$ **53.** $\frac{7}{10} \div \frac{5}{9}$ **54.** $\frac{8}{9} \div \frac{11}{15}$

55. $\frac{8}{15} \div \frac{2}{5}$ **56.** $\frac{5}{27} \div \frac{15}{54}$ VIDEO **57.** $\frac{8}{21} \div \frac{24}{35}$ **58.** $\frac{9}{28} \div \frac{27}{35}$

Find the least common denominator (LCD) for fractions with the given denominators.

59. 30 and 50 **60.** 36 and 48 **61.** 48 and 80 **62.** 60 and 84

63. 3, 4, and 5 **64.** 3, 4, and 6 **65.** 8, 10, and 15 **66.** 6, 22, and 33

67. 5, 10, and 25 **68.** 8, 24, and 48

< Objective 3 >

Add. Write each result in simplest form.

69. $\frac{2}{5} + \frac{1}{4}$ **70.** $\frac{2}{3} + \frac{3}{10}$ **71.** $\frac{2}{5} + \frac{7}{15}$ **72.** $\frac{2}{3} + \frac{4}{5}$

73. $\frac{3}{8} + \frac{5}{12}$ **74.** $\frac{5}{36} + \frac{7}{24}$ **75.** $\frac{7}{30} + \frac{5}{18}$ **76.** $\frac{9}{14} + \frac{10}{21}$

77. $\frac{7}{15} + \frac{13}{18}$ 

78. $\frac{12}{25} + \frac{19}{30}$ **79.** $\frac{1}{5} + \frac{1}{10} + \frac{1}{15}$ **80.** $\frac{1}{3} + \frac{1}{5} + \frac{1}{10}$

Subtract. Write each result in simplest form.

81. $\frac{8}{9} - \frac{3}{9}$ **82.** $\frac{9}{10} - \frac{6}{10}$ **83.** $\frac{6}{7} - \frac{2}{7}$ **84.** $\frac{11}{12} - \frac{7}{12}$

85. $\frac{7}{8} - \frac{2}{3}$ VIDEO **86.** $\frac{4}{9} - \frac{2}{5}$ **87.** $\frac{11}{18} - \frac{2}{9}$ **88.** $\frac{5}{6} - \frac{1}{4}$

89. $\frac{2}{3} - \frac{7}{11}$ **90.** $\frac{13}{18} - \frac{5}{12}$ **91.** $\frac{5}{42} - \frac{1}{36}$ **92.** $\frac{13}{18} - \frac{7}{15}$

93. **Crafts** If a pancake recipe calls for $\frac{1}{3}$ cup of white flour, $\frac{1}{3}$ cup of wheat flour, and $\frac{1}{2}$ cup of soy flour, how much flour is in the recipe?

94. **Business and Finance** Deductions from your paycheck are approximately $\frac{1}{8}$ for federal tax, $\frac{1}{20}$ for state tax, $\frac{1}{20}$ for Social Security, and $\frac{1}{40}$ for a savings withholding plan. What portion of your pay is deducted? VIDEO

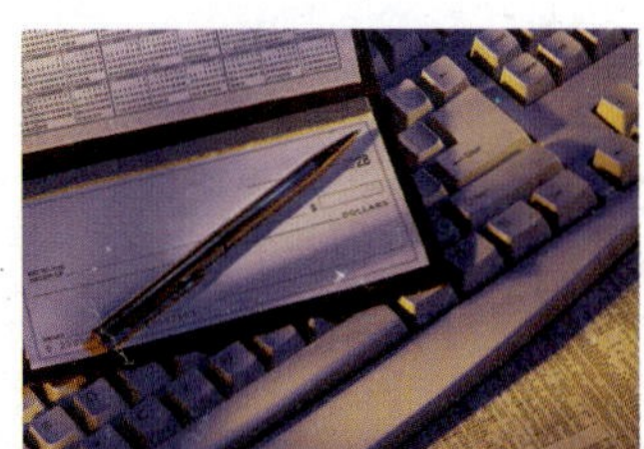

95. **Science and Medicine** Carol walked $\frac{3}{4}$ mile (mi) to the store, $\frac{1}{2}$ mi to a friend's house, and then $\frac{2}{3}$ mi home. How far did she walk?

96. **Geometry** Find the perimeter of, or the distance around, the figure by finding the sum of the lengths of the sides.

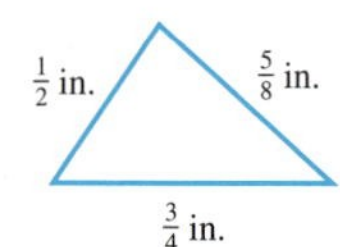

97. **Crafts** A hamburger that weighed $\frac{1}{4}$ pound (lb) before cooking weighed $\frac{3}{16}$ lb after cooking. How much weight was lost in cooking?

98. **Crafts** Geraldo has $\frac{3}{4}$ cup of flour. Biscuits use $\frac{5}{8}$ cup. Will he have enough left over for a small pie crust that requires $\frac{1}{4}$ cup? Explain.

Determine whether each statement is **true** *or* **false.**

99. When adding two fractions, we add the numerators together and we add the denominators together.

100. When multiplying two fractions, we multiply the numerators together and we multiply the denominators together.

Complete each statement with **always, sometimes,** *or* **never.**

101. The least common denominator of three fractions is __________ the product of the three denominators.

102. To add two fractions with different denominators, we __________ rewrite the fractions so that they have the same denominator.

Answers

1. $\frac{6}{14}, \frac{9}{21}, \frac{12}{28}$ **3.** $\frac{8}{18}, \frac{16}{36}, \frac{40}{90}$ **5.** $\frac{10}{12}, \frac{15}{18}, \frac{50}{60}$ **7.** $\frac{20}{34}, \frac{30}{51}, \frac{100}{170}$ **9.** $\frac{18}{32}, \frac{27}{48}, \frac{90}{160}$ **11.** $\frac{14}{18}, \frac{35}{45}, \frac{140}{180}$ **13.** $\frac{2}{3}$ **15.** $\frac{5}{7}$ **17.** $\frac{2}{3}$ **19.** $\frac{7}{8}$ **21.** $\frac{1}{4}$ **23.** $\frac{1}{3}$ **25.** $\frac{8}{9}$ **27.** $\frac{4}{5}$ **29.** $\frac{5}{7}$ **31.** $\frac{4}{5}$ **33.** $\frac{7}{9}$ **35.** $\frac{12}{35}$ **37.** $\frac{21}{20}$ **39.** $\frac{3}{7}$ **41.** $\frac{8}{39}$ **43.** $\frac{7}{33}$ **45.** $\frac{1}{9}$ **47.** $\frac{5}{21}$ **49.** $\frac{8}{15}$ **51.** $\frac{2}{3}$ **53.** $\frac{63}{50}$ **55.** $\frac{4}{3}$ **57.** $\frac{5}{9}$ **59.** 150 **61.** 240 **63.** 60 **65.** 120 **67.** 50 **69** $\frac{13}{20}$ **71.** $\frac{13}{15}$ **73.** $\frac{19}{24}$ **75.** $\frac{23}{45}$ **77.** $\frac{107}{90}$ **79.** $\frac{11}{30}$ **81.** $\frac{5}{9}$ **83.** $\frac{4}{7}$ **85.** $\frac{5}{24}$ **87.** $\frac{7}{18}$ **89.** $\frac{1}{33}$ **91.** $\frac{23}{252}$ **93.** $\frac{7}{6}$ cups or $1\frac{1}{6}$ cups **95.** $\frac{23}{12}$ mi or $1\frac{11}{12}$ mi **97.** $\frac{1}{16}$ lb **99.** False **101.** sometimes

0.2 Real Numbers

< 0.2 Objectives >

1 > Classify real numbers

2 > Plot numbers on a number line

3 > Find the opposite of a number

4 > Find the absolute value of a number

In Section 0.1, we said the numbers used to count things—1, 2, 3, 4, 5, and so on—are called the **natural** (or **counting**) **numbers.** The **whole numbers** consist of the natural numbers and zero—0, 1, 2, 3, 4, 5, and so on. They can be represented on a number line like the one shown. Zero (0) is called the origin.

The number line continues indefinitely in both directions.

When numbers are used to represent physical quantities (such as altitude, temperature, or an amount of money), it is often necessary to distinguish between *positive* and *negative* quantities. It is convenient to represent these quantities with plus (+) or minus (−) signs. For instance,

The Empire State building is 1,250 feet tall (+1,250).

The altitude at Badwater in Death Valley is 282 ft *below* sea level (−282).

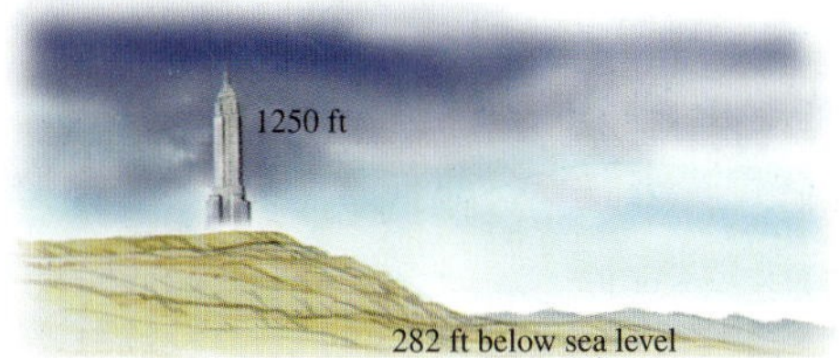

The temperature in Chicago might be 10° *below* zero (−10°).

An account could show a *gain* of \$100 (+100) or a *loss* of \$100 (−100).

These numbers suggest the need to extend the whole numbers to include both positive numbers (such as +100) and negative numbers (such as −282).

To represent the negative numbers, we extend the number line to the *left* of zero and name equally spaced points.

Numbers used to name points to the right of zero are positive numbers. They are written with a positive (+) sign or with no sign at all.

+6 and 9 are positive numbers

Numbers used to name points to the left of zero are negative numbers. They are always written with a negative (−) sign.

−3 and −20 are negative numbers

Read "negative 3."

Example 1 Classifying Real Numbers

< Objective 1 >

RECALL

If no sign appears, a nonzero number is positive.

+6 is a positive number.
−9 is a negative number.
5 is a positive number.
0 is neither positive nor negative.

Check Yourself 1

Classify each number as positive, negative, or neither.

(a) +3 **(b)** 7 **(c)** −5 **(d)** 0

Positive and negative numbers considered together (along with zero) are **real numbers.**

Here is a number line extended to include both positive and negative numbers.

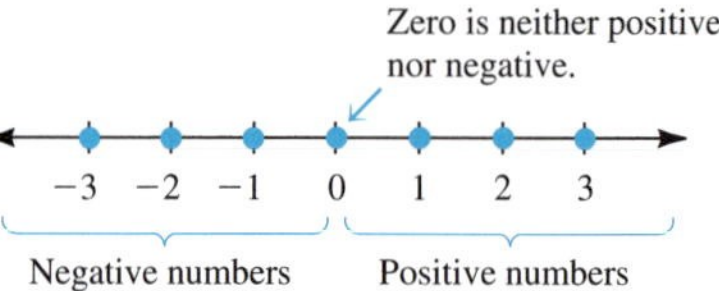

The numbers used to name the points shown on the number line are called the **integers.** The integers consist of the natural numbers, their negatives, and the number 0.

Definition

Integers

The **integers** consist of the natural numbers, their negatives, and zero. We can represent the set of integers by

{. . . , −3, −2, −1, 0, 1, 2, 3, . . .}

Example 2 Identifying Integers

Which numbers are integers?

$-3, \quad 5.3, \quad \frac{2}{3}, \quad 4$

Of these four numbers, only −3 and 4 are integers.

Check Yourself 2

Which numbers are integers?

$7, \quad 0, \quad \frac{4}{7}, \quad -5, \quad 0.2$

Definition

Rational Numbers

Any number that can be written as the ratio of two integers is called a **rational number.**

NOTE

6 is a rational number because it can be written as $\frac{6}{1}$.

Examples of rational numbers are $6, \frac{7}{3}, -\frac{15}{4}, 0, \frac{4}{1}$. You can estimate the location of a rational number on a number line, as Example 3 illustrates.

Example 3 Plotting Rational Numbers

< Objective 2 >

Plot each point on a number line.

$\frac{2}{3}$, $-3\frac{1}{4}$, $\frac{27}{5}$, -1.445

$\frac{2}{3}$ is between 0 and 1 (closer to one), so we plot that point on the number line as shown. $-3\frac{1}{4}$ is to the left of zero; it is $\frac{1}{4}$ farther than -3 from 0, so we plot this point, as well.

RECALL

Decimals are just a way of writing fractions when the denominator is a power of 10.

$-1.445 = -\frac{1,445}{1,000}$

To find $\frac{27}{5}$ on a number line, we can do division, $\frac{27}{5} = 27 \div 5 = 5.4$, or write it as a mixed number $\frac{27}{5} = 5\frac{2}{5}$. Either way, we find the same point, farther than 5 units from 0 on the number line.

Finally, the point -1.445 is nearly halfway between -1 and -2.

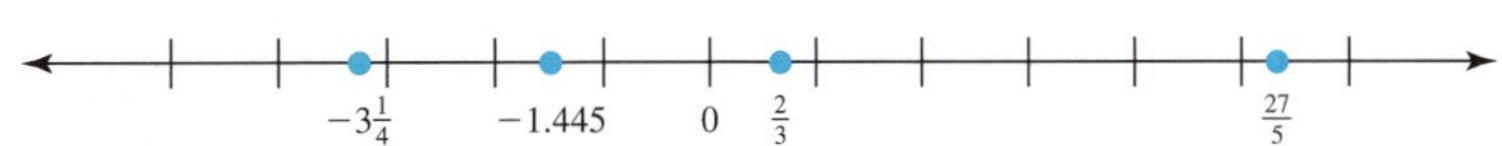

Check Yourself 3

Plot each point on the number line.

$-2\frac{1}{3}$, $\frac{37}{11}$, 5.66, $-\frac{1}{4}$

−6 −4 −2 0 2 4 6

One important property we can easily see on a number line is **order.** We say one number is **greater than** another if it is to the right on a number line. Similarly, the number on the left is **less than** the one on the right.

We use the symbols $>$ and $<$ to indicate order. The *inequality symbol* points to the smaller number. You should see how to use these symbols in the next example.

Example 4 Determining Order

> CAUTION

Because order is defined by position on a number line, you need to be careful when comparing two negative numbers.

(a) $6 > 3$

Six is *greater than* 3 because it is to the right of 3 on a number line.

(b) $2 < 5$

Two is *less than* 5; it is to the left of 5 on a number line.

(c) $-2 > -5$

-2 is to the right of -5 on a number line, so -2 is *greater than* -5.

Check Yourself 4

Fill in each blank with $>$, $<$, or $=$ to make a true statement.

(a) -7 ____ 4 **(b)** $\frac{13}{4}$ ____ 3.25 **(c)** -12.08 ____ -12.2

Numbers that cannot be written as the ratio of two integers are called **irrational numbers.** Some examples of irrational numbers are $\sqrt{3}$, $\sqrt{7}$, and π. We say more about these numbers later in the text. The next diagram illustrates the relationships among the various sets of numbers.

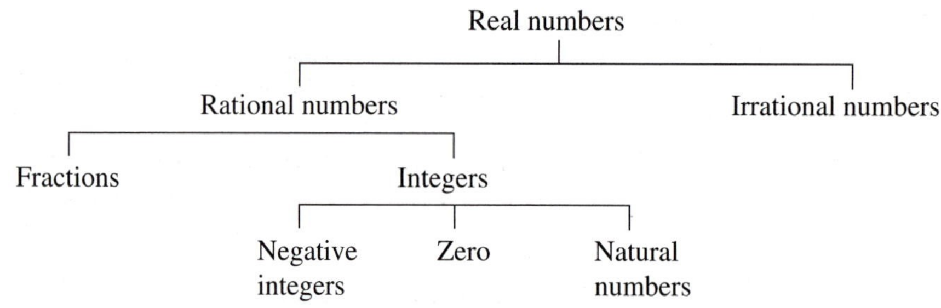

An important idea in our work with real numbers is the *opposite* of a number. Every number has an opposite.

Definition

Opposite of a Number

The **opposite** of a number corresponds to a point the same distance from 0 as the given number, but in the opposite direction.

Example 5 Writing the Opposite of a Real Number

< Objective 3 >

(a) The opposite of 5 is -5.

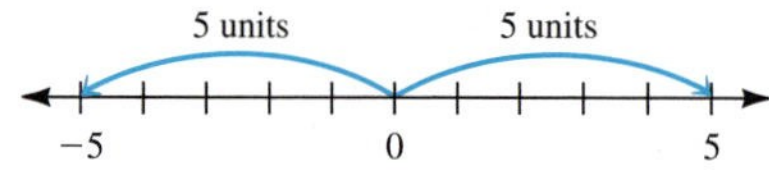

Both numbers are located 5 units from 0.

(b) The opposite of -3 is 3.

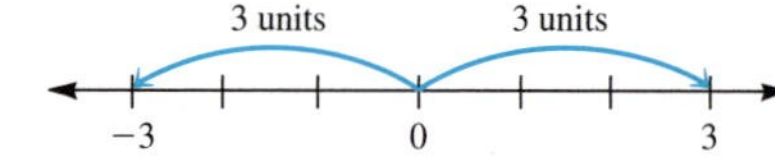

Both numbers correspond to points that are 3 units from 0.

Check Yourself 5

(a) What is the opposite of 8? **(b)** What is the opposite of -9?

NOTE

To represent the opposite of a number, place a minus sign in front of the number.

We write the opposite of 5 as -5. You can now think of -5 in two ways: as negative 5 and as the opposite of 5.

Using the same idea, we can write the opposite of a negative number. The opposite of -3 is $-(-3)$. Since we know from looking at a number line that the opposite of -3 is 3, this means that

$$-(-3) = 3$$

So the opposite of a negative number must be positive.

Property

The Opposite of a Real Number

1. The opposite of a positive number is negative.
2. The opposite of a negative number is positive.
3. The opposite of 0 is 0.

NOTES

The *magnitude* of a number is the same as its absolute value.

The absolute value of a number is always nonnegative.

An important property of a real number is its *magnitude*. The magnitude of a number is given by its distance from zero on a number line.

You can think of the magnitude of a number as the number without its sign. A more rigorous approach uses *absolute-value notation*. The magnitude of a number is the same as its absolute value.

Definition

Absolute Value

The **absolute value** of a real number is the distance (on a number line) between the number and 0.

To indicate the absolute value of a number, we enclose it with vertical bars. Given a number a, we write "the absolute value of a" as $|a|$.

Example 6 Finding Absolute Value

< Objective 4 >

NOTE

Both 5 and -5 have a magnitude of 5.

(a) The absolute value of 5 is 5.

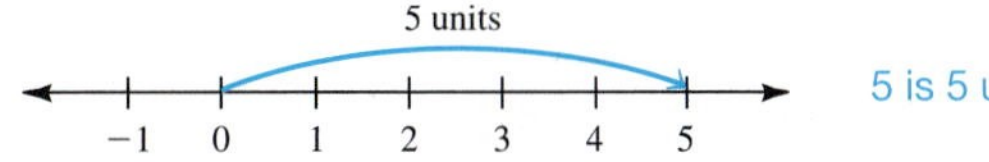

5 is 5 units from 0.

(b) The absolute value of -5 is 5.

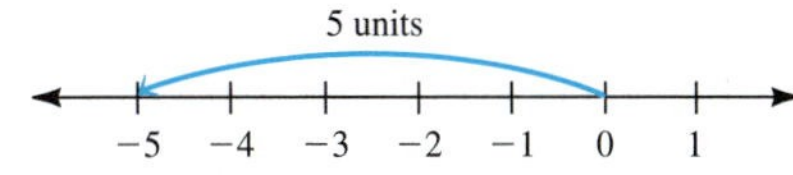

-5 is also 5 units from 0.

NOTE

$|5|$ is read "the absolute value of 5."

We write

$|5| = 5$ and $|-5| = 5$

Check Yourself 6

Complete each statement.

(a) The absolute value of 9 is __________.

(b) The absolute value of -12 is __________.

(c) $|-6| =$ **(d)** $|15| =$

Example 7 Applying Real Numbers

RECALL

To arrange a set of numbers in *ascending order,* list them from least to greatest.

The elevations, in inches, of several points on a job site are shown below.

-18 27 -84 37 59 -13 4 92 49 66 -45

Arrange the elevations in ascending order.

Step 1 The question asks us to arrange the given numbers from least to greatest.

NOTE

In this case, it makes sense to "combine" the remaining steps.

Steps 2 to 5 The number farthest left on a number line is -84, followed by -45, and so on.

$-84, -45, -18, -13, 4, 27, 37, 49, 59, 66, 92$

Check Yourself 7

Several resistors were tested using an ohmmeter. Their resistance levels were entered into a table indicating their variance from 10,000 ohms (Ω). For example, if a resistor were to measure 9,900 Ω, it would be listed at -100.

Use their measured resistances to list the resistors in ascending order.

Resistor	#1	#2	#3	#4	#5	#6	#7
Variance (10,000 Ω)	+175	−60	−188	+10	+218	−65	−302

Quite often, we refer to an unknown number by the set or group that it belongs to. We conclude this section by summarizing the sets of numbers discussed. When the instructions for an exercise refer to a specific set of numbers, you can refer back to this summary to remind you of which sets contain that number.

Definitions

Natural Numbers

The natural numbers are the counting numbers.
$\{1, 2, 3, \ldots\}$

Whole Numbers

The whole numbers are the natural numbers together with zero.
$\{0, 1, 2, 3, \ldots\}$

Integers

The integers consist of the natural numbers, their negatives, and zero.
$\{\ldots, -3, -2, -1, 0, 1, 2, 3, \ldots\}$

Rational Numbers

The rational numbers are the set of all numbers that can be written as the ratio of two integers.

When writing a ratio of two integers, the second integer cannot be zero. Therefore, a rational number can always be written as

$\frac{a}{b}$ where a is an integer and b is a nonzero integer.

Every decimal number that terminates or repeats is a rational number.

Real Numbers

Every point on a number line corresponds to a real number and every real number can be plotted on a number line.

The real numbers include the rational numbers as well as those numbers which cannot be written as fractions, such as $\sqrt{2}$, π, and $0.1011011101111\cdots$.

Check Yourself ANSWERS

1. **(a)** Positive; **(b)** positive; **(c)** negative; **(d)** neither **2.** 7, 0, -5 **3.**

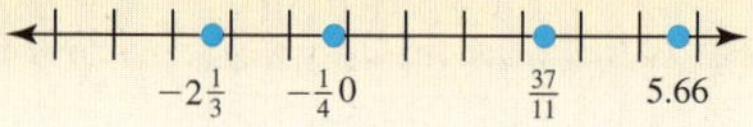

4. **(a)** $-7 < 4$; **(b)** $\frac{13}{4} = 3.25$; **(c)** $-12.08 > -12.2$ **5.** **(a)** -8; **(b)** 9 **6.** **(a)** 9; **(b)** 12; **(c)** 6; **(d)** 15

7. Resistors: #7, #3, #6, #2, #4, #1, and #5

Reading Your Text

These fill-in-the-blank exercises will help you understand some of the key vocabulary used in this section. The answers to these exercises are in the Answers Appendix in the back of the text.

(a) The ______________ numbers are the natural numbers together with zero.

(b) We indicate that a number is ______________ by placing a minus sign in front of the number.

(c) The set of ______________ consists of the natural numbers, their negatives, and zero.

(d) The ______________ of a number is its absolute value.

< Objectives 1, 3, and 4 >

Indicate whether each statement is **true** *or* **false.**

1. The opposite of -7 is 7.

2. The opposite of -10 is -10.

3. -9 is an integer.

4. 5 is an integer.

5. The opposite of -11 is 11.

6. The absolute value of -5 is 5.

7. $|-6| = -6$

8. $-(-30) = -30$

9. -12 is not an integer.

10. The opposite of -18 is 18.

11. $|-7| = -7$

12. The absolute value of -9 is -9.

13. $-(-8) = 8$

14. $\frac{2}{3}$ is not an integer.

15. $-|-15| = -15$

16. The absolute value of -3 is 3.

17. $\frac{3}{5}$ is an integer.

18. 0.7 is not an integer.

19. 0.15 is not an integer.

20. $|-9| = -9$

21. $\frac{5}{7}$ is not an integer.

22. 0.23 is not an integer.

23. $-(-7) = -7$

24. The opposite of 15 is -15.

Complete each statement.

25. The absolute value of -10 is ________.

26. $-(-12) =$ ________

27. $|-20| =$ ________ 

28. The absolute value of -12 is ________.

29. The absolute value of -7 is ________.

30. The opposite of -9 is ________.

31. The opposite of 30 is ________.

32. $-|-15| =$ ________

33. $-(-6) =$ ________

34. The absolute value of 0 is ________.

35. $|50| =$ ________

36. The opposite of 18 is ________.

37. The absolute value of the opposite of 3 is ________.

38. The opposite of the absolute value of 3 is ________.

39. The opposite of the absolute value of -7 is ________.

40. The absolute value of the opposite of -7 is ________.

Consider the numbers $-3, \frac{2}{3}, -1.5, 2,$ *and* 0.

41. Which of the numbers are integers? 

42. Which of the numbers are natural numbers?

43. Which of the numbers are whole numbers?

44. Which of the numbers are negative numbers?

Consider the numbers $-2, -\frac{4}{3}, 3.5, 0,$ *and* 1.

45. Which of the numbers are integers?

46. Which of the numbers are natural numbers?

47. Which of the numbers are whole numbers?

48. Which of the numbers are negative numbers?

< Objective 2 >

49. Plot each point on the number line.

$3, -3, -0.8, 4\frac{2}{5}, -\frac{9}{4}$

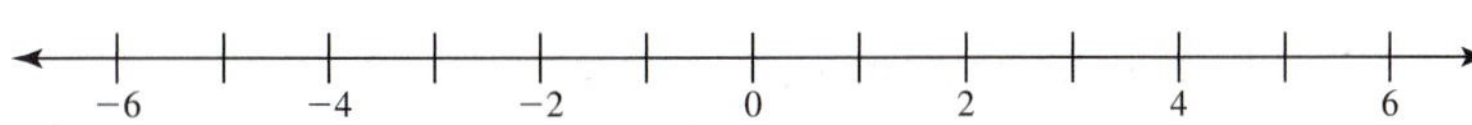

50. Plot each point on the number line.

$-1.5, \frac{3}{2}, 4, 0, -3\frac{3}{4}$

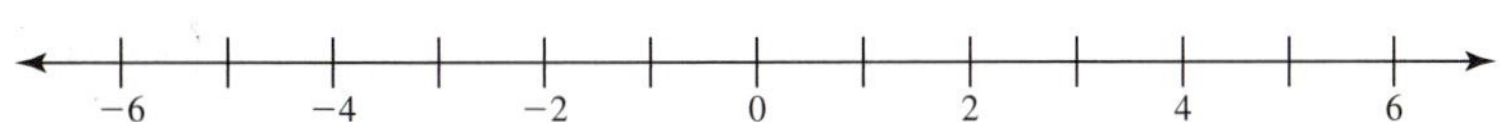

51. Plot each point on the number line.

$-5, 3.3, -2\frac{2}{3}, \frac{5}{2}, -0.85$

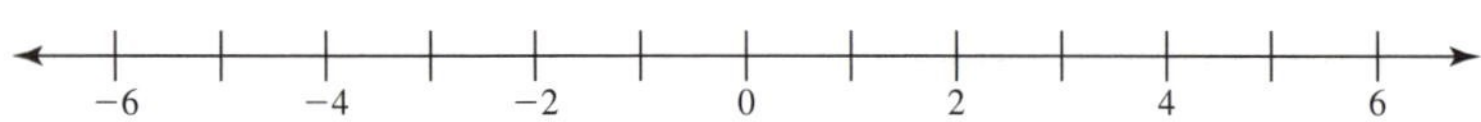

52. Plot each point on the number line.

$2, 5\frac{1}{3}, -\frac{6}{5}, -4.5, \frac{7}{8}$

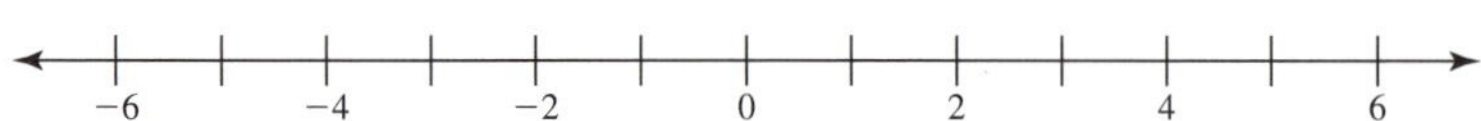

Fill in each blank with >, <, or = to make a true statement.

53. -5 ________ -9 

54. -15 ________ -10

55. -20 ________ -10

56. -15 ________ -14

57. $|3|$ ________ 3

58. $|-5|$ ________ $-(-5)$

59. -4 ________ $|-4|$

60. 7 ________ $|7|$

Use a real number to represent each quantity. Include the appropriate sign and unit with each answer.

61. **Business and Finance** A \$50 withdrawal from a checking account

62. **Business and Finance** A \$200 deposit into a savings account

63. **Science and Medicine** A 10°F temperature decrease in an hour

64. **Statistics** An eight-game losing streak by the local baseball team

65. **Social Science** A 25,000-person increase in a city's population

66. **Business and Finance** A country exported \$90,000,000 more than it imported, creating a positive trade balance.

Complete each statement with **always, sometimes,** *or* **never.**

67. The opposite of a negative number is __________ negative.

68. The absolute value of a nonzero number is __________ positive.

69. The absolute value of a number is __________ equal to that number. 

70. A rational number is __________ an integer.

Skills | Calculator/Computer | **Career Applications** | Above and Beyond

Use a real number to represent each quantity. Include the appropriate sign and unit with each answer.

71. **Agricultural Technology** The erosion of 4 in. of topsoil from an Iowa cornfield

72. **Agricultural Technology** The formation of 2.5 cm of new topsoil on the African savanna

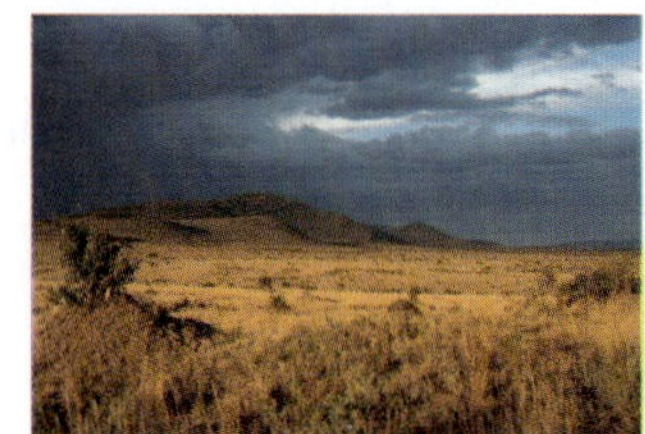

ELECTRICAL ENGINEERING Several 12-volt (V) batteries were tested using a voltmeter. The voltages were entered into a table indicating their variance from 12 V.

Battery	#1	#2	#3	#4	#5
Variance (12 V)	+1	0	−1	−3	+2

73. Use the voltages to list the batteries in ascending order.

74. Which battery had the highest voltage measurement? What was its voltage measurement?

Skills | Calculator/Computer | Career Applications | **Above and Beyond**

75. **(a)** Every number has an opposite. The opposite of 5 is −5. In English, a similar situation exists for words. For example, the opposite of *regular* is *irregular.* Write the opposite of each word.

irredeemable, uncomfortable, uninteresting, uninformed, irrelevant, immoral

(b) Note that the idea of an opposite is usually expressed by a prefix such as *un-* or *ir-*. What other prefixes can be used to negate or change the meaning of a word to its opposite? List four words using these prefixes, and use the words in a sentence.

76. **(a)** What is the difference between positive integers and nonnegative integers?

(b) What is the difference between negative and nonpositive integers?

77. Simplify each expression.

(a) $-(-3)$ **(b)** $-(-(-3))$ **(c)** $-(-(-(-3)))$

(d) Use the results of parts (a), (b), and (c) to create a rule for simplifying expressions of this type.

(e) Use the rule created in part (d) to simplify $-(-(-(-(-(-(-7))))))$.

Answers

1. True **3.** True **5.** True **7.** False **9.** False **11.** False **13.** True **15.** True **17.** False **19.** True **21.** True **23.** False **25.** 10 **27.** 20 **29.** 7 **31.** −30 **33.** 6 **35.** 50 **37.** 3 **39.** −7 **41.** −3, 2, 0 **43.** 0, 2 **45.** −2, 0, 1 **47.** 0, 1 **49.**

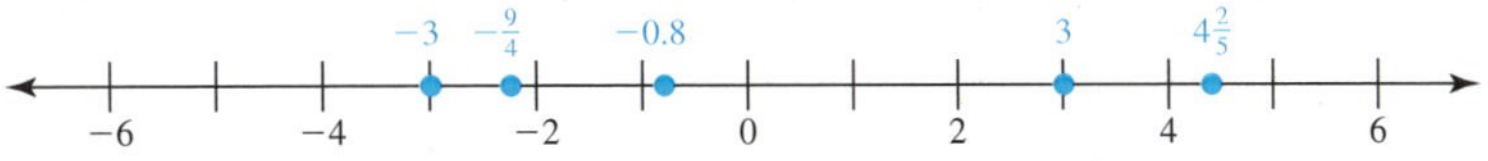

51.

53. > **55.** < **57.** = **59.** < **61.** −\$50 **63.** −10°F **65.** +25,000 people **67.** never **69.** sometimes **71.** −4 in. **73.** #4, #3, #2, #1, #5 **75.** Above and Beyond **77.** **(a)** 3; **(b)** −3; **(c)** 3; **(d)** Above and Beyond; **(e)** −7

0.3 Adding and Subtracting

< 0.3 Objectives >

1 > Add real numbers

2 > Use the properties of addition

3 > Subtract real numbers

We can use a number line to demonstrate how to add real numbers. To add a positive number, we move to the right; to add a negative number, we move to the left.

Example 1 Adding Real Numbers

< Objective 1 >

Find the sum $5 + (-2)$.

Begin 5 units *to the right* of 0. Then, to add -2, move 2 units *to the left*. We see that

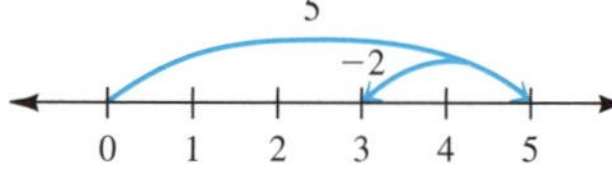

$5 + (-2) = 3$

Check Yourself 1

Find the sum.

$9 + (-7)$

We can also use a number line to picture adding two negative numbers. Example 2 illustrates this approach.

Example 2 Adding Real Numbers

NOTE

The sum of two positive numbers is positive, and the sum of two negative numbers is negative.

Find the sum $-2 + (-3)$.

Begin 2 units *to the left* of 0 (because the first number is -2). Then move 3 more units *to the left* to add negative 3. We see that

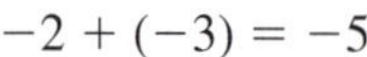

$-2 + (-3) = -5$

Check Yourself 2

Find the sum.

$-7 + (-5)$

You may have noticed some patterns from the examples. These patterns make it easier to do the work mentally.

Property

Adding Real Numbers

Case 1. If two numbers have the same sign, add their magnitudes. Give the sum the sign of the original numbers.

Case 2. If two numbers have different signs, subtract the smaller magnitude from the larger. Attach the sign of the number with the larger magnitude to the result.

RECALL

The magnitude of a number is given by its absolute value.

Example 3 Adding Real Numbers

Find each sum.

(a) $5 + 2 = 7$ The sum of two positive numbers is positive.

(b) $-2 + (-6) = -8$ Add the magnitudes of the two numbers ($2 + 6 = 8$). Give the sum the sign of the original numbers.

(c) $-4 + 7$

The numbers have different signs, so we subtract their magnitudes, the smaller from the larger.

$7 - 4 = 3$

The final result has the same sign as the number with the larger magnitude. In this case, the result is positive because 7 is positive and has a larger magnitude than -4.

$-4 + 7 = 3$

(d) $-12 + 7$

Again, we subtract their magnitudes,

$12 - 7 = 5$

This time, the result is negative because -12 has a larger magnitude than 7.

$-12 + 7 = -5$

Check Yourself 3

Find the sums.

(a) $6 + 7$ **(b)** $-8 + (-7)$

(c) $8 + (-15)$ **(d)** $-3 + 16$

There are three important parts to the study of algebra. The first is the set of numbers, which we discuss in this chapter. The second is the set of operations, such as addition and multiplication. The third is the set of rules, which we call **properties.** Example 4 enables us to look at an important property of addition.

Example 4 Adding Real Numbers

< Objective 2 >

Find each sum.

(a) $2 + (-7) = (-7) + 2 = -5$

(b) $-3 + (-4) = -4 + (-3) = -7$

In both cases the order in which we add the numbers does not affect the sum.

Check Yourself 4

Find the sums $-8 + 2$ and $2 + (-8)$. How do the results compare?

Property

The Commutative Property of Addition

The *order* in which we add two numbers does not change their sum. Addition is **commutative.** For any numbers a and b,

$$a + b = b + a$$

What if we want to add more than two numbers? Another property of addition is helpful.

Example 5 Adding Three Real Numbers

Find the sum $2 + (-3) + (-4)$. First,

$$\underbrace{[2 + (-3)]} + (-4) \quad \text{Add the first two numbers.}$$

$$= \quad -1 \quad + (-4) \quad \text{Then add the third to that sum.}$$

$$= \quad -5$$

Here is a second approach.

$$2 + \underbrace{[(-3) + (-4)]} \quad \text{This time, add the second and third numbers.}$$

$$= 2 + \quad (-7) \quad \text{Then add the first number to that sum.}$$

$$= -5$$

Check Yourself 5

Show that $-2 + (-3 + 5) = [-2 + (-3)] + 5$

Do you see that it makes no difference which way we group numbers in addition? The final sum is the same.

Property

The Associative Property of Addition

The way we *group* numbers does not change their sum. Addition is **associative.** For any numbers a, b, and c,

$$(a + b) + c = a + (b + c)$$

A number's opposite (or negative) is called its **additive inverse.** Use this rule to add opposite numbers.

Property

The Additive Inverse

The sum of any number and its additive inverse is 0. For any number a,

$$a + (-a) = 0$$

Example 6 Finding the Sum of Additive Inverses

Find each sum.

(a) $6 + (-6) = 0$

(b) $-8 + 8 = 0$

Check Yourself 6

Find the sum.

$9 + (-9)$

So far we have only added integers. The process is the same if we want to add other types of real numbers.

Example 7 Adding Real Numbers

NOTE

When your answer is a fraction, you should *always* write it in simplest terms.

Find each sum.

(a) $\frac{15}{4} + \left(-\frac{9}{4}\right) = \frac{6}{4} = \frac{3}{2}$ Subtract their magnitudes: $\frac{15}{4} - \frac{9}{4} = \frac{6}{4} = \frac{3}{2}$.

The sum is positive since $\frac{15}{4}$ has the larger magnitude.

(b) $-0.5 + (-0.2) = -0.7$ Add their magnitudes $(0.5 + 0.2 = 0.7)$. The sum is negative.

Check Yourself 7

Find each sum.

(a) $-\frac{5}{2} + \left(-\frac{7}{2}\right)$ **(b)** $5.3 + (-4.3)$

Now we turn our attention to subtraction. Subtraction is called the *inverse* operation to addition. This means that any subtraction problem can be written as a problem in addition.

Property

Subtracting Real Numbers

To subtract real numbers, add the first number and the *opposite* of the number being subtracted. By definition

$a - b = a + (-b)$

Example 8 illustrates this property.

Example 8 Subtracting Real Numbers

< Objective 3 >

(a) Subtract $5 - 3$.

$5 - 3 = 5 + (-3) = 2$ To subtract 3, we add the opposite of 3.

The opposite of 3

(b) Subtract $2 - 5$.

$2 - 5 = 2 + (-5) = -3$

The opposite of 5

NOTE

Use the subtraction property to add the opposite of 4, -4, to -3.

(c) Subtract $-3 - 4$.

$-3 - 4 = -3 + (-4) = -7$ — -4 is the opposite of 4.

(d) Subtract $-10 - 15$.

$-10 - 15 = -10 + (-15) = -25$ — -15 is the opposite of 15.

Check Yourself 8

Use the definition of subtraction to find each difference.

(a) $8 - 3$ (b) $7 - 9$
(c) $-5 - 9$ (d) $-12 - 6$

The subtraction rule works the same way when the number being subtracted is negative. Change the subtraction to addition and then replace the negative number being subtracted with its opposite, which is positive.

Example 9 Subtracting Real Numbers

Evaluate each expression.

> CAUTION

You can use a calculator to simplify the kinds of problems we encounter in this section. The negation key is [(−)] or [+/−] on the calculator. Do not confuse this with the subtraction key!

(a) $5 - (-2) = 5 + (+2) = 5 + 2 = 7$ — Change subtraction to addition and replace -2 with its opposite, $+2$ or 2.

(b) $7 - (-8) = 7 + (+8) = 7 + 8 = 15$

(c) $-9 - (-5) = -9 + 5 = -4$

(d) $-12.7 - (-3.7) = -12.7 + 3.7 = -9$

(e) $-\frac{3}{4} - \left(-\frac{7}{4}\right) = -\frac{3}{4} + \frac{7}{4} = \frac{4}{4} = 1$

(f) Subtract -4 from -5. We write

$-5 - (-4) = -5 + 4 = -1$

Check Yourself 9

Subtract.

(a) $8 - (-2)$ (b) $3 - (-10)$ (c) $-7 - (-2)$
(d) $-9.8 - (-5.8)$ (e) $7 - (-7)$

NOTE

If your calculator is different from either of the ones we describe, refer to your manual, or ask your instructor for assistance.

A calculator can be a useful tool for checking arithmetic or performing complicated computations. In order to master your calculator, you should become familiar with some of the keys.

The first key is the subtraction key, [−]. This key is usually found in the right column of calculator keys along with the other "operation" keys such as addition, multiplication, and division.

The second key to find is the one for negative numbers. On graphing calculators, it usually looks like [(−)], whereas on scientific calculators, the key usually looks like [+/−]. In either case, the negative number key is usually found in the bottom row.

One very important difference between the two types of calculators is that when using a graphing calculator, you input the negative sign before keying in the number (as it is written). When using a scientific calculator, you input the negative sign button after keying in the number.

In Example 10, we illustrate this difference, while showing that subtraction remains the same.

Example 10 Subtracting with a Calculator

Use a calculator to find each difference.

(a) $-12.43 - 3.516$

> **NOTE**
>
> Graphing calculators usually have an [ENTER] key, whereas scientific calculators have an [=] key.

Graphing Calculator

[(−)] 12.43 [−] 3.516 [ENTER] — The negative number sign comes before the number.

The display should read −15.946.

Scientific Calculator

12.43 [+/−] [−] 3.516 [=] — The negative number sign comes after the number.

The display should read −15.946.

```
-12.43-3.516
         -15.946
23.56--4.7
           28.26
■
```

(b) $23.56 - (-4.7)$

Graphing Calculator

23.56 [−] [(−)] 4.7 [ENTER] — Press the negative number key before the number.

The display should read 28.26.

Scientific Calculator

23.56 [−] 4.7 [+/−] [=] — Press the negative number key after the number.

The display should read 28.26.

Check Yourself 10

Use a calculator to find each difference.

(a) $-13.46 - 5.71$ **(b)** $-3.575 - (-6.825)$

Example 11 A Business and Finance Application

Oscar owns stock in four companies. This year, his holdings in Cisco went up $2,250; his holdings in AT&T went down $1,345; Chevron went down $5,215; and IBM went down $1,525.

How much did the total value of Oscar's holdings change during the year?

> **RECALL**
>
> We introduced this five-step problem-solving approach in Section 0.1.

Step 1 The question asks for the combined change in value of Oscar's holdings. We are given the amount each stock went up or down.

Step 2 To find the change in value, we add the increases and subtract the decreases.

Step 3 \$2,250 − \$1,345 − \$5,215 − \$1,525

Step 4 \$2,250 − \$1,345 − \$5,215 − \$1,525 = −\$5,835

Step 5 Oscar's holdings decreased in value by $5,835 during the year.

Reasonableness

Oscar lost money on three stocks including over $5,000 from one stock, so this answer seems reasonable.

Check Yourself 11

A bus with 15 people stopped at Avenue A. Nine people got off and 5 people got on. At Avenue B, 6 people got off and 8 people got on. At Avenue C, 4 people got off the bus and 6 people got on. How many people are now on the bus?

Check Yourself ANSWERS

1. 2 **2.** −12 **3.** **(a)** 13; **(b)** −15; **(c)** −7; **(d)** 13 **4.** −6 = −6 **5.** 0 = 0 **6.** 0 **7.** **(a)** −6; **(b)** 1 **8.** **(a)** 5; **(b)** −2; **(c)** −14; **(d)** −18 **9.** **(a)** 10; **(b)** 13; **(c)** −5; **(d)** −4; **(e)** 14 **10.** **(a)** −19.17; **(b)** 3.25 **11.** 15 people

Reading Your Text

These fill-in-the-blank exercises will help you understand some of the key vocabulary used in this section. The answers to these exercises are in the Answers Appendix in the back of the text.

(a) If two numbers have the same sign, add their ______________ and then give the sum the sign of the original numbers.

(b) The ______________ in which we add two numbers does not change their sum.

(c) Addition is ______________. For any numbers a, b, and c, $(a + b) + c = a + (b + c)$.

(d) The sum of any number and its additive inverse is ______________.

0.3 exercises

Skills | Calculator/Computer | Career Applications | Above and Beyond

< Objectives 1–3 >

Evaluate each expression.

1. $-6 + (-5)$

2. $3 + 9$

3. $11 + (-7)$

4. $-6 + (-7)$

5. $4 + (-6)$

6. $9 + (-2)$

7. $7 + 9$

8. $-7 + 11$

9. $(-11) + 5$

10. $5 + (-8)$

11. $-8 + (-7)$ VIDEO

12. $8 + (-7)$

13. $-12 + 4$

14. $7 + (-7)$

15. $-9 + 10$

16. $-6 + 8$

17. $-4 + 4$

18. $5 + (-20)$

19. $7 + (-13)$

20. $0 + (-10)$

21. $-8 + 5$

22. $-7 + 3$

23. $6 + (-6)$

24. $-9 + 9$

25. $\frac{45}{16} - \frac{9}{16}$

26. $-\frac{35}{16} + \frac{17}{16}$

27. $\frac{29}{8} + \left(-\frac{17}{8}\right)$

28. $-\frac{81}{20} + \left(-\frac{107}{20}\right)$

29. $-\frac{73}{16} + \frac{119}{16}$

30. $-\frac{13}{8} - \left(-\frac{15}{4}\right)$

31. $4 + (-7) + (-5)$

32. $-7 + 8 + (-6)$

33. $-2 + (-6) + (-4)$

34. $12 + (-6) + (-4)$

35. $-3 + (-7) + 5 + (-2)$

36. $7 + (-8) + (-9) + 10$

37. $11 - 13$ **38.** $7 - 5$ **39.** $9 - 3$ **40.** $4 - 9$

41. $-8 - 3$ VIDEO **42.** $-13 - 8$ **43.** $-12 - 8$ **44.** $9 - 15$

45. $-2 - (-3)$ **46.** $-9 - (-6)$ **47.** $-5 - (-5)$ **48.** $9 - (-7)$

49. $28 - (-22)$ **50.** $50 - (-25)$ **51.** $-15 - (-25)$ VIDEO **52.** $-20 - (-30)$

53. $-25 - (-15)$ **54.** $-30 - (-20)$ **55.** $-(-20) - (-15)$ **56.** $18 - (-12)$

57. $48 - (-15)$ **58.** $-25 - (-30)$ **59.** $\frac{10}{2} - \left(-\frac{7}{2}\right)$ **60.** $-\frac{4}{2} - \frac{3}{2}$

61. $-\frac{7}{8} - \left(-\frac{19}{8}\right)$ **62.** $\frac{13}{4} - \left(-\frac{7}{4}\right)$ **63.** $-7 - (-5) - 6$ **64.** $-5 - (-8) - 10$

65. $-10 - 8 - (-7)$ **66.** $-5 - 8 - (-15)$

Solve each application.

67. **SCIENCE AND MEDICINE** The temperature in Chicago dropped from 18°F at 4 P.M. to −9°F at midnight. What was the drop in temperature?

68. **BUSINESS AND FINANCE** Charley's checking account had $175 deposited at the beginning of the month. After he wrote checks for the month, the account was $95 *overdrawn.* What amount of checks did he write during the month?

69. **TECHNOLOGY** Micki entered the elevator on the 34th floor. From that point the elevator went up 12 floors, down 27 floors, down 6 floors, and up 15 floors before she got off. On what floor did she get off the elevator? VIDEO

70. **TECHNOLOGY** A submarine dives to a depth of 500 ft below the ocean's surface. It then dives another 217 ft before climbing 140 ft. What is the depth of the submarine?

71. **TECHNOLOGY** A helicopter is 600 ft above sea level, and a submarine directly below it is 325 ft below sea level. How far apart are they?

72. **BUSINESS AND FINANCE** Tom has received an overdraft notice from the bank telling him that his account is overdrawn by $142. How much must he deposit in order to have $625 in his account?

73. **SCIENCE AND MEDICINE** At 9:00 A.M., Jose had a temperature of 99.8°F. It rose another 2.5° before falling 3.7° by 1:00 P.M. What was his temperature at 1:00 P.M.?

74. **BUSINESS AND FINANCE** Olga has $250 in her checking account. She deposits $52 and then writes a check for $77. What is her new balance?

75. **STATISTICS** Ezra's scores on five tests taken in a mathematics class were 87, 71, 95, 81, and 90. What was the difference between the highest and the lowest of his scores?

76. **BUSINESS AND FINANCE** Aaron had $769 in his bank account on June 1. He deposited $125 and $986 during the month and wrote checks for $235, $529, and $712 during June. What was his balance at the end of the month?

Complete each statement with **always, sometimes,** *or* **never.**

77. The sum of two negative numbers is __________ negative.

78. The difference of two negative numbers is __________ negative.

79. The additive inverse of a negative number is __________ negative.

80. The sum of a number and its additive inverse is __________ zero.

Skills | **Calculator/Computer** | Career Applications | Above and Beyond

Use a calculator to find each difference.

81. $-11.392 - 13.491$

82. $-9.245 - 14.316$

83. $-7.259 - 4.235$

84. $-6.319 - 2.628$

85. $-18.271 - (-12.569)$

86. $-15.586 - (-9.874)$

87. $-17.346 - (-28.293)$

88. $-11.358 - (-23.145)$

Skills | Calculator/Computer | Career Applications | **Above and Beyond**

89. Complete the problem "4 − (−9) is the same as ____________." Write an application problem that might be answered using this subtraction.

90. Explain the difference between the two phrases "7 less than a number" and "a number subtracted from 7." Use both symbols and English to explain the meanings of these phrases. Write some other ways of expressing subtraction in English.

91. Construct an example to show that subtraction is *not* commutative.

92. Construct an example to show that subtraction is *not* associative.

93. **Calculators** If you use a scientific calculator, complete exercise (a); if you use a graphing calculator, complete exercise (b).

(a) **Scientific Calculator** Explain the difference between the [+/−] key and the [−] key on your calculator. Discuss what each key is used for and how to use the key.

(b) **Graphing Calculator** Explain the difference between the [(−)] key and the [−] key on your calculator. Discuss what each key is used for and how to use the key.

94. Do you think this statement is true?

$|a + b| = |a| + |b|$ for all numbers a and b

When we don't know whether such a statement is true, we refer to the statement as a **conjecture.** We may "test" the conjecture by substituting specific numbers for the letters.

Test the conjecture, using two positive numbers for a and b.

Test again, using a positive number for a and 0 for b.

Test again, using two negative numbers.

Now try using one positive number and one negative number.

Summarize your results in a rule that you feel is true.

95. If a represents a positive number and b represents a negative number, determine which expressions are positive and which are negative.

(a) $|b| + a$ **(b)** $b + (-a)$ **(c)** $(-b) + a$ **(d)** $-b + |-a|$

Answers

1. −11 **3.** 4 **5.** −2 **7.** 16 **9.** −6 **11.** −15 **13.** −8 **15.** 1 **17.** 0 **19.** −6 **21.** −3 **23.** 0 **25.** $\frac{9}{4}$ **27.** $\frac{3}{2}$ **29.** $\frac{23}{8}$ **31.** −8 **33.** −12 **35.** −7 **37.** −2 **39.** 6 **41.** −11 **43.** −20 **45.** 1 **47.** 0 **49.** 50 **51.** 10 **53.** −10 **55.** 35 **57.** 63 **59.** $\frac{17}{2}$ **61.** $\frac{3}{2}$ **63.** −8 **65.** −11 **67.** 27°F **69.** 28th floor **71.** 925 ft **73.** 98.6°F **75.** 24 points **77.** always **79.** never **81.** −24.883 **83.** −11.494 **85.** −5.702 **87.** 10.947 **89.** Above and Beyond **91.** Above and Beyond **93.** **(a)** Above and Beyond; **(b)** Above and Beyond **95.** **(a)** Positive; **(b)** negative; **(c)** positive; **(d)** positive

0.4 Multiplying and Dividing

< 0.4 Objectives >

1 > Multiply real numbers

2 > Use the properties of multiplication

3 > Divide real numbers

Multiplication can be thought of as repeated addition. That is, we can interpret

$3 \times 4 = 4 + 4 + 4 = 12$

We can use this interpretation, together with our work in Section 0.3, to find the product of two real numbers.

Example 1 Multiplying Real Numbers

< Objective 1 >

NOTE

We use parentheses ( ) to indicate multiplication when there are negative numbers.

Multiply.

(a) $(3)(-4) = (-4) + (-4) + (-4) = -12$

(b) $(4)\left(-\frac{1}{3}\right) = \left(-\frac{1}{3}\right) + \left(-\frac{1}{3}\right) + \left(-\frac{1}{3}\right) + \left(-\frac{1}{3}\right) = -\frac{4}{3}$

Check Yourself 1

Find the product by writing the expression as repeated addition.

$(4)(-3)$

Looking at the products we found by repeated addition in Example 1 should suggest our first rule for multiplying real numbers.

Property

Multiplying Real Numbers

Case 1 The product of two numbers with different signs is negative.

RECALL

The absolute value of a number is the same as its magnitude.

The rule is easy to use. To multiply two numbers with different signs, just multiply their absolute values and attach a negative sign to the product.

Example 2 Multiplying Real Numbers

Find each product.

$(5)(-6) = -30$

$(10)(-12) = -120$

$$(-7)\left(\frac{1}{8}\right) = -\frac{7}{8}$$

$$(1.5)(-0.3) = -0.45$$

$$\left(-\frac{5}{8}\right)\left(\frac{4}{15}\right) = -\left(\frac{\overset{1}{\cancel{5}}}{\underset{2}{\cancel{8}}} \times \frac{\overset{1}{\cancel{4}}}{\underset{3}{\cancel{15}}}\right)$$

$$= -\frac{1}{6}$$

RECALL

$$-\frac{5}{8} \times \frac{4}{15} = -\frac{\cancel{5} \times \cancel{2} \times \cancel{2}}{\cancel{2} \times \cancel{2} \times 2 \times 3 \times \cancel{5}}$$

$$= -\frac{1}{6}$$

Check Yourself 2

Find each product.

(a) $(15)(-5)$ **(b)** $(-0.8)(0.2)$ **(c)** $\left(-\frac{2}{3}\right)\left(\frac{6}{7}\right)$

The product of two negative numbers is harder to visualize. The pattern below may help you see how we can determine the sign of the product.

$$(3)(-2) = -6$$
$$(2)(-2) = -4$$
$$(1)(-2) = -2$$
$$(0)(-2) = 0$$
$$(-1)(-2) = 2$$
$$(-2)(-2) = 4$$

Do you see that the product *increases* by 2 each time the first factor *decreases* by 1?

RECALL

We already know that the product of two positive numbers is positive.

This suggests that the product of two negative numbers is positive, which is, in fact, the case.

Property

Multiplying Real Numbers

Case 2 The product of two numbers with the same sign is positive.

Example 3 Multiplying Real Numbers

> CAUTION

$(-9)(-6)$ tells you to multiply. The parentheses are *next to* one another. The expression $-9 - 6$ tells you to subtract. The numbers are *separated* by the operation sign.

Find each product.

$8 \times 7 = 56$ Each pair of numbers has the same sign, so each product is positive.

$(-9)(-6) = 54$

$(-0.5)(-2) = 1$

Check Yourself 3

Find each product.

(a) 5×7 **(b)** $(-8)(-6)$ **(c)** $(9)(-6)$ **(d)** $(-1.5)(-4)$

To multiply more than two real numbers, apply the multiplication rule repeatedly.

Example 4 Multiplying a Group of Real Numbers

NOTE

The original expression has an odd number of negative signs. Do you see that having an odd number of negative factors always results in a negative product?

Multiply.

$(5)(-7)(-3)(-2)$ $\quad (5)(-7) = -35$

$= (-35)(-3)(-2)$ $\quad (-35)(-3) = 105$

$= \quad (105)(-2)$

$= \quad -210$

Check Yourself 4

Find the product.

$(-4)(3)(-2)(5)$

In Section 0.3, we saw that the commutative and associative properties for addition can be extended to real numbers. The same is true for multiplication.

Example 5 Using the Commutative Property of Multiplication

< Objective 2 >

Find each product.

$(-5)(7) = (7)(-5) = -35$

$(-6)(-7) = (-7)(-6) = 42$

The order in which we multiply does not affect the product.

Check Yourself 5

Show that $(-8)(-5) = (-5)(-8)$.

Property

The Commutative Property of Multiplication

The order in which we multiply does not change the product. Multiplication is *commutative*. For any real numbers *a* and *b*,

$a \cdot b = b \cdot a$ $\quad$ The centered dot represents multiplication.

What about the way we group numbers in multiplication?

Example 6 Using the Associative Property of Multiplication

NOTE

The symbols [ ] are called *brackets* and are used to group numbers in the same way as parentheses.

Multiply.

$[(3)(-7)](-2)$	or	$(3)[(-7)(-2)]$
$= (-21)(-2)$		$= (3)(14)$
$= 42$		$= 42$

We group the first two numbers on the left and the second two numbers on the right. The product is the same in either case.

Check Yourself 6

Show that $[(2)(-6)](-3) = (2)[(-6)(-3)]$.

Property

The Associative Property of Multiplication

The way we *group* the numbers does not change the product. Multiplication is *associative*. For any real numbers *a*, *b*, and *c*,

$(a \cdot b) \cdot c = a \cdot (b \cdot c)$

RECALL

The area of a rectangle is the product of its length and width.

$A = L \cdot W$

Another important property in mathematics is the **distributive property.** The distributive property involves addition and multiplication together. We can illustrate the property with an application.

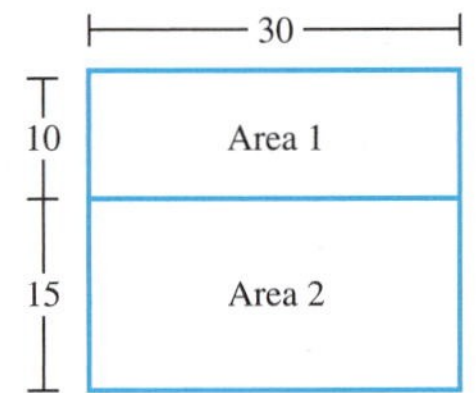

We can find the total area by multiplying the length by the overall width, which is found by adding the two widths.

or

We can find the total area as a sum of the two areas.

Length · Overall width

$30 \cdot (10 + 15)$

$= 30 \cdot 25$

$= 750$

(Area 1) Length · Width + (Area 2) Length · Width

$30 \cdot 10 + 30 \cdot 15$

$= 300 + 450$

$300 + 450 = 750$

So

$30 \cdot (10 + 15) = 30 \cdot 10 + 30 \cdot 15$

This leads us to the distributive property.

Property

The Distributive Property

If *a*, *b*, and *c* are any numbers,

$a \cdot (b + c) = a \cdot b + a \cdot b$ and $(a + b) \cdot c = a \cdot c + b \cdot c$

Example 7 Using the Distributive Property

NOTES

It is also true that

$5(3 + 4) = 5 \cdot (7) = 35$

It is also true that

$\frac{1}{3}(9 + 12) = \frac{1}{3}(21) = 7$

Use the distributive property to simplify (remove the parentheses).

(a) $5(3 + 4)$

$5(3 + 4) = 5 \cdot 3 + 5 \cdot 4 = 15 + 20 = 35$

(b) $\frac{1}{3}(9 + 12) = \frac{1}{3} \cdot 9 + \frac{1}{3} \cdot 12$

$= 3 + 4 = 7$

Check Yourself 7

Use the distributive property to simplify (remove the parentheses).

(a) $4(6 + 7)$ **(b)** $\frac{1}{5}(10 + 15)$

Many students have difficulty applying the distributive property when negative numbers are involved. One key to applying the property correctly is to remember that the sign of a number "travels" with that number.

Example 8 Applying the Distributive Property with Negative Numbers

RECALL

We usually enclose negative numbers in parentheses in the middle of an expression to avoid careless errors.

We use brackets rather than nesting parentheses to avoid careless errors.

Use the distributive property to evaluate each expression.

(a) $-7(3 + 6) = (-7) \cdot 3 + (-7) \cdot 6$ Apply the distributive property.

$= -21 + (-42)$ Multiply first, then add.

$= -63$

(b) $-3(5 - 6) = -3[5 + (-6)]$ First change the subtraction to addition.

$= (-3) \cdot 5 + (-3)(-6)$ Distribute the -3.

$= -15 + 18$ Multiply first, then add.

$= 3$

(c) $5(-2 - 6) = 5[-2 + (-6)]$

$= 5 \cdot (-2) + 5 \cdot (-6)$

$= -10 + (-30)$

$= -40$ The sum of two negative numbers is negative.

Check Yourself 8

Use the distributive property to evaluate each expression.

(a) $-2(-3 + 5)$ **(b)** $4(-3 + 6)$ **(c)** $-7(-3 - 8)$

We conclude our discussion of multiplication with a detailed explanation of why the product of two negative numbers must be positive.

Property

The Product of Two Negative Numbers

This argument shows why the product of two negative numbers is positive.

From our earlier work, we know that a number added to its opposite is 0. $5 + (-5) = 0$

Multiply both sides of the statement by -3. $(-3)[5 + (-5)] = (-3)(0)$

A number multiplied by 0 is 0, so on the right we have 0. $(-3)[5 + (-5)] = 0$

We can now use the distributive property on the left. $(-3)(5) + (-3)(-5) = 0$

We know that $(-3)(5) = -15$, so the statement becomes $-15 + (-3)(-5) = 0$

We now have a statement of the form $-15 + \square = 0$. This asks, What number must we add to -15 to get 0, where $\square$ is the value of $(-3)(-5)$? The answer is, of course, 15. This means that

$(-3)(-5) = 15$ The product must be positive.

No matter what numbers we use in the argument, the product of two negative numbers is always positive.

RECALL

We interpret a fraction as division.

$\frac{12}{3} = 12 \div 3$

Multiplication and division are related operations, so every division problem can be stated as an equivalent multiplication problem.

$8 \div 4 = 2$ because $8 = 4 \cdot 2$

$\frac{12}{3} = 4$ because $12 = 3 \cdot 4$

The operations are related, so the rules of signs for multiplication are also true for division.

Property

Dividing Real Numbers

Case 1 If two numbers have the same sign, their quotient is positive.

Case 2 If two numbers have different signs, their quotient is negative.

As you would expect, division with fractions or decimals uses the same rules for signs. Example 9 illustrates this.

Example 9 Dividing Real Numbers

< Objective 3 >

RECALL

$$\frac{3}{5} \cdot \frac{20}{9} = \frac{\cancel{3} \cdot 2 \cdot 2 \cdot \cancel{5}}{\cancel{5} \cdot \cancel{3} \cdot 3} = \frac{4}{3}$$

Divide.

$$\left(-\frac{3}{5}\right) \div \left(-\frac{9}{20}\right) = \frac{\overset{1}{\cancel{3}}}{\underset{1}{\cancel{5}}} \cdot \frac{\overset{4}{\cancel{20}}}{\underset{3}{\cancel{9}}} = \frac{4}{3}$$

The quotient of two negative numbers is positive, so we omit the negative signs and simply invert the divisor and multiply.

Check Yourself 9

Find each quotient.

(a) $-\frac{5}{8} \div \frac{3}{4}$ **(b)** $-4.2 \div (-0.6)$

We must be careful when 0 is involved in a division problem. Remember that 0 divided by any nonzero number is 0. However, division *by* 0 is not allowed and is described as *undefined*.

Example 10 Division and Zero

NOTE

An expression like $-9 \div 0$ has no meaning. There is no answer to the problem. Just write "undefined."

Divide.

(a) $0 \div 7 = 0$

(b) $\frac{0}{-4} = 0$

(c) $-9 \div 0$ is undefined.

(d) $\frac{-5}{0}$ is undefined.

Check Yourself 10

Find each quotient, if possible.

(a) $\frac{0}{-7}$ **(b)** $\frac{-12}{0}$

You can use a calculator to confirm your results from Example 10, as we do in Example 11.

Example 11 Dividing with a Calculator

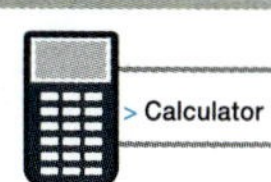


```
-12.567/0
```

Use a calculator to find each quotient.

(a) $\dfrac{-12.567}{0}$

The keystroke sequence on a graphing calculator

[(−)] 12.567 [÷] 0 [ENTER]

results in a "Divide by 0" error message. The calculator recognizes that it cannot divide by zero.

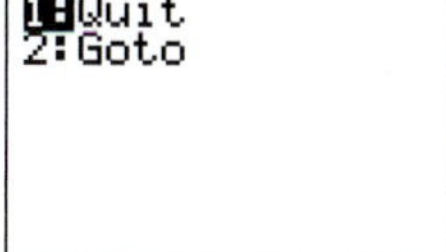

On a scientific calculator, 12.567 [+/−] [÷] 0 [=] results in an error message.

(b) $-10.992 \div -4.58$

The keystroke sequence

[(−)] 10.992 [÷] [(−)] 4.58 [ENTER]

or 10.992 [+/−] [÷] 4.58 [+/−] [=]

yields 2.4.

```
-12.567/0
-10.992/-4.58
              2.4
```

Check Yourself 11

Find each quotient.

(a) $\dfrac{-31.44}{6.55}$ **(b)** $-23.6 \div 0$

Keep in mind, a calculator is a useful tool when performing computations. However, it is only a tool. The real work should be yours. You should not rely only on a calculator.

You need to have a good sense of whether an answer is reasonable, especially a calculator-derived answer. We all commit "typos" by pressing the wrong button now and again. You need to be able to look at a calculator answer and determine when it is unreasonable, indicating that you made a mistake entering the operation.

We recommend that you perform all computations by hand and then use the calculator to check your work.

Recall that a negative sign indicates the opposite of the number that follows. For instance, the opposite of 5 is -5, whereas the opposite of -5 is 5. This last instance can be translated as $-(-5) = 5$.

Also recall that any number must correspond to some point on a number line. That is, any nonzero number is either positive or negative. No matter how many negative signs a quantity has, you can always simplify it so that it is represented by a positive or a negative number (zero or one negative sign).

Example 12 Simplifying Real Numbers

Simplify each expression.

(a) $-(-(-(-4)))$

The opposite of -4 is 4, so $-(-4) = 4$.

NOTES

You should see a pattern emerge. An even number of negative signs gives a positive number, whereas an odd number of negative signs produces a negative number.

In this text, we generally choose to write negative fractions with the sign outside the fraction, such as $-\frac{1}{2}$.

The opposite of 4 is -4, so $-(-(-4)) = -4$. The opposite of this last number, -4, is 4, so $-(-(-(-4))) = 4$

(b) $-\frac{-3}{4}$

This is the opposite of $\frac{-3}{4}$ which is $\frac{3}{4}$, a positive number.

Check Yourself 12

Simplify each expression.

(a) $-(-(-(-(-(-(-12))))))$ **(b)** $-\frac{-2}{-3}$

Example 13 A Finance Application

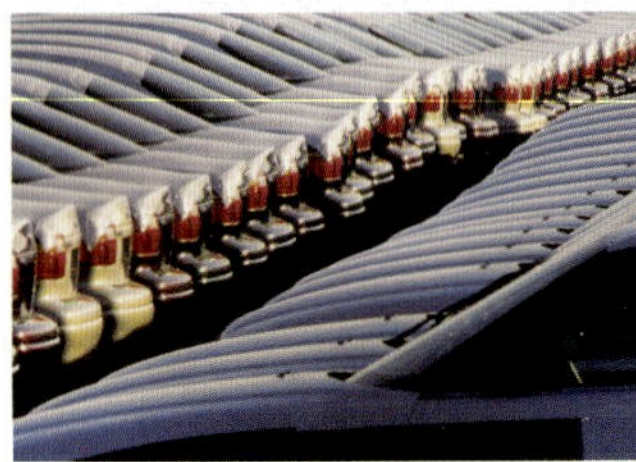

Bernal intends to purchase a new car for \$18,950. He will make a down payment of \$1,000 and agrees to make payments over a 48-month period. The total interest is \$8,546. What will his monthly payments be?

Step 1 We are trying to find the monthly payments.

Step 2 First, we subtract the down payment. Then we add the interest to that result. Finally, we divide that total by the 48 months.

Step 3 \$18,950 − \$1,000 = \$17,950 Subtract the down payment.

\$17,950 + \$8,546 = \$26,496 Add the interest.

Step 4 \$26,496 ÷ 48 = \$552 Divide that total by the 48 months.

Step 5 The monthly payments are \$552, which seems reasonable.

Check Yourself 13

One \$13 bag of fertilizer covers 310 sq ft. What does it cost to cover 7,130 sq ft?

Check Yourself ANSWERS

1. $(-3) + (-3) + (-3) + (-3) = -12$ **2.** **(a)** -75; **(b)** -0.16; **(c)** $-\frac{4}{7}$ **3.** **(a)** 35; **(b)** 48; **(c)** -54; **(d)** 6 **4.** 120 **5.** 40 = 40 **6.** 36 = 36 **7.** **(a)** 52; **(b)** 5 **8.** **(a)** -4; **(b)** 12; **(c)** 77 **9.** **(a)** $-\frac{5}{6}$; **(b)** 7 **10.** **(a)** 0; **(b)** undefined **11.** **(a)** -4.8; **(b)** undefined **12.** **(a)** -12; **(b)** $-\frac{2}{3}$ **13.** \$299

Reading Your Text

These fill-in-the-blank exercises will help you understand some of the key vocabulary used in this section. The answers to these exercises are in the Answers Appendix in the back of the text.

(a) ________ can be thought of as repeated addition.

(b) The product of two nonzero numbers with ________ signs is always negative.

(c) The product of two nonzero numbers with the same sign is ________.

(d) The ________ of a rectangle can be found by taking the product of its length and width.

Skills | Calculator/Computer | Career Applications | Above and Beyond

0.4 exercises

< Objectives 1 and 2 >

Multiply.

1. $7 \cdot 8$ **2.** $(6)(-12)$ **3.** $(4)(-3)$ **4.** $15 \cdot 5$

5. $(-8)(9)$ **6.** $(-8)(3)$ **7.** $(-5)\left(\frac{1}{3}\right)$ **8.** $(-12)(-2)$

9. $(-10)(0)$ **10.** $(10)(-10)$ **11.** $(-8)(-8)$ **12.** $(0)(-50)$

13. $(-4)\left(\frac{5}{7}\right)$ VIDEO **14.** $(-25)(-8)$ **15.** $(-9)(-12)$ VIDEO **16.** $(-3)(-27)$

17. $(-20)(1)$ **18.** $(1)(-30)$ **19.** $(-1.3)(6)$ **20.** $(-25)(5)$

21. $(-10)(-15)$ **22.** $(-2.4)(0.2)$ VIDEO **23.** $\left(-\frac{7}{10}\right)\left(-\frac{5}{14}\right)$ **24.** $\left(-\frac{7}{20}\right)\left(\frac{10}{21}\right)$

25. $\left(\frac{3}{5}\right)\left(-\frac{10}{27}\right)$ **26.** $\left(-\frac{15}{4}\right)(0)$ **27.** $\left(-\frac{5}{8}\right)\left(-\frac{4}{15}\right)$ **28.** $\left(-\frac{8}{21}\right)\left(-\frac{7}{4}\right)$

29. $(-5)(3)(-8)$ **30.** $(4)(-3)(-5)$ **31.** $(-5)(-9)(-3)$ **32.** $(-7)(-5)(-2)$

33. $(2)(-5)(-3)(-5)$ **34.** $(-2)(-5)(-5)(-6)$ **35.** $(-4)(-3)(-6)(-2)$ **36.** $(-8)(3)(-2)(5)$

Use the distributive property to remove parentheses and simplify.

37. $5(-6 + 9)$ **38.** $12(-5 + 9)$ **39.** $-8(-9 + 15)$ **40.** $-11(-8 + 3)$

41. $-4(-5 - 3)$ **42.** $-2(-7 - 11)$ **43.** $-4(-6 - 3)$ **44.** $-6(-3 - 2)$

< Objective 3 >

Divide.

45. $15 \div (-3)$ **46.** $\frac{90}{18}$ **47.** $\frac{54}{9}$ **48.** $-20 \div (-2)$

49. $\frac{-50}{5}$ **50.** $-36 \div 6$ **51.** $\frac{-24}{-3}$ **52.** $\frac{42}{-6}$

53. $\frac{90}{-6}$ VIDEO **54.** $70 \div (-10)$ **55.** $18 \div (-1)$ **56.** $\frac{-250}{-25}$

57. $\frac{0}{-9}$ **58.** $\frac{-12}{0}$ **59.** $-180 \div (-15)$ **60.** $\frac{0}{-10}$

61. $-7 \div 0$ **62.** $\frac{-25}{-1}$ **63.** $\frac{-150}{6}$ **64.** $\frac{-80}{-16}$

65. $-45 \div (-9)$ **66.** $-\frac{2}{3} \div \frac{4}{9}$ **67.** $-\frac{8}{11} \div \frac{18}{55}$ **68.** $(-8) \div (-4)$

69. $\frac{7}{10} \div \left(-\frac{14}{25}\right)$ **70.** $\frac{6}{13} \div \left(-\frac{18}{39}\right)$ **71.** $\frac{-75}{15}$ **72.** $-\frac{5}{8} \div \left(-\frac{5}{16}\right)$

Solve each application.

73. Science and Medicine A patient lost 42 pounds (lb). If he lost 3 lb each week, how long has he been dieting?

74. Business and Finance Patrick worked all day mowing lawns and was paid \$9 per hour. If he had \$125 at the end of a 9-hour day, how much did he have before he started working?

75. Business and Finance A 4.5-lb bag of greens costs \$8.91. What is the cost per pound?

76. Business and Finance Suppose that you and two friends bought equal shares of an investment for a total of \$20,000 and sold it later for \$16,232. How much did each person lose?

77. Science and Medicine Suppose that the temperature outside is dropping at a constant rate. At noon, the temperature is 70°F and it drops to 58°F at 5:00 P.M. How much did the temperature change each hour?

VIDEO

78. **Science and Medicine** A chemist has 84 ounces (oz) of a solution. She pours the solution into test tubes. Each test tube holds $\frac{2}{3}$ oz. How many test tubes can she fill?

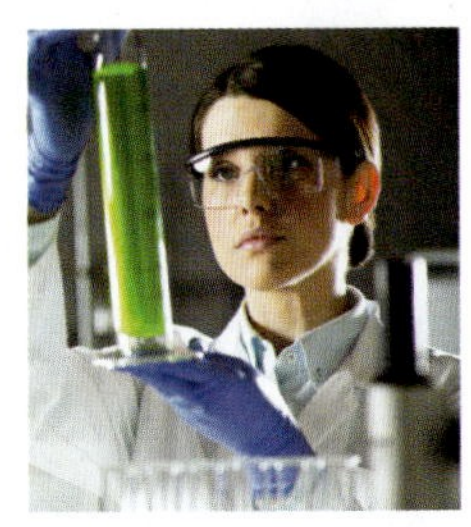

To evaluate an expression involving a fraction (indicating division), we evaluate the numerator and then the denominator. We then divide the numerator by the denominator as the last step. Using this approach, find the value of each expression.

79. $\frac{5 - 15}{2 + 3}$

80. $\frac{4 - (-8)}{2 - 5}$

81. $\frac{-6 + 18}{-2 - 4}$

82. $\frac{-4 - 21}{3 - 8}$

83. $\frac{(5)(-12)}{(-3)(5)}$

84. $\frac{(-8)(-3)}{(2)(-4)}$

Determine whether each statement is **true** *or* **false.**

85. A number divided by 0 is 0.

86. The product of 0 and any number is 0.

Complete each statement with **always, sometimes,** *or* **never.**

87. The product of three negative numbers is __________ positive.

88. The quotient of a positive number and a negative number is __________ negative.

Skills | **Calculator/Computer** | Career Applications | Above and Beyond

Use a calculator to divide. Round your answers to the nearest thousandth.

89. $-5.634 \div 2.398$

90. $-2.465 \div 7.329$

91. $-18.137 \div (-5.236)$

92. $-39.476 \div (-17.629)$

93. $32.245 \div (-48.298)$

94. $43.198 \div (-56.249)$

Skills | Calculator/Computer | Career Applications | **Above and Beyond**

95. Create an example to show that the division of real numbers is *not* commutative.

96. Create an example to show that the division of real numbers is *not* associative.

97. Here is another conjecture to consider:

$|ab| = |a|\,|b|$ for all numbers a and b

(See exercise 94 in Section 0.3 concerning testing a conjecture.) Test this conjecture for various values of a and b. Use positive numbers, negative numbers, and 0. Summarize your results in a rule.

98. Use a calculator (or mental calculations) to evaluate each expression.

$\frac{5}{0.1}, \frac{5}{0.01}, \frac{5}{0.001}, \frac{5}{0.0001}, \frac{5}{0.00001}$

In this series of problems, while the numerator is always 5, the denominator is getting smaller (and is getting closer to 0). As this happens, what is happening to the value of the fraction?

Write an argument that explains why $\frac{5}{0}$ could not have any finite value.

Answers

1. 56 **3.** −12 **5.** −72 **7.** $-\frac{5}{3}$ **9.** 0 **11.** 64 **13.** $-\frac{20}{7}$ **15.** 108 **17.** −20 **19.** −7.8 **21.** 150 **23.** $\frac{1}{4}$ **25.** $-\frac{2}{9}$ **27.** $\frac{1}{6}$ **29.** 120 **31.** −135 **33.** −150 **35.** 144 **37.** 15 **39.** −48 **41.** 32 **43.** 36 **45.** −5 **47.** 6 **49.** −10 **51.** 8 **53.** −15 **55.** −18 **57.** 0 **59.** 12 **61.** Undefined **63.** −25 **65.** 5 **67.** $-\frac{20}{9}$ **69.** $-\frac{5}{4}$ **71.** −5 **73.** 14 weeks **75.** $1.98 **77.** −2.4°F **79.** −2 **81.** −2 **83.** 4 **85.** False **87.** never **89.** −2.349 **91.** 3.464 **93.** −0.668 **95.** Above and Beyond **97.** Above and Beyond

0.5 Exponents and Order of Operations

< 0.5 Objectives >

1 > Write a product of like factors in exponential form

2 > Evaluate exponential expressions

3 > Use the order of operations to evaluate expressions

In Section 0.4, we noted that multiplication is a form for repeated addition. For example, an expression with repeated addition, such as

$3 + 3 + 3 + 3 + 3$

can be written as

$5 \cdot 3$

Thus, multiplication is "shorthand" for repeated addition.

In algebra, we frequently have a number or variable that is multiplied several times. For instance, we might have

$5 \cdot 5 \cdot 5$

To abbreviate this product, we write

$5 \cdot 5 \cdot 5 = 5^3$

> CAUTION

Be careful: 5^3 is *not* the same as $5 \cdot 3$.
$5^3 = 5 \cdot 5 \cdot 5 = 125$ and
$5 \cdot 3 = 15$

This is called **exponential notation** or an **exponential expression.** The exponent or power, here 3, indicates the number of times that the factor or base, here 5, appears in a product.

$5 \cdot 5 \cdot 5 = 5^3$

Exponent or power (the 3); Factor or base (the 5)

Example 1 Writing Exponential Expressions

< Objective 1 >

(a) Write $3 \cdot 3 \cdot 3 \cdot 3$ in exponential form.

The number 3 appears 4 times in the product, so

$3 \cdot 3 \cdot 3 \cdot 3 = 3^4$ ← Four factors of 3

This is read "3 to the fourth power."

(b) Write $10 \cdot 10 \cdot 10$ in exponential form.

Since 10 appears 3 times in the product, you can write

$10 \cdot 10 \cdot 10 = 10^3$

This is read "10 to the third power" or "10 cubed."

Check Yourself 1

Write in exponential form.

(a) $4 \cdot 4 \cdot 4 \cdot 4 \cdot 4 \cdot 4$ **(b)** $10 \cdot 10 \cdot 10 \cdot 10$

When evaluating a number raised to a power, it is important to note whether there is a sign attached to the number.

$(-2)^4 = (-2)(-2)(-2)(-2) = 16$

whereas,

$-2^4 = -(2)(2)(2)(2) = -16$

Example 2 Evaluating Exponential Expressions

< Objective 2 >

NOTE

$-3^4 = -1 \cdot 3^4$
$= -1 \cdot 81$
$= -81$

whereas,

$(-3)^4 = (-3)(-3)(-3)(-3)$
$= 81$

Evaluate each expression.

(a) $(-3)^3 = (-3)(-3)(-3) = -27$ **(b)** $-3^3 = -(3)(3)(3) = -27$

(c) $(-3)^4 = (-3)(-3)(-3)(-3) = 81$ **(d)** $-3^4 = -(3)(3)(3)(3) = -81$

Check Yourself 2

Evaluate each expression.

(a) $(-4)^3$ **(b)** -4^3 **(c)** $(-4)^4$ **(d)** -4^4

NOTE

Most computer software, such as Excel, use a caret ^ to indicate that an exponent follows.

You can use a calculator to help you evaluate expressions containing exponents. If you have a graphing calculator, you can use the caret key, [∧]. Enter the base, followed by the caret key, and then enter the exponent. Some calculators use a key labeled [y^x] instead of the caret key.

Remember, we use a calculator as an aid or tool, not a crutch. You should be able to evaluate each of these expressions by hand, if necessary.

Example 3 Evaluating Exponential Expressions

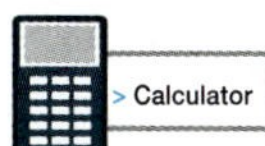

Use a calculator to evaluate each expression.

(a) $3^5 = 243$ Type 3 [∧] 5 [ENTER]

or 3 [y^x] 5 [=]

(b) $2^{10} = 1{,}024$ 2 [∧] 10 [ENTER]

or 2 [y^x] 10 [=]

Check Yourself 3

Use a calculator to evaluate each expression.

(a) 3^4 **(b)** 2^{16}

NOTE

To evaluate an expression, we find a number that is equal to the expression.

We used the word *expression* when discussing numbers taken to powers, such as 3^4. But what about something like $4 + 12 - 6$? We call *any* meaningful combination of numbers and operations an **expression.** When we evaluate an expression, we find a number that is equal to the expression. To evaluate an expression, we need to establish a set of rules that tell us the correct order in which to perform the operations. To see why, simplify the expression $5 + 2 \cdot 3$.

> CAUTION
>
> Only one of these results can be correct.

Method 1	or	Method 2
$\underbrace{5 + 2} \cdot 3$		$5 + \underbrace{2 \cdot 3}$
Add first		Multiply first
$= 7 \cdot 3$		$= 5 + 6$
$= 21$		$= 11$

Since we get different answers depending on how we do the problem, the language of algebra would not be clear if there were no agreement on which method is correct. We have rules that determine the order in which operations should be done.

Step by Step

The Order of Operations

Parentheses, brackets, and fraction bars are all examples of grouping symbols.

Step 1 Evaluate all expressions inside grouping symbols.

Step 2 Evaluate all expressions involving exponents.

Step 3 Do any multiplication or division in order, working from left to right.

Step 4 Do any addition or subtraction in order, working from left to right.

Example 4 Evaluating Expressions

< Objective 3 >

> **NOTE**
>
> Method 2 in the previous discussion is the correct one.

(a) Evaluate $5 + 2 \cdot 3$.

There are no parentheses or exponents, so start with step 3: First multiply and then add.

$5 + 2 \cdot 3$ ← Multiply first

$= 5 + 6$ ← Then add

$= 11$

(b) Evaluate $10 \cdot 4 \div 2 \cdot 5$.

Perform the multiplication and division from left to right.

$10 \cdot 4 \div 2 \cdot 5$

$= 40 \div 2 \cdot 5$

$= 20 \cdot 5$

$= 100$

Check Yourself 4

Evaluate each expression.

(a) $20 - 3 \cdot 4$ **(b)** $9 + 6 \div 3$ **(c)** $10 \cdot 6 \div 3 \cdot 2$

Example 5 Evaluating Expressions

Evaluate $-5 \cdot 3^2$.

$-5 \cdot 3^2$ ← Evaluate the exponential expression first.

$= -5 \cdot 9$

$= -45$

Check Yourself 5

Evaluate $-4 \cdot 2^4$.

Modern calculators correctly interpret the order of operations as demonstrated in Example 6.

Example 6 Using a Calculator to Evaluate Expressions

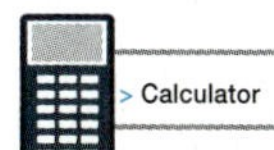

Use a calculator to evaluate each expression.

(a) $24.3 + 6.2 \cdot 3.5$

When evaluating expressions by hand, you must consider the order of operations. In this case, we multiply before adding. With a modern calculator, you need only enter the expression correctly. The calculator is programmed to follow the order of operations.

Entering 24.3 $\boxed{+}$ 6.2 $\boxed{\times}$ 3.5 $\boxed{\text{ENTER}}$

yields 46.

(b) $(2.45)^3 - 49 \div 8{,}000 + 12.2 \cdot 1.3$

As we mentioned earlier, some calculators use a caret ($\wedge$) to designate exponents. Others use the symbol x^y (or y^x).

Entering 2.45 $\boxed{\wedge}$ 3 $\boxed{-}$ 49 $\boxed{\div}$ 8000 $\boxed{+}$ 12.2 $\boxed{\times}$ 1.3 $\boxed{\text{ENTER}}$

or 2.45 $\boxed{\wedge}$ 3 $\boxed{\blacktriangleright}$ $\boxed{-}$ 49 $\boxed{\div}$ 8000 $\boxed{+}$ 12.2 $\boxed{\times}$ 1.3 $\boxed{\text{ENTER}}$

or 2.45 $\boxed{y^x}$ 3 $\boxed{-}$ 49 $\boxed{\div}$ 8000 $\boxed{+}$ 12.2 $\boxed{\times}$ 1.3 $\boxed{=}$

yields 30.56.

NOTE

Some graphing calculators display the caret symbol ^ and wrap long entries on multiple lines, as in the first image, below.

```
24.3+6.2*3.5
                 46
2.45^3-49/8000+1
2.2*1.3
              30.56
```

Other models display exponents as superscripts. Long entries may scroll off screen, as in the next image.

```
24.3+6.2*3.5
                 46
2.45³-49/8000+12▸
              30.56
```

If your calculator displays exponents as superscripts, you need to press the right arrow key $\boxed{\blacktriangleright}$ in order to enter in the remaining part of the expression.

Check Yourself 6

Use a calculator to evaluate each expression.

(a) $67.89 - 4.7 \cdot 12.7$ **(b)** $4.3 \cdot 55.5 - (3.75)^3 + 8{,}007 \div 1{,}600$

Operations inside grouping symbols are done first.

Example 7 Evaluating Expressions

Evaluate $(5 + 2) \cdot 3$.

Do the operation inside the parentheses as the first step.

$(5 + 2) \cdot 3 = 7 \cdot 3 = 21$

Add

Check Yourself 7

Evaluate $4(9 - 3)$.

The principle is the same when more than two “levels” of operations are involved.

Example 8 Evaluating Expressions

(a) Evaluate $4(-2 + 7)^3$.

Add inside the parentheses first.

$$4(-2 + 7)^3 = 4(5)^3$$

Evaluate the exponential expression.

$$= 4 \cdot 125$$

Multiply

$$= 500$$

(b) Evaluate $5(7 - 3)^2 - 10$.

Evaluate the expression inside the parentheses.

$$5(7 - 3)^2 - 10 = 5(4)^2 - 10$$

Evaluate the exponential expression.

$$= 5 \cdot 16 - 10$$

Multiply

$$= 80 - 10 = 70$$

Subtract

Check Yourself 8

Evaluate.

(a) $4 \cdot 3^3 + 8 \cdot (-11)$ **(b)** $12 + 4(2 + 3)^2$

Parentheses and brackets are not the only types of grouping symbols. Example 9 demonstrates the fraction bar as a grouping symbol.

Example 9 Using the Order of Operations with Grouping Symbols

> CAUTION

You may not "cancel" the 2's, because the numerator is being added, not multiplied.

$\frac{\cancel{2} + 14}{\cancel{2}}$ is incorrect!

Evaluate $3 + \frac{2 + 14}{2} \cdot 5$.

$$3 + \frac{2 + 14}{2} \cdot 5 = 3 + \frac{16}{2} \cdot 5$$

The fraction bar acts as a grouping symbol.

$$= 3 + 8 \cdot 5$$

We perform the division first because it precedes the multiplication.

$$= 3 + 40$$

$$= 43$$

Check Yourself 9

Evaluate $4 \cdot \frac{3^2 + 2 \cdot 3}{5} + 6$.

Many formulas require us to use the order of operations. We conclude this chapter with one such application.

For obvious ethical reasons, children are rarely subjects in medical research. Nonetheless, when children are ill, doctors sometimes determine that medication is necessary. Dosage recommendations for adults are based on research studies and the medical community believes that in most cases, children need smaller dosages than adults.

There are many formulas for determining the proper dosage for a child. The more complicated (and accurate) ones use a child's height, weight, or body mass. All of the formulas try to answer the question, "What fraction of an adult dosage should a child be given?"

Dr. Thomas Young constructed a conservative but simple model using only a child's age.

Example 10 An Allied Health Application

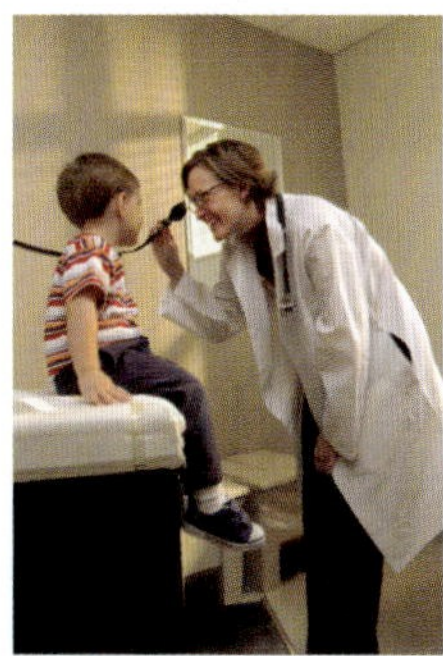

One formula for calculating the proper dosage of a medication for a child based on the recommended adult dosage and the child's age (in years) is Young's rule.

$$\text{Child's dose} = \left(\frac{\text{Age}}{\text{Age} + 12}\right) \times \text{Adult dose}$$

Use Young's rule to find the proper dose for a 3-year-old child if the recommended adult dose is 24 milligrams (mg).

Step 1 We are being asked to use the formula to find the proper dosage for a child who is 3, given that an adult should take 24 mg.

Step 2 We need to evaluate the expression formed when the age of the child and the adult dosage are taken into account.

NOTE

The child's age is 3; an adult should take a 24-mg dose.

Step 3

$$\text{Child's dose} = \left(\frac{\text{Age}}{\text{Age} + 12}\right) \times \text{Adult dose}$$

$$= \left(\frac{(3)}{(3) + 12}\right) \times (24 \text{ mg})$$

Step 4

$$\left(\frac{(3)}{(3) + 12}\right) \times (24 \text{ mg}) = \frac{3}{15} \times 24 \text{ mg}$$

$$= \frac{1}{5} \times \frac{24}{1}$$

$$= \frac{24}{5}$$

$$\frac{24}{5} = 4\frac{4}{5} = 4.8$$

The fraction bar is a grouping symbol, so we add the numbers in the denominator first, and then we continue simplifying the fraction.

RECALL

$\frac{24}{5} = 24 \div 5$

Step 5 According to Young's rule, the proper dose for a 3-year-old child is 4.8 mg.

Reasonableness

A 3-year-old is much younger than an adult, so we would expect the child's dose to be much smaller than the adult's dose.

Check Yourself 10

NOTE

This formula uses an approximation of the formula for the *circumference*, or distance around, a circle.

The approximate length of the belt pictured is given by

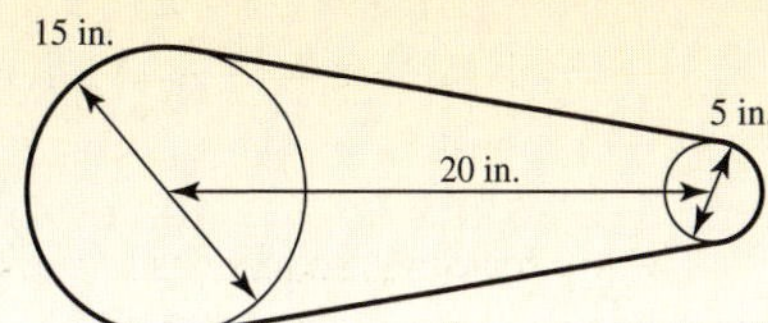

$$\frac{22}{7}\left(\frac{1}{2} \cdot 15 + \frac{1}{2} \cdot 5\right) + 2 \cdot 21$$

Find the length of the belt.

Check Yourself ANSWERS

1. **(a)** 4^6; **(b)** 10^4 **2.** **(a)** -64; **(b)** -64; **(c)** 256; **(d)** -256 **3.** **(a)** 81; **(b)** 65,536
4. **(a)** 8; **(b)** 11; **(c)** 40 **5.** -64 **6.** **(a)** 8.2; **(b)** 190.92 **7.** 24 **8.** **(a)** 20; **(b)** 112 **9.** 18
10. $73\frac{3}{7}$ in.

Reading Your Text

These fill-in-the-blank exercises will help you understand some of the key vocabulary used in this section. The answers to these exercises are in the Answers Appendix in the back of the text.

(a) A ______________ is a number or variable that is being multiplied by another number or variable.

(b) The first step in the order of operations is to evaluate all expressions inside ______________ symbols.

(c) When evaluating an expression, you should evaluate all ______________ expressions after evaluating any expressions inside grouping symbols.

(d) Some calculators use the caret key $\boxed{\wedge}$ to designate an ______________.

0.5 exercises

Skills | Calculator/Computer | Career Applications | Above and Beyond

< Objective 1 >

Use exponential notation to rewrite each expression.

1. $3 \cdot 3 \cdot 3 \cdot 3 \cdot 3$ **2.** $2 \cdot 2 \cdot 2 \cdot 2 \cdot 2 \cdot 2 \cdot 2$ **3.** $7 \cdot 7 \cdot 7 \cdot 7 \cdot 7$ 

4. $10 \cdot 10 \cdot 10 \cdot 10 \cdot 10$ **5.** $8 \cdot 8 \cdot 8 \cdot 8 \cdot 8 \cdot 8$ **6.** $5 \cdot 5 \cdot 5 \cdot 5 \cdot 5 \cdot 5$

7. $(-2)(-2)(-2)$ **8.** $(-4)(-4)(-4)(-4)$

< Objective 2 >

Evaluate each expression.

9. 3^2 **10.** 2^3 **11.** 2^4 **12.** 2^5

13. $(-8)^3$ **14.** 3^5 **15.** -8^3 **16.** 4^4

17. -5^2 **18.** $(-5)^2$ **19.** $(-4)^2$ **20.** $(-3)^4$

21. $-(-2)^5$ **22.** $-(-6)^4$ **23.** 10^3 **24.** 10^2

25. 10^6 **26.** 10^7

< Objective 3 >

27. 2×4^3 VIDEO
28. $(2 \times 4)^3$
29. 3×4^2
30. $(3 \times 4)^2$
31. $5 + 2^2$
32. $(5 + 2)^2$
33. $3^4 \times 2^4$
34. $(3 \times 2)^4$
35. $4 + 3 \cdot 5$
36. $10 - 4 \cdot 2$
37. $(7 + 2) \cdot 6$
38. $(9 - 5) \cdot 3$
39. $-12 - 8 \div 4$
40. $10 + 20 \div 5$
41. $(12 - 8) \div 4$ VIDEO
42. $(10 + 20) \div 5$
43. $8 \cdot 7 + 2 \cdot 2$
44. $48 \div 8 - 14 \div 2$
45. $(7 \cdot 5) + 3 \cdot 2$
46. $48 \div (8 - 4) \div 2$
47. $3 \cdot 5^2$
48. $5 \cdot 2^3$
49. $(3 \cdot 5)^2$
50. $(5 \cdot 2)^3$
51. $4 \cdot 3^2 - 2$ VIDEO
52. $3 \cdot 2^4 - 8$
53. $7(2^3 - 5)$
54. $3(7 - 3^2)$
55. $3 \cdot 2^4 - 26 \cdot 2$
56. $4 \cdot 2^3 - 15 \cdot 6$
57. $(2 \cdot 4)^2 - 8 \cdot 3$
58. $(3 \cdot 2)^3 - 7 \cdot 3$
59. $5(3 + 4)^2$
60. $3(8 - 4)^2$
61. $(5 \cdot 3 + 4)^2$
62. $(3 \cdot 8 - 4)^2$
63. $5[3(2 + 5) - 5]$
64. $\dfrac{11 - (-9) + 6(8 - 2)}{2 + 3 \cdot 4}$
65. $-2[(3 - 5)^2 - (-4 + 2)^3 \cdot (8 \div 4 \cdot 2)]$
66. $5 \cdot 4 - 2^3$
67. $4(2 + 3)^2 - 125$
68. $8 + 2(3 + 3)^2$
69. $(4 \cdot 2 + 3)^2 - 25$ VIDEO
70. $8 + (2 \cdot 3 + 3)^2$
71. $[-20 - 4^2 + (-4)^2 + 2] \div 9$
72. $14 + 3 \cdot 9 - 28 \div 7 \cdot 2$
73. $4 \cdot 8 \div 2 - 5^2$
74. $-12 - 8 \div 4 \cdot 2$
75. $15 + 5 - 3 \cdot 2 + (-2)^3$
76. $-8 + 14 \div 2 \cdot 4 - 3$

77. **Science and Medicine** Over the last 2,000 years, the earth's population has doubled approximately 5 times. Write the phrase "doubled 5 times" in exponential form.

78. **Geometry** The volume of a cube with each edge of length 9 inches (in.) is given by $9 \cdot 9 \cdot 9$. Write the volume, using exponential notation.

79. **Statistics** On an 8-hour trip, Jack drives $2\frac{3}{4}$ hr and Pat drives $2\frac{1}{2}$ hr. How much longer do they still need to drive?

80. **Statistics** A runner decides to run 20 miles (mi) each week. She runs $5\frac{1}{2}$ mi on Sunday, $4\frac{1}{4}$ mi on Tuesday, $4\frac{3}{4}$ mi on Wednesday, and $2\frac{1}{8}$ mi on Friday. How far does she need to run on Saturday to meet her goal?

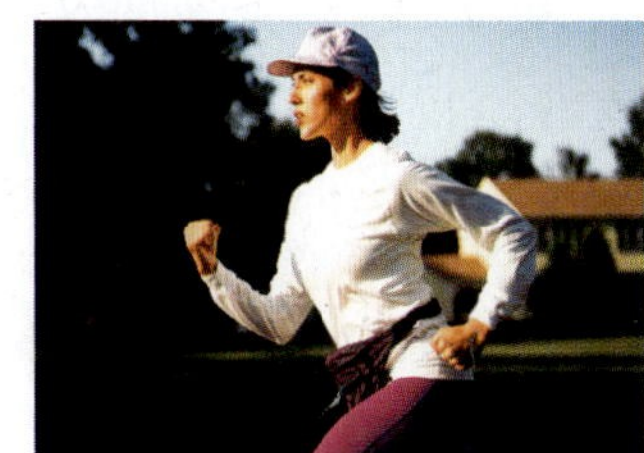

Determine whether each statement is **true** *or* **false.**

81. A negative number raised to an even power results in a positive number.

82. Exponential notation is shorthand for repeated addition.

Complete each statement with **always, sometimes,** *or* **never.**

83. Operations inside grouping symbols are ____________ done first.

84. In the order of operations, division is ____________ done before multiplication.

Skills	**Calculator/Computer**	Career Applications	Above and Beyond

Use a calculator to evaluate each expression. Round your answers to the nearest tenth.

85. $(1.2)^3 \div 2.0736 \cdot 2.4 + 1.6935 - 2.4896$

86. $(5.21 \cdot 3.14 - 6.2154) \div 5.12 - 0.45625$

87. $1.23 \cdot 3.169 - 2.05194 + (5.128 \cdot 3.15 - 10.1742)$

88. $4.56 + (2.34)^4 \div 4.7896 \cdot 6.93 \div 27.5625 - 3.1269 + (1.56)^2$

Skills	Calculator/Computer	**Career Applications**	Above and Beyond

89. **Business and Finance** The interest rate on an auto loan was $12\frac{3}{8}\%$ in May and $14\frac{1}{4}\%$ in September. By how many percentage points did the interest rate increase between May and September?

90. **Manufacturing Technology** A $3\frac{3}{8}$-in. cut needs to be made in a piece of material. The cut rate is $\frac{3}{4}$ in. per minute. How many minutes does it take to make this cut?

91. **Manufacturing Technology** Peer's Pipe Fitters started July with 1,789 gallons (gal) of liquefied petroleum gas (LP) in its tank. After 21 working days, there were 676 gal left in the tank. How much gas was used on each working day, on average?

92. **Business and Finance** Three friends bought equal shares in an investment. Between them, they paid $21,000 for the shares. Later, they were able to sell their shares for only $17,232. How much did each person lose on the investment?

Skills	Calculator/Computer	Career Applications	**Above and Beyond**

93. Insert grouping symbols in the proper place so that the value of the expression $36 \div 4 + 2 - 4$ is 2.

94. Work with a small group of students.

Part 1: Write the numbers 1 through 25 on slips of paper and put the slips in a pile, face down. Each of you randomly draws a slip of paper until each person has five slips. Turn the papers over and write down the five numbers. Put the five papers back in the pile, shuffle, and then draw one more. This last number is the answer. The first five numbers are the problem. Your task is to arrange the first five into a computation, using all you know about the order of operations, so that the answer is the last number. Each number must be used and may be used only once. If you cannot find a way to do this, pose it as a question to the whole class. Is this guaranteed to work?

Part 2: Use your five numbers in a problem, each number being used and used only once, for which the answer is 1. Try this 9 more times with the numbers 2 through 10. You may find more than one way to do each of these. Surprising, isn't it?

Part 3: Be sure that when you successfully find a way to get the desired answer by using the five numbers, you can then write your steps, using the correct order of operations. Write your 10 problems and exchange them with another group to see if they get these same answers when they do your problems.

Answers

1. 3^5 **3.** 7^5 **5.** 8^6 **7.** $(-2)^3$ **9.** 9 **11.** 16 **13.** −512 **15.** −512 **17.** −25 **19.** 16 **21.** 32 **23.** 1,000 **25.** 1,000,000 **27.** 128 **29.** 48 **31.** 9 **33.** 1,296 **35.** 19 **37.** 54 **39.** −14 **41.** 1 **43.** 60 **45.** 41 **47.** 75 **49.** 225 **51.** 34 **53.** 21 **55.** −4 **57.** 40 **59.** 245 **61.** 361 **63.** 80 **65.** −72 **67.** −25 **69.** 96 **71.** −2 **73.** −9 **75.** 6 **77.** 2^5 **79.** $2\frac{3}{4}$ hr **81.** True **83.** always **85.** 1.2 **87.** 7.8 **89.** $1\frac{7}{8}\%$ **91.** 53 gal/day **93.** $36 \div (4 + 2) - 4$

Definition/Procedure	Example	Reference
A Review of Fractions		**Section 0.1**
Equivalent Fractions If the numerator and denominator of a fraction are both multiplied by some nonzero number, the result is a fraction that is equivalent to the original fraction.	$\frac{4 \cdot 3}{5 \cdot 3} = \frac{12}{15}$	p. 3
Simplifying Fractions A fraction is in simplest terms when the numerator and denominator have no common factor.	$\frac{9}{21} = \frac{3 \cdot 3}{3 \cdot 7} = \frac{3}{7}$	p. 4
Multiplying Fractions To multiply two fractions, multiply the numerators, then multiply the denominators. Simplification can be done before or after the multiplication.	$\frac{2}{3} \cdot \frac{5}{6} = \frac{10}{18} = \frac{5}{9}$	p. 6
Dividing Fractions To divide two fractions, invert the divisor (the second fraction), then multiply the fractions.	$\frac{3}{5} \div \frac{2}{7} = \frac{3}{5} \cdot \frac{7}{2} = \frac{21}{10}$	p. 7
Adding Fractions To add two fractions, find the LCD (least common denominator), rewrite the fractions with this denominator, then add the numerators.	$\frac{2}{5} + \frac{5}{8} = \frac{16}{40} + \frac{25}{40} = \frac{41}{40}$	p. 8
Subtracting Fractions To subtract two fractions, find the LCD, rewrite the fractions with this denominator, then subtract the numerators.	$\frac{2}{3} - \frac{1}{4} = \frac{8}{12} - \frac{3}{12} = \frac{5}{12}$	p. 8
Solving Applications Follow this step-by-step approach when solving applications. **Step 1** Read the problem carefully to determine what you are being asked to find and what information is given in the application. **Step 2** Decide what you will do to solve the problem. **Step 3** Write down the complete (mathematical) statement necessary to solve the problem. **Step 4** Perform any calculations or other mathematics needed to solve the problem. **Step 5** Answer the question. Be sure to include units with your answer, when appropriate. Check to make certain that your answer is reasonable.	A foundation requires 2,668 blocks. If a contractor has 879 blocks on hand, how many more blocks need to be ordered? **Step 1** We want to find out how many more blocks the contractor needs. The contractor has 879 blocks, but needs a total of 2,668 blocks. **Step 2** This is a subtraction problem. **Step 3** 2,668 − 879 **Step 4** 2,668 − 879 = 1,789 **Step 5** The contractor needs to order 1,789 blocks. *Reasonableness Check* 1,789 + 879 = 2,668	p. 5
Real Numbers		**Section 0.2**
Positive Numbers Numbers used to name points to the right of 0 on a number line. **Negative Numbers** Numbers used to name points to the left of 0 on a number line.	Negative numbers / Positive numbers −3 −2 −1 0 1 2 3 Zero is neither positive nor negative.	p. 13

Continued

Definition/Procedure	Example	Reference
Natural Numbers The counting numbers	The natural numbers are {1, 2, 3, . . .}	*p.* 13
Integers The set consisting of the natural numbers, their opposites, and 0.	The integers are {. . .,−3, −2, −1, 0, 1, 2, 3, . . .}	*p.* 14
Rational Number Any number that can be expressed as the ratio of two integers.	Rational numbers are $\frac{2}{3}, \frac{5}{1}$, 0.234	*p.* 14
Irrational Number Any number that is not rational.	Irrational numbers include $\sqrt{2}$ and π	*p.* 15
Real Numbers Rational and irrational numbers together.	All the numbers listed are real numbers.	*p.* 16
Opposites Two numbers are opposites if the points name the same distance from 0 on a number line, but in opposite directions.	5 units 5 units −5 0 5 The opposite of 5 is −5.	*p.* 16
The opposite of a positive number is negative.		*p.* 16
The opposite of a negative number is positive.	3 units 3 units −3 0 3 The opposite of −3 is 3.	*p.* 16
0 is its own opposite.		*p.* 16
Absolute Value The distance on a number line between the point named by a number and 0. The absolute value of a number is always positive or 0. The absolute value of a number is called its **magnitude.**	The absolute value of a number a is written $\|a\|$. $\|7\| = 7 \quad \|-8\| = 8$	*p.* 17
Operations on Real Numbers *Adding Real Numbers*		Sections 0.3–0.4
1. If two numbers have the same sign, add their magnitudes. Give the sum the sign of the original numbers. **2.** If two numbers have different signs, subtract the smaller absolute value from the larger. Give the result the sign of the number with the larger magnitude.	$5 + 8 = 13$ $-3 + (-7) = -10$ $5 + (-3) = 2$ $7 + (-9) = -2$	*p.* 23
Subtracting Real Numbers To subtract real numbers, add the first number and the opposite of the number being subtracted.	$4 - (-2) = 4 + 2 = 6$ The opposite of −2	*p.* 25
Multiplying Real Numbers To multiply real numbers, multiply the absolute values of the numbers. Then attach a sign to the product according to the following rules: **1.** If the numbers have different signs, the product is negative **2.** If the numbers have the same sign, the product is positive.	$5 \cdot 7 = 35$ $(-4)(-6) = 24$ $(8)(-7) = -56$	*p.* 31

Continued

Definition/Procedure	Example	Reference
Dividing Real Numbers To divide real numbers, divide the absolute values of the numbers. Then attach a sign to the quotient according to the following rules: **1.** If the numbers have the same sign, the quotient is positive. **2.** If the numbers have different signs, the quotient is negative.	$\frac{-8}{-2} = 4$ $27 \div (-3) = -9$ $\frac{-16}{8} = -2$	*p.* 36
The Properties of Addition and Multiplication		
The Commutative Properties If a and b are any numbers, then **1.** $a + b = b + a$ **2.** $a \cdot b = b \cdot a$	$3 + 4 = 4 + 3$ $7 = 7$	*pp.* 24 and 33
The Associative Properties If a, b, and c are any numbers, then **1.** $a + (b + c) = (a + b) + c$ **2.** $a \cdot (b \cdot c) = (a \cdot b) \cdot c$	$3 \cdot (4 \cdot 5) = (3 \cdot 4) \cdot 5$ $3 \cdot (20) = (12) \cdot 5$ $60 = 60$	*pp.* 24 and 34
The Distributive Property If a, b, and c are any numbers, then $a \cdot (b + c) = a \cdot b + a \cdot c$ and $(a + b) \cdot c = a \cdot c + b \cdot c$	$2(5 + 3) = 2 \cdot 5 + 2 \cdot 3$ $2(8) = 10 + 6$ $16 = 16$	*p.* 34
Exponents and Order of Operations		Section 0.5
Notation Exponent ↓ $a^4 = \underbrace{a \cdot a \cdot a \cdot a}$ ↑ Base 4 factors The number or letter used as a factor, here a, is called the *base*. The *exponent*, which is written above and to the right of the *base*, tells us how many times the base is used as a factor.	$5^3 = 5 \cdot 5 \cdot 5$ $= 125$ $3^2 \cdot 7^3 = 3 \cdot 3 \cdot 7 \cdot 7 \cdot 7$	*p.* 42
The Order of Operations		
Step 1 Evaluate all expressions within grouping symbols. **Step 2** Evaluate all expressions containing exponents. **Step 3** Do any multiplication or division in order, working from left to right. **Step 4** Do any addition or subtraction in order, working from left to right.	Operate inside grouping symbols. $5 + 3(6 - 4)^2$ Evaluate the exponential expression $= 5 + 3 \cdot 2^2$ Multiply $= 5 + 3 \cdot 4$ Add $= 5 + 12$ $= 17$	*p.* 44

summary exercises :: chapter 0

This summary exercise set will help ensure that you have mastered each of the objectives of this chapter. The exercises are grouped by section. You should reread the material associated with any exercises that you find difficult. The answers to the odd-numbered exercises are in the Answers Appendix in the back of the text.

0.1 *Use the fundamental principle to find three fractions equivalent to each number.*

1. $\frac{5}{7}$ **2.** $\frac{3}{11}$ **3.** $\frac{4}{9}$ **4.** Simplify $\frac{24}{64}$.

Evaluate each expression. Write each answer in simplest form.

5. $\frac{7}{15} \times \frac{5}{21}$ **6.** $\frac{10}{27} \times \frac{9}{20}$ **7.** $\frac{5}{17} \div \frac{15}{34}$ **8.** $\frac{7}{15} \div \frac{14}{25}$

9. $\frac{7}{8} + \frac{15}{24}$ **10.** $\frac{5}{18} + \frac{7}{12}$ **11.** $\frac{11}{18} - \frac{2}{9}$ **12.** $\frac{11}{27} - \frac{5}{18}$

Solve each application.

13. CONSTRUCTION A kitchen measures $\frac{16}{3}$ by 4 yd. If you purchase linoleum that costs \$9 per square yard (you cannot purchase a partial square yard), how much will it cost to cover the floor?

14. SOCIAL SCIENCE The scale on a map uses 1 in. to represent 80 mi. If two cities are $\frac{11}{4}$ in. apart on the map, what is the actual distance between the cities?

15. CONSTRUCTION An 18-acre piece of land is to be subdivided into home lots that are each $\frac{3}{8}$ acre. How many lots can be formed?

16. GEOMETRY Find the perimeter of the given figure.

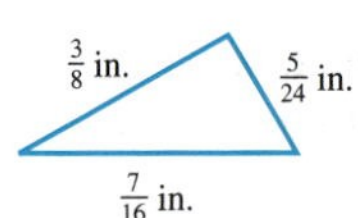

0.2 *Complete the statement.*

17. The absolute value of 12 is ________.

18. The opposite of -8 is ________.

19. $-|-3| =$ ________

20. $-(-20) =$ ________

21. $-|-4| =$ ________

22. $|-(-5)| =$ ________

23. The absolute value of -16 is ________.

24. The opposite of the absolute value of -9 is ________.

Complete the statement, using the symbol $<$, $>$, or $=$.

25. -3 ________ -1

26. -6 ________ $-|-6|$

27. $-|-7|$ ________ $-(-2)$

28. $-|-5|$ ________ $|-(-5)|$

0.3 *Simplify.*

29. $15 + (-7)$ **30.** $4 + (-9)$ **31.** $-23 - (-12)$ **32.** $\frac{5}{2} + \left(-\frac{4}{2}\right)$

33. $-\frac{9}{13} - \frac{4}{39}$ **34.** $5 + (-6) + (-3)$ **35.** $7 + (-4) + 8 + (-7)$ **36.** $-6 + 9 + 9 + (-5)$

37. $-35 + 30$ **38.** $-10 - 5$ **39.** $3 - (-2)$ **40.** $-7 - (-3)$

41. $\frac{23}{4} - \left(-\frac{3}{4}\right)$

42. $-3 - 2$

43. $8 - 12 - (-5)$

44. $-6 - 7 - (-18)$

45. $7 - (-4) - 7 - 4$

46. $-9 - (-6) - 8 - (-11)$

47. **Business and Finance** Jean deposited a check for \$625. She wrote two checks for \$69.74 and \$29.95, and used her debit card for a \$57.65 purchase. How much of her original deposit did she have left?

48. **Electrical Engineering** A certain electric motor spins at a rate of 5,400 rotations per minute (rpm). When a load is applied, the motor spins at 4,250 rpm. What is the change in rpm after loading?

0.4 *Multiply.*

49. $(-18)(-2)$

50. $(-10)(8)$

51. $(-5)(3)$

52. $\left(-\frac{3}{8}\right)\left(-\frac{4}{5}\right)$

53. $(-4)^2$

54. $(-2)(7)(-3)$

55. $(-6)(-5)(4)(-3)$

56. $(-9)(2)(-3)(1)$

Use the distributive property to remove parentheses and simplify.

57. $-4(8 - 7)$

58. $11(-15 + 4)$

59. $-8(5 - 2)$

60. $-4(-3 - 6)$

Divide.

61. $(-48) \div 12$

62. $\frac{-33}{-3}$

63. $-2 \div 0$

64. $-75 \div (-3)$

65. $-\frac{7}{9} \div \left(-\frac{2}{3}\right)$

66. $\left(-\frac{5}{11}\right) \div \frac{20}{33}$

67. $8 \div (-4)$

68. $(-12) \div (-1)$

69. **Business and Finance** An advertising agency lost a client who had been paying \$3,500 per month. How much revenue does the agency lose in a year?

70. **Social Science** A gambler lost \$180 over a 4-hr period. How much did the gambler lose per hour, on average?

0.5 *Write each expression in expanded form.*

71. 3^3

72. 5^4

73. 2^6

74. 4^5

Evaluate each expression.

75. $18 - 12 \div 2$

76. $(18 - 3) \cdot 5$

77. $6 \cdot 2^3$

78. $(5 \cdot 4)^2$

79. $5 \cdot 3^2 - 4$

80. $5(3^2 - 4)$

81. $5(4 - 2)^2$

82. $5 \cdot 4 - 2^2$

83. $(5 \cdot 4 - 2)^2$

84. $3(5 - 2)^2$

85. $3 \cdot 5 - 2^2$

86. $(3 \cdot 5 - 2)^2$

Statistics A professor grades a 20-question exam by awarding 5 points for each correct answer and subtracting 2 points for each incorrect answer. Points are neither added nor subtracted for answers left blank. Use this information to complete exercises 87 and 88.

87. Find the exam grade of a student who answers 14 questions correctly and 4 incorrectly, and leaves 2 questions blank.

88. Find the exam grade of a student who answers 17 questions correctly and 2 incorrectly, and leaves 1 question blank.

chapter test 0

Use this chapter test to assess your progress and to review for your next exam. Allow yourself about an hour to take this test. The answers to these exercises are in the Answers Appendix in the back of the text.

Use the fundamental principle to simplify each fraction.

1. $\frac{27}{99}$

2. $\frac{100}{64}$

Evaluate each expression. Write each answer in simplest form.

3. $\frac{4}{15} + \frac{7}{10}$

4. $\frac{9}{16} - \frac{3}{10}$

5. $13 + (-11) + (-5)$

6. $23 - 35$

7. $28 \div (-4)$

8. $(-44) \div (-11)$

9. $(-7)(5)$

10. $(-9)(-6)$

11. $-9 - 8 - (-5)$

12. $7 - 11 + 15$

13. $23 - 4 \times 12 \div 3 + |-4|$

14. $4 \cdot 5^2 - 35 + 21 - (-3)^3$

15. $\frac{3}{8} \times \frac{6}{11}$

16. $\frac{4}{5} \div \frac{2}{7}$

Fill in each blank with >, <, or = to make a true statement.

17. -7 ________ -5

18. $8 + (-3)^2$ ________ $8 - (-3)$

19. **Construction** A 14-acre piece of land is being developed into home lots. Each home site will be 0.35 acres and 2.8 acres will be used for roads. How many lots can be formed?

20. **Business and Finance** Michelle deposits $2,500 into her checking account each month. Each month, she pays her auto insurance ($200/mo), her auto loan ($250/mo), and her student loan ($275/mo). How much does she have left each month for other expenses?

CHAPTER 1

From Arithmetic to Algebra

INTRODUCTION

We expect to use mathematics both in our careers and when making financial decisions. But, there are many more opportunities to use math, even when enjoying life's pleasures. For instance, we use math regularly when traveling.

When traveling to another country, you need to be able to convert currency, temperature, and distance. Even figuring out when to call home so that you do not wake up family and friends during the night is a computation.

Equations are very old tools for solving problems and writing relationships clearly and accurately. In this chapter, we learn to solve linear equations and to write equations that accurately describe problem situations. Both of these skills are demonstrated in many settings, including international travel.

CHAPTER 1 OUTLINE

1.1 Transition to Algebra

< 1.1 Objectives >

1 > Use the symbols and language of algebra

2 > Identify algebraic expressions

3 > Use algebra to model an application

Tips for Student Success

Become Familiar with Your Syllabus

You probably received a syllabus in your first class meeting. You should add the important information to your calendar and address files.

1. Put all important dates into your *calendar*. These include homework due dates, quiz dates, test dates, and the date, time, and location of the final exam. Never be surprised by a deadline!
2. Put your instructor's name, contact number, and office number into your *address book*. Include your instructor's office hours, phone number, and email address. Make it a point to see your instructor early in the term. Although this is not the only person who can help you, your instructor is one of the most important resources available to you.
3. Familiarize yourself with other resources available to you. You can take advantage of the *Student's Solutions Manual* for your text as well as online resources offered through Connect Math Hosted by ALEKS Corp. and the ALEKS platform. Your college may also offer tutoring or other helpful services.

Given these resources, you have no reason to let confusion or frustration mount. If you can't "get it" from the text, try another resource. Take advantage of the resources available to you.

In arithmetic, you learned to calculate with numbers using addition, subtraction, multiplication, and division.

In algebra, we still use numbers and the same four operations. However, we also use letters to represent numbers. Letters such as x, y, L, and W are called **variables** when they represent numerical values.

Here we see two rectangles whose lengths and widths are labeled with numbers.

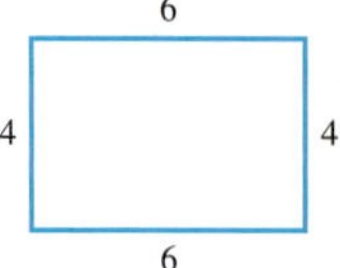

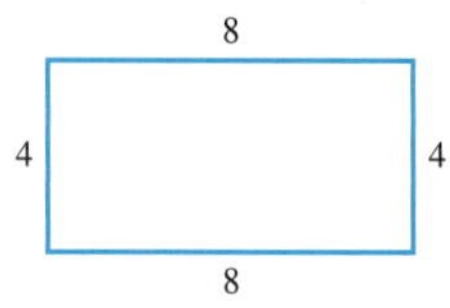

If we want to represent the length and width of *any* rectangle, we can use the variables L for length and W for width.

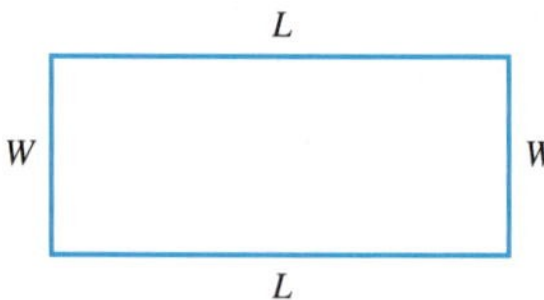

RECALL

In arithmetic
+ denotes addition
− denotes subtraction
× denotes multiplication
÷ denotes division

You are familiar with the four symbols (+, −, ×, ÷) used to indicate the fundamental operations of arithmetic.

To see how these operations are indicated in algebra, we begin by looking at addition.

Definition

Addition $x + y$ means the *sum* of x and y, or x *plus* y.

Example 1 Writing Expressions That Indicate Addition

< Objective 1 >

(a) The *sum* of a and 3 is written as $a + 3$.

(b) L *plus* W is written as $L + W$.

(c) 5 *more than* m is written as $m + 5$.

(d) x *increased by* 7 is written as $x + 7$.

Check Yourself 1

Write each phrase symbolically.

(a) The sum of y and 4 **(b)** a plus b
(c) 3 more than x **(d)** n increased by 6

Now look at how subtraction is indicated in algebra.

Definition

Subtraction $x - y$ means the *difference* of x and y, or x *minus* y.
Subtracting y is the same as adding its opposite, so
$x - y = x + (-y)$
$x - y$ is not the same as $y - x$.

Example 2 Writing Expressions That Indicate Subtraction

> CAUTION

"x minus y," "the difference of x and y," "x decreased by y," and "x take away y" are all written in the same order as the instructions are given, $x - y$.

However, we reverse the order when writing "x less than y" and "x subtracted from y." These two phrases are translated as $y - x$.

(a) r *minus* s is written as $r - s$.

(b) The *difference* of m and 5 is written as $m - 5$.

(c) x *decreased by* 8 is written as $x - 8$.

(d) 4 *less than* a is written as $a - 4$.

(e) x *subtracted from* 5 is written as $5 - x$.

(f) 7 *take away* y is written as $7 - y$.

Check Yourself 2

Write each phrase symbolically.

(a) w minus z **(b)** The difference of a and 7
(c) y decreased by 3 **(d)** 5 less than b
(e) b subtracted from 8 **(f)** 4 take away x

You have seen that the operations of addition and subtraction are written exactly the same way in algebra as in arithmetic. This is not true for multiplication because the symbol × looks like the letter x, so we use other symbols to show multiplication to avoid confusion. Here are some ways to write multiplication.

Definition

Multiplication

> **NOTE**
>
> x and y are called the factors of the product xy.

A centered dot	$x \cdot y$	All these indicate the *product* of x and y, or x *times* y.
Writing the letters next to each other or separated only by parentheses	xy $x(y)$ $(x)(y)$	

Example 3 Writing Expressions That Indicate Multiplication

> **NOTE**
>
> You can place letters next to each other or numbers and letters next to each other to show multiplication. But you cannot place numbers side by side to show multiplication: 37 means the number thirty-seven, not 3 times 7.

(a) The product of 5 and a is written as $5 \cdot a$, $(5)(a)$, or $5a$. The last expression, $5a$, is the shortest and most common way of writing the product.

(b) 3 times 7 can be written as $3 \cdot 7$ or $(3)(7)$.

(c) Twice z is written as $2z$.

(d) The product of 2, s, and t is written as $2st$.

(e) 4 more than the product of 6 and x is written as $6x + 4$.

Check Yourself 3

Write each phrase symbolically.

(a) m times n **(b)** The product of h and b
(c) The product of 8 and 9 **(d)** The product of 5, w, and y
(e) 3 more than the product of 8 and a

Before moving on to division, look at how we can combine the symbols we have learned so far.

Definition

Expression

An **expression** is a meaningful collection of numbers, variables, and operations.

Example 4 Identifying Expressions

< Objective 2 >

> **NOTE**
>
> Not every collection of symbols is an expression.

(a) $2m + 3$ is an expression. It means that we multiply 2 and m, then add 3.

(b) $x + \cdot + 3$ is not an expression. The three operations in a row have no meaning.

(c) $y = 2x - 1$ is not an expression, it is an *equation*. The equal sign is not an operation symbol.

(d) $3a + 5b - 4c$ is an expression.

Check Yourself 4

Identify the expressions.

(a) $7 - \cdot x$ **(b)** $6 + y = 9$
(c) $a + b - c$ **(d)** $3x - 5yz$

To write more complicated expressions in algebra, we need some "punctuation marks." Parentheses ( ) mean that an expression is to be thought of as a single quantity. Brackets [ ] are used in exactly the same way as parentheses in algebra. Example 5 shows expressions with grouping symbols.

Example 5 Writing Expressions

NOTES

This can be read as "3 times the quantity *a* plus *b*."

No parentheses are needed in part (*b*) since the 3 multiplies *only a*.

(a) 3 times the sum of a and b is written as

$3(a + b)$

The sum of a and b is a single quantity, so it is enclosed in parentheses.

(b) The sum of 3 times a and b is written as $3a + b$.

(c) 2 times the difference of m and n is written as $2(m - n)$.

(d) The product of s plus t and s minus t is written as $(s + t)(s - t)$.

(e) The product of b and 3 less than b is written as $b(b - 3)$.

Check Yourself 5

Write each phrase symbolically.

(a) Twice the sum of p and q
(b) The sum of twice p and q
(c) The product of a and the quantity $b - c$
(d) The product of x plus 2 and x minus 2
(e) The product of x and 4 more than x

NOTE

In algebra the fraction form is usually used.

Now we look at division. In arithmetic, you see the division sign ÷, the long division symbol $\overline{)\ \ }$, and fraction notation. For example, to indicate the quotient when 9 is divided by 3, you could write

$$9 \div 3 \qquad \text{or} \qquad 3\overline{)9} \qquad \text{or} \qquad \frac{9}{3}$$

Definition

Division $\frac{x}{y}$ means x *divided* by y or the *quotient* of x and y.

Example 6 Writing Expressions That Indicate Division

RECALL

The fraction bar is a grouping symbol.

(a) m divided by 3 is written as $\frac{m}{3}$.

(b) The sum of a and b, divided by 5, is written as $\frac{a + b}{5}$.

(c) The quantity p plus q divided by the quantity p minus q is written as $\frac{p + q}{p - q}$.

Check Yourself 6

Write each phrase symbolically.

(a) r divided by s
(b) The quotient when x minus y is divided by 7
(c) The quantity a minus 2 divided by the quantity a plus 2

We can use many different letters to represent variables. In Example 6, the letters m, a, b, p, and q represented different variables. We often choose a letter that reminds us of what it represents, for example, L for *length* or W for *width*. These variables may be uppercase or lowercase letters, although lowercase is used more often.

Example 7 Writing Geometric Expressions

(a) *Length* times *width* is written $L \cdot W$.

(b) One-half of *altitude* times *base* is written $\frac{1}{2}a \cdot b$.

(c) *Length* times *width* times *height* is written $L \cdot W \cdot H$.

(d) Pi (π) times *diameter* is written πd.

Check Yourself 7

Write each geometric expression symbolically.

(a) 2 times *length* plus two times *width*
(b) 2 times pi (π) times *radius*

Algebra can be used to model a variety of applications, such as the one shown in Example 8.

Example 8 Modeling Applications with Algebra

< Objective 3 >

NOTE

We are asked to describe her pay given that her hours may vary.

Carla earns \$10.25 per hour in her job. Write an expression that describes her weekly gross pay in terms of the number of hours she works.

We represent the number of hours she works in a week by the variable h. Carla's pay is figured by taking the product of her hourly wage and the number of hours she works.

So, the expression

$10.25h$

describes Carla's weekly gross pay.

NOTE

The words "twice" and "doubled" indicate multiplication by 2.

Check Yourself 8

The specifications for an engine cylinder call for the stroke length to be two more than twice the diameter of the cylinder. Write an expression for the stroke length of a cylinder based on its diameter.

We close this section by listing many of the common words used to indicate arithmetic operations.

Words Indicating Operations

The operations listed are usually indicated by the words shown.

Addition (+)	Plus, and, more than, increased by, sum
Subtraction (−)	Minus, from, less than, decreased by, difference, take away
Multiplication (·)	Times, of, by, product
Division (÷)	Divided, into, per, quotient

Check Yourself ANSWERS

1. **(a)** $y + 4$; **(b)** $a + b$; **(c)** $x + 3$; **(d)** $n + 6$ **2.** **(a)** $w - z$; **(b)** $a - 7$; **(c)** $y - 3$; **(d)** $b - 5$; **(e)** $8 - b$; **(f)** $4 - x$ **3.** **(a)** mn; **(b)** hb; **(c)** $8 \cdot 9$ or $(8)(9)$; **(d)** $5wy$; **(e)** $8a + 3$ **4.** **(a)** not an expression; **(b)** not an expression; **(c)** expression; **(d)** expression **5.** **(a)** $2(p + q)$; **(b)** $2p + q$; **(c)** $a(b - c)$; **(d)** $(x + 2)(x - 2)$; **(e)** $x(x + 4)$ **6.** **(a)** $\frac{r}{s}$; **(b)** $\frac{x - y}{7}$; **(c)** $\frac{a - 2}{a + 2}$ **7.** **(a)** $2L + 2W$; **(b)** $2\pi r$ **8.** $2d + 2$

Reading Your Text

These fill-in-the-blank exercises will help you understand some of the key vocabulary used in this section. The answers to these exercises are in the Answers Appendix in the back of the text.

(a) In algebra, we use letters to represent numbers. We call these letters ________.

(b) $x + y$ means the ________ of x and y.

(c) $x \cdot y$, $(x)(y)$, and xy are all ways of indicating ________ in algebra.

(d) An ________ is a meaningful collection of numbers, variables, and operations.

Skills Calculator/Computer Career Applications Above and Beyond

1.1 exercises

< Objective 1 >

Write each phrase symbolically.

1. The sum of c and d 

2. a plus 7

3. w plus z

4. The sum of m and n

5. x increased by 5

6. 3 more than b

7. 10 more than y 

8. m increased by 4

9. a minus b

10. s less than 5

11. 7 decreased by b

12. r minus 3

13. 6 less than r 

14. x decreased by 3

15. w times z

16. The product of 3 and c

17. The product of 5 and t 

18. 8 times a

19. The product of 8, m, and n

20. The product of 7, r, and s

21. The product of 8 and the quantity m plus n

22. The product of 5 and the sum of a and b

23. Twice the sum of x and y 

24. 3 times the sum of m and n

25. The sum of twice x and y

26. The sum of 3 times m and n

27. Twice the difference of x and y

28. 3 times the difference of c and d

29. The quantity a plus b times the quantity a minus b

30. The product of x plus y and x minus y

31. The product of m and 3 less than m 

32. The product of a and 7 less than a

33. 5 divided by x

34. The quotient when b is divided by 8

35. The sum of a and b, divided by 7

36. The quantity x minus y, divided by 9

37. The difference of p and q, divided by 4 

38. The sum of a and 5, divided by 9

39. The sum of a and 3, divided by the difference of a and 3

40. The difference of m and n, divided by the sum of m and n

Use x as the variable to write each phrase symbolically.

41. 5 more than a number

42. A number increased by 8

43. 7 less than a number

44. A number decreased by 8

45. 9 times a number

46. Twice a number

47. 6 more than 3 times a number

48. 5 times a number, decreased by the sum of the number and 3

49. Twice the sum of a number and 5

50. 3 times the difference of a number and 4 

51. The product of 2 more than a number and 2 less than that same number

52. The product of 5 less than a number and 5 more than that same number

53. The quotient of a number and 7

54. A number divided by the sum of the number and 7

55. The sum of a number and 5, divided by 8

56. The quotient when 7 less than a number is divided by 3

57. 6 more than a number divided by 6 less than that same number 

58. The quotient when 3 less than a number is divided by 3 more than that same number

Write each geometric expression symbolically.

59. Four times the length of a side s

60. $\frac{4}{3}$ times π times the cube of the radius r 

61. π times the radius r squared times the height h

62. Twice the length L plus twice the width W

63. One-half the product of the height h and the sum of two unequal sides b_1 and b_2

64. Six times the length of a side s squared

< Objective 2 >

Identify the expressions.

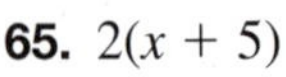

65. $2(x + 5)$ 

66. $4 - (x + 3)$

67. $4 + \div m$

68. $6 + a = 7$

69. $2b = 6$ 

70. $x(y + 3)$

71. $2a(3b + 5)$

72. $4x + \cdot 7$

< Objective 3 >

73. **Number Problem** Two numbers have a sum of 35. If one number is x, express the other number in terms of x.

74. **Science and Medicine** It is estimated that the earth is losing 4,000 species of plants and animals every year. If S represents the number of species living last year, how many species are on the earth this year? VIDEO

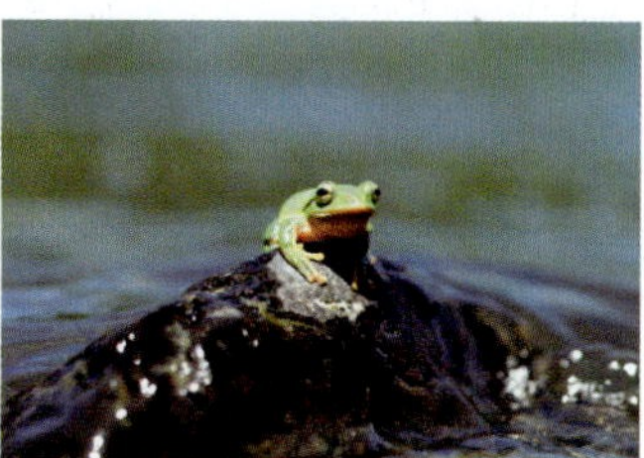

75. **Business and Finance** The simple interest earned when a principal P is invested at a rate r for a time t is calculated by multiplying the principal by the rate by the time. Write an expression for the interest earned.

76. **Science and Medicine** The kinetic energy of a particle of mass m is found by taking one-half of the product of the mass and the square of the velocity v. Write an expression for the kinetic energy of a particle.

77. **Business and Finance** Four hundred tickets were sold for a school play. The tickets were of two types: general admission and student. There were x general admission tickets sold. Write an expression for the number of student tickets sold.

78. **Business and Finance** Nate has $375 in his bank account. He wrote a check for x dollars for a concert ticket. Write an expression that represents the remaining money in his account.

Determine whether each statement is **true** *or* **false.**

79. The phrase "7 more than x" indicates addition.

80. A product is the result of dividing two numbers.

Complete each statement with **always, sometimes,** *or* **never.**

81. An expression is __________ an equation.

82. A number in front of a variable __________ indicates multiplication.

Skills | Calculator/Computer | **Career Applications** | Above and Beyond

83. **Construction Technology** K Jones Manufacturing produces hex bolts and carriage bolts. They sold 284 more hex bolts than carriage bolts last month. Write an expression that describes the number of carriage bolts they sold last month.

84. **Allied Health** The standard dosage given to a patient is equal to the product of the desired dose D and the available quantity Q divided by the available dose H. Write an expression for the standard dosage.

85. **Information Technology** Mindy is the manager of the help desk at a large cable company. She notices that, on average, her staff can handle 50 calls/hr. Last week, during a thunderstorm, the call volume increased from 65 calls/hr to 150 calls/hr.

    To determine the average number of customers in the system, she needs to take the quotient of the average rate of customer arrivals (the call volume) a and the difference of the average rate at which customers are served h and the average rate of customer arrivals a. Write an expression for the average number of customers in the system.

86. **Electrical Engineering** Electrical power is the product of voltage V and current I. Write an expression for the electrical power.

87. Rewrite each algebraic expression using English phrases. Exchange papers with another student to edit your writing. Be sure the meaning in English is the same as in algebra. These expressions are not complete sentences, so your English does not have to be in complete sentences. Here is an example.

Algebra: $2(x - 1)$

English: We could write "double 1 less than a number." Or we might write "a number diminished by 1 and then multiplied by 2."

(a) $n + 3$ **(b)** $\frac{x+2}{5}$ **(c)** $3(5 + a)$ **(d)** $3 - 4n$ **(e)** $\frac{x+6}{x-1}$

88. Use the Internet to find the origins of the symbols $+$, $-$, $\times$, and $\div$. Summarize your findings.

Answers

1. $c + d$ **3.** $w + z$ **5.** $x + 5$ **7.** $y + 10$ **9.** $a - b$ **11.** $7 - b$ **13.** $r - 6$ **15.** wz **17.** $5t$ **19.** $8mn$ **21.** $8(m + n)$ **23.** $2(x + y)$ **25.** $2x + y$ **27.** $2(x - y)$ **29.** $(a + b)(a - b)$ **31.** $m(m - 3)$ **33.** $\frac{5}{x}$ **35.** $\frac{a+b}{7}$ **37.** $\frac{p-q}{4}$ **39.** $\frac{a+3}{a-3}$ **41.** $x + 5$ **43.** $x - 7$ **45.** $9x$ **47.** $3x + 6$ **49.** $2(x + 5)$ **51.** $(x + 2)(x - 2)$ **53.** $\frac{x}{7}$ **55.** $\frac{x+5}{8}$ **57.** $\frac{x+6}{x-6}$ **59.** $4s$ **61.** $\pi r^2 h$ **63.** $\frac{1}{2}h(b_1 + b_2)$ **65.** Expression **67.** Not an expression **69.** Not an expression **71.** Expression **73.** $35 - x$ **75.** Prt **77.** $400 - x$ **79.** True **81.** never **83.** $H - 284$ **85.** $\frac{a}{h-a}$ **87.** Above and Beyond

Exchanging Money

In the opener to this chapter, we discussed international travel and using exchange rates when acquiring local currency. In this activity, we use exchange rates to explore the idea of variables. Recall that a **variable** is a symbol used to represent an unknown quantity or a quantity that varies.

Currency exchange rates are published on a daily basis by many sources such as *Yahoo!Finance* and the *Wall Street Journal*. For instance, on March 20, 2011, the exchange rate for trading US$ for CAN$ was 0.992. This means that US$1 is equivalent to CAN$0.992. That is, if you exchanged $100 of U.S. money, you would receive $99.20 in Canadian dollars.

CAN$ = Exchange rate × US$

Activity

1. Choose a country that you would like to visit. Use a search engine to find the exchange rate between US$ and the currency of your chosen country.
2. If you are visiting for only a short time, you may not need too much money. Determine how much of the local currency you will receive in exchange for US$250.
3. If you stay for an extended period, you will need more money. How much would you receive in exchange for US$900?

Here, we treated the amount (US$) as a *variable.* This quantity varied, depending on our needs. If we visit Canada and let x = the amount exchanged in US$ and y = the amount received in CAN$, then, using the exchange rate previously given, we have the equation

$$y = 0.992x$$

You may ask, "Isn't the amount of Canadian money received (y) a variable, too?" The answer is yes; in fact, all three quantities are variables. The exchange rate varies on a daily basis. For example, according to *Yahoo!Finance,* the exchange rate for US-CAN currency was 1.372 on December 14, 2001. If we let r = the exchange rate, then we can write the conversion equation as

$$y = rx$$

4. Consider the country you chose to visit above. Find the exchange rate for another date and repeat exercises 2 and 3 for this other exchange rate.
5. Choose another nation that you would like to visit. Repeat exercises 1–3 for this country.

This data set is provided for your convenience. We encourage you to find more current data on the Internet.

Data Set

Currency	US$	Yen (¥)	Euro (€)	CAN$	U.K. (£)	Aust$
1 US$	1	83.531	0.7557	0.992	0.6303	0.9534
1 Yen (¥)	0.012	1	0.009	0.0119	0.0075	0.0114
1 Euro (€)	1.3233	110.5365	1	1.3126	0.834	1.2616
1 CAN$	1.0081	84.2049	0.7618	1	0.6354	0.9611
1 U.K. (£)	1.5866	132.5327	1.199	1.5739	1	1.5127
1 Aust$	1.0489	87.6156	0.7926	1.0405	0.6611	1

Source: Yahoo!Finance; 3/20/11

1. We chose to visit Canada and will use the 3/20/11 exchange rate of 0.992 from the sample data set.
2. Exchange rate × US$ = CAN$

   (0.992) · (US$250) = CAN$248

   We would receive $248 in Canadian dollars for $250 in U.S. money.
3. (0.992) · (US$900) = CAN$892.80
4. Had we visited Canada on 12/14/01, we would have received an exchange rate of 1.372.

   (1.372) · (US$250) = CAN$343

   (1.372) · (US$900) = CAN$1,234.80
5. We choose to visit Japan. The 3/20/11 exchange rate was 83.531 Yen (¥) for each US$.

   (83.531) · (US$250) = ¥20,882.75

   (83.531) · (US$900) = ¥75,177.9

   We would receive 20,883 yen for US$250, and 75,178 yen for US$900.

1.2 Evaluating Algebraic Expressions

< 1.2 Objectives >

1 > Evaluate an algebraic expression

2 > Use a calculator to evaluate an expression

3 > Use expressions to solve applications

Tips for Student Success

Working Together

How many of your classmates do you know? Whether your nature is outgoing or shy, there are many benefits to getting to know your classmates.

1. It is important to have someone to call when you miss class or if you are unclear on an assignment.
2. Working with another person is almost always beneficial to both people. If you don't understand something, it helps to have someone to ask about it. If you do understand something, nothing cements that understanding better than explaining it to another person.
3. Sometimes we need to commiserate. If an assignment is particularly frustrating, it is reassuring to find that it is also frustrating for other students.
4. Have you ever thought you had the right answer, but it didn't match the answer in the text? Frequently the answers are equivalent, but that's not always easy to see. A different perspective can help you see that. Occasionally there is an error in a textbook. In such cases, it is wonderfully reassuring to find that someone else has the same answer as you do.

Algebra gives us a powerful tool to solve problems. Using algebra often requires us to *evaluate an algebraic expression*. This means that we have values or numbers to use in place of the variables in an expression and we use them to compute the overall value of the expression.

Step by Step

Evaluating an Algebraic Expression

Step 1 Replace each variable with its given number value.

Step 2 Compute, following the rules for order of operations.

Example 1 Evaluating Algebraic Expressions

< Objective 1 >

Let $a = 5$ and $b = 7$.

(a) To evaluate $a + b$, we replace a with 5 and b with 7.

$a + b = (5) + (7) = 12$

(b) To evaluate $3ab$, we again replace a with 5 and b with 7.

$3ab = 3 \cdot (5) \cdot (7) = 105$

Check Yourself 1

If $x = 6$ and $y = 7$, evaluate.

(a) $y - x$ **(b)** $5xy$

When necessary, we always follow the order of operations when evaluating an expression.

Example 2 Evaluating Algebraic Expressions

Evaluate each expression if $a = 2$, $b = 3$, $c = 4$, and $d = 5$.

(a) $5a + 7b$

$5a + 7b = 5(2) + 7(3)$ First substitute $a = 2$ and $b = 3$.

$= 10 + 21$ Multiply before adding.

$= 31$

> CAUTION

This is different from

$(3c)^2 = [3 \cdot (4)]^2$
$= 12^2 = 144$

(b) $3c^2$

$3c^2 = 3(4)^2$ Substitute $c = 4$.

$= 3 \cdot 16$ Apply the exponent before multiplying.

$= 48$

(c) $7(c + d)$

$7(c + d) = 7[(4) + (5)]$ Substitute $c = 4$ and $d = 5$.

$= 7 \cdot 9$ Do the work inside the grouping symbols first.

$= 63$ Multiply to complete the problem.

(d) $5a^4 - 2d^2$

$5a^4 - 2d^2 = 5(2)^4 - 2(5)^2$ Substitute $a = 2$ and $d = 5$.

$= 5 \cdot 16 - 2 \cdot 25$ Apply the exponents before multiplying.

$= 80 - 50$ Multiply before subtracting.

$= 30$ Finally, subtract.

Check Yourself 2

If $x = 3$, $y = 2$, $z = 4$, and $w = 5$, evaluate each expression.

(a) $4x^2 + 2$ **(b)** $5(z + w)$ **(c)** $7(z^2 - y^2)$

To evaluate an expression when there is a fraction bar, keep in mind that the fraction bar is a grouping symbol. First, complete all the computations in the numerator and denominator separately. Finally, divide the numerator by the denominator.

Example 3 Evaluating Algebraic Expressions

If $p = 2$, $q = 3$, and $r = 4$, evaluate.

(a) $\frac{8p}{r}$

Replace p with 2 and r with 4.

$\frac{8p}{r} = \frac{8 \cdot (2)}{(4)} = \frac{16}{4} = 4$ Divide as the last step.

(b) $\frac{7q + r}{p + q}$

$\frac{7q + r}{p + q} = \frac{7 \cdot (3) + (4)}{(2) + (3)}$

$$= \frac{21 + 4}{2 + 3}$$ Evaluate the top and bottom separately.

$$= \frac{25}{5} = 5$$

Check Yourself 3

Evaluate each expression if $c = 5$, $d = 8$, and $e = 3$.

(a) $\frac{6c}{e}$ **(b)** $\frac{4d + e}{c}$ **(c)** $\frac{10d - e}{d + e}$

As you might expect, we can use a calculator or computer to evaluate an algebraic expression.

Example 4 Using a Calculator to Evaluate an Expression

< Objective 2 >

RECALL

Graphing calculators use ENTER instead of an = key.

Use a calculator to evaluate each expression.

(a) $\frac{4x + y}{z}$ if $x = 2$, $y = 1$, and $z = 3$

Begin by writing the expression with the values substituted for the variables.

$$\frac{4x + y}{z} = \frac{4(2) + (1)}{(3)}$$

Then, enter the numerical expression into a calculator.

(4 × 2 + 1) ÷ 3 ENTER

Remember to enclose the entire numerator in parentheses.

The display should read 3.

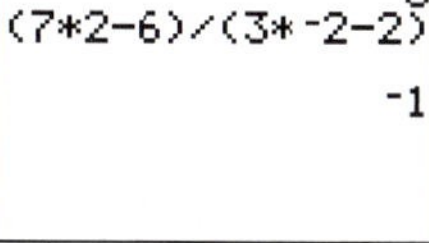

(b) $\frac{7x - y}{3z - x}$ if $x = 2$, $y = 6$, and $z = -2$

Again, we begin by substituting.

$$\frac{7x - y}{3z - x} = \frac{7(2) - (6)}{3(-2) - 2}$$

Then, we enter the expression into a calculator.

(7 × 2 − 6) ÷ (3 × (−) 2 − 2) ENTER

The display should read -1.

Check Yourself 4

Use a calculator to evaluate each expression if $x = 2$, $y = -6$, and $z = 5$. Round your results to the nearest hundredth, if necessary.

(a) $\frac{2x + y}{z}$ **(b)** $\frac{4y - 2z}{3x}$

NOTE

Does your calculator follow the order of operations? Enter

8 + 5 × 4

into your calculator.

If you get 28, then your calculator follows the order of operations.

If you get 52, then your calculator *does not* follow the order of operations.

Many calculators follow the correct order of operations when evaluating an expression. If we omit the parentheses in Example 4(b) and enter

7 × 2 − 6 ÷ 3 × (−) 2 − 2 ENTER

the calculator will interpret our input as $7 \cdot 2 - \frac{6}{3} \cdot (-2) - 2$, which is not what we wanted.

Whether working with a calculator or pencil and paper, you must remember to take care both with signs and with the order of operations.

Example 5 Evaluating Expressions

RECALL

Always follow the rules for the order of operations. Multiply first, then add.

Evaluate $5a + 4b$ if $a = -2$ and $b = \frac{3}{4}$.

Replace a with -2 and b with $\frac{3}{4}$.

$$5a + 4b = 5(-2) + 4\left(\frac{3}{4}\right)$$
$$= -10 + 3$$
$$= -7$$

Check Yourself 5

Evaluate $3x + 5y$ if $x = -2$ and $y = -\frac{4}{5}$.

We follow the same rules no matter how many variables are in the expression.

Example 6 Evaluating Expressions

Evaluate each expression if $a = -4$, $b = 2$, $c = -5$, and $d = 6$.

(a) $7a - 4c$

This becomes $-(-20)$, or $+20$.

$$7a - 4c = 7(-4) - 4(-5)$$
$$= -28 + 20$$
$$= -8$$

>CAUTION

When a squared variable is replaced by a negative number, square the negative.

$(-5)^2 = (-5)(-5) = 25$

The exponent applies to -5!

$-5^2 = -(5 \cdot 5) = -25$

The exponent applies only to 5!

(b) $7c^2$

Evaluate the power first, then multiply by 7.

$$7c^2 = 7(-5)^2 = 7 \cdot 25$$
$$= 175$$

(c) $b^2 - 4ac$

$$b^2 - 4ac = (2)^2 - 4(-4)(-5)$$
$$= 4 - 4(-4)(-5)$$
$$= 4 - 80$$
$$= -76$$

(d) $b(a + d)$

Add inside the brackets first.

$$b(a + d) = (2)[(-4) + (6)]$$
$$= 2(2)$$
$$= 4$$

Check Yourself 6

Evaluate if $p = -4$, $q = 3$, and $r = -2$.

(a) $5p - 3r$ **(b)** $2p^2 + q$ **(c)** $p(q + r)$
(d) $-q^2$ **(e)** $(-q)^2$

We look at another example with a fraction. Remember that the fraction bar is a grouping symbol. This means that you should do the required operations first in the numerator and then in the denominator. Divide as the last step.

Example 7 Evaluating Expressions

Evaluate each expression if $x = 4$, $y = -5$, $z = 2$, and $w = -3$.

(a) $\frac{z - 2y}{x}$

$$\frac{z - 2y}{x} = \frac{(2) - 2(-5)}{(4)} = \frac{2 + 10}{4}$$

$$= \frac{12}{4} = 3$$

(b) $\frac{3x - w}{2x + w}$

$$\frac{3x - w}{2x + w} = \frac{3(4) - (-3)}{2(4) + (-3)} = \frac{12 + 3}{8 + (-3)}$$

$$= \frac{15}{5} = 3$$

Check Yourself 7

Evaluate if $m = -6$, $n = 4$, and $p = -3$.

(a) $\frac{m + 3n}{p}$ **(b)** $\frac{4m + n}{m + 4n}$

The process of evaluating expressions has many common applications.

Example 8 An Application of Evaluating an Expression

< Objective 3 >

A car is advertised for rent at a cost of \$59 per day plus 20 cents per mile. The total cost can be found by evaluating the expression

$59d + 0.20m$

in which d represents the number of days and m the number of miles. Find the total cost for a 3-day rental if 250 miles are driven.

$59(3) + 0.20(250)$

$= 177 + 50$

$= 227$

The total cost is \$227.

Check Yourself 8

The cost to hold a wedding reception at a certain cultural arts center is \$195 per hour plus \$27.50 per guest. The total cost can be found by evaluating the expression

$195h + 27.50g$

in which h represents the number of hours and g the number of guests. Find the total cost for a 4-hour reception with 220 guests.

Check Yourself ANSWERS

1. **(a)** 1; **(b)** 210 **2.** **(a)** 38; **(b)** 45; **(c)** 84 **3.** **(a)** 10; **(b)** 7; **(c)** 7 **4.** **(a)** −0.4; **(b)** −5.67 **5.** −10 **6.** **(a)** −14; **(b)** 35; **(c)** −4; **(d)** −9; **(e)** 9 **7.** **(a)** −2; **(b)** −2 **8.** \$6,830

Reading Your Text

These fill-in-the-blank exercises will help you understand some of the key vocabulary used in this section. The answers to these exercises are in the Answers Appendix in the back of the text.

(a) Finding the value of an expression is called ________ the expression.

(b) If a squared variable is replaced by a negative number, the result is ________.

(c) Always follow the order of ________ when evaluating an algebraic expression.

(d) A fraction bar is a ________ symbol.

Graphing Calculator Option

Using the Memory Feature to Evaluate Expressions

The memory features of a graphing calculator are a great aid when you need to evaluate several expressions, using the same variables and the same values for those variables.

Your graphing calculator can store variable values for many different variables in different memory spaces. Using these memory spaces saves a great deal of time when evaluating expressions.

Note: We use the TI-84 Plus model graphing calculator throughout this text. If you have a different model, consult your instructor or the instruction manual. If you do not have the instruction manual, you can download it from the manufacturer's website.

Evaluate each expression if $a = 4.6$, $b = -\frac{2}{3}$, and $c = 8$. Round your results to the nearest hundredth.

(a) $a + \frac{b}{ac}$ **(b)** $b - b^2 + 3(a - c)$ **(c)** $bc - a^2 - \frac{ab}{c}$ **(d)** $a^2b^3c - ab^4c^2$

Begin by entering each variable's value into a calculator memory space. When possible, use the memory space that has the same name as the variable you are saving.

Step 1 Type the value associated with one variable.

Step 2 Press the store key, [STO▶], the green alphabet key to access the memory names, [ALPHA], and the key indicating which memory space you want to use.

Note: By pressing [ALPHA], you are accessing the green letters above selected keys. These letters name the variable spaces.

Step 3 Press [ENTER].

Step 4 Repeat until every variable value has been stored in an individual memory space.

To complete the example, we store 4.6 in **Memory A,** $-\frac{2}{3}$ in **Memory B,** and 8 in **Memory C.**

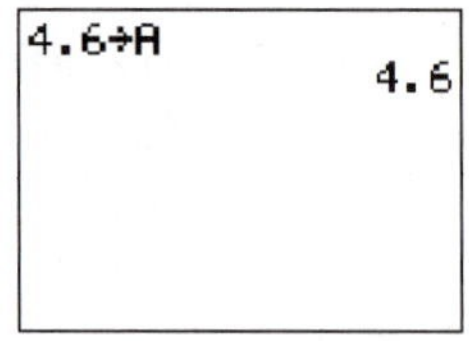

Memory A is with the [MATH] key.

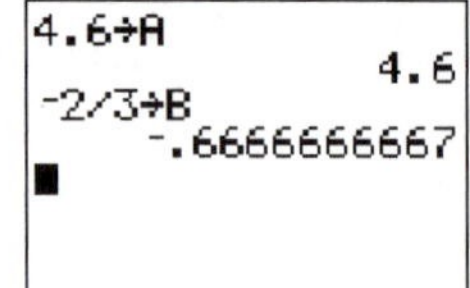

Memory B is with the [APPS] key.

Divide to form a fraction.

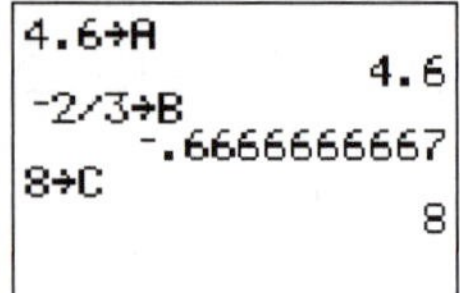

Memory C is with the [PRGM] key.

You can use the variables in the memory spaces rather than type in the numbers. Access the memory spaces by pressing [ALPHA] before pressing the key associated with the memory space. This will save time and make careless errors much less likely.

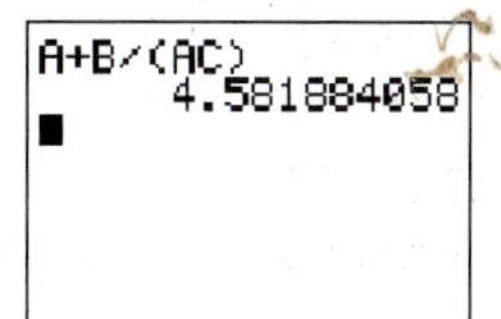

(a) $a + \dfrac{b}{ac}$

The keystrokes are [ALPHA], **A** (with [MATH]), [+], [ALPHA], **B** (with [APPS]), [÷], [(], [ALPHA], **A**, [ALPHA], **C**, [)], [ENTER].

$a + \dfrac{b}{ac} = 4.58$, to the nearest hundredth.

Note: Because the fraction bar is a grouping symbol, you must remember to enclose the denominator in parentheses.

(b) $b - b^2 + 3(a - c)$

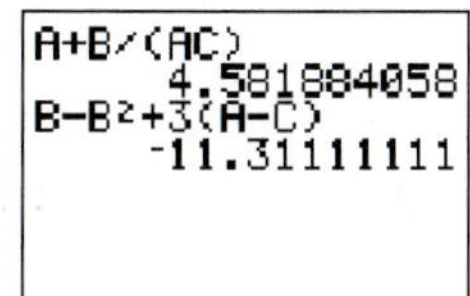

$b - b^2 + 3(a - c) = -11.31$

Use [x^2] to square a value.

(c) $bc - a^2 - \dfrac{ab}{c}$

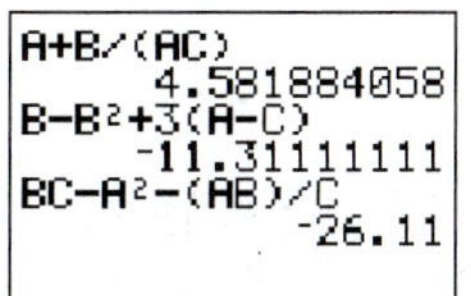

$bc - a^2 - \dfrac{ab}{c} = -26.11$

(d) $a^2b^3c - ab^4c^2$

```
          4.581884058
B-B²+3(A-C)
         -11.31111111
BC-A²-(AB)/C
               -26.11
A²B^3C-AB^4C²
         -108.3101235
```

$a^2b^3c - ab^4c^2 = -108.31$

Use the caret key, [^], for general exponents.

Graphing Calculator Check

Evaluate each expression if $x = -8.3$, $y = \dfrac{5}{4}$, and $z = -6$. Round your results to the nearest hundredth.

(a) $\dfrac{xy}{z} - xz$ **(b)** $5(z - y) + \dfrac{x}{x - z}$ **(c)** $x^2y^5z - (x + y)^2$ **(d)** $\dfrac{-2(x + z)^2}{y^3z}$

ANSWERS

(a) -48.07 **(b)** -32.64 **(c)** $-1{,}311.12$ **(d)** 34.9

Note: Throughout this text, we will provide additional graphing-calculator material. This material is optional. The authors will not assume that students have learned this, but we feel that students using a graphing calculator will benefit from these materials.

Skills Calculator/Computer Career Applications Above and Beyond

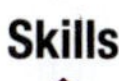

1.2 exercises

< Objective 1 >

Evaluate each expression if $a = -2$, $b = 5$, $c = -4$, and $d = 6$.

1. $3c - 2b$ 

2. $4c - 2b$

3. $7c + 6b$

4. $7a - 2c$

5. $-b^2 + b$

6. $(-c)^2 + 5c$

7. $3a^2$

8. $6c^2$

9. $c^2 - 2d$

10. $3a^2 + 4c$

11. $2a^2 + 3b^2$

12. $4b^2 - 2c^2$

13. $2(c - d)$ 14. $5(b - c)$ 15. $4(2a - d)$ 16. $6(3c - d)$

17. $a(b + 3c)$ 18. $c(3a - d)$ 19. $\frac{6d}{c}$ 20. $\frac{3a}{5b}$

21. $\frac{3d + 2c}{b}$ 22. $\frac{2b + 3d}{2a}$ 23. $\frac{2b - 3a}{c + 2d}$ VIDEO 24. $\frac{3d - 2b}{5a + d}$

25. $b^2 - d^2$ 26. $d^2 - b^2$ 27. $(b - d)^2$ 28. $(d - b)^2$

29. $(d - b)(d + b)$ 30. $(c - a)(c + a)$ 31. $c^3 - a^3$ 32. $c^3 + a^3$

33. $(c - a)^3$ 34. $(c + a)^3$ 35. $(d - b)(d^2 + db + b^2)$ 36. $(c + a)(c^2 - ac + a^2)$ VIDEO

37. $b^2 + a^2$ 38. $d^2 - a^2$ 39. $(b + a)^2$ 40. $(d - a)^2$

41. $a^2 + 2ad + d^2$ 42. $d^2 - 2ad + a^2$

Evaluate each expression if $x = -2$, $y = -3$, *and* $z = 4$.

43. $x^2 - 2y^2 + z^2$ 44. $4yz + 6xy$ 45. $2xy - (x^2 - 2yz)$ 46. $3yz - 6xyz + x^2y^2$

47. $2y(z^2 - 2xy) + yz^2$ 48. $-z - (-2x - yz)$

< Objective 3 >

49. **Electrical Engineering** The formula for the total resistance in a parallel circuit is $R_T = \frac{R_1R_2}{R_1 + R_2}$. Find the total resistance if $R_1 = 9$ ohms (Ω) and $R_2 = 15\ \Omega$.

50. **Geometry** The formula for the area of a triangle is given by $A = \frac{1}{2}ab$, where a is the altitude (or height) and b is the length of the base. Find the area of a triangle if $a = 4$ centimeters (cm) and $b = 8$ cm.

51. **Geometry** The perimeter of a rectangle of length L and width W is given by the formula $P = 2L + 2W$. Find the perimeter when $L = 10$ inches (in.) and $W = 5$ in.

52. **Business and Finance** The simple interest I on a principal of P dollars at interest rate r for time t, in years, is given by $I = Prt$. Find the simple interest on a principal of \$6,000 at 4% for 3 years.
Hint: 4% = 0.04.

53. **Business and Finance** Use the formula $P = \frac{I}{r \cdot t}$ to find the principal invested if the total interest earned was \$150 and the rate of interest was 4% for 2 years.

54. **Business and Finance** Use the formula $r = \frac{I}{P \cdot t}$ to find the interest rate if \$5,000 earns \$1,500 interest in 6 years.

55. **Science and Medicine** The formula that relates Celsius and Fahrenheit temperatures is $F = \frac{9}{5}C + 32$. If the temperature is $-10°C$, what is the Fahrenheit temperature?

56. **Geometry** If the area of a circle whose radius is r is given by $A = \pi r^2$, where $\pi = 3.14$, find the area when $r = 3$ meters (m).

57. **Business and Finance** A local telephone company offers a long-distance telephone plan that charges \$5.25 per month and \$0.08 per minute of calling time. The expression $0.08t + 5.25$ represents the monthly long-distance bill for a customer who makes t minutes (min) of long-distance calls on this plan. Find the monthly bill for a customer who makes 173 min of long-distance calls on this plan.

58. **Science and Medicine** The speed of a model car as it slows down is given by $v = 20 - 4t$, where v is the speed in meters per second (m/s) and t is the time in seconds (s) during which the car has slowed. Find the speed of the car 1.5 s after it has begun to slow.

Decide whether the given numbers make the statement **true** *or* **false.**

59. $x - 7 = 2y + 5;\ x = 22,\ y = 5$

60. $3(x - y) = 6;\ x = 5,\ y = -3$

61. $2(x + y) = 2x + y;\ x = -4,\ y = -2$ 

62. $x^2 - y^2 = x - y;\ x = 4,\ y = -3$

Determine whether each statement is **true** *or* **false.**

63. When evaluating an expression that has a fraction bar, dividing the numerator by the denominator is the first step.

64. The value of w^2 is always nonnegative.

Complete each statement with **always, sometimes,** *or* **never.**

65. When n is replaced with a number, the value of $-n^2$ is __________ positive.

66. When x is replaced with a number, the value of $-5x$ is __________ negative.

Skills | **Calculator/Computer** | Career Applications | Above and Beyond

< Objective 2 >

Use a calculator to evaluate each expression if $x = -2.34$, $y = -3.14$, *and* $z = 4.12$. *Round your answer to the nearest tenth.*

67. $x + yz$

68. $y - 2z$

69. $y^2 - 2x^2$

70. $x^2 + y^2$

71. $\dfrac{xy}{z - x}$

72. $\dfrac{y^2}{zy}$

73. $\dfrac{2x + y}{2x + z}$

74. $\dfrac{y^2z^2}{xy}$

Use a calculator to evaluate the expression $x^2 - 4x^3 + 3x$ *for each value.*

75. $x = 3$

76. $x = 12$

77. $x = 27$

78. $x = 48$

Use a calculator to evaluate each expression if $m = 232$, $n = -487$, *and* $p = 58$. *Round your results to the nearest tenth.*

79. $m + np^2$

80. $p - (m + 2n)$

81. $(p + n)^2 - m^2$

82. $\dfrac{pm - 2n}{n - 2m}$

83. $\dfrac{n^2 - p^2}{p^2 - m^2}$

84. $m^2 + (-n^2) + (-p^2)$

Skills | Calculator/Computer | **Career Applications** | Above and Beyond

85. **ALLIED HEALTH** The concentration, in micrograms per milliliter (μg/mL), of an antihistamine in a patient's bloodstream can be approximated using the expression $-2t^2 + 13t + 1$, in which t is the number of hours since the drug was administered. Approximate the concentration of the antihistamine 1 hour after being administered.

86. **ALLIED HEALTH** Use the expression given in exercise 85 to approximate the concentration of the antihistamine 3 hr after being administered.

87. **ELECTRICAL ENGINEERING** Evaluate $\dfrac{rT}{5{,}252}$ for $r = 1{,}180$ and $T = 3$ (round to the nearest thousandth).

88. **MECHANICAL ENGINEERING** The kinetic energy (in joules) of a particle is given by $\frac{1}{2}mv^2$. Find the kinetic energy of a particle if its mass is 60 kg and its velocity is 6 m/s.

89. Write an English interpretation of each algebraic expression or equation.

(a) $(2x^2 - y)^3$ **(b)** $3n = \frac{n-1}{2}$ **(c)** $(2n + 3)(n - 4)$

90. Is $a^n + b^n = (a + b)^n$? Try a few numbers and decide whether this is true for all numbers, for some numbers, or never true. Write an explanation of your findings and give examples.

91. **(a)** Evaluate the expression $4x(5 - x)(6 - x)$ for $x = 0, 1, 2, 3, 4$, and 5. Complete the table below.

Value of x	0	1	2	3	4	5
Value of expression						

(b) For which value of x does the expression value appear to be largest?

(c) Evaluate the expression for $x = 1.5, 1.6, 1.7, 1.8, 1.9, 2.0, 2.1, 2.2, 2.3, 2.4$, and 2.5. Complete the table.

Value of x	1.5	1.6	1.7	1.8	1.9	2.0	2.1	2.2	2.3	2.4	2.5
Value of expression											

(d) For which value of x does the expression value appear to be largest?

(e) Continue the search for the value of x that produces the greatest expression value. Determine this value of x to the nearest hundredth.

92. Work with other students on this exercise.

Part 1: Evaluate the three expressions $\frac{n^2 - 1}{2}$, n, $\frac{n^2 + 1}{2}$, using odd values of n: 1, 3, 5, 7, etc. Make a chart like the one below and complete it.

n	$a = \frac{n^2 - 1}{2}$	$b = n$	$c = \frac{n^2 + 1}{2}$	a^2	b^2	c^2
1						
3						
5						
7						
9						
11						
13						
15						

Part 2: The numbers a, b, and c that you get in each row have a surprising relationship to each other. Complete the last three columns and work together to discover this relationship. You may want to find out more about the history of this famous number pattern.

93. In exercise 92 you investigated the numbers obtained by evaluating the following expressions for odd positive integer values of n: $\frac{n^2 - 1}{2}$, n, $\frac{n^2 + 1}{2}$. Work with other students to investigate what three numbers you get when you evaluate for a *negative* odd value. Does the pattern you observed before still hold? Try several negative odd numbers to test the pattern.

Have no fear of fractions—does the pattern work with fractions? Try even integers. Is there a pattern for the three numbers obtained when you begin with even integers?

94. Enjoying patterns in art, music, and language is common to all cultures, and many delight in and draw spiritual significance from patterns in numbers. One such set of patterns is that of the "magic" square. One of these squares appears in a famous etching by Albrecht Dürer, who lived from 1471 to 1528 in Europe. He was one of the first artists in Europe to use geometry to give perspective, a feeling of three dimensions, in his work. The magic square in his work is this one:

16	3	2	13
5	10	11	8
9	6	7	12
4	15	14	1

Why is this square "magic"? It is magic because every row, every column, and both diagonals add to the same number. In this square there are 16 spaces for the numbers 1 through 16.

Part 1: What number does each row and column add to?

Write the square that you obtain by adding -17 to each number. Is this still a magic square? If so, what number does each column and row add to? If you add 5 to each number in the original magic square, do you still have a magic square? You have been studying the operations of addition, multiplication, subtraction, and division with integers and with rational numbers. What operations can you perform on this magic square and still have a magic square? Try to find something that will not work. Use algebra to help you decide what will work and what won't. Write a description of your work and explain your conclusions.

Part 2: Here is the oldest published magic square. It is from China, about 250 B.C. Legend has it that it was brought from the River Lo by a turtle to Emperor Yii, who was a hydraulic engineer.

4	9	2
3	5	7
8	1	6

Check to make sure that this is a magic square. Work together to decide what operation might be done to every number in the magic square to make the sum of each row, column, and diagonal the *opposite* of what it is now. What would you do to every number to cause the sum of each row, column, and diagonal to equal zero?

95. Use the Internet to research magic squares such as the one appearing in Dürer's work (see the previous exercise).

Answers

1. -22 **3.** 2 **5.** -20 **7.** 12 **9.** 4 **11.** 83 **13.** -20 **15.** -40 **17.** 14 **19.** -9 **21.** 2 **23.** 2 **25.** -11 **27.** 1 **29.** 11 **31.** -56 **33.** -8 **35.** 91 **37.** 29 **39.** 9 **41.** 16 **43.** 2 **45.** -16 **47.** -72 **49.** 5.625 Ω **51.** 30 in. **53.** $1,875 **55.** 14°F **57.** $19.09 **59.** True **61.** False **63.** False **65.** never **67.** -15.3 **69.** -1.1 **71.** 1.1 **73.** 14 **75.** -90 **77.** $-77{,}922$ **79.** $-1{,}638{,}036$ **81.** 130,217 **83.** -4.6 **85.** 12 μg/mL **87.** 0.674 **89.** Above and Beyond **91.** **(a)** 0, 80, 96, 72, 32, 0; **(b)** 2; **(c)** 94.5, 95.744, 96.492, 96.768, 96.596, 96, 95.004, 93.632, 91.908, 89.856, 87.5; **(d)** 1.8; **(e)** 1.81 **93.** Above and Beyond **95.** Above and Beyond

1.3 Simplifying Algebraic Expressions

< 1.3 Objectives >

1 > Use the vocabulary associated with algebraic expressions

2 > Combine like terms

3 > Add algebraic expressions

4 > Subtract algebraic expressions

RECALL

The perimeter of a figure is the distance around that figure.

To find the perimeter of a rectangle, we add 2 times the length and 2 times the width. In algebra, we write this as

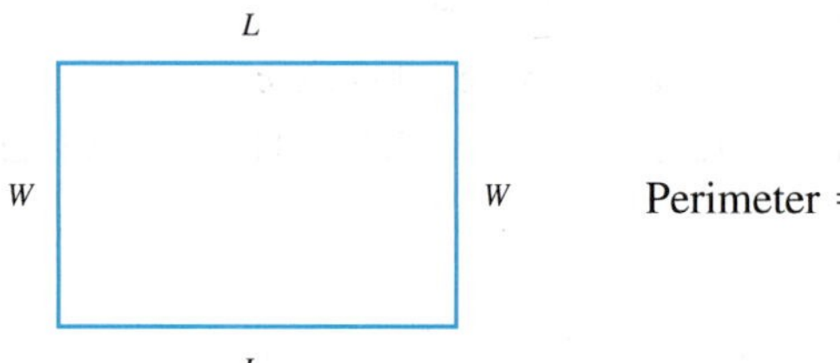

$$\text{Perimeter} = 2L + 2W$$

NOTE

If a variable has no exponent, it is raised to the power 1.

We call $2L + 2W$ an **algebraic expression,** or more simply an **expression.** Recall that an expression is a mathematical idea written symbolically. It is a meaningful collection of numbers, variables, and operations.

Some expressions are

$5x^2$ $\quad$ $3a + 2b$ $\quad$ $4x^3 - 2y + 1$ $\quad$ $3(x^2 + y^2)$

Addition and subtraction signs break expressions into smaller parts called *terms.*

Definition

Term — A **term** (more accurately, a *polynomial term*) can be written as a number or the product of a number and one or more variables and their whole-number exponents.

In an expression, each sign (+ or −) is a part of the term that follows the sign.

Example 1 Identifying Terms

< Objective 1 >

NOTE

Each term "owns" the sign that precedes it.

(a) $5x^2$ has one term.

(b) $\underbrace{3a}_{\text{Term}} + \underbrace{2b}_{\text{Term}}$ has two terms: $3a$ and $2b$.

(c) $\underbrace{4x^3}_{\text{Term}} \underbrace{- 2y}_{\text{Term}} \underbrace{+ 1}_{\text{Term}}$ has three terms: $4x^3$, $-2y$, and 1.

(d) $x - y$ has two terms: x and $-y$.

(e) (3)(2) is a term because we can write the product as the number 6.

Check Yourself 1

List the terms of each expression.

(a) $2b^4$ **(b)** $5m + 3n$ **(c)** $2s^2 - 3t - 6$

A term may have any number of factors. For instance, $5xy$ is a term. Its factors are 5, x, and y. The number factor of a term is called the **numerical coefficient.** For the term $5xy$, the numerical coefficient is 5.

Example 2 Identifying Numerical Coefficients

(a) $4a$ has the numerical coefficient 4.

(b) $6a^3b^4c^2$ has the numerical coefficient 6.

(c) $-7m^2n^3$ has the coefficient -7.

(d) x has the coefficient 1 since $x = 1 \cdot x$.

(e) $(4)(2)x^2$ has the coefficient 8 because we can write the expression as $8x^2$.

NOTE

We usually use *coefficient* instead of "numerical coefficient."

Check Yourself 2

Give the numerical coefficient for each term.

(a) $8a^2b$ **(b)** $-5m^3n^4$ **(c)** y

If terms contain exactly the *same variables* raised to the *same powers,* they are called **like terms.**

Example 3 Identifying Like Terms

(a) Each pair represents like terms.

$6a$ and $7a$

$5b^2$ and b^2

$10x^2y^3z$ and $-6x^2y^3z$

Each pair of terms has the same variables, with matching variables raised to the same power—the coefficients do not need to be the same.

(b) These are *not* like terms.

Different variables

$6a$ and $7b$

Different exponents

$5b^2$ and b^3

Different exponents

$3x^2y$ and $4xy^2$

Check Yourself 3

List the like terms.

$5a^2b$ ab^2 a^2b $-3a^2$ $4ab$ $3b^2$ $-7a^2b$

We can always combine like terms into a single term.

$$\underbrace{x + x}_{2x} + \underbrace{x + x + x + x + x}_{5x} = \underbrace{x + x + x + x + x + x + x}_{7x}$$

NOTES

Here we use the distributive property.

You don't have to write all this out—just do it mentally!

Rather than having to write out all those x's, try

$2x + 5x = (2 + 5)x = 7x$

In the same way,

$9b + 6b = (9 + 6)b = 15b$

and $10a - 4a = (10 - 4)a = 6a$

This leads us to a procedure for combining like terms.

Step by Step

Combining Like Terms

To combine like terms:

Step 1 Add or subtract the numerical coefficients.

Step 2 Attach the common variables.

Combining like terms is one of the elements of *simplifying an expression.*

Example 4 Combining Like Terms

< Objective 2 >

RECALL

When any factor is multiplied by 0, the product is 0.

Combine like terms.

(a) $8m + 5m = (8 + 5)m = 13m$

(b) $5pq^3 - 4pq^3 = 1pq^3 = pq^3$ We do not write the coefficient when it is 1.

(c) $7a^3b^2 - 7a^3b^2 = 0a^3b^2 = 0$

Check Yourself 4

Combine like terms.

(a) $6b + 8b$ **(b)** $12x^2 - 3x^2$
(c) $8xy^3 - 7xy^3$ **(d)** $9a^2b^4 - 9a^2b^4$

NOTE

When *simplifying an expression*, always combine like terms.

When no coefficient shown, it is understood to be one.

$x = 1 \cdot x$

We combine like terms in the same way, when we have fractions, decimals, and coefficients equal to one.

Example 5 Combining Like Terms

Combine like terms.

(a) $8x + x = (8 + 1)x$
$= 9x$

(b) $x - 0.65x = (1 - 0.65)x$
$= 0.35x$

NOTE

Because of the way we multiply fractions,

$\frac{x}{5} = \frac{1}{5}x$ and $\frac{3x}{2} = \frac{3}{2}x$.

We use both ways of writing such terms in this text.

(c) $\frac{2}{5}x - \frac{1}{5}x = \left(\frac{2}{5} - \frac{1}{5}\right)x$

$= \frac{x}{5}$

(d) $x + \frac{x}{2} = \left(1 + \frac{1}{2}\right)x$

$= \frac{3}{2}x$

Check Yourself 5

Combine like terms.

(a) $x - \frac{2}{3}x$ **(b)** $2.5x + 0.75x$ **(c)** $\frac{3x}{4} + \frac{3x}{2}$

Here are some expressions involving more than two terms. The idea is the same.

Example 6 Combining Like Terms

RECALL

The distributive property can be used with any number of like terms.

Combine like terms.

(a) $5ab - 2ab + 3ab$

$= (5 - 2 + 3)ab = 6ab$

(b) $8x - 2x + 5y$ Only like terms can be combined.

$= 6x + 5y$

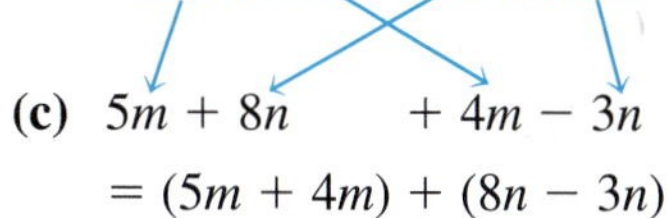

(c) $5m + 8n + 4m - 3n$ Rearrange the order of the terms using the associative and commutative properties of addition.

$= (5m + 4m) + (8n - 3n)$

$= 9m + 5n$

(d) $4x^2 + 2x - 3x^2 + x$

$= (4x^2 - 3x^2) + (2x + x)$

$= x^2 + 3x$

NOTE

With practice you will be doing this mentally rather than writing out these steps.

Check Yourself 6

Combine like terms.

(a) $4m^2 - 3m^2 + 8m^2$ **(b)** $9ab + 3a - 5ab$
(c) $4p + 7q + 5p - 3q$

> CAUTION

Be careful when moving terms. Remember that they own the signs in front of them.

RECALL

Like terms have exactly the same variables to the same powers.

As these examples illustrate, combining like terms often means changing the grouping and the order in which the terms are written. Again all this is possible because of the properties of addition.

We can remove the parentheses when adding two expressions.

$$(5x^2 + 3x + 4) + (4x^2 + 5x - 6) = 5x^2 + 3x + 4 + 4x^2 + 5x - 6$$

Because addition is *commutative,* we can add in any order we wish. In this case, we place *like terms* next to each other.

$$= \underbrace{5x^2 + 4x^2}_{\text{Like terms } (x^2)} + \underbrace{3x + 5x}_{\text{Like terms } (x)} + \underbrace{4 - 6}_{\text{Like terms (constants)}}$$

Addition is associative, so we can *group* the like terms together.

$= (5x^2 + 4x^2) + (3x + 5x) + (4 - 6)$

By combining like terms, we reach a final sum.

$= 9x^2 + 8x - 2$

Alternatively, we could perform the addition in a vertical format. When using this method, be certain to align like terms in each column. In a vertical format the same addition looks like this.

$$\begin{array}{r} 5x^2 + 3x + 4 \\ + \; 4x^2 + 5x - 6 \\ \hline 9x^2 + 8x - 2 \end{array}$$

Much of this work can be done mentally. You can then write the sum directly by locating like terms and combining.

Example 7 Combining Like Terms

< Objective 3 >

Add $3x - 5$ and $2x + 3$.

Write the sum.

$(3x - 5) + (2x + 3)$

$= 3x - 5 + 2x + 3 = 5x - 2$

Like terms   Like terms

NOTE

We combine like terms mentally.

$3x + 2x = 5x$ and

$-5 + 3 = -2$

Check Yourself 7

Add $6x^2 + 2x$ and $4x^2 - 7x$.

The distributive property provides us with a rule to remove grouping symbols when subtracting expressions.

Property

Removing Grouping Symbols When Subtracting

When subtracting expressions, if a minus sign (−) appears in front of a set of parentheses or other grouping symbols, the parentheses can be removed by changing the sign of each term inside the parentheses.

When applying this rule, we are actually distributing the negative. This is illustrated in Example 8.

Example 8 Removing Parentheses

In each case, remove the parentheses.

(a) $-(2x + 3y) = -2x - 3y$ Change each sign when removing the parentheses.

(b) $m - (5n - 3p) = m - 5n + 3p$

Sign changes

(c) $2x - (-3y + z) = 2x + 3y - z$

Sign changes

NOTE

This is the distributive property.

$-(2x + 3y) = (-1)(2x + 3y)$

$= -2x - 3y$

Check Yourself 8

Remove the parentheses.

(a) $-(3m + 5n)$ (b) $-(5w - 7z)$
(c) $3r - (2s - 5t)$ (d) $5a - (-3b - 2c)$

Subtracting expressions is now a matter of using the rule to remove the parentheses and then combining the like terms.

Example 9 Subtracting Expressions

< Objective 4 >

RECALL

The expression following *from* is written first in the problem.

(a) Subtract $5x - 3$ from $8x + 2$.

$(8x + 2) - (5x - 3)$

$= 8x + 2 - 5x + 3$

Sign changes

$= 3x + 5$ Combine like terms: $8x - 5x = 3x$ and $2 + 3 = 5$.

RECALL

Combine like terms.

$8x^2 - 4x^2 = 4x^2$

$5x + 8x = 13x$

$-3 - 3 = -6$

(b) Subtract $4x^2 - 8x + 3$ from $8x^2 + 5x - 3$.

$(8x^2 + 5x - 3) - (4x^2 - 8x + 3)$

$= 8x^2 + 5x - 3 - 4x^2 + 8x - 3$

Sign changes

$= 4x^2 + 13x - 6$

Check Yourself 9

(a) Subtract $7x + 3$ from $10x - 7$.
(b) Subtract $5x^2 - 3x + 2$ from $8x^2 - 3x - 6$.

NOTE

There are several other rules used to determine if a particular expression is in **simplest** form. You will learn more of them as we progress through this text.

Often, rather than use instructions such as "add" or "remove the parentheses," we just use the phrase "simplify." You have just learned two of the elements we use to determine whether an expression is **simplified.**

Property

Simplifying an Expression

1. Remove any grouping symbols that can be removed through addition or subtraction.
2. Combine all like terms.

You must always simplify an expression when giving your final answer.

In Example 10, we look at a business and finance application of some of the ideas presented in this section.

Example 10 A Business and Finance Application

S-Bar Electronics, Inc., sells a certain server for \$1,410. It pays the manufacturer \$849 for each server, and there are \$4,500 per week in fixed costs associated with the servers. Find an equation that represents the profit S-Bar Electronics earns by buying and selling these servers.

NOTE

A business can compute the profit it earns on a product by subtracting the costs associated with the product from the revenue earned by that product. We write $P = R - C$

Let x be the number of servers bought and sold during the week.

Then, the revenue earned by S-Bar from these servers can be modeled by the formula

$R = 1{,}410x$

The cost can be modeled with the formula

$C = 849x + 4{,}500$

The profit can be modeled by the difference between the revenue and the cost.

$P = 1{,}410x - (849 + 4{,}500)$

$P = 1{,}410x - 849x - 4{,}500$

NOTE

If the profit is negative, then the company suffered a loss.

Simplify the given profit formula.

The like terms are $1{,}410x$ and $-849x$. We combine these to give a simplified formula

$P = 561x - 4{,}500$

Check Yourself 10

S-Bar Electronics, Inc., also sells 27-in. LCD monitors for \$399. Each monitor costs them \$189. Additionally, there are weekly fixed costs of \$3,150 associated with the sale of the monitors. Model the profit earned on the sale of y monitors.

Check Yourself ANSWERS

1. **(a)** $2b^4$; **(b)** $5m$, $3n$; **(c)** $2s^2$, $-3t$, -6 **2.** **(a)** 8; **(b)** -5; **(c)** 1 **3.** The like terms are $5a^2b$, a^2b, and $-7a^2b$. **4.** **(a)** $14b$; **(b)** $9x^2$; **(c)** xy^3; **(d)** 0 **5.** **(a)** $\frac{x}{3}$; **(b)** $3.25x$; **(c)** $\frac{9}{4}x$ **6.** **(a)** $9m^2$; **(b)** $4ab + 3a$; **(c)** $9p + 4q$ **7.** $10x^2 - 5x$ **8.** **(a)** $-3m - 5n$; **(b)** $-5w + 7z$; **(c)** $3r - 2s + 5t$; **(d)** $5a + 3b + 2c$ **9.** **(a)** $3x - 10$; **(b)** $3x^2 - 8$ **10.** $P = 210y - 3{,}150$

Reading Your Text

These fill-in-the-blank exercises will help you understand some of the key vocabulary used in this section. The answers to these exercises are in the Answers Appendix in the back of the text.

(a) If a variable appears without an exponent, it is understood to be raised to the ________ power.

(b) A ________ can be written as a number or the product of a number and one or more variables and their exponents.

(c) A term may have any number of ________.

(d) In the term $5xy$, the factor 5 is called the ________.

Skills | Calculator/Computer | Career Applications | Above and Beyond

1.3 exercises

< Objective 1 >

List the terms of each expression.

1. $5a + 2$
2. $7a - 4b$
3. $5x^4$
4. $3x^2$
5. $3x^2 + 3x - 7$
6. $2a^3 - a^2 + a$

Identify the like terms in each list.

7. $5ab, 3b, 3a, 4ab$
8. $9m^2, 8mn, 5m^2, 7m$
9. $4xy^2, 2x^2y, 5x^2, -3x^2y, 5y, 6x^2y$ 
10. $8a^2b, 4a^2, 3ab^2, -5a^2b, 3ab, 5a^2b$

< Objective 2 >

Combine like terms.

11. $6p + 9p$
12. $6a^2 + 8a^2$
13. $7b^3 + 10b^3$ 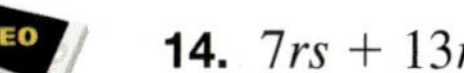
14. $7rs + 13rs$
15. $21xyz + 7xyz$
16. $4n^2m + 11n^2m$
17. $9z^2 - 3z^2$
18. $7m - 6m$
19. $5a^3 - 5a^3$
20. $9xy - 13xy$
21. $16p^2q - 17p^2q$
22. $7cd - 7cd$
23. $6p^2q - 21p^2q$
24. $8r^3s^2 - 17r^3s^2$
25. $10x^2 - 7x^2 + 3x^2$
26. $13uv + 5uv - 12uv$
27. $-6c + 3d + 5c$
28. $5m^2 - 3m + 6m^2$
29. $4x + 4y - 7x - 5y$
30. $7a - 4a^2 - 13a + 9a^2$
31. $2a - 7b - 3 + 2a - 3b + 2$
32. $5p^2 - 2p - 8 - 7p^2 + 5p + 6$

Remove the parentheses in each expression. Simplify where possible.

33. $-(2a + 3b)$
34. $-(7x - 4y)$
35. $5a - (2b - 3c)$
36. $7x - (4y + 3z)$
37. $3x - (4y + 5x)$
38. $10m - (3m - 2n)$
39. $5p - (-3p + 2q)$
40. $8d - (-7c - 2d)$

< Objective 3 >

Add.

41. $6a - 5$ and $3a + 9$
42. $9x + 3$ and $3x - 4$
43. $-7p^2 + 9p$ and $4p^2 - 5p$
44. $2m^2 + 3m$ and $6m^2 - 8m$
45. $3x^2 - 2x$ and $-5x^2 + 2x$
46. $3p^2 + 5p$ and $-7p^2 - 5p$
47. $2x^2 + 5x - 3$ and $3x^2 - 7x + 4$
48. $4d^2 - 8d + 7$ and $5d^2 - 6d - 9$
49. $(2b^2 + 8) + (5b + 8)$
50. $(5p - 2) + (4p^2 - 7p)$
51. $(8y^3 - 5y^2) + (5y^2 - 2y)$
52. $(9x^4 - 2x^2) + (2x^2 + 3)$
53. $(3x^2 - 7x^3) + (-5x^2 + 4x^3)$
54. $(9m^3 - 2m) + (-6m - 4m^3)$
55. $(4x^2 - 2 + 7x) + (5 - 8x - 6x^2)$
56. $(5b^3 - 8b + 2b^2) + (3b^2 - 7b^3 + 5b)$

< Objective 4 >

Subtract.

57. $x + 2$ from $3x - 5$
58. $x - 2$ from $3x + 5$
59. $3m^2 - 2m$ from $4m^2 - 5m$
60. $9a^2 - 5a$ from $11a^2 - 10a$
61. $6y^2 + 5y$ from $4y^2 + 5y$
62. $9x^2 - 2x$ from $6x^2 - 2x$
63. $x^2 - 4x - 3$ from $3x^2 - 5x - 2$
64. $3x^2 - 2x + 4$ from $5x^2 - 8x - 3$
65. $(8a^2 - 9a) - (3a + 7)$
66. $(4x^3 - 5x) - (3x^3 + x^2)$
67. $(-9p^2 + 4p) - (2p - 5p^2)$
68. $(3y^2 - 2y) - (7y - 3y^2)$
69. $(3x^2 - 8x + 7) - (x^2 - 5 - 8x)$
70. $(4x^3 + x - 3x^2) - (4x - 2x^2 + 4x^3)$

Perform the indicated operations.

71. Subtract $3b + 2$ from the sum of $4b - 2$ and $5b + 3$.

72. Subtract $5m - 7$ from the sum of $2m - 8$ and $9m - 2$.

73. Subtract $5x^2 + 7x - 6$ from the sum of $2x^2 - 3x + 5$ and $3x^2 + 5x - 7$.

74. Subtract $4x^2 - 5x - 3$ from the sum of $x^2 - 3x - 7$ and $2x^2 - 2x + 9$.

75. Subtract $2x^2 - 3x$ from the sum of $4x^2 - 5$ and $2x - 7$.

76. Subtract $5a^2 - 3a$ from the sum of $3a - 3$ and $5a^2 + 5$.

77. Subtract the sum of $3y^2 - 3y$ and $5y^2 + 3y$ from $2y^2 - 8y$.

78. Subtract the sum of $3y^3 + 7y^2$ and $5y^3 - 7y^2$ from $4y^3 - 5y^2$.

79. $[(9x^2 - 3x + 5) - (3x^2 + 2x - 1)] - (x^2 - 2x - 3)$

80. $[(5x^2 + 2x - 3) - (-2x^2 + x - 2)] - (2x^2 + 3x - 5)$

81. **Geometry** A rectangle has sides of $8x + 9$ and $6x - 7$. Find an expression that represents its perimeter.

82. **Geometry** A triangle has sides $4x + 7$, $6x + 3$, and $2x - 5$. Find an expression that represents its perimeter. 

83. **Business and Finance** The cost of producing x units of an item is $C = 150 + 25x$. The revenue for selling x units is $R = 90x - x^2$. The profit is given by the revenue minus the cost. Find an expression that represents profit.

84. **Business and Finance** The revenue for selling y units is $R = 3y^2 - 2y + 5$, and the cost of producing y units is $C = y^2 + y - 3$. Find an expression that represents profit.

85. **Construction** A wooden beam is $(3y^2 + 3y - 2)$ meters (m) long. If a piece $(y^2 - 8)$ m is cut, find an expression that represents the length of the remaining piece of beam.

86. **Construction** A steel girder is $(9y^2 + 6y - 4)$ m long. Two pieces are cut from the girder. One has length $(3y^2 + 2y - 1)$ m and the other has length $(4y^2 + 3y - 2)$ m. Find the length of the remaining piece.

87. **Geometry** Find an expression for the perimeter of the given triangle.

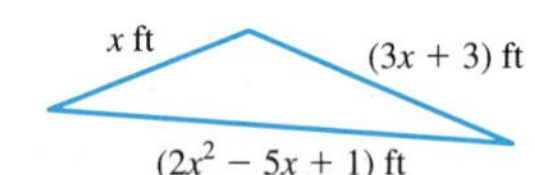

88. **Geometry** Find an expression for the perimeter of the given rectangle.

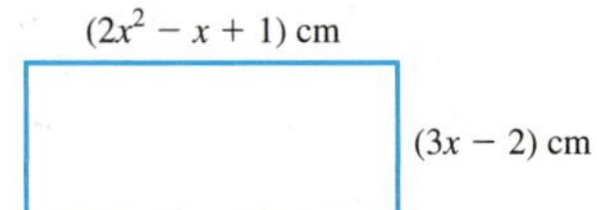

89. **Geometry** Find an expression for the perimeter of the given figure.

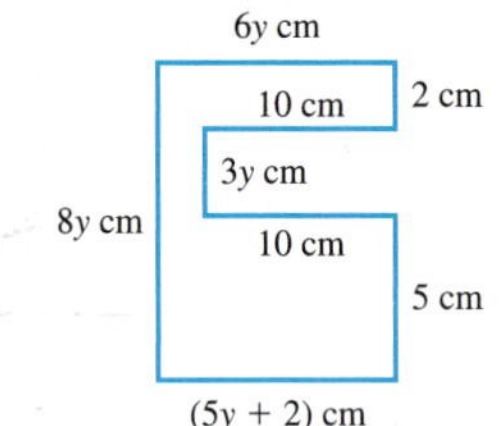

90. **Geometry** Find the perimeter of the accompanying figure.

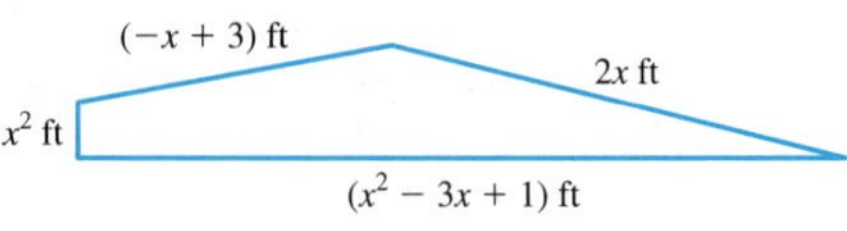

Determine whether each statement is **true** *or* **false.**

91. For two terms to be *like terms,* the numerical coefficients must match.

92. The key property that allows like terms to be combined is the distributive property.

Complete each statement with **always, sometimes,** *or* **never.**

93. Like terms can __________ be combined.

94. When adding two expressions, the terms can __________ be rearranged.

Skills | **Calculator/Computer** | Career Applications | Above and Beyond

Use a calculator to evaluate each expression. Round your answers to the nearest tenth.

95. $7x^2 - 5y^3$ for $x = 7.1695$ and $y = 3.128$

96. $2x^2 + 3y + 5x$ for $x = 3.61$ and $y = 7.91$

97. $4x^2y + 2xy^2 - 5x^3y$ for $x = 1.29$ and $y = 2.56$

98. $3x^3y - 4xy + 2x^2y^2$ for $x = 3.26$ and $y = 1.68$

Skills | Calculator/Computer | **Career Applications** | Above and Beyond

99. **Mechanical Engineering** A primary beam can support a load of $54p$. A second beam is added that can support a load of $32p$. What is the total load that the two beams can support? VIDEO

100. **Mechanical Engineering** Two objects are spinning on the same axis. The moment of inertia of the first object is $\frac{6^3}{12}b$. The moment of inertia of the second object is given by $\frac{30^3}{36}b$. The total moment of inertia is given by the sum of the moments of inertia of the two objects. Write a simplified expression for the total moment of inertia for the two objects described.

101. **Allied Health** A person's body mass index (BMI) can be calculated using their height h, in inches, and their weight w, in pounds, with the formula $\frac{703w}{h^2}$

Compute the BMI of a 69-inch, 190-pound man (to the nearest tenth).

102. **Allied Health** A person's body mass index (BMI) can be calculated using their height h, in centimeters, and their weight w, in kilograms, with the formula $\frac{10{,}000w}{h^2}$

Compute the BMI of a 160-cm, 70-kg woman (to the nearest tenth).

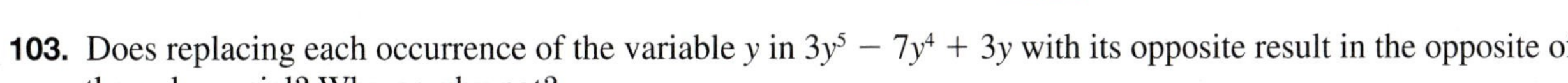

103. Does replacing each occurrence of the variable y in $3y^5 - 7y^4 + 3y$ with its opposite result in the opposite of the polynomial? Why or why not?

104. Write a paragraph explaining the difference between n^2 and $2n$.

105. Complete the explanation "x^3 and $3x$ are not the same because. . . ."

106. Complete the statement "$x + 2$ and $2x$ are different because. . . ."

107. Write an English phrase for each algebraic expression.

(a) $2x^3 + 5x$ **(b)** $(2x + 5)^3$ **(c)** $6(n + 4)^2$

108. Work with another student to complete this exercise. Place >, <, or = in the blanks in these statements.

1^2_____2^1

2^3_____3^2

3^4_____4^3

4^5_____5^4

Write an algebraic statement for the pattern of numbers. Do you think this is a pattern that continues? Add more examples and extend the pattern to the general case by writing the pattern in algebraic notation. Write a short paragraph stating your conjecture.

109. Compute and fill in the blanks.

Case 1: $1^2 - 0^2 =$ _____

Case 2: $2^2 - 1^2 =$ _____

Case 3: $3^2 - 2^2 =$ _____

Case 4: $4^2 - 3^2 =$ _____

Based on the pattern you see in these four cases, predict the value of case 5: $5^2 - 4^2$. Compute case 5 to check your prediction. Write an expression for case *n*. Describe in words the pattern that you see in this exercise.

Answers

1. $5a, 2$ **3.** $5x^4$ **5.** $3x^2, 3x, -7$ **7.** $5ab, 4ab$ **9.** $2x^2y, -3x^2y, 6x^2y$ **11.** $15p$ **13.** $17b^3$ **15.** $28xyz$ **17.** $6z^2$ **19.** 0 **21.** $-p^2q$ **23.** $-15p^2q$ **25.** $6x^2$ **27.** $-c + 3d$ **29.** $-3x - y$ **31.** $4a - 10b - 1$ **33.** $-2a - 3b$ **35.** $5a - 2b + 3c$ **37.** $-2x - 4y$ **39.** $8p - 2q$ **41.** $9a + 4$ **43.** $-3p^2 + 4p$ **45.** $-2x^2$ **47.** $5x^2 - 2x + 1$ **49.** $2b^2 + 5b + 16$ **51.** $8y^3 - 2y$ **53.** $-3x^3 - 2x^2$ **55.** $-2x^2 - x + 3$ **57.** $2x - 7$ **59.** $m^2 - 3m$ **61.** $-2y^2$ **63.** $2x^2 - x + 1$ **65.** $8a^2 - 12a - 7$ **67.** $-4p^2 + 2p$ **69.** $2x^2 + 12$ **71.** $6b - 1$ **73.** $-5x + 4$ **75.** $2x^2 + 5x - 12$ **77.** $-6y^2 - 8y$ **79.** $5x^2 - 3x + 9$ **81.** $28x + 4$ **83.** $-x^2 + 65x - 150$ **85.** $(2y^2 + 3y + 6)$ m **87.** $(2x^2 - x + 4)$ ft **89.** $(22y + 29)$ cm **91.** False **93.** always **95.** 206.8 **97.** 6.5 **99.** $86p$ **101.** 28.1 **103.** Above and Beyond **105.** Above and Beyond **107.** Above and Beyond **109.** Above and Beyond

1.4 Solving Equations with the Addition Property

< 1.4 Objectives >

1 > Determine whether a number is a solution to an equation

2 > Use the addition property to solve equations

3 > Use equations to solve applications

Tips for Student Success

Don't Procrastinate!

1. Complete your math homework while you are still fresh. Late at night, your mind is tired, making it difficult to understand new concepts.
2. Complete your homework the day it is assigned. The more recent the explanation, the easier it is to recall.
3. When you finish your homework, try reading the next section in the text. This will give you a sense of direction the next time you have class.

Remember, in a typical math class, you are expected to do 2 or 3 hours of homework for each hour of class time. This means 2 to 3 hours per day. Schedule the time and stay on schedule.

In this chapter, we work with one of the most important tools of mathematics—the equation. The ability to recognize and solve various types of equations is probably the most useful algebraic skill you will learn. We continue to build upon the methods of this chapter throughout the remainder of the text. To start, we define an *equation.*

Definition

Equation — An **equation** is a mathematical statement that two expressions are equal.

NOTE

An equation such as

$x + 3 = 5$

is called a **conditional equation** because it can be either true or false depending on the value of the variable.

Some examples are $3 + 4 = 7$, $x + 3 = 5$, $P = 2L + 2W$. As you can see, an equal sign (=) separates the two expressions. These expressions are often called the *left side* and the *right side* of the equation.

$$\underbrace{x + 3}_{\text{Left side}} \underset{\text{Equals}}{=} \underbrace{5}_{\text{Right side}}$$

An equation may be either true or false. For instance, $3 + 4 = 7$ is true because both sides name the same number. What about an equation such as $x + 3 = 5$ that has a variable on one side? Any number can replace x in the equation. However, only one number makes this equation a true statement.

$$\text{If} \quad x = \begin{cases} 1 & 1 + 3 = 5 \text{ is false} \\ 2 & 2 + 3 = 5 \text{ is true} \\ 3 & 3 + 3 = 5 \text{ is false} \end{cases}$$

The number 2 is called a *solution* (or *root*) of the equation $x + 3 = 5$ because substituting 2 for x gives a true statement. 2 is the only solution to this equation.

Definition

Solution

A **solution** to an equation is any value for the variable that makes the equation a true statement.

Example 1 Verifying a Solution

< Objective 1 >

NOTE

Until the left side equals the right side, we place a question mark over the equal sign.

(a) Is 3 a solution to the equation $2x + 4 = 10$?

To find out, replace x with 3 and evaluate $2x + 4$ on the left.

Left side		Right side
$2 \cdot (3) + 4$	$\stackrel{?}{=}$	10
$6 + 4$	$\stackrel{?}{=}$	10
10	$=$	10

Since $10 = 10$ is a true statement, 3 is a solution to the equation.

(b) Is $\frac{5}{3}$ a solution of the equation $3x - \frac{2}{3} = 2x + 1$?

To find out, replace x with $\frac{5}{3}$ and evaluate each side separately.

RECALL

Always apply the rules for the order of operations. Multiply first; then add or subtract.

Left side		Right side
$3 \cdot \left(\frac{5}{3}\right) - \frac{2}{3}$	$\stackrel{?}{=}$	$2 \cdot \left(\frac{5}{3}\right) + 1$
$\frac{15}{3} - \frac{2}{3}$	$\stackrel{?}{=}$	$\frac{10}{3} + \frac{3}{3}$
$\frac{13}{3}$	$=$	$\frac{13}{3}$

Because the two sides name the same number, we have a true statement, and $\frac{5}{3}$ is a solution.

Check Yourself 1

For the equation

$2x - 1 = x + 5$

(a) Is 6 a solution? **(b)** Is $\frac{8}{3}$ a solution?

NOTE

The equation $x^2 = 9$ is an example of a second-degree or **quadratic equation**. We will learn to solve them in Chapters 6 and 8.

You may be wondering whether an equation can have more than one solution. It certainly can. For instance,

$$x^2 = 9$$

has two solutions. They are 3 and -3 because

$$(3)^2 = 9 \qquad \text{and} \qquad (-3)^2 = 9$$

In this chapter, we generally work with *linear equations.* These are equations that can be put into the form

$$ax + b = 0$$

in which the variable is x, a and b are numbers, and a is not equal to 0. In a linear equation, the variable can only appear to the first power. No other power (x^2, x^3, etc.) can appear. Linear equations are also called **first-degree equations.** The *degree* of an equation in one variable is the highest power to which the variable is raised. In the equation $5x^4 - 9x^2 + 7x - 2 = 0$, the highest power to which the x is raised is four. Therefore, it is a fourth-degree equation.

Property

Solutions to Linear Equations

Linear equations in one variable are equations that can be written in the form

$ax + b = 0 \qquad a \neq 0$

Linear equations in one variable have exactly one solution.

Example 2 Identifying Expressions and Equations

Label each statement as an expression, a linear equation, or a nonlinear equation. Recall that an equation is a statement in which an equal sign separates two expressions.

(a) $4x + 5$ is an expression.

(b) $2x + 8 = 0$ is a linear equation.

(c) $3x^2 - 9 = 0$ is a nonlinear equation.

(d) $5x = 15$ is a linear equation.

(e) $\frac{3}{x} + 2 = 0$ is a nonlinear equation.

NOTE

Variables cannot be in the denominator of a linear equation.

Check Yourself 2

Label each as an expression, a linear equation, or a nonlinear equation.

(a) $2x^2 = 8$ **(b)** $2x - 3 = 0$
(c) $5x - 10$ **(d)** $2x + 1 = 7$
(e) $5 - \frac{6}{x} = 2x$

You can find the solution to an equation such as $x + 3 = 8$ by guessing the answer to the question "What plus 3 is 8?" Here the answer to the question is 5, which is also the solution to the equation. But for more complicated equations we need something more than guesswork. A better method is to transform the given equation to an *equivalent equation* whose solution can be found by inspection.

Definition

Equivalent Equations

Equations that have exactly the same solutions are called **equivalent equations.**

These are all equivalent equations.

$2x + 3 = 5 \qquad 2x = 2 \qquad \text{and} \qquad x = 1$

They all have the same solution, 1. We say that a linear equation is *solved* when it is transformed to an equivalent equation of the form

$x = \square$

The variable is alone on one side.

The other side is some number, the solution.

NOTE

In some cases we write the equation in the form

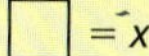

$\square = x$

The number is the solution whether the variable is isolated on either the left or the right.

The addition property of equality is the first property you need to transform an equation to an equivalent form.

Property

The Addition Property of Equality

If $\quad a = b$

then $\quad a + c = b + c$

Adding the same quantity to both sides of an equation gives an equivalent equation.

An equation is a statement that the two sides are equal. Adding the same quantity to both sides does not change the equality or "balance."

In Example 3 we apply this idea to solve an equation.

Example 3 Using the Addition Property to Solve an Equation

< Objective 2 >

> **NOTE**
>
> To check, replace x with 12 in the original equation.
>
> $x - 3 = 9$
>
> $(12) - 3 \stackrel{?}{=} 9$
>
> $9 = 9$ True
>
> Because we have a true statement, 12 is the solution.

Solve.

$x - 3 = 9$

Remember that our goal is to isolate x on one side of the equation. Because 3 is being subtracted from x, we can add 3 to remove it. We use the addition property to add 3 to both sides of the equation.

$$\begin{array}{rcl} x - 3 &=& 9 \\ +3 && +3 \\ \hline x &=& 12 \end{array}$$

Adding 3 "undoes" the subtraction and leaves x alone on the left.

Because 12 is the solution of the equivalent equation $x = 12$, it is the solution to our original equation.

Check Yourself 3

Solve and check.

$x - 5 = 4$

The addition property also allows us to add a negative number to both sides of an equation. This is really the same as subtracting the same quantity from both sides.

Example 4 Solving an Equation

Solve.

$x + 2 = \frac{11}{2}$

In this case, 2 is *added* to x on the left. We can use the addition property to subtract 2 from both sides. This "undoes" the addition and leaves the variable x alone on one side of the equation.

> **NOTE**
>
> Because subtraction is defined in terms of addition, we can also subtract the same quantity from both sides of an equation.

$$\begin{array}{rcl} x + 2 &=& \frac{11}{2} \\ -2 && -\frac{4}{2} \\ \hline x &=& \frac{7}{2} \end{array}$$

We subtract 2 from each side. $-\frac{4}{2} = -2$

The solution is $\frac{7}{2}$. To check, replace x with $\frac{7}{2}$.

$x + 2 = \frac{11}{2}$ The original equation

$\left(\frac{7}{2}\right) + 2 \stackrel{?}{=} \frac{11}{2}$ Substitute $x = \frac{7}{2}$.

$\frac{7}{2} + \frac{4}{2} \stackrel{?}{=} \frac{11}{2}$ $2 = \frac{4}{2}$

$\frac{11}{2} = \frac{11}{2}$ True

Check Yourself 4

Solve and check.

$x + 6 = \frac{11}{3}$

What if the equation has variable terms on both sides? You can use the addition property to add or subtract a term involving the variable to get the desired result.

Example 5 Solving an Equation

Solve.

$5x = 4x + 7$

We start by subtracting $4x$ from both sides of the equation. Do you see why? Remember that an equation is solved when we have an equivalent equation of the form $x = \square$.

$$\begin{array}{rrl} 5x = & 4x + 7 \\ -4x & -4x \\ \hline x = & 7 \end{array}$$

Subtracting $4x$ from both sides *removes* $4x$ from the right.

> **NOTE**
>
> Subtracting $4x$ is the same as adding $-4x$.

To check: Since 7 is a solution to the equivalent equation $x = 7$, it should be a solution of the original equation. To find out, replace x with 7.

$5 \cdot (7) \stackrel{?}{=} 4 \cdot (7) + 7$

$35 \stackrel{?}{=} 28 + 7$

$35 = 35$ True

Check Yourself 5

Solve and check.

$7x = 6x + 3$

Recall that addition can be set up either in a vertical format such as

$$\begin{array}{r} 256 \\ +192 \\ \hline 448 \end{array}$$

or in a horizontal format

$256 + 192 = 448$

When we use the addition property to solve an equation, the same choices are available. In our examples to this point we used the vertical format. In Example 6 we use the horizontal format. In the remainder of this text, we assume that you are familiar with both formats.

Example 6 Solving an Equation

Solve.

$7x - 8 = 6x$

We want all variables on *one* side of the equation. If we choose the left, we subtract $6x$ from both sides of the equation. This removes $6x$ from the right.

$7x - 8 - 6x = 6x - 6x$

$x - 8 = 0$

We want the variable alone, so we add 8 to both sides. This isolates x on the left.

$x - 8 + 8 = 0 + 8$

$x = 8$

We leave it to you to check that 8 is the solution.

Check Yourself 6

Solve and check.

$9x + 3 = 8x$

Often an equation has more than one variable term *and* more than one number, as in Example 7.

Example 7 Solving an Equation

Solve.

(a) $5x - 7 = 4x + 3$

We would like the variable terms on the left, so we start by subtracting $4x$ to remove that term from the right side of the equation.

$$\begin{array}{rr} 5x - 7 = & 4x + 3 \\ -4x \quad & -4x \\ \hline x - 7 = & 3 \end{array}$$

To isolate the variable, we add 7 to both sides to undo the subtraction on the left.

$$\begin{array}{rr} x - 7 = & 3 \\ +7 & +7 \\ \hline x \quad = & 10 \end{array}$$

The solution is 10. To check, replace x with 10 in the original equation.

$$5 \cdot (10) - 7 \stackrel{?}{=} 4 \cdot (10) + 3$$
$$43 = 43 \qquad \text{True}$$

(b) $12x + 6 = 11x + 6$

As before, we begin by bringing the variable terms to the same side. In this case, we subtract $11x$ from both sides to bring them to the left side.

$$\begin{array}{rr} 12x + 6 = & 11x + 6 \\ -11x \quad & -11x \\ \hline x + 6 = & 6 \end{array}$$

$12x - 11x = 1x = x$

To isolate the variable, we need to subtract 6 from the left side, so we subtract 6 from the right side as well.

$$\begin{array}{rr} x + 6 = & 6 \\ -6 & -6 \\ \hline x \quad = & 0 \end{array}$$

The solution is 0. To check, we replace every instance of x with 0 in the original equation.

Check

$$12x + 6 = 11x + 6$$
$$12(0) + 6 \stackrel{?}{=} 11(0) + 6$$
$$0 + 6 \stackrel{?}{=} 0 + 6$$
$$6 = 6 \qquad \text{True}$$

NOTE

You could just as easily have added 7 to both sides and *then* subtracted $4x$. The result is the same. In fact, some students prefer to combine the two steps.

NOTES

Zero is a perfectly legitimate solution. Zero is just a number, so

$x = 0$

is equivalent to

$12x + 6 = 11x + 6$

We check 0 as a solution, and it checks.

Check Yourself 7

Solve and check.

(a) $4x - 5 = 3x + 2$ (b) $4x - 5 = 5x - 5$
(c) $6x + 2 = 5x - 4$

RECALL

By *simplify,* we mean to combine all like terms.

When solving an equation, you should always simplify each side as much as possible before using the addition property.

Example 8 Simplifying an Equation

Solve $5 + 8x - 2 = 2x - 3 + 5x$.

Like terms Like terms

$$5 + 8x - 2 = 2x - 3 + 5x$$

Notice that like terms appear on both sides of the equation. We start by combining the numbers on the left (5 and -2). Then we combine the like terms ($2x$ and $5x$) on the right.

$$3 + 8x = 7x - 3$$

Now we can apply the addition property, as before.

$$\begin{aligned} 3 + 8x &= 7x - 3 \\ -7x &= -7x \quad \text{Subtract } 7x. \\ \hline 3 + x &= -3 \\ -3 \qquad &\quad\; -3 \quad \text{Subtract 3 to isolate } x. \\ \hline x &= -6 \end{aligned}$$

The solution is -6. To check, always return to the original equation. That catches any possible errors in simplifying. Replacing x with -6 gives

$$\begin{aligned} 5 + 8(-6) - 2 &\stackrel{?}{=} 2(-6) - 3 + 5(-6) \\ 5 - 48 - 2 &\stackrel{?}{=} -12 - 3 - 30 \\ -45 &= -45 \quad \text{True} \end{aligned}$$

Check Yourself 8

Solve and check.

(a) $3 + 6x + 4 = 8x - 3 - 3x$ (b) $5x + 21 + 3x = 20 + 7x - 2$

We may have to apply some other properties when solving equations. In Example 9, we use the distributive property to clear an equation of parentheses.

Example 9 Using the Distributive Property to Solve Equations

RECALL

$$\begin{aligned} 2(3x + 4) &= 2(3x) + 2(4) \\ &= 6x + 8 \end{aligned}$$

Solve.

$$2(3x + 4) = 5x - 6$$

Applying the distributive property on the left gives

$$6x + 8 = 5x - 6$$

We proceed as before.

$$\begin{aligned} 6x + 8 &= 5x - 6 \\ -5x \quad &= -5x \\ \hline x + 8 &= \quad - 6 \\ -8 \quad &\quad\; - 8 \\ \hline x &= -14 \end{aligned}$$

Subtract $5x$.

Subtract 8.

The solution is -14. We leave it to you to check this result.

Remember: Always return to the original equation to check.

Check Yourself 9

Solve and check each equation.

(a) $4(5x - 2) = 19x + 4$ **(b)** $3(5x + 1) = 2(7x - 3) - 4$

Given an expression such as

$-2(x - 5)$

we use the distributive property to create the equivalent expression

$-2x + 10$

Distributing negative numbers is shown in Example 10.

Example 10 Distributing a Negative Number

Solve each equation.

(a)

$$\begin{aligned} -2(x - 5) &= -3x + 2 \\ -2x + 10 &= -3x + 2 && \text{Distribute the } -2. \\ +3x \quad &\quad\; +3x && \text{Add } 3x. \\ \hline x + 10 &= 2 \\ -10 &= -10 && \text{Subtract 10.} \\ \hline x &= -8 && \text{The solution is } -8. \end{aligned}$$

(b)

$$\begin{aligned} -3(3x + 5) &= -5(2x - 2) \\ -9x - 15 &= -5(2x - 2) && \text{Distribute the } -3. \\ -9x - 15 &= -10x + 10 && \text{Distribute the } -5. \\ +10x \quad &\quad\; +10x && \text{Add } 10x. \\ \hline x - 15 &= 10 \\ +15 &\quad\; +15 && \text{Add 15.} \\ \hline x &= 25 && \text{The solution is 25.} \end{aligned}$$

Check:

$$\begin{aligned} -3[3(25) + 5] &\stackrel{?}{=} -5[2(25) - 2] \\ -3(75 + 5) &\stackrel{?}{=} -5(50 - 2) && \text{Follow the order of operations.} \\ -3(80) &\stackrel{?}{=} -5(48) \\ -240 &= -240 && \text{True} \end{aligned}$$

RECALL

Return to the original equation to check your solution.

Check Yourself 10

Solve each equation.

(a) $-2(x - 3) = -x + 5$ **(b)** $-4(2x - 1) = -3(3x + 2)$

The main reason for learning how to set up and solve algebraic equations is so that we can use them to solve word problems. In fact, algebraic equations were *invented* to make solving applications much easier. The first word problems that we know about are over 4,000 years old. They were literally "written in stone," on Babylonian tablets, about 500 years before the first algebraic equation made its appearance.

Before algebra, people solved word problems primarily by **substitution,** which is a method of finding unknown values by using trial and error in a logical way. Example 11 shows how to solve a word problem by using substitution.

Example 11 Solving a Word Problem

< Objective 3 >

NOTE

Consecutive integers are integers that follow each other, such as 8 and 9.

The sum of two consecutive integers is 37. Find the two integers.

We take a guess-and-check approach to this problem.

When we try 20 and 21, we get 41 as the sum, which is too large.

Trying smaller numbers, say 15 and 16, gives a sum of 31, which is too small.

We continue to guess until we come to 18 and 19. They are consecutive and their sum is 37. The answer is 18 and 19.

Check Yourself 11

The sum of two consecutive integers is 91. Find the two integers.

Most word problems are not quite so easy to solve. For more complicated word problems, we use a five-step procedure. Using this step-by-step approach helps to organize our work. Organization is a key to solving word problems.

Step by Step

To Solve Word Problems

Step 1 Read the problem carefully. Then reread it to decide what you are asked to find.

Step 2 Choose a variable to represent one of the unknowns in the problem. Then represent all other unknowns of the problem with expressions that use the same variable.

Step 3 Translate the problem to the language of algebra to form an equation.

Step 4 Solve the equation.

Step 5 Answer the question and include units in your answer, when appropriate. Check your solution by returning to the original problem.

The third step is usually the hardest. We must translate words to the language of algebra. Before we look at a complete example, this table may help you review that translation step.

RECALL

We discussed these translations in Section 1.1. You might find it helpful to review that section before going on.

Translating Words to Algebra

Words	Algebra
The sum of x and y	$x + y$
3 plus a	$3 + a$ or $a + 3$
5 more than m	$m + 5$
b increased by 7	$b + 7$
The difference of x and y	$x - y$
4 less than a	$a - 4$
s decreased by 8	$s - 8$
The product of x and y	$x \cdot y$ or xy
5 times a	$5 \cdot a$ or $5a$
Twice m	$2m$
The quotient of x and y	$\frac{x}{y}$
a divided by 6	$\frac{a}{6}$
One-half of b	$\frac{b}{2}$ or $\frac{1}{2}b$

Now let's look at some typical examples of translating phrases to algebra.

Example 12 Translating Statements

Translate each English expression to an algebraic expression.

(a) The sum of a and 2 times b

$a + 2b$

Sum — 2 times b

(b) 5 times m, increased by 1

$5m + 1$

5 times m — Increased by 1

(c) 5 less than 3 times x

$3x - 5$

3 times x — 5 less than

(d) The product of x and y, divided by 3

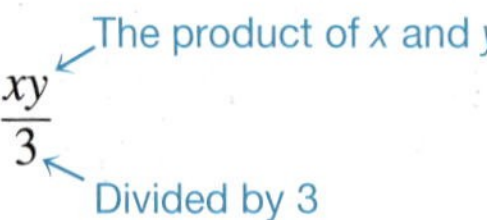

Check Yourself 12

Write each expression symbolically.

(a) 2 more than twice x **(b)** 4 less than 5 times n

(c) The product of twice a and b **(d)** The sum of s and t, divided by 5

Now we work through a complete example. Although this problem can be solved by guessing, we present it to help you practice the five-step approach.

Example 13 Solving an Application

The sum of a number and 5 is 17. What is the number?

Step 1 *Read carefully.* We must find the unknown number.

Step 2 *Choose letters or variables.* Let x represent the unknown number. There are no other unknowns.

Step 3 *Translate.*

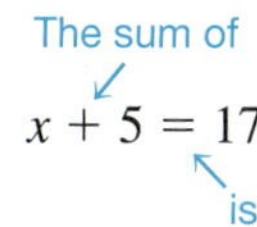

The sum of

$$x + 5 = 17$$

is

> CAUTION

Always return to the *original problem* to check your result and *not* to the equation in step 3. This helps prevent possible errors!

Step 4 *Solve.*

$$x + 5 = 17$$

$$x + 5 - 5 = 17 - 5 \quad \text{Subtract 5.}$$

$$x = 12$$

Step 5 *Answer.* The number is 12.

Check. Is the sum of 12 and 5 equal to 17? Yes, $12 + 5 = 17$.

Check Yourself 13

The sum of a number and 8 is 35. What is the number?

Of course, there are many applications requiring the addition property to solve an equation. Consider the consumer application in the next example.

Example 14 A Consumer Application

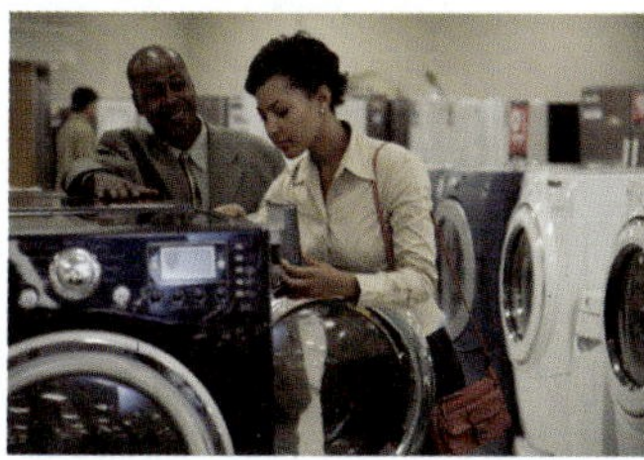

An appliance store is having a sale on washers and dryers. They are charging \$999 for a washer and dryer combination. If the washer sells for \$649, how much is someone paying for the dryer as part of the combination?

Step 1 *Read carefully.* We are asked to find the cost of a dryer in this application.

Step 2 *Choose letters or variables.* Let d represent the cost of a dryer as part of the washer-dryer combination. This is the only unknown quantity in the problem.

Step 3 *Translate.*

$$d + 649 = 999$$

The washer costs \$649.
Together, they cost \$999.

Step 4 *Solve.*

$$d + 649 = 999$$

$$d + 649 - 649 = 999 - 649 \quad \text{Subtract 649 to isolate the variable.}$$

$$d = 350$$

RECALL

Always answer an application with a full sentence.

Step 5 *Answer.* The dryer costs \$350 as part of this combination.

Check. A \$649 washer and a \$350 dryer cost a total of \$649 + \$350 = \$999.

Check Yourself 14

Of 18,540 votes cast in the school board election, 11,320 went to Carla. How many votes did her opponent Marco receive? Who won the election?

Check Yourself ANSWERS

1. **(a)** 6 is a solution; **(b)** $\frac{8}{3}$ is not a solution. **2.** **(a)** nonlinear equation; **(b)** linear equation; **(c)** expression; **(d)** linear equation; **(e)** nonlinear equation **3.** 9 **4.** $-\frac{7}{3}$ **5.** 3 **6.** -3
7. **(a)** 7; **(b)** 0; **(c)** -6 **8.** **(a)** -10; **(b)** -3 **9.** **(a)** 12; **(b)** -13 **10.** **(a)** 1; **(b)** -10
11. 45 and 46 **12.** **(a)** $2x + 2$; **(b)** $5n - 4$; **(c)** $2ab$; **(d)** $\frac{s+t}{5}$
13. The equation is $x + 8 = 35$. The number is 27.
14. Marco received 7,220 votes; Carla won the election.

Reading Your Text

These fill-in-the-blank exercises will help you understand some of the key vocabulary used in this section. The answers to these exercises are in the Answers Appendix in the back of the text.

(a) You should do your math homework while you are still _________.

(b) An equation is a mathematical statement that two _________ are equal.

(c) A _________ to an equation is any value for the variable that makes the equation a true statement.

(d) Linear equations in one variable have exactly _________ solution.

1.4 exercises

Skills | Calculator/Computer | Career Applications | Above and Beyond

< Objective 1 >

Is the number shown in parentheses a solution to the given equation?

1. $x + 7 = 12 \quad (5)$

2. $x + 2 = 11 \quad (8)$

3. $x - 15 = 6 \quad (-21)$ 

4. $x - 11 = 5 \quad (16)$

5. $5 - x = 2 \quad (4)$

6. $10 - x = 7 \quad (3)$

7. $8 - x = 5 \quad (-3)$

8. $5 - x = 6 \quad (-3)$

9. $3x + 4 = 13 \quad (8)$ 

10. $5x + 6 = 31 \quad (5)$

11. $4x - 5 = 7 \quad (3)$

12. $4x - 5 = 1 \quad \left(\frac{3}{2}\right)$

13. $5 - 2x = 10 \quad \left(-\frac{5}{2}\right)$

14. $4 - 5x = 9 \quad (-2)$

15. $6 - \frac{3}{4}x = 4 + \frac{1}{4}x \quad (2)$

16. $5x + 4 = 2x + 10 \quad (4)$

17. $x + 3 + 2x = 5 + x + 8 \quad (5)$ VIDEO

18. $5x - 3 + 2x = 3 + x - 12 \quad (-2)$

19. $\frac{3}{4}x = 18 \quad (20)$

20. $\frac{3}{5}x = 24 \quad (40)$

21. $\frac{3}{5}x + 5 = 11 \quad (10)$

22. $\frac{2}{3}x + 8 = -12 \quad (-6)$

Label each statement as an expression, a linear equation, or a nonlinear equation.

23. $2x + 1 = 9$

24. $5x - 11$

25. $7x + 2x + 8 - 3$

26. $x + 5 = 13$

27. $3x + 5 = 9$

28. $12x^2 - 5x + 2 = 5$

< Objective 2 >

Solve each equation and check your results.

29. $x + 7 = 9$

30. $x - 4 = 6$

31. $x - 8 = 3$

32. $x + 11 = 15$

33. $x - 8 = -10$

34. $x - 8 = -11$

35. $11 = x + 5$

36. $x + 7 = 0$

37. $4x = 3x + 4$

38. $7x = 6x - 8$

39. $11x = 10x - 10$

40. $2(x + 3) = x + 6$

41. $4x - 10 = 5(x - 2)$

42. $\frac{2}{3}x + \frac{7}{8} = \frac{5}{3}x + \frac{1}{8}$

43. $\frac{9}{5}x + \frac{5}{6} = \frac{4}{5}x + \frac{1}{6}$

44. $3x - 2 = 2x + 1$

45. $5x - 7 = 4x - 3$

46. $8x + 5 = 7x - 2$

47. $\frac{5}{3}x = 9 + \frac{2}{3}x$

48. $\frac{4}{7}x = 8 - \frac{3}{7}x$

49. $3 + 6x + 2 = 3x + 11 + 2x$

50. $6x - 3 + 2x = 7x + 8$

51. $4x + 7 + 3x = 5x + 13 + x$

52. $5x + 9 + 4x = 9 + 8x - 7$

53. $4(3x + 4) = 11x - 2$

54. $2(5x - 3) = 9x + 7$

55. $3(7x + 2) = 5(4x + 1) + 17$

56. $5(5x + 3) = 3(8x - 2) + 4$

57. $\frac{5}{4}x - 1 = \frac{1}{4}x + 7$

58. $\frac{7}{5}x + 3 = \frac{2}{5}x - 8$

59. $\frac{9}{2}x - \frac{3}{4} = \frac{7}{2}x + \frac{5}{4}$

60. $\frac{11}{3}x + \frac{1}{6} = \frac{8}{3}x + \frac{19}{6}$

61. $0.56x = 9 - 0.44x$

62. $8 - 0.37x = 5 + 0.63x$

63. $0.12x + 0.53x - 8 = -0.92x + 0.57x + 4$

64. $0.71x + 6 + 0.35x = 0.25x - 11 - 0.19x$

Translate each English statement to an algebraic equation. Use x as the variable in each case.

65. 3 more than a number is 7.

66. 5 less than a number is 12.

67. 7 less than 3 times a number is twice that same number.

68. 4 more than 5 times a number is 6 times that same number.

69. 2 times the sum of a number and 5 is 18 more than that same number.

70. 3 times the sum of a number and 7 is 4 times that same number.

< Objective 3 >

Solve each word problem. Show the equation you use to solve each problem.

71. **Number Problem** The sum of a number and 7 is 33. What is the number?

72. **Number Problem** The sum of a number and 15 is 22. What is the number?

73. **Number Problem** The difference between a number and 15 is 7. What is the number? 

74. **Number Problem** The difference between a number and 8 is 17. What is the number? VIDEO

75. **Social Science** In an election, the winning candidate has 1,840 votes. If the total number of votes cast was 3,260, how many votes did the losing candidate receive?

76. **Business and Finance** Mike and Stefanie work at the same company and make a total of $6,760 per month. If Stefanie makes $3,400 per month, how much does Mike earn every month?

77. **Business and Finance** A washer-dryer combination costs $650. If the washer costs $360, what does the dryer cost?

78. **Technology** You have $2,350 saved for the purchase of a new computer system that costs $3,675. How much more must you save?

79. Which equation is equivalent to $8x + 5 = 9x - 4$?

(a) $17x = -9$ (b) $x = -9$ (c) $8x + 9 = 9x$ (d) $9 = 17x$

80. Which equation is equivalent to $5x - 7 = 4x - 12$?

(a) $9x = 19$ (b) $9x - 7 = -12$ (c) $x = -18$ (d) $x - 7 = -12$

81. Which equation is equivalent to $12x - 6 = 8x + 14$?

(a) $4x - 6 = 14$ (b) $x = 20$ (c) $20x = 20$ (d) $4x = 8$

82. Which equation is equivalent to $7x + 5 = 12x - 10$?

(a) $5x = -15$ (b) $7x - 5 = 12x$ (c) $-5 = 5x$ (d) $7x + 15 = 12x$

Determine whether each statement is **true** *or* **false.**

83. Every linear equation with one variable has exactly one solution.

84. Isolating the variable on the right side of an equation results in a negative solution.

85. If we add the same number to both sides of an equation, we always obtain an equivalent equation.

86. The equations $x^2 = 9$ and $x = 3$ are equivalent equations.

Complete each statement with **always, sometimes,** *or* **never.**

87. An equation __________ has one solution.

88. If a first-degree equation has a variable term on both sides, we __________ use the addition property to solve the equation.

Skills | Calculator/Computer | **Career Applications** | Above and Beyond

89. **Construction Technology** K Jones Manufacturing produces hex bolts and carriage bolts. They sold 284 more hex bolts than carriage bolts last month. If they sold 2,680 carriage bolts, how many hex bolts did they sell?

90. **Electronics Technology** Berndt Electronics earns a marginal profit of $560 each on the sale of a particular server. If other costs involved amount to $4,500, will they earn a net profit of at least $5,000 on the sale of 15 servers? VIDEO

91. **Engineering Technology** The specifications for an engine cylinder of a particular ship call for the stroke length to be two more than twice the diameter of the cylinder. Write an expression for the required stroke length given a cylinder's diameter d.

92. **Engineering Technology** Use your answer to exercise 91 to determine the required stroke length if the cylinder has a diameter of 52 in.

Skills	Calculator/Computer	Career Applications	**Above and Beyond**

93. An algebraic equation is a complete sentence. It has a subject, a verb, and a predicate. For example, $x + 2 = 5$ can be written in English as "Two more than a number is five" or "A number added to two is five." Write an English version of each equation. Be sure to write complete sentences and that the sentences express the same idea as the equations. Exchange sentences with another student and see if your interpretations of each other's sentences result in the same equation.

(a) $2x - 5 = x + 1$ **(b)** $2(x + 2) = 14$

(c) $n + 5 = \frac{n}{2} - 6$ **(d)** $7 - 3a = 5 + a$

94. Complete the explanation in your own words: "The difference between $3(x - 1) + 4 - 2x$ and $3(x - 1) + 4 = 2x$ is . . . ."

95. "I make \$2.50 an hour more in my new job." If x = the amount I used to make per hour and y = the amount I now make, which of the equations say the same thing as the previous statement? Explain your choices by translating the equation to English and comparing with the original statement.

(a) $x + y = 2.50$ **(b)** $x - y = 2.50$ **(c)** $x + 2.50 = y$

(d) $2.50 + y = x$ **(e)** $y - x = 2.50$ **(f)** $2.50 - x = y$

96. "The river rose 4 feet above flood stage last night." If a = the river's height at flood stage and b = the river's height now (the morning after), which of the equations say the same thing as the previous statement? Explain your choices by translating the equations to English and comparing the meaning with the original statement.

(a) $a + b = 4$ **(b)** $b - 4 = a$ **(c)** $a - 4 = b$

(d) $a + 4 = b$ **(e)** $b + 4 = a$ **(f)** $b - a = 4$

97. "Surprising Results!" Work with other students to try this experiment. Each person should do the six steps mentally, not telling anyone else what his or her calculations are.

(a) Think of a number. **(b)** Add 7.

(c) Multiply by 3. **(d)** Add 3 more than the original number.

(e) Divide by 4. **(f)** Subtract the original number.

What number do you end up with? Compare your answer with everyone else's. Does everyone have the same answer? Make sure that everyone followed the directions accurately. How do you explain the results? Algebra makes the explanation clear. Work together to do the problem again, using a variable for the number. Make up another series of computations that give "surprising results."

98. **(a)** Do you think this is a linear equation in one variable?

$3(2x + 4) = 6(x + 2)$

(b) What happens when you use the properties of this section to solve the equation?

(c) Pick *any* number to substitute for x in this equation. Now try a different number to substitute for x in the equation. Try yet another number to substitute for x in the equation. Summarize your findings.

(d) Can this equation be called *linear in one variable?* Refer to the definition as you explain your answer.

99. **(a)** Do you think this is a linear equation in one variable?

$$4(3x - 5) = 2(6x - 8) - 3$$

(b) What happens when you use the properties of this section to solve the equation?

(c) Do you think it is possible to find a solution for this equation?

(d) Can this equation be called *linear in one variable?* Refer to the definition as you explain your answer.

Answers

1. Yes **3.** No **5.** No **7.** No **9.** No **11.** Yes **13.** Yes **15.** Yes **17.** Yes **19.** No **21.** Yes **23.** Linear equation **25.** Expression **27.** Linear equation **29.** 2 **31.** 11 **33.** -2 **35.** 6 **37.** 4 **39.** -10 **41.** 0 **43.** $-\frac{2}{3}$ **45.** 4 **47.** 9 **49.** 6 **51.** 6 **53.** -18 **55.** 16 **57.** 8 **59.** 2 **61.** 9 **63.** 12 **65.** $x + 3 = 7$ **67.** $3x - 7 = 2x$ **69.** $2(x + 5) = x + 18$ **71.** 26; $x + 7 = 33$ **73.** 22; $x - 15 = 7$ **75.** 1,420; $1,840 + x = 3,260$ **77.** \$290; $x + 360 = 650$ **79.** (c) **81.** (a) **83.** True **85.** True **87.** sometimes **89.** 2,964 hex bolts **91.** $2d + 2$ **93.** Above and Beyond **95.** Above and Beyond **97.** Above and Beyond **99.** Above and Beyond

1.5 Solving Equations with the Multiplication Property

< 1.5 Objectives >

1 > Use the multiplication property to solve equations

2 > Use the multiplication property to solve applications

Alone, the addition property is not enough to solve every equation. Consider the equation

$6x = 18$

Subtracting 6 from both sides of the equation yields

$6x - 6 = 18 - 6$

$6x - 6 = 12$

This made the equation more complicated, not less. To work with these equations, we need a second property.

Property

The Multiplication Property of Equality

If $a = b$ then $ac = bc$ where $c \neq 0$

Multiplying both sides of an equation by the same nonzero number produces an equivalent equation.

We work through some examples using our new rule.

Example 1 Using the Multiplication Property to Solve Equations

< Objective 1 >

NOTE

$\frac{1}{6}(6x) = \left(\frac{1}{6} \cdot 6\right)x$

$= 1 \cdot x$ or x

Solve.

$6x = 18$

The coefficient of the variable is 6. The reciprocal or multiplicative inverse of 6 is $\frac{1}{6}$.

Therefore, if we multiply the left side of the equation by $\frac{1}{6}$, we achieve the goal of producing an equation of the form $x = \square$.

Of course, what we do to one side, we must do to both sides.

$6x = 18$

$\frac{1}{6}(6x) = \frac{1}{6}(18)$ Multiply both sides by $\frac{1}{6}$.

$\left(\frac{1}{6} \cdot 6\right)x = 3$ Simplify.

$1x = 3$ or $x = 3$

The solution is 3. To check, replace x with 3.

$6 \cdot (3) \stackrel{?}{=} 18$

$18 = 18$ True

Check Yourself 1

Solve and check.

$8x = 32$

In Example 1 we solved the equation by multiplying both sides by the reciprocal of the coefficient of the variable. In Example 2, we use the multiplication property in a different way.

Example 2 Solving Equations

Solve.

$5x = -35$

The variable x is multiplied by 5. We *divide* both sides by 5 to "undo" that multiplication.

$$\frac{5x}{5} = \frac{-35}{5}$$ The right side simplifies to -7. Be careful with the rules for signs.

$$x = -7$$

The solution is -7.

We leave it to you to check the solution.

NOTE

Because division is defined in terms of multiplication, we can divide both sides of an equation by the same nonzero number.

Check Yourself 2

Solve and check.

$7x = -42$

Example 3 Solving Equations

Solve.

$-9x = 54$

In this case, x is multiplied by -9, so we divide both sides by -9 to isolate x on the left.

$$\frac{-9x}{-9} = \frac{54}{-9}$$

$$x = -6$$

The solution is -6.

Check

$(-9)(-6) \stackrel{?}{=} 54$

$54 = 54$ True

RECALL

Dividing by -9 and multiplying by $-\frac{1}{9}$ produce the same result—they are the same operation.

Check Yourself 3

Solve and check.

$-10x = -60$

Example 4 illustrates the multiplication property when there are fractions in an equation.

Example 4 Solving Equations

RECALL

$\frac{x}{3} = \frac{1}{3}x$

Solve each equation.

(a) $\frac{x}{3} = 6$

Here x is *divided* by 3, so we use multiplication to isolate x.

$3\left(\frac{x}{3}\right) = 3 \cdot (6)$ This leaves x alone on the left because $3\left(\frac{x}{3}\right) = \frac{3}{1} \cdot \frac{x}{3} = \frac{x}{1} = x$

$x = 18$

The solution is 18.

Check

$\frac{(18)}{3} \stackrel{?}{=} 6$

$6 = 6$ True

(b) $\frac{x}{5} = -9$

$5\left(\frac{x}{5}\right) = 5(-9)$ Because x is divided by 5, we multiply both sides by 5.

$x = -45$

The solution is -45. To check, we replace x with -45.

$\frac{(-45)}{5} \stackrel{?}{=} -9$

$-9 = -9$ True

The solution is verified.

RECALL

Remember to use the rules for signs when choosing a multiplication factor.

$-2\left(-\frac{1}{2}\right) = 1$

(c) $-\frac{x}{2} = 7$

We take the same approach when the fraction is negative. The reciprocal of $-\frac{1}{2}$ is -2, so we multiply both sides by -2.

$-\frac{x}{2} = 7$

$-2\left(-\frac{x}{2}\right) = -2(7)$ Multiply both sides by -2.

$x = -14$

The solution is -14.

Check

$-\frac{x}{2} = 7$ The original equation.

$-\frac{(-14)}{2} \stackrel{?}{=} 7$ Replace x with the solution -14.

$-(-7) \stackrel{?}{=} 7$ $\frac{-14}{2} = -7$.

$7 = 7$ Properly apply the rules of signs: $-(-7) = 7$.

Check Yourself 4

Solve and check.

(a) $\frac{x}{7} = 3$ **(b)** $\frac{x}{4} = -8$ **(c)** $-\frac{x}{4} = -11$

When the variable is multiplied by a fraction that has a numerator other than 1, there are two approaches to finding the solution.

Example 5 Using Reciprocals to Solve Equations

Solve each equation.

(a) $\frac{3}{5}x = 9$

One approach is to multiply by 5 as the first step. This clears the denominator from the left side of the equation.

$$5\left(\frac{3}{5}x\right) = 5 \cdot (9)$$
$$3x = 45$$

Now we divide by 3.

$$\frac{3x}{3} = \frac{45}{3}$$
$$x = 15$$

To check the solution 15, substitute 15 for x.

$$\frac{3}{5} \cdot (15) \stackrel{?}{=} 9$$
$$9 = 9 \qquad \text{True}$$

A second approach combines the multiplication and division steps and is more efficient. We multiply by $\frac{5}{3}$.

NOTE

We multiply by $\frac{5}{3}$ because it is the reciprocal of $\frac{3}{5}$, and the product of a number and its reciprocal is 1.

$\left(\frac{5}{3}\right)\left(\frac{3}{5}\right) = 1$

$$\frac{5}{3}\left(\frac{3}{5}x\right) = \frac{5}{3} \cdot (9)$$
$$x = \frac{5}{\cancel{3}_1} \cdot \frac{\cancel{9}^3}{1} = 15$$

So $x = 15$, as before.

(b) $-\frac{2}{3}x = 12$

RECALL

The product of two negative numbers is positive, so the reciprocal of a negative number is negative.

$\left(-\frac{2}{3}\right)\left(-\frac{3}{2}\right) = 1$

We take the same approach when the fraction is negative. The reciprocal of $-\frac{2}{3}$ is $-\frac{3}{2}$, so we multiply both sides by $-\frac{3}{2}$.

$$-\frac{2}{3}x = 12$$
$$\left(-\frac{3}{2}\right)\left(-\frac{2}{3}x\right) = \left(-\frac{3}{2}\right)(12) \qquad \text{Multiply both sides by } -\frac{3}{2}.$$
$$x = -18 \qquad -\frac{3}{\cancel{2}} \cdot \frac{\cancel{12}^6}{1} = -18.$$

The solution is –18.

Check

$$-\frac{2}{3}x = 12 \qquad \text{The original equation.}$$
$$-\frac{2}{3}(-18) \stackrel{?}{=} 12 \qquad \text{Replace } x \text{ with the solution } -18.$$
$$12 = 12 \qquad -\frac{2}{\cancel{3}} \cdot \left(-\frac{\cancel{18}^6}{1}\right) = 12$$

Check Yourself 5

Solve and check.

(a) $\frac{2}{3}x = 18$ (b) $-\frac{3}{4}x = 10$

You may sometimes have to simplify an equation before solving it. Example 6 illustrates this procedure.

Example 6 Combining Like Terms and Solving Equations

Solve and check.

$3x + 5x = 40$

Using the distributive property, we combine the like terms on the left to write

$8x = 40$

We now proceed as before.

$\frac{8x}{8} = \frac{40}{8}$ Divide by 8.

$x = 5$

The solution is 5. To check, we return to the original equation. Substituting 5 for x yields

$3 \cdot (5) + 5 \cdot (5) \stackrel{?}{=} 40$

$15 + 25 \stackrel{?}{=} 40$

$40 = 40$ True

The solution is verified.

Check Yourself 6

Solve and check.

$7x + 4x = -66$

As with the addition property, we can use the multiplication property to solve many applications.

Example 7 An Application Involving the Multiplication Property

< Objective 2 >

On her first day on the job in a photography lab, Samantha processed all of the film given to her. The next day, her boss gave her four times as much film to process. Over the two days, she processed 60 rolls of film. How many rolls did she process on the first day?

Step 1 We want to find the number of rolls Samantha processed on the first day.

Step 2 Let x be the number of rolls Samantha processed on her first day and solve the equation $x + 4x = 60$ to answer the question.

Step 3 $x + 4x = 60$

Step 4 $5x = 60$ Combine like terms first.

$$\frac{1}{5}(5x) = \frac{1}{5}(60)$$ Multiply by $\frac{1}{5}$ to isolate the variable.

$$x = 12$$

RECALL

Always use a sentence to answer an application.

Step 5 Samantha processed 12 rolls of film on her first day.

Check

$4 \times 12 = 48$; $12 + 48 = 60$

NOTE

The yen (¥) is the monetary unit of Japan.

Check Yourself 7

On a recent trip to Japan, Marilyn exchanged \$1,200 and received 97,428 yen. What exchange rate did she receive?

Check Yourself ANSWERS

1. 4 **2.** -6 **3.** 6 **4.** **(a)** 21; **(b)** -32; **(c)** 44 **5.** **(a)** 27; **(b)** $-\frac{40}{3}$ **6.** -6
7. She received 81.19 yen for each dollar.

Reading Your Text

These fill-in-the-blank exercises will help you understand some of the key vocabulary used in this section. The answers to these exercises are in the Answers Appendix in the back of the text.

(a) Multiplying both sides of an equation by the same ________ number yields an equivalent equation.

(b) $\frac{5}{3}$ is the ________ of $\frac{3}{5}$.

(c) To check a solution, we return to the ________ equation.

(d) The product of a number and its ________ is always 1.

1.5 exercises

Skills Calculator/Computer Career Applications Above and Beyond

< Objective 1 >

Solve and check.

1. $5x = 20$ 
2. $6x = 30$
3. $7x = 42$
4. $6x = -42$
5. $63 = 9x$
6. $66 = 6x$
7. $4x = -16$
8. $-3x = 27$
9. $-9x = 72$ 
10. $-9x = 90$
11. $6x = -54$
12. $-7x = 49$
13. $-4x = -12$ 
14. $15 = -9x$
15. $-21 = 24x$
16. $-7x = -35$

17. $-6x = -54$

18. $-4x = -24$

19. $\frac{x}{4} = 2$

20. $\frac{x}{3} = 2$

21. $\frac{x}{5} = 3$

22. $\frac{x}{8} = 5$

23. $6 = \frac{x}{7}$

24. $6 = \frac{x}{3}$

25. $\frac{x}{5} = -4$

26. $\frac{x}{5} = -7$

27. $-\frac{x}{3} = 8$ VIDEO

28. $-\frac{x}{4} = -3$

29. $\frac{2}{3}x = 6$ VIDEO

30. $\frac{4}{5}x = 10$

31. $\frac{3}{4}x = -16$

32. $\frac{7}{8}x = -21$

33. $-\frac{2}{5}x = 10$ VIDEO

34. $-\frac{5}{6}x = -15$

35. $5x + 4x = 36$

36. $8x - 3x = -50$

37. $\frac{7}{9}x - 5 = \frac{3}{9}x + 11$

38. $\frac{4}{11}x - 9 + \frac{2}{11}x = 18 - \frac{3}{11}x$

39. $4(x + 5) + 7x = -3(x - 2)$

40. $-2(x - 3) + 10 = -4(5 - 4x)$

We can also use the multiplication property to solve equations containing decimals. For instance, to solve $2.3x = 6.9$, we use the multiplication property to divide both sides of the equation by 2.3. This isolates x on the left as desired. Use this idea to solve each equation.

41. $3.2x = 12.8$

42. $5.1x = -15.3$

43. $-4.5x = 13.5$

44. $-8.2x = -32.8$

45. $1.3x + 2.8x = 12.3$

46. $2.7x + 5.4x = -16.2$

47. $9.3x - 6.2x = 12.4$

48. $12.5x - 7.2x = -21.2$

Translate each statement to an equation. Let x represent the number in each case.

49. 6 times a number is 72.

50. Twice a number is 36.

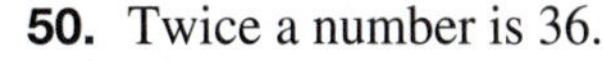

51. A number divided by 7 is equal to 6.

52. A number divided by 5 is equal to -4.

53. $\frac{1}{3}$ of a number is 8.

54. $\frac{1}{5}$ of a number is 10.

55. $\frac{3}{4}$ of a number is 18.

56. $\frac{2}{7}$ of a number is 8.

57. Twice a number, divided by 5, is 12.

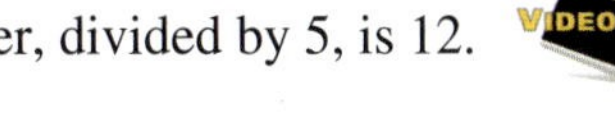

58. 3 times a number, divided by 4, is 36.

< Objective 2 >

Solve each application.

59. **Statistics** Three-fourths of the theater audience left in disgust. If 87 angry patrons walked out, how many were there originally?

60. **Number Problem** When a number is divided by -6, the result is 3. Find the number.

61. **Geometry** Suppose that the circumference of a tree measures 9 ft 2 in., or 110 in. To find the diameter of the tree at that point, we must solve the equation

$110 = 3.14d$

Find the diameter of the tree to the nearest inch.

Note: 3.14 is an approximation for π.

62. **Geometry** Suppose that the circumference of a circular swimming pool is 88 ft. Find the diameter of the pool to the nearest foot by solving the equation

$88 = 3.14d$

63. **Problem Solving** While traveling in Europe, Susan noticed that the distance to the city she was heading to was 200 kilometers (km). She knew that to estimate this distance in miles she could solve the equation

$$200 = \frac{8}{5}x$$

What was the equivalent distance in miles?  

64. **Problem Solving** Aaron was driving a rental car while traveling in France, and saw a sign indicating a speed limit of 95 km/hr. To approximate this speed in miles per hour, he used the equation

$$95 = \frac{8}{5}x$$

What is the corresponding speed, rounded to the nearest mile per hour? 

Determine whether each statement is **true** *or* **false.**

65. To isolate x in the equation $\frac{3}{4}x = 9$, we subtract $\frac{3}{4}$ from both sides.

66. Dividing both sides of an equation by 5 is the same as multiplying both sides by $\frac{1}{5}$.

Complete each statement with **always, sometimes,** *or* **never.**

67. To solve a linear equation, we __________ must use the multiplication property.

68. If we want to obtain an equivalent linear equation by multiplying both sides by a number, that number can __________ be zero.

Skills | Calculator/Computer | **Career Applications** | Above and Beyond

69. **Automotive Technology** One horsepower (hp) estimate of an engine is given by the formula

$$\text{hp} = \frac{d^2n}{2.5}$$

in which d is the diameter of the cylinder bore (in centimeters) and n is the number of cylinders.

Find the number of cylinders in a 194.4-hp engine if its cylinder bore has a 9-cm diameter.

70. **Automotive Technology** The horsepower (hp) of a diesel engine is calculated using the formula

$$\text{hp} = \frac{P \cdot L \cdot A \cdot N}{33{,}000}$$

in which P is the average pressure (in pounds per square inch), L is the length of the stroke (in feet), A is the area of the piston (in square inches), and N is the number of strokes per minute.

Determine the average pressure of a 144-hp diesel engine if its stroke length is $\frac{1}{3}$ ft, its piston area is 9 in.^{2}, and it completes 8,000 strokes per minute.

71. **Manufacturing Technology** The pitch of a gear is given by the number of teeth divided by the working diameter of the gear. Write an equation for the gear pitch p in terms of the number of teeth t and its diameter d.

72. **Manufacturing Technology** Use your answer to exercise 71 to determine the number of teeth needed for a gear with a working diameter of $6\frac{1}{4}$ in. to have a pitch of 4.

73. Describe the difference between the multiplication property and the addition property for solving equations. Give examples of when to use each property.

74. Describe when you should add a quantity to or subtract a quantity from both sides of an equation as opposed to when you should multiply or divide both sides by the same quantity.

Answers

1. 4 **3.** 6 **5.** 7 **7.** -4 **9.** -8 **11.** -9 **13.** 3 **15.** $-\frac{7}{8}$ **17.** 9 **19.** 8 **21.** 15 **23.** 42 **25.** -20 **27.** -24 **29.** 9 **31.** $-\frac{64}{3}$ **33.** -25 **35.** 4 **37.** 36 **39.** -1 **41.** 4 **43.** -3 **45.** 3 **47.** 4 **49.** $6x = 72$ **51.** $\frac{x}{7} = 6$ **53.** $\frac{1}{3}x = 8$ **55.** $\frac{3}{4}x = 18$ **57.** $\frac{2x}{5} = 12$ **59.** 116 patrons **61.** 35 in. **63.** 125 mi **65.** False **67.** sometimes **69.** 6 cylinders **71.** $p = \frac{t}{d}$ **73.** Above and Beyond

1.6 Combining the Rules to Solve Equations

< 1.6 Objectives >

1 > Combine the properties of equality to solve equations

2 > Recognize identities and contradictions

3 > Use linear equations to solve applications

NOTE

More generally, we reverse the order of operations to solve an equation so that we can *undo* each operation.

Up to this point, we solved equations by using the addition or multiplication properties of equality. Most of the time, we need to use both properties, along with the order of operations, to solve an equation.

When an equation requires both properties, we **always** apply the addition property before the multiplication property. We begin with some examples.

Example 1 Solving an Equation

< Objective 1 >

RECALL

Adding 5 *undoes* the subtraction on the left side of the equation.

The addition property requires us to add 5 to *both* sides of the equation.

Solve.

$3x - 5 = 4$

Our first goal is to get the variable term 3x all alone on one side. We call this *isolating the variable term*. We accomplish this by adding 5 to both sides of the equation.

$$3x - 5 = 4$$

$$3x - 5 + 5 = 4 + 5 \quad \text{Add 5 to both sides.}$$

$$3x = 9 \quad \text{This isolates the variable term } 3x.$$

We want the x-term alone on the left with a coefficient of 1 (we call this *isolating the variable*). To do this, we use the multiplication property and multiply both sides by $\frac{1}{3}$.

$$3x = 9$$

$$\frac{1}{3}(3x) = \frac{1}{3}(9) \quad \text{Multiply both sides by } \frac{1}{3}.$$

$$x = 3 \quad \frac{1}{3}(3x) = x \text{ because } \frac{1}{3} \cdot 3 = \frac{1}{3} \cdot \frac{3}{1} = 1.$$

Because any application of the addition or multiplication property leads to an equivalent equation, all of these equations have the same solution, 3.

To check this result, we replace x with 3 in the original equation.

$$3(3) - 5 \stackrel{?}{=} 4$$

$$9 - 5 \stackrel{?}{=} 4$$

$$4 = 4 \quad \text{True}$$

You may prefer a slightly different approach in the final step of this example. From the equation $3x = 9$, the multiplication property can be used to *divide* both sides of the equation by 3. Then

$$\frac{3x}{3} = \frac{9}{3}$$

$$x = 3$$

The result is the same.

Check Yourself 1

Solve and check.

$4x - 7 = 17$

We used both the addition and multiplication properties to solve the problem in Example 1. First, we used the addition property to *isolate the variable term*. Then, we used the multiplication property to *isolate the variable*.

Why did we do it in this order? Consider what happens when we multiply both sides by $\frac{1}{3}$ as the first step so that we have x instead of $3x$.

$3x - 5 = 4$

$\frac{1}{3}(3x - 5) = \frac{1}{3}(4)$ We must multiply the entire left side by $\frac{1}{3}$.

$\frac{1}{3}(3x) + \frac{1}{3}(-5) = \frac{4}{3}$ Distribute $\frac{1}{3}$ on the left side.

$x - \frac{5}{3} = \frac{4}{3}$ Now we are ready to add $\frac{5}{3}$ to both sides.

Since $\frac{4}{3} + \frac{5}{3} = 3$, we have the same result. We consider this approach to be more difficult and do not use it. Instead, we first get the variable term all alone with the addition property. Then, we use the multiplication property to get the variable all alone.

The steps involved in using the addition and multiplication properties to solve an equation are the same, even if there are more terms in the equation.

> CAUTION

When we think of dividing both sides by 3, we get

$\frac{3x - 5}{3} = \frac{4}{3}$

because we must divide the entire left side by 3.

Distribution requires us to divide 5 by 3 as well. The 3's do not cancel!

Example 2 Solving an Equation

Solve.

$5x - 11 = 2x - 7$

Our goal is to finish with an equivalent equation that looks like $x = \square$. To do this, we need to bring all of the x-terms to one side and the constant terms to the other side of the equation. We use the addition property to subtract $2x$ from both sides. This brings the variable terms to the left side. We then add 11 to both sides to bring the numbers to the right side.

$5x - 11 = 2x - 7$

$\underline{-2x \qquad\quad -2x}$ Subtract $2x$ from both sides.

$3x - 11 = -7$ Combine like terms to complete this step.

$\underline{\quad +11 \quad +11}$ Add 11 to both sides.

$3x = 4$

Now that the variable term is isolated on the left side, we use the multiplication rule to divide both sides by 3. This isolates the variable.

$\frac{\cancel{3}x}{\cancel{3}} = \frac{4}{3}$ Divide both sides by 3 and simplify where possible.

$x = \frac{4}{3}$

To check our result, we always return to the original equation.

$5x - 11 = 2x - 7$ The original equation.

$5\left(\frac{4}{3}\right) - 11 \stackrel{?}{=} 2\left(\frac{4}{3}\right) - 7$ Substitute the proposed solution for the variable.

$\frac{20}{3} - 11 \stackrel{?}{=} \frac{8}{3} - 7$ Multiply first, as required by the order of operations.

$\frac{20}{3} - \frac{33}{3} \stackrel{?}{=} \frac{8}{3} - \frac{21}{3}$ Rewrite 11 as $\frac{33}{3}$ and 7 as $\frac{21}{3}$ to combine the fractions.

$-\frac{13}{3} = -\frac{13}{3}$ True

Check Yourself 2

Solve and check.

$7x - 12 = 2x - 9$

If possible, you should always simplify each side of an equation before using the addition and multiplication properties. This might mean using the distributive property to remove parentheses or combining like terms. We illustrate both situations in Example 3.

Example 3 Solving Equations

> **NOTE**
>
> There are like terms on both sides of the equation.

Solve each equation.

(a) $8x + 2 - 3x = 8 + 3x + 2$

We begin by combining like terms on each side of the equation.

$8x + 2 - 3x = 8 + 3x + 2$

$5x + 2 = 3x + 10$ $\quad$ $8x - 3x = 5x$; $8 + 2 = 10$.

Now we have an equation that looks like the one we solved in Example 2. We can proceed as before.

We bring the x-terms to the left side by subtracting $3x$ and the numbers to the right side by subtracting 2 from both sides. We combine these two applications of the addition property into a single step. You will be able to do this with practice.

$$\begin{array}{rcl} 5x + 2 &=& 3x + 10 \\ -3x - 2 && -3x - 2 \\ \hline 2x &=& 8 \end{array}$$

Finally, we isolate the variable by dividing both sides by 2.

$$\frac{2x}{2} = \frac{8}{2}$$

$$x = 4$$

> **RECALL**
>
> Return to the original equation to check your solution.
>
> Always follow the order of operations when evaluating an expression.

Check

Return to the original equation to check your answer.

$8x + 2 - 3x = 8 + 3x + 2$

$8(4) + 2 - 3(4) \stackrel{?}{=} 8 + 3(4) + 2$ $\quad$ Substitute $x = 4$.

$32 + 2 - 12 \stackrel{?}{=} 8 + 12 + 2$ $\quad$ Follow the order of operations.

$22 = 22$ $\quad$ True

(b) $x + 3(3x - 1) = 4(x + 2) + 4$

We begin by using the distributive property to remove the parentheses.

$x + 3(3x - 1) = 4(x + 2) + 4$

$x + 3(3x) + 3(-1) = 4(x) + 4(2) + 4$ $\quad$ Use the distributive property to remove the parentheses.

$x + 9x - 3 = 4x + 8 + 4$ $\quad$ Simplify each term.

We now have an equation like the one in part (a). We simplify each side of the equation by combining like terms and then use the addition and multiplication properties to complete the solution.

$$x + 9x - 3 = 4x + 8 + 4$$

$$10x - 3 = 4x + 12 \quad \text{Combine like terms.}$$

$$\underline{-4x + 3 \quad -4x + 3} \quad \text{Subtract } 4x \text{ and add 3.}$$

$$\frac{6x}{6} = \frac{15}{6} \quad \text{Divide both sides by 6.}$$

$$x = \frac{5}{2} \quad \text{Simplify the fraction: } \frac{15}{6} = \frac{5}{2}.$$

The solution is $\frac{5}{2}$. Remember to return to the original equation to check this result.

Check Yourself 3

Solve each equation and check your results.

(a) $7x - 3 - 5x = 10 + 4x + 3$
(b) $x + 5(x + 2) = 3(3x - 2) + 18$

To solve an equation with fractions, a good first step is to multiply both sides of the equation by the **least common multiple (LCM)** of all denominators in the equation. This clears the equation of fractions, and allows us to proceed as before.

Example 4 Solving an Equation with Fractions

RECALL

The LCM of a set of denominators is also called the **least common denominator (LCD)**.

Solve.

$$\frac{x}{2} - \frac{2}{3} = \frac{5}{6}$$

First, multiply each side by 6, the least common multiple of 2, 3, and 6.

$$6\left(\frac{x}{2} - \frac{2}{3}\right) = 6\left(\frac{5}{6}\right)$$

$$6\left(\frac{x}{2}\right) - 6\left(\frac{2}{3}\right) = 6\left(\frac{5}{6}\right) \quad \text{Apply the distributive property.}$$

$$\overset{3}{\cancel{6}}\left(\frac{x}{\underset{1}{\cancel{2}}}\right) - \overset{2}{\cancel{6}}\left(\frac{2}{\underset{1}{\cancel{3}}}\right) = \overset{1}{\cancel{6}}\left(\frac{5}{\underset{1}{\cancel{6}}}\right) \quad \text{Simplify}$$

$$3x - 4 = 5$$

NOTE

We cleared the equation of fractions.

Next, isolate the variable x on the left side.

$$3x = 9$$

$$x = 3$$

The solution 3, should be checked as before by returning to the original equation.

Check Yourself 4

Solve.

$$\frac{x}{4} - \frac{4}{5} = \frac{19}{20}$$

Be sure that the distributive property is applied properly so that *every term* of the equation is multiplied by the LCM.

Example 5 Solving an Equation with Fractions

Solve.

$$\frac{2x-1}{5}+1=\frac{x}{2}$$

First, multiply each side by 10, the LCM of 5 and 2.

$$10\left(\frac{2x-1}{5}+1\right)=10\left(\frac{x}{2}\right)$$

$$\overset{2}{\cancel{10}}\left(\frac{2x-1}{\underset{1}{\cancel{5}}}\right)+10(1)=\overset{5}{\cancel{10}}\left(\frac{x}{\underset{1}{\cancel{2}}}\right)$$ Apply the distributive property on the left and simplify.

$$2(2x-1)+10=5x$$

$$4x-2+10=5x$$

$$4x+8=5x$$

$$8=x$$ Next, isolate x. Here we isolate x on the right side.

NOTE

We subtract $4x$ from both sides of the equation.

The solution is 8. We return to the original equation and follow the order of operations to check this result.

$$\frac{2(8)-1}{5}+1\stackrel{?}{=}\frac{(8)}{2}$$

$$\frac{16-1}{5}+1\stackrel{?}{=}4$$ The fraction bar is a grouping symbol.

$$\frac{15}{5}+1\stackrel{?}{=}4$$

$$3+1\stackrel{?}{=}4$$

$$4=4$$ True

Check Yourself 5

Solve and check.

$$\frac{3x+1}{4}-2=\frac{x+1}{3}$$

Conditional Equations, Identities, and Contradictions

1. An equation that is true for only particular values of the variable is called a **conditional equation.** For example, a linear equation that can be written in the form

   $ax+b=0$

   in which $a \neq 0$ is a conditional equation. We illustrated this case in our previous examples and exercises.
2. An equation that is true for all possible values of the variable is called an **identity.** This is equivalent to the equation $0=0$. This is the case if both sides of the equation reduce to the same expression (a true statement).
3. An equation that is never true, no matter what value the variable takes on, is called a **contradiction**. This is the case if the equation is equivalent to a false statement.

We have been practicing with the first case, conditional equations. We look at the second and third cases in Example 6.

Example 6 Identities and Contradictions

< Objective 2 >

NOTE

By adding 6 to both sides of this equation, we have 0 = 0.

(a) Solve $2(x - 3) - 2x = -6$.

Apply the distributive property to remove the parentheses.

$$2x - 6 - 2x = -6$$
$$-6 = -6 \quad \text{A } \textit{true} \text{ statement}$$

Because the two sides simplify to the true statement $-6 = -6$, the original equation is an *identity,* and the solution set is the set of all real numbers. This is sometimes written as $\mathbb{R}$, which is read, "the set of all real numbers."

NOTE

This agrees with the definition of a contradiction. Subtracting 3 from both sides yields 0 = 1.

(b) Solve $3(x + 1) - 2x = x + 4$.

Again, apply the distributive property.

$$3x + 3 - 2x = x + 4$$
$$x + 3 = x + 4$$
$$3 = 4 \quad \text{A } \textit{false} \text{ statement}$$

Because the two sides reduce to the false statement $3 = 4$, the original equation is a contradiction. There are no values of the variable that can satisfy the equation. There is no solution. We sometimes use **empty set** or **null set** notation to write this: $\varnothing$ or $\{ \}$.

Check Yourself 6

Determine whether each equation is a conditional equation, an identity, or a contradiction.

(a) $2(x + 1) - 3 = x$ **(b)** $2(x + 1) - 3 = 2x + 1$
(c) $2(x + 1) - 3 = 2x - 1$

NOTE

An algorithm is a step-by-step process for problem solving.

An organized step-by-step procedure is the key to an effective equation-solving strategy. This algorithm summarizes our work in this section and gives you guidance in approaching problems throughout this text.

Step by Step

Solving Linear Equations in One Variable

Step 1 Remove any grouping symbols by applying the distributive property.

Step 2 Multiply both sides of the equation by the LCM to clear the equation of fractions or decimals

Step 3 Combine any like terms that appear on either side of the equation.

Step 4 Apply the addition property of equality to write an equivalent equation with the variable term on *one side* of the equation and the constant term on the *other side*.

Step 5 Apply the multiplication property of equality to write an equivalent equation with the variable isolated on one side of the equation with coefficient 1.

Step 6 State the answer and check the solution in the *original* equation.

Note: If the equation derived in step 5 is always true, the original equation is an *identity*. If the equation is always false, the original equation is a *contradiction*. If you find a unique solution, the equation is *conditional*.

When you are solving an equation with a calculator, it is often easiest to do all calculations as the last step. For more complex equations, it may be best to calculate at each step.

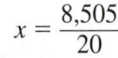

Example 7 Solving Equations with a Calculator

Solve.

$$5(x - 3.25) + \frac{3}{4} = 2{,}110.75$$

Following the steps of the algorithm, we get

$$5x - 16.25 + \frac{3}{4} = 2{,}110.75 \qquad \text{Remove parentheses.}$$

$$20x - 65 + 3 = 8{,}443 \qquad \text{Multiply by the LCD, 4.}$$

$$20x = 8{,}443 + 62$$

$$x = \frac{8{,}505}{20} \qquad \text{Isolate the variable.}$$

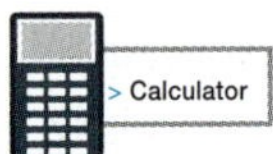

Now, we use a calculator to simplify the expression on the right.

$x = 425.25$

Check Yourself 7

Solve.

$$7(x + 4.3) - \frac{3}{5} = 467$$

We first solved problems involving consecutive integers in Section 1.4. We use these properties to solve these problems algebraically.

Property

Consecutive Integers

If x is an integer, then $x + 1$ is the next **consecutive integer,** $x + 2$ is the next, and so on.

If x is an odd integer, the next **consecutive odd integer** is $x + 2$, and the next is $x + 4$.

If x is an even integer, the next **consecutive even integer** is $x + 2$, and the next is $x + 4$.

NOTE

Consecutive integers are integers that follow one another, such as 10, 11, and 12.

We use this idea in Example 8.

Example 8 Solving an Application

< Objective 3 >

RECALL

We use the five-step method to solve word problems that we introduced in Section 1.4.

The sum of two consecutive integers is 41. What are the two integers?

Step 1 We want to find the two consecutive integers.

Step 2 Let x be the first integer. Then $x + 1$ must be the next.

Step 3

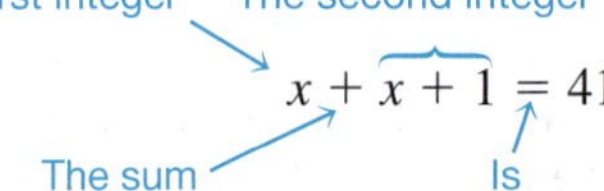

Step 4 $x + x + 1 = 41$

$2x + 1 = 41$

$2x = 40$

$x = 20$

Step 5 The first integer is 20 and the next integer $x + 1$ is 21.

The sum of the two integers 20 and 21 is 41.

Check Yourself 8

The sum of three consecutive integers is 51. What are the three integers?

Sometimes we use algebra to reconstruct missing information. Example 9 does just that with some election information.

Example 9 Solving an Application

There were 55 more yes votes than no votes on an election measure. If 735 votes were cast in all, how many yes votes were there? How many no votes?

Step 1 We want to find the number of yes votes and the number of no votes.

Step 2 Let x be the number of no votes. Then

$x + 55$

55 more than x

is the number of yes votes.

Step 3 $x + x + 55 = 735$

No votes Yes votes Total votes cast

Step 4 $x + x + 55 = 735$

$2x + 55 = 735$

$2x = 680$

$x = 340$

Step 5 No votes $= 340$

Yes votes $x + 55 = 395$

Since 340 no votes plus 395 yes votes equals 735 total votes, the solution checks.

Check Yourself 9

Francine earns $240 per month more than Rob. If they earn a total of $5,360 per month, what are their monthly salaries?

Similar methods allow you to solve a variety of word problems. Example 10 includes three unknown quantities but uses the same basic solution steps.

Example 10 Solving an Application

Juan worked twice as many hours as Jerry. Marcia worked 3 hr more than Jerry. If they worked a total of 31 hr, find out how many hours each worked.

NOTE

There are other choices for x, but choosing the smallest quantity usually gives the easiest equation to write and solve.

Step 1 We want to find the hours each worked, so there are three unknowns.

Step 2 Let x be the hours that Jerry worked.

Juan worked twice Jerry's hours.

Then $2x$ is Juan's hours worked.

Marcia worked 3 hr more than Jerry worked.

And $x + 3$ is Marcia's hours.

Step 3 Jerry, Juan, Marcia

$$x + 2x + x + 3 = 31$$

Sum of their hours

Step 4

$$\begin{aligned} x + 2x + x + 3 &= 31 \\ 4x + 3 &= 31 \\ 4x &= 28 \\ x &= 7 \end{aligned}$$

Step 5

Jerry's hours $x = 7$

Juan's hours $2x = 14$

Marcia's hours $x + 3 = 10$

The sum of their hours $7 + 14 + 10$ is 31, and the solution is verified.

Check Yourself 10

Lucy jogged twice as many miles as Paul but 3 mi less than Isaac. If the three ran a total of 23 mi, how far did each person run?

Many applied problems involve *percents.* Percents are a useful way of naming parts of a whole. We can think of a percent as a fraction whose denominator is 100. Thus, 15% is equal to $\frac{15}{100}$ and represents 15 parts out of 100. A percentage can also be expressed as a decimal by converting the fraction to a decimal. So 15% is 0.15. Examples 11 and 12 illustrate percents in applications.

Example 11 Solving an Application

Marzenna inherits \$5,000 and invests part of her money in bonds at 4% and the remaining in savings at 3%. What amount has she invested at each rate if she receives \$180 in interest for 1 year?

NOTE

All of the \$5,000 was invested at 3% except the portion that was invested at 4%. Subtract x from \$5,000 to get the amount invested at 3%.

Step 1 We want to find the amount invested at each rate, so there are two unknowns.

Step 2 Let x be the amount invested at 4%.

\$5,000 was the total amount of money invested.

So $5{,}000 - x$ is the amount invested at 3%.

$0.04x$ is the amount of interest from the 4% investment.

$0.03(5{,}000 - x)$ is the amount of interest from the 3% investment.

\$180 is the total interest for the year.

Step 3 $0.04x + 0.03(5{,}000 - x) = 180$

Step 4

$$0.04x + 0.03(5{,}000) - 0.03x = 180$$
$$0.04x + 150 - 0.03x = 180$$
$$0.04x - 0.03x = 180 - 150$$
$$0.01x = 30$$
$$x = \frac{30}{0.01}$$
$$x = 3{,}000$$

NOTE

We leave the check to you.

Step 5 Amount invested at 4%: $x = \$3{,}000$

Amount invested at 3%: $5{,}000 - x = \$2{,}000$

Check Yourself 11

Greg received an \$8,000 bonus. He invested some of it in bonds at 2% and the rest in savings at 5%. If he received \$295 interest for 1 year, how much was invested at each rate?

Example 12 Solving an Application

Tony earns a take-home pay of \$592 per week. If his deductions for taxes, retirement, union dues, and a medical plan amount to 26% of his wages, what is his weekly pay before the deductions?

Step 1 We want to find his weekly pay before deductions (gross pay).

Step 2 Let x = gross pay.

Since 26% of his gross pay is deducted from his weekly salary, the amount deducted is $0.26x$.

\$592 is Tony's take-home pay (net pay).

Step 3 Net pay = Gross pay − Deductions

$$\$592 = x - 0.26x$$

Step 4

$$592 = 0.74x$$
$$\frac{592}{0.74} = x$$
$$800 = x$$

Step 5 So Tony's weekly pay before deductions is \$800.

Check Yourself 12

Joan gives 10% of her take-home pay to her church. This amounts to \$270 per month. In addition, her paycheck deductions are 25% of her gross monthly income. What is her gross monthly income?

Check Yourself ANSWERS

1. 6 **2.** $\frac{3}{5}$ **3.** **(a)** -8; **(b)** $-\frac{2}{3}$ **4.** 7 **5.** 5 **6.** **(a)** Conditional equation; **(b)** contradiction; **(c)** identity **7.** 62.5 **8.** 16, 17, 18 **9.** Francine: \$2,800; Rob: \$2,560 **10.** Paul: 4 mi; Lucy: 8 mi; Isaac: 11 mi **11.** \$3,500 at 2%; \$4,500 at 5% **12.** \$3,600

Reading Your Text

These fill-in-the-blank exercises will help you understand some of the key vocabulary used in this section. The answers to these exercises are in the Answers Appendix in the back of the text.

(a) Given an equation such as $3x = 9$, the multiplication property can be used to ________ both sides of the equation by 3.

(b) ________ by $\frac{1}{3}$ is the same as dividing by 3.

(c) If we need to use both properties, always apply the addition property ________ applying the multiplication property when solving an equation.

(d) Both sides of an equation should be ________ as much as possible before using the addition and multiplication properties.

1.6 exercises

Skills | Calculator/Computer | Career Applications | Above and Beyond

< Objective 1 >

Solve and check each equation.

1. $3x + 1 = 13$
2. $3x - 1 = 17$
3. $3x - 2 = 7$ VIDEO
4. $5x + 3 = 23$
5. $4 - 7x = 18$
6. $7 - 4x = -5$
7. $3 - 4x = -9$ VIDEO
8. $5 - 4x = 25$ VIDEO
9. $\frac{x}{2} + 1 = 5$
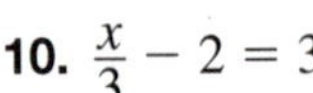
10. $\frac{x}{3} - 2 = 3$
11. $\frac{3}{4}x + 8 = 32$
12. $\frac{5}{6}x - 9 = 16$
13. $5x = 2x + 9$
14. $7x = 18 - 2x$
15. $9x + 2 = 3x + 38$
16. $4(2x - 1) = 2(3x - 2)$
17. $4x - 8 = x - 14$
18. $6x - 5 = 3x - 29$
19. $5(3x + 4) = 10(x + 2)$
20. $\frac{4}{3}x - 7 = 11 - \frac{5}{3}x$
21. $5x + 4 = 7x - 8$
22. $2x + 23 = 6x - 5$
23. $6x + 7 - 4x = 8 + 7x - 26$ VIDEO
24. $7x - 2 - 3x = 5 + 8x + 13$
25. $6x - 3 + 5x + 11 = 8x - 12$
26. $3x + 3 + 8x - 9 = 7x + 5$
27. $5(8 - x) = 3x$
28. $7x = 7(6 - x)$
29. $7(2x - 1) - 5x = x + 25$
30. $9(3x + 2) - 10x = 12x - 7$
31. $2(2x - 1) = 3(x + 1)$
32. $3(3x - 1) = 4(3x + 1)$
33. $8x - 3(2x - 4) = 17$
34. $7x - 4(3x + 4) = 9$
35. $7(3x + 4) = 8(2x + 5) + 13$
36. $-4(2x - 1) + 3(3x + 1) = 9$

37. $9 - 4(3x + 1) = 3(6 - 3x) - 9$
38. $13 - 4(5x + 1) = 3(7 - 5x) - 15$
39. $5.3x - 7 = 2.3x + 5$ VIDEO

40. $9.8x + 2 = 3.8x + 20$

Solve each equation.

41. $\frac{2x}{3} - \frac{5}{3} = 3$
42. $\frac{3x}{4} + \frac{1}{4} = 4$
43. $\frac{x}{6} + \frac{x}{5} = 11$
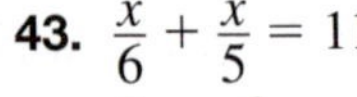
44. $\frac{x}{6} - \frac{x}{8} = 1$
45. $\frac{2x}{5} - \frac{x}{3} = \frac{7}{15}$
46. $\frac{2x}{7} - \frac{3x}{5} = \frac{6}{35}$
47. $\frac{x}{5} - \frac{x - 7}{3} = \frac{1}{3}$
48. $\frac{x}{6} + \frac{3}{4} = \frac{x - 1}{4}$
49. $\frac{5x - 3}{4} - 2 = \frac{x}{3}$
50. $\frac{6x - 1}{5} - \frac{2x}{3} = 3$
51. $\frac{3x - 2}{3} - \frac{2x - 5}{5} = \frac{7}{15}$ VIDEO
52. $\frac{4x}{7} - \frac{2x - 3}{3} = \frac{19}{21}$

< Objective 2 >

Classify each equation as a conditional equation, an identity, or a contradiction and give the solution.

53. $3(x - 1) = 2x + 3$

54. $2(x + 3) = 2x + 6$

55. $3(x - 1) = 3x + 3$

56. $2(x + 3) = x + 5$

57. $3(x - 1) = 3x - 3$

58. $2(2x - 1) = 3x - 4$

59. $3x - (x - 3) = 2(x + 1) + 2$

60. $5x - (x + 4) = 4(x - 2) + 4$

61. $\frac{x}{2} - \frac{x}{3} = \frac{x}{6}$

62. $\frac{3x}{4} - \frac{2x}{3} = \frac{1}{6}$

Translate each statement to an equation. Let x represent the number in each case.

63. 3 more than twice a number is 7.

64. 5 less than 3 times a number is 25.

65. 7 less than 4 times a number is 41.

66. 10 more than twice a number is 44.

67. 5 more than two-thirds of a number is 21.

68. 3 less than three-fourths of a number is 24.

69. 3 times a number is 12 more than that number.

70. 5 times a number is 8 less than that number.

< Objective 3 >

Solve each problem.

71. **Number Problem** The sum of twice a number and 16 is 24. What is the number?

72. **Number Problem** 3 times a number, increased by 8, is 50. Find the number.

73. **Number Problem** 5 times a number, minus 12, is 78. Find the number.

74. **Number Problem** 4 times a number, decreased by 20, is 44. What is the number?

75. **Number Problem** The sum of two consecutive integers is 71. Find the two integers.

76. **Number Problem** The sum of two consecutive integers is 145. Find the two integers.

77. **Number Problem** The sum of three consecutive integers is 90. What are the three integers?

78. **Number Problem** If the sum of three consecutive integers is 93, find the three integers.

79. **Number Problem** The sum of two consecutive even integers is 66. What are the two integers?
Hint: Consecutive even integers such as 10, 12, and 14 can be represented by x, $x + 2$, $x + 4$, and so on.

80. **Number Problem** If the sum of two consecutive even integers is 110, find the two integers.

81. **Number Problem** If the sum of two consecutive odd integers is 52, what are the two integers?
Hint: Consecutive odd integers such as 21, 23, and 25 can be represented by x, $x + 2$, $x + 4$, and so on.

82. **Number Problem** The sum of two consecutive odd integers is 88. Find the two integers.

83. **Number Problem** 4 times an integer is 9 more than 3 times the next consecutive integer. What are the two integers?

84. **Number Problem** 4 times an even integer is 30 less than 5 times the next consecutive even integer. Find the two integers.

85. **Social Science** In an election, the winning candidate had 160 more votes than the loser. If the total number of votes cast was 3,260, how many votes did each candidate receive?

86. **Business and Finance** Jody earns $280 more per month than Frank. If their monthly salaries total $5,520, what amount does each earn?

87. **Business and Finance** A washer-dryer combination costs \$950. If the washer costs \$90 more than the dryer, what does each appliance cost?

88. **Problem Solving** Yan Ling is 1 year less than twice as old as his sister. If the sum of their ages is 14 years, how old is Yan Ling?

89. **Problem Solving** Diane is twice as old as her brother Dan. If the sum of their ages is 27 years, how old are Diane and her brother?

90. **Business and Finance** Patrick invested \$15,000 in two bonds; one bond yields 4% annual interest, and the other yields 3% annual interest. How much is invested in each bond if the combined yearly interest from both bonds is \$545?

91. **Business and Finance** Johanna deposited \$21,000 in two banks. One bank gives $2\frac{1}{2}\%$ annual interest, and the other gives $3\frac{1}{4}\%$ annual interest. How much did she deposit in each bank if she received a total of \$615 in annual interest?

92. **Business and Finance** Tonya takes home \$1,080 per week. If her deductions amount to 28% of her wages, what is her weekly pay before deductions?

93. **Business and Finance** Sam donates 5% of his net income to charity. This amounts to \$190 per month. His payroll deductions are 24% of his gross monthly income. What is Sam's gross monthly income?

94. **Business and Finance** The Randolphs used 12 more gallons (gal) of fuel oil in October than in September and twice as much oil in November as in September. If they used 132 gal for the 3 months, how much was used during each month?

95. **Science and Medicine** While traveling in South America, Richard noted that temperatures were given in degrees Celsius. Wondering what the temperature 95°F would correspond to, he found that he could answer this if he could solve the equation

$$95 = \frac{9}{5}C + 32$$

What was the corresponding temperature? 

96. **Science and Medicine** While traveling in England, Marissa noted an outdoor thermometer showing 20°C. To convert this to degrees Fahrenheit, she solved the equation

$$20 = \frac{5}{9}(F - 32)$$

What was the Fahrenheit temperature? 

Determine whether each statement is **true** *or* **false.**

97. An equation that is never true, no matter what value is substituted, is called an identity.

98. A conditional equation can be an identity.

Complete each statement with **always, sometimes,** *or* **never.**

99. To solve a linear equation, we __________ use both the addition and multiplication properties.

100. We should __________ check a possible solution by substituting it into the original equation.

Use a calculator to solve each equation. Round your answers to the nearest hundredth.

101. $230x - 52 = 191$

102. $321 - 45x = 1{,}021x + 658$

103. $360 - 29(2x + 1) = 2{,}464$

104. $81(x + 26) = 35(86 - 4x)$

105. $23.12x - 34.2 = 34.06$

106. $46.1x + 5.78 = x - 12$

107. $3.2(0.5x - 5.1) = -6.4(9.7x + 15.8)$

108. $x - 11.304(2 - 1.8x) = 2.4x + 3.7$

Skills | Calculator/Computer | **Career Applications** | Above and Beyond

109. **Allied Health** The internal diameter d (in millimeters) of an endotracheal tube for a child is calculated using the formula

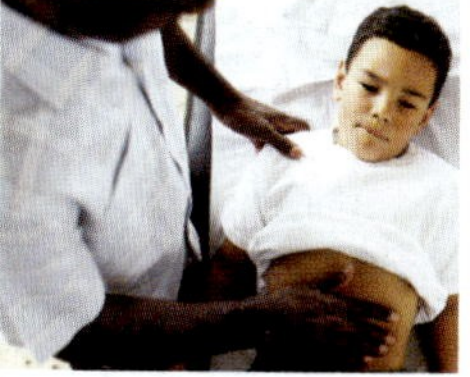

$$d = \frac{t + 16}{4}$$

in which t is the child's age (in years).

How old is a child who requires an endotracheal tube with an internal diameter of 7 mm?

110. **Construction Technology** The number of studs, s, required to build a wall (with studs spaced 16 inches on center) is equal to one more than $\frac{3}{4}$ times the length of the wall, w, in feet. We model this with the formula

$$s = \frac{3}{4}w + 1$$

If a contractor uses 22 studs to build a wall, how long is the wall?

111. **Information Technology** A compression program reduces the size of files by 36%. If a compressed folder has a size of 11.2 MB, how large was it before compressing? VIDEO

112. **Agricultural Technology** A farmer harvested 2,068 bushels of barley. This amounted to 94% of his bid on the futures market. How many bushels did he bid to sell on the futures market?

Skills | Calculator/Computer | Career Applications | **Above and Beyond**

113. Complete this statement in your own words: "You can tell that an equation is a linear equation when. . . ."

114. What is the common characteristic of equivalent equations?

115. What is meant by a *solution* to a linear equation?

116. Define **(a)** identity and **(b)** contradiction.

117. Why does the multiplication property of equality not include multiplying both sides of the equation by 0?

118. Maxine lives in Pittsburgh, Pennsylvania, and pays 8.33 cents per kilowatt-hour (kWh) for electricity. During the 6 months of cold winter weather, her household uses about 1,500 kWh of electric power per month. During the two hottest summer months, the usage is also high because the family uses electricity to run an air conditioner. During these summer months, the usage is 1,200 kWh per month; the rest of the year, usage averages 900 kWh per month.

(a) Write an expression for the total yearly electric bill.

(b) Maxine is considering spending \$2,000 for more insulation for her home so that it is less expensive to heat and to cool. The insulation company claims that "with proper installation the insulation will reduce your heating and cooling bills by 25%." If Maxine invests the money in insulation, how long will it take her to get her money back in savings on her electric bill? Write to her about what information she needs to answer this question. Give her your opinion about how long it will take to save \$2,000 on heating bills, and explain your reasoning. What is your advice to Maxine?

119. Solve each equation. Express each solution as a fraction.

(a) $2x + 3 = 0$ **(b)** $4x + 7 = 0$ **(c)** $6x - 1 = 0$

(d) $5x - 2 = 0$ **(e)** $-3x + 8 = 0$ **(f)** $-5x - 9 = 0$

(g) Based on these problems, express the solution to the equation

$$ax + b = 0$$

where a and b represent real numbers and $a \neq 0$.

120. You are asked to solve an equation, but one number is missing. It reads

$$\frac{5x - ?}{4} = \frac{9}{2}$$

The solution to the equation is 4. What is the missing number?

Answers

1. 4 **3.** 3 **5.** -2 **7.** 3 **9.** 8 **11.** 32 **13.** 3 **15.** 6 **17.** -2 **19.** 0 **21.** 6 **23.** 5 **25.** $-\frac{20}{3}$ **27.** 5 **29.** 4 **31.** 5 **33.** $\frac{5}{2}$ **35.** 5 **37.** $-\frac{4}{3}$ **39.** 4 **41.** 7 **43.** 30 **45.** 7 **47.** 15 **49.** 3 **51.** $\frac{2}{9}$ **53.** Conditional; 6 **55.** Contradiction; { } **57.** Identity; $\mathbb{R}$ **59.** Contradiction; { } **61.** Identity; $\mathbb{R}$ **63.** $2x + 3 = 7$ **65.** $4x - 7 = 41$ **67.** $\frac{2}{3}x + 5 = 21$ **69.** $3x = x + 12$ **71.** 4 **73.** 18 **75.** 35, 36 **77.** 29, 30, 31 **79.** 32, 34 **81.** 25, 27 **83.** 12, 13 **85.** 1,550 votes, 1,710 votes **87.** Washer: \$520; dryer: \$430 **89.** 18 yr old, 9 yr old **91.** \$12,000 at $3\frac{1}{4}$%; \$9,000 at $2\frac{1}{2}$% **93.** \$5,000 **95.** 35°C **97.** False **99.** sometimes **101.** 1.06 **103.** -36.78 **105.** 2.95 **107.** -1.33 **109.** 12 yr old **111.** 17.5 MB **113.** Above and Beyond **115.** A value for which the original equation is true **117.** Multiplying by 0 would always give $0 = 0$. **119.** (a) $-\frac{3}{2}$; (b) $-\frac{7}{4}$; (c) $\frac{1}{6}$; (d) $\frac{2}{5}$; (e) $\frac{8}{3}$; (f) $-\frac{9}{5}$; (g) $-\frac{b}{a}$

1.7 Linear Inequalities

< 1.7 Objectives >

1 > Use inequality notation

2 > Graph the solution set of a linear inequality

3 > Solve a linear inequality

Tips for Student Success

Preparing for a test

Test prep begins on the first day of class. Everything you do in class and at home is part of that preparation. In fact, if you attend class every day, take good notes, and keep up with the homework, then you will already be prepared and will not need to "cram" for your exam.

Instead of cramming, here are a few things to focus on in the days before an exam.

1. Study for your exam, but finish studying 24 hours before the test. Make certain to get some good rest before taking a test.
2. Study for the exam by going over the homework and class notes. Write down all of the problem types, formulas, and definitions that you think might give you trouble on the test.
3. The last item before you finish studying is to take the notes you made in step 2 and transfer the most important ideas to a 3×5 (index) card. You should complete this step a full 24 hours before your exam.
4. One hour before your exam, review the information on the 3×5 card you made in step 3. You will be surprised at how much you remember about each concept.
5. The biggest obstacle for many students is to believe that they can be successful on a test. You can overcome this obstacle easily enough. If you have been completing the homework and keeping up with the classwork, then you should perform quite well on the test. Truly anxious students are often surprised to score well on an exam. These students attribute a good test score to blind luck when it is not luck at all. This is the first sign that you "get it." Enjoy the success!

We know that an equation is a statement that two expressions are equal. In algebra, an **inequality** is a statement that one expression is less than or greater than another. We use inequality symbols to write inequalities.

Example 1 Inequality Notation

< Objective 1 >

$5 < 8$ is an inequality and says "5 is less than 8."

$9 > 6$ says "9 is greater than 6."

RECALL

The "arrowhead" always points toward the smaller quantity.

Check Yourself 1

Fill in each blank with the symbol < or >.

(a) 12 ______ 8 **(b)** 20 ______ 25

As is the case with equations, inequalities with variables may be either true or false depending on the value that we give to the variable. For instance, consider the inequality

$x < 6$

$$\text{If } x = \begin{cases} 3 & 3 < 6 \text{ is true} \\ 6 & 6 < 6 \text{ is false} \\ -10 & -10 < 6 \text{ is true} \\ 8 & 8 < 6 \text{ is false} \end{cases}$$

Therefore, 3 and -10 are both *solutions* to the inequality $x < 6$; they make the inequality a true statement. You should see that 6 and 8 are *not* solutions.

Recall from Section 1.4 that a *solution to an equation* is any value for the variable that makes the equation a true statement. Similarly, a **solution to an inequality** is a value for the variable that makes the inequality a true statement.

RECALL

The equation $x^2 = 9$ has two solutions.

Identities have an infinite number of solutions.

There is more than one solution to the inequality $x < 6$. We have also seen equations with more than one solution. To talk clearly about this type of problem, we define a term for all of the solutions of an equation or inequality in one variable. In Chapter 2, we expand this definition to include equations and inequalities with more than one variable.

Definition

Solution Set

The **solution set** of an equation or inequality in one variable is the set of all values for the variable that make the equation or inequality a true statement.

That is, the solution set is the set of all solutions to an equation or inequality.

Example 2 Graphing Inequalities

< Objective 2 >

To graph the solution set of the inequality $x < 6$, we want to include all real numbers that are "less than" 6. This means all numbers *to the left* of 6 on a number line.

We start at 6 and draw an arrow extending left, as shown.

NOTE

Since there are so many solutions (an infinite number, in fact), we certainly do not want to try to list them all! A convenient way to show the solutions of an inequality is with a number line.

Note: The parenthesis at 6 means that we do not include 6 in the solution set (6 is not less than itself). The colored arrow shows all the numbers in the solution set, with the arrowhead indicating that the solution set continues to the left indefinitely.

Check Yourself 2

Graph the solution set of $x < -2$.

There are two other inequality symbols that we use. They appear in inequalities such as

$x \geq 5$ and $x \leq 2$

$x \geq 5$ is a combination of the two statements $x > 5$ and $x = 5$. It is read "x is greater than or equal to 5." The solution set includes 5 in this case.

The inequality $x \leq 2$ combines the statements $x < 2$ and $x = 2$. It is read "x is less than or equal to 2."

Example 3 Graphing Inequalities

NOTE

The bracket means that we include 5 in the solution set.

The solution set of $x \geq 5$ is graphed as

Check Yourself 3

Graph each solution set.

(a) $x \leq -4$ **(b)** $x \geq 3$

We looked at graphs of the solution sets of some simple inequalities, such as $x < 6$ or $x \geq 5$. Now we look at more complicated inequalities, such as

$2x - 3 < x + 4$

Fortunately, the methods used to solve this type of inequality are very similar to those used to solve linear equations. Here is our first property for inequalities.

Property

The Addition Property of Inequality

If $a < b$ then $a + c < b + c$

Adding the same quantity to both sides of an inequality gives an **equivalent inequality.**

Equivalent inequalities have the same solution set.

The addition property and other rules we use to solve inequalities hold true for any of the four inequality symbols.

If $a > b$, then $a + c > b + c$. If $a < b$, then $a + c < b + c$.

If $a \geq b$, then $a + c \geq b + c$. If $a \leq b$, then $a + c \leq b + c$.

Example 4 Solving Inequalities

< Objective 3 >

NOTE

The inequality is solved when an equivalent inequality has the form

$x < \square$ or $x > \square$

Solve and graph the solution set of $x - 8 < 7$.

To solve $x - 8 < 7$, use the addition property to add 8 to both sides of the inequality.

$x - 8 < 7$

$x - 8 + 8 < 7 + 8$ Add 8 to both sides.

$x < 15$ The inequality is solved.

The graph of the solution set is

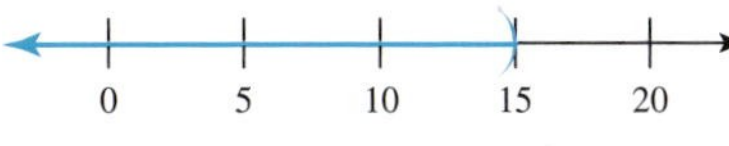

Check Yourself 4

Solve and graph the solution set of

$x - 9 > -3$

As with equations, the addition property allows us to *subtract* the same quantity from both sides of an inequality.

Example 5 Solving Inequalities

NOTES

We subtracted $3x$ and then added 2 to both sides. If these steps are done in the other order, the result is the same.

As with equations, we can combine these into a single step.

Solve and graph the solution set of $4x - 2 \geq 3x + 5$.

First, we subtract $3x$ from both sides of the inequality.

$$4x - 2 \geq 3x + 5$$

$$4x - 3x - 2 \geq 3x - 3x + 5 \quad \text{Subtract } 3x \text{ from both sides.}$$

$$x - 2 \geq 5$$

$$x - 2 + 2 \geq 5 + 2 \quad \text{Add 2 to both sides.}$$

$$x \geq 7$$

The graph of the solution set is

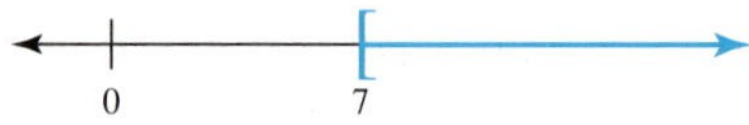

Check Yourself 5

Solve and graph the solution set.

$7x - 8 \leq 6x + 2$

Note that $x < 3$ is the same as $3 > x$. In our next example, we graph an inequality in which the variable is on the right side.

Example 6 Solving an Inequality

Solve and graph the solution set of the inequality

$2x + 3 < 3x + 6$

The coefficient of x is larger on the right side of the inequality than on the left side. Therefore, we isolate the variable on the right side.

$$2x - 2x + 3 < 3x - 2x + 6 \quad \text{Subtract } 2x \text{ from both sides.}$$

$$3 < x + 6$$

$$3 - 6 < x + 6 - 6 \quad \text{Subtract 6 from both sides.}$$

$$-3 < x$$

The graph of the solution set is

Check Yourself 6

Solve and graph the solution set of the inequality

$4x - 5 < 5x - 9$

As with equations, we need a rule for multiplying both sides of an inequality. Here we have to be a bit careful. There is a difference between the multiplication property for inequalities and the one for equations.

$2 < 7$ A true inequality

Multiply both sides by 3.

$$2 < 7$$
$$3 \cdot 2 < 3 \cdot 7$$
$$6 < 21 \qquad \text{A true inequality}$$

Start again, but multiply both sides by -3.

$$2 < 7 \qquad \text{The original inequality}$$
$$(-3)(2) < (-3)(7)$$
$$-6 < -21 \qquad \textit{Not} \text{ a true inequality}$$

Let's try something different.

$$2 < 7 \qquad \text{Change the direction of the inequality.}$$
$$(-3)(2) > (-3)(7) \qquad < \text{ becomes } >.$$
$$-6 > -21 \qquad \text{This is now a true inequality.}$$

NOTE

When we multiply both sides of an inequality by a negative number, we *must* reverse the direction of the inequality sign.

This suggests that multiplying both sides of an inequality by a negative number changes the direction of the inequality.

Property

The Multiplication Property of Inequality

If $a < b$ then $ac < bc$ if $c > 0$

and $ac > bc$ if $c < 0$

Multiplying both sides of an inequality by a *positive* number gives an equivalent inequality.

Multiplying both sides of an inequality by a *negative* number gives an equivalent inequality if we also reverse the direction of the inequality sign.

As with equations, this rule applies to division, as well.

- Dividing both sides of an inequality by the same *positive* number gives an equivalent inequality.

  If $a < b$, then $\frac{a}{c} < \frac{b}{c}$ if $c > 0$.

- When dividing both sides of an inequality by the same *negative* number we must reverse the direction of the inequality sign to get an equivalent inequality.

  If $a < b$, then $\frac{a}{c} > \frac{b}{c}$ if $c < 0$.

Example 7 Solving and Graphing Inequalities

(a) Solve and graph the solution set of $5x < 30$.

Multiplying both sides of the inequality by $\frac{1}{5}$ gives

$$\frac{1}{5}(5x) < \frac{1}{5}(30)$$

Simplifying, we have

$$x < 6$$

The graph of the solution set is

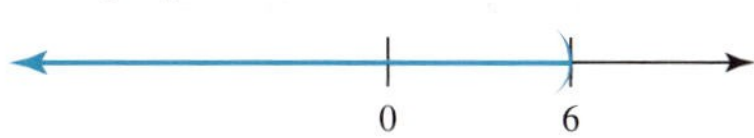

NOTE

Multiplying both sides of the inequality by $\frac{1}{5}$ is the same as dividing both sides by 5.

$$\frac{5x}{5} < \frac{30}{5}$$

(b) Solve and graph the solution set of $-4x \geq 28$.

To isolate the variable, we need to remove the coefficient. In this case, we divide both sides of the inequality by -4.

$$-4x \geq 28$$

$$\frac{-4x}{-4} \leq \frac{28}{-4}$$

Reverse the direction of the inequality symbol when dividing by a negative number!

$$x \leq -7$$

The graph of the solution set is

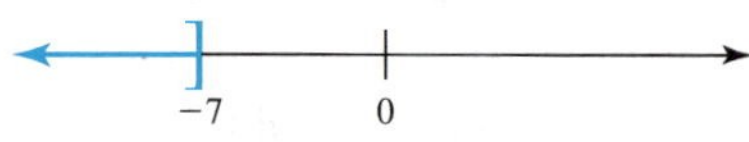

Check Yourself 7

Solve and graph each solution set.

(a) $7x > 35$ **(b)** $-8x \leq 48$

Example 8 illustrates the multiplication property when there are fractions in an inequality.

Example 8 Solving and Graphing Inequalities

(a) Solve and graph the solution set of

$$\frac{x}{4} > 3$$

Here we multiply both sides of the inequality by 4. This isolates x on the left.

$$4\left(\frac{x}{4}\right) > 4(3)$$

$$x > 12$$

The graph of the solution set is

(b) Solve and graph the solution set of

$$-\frac{x}{6} \geq -3$$

In this case, we multiply both sides of the inequality by -6.

RECALL

We reverse the direction of the inequality because we are multiplying by a negative number.

$$(-6)\left(-\frac{x}{6}\right) \leq (-6)(-3)$$

$$x \leq 18$$

The graph of the solution set is

Check Yourself 8

Solve and graph the solution set of each inequality.

(a) $\frac{x}{5} \leq 4$ **(b)** $-\frac{x}{3} < -7$

We summarize our work by looking at the step-by-step procedure for solving an inequality in one variable. The steps are nearly identical to those given to solve an equation in Section 1.6.

Step by Step

Solving a Linear Inequality in One Variable

Step 1 Remove any grouping symbols by applying the distributive property.

Step 2 Multiply both sides of the equation by the LCM to clear the inequality of fractions or decimals.

Step 3 Combine any like terms that appear on either side of the inequality.

Step 4 Apply the addition property of inequalities to write an equivalent inequality with the variable term on one side of the inequality and the constant term on the other.

Step 5 Apply the multiplication property to write an equivalent inequality with the variable isolated on one side of the inequality. Be sure to reverse the direction of the inequality if you multiply or divide by a negative number.

You should see the similarities and differences between equations and inequalities from the problems in the next example. Study them carefully and then complete Check Yourself 9 on your own.

Example 9 Solving and Graphing Inequalities

(a) Solve and graph the solution set of $5x - 3 < 2x$.

First, add 3 to both sides to undo the subtraction on the left.

$$5x - 3 < 2x$$

$$5x - 3 + 3 < 2x + 3$$ Add 3 to both sides to undo the subtraction.

$$5x < 2x + 3$$

Now subtract $2x$, so that only the number remains on the right.

$$5x < 2x + 3$$

$$5x - 2x < 2x - 2x + 3$$ Subtract $2x$ to isolate the number on the right.

$$3x < 3$$

Next *divide* both sides by 3. Because 3 is positive, we leave the inequality symbol as is.

$$\frac{3x}{3} < \frac{3}{3}$$

$$x < 1$$

The graph of the solution set is

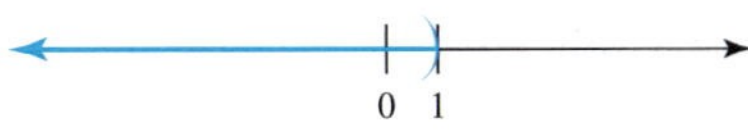

(b) Solve and graph the solution set of $2 - 5x < 7$.

$$2 - 5x < 7$$

$$2 - 2 - 5x < 7 - 2$$ Subtract 2.

$$-5x < 5$$

$$\frac{-5x}{-5} > \frac{5}{-5}$$ Divide by -5. Be sure to reverse the direction of the inequality.

or $$x > -1$$

RECALL

The multiplication property also allows us to divide both sides by a nonzero number.

The graph of the solution set is

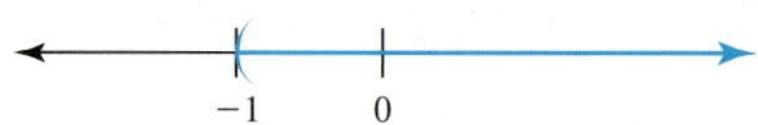

(c) Solve and graph the solution set of $5x - 5 \geq 3x + 4$.

$$5x - 5 \geq 3x + 4$$

$$5x - 5 + 5 \geq 3x + 4 + 5 \qquad \text{Add 5.}$$

$$5x \geq 3x + 9$$

$$5x - 3x \geq 3x - 3x + 9 \qquad \text{Subtract } 3x.$$

$$2x \geq 9$$

$$\frac{2x}{2} \geq \frac{9}{2} \qquad \text{Divide by 2.}$$

$$x \geq \frac{9}{2}$$

RECALL

Place $\frac{9}{2}$ on a number line in between 4 and 5.

The graph of the solution set is

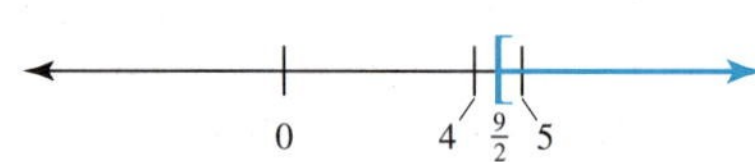

(d) Solve and graph the solution set of $x + 2 < \frac{5}{2}x - 1$.

$$2(x + 2) < 2\left(\frac{5}{2}x - 1\right) \qquad \text{Multiply by the LCD.}$$

$$2x + 4 < 5x - 2$$

$$2x + 4 - 4 < 5x - 2 - 4 \qquad \text{Subtract 4.}$$

$$2x < 5x - 6$$

$$2x - 5x < 5x - 5x - 6 \qquad \text{Subtract } 5x.$$

$$-3x < -6$$

$$\frac{-3x}{-3} > \frac{-6}{-3} \qquad \text{Divide by } -3 \text{ and reverse the direction of the inequality.}$$

$$x > 2$$

The graph of the solution set is

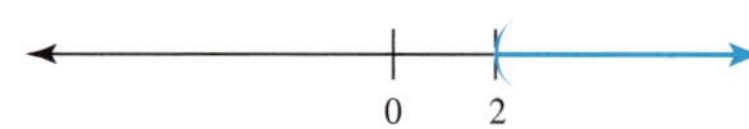

Check Yourself 9

Solve each inequality and graph each solution set.

(a) $4x + 9 \geq x$ **(b)** $5 - 6x < 41$
(c) $8x + 3 < 4x - 13$ **(d)** $5x + 12 \geq 10x - 8$

Many applications are better modeled with inequalities rather than equations. However, the English language sometimes makes it difficult to know how to translate sentences to algebra with inequalities. We build some inequalities in the next example.

Example 10 Writing Linear Inequalities

Write each statement as an inequality.

(a) All numbers greater than 12.

$x > 12$ 12 is not part of the set.

(b) The numbers that are not less than 12.

$x \geq 12$ 12 is part of this set.

(c) Numbers at least as large as 12.

$x \geq 12$ 12 is part of this set.

Check Yourself 10

Write each statement as an inequality. Use x as the variable.

(a) The numbers that are not more than 4.
(b) All numbers that do not exceed 4.
(c) All numbers less than 4.

We are ready to use inequalities to model applications.

Example 11 Solving an Inequality Application

Mohammed needs a mean score of 92 or higher on four tests to earn an A. So far his scores are 94, 89, and 88. What scores on the fourth test will get him an A?

NOTES

What do you need to find?

Assign a letter to the unknown.

Write an inequality.

Solve the inequality.

Step 1 We are looking for the scores that will, when combined with the other scores, give Mohammed an A.

Step 2 Let x represent a fourth-test score that will earn him an A.

Step 3 The inequality has the mean on the left side, which must be greater than or equal to the 92 on the right.

$$\frac{94 + 89 + 88 + x}{4} \geq 92$$

Step 4 First, multiply both sides by 4.

$$94 + 89 + 88 + x \geq 368$$

Then add the test scores.

$$271 + x \geq 368$$

Subtract 271 from both sides.

$$x \geq 97$$

Step 5 Mohammed needs a 97 or above to earn an A.

To check the solution, we find the mean of the four test scores, 94, 89, 88, and 97.

$$\frac{94 + 89 + 88 + (97)}{4} = \frac{368}{4} = 92$$

Check Yourself 11

Felicia needs a mean score of at least 75 on five tests to earn a passing grade in her health class. On her first four tests she has scores of 68, 79, 71, and 70. What scores on the fifth test will give her a passing grade?

So far, we have represented our solution sets by graphing them on a number line. In Chapter 2, you will learn to present these solution sets algebraically by using *set-builder* and *interval* notations.

Check Yourself ANSWERS

1. **(a)** >; **(b)** < **2.** 

3. **(a)** 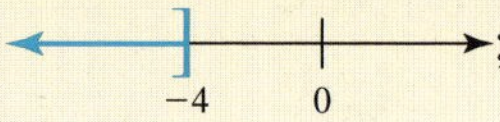; **(b)** 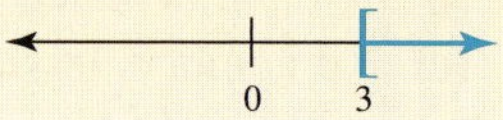 **4.** $x > 6$ 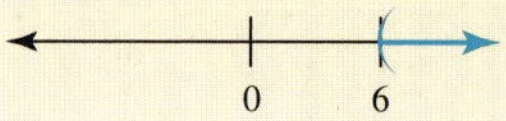

5. $x \le 10$ 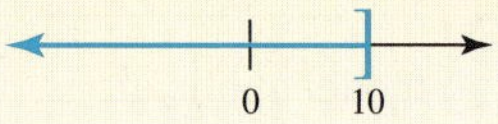 **6.** $x > 4$ 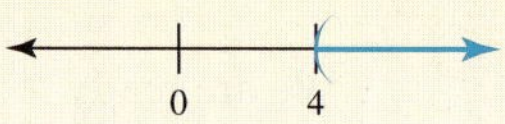

7. **(a)** $x > 5$ 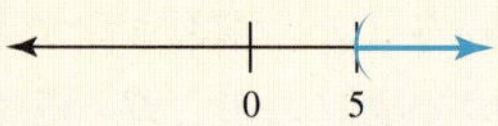 **(b)** $x \ge -6$ 

8. **(a)** $x \le 20$ 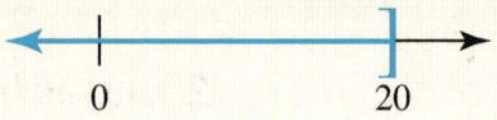 **(b)** $x > 21$ 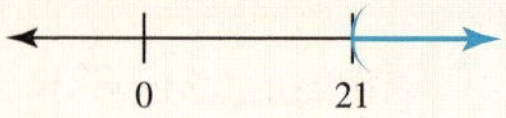

9. **(a)** $x \ge -3$  **(b)** $x > -6$ 

(c) $x < -4$ 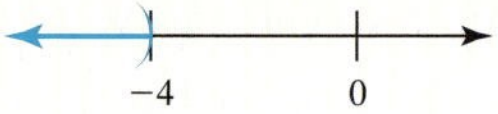 **(d)** $x \le 4$ 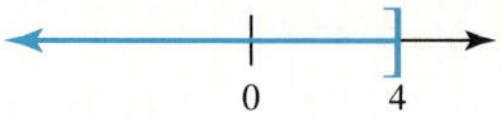

10. **(a)** $x \le 4$; **(b)** $x \le 4$; **(c)** $x < 4$ **11.** She needs to score 87 or better.

Reading Your Text

These fill-in-the-blank exercises will help you understand some of the key vocabulary used in this section. The answers to these exercises are in the Answers Appendix in the back of the text.

(a) $9 > 6$ is read "9 is ________ than 6."

(b) Adding the same quantity to both sides of an inequality yields an ________ inequality.

(c) Multiplying both sides of an inequality by a ________ number yields an equivalent inequality.

(d) Multiplying both sides of an inequality by a ________ number yields an equivalent inequality only if we also reverse the direction of the inequality sign.

1.7 exercises

< Objective 1 >

Complete the statements, using the symbol $<$ or $>$.

1. 5 ______ 10 **2.** 9 ______ 8 **3.** 7 ______ -2 **4.** 0 ______ -5

5. 0 ______ 4 **6.** -10 ______ -5 **7.** -2 ______ -5 VIDEO **8.** -4 ______ -11

Write each inequality in words.

9. $x < 3$ **10.** $x \le -5$ **11.** $x \ge -4$ **12.** $x < -2$

13. $-5 \le x$ **14.** $2 < x$

< Objective 2 >

Graph the solution set of each inequality.

15. $x > 2$ **16.** $x < -3$ **17.** $x < 6$

18. $x > 4$ **19.** $x > 1$ **20.** $x < -2$

21. $x < 8$ **22.** $x > 3$ **23.** $x > -5$

24. $x < -2$ **25.** $x \ge 9$ **26.** $x \ge 0$ VIDEO

27. $x < 0$ **28.** $x \le -3$

< Objective 3 >

Graph the solution set of each inequality.

29. $x - 8 \le 3$ **30.** $x + 5 \le 4$ **31.** $x + 8 \ge 10$

32. $x - 11 > -14$ **33.** $5x < 4x + 7$ VIDEO **34.** $8x \ge 7x - 4$

35. $6x - 8 \le 5x$ **36.** $3x + 2 > 2x$ **37.** $8x + 1 \ge 7x + 9$

38. $5x + 2 \le 4x - 6$

39. $7x + 5 < 6x - 4$

40. $8x - 7 > 7x + 3$

41. $\frac{3}{4}x - 5 \ge 7 - \frac{1}{4}x$

42. $\frac{7}{8}x + 6 < 3 - \frac{1}{8}x$

43. $11 + 0.63x > 9 - 0.37x$

44. $0.54x + 0.12x + 9 \le 19 - 0.34x$

45. $3x \le 9$

46. $5x > 20$

47. $5x > -35$

48. $6x \le -18$

49. $-6x \ge 18$

50. $-9x < 45$

51. $-2x < -12$

52. $-12x \ge -48$

53. $\frac{x}{4} > 5$

54. $\frac{x}{3} \le -3$

55. $-\frac{x}{2} \ge -3$

56. $-\frac{x}{4} < 5$

57. $\frac{2x}{3} < 6$

58. $\frac{3x}{4} \ge -9$

59. $5x > 3x + 8$

60. $4x \le x - 9$

61. $5x - 2 < 3x$ 

62. $7x + 3 \ge 2x$

63. $3 - 2x > 5$

64. $5 - 3x \le 17$

65. $2x \ge 5x + 18$

66. $3x < 7x - 28$

67. $\frac{1}{3}x - 5 \le \frac{5}{3}x + 11$

68. $\frac{3}{7}x + 6 \ge -\frac{12}{7}x - 9$

69. $0.34x + 21 \ge 19 - 1.66x$

70. $-1.57x - 15 \ge 1.43x + 18$

71. $7x - 5 < 3x + 2$

72. $5x - 2 \ge 2x - 7$

73. $5x + 7 > 8x - 17$

74. $4x - 3 \le 9x + 27$

75. $3x - 2 \le 5x + 3$ 

76. $2x + 3 > 8x - 2$

Translate each statement to an inequality. Let x represent the variable.

77. 5 more than a number is greater than 3.

78. 3 less than a number is less than or equal to 5.

79. 4 less than twice a number is less than or equal to 7. 

80. 10 more than a number is greater than negative 2.

81. 4 times a number, decreased by 15, is greater than that number.

82. 2 times a number, increased by 28, is less than or equal to 6 times that number.

83. **Statistics** There are fewer than 1,000 wild giant pandas left in the bamboo forests of China. Write an inequality expressing this relationship.

84. **Statistics** Let C represent the amount of Canadian forest and M represent the amount of Mexican forest. Write an inequality showing the relationship of the forests of Mexico and Canada if Canada contains at least 9 times as much forest as Mexico.

85. **Statistics** To pass a course with a grade of B or better, Liza must have an average of 80 or more. Her grades on three tests are 72, 81, and 79. Write an inequality representing the scores that Liza can receive on the fourth test to obtain a B average or better for the course.

86. **Statistics** Sam must average 70 or more in his summer course in order to obtain a grade of C. His first three test grades were 75, 63, and 68. Write an inequality representing the scores that Sam can receive on the last test in order to earn a C grade.

87. **Business and Finance** Juanita is a salesperson for a manufacturing company. She may choose to receive $500 or 5% commission on her sales as a bonus. Write an inequality representing the amounts she needs to sell to make the 5% offer a better deal.

88. **Business and Finance** The cost for a long-distance telephone call from a pay phone is $0.24 for the first minute and $0.11 for each additional minute or portion thereof. The total cost of the call cannot exceed $3. Write an inequality representing the number of minutes a person could talk without exceeding $3.

89. **Business and Finance** Samantha's financial aid stipulates that her tuition not exceed $1,500 per semester. If her local community college charges a $45 service fee plus $290 per course, what is the greatest number of courses for which Samantha can register?

90. **Statistics** Nadia is taking a mathematics course in which five tests are given. To earn a B, a student must average at least 80 on the five tests. Nadia scored 78, 81, 76, and 84 on the first four tests. What score on the last test will earn her at least a B?

91. **Geometry** The width of a rectangle is fixed at 40 cm, and the perimeter can be no greater than 180 cm. Find the maximum length of the rectangle.

92. **Business and Finance** The women's soccer team can spend at most $900 for its annual awards banquet. If the restaurant charges a $75 setup fee and $24 per person, at most how many people can attend?

93. **Business and Finance** Joyce is determined to spend no more than $125 on clothes. She wants to buy two pairs of identical jeans and a blouse. If she spends $29 on the blouse, what is the maximum amount she can spend on each pair of jeans?

94. **Business and Finance** Ben earns $750 per month plus 4% commission on all his sales over $900. Find the minimum sales that will allow Ben to earn at least $2,500 per month.

Match each inequality on the right with a statement on the left.

95. x is nonnegative. **(a)** $x \geq 0$

96. x is negative. **(b)** $x \geq 5$

97. x is no more than 5. **(c)** $x \leq 5$

98. x is positive. **(d)** $x > 0$

99. x is at least 5. **(e)** $x < 5$

100. x is less than 5. **(f)** $x < 0$

Determine whether each statement is **true** *or* **false.**

101. A linear inequality in one variable can have an infinite number of solutions.

102. The statement $x < 5$ has the same solution set as the statement $5 < x$.

103. The solution set of $3 \geq x$ is the same as the solution set of $x \leq 3$.

104. If we add a negative number to both sides of an inequality, we must reverse the direction of the inequality symbol.

Complete each statement with **always, sometimes,** *or* **never.**

105. Adding the same quantity to both sides of an inequality __________ gives an equivalent inequality. 

106. We can __________ solve an inequality just by using the addition property of inequality.

107. When both sides of an inequality are multiplied by a negative number, the direction of the inequality symbol is __________ reversed.

108. If the graph of the solution set for an inequality extends infinitely to the right, the solution set __________ includes the number 0.

Skills | Calculator/Computer | **Career Applications** | Above and Beyond

109. **Construction Technology** Pressure-treated wooden studs can be purchased for \$7.19 each. How many studs can be bought if a project's budget allots no more than \$450 for studs?

110. **Electronics Technology** Berndt Electronics earns a marginal profit of \$560 each on the sale of a particular server. If other costs involved amount to \$4,500, then how many servers does the company need to sell in order to earn a net profit of at least \$12,000?

Skills | Calculator/Computer | Career Applications | **Above and Beyond**

111. If an inequality simplifies to $7 > -5$, what is the solution set and why?

112. If an inequality simplifies to $7 < -5$, what is the solution set and why?

113. You are the office manager for a small company and need to acquire a new copier for the office. You find a suitable one that leases for \$250 per month. It costs 2.5¢ per copy to run the machine. You purchase paper for \$3.50 per ream (500 sheets). If your copying budget is no more than \$950 per month, is this machine a good choice? Write a brief recommendation to the purchasing department. Use equations and inequalities to explain your recommendation.

114. Nutritionists recommend that, for good health, no more than 30% of our daily intake of calories come from fat. Algebraically, we can write this as $f \le 0.30c$, where f = calories from fat and c = total calories for the day. But this does not mean that everything we eat must meet this requirement.

For example, if you eat $\frac{1}{2}$ cup of Ben and Jerry's vanilla ice cream for dessert after lunch, you are eating a total of 250 calories, of which 150 are from fat. This amount is considerably more than 30% from fat, but if you are careful about what you eat the rest of the day, you can stay within the guidelines.

Set up an inequality based on your normal caloric intake. Solve the inequality to find how many calories in fat you could eat over the day and still have no more than 30% of your daily calories from fat. The American Heart Association says that to maintain your weight, your daily caloric intake should be 15 calories for every pound. You can compute this number to estimate the number of calories a day you normally eat. Do some research in your grocery store or on the Internet to determine what foods satisfy the requirements for your diet for the rest of the day. There are 9 calories in every gram of fat; many food labels give the amount of fat only in grams.

115. Your aunt calls to ask your help in making a decision about buying a new refrigerator. She says that she found two that seem to fit her needs, and both are supposed to last at least 14 years, according to *Consumer Reports*. The initial cost for one refrigerator is \$712, but it uses only 88 kilowatt-hours (kWh) per month. The other refrigerator costs \$519 and uses an estimated 100 kWh/month. You do not know the price of electricity per kilowatt-hour where your aunt lives, so you will have to decide what, in cents per kilowatt-hour, will make the first refrigerator cheaper to run for its 14 years of expected usefulness. Write your aunt a letter, explaining what you did to calculate this cost, and tell her to make her decision based on how the kilowatt-hour rate she has to pay in her area compares with your estimation.

Answers

1. $<$ **3.** $>$ **5.** $<$ **7.** $>$ **9.** x is less than 3 **11.** x is greater than or equal to -4 **13.** -5 is less than or equal to x

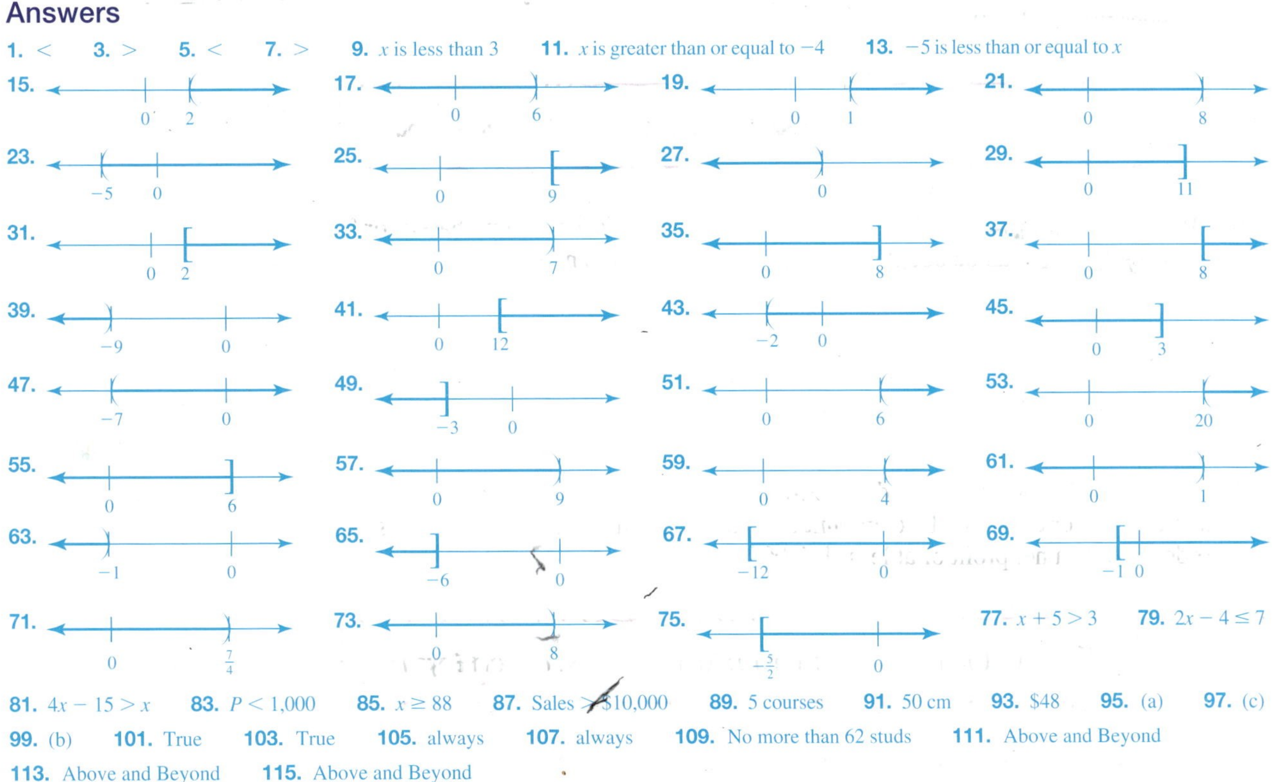

77. $x + 5 > 3$ **79.** $2x - 4 \le 7$ **81.** $4x - 15 > x$ **83.** $P < 1{,}000$ **85.** $x \ge 88$ **87.** Sales $> \$10{,}000$ **89.** 5 courses **91.** 50 cm **93.** \$48 **95.** (a) **97.** (c) **99.** (b) **101.** True **103.** True **105.** always **107.** always **109.** No more than 62 studs **111.** Above and Beyond **113.** Above and Beyond **115.** Above and Beyond

summary :: chapter 1

Definition/Procedure	Example	Reference
Transition to Algebra		**Section 1.1**
Addition $x + y$ means the **sum** of x **and** y or x **plus** y. Some other words indicating addition are *more than* and *increased by*.	The sum of x and 5 is $x + 5$. 7 more than a is $a + 7$. b increased by 3 is $b + 3$.	p. 59
Subtraction $x - y$ means the **difference** of x **and** y or x **minus** y. Some other words indicating subtraction are *less than* and *decreased by*.	The difference of x and 3 is $x - 3$. 5 less than p is $p - 5$. a decreased by 4 is $a - 4$.	p. 59
Multiplication $x \cdot y$, $(x)(y)$, xy — All these mean the **product** of x and y or x **times** y.	The product of m and n is mn. The product of 2 and the sum of a and b is $2(a + b)$.	p. 60
Division $\frac{x}{y}$ means x **divided by** y or the **quotient** when x is divided by y.	n divided by 5 is $\frac{n}{5}$. The sum of a and b, divided by 3, is $\frac{a+b}{3}$.	p. 61
Evaluating Algebraic Expressions		**Section 1.2**
Evaluating Algebraic Expressions **Step 1** Replace each variable by the given number value. **Step 2** Compute. (Be sure to follow the rules for the order of operations.)	Evaluate $\frac{4a - b}{2c}$ if $a = -6$, $b = 8$, and $c = -4$. $\frac{4a - b}{2c} = \frac{4(-6) - 8}{2(-4)}$ $= \frac{-24 - 8}{-8}$ $= \frac{-32}{-8} = 4$	p. 69
Simplifying Algebraic Expressions		**Section 1.3**
Term A number or the product of a number and one or more variables and their exponents.	$3x^2y$ is a term.	p. 80
Like Terms Terms that contain exactly the same variables raised to the same powers.	$4a^2$ and $3a^2$ are like terms. $5x^2$ and $2xy^2$ are not like terms.	p. 81
Combining Like Terms **Step 1** Add or subtract the numerical coefficients. **Step 2** Attach the common variables.	$5a + 3a = 8a$ $7xy - 3xy = 4xy$	p. 82
Solving Algebraic Equations		**Sections 1.4–1.6**
Equation A statement that two expressions are equal.	$3x - 5 = 7$ is an equation.	p. 91
Solution A value for the variable that will make an equation a true statement.	4 is a solution to the equation because $3 \cdot 4 - 5 \stackrel{?}{=} 7$ $12 - 5 \stackrel{?}{=} 7$ $7 = 7$ True	p. 92
Equivalent Equations Equations that have exactly the same solutions.		p 93

Continued

Definition/Procedure	Example	Reference
Writing Equivalent Equations There are two basic properties that will yield equivalent equations.		
Addition Property If $a = b$, then $a + c = b + c$. Adding (or subtracting) the same quantity on each side of an equation gives an equivalent equation.	If $x = y + 3$, then $x + 2 = y + 5$.	*p.* 93
Multiplication Property If $a = b$, then $ac = bc, c \neq 0$. Multiplying (or dividing) both sides of an equation by the same nonzero number gives an equivalent equation.	$5x = 20$ and $x = 4$ are equivalent equations.	*p.* 107
Solving Linear Equations We say that an equation is "solved" when we have an equivalent equation of the form $x = \square$ or $\square = x$ where $\square$ is some number The steps of solving a linear equation are **Step 1** Remove any grouping symbols by applying the distributive property. **Step 2** Multiply both sides of the equation by the LCM to clear the equation of fractions or decimals. **Step 3** Combine any like terms that appear on either side of the equation. **Step 4** Apply the addition property of equality to write an equivalent equation with the variable term on *one side* of the equation and the constant term on the *other side*. **Step 5** Apply the multiplication property of equality to write an equivalent equation with the variable isolated on one side of the equation with coefficient 1. **Step 6** State the answer and check the solution in the *original* equation.	Solve. $3(x - 2) + 4x = 3x + 14$ $3x - 6 + 4x = 3x + 14$ $7x - 6 = 3x + 14$ $+6 \quad\quad +6$ $7x = 3x + 20$ $-3x \quad\quad -3x$ $4x = 20$ $\frac{4x}{4} = \frac{20}{4}$ $x = 5$	*p.* 121
Linear Inequalities		**Section 1.7**
Inequality A statement that one quantity is less than (or greater than) another. Four symbols are used: $a < b$ ↑ *a* is less than *b*. $a > b$ ↑ *a* is greater than *b*. $a \le b$ ↑ *a* is less than or equal to *b*. $a \ge b$ ↑ *a* is greater than or equal to *b*.	$4 < 9$ $-1 > -6$ $2 \le 2$ $3 \ge -4$	*p.* 131
Graphing Inequalities To graph $x < a$, we use a parenthesis and an arrow pointing left.  To graph $x \ge b$, we use a bracket and an arrow pointing right.  	$x < 6$ 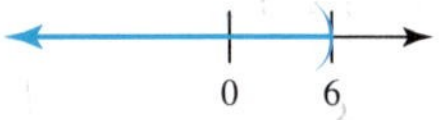 $x \ge 5$ 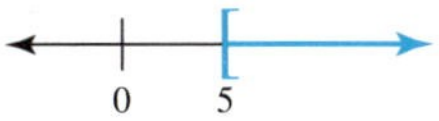 	*p.* 132

Continued

Definition/Procedure	Example	Reference
Solving Inequalities An inequality is "solved" when it is in the form $x < \square$ or $x > \square$. **Addition Property** If $a < b$, then $a + c < b + c$. Adding (or subtracting) the same quantity to both sides of an inequality gives an equivalent inequality.	$2x - 3 > 5x + 6$ $+3 \quad +3$ $2x > 5x + 9$ $-5x \quad -5x$ $-3x > 9$	p. 133
Multiplication Property If $a < b$, then $ac < bc$ when $c > 0$ and $ac > bc$ when $c < 0$. Multiplying both sides of an inequality by the same *positive number* gives an equivalent inequality. When both sides of an inequality are multiplied by the same *negative number, you must reverse the direction* of the inequality to give an equivalent inequality.	$\frac{-3x}{-3} < \frac{9}{-3}$ $x < -3$ 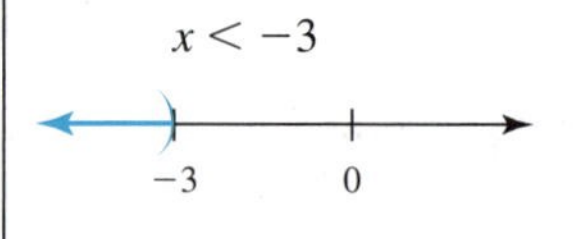	p. 135

summary exercises :: chapter 1

This summary exercise set will help ensure that you have mastered each of the objectives of this chapter. The exercises are grouped by section. You should reread the material associated with any exercises that you find difficult. The answers to the odd-numbered exercises are in the Answers Appendix in the back of the text.

1.1 *Express each phrase symbolically.*

1. 8 more than y
2. c decreased by 10
3. The product of 8 and a
4. 5 times the product of m and n
5. The product of x and 7 less than x
6. 3 more than the product of 17 and x
7. The quotient when a plus 2 is divided by a minus 2
8. The product of 6 more than a number and 6 less than the same number
9. The quotient of 9 and a number
10. The product of a number and 3 more than twice the same number

1.2 *Evaluate each expression if $x = -3$, $y = 6$, $z = -4$, and $w = 2$.*

11. $3x + w$
12. $5y - 4z$
13. $x + y - 3z$
14. $5z^2$
15. $5(x^2 - w^2)$
16. $\frac{6z}{2w}$
17. $\frac{2x - 4z}{y - z}$
18. $\frac{y(x - w)^2}{x^2 - 2xw + w^2}$
19. $-4x^2 - 2zw^2 + 4z$
20. $3x^3w^2 + xy^2$

1.3 *List the terms of each expression.*

21. $4a^3 - 3a^2$
22. $5x^2 - 7x + 3$

List the like terms.

23. $5m^2, -3m, -4m^2, 5m^3, m^2$

24. $4ab^2, 3b^2, -5a, ab^2, 7a^2, -3ab^2, 4a^2b$

Combine like terms.

25. $9x + 7x$

26. $2x + 5x$

27. $9xy - 6xy$

28. $5ab^2 + 2ab^2$

29. $7a + 3b + 12a - 2b$

30. $-3x + 2y - 5x - 7y$

31. $5x^3 + 17x^2 - 2x^3 - 8x^2$

32. $3a^3 + 5a^2 + 4a - 2a^3 - 3a^2 - a$

33. Subtract $4a^3$ from the sum of $2a^3$ and $12a^3$.

34. Subtract the sum of $3x^2$ and $5x^2$ from $15x^2$.

Write an expression for each exercise.

35. **Construction** If x feet (ft) is cut off the end of a board that is 37 ft long, how much is left?

36. **Business and Finance** Sergei has 25 nickels and dimes in his pocket. If x of these are dimes, how many of the coins are nickels?

37. **Geometry** The length of a rectangle is 4 meters (m) more than the width. Write an expression for the length of the rectangle.

38. **Number Problem** A number is 7 less than 6 times the number n. Write an expression for the number.

39. **Construction** A 25-ft plank is cut into two pieces. Write expressions for the length of each piece.

40. **Business and Finance** Bernie has d dimes and q quarters in his pocket. Write an expression for the amount of money (in dollars) that Bernie has in his pocket.

41. **Geometry** Find the perimeter of the given rectangle.

$2x$ m

$(-x + 4)$ m

42. **Geometry** If the length of a building is x m and the width is $\frac{x}{6}$ m, what is the perimeter of the building?

1.4 *Determine whether the number shown in parentheses is a solution to the given equation.*

43. $5x - 3 = 7$ (2)

44. $5x - 8 = 3x + 2$ (4)

45. $7x - 2 = 2x + 8$ (2)

46. $\frac{2}{3}x - 2 = 10$ (21)

Solve each equation and check your results.

47. $x + 3 = 5$

48. $x - 9 = 3$

49. $5x = 4x - 5$

50. $4x - 9 = 3x$

51. $9x - 7 = 8x - 6$

52. $3 + 4x - 1 = x - 7 + 2x$

53. $4(2x + 3) = 7x + 5$

54. $5(5x - 3) = 6(4x + 1)$

1.5 and 1.6

55. $5x = 35$

56. $7x = -28$

57. $-9x = 36$

58. $-9x = -63$

59. $\frac{2}{3}x = 18$

60. $\frac{7}{8}x = 28$

61. $7x + 8 = 3x$

62. $3 - 5x = -17$

63. $4x - 7 = 2x$

64. $2 - 4x = 5$

65. $\frac{x}{3} - 5 = 1$

66. $\frac{3}{4}x - 2 = 7$

67. $7x + 4 = 2x + 6$

68. $9x - 8 = 7x - 3$

69. $2x + 7 = 4x - 5$

70. $3x - 15 = 7x - 10$

71. $\frac{10}{3}x - 5 = \frac{4}{3}x + 7$

72. $\frac{11}{4}x - 15 = 5 - \frac{5}{4}x$

73. $3.7x + 8 = 1.7x + 16$

74. $2.4x + 6 - 1.2x = 9 - 1.8x + 12$

75. $5(3x - 1) - 6x = 3x - 2$

76. $5x + 2(3x - 4) = 14x - 7$

77. $8x - 5(x + 3) = -10$

78. $3(2x - 5) - 2(x - 3) = 11$

79. $\frac{2x}{3} - \frac{x}{4} = 5$

80. $\frac{3x}{4} - \frac{2x}{5} = 7$

81. $\frac{x}{2} - \frac{x + 1}{3} = \frac{1}{6}$

82. $\frac{x + 1}{5} - \frac{x - 6}{3} = \frac{1}{3}$

Solve each application.

83. **Business and Finance** A mechanic charged \$75 an hour plus \$225 for parts to replace the ignition coil on a car. If the total bill was \$450, how many hours did the repair job take?

84. **Business and Finance** A call to Phoenix, Arizona, from Dubuque, Iowa, costs 55 cents for the first minute and 23 cents for each additional minute or portion of a minute. If Barry has \$6.30 in change, how long can he talk?

85. **Number Problem** The sum of 4 times a number and 14 is 34. Find the number.

86. **Number Problem** If 6 times a number is subtracted from 42, the result is 24. Find the number.

1.7 *Graph the solution set for each inequality.*

87. $x > 5$

88. $x \le -4$

89. $x \ge 9$

90. $x < 0$

91. $x - 2 \le 9$

92. $5x > 4x - 3$

93. $4x \ge -12$

94. $-\frac{x}{3} \ge 5$

95. $2x \le 8x - 3$

96. $7 - 6x > 15$

97. $5x - 2 \le 4x + 5$

98. $4x - 2 < 7x + 16$

Use this chapter test to assess your progress and to review for your next exam. Allow yourself about an hour to take this test. The answers to these exercises are in the Answers Appendix in the back of the text.

Express each phrase symbolically.

1. The sum of x and y

2. The difference m minus n

3. The product of a and b

4. The quotient when p is divided by 3 less than q

5. 5 less than c

6. The product of 3 and the quantity $2x$ minus $3y$

7. 3 times the difference of m and n

8. All numbers not more than 8.

Evaluate when $x = -4$.

9. $-4x - 12$

10. $3x^2 + 2x - 4$

Evaluate each expression if $a = -2$, $b = 6$, and $c = -4$.

11. $4a - c$

12. $\dfrac{3a - 4b}{a + c}$

Combine like terms.

13. $8a - 3b - 5a + 2b$

14. $7x^2 - 3x + 2 - (5x^2 - 3x - 6)$

Determine whether the number shown in parentheses is a solution to each equation.

15. $7x - 3 = 25$ (5)

16. $8x - 3 = 5x + 9$ (4)

Solve each equation and check your results.

17. $7x - 12 = 6x$

18. $\frac{4}{5}x = 24$

19. $5x - 3(x - 5) = 19$

20. $\dfrac{x - 5}{3} = \dfrac{5}{4}$

Solve and graph the solution set of each inequality.

21. $x - 5 \le 9$

22. $5 - 3x > 17$

Solve each application.

23. **Number Problem** 5 times a number, decreased by 7, is 28. What is the number?

24. **Problem Solving** Jan is twice as old as Juwan, while Rick is 5 years older than Jan. If the sum of their ages is 35 years, find each of their ages.

25. **Business and Finance** A grocery store adds a 25% markup to the wholesale price of goods to determine their retail price. Find the wholesale price of a gallon of milk that retails for $2.59.

CHAPTER

2

Functions and Graphs

CHAPTER 2 OUTLINE

INTRODUCTION

Math is used in so many places that, although we try to provide our readers with a variety of applications, we can touch on only a few of the settings and fields in which mathematics is applied.

Though the methods learned in introductory algebra have not changed, the technology associated with "doing mathematics" is different. Today, the power of math comes from the use of functions to model applications. We can concentrate on understanding the function model precisely because the tools and technology enhance our experience with "doing mathematics."

In Activity 2, we introduce you to many of the features of graphing calculators. If you have not had the opportunity to use a graphing calculator, we suggest that you work through the activity in this chapter. If you have had experience with a graphing calculator, you will undoubtedly agree that it is a very helpful tool for examining and understanding the function model.

2.1 Formulas and Problem Solving

< 2.1 Objectives >

1 > Solve a formula for any variable

2 > Solve applications involving geometric figures

3 > Solve motion problems

Formulas are extremely useful tools in many fields. Formulas are simply equations that express relationships between two or more variables. You are familiar with many formulas, such as

$A = \frac{1}{2}bh$ The area of a triangle

$I = Prt$ Interest

$V = \pi r^2 h$ The volume of a cylinder

A formula is also called a literal equation because it involves several letters or variables. For instance, our first formula or literal equation, $A = \frac{1}{2}bh$, involves the three variables A (for area), b (for base), and h (for height).

Unfortunately, formulas are not always given in the form we need to solve a particular problem. In such cases, we use algebra to change the formula to a more useful equivalent equation, solved for a particular variable. The steps used in the process are the same as those used to solve linear equations. Consider an example.

Example 1 Solving a Formula for a Variable

< Objective 1 >

Suppose we know the area A and the base b of a triangle and want to find its height h.

We are given

$$A = \frac{1}{2}bh$$

We need to find an equivalent equation with h, the unknown, by itself on one side and everything else on the other side. We can think of $\frac{1}{2}b$ as the **coefficient** of h.

We can remove the two *factors* of that coefficient, $\frac{1}{2}$ and b, separately.

$$2A = 2\left(\frac{1}{2}bh\right)$$ Multiply both sides by 2 to clear the equation of fractions.

or

$$2A = bh$$

$$\frac{2A}{b} = \frac{bh}{b}$$ Divide by b to isolate h.

$$\frac{2A}{b} = h$$

or

$$h = \frac{2A}{b}$$ Reverse the sides to write h on the left.

We now have the height h in terms of the area A and the base b. This is called **solving the equation for** h and means that we are rewriting the formula as an equivalent equation of the form

$$h = \square$$

RECALL

A coefficient is the factor by which a variable is multiplied.

NOTE

$$2\left(\frac{1}{2}bh\right) = \left(2 \cdot \frac{1}{2}\right)(bh)$$
$$= 1 \cdot bh$$
$$= bh$$

NOTE

Here, $\square$ means an expression containing all the numbers and variables *other than h.*

Check Yourself 1

Solve $V = \frac{1}{3}Bh$ for h.

You learned the methods needed to solve most formulas for some specified variable. As Example 1 illustrates, we apply the rules you learned in Section 1.6 in exactly the same way as we did with equations with one variable.

You may have to apply both the addition and the multiplication properties when solving a formula for a specified variable. Example 2 illustrates this situation.

Example 2 Solving a Formula

NOTE

This is a linear equation in two variables. You will see this again.

(a) Solve $y = mx + b$ for x.

Remember that we want to end up with x alone on one side of the equation. Start by subtracting b from both sides to undo the addition on the right.

$$y = mx + b$$
$$y - b = mx + b - b$$
$$y - b = mx$$

If we divide both sides by m, then x will be alone on the right side.

$$\frac{y-b}{m} = \frac{mx}{m}$$
$$\frac{y-b}{m} = x$$

or

$$x = \frac{y-b}{m}$$

(b) Solve $3x + 2y = 12$ for y.

Begin by isolating the y-term.

$$\begin{aligned} 3x + 2y &= \quad 12 \\ -3x \qquad &\quad -3x \\ \hline 2y &= -3x + 12 \end{aligned}$$

Then, isolate y by dividing by its coefficient.

RECALL

Dividing by 2 is the same as multiplying by $\frac{1}{2}$.

$$\frac{2y}{2} = \frac{-3x+12}{2}$$
$$y = \frac{-3x+12}{2}$$

Often, in a situation like this, we use the distributive property to separate the terms on the right side of the equation.

$$\begin{aligned} y &= \frac{-3x+12}{2} \\ &= \frac{-3x}{2} + \frac{12}{2} \\ &= -\frac{3}{2}x + 6 \end{aligned}$$

$\frac{-3x}{2} = -\frac{3}{2}x$

NOTE

v and v_0 represent distinct quantities.

Check Yourself 2

(a) Solve $v = v_0 + gt$ for t.
(b) Solve $4x - 3y = 8$ for x.

We summarize the steps illustrated by our examples.

Step by Step

Solving a Formula or Literal Equation

NOTE

These are the same basic steps used to solve **any** linear equation.

Step 1 Remove any grouping symbols by applying the distributive or multiplication property.

Step 2 Multiply both sides of the equation by the LCM to clear the equation of fractions or decimals.

Step 3 Combine any like terms that appear on either side of the equation.

Step 4 Use the addition property of equality to write an equivalent equation with the term containing the desired variable on one side of the equation and all other terms on the other side.

Step 5 Use the multiplication property to write an equivalent equation with the desired variable isolated on one side and its coefficient equal to 1.

Here is one more example, using these steps.

Example 3 Solving a Formula for a Variable

NOTE

$A = P + Prt$ is a formula for the *amount* of money in an account after earning interest.

Solve $A = P + Prt$ for r.

$$A = P + Prt$$

$$A - P = P - P + Prt$$

Subtracting P from both sides leaves the term involving r alone on the right.

$$A - P = Prt$$

$$\frac{A - P}{Pt} = \frac{Prt}{Pt}$$

Dividing both sides by Pt isolates r on the right.

$$\frac{A - P}{Pt} = r$$

or

$$r = \frac{A - P}{Pt}$$

Check Yourself 3

Solve $2x + 3y = 6$ for y.

Now we look at an application requiring us to solve a formula.

Example 4 Using a Formula

Suppose that the amount in an account 3 years after a principal of $5,000 was invested is $6,050. What was the average interest rate?

From Example 3,

$$A = P + Prt$$

NOTE

Do you see the advantage of having the equation solved for the desired variable?

in which A is the amount in the account, P is the principal, r is the interest rate, and t is the time in years that the money has been invested. By the result of Example 3 we have

$$r = \frac{A - P}{Pt}$$

and we can substitute the known values in the second equation:

$$r = \frac{(6{,}050) - (5{,}000)}{(5{,}000)(3)}$$

$$= \frac{1{,}050}{15{,}000} = 0.07 = 7\%$$

The interest rate was 7%.

Check Yourself 4

Suppose that the amount in an account 4 years after a principal of \$3,000 was invested is \$3,720. What was the interest rate?

In subsequent applications, we use the five-step process first described in Section 1.4. As a reminder, here are those steps.

Step by Step

Solving Applications

Step 1 Read the problem carefully. Then reread it to decide what you are asked to find.

Step 2 Choose a variable to represent one of the unknowns in the problem. Then represent all other unknowns of the problem with expressions that use the same variable.

Step 3 Translate the problem to the language of algebra to form an equation.

Step 4 Solve the equation.

Step 5 Answer the question and include units in your answer, when appropriate. Check your solution by returning to the original problem.

NOTE

Part of checking a solution is making certain that it is reasonable.

Example 5 Solving a Geometry Application

< Objective 2 >

The length of a rectangle is 1 centimeter (cm) less than 3 times the width. If the perimeter is 54 cm, find the dimensions of the rectangle.

Step 1 You want to find the dimensions (the width and length).

Step 2 Let x be the width.

Then $3x - 1$ is the length.

(3 times the width: $3x$; 1 less than: -1)

Step 3 To write an equation, we use the formula for the perimeter of a rectangle.

$$P = 2W + 2L \qquad \text{or} \qquad 2W + 2L = P$$

So

$$2x + 2(3x - 1) = 54$$

(Twice the width: $2x$; Twice the length: $2(3x - 1)$; Perimeter: 54)

NOTE

When an application involves geometric figures, draw a sketch of the problem, including the labels you assigned in step 2.

Length $3x - 1$

Width x

Step 4 Solve the equation.

$$2x + 2(3x - 1) = 54$$
$$2x + 6x - 2 = 54$$
$$8x = 56$$
$$x = 7$$

NOTE

Be sure to return to the original statement of the problem when checking your result.

Step 5 The width x is 7 cm, and the length, $3x - 1$, is 20 cm.

Check: We look at the two conditions specified in this problem.

The relationship between the length and the width
20 is 1 less than 3 times 7, so this condition is met.

The perimeter of a rectangle
The sum of twice the width and twice the length is
$2(7) + 2(20) = 14 + 40 = 54$, which checks.

Check Yourself 5

The length of a rectangle is 5 inches (in.) more than twice the width. If the perimeter of the rectangle is 76 in., what are the dimensions of the rectangle?

RECALL

The circumference of a circle is the distance around the circle.

π is used to represent an irrational number.

$\pi \approx 3.14$

One reason you might need to *manipulate* a geometric formula is because it is sometimes easier to measure the *output* of a formula.

For instance, the formula for the circumference of a circle is

$$C = 2\pi r$$

However, in practice, we might be able to measure the circumference of a round object directly, but not its radius. But if we wanted to compute the area (or volume) of this object, we need to know its radius.

Example 6 Solving a Geometry Application

Poplar trees often have round trunks. You use a tape measure to find the circumference of one poplar tree. Its circumference is approximately 8.8 in.

(a) Find the radius of the trunk, to the nearest tenth of an inch.

We are asked to find the radius of this tree trunk.

We begin with the formula for the circumference of a circle and solve for the radius, r.

$$C = 2\pi r$$

Now we divide both sides by 2π. This is the "coefficient" of r and results in r being isolated on one side.

$$\frac{C}{2\pi} = \frac{2\pi r}{2\pi}$$ Divide both sides by 2π and simplify.

$$\frac{C}{2\pi} = r \quad \text{or} \quad r = \frac{C}{2\pi}$$

Now we can substitute the circumference, 8.8 in.

$$r = \frac{C}{2\pi}$$
$$= \frac{(8.8)}{2\pi}$$
$$\approx 1.4$$

The radius is approximately 1.4 in.

NOTE

```
8.8/(2π)
        1.400563499
Ans→R
        1.400563499
```

Be sure to place parentheses around the denominator.

Recall that you can store this value if you want to use it later.

```
π(1.4)²(420)
        2586.159072
πR²*420
        2588.241347
```

If you use the stored value for the radius, you get 2,588 in.^{3}, which is more precise.

(b) The trunk of this particular poplar tree is 35 ft tall (420 in.). The volume of the trunk, in cubic inches, is given by the formula

$$V = \pi r^2 h$$

in which r is the radius and h is the height (both in inches).

Find the volume of this poplar trunk, to the nearest cubic inch.

We use the radius found in part (a) along with the height, in inches.

$$\begin{aligned} V &= \pi r^2 h \\ &= \pi(1.4)^2(420) \\ &\approx 2{,}586 \end{aligned}$$

The volume is approximately 2,586 in.^{3}.

Check Yourself 6

The circumference of a telephone pole measures approximately 31.4 in.

(a) Find the radius of a telephone pole, to the nearest inch.
(b) Find the volume, to the nearest cubic inch, if the telephone pole is 40 ft (480 in.) tall.

> CAUTION

Be careful to make your units consistent. If a rate is given in *miles per hour*, then the time must be given in *hours* and the distance in *miles*.

One common application is the *motion problem*. Motion problems involve a distance traveled, a rate (or speed), and an amount of time. To solve a motion problem, we need a relationship between these three quantities.

Suppose you travel at a rate of 50 miles per hour (mi/hr) on a highway for 6 hr. How far (what distance) will you have gone? To find the distance, multiply.

(50 mi/hr)(6 hr) = 300 mi

↑ Speed or rate ↑ Time ↑ Distance

Property

Motion Problems

If r is the rate, t is the time, and d is the distance traveled, then

$$d = r \cdot t$$

We apply this relationship in Example 7.

Example 7 Solving a Motion Problem

< Objective 3 >

On Friday morning Ricardo drove from his house to the beach in 4 hr. When coming back Sunday afternoon, heavy traffic slowed his speed by 10 mi/hr, and the trip took 5 hr. What was his average speed (rate) in each direction?

Step 1 We want the speed or rate in each direction.

Step 2 Let x be Ricardo's speed to the beach. Then $x - 10$ is his return speed.

It is always a good idea to sketch the given information in a motion problem. Here we have

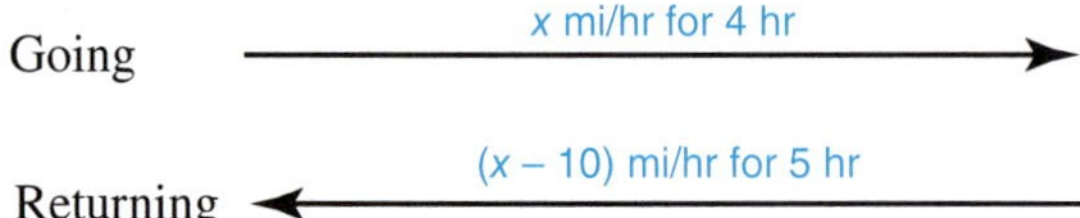

Step 3 A chart or table can help summarize the given information, especially when stumped about how to proceed. We begin with an "empty" table.

	Rate	Time	Distance
Going			
Returning			

Next, we fill the table with the information given in the problem.

	Rate	Time	Distance
Going	x	4	
Returning	$x - 10$	5	

Now we fill in the missing information. Here we use the fact that $d = rt$ to complete the table.

	Rate	Time	Distance
Going	x	4	$4x$
Returning	$x - 10$	5	$5(x - 10)$

Since we know that the distance is the same each way, we can write an equation using the fact that the product of the rate and the time each way must be the same.

Distance (going) = Distance (returning)

Time · rate (going) = Time · rate (returning)

$$\underbrace{4x}_{\text{Time} \cdot \text{rate (going)}} = \underbrace{5(x - 10)}_{\text{Time} \cdot \text{rate (returning)}}$$

Step 4 Solve.

$4x = 5(x - 10)$

$4x = 5x - 50$ Use the distributive property to remove the parentheses.

$-x = -50$ Subtract $5x$ from both sides to isolate the x-term.

$x = 50$ Multiply both sides by -1 to isolate the variable.

NOTE

x was his rate going; $x - 10$, his rate returning.

Step 5 So Ricardo's rate going to the beach was 50 mi/hr, and his rate returning was 40 mi/hr.

To check, you should verify that the product of the time and the rate is the same in each direction.

Check Yourself 7

A plane made a flight (with the wind) between two towns in 2 hr. Returning against the wind, the plane's speed was 60 mi/hr slower, and the flight took 3 hr. What was the plane's speed in each direction?

Example 8 illustrates another way to use the distance relationship.

Example 8 Solving a Motion Problem

Katy leaves Las Vegas, Nevada, for Los Angeles, California, at 10 A.M., driving 50 mi/hr. At 11 A.M. Jensen leaves Los Angeles for Las Vegas, driving 55 mi/hr along the same route. If the cities are 260 mi apart, at what time will they meet?

Step 1 Let's find the time that Katy travels until they meet.

Step 2 Let x be Katy's time.

Then $x - 1$ is Jensen's time. Jensen left 1 hr later!

Again, you should draw a sketch of the given information.

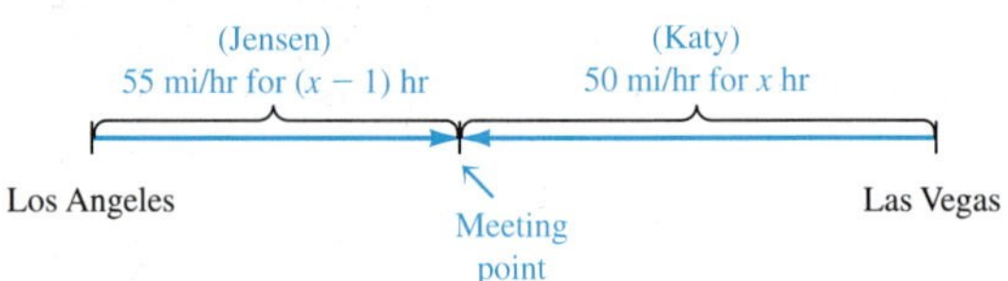

Step 3 As before, we can use a table to solve.

	Rate	Time	Distance
Katy	50	x	$50x$
Jensen	55	$x - 1$	$55(x - 1)$

To write an equation, we again need the relationship $d = rt$. From this equation, we write

Katy's distance $= 50x$

Jensen's distance $= 55(x - 1)$

From the original problem, the sum of those distances is 260 mi, so

$$50x + 55(x - 1) = 260$$

Step 4

$$50x + 55(x - 1) = 260$$
$$50x + 55x - 55 = 260$$
$$105x - 55 = 260$$
$$105x = 315$$
$$x = 3$$

NOTE

Be sure to answer the question asked in the problem.

Step 5 Finally, since Katy left at 10 A.M., the two will meet at 1 P.M. We leave the check of this result to you.

Check Yourself 8

At noon a jogger leaves one point, running 8 mi/hr. One hour later a bicyclist leaves the same point, traveling 20 mi/hr in the opposite direction. At what time will they be 36 mi apart?

Check Yourself ANSWERS

1. $h = \frac{3V}{B}$ **2.** **(a)** $t = \frac{v - v_0}{g}$; **(b)** $x = \frac{3}{4}y + 2$ **3.** $y = \frac{6 - 2x}{3}$ or $y = -\frac{2}{3}x + 2$
4. The interest rate was 6%. **5.** The width is 11 in.; the length is 27 in.
6. **(a)** 5 in.; **(b)** 37,699 in.^{3} **7.** 180 mi/hr with the wind and 120 mi/hr against the wind **8.** At 2 P.M.

Reading Your Text

These fill-in-the-blank exercises will help you understand some of the key vocabulary used in this section. The answers to these exercises are in the Answers Appendix in the back of the text.

(a) A ________ is also called a literal equation because it involves several letters or variables.

(b) A ________ is the factor by which a variable is multiplied.

(c) Always return to the ________ equation or statement when checking your result.

(d) In a motion problem, the ________ traveled is found by taking the product of the rate of travel (speed) and the time traveled.

Skills | Calculator/Computer | Career Applications | Above and Beyond

2.1 exercises

< Objective 1 >

Solve each formula for the indicated variable.

1. $P = 4s$ (for s)
Perimeter of a square

2. $V = Bh$ (for B)
Volume of a prism

3. $E = IR$ (for R)
Voltage in an electric circuit

4. $I = Prt$ (for r)
Simple interest

5. $V = LWH$ (for H)
Volume of a rectangular solid

6. $V = \pi r^2 h$ (for h)
Volume of a cylinder

7. $A + B + C = 180$ (for B)
Measure of angles in a triangle

8. $P = I^2R$ (for R)
Power in an electric circuit

9. $ax + b = 0$ (for x)
Linear equation in one variable

10. $y = mx + b$ (for m) VIDEO
Slope-intercept form for a line

11. $s = \frac{1}{2}gt^2$ (for g)
Distance

12. $K = \frac{1}{2}mv^2$ (for m)
Energy

13. $x + 5y = 15$ (for y)
Linear equation in two variables

14. $2x + 3y = 6$ (for x)
Linear equation in two variables

15. $P = 2L + 2W$ (for L)
Perimeter of a rectangle

16. $ax + by = c$ (for y)
Linear equation in two variables

17. $V = \frac{KT}{P}$ (for T)
Volume of a gas

18. $V = \frac{1}{3}\pi r^2 h$ (for h)
Volume of a cone

19. $x = \frac{a + b}{2}$ (for b)
Average of two numbers

20. $D = \frac{C - s}{n}$ (for s)
Depreciation

21. $F = \frac{9}{5}C + 32$ (for C)
Celsius/Fahrenheit conversion

22. $A = P + Prt$ (for t)
Amount at simple interest

23. $S = 2\pi r^2 + 2\pi rh$ (for h)
Total surface area of a cylinder

24. $A = \frac{1}{2}h(B + b)$ (for b)
Area of a trapezoid

Write each statement as an equation with x as the variable.

25. Twice the sum of a number and 6 is 18.

26. The sum of twice a number and 4 is 20.

27. 3 times the difference of a number and 5 is 21.

28. The difference of 3 times a number and 5 is 21.

29. The sum of twice an integer and 3 times the next consecutive integer is 48.

30. The sum of 4 times an odd integer and twice the next consecutive odd integer is 46.

< Objectives 2 and 3 >

Solve each application.

31. **Geometry** A rectangular solid has a base with length 8 cm and width 5 cm. If the volume of the solid is 120 cm^3, find the height of the solid. (See exercise 5.) VIDEO

32. **Geometry** A cylinder has a radius of 4 in. If the volume of the cylinder is 144π $in.^3$, what is the height of the cylinder? (See exercise 6.)

33. **Business and Finance** A principal of $2,000 was invested in a savings account for 4 years. If the interest earned for the period was $240, what was the interest rate? (See exercise 4.)

34. **Geometry** If the perimeter of a rectangle is 60 ft and the width is 12 ft, find its length. (See exercise 15.)

35. **Statistics** The high temperature in New York for a particular day was reported at 77°F. How would the same temperature have been given in degrees Celsius? (See exercise 21.) chapter 2 > Make the Connection

36. **Geometry** Rose's garden is in the shape of a trapezoid. If the height of the trapezoid is 16 m, one base is 20 m, and the area is 224 m^2, find the length of the other base. (See exercise 24.)

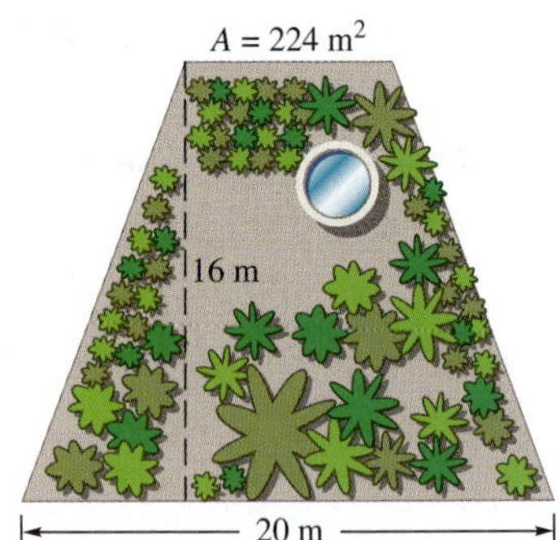

37. **Number Problem** One number is 8 more than another. If the sum of the smaller number and twice the larger number is 46, find the two numbers.

38. **Number Problem** One number is 3 less than another. If 4 times the smaller number minus 3 times the larger number is 4, find the two numbers.

39. **Number Problem** One number is 7 less than another. If 4 times the smaller number plus 2 times the larger number is 62, find the two numbers.

40. **Number Problem** One number is 10 more than another. If the sum of twice the smaller number and 3 times the larger number is 55, find the two numbers. VIDEO

41. **Number Problem** Find two consecutive integers such that the sum of twice the first integer and 3 times the second integer is 28.
Hint: If x represents the first integer, $x + 1$ represents the next consecutive integer.

42. **Number Problem** Find two consecutive odd integers such that 3 times the first integer is 5 more than twice the second.
Hint: If x represents the first integer, $x + 2$ represents the next consecutive odd integer.

43. **Geometry** The length of a rectangle is 1 in. more than twice its width. If the perimeter of the rectangle is 74 in., find the dimensions of the rectangle.

44. **Geometry** The length of a rectangle is 5 cm less than 3 times its width. If the perimeter of the rectangle is 46 cm, find the dimensions of the rectangle. VIDEO

45. **Geometry** The length of a rectangular garden is 5 m more than 3 times its width. The perimeter of the garden is 74 m. What are the dimensions of the garden?

46. **Geometry** The length of a rectangular playing field is 5 ft less than twice its width. If the perimeter of the playing field is 230 ft, find the length and width of the field.

47. **Geometry** The base of an isosceles triangle is 3 cm less than the length of the equal sides. If the perimeter of the triangle is 36 cm, find the length of each side.

48. **Geometry** The length of one of the equal legs of an isosceles triangle is 3 in. less than twice the length of the base. If the perimeter is 29 in., find the length of each side.

49. **Science and Medicine** Patrick drove 3 hr to attend a meeting. On the return trip, his speed was 10 mi/hr less, and the trip took 4 hr. What was his speed each way?

50. **Science and Medicine** A bicyclist rode into the country for 5 hr. In returning, her speed was 5 mi/hr faster and the trip took 4 hr. What was her speed each way?

51. **Science and Medicine** A car leaves a city at 2 P.M. and goes north at a rate of 50 mi/hr. One hour later a second car leaves, traveling south at a rate of 40 mi/hr. At what time will the two cars be 320 mi apart? VIDEO

52. **Science and Medicine** A bus leaves a station at 1 P.M., traveling west at an average rate of 44 mi/hr. One hour later a second bus leaves the same station, traveling east at a rate of 48 mi/hr. At what time will the two buses be 274 mi apart?

53. **Science and Medicine** At 8:00 A.M., Catherine leaves on a trip at 45 mi/hr. One hour later, Max decides to join her and leaves along the same route, traveling at 54 mi/hr. When will Max catch up with Catherine?

54. **Science and Medicine** Martina leaves home at 9 A.M., bicycling at a rate of 24 mi/hr. Two hours later, John leaves, driving at the rate of 48 mi/hr. At what time will John catch up with Martina?

55. **Science and Medicine** If the temperature in Madrid is given as 35°C, what is the corresponding temperature in degrees Fahrenheit? chapter 2 > Make the Connection

56. **Science and Medicine** What temperature in degrees Celsius is equivalent to 59°F? 

57. **Science and Medicine** Mika leaves Boston for Baltimore at 10:00 A.M., traveling at 45 mi/hr. One hour later, Hiroko leaves Baltimore for Boston on the same route, traveling at 50 mi/hr. If the two cities are 425 mi apart, when will Mika and Hiroko meet?

58. **Science and Medicine** A train leaves town A for town B, traveling at 35 mi/hr. At the same time, a second train leaves town B for town A at 45 mi/hr. If the two towns are 320 mi apart, how long will it take for the two trains to meet?

Determine whether each statement is **true** *or* **false.**

59. Another name for a *formula* is *literal equation.*

60. The formula for the area of a rectangle is $P = 2L + 2W$.

61. The key relationship in motion problems is $d = rt$.

62. When solving for a variable in a formula, we use the same steps used in solving linear equations.

Skills | Calculator/Computer | **Career Applications** | Above and Beyond

63. **Electrical Engineering** Resistance R (in ohms, Ω) is given by the formula

$$R = \frac{V^2}{D}$$

in which D is the power dissipation (in watts, W) and V is the voltage. Determine the power dissipation when 13.2 volts pass through a 220-Ω resistor.

64. **Mechanical Engineering** In a planetary gear, the size and number of teeth must satisfy the equation

$$Cx = By(F - 1)$$

Calculate the number of teeth y needed if $C = 9$ in., $x = 14$ teeth, $B = 2$ in., and $F = 8$.

65. **Allied Health** Yohimbine is used to reverse the effects of xylazine in deer. The recommended dose is 0.125 mg per kilogram of a deer's weight.
   (a) Write a formula that expresses the required dosage level d for a deer of weight w.
   (b) How much yohimbine should be administered to a 15-kg fawn?
   (c) What size deer requires a 5.0-mg dosage?

66. **Electronics Technology** Temperature sensors output voltage, which varies with respect to temperature. For a particular sensor, the output voltage V for a given Celsius temperature C is given by

$$V = 0.28C + 2.2$$

   (a) Determine the output voltage at 0°C.
   (b) Determine the output voltage at 22°C.
   (c) Determine the temperature if the sensor outputs 14.8 V.
   (d) At what temperature is there no voltage output (two decimal places)?

Skills	Calculator/Computer	Career Applications	**Above and Beyond**

67. Here is a common mistake in solving equations.

The equation: $2(x - 2) = x + 3$
First step in solving: $2x - 2 = x + 3$

Write a clear explanation of what error has been made. What could be done to avoid this error?

68. Here is another very common mistake.

The equation: $6x - (x + 3) = 5 + 2x$
First step in solving: $6x - x + 3 = 5 + 2x$

Write a clear explanation of what error has been made and what could be done to avoid the mistake.

69. There is a universally agreed on *order of operations* used to simplify expressions. Explain how the order of operations is used in solving equations. Be sure to use complete sentences.

70. Write an algebraic equation for the English statement "Subtract 5 from the sum of x and 7 times 3 and the result is 20." Compare your equation with those of other students. Did you all write the same equation? Are all the equations correct even though they don't look alike? Do all the equations have the same solution? What is wrong? The English statement is *ambiguous.* Write another English statement that leads correctly to more than one algebraic equation. Exchange with another student and see if she or he thinks the statement is ambiguous. Notice that the algebra is *not* ambiguous!

Answers

1. $s = \frac{P}{4}$ **3.** $R = \frac{E}{I}$ **5.** $H = \frac{V}{LW}$ **7.** $B = 180 - A - C$ **9.** $x = -\frac{b}{a}$ **11.** $g = \frac{2s}{t^2}$ **13.** $y = \frac{15 - x}{5}$ or $y = -\frac{1}{5}x + 3$ **15.** $L = \frac{P - 2W}{2}$ or $L = \frac{P}{2} - W$ **17.** $T = \frac{PV}{K}$ **19.** $b = 2x - a$ **21.** $C = \frac{5}{9}(F - 32)$ or $C = \frac{5(F - 32)}{9}$ **23.** $h = \frac{S - 2\pi r^2}{2\pi r}$ or $h = \frac{S}{2\pi r} - r$ **25.** $2(x + 6) = 18$ **27.** $3(x - 5) = 21$ **29.** $2x + 3(x + 1) = 48$ **31.** 3 cm **33.** 3% **35.** 25°C **37.** 10, 18 **39.** 8, 15 **41.** 5, 6 **43.** 12 in., 25 in. **45.** 8 m, 29 m **47.** Legs: 13 cm; base: 10 cm **49.** going 40 mi/hr, returning 30 mi/hr **51.** 6 P.M. **53.** 2 P.M. **55.** 95°F **57.** 3 P.M. **59.** True **61.** True **63.** 0.792 W **65.** **(a)** $d = 0.125w$; **(b)** 1.875 mg; **(c)** 40 kg **67.** Above and Beyond **69.** Above and Beyond

chapter 2 > Make the Connection

Activity 2 ::

Graphing with a Calculator

A graphing calculator is a tool that can be used to help you solve many different kinds of problems. This activity walks you through several features of the TI-84 Plus. By the time you complete this activity, you will be able to graph equations, change the viewing window to better accommodate a graph, or look at a table of values that represent some of the solutions to an equation. The first portion of this activity demonstrates how you can create the graph of an equation. The features described here can be found on most graphing calculators. See your calculator manual to learn how to get your particular calculator model to perform this activity.

Menus and Graphing

1. To graph the equation $y = 2x + 3$ on a graphing calculator, follow these steps.

a. Press the [Y =] key.

```
Plot1 Plot2 Plot3
\Y1=
\Y2=
\Y3=
\Y4=
\Y5=
\Y6=
\Y7=
```

b. Type $2x + 3$ at the Y_1 prompt. (This represents the first equation. You can type up to 10 separate equations.) Use the [X, T, θ, *n*] key for the variable.

```
Plot1 Plot2 Plot3
\Y1=2X+3
\Y2=
\Y3=
\Y4=
\Y5=
\Y6=
\Y7=
```

c. Press the [GRAPH] key to see the graph.

d. Press the [TRACE] key to display the equation. **Once** you have selected the [TRACE] key, you can use the left and right arrows of the calculator to move the cursor along the line. Experiment with this movement. Look at the coordinates at the bottom of the display screen as you move along the line.

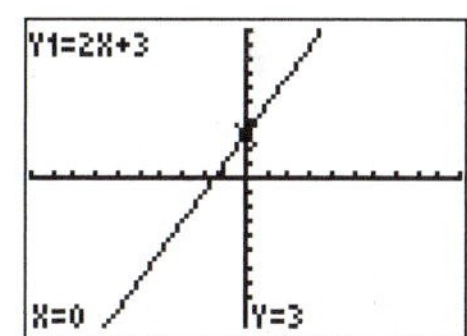

NOTE

Be sure the window is the standard window to see the same graph displayed.

Frequently, we can learn more about an equation if we look at a different section of the graph than the one offered on the display screen. The portion of the graph displayed is called the **window.** The second portion of the activity explains how this window can be changed.

2. Press the [WINDOW] key. The **standard** graphing screen is shown.

Xmin = left edge of screen
Xmax = right edge of screen
Xscl = scale given by each tick mark on *x*-axis
Ymin = bottom edge of screen
Ymax = top edge of screen
Yscl = scale given by each tick mark on *y*-axis
Xres = resolution (do not alter this)

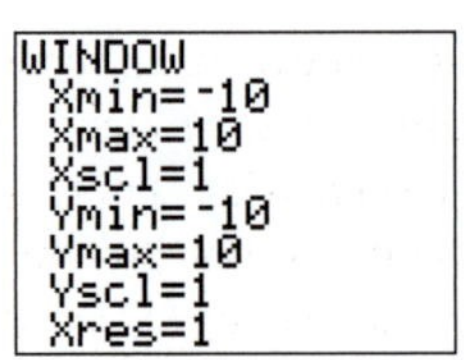

Note: To turn the scales off, enter a 0 for Xscl or Yscl. Do this when the intervals used are very large.

By changing the values for Xmin, Xmax, Ymin, and Ymax, you can adjust the viewing window. Change the viewing window so that Xmin = 0, Xmax = 40, Ymin = 0, and Ymax = 10. Again, press GRAPH. Notice that the tick marks along the x-axis are now much closer together. Changing Xscl from 1 to 5 will improve the display. Try it.

Sometimes we can learn something important about a graph by zooming in or zooming out. The third portion of this activity discusses this calculator feature.

3. a. Press the ZOOM key. Use the ▼ key to scroll down.

b. Selecting the first option, ZBox, allows the user to enlarge the graph within a specified rectangle.

i. Graph the equation $y = x^2 + x - 1$ in the standard window.
Note: To type in the exponent, use the x^2 key or the ∧ key.

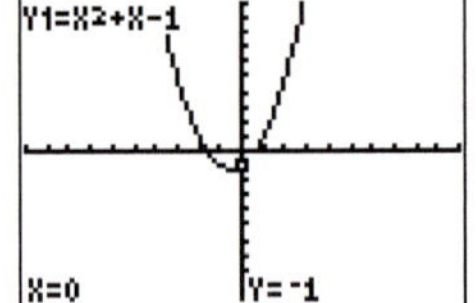

ii. When ZBox is selected, a blinking "+" cursor will appear in the graph window. Use the arrow keys to move the cursor to where you would like a corner of the screen to be; then press the ENTER key.

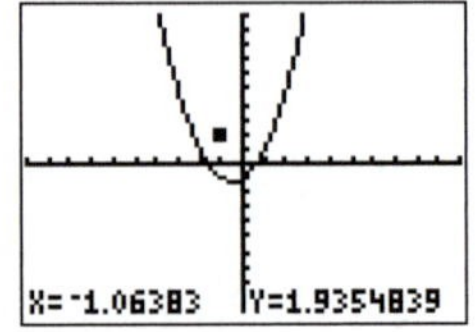

iii. Use the arrow keys to trace out the box containing the desired portion of the graph. Do not press the ENTER key until you have reached the diagonal corner and a full box is on your screen.

After using the down arrow

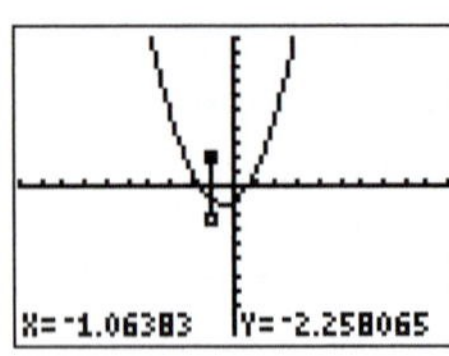

After using the right arrow

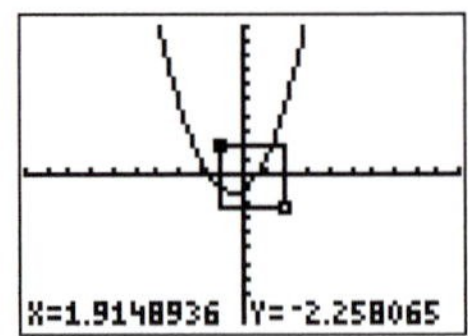

After pressing the ENTER key a second time

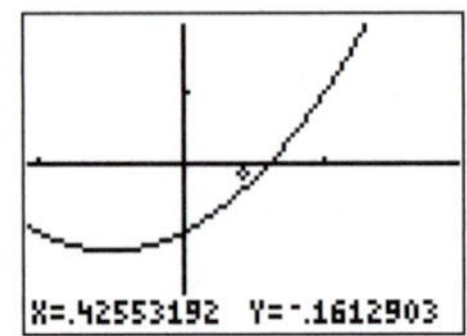

Now the desired portion of a graph can be seen more clearly.

The Zbox feature is especially useful when analyzing the roots (x-intercepts) of an equation.

c. Another feature that allows us to focus is Zoom In. Select the Zoom In option on the Zoom menu. Place the cursor in the center of the portion of the graph you are interested in and press the ENTER key. The window will reset with the cursor at the center of a zoomed-in view.
d. Zoom Out works like Zoom In, except that it sets the view larger (that is, it zooms out) to enable you to see a larger portion of the graph.
e. ZStandard sets the window to the standard window. This is a quick and convenient way to reset the viewing window.
f. ZSquare recalculates the view so that one horizontal unit is the same length as one vertical unit. This is sometimes necessary to get an accurate view of a graph because the width of the calculator screen is greater than its height.

4. **Home Screen** This is where all the basic computations take place. To get to the home screen from any other screen, press 2nd Mode. This accesses the QUIT feature. To clear the home screen of calculations, press the CLEAR key (once or twice).

5. **Tables** The final feature that we look at here is the TABLE. Enter the equation $y = 2x + 3$ into the Y = menu. Then press 2nd WINDOW to access the TBLSET menu. Set the table as shown here and press 2nd GRAPH to access the TABLE feature. You will see the screens shown here.

```
TABLE SETUP
 TblStart=0
 ΔTbl=1
Indpnt: Auto Ask
Depend: Auto Ask
```

X	Y1
0	3
1	5
2	7
3	9
4	11
5	13
6	15

X=0

2.2 Sets and Set Notation

< 2.2 Objectives >

1 > List the elements of a set in roster form

2 > Use set-builder notation to describe a set

3 > Use interval notation to describe a set

4 > Find the union and intersection of sets

Jacob enrolled in his local community college. He registered for intermediate algebra, English composition, small business management I, and photography I. We can call this collection "Jacob's classes." Such a collection is called a **set**. The items in the set are called **elements** of the set. We can write the set as {intermediate algebra, English composition, small business management I, photography I}. The braces tell us where the set begins and ends. Every student has a set describing the classes they are taking.

What if a student takes a term off and does not enroll in any classes? What would their set look like? It would be the set { }, which we call the **empty set.** Sometimes we use the symbol ∅ to indicate the empty set.

RECALL

We introduced *empty sets* in Section 1.6.

Many sets can be written in *roster form,* as was the case with Jacob's classes. The set of prime numbers less than 15 can be written in roster form as $\{2, 3, 5, 7, 11, 13\}$. In Example 1, we list some sets in roster form. **Roster form** is a list enclosed in braces.

Example 1 Listing the Elements of a Set

< Objective 1 >

List the elements of each set in roster form.

(a) The set of all factors of 12

The set of factors is $\{1, 2, 3, 4, 6, 12\}$.

(b) The set of all integers with an absolute value less than 4

The set of integers is $\{-3, -2, -1, 0, 1, 2, 3\}$.

(c) The set of integers between 5 and 8.

The word "between" indicates that we should omit 5 and 8.

$\{6, 7\}$

Check Yourself 1

List the elements of each set in roster form.

(a) The set of factors of 18
(b) The set of even prime numbers
(c) The set of integers between –10 and –5

In this text, we write *between* to mean that the endpoints are not included. So the integers between 1 and 5 are $\{2, 3, 4\}$; the set does not include 1 or 5.

If we want to include the endpoints, we explicitly say so. For instance, we might write "the integers between 1 and 5 (inclusive)," or "the integers between 1 and 5, including the endpoints" for $\{1, 2, 3, 4, 5\}$.

Each set that we examined had a limited number of elements. If we need to indicate that a set continues in some pattern, we use three dots, called an *ellipsis,* to indicate that the set continues with the pattern it started.

Example 2 Listing the Elements of a Set

List the elements of each set in roster form.

(a) The set of natural numbers less than 100

The set $\{1, 2, 3, \ldots, 98, 99\}$ indicates that we continue increasing the numbers by 1 until we get to 99.

(b) The set of positive multiples of 4

The set $\{4, 8, 12, 16, \ldots\}$ indicates that we continue counting by fours forever. (There is no stopping point.)

(c) The set of integers

$\{\ldots, -2, -1, 0, 1, 2, \ldots\}$ indicates that we continue forever in both directions.

Check Yourself 2

List the elements of each set in roster form.

(a) The set of natural numbers between 200 and 300
(b) The set of positive multiples of 3
(c) The set of even integers

Not all sets can be described using the roster form. For instance, if we wanted to describe the set of all real numbers between 1 and 2, we would be unable to list them all, even with ellipses.

NOTE

Some texts use a colon : or the abbreviation s.t. (such that) instead of the vertical bar |.

A statement such as $1 < x < 2$ is called a *compound inequality*.

The most powerful method for describing sets is set-builder notation. Just about every set can be described using this method. For the set of real numbers between 1 and 2, we write

$\{x \mid 1 < x < 2\}$ The inequality symbols are in the same direction.

We read this as "the set of all real numbers with the property that the numbers are between 1 and 2." In this case, we call x a *dummy variable*. We can use any variable in this situation, so $\{y \mid 1 < y < 2\}$ and {☺|1 < ☺ < 2} both represent the same set as $\{x \mid 1 < x < 2\}$.

To the right of the vertical bar, we write the properties that make an object a member of the set. We use the dummy variable to refer to a general element of the set.

Example 3 Using Set-Builder Notation

< Objective 2 >

Use set-builder notation to describe each set.

(a) The set of real numbers less than 100

We write $\{x \mid x < 100\}$.

(b) The set of real numbers greater than -4 but less than or equal to 9

$\{x \mid -4 < x \leq 9\}$

The symbol $\leq$ is a combination of the symbols $<$ and $=$. When we write $x \leq 9$, we are indicating that either x is equal to 9 or it is less than 9.

Check Yourself 3

Use set-builder notation to describe each set.

(a) The set of real numbers greater than -2
(b) The set of real numbers between 3 and 10 (inclusive)

Interval notation is another way to describe a set. For example, all the real numbers between 1 and 2 can be written as (1, 2). We use parentheses to indicate that neither 1 nor 2 is included in the set.

Interval notation should feel familiar based on your work graphing the solution set of an inequality on a number line in Section 1.7. You are simply "removing" the number line from the notation.

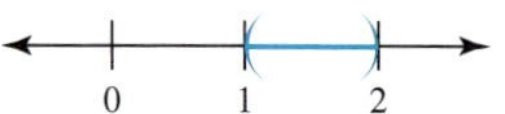

(1, 2)

With number line Without number line

When we describe an interval and want to include the endpoints, we use brackets, just like we do when graphing on a number line. So, the set of numbers between 1 and 2, including the endpoints 1 and 2 looks like [1, 2] in interval notation.

Example 4 Using Interval Notation

< Objective 3 >

Use interval notation to describe each set.

(a) The set of real numbers between 4 and 5

We write (4, 5).

(b) The set of real numbers greater than -3 but less than or equal to 9

We write $(-3, 9]$.

We use a square bracket with 9 to indicate that 9 is included in the interval and a parenthesis with -3 because -3 is not part of the interval.

(c) The set of real numbers greater than or equal to 45

We write $[45, \infty)$.

The positive infinity symbol ∞ does not indicate a number. It shows that the interval includes all real numbers greater than or equal to 45.

(d) The set of real numbers less than 15

We write $(-\infty, 15)$.

The negative infinity symbol $-\infty$ is used to show that the interval includes all real numbers less than 15.

NOTES

Looking at interval notation in terms of a number line, ∞ means that we would shade in the number line as far as it goes: $[45, \infty)$.

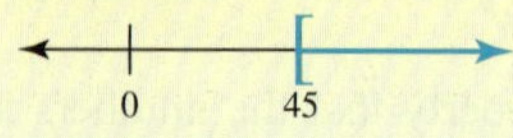

We *always* use a parenthesis with an *infinity symbol.* A square bracket *never* accompanies an infinity symbol.

Check Yourself 4

Use interval notation to describe each set.

(a) The set of real numbers less than 75
(b) The set of real numbers between 5 and 10
(c) The set of real numbers greater than 60
(d) The set of real numbers greater than or equal to 23 but less than or equal to 38

RECALL

"Between" does not include the endpoints, unless otherwise directed.

We can also use a graph to represent a set of numbers. In Example 5, we look at the connection between sets and their graphs.

Example 5 Using a Number Line

Plot the elements of each set on a number line.

(a) $\{-2, 1, 5\}$

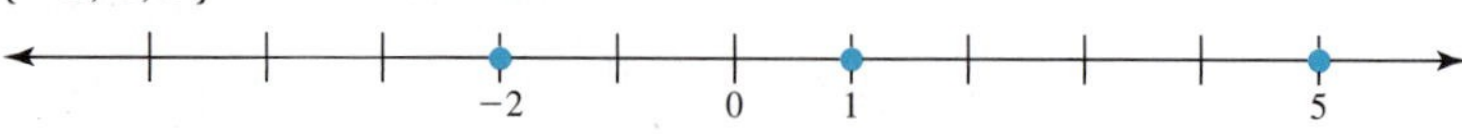

> **NOTE**
>
> Just like interval notation, we use parentheses (,) when the endpoints are not included.
>
> We use brackets [,] when we want to include endpoints.

(b) $\{x \mid x < 3\}$

The blue line and arrow indicate that we continue forever in the negative direction. The parenthesis at 3 indicates that the 3 is not part of the graph.

(c) $\{x \mid -2 < x < 5\}$

The parentheses indicate that the numbers -2 and 5 are not part of the set.

(d) $\{x \mid x \geq -2\}$

The bracket indicates that -2 is part of the set.

Check Yourself 5

Plot the elements of each set on a number line.

(a) $\{-5, -3, 0\}$ **(b)** $\{x \mid -3 < x < -1\}$ **(c)** $\{x \mid x \leq 5\}$

This table summarizes the different ways of describing a set.

> **RECALL**
>
> When using interval notation to describe a set, the infinity symbol is always accompanied by a parenthesis, never a bracket.

Basic Set Notation (a and b represent any real numbers)

Set	Set-Builder Notation	Interval Notation	Graph
All real numbers greater than b	$\{x \mid x > b\}$	(b, ∞)	b
All real numbers less than or equal to b	$\{x \mid x \leq b\}$	$(-\infty, b]$	b
All real numbers greater than a and less than b	$\{x \mid a < x < b\}$	(a, b)	a b
All real numbers greater than or equal to a and less than b	$\{x \mid a \leq x < b\}$	$[a, b)$	a b

The enclosures are important. When describing a set, you are communicating mathematical and technical information. If someone else is going to understand what you mean, then you need to use notation correctly.

Roster form uses curly braces: $\{1, 2, 3\}$.

Set-builder notation uses curly braces: $\{x \mid x \geq 4\}$.

Interval notation uses parentheses to indicate an endpoint is not in the set and brackets when the endpoint is in the set: $(2, 5]$ indicates that 2 is not part of the set, but 5 is in it.

An infinity symbol is always accompanied by a parenthesis in interval notation.

Example 6 Using Set Notation

Express each set in both set-builder and interval notation.

(a)

In set-builder notation the set is $\{x \mid x > -5\}$.

In interval notation it is $(-5, \infty)$.

(b) 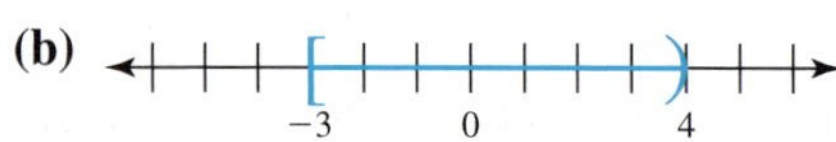

In set-builder notation the set is $\{x \mid -3 \le x < 4\}$.

In interval notation it is $[-3, 4)$.

Check Yourself 6

Express each set in both set-builder and interval notation.

(a) 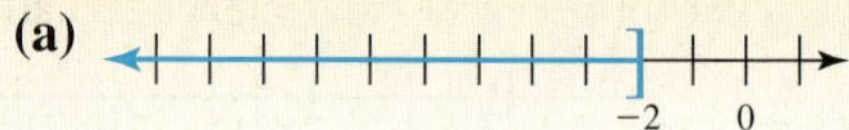

(b) 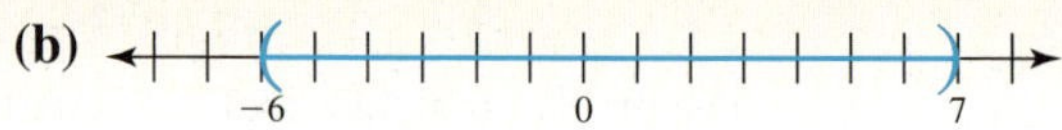

In Section 1.7, you learned to solve inequalities and graph their *solution sets.* The language and notation of sets allow us to present the solution set of an inequality in other ways.

Example 7 Solving and Graphing Inequalities

Solve each inequality. Represent each solution set using set-builder notation, interval notation, and with a graph, as appropriate.

(a) $5(x - 2) \ge -8$

Applying the distributive property on the left yields

$5x - 10 \ge -8$

Solving yields

$$5x - 10 + 10 \ge -8 + 10 \qquad \text{Add 10.}$$

$$5x \ge 2$$

$$x \ge \frac{2}{5} \qquad \text{Divide by 5.}$$

The graph of the solution set, $\left\{x \mid x \ge \frac{2}{5}\right\}$, is

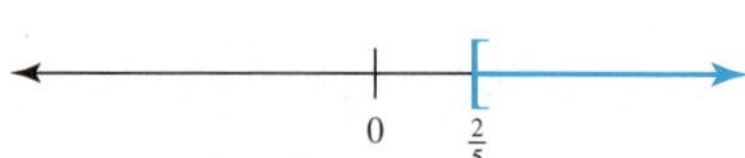

We write the solution set using interval notation as $\left[\frac{2}{5}, \infty\right)$.

(b) $-3(x + 2) \ge 5 - 3x$

$$-3x - 6 \ge 5 - 3x \qquad \text{Apply the distributive property.}$$

$$-6 \ge 5 \qquad \text{Add } 3x \text{ to both sides.}$$

$$0 \ge 11 \qquad \text{Add 6 to both sides.}$$

This is a false statement, so no real number satisfies this inequality or the original inequality. Thus, the solution set is the empty set, written { }. The graph of the solution set contains no points at all. (You might say that it is pointless!)

NOTE

When an answer is the empty set, we neither graph the solution nor use interval notation.

NOTE

Interval notation for the set of all real numbers is $(-\infty, \infty)$. We do not usually graph the solution set if it is the entire number line.

(c) $3(x + 2) > 3x - 4$

$3x + 6 > 3x - 4$ Apply the distributive property.

$6 > -4$ Subtract $3x$ from both sides.

This is a true statement for all values of x, so this inequality and the original inequality are true for all real numbers.

The graph of the solution set $\{x | x \in \mathbb{R}\}$ is every point on the number line.

Check Yourself 7

Solve each inequality. Represent each solution set using set-builder notation, interval notation, and with a graph, as appropriate.

(a) $4(x + 3) < 9$ **(b)** $-2(4 - x) \leq 5 + 2x$

(c) $-4(x - 1) < 3 - 4x$

We used the *element* symbol $\in$ in the previous example to indicate that our answer is the set of all real numbers $\{x | x \in \mathbb{R}\}$. More generally, we can use this symbol to make a statement about any element and any set.

$2 \in \{1, 2, 3\}$ 2 is in the set {1, 2, 3}.

$4 \notin \{1, 2, 3\}$ 4 is not in the set {1, 2, 3}.

We also used the standard symbol for the set of real numbers $\mathbb{R}$. Some other sets of numbers have agreed upon symbols, as well.

$\mathbb{N}$ symbolizes the set of natural (or counting) numbers, $\{1, 2, 3, \ldots\}$.

$\mathbb{Z}$ indicates the set of integers.

$\mathbb{Q}$ represents the set of rational numbers.

$\mathbb{R}$ is the set of real numbers.

There are occasions when we need to combine sets. Two commonly used operations are *union* and *intersection.*

Definition

Union and Intersection of Sets

The **union** of two sets A and B, written $A \cup B$, is the set of all elements that belong to either A or B or both.

The **intersection** of two sets A and B, written $A \cap B$, is the set of all elements that belong to both A and B.

Example 8 Union and Intersection

< Objective 4 >

Let $A = \{1, 3, 5\}$, $B = \{3, 5, 9\}$, and $C = \{9, 11\}$. Describe each set.

(a) $A \cup B$

This is the set of elements that are in A or B or both.

$A \cup B = \{1, 3, 5, 9\}$

(b) $A \cap B$

This is the set of elements common to A and B.

$A \cap B = \{3, 5\}$

(c) $A \cup C$

This is the set of elements in A or C or both.

$A \cup C = \{1, 3, 5, 9, 11\}$

(d) $A \cap C$

This is the set of elements common to A and C.

$A \cap C = \{\ \}$ or $\varnothing$ since there are no elements in common.

RECALL

$\{\ \}$ or $\varnothing$ is the symbol for the empty set.

Check Yourself 8

Let $A = \{2, 4, 7\}$, $B = \{4, 7, 10\}$, and $C = \{8, 12\}$. Find

(a) $A \cup B$ **(b)** $A \cap B$ **(c)** $A \cup C$ **(d)** $A \cap C$

Check Yourself ANSWERS

1. (a) $\{1, 2, 3, 6, 9, 18\}$; **(b)** $\{2\}$; **(c)** $\{-9, -8, -7, -6\}$ **2. (a)** $\{201, 202, 203, \ldots, 298, 299\}$; **(b)** $\{3, 6, 9, 12, \ldots\}$; **(c)** $\{\ldots, -6, -4, -2, 0, 2, 4, 6, \ldots\}$ **3. (a)** $\{x \mid x > -2\}$; **(b)** $\{x \mid 3 \le x \le 10\}$
4. (a) $(-\infty, 75)$; **(b)** $(5, 10)$; **(c)** $(60, \infty)$; **(d)** $[23, 38]$ **5. (a)** 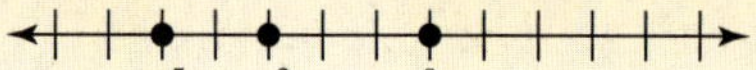

(b) 

(c) 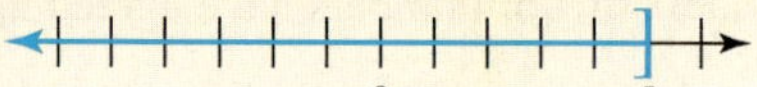

6. (a) $\{x \mid x \le -2\}$, $(-\infty, -2]$; **(b)** $\{x \mid -6 < x < 7\}$, $(-6, 7)$
7. (a) $\left\{x \mid x < -\frac{3}{4}\right\}$, $\left(-\infty, -\frac{3}{4}\right)$, 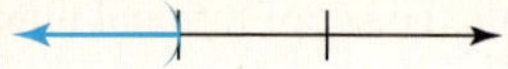

; **(b)** $\{x \mid x \in \mathbb{R}\}$, $(-\infty, \infty)$; **(c)** $\{\ \}$
8. (a) $\{2, 4, 7, 10\}$; **(b)** $\{4, 7\}$; **(c)** $\{2, 4, 7, 8, 12\}$; **(d)** $\varnothing$

Reading Your Text

These fill-in-the-blank exercises will help you understand some of the key vocabulary used in this section. The answers to these exercises are in the Answers Appendix in the back of the text.

(a) The objects in a set are called the __________ of the set.

(b) The symbol $\varnothing$ is used to represent the __________ set.

(c) The notation $\{x \mid x < 0\}$ is an example of __________ notation.

(d) The notation in which the set of real numbers between 0 and 1 is written as (0, 1) is called __________ notation.

Skills | Calculator/Computer | Career Applications | Above and Beyond

2.2 exercises

< Objective 1 >

Use the roster method to list the elements of each set.

1. The set of all the days of the week

2. The set of all months of the year that have 31 days

3. The set of all factors of 18

4. The set of all factors of 24

5. The set of all prime numbers less than 30

6. The set of all prime numbers between 20 and 40

7. The set of all negative integers greater than -6

8. The set of all positive integers less than 6

9. The set of all even whole numbers less than 13

10. The set of all odd whole numbers less than 14

11. The set of integers greater than 2 and less than 7

12. The set of integers greater than 5 and less than 10

13. The set of integers greater than -4 and less than -1

14. The set of integers greater than -8 and less than -3

15. The set of integers between -5 and 2, inclusive

16. The set of integers between -1 and 4

17. The set of odd whole numbers

18. The set of even whole numbers

19. The set of all even whole numbers less than 100

20. The set of all odd whole numbers less than 100

21. The set of all positive multiples of 5

22. The set of all positive multiples of 6

< Objectives 2 and 3 >

Use set-builder notation and interval notation to describe each set.

23. The set of all real numbers greater than 10

24. The set of all real numbers less than 25

25. The set of all real numbers greater than or equal to -5

26. The set of all real numbers less than or equal to -3 

27. The set of all real numbers greater than or equal to 2 and less than or equal to 7

28. The set of all real numbers greater than -3 and less than -1

29. The set of all real numbers between -4 and 4, inclusive

30. The set of all real numbers between -8 and 3, inclusive

Plot the elements of each set on a number line.

31. $\{-2, -1, 0, 4\}$

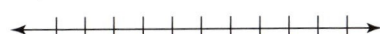

32. $\{-5, -1, 2, 3, 5\}$

33. $\{x \mid x > 4\}$

34. $\{x \mid x > -1\}$

35. $\{x \mid x \geq -3\}$

36. $\{x \mid x \leq 6\}$

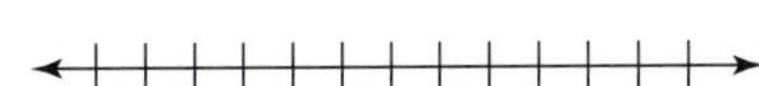

37. $\{x \mid 2 < x < 7\}$

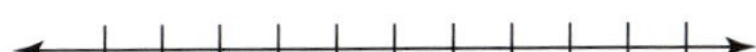

38. $\{x \mid 4 < x < 8\}$

39. $\{x \mid -3 < x \le 5\}$

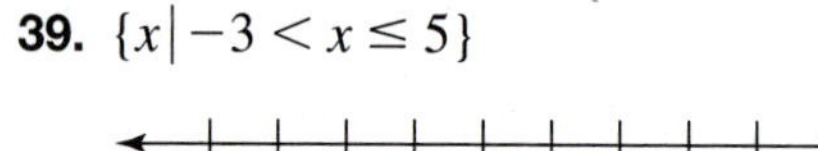

40. $\{x \mid -6 < x \le 1\}$

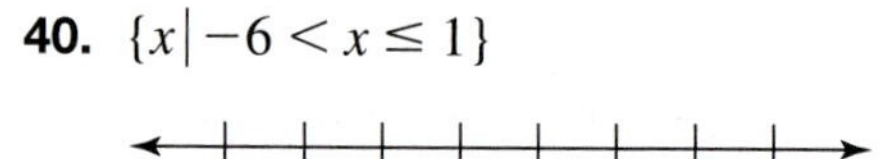

41. $\{x \mid -4 \le x < 0\}$

42. $\{x \mid -5 \le x < 2\}$ VIDEO

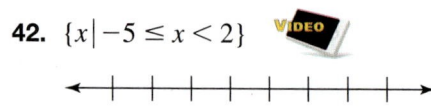

43. $\{x \mid -7 \le x \le -3\}$

44. $\{x \mid -1 \le x \le 4\}$

45. The set of all integers between -7 and -3, inclusive

46. The set of all integers between -1 and 4, inclusive

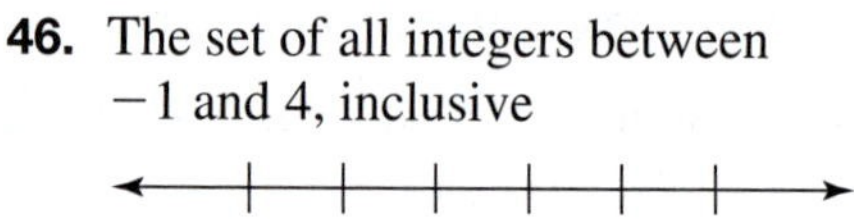

Solve each inequality. Represent each solution set using set-builder notation, interval notation, and a graph, as appropriate.

47. $3(x - 3) < 3$

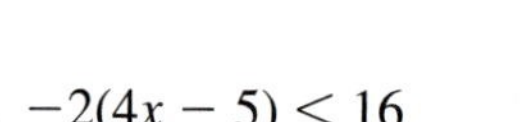

48. $3 - 2x \le 2$

49. $-2(4x - 5) < 16$

50. $5x + 4 \le 2x + 4$

51. $-2(4 - x) \le 7 + 2x$

52. $6(x + 3) \ge 4 + 6x$

53. $-2(5 - x) \le -3(x + 2) + 5x$

54. $3(x + 5) \le 6(x + 2) - 3x$

Use set-builder notation and interval notation to describe each set.

55.

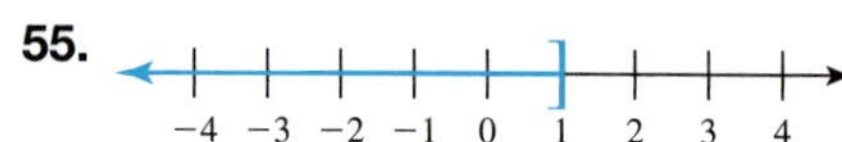

56.

57.

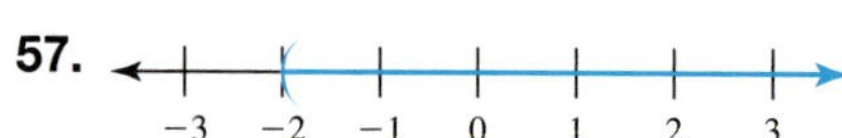

58.

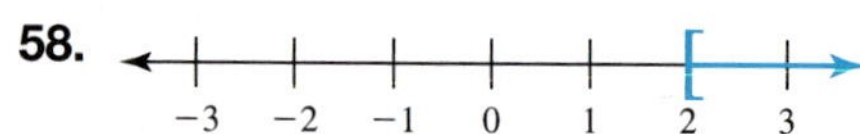

59.

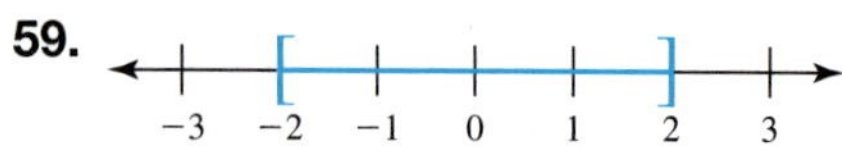

60.

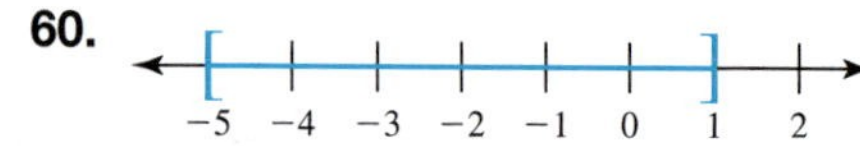

61.

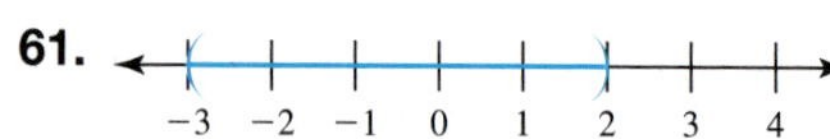

62.

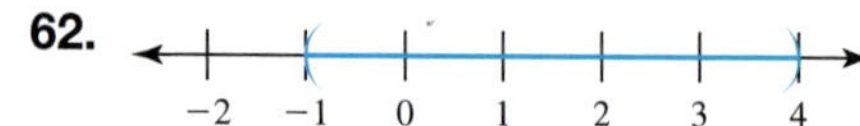

63.

64.

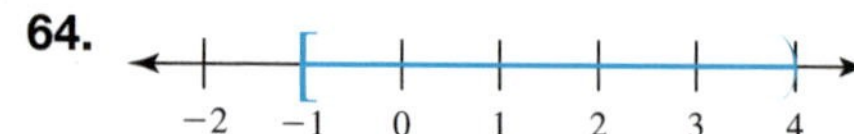

65.

66.

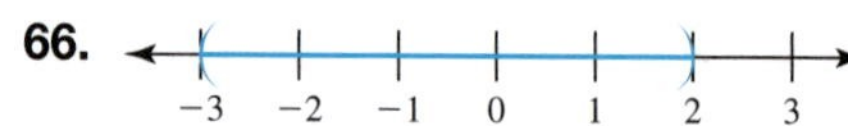

67.

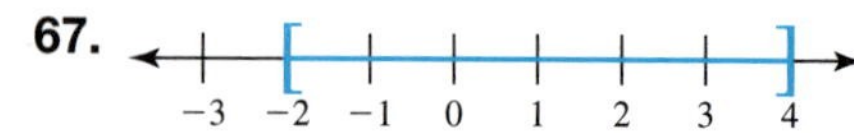

68.

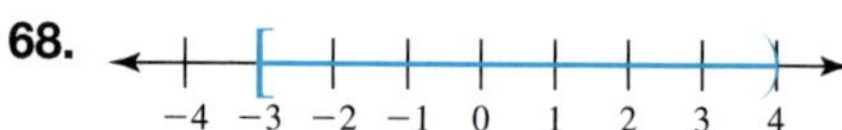

< Objective 4 >

Let $A = \{x|x$ *is an even natural number less than* $10\}$, $B = \{1, 3, 5, 7, 9\}$, *and* $C = \{1, 2, 3, 4, 5\}$. *List the elements in each set.*

69. $A \cup B$

70. $A \cap B$

71. $B \cap \varnothing$

72. $C \cup A$

73. $A \cup \varnothing$

74. $B \cup C$

75. $B \cap C$

76. $C \cap A$

77. $(A \cup C) \cap B$

78. $A \cup (C \cap B)$

Determine whether each statement is **true** *or* **false**.

79. The set of all even primes is finite.

80. We can list all the real numbers between 3 and 4 in roster form.

Complete each statement with **always, sometimes,** *or* **never.**

81. The intersection of two nonempty sets is __________ empty.

82. The union of two nonempty sets is __________ empty.

Skills | Calculator/Computer | Career Applications | **Above and Beyond**

83. Use the Internet to research the history of sets in mathematics.

Answers

1. {Monday, Tuesday, Wednesday, Thursday, Friday, Saturday, Sunday} **3.** {1, 2, 3, 6, 9, 18} **5.** {2, 3, 5, 7, 11, 13, 17, 19, 23, 29} **7.** {−5, −4, −3, −2, −1} **9.** {0, 2, 4, 6, 8, 10, 12} **11.** {3, 4, 5, 6} **13.** {−3, −2} **15.** {−5, −4, −3, −2, −1, 0, 1, 2} **17.** {1, 3, 5, 7, . . .} **19.** {0, 2, 4, 6, . . . , 96, 98} **21.** {5, 10, 15, 20, . . .} **23.** $\{x \mid x > 10\}$; $(10, \infty)$ **25.** $\{x \mid x \geq -5\}$; $[-5, \infty)$ **27.** $\{x \mid 2 \leq x \leq 7\}$; $[2, 7]$ **29.** $\{x \mid -4 \leq x \leq 4\}$; $[-4, 4]$

31.
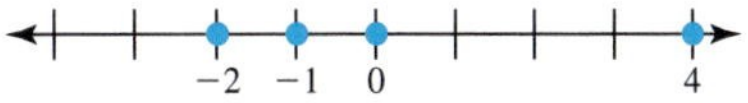

33.
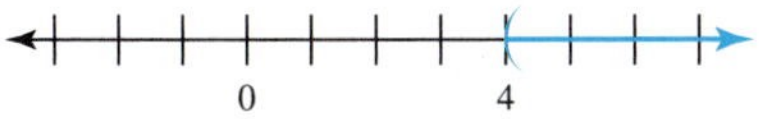

35.
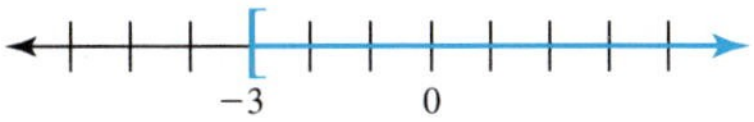

37.
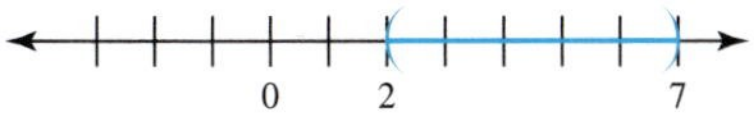

39.

41.

43.
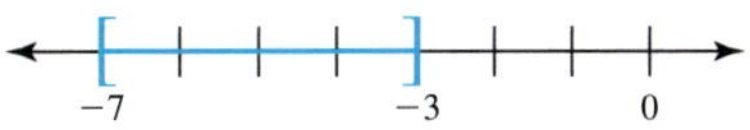

45.
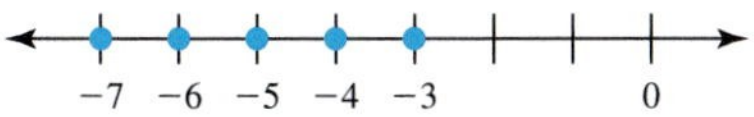

47. $\{x \mid x < 4\}$; $(-\infty, 4)$;
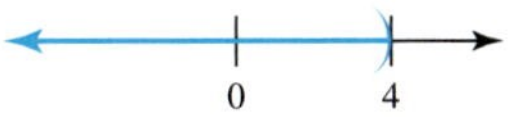

49. $\left\{x \mid x > -\frac{3}{4}\right\}$; $\left(-\frac{3}{4}, \infty\right)$;
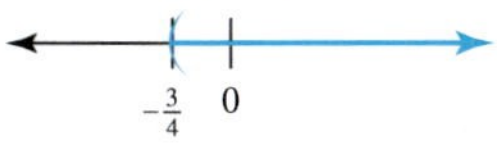

51. $\{x \mid x \in \mathbb{R}\}$; $(-\infty, \infty)$ **53.** $\{x \mid x \in \mathbb{R}\}$; $(-\infty, \infty)$ **55.** $\{x \mid x \leq 1\}$; $(-\infty, 1]$ **57.** $\{x \mid x > -2\}$; $(-2, \infty)$ **59.** $\{x \mid -2 \leq x \leq 2\}$; $[-2, 2]$ **61.** $\{x \mid -3 < x < 2\}$; $(-3, 2)$ **63.** $\{x \mid -2 \leq x < 4\}$; $[-2, 4)$ **65.** $\{x \mid -2 < x \leq 4\}$; $(-2, 4]$ **67.** $\{x \mid -2 \leq x \leq 4\}$; $[-2, 4]$ **69.** {1, 2, 3, 4, 5, 6, 7, 8, 9} **71.** ∅ **73.** {2, 4, 6, 8} **75.** {1, 3, 5} **77.** {1, 3, 5} **79.** True **81.** sometimes **83.** Above and Beyond

2.3 Two-Variable Equations

< **2.3 Objectives** >

1 > Identify solutions of an equation in two variables

2 > Use ordered-pair notation to write solutions to equations in two variables

3 > Use two-variable equations in applications

RECALL

An equation is a statement that two expressions are equal.

Recall that a solution to an equation is a value for the variable that "satisfies" the equation, or makes the equation a true statement. For example, we know that 4 is a solution to the equation

$2x + 5 = 13$

because when we replace x with 4, we have

$$2(4) + 5 \stackrel{?}{=} 13$$
$$8 + 5 \stackrel{?}{=} 13$$
$$13 = 13 \quad \text{A true statement}$$

We now want to consider **equations in two variables.** In fact, in this chapter we study equations of the form $Ax + By = C$, where A and B are not both 0. Such equations are called **linear equations in two variables,** and are said to be in **standard form.** An example is

$x + y = 5$

What does a solution look like? It is not going to be a single number, because there are two variables. Here a solution is a pair of numbers—one value for each of the variables x and y. Suppose that x has the value 3. In the equation $x + y = 5$, you can substitute 3 for x.

NOTE

An equation in two variables "pairs" two numbers, one for x and one for y.

$(3) + y = 5$

Solving for y gives

$y = 2$

So the pair of values $x = 3$ and $y = 2$ satisfies the equation because

$(3) + (2) = 5$

That pair of numbers is a *solution* to the equation in two variables.

Property

Equation in Two Variables

An **equation in two variables** is an equation for which *every* solution is a pair of values.

How many such pairs are there? Given $Ax + By = C$, if neither A nor B is 0, then we can choose any value for x (or for y) and find the other *paired* or *corresponding* value to form a solution.

Think about this. There are infinitely many solutions to this type of equation, one corresponding to each real-number choice of x (or of y).

In fact, as long as either $A \neq 0$ or $B \neq 0$ in $Ax + By = C$, we can always find an infinite number of pairs of numbers that are solutions.

In Example 1, we find some other solutions to the equation $x + y = 5$.

Example 1 Solving for Corresponding Values

< Objective 1 >

For the equation $x + y = 5$, find **(a)** y if $x = 5$ and **(b)** x if $y = 4$.

(a) If $x = 5$,

$(5) + y = 5, \quad$ so $\quad y = 0$

(b) If $y = 4$,

$x + (4) = 5, \quad$ so $\quad x = 1$

So the pairs $x = 5$, $y = 0$ and $x = 1$, $y = 4$ are both solutions.

Check Yourself 1

You are given the equation $2x + 3y = 26$.

(a) If $x = 4$, $y = ?$ **(b)** If $y = 0$, $x = ?$

To simplify writing the pairs that satisfy an equation, we use **ordered-pair notation.** The numbers are written in parentheses and are separated by a comma. For example, we know that the values $x = 3$ and $y = 2$ satisfy the equation $x + y = 5$. So we write the pair as

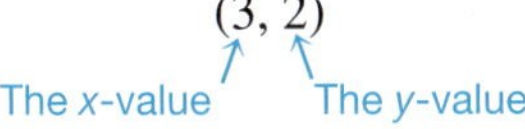

> CAUTION

(3, 2) means $x = 3$ and $y = 2$. (2, 3) means $x = 2$ and $y = 3$. (3, 2) and (2, 3) are entirely different. That's why we call them *ordered pairs*.

The first number of the pair is *always* the value for x and is called the ***x*-coordinate.** The second number of the pair is *always* the value for y and is the ***y*-coordinate.**

Using ordered-pair notation, we can say that (3, 2), (5, 0), and (1, 4) are all *solutions* to the equation $x + y = 5$. Each pair gives values for x and y that satisfy the equation.

Example 2 Identifying Solutions of Two-Variable Equations

< Objective 2 >

Which of the ordered pairs (2, 5), (5, −1), and (3, 4) are solutions to the equation $2x + y = 9$?

(a) To check whether (2, 5) is a solution, let $x = 2$ and $y = 5$ and see if the equation is satisfied.

$2x + y = 9$

$$2(2) + (5) \stackrel{?}{=} 9 \qquad \text{Substitute 2 for } x \text{ and 5 for } y.$$

$$4 + 5 \stackrel{?}{=} 9$$

$$9 = 9 \qquad \text{A true statement}$$

So (2, 5) is a solution to the equation $2x + y = 9$.

NOTE

(2, 5) is a solution because a *true statement* results.

(b) For $(5, -1)$, let $x = 5$ and $y = -1$.

$$2(5) + (-1) \stackrel{?}{=} 9$$
$$10 - 1 \stackrel{?}{=} 9$$
$$9 = 9 \quad \text{A true statement}$$

So $(5, -1)$ is a solution to $2x + y = 9$.

(c) For $(3, 4)$, let $x = 3$ and $y = 4$. Then

$$2(3) + (4) \stackrel{?}{=} 9$$
$$6 + 4 \stackrel{?}{=} 9$$
$$10 = 9 \quad \textit{Not} \text{ a true statement}$$

So $(3, 4)$ is *not* a solution to the equation.

Check Yourself 2

Which of the ordered pairs $(3, 4)$, $(4, 3)$, $(1, -2)$, and $(0, -5)$ are solutions to the equation.

$$3x - y = 5$$

Equations such as those seen in Examples 1 and 2 are said to be in **standard form.**

Definition

Standard Form of Linear Equation

A linear equation in two variables is in **standard form** if it is written as

$Ax + By = C$ in which A and B are not both 0.

For example, if $A = 1$, $B = 1$, and $C = 5$, we have

$$(1)x + (1)y = (5)$$
$$x + y = 5$$

which is the equation in Example 1.

It is possible to view an equation in one variable as a two-variable equation. For example, if we have the equation $x = 2$, we can view this in standard form as

$$1x + 0y = 2$$

and we may search for ordered-pair solutions. The key is this: If the equation contains only one variable (in this case x), then the missing variable (in this case y) can take on any value. Consider Example 3.

Example 3 Identifying Solutions of One-Variable Equations

Which of the ordered pairs $(2, 0)$, $(0, 2)$, $(5, 2)$, $(2, 5)$, and $(2, -1)$ are solutions to the equation $x = 2$?

An ordered pair is a solution if substituting the x- and y-coordinates wherever they appear in the equation results in a true statement. In the case of the equation $x = 2$, we only need to substitute the x-coordinate since y does not appear in the equation.

Looking at the ordered pair $(2, 0)$, we substitute

$$x = 2 \quad \text{The original equation}$$
$$(2) \stackrel{?}{=} 2 \quad \text{Substitute } x = 2.$$
$$2 = 2 \quad \text{True}$$

NOTE

What about the y-coordinate? Think of it this way, we are substituting $y = 0$ everywhere that y appears in the equation.

So, (2, 0) is a solution to the equation $x = 2$.

Next, look at the ordered pair (0, 2).

$$x = 2 \quad \text{The original equation}$$
$$(0) \stackrel{?}{=} 2 \quad \text{Substitute } x = 0.$$
$$0 = 2 \quad \text{False}$$

Since a false statement results, we know that (0, 2) is *not* a solution to $x = 2$.

It should be clear that in order for an ordered pair to be a solution to the equation $x = 2$, the x-coordinate must be 2. Therefore, (2, 5) and (2, −1) are also solutions to the equation $x = 2$. On the other hand, (5, 2) is not a solution because the x-coordinate is not 2.

Check Yourself 3

Which of the ordered pairs (3, 0), (0, 3), (3, 3), (−1, 3), and (3, −1) are solutions to the equation $y = 3$?

Remember that when an ordered pair is presented, the first number is always the x-coordinate and the second number is always the y-coordinate.

Example 4 Completing Ordered-Pair Solutions

NOTE

When we subtitute a number for one of the variables, the result is a linear equation in one variable. In Chapter 1, you learned that all such equations have exactly one solution. When we choose a value for x, we can always find a corresponding value for y that results in a true statement.

Complete the ordered pairs (9,), (, −1), (0,), and (, 0) so that each is a solution to the equation $x - 3y = 6$.

(a) The first number, 9, appearing in (9,) represents the x-value. To complete the pair (9,), substitute 9 for x and then solve for y.

$$(9) - 3y = 6$$
$$-3y = -3$$
$$y = 1$$

The ordered pair (9, 1) is a solution of $x - 3y = 6$.

(b) To complete the pair (, −1), let y be −1 and solve for x.

$$x - 3(-1) = 6$$
$$x + 3 = 6$$
$$x = 3$$

The ordered pair (3, −1) is a solution of the equation $x - 3y = 6$.

(c) To complete the pair (0,), let x be 0.

$$(0) - 3y = 6$$
$$-3y = 6$$
$$y = -2$$

So (0, −2) is a solution.

(d) To complete the pair (, 0), let y be 0.

$$x - 3(0) = 6$$
$$x - 0 = 6$$
$$x = 6$$

Then (6, 0) is a solution.

Check Yourself 4

Complete the ordered pairs so that each is a solution to the equation $2x + 5y = 10$.

(10,), (, 4), (0,), and (, 0)

Often, we encounter fractions when finding ordered-pair solutions to an equation.

Example 5 Completing Ordered-Pair Solutions

Complete the ordered pairs so that each is a solution to the equation

$2x + 3y = 18$

(a) $\left(\frac{1}{2}, \quad\right)$

We let $x = \frac{1}{2}$ and solve for y.

$$2\left(\frac{1}{2}\right) + 3y = 18$$

$$1 + 3y = 18$$

$$3y = 17 \qquad \text{Subtract 1 from both sides.}$$

$$y = \frac{17}{3} \qquad \text{Divide both sides by 3.}$$

The ordered pair $\left(\frac{1}{2}, \frac{17}{3}\right)$ is a solution.

(b) (, 3)

We let $y = 3$ and solve for x.

$$2x + 3(3) = 18$$

$$2x + 9 = 18$$

$$2x = 9$$

$$x = \frac{9}{2}$$

The ordered pair $\left(\frac{9}{2}, 3\right)$ is a solution.

Check Yourself 5

Complete the ordered pairs so that each is a solution to the equation

$3x - 4y = 16$

(a) (2,) **(b)** $\left(\quad, \frac{3}{2}\right)$ **(c)** (8,)

We can choose any real number for one of the variables and find the corresponding value for the other variable. When finding solutions, we choose numbers that make it easy to find the other value.

Example 6 Finding Some Solutions of a Two-Variable Equation

Find four solutions to the equation

$2x + y = 8$

> **NOTE**
>
> Generally, you want to pick values for x (or for y) so that the resulting equation in one variable is easy to solve.

In this case the values used to form the solutions are *up to you.* You can assign any value for x (or for y). We demonstrate with some possible choices.

Solution with $x = 2$:

$$\begin{aligned} 2x + y &= 8 \\ 2(2) + y &= 8 \\ 4 + y &= 8 \\ y &= 4 \end{aligned}$$

The ordered pair (2, 4) is a solution of $2x + y = 8$.

Solution with $y = 6$:

$$\begin{aligned} 2x + y &= 8 \\ 2x + (6) &= 8 \\ 2x &= 2 \\ x &= 1 \end{aligned}$$

So (1, 6) is also a solution of $2x + y = 8$.

Solution with $x = 0$:

$$\begin{aligned} 2x + y &= 8 \\ 2(0) + y &= 8 \\ y &= 8 \end{aligned}$$

And (0, 8) is a solution.

> **NOTE**
>
> The solutions (0, 8) and (4, 0) have special significance when graphing. They are also easy to find!

Solution with $y = 0$:

$$\begin{aligned} 2x + y &= 8 \\ 2x + (0) &= 8 \\ 2x &= 8 \\ x &= 4 \end{aligned}$$

So (4, 0) is a solution.

Check Yourself 6

Find four solutions to $x - 3y = 12$.

When finding your own ordered-pair solutions to an equation, you can usually choose x- or y-values that allow you to avoid fractions. Consider Example 7.

Example 7 Finding Solutions to an Equation

Find four solutions to the equation $5x - 2y = 20$.

We can choose x-values and then solve to find the corresponding y-values or we can choose y-values and solve for x. It does not matter which approach we take.

More important to consider is choosing values that lead to ordered pairs of integers, rather than fractions. Consider our first choice of $x = 1$.

If $x = 1$, then

$$\begin{aligned} 5(1) - 2y &= 20 && \text{Substitute } x = 1. \\ 5 - 2y &= 20 \\ -2y &= 15 && \text{Subtract 5 from both sides.} \\ y &= -\frac{15}{2} && \text{Divide by } -2. \end{aligned}$$

The ordered pair $\left(1, -\frac{15}{2}\right)$ is a solution.

What if we wanted to avoid fractions? In that case, we need $5x$ to be an even number. Do you see why?

Choose $x = 2$.

$5(2) - 2y = 20$

$10 - 2y = 20$

$-2y = 10$ Subtract 10.

$y = -5$ Divide by -2: $\frac{10}{-2} = -5$.

This gives the solution $(2, -5)$.

If we make a choice for y, then we need to solve for x.

Choose $y = 1$.

$5x - 2(1) = 20$

$5x - 2 = 20$

$5x = 22$

$x = \frac{22}{5}$

So, $\left(\frac{22}{5}, 1\right)$ is a solution.

Notice that to isolate x, we had to divide by 5 because the x-term is $5x$. Since 20 is divisible by 5, we need $2y$ to be divisible by 5 if we want to avoid fractions. In this case, it means that our choice of y must be divisible by 5.

Choose $y = 5$.

$5x - 2(5) = 20$

$5x - 10 = 20$

$5x = 30$

$x = 6$

$(6, 5)$ is a solution.

Check Yourself 7

Find four solutions to the equation $4x + 5y = 40$. Choose variable values so that at least two of your solutions are ordered pairs of integers.

Each variable in a two-variable equation plays a different role. The variable for which the equation is solved is called the **dependent** or **output variable** because its value depends on what value is given the other variable, which is called the **independent** or **input variable.** Generally we use x for the independent variable and y for the dependent variable.

In applications, different letters tend to be used for the variables. These letters are selected to help us see what they stand for, so h is used for height, A is used for area, and so on.

Example 8 Applications of Two-Variable Equations

< Objective 3 >

Suppose that it costs the manufacturer \$1.25 for each stapler that is produced. In addition, fixed costs (related to staplers) are \$110 per day.

(a) Write an equation relating the total daily costs C to the number x of staplers produced in a day.

Because each stapler costs \$1.25 to produce, the cost of producing staplers is $1.25x$. Adding the fixed cost to this gives us an equation for the total daily costs.

$$C = 1.25x + 110$$

(b) What is the total cost of producing 500 staplers in a day?

We substitute 500 for x in the equation from part (a) and calculate the total cost.

$$\begin{aligned} C &= 1.25(500) + 110 \\ &= 625 + 110 \\ &= 735 \end{aligned}$$

It costs the manufacturer a total of \$735 to produce 500 staplers in one day.

RECALL

Divide both sides by 1.25.

$$\frac{1,000}{1.25} = \frac{\cancel{1.25}x}{\cancel{1.25}}$$

$$800 = x$$

(c) How many staplers can be produced for \$1,110?

In this case, we substitute 1,110 for C in the equation from part (a) and solve for x.

$$(1,110) = 1.25x + 110$$

$$1,000 = 1.25x \quad \text{Subtract 110 from both sides.}$$

$$800 = x \quad \text{Divide both sides by 1.25.}$$

800 staplers can be produced at a cost of \$1,110.

Check Yourself 8

Suppose that the stapler manufacturer earns a profit of \$1.80 on each stapler shipped. However, it costs \$120 to operate each day.

(a) Write an equation relating the daily profit P to the number x of staplers shipped in a day.
(b) What is the total profit of shipping 500 staplers in a day?
(c) How many staplers need to be shipped to produce a profit of \$1,500?

We close this section with an application from the field of medicine.

Example 9 An Allied Health Application

NOTE

The value of the independent variable d, which represents the number of days, can only be a positive integer. We call the set of all possible values for the independent variable the *domain*.

For a particular patient, the weight w, in grams, of a uterine tumor is related to the number of days d of chemotherapy treatment by the equation

$$w = -1.75d + 25$$

(a) What was the original size of the tumor?

The original size of the tumor is the value of w when $d = 0$. Substituting 0 for d in the equation gives

$$\begin{aligned} w &= -1.75(0) + 25 \\ &= 25 \end{aligned}$$

The tumor was originally 25 grams.

NOTE

We round **up** in this case, because the tumor will be eliminated on the 15th day.

(b) How many days of chemotherapy are required to eliminate the tumor?

The tumor is eliminated when its weight is 0. So

$$\begin{aligned} (0) &= -1.75d + 25 \\ -25 &= -1.75d \\ d &\approx 14.3 \end{aligned}$$

It will take about 14.3 days to eliminate the tumor. Because d is limited to whole numbers, we answer the original question by saying it will take 15 days to eliminate the tumor.

Check Yourself 9

For a particular patient, the weight w, in grams, of a uterine tumor is related to the number of days d of chemotherapy treatment by the equation

$$w = -1.6d + 32$$

(a) Find the original size of the tumor.
(b) Determine the number of days of chemotherapy required to eliminate the tumor.

Check Yourself ANSWERS

1. **(a)** $y = 6$; **(b)** $x = 13$ **2.** (3, 4), (1, −2), and (0, −5) are solutions. **3.** (0, 3), (3, 3), and (−1, 3) are solutions. **4.** (10, −2), (−5, 4), (0, 2), and (5, 0) **5.** **(a)** $\left(2, -\frac{5}{2}\right)$; **(b)** $\left(\frac{22}{3}, \frac{3}{2}\right)$; **(c)** (8, 2) **6.** (6, −2), (3, −3), (0, −4), and (12, 0) are four possibilities. **7.** (0, 8), (10, 0), (5, 4), and (−5, 12) are four possibilities. **8.** **(a)** $P = 1.80x - 120$; **(b)** \$780; **(c)** 900 staplers **9.** **(a)** 32 grams; **(b)** 20 days

Reading Your Text

These fill-in-the-blank exercises will help you understand some of the key vocabulary used in this section. The answers to these exercises are in the Answers Appendix in the back of the text.

(a) An equation in two variables is an equation for which every ________ is a pair of values.

(b) Given an equation such as $x + y = 5$, there is an ________ number of solutions.

(c) To simplify writing the pairs that satisfy an equation, we use ________ notation.

(d) When an equation in two variables is solved for y, we say that y is the ________ variable.

2.3 exercises

Skills | Calculator/Computer | Career Applications | Above and Beyond

< Objectives 1 and 2 >

Determine which of the given ordered pairs are solutions to each equation.

1. $x + y = 6$ (4, 2), (−2, 4), (0, 6), (−3, 9)

2. $x - y = 10$ (11, 1), (11, −1), (10, 0), (5, 7)

3. $2x - y = 8$ (5, 2), (4, 0), (0, 8), (6, 4)

4. $x + 5y = 20$ (10, −2), (10, 2), (20, 0), (25, −1)

5. $4x + y = 8$ (2, 0), (2, 3), (0, 2), (1, 4)

6. $x - 2y = 8$ (8, 0), (0, 4), (5, −1), (10, −1)

7. $2x - 3y = 6$ (0, 2), (3, 0), (6, 2), (0, −2)

8. $6x + 2y = 12$ (0, 6), (2, 6), (2, 0), (1, 3)

9. $3x - 2y = 12$ $(4, 0), \left(\frac{2}{3}, -5\right), (0, 6), \left(5, \frac{3}{2}\right)$

10. $3x + 4y = 12$ $(-4, 0), \left(\frac{2}{3}, \frac{5}{2}\right), (0, 3), \left(\frac{2}{3}, 2\right)$

11. $3x + 5y = 15$ $(0, 3), \left(1, \frac{12}{5}\right), (5, 3)$

12. $y = 2x - 1$ $(0, -2), (0, -1), \left(\frac{1}{2}, 0\right), (3, -5)$

13. $x = 3$ $(3, 5), (0, 3), (3, 0), (3, 7)$ 

14. $y = 7$ $(0, 7), (3, 7), (-1, -4), (7, 7)$

Complete the ordered pairs so that each is a solution to the given equation.

15. $x + y = 12$ $(4, \ ), (\ , 5), (0, \ ), (\ , 0)$

16. $x - y = 7$ $(\ , 4), (15, \ ), (0, \ ), (\ , 0)$

17. $3x + y = 9$ $(3, \ ), (\ , 9), (\ , -3), (0, \ )$

18. $x + 4y = 12$ $(0, \ ), (\ , 2), (8, \ ), (\ , 0)$

19. $5x - y = 15$ $(\ , 0), (2, \ ), (4, \ ), (\ , -5)$

20. $x - 3y = 9$ $(0, \ ), (12, \ ), (\ , 0), (\ , -2)$

21. $4x - 2y = 16$ $(\ , 0), (\ , -6), (2, \ ), (\ , 6)$

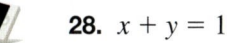

22. $2x + 5y = 20$ $(0, \ ), (5, \ ), (\ , 0), (\ , 6)$

23. $y = 3x + 9$ $(\ , 0), \left(\frac{2}{3}, \ \right), (0, \ ), \left(-\frac{2}{3}, \ \right)$

24. $6x + 8y = 24$ $(0, \ ), \left(\ , \frac{3}{4}\right), (\ , 0), \left(-\frac{2}{3}, \ \right)$

25. $y = 3x - 4$ $(0, \ ), (\ , 5), (\ , 0), \left(\frac{5}{3}, \ \right)$

26. $y = -2x + 5$ $(0, \ ), (\ , 5), \left(\frac{3}{2}, \ \right), (\ , 1)$

Find four solutions for each equation.
Note: *Your answers may vary from those shown in the answer section.*

27. $x - y = 10$

28. $x + y = 18$

29. $2x - y = 6$

30. $4x - 2y = 8$

31. $x + 4y = 8$

32. $x + 3y = 12$

33. $5x - 2y = 10$

34. $2x + 7y = 14$

35. $y = 2x + 3$

36. $y = 5x - 8$

37. $x = -5$

38. $y = 8$

Find two solutions for each equation.
Note: *Your answers may vary from those shown in the answer section.*

39. $\frac{1}{2}x + \frac{1}{3}y = 1$

40. $\frac{1}{3}x - \frac{1}{4}y = 1$

41. $0.3x + 0.5y = 2$

42. $0.6x - 0.2y = 5$

43. $\frac{3}{4}x - \frac{2}{5}y = 6$

44. $\frac{4}{5}x + \frac{2}{3}y = 8$

45. $0.4x - 0.7y = 3$

46. $0.8x + 0.9y = 2$

< Objective 3 >

47. **Business and Finance** When an employee produces x units per hour, the hourly wage in dollars is given by $y = 0.75x + 8$. What are the hourly wages for the following number of units: 2, 5, 10, 15, and 20?

48. **Science and Medicine** Celsius temperature readings can be converted to Fahrenheit readings by using the formula $F = \frac{9}{5}C + 32$. What is the Fahrenheit temperature that corresponds to each of the following Celsius temperatures: -10, 0, 15, 100?

49. **Geometry** The perimeter of a square is given by $P = 4s$. What are the perimeters of the squares whose sides are 5, 10, 12, and 15 cm?

50. **Geometry** The area of a square is given by $A = s^2$. What is the area of the squares whose sides are 4 cm, 11 cm, 14 cm, and 17 cm?

51. **Business and Finance** When x units are sold, the price of each unit is given by $p = 4x + 15$. Find the unit price in dollars when the following quantities are sold: 1, 5, 10, 12.

52. **Business and Finance** When x units are sold, the price of each unit (in dollars) is given by $p = -\frac{x}{2} + 75$. Find the unit price when each quantity is sold: 2, 7, 9, 11.

53. **Statistics** The number of programs for people with disabilities in the United States for a 5-year period is approximated by the equation $y = 162x + 4{,}365$, where x represents particular years. Complete the table.

x	1	2	3	4	6
y					

54. **Business and Finance** Your monthly pay as a car sales associate is determined using the equation $S = 200x + 1{,}500$ in which x is the number of cars you can sell each month.

(a) Complete the table.

x	12	15	17	18
S				

(b) You are offered a job at a salary of $56,400 per year. How many cars would you have to sell per month to equal this salary?

Determine whether each statement is **true** *or* **false.**

55. The ordered pair (a, b) means the same thing as the ordered pair (b, a).

56. An equation in two variables has exactly two solutions.

57. For any number k, $(0, k)$ is a solution for the equation $y = k$.

58. For any number h, $(0, h)$ is a solution for the equation $x = h$.

59. When finding solutions for the equation $1 \cdot x + 0 \cdot y = 5$, you can choose any number for x.

60. When finding solutions for the equation $1 \cdot x + 1 \cdot y = 5$, you can choose any number for x.

Complete each statement with **always, sometimes,** *or* **never.**

61. If (a, b) is a solution to a particular two-variable equation, then (b, a) is __________ a solution to the same equation.

62. There are __________ infinitely many solutions to a two-variable equation in standard form.

Skills | **Calculator/Computer** | Career Applications | Above and Beyond

63. Given $y = 3.12x - 14.79$, use the TABLE utility on a graphing calculator to complete the ordered pairs.

(10,), (20,), (30,), (40,), (50,) chapter 2 > Make the Connection VIDEO

64. Given $y = -16x^2 + 90x + 23$, use the TABLE utility on a graphing calculator to complete the ordered pairs.

(1.5,), (2.5,), (3.5,), (4.5,), (5.5,) chapter 2 > Make the Connection

65. **Construction Technology** The number of studs s (16 in. on center) required to build a wall that is L ft long is given by the formula

$$s = \frac{3}{4}L + 1$$

Determine the number of studs required to build walls of length 12 ft, 20 ft, and 24 ft.

66. **Manufacturing Technology** The number of board feet b of lumber in a 2″ × 6″ board of length L (in feet) is given by the equation

$$b = \frac{8.25}{144}L$$

Determine the number of board feet in 2″ × 6″ boards of length 12 ft, 16 ft, and 20 ft. Round your results to the nearest hundredth bd ft.

67. **Allied Health** The recommended dosage d (in mg) of the antibiotic ampicillin sodium for children weighing less than 40 kg is given by the linear equation $d = 7.5w$, in which w represents the child's weight (in kg). VIDEO

(a) Determine the dosage for a 30-kg child.

(b) What is the weight of a child who requires a 150-mg dose?

68. **Allied Health** The recommended dosage d (in μg) of neupogen (medication given to bone-marrow transplant patients) is given by the linear equation $d = 8w$, in which w is the patient's weight (in kg).

(a) Determine the dosage for a 92-kg patient.

(b) What is the weight of a patient who requires a 250-μg dose?

69. **Mechanical Engineering** The force that a coil exerts on an object is related to the distance that the coil is pulled from its natural position. The formula to describe this is $F = kx$. If $k = 72$ lb/ft for a certain coil, determine the force exerted if $x = 3$ ft or $x = 5$ ft.

70. **Mechanical Engineering** If a particular machine is to operate under water, it must be designed to handle the pressure p, measured in pounds, which depends on the depth d, measured in feet, of the water. The relationship for this machine is approximated by the formula $p = 59d + 13$. Determine the pressure at depths of 10 ft, 20 ft, and 30 ft.

Skills Calculator/Computer Career Applications **Above and Beyond**

An equation in three variables has an ordered triple as a solution. For example, (1, 2, 2) *is a solution to the equation* $x + 2y - z = 3$. *Complete the ordered-triple solutions for each equation.*

71. $x + y + z = 0$ (2, −3,)

72. $2x + y + z = 2$ (, −1, 3)

73. $x + y + z = 0$ (1, , 5)

74. $x + y - z = 1$ (4, , 3)

75. $2x + y + z = 2$ (−2, , 1)

76. $x + y - z = 1$ (−2, 1,)

77. You now have had practice solving equations with one variable and equations with two variables. Compare equations with one variable to equations with two variables. How are they alike? How are they different?

78. Each sentence describes pairs of numbers that are related. After completing the sentences in parts (a) to (g), write two of your own sentences in (h) and (i).

(a) The *number of hours you work* determines the *amount you are* __________.

(b) The *number of gallons of gasoline* you put in your car determines *the amount you* __________.

(c) The *amount of the* __________ in a restaurant is related to *the amount of the tip.*

(d) The *sales amount of a purchase in a store* determines __________.

(e) The *age of an automobile* is related to __________.

(f) The *amount of electricity you use in a month* determines __________.

(g) The *cost of food for a family* is related to __________.

Think of two more:

(h) ____.

(i) ____.

Answers

1. (4, 2), (0, 6), (−3, 9) **3.** (5, 2), (4, 0), (6, 4) **5.** (2, 0), (1, 4) **7.** (3, 0), (6, 2), (0, −2) **9.** (4, 0), $\left(\frac{2}{3}, -5\right)$, $\left(5, \frac{3}{2}\right)$ **11.** (0, 3), $\left(1, \frac{12}{5}\right)$ **13.** (3, 5), (3, 0), (3, 7) **15.** 8, 7, 12, 12 **17.** 0, 0, 4, 9 **19.** 3, −5, 5, 2 **21.** 4, 1, −4, 7 **23.** −3, 11, 9, 7 **25.** −4, 3, $\frac{4}{3}$, 1 **27.** (0, −10), (10, 0), (5, −5), (12, 2) **29.** (0, −6), (3, 0), (6, 6), (9, 12) **31.** (8, 0), (−4, 3), (0, 2), (4, 1) **33.** (0, −5), (4, 5), (−6, −20), (2, 0) **35.** (0, 3), (1, 5), (2, 7), (3, 9) **37.** (−5, 0), (−5, 1), (−5, 2), (−5, 3) **39.** (2, 0), (0, 3) **41.** (0, 4), $\left(\frac{20}{3}, 0\right)$ **43.** (8, 0), (0, −15) **45.** $\left(\frac{15}{2}, 0\right)$, $\left(0, -\frac{30}{7}\right)$ **47.** \$9.50, \$11.75, \$15.50, \$19.25, \$23 **49.** 20 cm, 40 cm, 48 cm, 60 cm **51.** \$19, \$35, \$55, \$63 **53.** 4,527, 4,689, 4,851, 5,013, 5,337 **55.** False **57.** True **59.** False **61.** sometimes **63.** 16.41, 47.61, 78.81, 110.01, 141.21 **65.** 10 studs, 16 studs, 19 studs **67.** **(a)** 225 mg; **(b)** 20 kg **69.** 216 lb, 360 lb **71.** 1 **73.** −6 **75.** 5 **77.** Above and Beyond

2.4

The Cartesian Coordinate System

< **2.4 Objectives** >

1 > Determine the coordinates of a plotted point

2 > Plot ordered pairs

3 > Scale the axes of a graph

In Section 2.3, we used ordered pairs to write solutions to equations in two variables. The next step is to graph those ordered pairs as points in a plane.

Since there are two numbers (one for x and one for y), we need two number lines. We draw one line horizontally and the other vertically. Their point of intersection (at their respective zero points) is called the **origin.** The horizontal line is called the ***x*-axis,** and the vertical line is called the ***y*-axis.** Together the lines form the **rectangular** or **Cartesian coordinate system.**

> **NOTE**
>
> This system is called the **Cartesian coordinate system** in honor of its inventor, René Descartes (1596–1650), a French mathematician and philosopher.

The axes (pronounced "axees") divide the plane into four regions called **quadrants,** which are numbered (with Roman numerals) counterclockwise from the upper right.

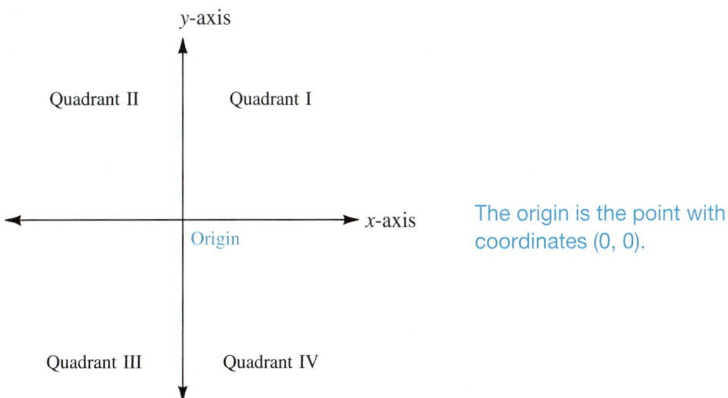

The origin is the point with coordinates (0, 0).

We want to establish correspondences between ordered pairs of numbers (x, y) and points in the plane.

For any ordered pair,

(x, y)

x-coordinate   y-coordinate

1. If the x-coordinate is

   Positive, the point corresponding to that pair is located x units to the *right* of the y-axis.

   Negative, the point is x units to the *left* of the y-axis.

   Zero, the point is on the y-axis.

2. If the y-coordinate is

   Positive, the point is y units *above* the x-axis.

   Negative, the point is y units *below* the x-axis.

   Zero, the point is on the x-axis.

y
x is negative
x is positive
x

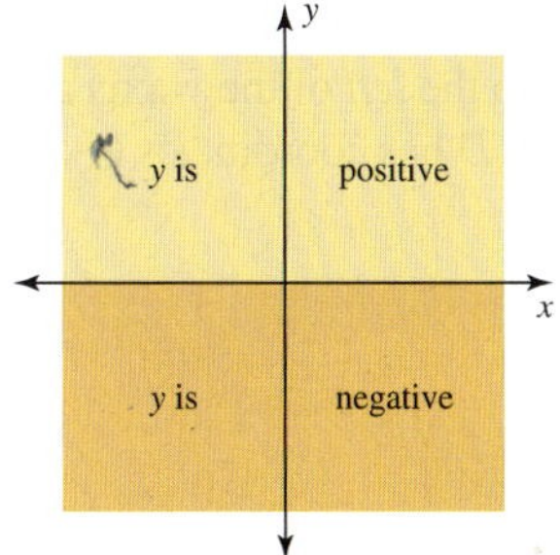

Example 1 illustrates how to use these guidelines to match coordinates with points in the plane.

Example 1 Identifying the Coordinates of a Point

< Objective 1 >

Give the coordinates of each point shown. Assume that each tick mark represents 1 unit.

NOTE

The *x*-coordinate gives the *horizontal* distance from the *y*-axis. The *y*-coordinate gives the *vertical* distance from the *x*-axis.

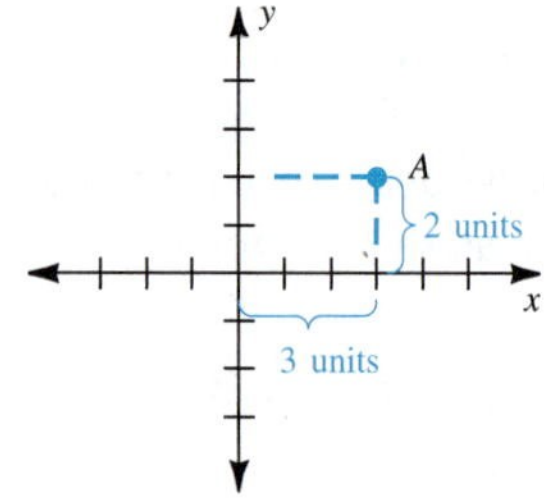

(a) Point *A* is 3 units to the *right* of the *y*-axis and 2 units *above* the *x*-axis. Point *A* has coordinates (3, 2).

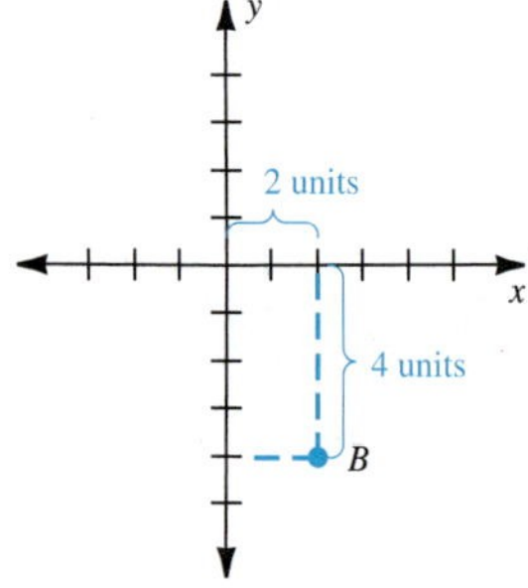

(b) Point *B* is 2 units to the *right* of the *y*-axis and 4 units *below* the *x*-axis. Point *B* has coordinates (2, -4).

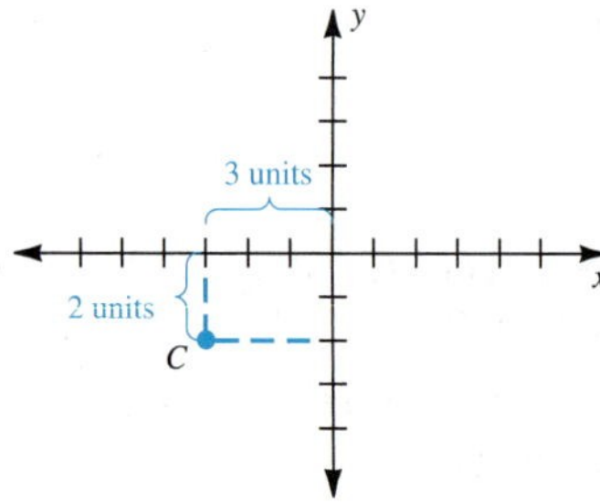

(c) Point *C* is 3 units to the *left* of the *y*-axis and 2 units *below* the *x*-axis. Point *C* has coordinates (-3, -2).

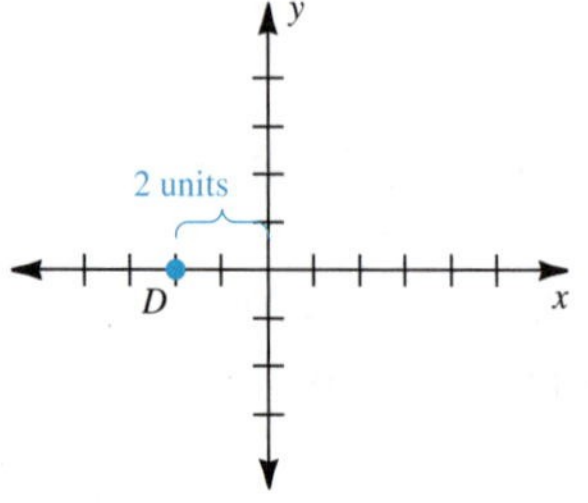

(d) Point *D* is 2 units to the *left* of the *y*-axis and *on* the *x*-axis. Point *D* has coordinates (-2, 0).

Check Yourself 1

Give the coordinates of points P, Q, R, and S.

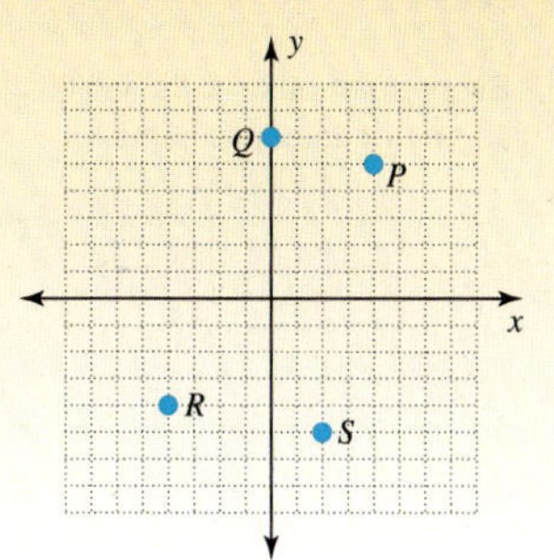

NOTE

Graphing individual points is sometimes called **point plotting.**

Reversing the process used in Example 1 allows us to graph (or plot) a point in the plane, given the coordinates of the point.

Step by Step

Graphing a Point in the Plane

Step 1	Start at the origin.
Step 2	Move right or left according to the value of the x-coordinate.
Step 3	Move up or down according to the value of the y-coordinate.

Example 2 Graphing Points

< Objective 2 >

(a) Graph the point corresponding to the ordered pair (4, 3).

Move 4 units to the right on the x-axis. Then move 3 units up from the point where you stopped on the x-axis. This locates the point corresponding to (4, 3).

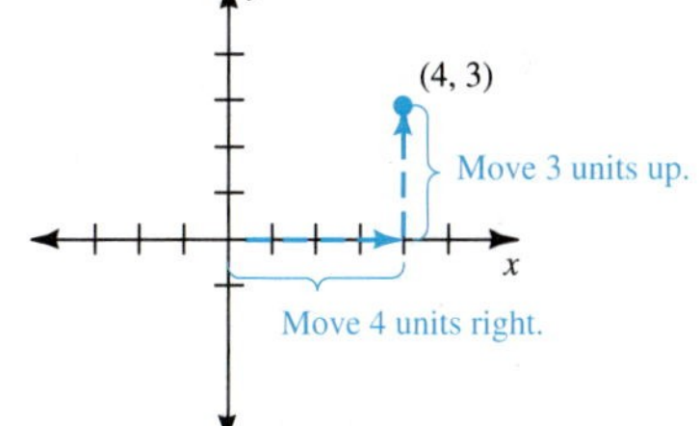

(b) Graph the point corresponding to the ordered pair (−5, 2).

In this case move 5 units *left* (because the x-coordinate is negative) and then 2 units *up*.

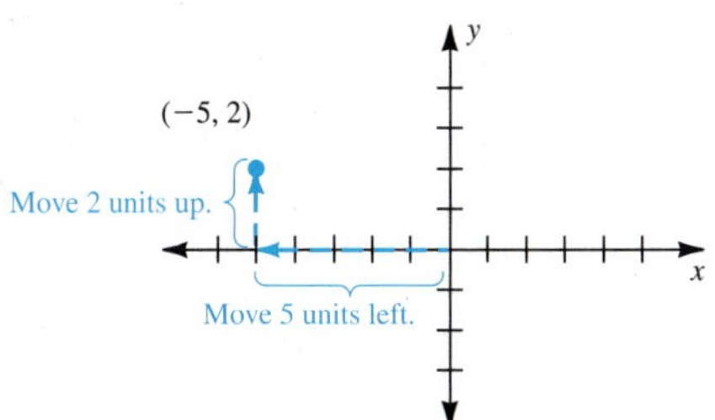

(c) Graph the point corresponding to (−4, −2).

Here move 4 units *left* and then 2 units *down* (the y-coordinate is also negative).

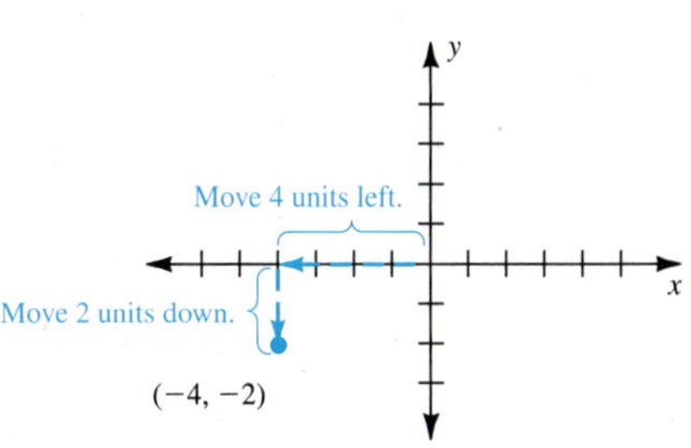

NOTE

Every point on an axis has 0 as one of its coordinates.

(d) Graph the point corresponding to $(0, -3)$.

There is *no* horizontal movement because the x-coordinate is 0. Move 3 units *down*.

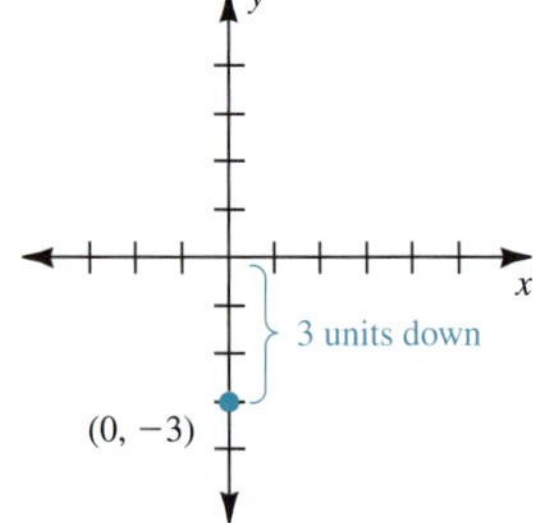

(e) Graph the point corresponding to $(5, 0)$.

Move 5 units *right*. The desired point is on the x-axis because the y-coordinate is 0.

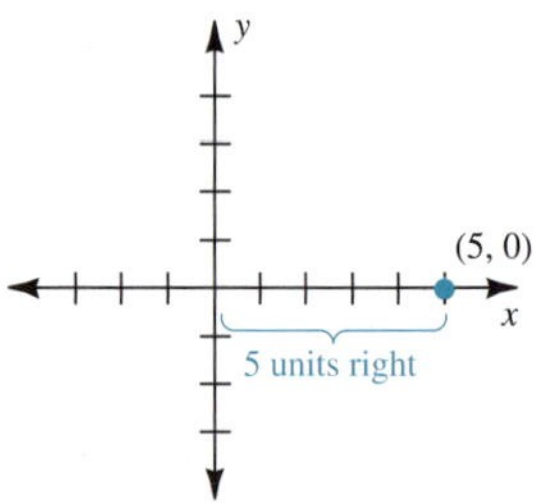

Check Yourself 2

Graph the points corresponding to $M(4, 3)$, $N(-2, 4)$, $P(-5, -3)$, and $Q(0, -3)$.

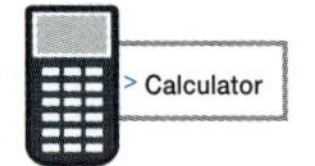

It is not necessary, or even desirable, to always use the same scale on both the x- and y-axes. For example, if we were plotting ordered pairs in which the first value represented the age of a used car and the second value represented the number of miles driven, it would be necessary to have a different scale on the two axes. If not, extreme cases could happen.

NOTE

The same decisions must be made when you use a graphing calculator. When graphing this kind of relation on a calculator, you must decide on an appropriate **viewing window.**

Assume that the cars range in age from 1 to 15 years. The cars have mileages from 2,000 to 150,000 miles (mi). If we used the same scale on both axes, 0.5 in. between each two counting numbers, how large would the paper have to be on which the points were plotted? The horizontal axis would have to be $15(0.5) = 7.5$ in. The vertical axis would have to be $150{,}000(0.5) = 75{,}000$ in. $= 6{,}250$ feet (ft) $=$ almost 1.2 mi long!

NOTE

The age of a car is nonnegative, as is its mileage. Therefore, we only need the graph to show the first quadrant.

So what do we do? We simply use a different, but clearly marked, scale on each axis. In this case, we mark the horizontal axis in 5's with gridlines every unit, and we mark the vertical axis in 50,000's with gridlines every 10,000 units. Additionally, all the numbers are positive, so we only need the first quadrant, in which x and y are both always positive.

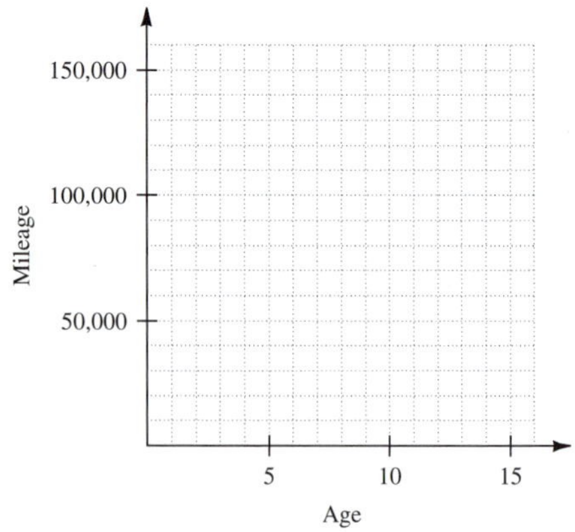

Example 3 Scaling the Axes

< Objective 3 >

A survey of residents in a large apartment building was recently taken. The plotted points represent ordered pairs in which the first number is the number of years of education a person has and the second number is his or her year 2012 income (in thousands of dollars). Estimate and interpret each ordered pair represented.

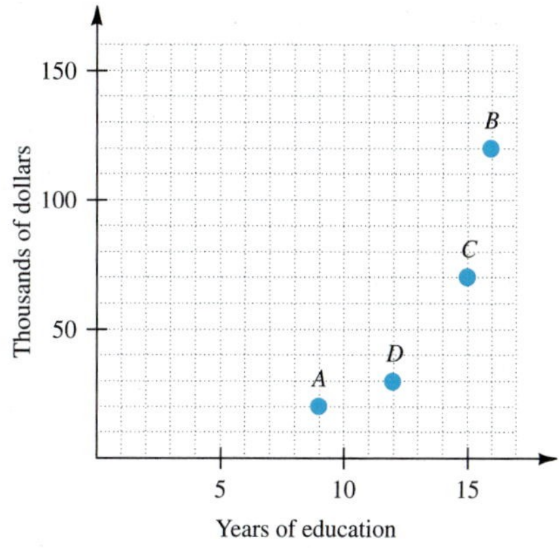

Point A is (9, 20), B is (16, 120), C is (15, 70), and D is (12, 30). Person A completed 9 years of education and made \$20,000 in 2012. Person B completed 16 years of education and made \$120,000 in 2012. Person C had 15 years of education and made \$70,000. Person D had 12 years and made \$30,000.

Note that there is no obvious "relation" that would allow one to predict income from years of education, but you might suspect that in most cases, more education results in more income.

Check Yourself 3

Every 6 months, Armand records his son's weight. The plotted points represent ordered pairs in which the first number represents his son's age and the second number represents his son's weight. For example, point A indicates that when his son was 1 year old, he weighed 14 lb. Estimate and interpret each ordered pair represented.

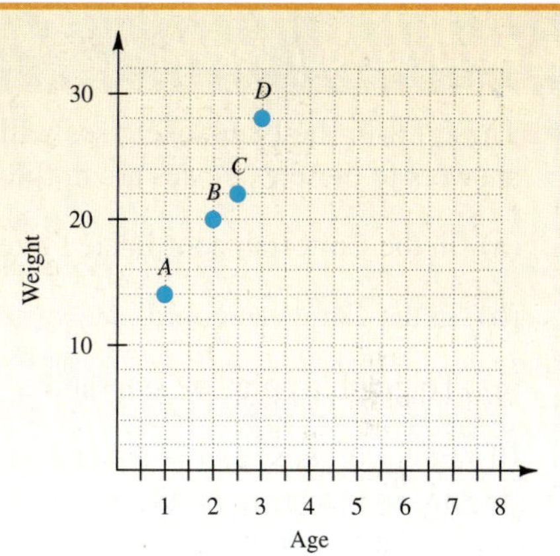

Here is an application from the field of manufacturing.

Example 4 A Graphing Application

A computer-aided design (CAD) operator located three corners of a rectangle at (5, 9), (−2, 9), and (5, 2). Find the location of the fourth corner.

We plot the three points.

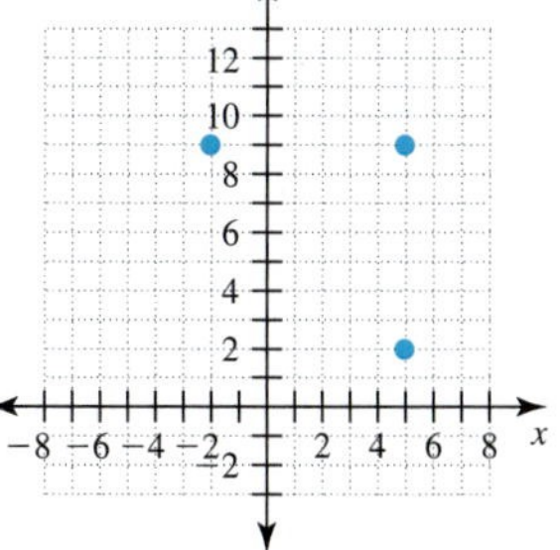

The fourth corner must lie directly underneath the point (−2, 9), so the x-coordinate must be −2. The corner must lie on the same horizontal as the point (5, 2), so the y-coordinate must be 2. Therefore, the coordinates of the fourth corner must be (−2, 2).

Check Yourself 4

A CAD operator has located three corners of a rectangle. The corners are at (−3, 4), (6, 4), and (−3, −7). Find the location of the fourth corner.

Check Yourself ANSWERS

1. $P(4, 5)$, $Q(0, 6)$, $R(-4, -4)$, and $S(2, -5)$ **2.**

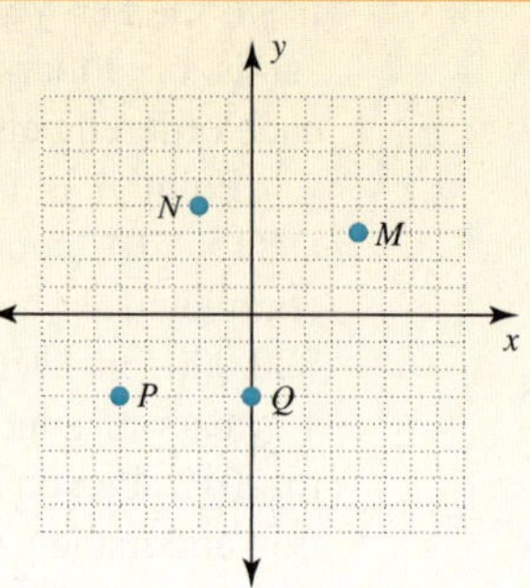

3. $A(1, 14)$, $B(2, 20)$, $C\left(\frac{5}{2}, 22\right)$, and $D(3, 28)$; The first number in each ordered pair represents the age, in years. The second number represents the weight, in pounds. **4.** $(6, -7)$

Reading Your Text

These fill-in-the-blank exercises will help you understand some of the key vocabulary used in this section. The answers to these exercises are in the Answers Appendix in the back of the text.

(a) In the Cartesian coordinate system the horizontal line is called the ________.

(b) In the Cartesian coordinate system the vertical line is called the ________.

(c) To graph a point we start at the ________.

(d) Every ordered pair is either in one of the ________ or on one of the axes.

2.4 exercises

Skills Calculator/Computer Career Applications Above and Beyond

< Objective 1 >

Give the coordinates of each plotted point.

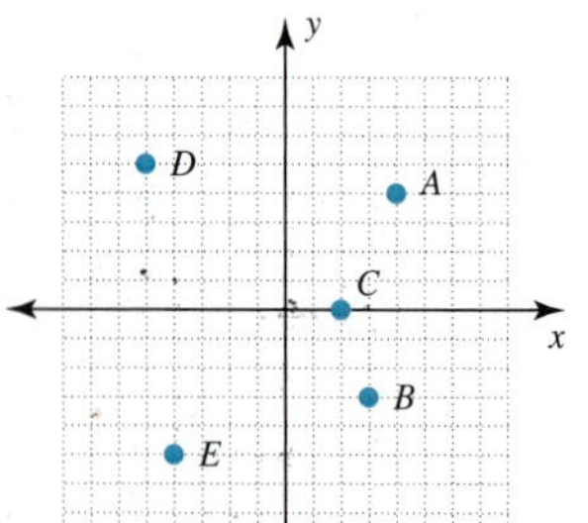

1. *A*

2. *B*

3. *C* VIDEO

4. *D*

5. *E* VIDEO

Give the coordinates of each plotted point.

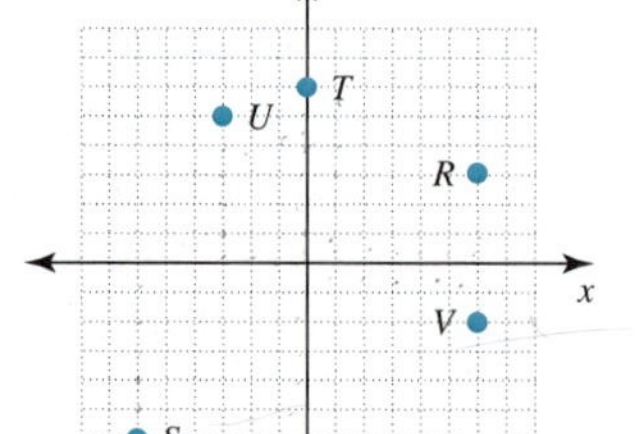

6. *R*

7. *S*

8. *T*

9. *U*

10. *V*

< Objective 2 >

Plot each point on a rectangular coordinate system.

11. $M(5, 3)$ **12.** $N(0, -3)$ **13.** $P(-4, 5)$ **14.** $Q(5, 0)$

15. $R(-4, -6)$ **16.** $S(-4, -3)$ **17.** $F(-3, -1)$ **18.** $G(4, 3)$

19. $H(4, -3)$ **20.** $I(-3, 0)$ **21.** $J(-5, 3)$ 

22. $K(0, 4)$

Give the quadrant in which each point is located or the axis on which the point lies.

23. $(4, 5)$ **24.** $(-3, 2)$ **25.** $(-6, -8)$ **26.** $(2, -4)$

27. $(5, 0)$ **28.** $(-1, 11)$ **29.** $(-4, 7)$ **30.** $(-3, -7)$

31. $(0, -4)$ **32.** $(-3, 0)$ **33.** $\left(5\frac{3}{4}, -3\right)$ **34.** $\left(-3, 4\frac{2}{3}\right)$

< Objective 3 >

35. A company kept a record of the number of items produced by an employee as the number of days on the job increases. In the graph, points correspond to an ordered-pair relationship in which the first number represents days on the job and the second number represents the number of items produced. Estimate each ordered pair represented.

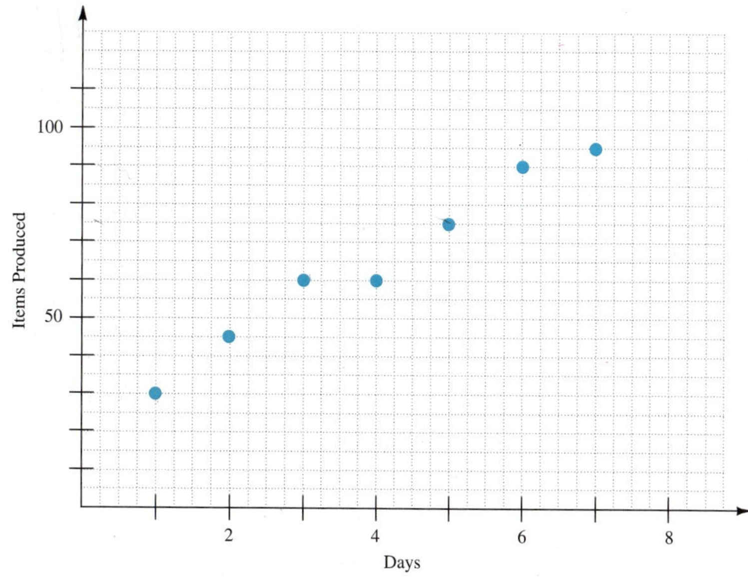

36. In the graph, points correspond to an ordered-pair relationship between height and age in which the first number represents age and the second number represents height. Estimate each ordered pair represented.

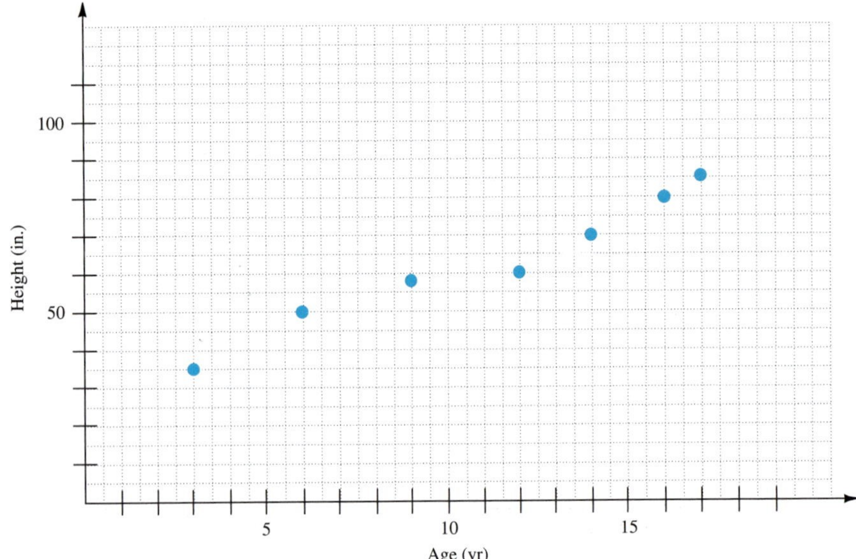

37. An unidentified company kept a record of the number of hours devoted to safety training and the number of work hours lost due to on-the-job accidents. In the graph, the points correspond to an ordered-pair relationship in which the first number represents hours in safety training and the second number represents hours lost by accidents. Estimate each ordered pair represented.

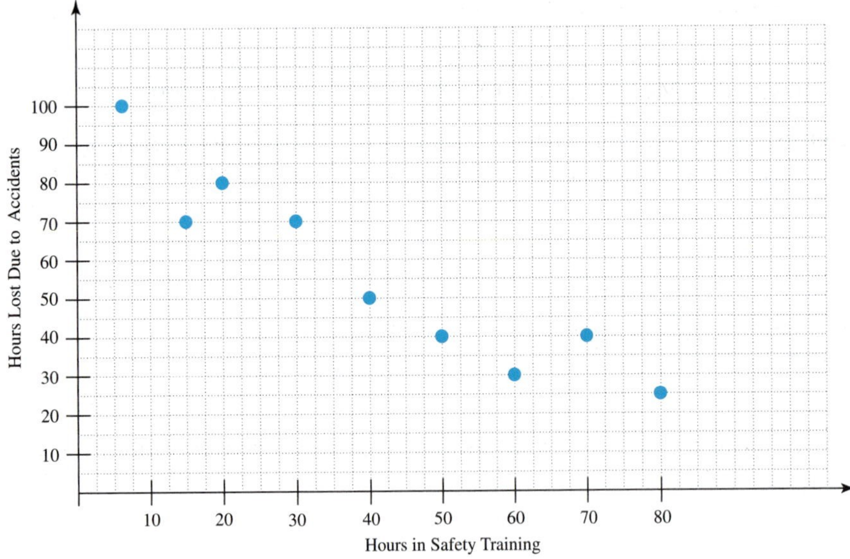

38. In the graph, points correspond to an ordered-pair relationship between the age of a person and the annual average number of visits to doctors and dentists for a person that age. The first number represents the age, and the second number represents the number of visits. Estimate each ordered pair represented.

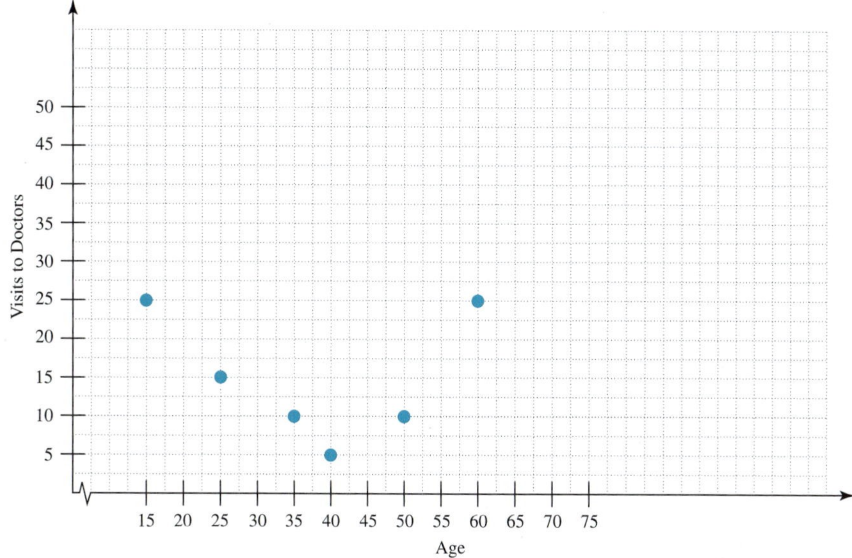

39. **Business and Finance** A plastics company is sponsoring a plastics recycling contest for the local community. The focus of the contest is on collecting plastic milk, juice, and water jugs. The company will award \$200, plus the current market price of the jugs collected, to the group that collects the most jugs in a single month. The number of jugs collected and the amount of money won can be represented as an ordered pair.

(a) In April, group A collected 1,500 lb of jugs to win first place. The prize for the month was \$350. If x represents the pounds of jugs and y represents the amount of money that the group won, graph the point that represents the winner for April.

(b) In May, group B collected 2,300 lb of jugs to win first place. The prize for the month was \$430. Graph the point that represents the May winner on the same grid you used in part (a).

40. **Science and Medicine** The table gives the average temperature y (in degrees Fahrenheit) for a sequence of months x. The months are numbered 1 through 6, with 1 corresponding to January. Plot the data given in the table. VIDEO

x	1	2	3	4	5	6
y	4	14	26	33	42	51

41. **Business and Finance** The table gives the total salary of a salesperson y for each quarter x. Plot the data given in the table.

x	1	2	3	4
y	\$6,000	\$5,000	\$8,000	\$9,000

42. **Business and Finance** The Center for Economic Transformation compared the mean marginal tax rate for the highest income bracket to the mean annual growth in the gross domestic product (GDP) in the United States by

decade. Plot the data points on a graph with the x-coordinate giving the marginal tax rates and the y-coordinate giving the GDP growth rate.

Period	Mean Marginal Tax Rate (%)	Mean Real GDP Growth Rate (%)
1950–1959	91	4.4
1960–1969	80	4.4
1970–1979	70	3.4
1980–1989	48	3.1
1990–1999	36	3.4
2000–2009	36	1.6

Complete each statement with **always, sometimes,** *or* **never.**

43. In the plane, a point on an axis __________ has a coordinate equal to zero.

44. The ordered pair (a, b) is __________ equal to the ordered pair (b, a).

45. If, in the ordered pair (a, b), a and b have different signs, then the point (a, b) is __________ in the second quadrant.

46. If $a \neq b$, then the ordered pair (a, b) is __________ equal to the ordered pair (b, a).

Skills | Calculator/Computer | **Career Applications** | Above and Beyond

47. **Mechanical Engineering** Plot the temperature and pressure relationship of a coolant as described in the table.

Temperature (°F)	−10	10	30	50	70	90
Pressure (psi)	4.6	14.9	28.3	47.1	71.1	99.2

48. **Mechanical Engineering** Use the graph in exercise 47 to answer each question.

(a) Predict the pressure when the temperature is 60°F.

(b) At what temperature would you expect the coolant to be if the pressure reads 37 psi?

49. **Allied Health** Plot the baby's weight w (in pounds) recorded at well-baby checkups at the ages x (in months), as described in the table.

Age (months)	0	0.5	1	2	7	9
Weight (pounds)	7.8	7.14	9.25	12.5	20.25	21.25

50. **Electronics** A solenoid uses an applied electromagnetic force to cause mechanical force. Typically, a wire conductor is coiled and current is applied, creating an electromagnet. The magnetic field induced by the energized coil attracts a piece of iron, creating mechanical movement.

Plot the force y (in newtons) for each applied voltage x (in volts) of a solenoid shown in the table. VIDEO

x	5	10	15	20
y	0.12	0.24	0.36	0.49

51. Graph points with coordinates $(-1, 3)$, $(0, 0)$, and $(1, -3)$. What do you observe? Can you give the coordinates of another point with the same property?

52. Graph points with coordinates $(1, 5)$, $(-1, 3)$, and $(-3, 1)$. What do you observe? Can you give the coordinates of another point with the same property?

53. Although high employment is a measure of a country's economic vitality, economists worry that periods of low unemployment lead to inflation. Consider the table.

Year	Unemployment Rate (%)	Inflation Rate (%)
1975	4.9	5.8
1980	8.5	9.2
1985	7.1	13.6
1990	5.6	3.6
1995	5.6	5.4
2000	4.0	3.4
2005	5.1	3.4
2010	9.6	1.6

Source: Bureau of Labor Statistics

Plot the figures in the table with unemployment rates on the x-axis and inflation rates on the y-axis. What does the plot tell you? Do higher inflation rates seem to be associated with lower unemployment rates? Explain.

54. The Cartesian coordinate system was named for the French philosopher and mathematician René Descartes. What philosophy book is Descartes most famous for?

55. How would you describe a rectangular coordinate system? Explain what information is needed to locate a point in a coordinate system.

56. What characteristic is common to all points on the x-axis? On the y-axis?

57. Some newspapers have a special day that they devote to automobile want ads. Use this special section or the Sunday classified ads from your local newspaper to find all the want ads for a particular automobile model. Make a list of the model year and asking price for 10 ads, being sure to get a variety of ages for this model. After collecting the information, make a plot of the age and the asking price for the car.

Describe your graph, including an explanation of how you decided which variable to put on the vertical axis and which on the horizontal axis. What trends or other information does the graph portray?

Answers

1. (4, 4) **3.** (2, 0) **5.** $(-4, -5)$ **7.** $(-6, -6)$ **9.** $(-3, 5)$

11.

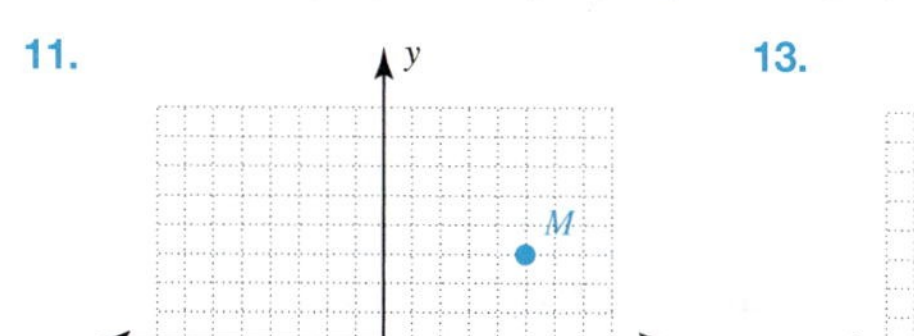

13.

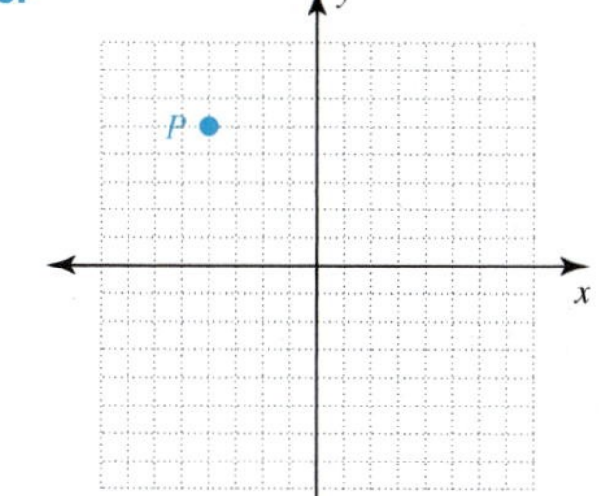

15.

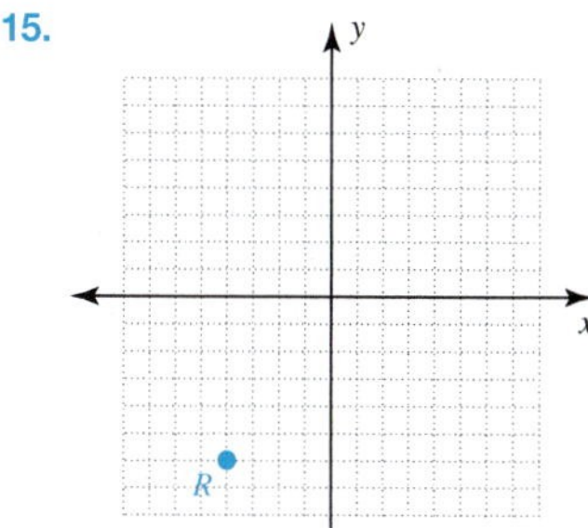

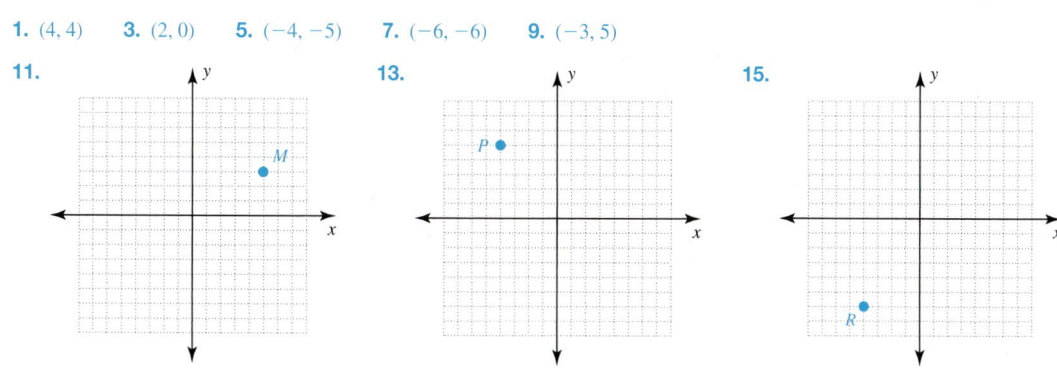

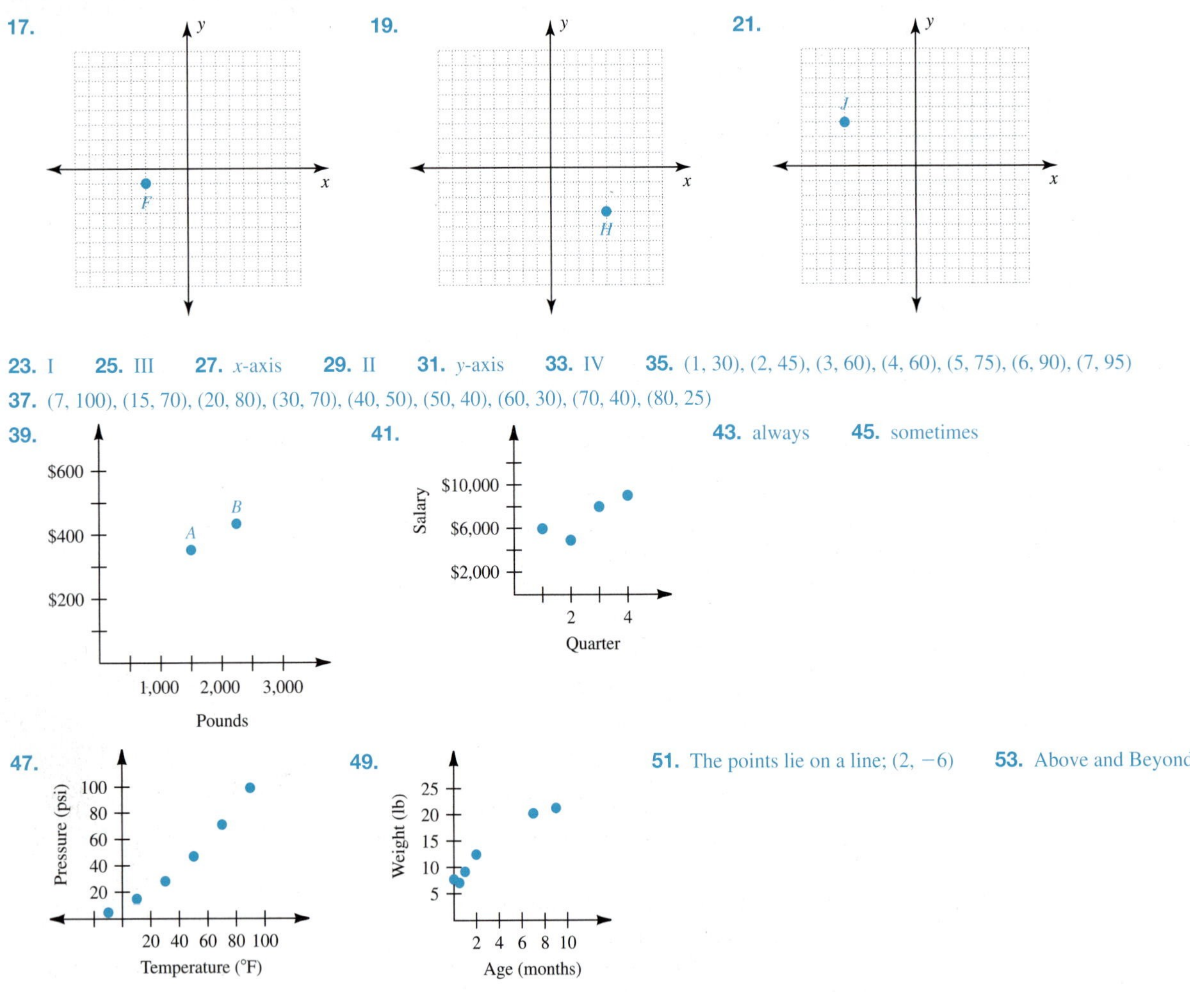

23. I **25.** III **27.** *x*-axis **29.** II **31.** *y*-axis **33.** IV **35.** (1, 30), (2, 45), (3, 60), (4, 60), (5, 75), (6, 90), (7, 95)

37. (7, 100), (15, 70), (20, 80), (30, 70), (40, 50), (50, 40), (60, 30), (70, 40), (80, 25)

43. always **45.** sometimes

51. The points lie on a line; (2, −6) **53.** Above and Beyond

55. Above and Beyond **57.** Above and Beyond

2.5 Relations and Functions

< 2.5 Objectives >

1 > Identify the domain and range of a relation

2 > Determine whether a relation is a function

3 > Evaluate a function

4 > Use function notation to write an equation

In Section 2.3, we introduced you to ordered pairs. We now consider sets of ordered pairs in this section.

Definition

Relation A set of ordered pairs is called a **relation.**

We usually use capital letters to name relations. Given

$A = \{$(Jane Trudameier, 123-45-6789),
(Jacob Smith, 987-65-4321),
(Julia Jones, 111-22-3333)$\}$

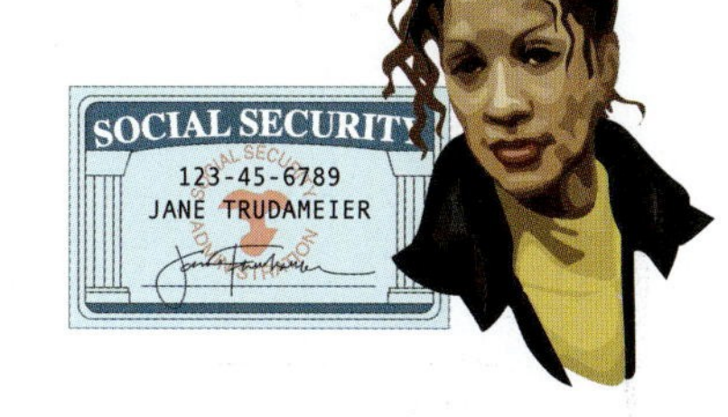

we have a relation, which we call A. In this case, there are three ordered pairs in the relation A.

Within this relation, there are two interesting sets. The first is the set of names, which is the set of first elements. The second is the set of Social Security numbers, which is the set of second elements. Each of these sets has a name.

Definition

Domain The set of first elements in a relation is called the domain of the relation.

Example 1 Finding the Domain of a Relation

< Objective 1 >

Find the domain of each relation.

(a) $A = \{$(Ben Bender, 58), (Carol Clairol, 32), (David Dorin, 29)$\}$

The domain of A is $\{$Ben Bender, Carol Clairol, David Dorin$\}$.

(b) $B = \left\{\left(5, \frac{1}{2}\right), (-4, -5), (-12, 10), (-16, \pi)\right\}$

The domain of B is $\{5, -4, -12, -16\}$.

Check Yourself 1

Find the domain of each relation.

(a) $A = \{$(Brevard, 10), (Lee, 8), (Orange, 5), (Duval, 7)$\}$

(b) $B = \left\{\left(-\frac{1}{2}, \frac{3}{4}\right), (0, 0), (1, 5), (\pi, \pi)\right\}$

Definition

Range The set of second elements in a relation is called the **range** of the relation.

Example 2 Finding the Range of a Relation

Find the range of each relation.

(a) $A = \{(\text{Ben Bender}, 58), (\text{Carol Clairol}, 32), (\text{David Dorin}, 29)\}$

The range of A is $\{58, 32, 29\}$.

(b) $B = \left\{\left(5, \frac{1}{2}\right), (-4, -5), (-12, 10), (-16, \pi), (-16, 1)\right\}$

The range of B is $\left\{\frac{1}{2}, -5, 10, \pi, 1\right\}$.

Check Yourself 2

Find the range of each relation.

(a) $A = \{(\text{Brevard}, 10), (\text{Lee}, 8), (\text{Orange}, 5), (\text{Duval}, 7)\}$

(b) $B = \left\{\left(-\frac{1}{2}, \frac{3}{4}\right), (0, 0), (1, 5), (\pi, \pi)\right\}$

We can represent the set of ordered pairs $B = \{(-2, 1), (-1, 1), (0, 3), (4, 3)\}$ with a table.

x	y
−2	1
−1	1
0	3
4	3

The same set of ordered pairs can also be presented as a mapping.

Note that, in this mapping, no x-value (domain element) is mapped to two different y-values (range elements). That leads to our definition of a function.

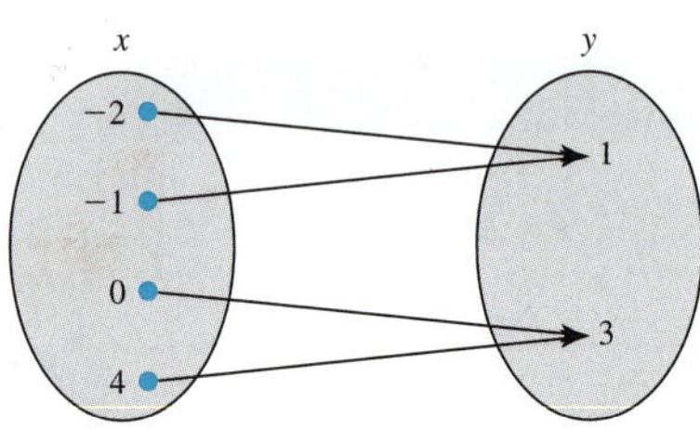

Definition

Function A **function** is a set of ordered pairs in which no element of the domain is paired with more than one element of the range.

Example 3 Identifying a Function

< Objective 2 >

For each table of values, decide whether the relation is a function.

(a)

x	y
−2	1
−1	1
1	3
2	3

(b)

x	y
−5	−2
−1	3
−1	6
2	8

(c)

x	y
−3	1
−1	0
0	2
2	4

In part (a), no two first elements in the table are equal, so we have a function.

You should see that in two cases, two distinct domain elements are mapped to a single range element, but this does not prevent the set of ordered pairs represented by the table from being a function.

In part (b), we see that -1 maps to both 3 and 6. We have two distinct range elements corresponding to a single domain element. This does not represent a function.

In (c), neither of these situations occur. The table in part (c) gives a set of ordered pairs that represent a function.

Because no range elements repeat, the relation in (c) represents a special type of function. You will learn more about these functions in Chapter 10.

Check Yourself 3

For each table of values below, decide whether the relation is a function.

(a)

x	y
-3	0
-1	1
1	2
3	3

(b)

x	y
-2	-2
-1	-2
1	3
2	3

(c)

x	y
-2	0
-1	1
0	2
0	3

Next we look at another way to represent functions. Rather than being given a set of ordered pairs or a table, we may instead give a rule or equation that we can use to generate ordered pairs. To generate ordered pairs, we need to recall how to evaluate an expression, first introduced in Section 1.2, and apply the order-of-operations.

In Section 1.2 you used addition, subtraction, multiplication, and division to evaluate expressions such as

$$3 + 5 \qquad 7x - 4 \qquad x^2 - 3x - 4 \qquad x^4 - x^2 + 2$$

Example 4 Evaluating Expressions

Evaluate the expression $x^4 - 2x^2 + 3x + 4$ as indicated.

(a) $x = 0$

Substituting 0 for x in the expression yields

$$\begin{aligned}(0)^4 - 2(0)^2 + 3(0) + 4 &= 0 - 0 + 0 + 4 \\ &= 4\end{aligned}$$

(b) $x = 2$

Substituting 2 for x in the expression yields

$$\begin{aligned}(2)^4 - 2(2)^2 + 3(2) + 4 &= 16 - 8 + 6 + 4 \\ &= 18\end{aligned}$$

(c) $x = -1$

Substituting -1 for x in the expression yields

$$\begin{aligned}(-1)^4 - 2(-1)^2 + 3(-1) + 4 &= 1 - 2 - 3 + 4 \\ &= 0\end{aligned}$$

Check Yourself 4

Evaluate the expression $2x^3 - 3x^2 + 3x + 1$ as indicated.

(a) $x = 0$ **(b)** $x = 1$ **(c)** $x = -2$

We could design a machine whose purpose would be to crank out the value of an expression for each given value of x. We could call this machine something simple such as f, our **function machine.** Our machine might look like this.

For example, if we put -1 into the machine, the machine would substitute -1 for x in the expression, and 5 would come out the other end because

$$2(-1)^3 + 3(-1)^2 - 5(-1) - 1 = -2 + 3 + 5 - 1 = 5$$

Note that, with this function machine, an input of -1 always results in an output of 5. One of the most important aspects of a function machine is that each input has a unique output.

In fact, the idea of the function machine is very useful in mathematics. Your graphing calculator can be used as a function machine. You can enter the expression into the calculator as Y_1 and then evaluate Y_1 for different values of x.

Generally, in mathematics, we do not write $Y_1 = 2x^3 + 3x^2 - 5x - 1$. Instead, we write $f(x) = 2x^3 + 3x^2 - 5x - 1$, which is read "$f$ of x is equal to. . . ." Instead of calling f a function machine, we say that f is a function of x. The greatest benefit of this notation is that it lets us easily note the input value of x along with the output of the function. Instead of "the value of Y_1 is 155 when $x = 4$," we can write $f(4) = 155$.

NOTE

Two distinct input elements can have the same output. However, each input element can only be associated with exactly one output element.

Example 5 Evaluating a Function

< Objective 3 >

Given $f(x) = x^3 + 3x^2 - x + 5$, find

(a) $f(0)$

Substituting 0 for x in the above expression, we get

$$(0)^3 + 3(0)^2 - (0) + 5 = 5$$

So we write

$f(0) = 5$ The input or x-value 0 replaces x in this notation.

(b) $f(-3)$

Substituting -3 for x in the expression, we get

$$(-3)^3 + 3(-3)^2 - (-3) + 5 = -27 + 27 + 3 + 5$$
$$= 8$$

We write

$f(-3) = 8$ The function at $x = -3$ equals the output or y-value 8.

NOTE

$f(x)$ is just another name for y. The advantage of the $f(x)$ notation is seen here. It allows us to indicate the value we are using to evaluate the function.

(c) $f\left(\frac{1}{2}\right)$

Substituting $\frac{1}{2}$ for x in the earlier expression, we get

$$\left(\frac{1}{2}\right)^3 + 3\left(\frac{1}{2}\right)^2 - \left(\frac{1}{2}\right) + 5 = \frac{1}{8} + 3\left(\frac{1}{4}\right) - \frac{1}{2} + 5$$
$$= \frac{1}{8} + \frac{3}{4} - \frac{1}{2} + 5$$
$$= \frac{1}{8} + \frac{6}{8} - \frac{4}{8} + 5$$
$$= \frac{3}{8} + 5$$
$$= 5\frac{3}{8} \quad \text{or} \quad \frac{43}{8}$$

$f\left(\frac{1}{2}\right) = \frac{43}{8}$ At $x = \frac{1}{2}$, $y = \frac{43}{8}$.

Check Yourself 5

Given $f(x) = 2x^3 - x^2 + 3x - 2$, find

(a) $f(0)$ **(b)** $f(3)$ **(c)** $f\left(-\frac{1}{2}\right)$

We can rewrite the relationship between x and $f(x)$ in Example 5 as a series of ordered pairs.

$f(x) = x^3 + 3x^2 - x + 5$

From this we found that

$f(0) = 5,\quad f(-3) = 8,\quad$ and $\quad f\left(\frac{1}{2}\right) = \frac{43}{8}$

NOTE

Because $y = f(x)$, $(x, f(x))$ is another way of writing (x, y).

There is an ordered pair, which we could write as $(x, f(x))$, associated with each of these. Those three ordered pairs are

$(0, 5),\quad (-3, 8),\quad$ and $\quad \left(\frac{1}{2}, \frac{43}{8}\right)$

Example 6 Finding Ordered Pairs

Given the function $f(x) = 2x^2 - 3x + 5$, find the ordered pair $(x, f(x))$ associated with each value for x.

(a) $x = 0$

$f(0) = 2(0)^2 - 3(0) + 5 = 5$

The ordered pair is $(0, 5)$.

(b) $x = -1$

$f(-1) = 2(-1)^2 - 3(-1) + 5 = 10$

The ordered pair is $(-1, 10)$.

(c) $x = \frac{1}{4}$

$f\left(\frac{1}{4}\right) = 2\left(\frac{1}{4}\right)^2 - 3\left(\frac{1}{4}\right) + 5 = \frac{35}{8}$

The ordered pair is $\left(\frac{1}{4}, \frac{35}{8}\right)$.

Check Yourself 6

Given $f(x) = 2x^3 - x^2 + 3x - 2$, find the ordered pair associated with each value of x.

(a) $x = 0$ **(b)** $x = 3$ **(c)** $x = -\frac{1}{2}$

We began this section by defining a relation as a set of ordered pairs. In Example 7, we determine if a relation is also a function.

Example 7 Modeling with a Function Machine

Determine which relations are also functions.

(a) The set of all possible ordered pairs in which the first element is a U.S. state and the second element is a U.S. Senator from that state.

This relation is not a function. Because there are two senators from each state, each input does not have a unique output. In the picture, New Jersey is the input, but New Jersey has two different senators.

(b) The set of all ordered pairs in which the input is the year and the output is the U.S. Open golf champion of that year.

This relation is a function. Each input has a unique output. In the picture, an input of 2010 gives an output of Graeme McDowell. For any input year, there is exactly one U.S. Open golf champion.

(c) The set of all ordered pairs in the relation R, when

$R = \{(1, 3), (2, 5), (2, 7), (3\ -4)\}$

This relation is not a function. An input of 2 results in two different outputs, 5 and 7.

(d) The set of all ordered pairs in the relation S, when

$S = \{(-1, 3), (0, 3), (3, 5), (5, -2)\}$

This relation is a function. Each input has a unique output.

Check Yourself 7

Determine which relations are also functions.

(a) The set of all ordered pairs in which the first element is a U.S. city and the second element is the mayor of that city
(b) The set of all ordered pairs in which the first element is a street name and the second element is a U.S. city in which a street of that name is found
(c) The relation $A = \{(-2, 3), (-4, 9), (9, -4)\}$
(d) The relation $B = \{(1, 2), (3, 4), (3, 5)\}$

NOTE

We begin graphing functions in Section 2.6 and continue in Chapter 3.

If we are working with an equation in x and y, we may wish to rewrite the equation as a function of x. This is particularly useful if we want to use a graphing calculator to find y for a given x or to view a graph of the equation.

Example 8 Writing Equations as Functions

< Objective 4 >

Rewrite each linear equation as a function of x. Use $f(x)$ notation in the final result.

(a) $y = 3x - 4$

We note that y is already isolated. Simply replace y with $f(x)$.

$f(x) = 3x - 4$

(b) $2x - 3y = 6$

First we solve for y.

$-3y = -2x + 6$

$y = \dfrac{-2x + 6}{-3}$ y is isolated.

$y = \frac{2}{3}x - 2$

$f(x) = \frac{2}{3}x - 2$ Now replace y with $f(x)$.

Check Yourself 8

Rewrite each linear equation as a function of x. Use $f(x)$ notation in the final result.

(a) $y = -2x + 5$ **(b)** $3x + 5y = 15$

One benefit of having a function written in $f(x)$ form is that it makes it fairly easy to substitute values for x. Sometimes it is useful to substitute nonnumeric values for x.

Example 9 Substituting Nonnumeric Values for x

Let $f(x) = 2x + 3$. Evaluate f as indicated.

(a) $f(a)$

Substituting a for x in the equation, we see that

$f(a) = 2a + 3$

NOTE

It is convenient to think of x as a dummy variable in the function $f(x) = 2x + 3$. The **rule** is to multiply the input by 2, and then add 3 to get a final result. $f(t) = 2t + 3$ describes the same rule, so it represents the same function.

(b) $f(2 + h)$

Substituting $2 + h$ for x in the equation, we get

$f(2 + h) = 2(2 + h) + 3$

Distributing the 2 and then simplifying, we have

$$\begin{aligned} f(2 + h) &= 4 + 2h + 3 \\ &= 2h + 7 \end{aligned}$$

Check Yourself 9

Let $f(x) = 4x - 2$. Evaluate f as indicated.

(a) $f(b)$

(b) $f(4 + h)$

The TABLE utility on a graphing calculator can also be used to evaluate a function.

Example 10 Using a Graphing Calculator to Evaluate a Function

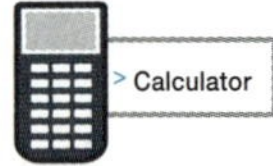

NOTE

Although we use a Texas Instruments calculator, most graphing calculators have similar capabilities.

Evaluate the function $f(x) = 3x^3 + x^2 - 2x - 5$ for each x in the set $\{-6, -5, -4, -3, -2\}$.

1. Enter the function into the Y= menu.
2. Find the table setup screen.
3. Start the table at -6 with a change of 1.
4. View the table.

Plot1 Plot2 Plot3
\Y1=3X³+X²-2X-5
\Y2=
\Y3=
\Y4=
\Y5=
\Y6=
\Y7=

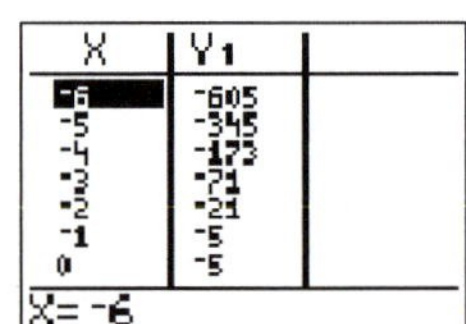

The x-values are the inputs. The output values are in the Y_1 column.

Check Yourself 10

Evaluate the function $f(x) = 2x^3 - 3x^2 - x + 2$ for each x in the set $\{-5, -4, -3, -2, -1, 0, 1\}$.

Check Yourself ANSWERS

1. **(a)** The domain of A is {Brevard, Lee, Orange, Duval}; **(b)** the domain of B is $\left\{-\frac{1}{2}, 0, 1, \pi\right\}$.
2. **(a)** The range of A is {10, 8, 5, 7}; **(b)** the range of B is $\left\{\frac{3}{4}, 0, 5, \pi\right\}$.
3. **(a)** Function; **(b)** function; **(c)** not a function 4. **(a)** 1; **(b)** 3; **(c)** -33 5. **(a)** -2; **(b)** 52; **(c)** -4
6. **(a)** $(0, -2)$; **(b)** $(3, 52)$; **(c)** $\left(-\frac{1}{2}, -4\right)$
7. **(a)** Function; **(b)** not a function; **(c)** function; **(d)** not a function
8. **(a)** $f(x) = -2x + 5$; **(b)** $f(x) = -\frac{3}{5}x + 3$ 9. **(a)** $4b - 2$; **(b)** $4h + 14$ 10.

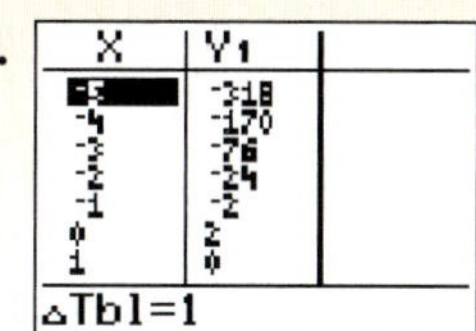

Reading Your Text

These fill-in-the-blank exercises will help you understand some of the key vocabulary used in this section. The answers to these exercises are in the Answers Appendix in the back of the text.

(a) A set of ordered pairs is called a ________.

(b) The set of all first elements in a relation is called the ________ of the relation.

(c) The set of second elements of a relation is called the ________ of the relation.

(d) In a function of the form $y = f(x)$, x is called the ________ variable, and y is called the dependent variable.

Skills Calculator/Computer Career Applications Above and Beyond

2.5 exercises

< Objective 1 >

Find the domain and range of each relation.

1. $A = \{$(Colorado, 21), (Edmonton, 5), (Calgary, 18), (Vancouver, 17)$\}$

2. $F = \left\{\left(\text{St. Louis}, \frac{1}{2}\right), \left(\text{Denver}, -\frac{3}{4}\right), \left(\text{Green Bay}, \frac{7}{8}\right), \left(\text{Dallas}, -\frac{4}{5}\right)\right\}$

3. $G = \left\{(\text{Chamber}, \pi), (\text{Testament}, 2\pi), \left(\text{Rainmaker}, \frac{1}{2}\right), (\text{Street Lawyer}, 6)\right\}$

4. $C = \{$(John Adams, -16), (John Kennedy, -23), (Richard Nixon, -5), (Harry Truman, -11)$\}$

5. $\{(1, 2), (3, 4), (5, 6), (7, 8), (9, 10)\}$

6. $\{(2, 3), (3, 5), (4, 7), (5, 9), (6, 11)\}$

7. $\{(1, 2), (1, 3), (1, 4), (1, 5), (1, 6)\}$

8. $\{(3, 4), (3, 6), (3, 8), (3, 9), (3, 10)\}$

9. $\{(-1, 3), (-2, 4), (-3, 5), (4, 4), (5, 6)\}$ 

10. $\{(-2, 4), (1, 4), (-3, 4), (5, 4), (7, 4)\}$

11. **Business and Finance** The Dow Jones Industrial Average (DJIA) is a common index used to gauge the health and performance of the stock market and the American economy. The DJIA at close over a recent 5-day period are shown in the table. Give this information as a set of ordered pairs, with the days as the domain and the averages as the range.

Day	1	2	3	4	5
Average	12,961.45	13,178.25	13,195.16	13,251.28	13,232.21

12. **Business and Finance** In the snack department of the local supermarket, candy costs \$2.16 per pound. For 1 to 5 lb, write the cost of candy as a set of ordered pairs.

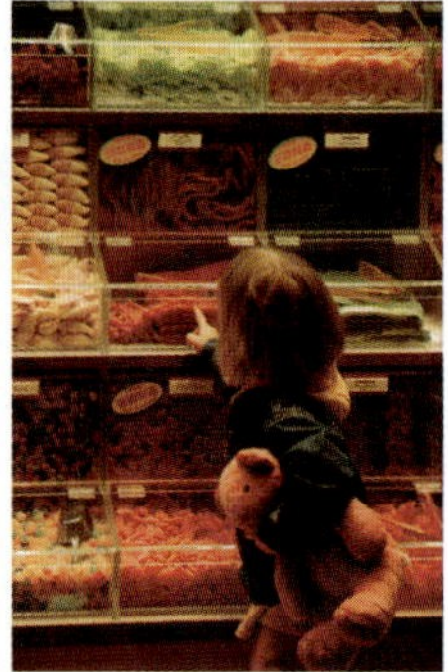

Write a set of ordered pairs that describes each situation. Give the domain and range of each relation.

13. The first element is an integer between -3 and 3. The second coordinate is the cube of the first coordinate.

14. The first element is a positive integer less than 6. The second coordinate is the sum of the first coordinate and -2.

15. The first element is the number of hours worked—10, 20, 30, 40; the second coordinate is the salary at \$9 per hour.

16. The first coordinate is the number of toppings on a pizza (up to four); the second coordinate is the price of the pizza, which is \$9 plus \$1 per topping.

< Objective 2 >

Determine which relations are functions.

17. $\{(1, 6), (2, 8), (3, 9)\}$

18. $\{(2, 3), (3, 4), (5, 9)\}$

19. $\{(-1, 4), (-2, 5), (-3, 7)\}$

20. $\{(-2, 1), (-3, 4), (-4, 6)\}$ 

21. $\{(1, 3), (1, 2), (1, 1)\}$

22. $\{(2, 4), (2, 5), (3, 6)\}$

23. $\{(-3, 5), (6, 3), (6, 9)\}$

24. $\{(4, -4), (2, 8), (4, 8)\}$

25.

x	y
3	1
-2	4
5	3
-7	4

26.

x	y
-2	3
1	4
5	6
2	-1

27.

x	y
2	3
4	2
2	-5
-6	-3

28.

x	y
1	5
3	-6
1	-5
-2	-9

29.

x	y
-1	2
3	6
6	2
-9	4

30.

x	y
4	-6
2	3
-7	1
-3	-6

< Objective 3 >

Evaluate each function as indicated.

31. $f(x) = x^2 - x - 2$; find **(a)** $f(0)$, **(b)** $f(-2)$, and **(c)** $f(1)$.

32. $f(x) = x^2 - 7x + 10$; find **(a)** $f(0)$, **(b)** $f(5)$, and **(c)** $f(-2)$.

33. $f(x) = 3x^2 + x - 1$; find **(a)** $f(-2)$, **(b)** $f(0)$, and **(c)** $f(1)$.

34. $f(x) = -x^2 - x - 2$; find **(a)** $f(-1)$, **(b)** $f(0)$, and **(c)** $f(2)$.

35. $f(x) = x^3 - 2x^2 + 5x - 2$; find **(a)** $f(-3)$, **(b)** $f(0)$, and **(c)** $f(1)$.

36. $f(x) = -2x^3 + 5x^2 - x - 1$; find **(a)** $f(-1)$, **(b)** $f(0)$, and **(c)** $f(2)$.

37. $f(x) = -3x^3 + 2x^2 - 5x + 3$; find **(a)** $f(-2)$, **(b)** $f(0)$, and **(c)** $f(3)$. 

38. $f(x) = -x^3 + 5x^2 - 7x - 8$; find **(a)** $f(-3)$, **(b)** $f(0)$, and **(c)** $f(2)$.

39. $f(x) = 2x^3 + 4x^2 + 5x + 2$; find **(a)** $f(-1)$, **(b)** $f(0)$, and **(c)** $f(1)$.

40. $f(x) = -x^3 + 2x^2 - 7x + 9$; find **(a)** $f(-2)$, **(b)** $f(0)$, and **(c)** $f(2)$.

< Objective 4 >

Rewrite each equation as a function of x. Use f(x) notation in the final result.

41. $y = -3x + 2$

42. $y = 5x + 7$

43. $y = 4x - 8$

44. $y = -7x - 9$

45. $3x + 2y = 6$

46. $4x + 3y = 12$

47. $-2x + 6y = 9$

48. $-3x + 4y = 11$

49. $-5x - 8y = -9$

50. $4x - 7y = -10$

If $f(x) = 5x - 1$, *find*

51. $f(a)$

52. $f(2r)$

53. $f(x + 1)$ 

54. $f(a - 2)$

55. $f(x + h)$

56. $\dfrac{f(x + h) - f(x)}{h}$

If $g(x) = -3x + 2$, *find*

57. $g(m)$

58. $g(5n)$

59. $g(x + 2)$

60. $g(s - 1)$

61. $g(x + h)$

62. $\dfrac{g(x + h) - g(x)}{h}$

If $k(x) = \frac{1}{2}x + 3$, *find*

63. $k(t)$

64. $k(2t)$

65. $k(x - 6)$

66. $k(2x + 6)$

67. $k(x + h)$

68. $\dfrac{k(x + h) - k(x)}{h}$

Solve each application.

69. **Business and Finance** A manufacturer determines that their profit from the sale of x espresso machines can be modeled by the function

$P(x) = 50x - 15{,}000$

Find the profit if they sell 2,500 machines.

70. **Business and Finance** The inventor of a new product believes that the cost of producing the product is given by the function

$C(x) = 1.75x + 7{,}000$ where x = units produced

What does it cost to produce 2,000 units?

71. **Business and Finance** A phone company has two different rates for calls made at different times of the day. These rates are given by the function

$$C(x) = \begin{cases} 24x + 33 & \text{between 5 P.M. and 11 P.M.} \\ 36x + 52 & \text{between 8 A.M. and 5 P.M.} \end{cases}$$

where x is the number of minutes of a call and C is the cost of a call in cents.

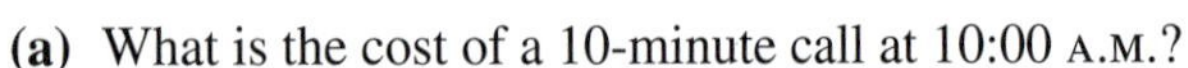

(a) What is the cost of a 10-minute call at 10:00 A.M.?

(b) What is the cost of a 10-minute call at 10:00 P.M.?

72. **Statistics** The number of accidents in 1 month involving drivers x years of age can be approximated by the function

$f(x) = 2x^2 - 125x + 3{,}000$

Find the number of accidents in 1 month that involved **(a)** 17-year-olds and **(b)** 25-year-olds.

73. **Science and Medicine** The distance x (in feet) that a car will skid on a certain road surface after the brakes are applied is a function of the car's velocity v (in miles per hour). The function can be approximated by

$f(v) = 0.017v^2$

How far will the car skid if the brakes are applied at **(a)** 55 mi/hr? **(b)** 70 mi/hr?

74. **Science and Medicine** An object is thrown upward with an initial velocity of 128 ft/s. Its height h in feet after t seconds is given by the function

$h(t) = -16t^2 + 128t$

What is the height of the object at **(a)** 2 s? **(b)** 4 s? **(c)** 6 s?

75. **Science and Medicine** Suppose that the weight (in pounds) of a baby boy x months old is predicted, for his first 10 months, by the function

$f(x) = 1.5x + 8.3$

(a) Find the predicted weight at the age of 4 months.

(b) Find the predicted weight at 8 months.

76. **Science and Medicine** Suppose that the height (in centimeters) of a baby girl x months old is predicted, for her first year, by the function

$f(x) = 1.9x + 47$

(a) Find the predicted height at the age of 4 months.

(b) Find the predicted height at 8 months.

Complete each statement with **always, sometimes,** *or* **never.**

77. The domain of a relation __________ consists of the set of all first coordinates of the ordered pairs of the relation.

78. When evaluating a function at a particular x-value, we __________ obtain two y-values.

Skills | **Calculator/Computer** | Career Applications | Above and Beyond

Use a graphing calculator to evaluate each function for the values in each set.

79. $f(x) = 3x^2 - 5x + 7$; $\{-5, -4, -3, -2, -1, 0, 1, 2, 3, 4, 5\}$

80. $f(x) = 4x^3 - 7x^2 + 9$; $\{-3, -2, -1, 0, 1, 2, 3\}$

81. $f(x) = 2x^3 - 4x^2 + 5x - 9$; $\{-4, -3, -2, -1, 0, 1, 2, 3, 4\}$

82. $f(x) = -3x^4 + 5x^2 - 7x - 15$; $\{-3, -2, -1, 0, 1, 2, 3\}$

Skills | Calculator/Computer | **Career Applications** | Above and Beyond

83. **Manufacturing Technology** The pitch of a 6-in. gear is given by the number of teeth divided by 6.

(a) Write a function to describe this relationship.

(b) What is the pitch of a 6-in. gear with 30 teeth?

84. **Allied Health** Dimercaprol (BAL) is used to treat arsenic poisoning in mammals. The recommended dose is 4 mg per kg of the animal's weight.

(a) Construct a function for the dosage in terms of an animal's weight.

(b) How much BAL must be administered to a 5-kg cat?

(c) What size cow requires a 1,450-mg dose of BAL?

85. **Construction Technology** The cost of building a house is \$90 per square foot plus \$12,000 for the foundation.

(a) Give the cost of building a house as a function of the area of the house.

(b) How much does it cost to build an 1,800-ft^2 house?

86. **Mechanical Engineering** A computer-aided design (CAD) operator has located 3 corners of a rectangle, at (5, 9), (−2, 9), and (5, 2). Give the coordinates of the fourth corner.

Answers

1. Domain: {Colorado, Edmonton, Calgary, Vancouver}; Range: {21, 5, 18, 17}
3. Domain: {Chamber, Testament, Rainmaker, Street Lawyer}; Range: $\{\pi, 2\pi, \frac{1}{2}, 6\}$ **5.** Domain: {1, 3, 5, 7, 9}; Range: {2, 4, 6, 8, 10}
7. Domain: {1}; Range: {2, 3, 4, 5, 6} **9.** Domain: {−3, −2, −1, 4, 5}; Range: {3, 4, 5, 6}
11. {(1, 12,961.45), (2, 13,178.25), (3, 13,195.16), (4, 13,251.28), (5, 13,232.21)}
13. {(−2, −8), (−1, −1), (0, 0), (1, 1), (2, 8)}; Domain: {−2, −1, 0, 1, 2}; Range: {−8, −1, 0, 1, 8}
15. {(10, 90), (20, 180), (30, 270), (40, 360)}; Domain: {10, 20, 30, 40}; Range: {90, 180, 270, 360} **17.** Function **19.** Function
21. Not a function **23.** Not a function **25.** Function **27.** Not a function **29.** Function **31.** (a) −2; (b) 4; (c) −2
33. (a) 9; (b) −1; (c) 3 **35.** (a) −62; (b) −2; (c) 2 **37.** (a) 45; (b) 3; (c) −75 **39.** (a) −1; (b) 2; (c) 13 **41.** $f(x) = -3x + 2$
43. $f(x) = 4x - 8$ **45.** $f(x) = -\frac{3}{2}x + 3$ **47.** $f(x) = \frac{1}{3}x + \frac{3}{2}$ **49.** $f(x) = -\frac{5}{8}x + \frac{9}{8}$ **51.** $5a - 1$ **53.** $5x + 4$ **55.** $5x + 5h - 1$
57. $-3m + 2$ **59.** $-3x - 4$ **61.** $-3x - 3h + 2$ **63.** $\frac{1}{2}t + 3$ **65.** $\frac{1}{2}x$ **67.** $\frac{1}{2}x + \frac{1}{2}h + 3$ **69.** \$110,000
71. (a) \$4.12; (b) \$2.73 **73.** (a) 51.425 ft; (b) 83.3 ft **75.** (a) 14.3 lb; (b) 20.3 lb **77.** always **79.** 107, 75, 49, 29, 15, 7, 5, 9, 19, 35, 57
81. −221, −114, −51, −20, −9, −6, 1, 24, 75 **83.** (a) $P(t) = \frac{t}{6}$; (b) 5 **85.** (a) $C(x) = 90x + 12{,}000$; (b) \$174,000

2.6 Tables and Graphs

< 2.6 Objectives >

1 > Use the vertical line test

2 > Identify the domain and range of a relation from its graph

3 > Read function values from a table

4 > Read function values from a graph

In Section 2.5, we defined a function in terms of ordered pairs. A set of ordered pairs can be specified in several ways; here are the most common.

Property

Ordered Pairs

1. We can present ordered pairs in a list or table.
2. We can give a rule or equation that generates ordered pairs.
3. We can use a graph to indicate ordered pairs. The graph can show distinct ordered pairs or it can show all the ordered pairs on a line or curve.

We have already seen functions presented as lists of ordered pairs, in tables, and as rules or equations. We now look at graphs of the ordered pairs from Example 3 in Section 2.5 to introduce the **vertical line test,** which is a graphical test for identifying a function.

RECALL

If a grid has no numeric labels, each mark represents one unit. In this text, each such grid represents x- and y-values from -8 to 8.

(a) As a set of ordered pairs, the relation is $\{(-2, 1), (-1, 1), (1, 3), (2, 3)\}$. Recall that this relation represents a function.

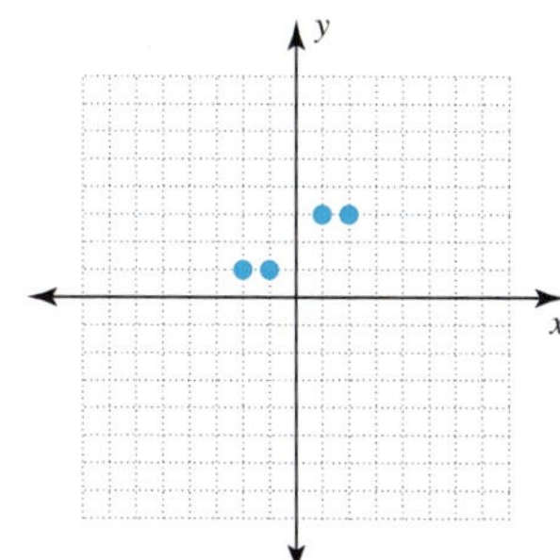

(b) As a set of ordered pairs, the relation is $\{(-5, -2), (-1, 3), (-1, 6), (2, 8)\}$. Recall that this relation does *not* represent a function.

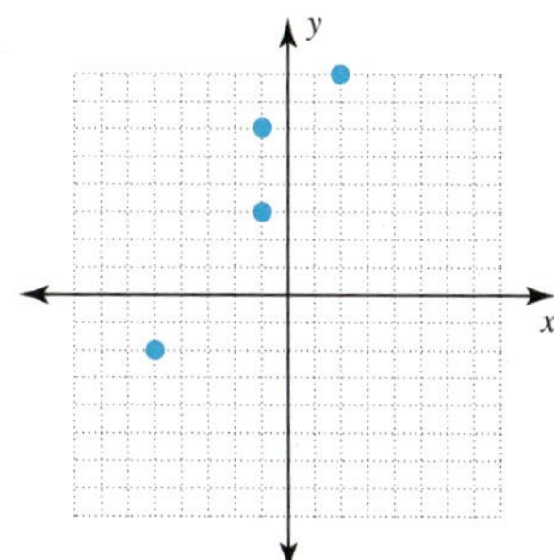

(c) As a set of ordered pairs, the relation is $\{(-3, 1), (-1, 0), (0, 2), (2, 4)\}$. Recall that this relation represents a function.

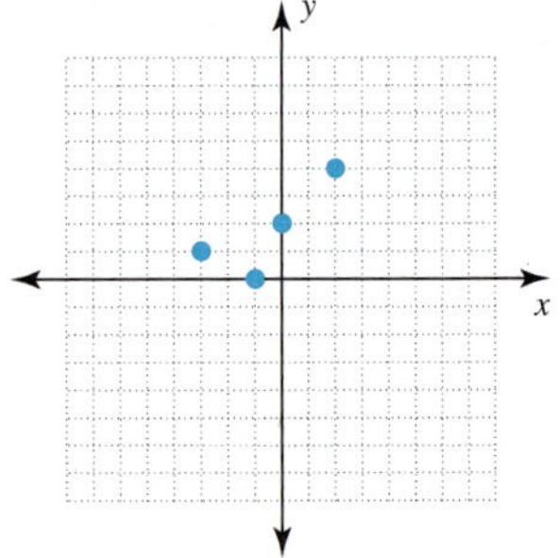

Notice that in the graphs of relations (a) and (c), there is no vertical line that can pass through two different points of the graph. In relation (b), a vertical line can pass through the two points that represent the ordered pairs $(-1, 3)$ and $(-1, 6)$. We call this the **vertical line test.**

Property

Vertical Line Test A relation is a function if no vertical line can pass through two or more points on its graph.

Example 1 Identifying a Function

< Objective 1 >

For each set of ordered pairs, plot the points and use the vertical line test to determine if it is a function.

(a) $\{(0, -1), (2, 3), (2, 6), (4, 2), (6, 3)\}$

Because a vertical line can be drawn through the points (2, 3) and (2, 6), the relation does not pass the vertical line test. That is, if the input is 2, the output is *both* 3 and 6. This is not a function.

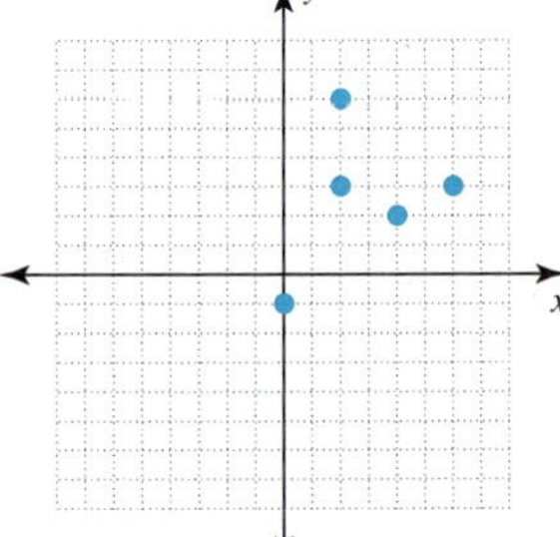

(b) $\{(1, 1), (2, 0), (3, 3), (4, 3), (5, 3)\}$

This is a function. Although a horizontal line can be drawn through several points, no vertical line passes through more than one point.

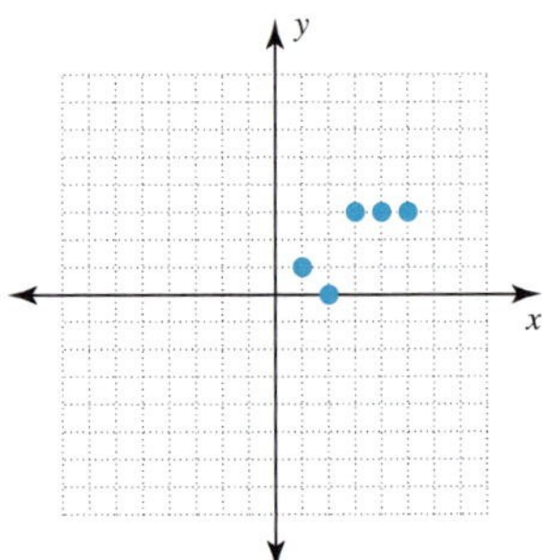

Check Yourself 1

For each set of ordered pairs, plot the points and use the vertical line test to determine if it is a function.

(a) $\{(-2, 4), (-1, 4), (0, 4), (1, 3), (5, 5)\}$
(b) $\{(-3, -1), (-1, -3), (1, -3), (1, 3)\}$

Consider why the vertical line test works. The points on a vertical line all have the same x-coordinate. Consider the graph shown.

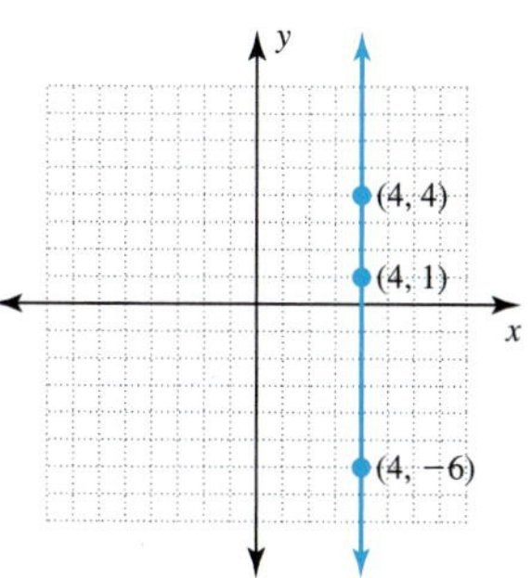

Notice that in this case, all of the points on the line have 4 as the first coordinate or x-value.

For a relation to be a function, no two distinct ordered pairs can have the same first coordinate. If a vertical line intersects the graph of a relation more than once, that relation has more than one point with the same first coordinate so it is not a function.

We can also use the graph of a relation to determine its domain and range. Remember that the domain is the set of x-values in the ordered pairs and the range is given by the y-values of the relation. Consider Example 2.

Example 2 Identifying Functions, Domain, and Range

< Objective 2 >

Determine whether each graph represents a function. Provide the domain and range in each case.

(a)

This is not a function. The vertical line at $x = 4$ passes through three points. The domain D of this relation is

$$D = \{-5, -2, 2, 4\}$$

The range R is

$$R = \{-2, 0, 1, 2, 3, 5\}$$

(b)

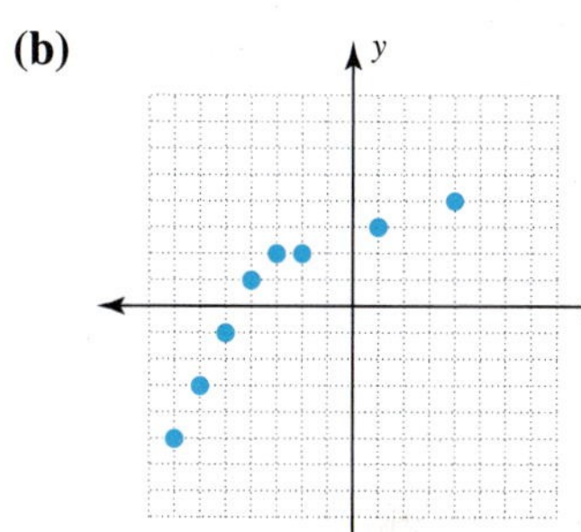

This is a function. No vertical line passes through more than one point. The domain is

$$D = \{-7, -6, -5, -4, -3, -2, 1, 4\}$$

The range is

$$R = \{-5, -3, -1, 1, 2, 3, 4\}$$

Check Yourself 2

Determine whether each graph represents a function. Provide the domain and range in each case.

(a)

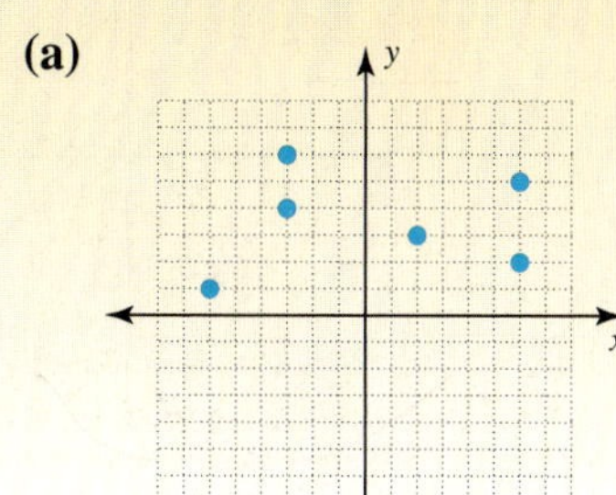

(b)

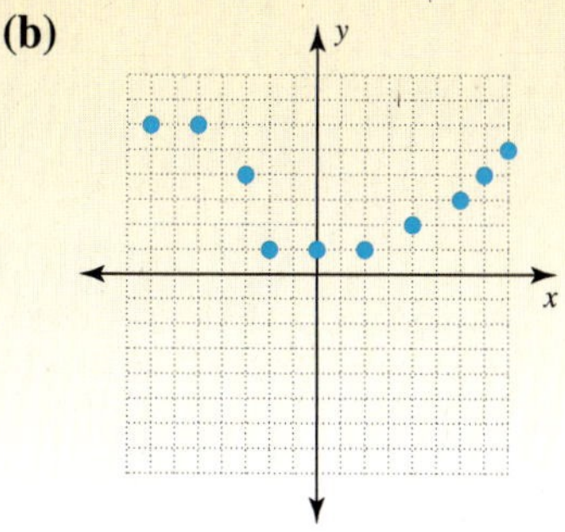

So far, we have looked at **finite** sets of ordered pairs. Now consider graphs composed of line segments, lines, or curves. Each such graph represents an **infinite** collection of points. We can still use the vertical line test to decide whether a relation is a function and we use the graph to find its domain and range.

Example 3 Identifying Functions, Domain, and Range

Determine whether each graph represents a function. Provide the domain and range in each case.

RECALL

When you see the statement $-2 \le x \le 4$, think "all real numbers between -2 and 4, including -2 and 4."

(a)

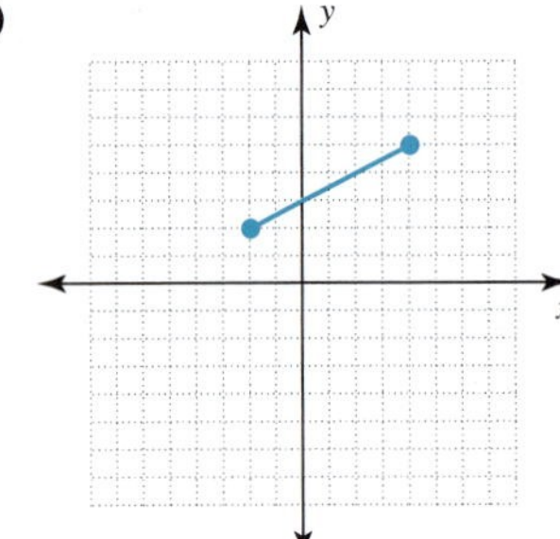

Because no vertical line passes through more than one point, this is a function. The x-values in the ordered pairs go from -2 to 4, inclusive. In set-builder notation, we write the domain as

$D = \{x \mid -2 \le x \le 4\}$

In interval notation, we have

$D = [-2, 4]$

The y-values go from 2 to 5, inclusive. The range is

$R = \{y \mid 2 \le y \le 5\}$

In interval notation, we have

$R = [2, 5]$

RECALL

When the endpoints are included, we use the "less than or equal to" symbol $\le$.

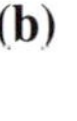

(b)

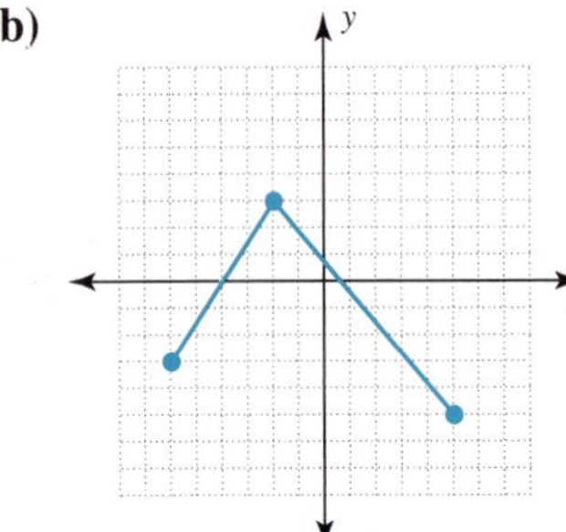

The relation graphed here is a function. The x-values run from -6 to 5, so

$D = \{x \mid -6 \le x \le 5\}$ or $D = [-6, 5]$

The y-values go from -5 to 3, so

$R = \{y \mid -5 \le y \le 3\}$ or $R = [-5, 3]$

Check Yourself 3

Determine whether each graph represents a function. Provide the domain and range in each case.

(a)

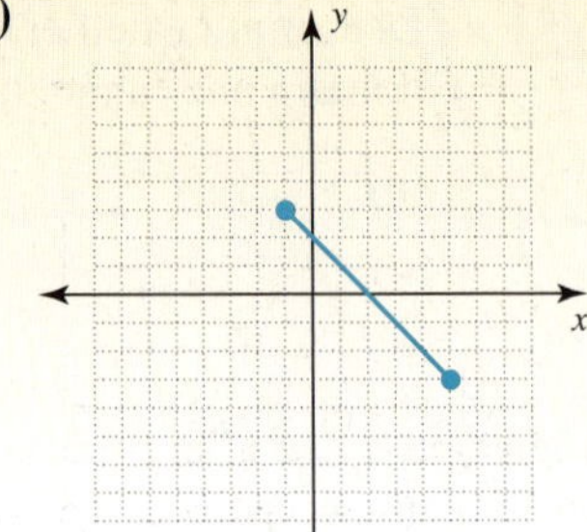

(b)

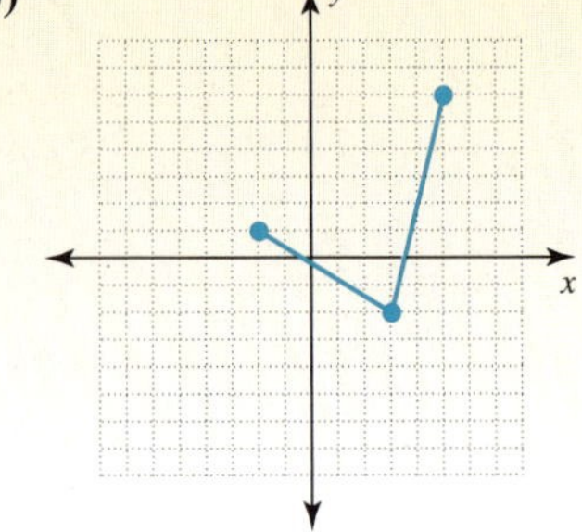

In Example 4, we consider the graphs of some common curves.

Example 4 Identifying Functions, Domain, and Range

Determine whether each graph represents a function. Provide the domain and range in each case.

NOTE

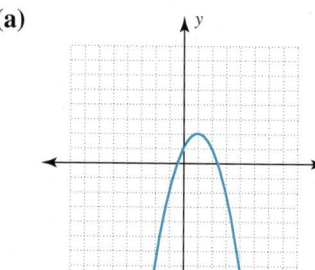

This curve is called a *parabola.*

RECALL

$\mathbb{R}$ is the symbol for the set of all real numbers.

(a)

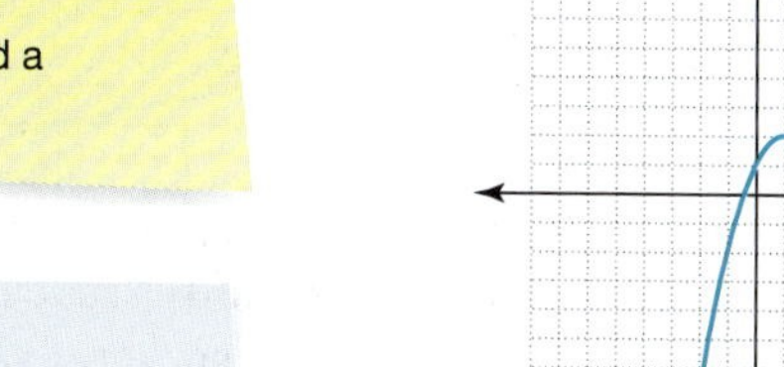

Since no vertical line passes through more than one point, this is a function. Note that the arrows on the ends of the graph indicate that the pattern continues indefinitely. The x-values that are used in this graph therefore consist of all real numbers. The domain is

$D = \{x \mid x \text{ is a real number}\}$

or simply $D = \mathbb{R}$.

The y-values, however, are never higher than 2. The range is the set of all real numbers less than or equal to 2.

$R = \{y \mid y \leq 2\} = (-\infty, 2]$ The infinity symbol always gets a parenthesis.

NOTE

This curve is also a parabola.

(b)

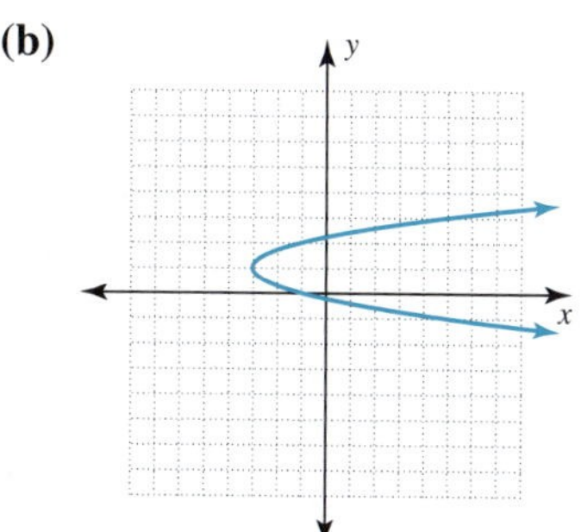

This relation is not a function. A vertical line drawn anywhere to the right of -3 passes through two points. The x-values begin at -3 and continue indefinitely to the right, so

$D = \{x \mid x \geq -3\} = [-3, \infty)$

The y-values consist of all real numbers, so

$R = \{y \mid y \text{ is a real number}\}$

or simply $R = \mathbb{R}$.

NOTE

This curve is called an *ellipse.*

(c)

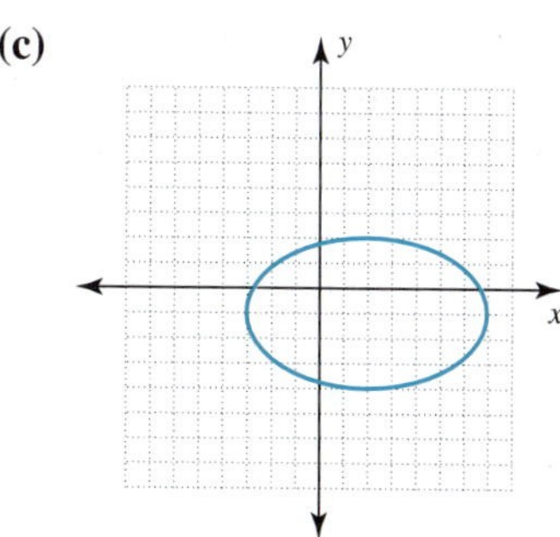

This relation is not a function. A vertical line drawn anywhere between -3 and 7 passes through two points. The x-values run from -3 to 7, inclusive. Thus,

$D = \{x \mid -3 \leq x \leq 7\} = [-3, 7]$

The y-values range from -4 to 2, inclusive, so

$R = \{y \mid -4 \leq y \leq 2\} = [-4, 2]$

Check Yourself 4

Determine whether each graph represents a function. Provide the domain and range in each case.

(a)

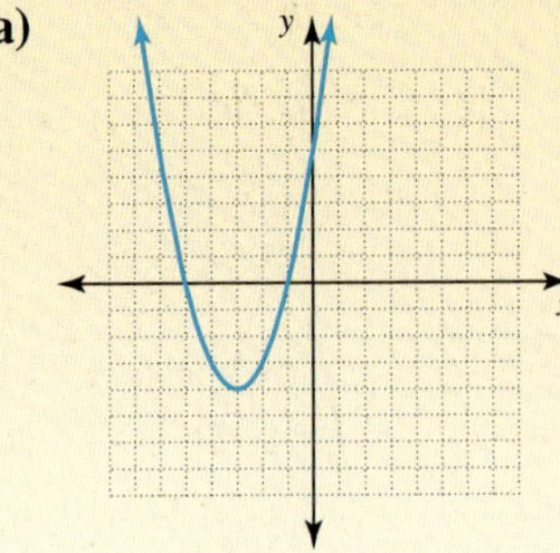

(b)

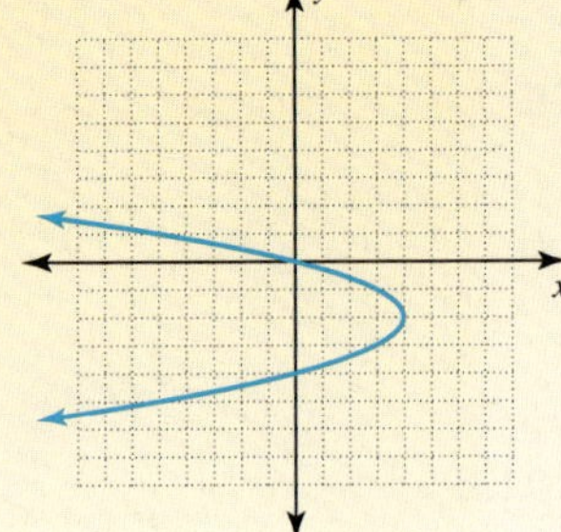

(c)

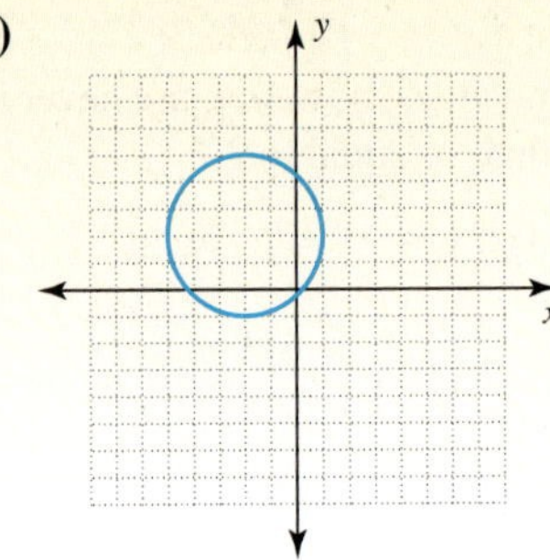

Reading tables and graphs are important skills when working with functions. If we are given a function f in either of these forms, we usually have one of two goals.

1. Given x, we want to find $f(x)$.
2. Given $f(x)$, we want to find x.

Example 5 illustrates.

Example 5 Reading Values from a Table

< Objective 3 >

Suppose we have the functions f and g, as shown.

x	$f(x)$
-4	8
0	6
2	-4
1	-2

x	$g(x)$
-2	5
1	0
4	-4
8	-2

NOTE

Think of the x-values as "input" values and the $f(x)$ values as "outputs."

(a) Find $f(0)$.

This means that 0 is the input value (a value for x). We want to know what f does to 0. Looking in the table, we see that the output value is 6. So $f(0) = 6$.

(b) Find $g(4)$.

We are given $x = 4$, and we want $g(x)$. In the table we find $g(4) = -4$.

(c) Find x, given that $f(x) = -4$.

Now we are given the output value of -4. We ask, what x-value results in an output value of -4? The answer is 2. So $x = 2$.

(d) Find x, given that $g(x) = -2$.

Since the output is given as -2, we look in the table to find that when $x = 8$, $g(x) = -2$. So $x = 8$.

Check Yourself 5

Use the functions in Example 5 to find

(a) $f(1)$
(b) $g(-2)$
(c) x, given that $f(x) = 8$
(d) x, given that $g(x) = 5$

In Example 6 we consider the same goals, given the graph of a function: (1) given x, find $f(x)$; and (2) given $f(x)$, find x.

Example 6 Reading Values from a Graph

< Objective 4 >

Given the graph of f shown, find the desired values.

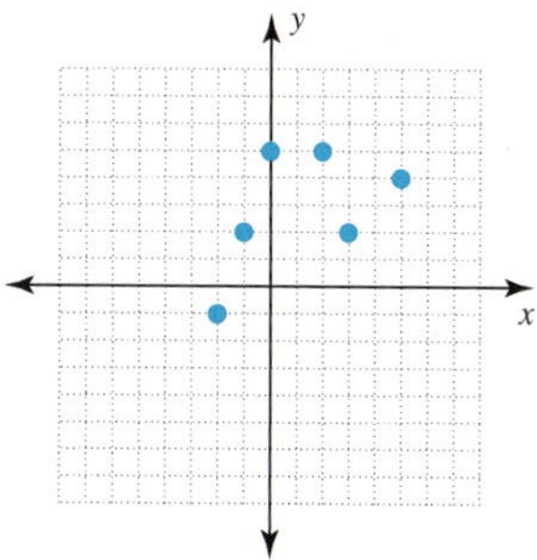

(a) Find $f(2)$.

Since 2 is the x-value, we move to 2 on the x-axis and then search vertically for a plotted point. We find (2, 5), which tells us that an input of 2 results in an output of 5. Thus, $f(2) = 5$.

(b) Find $f(-1)$.

Since $x = -1$, we move to -1 on the x-axis. We note the point $(-1, 2)$, so $f(-1) = 2$.

(c) Find all x such that $f(x) = 2$.

Now we are told that the output value is 2, so we move up to 2 on the y-axis and search horizontally for plotted points. There are two: $(-1, 2)$ and (3, 2). So the desired x-values are -1 and 3.

(d) Find all x such that $f(x) = 4$.

We move to 4 on the y-axis and search horizontally. We find one point: (5, 4). So $x = 5$.

Check Yourself 6

Given the graph of f shown, find the desired values.

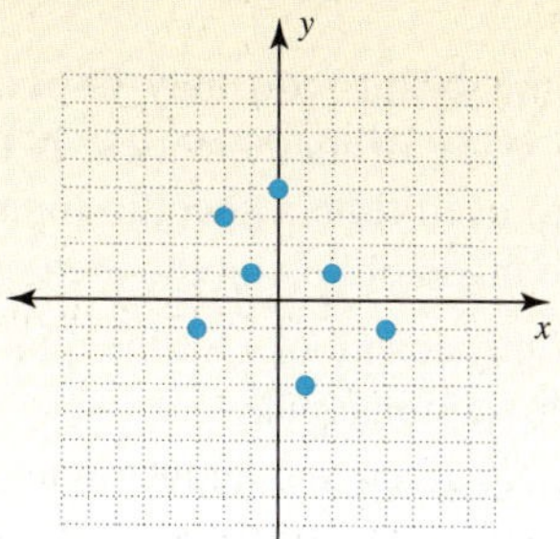

(a) Find $f(1)$.
(b) Find $f(-3)$.
(c) Find all x such that $f(x) = 1$.
(d) Find all x such that $f(x) = 4$.

Example 7 considers graphs that represent **infinite** collections of points.

Example 7 Reading Values from a Graph

(a) Given the graph of f shown, find the desired values.

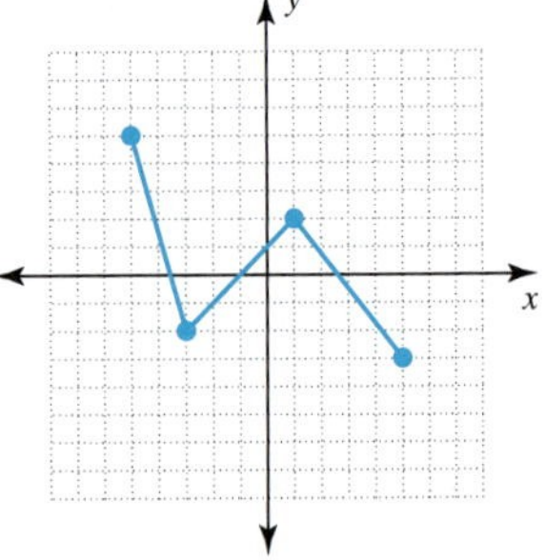

(i) Find $f(-1)$.

Since $x = -1$, we move to -1 on the x-axis. There we find the point $(-1, 0)$. So $f(-1) = 0$.

(ii) Find all x such that $f(x) = -1$.

We are given the output -1, so we move to -1 on the y-axis and search horizontally for plotted points. There are three of them, and we must estimate the coordinates for a couple of these. One point is exactly $(-2, -1)$, one is approximately $(-3.3, -1)$, and one is approximately $(3.5, -1)$. So the desired x-values are -3.3, -2, and 3.5.

NOTE

Often, we need to approximate a value when working with graphs.

(b) Given the graph of f shown, find the desired values.

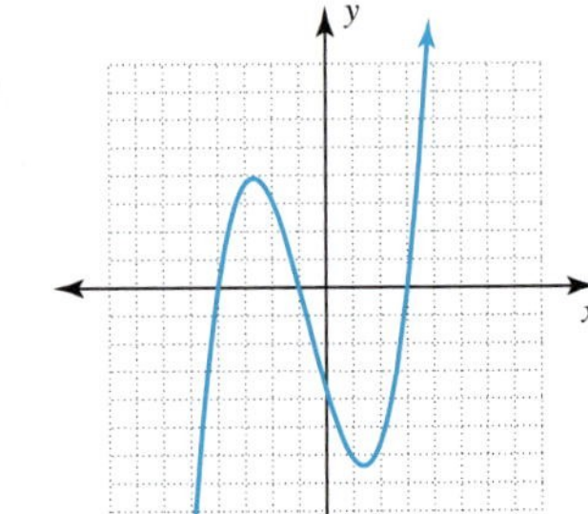

(i) Find $f(-3)$.

Since $x = -3$, we move to -3 on the x-axis. We search vertically and estimate a plotted point at approximately $(-3, 3.7)$. So $f(-3) \approx 3.7$.

(ii) Find all x such that $f(x) = 0$.

Since the output (y-value) is 0, we look for points with a y-coordinate of 0. There are three: $(-4, 0)$, $(-1, 0)$, and $(3, 0)$. So the desired x-values are -4, -1, and 3.

Check Yourself 7

Given the graph of f shown, find the desired values.

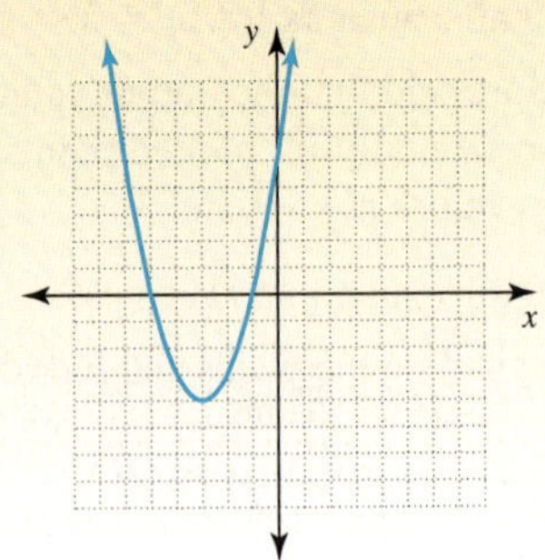

(a) Find $f(-4)$.
(b) Find all x such that $f(x) = 0$.

At this point, you may be wondering how functions relate to anything outside the study of mathematics. A function is a relation that yields a single output (y-value) each time a specific input (x-value) is given. Any field in which predictions are made builds on the idea of functions.

- A physicist looks for the relationship that uses a planet's mass to predict its gravitational pull.
- An economist looks for the relationship that uses the tax rate to predict the employment rate.

- A business marketer looks for the relationship that uses an item's price to predict the number that will be sold.
- A college board looks for the relationship between tuition costs and the number of students enrolled at the college.
- A biologist looks for the relationship that uses temperature to predict a body of water's nutrient level.

In your future study of mathematics, you will see functions applied in areas such as these. In those applications, you should find that you put to good use the basic skills developed here: (1) given x, find $f(x)$; and (2) given $f(x)$, find x.

Check Yourself ANSWERS

1. **(a)** Function; **(b)** not a function **2.** **(a)** Not a function; $D = \{-6, -3, 2, 6\}$; $R = \{1, 2, 3, 4, 5, 6\}$; **(b)** function; $D = \{-7, -5, -3, -2, 0, 2, 4, 6, 7, 8\}$; $R = \{1, 2, 3, 4, 5, 6\}$
3. **(a)** Function; $D = \{x \mid -1 \le x \le 5\} = [-1, 5]$; $R = \{y \mid -3 \le y \le 3\} = [-3, 3]$; **(b)** function; $D = \{x \mid -2 \le x \le 5\} = [-2, 5]$; $R = \{y \mid -2 \le y \le 6\} = [-2, 6]$
4. **(a)** Function; $D = \mathbb{R}$; $R = \{y \mid y \ge -4\} = [-4, \infty)$; **(b)** not a function; $D = \{x \mid x \le 4\} = (-\infty, 4]$; $R = \mathbb{R}$; **(c)** not a function; $D = \{x \mid -5 \le x \le 1\} = [-5, 1]$; $R = \{y \mid -1 \le y \le 5\} = [-1, 5]$
5. **(a)** -2; **(b)** 5; **(c)** -4; **(d)** -2 **6.** **(a)** -3; **(b)** -1; **(c)** -1 and 2; **(d)** 0 **7.** **(a)** -3; **(b)** -5 and -1

Reading Your Text

These fill-in-the-blank exercises will help you understand some of the key vocabulary used in this section. The answers to these exercises are in the Answers Appendix in the back of the text.

(a) The vertical line test is a graphical test for identifying a ________.

(b) A ________ is a function if no vertical line passes through two or more points on its graph.

(c) The ________ of a function is the set of inputs that can be substituted for the independent variable.

(d) The range of a function is the set of ________ or y-values.

Skills | Calculator/Computer | Career Applications | Above and Beyond

2.6 exercises

< Objective 1 >

Plot each set of ordered pairs and use the vertical line test to determine if it is a function.

1. $\{(-3, 1), (-1, 2), (-2, 3), (1, 4)\}$

2. $\{(2, 2), (1, 1), (3, 3), (4, 5)\}$

3. $\{(-1, 1), (2, 2), (3, 4), (5, 6)\}$

4. $\{(1, 4), (-1, 5), (0, 2), (2, 3)\}$ 

5. $\{(1, 2), (1, 3), (2, 1), (3, 1)\}$

6. $\{(-1, 1), (3, 4), (-1, 2), (5, 3)\}$

< Objective 2 >

Determine whether the relation is a function. Provide the domain and range.

7.

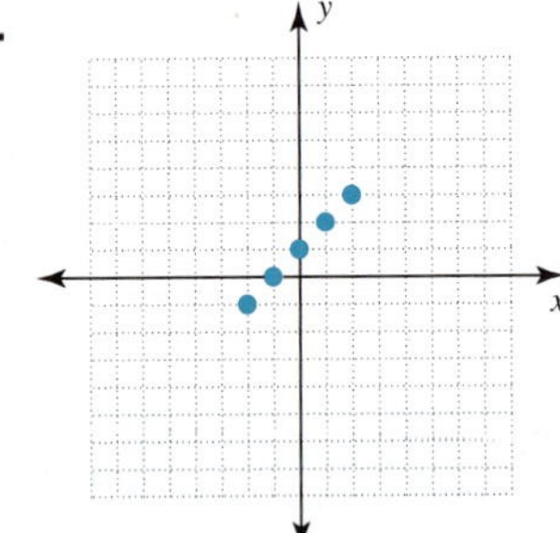

8.

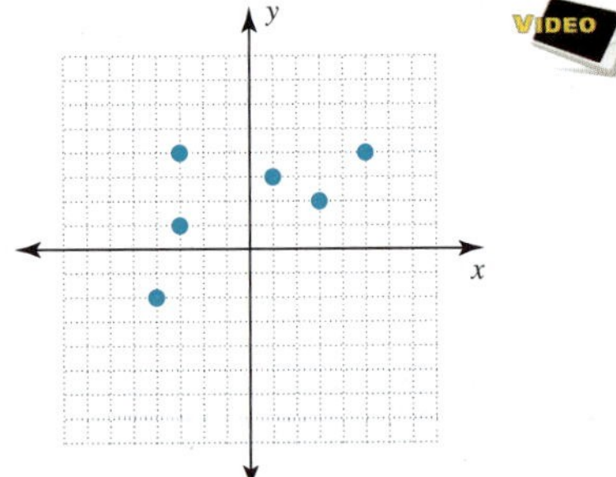

9.

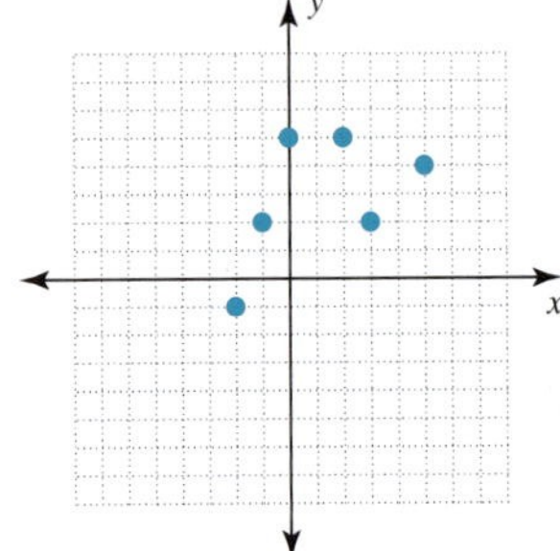

10.

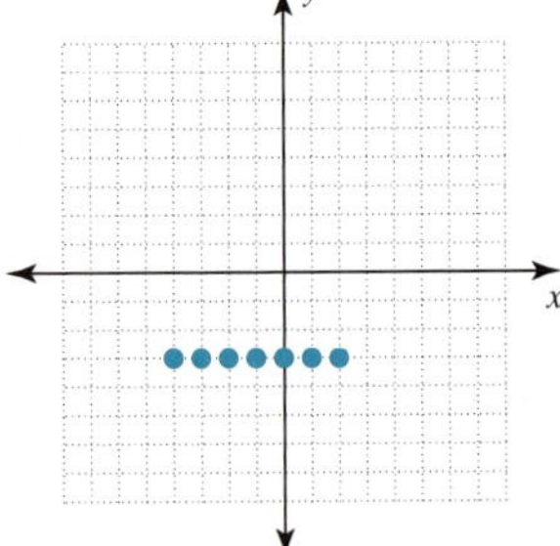

11.

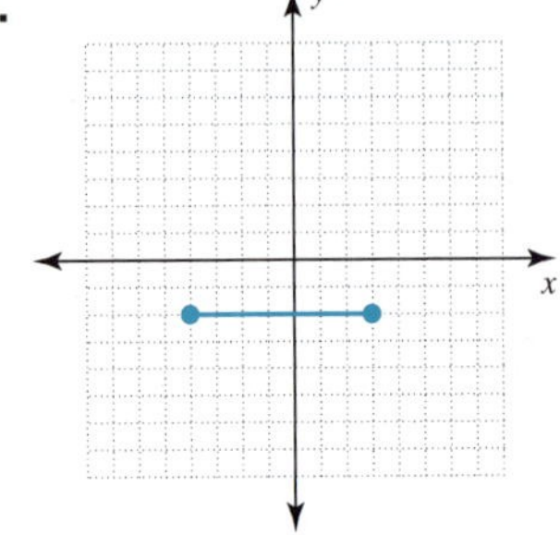

12.

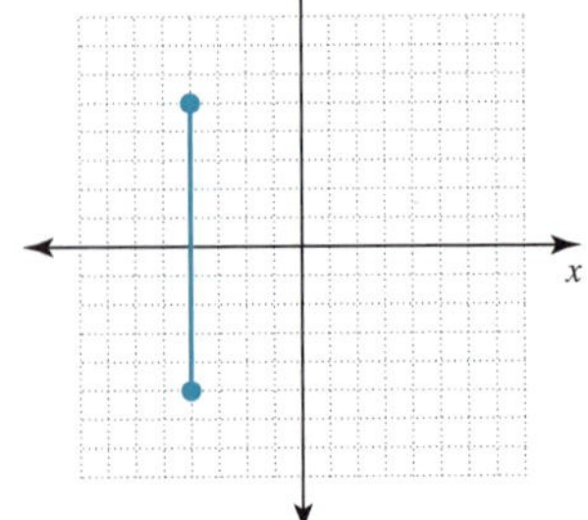

13.

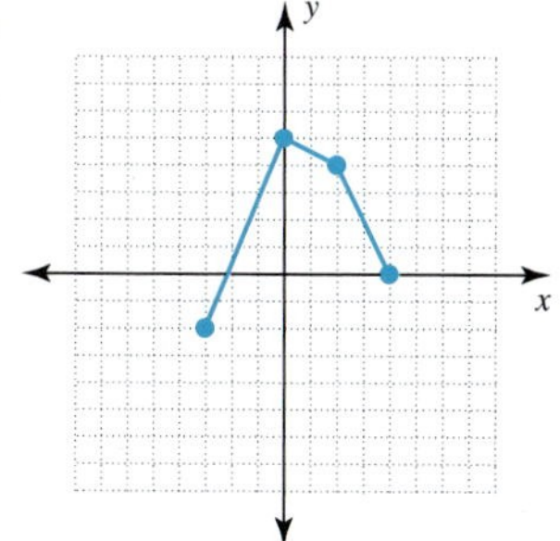

14.

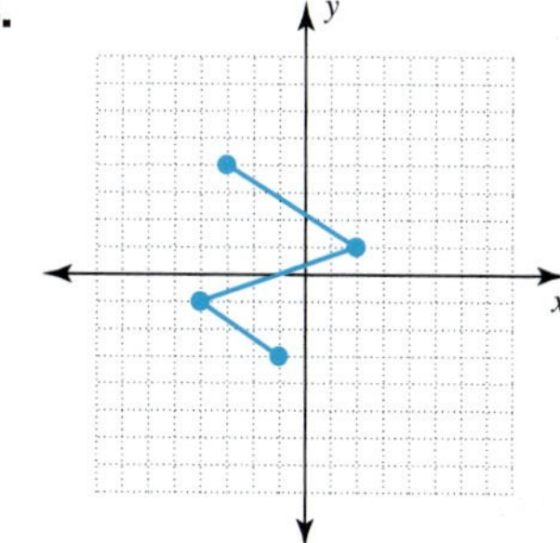

15.

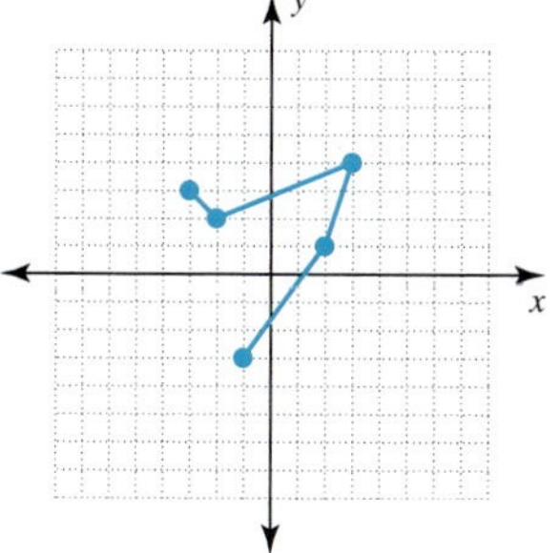

16.

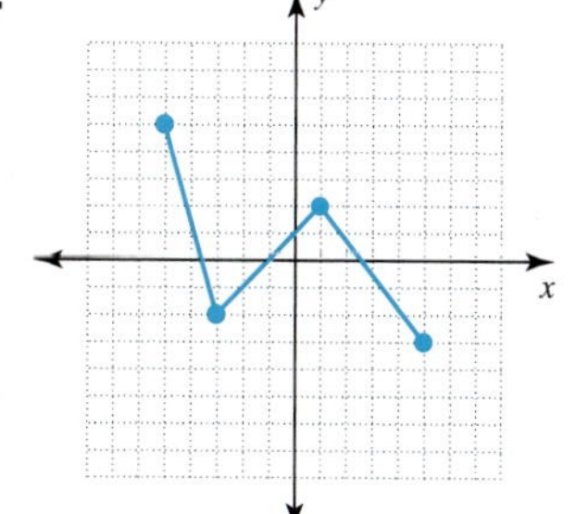

17.

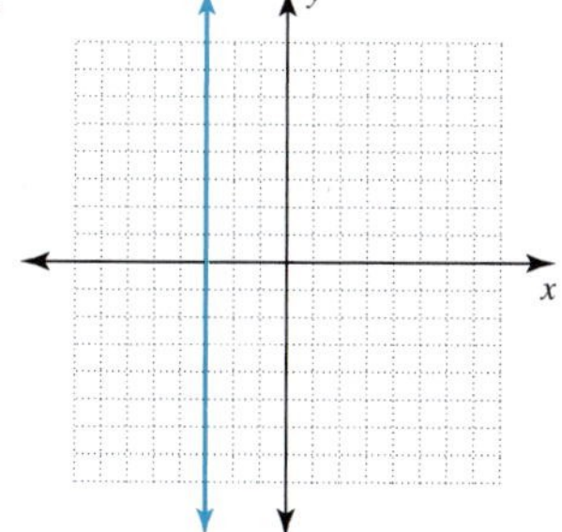

18.

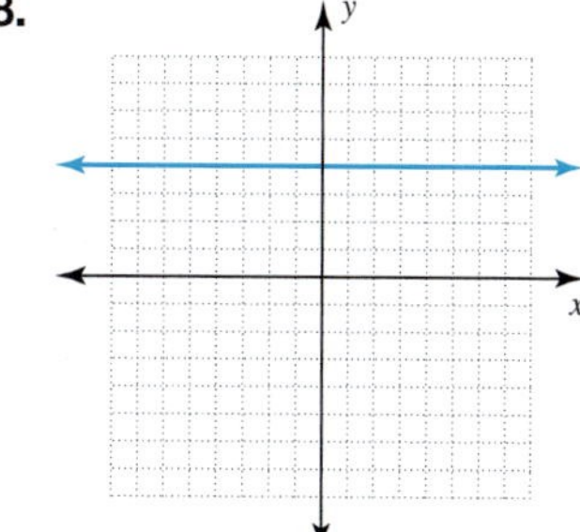

19.

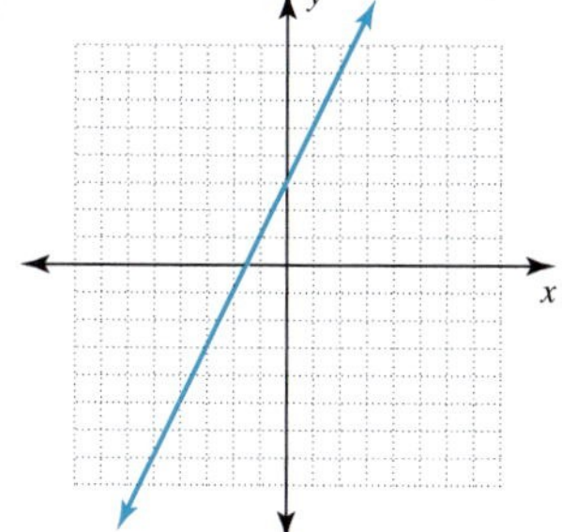

20.

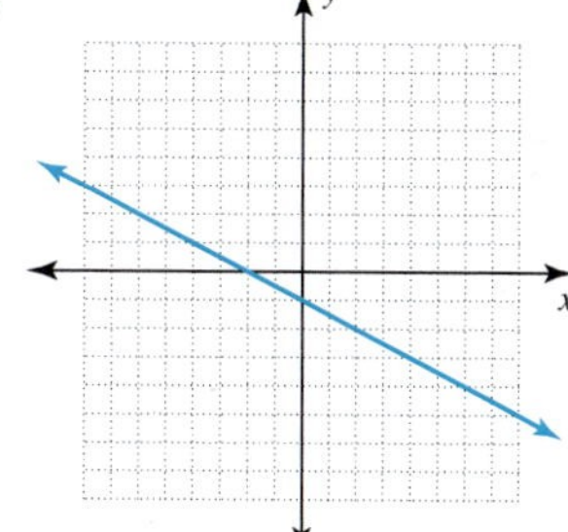

21.

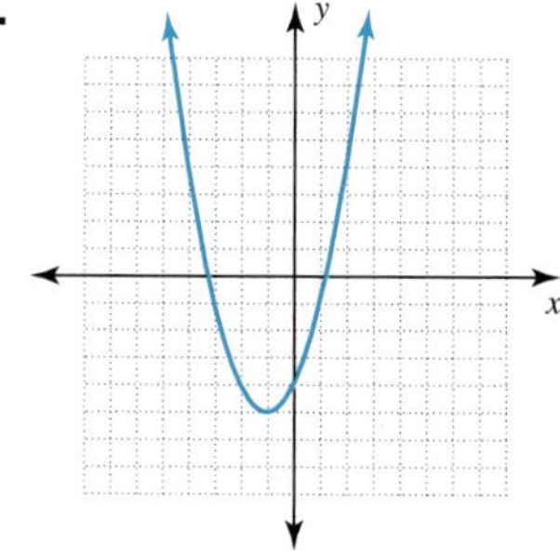

22.

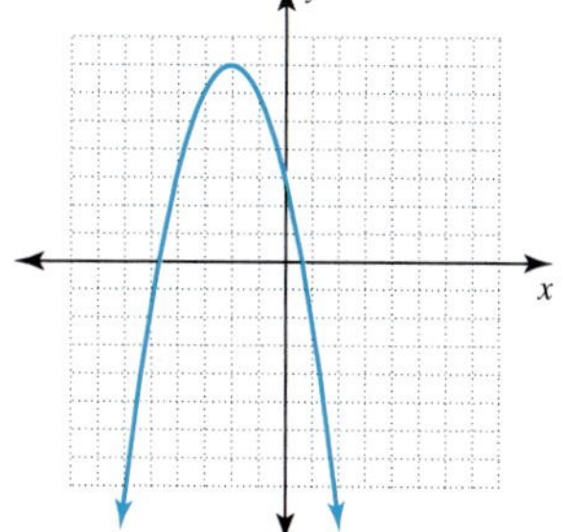

23.

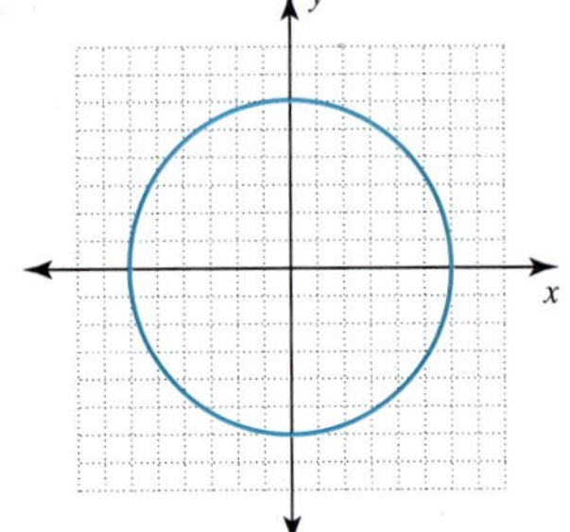

24.

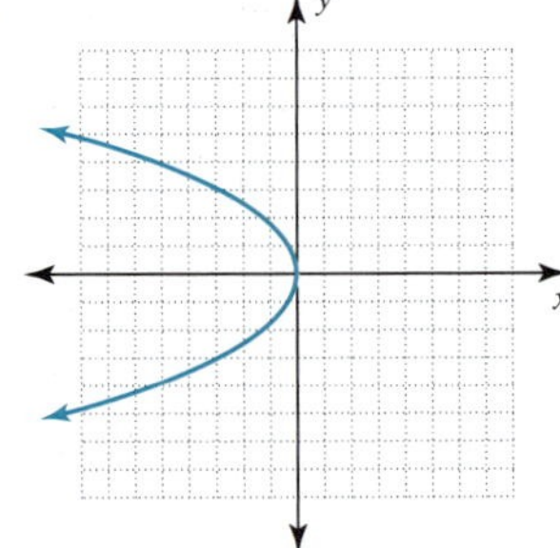

25.

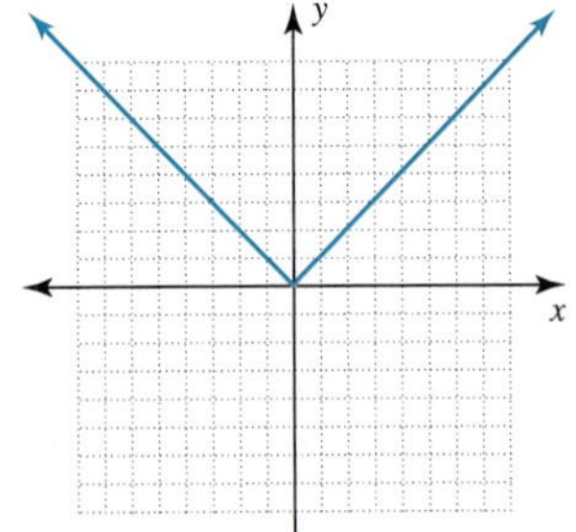

26.

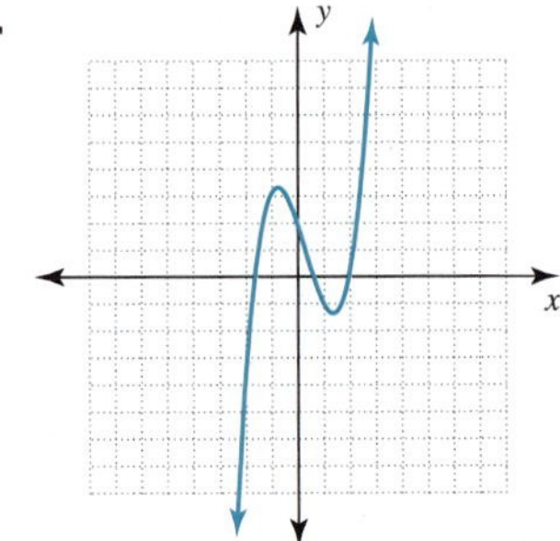

27.

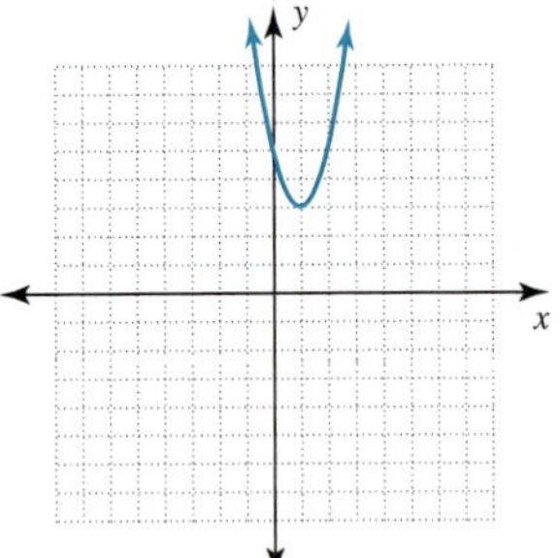

28.

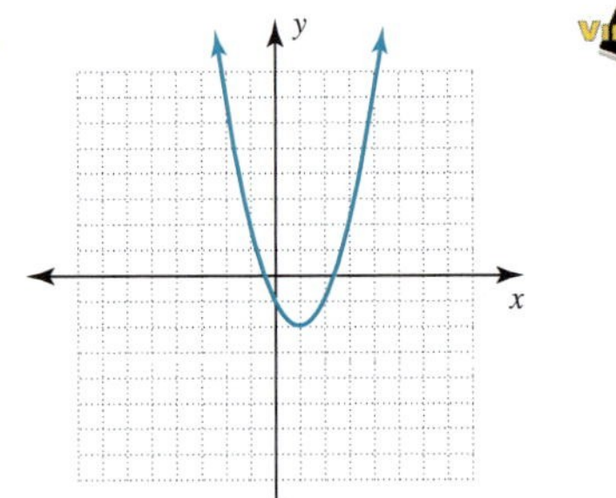

29.

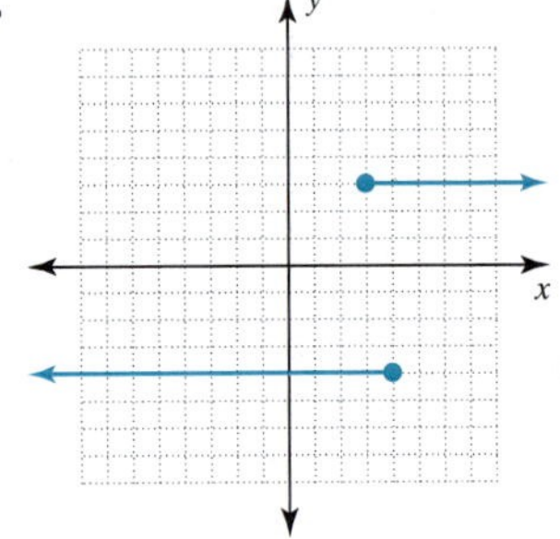

30.

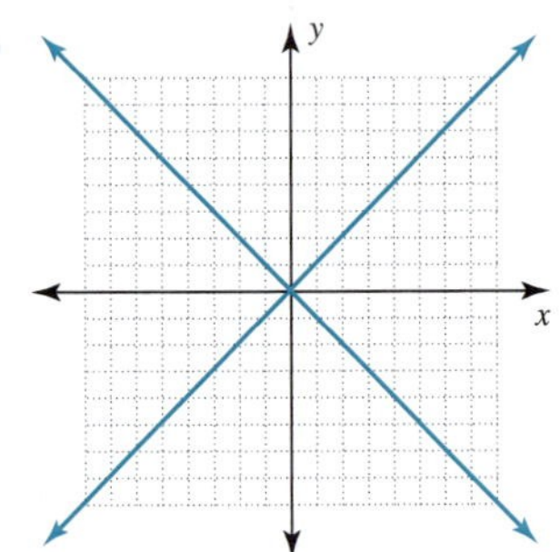

< Objective 3 >

Use the tables to find the desired values.

x	$f(x)$
−3	−8
−1	2
2	4
5	−3

x	$g(x)$
−6	1
0	3
1	3
4	5

x	$h(x)$
−4	8
−2	7
3	5
7	−4

x	$k(x)$
−5	2
−3	−4
0	2
6	−4

31. $f(5)$

32. $g(-6)$

33. $h(3)$

34. $k(-5)$

35. All x such that $f(x) = -8$

36. All x such that $g(x) = 1$

37. All x such that $g(x) = 3$

38. All x such that $k(x) = 2$

39. $k(0)$

40. $g(4)$

41. $g(1)$

42. $h(-4)$

43. All x such that $k(x) = -4$

44. All x such that $h(x) = 3$

< Objective 4 >

Use the graphs to find, or estimate, the desired values.

45.

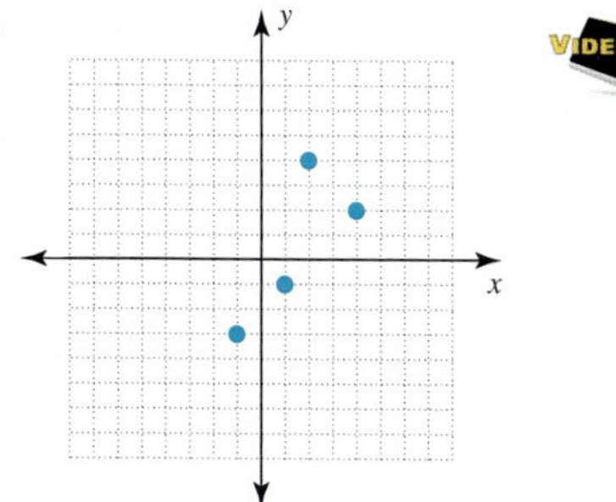

(a) Find $f(2)$.
(b) Find $f(-1)$.
(c) Find all x such that $f(x) = 2$.
(d) Find all x such that $f(x) = -1$.

46.

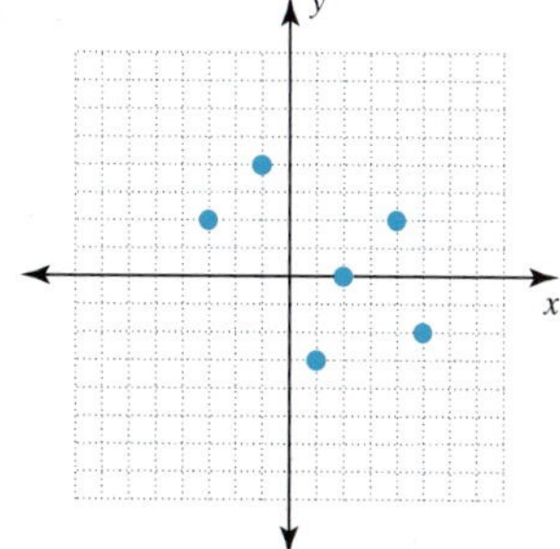

(a) Find $f(2)$.
(b) Find $f(-1)$.
(c) Find all x such that $f(x) = 2$.
(d) Find all x such that $f(x) = -3$.

47.

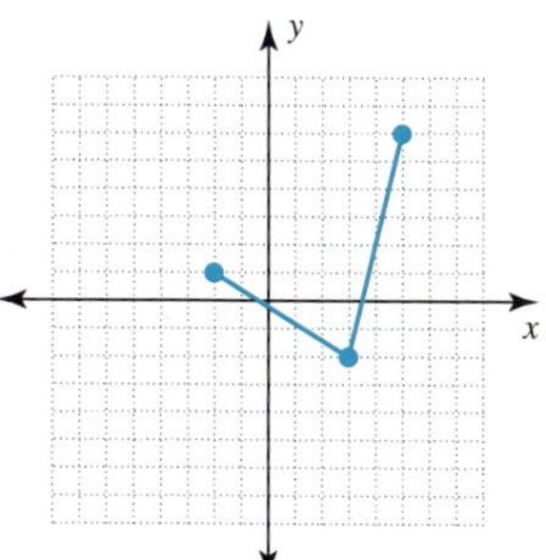

(a) Find $f(3)$.
(b) Find $f(4)$.
(c) Find all x such that $f(x) = 1$.
(d) Find all x such that $f(x) = 4$.

48.

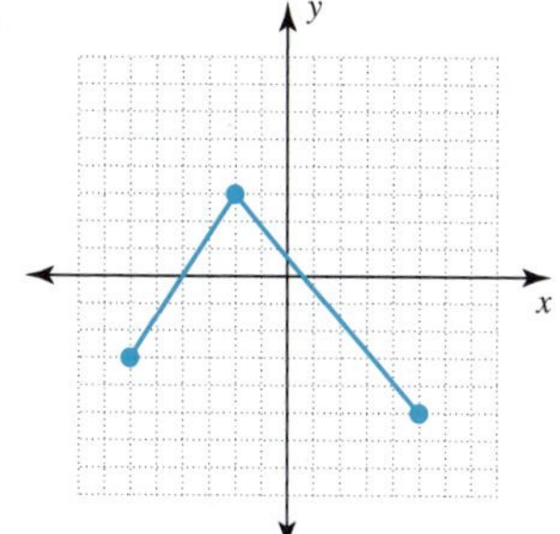

(a) Find $f(-2)$.
(b) Find $f(-5)$.
(c) Find all x such that $f(x) = 0$.
(d) Find all x such that $f(x) = -2$.

49.

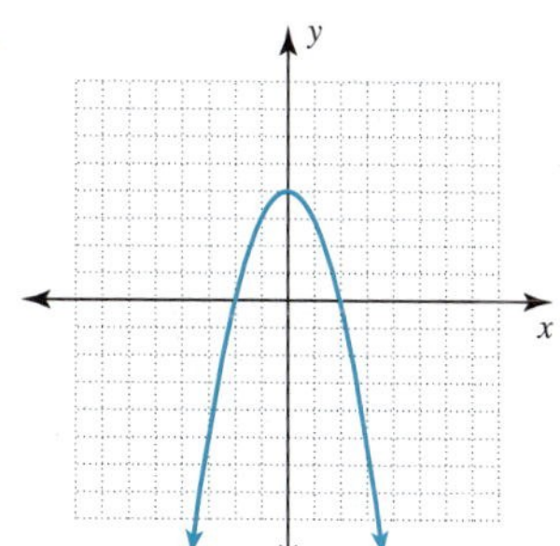

(a) Find $f(3)$.
(b) Find $f(0)$.
(c) Find all x such that $f(x) = 0$.
(d) Find all x such that $f(x) = -2$.

50.

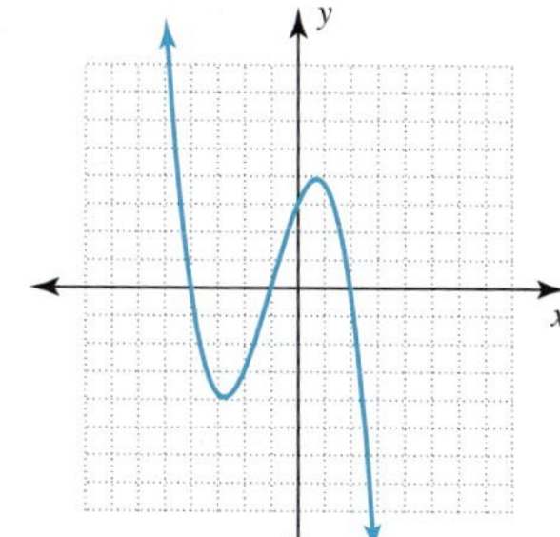

(a) Find $f(2)$.
(b) Find $f(-2)$.
(c) Find all x such that $f(x) = 3$.
(d) Find all x such that $f(x) = 5$.

Complete each statement with **always, sometimes,** *or* **never.**

51. If a vertical line passes through two points on the graph of a relation, the relation is __________ a function.

52. If a horizontal line passes through two points on the graph of a relation, the relation is __________ a function.

53. If the graph of a relation is a line that is not vertical, the relation is __________ a function.

54. If the graph of a relation is a circle, the relation is __________ a function.

Answers

1. 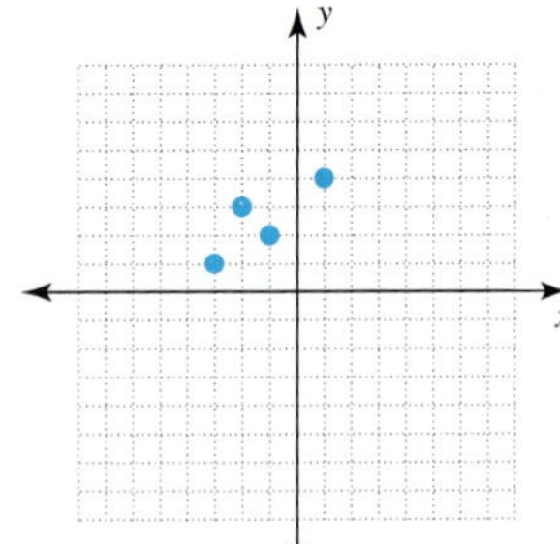

Function

3. 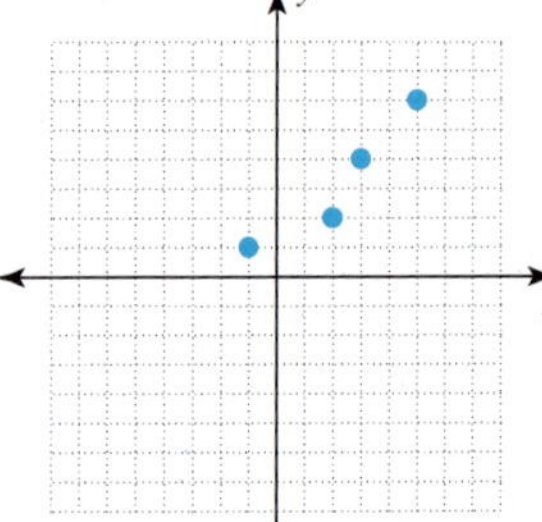

Function

5. 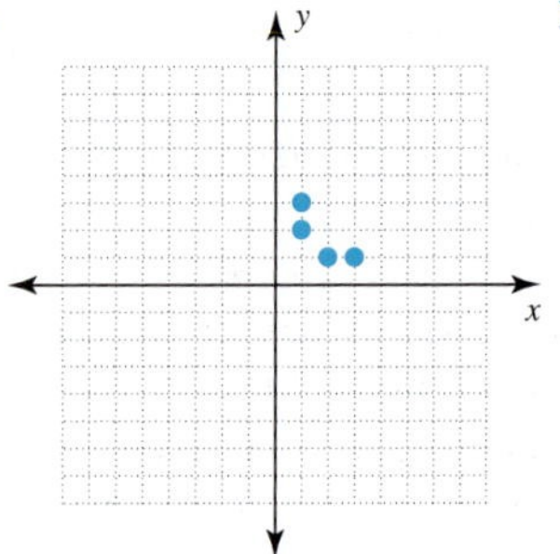

Not a function

7. Function; $D = \{-2, -1, 0, 1, 2\}$; $R = \{-1, 0, 1, 2, 3\}$ **9.** Function; $D = \{-2, -1, 0, 2, 3, 5\}$; $R = \{-1, 2, 4, 5\}$
11. Function; $D = \{x \mid -4 \le x \le 3\} = [-4, 3]$; $R = \{-2\}$ **13.** Function; $D = \{x \mid -3 \le x \le 4\} = [-3, 4]$; $R = \{y \mid -2 \le y \le 5\} = [-2, 5]$
15. Not a function; $D = \{x \mid -3 \le x \le 3\} = [-3, 3]$; $R = \{y \mid -3 \le y \le 4\} = [-3, 4]$ **17.** Not a function; $D = \{-3\}$; $R = \mathbb{R}$
19. Function; $D = \mathbb{R}$; $R = \mathbb{R}$ **21.** Function; $D = \mathbb{R}$; $R = \{y \mid y \ge -5\} = [-5, \infty)$
23. Not a function; $D = \{x \mid -6 \le x \le 6\} = [-6, 6]$; $R = \{y \mid -6 \le y \le 6\} = [-6, 6]$ **25.** Function; $D = \mathbb{R}$; $R = \{y \mid y \ge 0\} = [0, \infty)$
27. Function; $D = \mathbb{R}$; $R = \{y \mid y \ge 3\} = [3, \infty)$ **29.** Not a function; $D = \mathbb{R}$; $R = \{-4, 3\}$ **31.** -3 **33.** 5 **35.** -3 **37.** 0, 1
39. 2 **41.** 3 **43.** $-3, 6$ **45.** (a) 4; (b) -3; (c) 4; (d) 1 **47.** (a) -2; (b) 2; (c) $-2, 3.7$; (d) 4.5
49. (a) -5; (b) 4; (c) $-2, 2$; (d) $-2.5, 2.5$ **51.** never **53.** always

Definition/Procedure	Example	Reference
Formulas and Problem Solving		**Section 2.1**
Formula or **Literal Equation** An equation that expresses a relationship between two or more variables.	$a = \frac{2b + c}{3}$ is a formula or literal equation.	p. 153
Solving Formulas **Step 1** Remove any grouping symbols by applying the distributive or multiplication property. **Step 2** Multiply both sides of the equation by the LCM to clear the equation of fractions or decimals. **Step 3** Combine any like terms that appear on either side of the equation. **Step 4** Use the addition property of equality to write an equivalent equation with the term containing the desired variable on one side of the equation and all other terms on the other side. **Step 5** Use the multiplication property to write an equivalent equation with the desired variable isolated on one side and its coefficient equal to 1.	Solve for b: $a = \frac{2b + c}{3}$ $3a = 3\left(\frac{2b + c}{3}\right)$ $3a = 2b + c$ $\underline{\ -c \qquad -c}$ $3a - c = 2b$ $\frac{3a - c}{2} = b$	p. 155
Solving Applications **Step 1** Read the problem carefully. Then reread it to decide what you are asked to find. **Step 2** Choose a variable to represent one of the unknowns in the problem. Then represent each of the unknowns with an expression that uses the same variable. **Step 3** Translate the problem to the language of algebra to form an equation. **Step 4** Solve the equation. **Step 5** Answer the question and include units in your answer, when appropriate. Check your solution by returning to the original problem.		p. 156
Sets and Set Notation		**Section 2.2**
Set A set is a collection of objects.	$A = \{2, 3, 4, 5\}$ is a set.	p. 168
Elements The elements are the objects in a set.	2 is an element of set A.	p. 168
Roster Form A set is said to be in roster form if the elements are listed and enclosed in braces.	$S = \{2, 4, 6, 8\}$ is in roster form.	p. 168
Set-Builder Notation $\{x \mid x > a\}$ is read "the set of all x, where x is greater than a." $\{x \mid x < a\}$ is read "the set of all x, where x is less than a." $\{x \mid a < x < b\}$ is read "the set of all x, where x is greater than a and less than b."	$\{x \mid x < 4\}$ is written in set-builder notation.	p. 169
Interval Notation (a, ∞) is read "all real numbers greater than a." $(-\infty, b)$ is read "all real numbers less than b." (a, b) is read "all real numbers greater than a and less than b." $[a, b]$ is read "all real numbers greater than or equal to a and less than or equal to b."	$(-4, 5]$ is written in interval notation.	p. 170

Continued

Definition/Procedure	Example	Reference
Plotting the Elements of a Set on a Number Line $\{x \mid x < a\}$ indicates the set of all points on the number line to the left of a. We plot those points by using a parenthesis at a (indicating that a is not included), then a bold line to the left.	$\{x \mid x < 4\}$  The parenthesis indicates every number below the marked value (here it is 4).	*p.* 171
$\{x \mid x \geq a\}$ indicates the set of all points on the number line to the right of, and including, a. We plot those points by using a bracket at a (indicating that a is included), then a bold line to the right.	$\{x \mid x \geq -3\}$  The bracket indicates every number at or above the indicated value (-3).	*p.* 171
$\{x \mid a \leq x < b\}$ indicates the set of all points on the number line between a and b, including a. We plot those points by using an opening bracket at a and a closing parenthesis at b, then a bold line in between.	$\{x \mid 3 \leq x < 10\}$ 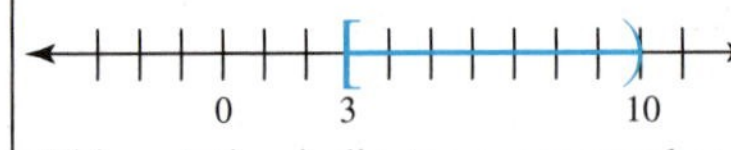 This notation indicates every number between 3 and 10, including 3 but not including 10.	*p.* 171
Set Operations **Union** $A \cup B$ is the set of elements in A or B or both. **Intersection** $A \cap B$ is the set of elements in both A and B.	$A = \{2, 3, 4, 6\}$ and $B = \{3, 4, 7, 8\}$ $A \cup B = \{2, 3, 4, 6, 7, 8\}$ $A \cap B = \{3, 4\}$	*p.* 173
Two-Variable Equations		**Section 2.3**
Solutions of Linear Equations Pairs of values that satisfy the equation. Solutions to linear equations in two variables are written as *ordered pairs*. An ordered pair has the form 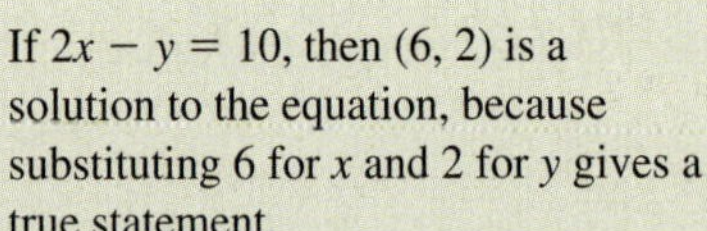 	If $2x - y = 10$, then (6, 2) is a solution to the equation, because substituting 6 for x and 2 for y gives a true statement.	*p.* 178
The Cartesian Coordinate System		**Section 2.4**
The Rectangular Coordinate System A system formed by two perpendicular axes that intersect at a point called the **origin**. The horizontal line is called the ***x*-axis.** The vertical line is called the ***y*-axis.**	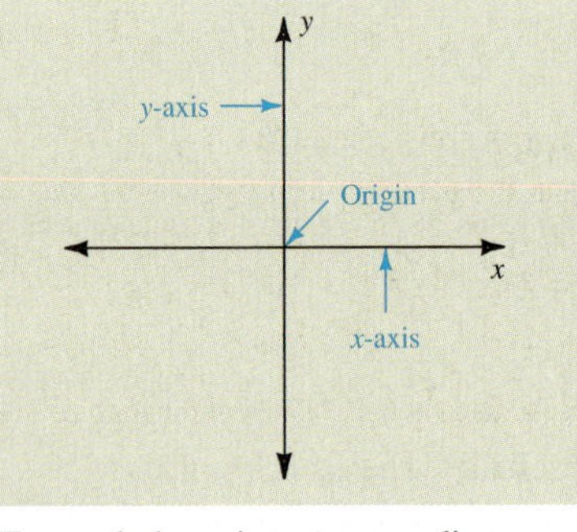 	*p.* 191
Graphing Points from Ordered Pairs The coordinates of an ordered pair allow you to associate a point in the plane with every ordered pair. **Step 1** Start at the origin. **Step 2** Move right or left according to the value of the x-coordinate: to the right if x is positive or to the left if x is negative. **Step 3** Then move up or down according to the value of the y-coordinate: up if y is positive or down if y is negative.	To graph the point corresponding to (2, 3): 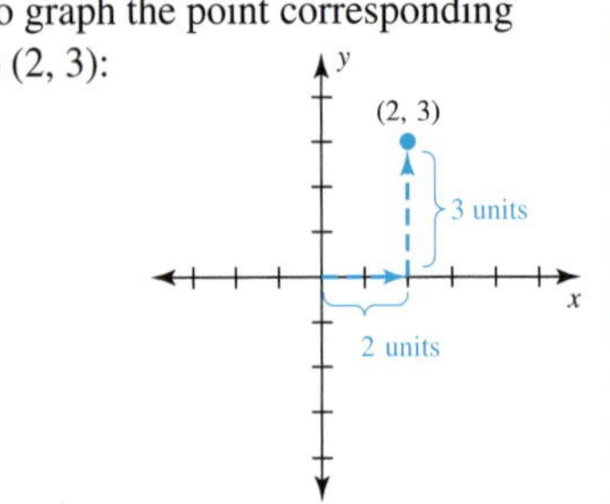 	*p.* 193

Continued

Definition/Procedure	Example	Reference
Relations and Functions		**Section 2.5**
Relation A relation is a set of ordered pairs.	The set {(1, 4), (2, 5), (1, 6)} is a relation.	*p.* 203
Domain The domain is the set of all first elements of a relation.	The domain is {1, 2}.	*p.* 203
Range The range is the set of all second elements of a relation.	The range is {4, 5, 6}.	*p.* 204
Function A function is a set of ordered pairs (a relation) in which no two first elements are equal.	{(1, 2), (2, 3), (3, 4)} is a function. {(1, 2), (2, 3), (2, 4)} is *not* a function.	*p.* 204
Tables and Graphs		**Section 2.6**
Graph The graph of a relation is the set of points in the plane that correspond to the ordered pairs of the relation.	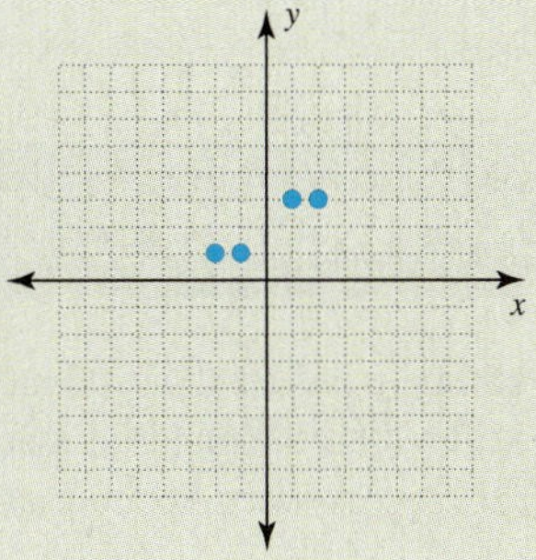	*p.* 217
Vertical Line Test The vertical line test is used to determine, from the graph, whether a relation is a function. If a vertical line meets the graph of a relation in two or more points, the relation is *not* a function. If no vertical line passes through two or more points on the graph of a relation, it is the graph of a function.	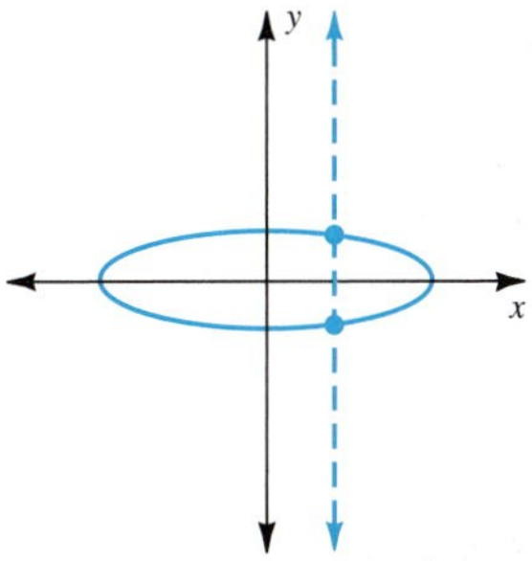 A relation—*not* a function	*p.* 218
Reading Values from Graphs We ask two questions from the graph of a function. • Given a value for x, find $f(x)$. • Find all x-values that produce a given value for $f(x)$. Given a value for x, find $f(x)$. **1.** Locate the x-value on the x-axis. **2.** Move vertically until reaching the graph. **3.** The y-value at this point is the value for $f(x)$.	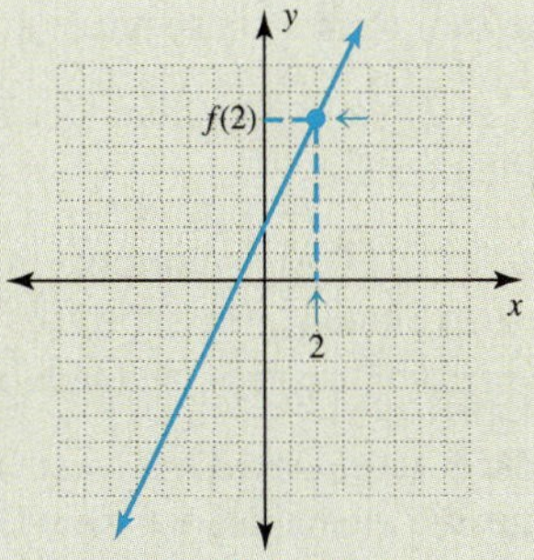 If $x = 2$, find $f(2)$. $f(2) = 6$.	*p.* 222

Continued

Definition/Procedure	Example	Reference
Find all x-values that produce a given value for $f(x)$. **1.** Locate the value for $f(x)$ on the y-axis. **2.** Move horizontally to reach each point on the graph corresponding to the y-value. **3.** The x-value of each intersection point produces the given $f(x)$.	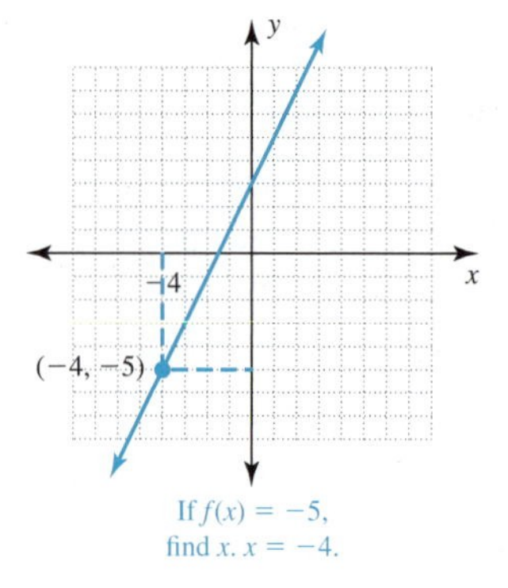 	p. 222

summary exercises :: chapter 2

This summary exercise set will help ensure that you have mastered each of the objectives of this chapter. The exercises are grouped by section. You should reread the material associated with any exercises that you find difficult. The answers to the odd-numbered exercises are in the Answers Appendix in the back of the text.

2.1 *Solve for the indicated variable.*

1. $V = LWH$ (for W)

2. $P = 2L + 2W$ (for L)

3. $ax + by = c$ (for y)

4. $A = \frac{1}{2}bh$ (for h)

5. $A = P + Prt$ (for t)

6. $m = \frac{n - p}{q}$ (for p)

Solve each application.

7. **Number Problem** The sum of 3 times a number and 7 is 25. What is the number?

8. **Number Problem** 5 times a number, decreased by 8, is 32. Find the number.

9. **Number Problem** If the sum of two consecutive integers is 85, find the two integers.

10. **Problem Solving** Larry is 2 years older than Susan, while Nathan is twice as old as Susan. If the sum of their ages is 30 years, find each of their ages.

11. **Science and Medicine** Lisa left Friday morning, driving on the freeway to visit friends for the weekend. Her trip took 4 hr. When she returned on Sunday, heavier traffic slowed her average speed by 6 mi/hr, and the trip took $4\frac{1}{2}$ hr. What was her average speed in each direction, and how far did she travel each way?

12. **Science and Medicine** At 9 A.M., David left New Orleans, Louisiana, for Tallahassee, Florida, averaging 47 mi/hr. Two hours later, Gloria left Tallahassee for New Orleans along the same route, driving 5 mi/hr faster than David. If the two cities are 391 mi apart, at what time will David and Gloria meet?

13. **Business and Finance** A firm producing running shoes finds that its fixed costs are \$3,900 per week, and its variable cost is \$21 per pair of shoes. If the firm can sell the shoes for \$47 per pair, how many pairs of shoes must be produced and sold each week for the company to break even?

14. **Business and Finance** Water Features Plus sells an indoor waterfall for \$349. Each unit costs them \$125 and the fixed costs associated with this particular waterfall come to \$4,300 per week. Find their break even point for this waterfall.

2.2 *Use the roster method to list the elements of each set.*

15. The set of all factors of 3

16. The set of all positive integers less than 7

17. The set of integers greater than -2 and less than 4

18. The set of integers between -4 and 3, inclusive

19. The set of all odd whole numbers less than 4

20. The set of all integers greater than -2 and less than 3

Plot each set on a number line.

21. $\{x|x \geq 1\}$

22. $\{x|x \leq -2\}$

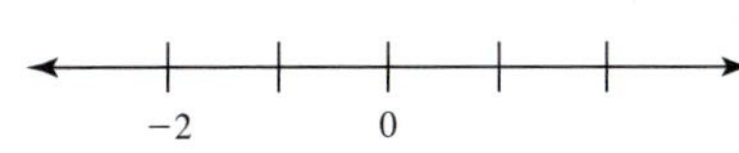

23. $\{x|-2 \leq x < 3\}$

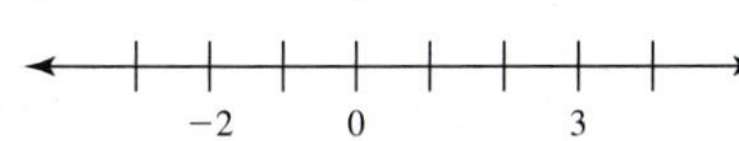

24. $\{x|-7 < x \leq -1\}$

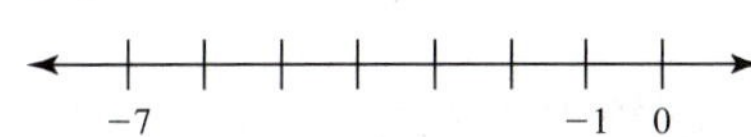

Use set-builder notation and interval notation to describe each set.

25. The set of all real numbers greater than 9

26. The set of all real numbers greater than -2 and less than 4

27. The set of all real numbers less than or equal to -5

28. The set of all real numbers between -4 and 3, inclusive

29. 

30. 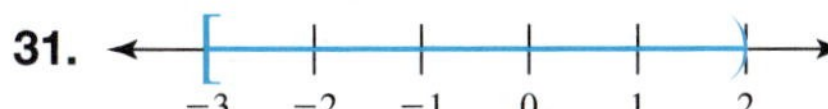

31. 

32. 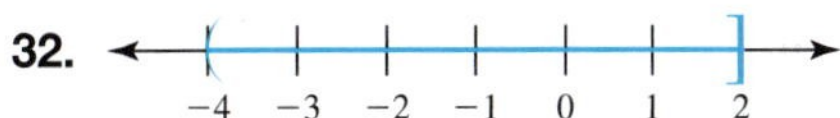

33. 

34.

Let $A = \{1, 5, 7, 9\}$ and $B = \{2, 5, 9, 11, 15\}$. List the elements in each set.

35. $A \cup B$

36. $A \cap B$

37. $B \cup \varnothing$

38. $A \cap \varnothing$

2.3 *Determine which ordered pairs are solutions to the given equations.*

39. $x - y = 6$ (6, 0), (3, 3), (3, -3), (0, -6)

40. $2x + 3y = 6$ (3, 0), (6, 2), (-3, 4), (0, 2)

2.4 *Give the coordinates of each plotted point.*

41. A

42. B

43. E

44. F

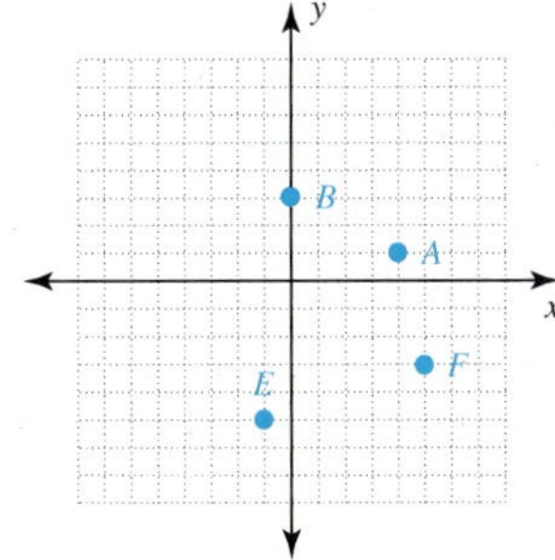

Plot each point.

45. $P(4, 0)$ **46.** $Q(5, 4)$ **47.** $T(-2, 4)$ **48.** $U(4, -2)$

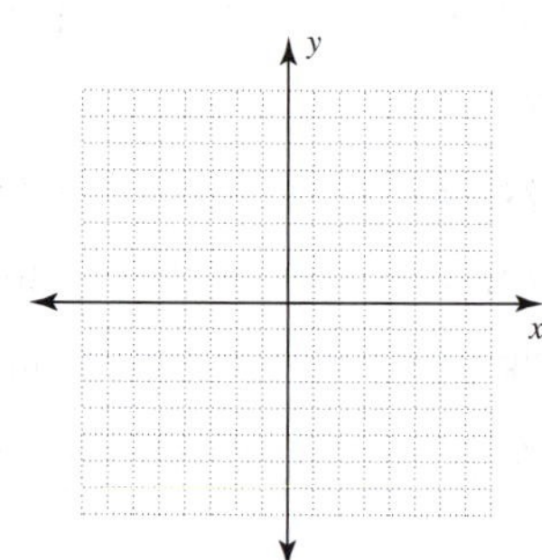

49. $(-1, 4)$ **50.** $(-1.25, 3.5)$ **51.** $(6, 3)$ **52.** $(-5, -2)$

Give the quadrant in which each point is located or the axis on which the point lies.

53. $(3, 6)$ **54.** $(-7, 5)$ **55.** $(-1, -6)$

56. $(-7, 8)$ **57.** $(-5, 0)$ **58.** $(0, -5)$

2.5 *Find the domain and range of each relation.*

59. A = {(2001, Halle Berry), (2004, Hilary Swank), (2007, Marion Cotillard), (2010, Natalie Portman)}

60. B = {(New Mexico, 5), (Nebraska, 5), (Nevada, 6), (Virginia, 13)}

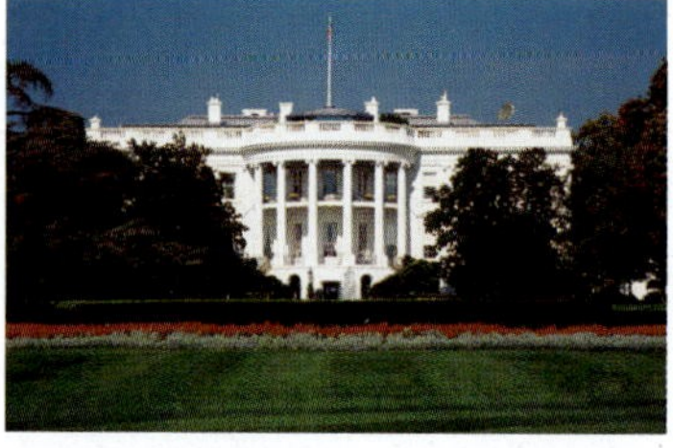

61. C = {(Brushed Nickel, \$199.99), (Jeweled Bronze, \$399.99), (Iron Amber, \$349.99), (Walnut Silver, \$129.99)}

62. E = {(2003, Jean-Pierre Serre), (2005, Peter Lax), (2009, Mikhail Gromov), (2011, John Milnor)}

63. $\{(3, 5), (4, 6), (1, 2), (8, 1), (7, 3)\}$

64. $\{(-1, 3), (-2, 5), (3, 7), (1, 4), (2, -2)\}$

65. $\{(1, 3), (1, 5), (1, 7), (1, 9), (1, 10)\}$

66. $\{(2, 4), (-1, 4), (-3, 4), (1, 4), (6, 4)\}$

Determine which relations are functions.

67. $\{(1, 3), (2, 4), (5, -1), (-1, 3)\}$

68. $\{(-2, 4), (3, 6), (1, 5), (0, 1)\}$

69. $\{(1, 2), (0, 4), (1, 3), (2, 5)\}$

70. $\{(1, 3), (2, 3), (3, 3), (4, 3)\}$

71.

x	y
−3	2
−1	1
0	3
1	4
3	5

72.

x	y
−1	3
0	2
1	3
2	4
3	5

73.

x	y
−2	3
−1	4
0	−1
1	5
−2	13

74.

x	y
−3	−4
1	0
2	3
1	5
5	2

Evaluate each function for the value specified.

75. $f(x) = x^2 - 3x + 5$; find **(a)** $f(0)$, **(b)** $f(-1)$, and **(c)** $f(1)$.

76. $f(x) = -2x^2 + x - 7$; find **(a)** $f(0)$, **(b)** $f(2)$, and **(c)** $f(-2)$.

77. $f(x) = x^3 - x^2 - 2x + 5$; find **(a)** $f(-1)$, **(b)** $f(0)$, and **(c)** $f(2)$.

78. $f(x) = -x^2 + 7x - 9$; find **(a)** $f(-3)$, **(b)** $f(0)$, and **(c)** $f(1)$.

79. $f(x) = 3x^2 - 5x + 1$; find **(a)** $f(-1)$, **(b)** $f(0)$, and **(c)** $f(2)$.

80. $f(x) = -x^3 + 3x - 5$; find **(a)** $f(2)$, **(b)** $f(0)$, and **(c)** $f(1)$.

Rewrite each equation as a function of x. Use f(x) notation in the final result.

81. $y = -2x + 5$

82. $y = 3x + 2$

83. $2x + 3y = 6$

84. $4x + 2y = 8$

85. $-3x + 4y = 12$

86. $-2x - 5y = -10$

Let $f(x) = -\frac{3}{4}x + 2$. *Evaluate, as indicated.*

87. $f(t)$

88. $f(x + 4)$

89. $f(x + h)$

90. $\dfrac{f(x + h) - f(x)}{h}$

If $f(x) = 3x - 2$, *find*

91. $f(a)$

92. $f(x - 1)$

93. $f(x + h)$

94. $\dfrac{f(x + h) - f(x)}{h}$

2.6 *Plot each set of points. Use the vertical line test to determine which sets are functions.*

95. $\{(-3, -3), (-2, -2), (2, 2), (3, 3)\}$

96. $\{(-4, -4), (-2, 4), (2, 3), (0, 4)\}$

97. $\{(-2, 1), (-2, 3), (0, 1), (1, 2)\}$

98. $\{(0, 5), (1, 6), (1, -2), (3, 4)\}$

Use the vertical line test to determine whether each graph represents a function. Find the domain and range of the relation.

99.

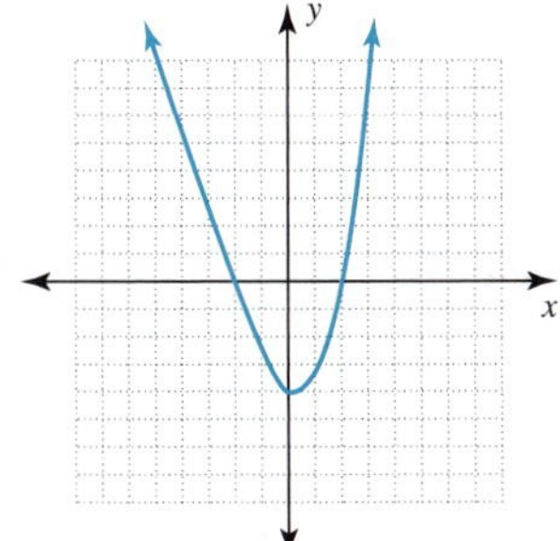

100.

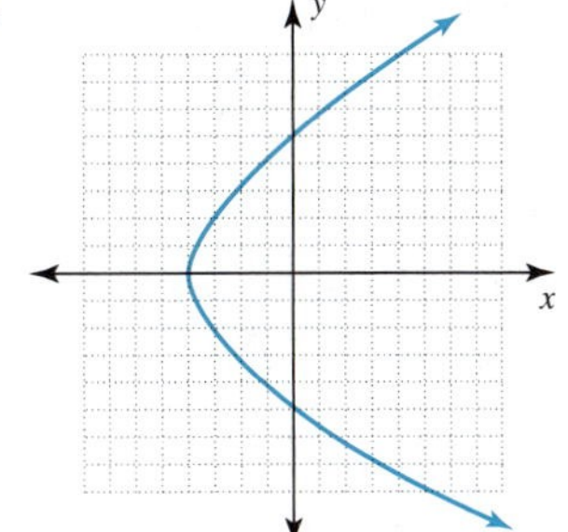

101.

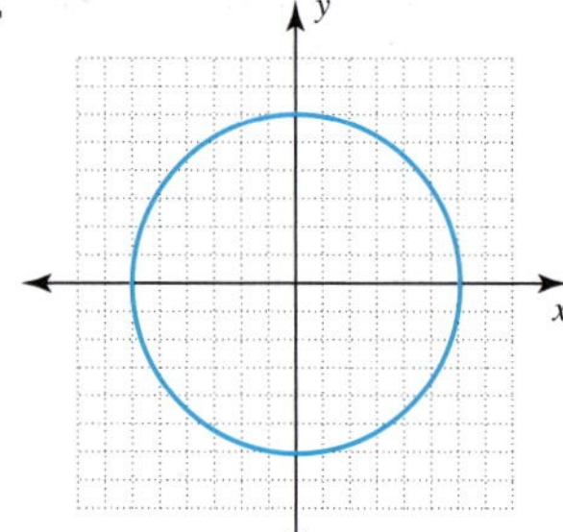

102.

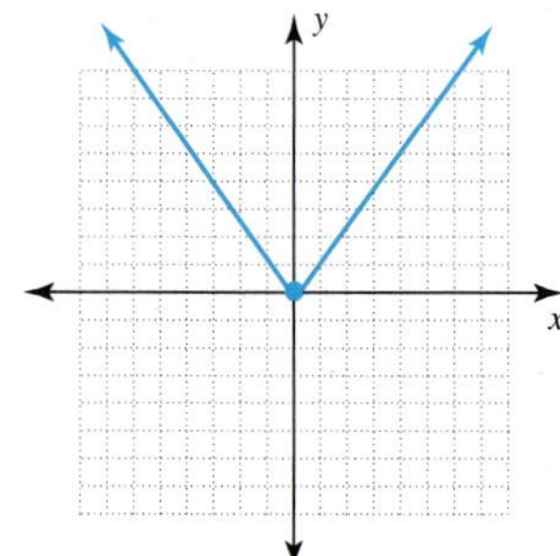

103. Complete each exercise. Estimate values where necessary.

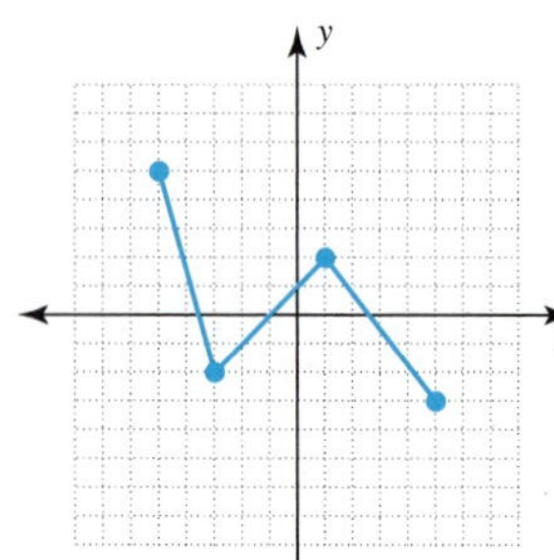

(a) Find $f(-1)$.
(b) Find $f(5)$.
(c) Find all x such that $f(x) = 5$.
(d) Find all x such that $f(x) = -3$.
(e) Find all x such that $f(x) = -1$.
(f) Find all x such that $f(x) = 2$.

104. Complete each exercise. Estimate values where necessary.

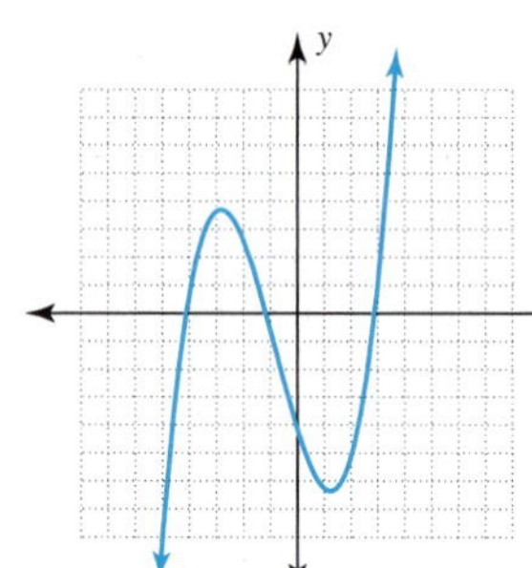

(a) Find $f(-3)$.
(b) Find $f(2)$.
(c) Find all x such that $f(x) = 7$.
(d) Find all x such that $f(x) = -8$.
(e) Find all x such that $f(x) = 3$.
(f) Find all x such that $f(x) = -4$.

105. Complete each exercise.

x	$f(x)$
−5	−2
−2	0
4	−2
8	5
12	12

(a) Find $f(-5)$.
(b) Find $f(12)$.
(c) Find all x such that $f(x) = -2$.
(d) Find all x such that $f(x) = 0$.
(e) Find all x such that $f(x) = 5$.
(f) Find all x such that $f(x) = 12$.

106. Complete each exercise.

x	$f(x)$
−3	0
−2	−3
0	−3
3	−2
7	0

(a) Find $f(-3)$.
(b) Find $f(0)$.
(c) Find $f(3)$.
(d) Find all x such that $f(x) = -3$.
(e) Find all x such that $f(x) = -2$.
(f) Find all x such that $f(x) = 0$.

chapter test 2

Use this chapter test to assess your progress and to review for your next exam. Allow yourself about an hour to take this test. The answers to these exercises are in the Answers Appendix in the back of the text.

1. Plot the set $\{x \mid -5 < x \le 3\}$ on a number line.

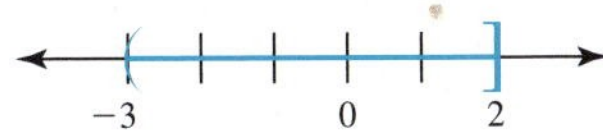

2. (a) Use set-builder notation to describe the set pictured below.
   (b) Describe the set using interval notation.

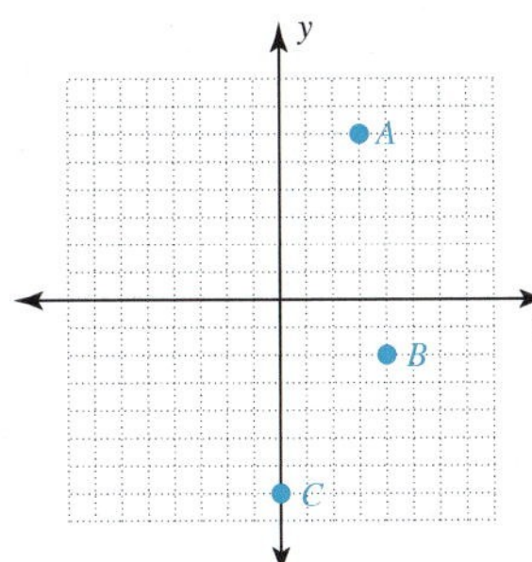

Give the coordinates of the plotted points.

3. A

4. B

5. C

Plot each point.

6. $S(1, -2)$

7. $T(0, 3)$

8. $U(-4, 5)$

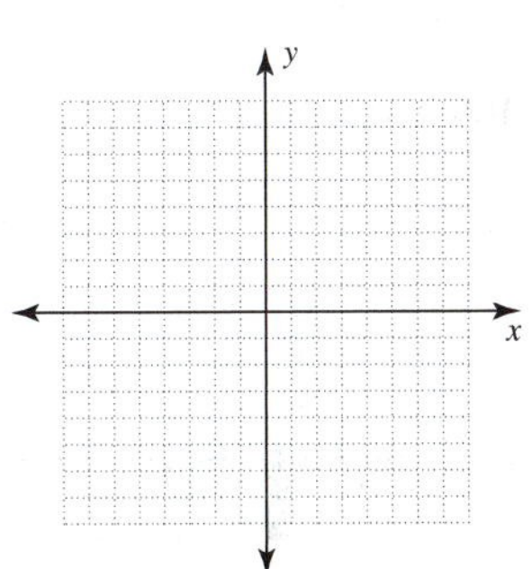

9. Plot the points on a graph and use the vertical line test to determine whether the graph represents a function.

   $\{(-1, 2), (0, 1), (2, 2), (3, -4)\}$

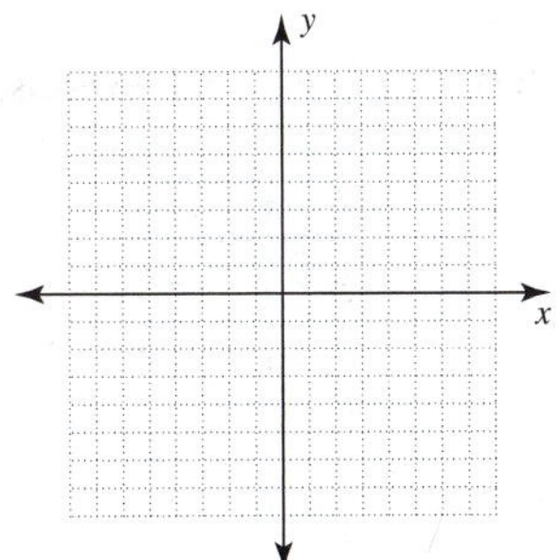

10. Determine which of the ordered pairs are solutions to the given equation.

    $4x - y = 16 \quad (4, 0), (3, -1), (5, 4)$

11. Complete each ordered pair so that it is a solution to the equation.

    $4x + 3y = 12 \quad (3, \ ), (\ , 4), (\ , 3)$

12. If $f(x) = x^2 - 5x + 6$, find **(a)** $f(0)$; **(b)** $f(-1)$; and **(c)** $f(1)$.

13. If $f(x) = -3x^2 - 2x + 3$, find **(a)** $f(-1)$; and **(b)** $f(-2)$.

14. Use the table to find the desired values.

x	y
-5	3
-3	5
0	-1
1	9
4	2
5	3

(a) $f(-5)$

(b) $f(4)$

(c) Values of x such that $f(x) = 9$

(d) Values of x such that $f(x) = 3$

15. For each set of ordered pairs, identify the domain and range.

(a) $\{(1, 6), (-3, 5), (2, 1), (4, -2), (3, 0)\}$

(b) {(United States, 101), (Germany, 65), (Russia, 63), (China, 50)}

16. Determine whether each relation is a function and identify its domain and range.

(a) $\{(2, 5), (-1, 6), (0, 2), (-4, 5)\}$

(b)

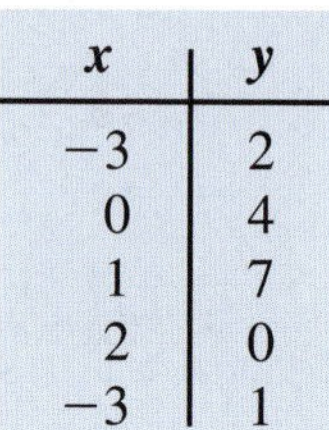

x	y
-3	2
0	4
1	7
2	0
-3	1

Determine whether each graph represents a function.

17.

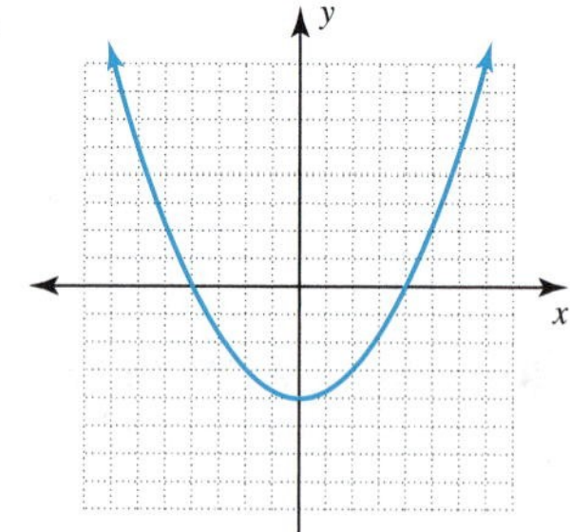

18.

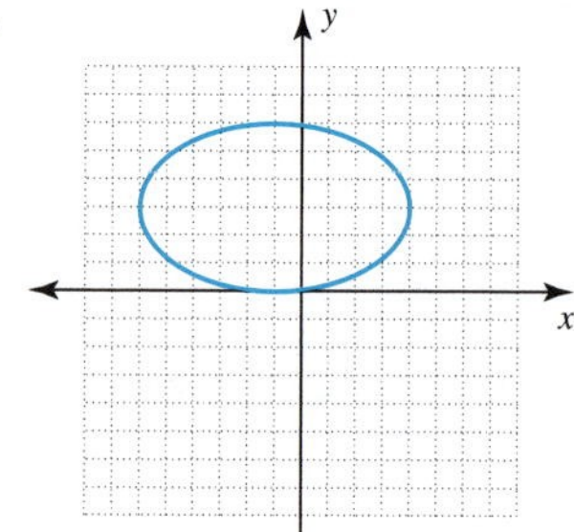

19. If $A = \{1, 2, 5\}$, and $B = \{3, 5, 7\}$, find

(a) $A \cup B$

(b) $A \cap B$

20. Rewrite the equation

$3x - 7y = 28$

as a function of x. Use $f(x)$ notation in the final result.

Solve for the indicated variable.

21. $V = \frac{1}{3}Bh$ (for B)

Solve each word problem.

22. 5 times a number, decreased by 7, is 28. What is the number?

23. Jan is twice as old as Juwan, while Rick is 5 years older than Jan. If the sum of their ages is 35 years, find each of their ages.

24. At 10 A.M., Sandra left her house on a business trip and drove an average of 45 mi/hr. One hour later, Adam discovered that Sandra had left her briefcase behind, and he began driving at 55 mi/hr along the same route. When will Adam catch up with Sandra?

25. A manufacturer produces notepads at a cost of \$0.20 each with a fixed cost of \$450 per week. If they sell the notepads for \$0.95 each, how many do they need to sell to break even?

cumulative review chapters 0–2

Use this exercise set to review concepts from earlier chapters. While it is not a comprehensive exam, it will help you identify any material that you need to review before moving on to the next chapter. The answers to these exercises are in the Answers Appendix in the back of the text.

Use the fundamental principle of fractions to simplify each fraction.

1. $\frac{56}{88}$

2. $\frac{132}{110}$

Evaluate each expression. Write your answers in simplest form.

3. $2 \cdot 3^2 - 8 \cdot 2$

4. $5(7 - 3)^2$

5. $|12 - 5|$

6. $|12| - |5|$

7. $(-7) + (-9)$

8. $\frac{17}{3} + \left(-\frac{5}{3}\right)$

9. $(-7)(-9)$

10. $(-3.2)(5)$

11. $\frac{0}{-13}$

12. $8 - 12 \div 2 \cdot 3 + 5$

13. $5 - 4^2 \div (-8) \cdot 2$

14. $\frac{4}{9} \times \frac{27}{36}$

15. $\frac{3}{4} + \frac{5}{6}$

16. $\frac{5}{6} \div \frac{25}{21}$

Evaluate each expression if $x = -2$, $y = 3$, and $z = 5$.

17. $3x - y$

18. $4x^2 - y$

19. $\frac{5z - 4x}{2y + z}$

20. $-y^2 - 8x$

Simplify and combine like terms.

21. $7x - 3y + 2(4x - 3y)$

22. $6x^2 - (5x - 4x^2 + 7) - 8x + 9$

Solve each equation.

23. $12x - 3 = 10x + 5$

24. $\frac{x - 2}{3} - \frac{x + 1}{4} = 5$

25. $4(x - 1) - 2(x - 5) = 14$

Solve each inequality.

26. $7x + 5 \le 4x - 7$

27. $-5 \le 2x + 1 < 7$

Solve each equation for the indicated variable.

28. $I = Prt$ (for r)

29. $A = \frac{1}{2}bh$ (for h)

30. $ax + by = c$ (for y)

31. $P = 2L + 2W$ (for W)

Use the graph to complete each exercise.

32. $f(-3)$

33. $f(0)$

34. Value of x for which $f(x) = 3$

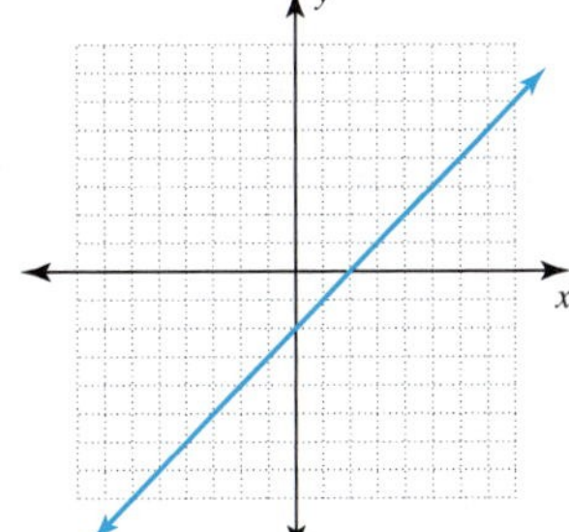

Solve each application. Be sure to show the equation used for the solution.

35. If 4 times a number decreased by 7 is 45, find that number.

36. The sum of two consecutive integers is 85. What are those two integers?

37. If 3 times an odd integer is 12 more than the next consecutive odd integer, what is that integer?

38. Michelle earns \$120 more per week than Dmitri. If their weekly salaries total \$720, how much does Michelle earn?

39. The length of a rectangle is 2 centimeters (cm) more than 3 times its width. If the perimeter of the rectangle is 44 cm, what are the dimensions of the rectangle?

40. One side of a triangle is 5 in. longer than the shortest side. The third side is twice the length of the shortest side. If the triangle's perimeter is 37 in., find the length of each leg.

Graphing Linear Functions

INTRODUCTION

Linear models describe many situations that we encounter in our classes and careers. For instance, many people earn a paycheck based on the number of hours worked. In fact, many business and finance applications are best modeled with linear functions.

In this chapter, we learn to build, graph, and describe linear functions. We use properties such as *rate-of-change* to describe important ideas like *marginal profit*. Using technology and real-world data, we look to you to make these powerful models your own. Doing so will ensure that you can use the math you learn in later settings.

CHAPTER 3 OUTLINE

3.1 Graphing Linear Functions

< 3.1 Objectives >

1 > Graph a linear equation by plotting points

2 > Graph horizontal and vertical lines

3 > Use the intercept method to graph a linear equation

4 > Use function notation to write a linear equation

In Section 2.3, we used ordered pairs to write the solutions to equations in two variables. In Section 2.4, we graphed ordered pairs in the Cartesian plane. Putting these ideas together helps us graph certain equations. Example 1 illustrates one approach to finding the graph of a linear equation.

Example 1 Graphing a Linear Equation

< Objective 1 >

Graph $x + 2y = 4$.

Step 1 Find some solutions to $x + 2y = 4$. To find solutions, we choose any convenient values for x, say $x = 0$, $x = 2$, and $x = 4$. Given these values for x, we can substitute and then solve for the corresponding value for y.

When $x = 0$, we have

$x + 2y = 4$

$(0) + 2y = 4$ Substitute $x = 0$.

$2y = 4$

$y = 2$ Divide both sides by 2.

Therefore, (0, 2) is a solution.

When $x = 2$, we have

$(2) + 2y = 4$ Substitute $x = 2$.

$2y = 2$ Subtract 2 from both sides.

$y = 1$ Divide both sides by 2.

So, (2, 1) is a solution.

When $x = 4$, $y = 0$, so (4, 0) is a solution.

NOTES

We find *three* solutions for the equation.

A table is a convenient way to display the information. It is the same as writing (0, 2), (2, 1), and (4, 0).

A handy way to show this information is in a table.

x	y
0	2
2	1
4	0

Step 2 We now plot the solutions found in step 1.

$x + 2y = 4$

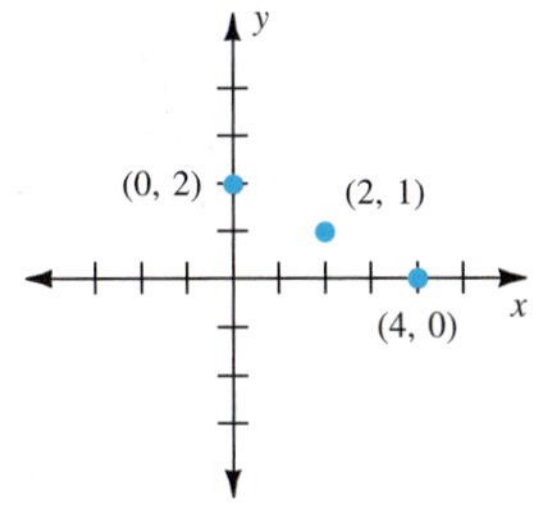

What pattern do you see? It appears that the three points lie on a straight line, which is the case.

Step 3 Draw a straight line through the three points graphed in step 2.

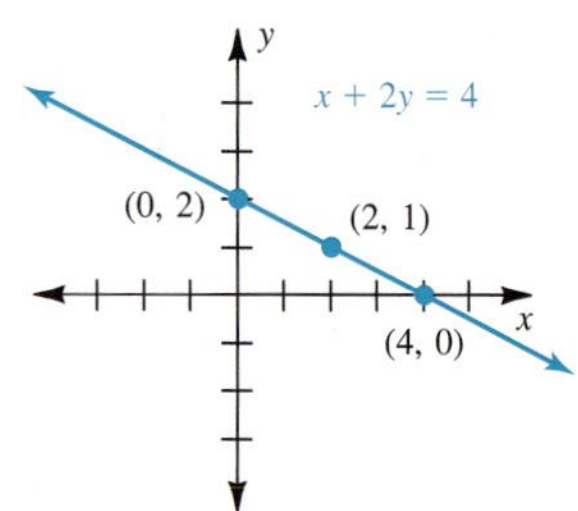

NOTE

Arrowheads on the end of the line mean that the line extends infinitely in each direction.

The line shown is the **graph** of the equation $x + 2y = 4$. It represents *all* the ordered pairs that are solutions (an infinite number) to that equation.

Every ordered pair that is a solution is a point on this line. Any point on the line represents a pair of numbers that is a solution to the equation.

NOTE

A graph is a "picture" of the solutions of an equation.

Note: Why did we find *three* solutions in step 1? Two points determine a line, so technically you need only two. The third point is a check to catch any possible errors.

Check Yourself 1

Graph $2x - y = 6$.

As mentioned in Section 2.3, an equation that can be written in the form

$$Ax + By = C \qquad \text{where } A \text{ and } B \text{ are not both } 0$$

is called a linear equation in two variables **in standard form.** The graph of this equation is a *line.* That is why we call it a *linear* equation.

Step by Step

Graphing a Linear Equation

Step 1 Find at least three solutions to the equation and put your results into a table.

Step 2 Plot the solutions found in step 1.

Step 3 Draw a straight line through the points found in step 2 to form the graph of the equation.

Example 2 Graphing a Linear Equation

Graph $y = 3x$.

NOTE

Let $x = 0$, 1, and 2, and substitute to determine the corresponding y-values. Again the choices for x are simply convenient. Other values for x would serve the same purpose.

Step 1 Some solutions are

x	y
0	0
1	3
2	6

Step 2 Plot the points.

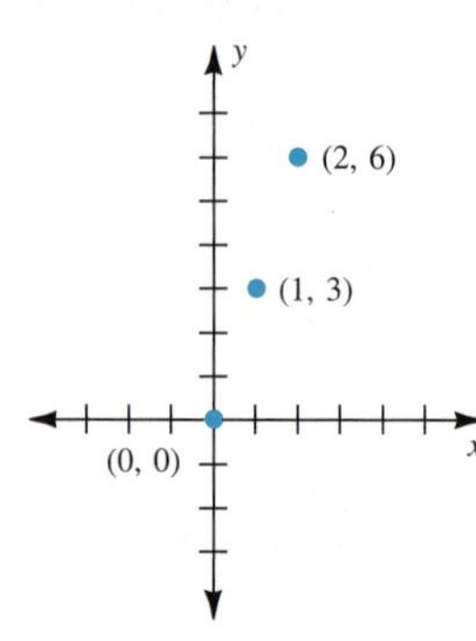

NOTE

Connecting any two of these points produces the same line.

Step 3 Draw a line through the points.

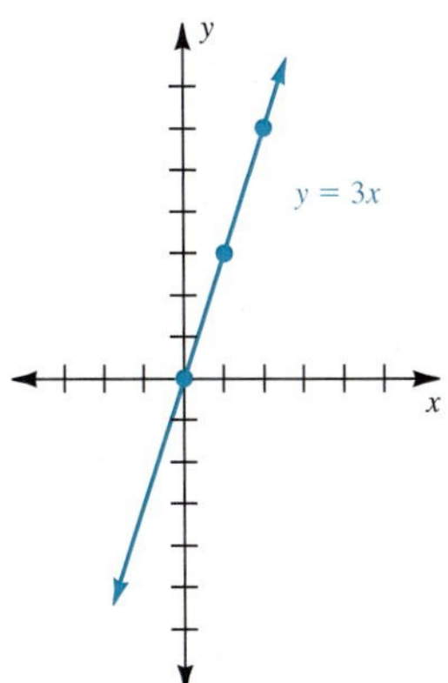

Check Yourself 2

Graph the equation $y = -2x$ after completing the table of values.

x	y
0	
1	
2	

We work through another example of graphing a line from its equation.

Example 3 Graphing a Linear Equation

Graph $y = 2x + 3$.

Step 1 Some solutions are

x	y
0	3
1	5
2	7

Step 2 Plot the points corresponding to these values.

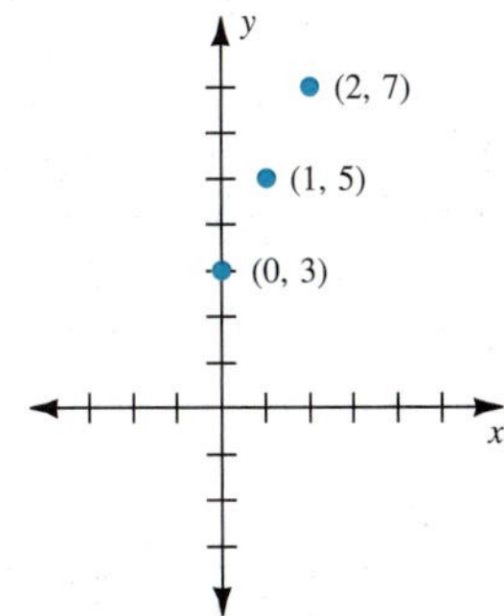

Step 3 Draw a line through the points.

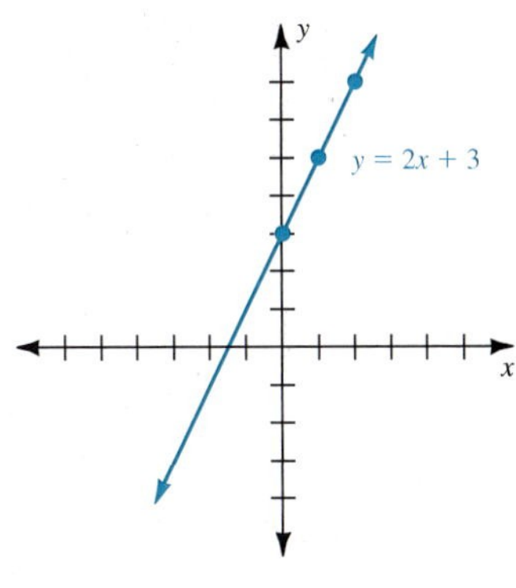

Check Yourself 3

Graph the equation $y = 3x - 2$ after completing the table of values.

x	y
0	
1	
2	

When there are fractions, you can find easy points by choosing x-values carefully. Consider Example 4.

Example 4 Graphing a Linear Equation

Graph

$$y = \frac{3}{2}x - 2$$

As before, we want to find solutions to the equation by picking convenient values for x. Because the denominator of the coefficient is 2, we should choose multiples of 2 for our x-values. For instance, here we might choose values of -2, 0, and 2 for x.

Step 1

If $x = -2$:

$$y = \frac{3}{2}x - 2$$

$$= \frac{3}{2}(-2) - 2$$

$$= -3 - 2 = -5$$

So, $(-2, -5)$ is a solution.

If $x = 0$:

$$y = \frac{3}{2}x - 2$$

$$= \frac{3}{2}(0) - 2$$

$$= 0 - 2 = -2$$

$(0, -2)$ is a solution.

If $x = 2$:

$$y = \frac{3}{2}x - 2$$

$$= \frac{3}{2}(2) - 2$$

$$= 3 - 2 = 1$$

$(2, 1)$ is a solution.

NOTE

Suppose we do *not* choose a multiple of 2, say, $x = 3$. Then

$$y = \frac{3}{2}(3) - 2$$

$$= \frac{9}{2} - 2$$

$$= \frac{5}{2}$$

$\left(3, \frac{5}{2}\right)$ is still a valid solution, but graphing a point with fractions as coordinates is more difficult.

We write these solutions in a table.

x	y
−2	−5
0	−2
2	1

Step 2 Plot the points found in step 1.

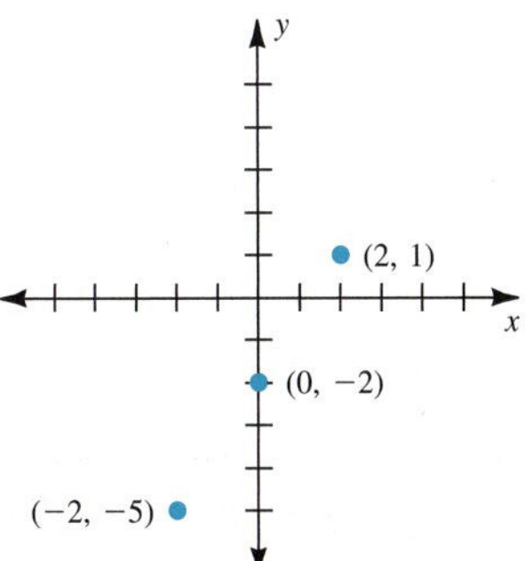

Step 3 Draw a line through the points.

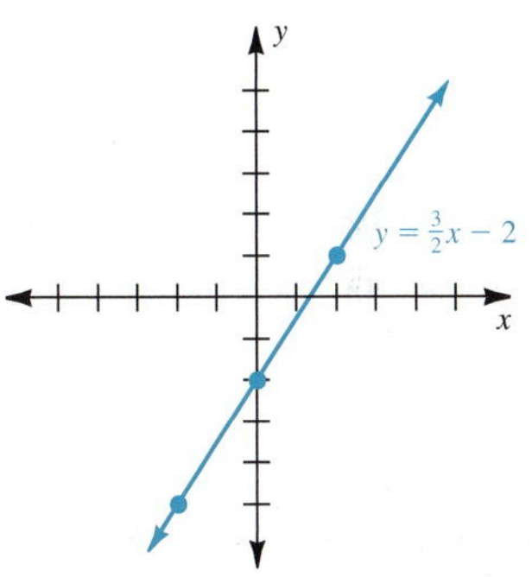

Check Yourself 4

Graph the equation $y = -\frac{1}{3}x + 3$ after completing the table of values.

x	y
−3	
0	
3	

We consider some special cases of linear equations in Example 5.

Example 5 Graphing Special Equations

< Objective 2 >

NOTE

We cannot write $x = 3$ so that y is a function of x. Therefore, this equation does not represent a function.

(a) Graph $x = 3$.

The equation $x = 3$ is equivalent to $1 \cdot x + 0 \cdot y = 3$. Let's look at some solutions.

If $y = 1$: $x + 0 \cdot (1) = 3$, $x = 3$

If $y = 4$: $x + 0 \cdot (4) = 3$, $x = 3$

If $y = -2$: $x + 0(-2) = 3$, $x = 3$

In a table,

x	y
3	1
3	4
3	−2

What do you observe? The variable x has the value 3, regardless of the value of y. Look at the graph.

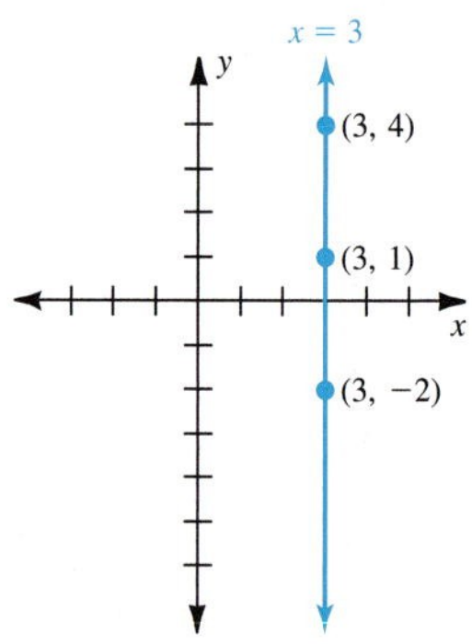

The graph of $x = 3$ is a vertical line crossing the x-axis at (3, 0).

Note that graphing (or plotting) points in this case is not really necessary. Simply recognize that the graph of $x = 3$ *must* be a vertical line (parallel to the y-axis) that intersects the x-axis at (3, 0).

(b) Graph $y = 4$.

Since $y = 4$ is equivalent to $0 \cdot x + 1 \cdot y = 4$, any value for x paired with 4 for y forms a solution. A table of values might be

x	y
−2	4
0	4
2	4

NOTE

A horizontal line represents the graph of a *constant function.* In this case, the function is written as $f(x) = 4$.

Here is the graph.

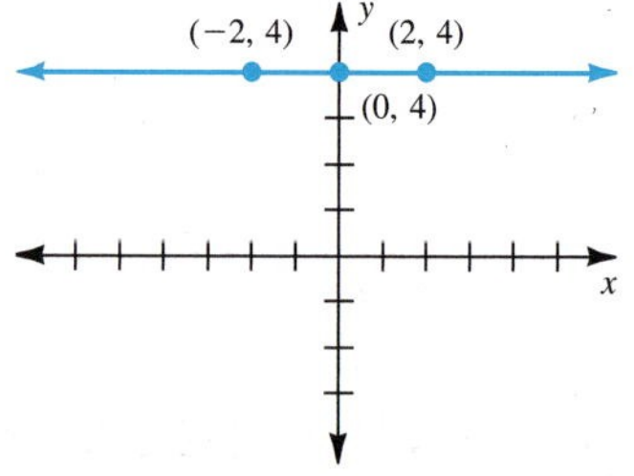

This time the graph is a horizontal line that crosses the y-axis at (0, 4). Again graphing the points is not required. The graph of $y = 4$ *must* be horizontal (parallel to the x-axis) and intersect the y-axis at (0, 4).

Check Yourself 5

(a) Graph the equation $x = -2$.
(b) Graph the equation $y = -3$.

We call the function $f(x) = b$ a constant function because the y-value does not change, even as the input x changes. On the graph, the height of the line does not change so we think of it as constant.

The vertical line produced by the linear equation $x = a$ does not represent a function. We cannot write this equation so that y is a function of x because the one x-value is a and this "maps" to every real number y.

The property box summarizes our work in Example 5.

Property

Vertical and Horizontal Lines

1. The graph of $x = a$ is a *vertical line* crossing the x-axis at $(a, 0)$.
2. The graph of $y = b$ is a *horizontal line* crossing the y-axis at $(0, b)$.

Some students prefer the **intercept method** when graphing linear equations. This method makes use of the fact that the solutions that are easiest to find are those with an x-coordinate or a y-coordinate of 0. For instance, let's graph the equation

$4x + 3y = 12$

NOTE

With practice, this all can be done mentally, which is the big advantage of this method.

First, let $x = 0$ and solve for y.

$$4x + 3y = 12$$
$$4(0) + 3y = 12$$
$$3y = 12$$
$$y = 4$$

So (0, 4) is one solution. Now let $y = 0$ and solve for x.

RECALL

Only two points are needed to graph a line. A third point is used only as a check.

$$4x + 3y = 12$$
$$4x + 3(0) = 12$$
$$4x = 12$$
$$x = 3$$

A second solution is (3, 0).

The two points corresponding to these solutions can now be used to graph the equation.

NOTE

The **intercepts** are the points where the line intersects the x- and y-axes. Here, the x-intercept has coordinates (3, 0), and the y-intercept has coordinates (0, 4).

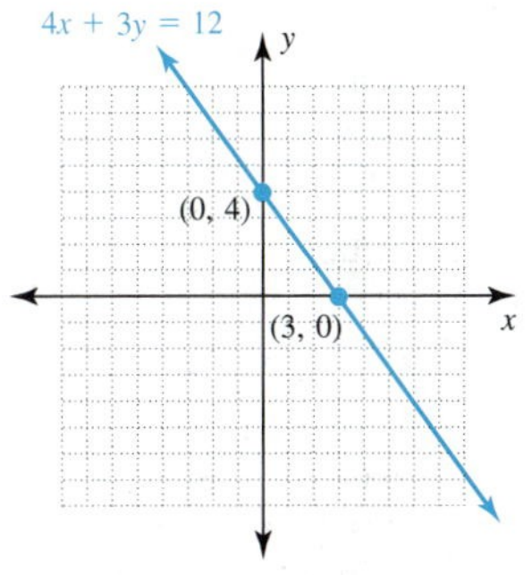

The point (3, 0) is called the ***x*-intercept,** and the point (0, 4) is the ***y*-intercept** of the graph. Using these points to draw the graph gives the name to this method. Here is another example of graphing by the intercept method.

Example 6 Using the Intercept Method to Graph a Line

< Objective 3 >

Use the intercept method to graph $3x - 5y = 15$.

To find the x-intercept, let $y = 0$.

$$3x - 5 \cdot (0) = 15$$
$$x = 5$$

The x-value of the intercept.

To find the y-intercept, let $x = 0$.

$$3 \cdot (0) - 5y = 15$$
$$y = -3$$

The y-value of the intercept.

So (5, 0) and (0, −3) are solutions to the equation, and we use the corresponding points to graph the equation.

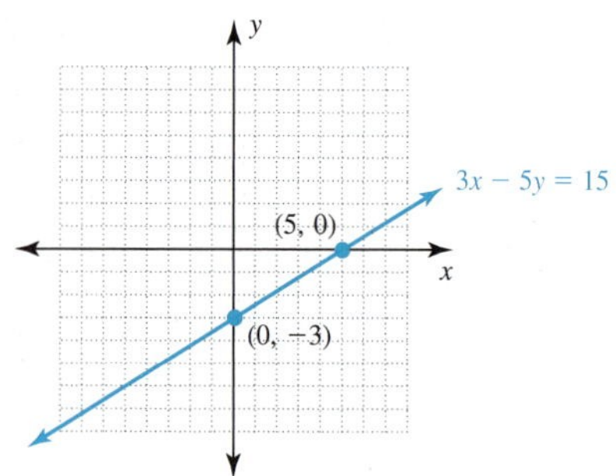

Check Yourself 6

Use the intercept method to graph $4x + 5y = 20$.

NOTE

Finding a third "checkpoint" is always a good idea.

This all looks quite easy, and for many equations it is. What are the drawbacks? For one, you don't have a third checkpoint and it is possible for errors to occur. You can, of course, still find a third point (other than the two intercepts) to be sure your graph is correct. A second difficulty arises when the x- and y-intercepts are very close to each other (or are actually the same point—the origin). For instance, if we have the equation

$$3x + 2y = 1$$

the intercepts are $\left(\frac{1}{3}, 0\right)$ and $\left(0, \frac{1}{2}\right)$. It is hard to draw a line accurately through these intercepts, so choose other solutions farther away from the origin for your points.

We summarize the steps of graphing by the intercept method for appropriate equations.

Step by Step

Graphing a Line by the Intercept Method

Step 1 To find the x-intercept: Let $y = 0$, then solve for x.

Step 2 To find the y-intercept: Let $x = 0$, then solve for y.

Step 3 Graph the x- and y-intercepts.

Step 4 Draw a straight line through the intercepts.

A third method of graphing linear equations involves **solving the equation for y.** The reason we use this extra step is that it often makes it much easier to find solutions to the equation.

Example 7 Graphing a Linear Equation

Graph $2x + 3y = 6$.

RECALL

Solving for y means that we want y isolated on one side.

Rather than finding solutions for the equation in this form, we solve for y.

$$2x + 3y = 6$$

$$3y = 6 - 2x \qquad \text{Subtract } 2x.$$

$$y = \frac{6 - 2x}{3} \qquad \text{Divide by 3.}$$

RECALL

We can write this equation in function form,

$f(x) = -\frac{2}{3}x + 2$

We solved the equation for y (or isolated y on one side). For reasons that will become apparent as you work through this chapter, we use distribution to rewrite the equation.

$$y = \frac{6 - 2x}{3}$$

$$y = \frac{6}{3} - \frac{2x}{3} \qquad \text{Distribute the denominator 3.}$$

$$y = 2 - \frac{2}{3}x \qquad \text{Simplify.}$$

$$y = -\frac{2}{3}x + 2 \qquad \text{Use the commutative property.}$$

Now find your solutions by picking convenient x-values.

NOTE

Again, to pick convenient values for x, we suggest you look at the equation carefully. Here, picking multiples of 3 for x makes the work much easier.

If $x = -3$:

$$\begin{aligned} y &= -\frac{2}{3}x + 2 \\ &= -\frac{2}{3}(-3) + 2 \\ &= 2 + 2 = 4 \end{aligned}$$

So, $(-3, 4)$ is a solution.

If $x = 0$:

$$\begin{aligned} y &= -\frac{2}{3}x + 2 \\ &= -\frac{2}{3}(0) + 2 \\ &= 0 + 2 = 2 \end{aligned}$$

$(0, 2)$ is a solution.

If $x = 3$:

$$\begin{aligned} y &= -\frac{2}{3}x + 2 \\ &= -\frac{2}{3}(3) + 2 \\ &= -2 + 2 = 0 \end{aligned}$$

$(3, 0)$ is a solution.

We can now plot the points that correspond to these solutions and graph the equation.

x	y
-3	4
0	2
3	0

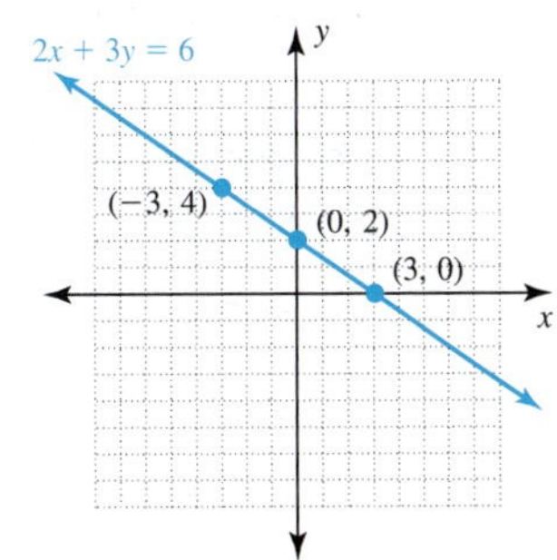

Check Yourself 7

Graph the equation $5x + 2y = 10$. Solve for y to determine solutions.

x	y
0	
2	
4	

Many students find it easier to keep themselves organized by using function notation when working with linear equations. In Chapter 2, you learned that when we solve a two-variable equation for y, we can write y as a function of x. In this case, we write

$$y = f(x)$$

One advantage to writing an equation with function notation is that it allows us to see the value we are using for x when evaluating the function.

Example 8 Graphing a Linear Function

< Objective 4 >

RECALL

To write y as a function of x, solve the equation for y and replace y with $f(x)$.

Rewrite the equation shown so that y is a function of x. Graph the function.

$$x - 2y = 6$$

Begin by solving the equation for y.

$$x - 2y = 6$$

$$x - 6 = 2y \quad \text{Subtract 6 and add } 2y \text{ to both sides.}$$

$$2y = x - 6 \quad \text{Switch sides so that the } y\text{-term is on the left.}$$

$$y = \frac{x-6}{2} \quad \text{Divide both sides by 2.}$$

$$y = \frac{1}{2}x - 3 \quad \text{Remember to distribute.}$$

$$f(x) = \frac{1}{2}x - 3$$

NOTE

Choosing even numbers for your inputs (or x-values) guarantees whole-number outputs.

Now we evaluate the function at three points to graph it.

$$f(0) = \frac{1}{2}(0) - 3 = -3 \qquad (0, -3)$$

$$f(2) = \frac{1}{2}(2) - 3 = 1 - 3 = -2 \qquad (2, -2)$$

$$f(4) = \frac{1}{2}(4) - 3 = 2 - 3 = -1 \qquad (4, -1)$$

Finally, we plot the three points and draw the line through them.

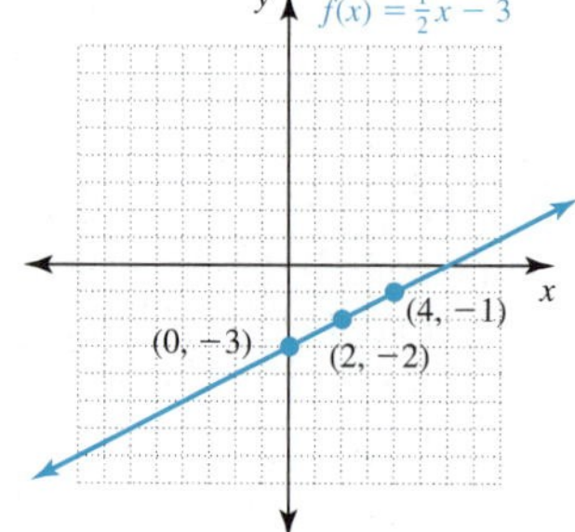

Check Yourself 8

Rewrite the equation shown so that y is a function of x and graph the function.

$6x + 2y = 4$

One important reason to solve a linear equation for y is so that we can analyze it with a graphing calculator. In order to enter an equation into the [Y=] menu, we need to isolate y because your calculator needs to work with functions.

Example 9 Using a Graphing Calculator

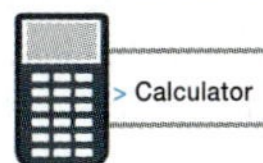

NOTES

A graphing calculator needs to "think" of the equation as a function so it must look like "*y* is a function of *x*."

A good way to enter fractions is to enclose them in parentheses.

Use a graphing calculator to graph the equation

$2x + 3y = 6$

In Example 7, we solved this equation for y to form the equivalent equation

$$y = -\frac{2}{3}x + 2$$

Enter the right side of the equation into the [Y=] menu in the Y_1 field, and then press the [GRAPH] key.

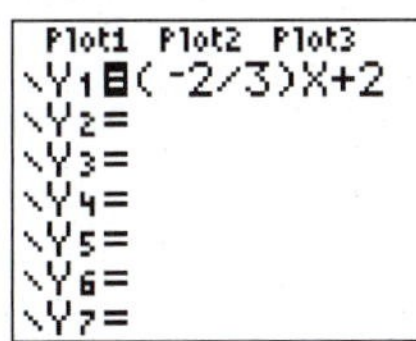

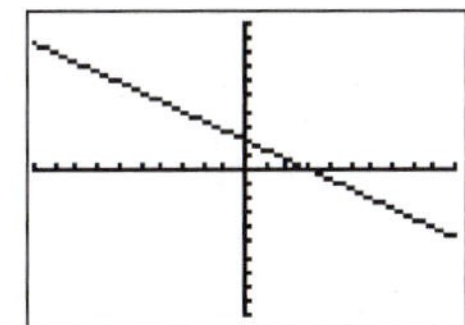

Check Yourself 9

Use a graphing calculator to graph the equation

$5x + 2y = 10$

When we scale the axes, it is important to include numbers on the axes at convenient grid lines. If we set the axes so that part of an axis is removed, we may include a mark to indicate this. Both of these situations are illustrated in Example 10.

Example 10 Graphing in Nonstandard Windows

NOTE

In business, the constant, 2,500, is called the **fixed cost.** The slope, 45, is referred to as the **marginal cost.**

The cost y to produce x Blu-ray disc players is given by the equation $y = 45x + 2{,}500$. Graph the cost equation, with appropriately scaled and set axes.

The y-intercept is (0, 2,500). We create a table to find more points.

x	10	20	30	40	50
y	2,950	3,400	3,850	4,300	4,750

NOTE

We can also graph this with a graphing calculator.

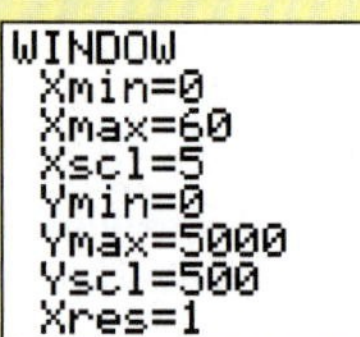

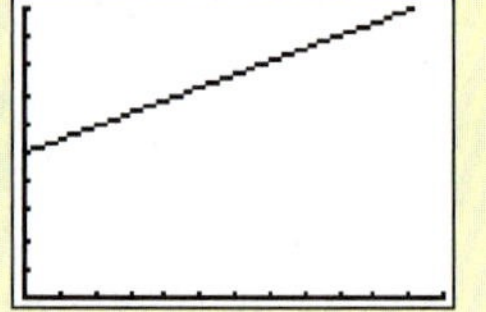

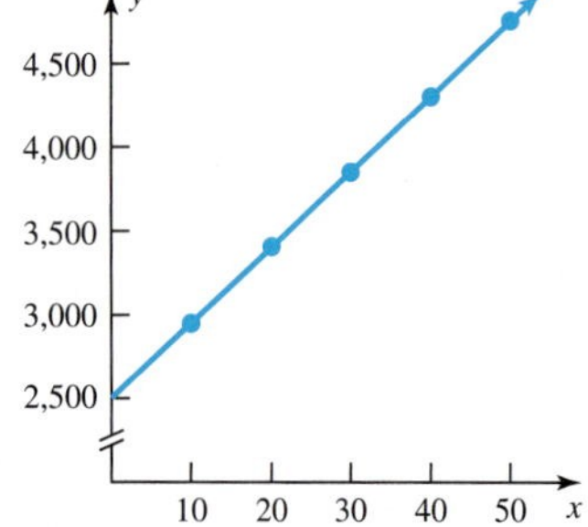

We removed part of the y-axis.

Check Yourself 10

Graph the cost equation given by $y = 60x + 1{,}200$, with appropriately scaled and set axes.

Here is an application from the field of medicine.

Example 11 A Health Sciences Application

NOTE

The domain (the set of possible values for A) is restricted to positive integers.

The arterial oxygen tension P_aO_2, in millimeters of mercury (mm Hg), of a patient can be estimated based on the patient's age A, in years. If the patient is lying down, the equation $P_aO_2 = 103.5 - 0.42A$ is used to determine arterial oxygen tension. Graph this equation, using appropriately scaled and set axes.

We begin by creating a table. Using a calculator here is very helpful.

A	0	10	20	30	40	50	60	70	80
P_aO_2	103.5	99.3	95.1	90.9	86.7	82.5	78.3	74.1	69.9

RECALL

We can use the table feature to determine a reasonable viewing window.

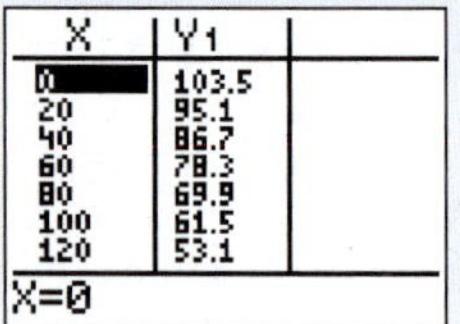

We can also graph this with a graphing calculator.

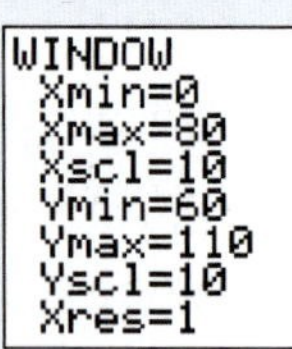

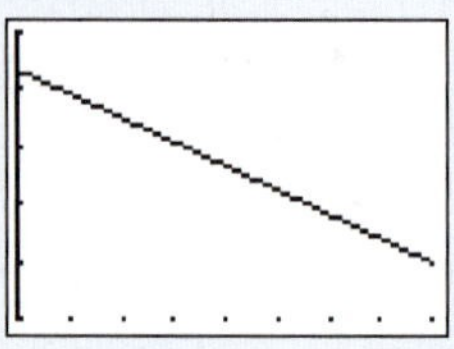

Seeing these values allows us to decide on the vertical axis scaling. We scale from 60 to 110, and include a mark to show a break in the axis. We estimate the locations of these coordinates, and draw the line.

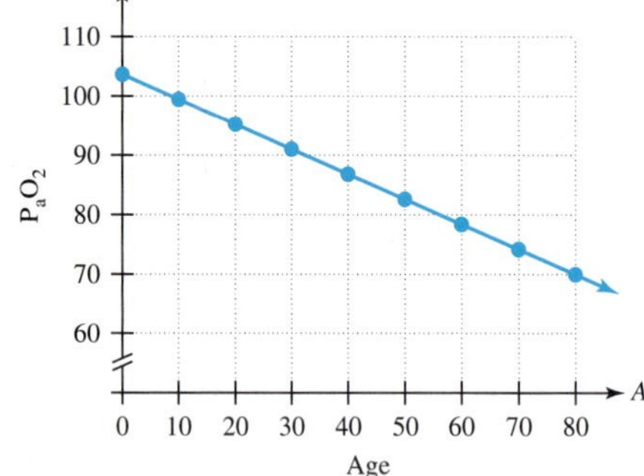

Check Yourself 11

The arterial oxygen tension P_aO_2, in millimeters of mercury (mm Hg), of a patient can be estimated based on the patient's age A, in years. If the patient is seated, the equation $P_aO_2 = 104.2 - 0.27A$ is used to approximate arterial oxygen tension. Graph this equation, using appropriately scaled and set axes.

Check Yourself ANSWERS

1.

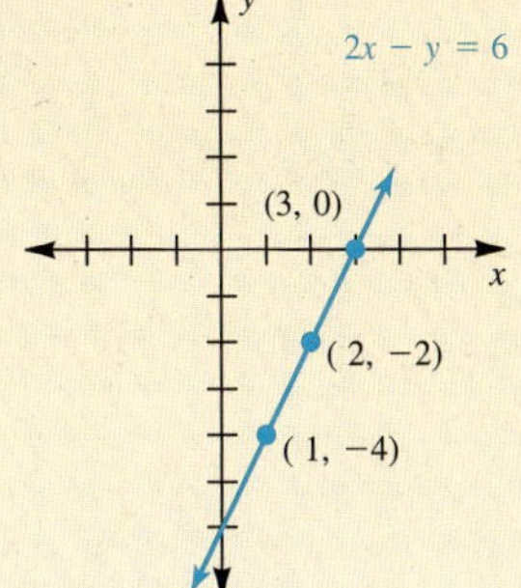

2.

x	y
0	0
1	−2
2	−4

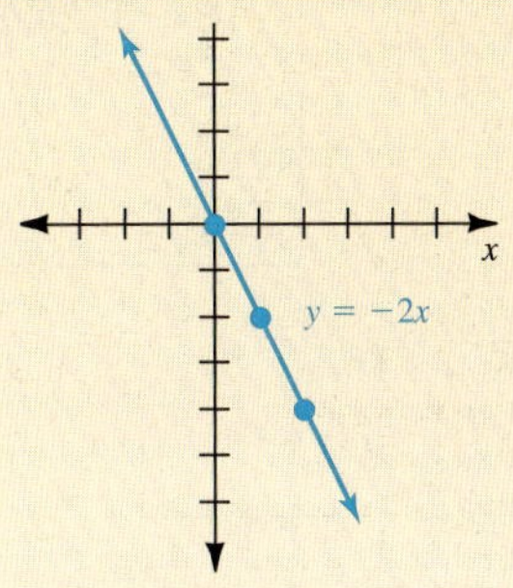

3.

x	y
0	−2
1	1
2	4

4.

x	y
−3	4
0	3
3	2

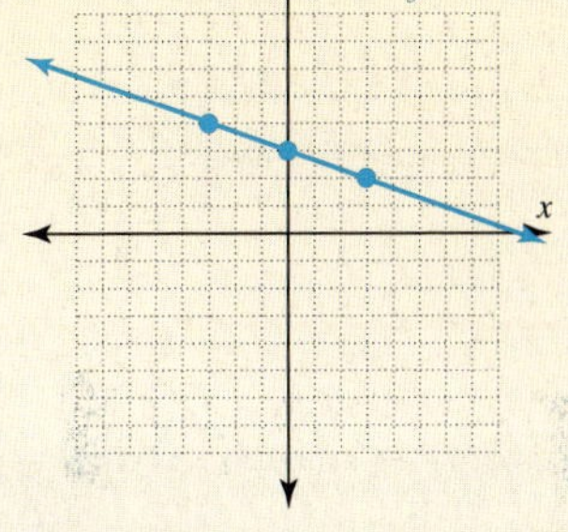

5. (a)

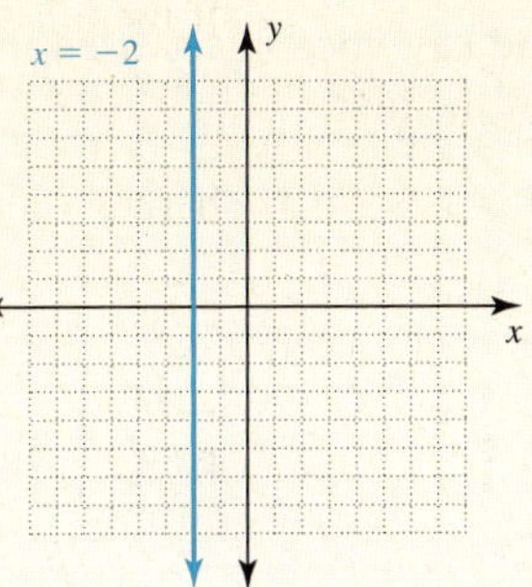

(b)

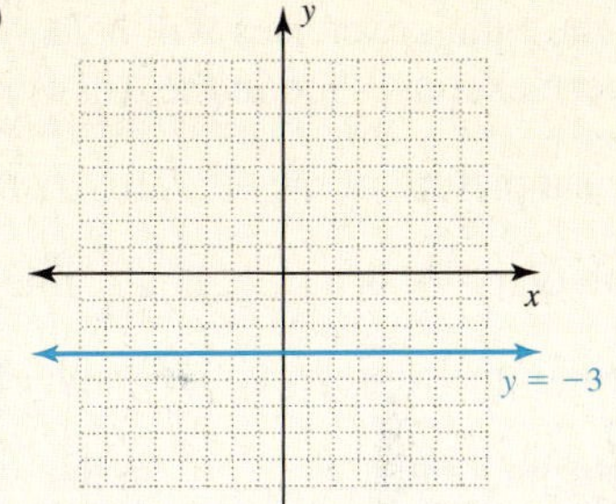

6.

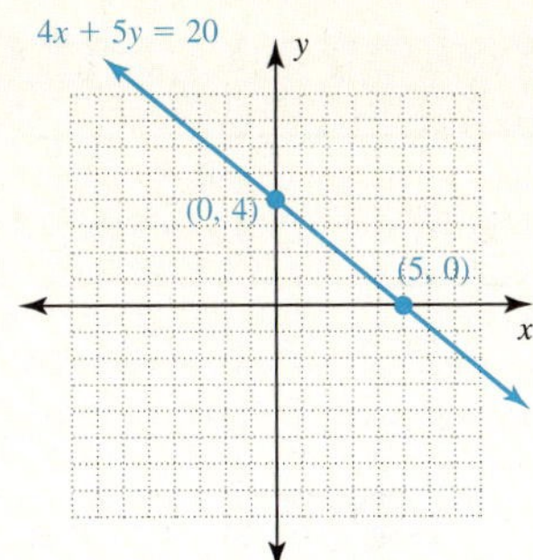

7.

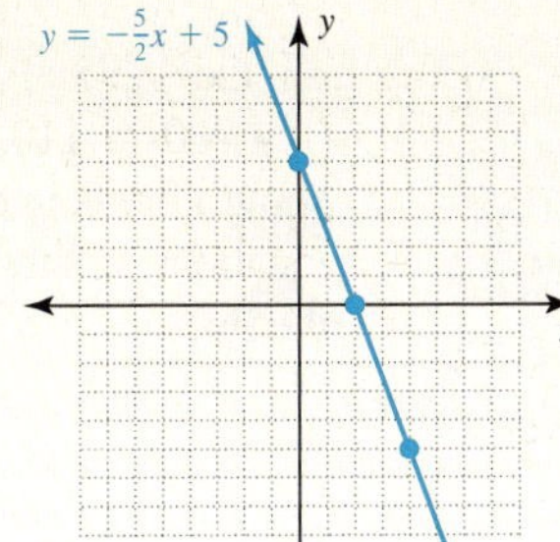

x	y
0	5
2	0
4	−5

8. $f(x) = -3x + 2$

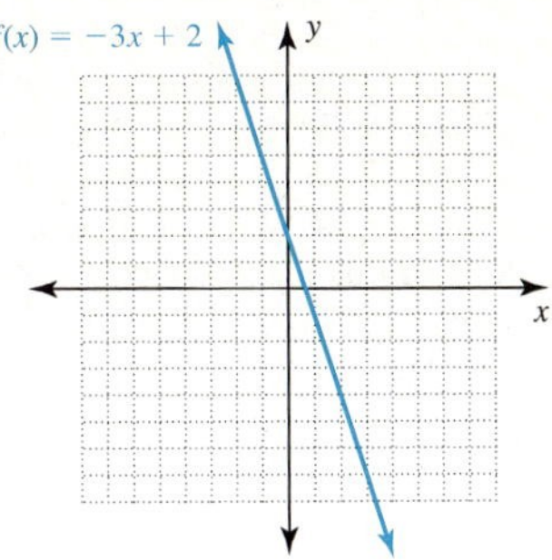

9.

```
Plot1 Plot2 Plot3
\Y1■(-5/2)X+5
\Y2=
\Y3=
\Y4=
\Y5=
\Y6=
\Y7=
```

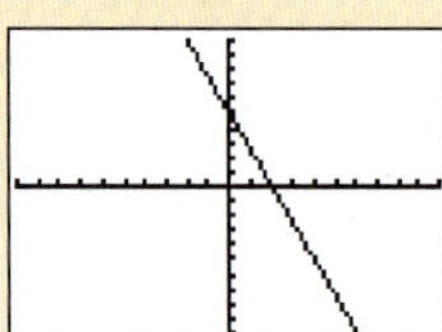

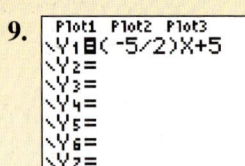

10.

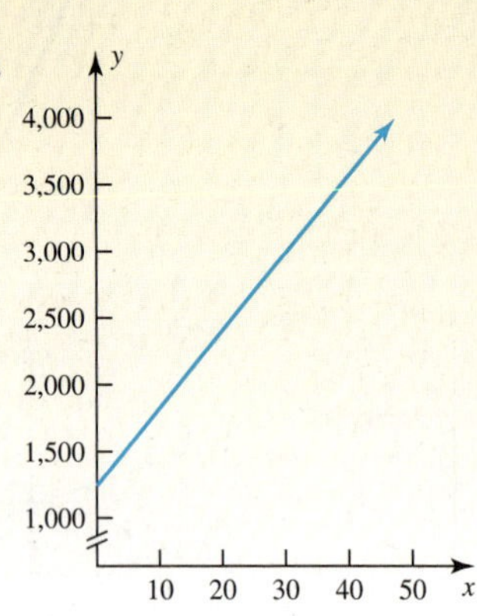

11.

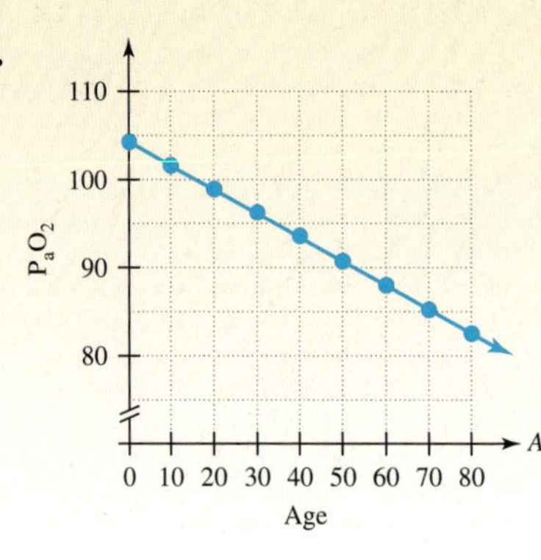

Reading Your Text

These fill-in-the-blank exercises will help you understand some of the key vocabulary used in this section. The answers to these exercises are in the Answers Appendix in the back of the text.

(a) A graph is a picture of the ________ to an equation.

(b) The graph of $x = a$ is a ________ line.

(c) A horizontal line represents the graph of a ________ function.

(d) The x-coordinate is ________ at the y-intercept of a graph.

3.1 exercises

Skills Calculator/Computer Career Applications Above and Beyond

< Objectives 1–3 >

Graph each equation.

1. $x + y = 6$

2. $x - y = 5$

3. $x - y = -3$ 

4. $x + y = -3$

5. $3x + y = 6$

6. $x - 2y = 6$

7. $3x + y = 0$

8. $2x - y = 4$

9. $x + 4y = 8$

10. $2x - 3y = 6$

11. $y = 3x$

12. $y = -4x$

13. $y = 2x - 1$

14. $y = 2x + 5$

15. $y = -3x + 1$

16. $y = -3x - 3$

17. $y = \frac{1}{5}x$

18. $y = -\frac{1}{4}x$

19. $y = \frac{2}{3}x - 3$

20. $y = \frac{3}{4}x + 2$

21. $x = -3$

22. $y = -3$

23. $y = 1$

24. $x = 4$

25. $x - 2y = 4$

26. $6x + y = 6$

27. $5x + 2y = 10$

28. $2x + 3y = 6$

29. $3x + 5y = 15$

30. $4x + 3y = 12$

< Objective 4 >

Solve each equation for y, write the equation in function form, and graph the function.

31. $x + 3y = 6$

32. $x - 2y = 6$

33. $3x + 4y = 12$

34. $2x - 3y = 12$

35. $5x - 4y = 20$

36. $7x + 3y = 21$

Write an equation that describes each relationship between x and y.

37. y is twice x.

38. y is 3 times x.

39. y is 3 more than x.

40. y is 2 less than x.

41. y is 3 less than 3 times x.

42. y is 4 more than twice x. 

43. The difference of x and the product of 4 and y is 12.

44. The difference of twice x and y is 6.

Graph each pair of equations on the same grid. Give the coordinates of the point where the lines intersect.

45. $x + y = 4$
$x - y = 2$

46. $x - y = 3$
$x + y = 5$

47. **Business and Finance** The function $f(x) = 0.10x + 200$ describes the amount of winnings a group earns for collecting plastic bottles in a recycling contest. Sketch the graph of the line.

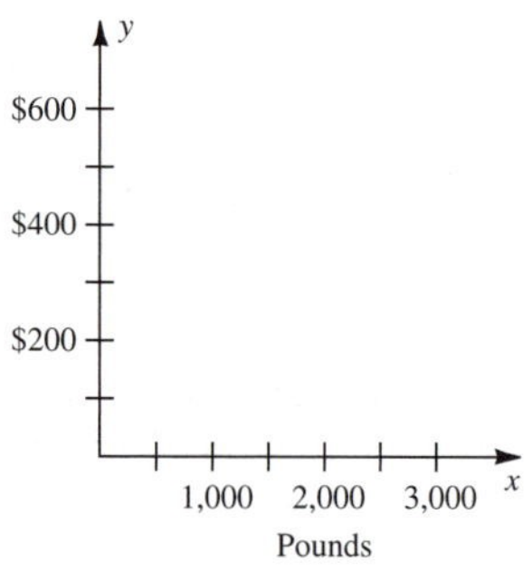

48. **Business and Finance** In exercise 47, the contest sponsor will award a prize only if the winning group in the contest collects 100 lb of bottles or more. Use your graph to determine the minimum prize possible.

49. **Business and Finance** A high school class wants to raise some money by recycling newspapers. They decide to rent a truck for a weekend and to collect the newspapers from homes in the neighborhood. At the time, the market price for recycled newsprint is $15 per ton. The function $f(x) = 15x - 100$ describes the amount of money the class will make, where $f(x)$ is the amount of money made in dollars, x is the number of tons of newsprint collected, and $100 is the cost to rent the truck.

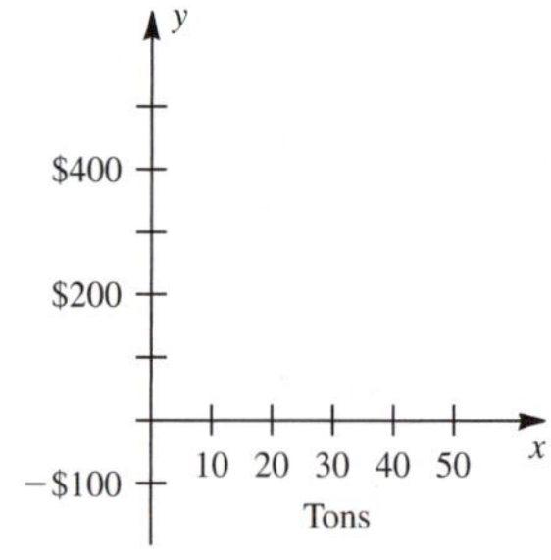

(a) Draw a graph that represents the relationship between newsprint collected and money earned.

(b) The truck costs the class $100. How many tons of newspapers must the class collect to break even on this project?

(c) If the class members collect 16 tons of newsprint, how much money will they earn?

(d) Six months later the price of newsprint is $17 dollars per ton, and the cost to rent the truck is $125. Construct a function describing the amount of money the class might make at that time.

50. **Business and Finance** The cost of producing x items is given by $C(x) = mx + b$, where b is the fixed cost and m is the marginal cost (the cost of producing one additional item).

(a) If the fixed cost is $40 and the marginal cost is $10, write the cost function.

(b) Graph the cost function.

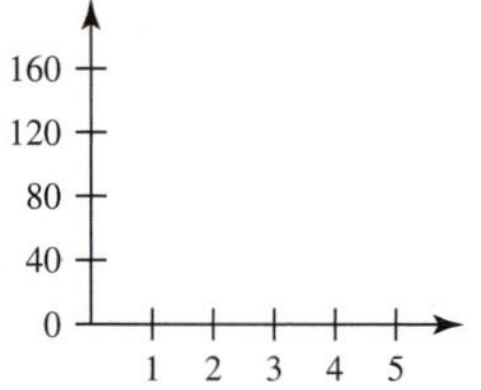

(c) The revenue generated from the sale of x items is given by $R(x) = 50x$. Graph the revenue function on the same set of axes as the cost function.

(d) How many items must be produced for the revenue to equal the cost (the break-even point)?

51. **Business and Finance** The insurance to rent a compact automobile comes to \$12 per day plus 8¢ per mile. The cost of the insurance C and the number of miles driven per day s are related by the equation

$C = 0.08s + 12$

Graph the relationship between C and s. Be sure to select appropriate scaling for the C and s axes.

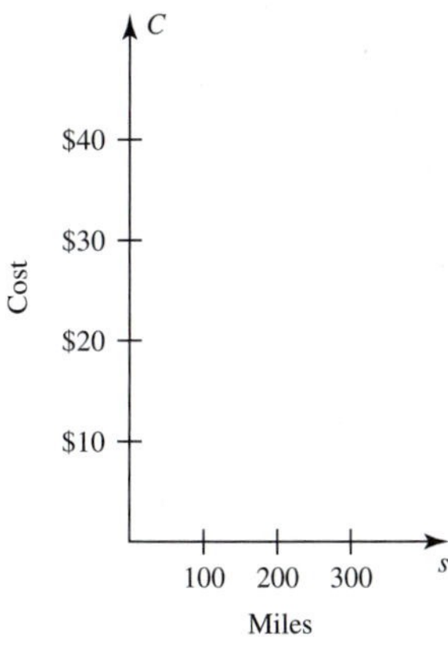

52. **Business and Finance** A bank charges \$8 per month to maintain a checking account. Each check costs 4¢. The monthly cost of an account C and the number of checks written per month n are related by the equation

$C = 0.04n + 8$

Graph the relationship between C and n. 

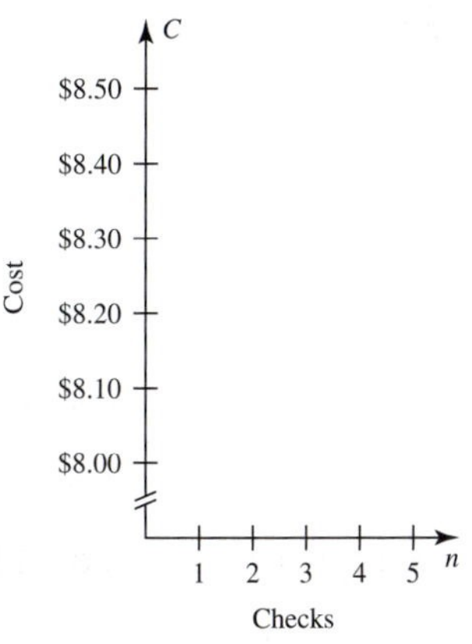

53. **Business and Finance** A college charges tuition based on a pattern. Tuition is \$35 per credit-hour plus a fixed student fee of \$75.

(a) Write a linear function describing the relationship between the total tuition charge T and the number of credit-hours taken h.

(b) Graph the relationship between T and h.

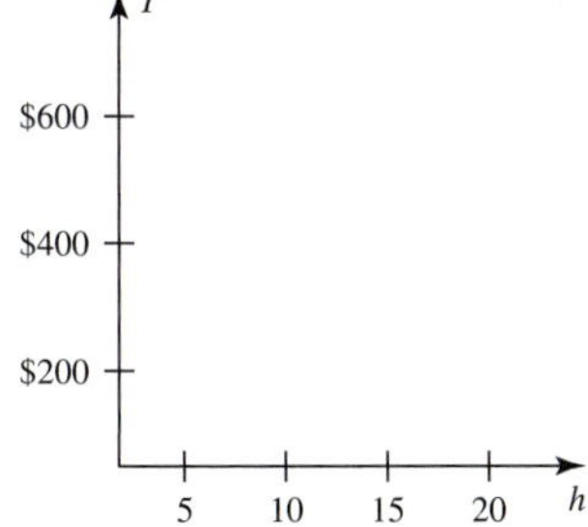

54. **Business and Finance** A sales associate's weekly salary is based on a fixed amount of \$200 plus 10% of the total amount of weekly sales.

(a) Write an equation that shows the relationship between the weekly salary S and the amount of weekly sales x (in dollars).

(b) Graph the relationship between S and x.

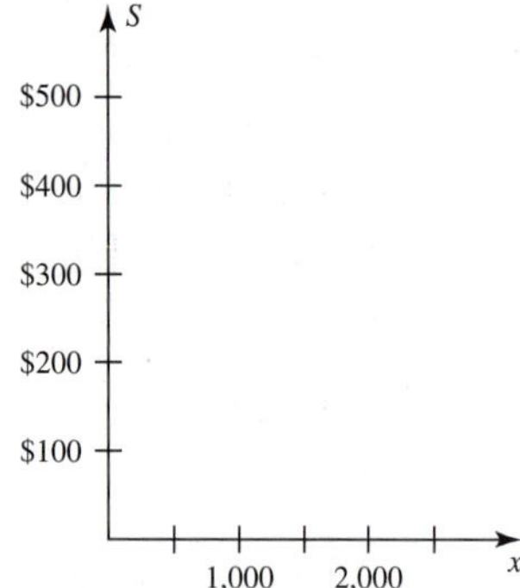

Complete each statement with **always, sometimes,** *or* **never.**

55. If the ordered pair (x, y) is a solution to an equation in two variables, then the point (x, y) is __________ on the graph of the equation.

56. If the graph of a linear equation $Ax + By = C$ passes through the origin, then C __________ equals zero.

57. If the ordered pair (x, y) is *not* a solution to an equation in two variables, then the point (x, y) is __________ on the graph of the equation.

58. The graph of a horizontal line __________ passes through the origin.

Skills | **Calculator/Computer** | Career Applications | Above and Beyond

Use a graphing calculator to graph each equation in the standard viewing window.

59. $y = 3x - 5$ chapter 3 > Make the Connection

60. $y = 2x + 3$ chapter 3 > Make the Connection

61. $y = -2x + 4$ chapter 3 > Make the Connection

62. $y = -4x - 1$ chapter 3 > Make the Connection

63. $y = \frac{2}{3}x + 3$ chapter 3 > Make the Connection

64. $y = -\frac{x}{2} + 2$ chapter 3 > Make the Connection

65. $3x + 4y = 8$ 

Hint: First solve for y.

66. $5x - 2y = -3$ 

Hint: First solve for y.

67. Use a graphing calculator to draw the graph for the equation you created in exercise 49, part (d). Choose a window that shows results from $x = 0$ to $x = 20$. Sketch the graph you see on your screen, and indicate the viewing window that you chose.

68. Use a graphing calculator to draw the graphs of the equations you created in exercise 50, parts (a) and (c). Choose a window that shows results from $x = 0$ to $x = 4$. Sketch what you see on your screen, and indicate the viewing window that you chose.

69. Use a graphing calculator to draw the graph of the equation given in exercise 51. Choose a window that shows results from $s = 0$ to $s = 300$. Sketch the graph you see on your screen, and indicate the viewing window that you chose.

70. Use a graphing calculator to draw the graph for the equation you created in exercise 54. Choose a window that shows results from $x = 0$ to $x = 3{,}000$. Sketch the graph you see on your screen, and indicate the viewing window that you chose.

71. **Allied Health** The weight w (in kilograms) of a uterine tumor is related to the number of days d of chemotherapy treatment by the function $w(d) = -1.75d + 25$. Sketch a graph of the weight of a tumor in terms of the number of days of treatment.

72. **Mechanical Engineering** The force that a coil exerts on an object is related to the distance that the coil is pulled from its natural (at rest) position. The formula to describe this is $F = kx$. Graph this relationship for a coil for which $k = 72$ pounds per foot.

73. **Construction Technology** The number of studs s (16 in. on center) required to build a wall that is L ft long is given by the formula

$$s = \frac{3}{4}L + 1$$

Graph the equation with appropriately scaled axes. 

74. **Manufacturing Technology** The number of board feet b of lumber in a 2" × 6" board of length L (in feet) is given by the equation

$$b = \frac{8.25}{144}L$$

Graph the equation with appropriately scaled axes.

Skills | Calculator/Computer | Career Applications | **Above and Beyond**

Graph both functions on the same set of axes and report what you observe about the graphs.

75. $f(x) = 2x$ and $g(x) = 2x + 1$

76. $f(x) = 3x + 1$ and $g(x) = 3x - 1$

77. $f(x) = 2x$ and $g(x) = -\frac{1}{2}x$

78. $f(x) = \frac{1}{3}x + \frac{7}{3}$ and $g(x) = -3x + 2$

79. Consider the equation $y = 2x + 3$.

(a) Complete the table of values and plot the points.

Point	x	y
A	5	
B	6	
C	7	
D	8	
E	9	

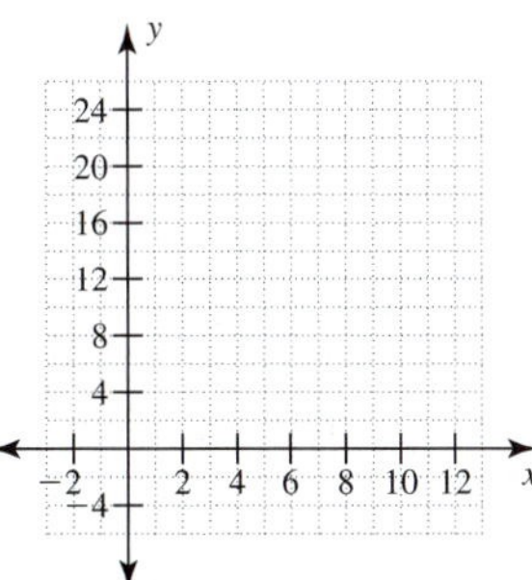

(b) As the x-coordinate changes by 1 (for example, as you move from point A to point B), how much does the corresponding y-coordinate change?

(c) Is your answer to part (b) the same if you move from B to C? from C to D? from D to E?

(d) Describe the "growth rate" of the line, using these observations. Complete the statement: When the x-value grows by 1 unit, the y-value ___________.

80. Describe how answers to parts (b), (c), and (d) change if you repeat exercise 79 using $y = 2x + 5$.

81. Describe how answers to parts (b), (c), and (d) change if you repeat exercise 79 using $y = 3x - 2$.

82. Describe how answers to parts (b), (c), and (d) change if you repeat exercise 79 using $y = 3x - 4$.

83. Describe how answers to parts (b), (c), and (d) change if you repeat exercise 79 using $y = -4x + 50$.

84. Describe how answers to parts (b), (c), and (d) change if you repeat exercise 79 using $y = -4x + 40$.

Answers

1.

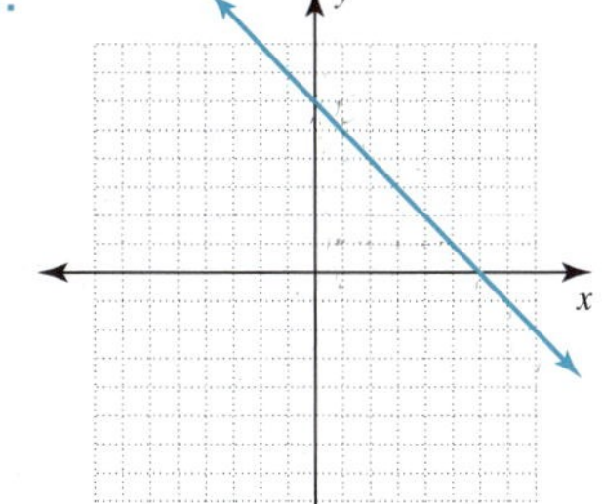

3.

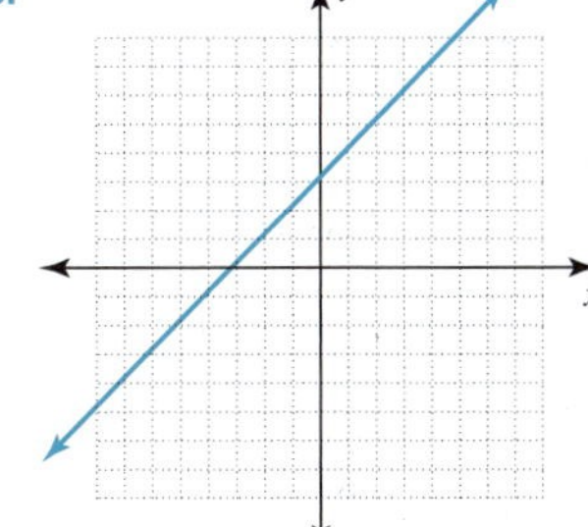

5.

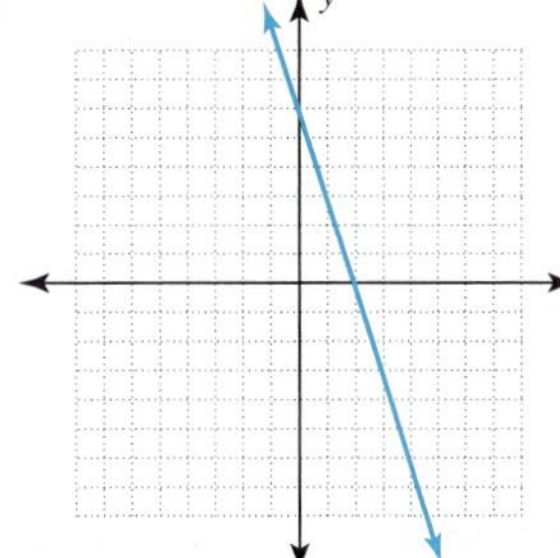

7.

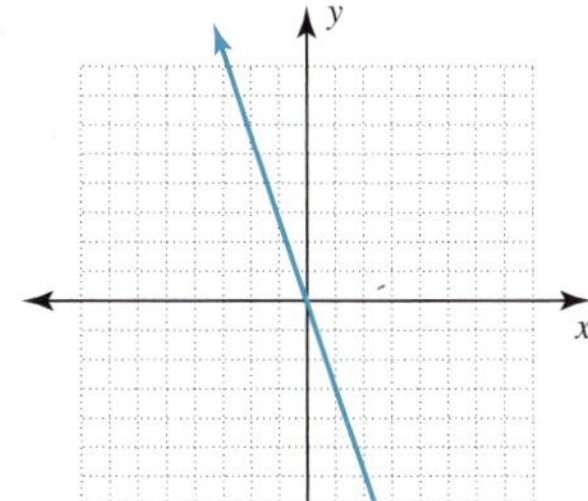

9.

11.

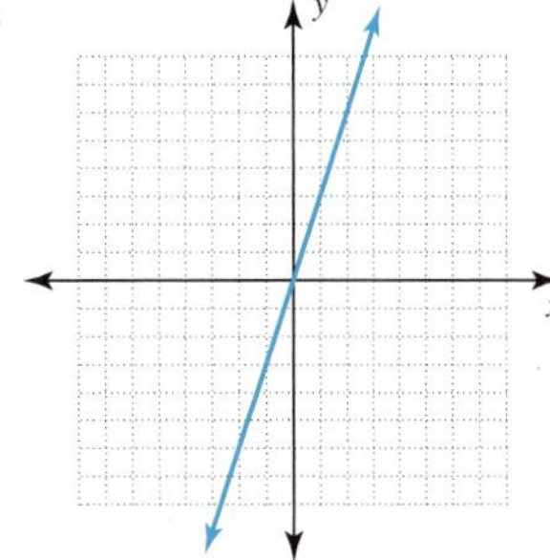

13.

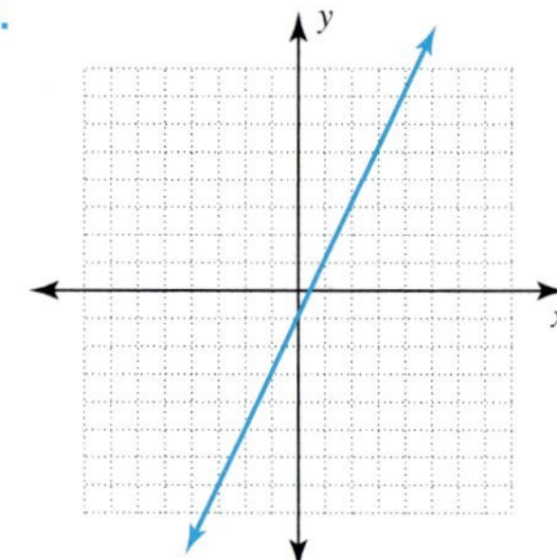

15.

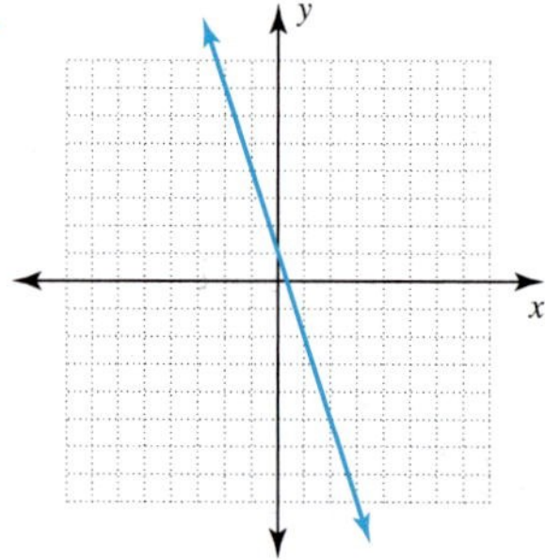

17.

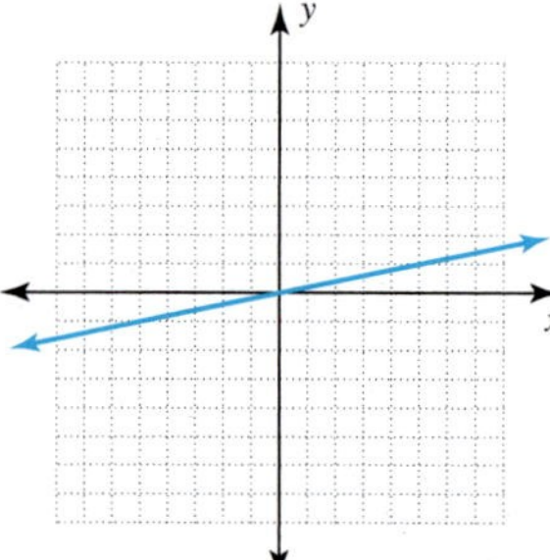

19.

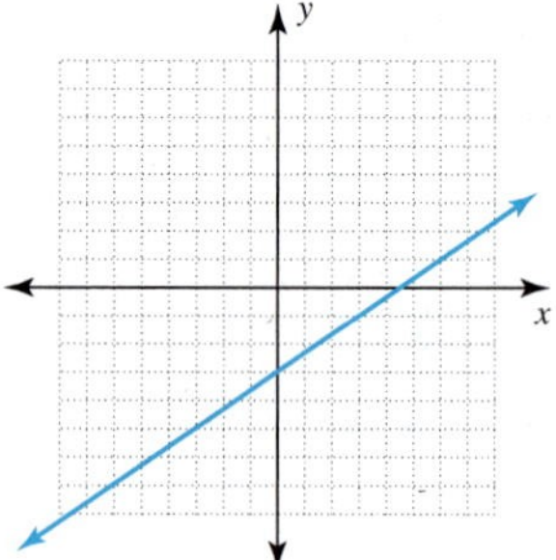

21.

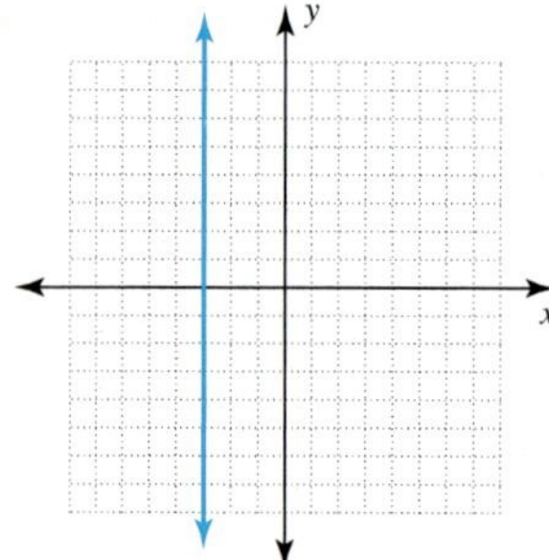

23.

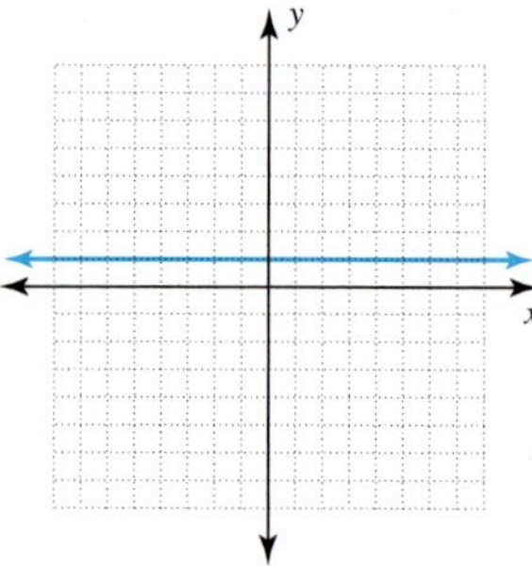

25.

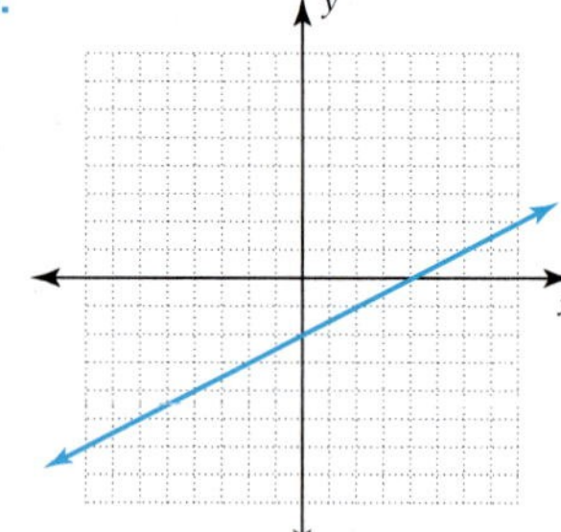

27.

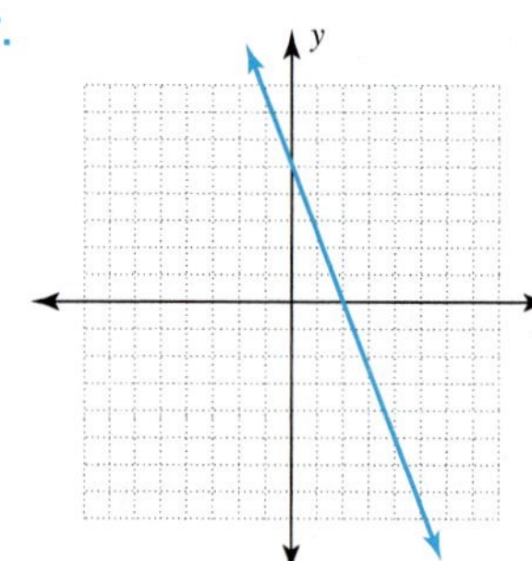

29.

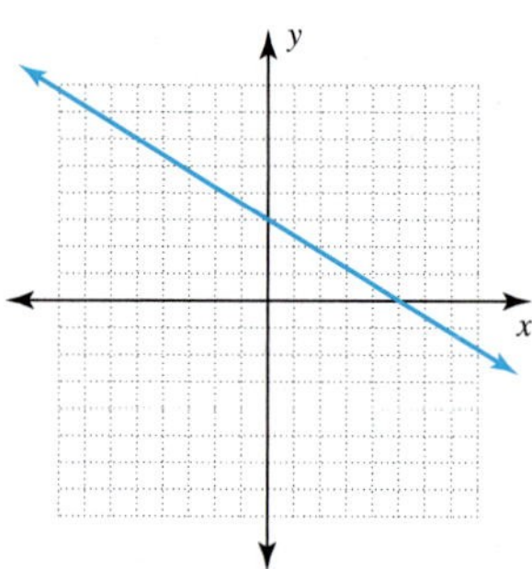

31.

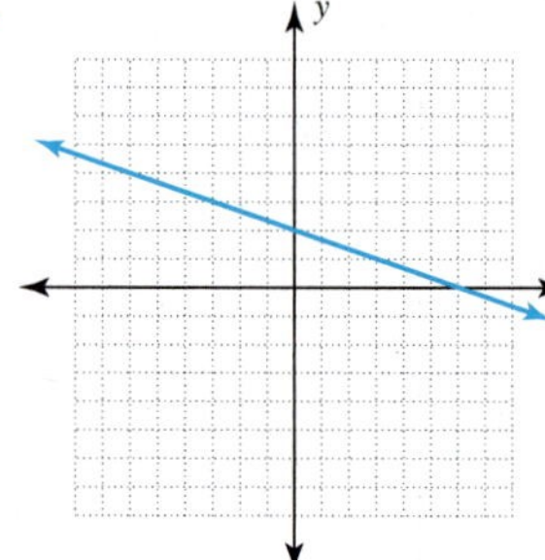

$f(x) = -\frac{1}{3}x + 2$

33.

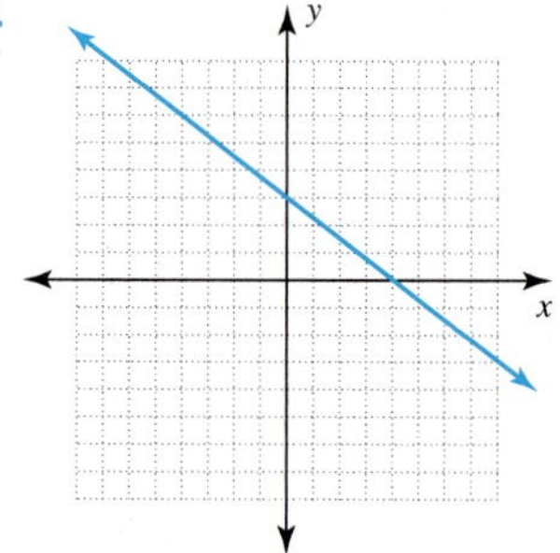

$f(x) = -\frac{3}{4}x + 3$

35.

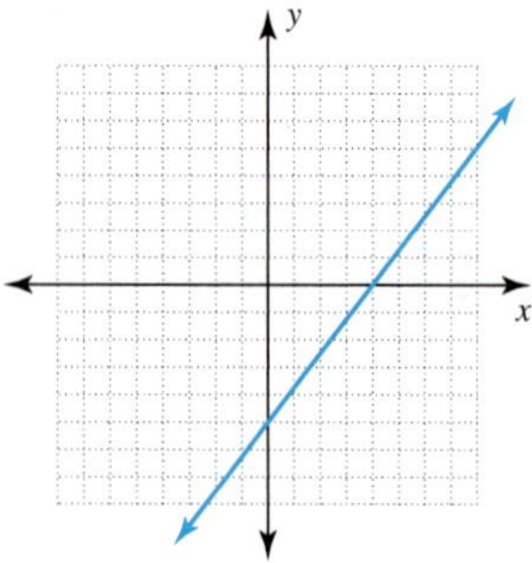

$f(x) = \frac{5}{4}x - 5$

37. $y = 2x$ **39.** $y = x + 3$ **41.** $y = 3x - 3$ **43.** $x - 4y = 12$ **45.** $(3, 1)$

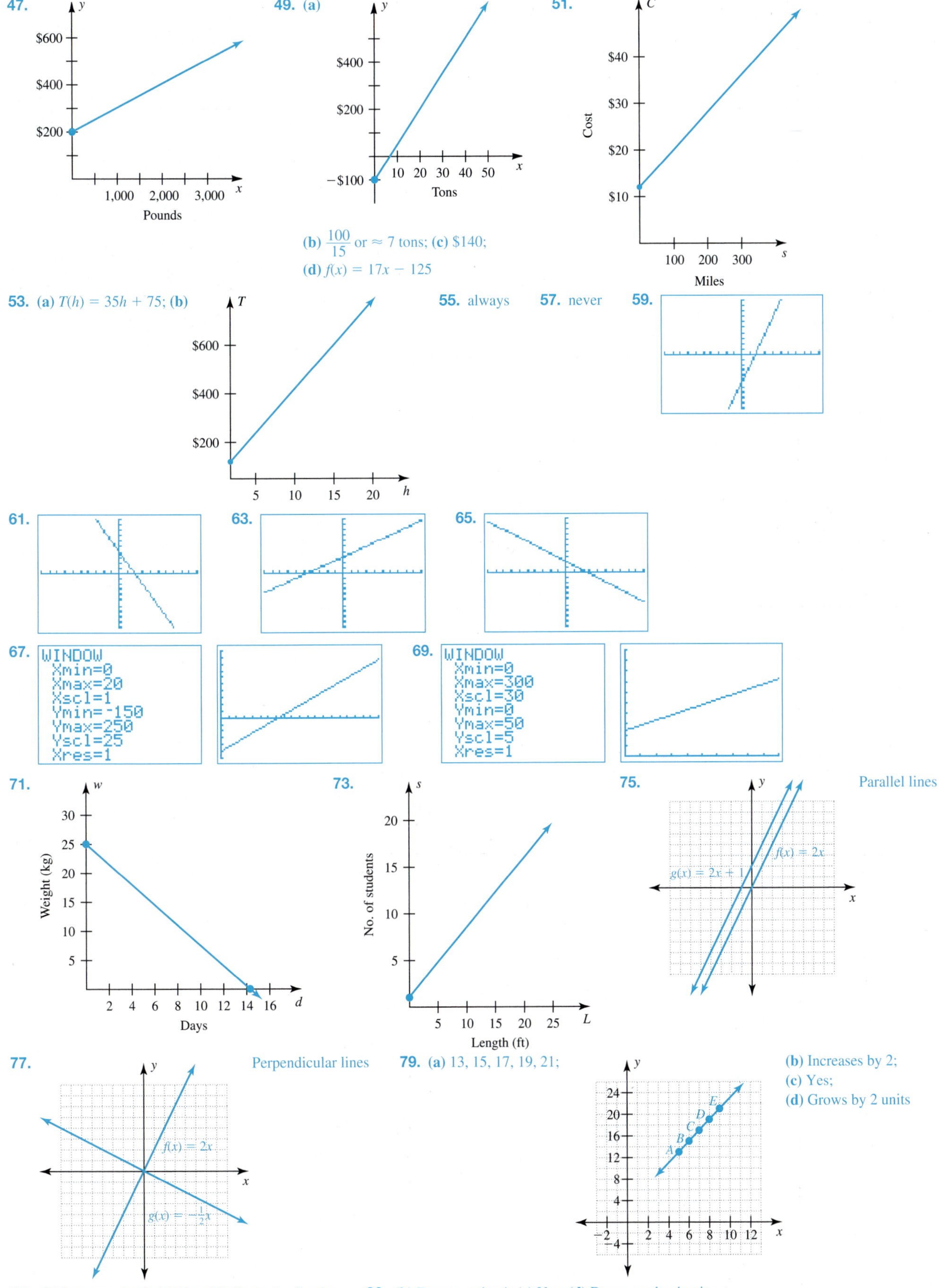

47.
y
$600
$400
$200
1,000 2,000 3,000 x
Pounds
49. (a)
y
$400
$200
−$100
10 20 30 40 50 x
Tons
(b) $\frac{100}{15}$ or ≈ 7 tons; (c) $140;
(d) $f(x) = 17x - 125$
51.
C
$40
$30
$20
$10
Cost
100 200 300 s
Miles
53. (a) $T(h) = 35h + 75$; (b)
T
$600
$400
$200
5 10 15 20 h
55. always
57. never
59.
61.
63.
65.
67.
WINDOW
Xmin=0
Xmax=20
Xscl=1
Ymin=-150
Ymax=250
Yscl=25
Xres=1
69.
WINDOW
Xmin=0
Xmax=300
Xscl=30
Ymin=0
Ymax=50
Yscl=5
Xres=1
71.
w
30
25
20
15
10
5
Weight (kg)
2 4 6 8 10 12 14 16 d
Days
73.
s
20
15
10
5
No. of students
5 10 15 20 25 L
Length (ft)
75.
Parallel lines
y
$f(x) = 2x$
$g(x) = 2x + 1$
x
77.
Perpendicular lines
y
$f(x) = 2x$
$g(x) = -\frac{1}{2}x$
x
79. (a) 13, 15, 17, 19, 21;
y
24
20
16
12
8
4
−2 −4
2 4 6 8 10 12 x
A B C D E
(b) Increases by 2;
(c) Yes;
(d) Grows by 2 units
81. (b) Increases by 3; (c) Yes; (d) Grows by 3 units
83. (b) Decreases by 4; (c) Yes; (d) Decreases by 4 units

Activity 3 ::

Linear Regression: A Graphing Calculator Activity

In Section 3.4, you will learn to build functions that model real-world phenomena. One of the more powerful features of graphing calculators is that they can create *regression equations* to approximate a data set. We will describe how to use Texas Instruments calculators, the TI-84 Plus, to plot data, find a linear regression equation, and use that equation. See your instructor or your calculator manual to learn the steps necessary to get your particular calculator model to perform these functions.

Scatter Plots

We begin by putting together a *scatter plot.* At its most basic level, a scatter plot is simply a set of points on the same graph. Of course, in order to be useful, the points should all be related in some way.

The data that we use relate the amount of time (in hours) each of 13 students spent studying for an exam and their grades on the exam.

Study Time, *x*	0	1	2	4	4	5	5	5	6	6	7	7	8
Exam Grade, *y*	50	51	72	52	74	78	81	74	86	93	84	92	94

We want to enter the data into our calculator so that we can create a scatter plot.

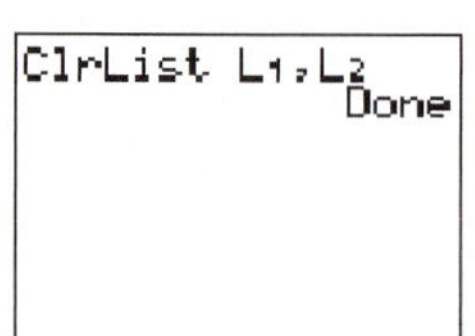

1. Clear any existing data from the lists you will use.
   We will use Lists 1 and 2, so our first step is to make sure these lists are empty. We do this by accessing the statistics menu and clearing the lists.
   STAT **4:ClrList**
   "ClrList" will appear on the home screen.
   Then, tell the calculator to clear Lists 1 and 2.
   2nd [L1] , 2nd [L2] ENTER
   Note: On the Texas Instruments models, [L1] and [L2] are the second functions of the **1** and **2** number keys.
2. Enter the data into the lists.
   Access the lists by choosing the edit option from the statistics menu.
   STAT **1:Edit**
   Then, enter the x-values in the first list, pressing ENTER after each one.
   After entering the x-values, use the right-arrow key ▶ to move to the second list and enter the y-values.
   Note: It is surprisingly easy to make a mistake when entering data into the lists. You should double-check that you entered the data correctly and that the y-values that you enter are on the same line as the corresponding x-values.

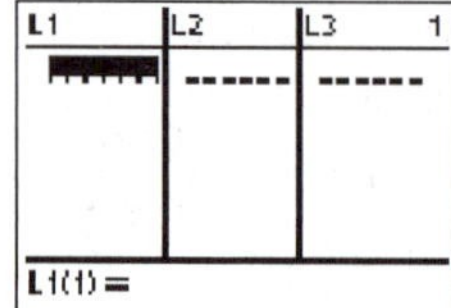

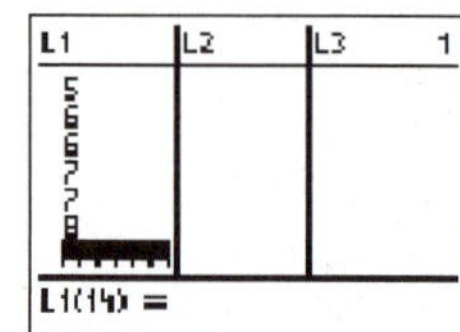

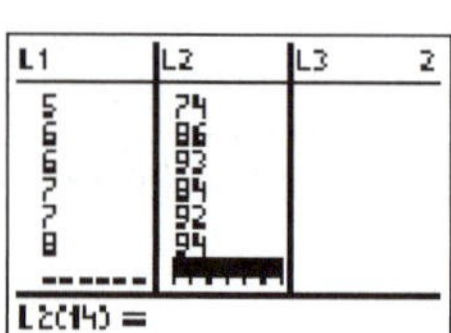

3. Create a scatter plot from the data.
Clear any equations from the function, or [Y=], menu.
Then, access the StatPlot menu, it is the second function of the [Y=] key. Select the first plot.
[2nd] [STAT PLOT] **1:Plot 1**
Select the **On** option and make sure the **Type** selected is the scatter plot, as shown in the figure to the right.

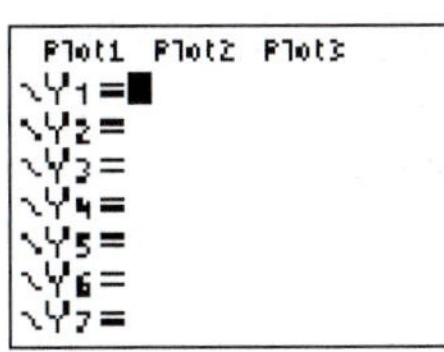

4. View the scatter plot.
Let the calculator choose an appropriate viewing window by using the ZoomStat feature.
[ZOOM] **9:ZoomStat**
You can use the window menu, [WINDOW], to modify the viewing window to improve your graph, if you wish. Press [GRAPH] to see the scatter plot when you are done.

NOTE

In the window menu, we increased the Yscl value to 10. Yscl gives the space between "tick marks" on the *y*-axis.

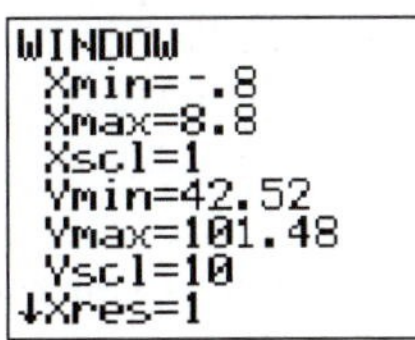
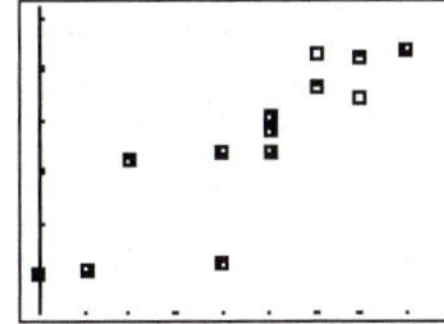

Regression Analysis

In this chapter, you will learn to construct a linear equation based on two points. Your calculator can accomplish the much more intense task of creating the *best* linear function to fit a larger set of data points.

1. Set your calculator to perform data analysis.
Access the statistics menu and set up the editor; you will need to enter the command in the home screen when it comes up:
[STAT] **5:SetUpEditor** [ENTER]
You also need to turn the calculator's diagnostics program on. You can do this by going to the *catalog* menu. The catalog menu is a complete listing of every function programmed into your calculator. The catalog menu is the second function of the **0** key.
[2nd] [CATALOG]
Move down the list until you reach **DiagnosticOn.** Press [ENTER] to send it to the home screen and press [ENTER] again to load the app.
We are now ready to perform a regression analysis.

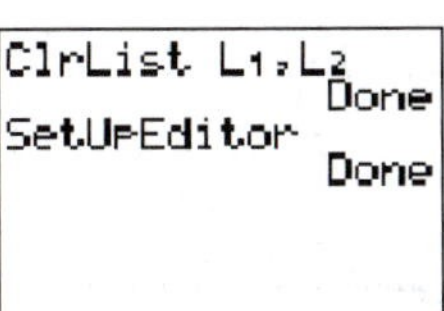

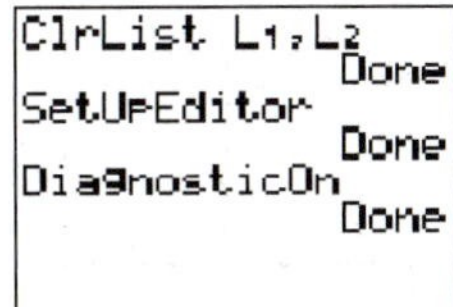

2. Perform a regression analysis on the data.
   Access the regression options by moving to the **CALC** submenu of the statistics menu. Then select the linear regression model.
   STAT ▶
   Note: This brings you to the **CALC** submenu of the statistics menu.
   4:LinReg(ax+b) ENTER
   If you are using an older calculator model, this is sufficient to perform the regression analysis. On newer models, you also need to tell the calculator to complete the analysis. After selecting the linear regression, you see the middle screen shown. Use the down arrow key to reach **Calculate** and press ENTER.

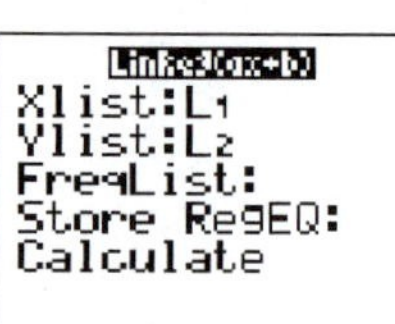

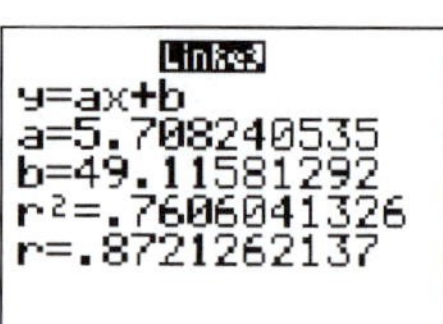

NOTE

Your calculator constructs a linear regression model by finding the line that minimizes the vertical distance between that line and the data set's y-values.

We will learn about the slope of a line beginning with the next section. After completing Sections 3.2 and 3.3, you should read this activity again. We briefly describe the information your calculator gives you.

The calculator is modeling a linear equation $y = ax + b$, so it is calling the slope a and the y-intercept is $(0, b)$ (in this equation, the number that multiplies x is the slope). In this case, the calculator is showing the line that best fits the student study data as

$y = 5.7x + 49.1$ (to one decimal place)

In the context of this application, a slope of 5.7 indicates that each additional hour of studying increased a student's exam score by 5.7 points. The y-intercept tells us that a student who did not study at all could expect to receive a 49.1 on the exam. **Note:** $\mathbf{r}^2$ and **r** are used to measure the validity of the model. The closer **r** is to 1 or –1, the better the model; the closer **r** is to 0, the worse the model.

3. Graph the linear regression model on the scatter plot.
   We command the calculator to paste the linear regression model into the function menu, Y=. The calculator has saved the regression model in a variables menu.
   Y= VARS **5:Statistics . . .** ▶ ▶ **1:RegEQ** GRAPH

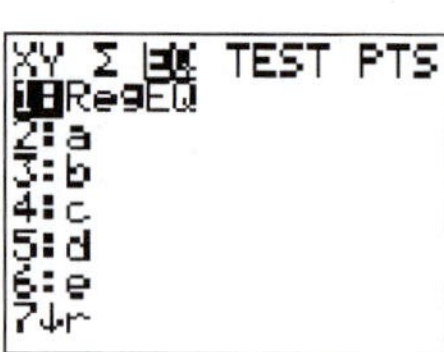

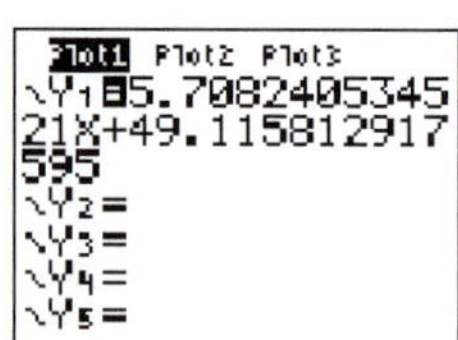

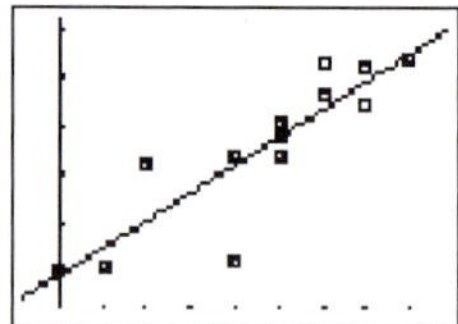

Exercise

The table gives the total acreage devoted to wheat in the United States over a recent 5-year period (all figures are in millions of acres).

Year	1	2	3	4	5
Acreage Planted, x	65.8	62.7	62.6	59.6	60.4
Acreage Harvested, y	59.0	53.8	53.1	48.6	45.8

Source: Farm Service Agency; U.S. Department of Agriculture

(a) Create a scatter plot relating the acres planted and harvested.
(b) Perform a regression analysis on the data.
(c) Give the slope and y-intercept (one decimal place of accuracy) and interpret them in the context of this application.
(d) Graph the regression equation in the same window with the scatter plot.

Answers

(a)

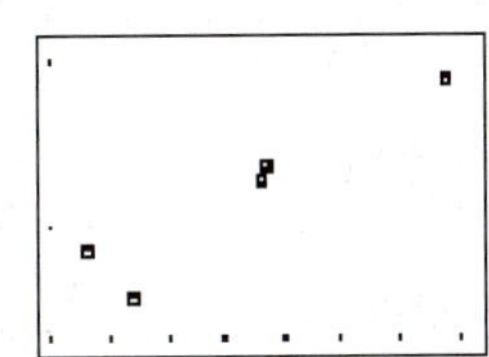

(b)

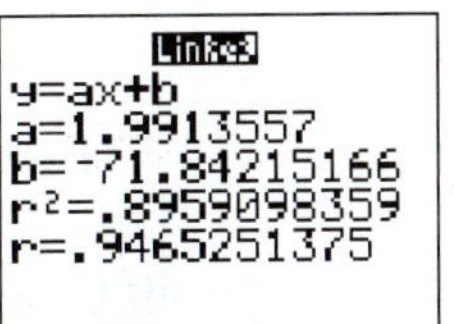

(c) The slope is approximately 2, which means that for each additional acre planted, we expect to harvest two additional acres of wheat. The y-intercept is $(0, -71.8)$, which claims that if we planted no wheat, we would harvest a negative amount of wheat. This is, of course, not true. This means that our model is not valid near $x = 0$.

(d)

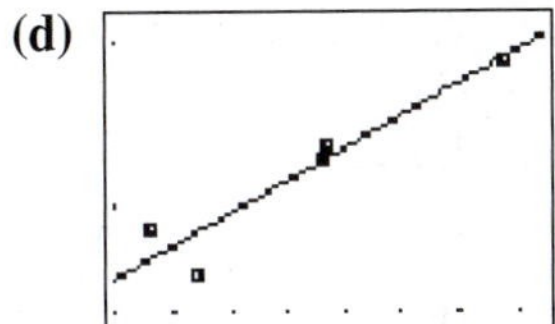

3.2 The Slope of a Line

< 3.2 Objectives >

1 > Find the slope of a line

2 > Find the slope and y-intercept of a line from an equation

3 > Write the equation of a line given its slope and y-intercept

4 > Graph a linear equation

On the coordinate system, plot a random point.

How many different lines can you draw through that point? Hundreds? Thousands? Millions? Actually, there is no limit to the number of different lines that pass through that point.

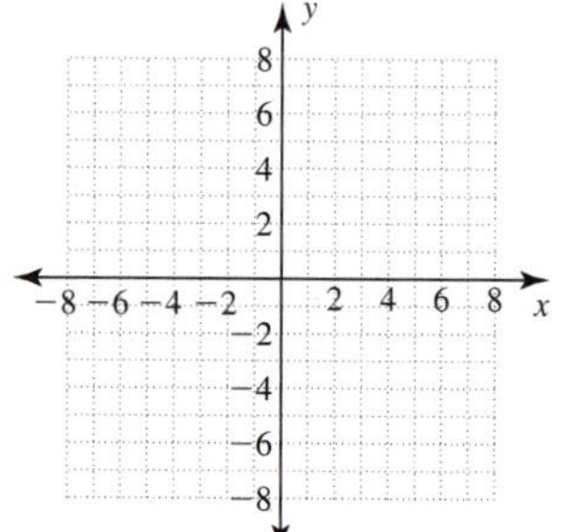

On the coordinate system, plot two distinct points.

Now, how many different (straight) lines can you draw through those points? Only one! Two points are enough to define the line.

In Section 3.3, we will see how to find the equation of a line given two points. The first part of finding that equation is finding the **slope** of the line, which is a way of describing the *steepness* of a line.

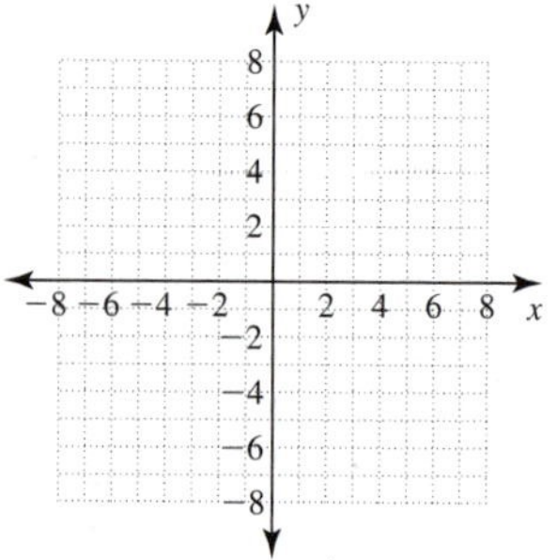

Let us assume that the two points selected were $(-2, -3)$ and $(3, 7)$.

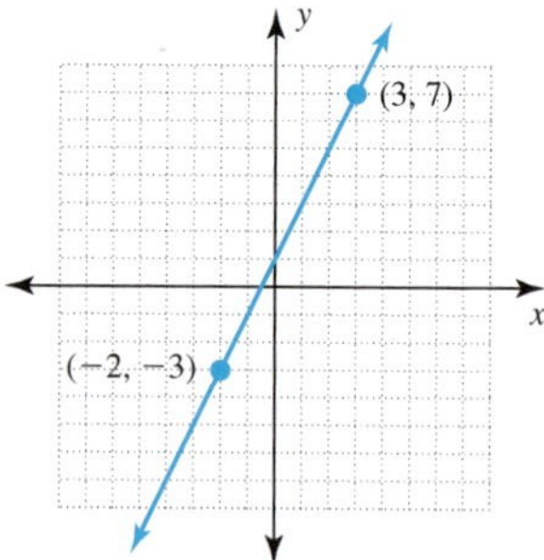

When moving between these two points, we go up 10 units and over 5 units.

We refer to the 10 units as the *rise*. The 5 units is called the *run*. The slope is found by dividing the rise by the run. In this case, we have

$$\frac{\text{Rise}}{\text{Run}} = \frac{10}{5} = 2$$

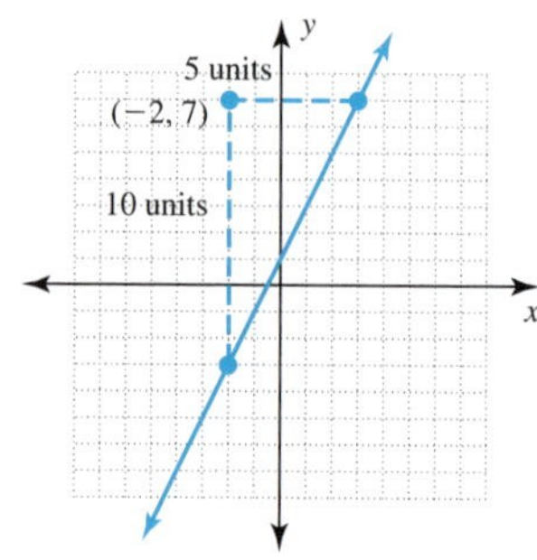

The slope of this line is 2. This means that for any two points on the line, the rise (the change in the y-value) is twice as much as the run (the change in the x-value).

We now proceed to a more formal look at the process of finding the slope of the line through two points.

To define a formula for slope, choose any two distinct points on the line, say, P with coordinates (x_1, y_1) and Q with coordinates (x_2, y_2). As we move along the line from P to Q, the x-value, or coordinate, changes from x_1 to x_2. That change in x, also called the **horizontal change,** is $x_2 - x_1$. Similarly, as we move from P to Q, the corresponding change in y, called the **vertical change,** is $y_2 - y_1$. The *slope* is defined as the ratio of the vertical change to the horizontal change. The letter m is used to represent the slope, which we now define.

NOTE

We read x_1 as "x sub 1," x_2 as "x sub 2," and so on. The 1 and 2 in x_1 and x_2 are called **subscripts.**

Definition

Slope of a Line

The **slope** of the line through two distinct points $P(x_1, y_1)$ and $Q(x_2, y_2)$ is given by

$$m = \frac{\text{Change in } y}{\text{Change in } x} = \frac{y_2 - y_1}{x_2 - x_1}$$

where $x_1 \neq x_2$.

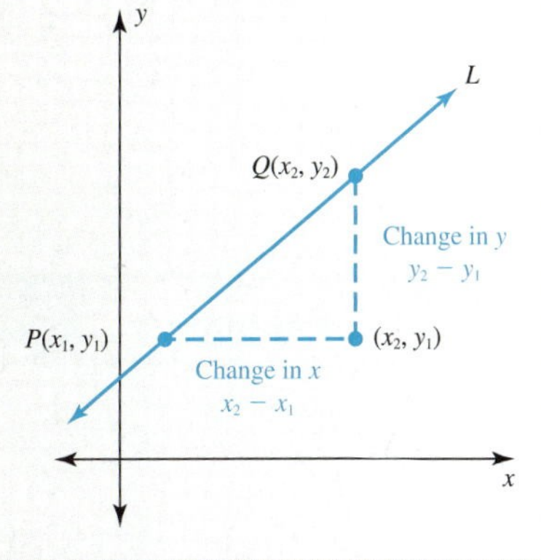

NOTE

The difference $x_2 - x_1$ is called the **run.** The difference $y_2 - y_1$ is the **rise.**

Note that $x_1 \neq x_2$, or $x_2 - x_1 \neq 0$, ensures that the denominator is nonzero, so that the slope is defined.

This definition provides the numerical measure of "steepness" that we want. If a line "rises" as we move from left to right, its slope is positive—the steeper the line, the larger the numerical value of the slope. If the line "falls" from left to right, its slope is negative.

Example 1 Finding Slope

< Objective 1 >

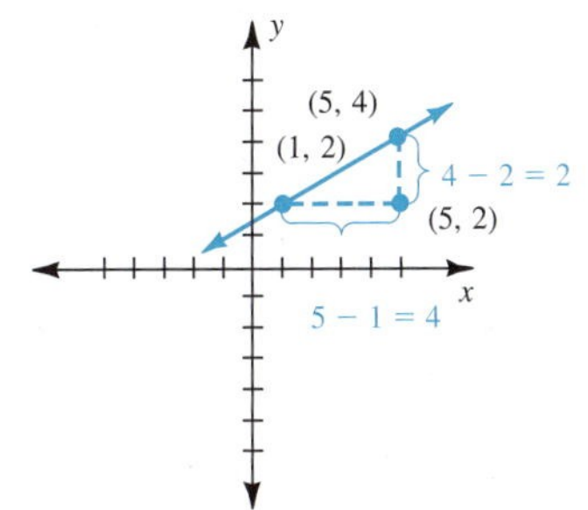

(a) Find the slope of the line through the points (1, 2) and (5, 4).

Let $P(x_1, y_1) = (1, 2)$ and $Q(x_2, y_2) = (5, 4)$. The formula for the slope of a line gives

$$m = \frac{y_2 - y_1}{x_2 - x_1} = \frac{(4) - (2)}{(5) - (1)} = \frac{2}{4} = \frac{1}{2}$$

Note: We have the same slope if we reverse P and Q and subtract in the other order. In that case, $P(x_1, y_1) = (5, 4)$ and $Q(x_2, y_2) = (1, 2)$, so

$$m = \frac{(2) - (4)}{(1) - (5)} = \frac{-2}{-4} = \frac{1}{2}$$

It makes no difference which point is labeled (x_1, y_1) and which is (x_2, y_2)—the slope is the same. You must simply stay with your choice once it is made and *not* reverse the order of the subtraction in your calculations.

(b) Find the slope of the line through the points $(-1, -2)$ and $(3, 6)$.

Again, applying the definition, we have

$$m = \frac{(6) - (-2)}{(3) - (-1)} = \frac{6 + 2}{3 + 1} = \frac{8}{4} = 2$$

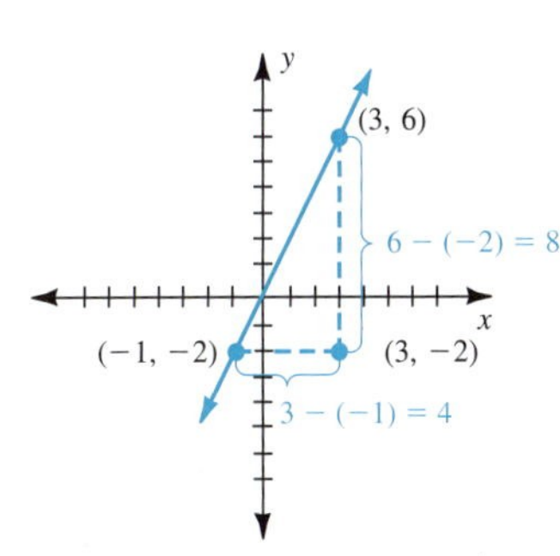

The figure compares the slopes found in parts (a) and (b). Line l_1, from part (a), had slope $\frac{1}{2}$. Line l_2, from part (b), had slope 2. Do you see the idea of slope measuring steepness? The greater the value of a positive slope, the more steeply the line is inclined upward.

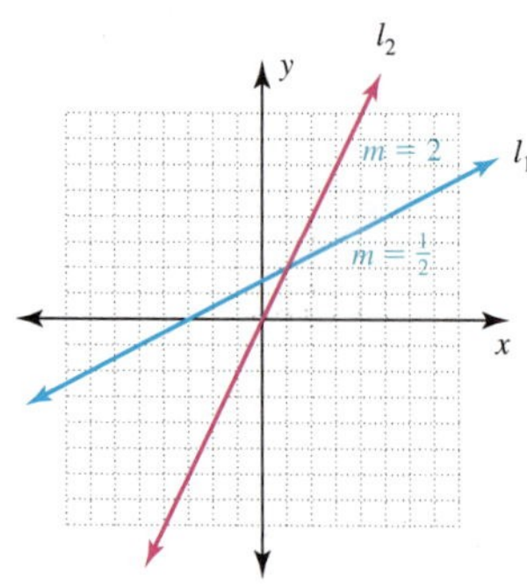

Check Yourself 1

(a) Find the slope of the line through the points (2, 3) and (5, 5).
(b) Find the slope of the line through the points (−1, 2) and (2, 7).
(c) Graph both lines on the same set of axes. Compare the lines and their slopes.

We now look at lines with a negative slope.

Example 2 Finding Slope

Find the slope of the line through the points (−2, 3) and (1, −3).

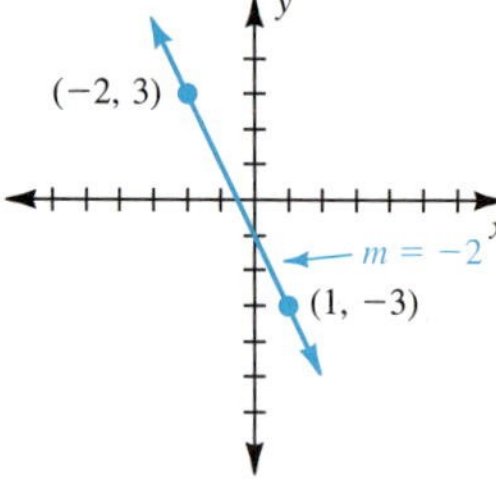

Using the slope formula, we have

$$m = \frac{(-3) - (3)}{(1) - (-2)} = \frac{-6}{3} = -2$$

This line has a *negative* slope. The line *falls* as we move from left to right.

Check Yourself 2

Find the slope of the line through the points (−1, 3) and (1, −3).

Lines with positive slope rise from left to right; lines with negative slope fall from left to right. What about lines with a slope of 0? A line with a slope of 0 is especially important in mathematics.

Example 3 Finding Slope

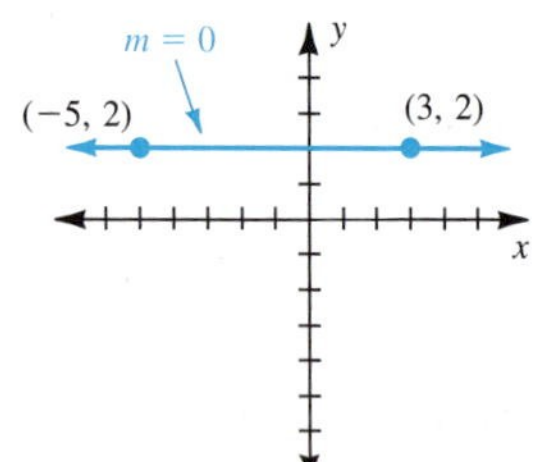

Find the slope of the line through the points (−5, 2) and (3, 2).

From the formula,

$$m = \frac{(2) - (2)}{(3) - (-5)} = \frac{0}{8} = 0$$

The slope of the line is 0. This is true for any horizontal line. Since any two points on the line have the same *y*-coordinate, the vertical change $y_2 - y_1$ is always 0, and so the resulting slope is 0.

You should recall that this is the graph of $y = 2$.

Check Yourself 3

Find the slope of the line through the points (−2, −4) and (3, −4).

Since division by 0 is undefined, it is possible to have a line with an undefined slope.

Example 4 Finding Slope

Find the slope of the line through the points (2, −5) and (2, 5).

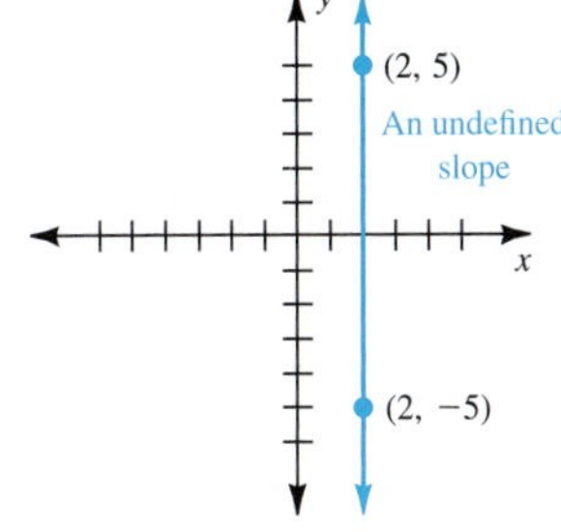

By the definition,

$$m = \frac{(5) - (-5)}{(2) - (2)} = \frac{10}{0}$$

Remember that division by 0 is undefined.

We say the vertical line has an undefined slope. On a vertical line, any two points have the same x-coordinate. This means that the horizontal change $x_2 - x_1$ is 0, and since division by 0 is undefined, the slope of a vertical line is always undefined.

You should recall that this is the graph of $x = 2$.

Check Yourself 4

Find the slope of the line through the points (−3, −5) and (−3, 2).

NOTE

As the slope gets closer to 0, the line gets "flatter."

This sketch summarizes our results from Examples 1 through 4.

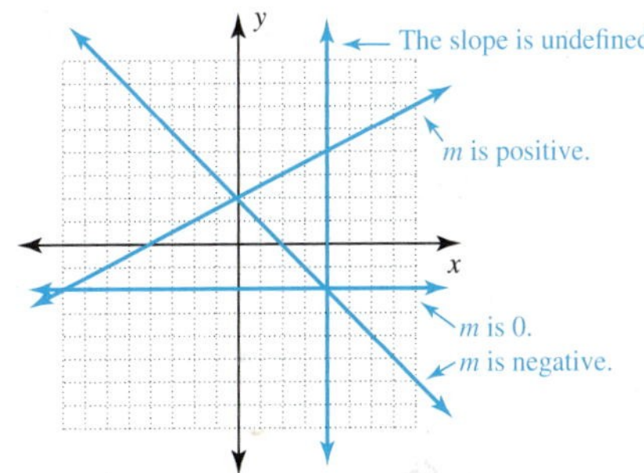

Four lines are illustrated in the figure.

1. The slope of a line that rises from left to right is positive.
2. The slope of a line that falls from left to right is negative.
3. The slope of a horizontal line is 0.
4. A vertical line has an undefined slope.

Finding an equation of a line when we know its slope and y-intercept is very straightforward. Suppose that the y-intercept is $(0, b)$. That is, the point at which the line crosses the y-axis has coordinates $(0, b)$.

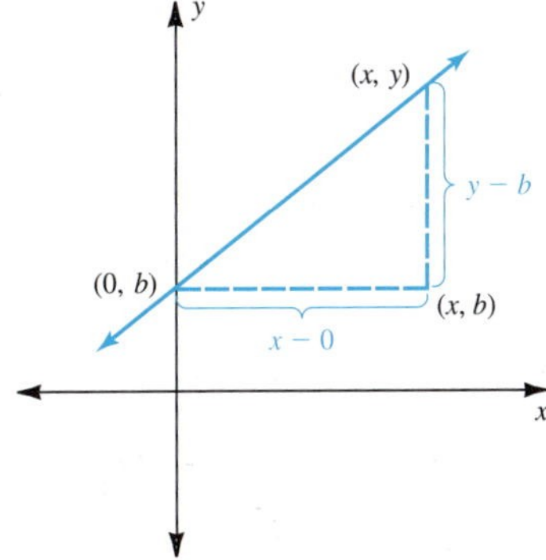

Now, using any other point (x, y) on the line and the slope formula, we write

$$m = \frac{\overbrace{y - b}^{\text{Change in } y}}{\underbrace{x - 0}_{\text{Change in } x}}$$

or $$m = \frac{y - b}{x}$$

Multiplying both sides by x, we have

$mx = y - b$

Finally, adding b to both sides gives

$$mx + b = y$$

or $$y = mx + b$$

We summarize this discussion with a property.

Property

The Slope-Intercept Equation for a Line

A linear function with slope m and y-intercept $(0, b)$ is expressed in *slope-intercept form* as

$y = mx + b$ or $f(x) = mx + b$ (using function notation)

In this form, the equation is *solved* for y. The coefficient of x gives you the slope of the line, and the constant term gives the y-intercept.

Example 5 Finding the Slope and y-Intercept

< Objective 2 >

(a) Find the slope and y-intercept for the graph of the equation

$$y = \underset{\uparrow\, m}{3}x + \underset{\uparrow\, b}{4}$$

The graph has slope 3 and y-intercept $(0, 4)$.

(b) Find the slope and y-intercept for the graph of the equation

$$y = \underset{\uparrow\, m}{-\frac{2}{3}}x - \underset{\uparrow\, b}{5}$$

The slope of the line is $-\frac{2}{3}$; the y-intercept is $(0, -5)$.

NOTES

You briefly encountered this idea in Activity 3. You might want to review that activity after completing this section.

The y-intercept of a graph is a point not a number; it has an x- and y-coordinate. Therefore, we always write the y-intercept as an ordered pair. In part (b), while $b = -5$, the y-intercept is written $(0, -5)$.

Check Yourself 5

Find the slope and y-intercept for the graph of each equation.

(a) $y = -3x - 7$ **(b)** $y = \frac{3}{4}x + 5$

As Example 6 illustrates, we may have to solve for y as the first step in determining the slope and the y-intercept for the graph of an equation.

Example 6 Finding the Slope and y-Intercept

Find the slope and y-intercept for the graph of the equation

$3x + 2y = 6$

First, we solve the equation for y.

$$3x + 2y = 6$$

$$2y = -3x + 6 \quad \text{Subtract } 3x \text{ from both sides.}$$

$$y = -\frac{3}{2}x + 3 \quad \text{Divide each term by 2.}$$

The equation is now in slope-intercept form. The slope is $-\frac{3}{2}$, and the y-intercept is $(0, 3)$.

NOTES

If we write the equation as

$y = \frac{-3x + 6}{2}$

it is more difficult to identify the slope and the y-intercept.

Remember to write the y-intercept as the ordered pair $(0, 3)$, not as a single number.

Check Yourself 6

Find the slope and y-intercept for the graph of the equation

$$2x - 5y = 10$$

Knowing certain properties of a line (namely, its slope and y-intercept) allows us to write the equation of the line by using the slope-intercept equation. Example 7 illustrates this approach.

Example 7 Writing an Equation of a Line

< Objective 3 >

(a) Write an equation of the line with slope 3 and y-intercept $(0, 5)$.

We know that $m = 3$ and $b = 5$. Using the slope-intercept equation, we have

$$y = \underset{m}{3}x + \underset{b}{5}$$

(b) Write an equation of the line with slope $-\frac{3}{4}$ and y-intercept $(0, -3)$.

We know that $m = -\frac{3}{4}$ and $b = -3$. In this case,

$$y = \overset{m}{-\frac{3}{4}}x + \overset{b}{(-3)}$$

or $y = -\frac{3}{4}x - 3$

Check Yourself 7

Write an equation of the line with each set of properties.

(a) Slope -2 and y-intercept $(0, 7)$

(b) Slope $\frac{2}{3}$ and y-intercept $(0, -3)$

We can also use the slope and y-intercept of a line to sketch its graph.

Example 8 Graphing a Line

< Objective 4 >

> **NOTE**
>
> $m = \frac{2}{3} = \frac{\text{Rise}}{\text{Run}}$
>
> The line rises from left to right because the slope is positive.

(a) Graph the line with slope $\frac{2}{3}$ and y-intercept $(0, 2)$.

Because the y-intercept is $(0, 2)$, we begin by plotting this point. The horizontal change (or run) is 3, so we move 3 units to the right *from that y-intercept.* The vertical change (or rise) is 2, so we move 2 units up to locate another point on the graph. This locates a second point at $(3, 4)$. The final step is to draw a line through that point and the y-intercept.

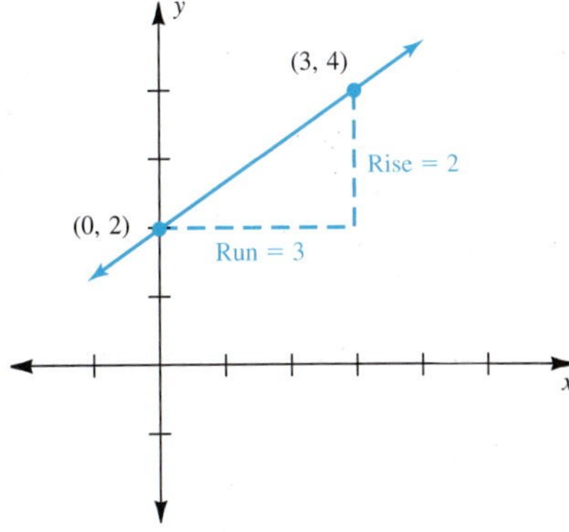

The equation of this line is $y = \frac{2}{3}x + 2$.

(b) Graph the line with slope -3 and y-intercept $(0, 3)$.

As before, first we plot the intercept point. In this case, we plot $(0, 3)$. The slope is -3, which we interpret as $\frac{-3}{1}$. Because the rise is negative, we go down rather than up. We move 1 unit in the horizontal direction, then 3 units down in the vertical direction. We plot this second point, $(1, 0)$, and connect the two points to form the line. The equation of this line is $y = -3x + 3$.

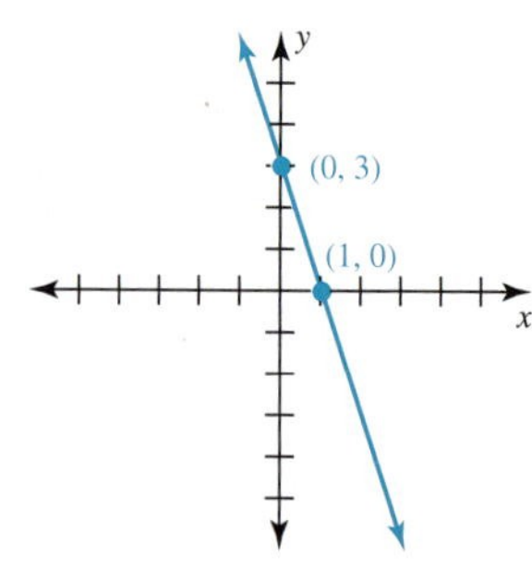

Check Yourself 8

Graph the line with slope $\frac{3}{5}$ and y-intercept $(0, -2)$.

A line can certainly pass through the origin, as Example 9 demonstrates. In such cases, the y-intercept is $(0, 0)$.

Example 9 Graphing a Line

Graph the lines given by $y = -3x$ and $y = -3x - 3$.

In the first case, the slope is -3 and the y-intercept is $(0, 0)$. We begin with the point $(0, 0)$. From there, we move down 3 units and to the right 1 unit, arriving at the point $(1, -3)$. Now we draw a line through those two points. On the same axes, we draw the line with slope -3 through the intercept $(0, -3)$. The two lines are parallel to each other.

NOTE

Nonvertical parallel lines have the same slope.

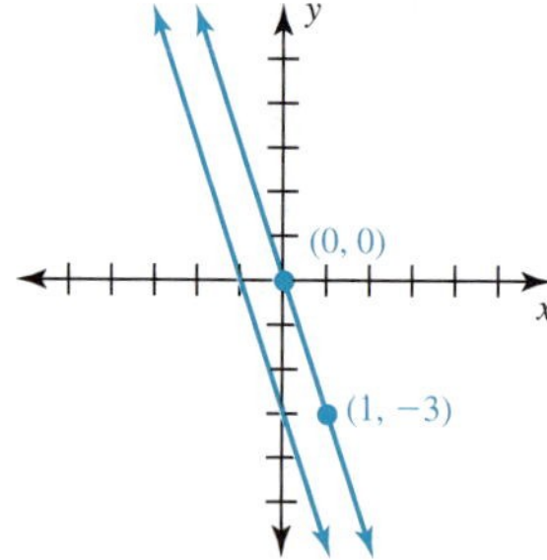

Check Yourself 9

Graph the line given by $y = \frac{7}{2}x$.

Our work graphing a line from its equation leads us to this algorithm.

Step by Step

Graphing by Using the Slope-Intercept Form

Step 1 Write the equation of the line in slope-intercept form $y = mx + b$.

Step 2 Determine the slope m and the y-intercept $(0, b)$.

Step 3 Plot the y-intercept at $(0, b)$.

Step 4 Use m (the change in y over the change in x) to determine a second point on the line.

Step 5 Draw a line through the two points to complete the graph.

You have seen two methods for graphing a line from its equation: the slope-intercept method (this section) and the intercept method (Section 3.1). When you graph a linear equation, you should first decide which method is easiest.

Example 10 Selecting an Appropriate Graphing Method

Decide which of the two methods for graphing lines—the intercept method or the slope-intercept method—is easier to graph each equation.

(a) $2x - 5y = 10$

Because both intercepts are easy to find, you should choose the intercept method to graph this equation.

(b) $2x + y = 6$

This equation can be quickly graphed by either method. As it is written, you might choose the intercept method. It can, however, be rewritten as $y = -2x + 6$, in which case the slope-intercept method is more appropriate.

(c) $y = \frac{1}{4}x - 4$

Since the equation is in slope-intercept form, that is the more appropriate method to choose.

Check Yourself 10

Which is more appropriate for graphing each equation, the intercept method or the slope-intercept method?

(a) $x + y = -2$ **(b)** $3x - 2y = 12$ **(c)** $y = -\frac{1}{2}x - 6$

When working with applications, we are frequently asked to interpret the slope of a function as its *rate of change.* We explore this more fully in Sections 3.3 and 3.4.

In short, the slope represents the change in the output, y or $f(x)$, when the input x is increased by one unit.

Graphically, the slope of a line is the change in the line's height when x increases by one unit. To remind you, a constant function has a slope equal to zero because the height of a horizontal line does not change when the input x increases by one unit.

In business applications, the slope of a linear function often correlates to the idea of *margin.* We learned about *marginal revenue, marginal cost,* and *marginal profit* in Chapter 2 and again in Section 3.1.

We conclude this section with an application from the field of electronics.

Example 11 An Electronics Application

The accompanying graph depicts the relationship between the position of a linear potentiometer (variable resistor) and the output voltage of some DC source. Consider the potentiometer to be a slider control, possibly to control volume of a speaker or the speed of a motor.

The linear position of the potentiometer is represented on the x-axis, and the resulting output voltage is represented on the y-axis. At the 2-cm position, the output voltage measured with a voltmeter is 16 VDC. At a position of 3.5 cm, the measured output was 10 VDC.

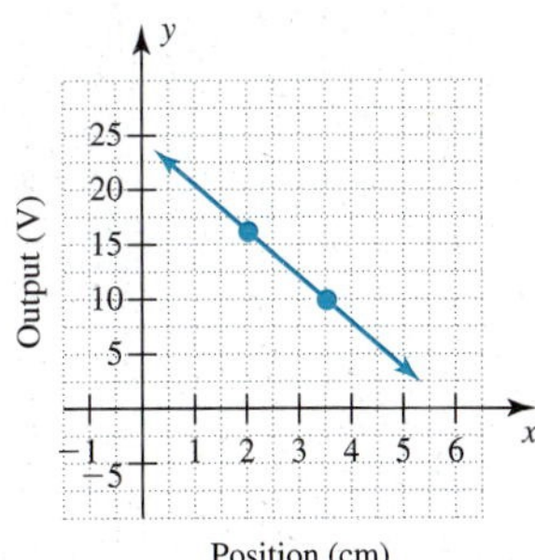

What is the slope of the resulting line?

We see that we have two ordered pairs: (2, 16) and (3.5, 10). Using the slope formula gives

$$m = \frac{(10) - (16)}{(3.5) - (2)} = \frac{-6}{1.5} = -4$$

The slope is -4.

Interpret the slope in the context of this application.

Because the x-values correspond to the slider's position and the y-values correspond to the output voltage, the slope refers to the change in output voltage if the slider's position increases by one unit.

In this case, increasing the slider's position 1 cm decreases the output voltage by 4 VDC.

Check Yourself 11

The same potentiometer described in Example 11 is used in another circuit. This time, though, when at position 0 cm, the output voltage is 12 volts. At position 5 cm, the output voltage is 3 volts. Draw a graph using the new data, determine the slope, and interpret the slope in the context of this application.

Check Yourself ANSWERS

1. (a) $m = \frac{2}{3}$; **(b)** $m = \frac{5}{3}$; **(c)**

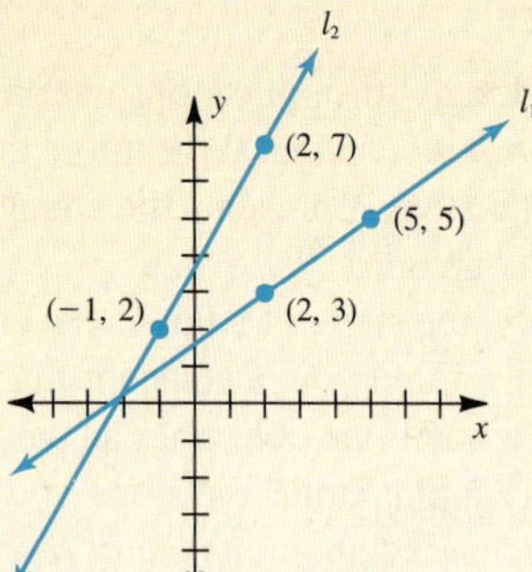

2. $m = -3$ **3.** $m = 0$

4. m is undefined **5. (a)** $m = -3$, y-intercept: $(0, -7)$; **(b)** $m = \frac{3}{4}$, y-intercept: $(0, 5)$

6. $y = \frac{2}{5}x - 2$; $m = \frac{2}{5}$; y-intercept: $(0, -2)$ **7. (a)** $y = -2x + 7$; **(b)** $y = \frac{2}{3}x - 3$

8.

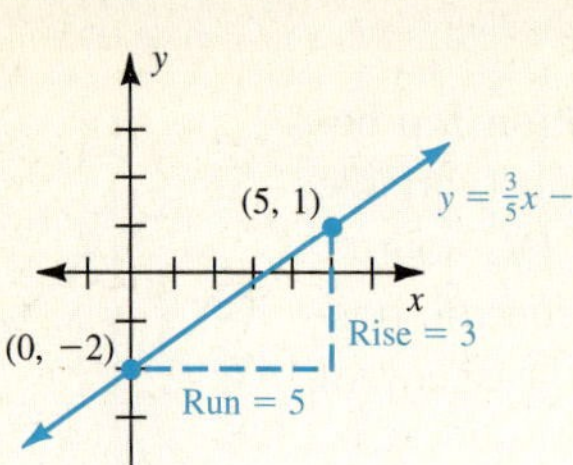

9.

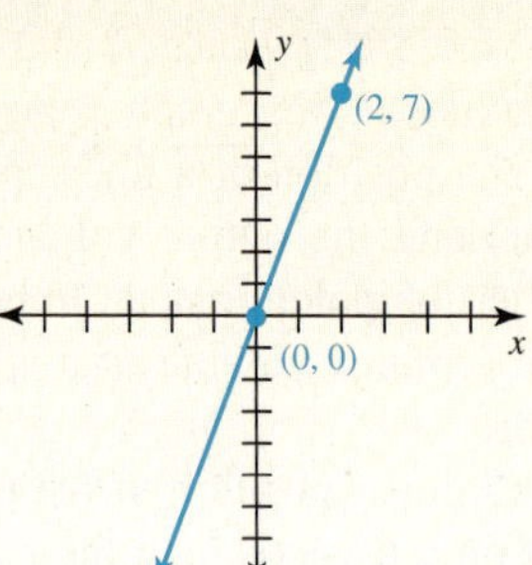

10. (a) either; **(b)** intercept; **(c)** slope-intercept

11. The slope is $-\frac{9}{5}$. Moving the slider one centimeter decreases the output voltage by $\frac{9}{5}$ or 1.8 VDC.

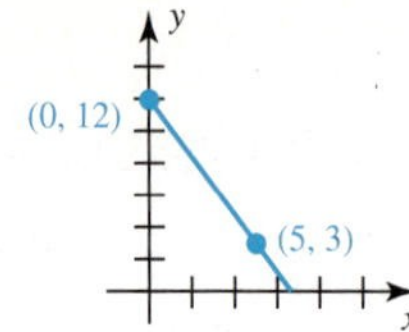

Reading Your Text

These fill-in-the-blank exercises will help you understand some of the key vocabulary used in this section. The answers to these exercises are in the Answers Appendix in the back of the text.

(a) The __________ of a line describes its steepness.

(b) The slope is defined as the ratio of the vertical change to the __________ change.

(c) The change in the x-values between two points is called the run. The change in the y-values is called the __________.

(d) Lines with __________ slope fall from left to right.

Skills Calculator/Computer Career Applications Above and Beyond

3.2 exercises

< Objective 1 >

Find the slope of the line through each pair of points.

1. $(5, 7)$ and $(9, 11)$
2. $(4, 9)$ and $(8, 17)$
3. $(-3, -1)$ and $(2, 3)$
4. $(-3, 2)$ and $(0, 17)$
5. $(-2, 3)$ and $(3, 7)$
6. $(-2, -5)$ and $(1, -4)$
7. $(-3, 2)$ and $(2, -8)$
8. $(-6, 1)$ and $(2, -7)$
9. $(3, -2)$ and $(5, -5)$ VIDEO
10. $(-2, 4)$ and $(3, 1)$
11. $(5, -4)$ and $(5, 2)$
12. $(-2, 8)$ and $(6, 8)$
13. $(-4, -2)$ and $(3, 3)$
14. $(-5, -3)$ and $(-5, 2)$
15. $(-2, -6)$ and $(8, -6)$
16. $(-5, 7)$ and $(2, -2)$
17. $(-1, 7)$ and $(2, 3)$
18. $(-3, -5)$ and $(2, -2)$

< Objective 2 >

Find the slope and y-intercept of the line represented by each equation.

19. $y = 3x + 5$
20. $y = -7x + 3$
21. $y = -3x - 6$
22. $y = 5x - 2$
23. $y = \frac{3}{4}x + 1$
24. $y = -5x$
25. $y = \frac{2}{3}x$
26. $y = -\frac{3}{5}x - 2$

Write each equation in function form. Give the slope and y-intercept of each function.

27. $4x + 3y = 12$
28. $5x + 2y = 10$
29. $y = 9$
30. $2x - 3y = 6$
31. $3x - 2y = 8$

32. $x = -3$

< Objective 3 >

Write an equation of the line with given slope and y-intercept. Then graph each line.

33. Slope 3; y-intercept: $(0, 5)$

34. Slope -2; y-intercept: $(0, 4)$
35. Slope -4; y-intercept: $(0, 5)$

36. Slope 5; y-intercept: $(0, -2)$

37. Slope $\frac{1}{2}$; y-intercept: $(0, -2)$

38. Slope $-\frac{2}{5}$; y-intercept: $(0, 6)$

39. Slope $\frac{4}{3}$; y-intercept: $(0, 0)$

40. Slope $\frac{2}{3}$; y-intercept: $(0, -2)$

41. Slope $\frac{3}{4}$; y-intercept: $(0, 3)$

42. Slope -3; y-intercept: $(0, 0)$

< Objective 4 >

Solve each equation for y and graph it.

43. $2x + 5y = 10$ 

44. $5x - 3y = 12$

45. $x + 7y = 14$

46. $-2x - 3y = 9$

Use the graph to determine the slope of each line.

47.

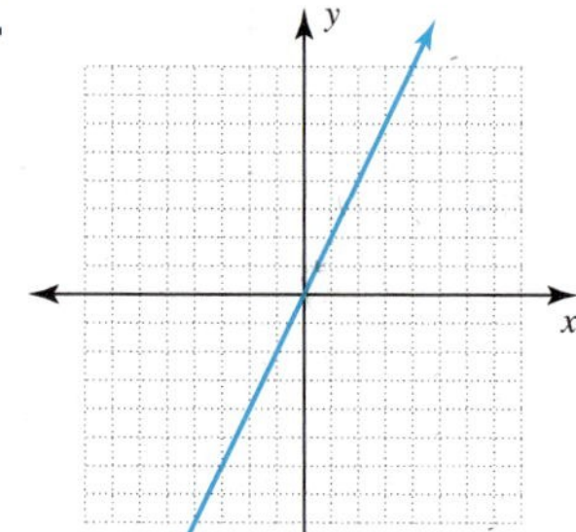

48.

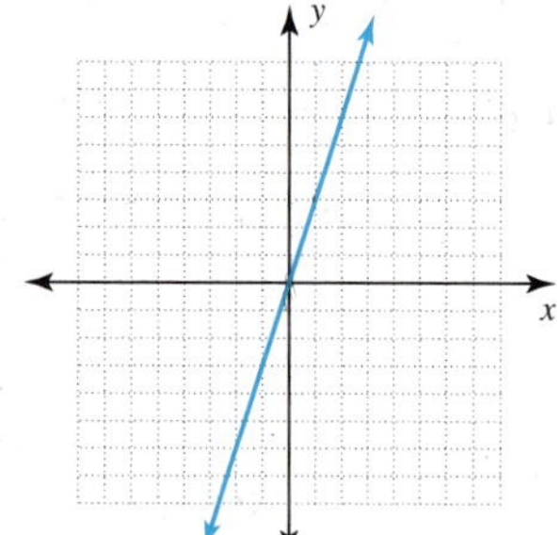

49.

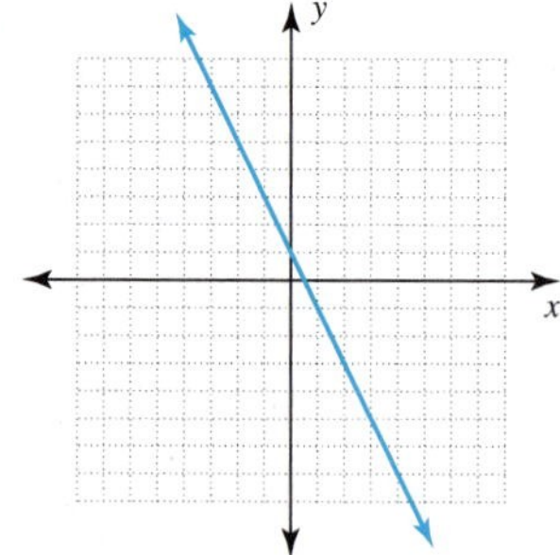

50.

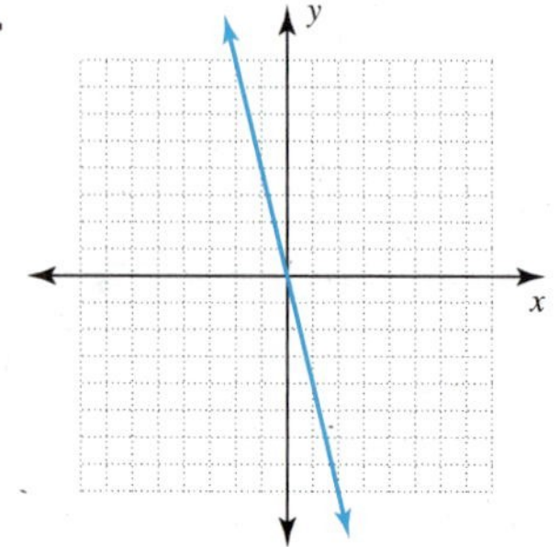

51.

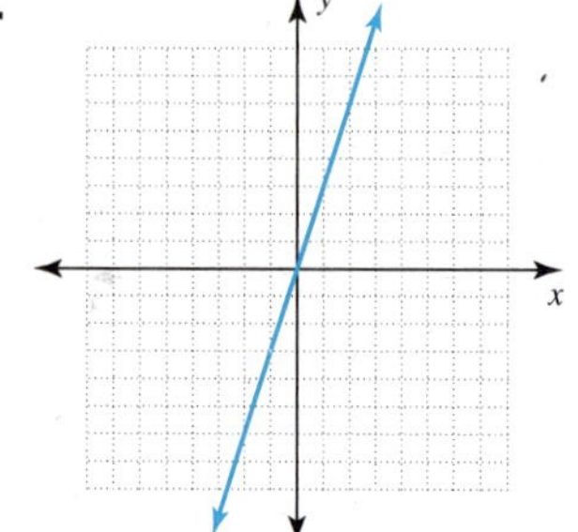

52.

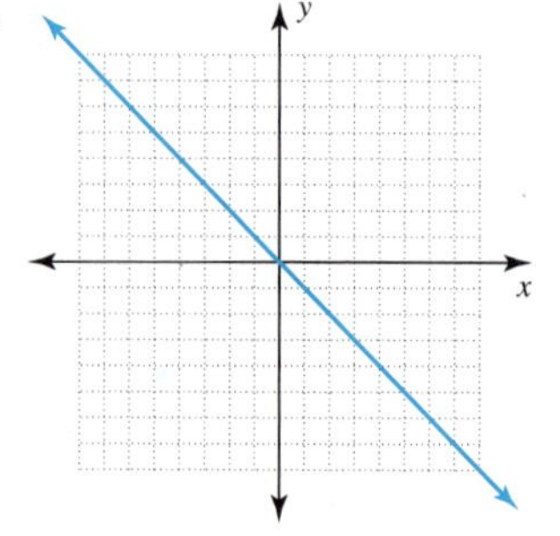

53. 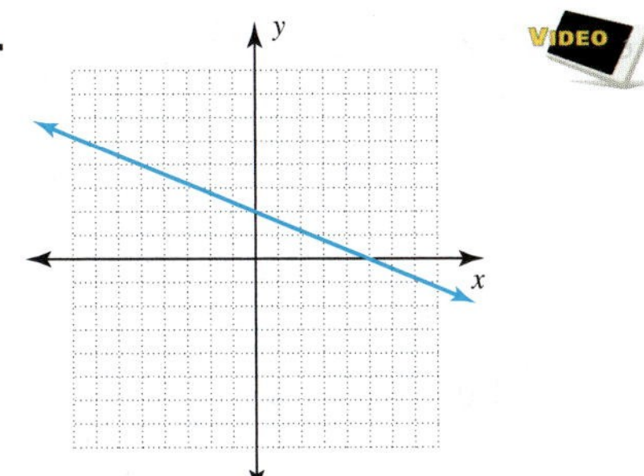

54. 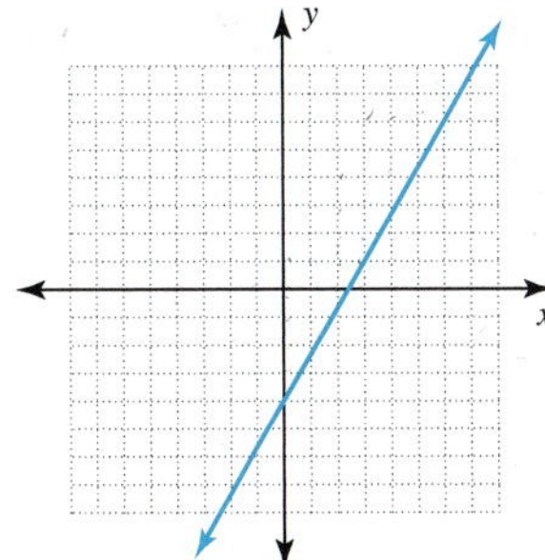

55. **Business and Finance** The equation $y = 0.10x + 200$ describes the award money in a recycling contest. What is the slope and y-intercept for this equation? What does the slope of the line represent in the equation? What does the y-intercept represent?

56. **Business and Finance** The equation $y = 15x - 100$ describes the amount of money a high school class might earn from a paper drive. What are the slope and y-intercept for this equation?

57. **Business and Finance** In the equation in exercise 56, what does the slope of the line represent? What does the y-intercept represent?

58. **Construction** A roof rises 8.75 ft in a horizontal distance of 15.09 ft. Find the slope of the roof to the nearest hundredth.

59. **Science and Medicine** An airplane covered 15 mi of its route while decreasing its altitude by 24,000 ft. Find the slope of the line of descent that was followed. (1 mi = 5,280 ft) Round to the nearest hundredth.

60. **Science and Medicine** Driving down a mountain, Tom finds that he has descended 1,800 ft in elevation by the time he is 3.25 mi horizontally away from the top of the mountain. Find the slope of his descent to the nearest hundredth.

61. **Business and Finance** In 1970, the cost of a soft drink was 50¢. By 2012, the cost of the same soft drink had risen to \$1.80. During this time period, what was the annual rate of change of the cost of the soft drink?

62. **Science and Medicine** On a February day in Philadelphia, the temperature at 6:00 A.M. was 10°F. By 2:00 P.M. the temperature was up to 26°F. What was the hourly rate of temperature change?

In which quadrant(s) are there no solutions for each equation?

63. $y = 4x + 5$

64. $y = 3x + 2$

65. $y = -x + 1$

66. $y = -7x + 3$

67. $y = -2x - 5$

68. $y = -5x - 7$

69. $y = 5$

70. $x = -2$

Match each graph with one of the equations.

(a) $y = 2x$, (b) $y = x + 1$, (c) $y = -x + 3$, (d) $y = 2x + 1$,

(e) $y = -3x - 2$, (f) $y = \frac{2}{3}x + 1$, (g) $y = -\frac{3}{4}x + 1$, (h) $y = -4x$

71. 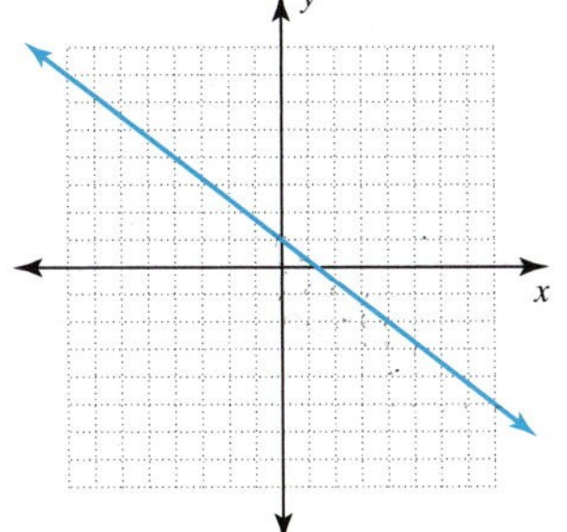

72. 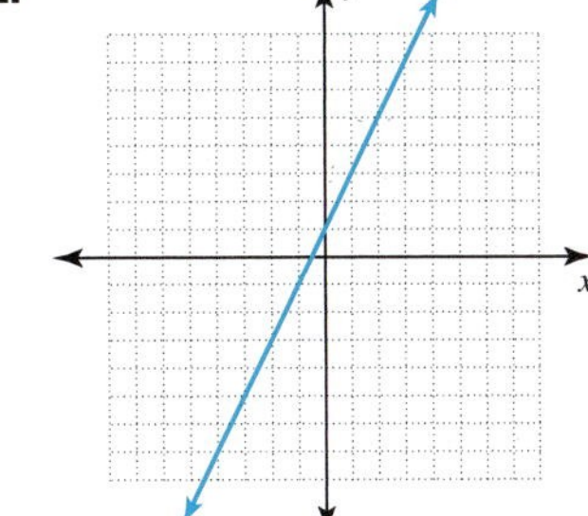

73. 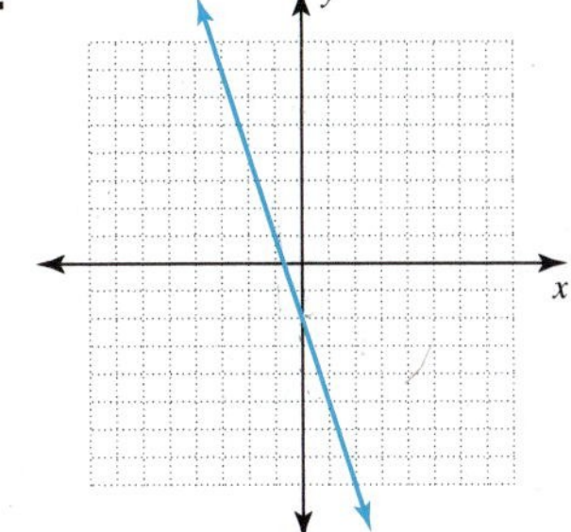

74.

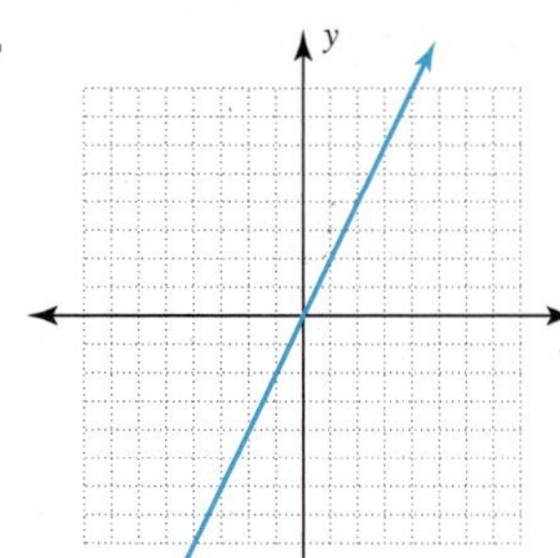

75.

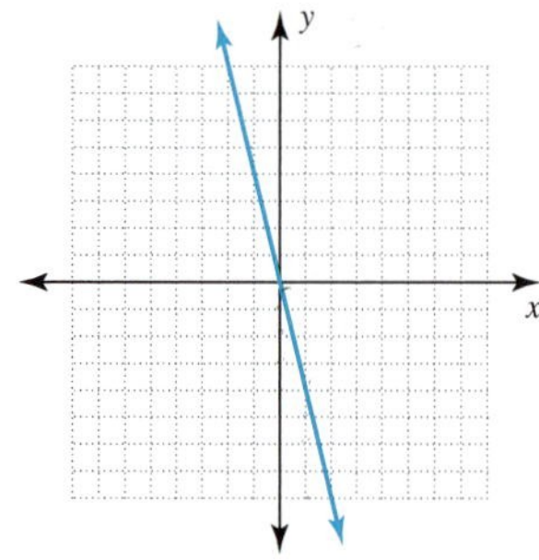

76.

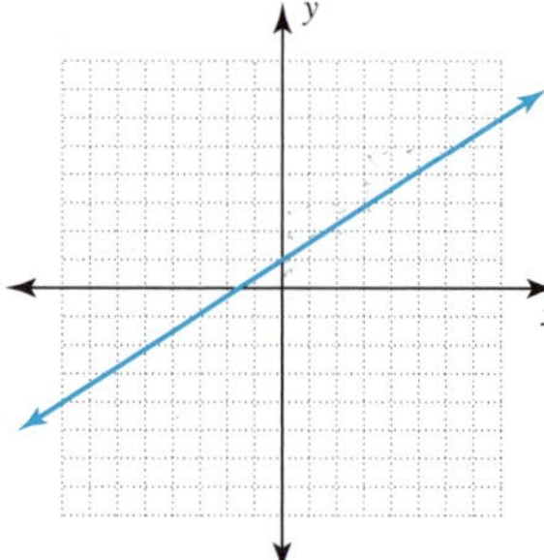

77.

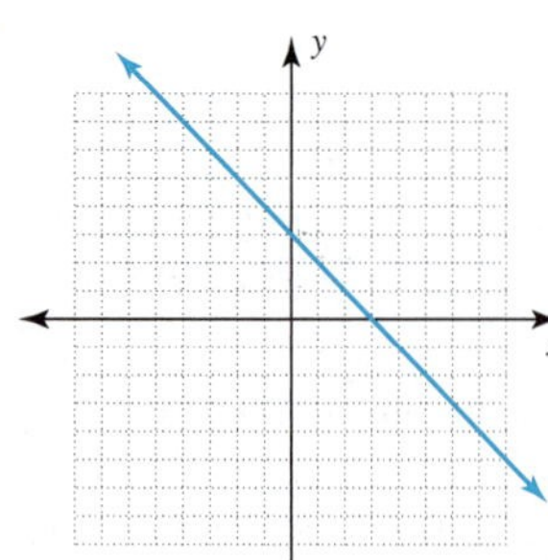

78.

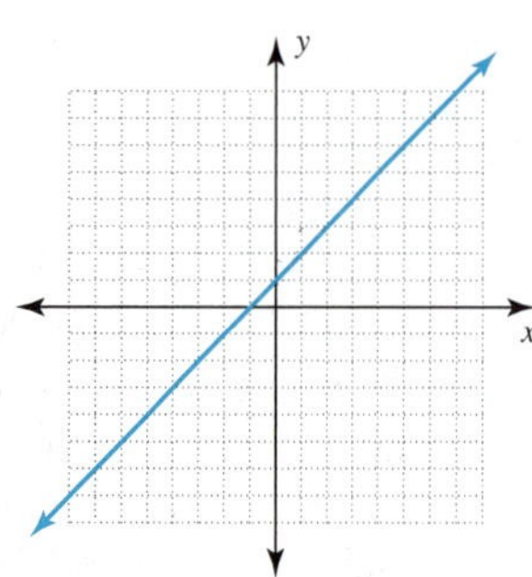

Complete each statement with **always, sometimes,** *or* **never.**

79. The slope of a line through the origin is __________ zero.

80. A line with an undefined slope is __________ the same as a line with a slope of zero.

81. Lines __________ have exactly one x-intercept.

82. The y-intercept of a nonvertical line through the origin is __________ (0, 0).

Skills | Calculator/Computer | **Career Applications** | Above and Beyond

83. **Allied Health** The recommended dosage d (in milligrams) of the antibiotic ampicillin sodium for children weighing less than 40 kg is given by the linear equation $d = 7.5w$, in which w represents the child's weight (in kilograms). Sketch a graph of this equation.

84. **Allied Health** The recommended dosage d (in micrograms) of neupogen (medication given to bone-marrow transplant patients) is given by the linear equation $d = 8w$, in which w is the patient's weight (in kilograms). Sketch a graph of this equation.

Mechanical Engineering The graph shows the bending moment of a wood beam at various points x feet from the left end of the beam. Use the graph to complete exercises 85 and 86.

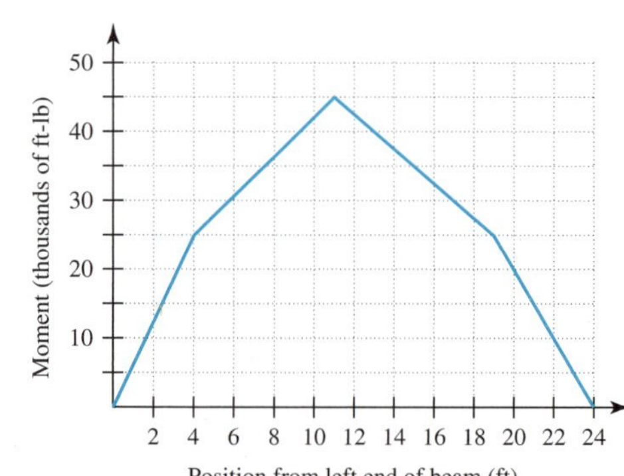

85. Determine the slope of the moment graph for points between 0 and 4 ft from the left end of the beam.

86. Determine the slope of the moment graph for points between 4 and 11 ft and between 11 and 19 ft.

Skills	Calculator/Computer	Career Applications	**Above and Beyond**

87. Complete the statement: "The difference between an undefined slope and a zero slope is. . . ."

88. Complete the statement: "The slope of a line tells you. . . ."

89. On two occasions last month, Sam Johnson rented a car on a business trip. Both times it was the same model from the same company, and both times it was in San Francisco. On both occasions he dropped the car at the airport booth and just got the total charge, not the details. Sam now has to fill out an expense account form and needs to know how much he was charged per mile and the base rate. All Sam knows is that he was charged \$210 for 625 mi on the first occasion and \$133.50 for 370 mi on the second trip. Sam called accounting to ask for help. Plot these two points on a graph, and draw the line that goes through them. What question does the slope of the line answer for Sam? How does the y-intercept help? Write a memo to Sam, explaining the answers to his questions and how a knowledge of algebra and graphing has helped you find the answers.

90. On the same graph, sketch each line.

$y = 2x - 1$ and $y = 2x + 3$

What do you observe about these graphs? Will the lines intersect?

91. Repeat Exercise 90, using

$y = -2x + 4$ and $y = -2x + 1$

92. On the same graph, sketch each line.

$y = \frac{2}{3}x$ and $y = -\frac{3}{2}x$

What do you observe concerning these graphs? Find the product of the slopes of these two lines.

93. Repeat Exercise 92, using

$y = \frac{4}{3}x$ and $y = -\frac{3}{4}x$

94. Based on Exercises 92 and 93, write an equation of a line that is perpendicular to

$y = \frac{3}{5}x$

Answers

1. 1 **3.** $\frac{4}{5}$ **5.** $\frac{4}{5}$ **7.** -2 **9.** $-\frac{3}{2}$ **11.** Undefined **13.** $\frac{5}{7}$ **15.** 0 **17.** $-\frac{4}{3}$ **19.** Slope 3; y-intercept: (0, 5) **21.** Slope -3; y-intercept: (0, -6) **23.** Slope $\frac{3}{4}$; y-intercept: (0, 1) **25.** Slope $\frac{2}{3}$; y-intercept: (0, 0) **27.** $f(x) = -\frac{4}{3}x + 4$; slope: $-\frac{4}{3}$; y-intercept: (0, 4) **29.** $f(x) = 9$; slope: 0; y-intercept: (0, 9) **31.** $f(x) = \frac{3}{2}x - 4$; slope: $\frac{3}{2}$; y-intercept: (0, -4)

33.

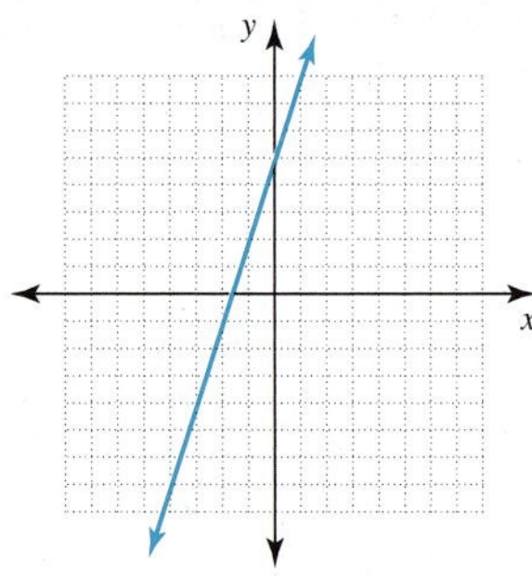

$y = 3x + 5$

35.

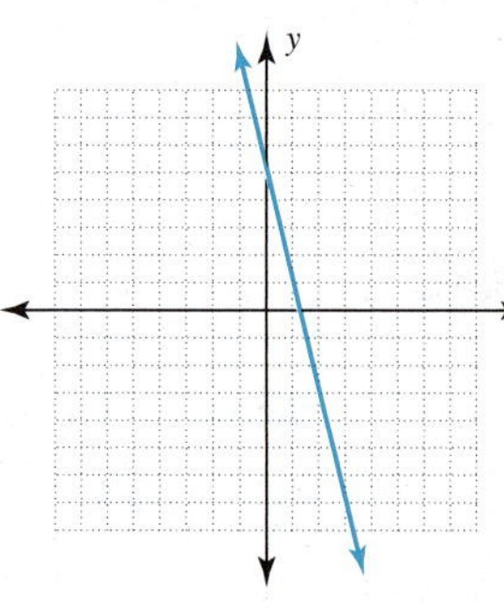

$y = -4x + 5$

37.

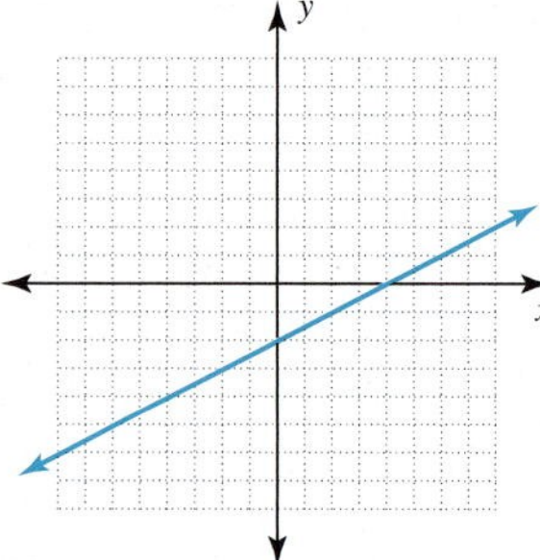

$y = \frac{1}{2}x - 2$

39.

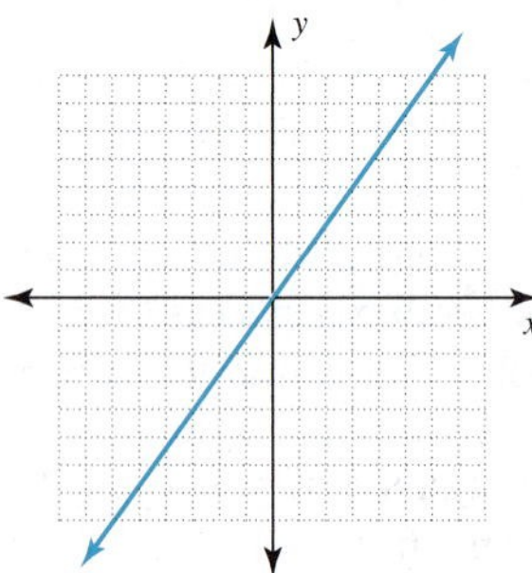

$y = \frac{4}{3}x$

41.

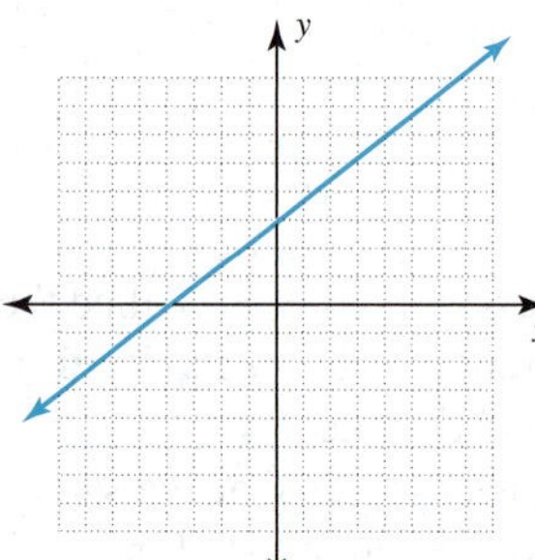

$y = \frac{3}{4}x + 3$

43.

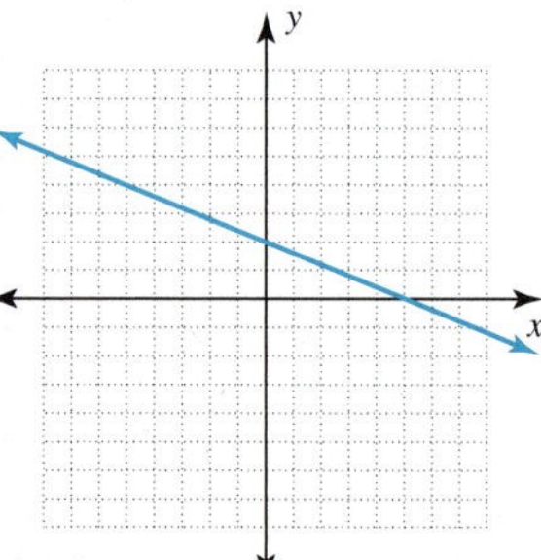

$y = -\frac{2}{5}x + 2$

45.

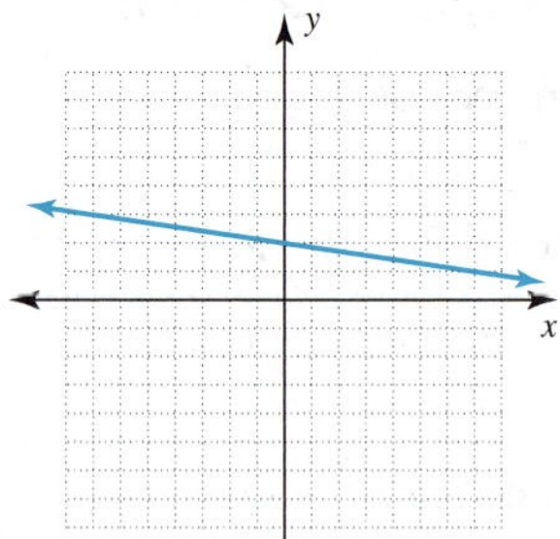

$y = -\frac{1}{7}x + 2$ **47.** 2 **49.** -2 **51.** 3 **53.** $-\frac{2}{5}$

55. Slope: 0.10, market price per pound; y-intercept: (0, 200), the minimum \$200 award

57. Slope represents price of newsprint; y-intercept represents the fixed cost **59.** -0.30 **61.** 3.1¢/yr **63.** IV **65.** III **67.** I

69. III and IV **71.** (g) **73.** (e) **75.** (h) **77.** (c) **79.** sometimes **81.** sometimes **83.**

d: 300, 250, 200, 150, 100, 50; w: 10, 20, 30, 40

85. 6,250 ft-lb per foot **87.** Above and Beyond **89.** Above and Beyond

91.

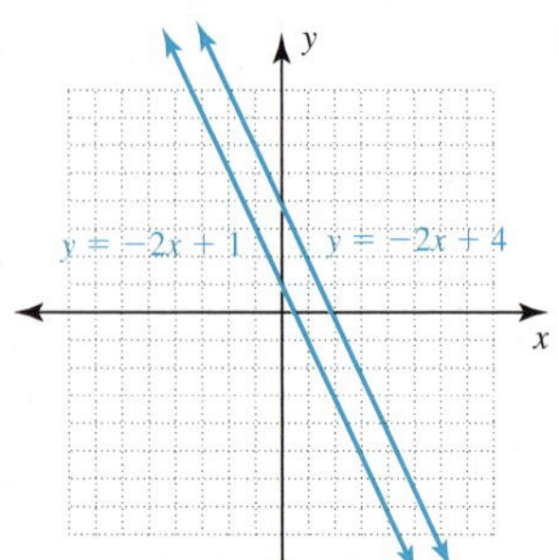

Parallel lines; no

93.

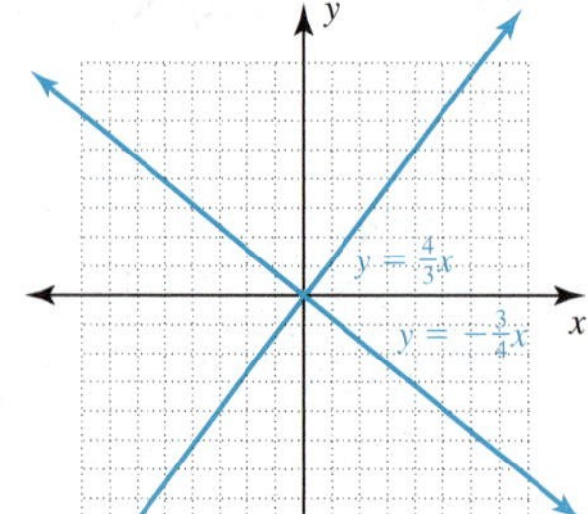

Perpendicular lines; -1

3.3 Linear Equations

< 3.3 Objectives >

1 > Determine whether lines are parallel or perpendicular

2 > Write an equation for a line from its slope and a point

3 > Write an equation of the line through two points

4 > Write an equation of a line based on geometric conditions

Recall that the form

$Ax + By = C$

in which A and B are not both zero, is called the **standard form for a linear equation.** In Section 3.2 we found the slope of a line through two points. We then used the slope to write an equation of the line.

In this section, we will see that the slope-intercept equation makes it easy to determine whether two lines are parallel, perpendicular, or neither.

Definition

Parallel Lines and Perpendicular Lines

When two nonvertical lines have the same slope, we say they are **parallel lines.**

When two lines meet at right angles, we say they are **perpendicular lines.**

We write the slopes of two nonvertical lines as m_1 and m_2. For parallel lines, it is always the case that

$m_1 = m_2$

NOTE

We assume that neither line is vertical. We discuss the special case involving vertical lines shortly.

For perpendicular lines, it is always true that the slopes are negative reciprocals. Algebraically, we write

$m_1 = -\frac{1}{m_2}$

Multiplying both sides by m_2 allows us to write this as

$m_1 \cdot m_2 = -1$

Example 1 illustrates this concept.

Example 1 Verifying That Two Lines Are Perpendicular

< Objective 1 >

Show that the graphs of $3x + 4y = 4$ and $-4x + 3y = 12$ are perpendicular lines.

First, we solve each equation for y.

$$
\begin{aligned}
3x + 4y &= 4 \\
4y &= -3x + 4 \\
y &= -\frac{3}{4}x + 1
\end{aligned}
$$

The slope of the line is $m_1 = -\frac{3}{4}$.

$$
\begin{aligned}
-4x + 3y &= 12 \\
3y &= 4x + 12 \\
y &= \frac{4}{3}x + 4
\end{aligned}
$$

The slope of the line is $m_2 = \frac{4}{3}$.

We now look at the product of the two slopes: $-\frac{3}{4} \cdot \frac{4}{3} = -1$. Any two lines whose slopes have a product of -1 are perpendicular lines. These two lines are perpendicular.

Check Yourself 1

Show that the graphs of the equations

$-3x + 2y = 4$ and $2x + 3y = 9$

are perpendicular lines.

In Example 2, we review how to use the slope-intercept equation to graph a line.

Example 2 Graphing a Line

Graph the line $2x + 3y = 3$.

Solving for y, we find the slope-intercept form for this equation is

NOTE

We treat $-\frac{2}{3}$ as $\frac{-2}{+3}$ to move to the right 3 units and down 2 units.

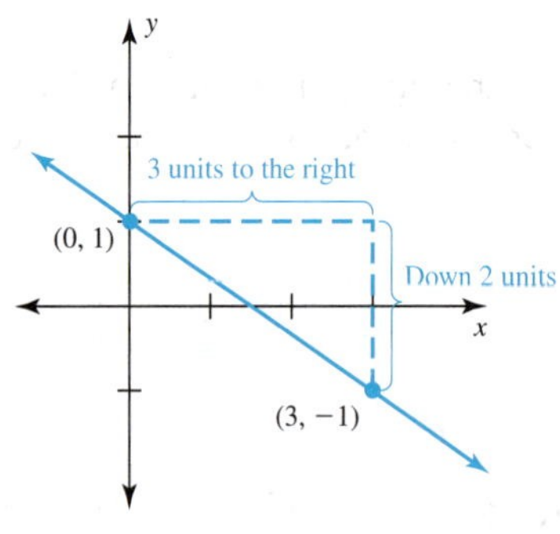

$$y = -\frac{2}{3}x + 1$$

To graph the line, plot the y-intercept at (0, 1). Because the slope m is equal to $-\frac{2}{3}$, we move from (0, 1) to the right 3 units and then *down* 2 units, to locate a second point on the graph of the line, here $(3, -1)$. We draw a line through the two points to complete the graph.

Check Yourself 2

Graph the line with equation

$3x - 4y = 8$

Hint: First write the equation in slope-intercept form.

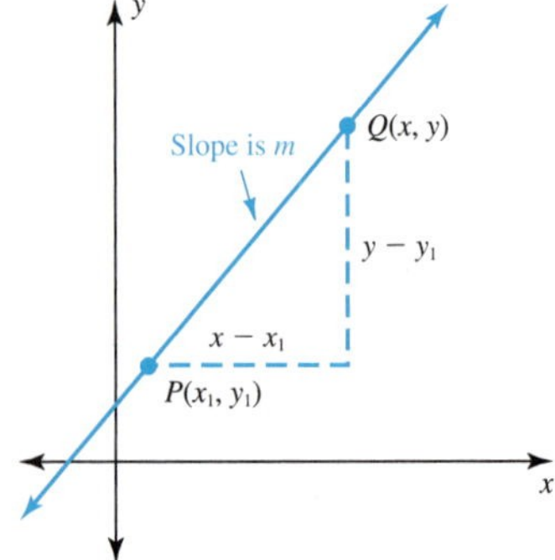

We know how to find an equation for a line from its slope and y-intercept. But, what if we have some point besides the y-intercept? Given any point and a slope, we can graph a line. Just plot the given point and then use the slope to find a second point by moving horizontally and vertically according to the slope.

However, we can use the definition of slope to find another useful form for the equation of a line. The **point-slope form** produces an equation for a line through any given point, not just the intercept.

Consider the line with slope m that passes through the point $P(x_1, y_1)$. We consider some unknown second point on the line $Q(x, y)$. From the slope formula, we have

$$m = \frac{y - y_1}{x - x_1}$$

We multiply both sides by $x - x_1$ to remove the fraction.

$$m = \frac{y - y_1}{x - x_1}$$

$m(x - x_1) = y - y_1$ Multiply both sides by $x - x_1$.

or $y - y_1 = m(x - x_1)$

This last equation is called the **point-slope form** of the equation of a line. All points lying on the line satisfy this equation.

Property

Point-Slope Form of the Equation of a Line

The point-slope equation of the line with slope m that passes through (x_1, y_1) is given by

$$y - y_1 = m(x - x_1)$$

Example 3 Finding the Equation of a Line

< Objective 2 >

Write an equation for the line that passes through (3, −1) with a slope of 3.

Letting $(x_1, y_1) = (3, -1)$ and $m = 3$, we use the point-slope form to get

$$y - (-1) = 3[x - (3)]$$

or $$y + 1 = 3(x - 3)$$

Usually, we write the equation of a line in slope-intercept form. To rewrite the equation in slope-intercept form, solve for y and simplify.

$y + 1 = 3(x - 3)$

$y + 1 = 3x - 9$ Distribute to remove the parentheses.

$y = 3x - 10$ Subtract 1 from both sides.

Check Yourself 3

Write the slope-intercept equation of the line that passes through $(-2, 4)$ with a slope of $\frac{3}{2}$.

Since we know that two points determine a line, it is natural that we should be able to write an equation for a line passing through two points. Using the point-slope equation together with the slope formula allows us to write such an equation.

Example 4 Finding the Equation of a Line

< Objective 3 >

Write the slope-intercept equation of the line passing through (2, 4) and (4, 7).

First, we find m, the slope of the line. Here

$$m = \frac{7 - 4}{4 - 2} = \frac{3}{2}$$

Now we apply the point-slope form with $m = \frac{3}{2}$ and $(x_1, y_1) = (2, 4)$.

$$\underset{y_1}{y - (4)} = \underset{m}{\left(\frac{3}{2}\right)}[x - \underset{x_1}{(2)}]$$

$$y - 4 = \frac{3}{2}(x - 2)$$

NOTE

We could just as well choose to let

$(x_1, y_1) = (4, 7)$

The slope-intercept form is the same in either case. Take time to verify this for yourself.

We finish by writing the equation in slope-intercept form.

$$y - 4 = \frac{3}{2}(x - 2)$$

$$y - 4 = \frac{3}{2}x - 3$$ Distribute to remove the parentheses: $\frac{3}{2} \cdot 2 = 3$.

$$y = \frac{3}{2}x + 1$$ Add 4 to both sides.

Check Yourself 4

Write the slope-intercept equation of the line passing through $(-2, 5)$ and $(1, 3)$.

A line with slope zero is a horizontal line. A line with an undefined slope is vertical. Example 5 illustrates the equations of such lines.

Example 5 Finding an Equation of a Line

< Objective 4 >

(a) Find an equation of the line passing through $(7, -2)$ with a slope of 0.

We could find the equation by letting $m = 0$. Substituting into the slope-intercept formula, we solve for b.

$$y = mx + b$$

$$-2 = 0(7) + b$$

$$-2 = b$$

So $\quad y = 0x - 2, \quad$ or $\quad y = -2$

It is far easier to remember that any line with a zero slope is a horizontal line and has the form

$$y = b$$

The value for b is always the y-coordinate of the given point.

Note that, for any horizontal line, all of the points have the same y-value. Look at the graph of the line $y = -2$. We labeled three points on the line. All of them have the same y-coordinate.

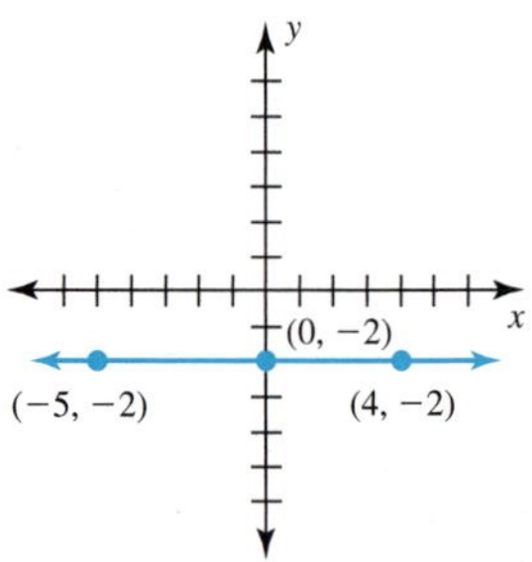

(b) Find an equation of the line with undefined slope passing through $(4, -5)$.

A line with undefined slope is vertical. It always has the form $x = a$, where a is the x-coordinate of the given point. The equation is

$$x = 4$$

Note that, for any vertical line, all of the points have the same x-value. Look at the graph of the line $x = 4$. All three labeled points have the same x-coordinate.

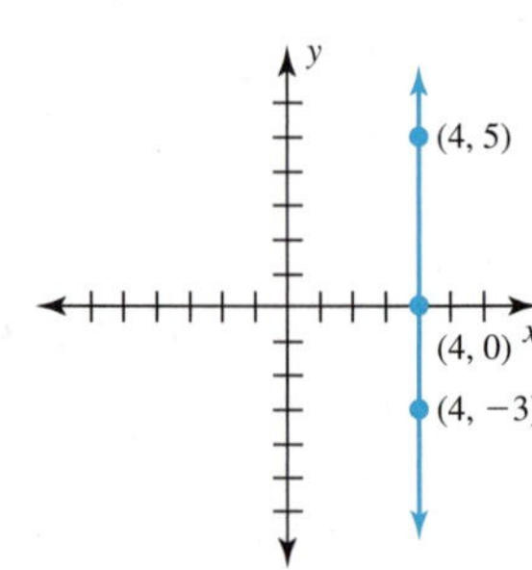

Check Yourself 5

(a) Find an equation of the line with zero slope that passes through $(-3, 5)$.

(b) Find an equation of the line passing through $(-3, -6)$ with undefined slope.

There are other ways of finding an equation of the line through two points. Example 6 shows such an approach.

Example 6 Finding the Equation of a Line

Write the slope-intercept equation of the line through $(-2, 3)$ and $(4, 5)$.

First, we find m.

$$m = \frac{(5) - (3)}{(4) - (-2)} = \frac{2}{6} = \frac{1}{3}$$

NOTE

We could, of course, use the point-slope form instead.

We now make use of the slope-intercept form, but in a different manner.

Using $y = mx + b$, and the slope just calculated, we can immediately write

$$y = \frac{1}{3}x + b$$

NOTE

We substitute these values because the line must pass through (−2, 3).

Now, if we substitute a known point for x and y, we can solve for b. We may choose either of the two given points. Using $(-2, 3)$, we have

$$3 = \frac{1}{3}(-2) + b$$

$$3 = -\frac{2}{3} + b$$

$$3 + \frac{2}{3} = b$$

$$b = \frac{11}{3}$$

Therefore, the equation of the line is

$$y = \frac{1}{3}x + \frac{11}{3}$$

Check Yourself 6

Repeat Check Yourself 4, using the technique illustrated in Example 6.

We now know that we can write an equation of a line once we have been given a point on the line and the slope of that line. In some applications, we need to use specified geometric conditions to find the slope of a line.

Example 7 Finding an Equation of a Parallel Line

Find the slope-intercept equation of the line passing through $(-4, -3)$ and parallel to the line determined by $3x + 4y = 12$.

First, we find the slope of the given parallel line.

$$3x + 4y = 12$$
$$4y = -3x + 12$$
$$y = -\frac{3}{4}x + 3$$

NOTES

The slope of the given line is $-\frac{3}{4}$, the coefficient of x.

The line must pass through $(-4, -3)$, so let $(x_1, y_1) = (-4, -3)$

The slopes of parallel lines are the same. Because the slope of the desired line must also be $-\frac{3}{4}$, we can use the point-slope form to write the required equation.

$$y - y_1 = m(x - x_1)$$

$$y - (-3) = -\frac{3}{4}[x - (-4)]$$ $m = -\frac{3}{4}$ is the slope;

$$y + 3 = -\frac{3}{4}(x + 4)$$ $(-4, -3)$ is a point on the line.

We simplify this to its slope-intercept form, $y = mx + b$.

$$y + 3 = -\frac{3}{4}(x + 4)$$

$$y + 3 = -\frac{3}{4}x - 3$$ Distribute to remove the parentheses.

$$y = -\frac{3}{4}x - 6$$ Subtract 3 from both sides.

Check Yourself 7

Find the slope-intercept equation of the line passing through $(2, -5)$ and parallel to the line determined by $4x - y = 9$.

Example 8 Finding an Equation of a Perpendicular Line

Find the slope-intercept equation of the line passing through $(3, -1)$ and perpendicular to the line $3x - 5y = 2$.

First, find the slope of the perpendicular line.

$$3x - 5y = 2$$
$$-5y = -3x + 2$$
$$y = \frac{3}{5}x - \frac{2}{5}$$

The slope of the perpendicular line is $\frac{3}{5}$. Recall that the slopes of perpendicular lines are negative reciprocals. The slope of our line is the negative reciprocal of $\frac{3}{5}$ or $-\frac{5}{3}$.

Using the point-slope equation, we have

$$y - (-1) = -\frac{5}{3}[x - (3)]$$
$$y + 1 = -\frac{5}{3}x + 5$$
$$y = -\frac{5}{3}x + 4$$

Check Yourself 8

Find the slope-intercept equation of the line passing through $(5, 4)$ and perpendicular to the line with equation $2x - 5y = 10$.

Before looking at some applications, we summarize the various forms a linear equation may take. The graph of each equation is a straight line.

Form	Equation for Line L	Conditions
Standard	$Ax + By = C$	Constants A and B cannot both be zero.
Slope-intercept	$y = mx + b$	Line has y-intercept $(0, b)$ and slope m.
Point-slope	$y - y_1 = m(x - x_1)$	Line passes through (x_1, y_1) with slope m.
Horizontal	$y = a$	Slope is zero.
Vertical	$x = b$	Slope is undefined.

In real-world applications, we rarely begin with the equation of a line. Rather, we usually have points relating two variables based on actual data. If we believe that the two variables are linearly related, we build the equation of the line through the data points. This line can then be used to determine other points and make predictions.

Example 9 A Business and Finance Application

NOTE

In applications, it is common to use letters other than x and y. In this case, we use C to represent the **cost**.

In producing a new line of lamps, a manufacturer predicts that the number of lamps produced x and the cost in dollars C of producing them are related by a linear equation.

Suppose that the cost of producing 100 lamps is \$5,000 and the cost of producing 500 lamps is \$15,000. Find the linear equation relating x and C.

To solve this problem, we must find an equation of the line passing through points (100, 5,000) and (500, 15,000). Even though the numbers are considerably larger than we have encountered thus far in this section, the process is exactly the same.

First, we find the slope:

$$m = \frac{15{,}000 - 5{,}000}{500 - 100} = \frac{10{,}000}{400} = 25$$

We use the point-slope formula to find the equation.

$$C - 5{,}000 = 25(x - 100)$$
$$C - 5{,}000 = 25x - 2{,}500$$
$$C = 25x + 2{,}500$$

NOTE

The scaling "distorts" the slope of the line.

To graph the equation, we must choose the scaling on the x- and C-axes carefully to get a "reasonable" picture. Here we choose increments of 100 on the x-axis and 2,500 on the C-axis since those seem appropriate for the given information.

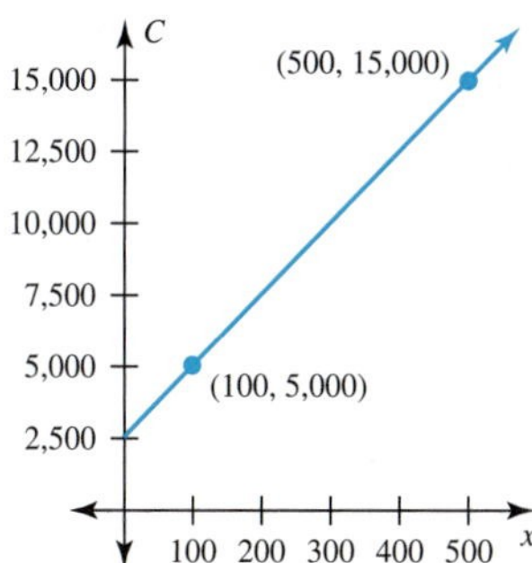

Check Yourself 9

A technical diving company predicts that the value in dollars V, and the time that a front-entry drysuit has been in use t, are related by a linear equation. If the drysuit is valued at \$1,500 after 2 years and at \$300 after 10 years, find the linear equation relating t and V.

Earlier, we mentioned that when working with applications, we are frequently asked to interpret the slope of a function as its *rate of change*.

Recall that the slope represents the change in the output, y or $f(x)$, when the input x is increased by one unit. We ask for such an interpretation in the next example, from the health sciences field.

Example 10 An Allied Health Application

A person's body mass index (BMI) can be calculated using his or her height h, in inches, and weight w, in pounds, with the formula

$$\text{BMI} = \frac{703w}{h^2}$$

NOTE

$69^2 = 4{,}761$

In the case of a 69-in. man, his height remains constant over many years, but his weight might vary, so we can model his body mass index as a function of his weight w.

$$B(w) = \frac{703}{4{,}761}w$$

(a) Find the body mass index of a 190-lb, 69-in. man (to the nearest tenth). We use the model above with $w = 190$.

$$\begin{aligned} B(190) &= \frac{703}{4{,}761}(190) \\ &= \frac{133{,}570}{4{,}761} \\ &\approx 28.1 \end{aligned}$$

(b) Determine the slope of this function.

The slope is $\dfrac{703}{4{,}761} \approx 0.15$

(c) Interpret the slope of this function in the context of the application.

The input of this function is the man's weight, which is given in pounds. Therefore, the slope can be interpreted as "for each additional pound that the man weighs, his body mass index increases by 0.15."

Check Yourself 10

Using the metric system, a person's body mass index (BMI) can be calculated using his or her height h, in centimeters, and weight w, in kilograms, with the formula

$$\text{BMI} = \frac{10{,}000w}{h^2}$$

In the case of a 160-cm woman, we can model her body mass index as a function of her weight w.

$$B(w) = \frac{25}{64}w$$

(a) Find the body mass index of a 160-cm, 70-kg woman (to the nearest tenth).
(b) Determine the slope of this function.
(c) Interpret the slope of this function in the context of the application.

NOTE

$$\frac{10{,}000}{h^2} = \frac{10{,}000}{160^2} = \frac{10{,}000}{25{,}600} = \frac{25}{64}$$

Check Yourself ANSWERS

1. $m_1 = \frac{3}{2}$ and $m_2 = -\frac{2}{3}$; $(m_1)(m_2) = -1$ **2.**

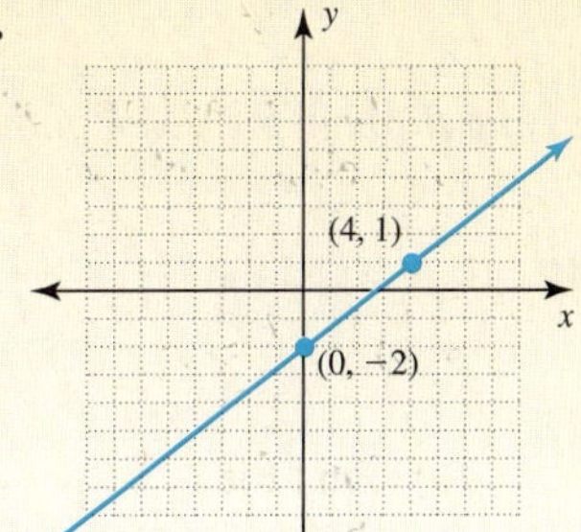

3. $y = \frac{3}{2}x + 7$

4. $y = -\frac{2}{3}x + \frac{11}{3}$ **5.** **(a)** $y = 5$; **(b)** $x = -3$ **6.** $y = -\frac{2}{3}x + \frac{11}{3}$ **7.** $y = 4x - 13$

8. $y = -\frac{5}{2}x + \frac{33}{2}$ **9.** $V = -150t + 1{,}800$ **10.** **(a)** $\frac{875}{32} \approx 27.3$; **(b)** $\frac{25}{64} \approx 0.39$;

(c) Each additional kilogram increases her BMI by 0.39.

Reading Your Text

These fill-in-the-blank exercises will help you understand some of the key vocabulary used in this section. The answers to these exercises are in the Answers Appendix in the back of the text.

(a) Two nonvertical lines are __________ if, and only if, their slopes are equal.

(b) Two nonvertical lines are __________ if, and only if, their slopes are negative reciprocals.

(c) A vertical line has a slope that is __________.

(d) A horizontal line has a slope that is __________.

3.3 exercises

Skills | Calculator/Computer | Career Applications | Above and Beyond

< Objective 1 >

Determine whether each pair of lines is **parallel, perpendicular,** *or* **neither.**

1. L_1 through $(-2, -3)$ and $(4, 3)$; L_2 through $(3, 5)$ and $(5, 7)$

2. L_1 through $(-2, 4)$ and $(1, 8)$; L_2 through $(-1, -1)$ and $(-5, 2)$

3. L_1 through $(7, 4)$ and $(5, -1)$; L_2 through $(8, 1)$ and $(-3, -2)$

4. L_1 through $(-2, -3)$ and $(3, -1)$; L_2 through $(-3, 1)$ and $(7, 5)$

5. L_1 with equation $x - 3y = 6$; L_2 with equation $3x + y = 3$

6. L_1 with equation $2x + 4y = 8$; L_2 with equation $4x + 8y = 10$

7. Find the slope of any line parallel to the line through points $(-2, 3)$ and $(4, 5)$.

8. Find the slope of any line perpendicular to the line through points $(0, 5)$ and $(-3, -4)$.

9. A line passing through $(-1, 2)$ and $(4, y)$ is parallel to a line with slope 2. What is the value of y?

10. A line passing through $(2, 3)$ and $(5, y)$ is perpendicular to a line with slope $\frac{3}{4}$. What is the value of y?

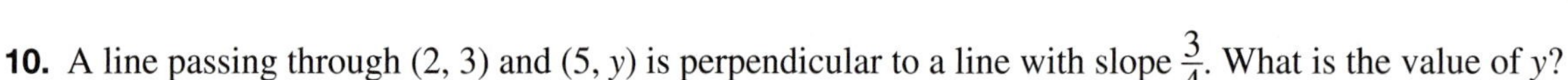

< Objective 2 >

Write the equation of the line passing through the given point with the indicated slope. Give your results in slope-intercept form, where possible.

11. $(0, -5)$, slope $\frac{5}{4}$

12. $(0, -4)$, slope $-\frac{3}{4}$

13. $(1, 3)$, slope 5

14. $(-1, 2)$, slope 3

15. $(-2, -3)$, slope -3

16. $(1, 3)$, slope -2

17. $(5, -3)$, slope $\frac{2}{5}$

18. $(4, 3)$, slope 0

19. $(1, -4)$, slope undefined

20. $(2, -5)$, slope $\frac{1}{4}$

21. $(5, 0)$, slope $-\frac{4}{5}$

22. $(-3, 4)$, slope undefined

< Objective 3 >

Write the equation for the line passing through each pair of points. Write your result in slope-intercept form, where possible.

23. $(2, 3)$ and $(5, 6)$

24. $(3, -2)$ and $(6, 4)$

25. $(-2, -3)$ and $(2, 0)$

26. $(-1, 3)$ and $(4, -2)$

27. $(-3, 2)$ and $(4, 2)$

28. $(-5, 3)$ and $(4, 1)$

29. $(2, 0)$ and $(0, -3)$

30. $(2, -3)$ and $(2, 4)$

31. $(0, 4)$ and $(-2, -1)$

32. $(-4, 1)$ and $(3, 1)$

< Objective 4 >

Write an equation for the line L satisfying the given geometric conditions.

33. L has slope 4 and y-intercept $(0, -2)$.

34. L has slope $-\frac{2}{3}$ and y-intercept $(0, 4)$.

35. L has x-intercept $(4, 0)$ and y-intercept $(0, 2)$.

36. L has x-intercept $(-2, 0)$ and slope $\frac{3}{4}$.

37. L has y-intercept $(0, 4)$ and a 0 slope.

38. L has x-intercept $(-2, 0)$ and an undefined slope.

39. L passes through $(2, 3)$ with a slope of -2.

40. L passes through $(-2, -4)$ with a slope of $-\frac{3}{2}$.

41. L has y-intercept $(0, 3)$ and is parallel to the line with equation $y = 3x - 5$.

42. L has y-intercept $(0, -3)$ and is parallel to the line with equation $y = \frac{2}{3}x + 1$.

43. L has y-intercept $(0, 4)$ and is perpendicular to the line with equation $y = -2x + 1$.

44. L has y-intercept $(0, 2)$ and is parallel to the line with equation $y = -1$.

45. L has y-intercept $(0, 3)$ and is parallel to the line with equation $y = 2$.

46. L has y-intercept $(0, 2)$ and is perpendicular to the line with equation $2x - 3y = 6$.

47. L passes through $(-4, 5)$ and is parallel to the line $y = -4x + 5$.

48. L passes through $(-4, 3)$ and is parallel to the line with equation $y = -2x + 1$.

49. L passes through $(3, 2)$ and is parallel to the line with equation $y = \frac{4}{3}x + 4$.

50. L passes through $(-2, -1)$ and is perpendicular to the line with equation $y = 3x + 1$.

51. L passes through $(3, -1)$ and is perpendicular to the line with equation $y = -\frac{2}{3}x + 5$.

52. L passes through $(-4, 2)$ and is perpendicular to the line with equation $y = 4x + 5$.

53. L passes through $(-2, 1)$ and is parallel to the line with equation $x + 2y = 4$.

54. L passes through $(-3, 5)$ and is parallel to the x-axis.

55. **Science and Medicine** A temperature of 10°C corresponds to a temperature of 50°F. 40°C corresponds to 104°F. Find the linear equation relating F and C.

56. **Business and Finance** In planning for a new line of cordless drills, a manufacturer assumes that the number of drills produced x and the cost in dollars C of producing these items are related by a linear equation. Projections are that 100 drills will cost \$10,000 to produce and that 300 drills will cost \$22,000 to produce. Find the equation that relates C and x.

57. **Business and Finance** Mike bills a customer \$65 per hour plus a fixed service call charge of \$75.

    (a) Write an equation to compute the total bill for any number of hours x that it takes to complete a job.

    (b) How much does a 3.5-hr job cost?

    (c) How many hours did a job have to take if the bill was \$247.25?

58. **Business and Finance** Two years after an expansion, a company had monthly sales of \$42,000. Four years later (6 years after the expansion) the monthly sales came to \$102,000. Assuming that the monthly sales in dollars S and the time t in years are related by a linear equation, find the equation relating S and t.

Use the graph to find the slope and y-intercept of each line.

59.

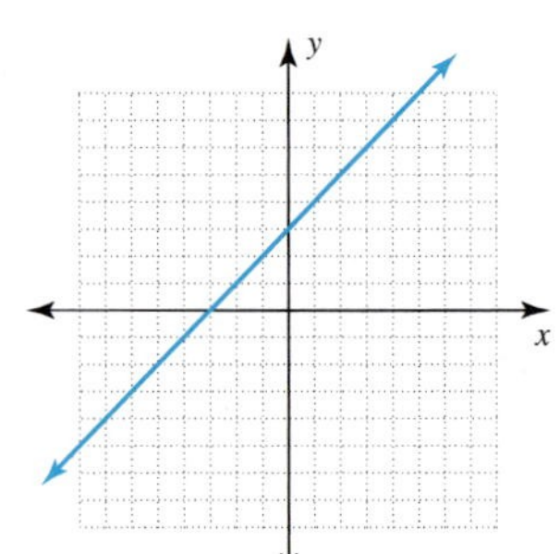

VIDEO

60.

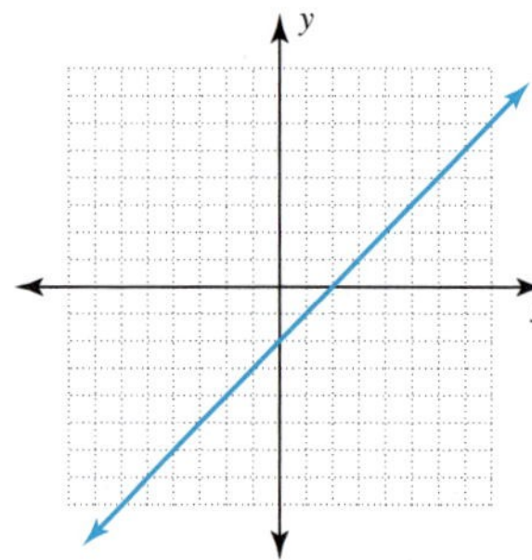

61.

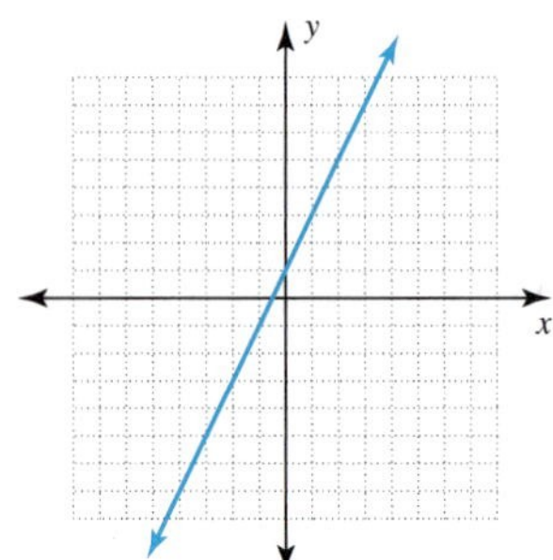

62.

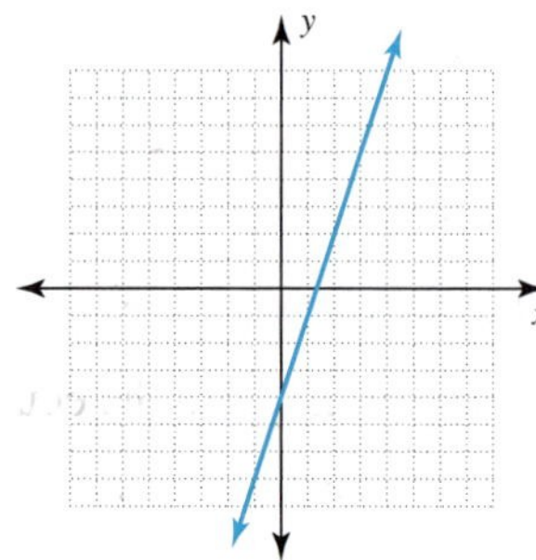

63.

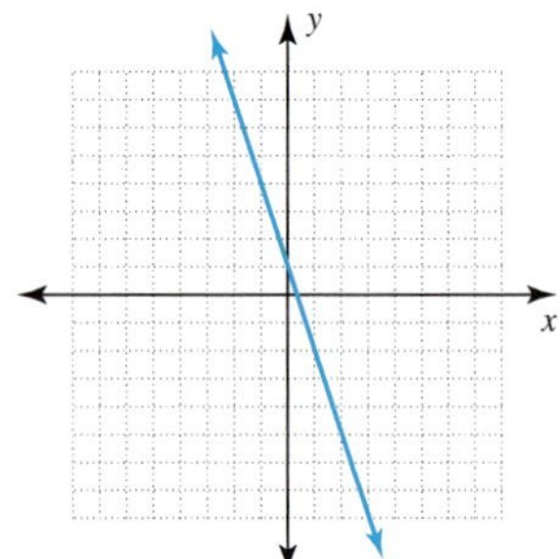

64.

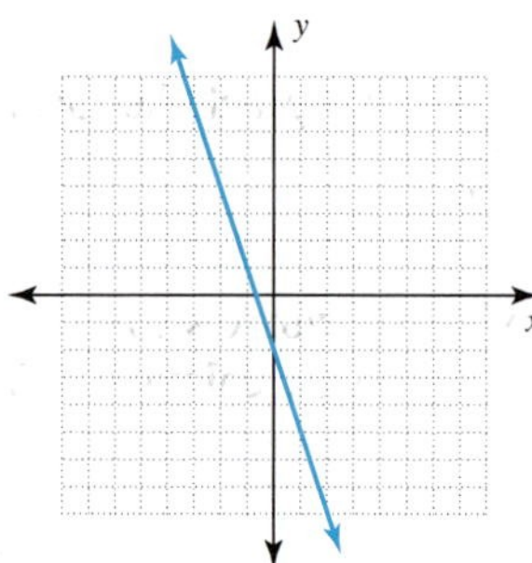

65.

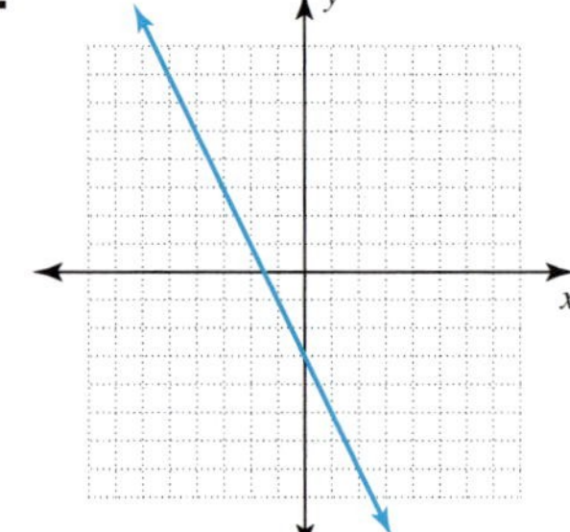

66.

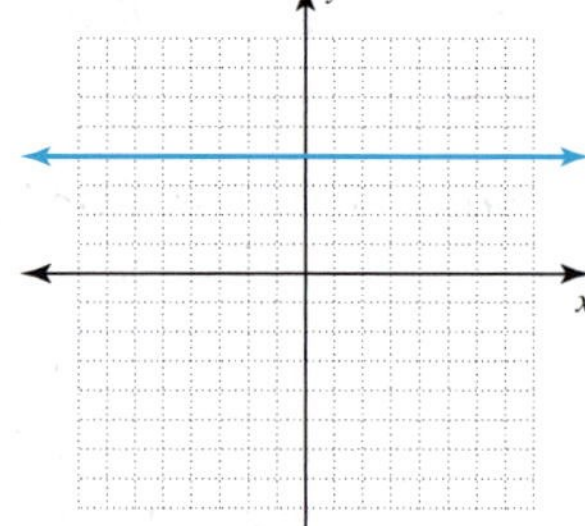

A four-sided figure (quadrilateral) is a parallelogram if the opposite sides have the same slope. If the adjacent sides are perpendicular, the figure is a rectangle. In exercises 67 to 70, for each quadrilateral ABCD, determine whether it is a parallelogram; then determine whether it is a rectangle.

67. $A(0, 0)$, $B(2, 0)$, $C(2, 3)$, $D(0, 3)$

68. $A(-3, 2)$, $B(1, -7)$, $C(3, -4)$, $D(-1, 5)$

69. $A(0, 0), B(4, 0), C(5, 2), D(1, 2)$

70. $A(-3, -5), B(2, 1), C(-4, 6), D(-9, 0)$

Determine whether each statement is **true** *or* **false.**

71. If two nonvertical lines are parallel, then they have the same slope.

72. If two lines are perpendicular, with slopes m_1 and m_2, then the product of the slopes is 1.

Complete each statement with **always, sometimes,** *or* **never.**

73. Given two points of a line, we can __________ find the equation of the line.

74. Given a nonvertical line, the slope of a line perpendicular to it is __________ zero.

Skills **Calculator/Computer** Career Applications Above and Beyond

75. Use a graphing calculator to graph the equations on the same screen.

$y = 0.5x + 7 \qquad y = 0.5x + 3 \qquad y = 0.5x - 1 \qquad y = 0.5x - 5$

Use the standard viewing window. Describe the results. 

76. Use a graphing calculator to graph the equations on the same screen.

$y = \frac{2}{3}x \qquad y = -\frac{3}{2}x$

Use the standard viewing window first, and then regraph using a Zsquare utility on the calculator. Describe the results.

Skills Calculator/Computer **Career Applications** Above and Beyond

77. **Agricultural Technology** The yield Y (in bushels per acre) for a cornfield is estimated from the amount of rainfall R (in inches) using the formula

$Y = \frac{43{,}560}{8000}R$ 

(a) Find the slope of the line described by this equation (to the nearest tenth).

(b) Interpret the slope in the context of this application.

78. **Agricultural Technology** During one summer period, the growth of corn plants followed a linear pattern approximated by the equation

$h = 1.77d + 24.92$

in which h was the height (in inches) of the corn plants and d was the number of days that passed.

(a) Find the slope of the line described by this equation.

(b) Interpret the slope in the context of this application.

Allied Health The arterial oxygen tension (P_aO_2, in mm Hg) of a patient can be estimated based on the patient's age *A* (in years). The equation used depends on the position of the patient. Use this information to complete exercises 79 and 80.

79. If a patient is lying down, the arterial oxygen tension can be approximated using the formula

$$P_aO_2 = 103.5 - 0.42A$$

(a) Determine the slope of this formula.

(b) Interpret the slope in the context of this application.

80. If a patient is seated, the arterial oxygen tension can be approximated using the formula

$$P_aO_2 = 104.2 - 0.27A$$

(a) Determine the slope of this formula.

(b) Interpret the slope in the context of this application.

Skills | Calculator/Computer | Career Applications | **Above and Beyond**

81. Describe the process for finding an equation for a line if you are given two points on the line.

82. How would you find an equation for a line if you were given the slope and the *x-intercept?*

Answers

1. Parallel **3.** Neither **5.** Perpendicular **7.** $\frac{1}{3}$ **9.** 12 **11.** $y = \frac{5}{4}x - 5$ **13.** $y = 5x - 2$ **15.** $y = -3x - 9$
17. $y = \frac{2}{5}x - 5$ **19.** $x = 1$ **21.** $y = -\frac{4}{5}x + 4$ **23.** $y = x + 1$ **25.** $y = \frac{3}{4}x - \frac{3}{2}$ **27.** $y = 2$ **29.** $y = \frac{3}{2}x - 3$
31. $y = \frac{5}{2}x + 4$ **33.** $y = 4x - 2$ **35.** $y = -\frac{1}{2}x + 2$ **37.** $y = 4$ **39.** $y = -2x + 7$ **41.** $y = 3x + 3$ **43.** $y = \frac{1}{2}x + 4$
45. $y = 3$ **47.** $y = -4x - 11$ **49.** $y = \frac{4}{3}x - 2$ **51.** $y = \frac{3}{2}x - \frac{11}{2}$ **53.** $y = -\frac{1}{2}x$ **55.** $F = \frac{9}{5}C + 32$
57. **(a)** $C = 65x + 75$; **(b)** \$302.50; **(c)** 2.65 hr **59.** Slope 1, *y*-intercept (0, 3) **61.** Slope 2, *y*-intercept (0, 1)
63. Slope −3, *y*-intercept (0, 1) **65.** Slope −2, *y*-intercept (0, −3) **67.** Yes; yes **69.** Yes; no **71.** True **73.** always
75. The lines are parallel.

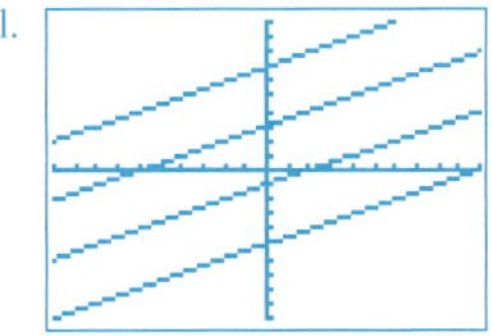

77. **(a)** 5.4; **(b)** Each additional inch of rainfall yields an additional 5.4 bushels per acre.

79. **(a)** −0.42; **(b)** Each additional year of age reduces the arterial oxygen tension by 0.42 mm Hg. **81.** Above and Beyond

3.4 Rate of Change and Linear Regression

< 3.4 Objectives >

1 > Construct a linear function to model an application

2 > Construct a linear function based on two data points

3 > Solve a function for an output value

4 > Use regression analysis to produce a linear model and make predictions

5 > Use a graphing calculator to perform a regression analysis

In this section, we bring together the two main ideas of Chapters 2 and 3: functions and linear equations. This will allow us to understand these powerful tools in real-world settings.

To begin, we look to gain a more complete understanding of the phenomena described by slope. There are several ways of interpreting the slope of a linear equation and the relationship between x and y. When graphing, we use slope to describe the steepness of a line and use

$$m = \frac{\text{change in } y}{\text{change in } x} = \frac{\text{rise}}{\text{run}}$$

to compute and interpret the slope.

This means that for an equation that has a slope of $\frac{2}{3}$ such as $y = \frac{2}{3}x - 3$, we interpret the slope to mean that y increases by 2 when x increases by 3. If the slope were -3, such as in the equation $y = -3x + 2$, we interpret the slope $m = -3 = \frac{-3}{1}$ to mean that y decreases by 3 when x increases by 1.

Another way of interpreting the slope of a line that is useful in applications uses the idea of *margin*. This interpretation always considers the change in y as x increases by 1.

In the example $y = \frac{2}{3}x - 3$, we think of the slope $m = \frac{2}{3}$ as a complex fraction.

$$m = \frac{\frac{2}{3}}{1}$$

In this case, we interpret the slope to mean that y increases by $\frac{2}{3}$ when x increases by 1. We call this the **rate of change of the function.** It is an especially useful interpretation when working with business applications and functions and is a powerful way of using the slope of a linear model.

We look at these two ways of interpreting the slope of a linear equation in Example 1. We then move on to understanding the properties the rate of change represents in applications.

RECALL

When the slope is negative, y *decreases* as x increases.

NOTE

These two interpretations are the same.

If y increases by $\frac{2}{3}$ when x increases by 1, then over three iterations, y increases by

$\frac{2}{3} + \frac{2}{3} + \frac{2}{3} = 2$

when x increases by

$1 + 1 + 1 = 3$.

Example 1 Interpreting the Slope

Interpret the slope of each equation.

(a) $y = 2x + 5$

Since $2 = \frac{2}{1}$, both interpretations are the same. In this case, y increases by 2 units when x increases by 1 unit.

(b) $y = -\frac{3}{4}x + 4$

We usually consider the sign in terms of y, so one interpretation reads y decreases by 3 units when x increases by 4 units.

In this case, the second interpretation says that y decreases by $\frac{3}{4}$ when x increases by 1.

NOTES

We also can use the fraction to write that y increases by $\frac{4}{5}$ when x increases by 1.

In part (d), n takes the place of x and C acts as y when formatted as

$C = mn + b$

(c) $y = 0.8x + 6.2$

We can use the decimal to write that y increases by 0.8 as x increases by 1. We could also rewrite it as a fraction, $m = 0.8 = \frac{4}{5}$, in which case, we would say that y increases by 4 when x increases by 5.

(d) $C = 25n + 120$

The slope $m = 25$ indicates that C increases by 25 when n increases by 1.

Check Yourself 1

Interpret the slope of each equation.

(a) $y = -x + 5$ **(b)** $y = -0.25x - 4$

(c) $y = \frac{x}{5} - 3$ **(d)** $P = 12x - 350$

In Example 2, we construct a linear function to model an application. Inherent in this modeling is an understanding of the slope as the rate of change of a function.

Example 2 Constructing a Function

< Objective 1 >

A store charges \$99.95 for a certain calculator. If we are interested in the revenue from the sales of this calculator, then the quantity that varies is the number of calculators sold.

We begin by identifying this quantity and representing it with a variable.

Let x be the number of calculators sold.

Because the store's revenue depends on the number of calculators sold and is computed by multiplying the number of calculators sold by the price of each, we identify, and name, a function to describe this relationship.

Let R represent the revenue from the sale of x calculators.

We use function notation to model this relationship.

RECALL

This does not mean R times x.

$R(x) = 99.95x$ Revenue is determined by multiplying the number of calculators sold by the price of each one.

NOTE

The input, or independent variable, represents the number of calculators sold.

The output, or function value, gives the revenue.

This function is read, "R of x is 99.95 times x." We say "R is a function of x."

The function above is a linear function. The y-intercept is (0, 0) because the store earns no revenue if it does not sell any calculators.

The slope of the function is 99.95. This means that each additional calculator sold increases the revenue \$99.95.

Consider the question, "How much is the store's revenue if it sells 10 calculators?" We are trying to find $R(10)$.

$R(10) = 99.95(10) = 999.50$ We replace x with 10 everywhere it appears.

The store earns \$999.50 in revenue from the sale of 10 calculators.

Check Yourself 2

A store sells coffee by the pound. It charges $6.99 for each pound of coffee beans.

(a) Construct a function that models its revenue from the sale of x pounds of coffee.
(b) Use the function created in (a) to determine its revenue if it sells 32.5 pounds of coffee.

In retail applications, it is common for revenue models to have the origin as the y-intercept because a business needs to make a sale in order to earn revenue.

On the other hand, most businesses incur costs independent of how much they sell. There are two types of costs that we focus on: *fixed cost* and *variable cost*.

The fixed cost represents the cost of running a business and having that business available. Fixed cost might include the cost of a lease, insurance costs, energy costs, salaried labor costs, and other costs.

Marginal cost represents the cost of each item being sold. For a retail store, this is usually the *wholesale price* of an item. For instance, if the store in Example 2 bought the calculators from the manufacturer for $64.95 each, then this is their wholesale price and represents the marginal cost associated with the calculators.

The variable cost of an item is the product of the marginal cost and the number of items sold.

Example 3 Modeling a Cost Function

A store purchases a graphing-calculator model at a wholesale price of $64.95 each. Additionally, the store has a weekly fixed cost of $450 associated with the sale of these graphing calculators.

> **NOTE**
>
> The slope of the function is given by the marginal cost, 64.95, because each additional calculator increases the cost $64.95.

(a) Construct a function to model the cost of selling these calculators.

Let x represent the number of these graphing calculators that the store buys.
Let C represent the cost of purchasing x calculators.

Then, we construct the cost function.

$$C(x) = \underbrace{64.95}_{\text{Marginal cost}}\,x + \underbrace{450}_{\text{Fixed cost}}$$

(b) Find the cost to the store if it sells 35 calculators in one week.

$$\begin{aligned} C(35) &= 64.95(35) + 450 \\ &= 2{,}723.25 \end{aligned}$$

It costs the store $2,723.25 to sell 35 calculators in one week.

(c) What is the additional cost if the store were to sell 36 calculators?

The slope gives the cost of selling one more unit. Therefore, selling one more calculator costs the store an additional $64.95.

Check Yourself 3

A store purchases coffee beans at a wholesale price of $4.50 per pound (lb). Its daily fixed cost, associated with the sale of coffee beans, is $60.

(a) Construct a function to model the cost of coffee-bean sales.
(b) Find the cost if the store sells 40 lb of coffee beans in a day.
(c) What is the additional cost to the store if it were to sell 41 lb of coffee?

In many cases, there isn't an explicit or obvious function to use. For instance, consider the question, "How will an increase in spending on advertising affect sales?" We might think that sales will increase, but we do not know by what amount.

We saw in Section 3.3 that we can construct a linear model if we have two points. Before moving to more complicated examples in which we need to use technology, we review the techniques for building a linear model from two points.

Example 4 Building a Linear Model

< Objective 2 >

A small financial services company spent \$40,000 on advertising one month. It earned \$325,000 in profit that month. The following month, the company increased its advertising budget to \$55,000 and saw its profit increase to \$445,000.

NOTE

Forty thousand dollars spent on advertising led to a profit of 325 thousand dollars.

Similarly, 55 thousand dollars in advertising gave the company 445 thousand dollars in profit.

(a) Use this information to determine two points and model the profit as a linear function of the advertising budget.

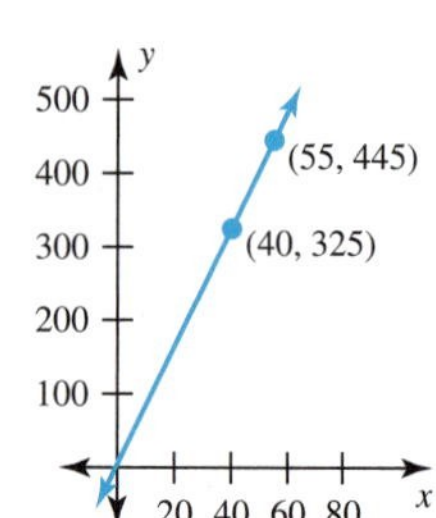

The company is interested in how its advertising spending affects its profit. As such, the independent, or input, variable is the amount of the advertising budget.

Let x be the amount spent on advertising in a month and let P be the profit that month.

We use thousands, as our unit, to keep the numbers reasonably small. This gives the points (40, 325) and (55, 445), because $P(40) = 325$ and $P(55) = 445$.

To construct a model, we find the slope of the line through our two points. We then use the point-slope equation for a line to construct the function.

$$m = \frac{y_2 - y_1}{x_2 - x_1} = \frac{445 - 325}{55 - 40} = \frac{120}{15} = 8$$

We choose to use the first point, (40, 325), in the point-slope formula with the slope $m = 8$. Recall that the point-slope formula for a line, given a point and the slope, is

$$y - y_1 = m(x - x_1)$$

Substitute into this formula:

$$y - (325) = (8)(x - 40)$$

$$y - 325 = 8x - 320 \quad \text{Distribute the 8 to remove the parentheses on the right.}$$

$$y = 8x + 5 \quad \text{Add 325 to both sides.}$$

$$P(x) = 8x + 5 \quad \text{Write the model using function notation.}$$

NOTE

Choosing the other point, (55, 445), leads to the same function.

$y - 445 = 8(x - 55)$

$y - 445 = 8x - 440$

$y = 8x + 5$

(b) Interpret the slope as the rate of change of this function in the context of this application.

The slope is 8. This means that each time x increases by 1, y increases by 8. In the context of this application, the company predicts that profit increases \$8,000 for each \$1,000 increase in advertising spending.

RECALL

The y-intercept occurs where the x-value is 0.

In this case, $x = 0$ corresponds to the company spending \$0 on advertising.

(c) Interpret the y-intercept in the context of this application.

The y-intercept is (0, 5). This means that the model predicts that the company would earn \$5,000 in profit if it did not spend anything on advertising.

Check Yourself 4

At an underwater depth of 30 ft, the atmospheric pressure is approximately 28.2 pounds per square inch (psi). At 80 feet, the pressure increases to approximately 50.7 psi.

(a) Construct a linear function modeling the pressure underwater as a function of the depth.
(b) Give the slope of the function. Interpret the slope in the context of this application.
(c) Give the y-intercept and interpret it in the context of this application.

Once we build a model to describe a situation, we can use the model in several different ways. We can examine different outputs based on changes to the input. Or, we can predict an input that would lead to a desired output. We do both in Example 5.

Example 5 A Business and Finance Model

< Objective 3 >

In Example 4, we modeled the profit earned by a financial services company as a function of its advertising budget: $P(x) = 8x + 5$, in thousands.

(a) Use this model to predict the company's profit if it budgets \$50,000 to advertising.

Because our units are thousands of dollars, we are being asked to find $P(50)$:

$P(50) = 8(50) + 5$ Replace x with 50 and evaluate.

$= 405$ Remember to follow the order of operations.

The company expects to earn \$405,000 in profit if it budgets \$50,000 to advertising.

(b) How much would it need to budget to advertising in order for their profit to reach \$500,000?

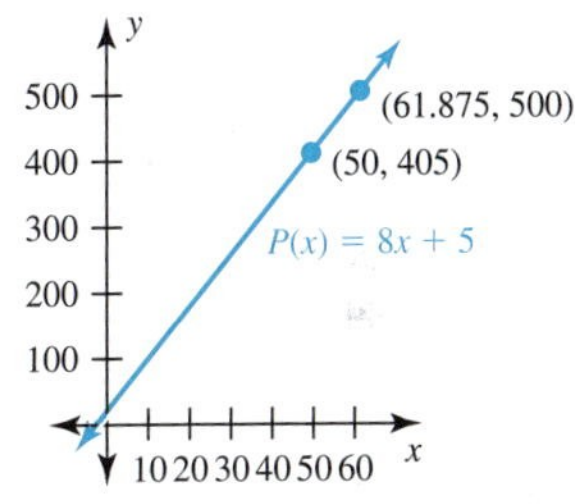

Do you see how this differs from the problem in (a)? This time, we are being given the profit, or output, of the function and being asked to find the appropriate input.

That is, we want to find x so that $P(x) = 500$.

We need to solve the equation

$8x + 5 = 500$ Because $P(x) = 500$, we set the expression $8x + 5$ equal to 500.

$8x = 495$ Subtract 5 from both sides.

$x = \frac{495}{8}$ Divide both sides by 8.

$= 61.875$

NOTE

Because x is in thousands, we have

$61.875 \times 1{,}000 = 61{,}875$

Therefore, in order to earn a \$500,000 profit, the company should invest \$61,875 in advertising.

Check Yourself 5

In Check Yourself 4, you modeled atmospheric pressure as a function of underwater depth: $P(x) = 0.45x + 14.7$.

(a) What is the approximate pressure felt by a diver at a depth of 130 ft?
(b) At what depth is the pressure 60 psi (to the nearest foot)?

Of course, a company does not usually have a nice model demonstrating its profit as a function of its advertising. More likely, a company might spend \$40,000 one month and see a \$325,000 return, but might earn \$300,000 the next time they spend that much in advertising. There are many factors that influence a company's profit, and advertising is just one of them.

When modeling an application, it is much more likely that the data set does not form a straight line, even when the underlying phenomenon is basically linear. This is especially true when the data being studied relate to human health, such as children's heights or drug studies. In these situations, each subject is unique because each is a person. No two people respond exactly the same to similar conditions or medication dosages.

NOTE

You will learn to construct regression models by hand when you study calculus. Until then, we use technology to do the computations.

We use the *regression analysis* techniques you learned in Activity 3 following Section 3.1. In that activity, you learned to enter data into a graphing calculator to create a scatter plot and to find the linear function that *best fits* the data.

Before performing a complete regression analysis, we look to interpret the output from a graphing calculator.

Example 6 Interpreting a Regression Analysis

< Objective 4 >

The average maximum temperature (in degrees Fahrenheit) and rainfall (in inches) for selected months over a 30-year period is shown.

Month	Jan.	Mar.	May	Jul.	Sep.	Nov.
Temperature	45.3°	55.9°	67.1°	79.9°	74.5°	52.5°
Rainfall	5.3	3.6	2.1	0.6	1.7	5.3

Source: World Climate; Portland International Airport

If we follow the methods described in Activity 3, we have

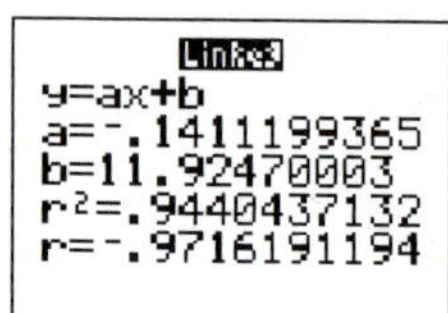
LinReg
y=ax+b
a=-.1411199365
b=11.92470003
r²=.9440437132
r=-.9716191194

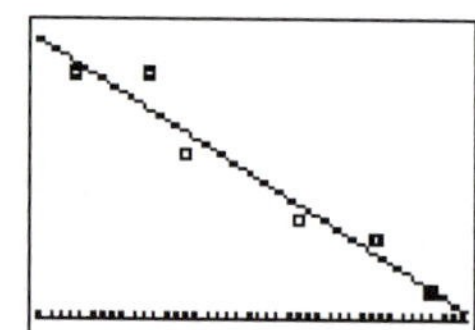

NOTE

r^2 measures whether a line is appropriate when modeling a data set. The closer r^2 is to 1, the better the model.

In this case, the model is very strong.

(a) Use function notation to write the equation of the best-fit line.

The regression analysis tells us that the slope of the best-fit line is approximately –0.14 and the y-intercept is approximately 11.92. The data can be modeled by the function

$f(x) = -0.14x + 11.92$

(b) Interpret the slope and y-intercept of the model in the context of this application.

The slope (or rate of change) tells us that for each degree increase in the average monthly high temperature, we would expect a decrease of 0.14 in. of rainfall that month.

The y-intercept tells us that in a month with a high temperature of 0°F, we should expect there to be 11.92 in. of rainfall.

(c) We can evaluate the function at various points in order to predict the rainfall or solve the function for a given amount of rainfall to determine the likely high temperature.

Use the model to predict the amount of rainfall in a month with a high temperature of 60ºF.

We are looking for the function when $x = 60$.

$$f(60) = -0.14(60) + 11.92$$
$$= 3.52$$

We would expect there to be 3.52 in. of rainfall if the monthly high temperature were 60°F.

(d) What monthly high temperature correlates to 5 in. of rain?

Now we are asked to solve the function for $f(x) = 5$.

$$-0.14x + 11.92 = 5$$
$$-0.14x = -6.92 \quad \text{Subtract 11.92 from both sides.}$$
$$x = 49.4 \quad \text{Divide by } -0.14 \text{ and round to one decimal place.}$$

We would expect 5 in. of rain in a month with a high temperature of 49.4°F.

Check Yourself 6

A company compares the number of visitors to its website to the number of online sales completed over a 6-week period.

Week	1	2	3	4	5	6
Visitors	120	105	146	168	140	132
Sales	41	35	48	52	50	44

If we plot this data and perform a regression analysis, we have

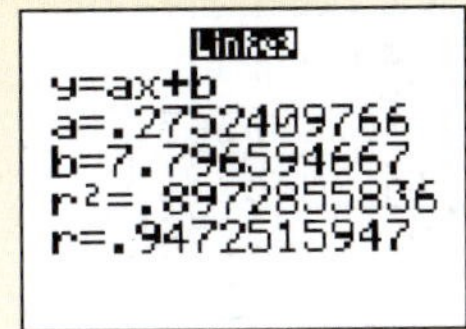

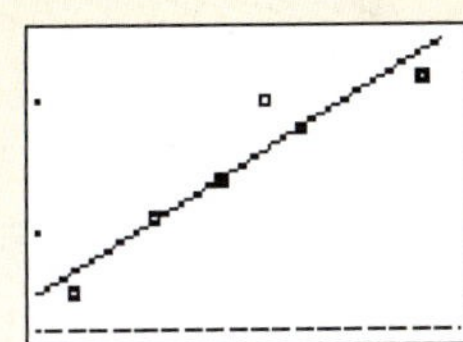

(a) Use function notation to write the line of best fit. Round the slope and y-intercept to the nearest tenth.

(b) Interpret the slope and y-intercept in the context of this application.

(c) How many sales would they expect if their website received 150 visitors in a week?

(d) How many visitors would they need in order to make 60 sales?

We are ready to put all of this together. In Examples 7 and 8, we use a graphing calculator to build regression models and then interpret the results.

Example 7 Linear Regression

< Objective 5 >

The table gives the recent population (in millions) and CO_2 emissions (in teragrams) for selected nations.

Nation	Population	Emissions
Australia	20	384
Austria	8	80
Belarus	10	55
Bulgaria	8	55
Canada	32	583
Czech Republic	10	126
Denmark	5	52
Finland	5	57
France	61	417
Germany	82	873
Italy	59	493
United Kingdom	60	558

Source: Statistics Division; United Nations

(a) Use a graphing calculator to create a scatter plot, perform a regression analysis, and graph the best-fit linear model on the scatter plot.

We follow the techniques learned in Activity 3 to create each screen.

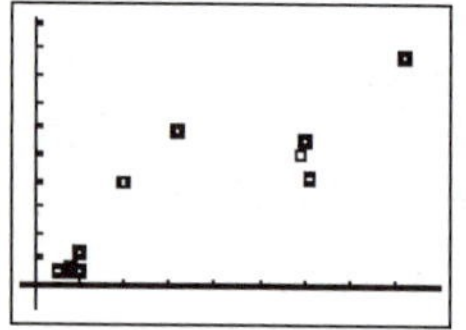

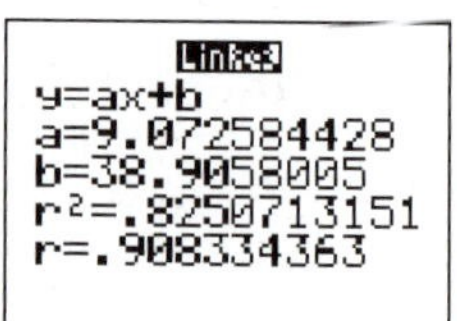

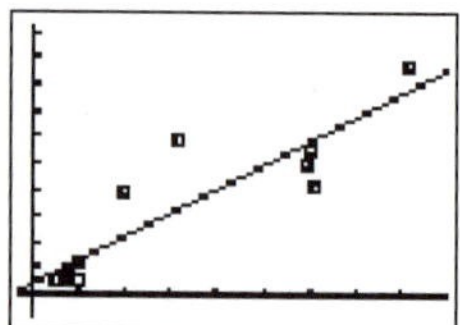

(b) Write the equation of the line-of-best-fit for the data, the linear regression model (round to the nearest tenth). What is the slope? Interpret the slope in the context of this application.

On the regression screen, a gives the slope of the line.

$y = 9.1x + 38.9$

The slope is approximately 9.1. We interpret this to mean that an increase of one million people leads to an increase of 9.1 teragrams of CO_2 emissions.

Check Yourself 7

The table gives the age (in months) and weight (in pounds) for a set of 10 boys.

Age x	12	12	13	15	15	16	19	20	20	24
Weight y	19	25	24	21	26	26	29	26	31	33

Source: Adapted from U.S. Center for Disease Control and Prevention data

(a) Use a graphing calculator to create a scatter plot, perform a regression analysis, and graph the best-fit linear model on the scatter plot.
(b) Write the equation of the line-of-best-fit for the data, the linear regression model (round to the nearest tenth). What is the slope? Interpret the slope in the context of this application.

Looking at Example 5 suggests how we can expand on Example 7. In Example 5, you evaluated the linear function at important points that were not part of the original model. We can use the line-of-best-fit in the same way.

Example 8 Using the Line-of-Best-Fit

Use the linear function constructed in Example 7 to answer each question. Use the model rounded to the nearest tenth.

(a) Estimate the CO_2 emissions (in teragrams) of a country if its population is 50 million people.

Use the function $f(x) = 9.1x + 38.9$, from Example 7, with $x = 50$.

$$f(50) = 9.1(50) + 38.9$$
$$= 493.9$$

We expect a country of 50 million people to emit approximately 493.9 teragrams of CO_2.

(b) That year, the United States emitted approximately 6,064 teragrams of CO_2. Use the line-of-best-fit from Example 7 (to the nearest tenth) to estimate the population of a nation that emits 6,064 teragrams of CO_2.

This time, we are being given the output, $f(x) = 6{,}064$, and asked to find the input that produces that result.

NOTE

The U.S. Census Bureau estimated the nation's 2010 population at 309 million.

$$9.1x + 38.9 = 6{,}064$$
$$9.1x = 6{,}025.1 \quad \text{Subtract 38.9 from both sides.}$$
$$x = 662.1 \quad \text{Divide both sides by 9.1; round to one decimal place.}$$

We would expect a nation that emits 6,064 teragrams of CO_2 to have a population about 662 million people.

Check Yourself 8

In Check Yourself 7, you constructed a model for a boy's weight, in pounds, based on his age, in months. Use that model, rounded to one decimal place, to answer each question.

(a) Estimate the weight of an 18-month-old boy.
(b) Estimate the age (to the nearest month) of a boy who weighs 25 lb.

Check Yourself ANSWERS

1. **(a)** y decreases by 1 when x increases by 1; **(b)** y decreases by 0.25 or $\frac{1}{4}$ when x increases by 1 or y decreases by 1 when x increases by 4; **(c)** y increases by 1 when x increases by 5 or y increases by $\frac{1}{5}$ when x increases by 1; **(d)** P increases by 12 when x increases by 1.
2. **(a)** $R(x) = 6.99x$; **(b)** \$227.18 **3.** **(a)** $C(x) = 4.5x + 60$; **(b)** \$240; **(c)** \$4.50
4. **(a)** $P(x) = 0.45x + 14.7$; **(b)** The slope is approximately 0.45; the pressure increases approximately 0.45 psi for each additional foot of depth underwater;
   (c) The y-intercept is approximately (0, 14.7). At the surface (depth is 0 ft), the atmospheric pressure is approximately 14.7 psi.
5. **(a)** 73.2 psi; **(b)** 101 ft
6. **(a)** $f(x) = 0.3x + 7.8$; **(b)** The slope is 0.3. There is an additional 0.3 sale for each additional visitor to the company's website. The y-intercept is (0, 7.8). The company would expect to make 7.8 sales in a week in which no one visited their website. **(c)** 52.8 sales; **(d)** 174 visitors
7. **(a)**

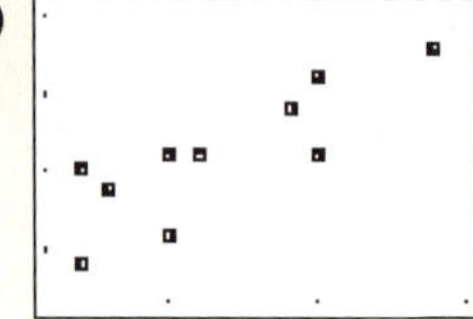

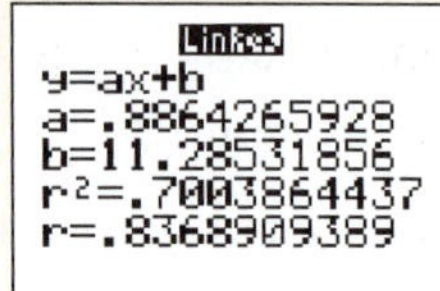

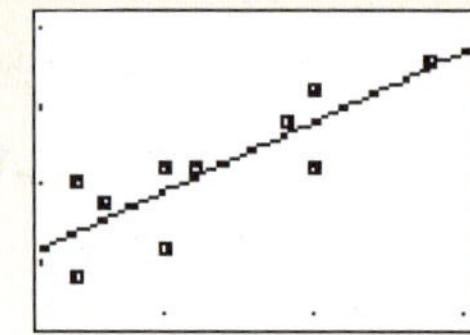

   (b) $y = 0.9x + 11.3$; the slope is approximately 0.9, which means that for each month that a boy ages, we expect him to gain an additional 0.9 lb.
8. **(a)** 27.5 lb; **(b)** 15 months

Reading Your Text

These fill-in-the-blank exercises will help you understand some of the key vocabulary used in this section. The answers to these exercises are in the Answers Appendix in the back of the text.

(a) The slope of a line represents the change in y when x is increased by ________ unit.

(b) The ________ of a linear function is called the rate of change of the function.

(c) The y-intercept occurs where the x-coordinate is ________.

(d) We use regression analysis to find the ________ that best fits a data set.

3.4 exercises

Skills | Calculator/Computer | Career Applications | Above and Beyond

Interpret the slope of each equation.

1. $y = 3x$ **2.** $y = -4x$ **3.** $y = -x$

4. $y = x + 5$ **5.** $y = \frac{3}{2}x - 1$ **6.** $y = -\frac{2x}{5} - 3$

7. $y = -0.4x + 11$ **8.** $y = 62.5x + 135$

< Objective 1 >

Model each relationship with a linear function.

9. In the snack department of a local supermarket, candy costs \$1.58 per pound.

10. A cheese pizza costs \$11.50. Each topping costs an additional \$1.25.

11. The perimeter of a square is a function of the length of a side.

12. The temperature in degrees Celsius is a function of the temperature in degrees Fahrenheit.

Hint: You can find this on the Internet, or look in some cookbooks or science books. You can even build it using

0°C = 32°F; 100°C = 212°F.

Use the functions constructed in exercises 9 to 12 and function notation to answer each question.

13. How much does it cost to purchase 7 pounds of candy (exercise 9)?

14. How much does a pizza with 3 toppings cost (exercise 10)?

15. What is the perimeter of a square if the length of a side is 18 cm (exercise 11)?

16. What is the Celsius equivalent of 65°F (exercise 12)?

< Objective 3 >

17. How much candy can be purchased for \$12 (exercise 9)?

18. How many pizza toppings can you get for \$14 (exercise 10)?

19. What is the length of a side of a square if its perimeter is 42 ft (exercise 11)?

20. What is the Fahrenheit equivalent of 13°C (exercise 12)?

< Objective 2 >

21. **Science and Medicine** At 3 months old, a kitten weighed 4 lb. It reached 9 lb by the time it was 8 months old.

(a) Construct a linear function modeling the kitten's weight as a function of its age.

(b) Give the slope of the function. Interpret the slope in the context of this application.

22. **Science and Medicine** A young girl weighed 25 lb at 24 months old. When she reached 30 months old, she weighed 27 lb.

(a) Construct a linear function modeling the girl's weight as a function of her age.

(b) Give the slope of the function. Interpret the slope in the context of this application.

23. **Information Technology** In AAC audio format, one song measured 3:13 (3 minutes 13 seconds or 193 seconds) and was 3.0 megabytes (MB) in size. A second song was 4:53 (293 seconds) long and took 4.2 MB.

(a) Construct a linear function modeling the size of a song as a function of length.

(b) Give the slope of the function. Interpret the slope in the context of this application.

(c) How much space would be required to store a 6:22 song (one decimal place)?

(d) How long would a song be if it required 4.6 MB (to the nearest second)?

24. **Social Science** A driver used 10.3 gal of gas driving 327 mi. The same driver drove 152 mi and used 5.4 gal.

(a) Construct a linear function modeling the gas used as a function of miles driven.

(b) Give the slope of the function. Interpret the slope in the context of this application.

(c) How much gas would be required to drive 225 mi (one decimal place)?

(d) How far can the driver go on 12 gal of gas (to the nearest mile)?

< Objective 4 >

25. TECHNOLOGY The table gives the highway fuel efficiency (in miles per gallon) of a particular automobile based on its speed (in miles per hour). Using a calculator to perform a regression analysis and a graph of this data gives the calculator screens shown.

Fuel Efficiency vs Speed

Speed (mi/hr)	Fuel Efficiency (mi/gal)
55	30.0
60	29.1
65	27.6
70	24.9
75	23.1

Source: U.S. Department of Energy

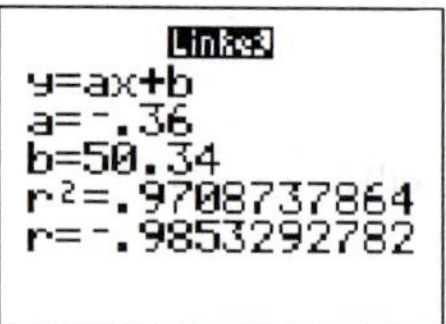

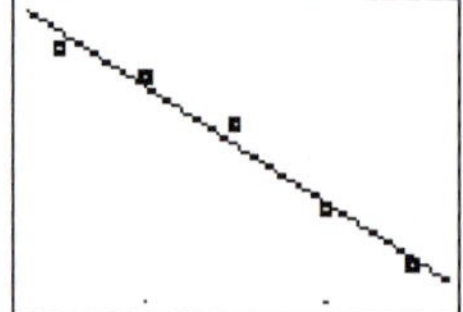

(a) Use function notation to write the equation of the best-fit line.

(b) Interpret the slope and y-intercept of the model in the context of this application.

(c) According to the model, what fuel efficiency would you expect if you drove 80 mi/hr? Round your result to one decimal place.

(d) What speed correlates to 25.0 mi/gal? Round your result to one decimal place.

26. BUSINESS AND FINANCE A college publishes its salary scale for one category of professor, based on the number of years of service. Using a calculator to perform a regression analysis and a graph of this data gives the calculator screens shown.

Salary (in thousands of dollars)

Years	Salary
0	$41.9
5	47.4
10	53.6
15	60.7
20	68.7

Source: College of Southern Nevada

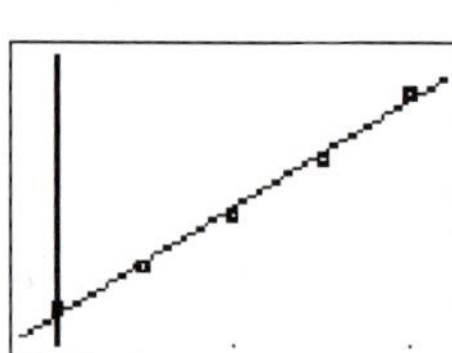

(a) Use function notation to write the equation of the best-fit line. Round the slope and y-intercept to one decimal place.

(b) Interpret the slope and y-intercept of the model in the context of this application.

(c) According to the model, how much would a professor with 17 years of service earn?

(d) If a professor earns $50,000, how many years have they been at the college? Round your results to the nearest year.

27. Business and Finance Median annual earnings for full-time employees, by education levels, are shown in the table. Using a calculator to perform a regression analysis and a graph of this data gives the calculator screens shown.

Earnings (in thousands of dollars)

Years	Earnings
12	$34.2
14	44.1
16	57.0
18	70.0
20	88.9

Source: U.S. Census Bureau

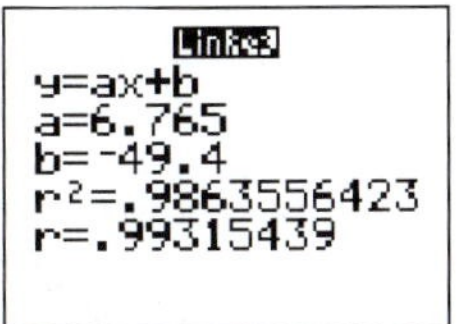

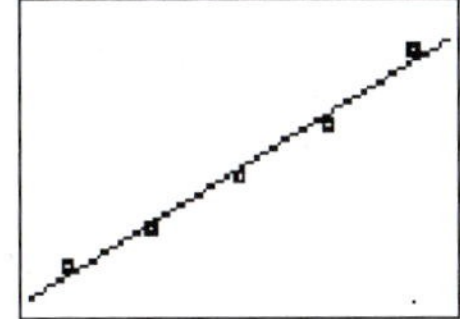

(a) Use function notation to write the equation of the best-fit line. Round the slope and y-intercept to one decimal place.

(b) Interpret the slope and y-intercept of the model in the context of this application.

(c) According to the model, how much would a person earn if they had 15 years of education?

(d) According to the model, how many years of education does a person need in order to earn $75,000 per year? Round your results to the nearest year.

28. Business and Finance The table shows the total annual amount spent on health care (in billions of dollars) in the United States over a series of years. Year 0 corresponds to 1990, year 5 corresponds to 1995, and so on. Using a calculator to perform a regression analysis and a graph of this data gives the calculator screens shown.

Years	Expenditures (billions)
0	$ 724
5	1,028
10	1,377
15	2,029
20	2,594

Source: Center for Medicare and Medicaid Services

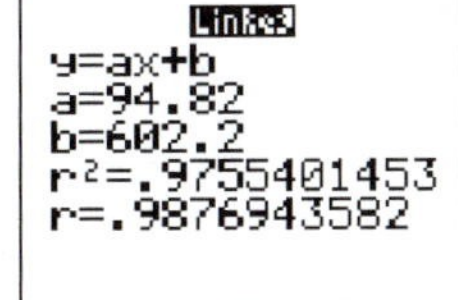

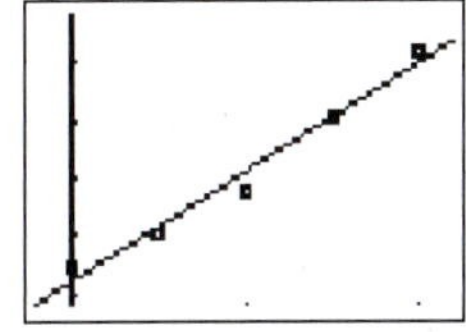

(a) Use function notation to write the equation of the best-fit line. Round the slope and y-intercept to one decimal place.

(b) Interpret the slope and y-intercept of the model in the context of this application.

(c) According to the model, how much was spent on health care in 2008?

(d) According to the model, when did health care spending surpass $2,000 billion? Round your results to the nearest year.

< Objective 5 >

29. Science and Medicine The table gives the age (in months) and weight (in pounds) for a set of 10 girls.

Age	24	24	25	26	26	28	32	32	33	36
Weight	23	29	28	32	30	26	27	35	33	35

Source: Adapted from U.S. Center for Disease Control and Prevention data

(a) Use a graphing calculator to create a scatter plot, perform a regression analysis, and graph the best-fit linear model on the scatter plot.

(b) Write the equation of the line-of-best-fit for the data, the linear regression model (round to the nearest tenth).

(c) What is the slope? Interpret the slope in the context of this application.

30. **Social Science** The table gives recent population (in millions) and CO_2 emissions (in teragrams) for selected nations.

Nation	Population	Emissions
Croatia	4	23
Greece	11	110
Ireland	4	46
Japan	128	1,288
Netherlands	16	181
Portugal	11	66
Russian Federation	143	1,698
Spain	44	352
Turkey	72	242
United States	296	6,064

Source: Statistics Division; United Nations

(a) Use a graphing calculator to create a scatter plot, perform a regression analysis, and graph the best-fit linear model on the scatter plot.

(b) Write the equation of the line-of-best-fit for the data, the linear regression model (round to the nearest tenth).

(c) What is the slope? Interpret the slope in the context of this application.

(d) How does the slope compare to that found in Example 7? Provide a reason for this discrepancy.

31. **Statistics** A brief review of ten syndicated news columns showed the number of words and the number of characters (including punctuation but not spaces) in the fourth paragraph of each column.

Words	53	90	52	27	22	49	25	44	87	98
Characters	281	510	324	142	119	233	128	225	435	417

(a) Use a graphing calculator to create a scatter plot, perform a regression analysis, and graph the best-fit linear model on the scatter plot.

(b) Write the equation of the line-of-best-fit for the data, the linear regression model (round to the nearest tenth).

(c) What is the slope? Interpret the slope in the context of this application.

(d) How many characters would you expect if a paragraph had 75 words?

(e) If a paragraph required 200 characters, how many words would you expect it to have?

32. **Science and Medicine** Each of the Great Lakes contains many islands. The table below compares the number of islands in each lake to the total area of the lake's islands (in thousands of acres).

Lake	Islands	Area
Superior	41	390
Michigan	21	96
Huron	66	979
Erie	7	25
Ontario	16	82

Source: U.S. National Oceanic and Atmospheric Administration

(a) Use a graphing calculator to create a scatter plot, perform a regression analysis, and graph the best-fit linear model on the scatter plot.

(b) Write the equation of the line-of-best-fit for the data, the linear regression model (round to the nearest tenth).

(c) What is the slope? Interpret the slope in the context of this application.

(d) How much area would you expect 30 islands to require in a lake similar to a Great Lake?

(e) In a lake similar to a Great Lake, if islands made up 500,000 acres, how many islands would you expect?

Skills | Calculator/Computer | **Career Applications** | Above and Beyond

33. **Allied Health** Dimercaprol (BAL) is used to treat arsenic poisoning in mammals. The recommended dose is 4 mg per kilogram of the animal's weight.

(a) Construct a linear function describing the relationship between the recommended dose and the animal's weight.

(b) How much BAL must be administered to a 5-kg cat?

(c) What size cow requires a 1,450-mg dose of BAL?

34. **Allied Health** Yohimbine is used to reverse the effects of xylazine in deer. The recommended dose is 0.125 mg per kilogram of the deer's weight.

(a) Express the recommended dosage as a linear function of a deer's weight.

(b) How much yohimbine should be administered to a 15-kg fawn?

(c) What size deer requires a 5.0-mg dose of yohimbine?

35. **Allied Health** An abdominal tumor originally weighed 32 g. Every day, chemotherapy treatment reduces the size of the tumor by 2.33 g.

(a) Express the size of the tumor as a linear function of the number of days spent in chemotherapy.

(b) How much does the tumor weigh after 5 days of treatment?

(c) How many days of chemotherapy are required to eliminate the tumor?

36. **Allied Health** A brain tumor originally weighs 41 g. Every day of chemotherapy treatment reduces the size of the tumor by 0.83 g.

(a) Express the size of the tumor as a linear function of the number of days spent in chemotherapy.

(b) How much does the tumor weigh after 2 weeks of treatment?

(c) How many days of chemotherapy are required to eliminate the tumor?

37. **Mechanical Engineering** The input force required to lift an object with a two-pulley system is equal to one-half the object's weight plus eight pounds (to overcome friction).

(a) Express the input force required as a linear function of an object's weight.

(b) Report the input force required to lift a 300-lb object.

(c) How much weight can be lifted with an input force of 650 lb?

38. **Mechanical Engineering** The pitch of a 6-in. gear is given by the number of teeth the gear has divided by six.

(a) Express the pitch as a linear function of the number of teeth.

(b) Report the pitch of a 6-in. gear with 30 teeth.

(c) How many teeth does a 6-in. gear have if its pitch is 8?

39. **Mechanical Engineering** The working depth of a gear (in inches) is given by 2.157 divided by the pitch of the gear.

(a) Express the depth of a gear as a function of its pitch. (This function is not linear.)

(b) What is the working depth of a gear that has a pitch of 3.5 (round your result to the nearest hundredth of an inch)?

40. **Mechanical Engineering** Use exercises 38 and 39 to determine the working depth of a 6-in. gear with 42 teeth (to the nearest hundredth of an inch).

Electronics A temperature sensor outputs voltage at a certain temperature. The output voltage varies linearly with respect to temperature. For a particular sensor, the function describing the voltage output V for a given Celsius temperature x is given by

$V(x) = 0.28x + 2.2$

41. Determine the output voltage if $x = 0°C$.

42. Evaluate $V(22°C)$.

43. Determine the temperature if the sensor puts out 7.8 V.

44. Interpret the slope of the model in the context of this application.

Extrapolation and Interpolation In Exercise 21, you modeled a kitten's weight (in pounds) based on its age (in months).

$W(x) = x + 1$

This model gives the weight of a kitten as 1 more than its age.

45. How much should a 7-month-old kitten weigh?

46. According to the model, how much should the kitten weigh when it is 5 years old (60 months)?

47. Write a paragraph giving your interpretations of the answers to exercises 45 and 46.

The extrapolation problem, above, has difficulty with making predictions based on data-derived models. Every model should be accompanied by a *domain* stating the input values for which the model is valid. Making predictions outside the given data is called *extrapolation*.

For instance, in the kitten model, the domain might be $3 \le x \le 12$, which means that the model could be used on kittens at least 3 months old but not older than a year. This would make sense because as they become cats, their growth rates (and weight gain) slows.

On the other hand, exercise 45 asks you to *interpolate*. This means that you are making a prediction based on an input (7 months) that is between the extremes of your data. That is, your data points were for a 3-month-old and an 8-month-old kitten.

We can usually extrapolate near to our data. For instance, it might be safe to predict the weight of a 10-month-old kitten, but a 5-year-old cat will not weigh 61 lb!

In exercise 29, you modeled a young girl's weight as a function of her age, based on 10 girls between 24 months old and 36 months old.

$W(x) = 0.6x + 12.9$

48. According to the model, how much should a 32-month-old girl weigh?

49. According to the model, how much should a 40-month-old girl weigh?

50. According to the model, how much should a 50-year-old (600 months) woman weigh?

51. Which of the predictions above are interpolations and which are extrapolations?

52. Write a paragraph interpreting your predictions in exercises 48–50.

Answers

1. y increases by 3 units when x increases by 1 unit. **3.** y decreases by 1 unit when x increases by 1 unit. **5.** y increases by 3 units when x increases by 2 units. y increases by $\frac{3}{2}$ when x increases by 1 unit. **7.** y decreases by 0.4 unit when x increases by 1 unit.
9. $C(x) = 1.58x$ **11.** $P(s) = 4s$ **13.** $C(7) = 11.06$; $11.06 **15.** $P(18) = 72$; 72 cm **17.** $C(7.59) \approx 12$; 7.59 lb
19. $P(10.5) = 42$; 10.5 ft **21.** **(a)** $W(x) = x + 1$; **(b)** 1; the kitten gained one pound per month.
23. **(a)** $f(x) = 0.012x + 0.684$; **(b)** 0.012; each second requires 0.012 MB of space. **(c)** 5.3 MB; **(d)** 5:26 or 326 s
25. **(a)** $f(x) = -0.36x + 50.34$; **(b)** The slope is -0.36; each additional mi/hr of speed reduces fuel efficiency by 0.36 mi/gal. The y-intercept is $(0, 50.34)$; if the car were driven at 0 mi/hr, it would get 50.34 mi/gal. **Note:** In this case, the y-intercept does not apply to any realistic phenomenon. **(c)** 21.5 mi/gal; **(d)** 70.4 mi/hr.
27. **(a)** $f(x) = 6.8x - 49.4$; **(b)** The slope is 6.8; a person's annual earnings increase $6,800 with each additional year of education. The y-intercept is $(0, -49.4)$; a person with no education is predicted to earn a negative salary. **Note:** In this case, the y-intercept does not apply to any realistic phenomenon. **(c)** $52,600; **(d)** 18 yr.
29. **(a)**

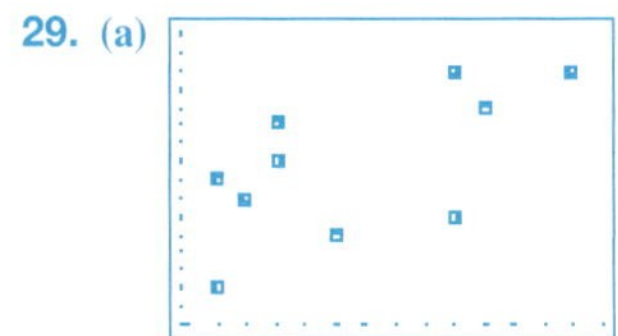

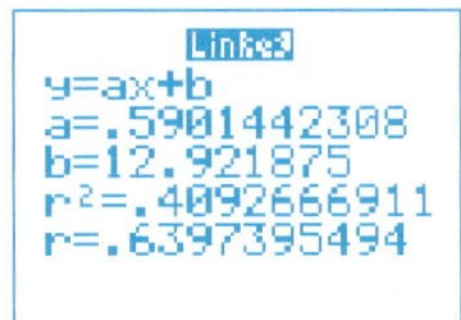

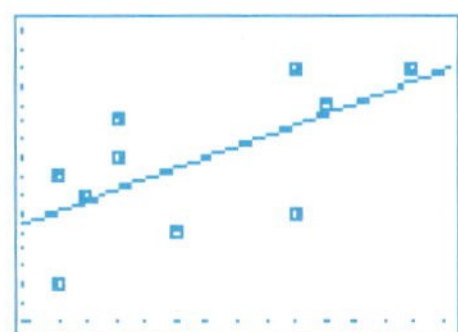

(b) $y = 0.6x + 12.9$; **(c)** 0.6; a young girl's weight increases about 0.6 lb for every month she ages.

31. **(a)**

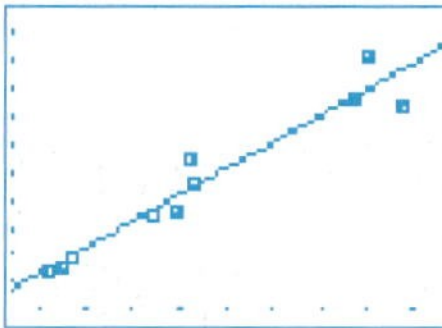

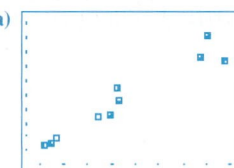

(b) $y = 4.7x + 21.9$; **(c)** 4.7; each additional word leads to approximately 4.7 additional characters in a paragraph; **(d)** 374.4 characters; **(e)** 37.9 words

33. **(a)** $d(x) = 4x$; **(b)** 20 mg; **(c)** 362.5 kg **35.** **(a)** $W(x) = -2.33x + 32$; **(b)** 20.35 g; **(c)** 14 days

37. **(a)** $f(x) = \frac{1}{2}x + 8$; **(b)** 158 lb; **(c)** 1,284 lb **39.** **(a)** $D(p) = \frac{2.157}{p}$; **(b)** 0.62 in. **41.** 2.2 V **43.** 20ºC **45.** 8 lb

47. Above and Beyond **49.** 36.9 lb **51.** Interpolation: Exercise 48; Extrapolation: Exercises 49–50

3.5 Linear Inequalities in Two Variables

< 3.5 Objectives >

1 > Graph a linear inequality in two variables

2 > Graph a region defined by linear inequalities

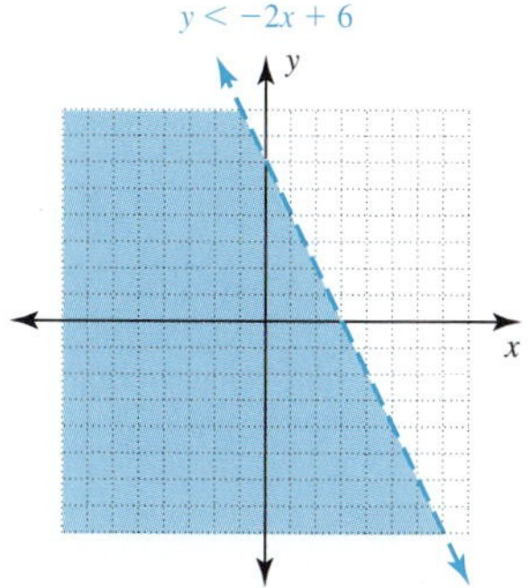

What does the solution set of an inequality in two variables look like? We will see that it is a set of ordered pairs best represented by a shaded region. The general form for a linear inequality in two variables is

$Ax + By < C$

in which A and B cannot both be 0. The symbol $<$ can be replaced with $>$, $\le$, or $\ge$. Some examples are

$y < -2x + 6 \qquad x - 2y \le 4 \qquad \text{and} \qquad 2x - 3y \ge x + 5y$

As was the case with equations, the solution set of a linear inequality is a set of ordered pairs of real numbers. However, the solution set of a linear inequality consists of an entire region in the plane. We call this region a **half-plane.**

To graph the solution set of

$y < -2x + 6$

we begin by writing the corresponding linear equation

$y = -2x + 6$

Note that the graph of $y = -2x + 6$ is simply a straight line.

To graph the solution set of $y < -2x + 6$, we must include all ordered pairs that satisfy that inequality. For instance if $x = 1$, we have

$y < -2(1) + 6$

$y < 4$

So we want to include all points of the form $(1, y)$, where $y < 4$. Of course, since $(1, 4)$ is *on* the corresponding line, this means that we want all points *below* the line along the vertical line $x = 1$. The result is similar for any choice of x, and our solution set contains all of the points below the line $y = -2x + 6$. We can graph the solution set as the shaded region shown.

NOTES

The line is dashed to indicate that points on the line are *not* included.

We call the graph of the equation

$Ax + By = C$

the **boundary line** of the half-planes.

Definition

Solution Set of an Inequality

The solution set of an inequality of the form

$Ax + By < C \quad \text{or} \quad Ax + By > C \quad \text{with} \quad B \ne 0$

can be represented by a half-plane either above or below the corresponding line determined by

$Ax + By = C$

NOTE

If the inequality is $\le$ or $\ge$, then the boundary line is part of the solution set.

How do we decide which half-plane represents the desired solution set? A **test point** provides an easy answer. Choose any point *not* on the line. Then substitute the coordinates of that point into the given inequality. If the coordinates satisfy the inequality (result in a true statement), then shade the region or half-plane that includes the test point; if not, shade the opposite half-plane. Example 1 illustrates the process.

Example 1 Graphing a Linear Inequality

< Objective 1 >

RECALL

The graph of $x - 2y = 4$ is shown below.

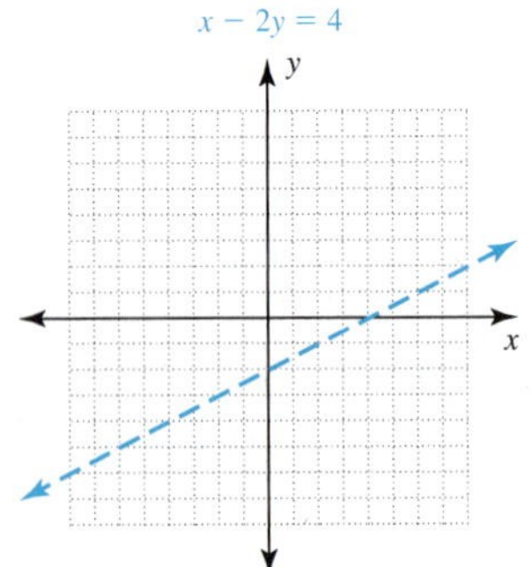

Graph the linear inequality

$x - 2y < 4$

First, we graph the corresponding equation

$x - 2y = 4$

to find the boundary line. To determine which half-plane is part of the solution set, we need a test point *not* on the line. As long as the line *does not pass through the origin,* we can use (0, 0) as a test point. It provides the easiest computation.

Letting $x = 0$ and $y = 0$, we have

$$(0) - 2(0) \overset{?}{<} 4$$
$$0 < 4$$

Because this is a true statement, we shade the half-plane that includes the origin (the test point), as shown.

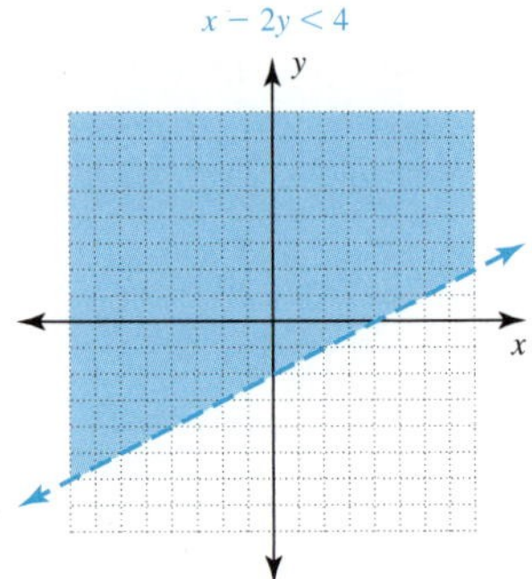

NOTE

Because we have a strict inequality, $x - 2y < 4$, the boundary line does not include solutions. In this case, we use a dashed line.

Check Yourself 1

Graph the solution set of $3x + 4y > 12$.

The graphs of some linear inequalities include the boundary line. That is the case whenever equality is included with the inequality statement, as illustrated in Example 2.

Example 2 Graphing a Linear Inequality

Graph the inequality

$2x + 3y \geq 6$

First, we graph the boundary line, here corresponding to $2x + 3y = 6$. This time we use a solid line because equality is included in the original statement.

Again, we choose a convenient test point not on the line. As before, the origin provides the simplest computation.

Substituting $x = 0$ and $y = 0$, we have

$$2(0) + 3(0) \overset{?}{\geq} 6$$
$$0 \geq 6$$

NOTE

A solid boundary line means that points on the line are solutions.

This occurs when the inequality symbol is either $\geq$ or $\leq$.

This is a *false* statement. Hence, the graph consists of all points on the *opposite* side of the boundary line. The graph is the upper half-plane shown.

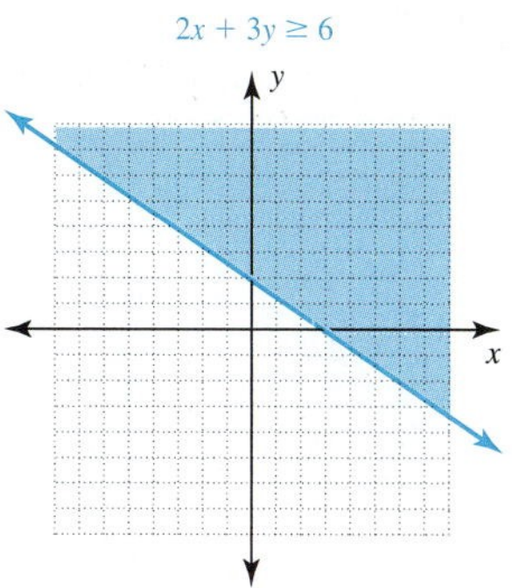

Check Yourself 2

Graph the solution set of $x - 3y \leq 6$.

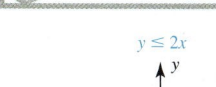

Example 3 Graphing a Linear Inequality

$y \leq 2x$

NOTE

The choice of (1, 1) is arbitrary. We simply want *any* point *not* on the line.

Graph the solution set of

$y \leq 2x$

We proceed as before by graphing the boundary line (it is solid since equality is included). The only difference between this and previous examples is that we *cannot use the origin* as a test point. Do you see why?

Choosing (1, 1) as our test point gives the statement

$$(1) \stackrel{?}{\leq} 2(1)$$
$$1 \leq 2$$

Because the statement is *true,* we shade the half-plane that *includes* the test point (1, 1).

Check Yourself 3

Graph the solution set of $3x + y > 0$.

We now consider a special case of graphing linear inequalities in the rectangular coordinate system.

Example 4 Graphing a Linear Inequality

NOTE

Here we specify the rectangular coordinate system to indicate we want a two-dimensional graph.

Graph the solution set of $x > 3$ in the rectangular coordinate system.

First, we draw the boundary line (a dashed line because equality is not included) corresponding to

$x = 3$

We can choose the origin as a test point in this case. It results in the false statement

$0 > 3$

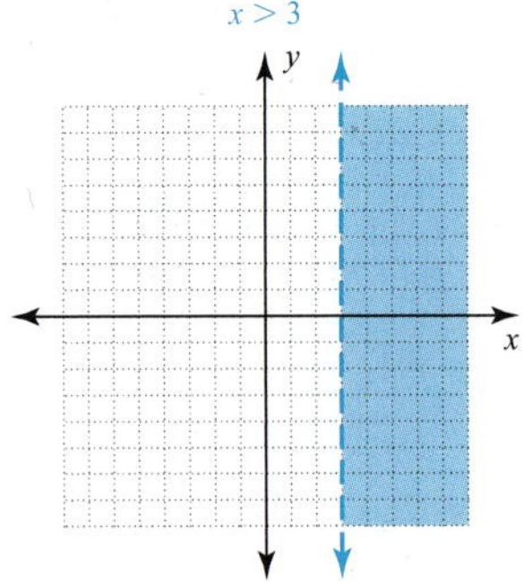

We then shade the half-plane *not* including the origin. In this case, the solution set is represented by the half-plane to the right of the vertical boundary line.

As you may have observed, in this special case choosing a test point is not really necessary. Because we want values of x that are *greater than* 3, we want those ordered pairs that are to the *right* of the boundary line.

Check Yourself 4

Graph the solution set of

$$y \leq 2$$

in the rectangular coordinate system.

Applications of linear inequalities often involve more than one inequality condition. Consider Example 5.

Example 5 Graphing a Region Defined by Linear Inequalities

< Objective 2 >

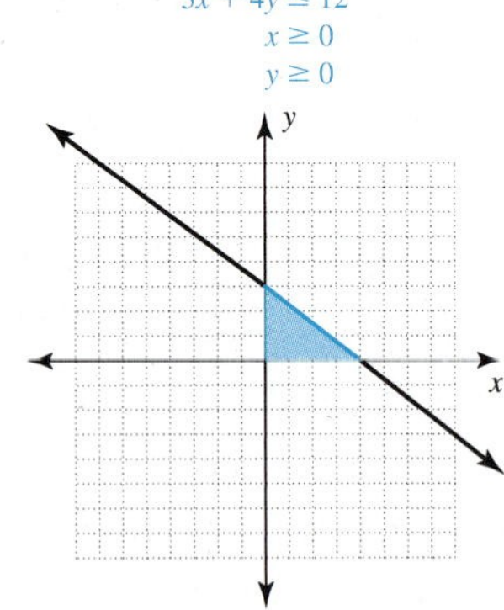

Graph the region satisfying the conditions.

$$3x + 4y \leq 12$$
$$x \geq 0$$
$$y \geq 0$$

The solution set in this case must satisfy *all three conditions.* As before, the solution set of the first inequality is graphed as the half-plane *below* the boundary line. The second and third inequalities mean that x and y must also be nonnegative. Therefore, our solution set is restricted to the first quadrant (and the appropriate segments of the x- and y-axes), as shown.

Check Yourself 5

Graph the region satisfying the conditions.

$$3x + 4y < 12$$
$$x \geq 0$$
$$y \geq 0$$

Here is an algorithm summarizing our work in graphing linear inequalities in two variables.

Step by Step

Graphing a Linear Inequality

Step 1 Replace the inequality symbol with an equal sign to form the equation of the boundary line of the solution set.

Step 2 Graph the boundary line. Use a dashed line if equality is not included ($<$ or $>$). Use a solid line if equality is included ($\leq$ or $\geq$).

Step 3 Choose any convenient test point *not* on the boundary line.

Step 4 If the inequality is *true* for the test point, shade the half-plane that *contains* the test point. If the inequality is *false* for the test point, shade the half-plane that does *not contain* the test point.

Check Yourself ANSWERS

1. 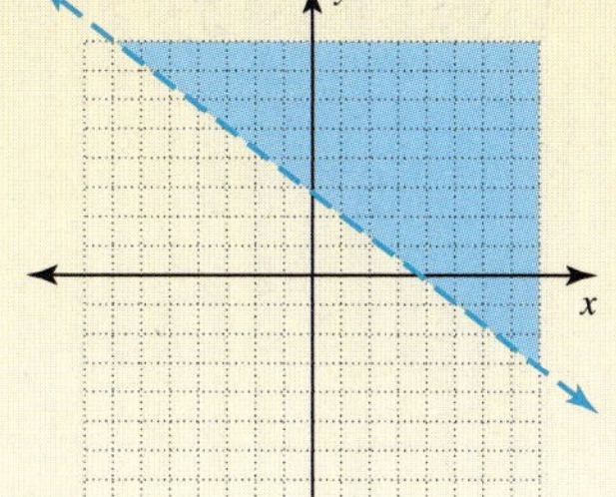

2. 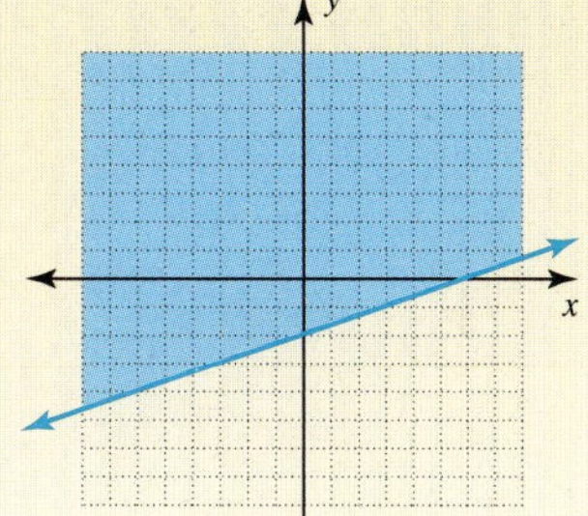

3. 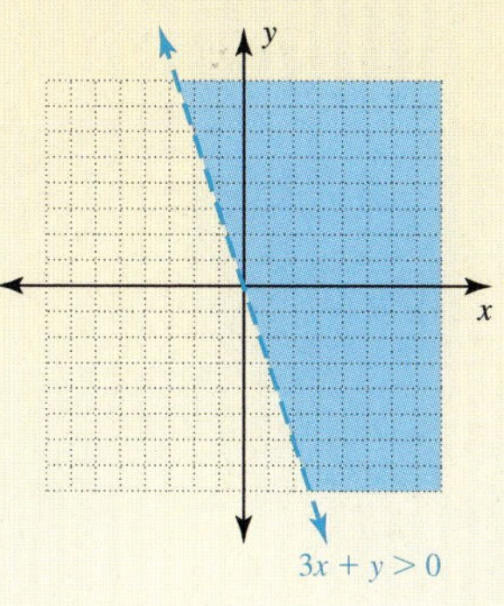

4. 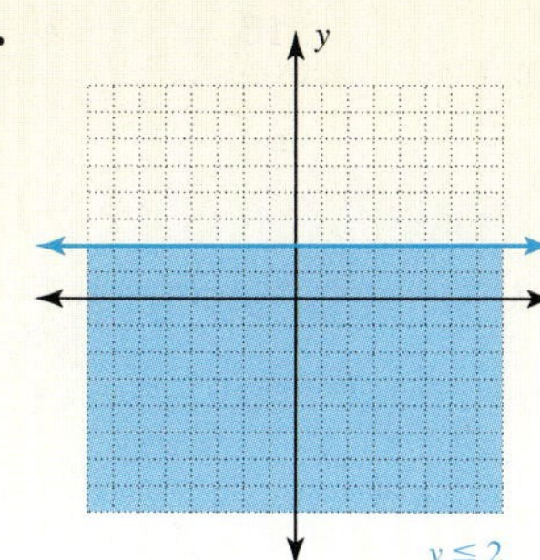

5. 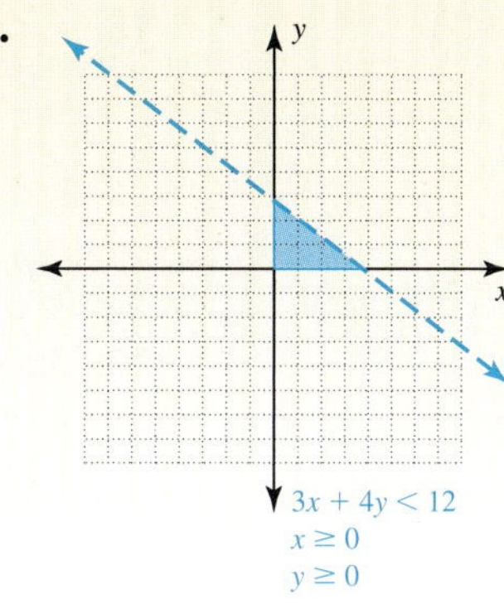

Reading Your Text

These fill-in-the-blank exercises will help you understand some of the key vocabulary used in this section. The answers to these exercises are in the Answers Appendix in the back of the text.

(a) In the case of linear inequalities, the solution set consists of all the points in an entire region of the plane, called a __________.

(b) To decide which region represents the solution set for an inequality, we use a __________ point.

(c) A __________ boundary line means that the points on the line are solutions to the inequality.

(d) If the inequality is __________ for the test point, shade the half-plane that does *not* include the test point.

Skills | Calculator/Computer | Career Applications | Above and Beyond

3.5 exercises

< Objective 1 >

Graph the solution set of each linear inequality.

1. $x + y < 4$ 

2. $x + y \geq 6$

3. $x - y \geq 3$

4. $x - y < 5$

5. $y \geq 2x + 1$

6. $y < 3x - 4$

7. $2x + 3y < 6$ VIDEO

8. $3x - 4y \geq 12$

9. $x - 4y > 8$

10. $2x + 5y \leq 10$

11. $y \geq 3x$

12. $y \leq -2x$

13. $x - 2y > 0$

14. $x + 4y \leq 0$

15. $x < 3$

16. $y < -2$

17. $y > 3$

18. $x \leq -4$

19. $3x - 6 \leq 0$

20. $-2y > 6$

< Objective 2 >

Graph the region satisfying each set of conditions.

21. $0 < x < 1$

22. $-2 \leq y \leq 1$

23. $1 \leq x \leq 3$

24. $1 < y < 5$

25. $0 \leq x \leq 3$
$2 \leq y \leq 4$

26. $1 \leq x \leq 5$
$0 \leq y \leq 3$

27. $x + 2y \leq 4$
$x \geq 0$
$y \geq 0$

28. $2x + 3y \leq 6$
$x \geq 0$
$y \geq 0$

29. **Business and Finance** A manufacturer produces LCD and plasma models of television sets. The LCD models require 12 hr to produce, while the plasma models require 18 hr. The labor available is limited to 360 hr per week.

If x represents the number of LCD sets produced per week and y represents the number of plasma models, draw a graph of the region representing the feasible values for x and y. Keep in mind that the values for x and y must be nonnegative since they represent a quantity of items. (This is the solution set for the system of inequalities.)

30. **Business and Finance** A manufacturer produces DVD and Blu-ray players. The DVD players require 10 hr of labor to produce while Blu-ray players require 20 hr. Let x represent the number of DVD players produced and y the number of Blu-ray players.

If the labor hours available are limited to 300 hr per week, graph the region representing the feasible values for x and y.

31. **Business and Finance** A hospital food service department can serve at most 1,000 meals per day. Patients on a normal diet receive 3 meals per day, and patients on a special diet receive 4 meals per day. Write a linear inequality that describes the number of patients that can be served per day and draw its graph.

32. **Business and Finance** The movie and TV critic for the local radio station spends 3 to 7 hr daily reviewing movies and less than 4 hr reviewing TV shows. Let x represent the time (in hours) watching movies and y represent the time spent watching TV. Write two inequalities that model the situation, and graph their intersection.

Write an inequality for the shaded region shown in each figure.

33.

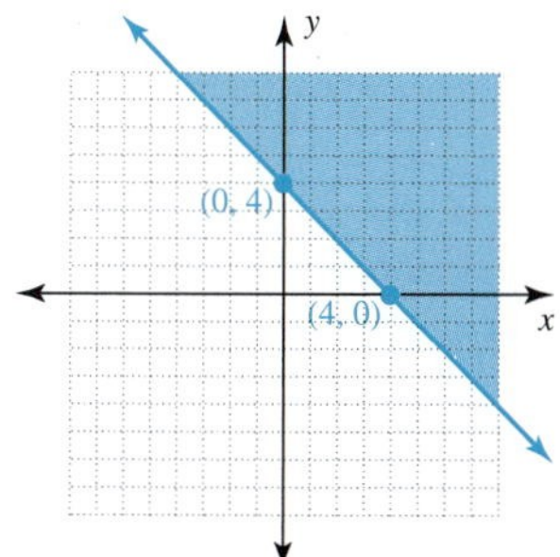

34.

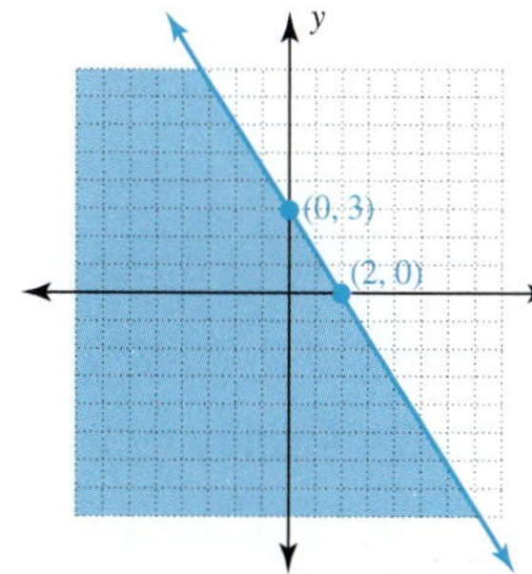

35.

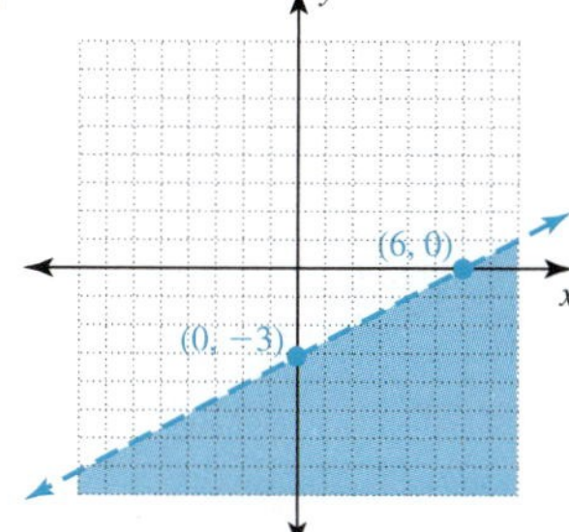

36.

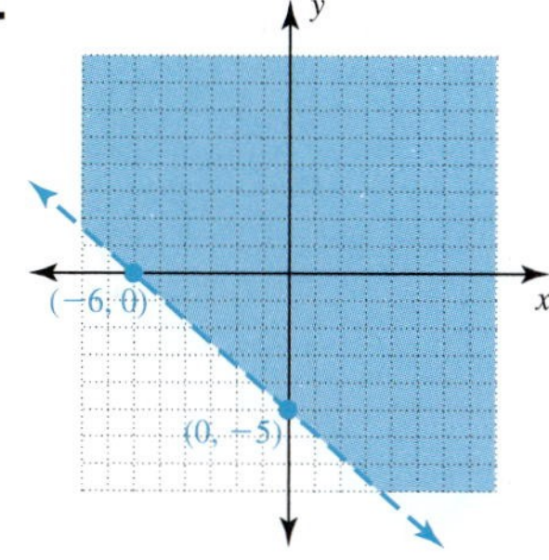

Determine whether each statement is **true** *or* **false**.

37. If a test point satisfies a linear inequality, then we shade the half-plane that contains the test point.

38. A dashed boundary line means that the points on that line are solutions for the inequality.

Complete each statement with **always, sometimes,** *or* **never.**

39. When graphing a linear inequality, there is __________ a straight-line boundary.

40. When graphing a linear inequality, the point (0, 0) is __________ on the boundary line.

41. **Manufacturing Technology** A manufacturer produces two-slice toasters and four-slice toasters. The two-slice toasters require 8 hr to produce, and the four-slice toasters require 10 hr to produce. The manufacturer has 400 hr of labor available each week.

(a) Write a linear inequality to represent the number of each type of toaster the manufacturer can produce in a week (use x for the two-slice toasters and y for the four-slice toasters).

(b) Graph the inequality (in the first quadrant).

(c) Is it feasible to produce 20 two-slice toasters and 30 four-slice toasters in the same week?

42. **Manufacturing Technology** A company produces standard clock radios and deluxe clock radios. It costs the company \$15 to produce each standard clock radio and \$20 to produce each deluxe model. The company's budget limits production costs to \$3,000 per day.

(a) Write a linear inequality to represent the number of each type of clock radio that the company can produce in a day (use x for the standard model and y for the deluxe model).

(b) Graph the inequality (in the first quadrant).

(c) Is it feasible to produce 80 of each type of clock radio in the same day?

43. Assume that you are working only with the variable x. Describe the set of solutions for the statement $x > -1$.

44. Assume that you are working in two variables x and y. Describe the set of solutions for the statement $x > -1$.

Answers

1.

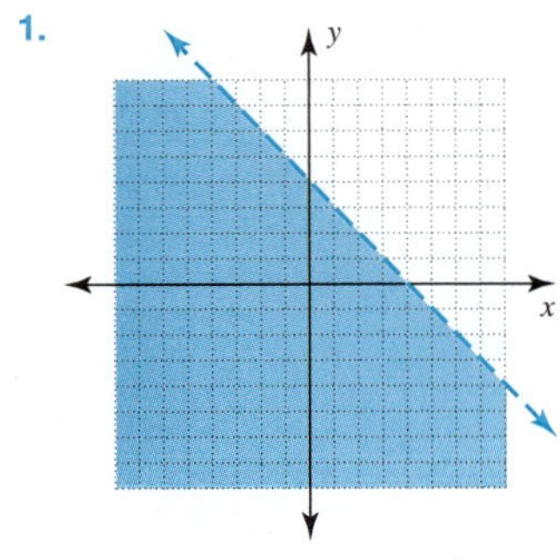

3.

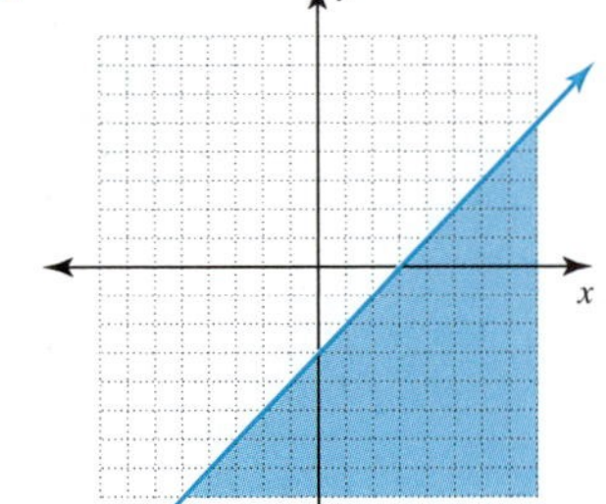

5.

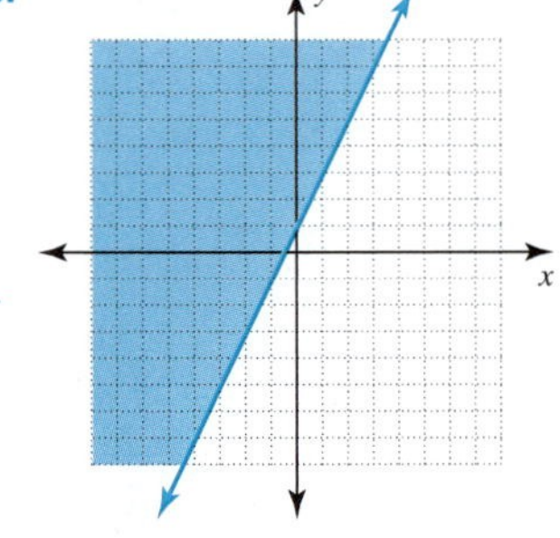

7.

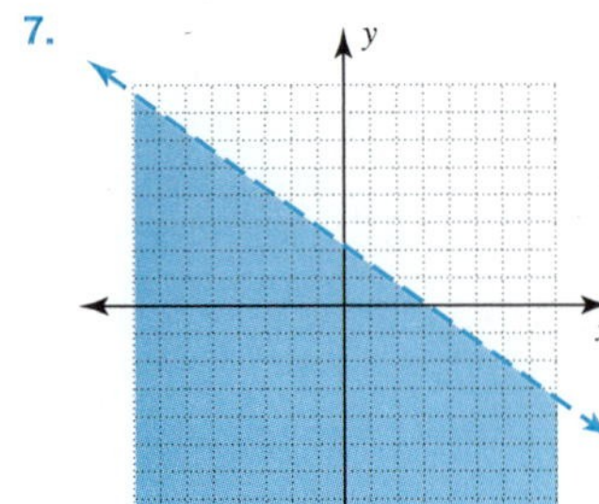

9.

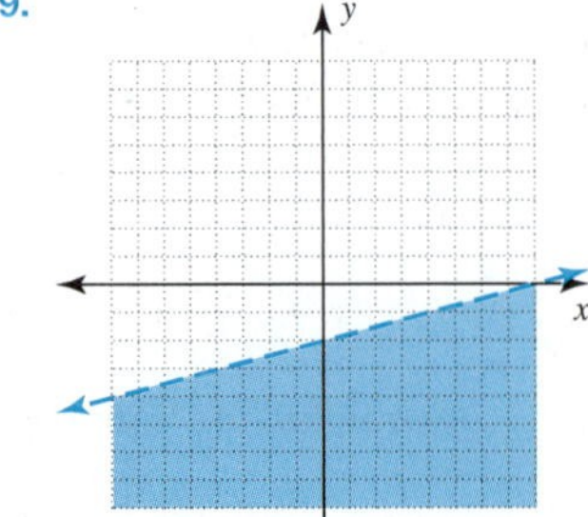

11.

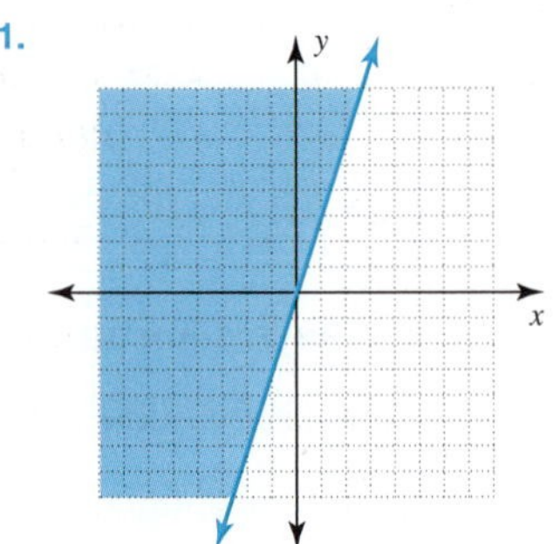

13.

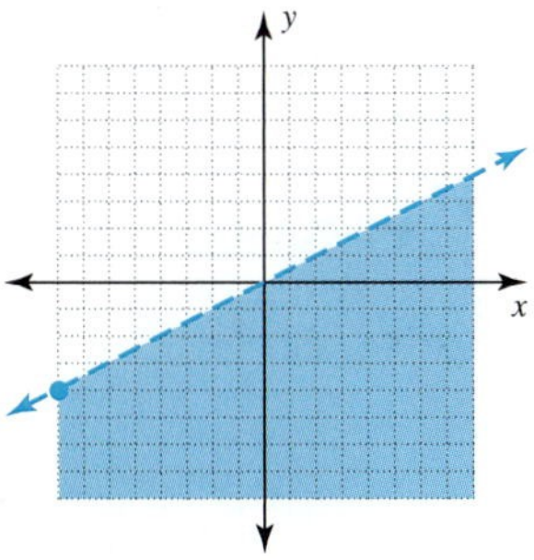

15.

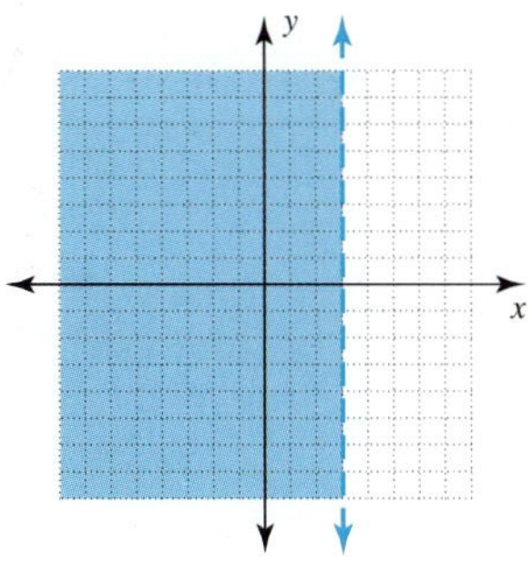

17.

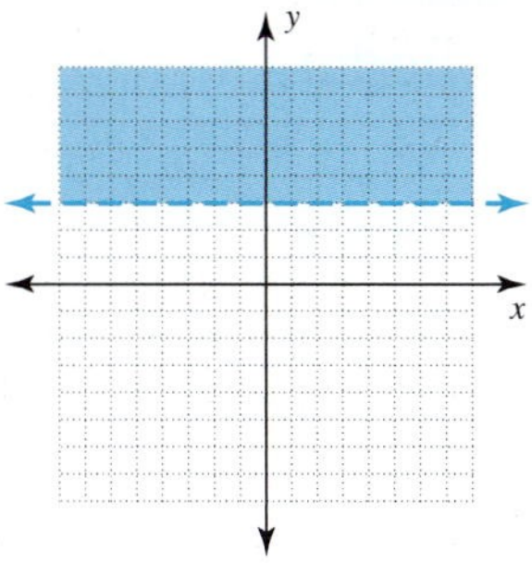

19.

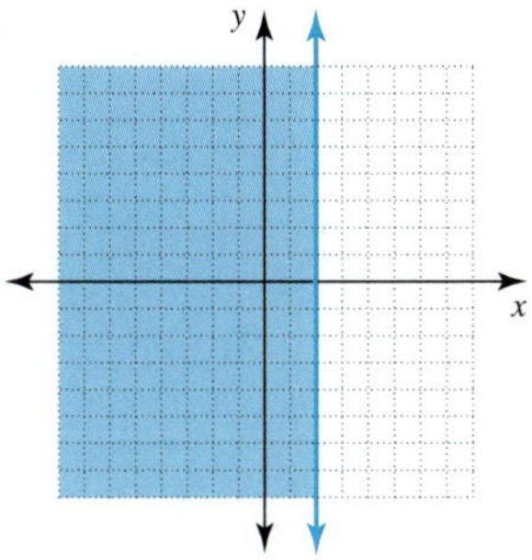

21.

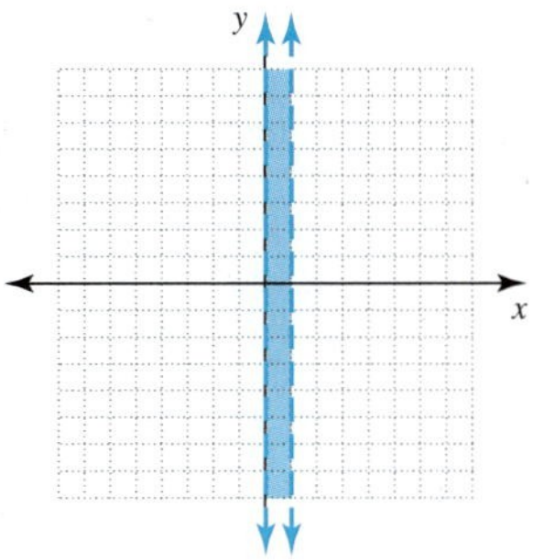

23.

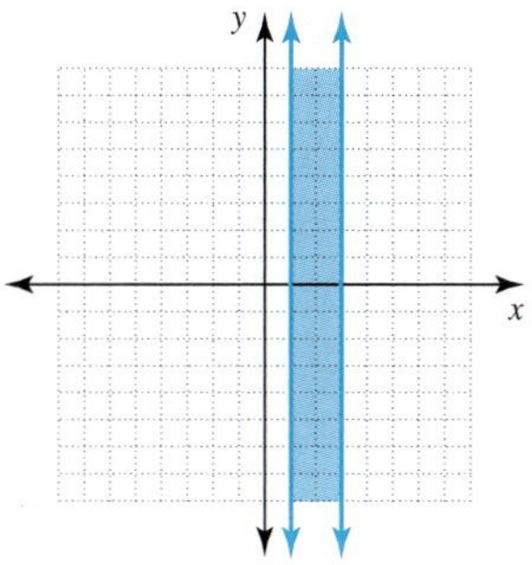

25.

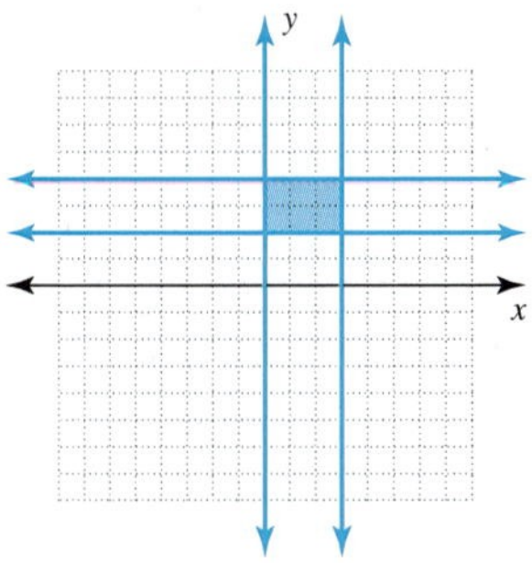

27.

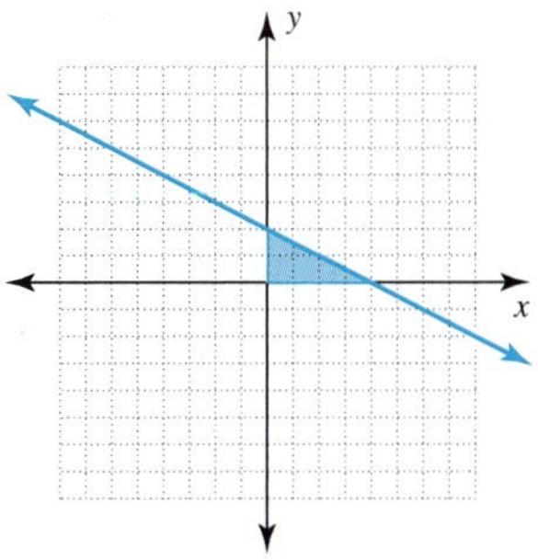

29. 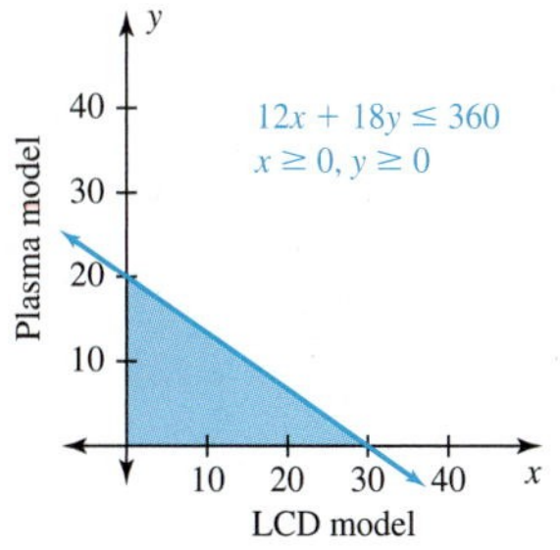 $12x + 18y \le 360$, $x \ge 0, y \ge 0$

31. 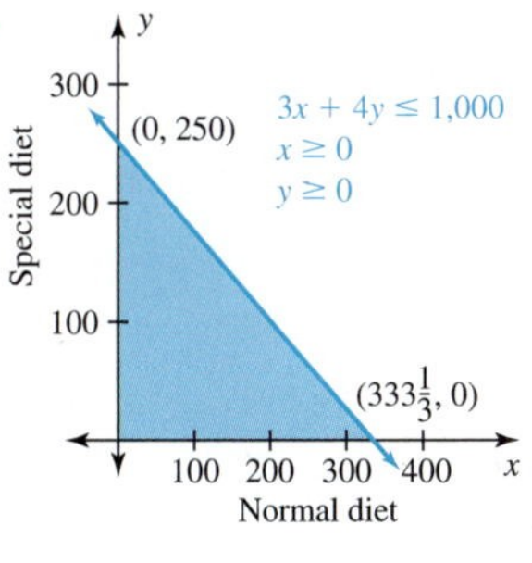 $3x + 4y \le 1{,}000$, $x \ge 0$, $y \ge 0$; (0, 250), $(333\frac{1}{3}, 0)$

33. $y \ge -x + 4$ 35. $y < \frac{1}{2}x - 3$ 37. True 39. always

41. (a) $8x\ 110y \le 400$; (b) (c) No 43. Above and Beyond

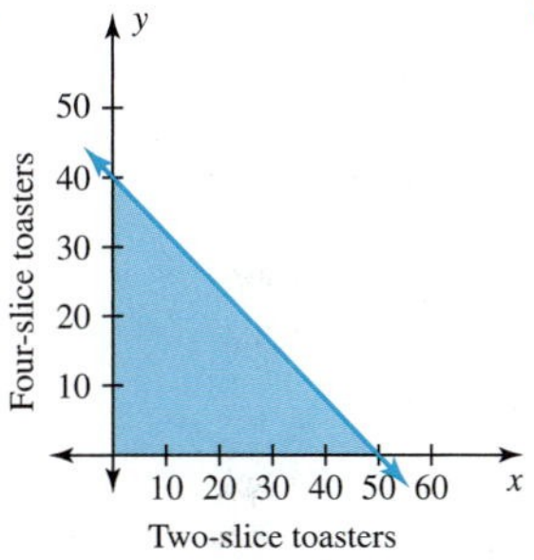

summary :: chapter 3

Definition/Procedure	Example	Reference
Graphing Linear Functions		**Section 3.1**
Linear Equation An equation that can be written in the form $Ax + By = C$ in which A and B are not both 0.	$2x - 3y = 4$ is a linear equation.	p. 244
Graphing Linear Equations **Step 1** Find at least three solutions to the equation, and put your results in a table. **Step 2** Graph the solutions found in step 1. **Step 3** Draw a straight line through the points found in step 2 to form the graph of the equation.	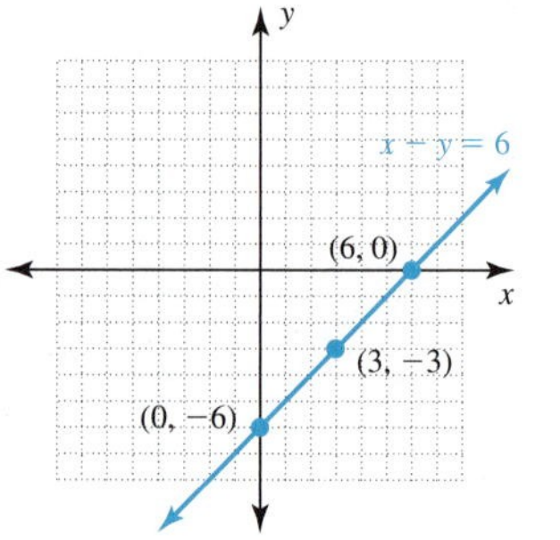 $x - y = 6$ (6, 0), (3, −3), (0, −6) x: 0, 3, 6; y: −6, −3, 0	p. 244
Writing Linear Equations as Functions **Step 1** Solve the equation for the dependent variable y. **Step 2** Replace y with $f(x)$.	$2x + 3y = 6$ **Step 1** $3y = -2x + 6$ $y = -\frac{2}{3}x + 2$ **Step 2** $f(x) = -\frac{2}{3}x + 2$	p. 252
The Slope of a Line		**Section 3.2**
Slope The slope of a line gives a numerical measure of the steepness of the line. The slope m of a line containing the distinct points in the plane $P(x_1, y_1)$ and $Q(x_2, y_2)$ is given by $m = \frac{y_2 - y_1}{x_2 - x_1}$ where $x_2 \neq x_1$.	To find the slope of the line through $(-2, -3)$ and $(4, 6)$, $m = \frac{(6) - (-3)}{(4) - (-2)}$ $= \frac{6 + 3}{4 + 2}$ $= \frac{9}{6} = \frac{3}{2}$	p. 271

Slopes and Lines
• The slope of a line that rises from left to right is positive.
• The slope of a line that falls from left to right is negative.
• The slope of a horizontal line is 0.
  • The equation of the horizontal line with y-intercept $(0, b)$ is $y = b$.
• The slope of a vertical line is undefined.
  • The equation of the vertical line with x-intercept $(a, 0)$ is $x = a$.
	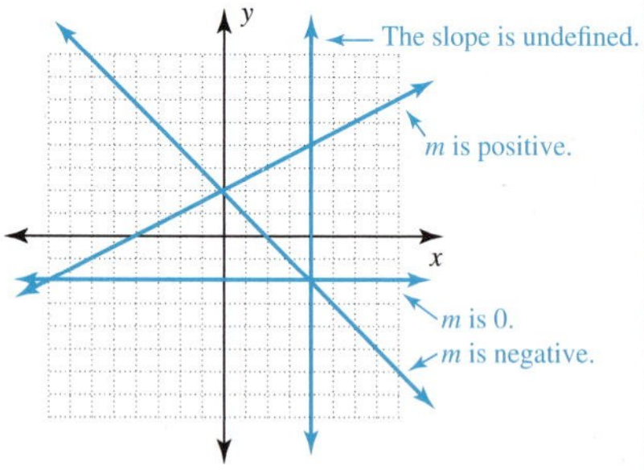 The slope is undefined.
	m is positive.
	m is 0.
	m is negative.
		p. 273

Slope-Intercept Form The slope-intercept form for the equation of a line is $y = mx + b$ in which the line has slope m and y-intercept $(0, b)$.	For the equation $y = \frac{2}{3}x - 3$ the slope m is $\frac{2}{3}$ and b, which determines the y-intercept, is -3.	p. 274

Continued

Definition/Procedure	Example	Reference

Slope-Intercept and Graphing

Step 1 Write the equation of the line in slope-intercept form.

Step 2 Find the slope and y-intercept.

Step 3 Plot the y-intercept.

Step 4 Plot a second point based on the slope of the line.

Step 5 Draw a line through the two points.

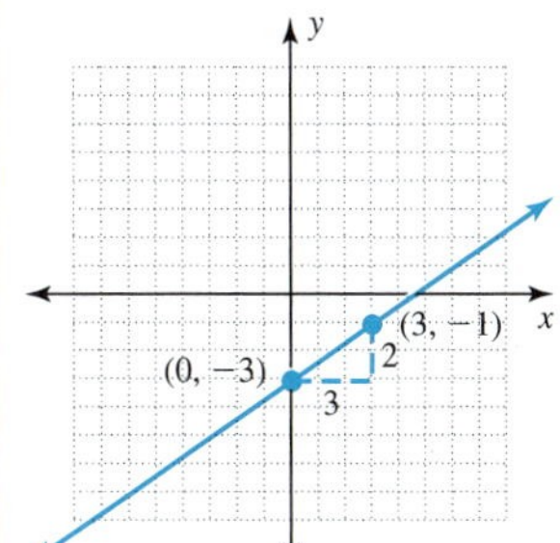

p. 276

Linear Equations

Section 3.3

Parallel Lines Two nonvertical lines are **parallel** if, and only if, they have the same slope, so

$m_1 = m_2$

or both are vertical.

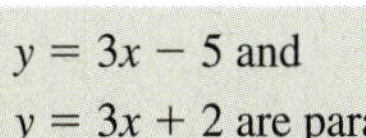

$y = 3x - 5$ and

$y = 3x + 2$ are parallel

Parallel lines

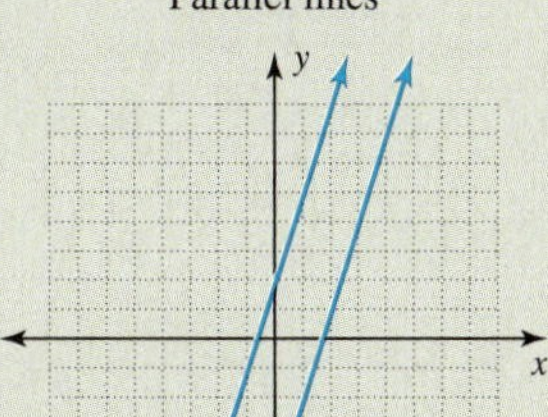

p. 285

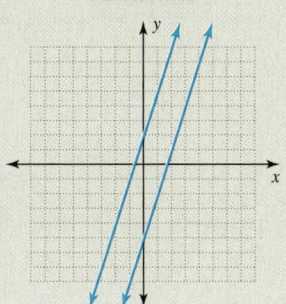

Perpendicular Lines Two nonvertical lines are **perpendicular** if, and only if, their slopes are negative reciprocals, that is, when

$m_1 \cdot m_2 = -1$

or if one is vertical and the other horizontal.

$y = 5x + 2$ and

$y = -\frac{1}{5}x - 3$ are perpendicular.

Perpendicular lines

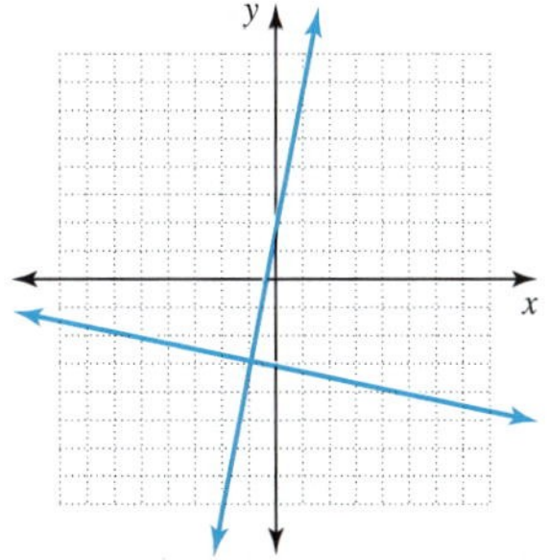

p. 285

Point-Slope Form The point-slope equation of the line with slope m that passes through the point (x_1, y_1) is

$y - y_1 = m(x - x_1)$

The line with slope $\frac{1}{3}$ passing through (4, 3) has the equation

$y - 3 = \frac{1}{3}(x - 4)$

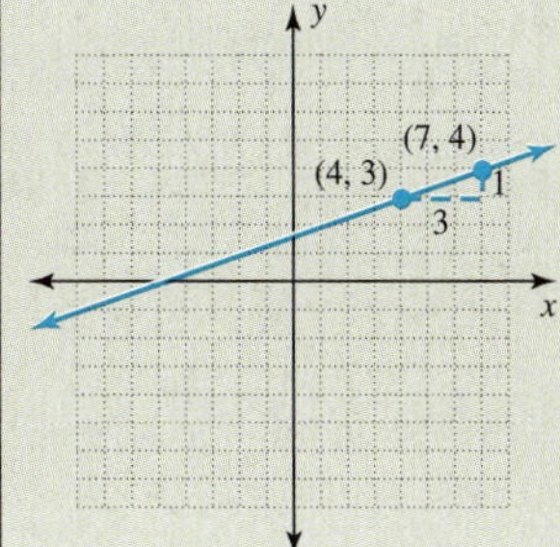

p. 287

Continued

Definition/Procedure	Example	Reference

Rate of Change and Linear Regression

Section 3.4

Rate of Change The rate of change of a linear function is equal to its slope. It represents the change in the output when the input is increased by 1.

Consider the cost model,

$C(x) = 12x + 250$

The rate of change of this function is 12, which means that the cost increases by $12 for each additional unit produced.

p. 299

Linear Regression

Step 1 Enter the x- and y-values into a graphing calculator's lists.

Step 2 Create a scatter plot from the data.

Step 3 Perform a regression analysis on the data.

Step 4 Graph the line-of-best-fit on the scatter plot.

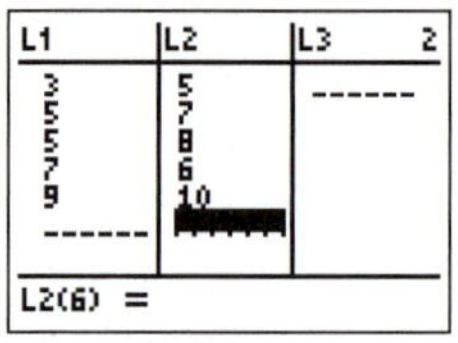

```
L1      L2      L3     2
3       5       ------
5       7
5       8
7       6
9       10
------
L2(6) =
```

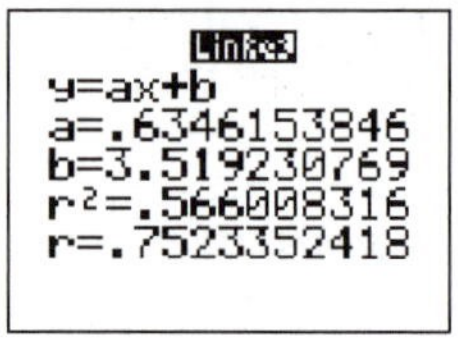

```
LinReg
y=ax+b
a=.6346153846
b=3.519230769
r²=.566008316
r=.7523352418
```

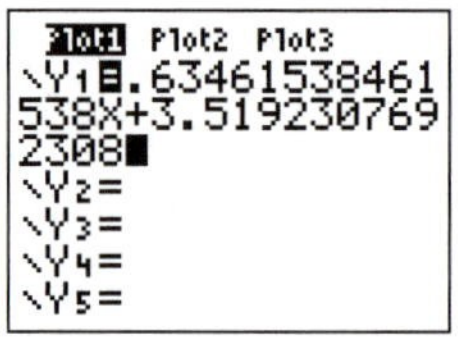

```
Plot1 Plot2 Plot3
\Y1=.63461538461
538X+3.519230769
2308
\Y2=
\Y3=
\Y4=
\Y5=
```

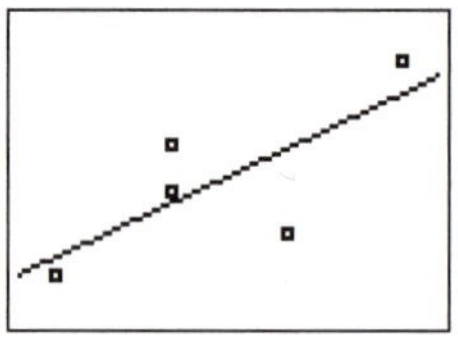

p. 304

Linear Inequalities in Two Variables

Section 3.5

In general, the solution set of an inequality of the form

$Ax + By < C$ or $Ax + By > C$ where $B \neq 0$

will be a **half-plane** either above or below the **boundary line** determined by

$Ax + By = C$

The boundary line is included in the graph if equality is included in the statement of the original inequality. Such a line is solid. The boundary line is dashed if it is not included in the graph.

To graph

$x - 2y < 4$

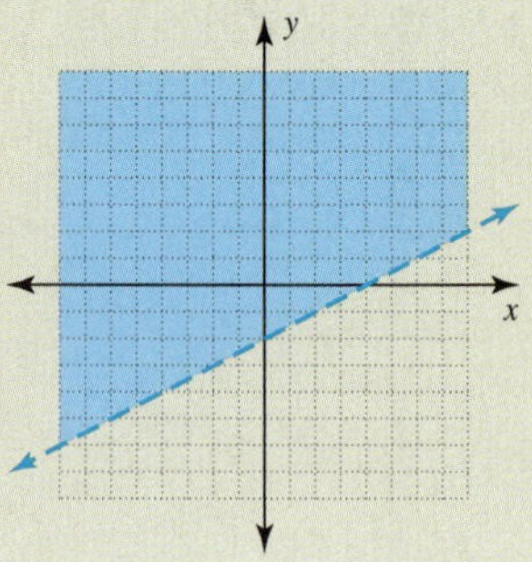

p. 317

Continued

Definition/Procedure	Example	Reference
Graphing Linear Inequalities **Step 1** Replace the inequality symbol with an equal sign to form the equation of the boundary line of the solution set. **Step 2** Graph the boundary line. Use a dashed line if equality is not included ($<$ or $>$). Use a solid line if equality is included ($\le$ or $\ge$). **Step 3** Choose any convenient test point *not* on the boundary line. **Step 4** If the inequality is *true* for the test point, shade the half-plane that *contains* the test point. If the inequality is *false* for the test point, shade the half-plane that does *not contain* the test point.		*p.* 318

summary exercises :: chapter 3

This summary exercise set will help ensure that you have mastered each of the objectives of this chapter. The exercises are grouped by section. You should reread the material associated with any exercises that you find difficult. The answers to the odd-numbered exercises are in the Answers Appendix in the back of the text.

3.1 *Graph each equation.*

1. $x + y = 5$

2. $x - y = 6$

3. $y = 5x$

4. $y = -3x$

5. $y = \frac{3}{2}x$

6. $y = 3x + 2$

7. $y = -2x + 4$

8. $y = -3x + 4$

9. $y = \frac{2}{3}x + 2$

10. $3x - y = 3$

11. $2x + y = 6$

12. $3x + 2y = 12$

13. $3x - 4y = 12$

14. $x = -5$

15. $y = -2$

16. $5x - 3y = 15$

17. $3x + 4y = 12$

18. $2x + y = 6$

19. $3x + 2y = 6$

20. $-4x - 5y = 20$

3.2 *Find the slope of the line through each pair of points.*

21. $(3, 4)$ and $(5, 8)$

22. $(-2, 3)$ and $(1, -6)$

23. $(-2, 5)$ and $(2, 3)$

24. $(-5, -2)$ and $(1, 2)$

25. $(-2, 6)$ and $(5, 6)$

26. $(-3, 2)$ and $(-1, -3)$

27. $(-3, -6)$ and $(5, -2)$

28. $(-6, -2)$ and $(-6, 3)$

Find the slope and y-intercept of the line represented by each equation.

29. $y = 2x + 5$

30. $y = -4x - 3$

31. $y = -\frac{3}{4}x$

32. $y = \frac{2}{3}x + 3$

33. $2x + 3y = 6$

34. $5x - 2y = 10$

35. $y = -3$

36. $x = 2$

Write an equation of the line with the given slope and y-intercept.

37. Slope 2, y-intercept $(0, 3)$

38. Slope $\frac{3}{4}$, y-intercept $(0, -2)$

39. Slope $-\frac{2}{3}$, y-intercept $(0, 2)$

40. Slope 0, y-intercept $(0, -5)$

3.3 *Determine whether each pair of lines is* **parallel, perpendicular,** *or* **neither.**

41. L_1 through $(-3, -2)$ and $(1, 3)$
L_2 through $(0, 3)$ and $(4, 8)$

42. L_1 through $(-4, 1)$ and $(2, -3)$
L_2 through $(0, -3)$ and $(2, 0)$

43. L_1 with equation $x + 2y = 6$
L_2 with equation $x + 3y = 9$

44. L_1 with equation $4x - 6y = 18$
L_2 with equation $2x - 3y = 6$

Write an equation of the line passing through each point with the indicated slope. Give your result in slope-intercept form, where possible.

45. $(0, -5)$, $m = \frac{2}{3}$

46. $(0, -3)$, $m = 0$

47. $(2, 3)$, $m = 3$

48. $(4, 3)$, m is undefined

49. $(3, -2)$, $m = \frac{5}{3}$

50. $(-2, -3)$, $m = 0$

51. $(-2, -4)$, $m = -\frac{5}{2}$

52. $(-3, 2)$, $m = -\frac{4}{3}$

53. $\left(\frac{2}{3}, -5\right)$, $m = 0$

54. $\left(-\frac{5}{2}, -1\right)$, m is undefined

Write an equation of the line L satisfying each set of conditions.

55. L passes through $(-3, -1)$ and $(3, 3)$.

56. L passes through $(2, 3)$ and $(-2, -5)$.

57. L has slope $\frac{3}{4}$ and y-intercept $(0, 3)$.

58. L passes through $(4, -3)$ with a slope of $-\frac{5}{4}$.

59. L has y-intercept $(0, -4)$ and is parallel to the line with equation $3x - y = 6$.

60. L passes through $(-5, 2)$ and is perpendicular to the line with equation $5x - 3y = 15$.

61. L passes through $(2, -1)$ and is perpendicular to the line with equation $3x - 2y = 5$.

62. L passes through the point $(-5, -2)$ and is parallel to the line with equation $4x - 3y = 9$.

3.4

Business and Finance It costs a lunch cart \$1.75 to make each gyro. The portion of the cart's fixed cost attributable to gyros comes to \$30 per day.

63. Construct a linear function to model the cart's gyro costs.

64. How much does it cost to make 35 gyros in one day?

65. How many gyros can the cart make if it can spend \$150 making gyros?

Business and Finance The lunch cart earns a profit of \$2.75 on each gyro by selling them for \$4.50. The fixed cost associated with gyros reduces profits by \$30 per day.

66. Construct a linear function to model the cart's gyro profits.

67. How much profit does the cart make by selling 35 gyros in one day?

68. How many gyros do they need to sell if they want to earn \$100 in gyro profits?

Statistics On a 63-mi trip, a driver used two gallons of gas. The same driver used 8 gal on a 252-mi trip.

69. Construct a linear model for the gas used as a function of the miles driven.

70. What is the rate of change of the function constructed in exercise 69?

71. Interpret the rate of change in the context of this application

72. What is the y-intercept of the function constructed in exercise 69?

73. Interpret the y-intercept in the context of this application.

74. (a) How much gas is needed for a 100-mi trip? Round your result to one decimal place.
    (b) How many miles can be driven on a full 12-gal tank of gas?

Social Science A survey of public school libraries and media centers provided data comparing the state's expenditures for library materials (per student) to the number of books acquired during the year (per 100 students). The data for five states in a recent year are shown in the table.

State	Expenditures	Acquisitions
Arizona	\$15.30	121
Georgia	\$14.20	76
Minnesota	\$15.20	111
Ohio	\$10.90	75
Virginia	\$16.20	88

Source: National Center for Education Statistics

75. Create a scatter plot of the data and include the line-of-best-fit on your graph.

76. What is the equation of the best-fit line (two decimal places of accuracy)?

77. What is the slope of the best-fit line?

78. Interpret the slope in the context of this application.

79. How many books would you expect to be acquired (per 100 students) if a state's per-student expenditures were \$17 (to the nearest whole number)?

80. What per-student expenditures should policy makers approve if they wanted their state's libraries to acquire 100 books (per 100 students) in a given year (to the nearest cent)?

3.5 *Graph the solution set for each linear inequality.*

81. $y < 2x + 1$

82. $y \geq -2x + 3$

83. $3x + 2y \geq 6$

84. $3x - 5y < 15$

85. $y < -2x$

86. $4x - y \geq 0$

87. $y \geq -3$

88. $x < 4$

chapter test 3

CHAPTER 3

Use this chapter test to assess your progress and to review for your next exam. Allow yourself about an hour to take this test. The answers to these exercises are in the Answers Appendix in the back of the text.

Find the slope of the line through each pair of points.

1. $(-3, 5)$ and $(2, 10)$

2. $(4, 9)$ and $(-3, 6)$

Find the slope and y-intercept of the line represented by each equation.

3. $y = -5x - 9$

4. $6x + 5y = 30$

5. $y = 5$

6. $x = -3$

Write the equation of the line with the given slope and y-intercept. Then graph each line.

7. Slope -3; y-intercept $(0, 6)$

8. 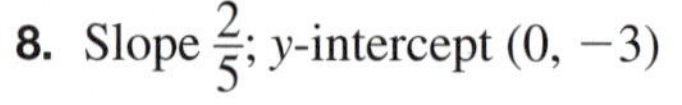
Slope $\frac{2}{5}$; y-intercept $(0, -3)$

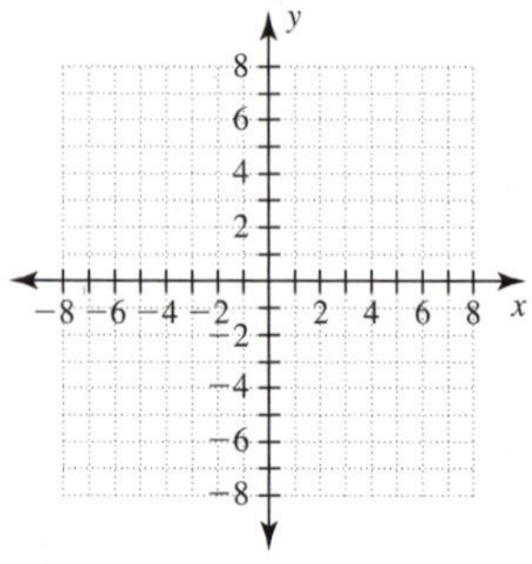

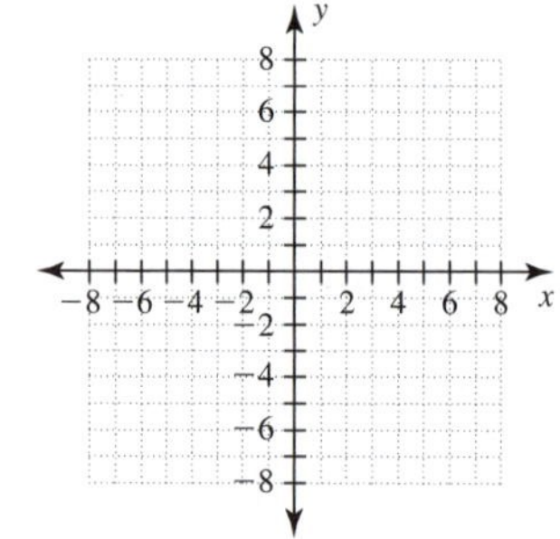

Write an equation of the line L satisfying the given set of conditions.

9. L has slope 5 and y-intercept $(0, -2)$.

10. L passes through $(-5, 4)$ and $(-2, -8)$.

11. L has y-intercept $(0, 3)$ and is parallel to the line given by $4x - y = 9$.

12. L passes through the point $(-6, -2)$ and is perpendicular to the line given by $2x - 5y = 10$.

Graph each equation.

13. $x + y = 4$

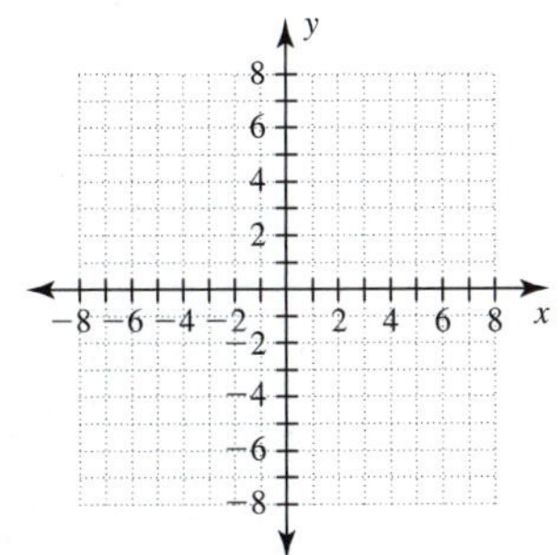

14. $y = 3x$

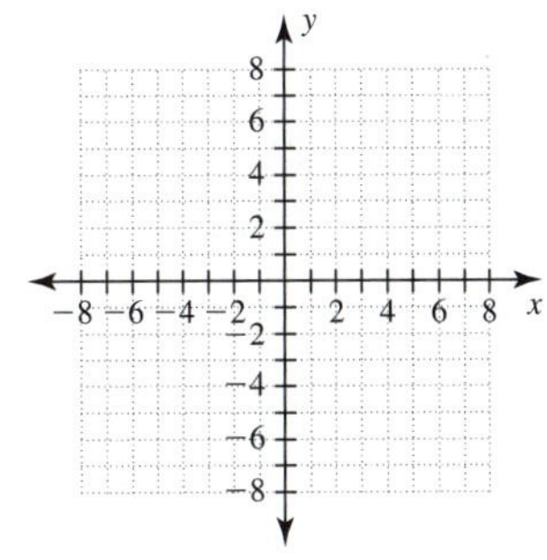

15. $y = \frac{3}{4}x - 4$

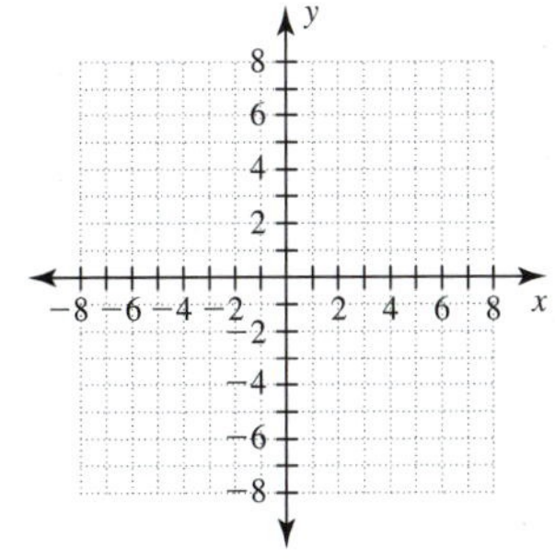

16. $x + 3y = 6$

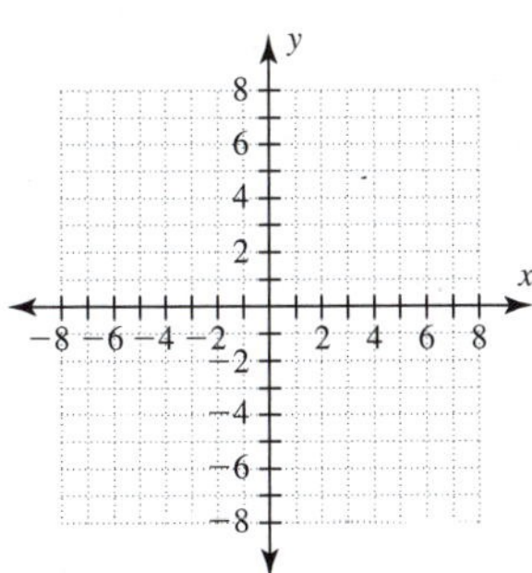

17. $2x + 5y = 10$

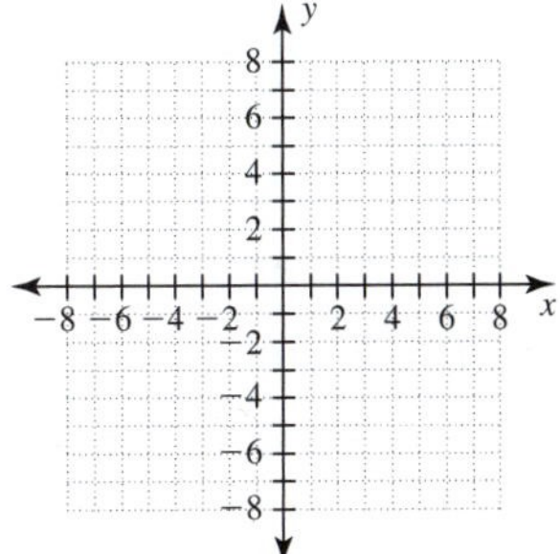

18. $y = -4$

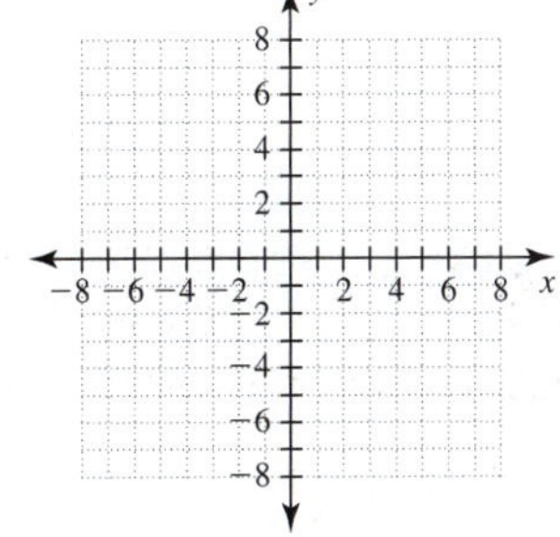

Graph each inequality.

19. $5x + 6y \leq 30$

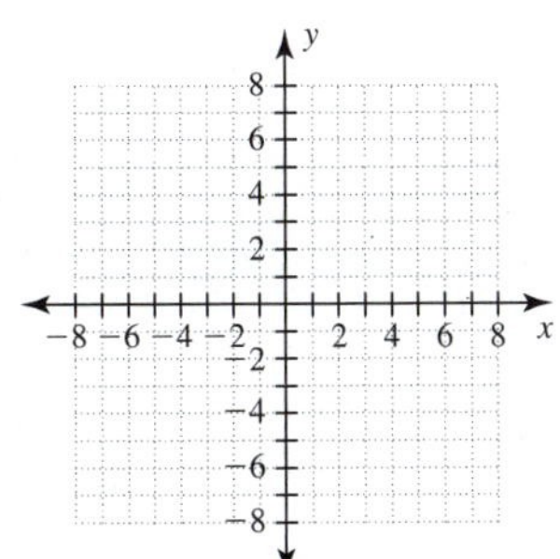

20. $x + 3y > 6$

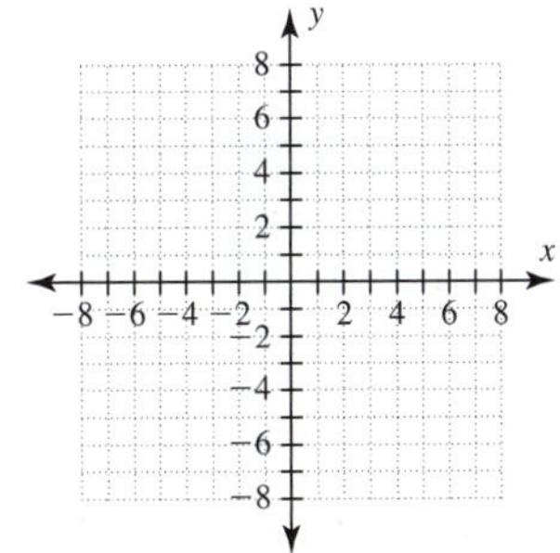

21. $4x - 8 \le 0$

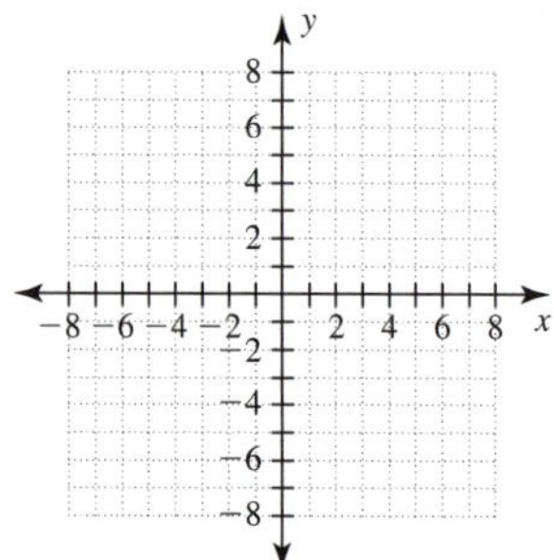

22. $2y + 4 > 0$

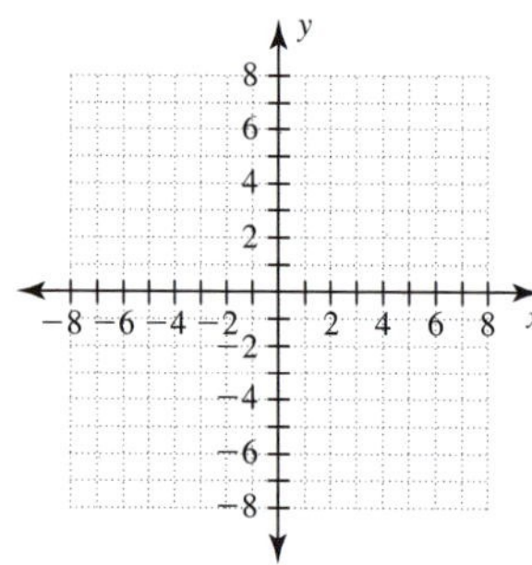

Crafts A cookbook recommends that you should roast a 10-lb stuffed turkey for 4 hr and an 18-lb stuffed bird for 6 hr.

23. Construct a linear model for roasting times as a function of the size of a stuffed turkey.

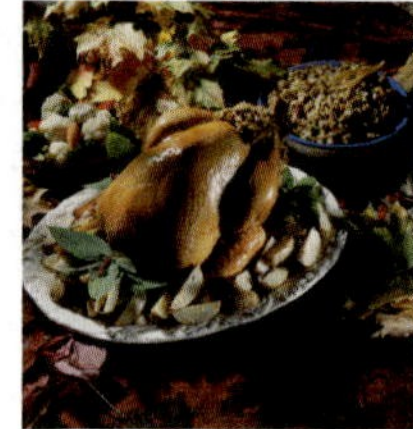

24. According to the model, for how long should you roast a 16-lb stuffed turkey?

Science and Medicine The high and low temperatures at five locations were recorded one day.

Low	39°F	42°F	54°F	64°F	66°F
High	70°F	77°F	77°F	79°F	75°F

Source: National Weather Service

25. Construct a scatter plot of the data and include the line-of-best-fit.

26. Find the equation of the line-of-best-fit (round to two decimal places).

cumulative review chapters 0–3

Use this exercise set to review concepts from earlier chapters. While it is not a comprehensive exam, it will help you identify any material that you need to review before moving on to the next chapter. The answers to these exercises are in the Answers Appendix in the back of the text.

Perform the indicated operations. Write your results in simplest form.

1. $\frac{5}{6} - \left(\frac{2}{3} + \frac{1}{2}\right)$

2. $\frac{7}{15} \times \left(\frac{5}{6} \div \frac{7}{12}\right)$

3. $2^3 - |-8| \div (-4) \cdot 2 + 5$

4. $4^2 + (-16 \div 4 \cdot 2)$

Evaluate each expression if $x = -1$, $y = 3$, and $z = -2$.

5. $-4x^2 + 3y + 2z$

6. $\dfrac{2z - 3y^2 - 2}{y^2 + 2z}$

Simplify each expression.

7. $9x - 5y - (3x - 8y)$

8. $2x^2 - 4x - (-3x + x^2) - (4 + x^2)$

9. $-4x^2 + 7x - 4 - (-7x^2 + 11x) - (9x^2 - 5x - 6)$

10. $7 - 5x + 2x^2 + 2(9 - 5x^2)$

Solve each equation.

11. $5x - 3(2x - 6) + 9 = -2(x - 5) + 6$

12. $\frac{4}{5}x - 2 = 3 + \frac{3}{4}x$

13. $\frac{x + 1}{3} - \frac{2x + 3}{4} = \frac{1}{6}$

14. $2x(x - 3) - 9 = 2x^2$

Solve and graph the solution set of each inequality.

15. $4x - 7 < 9$

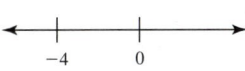

0 4

16. $6x + 4 > 3x - 8$

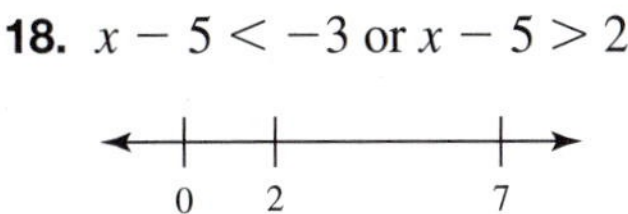

17. $4 \le 2x - 6 \le 12$

0 5 9

18. $x - 5 < -3$ or $x - 5 > 2$

0 2 7

Solve each equation for the indicated variable.

19. $F = \frac{9}{5}C + 32$ (for C)

20. $V = \frac{1}{3}\pi r^2 h$ (for h)

Find the slope of the line through each pair of points.

21. $(6, -4)$ and $(-2, 12)$

22. $(4, 5)$ and $(7, 5)$

Find the slope and the y-intercept of the line represented by each equation.

23. $y = -4x + 9$

24. $2x - 5y = 10$

25. $y = 9$

26. $x = 7$

Write an equation of the line L that satisfies the given conditions.

27. L has slope 5 and y-intercept of $(0, -6)$.

28. L passes through $(-4, 9)$ and $(6, 8)$.

29. L has y-intercept $(0, 6)$ and is parallel to the line with the equation $2x + 3y = 6$.

30. L passes through the point $(2, 4)$ and is perpendicular to the line with the equation $4x - 5y = 20$.

31. L has x-intercept $(2, 0)$ and y-intercept $(0, -3)$.

32. L has slope 3 and passes through the point $(-2, 4)$.

Solve each problem.

33. Number Problem If one-third of a number is added to 3 times the number, the result is 30. Find the number.

34. Number Problem Two more than 4 times a number is 30. Find the number.

35. Business and Finance On a particular flight, the cost of a coach ticket is one-half the cost of a first-class ticket. If the total cost of the tickets is \$1,350, how much does each ticket cost?

36. **Geometry** The length of one side of a triangle is twice that of the second and 4 less than that of the third. If the perimeter is 64 meters (m), find the length of each of the sides.

37. Graph the solution set for the inequality

$2x + 3y < 6$

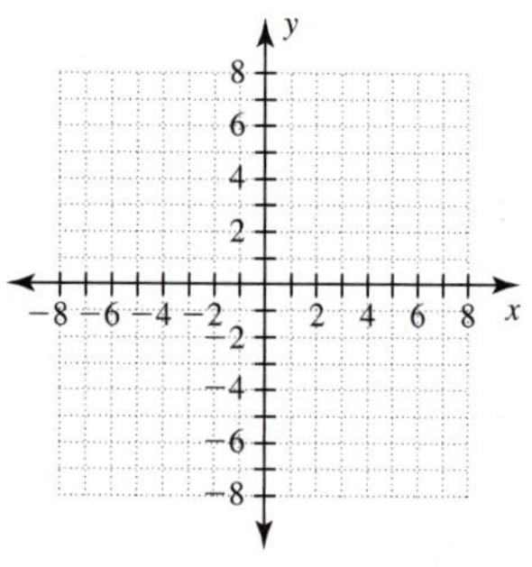

Business and Finance A shipping company charged \$50.52 to ship a 5-lb package across the country overnight. It charged \$70.27 to ship a 10-lb package overnight between the same addresses.

38. Construct a linear model for the cost of shipping as a function of a package's weight.

39. How much would you expect it to cost to ship a 12-lb package?

40. Interpret the slope of the model in the context of the application.

CHAPTER

4

Systems of Linear Equations

INTRODUCTION

Although agriculture is not typically thought of as a high-tech industry, technology has long been an important element of farming. In the industrial revolution, a lot of time and energy was spent assuring that farms were supplied with equipment to increase productivity.

In the computer information era, agriculture has again benefited greatly. Whether it is computer-operated watering systems or market analysis, computers and mathematics play an important role in agronomy.

We explore how we use systems of equations to determine appropriate levels of nutrients for fertilizers in Activity 4.

CHAPTER 4 OUTLINE

4.1 Graphing Systems of Linear Equations

< 4.1 Objectives >

1 > Solve a system of equations by graphing

2 > Classify systems of equations

In Section 1.7, we defined a solution set as "the set of all values for the variable that make the equation or inequality a true statement." For the equation

$$2x - 3x + 5 = x - 7$$

the solution set is $\{6\}$. This tells us that 6 is the only value for the variable x that makes the equation a true statement.

When we studied equations in two variables in Chapter 2, we found that a solution to a two-variable equation is an ordered pair. Given the equation

$$y = 2x - 5$$

one possible solution to the equation is the ordered pair $(1, -3)$. In fact, there are infinitely many solutions which form a line when graphed.

In this chapter, we introduce a topic that has many applications in chemistry, business, economics, and physics. Each of these fields has occasion to solve systems of equations.

Definition

Systems of Equations

A **system of equations** is a set of two or more related equations.

Our goal in this chapter is to solve linear systems of equations.

Definition

Solutions to Systems of Equations

A **solution** to a system of equations in two variables is an ordered pair of real numbers (x, y) that satisfies all of the equations in the system.

Over the course of this chapter, we look at different ways in which a linear system of equations can be used and solved. Our first method is a graphical method for solving a system.

Example 1 Solving a System by Graphing

< Objective 1 >

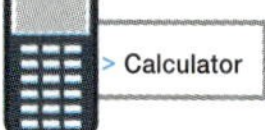

Solve the system by graphing.

$$2x + y = 4$$
$$x - y = 5$$

NOTES

If we solve the equations for y, we can use a graphing calculator to solve the system.

$Y_1 = -2x + 4$

and

$Y_2 = x - 5$

We *approximate* the solution by tracing the curves near their intersection.

Because there are two variables in the equations, we are searching for ordered pairs. We are looking for all of the ordered pairs that make *both* equations true.

We graph the lines corresponding to the two equations of the system.

Each equation has an infinite number of solutions (ordered pairs) corresponding to the points on its line. The point of intersection $(3, -2)$ is the *only* point lying on both lines, so $(3, -2)$ is the only ordered pair satisfying both equations and $(3, -2)$ is the solution to the system. The solution set is $\{(3, -2)\}$.

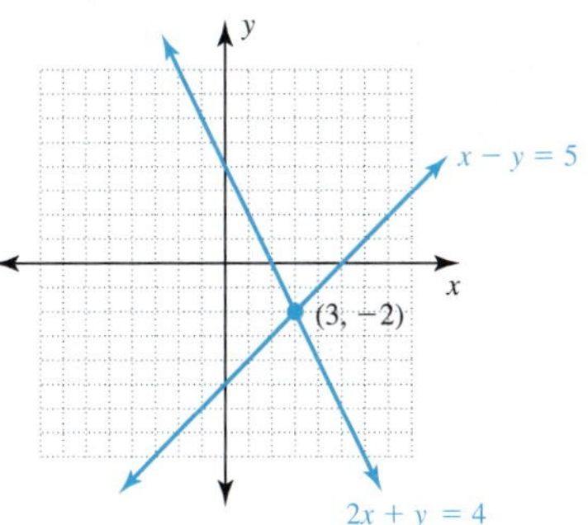

Check Yourself 1

Solve the system by graphing.

$$3x - y = 2$$
$$x + y = 6$$

We said that a *solution to a system of equations* must satisfy all of the equations in the system. It is always a good idea to check a solution to a system, but it is especially important to do so when using a graphing approach.

Example 2 Checking the Solution to a System of Equations

> CAUTION

The difficulty with determining a solution exactly by graphing makes it especially important that you check solutions found using this method.

NOTE

Remember to check the solution in *both* equations.

In Example 1, we found that $(3, -2)$ is a solution to the system of equations

$$2x + y = 4$$
$$x - y = 5$$

Check this result.

We check the solution to the system by checking that it is a solution to each equation, individually. Begin by substituting 3 for x and -2 for y into the first equation and see if the result is true.

$2x + y = 4$ Always use the original equation to check a result.

$2(3) + (-2) \stackrel{?}{=} 4$ Substitute $x = 3$ and $y = -2$.

$6 - 2 \stackrel{?}{=} 4$

$2 = 4$ True

Then, check the result using the second equation.

$x - y = 5$

$(3) - (-2) \stackrel{?}{=} 5$ Substitute into the second equation.

$5 = 5$ True

Because $(3, -2)$ checks as a solution in both equations, it is a solution to the system of equations.

Check Yourself 2

Check that $(2, 4)$ is a solution to the system of equations from Check Yourself 1.

$$3x - y = 2$$
$$x + y = 6$$

We put these last two ideas together into a single example.

Example 3 Solving a System of Equations

RECALL

Solve each equation for y to graph it. $5x + 2y = 5$ can be rewritten as

$y = -\frac{5}{2}x + \frac{5}{2}$

$3x + y = 2$ is equivalent to $y = -3x + 2$.

Solve the system by graphing and check the solution.

$$5x + 2y = 5$$
$$3x + y = 2$$

We begin by graphing the equations.

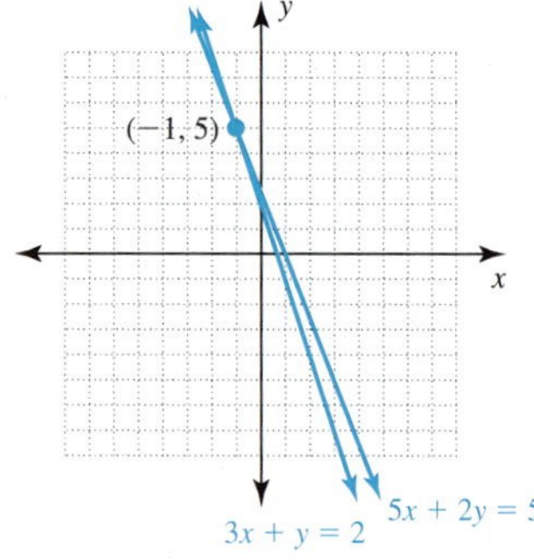

It looks like the graphed lines intersect at $(-1, 5)$. To be certain, we check that this is a solution to the system. We check the solution by substituting the x- and y-value in each equation.

First Equation

$5x + 2y = 5$ The first equation

$5(-1) + 2(5) \stackrel{?}{=} 5$ Substitute $x = -1$ and $y = 5$.

$-5 + 10 \stackrel{?}{=} 5$ Follow the order of operations.

$5 = 5$ True

Second Equation

$3x + y = 2$ The second equation

$3(-1) + (5) \stackrel{?}{=} 2$ Substitute $x = -1$ and $y = 5$.

$-3 + 5 \stackrel{?}{=} 2$ Follow the order of operations.

$2 = 2$ True

The solution $(-1, 5)$ checks in both equations so the solution set for the given system of equations is $\{(-1, 5)\}$.

Check Yourself 3

Solve the system by graphing and check your solution.

$$5x - 2y = 7$$
$$x + y = 7$$

In the previous examples, the two lines are nonparallel and intersect at only one point. Each system has a unique solution corresponding to that point. Such a system is called a **consistent system.** In Example 4, we examine a system in which the lines do not intersect.

Example 4 Graphing a System

Solve the system by graphing.

$$2x - y = 4$$
$$6x - 3y = 18$$

The lines corresponding to the two equations are graphed here.

The lines are distinct and parallel. There is no point at which they intersect, so the system has **no solution.** We call such a system an **inconsistent system.**

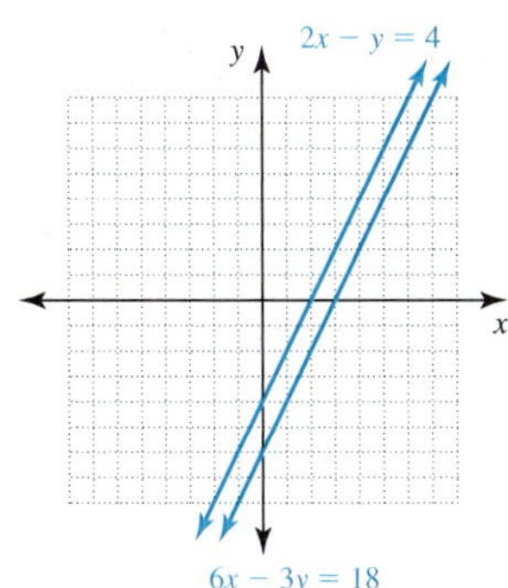

Check Yourself 4

Solve the system, if possible.

$$3x - y = 1$$
$$6x - 2y = 13$$

Sometimes the equations in a system have the same graph.

Example 5 Graphing a System

Solve the system by graphing.

$$2x - y = 2$$
$$4x - 2y = 4$$

We begin by graphing the equations.

> **NOTE**
>
> Every point on the line represents a solution to the system.

Both equations graph the same line, so they have an **infinite number of solutions** in common. We call such a system a **dependent system.**

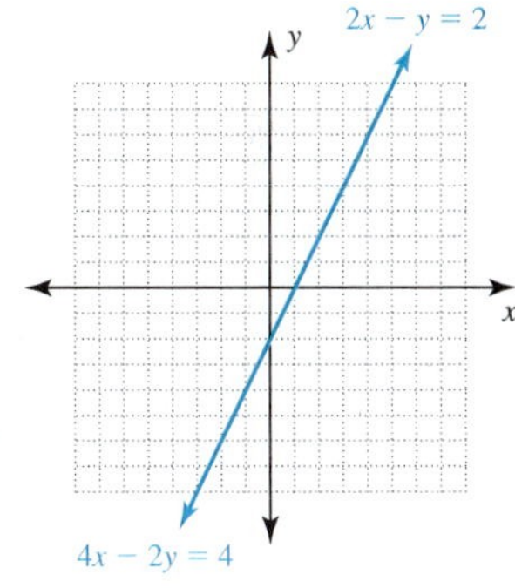

Check Yourself 5

Solve the system by graphing.

$$6x - 3y = 12$$
$$y = 2x - 4$$

You have now seen the three possible types of solutions to a system of two linear equations. There is a single solution (a consistent system), an infinite number of solutions (a dependent system), or no solution (an inconsistent system).

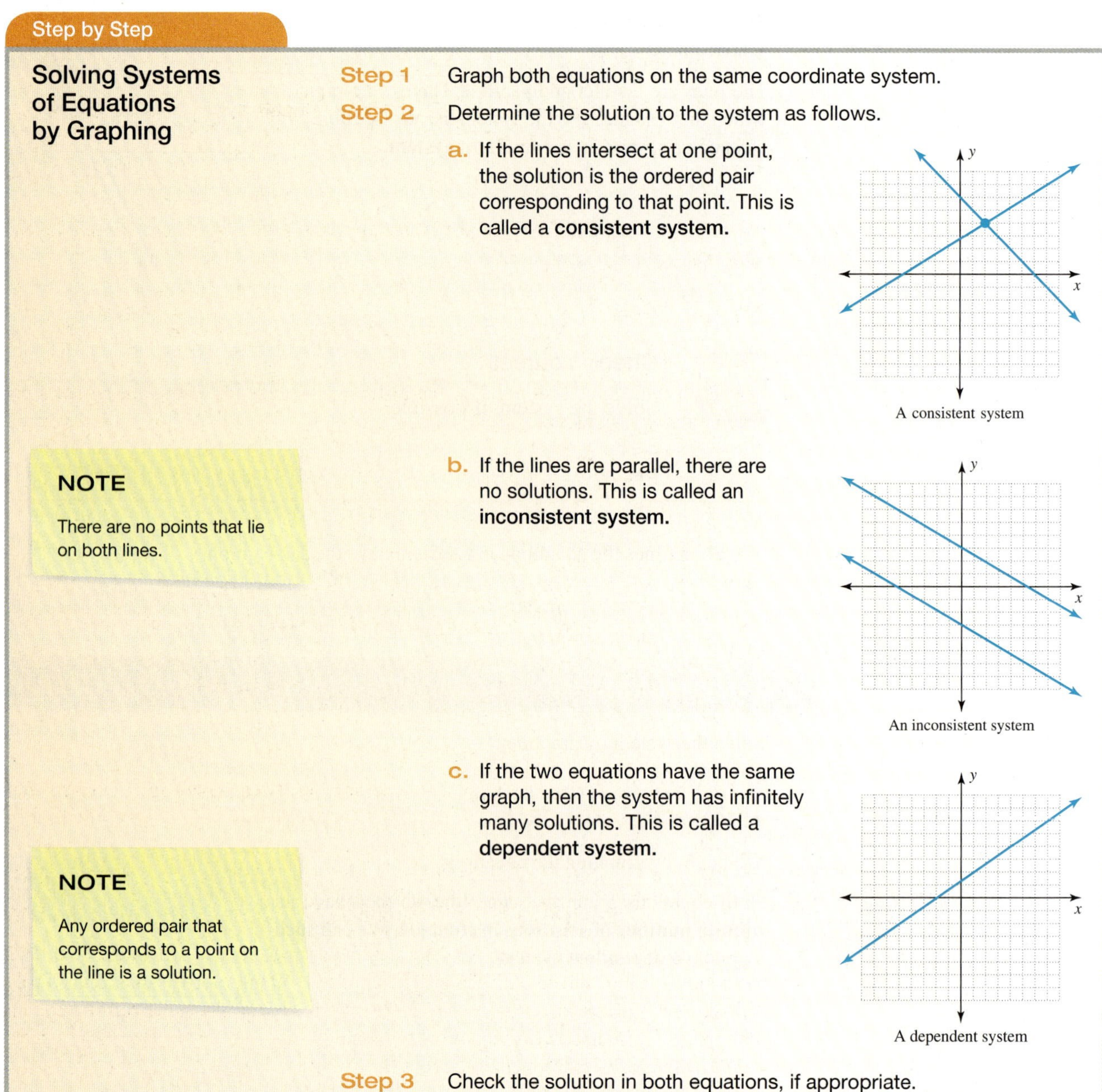

Step by Step
Solving Systems of Equations by Graphing
Step 1 Graph both equations on the same coordinate system.
Step 2 Determine the solution to the system as follows.
a. If the lines intersect at one point, the solution is the ordered pair corresponding to that point. This is called a consistent system.
y
x
A consistent system
NOTE
There are no points that lie on both lines.
b. If the lines are parallel, there are no solutions. This is called an inconsistent system.
y
x
An inconsistent system
c. If the two equations have the same graph, then the system has infinitely many solutions. This is called a dependent system.
NOTE
Any ordered pair that corresponds to a point on the line is a solution.
y
x
A dependent system
Step 3 Check the solution in both equations, if appropriate.

In both dependent and inconsistent systems, the slopes of the two lines in the system must be the same. (Do you see why that is true?) Two lines with different slopes *always* intersect at exactly one point.

Example 6 Classifying Systems

< Objective 2 >

For each system, determine the number of solutions and identify the type of system.

(a) $y = 2x - 5$
$y = 2x + 9$

Both lines have a slope of 2, but different y-intercepts. We have two distinct parallel lines, so there are no solutions. The system is inconsistent.

(b) $y = 3x + 7$

$y = -\frac{1}{3}x + 2$

These lines are perpendicular. There is one solution. The system is consistent.

(c) $2x - 3y = 7$

$3x + 5y = 2$

The lines have different slopes. The slopes are $\frac{2}{3}$ and $-\frac{3}{5}$. There is a single solution. The system is consistent.

NOTE

Solving $2x - 3y = 12$ for y gives $y = \frac{2}{3}x - 4$.

(d) $y = \frac{2}{3}x - 6$

$2x - 3y = 12$

Both lines have a slope of $\frac{2}{3}$, but different y-intercepts. There are no solutions. The system is inconsistent.

Check Yourself 6

For each system, determine the number of solutions and identify the type of system.

(a) $y = 2x - 1$
$y = 3x + 7$

(b) $y = -3x - 2$
$y = -\frac{1}{3}x + 4$

(c) $6x - 3y = 4$
$-2x + y = 9$

(d) $y = \frac{1}{2}x - 4$
$x - y = 6$

In Example 7, we use a graphical approach to solve an application from the field of medicine.

Example 7 A Science Application

A medical lab technician needs to determine how much 15% hydrochloric acid (HCl) solution x must be mixed with 5% HCl y to produce 50 mL of a 9% solution. To solve this problem, graph $x + y = 50$ and $15x + 5y = 450$ on the same set of axes, and determine the intersection point of the two lines.

The graphs of the two equations are shown here.

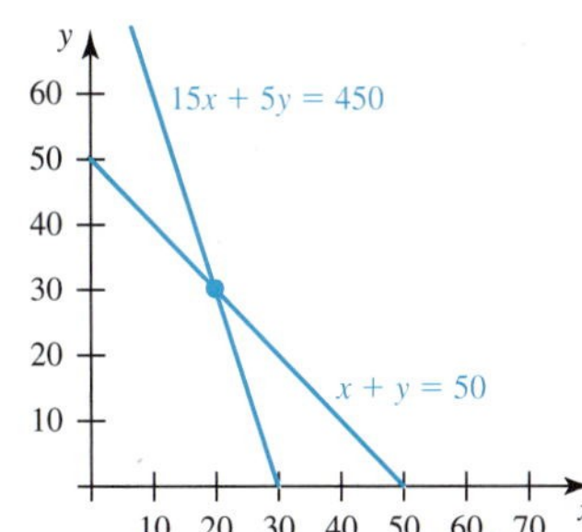

The intersection point appears to be (20, 30). Substituting into each equation verifies this. So, 20 mL of 15% HCl should be mixed with 30 mL of 5% HCl to obtain 50 mL of 9% HCl.

NOTE

If you have a graphing calculator, try solving this application on it. Use an "intersect" utility.

Check Yourself 7

A medical lab technician needs to determine how much 6-molar (M) copper sulfate ($CuSO_4$) solution x must be mixed with 2 M $CuSO_4$ y to produce 200 mL of a 3-M solution. To solve this problem, graph $x + y = 200$ and $6x + 2y = 600$ on the same set of axes, and determine the intersection point of the two lines.

Check Yourself ANSWERS

1.

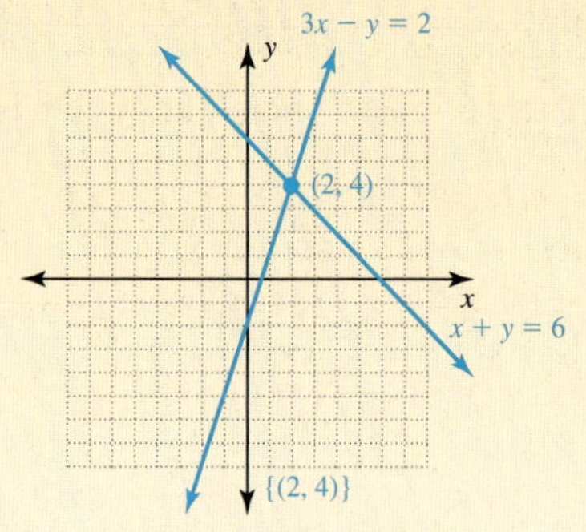

2. $3x - y = 2$
$3(2) - (4) \stackrel{?}{=} 2$
$6 - 4 \stackrel{?}{=} 2$
$2 = 2$ True

$x + y = 6$
$(2) + (4) \stackrel{?}{=} 6$
$6 = 6$ True

3. $\{(3, 4)\}$

4.

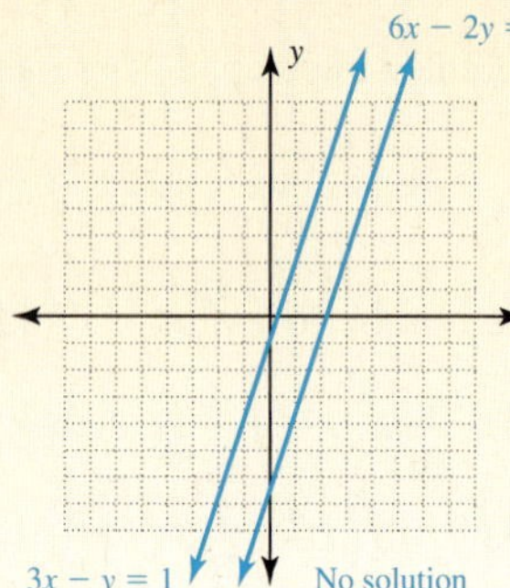

5.

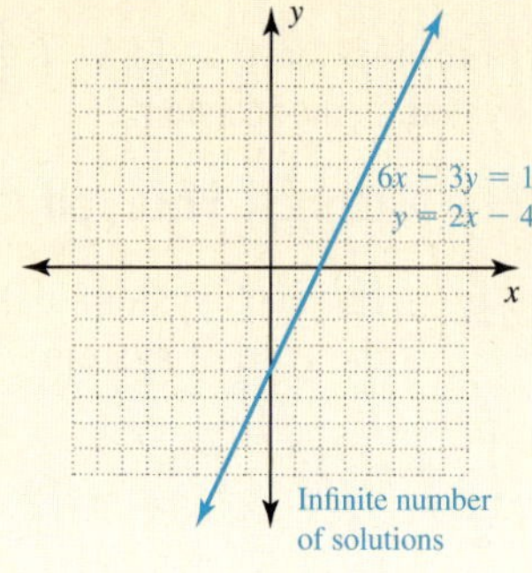

6. **(a)** one solution, consistent; **(b)** one solution, consistent; **(c)** no solutions, inconsistent; **(d)** one solution, consistent

7.

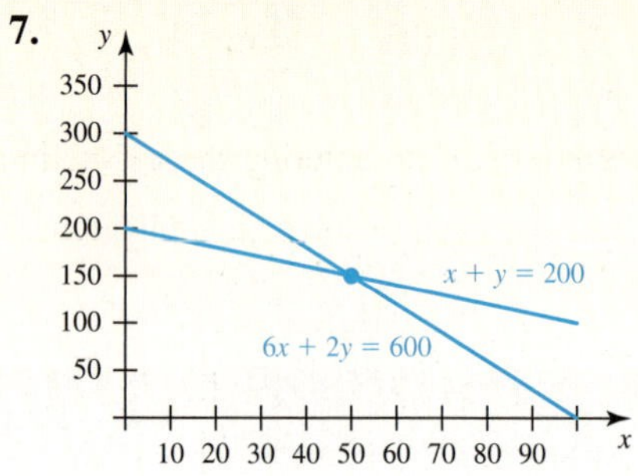

Solution: (50, 150); 50 mL of 6-M copper sulfate should be mixed with 150 mL of 2-M copper sulfate.

Reading Your Text

These fill-in-the-blank exercises will help you understand some of the key vocabulary used in this section. The answers to these exercises are in the Answers Appendix in the back of the text.

(a) A system of equations is a set of two or more ________ equations.

(b) A solution to a system of equations in two variables is an ________ of real numbers that satisfies all of the equations in the system.

(c) A system that has a unique solution corresponding to only one point is called a ________ system.

(d) A system having no solution is called an ________ system.

Graphing Calculator Option

Solving a System of Equations

A graphing calculator can help us solve a system of equations. In order to use a graphing calculator, we must first solve each equation for y. Note that we do not actually need to put the equations into slope-intercept form.

After graphing both lines in a system, we find a good viewing window and use the calculator's **intersect** utility to find the point of intersection. If the lines do not intersect at a *nice* point, the calculator gives us an estimate of the coordinates.

Consider the system

$$37x + 15y = 2{,}531$$
$$45x + 29y = 3{,}946$$

We begin by solving each equation for y.

$$37x + 15y = 2{,}531$$
$$15y = 2{,}531 - 37x$$
$$y = \frac{2{,}531 - 37x}{15}$$

There is no need to write the equation in slope-intercept form.

$$45x + 29y = 3{,}946$$
$$29y = 3{,}946 - 45x$$
$$y = \frac{3{,}946 - 45x}{29}$$

It is important to remember to place the entire numerator in parentheses when entering these functions into a calculator.

Enter the functions into a graphing calculator and graph them on the same set of axes.

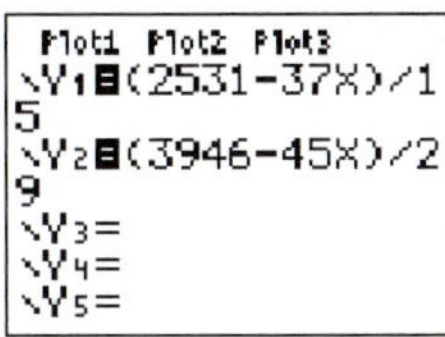

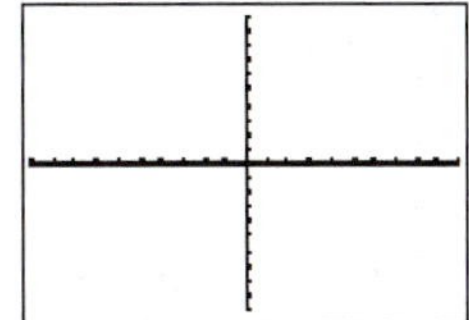

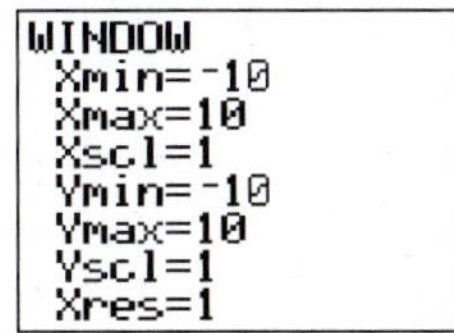

The graphs do not show on the standard (default) graphing window. This is because the graphs are outside this small range. That is, the y-values are not between -10 and 10 when x is in that range.

We can use the **TABLE** utility to find an appropriate viewing window.

X	Y1	Y2
0	168.73	136.07
10	144.07	120.55
20	119.4	105.03
30	94.733	89.517
40	70.067	74
50	45.4	58.483
60	20.733	42.966

X=30

When $x = 0$, the y-value of the first equation is approximately 169 and 136 in the second equation. Therefore, we need our window to include these y-values in order to see the graphs if the y-axis is part of our viewing window. To simplify our tasks, we set the graphs in the first quadrant and see what happens.

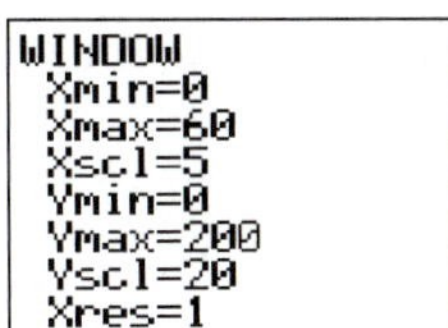

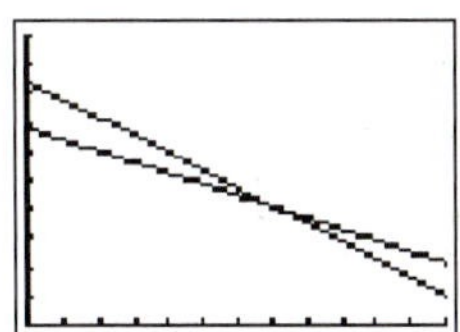

We can see the point of intersection on the screen, so there is no reason to modify the viewing window.

Next, we look for the point of intersection.

On the TI-84 Plus, we begin by opening the **CALC** menu. It is the second function above the TRACE key. Then, select the **intersect** utility.

2nd [CALC] **5:intersect**

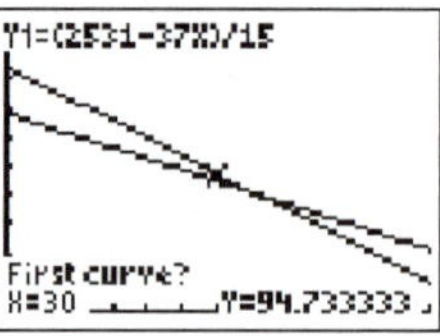

You must tell the calculator which graphs to examine and you must provide a guess. Simply press ENTER for each curve and for the guess and the calculator approximates the intersection point.

ENTER ENTER ENTER

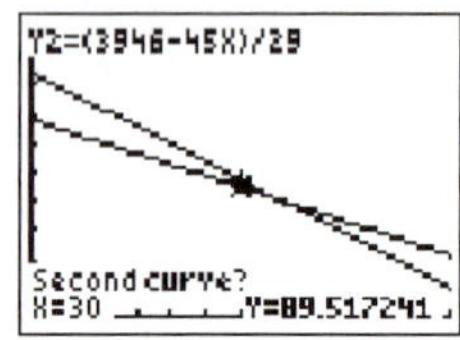

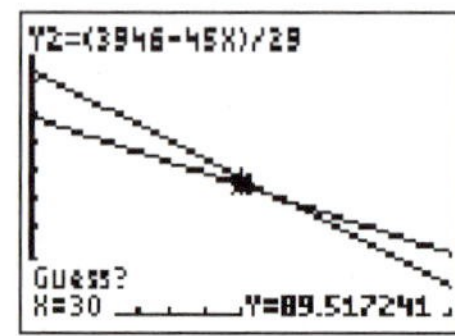

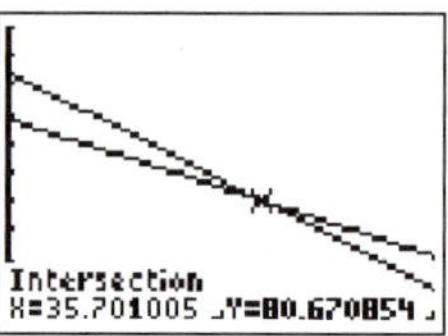

Note: You need to use the left/right arrows to move the cursor near the point of intersection when responding to the **Guess?** prompt if there is more than one intersection point on the screen. Similarly, you would need to use the up/down arrows to cycle to the correct curves if there are more than two functions graphed on your screen.

The final window gives the intersection point. We see that the solution to the system, to the nearest hundredth, is (35.70, 80.67).

Graphing Calculator Check

Use a graphing calculator to solve each system (round your results to the nearest hundredth).

(a) $19x + 83y = 4{,}587$
$36x + 51y = 4{,}229$

(b) $28x + 14y = 3{,}757$
$8x - 7y = -91$

(c) $3x + 5y = 10\sqrt{2}$
$x - 3y = 15$

(d) $x^2 - y = 6$
$x + 2y = 3\pi$

Note: This is not a linear system, but the methods are the same. In this case, there are two solutions.

ANSWERS

(a) $\{(57.98, 41.99)\}$

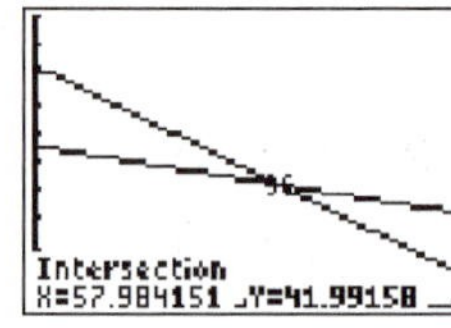

(b) $\{(81.25, 105.86)\}$

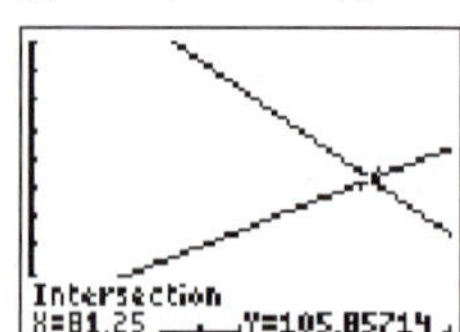

(c) $\{(8.39, -2.20)\}$

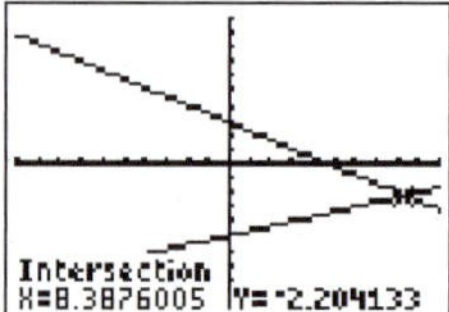

(d) $\{(-3.53, 6.48), (3.03, 3.20)\}$

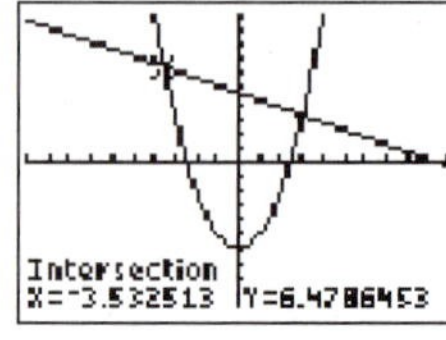

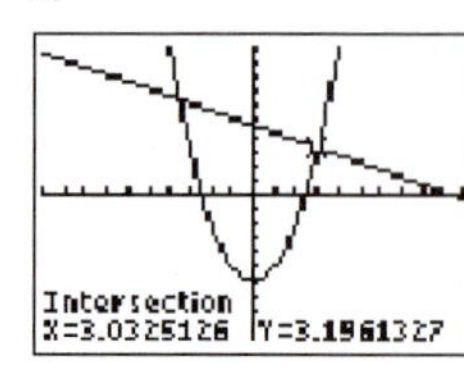

< Objective 1 >

Graph each system of equations and then solve the system.

1. $x + y = 6$
$x - y = 4$

2. $x - y = 8$
$x + y = 2$

3. $x + y = 5$
$-x + y = 7$

4. $x + y = 7$
$-x + y = 3$

5. $x + 2y = 4$
$x - y = 1$

6. $3x + y = 6$
$x + y = 4$

7. $3x - y = 21$
$3x + y = 15$

8. $x - 2y = -2$
$x + 2y = 6$

9. $x + 3y = 12$
$2x - 3y = 6$

10. $2x - y = 4$
$2x - y = 6$

11. $3x + 2y = 12$
$y = 3$

12. $5x - y = 11$
$2x - y = 8$

13. $x - y = 4$
$2x - 2y = 8$

14. $2x - y = 8$
$x = 2$

15. $x - 4y = -4$
$x + 2y = 8$

16. $4x + y = -7$
$-2x + y = 5$

17. $3x - 2y = 6$
$2x - y = 5$

18. $4x + 3y = 12$
$x + y = 2$

19. $3x - y = 3$
$3x - y = 6$

20. $3x - 6y = 9$
$x - 2y = 3$

21. $2y = 3$
$x - 2y = -3$

22. $x + y = -6$
$-x + 2y = 6$

23. $x = -5$
$y = 3$

24. $x = -3$
$y = 5$

< Objective 2 >

Determine whether each system is **consistent, inconsistent,** *or* **dependent.**

25. $y = 3x + 7$
$y = 7x - 2$

26. $y = 2x - 5$
$y = -2x + 9$

27. $y = 7x - 1$
$y = 7x + 8$

28. $y = -5x + 9$
$y = -5x - 11$

29. $3x + 4y = 12$
$9x - 5y = 10$

30. $2x - 4y = 11$
$-8x + 16y = 15$

31. $7x - 2y = 5$
$14x - 4y = 10$

32. $3x + 2y = 8$
$6x - 4y = -12$

Business and Finance *Two construction companies offer different pricing structures for jobs that they accept.*

Company A: $5,000 plus $15 per square foot of building

Company B: $7,500 plus $12.50 per square foot of building

Use this information to complete exercises 33 to 38.

33. Construct an equation to model the pricing structure used by company A.

34. Construct an equation to model the pricing structure used by company B.

35. Graph the equations constructed in exercises 33 and 34 on the same set of axes.

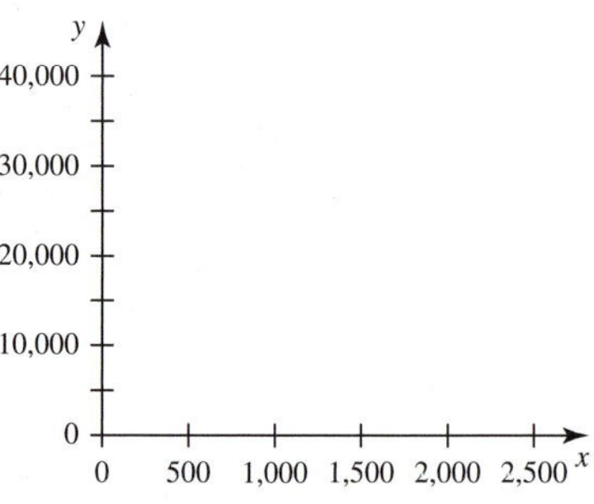

36. Assuming both companies do comparable work, which company offers the lowest price to build an 800-ft^2 shop? What is their price?

37. Which company offers the lower price to build an 1,800-ft^2 shop? What is their price?

38. For what building size will the two companies charge the same price? What is that price?

Complete each statement with **always, sometimes,** *or* **never.**

39. A linear system __________ has at least one solution.

40. If the graphs of two linear equations in a system have different slopes, the system __________ has exactly one solution.

41. If the graphs of two linear equations in a system have equal slopes, the system __________ has exactly one solution.

42. If the graphs of two linear equations in a system have equal slopes and equal y-intercepts, the system __________ has an infinite number of solutions.

Skills | **Calculator/Computer** | Career Applications | Above and Beyond

Use a graphing calculator to solve each exercise. Estimate your answers to the nearest hundredth. You may need to adjust the viewing window to see the point of intersection.

43. $88x + 57y = 1{,}909$
$95x + 48y = 1{,}674$

44. $32x + 45y = 2{,}303$
$29x - 38y = 1{,}509$

45. $25x - 65y = 5{,}312$
$-21x + 32y = 1{,}256$

46. $-27x + 76y = 1{,}676$
$56x - 2y = -678$

47. $15x + 20y = 79$
$7x + 5y = 115$

48. $23x - 31y = 1{,}915$
$15x + 42y = 1{,}107$

Skills | Calculator/Computer | **Career Applications** | Above and Beyond

49. Construction Technology The beam shown in the figure is 15 ft long and has the load indicated on each end. Graphically solve the system of equations shown in order to determine the point at which the beam balances.

$$x + y = 15$$
$$80x = 120y$$

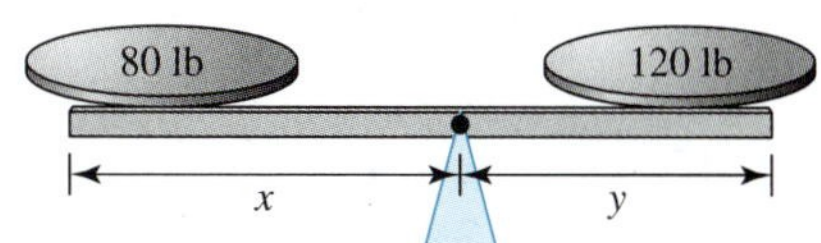

50. Mechanical Engineering For a plating bath, 10,000 L of 13% electrolyte solution is required. You have 8% and 16% solutions in stock. Solve the system of equations graphically, in which x represents the amount of 8% solution to use, to solve the application.

$$x + y = 10{,}000$$
$$0.08x + 0.16y = 1{,}300$$

51. Mechanical Engineering At 2,100°C, a 60% aluminum oxide and 40% chromium oxide alloy separates into two different alloys. The first alloy is 78% Al_2O_3 and 22% Cr_2O_3, and the second is 50% Al_2O_3 and 50% Cr_2O_3. If the total amount of the original alloy present is 7,000 grams, use a system of equations and solve by graphing to find out how many grams of each type the original alloy separates into.

Use the system $0.78x + 0.50y = 4{,}200$ and $0.22x + 0.50y = 2{,}800$

where x is the amount of the 78%/22% alloy and y is the amount of the 50%/50% alloy.

52. Manufacturing A manufacturer has two machines that produce door handles. On Monday, machine A operates for 10 hr and machine B operates for 7 hr, and 290 door handles are produced. On Tuesday, machine A operates for 6 hours and machine B operates for 12 hr, and 330 door handles are produced.

Use the system

$$10x + 7y = 290$$
$$6x + 12y = 330$$

where x is the number of handles produced by machine A in an hour, and y is the number of handles produced by machine B in an hour.

Solve the system graphically.

53. Construction The symmetric gambrel roof pictured here has several missing dimensions. These dimensions can be calculated using a system of equations.

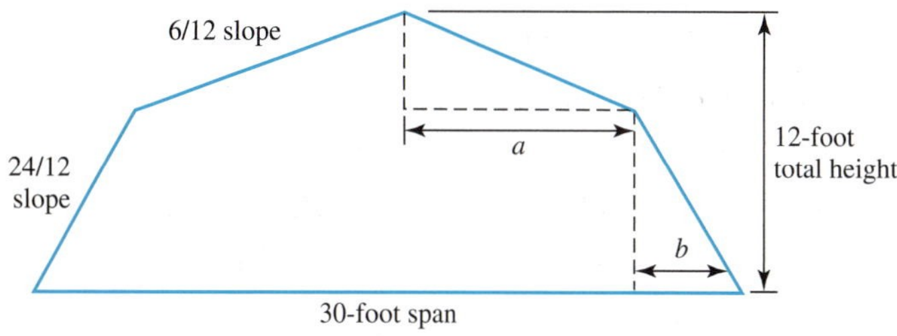

The system of equations is

$$2a + 2b = 30$$

$$\frac{24}{12}a + \frac{6}{12}b = 12$$

Solve this system of equations graphically. 

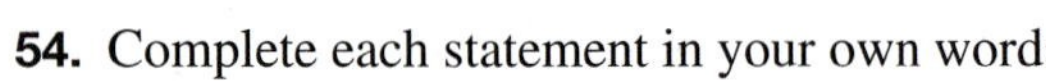

54. Complete each statement in your own words.
"To solve an equation means to . . . ."
"To solve a system of equations means to . . . ."

55. Find values for a and b so that (1, 2) is the solution to the system.

$$ax + 3y = 8$$
$$-3x + 4y = b$$

56. Find values for a and b so that (−3, 4) is the solution to the system.

$$5x + 7y = b$$
$$ax + y = 22$$

57. A system of equations such as the one below is sometimes called a *2-by-2* system of linear equations.

$$3x + 4y = 1$$
$$x - 2y = 6$$

Explain this term.

58. Complete this statement in your own words: "All the points on the graph of the equation $2x + 3y = 6$ . . . ." Exchange statements with other students. Do you agree with other students' statements?

59. Does a system of linear equations always have a solution? How can you tell without graphing that a system of two equations will be graphed as two parallel lines? Give some examples to explain your reasoning.

60. Suppose we have the linear system

$$Ax + By = C$$
$$Dx + Ey = F$$

(a) Write the slope of the line determined by the first equation.
(b) Write the slope of the line determined by the second equation.
(c) What must be true about the given coefficients in order to guarantee that the system is consistent?

Answers

1. 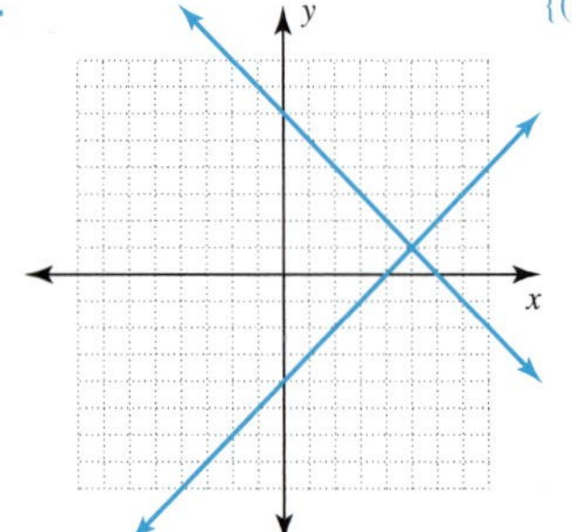

3. 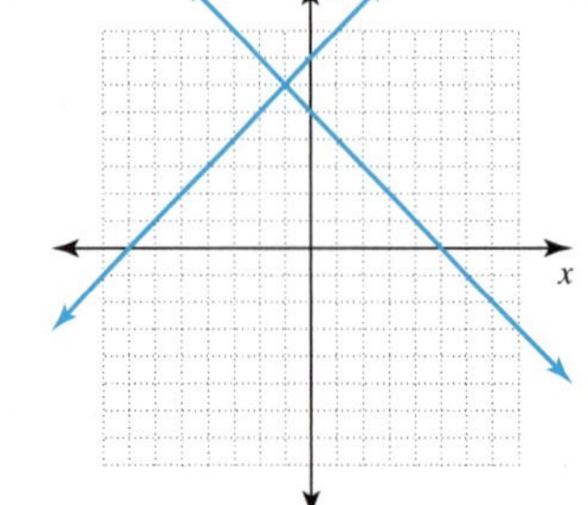

5. 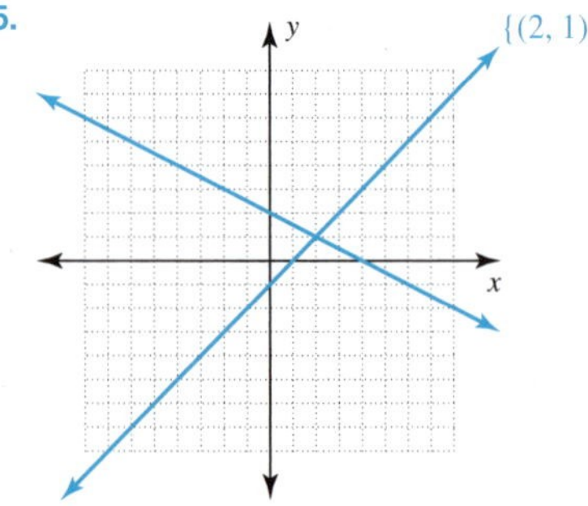

7. 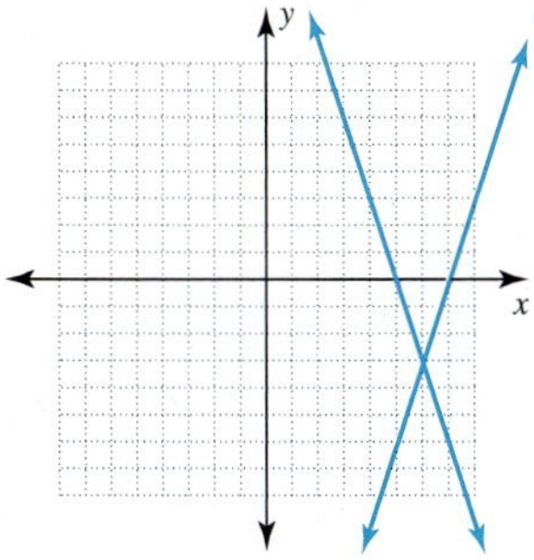 $\{(6, -3)\}$

9. 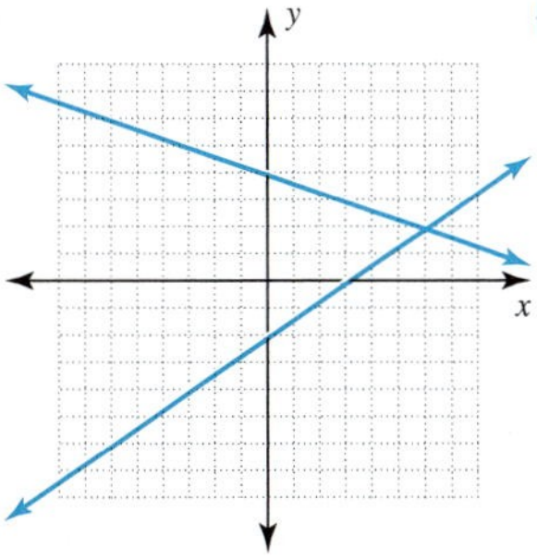 $\{(6, 2)\}$

11. 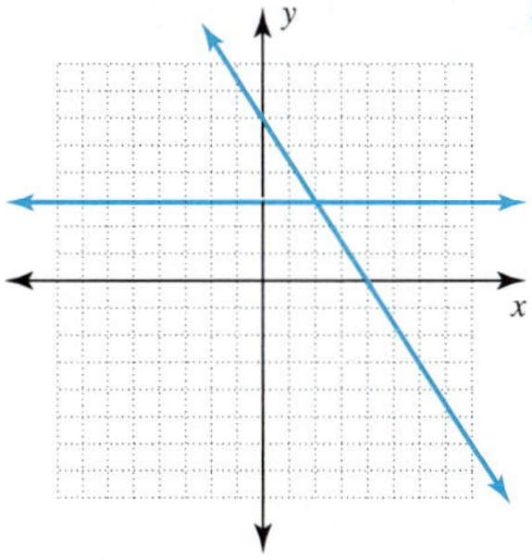 $\{(2, 3)\}$

13.

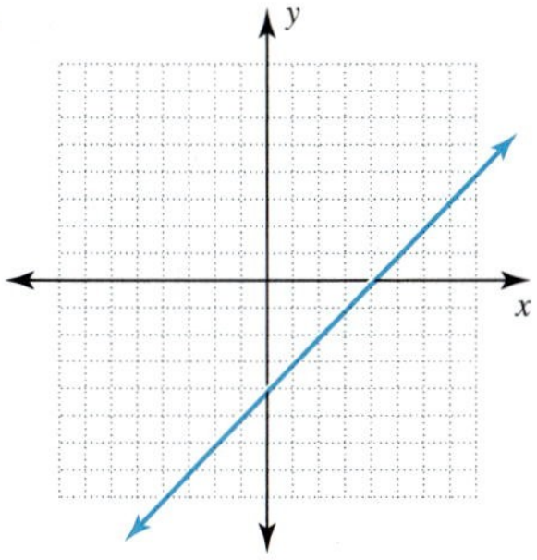

Infinite number of solutions, dependent system

15. 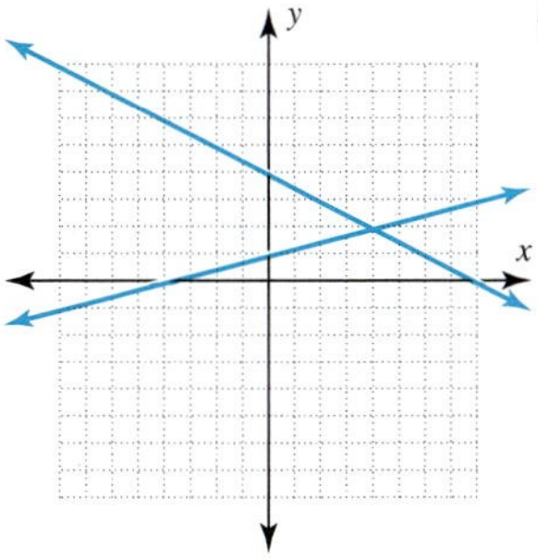 $\{(4, 2)\}$

17. 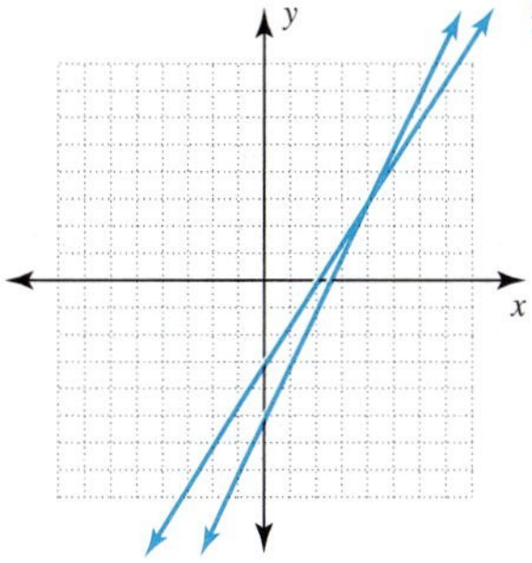 $\{(4, 3)\}$

19.

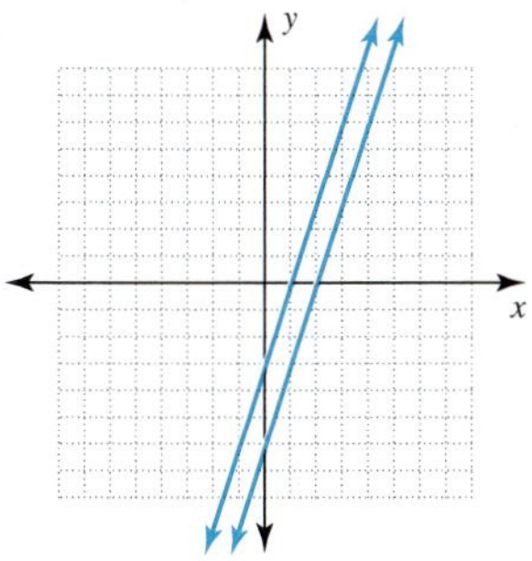

No solutions, inconsistent system

21. 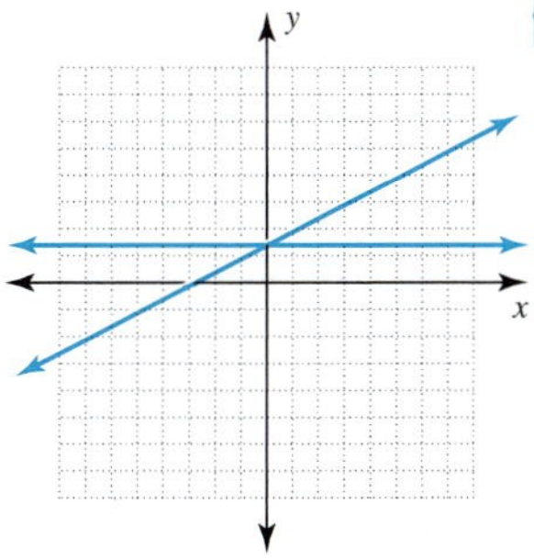 $\left\{\left(0, \frac{3}{2}\right)\right\}$

23. 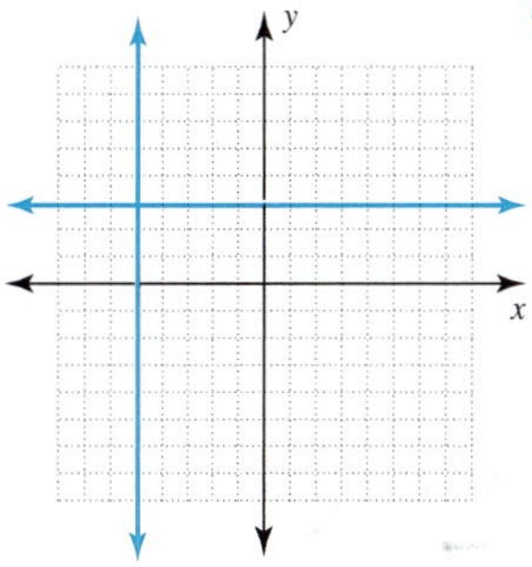 $\{(-5, 3)\}$

25. Consistent **27.** Inconsistent **29.** Consistent **31.** Dependent **33.** $A = 15x + 5{,}000$

35.

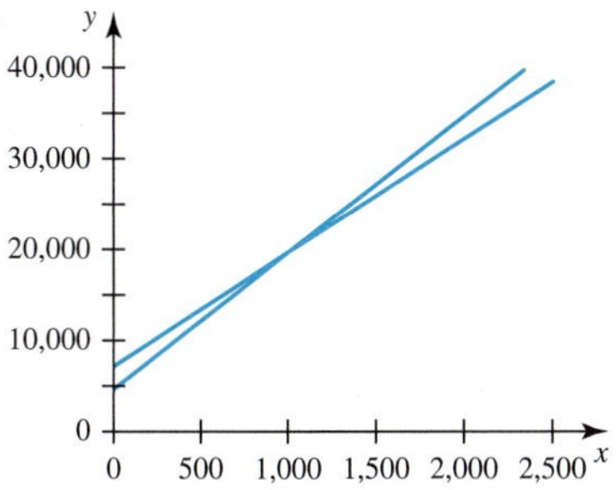

37. Company B at \$30,000 **39.** sometimes **41.** never **43.** $\{(3.18, 28.58)\}$

45. $\{(-445.35, -253.01)\}$ **47.** $\{(29.31, -18.03)\}$ **49.** $x = 9$ ft, $y = 6$ ft

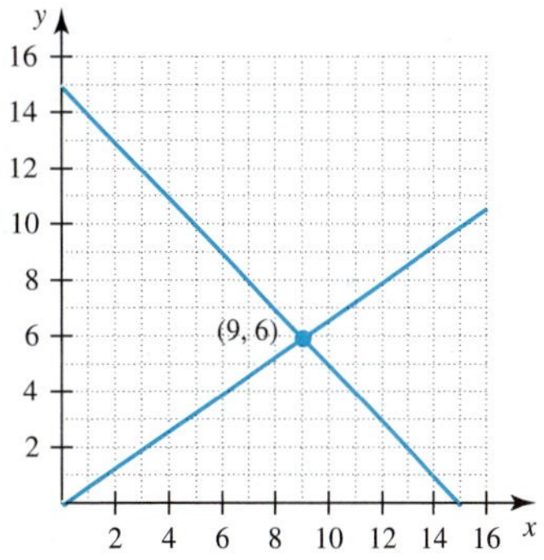

51. 78%/22%: 2,500 g; 50%/50%: 4,500 g **53.** $a = 3$; $b = 12$ **55.** $a = 2$; $b = 5$ **57.** Above and Beyond **59.** Above and Beyond

Activity 4 ::

Agricultural Technology

Nutrients and Fertilizers

When growing crops, it is not enough just to till the soil and plant seeds. The soil must be properly prepared before planting. Each crop takes nutrients out of the soil that must be replenished. Some of this is done with crop rotations (each crop takes some nutrients out of the soil while replenishing other nutrients), but maintaining proper nutrient levels often requires that some additional nutrients be added. This may be accomplished with fertilizers.

The three most vital nutrients are nitrogen, phosphorus, and potassium. Three different fertilizer mixes are available:

Urea: Contains 46% nitrogen

Growth: Contains 16% nitrogen, 48% phosphorus, and 12% potassium

Restorer: Contains 21% phosphorus and 62% potassium

A soil test shows that a field requires 115 lb of nitrogen, 78 lb of phosphorus, and 61 lb of potassium per acre. We need to determine how many pounds of each type of fertilizer to use on the field.

1. Let x equal the number of pounds of urea used. How many pounds of each nutrient are in a batch of urea?
2. Let y equal the number of pounds of the growth blend used. How many pounds of each nutrient are in a batch?
3. Let z equal the number of pounds of the soil restorer used. How many pounds of each nutrient are in a batch?
4. Create an expression for the amount of nitrogen in x pounds of urea, y pounds of growth blend, and z pounds of soil restorer.
5. Create similar expressions for phosphorus and potassium.
6. Construct and solve a system of equations to find the amount of each type of fertilizer required.

4.2 Solving Equations in One Variable Graphically

< 4.2 Objectives >

1 > Rewrite a linear equation in one variable as $f(x) = g(x)$

2 > Find and interpret the point of intersection of $f(x)$ and $g(x)$

We solved linear equations in one variable algebraically in Chapter 1. Our work in Section 4.1 provides us with a more visual way of solving equations in one variable. These techniques do not replace algebraic methods, but they are a powerful alternative approach to solving equations.

Our approach is to create a system of equations based on the two sides of an equation. We then graph the system and find its point of intersection.

We begin with a linear equation in one variable in standard form.

Example 1 Solving a Linear Equation Graphically

< Objective 1 >

Graphically solve the equation.

$2x - 6 = 0$

Step 1 Let each side of the equation represent a function of x.

$f(x) = 2x - 6$

$g(x) = 0$

Step 2 Graph the two functions on the same set of axes.

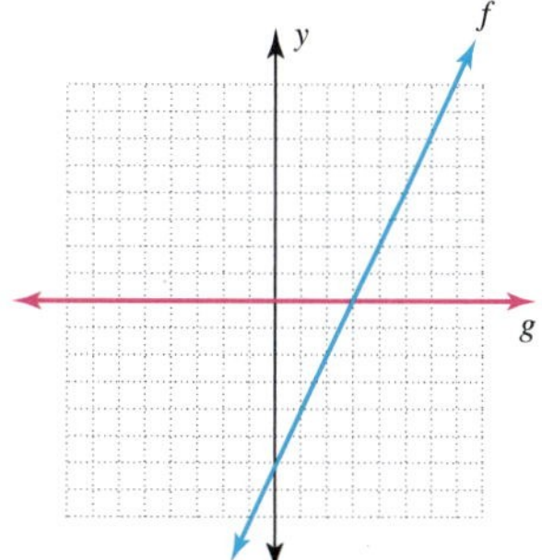

The graph of $y = g(x)$ is simply the x-axis.

NOTE

We ask the question, When is the graph of f equal to the graph of g? Specifically, for what values of x does this occur?

Step 3 Find the point of intersection of the two graphs. The x-coordinate of this point represents the solution to the original equation.

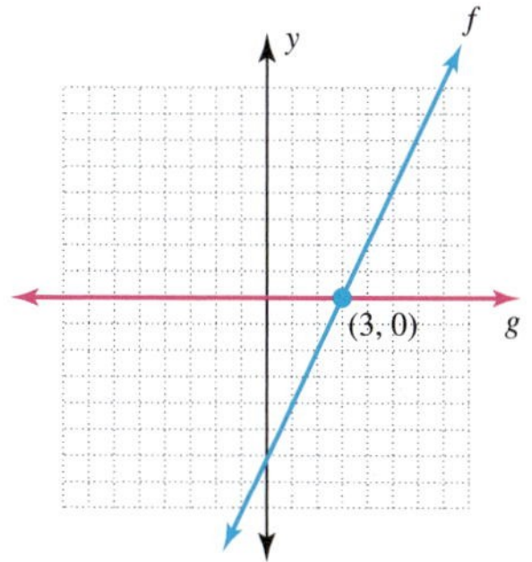

The two lines intersect on the x-axis at the point (3, 0). Again, because we are solving an equation in one variable x, we are interested only in x-values. Thus, the solution is $x = 3$ and the solution set is $\{3\}$.

It is always a good idea to check your work, and it is especially important when you solve a problem graphically.

We check our solution by substituting it back into the original equation.

Check

$2x - 6 = 0$ The original equation

$2(3) - 6 \stackrel{?}{=} 0$ Substitute $x = 3$ into the original equation.

$6 - 6 \stackrel{?}{=} 0$

$0 = 0$ True!

Check Yourself 1

Graphically solve the equation.

$-3x + 6 = 0$

We use the same three-step process to solve any equation. In Example 2, we look for a point of intersection that is *not* on the x-axis.

Example 2 Solving a Linear Equation Graphically

< Objective 2 >

Graphically solve the equation.

$2x - 6 = -3x + 4$

Step 1 Let each side of the equation represent a function of x.

$f(x) = 2x - 6$

$g(x) = -3x + 4$

Step 2 Graph the two functions on the same set of axes.

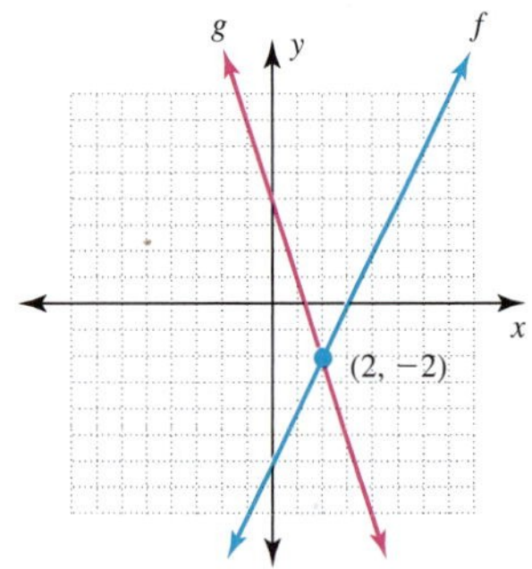

Step 3 Find the point of intersection of the two graphs. Here, they intersect at $(2, -2)$. Remember, we are only interested in the x-coordinate of their intersection, so the solution to our original equation is 2; the solution set is $\{2\}$.

We check our work by returning to the original equation.

Check

$2x - 6 = -3x + 4$ Return to the original equation to check.

$2(2) - 6 \overset{?}{=} -3(2) + 4$ Substitute 2 for x.

$4 - 6 \overset{?}{=} -6 + 4$

$-2 = -2$ True.

The x-coordinate at the point of intersection gives the solution to the original equation. But what does the y-coordinate represent?

The y-value gives the left and right sides of the equation after plugging in the x-value. If you look at the check in the previous example, you will notice that the last line is $-2 = -2$ and the y-value of the intersection point is -2.

Check Yourself 2

Graphically solve the equation.

$-2x - 5 = x - 2$

This algorithm summarizes our work solving linear equations graphically.

Step by Step

Solving a Linear Equation Graphically

Step 1 Let each side of the equation represent a function of x.

Step 2 Graph the two functions on the same set of axes.

Step 3 Find the point of intersection of the two graphs. The x-coordinate of the intersection is the solution to the original equation.

We often apply some algebra even when we take a graphical approach. Consider Example 3.

Example 3 Solving a Linear Equation Graphically

Solve the equation graphically.

$2(x + 3) = -3x - 4$

Use the distributive property to remove the parentheses from the left side.

$2x + 6 = -3x - 4$

Now let $f(x) = 2x + 6$

$g(x) = -3x - 4$

Graphing both lines, we get

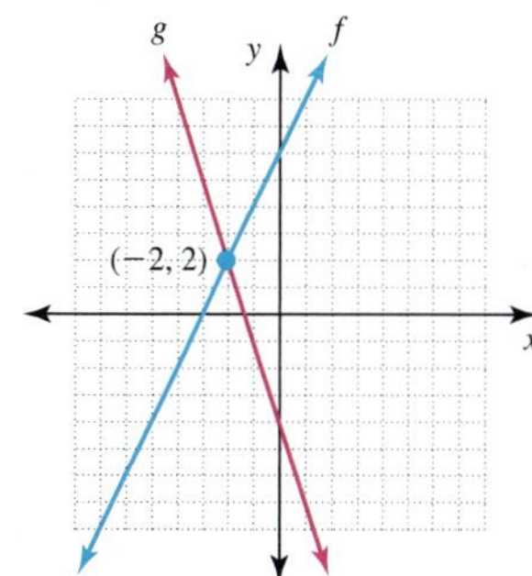

The point of intersection is $(-2, 2)$. The x-coordinate is -2. The solution set for the original equation is $\{-2\}$.

As before, we should check that our proposed solution is correct.

Check

$2(x + 3) = -3x - 4$ The original equation

$2[(-2) + 3] \stackrel{?}{=} -3(-2) - 4$ Substitute $x = -2$ into the equation.

$2(1) \stackrel{?}{=} 6 - 4$ Follow the order of operations.

$2 = 2$ True!

Check Yourself 3

Graphically solve the equation and check your result.

$3(x - 2) = -4x + 1$

We can use a graphing calculator to solve equations in this manner. We now demonstrate this with the same equation seen in Example 3.

Example 4 Solving a Linear Equation with a Graphing Calculator

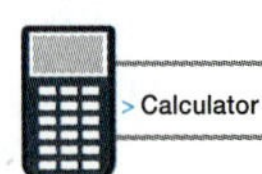

Use a graphing calculator to solve the equation.

$2(x + 3) = -3x - 4$

As before, let each side define a function.

$Y_1 = 2(x + 3)$
$Y_2 = -3x - 4$

When we graph these in the "standard viewing" window, we see

NOTE

This window typically shows x-values from -10 to 10 and y-values from -10 to 10.

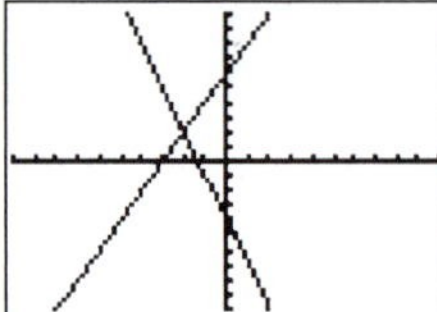

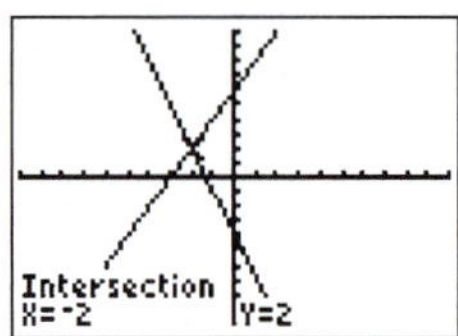

Using the INTERSECT utility, we then see the view shown to the right.

The calculator reports the intersection point as $(-2, 2)$. Since we are interested only in the x-value, the solution is $x = -2$. The solution set is $\{-2\}$.

RECALL

We introduced the intersect utility in the Graphing Calculator Option segment at the end of Section 4.1.

Check Yourself 4

Use a graphing calculator to solve the equation.

$-2x - 5 = x - 2$

Graphing calculators are particularly effective when solving equations with "messy" coefficients. With technology, we can obtain a solution to any desired level of precision.

Example 5 Solving a Linear Equation with a Graphing Calculator

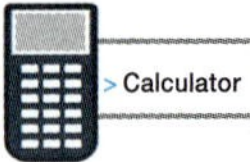

Use a graphing calculator to solve the equation. Round your result to the nearest hundredth.

$2.05(x - 4.83) = -3.17(x + 0.29)$

In the calculator we define

$Y_1 = 2.05(x - 4.83)$
$Y_2 = -3.17(x + 0.29)$

In the standard viewing window, we see

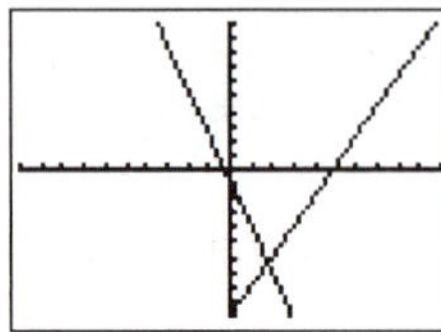

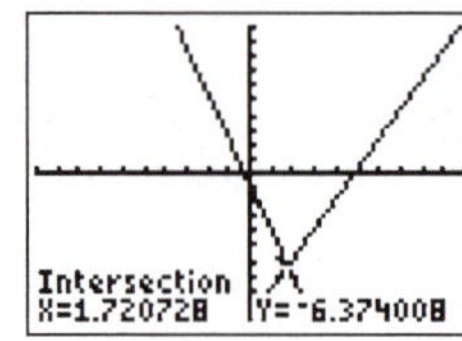

With the INTERSECT utility, we find the intersection point to be (1.720728, −6.374008). Because we want only the x-value, the solution (rounded to the nearest hundredth) is 1.72. The solution set is $\{1.72\}$.

Check Yourself 5

Use a graphing calculator to solve the equation. Round your result to the nearest hundredth.

$$-0.87x + 1.14 = -2.69(x + 4.05)$$

In Example 6, we turn to a business application.

Example 6 A Business and Finance Application

A manufacturer can produce and sell x chairs per day at a cost, in dollars, given by

$$C(x) = 30x + 800$$

The revenue from selling chairs is given by

$$R(x) = 110x$$

Use a graphical approach to find the break-even point, which is the number of chairs at which the revenue equals the cost. That is, we wish to graphically solve the equation

$$110x = 30x + 800$$

Graphing the two functions, we have

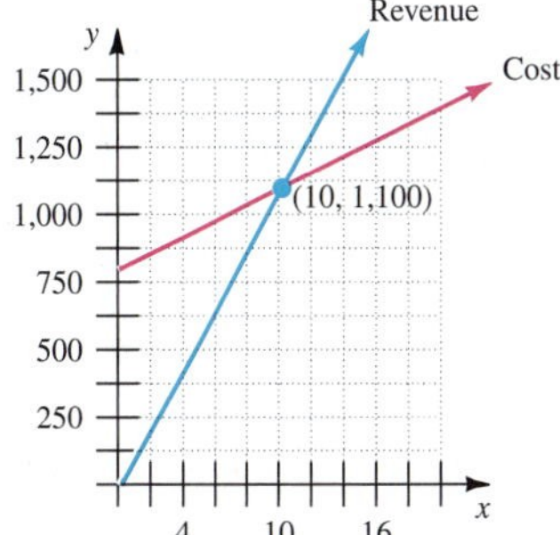

NOTE

Try graphing these functions on your graphing calculator.

We see that the desired x-value (the break-even point) is 10 chairs per day. Note that if the company sells more than 10 chairs, it makes a profit since the revenue exceeds the cost.

In this case, we can interpret the y-coordinate of the intersection point, as well. The intersection point is (10, 1,100). This tells us that when the company sells 10 chairs, their revenue and cost are both $1,100.

Check Yourself 6

A manufacturer can produce and sell x tables per day at a cost of

$$C(x) = 30x + 1{,}800$$

The revenue from selling tables is given by

$$R(x) = 120x$$

Use a graphical approach to find the break-even point.

Check Yourself ANSWERS

1. $f(x) = -3x + 6$
   $g(x) = 0$

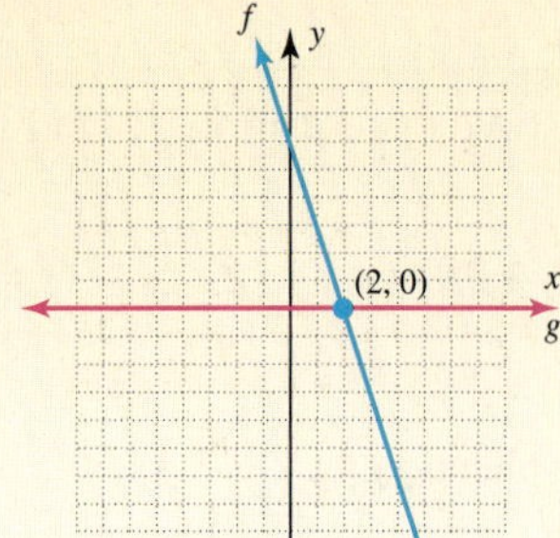

Solution set: {2}

2. $f(x) = -2x - 5$
   $g(x) = x - 2$

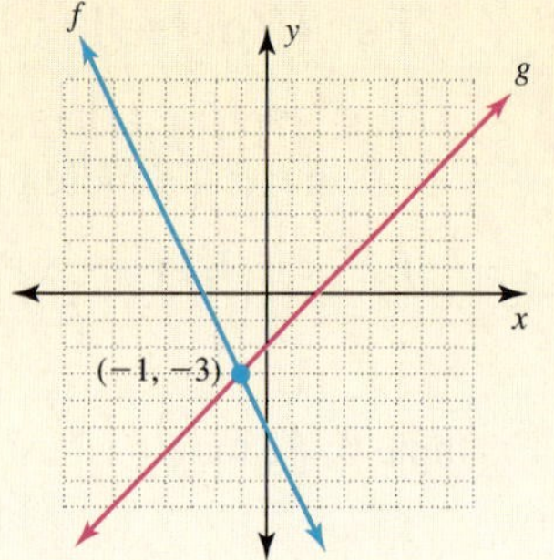

Solution set: {−1}

3. $f(x) = 3(x - 2) = 3x - 6$
   $g(x) = -4x + 1$

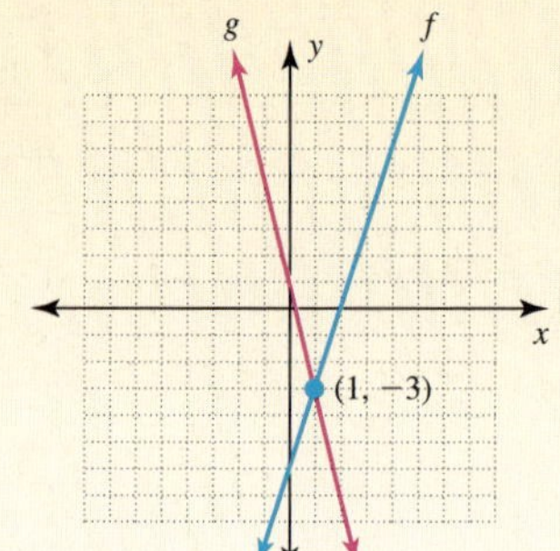

Solution set: {1}

4. $Y_1 = -2x - 5$
   $Y_2 = x - 2$

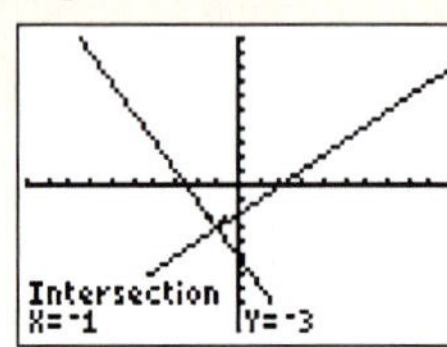

Solution set: {−1}

5. $Y_1 = -0.87x + 1.14$
   $Y_2 = -2.69(x + 4.05)$

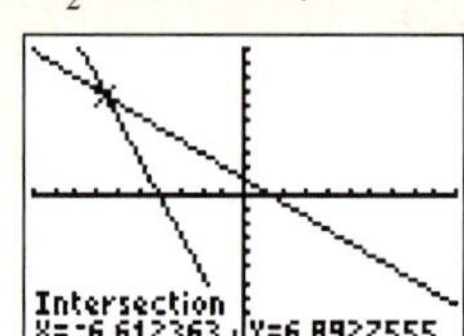

Solution set: {−6.61}

6. 20 tables

Reading Your Text

These fill-in-the-blank exercises will help you understand some of the key vocabulary used in this section. The answers to these exercises are in the Answers Appendix in the back of the text.

(a) When taking a graphical approach to solving a linear equation in one variable, the x-value at the point of intersection gives the ________ to the equation.

(b) It is especially important to ________ your work when solving an equation by graphing.

(c) You can use the ________ utility of a graphing calculator to find the point where two curves intersect.

(d) Always use the ________ equation to check a solution.

4.2 exercises

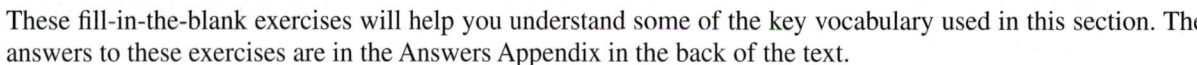

Skills | Calculator/Computer | Career Applications | Above and Beyond

< Objectives 1 and 2 >

Solve each equation graphically. Do not use a calculator.

1. $2x - 8 = 0$

2. $4x + 12 = 0$ VIDEO

3. $7x - 7 = 0$

4. $2x - 6 = 0$

5. $5x - 8 = 2$

6. $4x + 5 = -3$

7. $2x - 3 = 7$

8. $5x + 9 = 4$

9. $4x - 2 = 3x + 1$

10. $6x + 1 = x + 6$

11. $\frac{7}{5}x - 3 = \frac{3}{10}x + \frac{5}{2}$

12. $2x - 3 = 3x - 2$

13. $3(x - 1) = 4x - 5$

14. $2(x + 1) = 5x - 7$

15. $7\left(\frac{1}{5}x - \frac{1}{7}\right) = x + 1$

16. $2(3x - 1) = 12x + 4$

17. **Business and Finance** A firm producing flashlights finds that its fixed cost is \$2,400 per week and its marginal cost is \$4.50 per flashlight. The revenue is \$7.50 per flashlight, so the cost and revenue equations are, respectively,

$C(x) = 4.50x + 2{,}400$ and $R(x) = 7.50x$

where x represents the number of flashlights produced and sold. Find the break-even point for the firm (the point at which the revenue equals the cost). Use a graphical approach.

18. **Business and Finance** A company that produces portable television sets determines that its fixed cost is \$8,750 per month. The marginal cost is \$70 per set, and the revenue is \$105 per set. The cost and revenue equations, respectively, are

$C(x) = 70x + 8{,}750$ and $R(x) = 105x$

where x represents the number of TVs produced and sold. Find the number of sets the company must produce and sell in order to break even. Use a graphical approach.

Determine whether each statement is **true** *or* **false.**

19. When we solve an equation graphically, we let each side of the equation define a function.

20. When solving an equation graphically, the y-coordinate of the point of intersection is the solution to the equation.

Complete each statement with **always, sometimes,** *or* **never.**

21. If each side of an equation is used to define a linear function, there will __________ be exactly two solutions to the equation.

22. If we have a zero on one side of an equation and an expression defining a linear function (with nonzero slope) on the other, the solution for the equation will __________ be the x-coordinate of the x-intercept.

Skills | **Calculator/Computer** | Career Applications | Above and Beyond

Solve each equation with a graphing calculator. Round your results to the nearest hundredth.

23. $4.17(x + 3.56) = 2.89(x + 0.35)$

24. $3.10(x - 2.57) = -4.15(x + 0.28)$ 

25. $3.61(x + 4.13) = 2.31(x - 2.59)$

26. $5.67(x - 2.13) = 1.14(x - 1.23)$

Skills | Calculator/Computer | **Career Applications** | Above and Beyond

ELECTRONICS TECHNOLOGY *Temperature sensors output voltage at a certain temperature. The output voltage varies with respect to temperature. For a particular sensor, the output voltage V for a given Celsius temperature C is given by*

$V = 0.28C + 2.2$

27. Determine the temperature (to the nearest tenth) if the sensor outputs 12.5 V. 

28. Determine the output voltage at 37°C.

ALLIED HEALTH *Yohimbine is used to reverse the effects of xylazine in deer. The recommended dose is 0.125 mg per kilogram of a deer's weight. We model the recommended dosage in terms of a deer's weight with the equation* $d = 0.125w$.

29. What size fawn requires a 2.4-mg dose?

30. How much yohimbine should be administered to a 60-kg buck?

Skills | Calculator/Computer | Career Applications | **Above and Beyond**

31. The graph represents the rates that two different car rental agencies charge. The x-axis represents the number of miles driven (in hundreds of miles), and the y-axis represents the total charge. How would you use this graph to decide which agency to use?

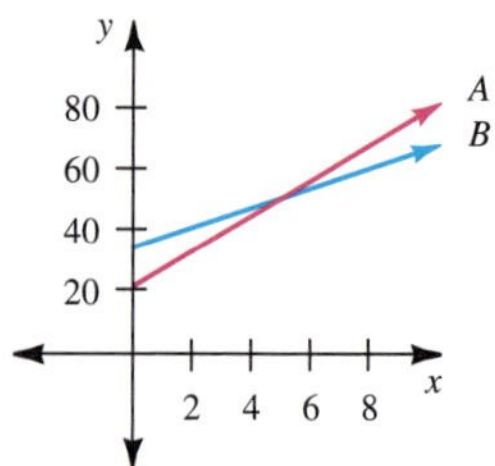

32. Graphs can be used to solve distance, time, and rate problems because graphs make pictures of the action.

(a) Consider this exercise: "Robert left on a trip, traveling at 45 mi/hr. One-half hour later, Laura discovered that Robert forgot his luggage, and so she left along the same route, traveling at 54 mi/hr, to catch up with him. When did Laura catch up with Robert?" How could drawing a graph help solve this problem? If you graph Robert's distance as a function of time and Laura's distance as a function of time, what does the slope of each line correspond to in the problem?

(b) Use a graph to solve this problem: Marybeth and Sam left her mother's house to drive home to Minneapolis along the interstate. They drove an average of 60 mi/hr. After they had been gone for $\frac{1}{2}$ hr, Marybeth's mother realized they had left their laptop computer. She grabbed it, jumped into her car, and pursued the two at 70 mi/hr. Marybeth and Sam also noticed the missing computer, but not until 1 hr after they had left. When they noticed that it was missing, they slowed to 45 mi/hr while they considered what to do. After driving for another $\frac{1}{2}$ hr, they turned around and drove back toward the home of Marybeth's mother at 65 mi/hr. Where did they pass each other? How long had Marybeth's mother been driving when they met?

(c) Now that you have become an expert at this, try solving this problem by drawing a graph. It requires you to think about the slope and perhaps make several guesses when drawing the graphs. If you ride your new bicycle to class, it takes you 1.2 hr. If you drive, it takes you 40 min. If you drive in traffic an average of 15 mi/hr faster than you can bike, how far away from school do you live? Write an explanation of how you solved this problem by using a graph.

33. **Business and Finance** The family next door to you is trying to decide which health maintenance organization (HMO) to join. One parent has a job with health benefits for the employee only, but the rest of the family can be covered if the employee agrees to a payroll deduction. The choice is between The Empire Group, which would cost the family \$385 per month for coverage and \$25.50 for each office visit, and Group Vitality, which costs \$460 per month and \$4.00 for each office visit.

(a) Write an equation showing total yearly costs for each HMO. Graph the cost per year as a function of the number of visits, and put both graphs on the same axes.

(b) Write a note to the family explaining when The Empire Group would be better and when Group Vitality would be better. Explain how they can use your data and graph to help make a good decision. What other issues might be of concern to them?

Answers

1.

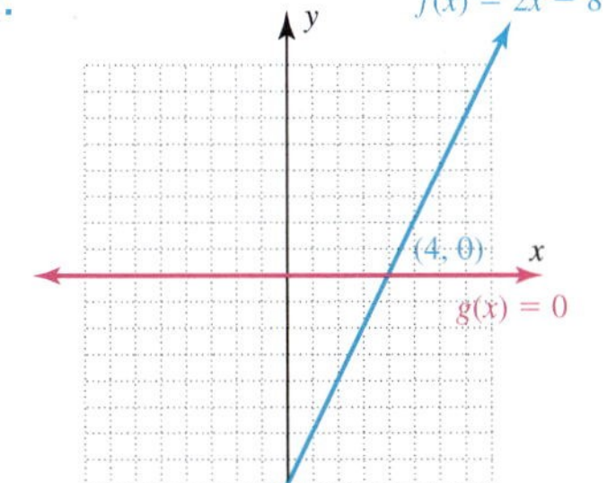

3.

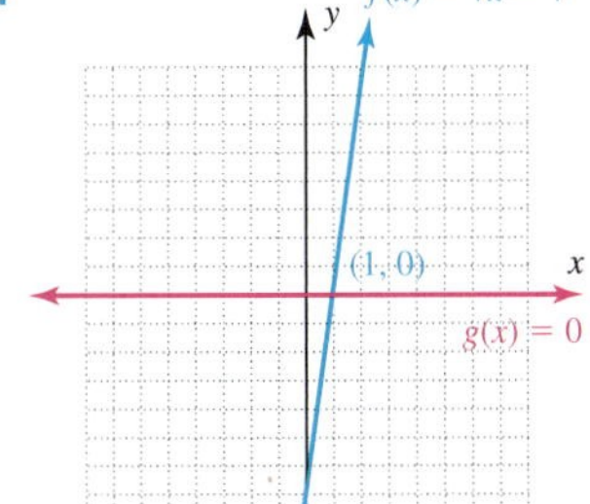

5.

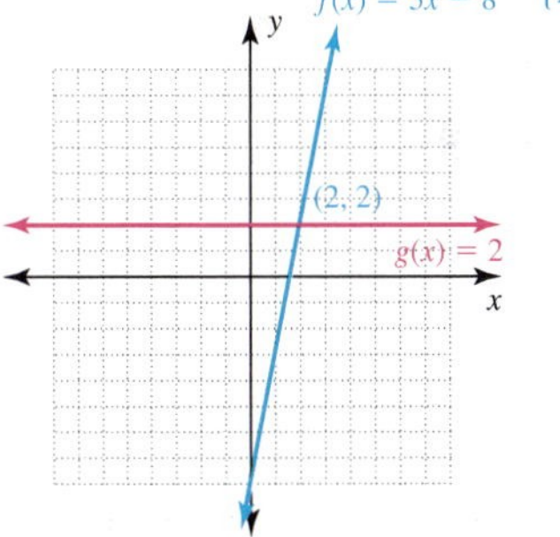

7.

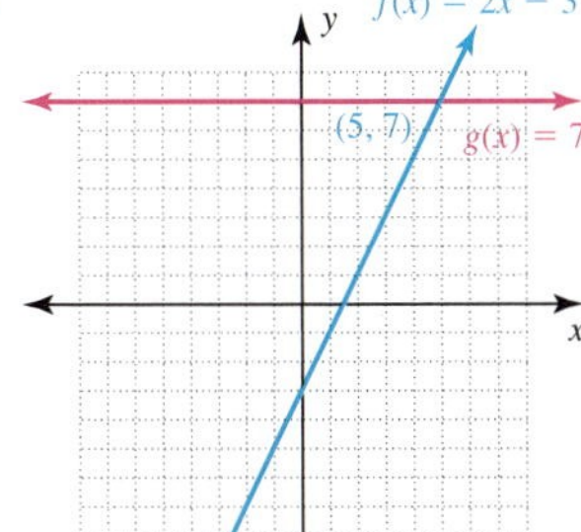

9.

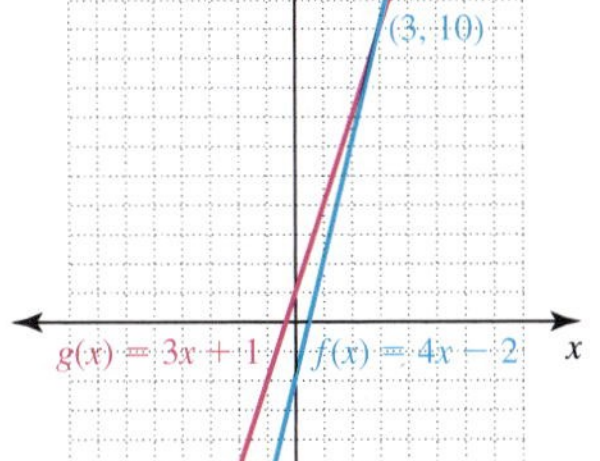

11.

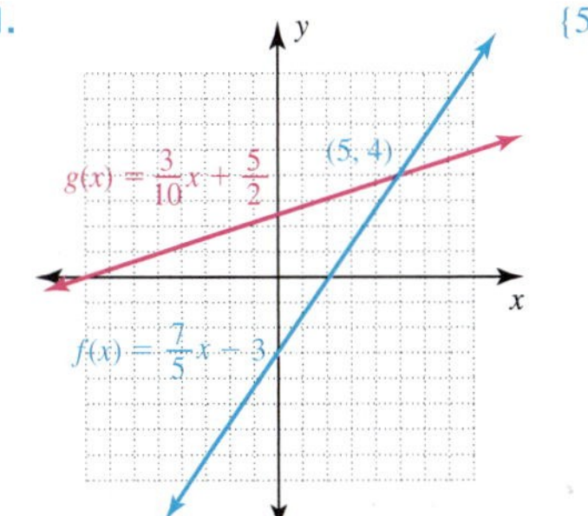

13.

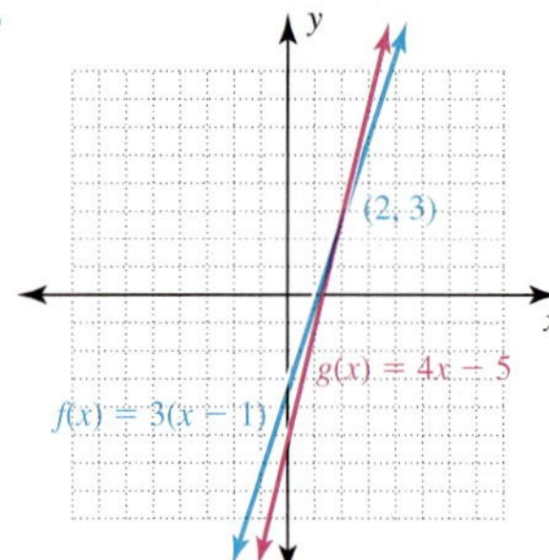

{2}

15.

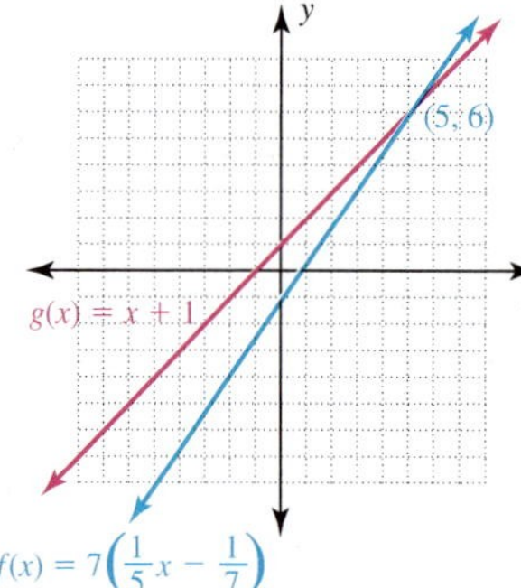

{5}

17. 800 flashlights **19.** True **21.** never

23. {−10.81}

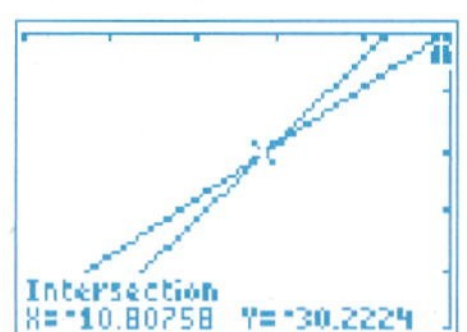

25. {−16.07}

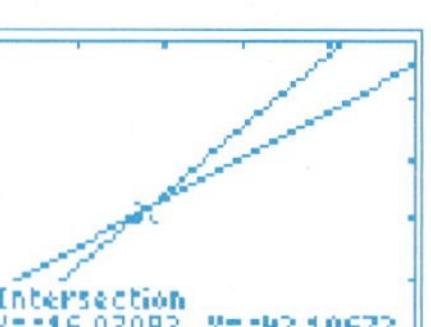

27. 36.8°C **29.** 19.2 kg

31. For a given number of miles, the lower graph gives the cheaper cost. **33.** Above and Beyond

4.3 Systems of Equations in Two Variables

< 4.3 Objectives >

1 > Use the addition method to solve a system of equations

2 > Use the substitution method to solve a system of equations

3 > Use systems of equations to solve applications

Graphing linear systems to find their intersection point is excellent for seeing and estimating solutions. The drawback comes in precision. No matter how carefully one graphs the lines, the displayed solution rarely leads to an exact value. This problem is exaggerated when the solution includes fractions. In this section, we look at some methods that result in exact solutions.

NOTE

The *addition method* is also called the **elimination method.**

One algebraic approach to solving a system of linear equations in two variables is the **addition method.** The basic idea to the addition method is to add the equations together so that one variable is eliminated.

Example 1 Using the Addition Method to Solve a System

< Objective 1 >

Use the addition method to solve the system.

$$5x - 2y = 12$$
$$3x + 2y = 12$$

In this case, adding the equations eliminates the y-variable.

$8x = 24$ Remember to add the right sides of the equations together, as well.

Now, solve this last equation for x by dividing both sides by 8.

$$\frac{8x}{8} = \frac{24}{8}$$
$$x = 3$$

This last equation gives the x-value of the solution to the system of equations. We can take this value and substitute it into either of the original equations to find the y-value of the system's solution.

We substitute $x = 3$ into the first of the original equations in the system.

$5x - 2y = 12$ The first equation in the original system.

$5(3) - 2y = 12$ Substitute $x = 3$ to find the y-value.

$15 - 2y = 12$

$-2y = -3$ Subtract 15 from both sides.

$y = \frac{3}{2}$ Divide both sides by -2: $\frac{-3}{-2} = \frac{3}{2}$.

Therefore, $\left(3, \frac{3}{2}\right)$ is the solution to the system of equations.

NOTE

Using the other equation gives the same result.

$$3x + 2y = 12$$
$$3(3) + 2y = 12$$
$$9 + 2y = 12$$
$$2y = 3$$
$$y = \frac{3}{2}$$

Check Yourself 1

Use the addition method to solve the system.

$$4x - 3y = 19$$
$$-4x + 5y = -25$$

RECALL

Multiplying both sides of an equation by the same nonzero number results in an equivalent equation.

Example 1 and the accompanying Check Yourself exercise were straightforward, in that adding the equations together eliminated one of the variables.

However, we may need to multiply both sides of an equation by some nonzero number in order to eliminate a variable when we add the equations together. In fact, we may need to multiply both equations by (different) numbers to eliminate a variable. We see this in Example 2.

Example 2 Using the Addition Method to Solve a System

Use the addition method to solve the given system.

$$3x - 5y = 19$$
$$5x + 2y = 11$$

NOTE

We could multiply the first equation by 5 and the second equation by -3 to eliminate the x-variable.

It should be clear that adding the two equations does not eliminate either variable. In this case, we decide which variable to eliminate and form an equivalent system by multiplying each equation by a constant.

We choose to eliminate the y-variable because the y-terms have different signs in the system. The least common multiple of 5 and 2 is 10, so we multiply the first equation by 2 and the second equation by 5.

$$3x - 5y = 19 \xrightarrow{\times 2} 6x - 10y = 38$$
$$5x + 2y = 11 \xrightarrow{\times 5} 25x + 10y = 55$$

Remember to multiply both sides of the equations.

This gives an equivalent system of equations. We can now eliminate a variable by adding the equations together.

$$\begin{array}{r} 6x - 10y = 38 \\ 25x + 10y = 55 \\ \hline 31x \qquad = 93 \end{array}$$

Divide both sides of this last equation by 31 to find the x-value of the solution.

$$\frac{31x}{31} = \frac{93}{31}$$
$$x = 3$$

Next, we substitute $x = 3$ into either of the original equations to find the y-value of the solution. We choose to substitute into the first equation.

NOTE

The solution is unique. Because the lines have different slopes, there is a single point of intersection.

$$3x - 5y = 19 \qquad \text{Use an equation from the } \textit{original} \text{ system.}$$
$$3(3) - 5y = 19 \qquad \text{Substitute } x = 3 \text{ into this equation.}$$
$$9 - 5y = 19 \qquad \text{Solve for } y.$$
$$-5y = 10$$
$$y = -2$$

$\{(3, -2)\}$ is the solution set for the system of equations.

You should recall from Section 4.1 that we can check our solution by showing that it is a solution to each equation in the system.

Check

$$3x - 5y = 19$$
$$3(3) - 5(-2) \stackrel{?}{=} 19$$
$$9 + 10 \stackrel{?}{=} 19$$
$$19 = 19 \qquad \text{True}$$

$$5x + 2y = 11$$
$$5(3) + 2(-2) \stackrel{?}{=} 11$$
$$15 - 4 \stackrel{?}{=} 11$$
$$11 = 11 \qquad \text{True}$$

Check Yourself 2

Use the addition method to solve the system. Check your work.

$$\begin{aligned} 2x + 3y &= -18 \\ 3x - 5y &= 11 \end{aligned}$$

Here is the addition method for solving a system of linear equations put together as an algorithm.

Step by Step

Solving by the Addition Method

Step 1 If necessary, multiply one or both of the equations by a constant so that one of the variables can be eliminated by addition.

Step 2 Add the equations of the equivalent system formed in step 1.

Step 3 Solve the equation found in step 2.

Step 4 Substitute the value found in step 3 into either of the equations of the original system to find the corresponding value of the remaining variable.

Step 5 The ordered pair found in step 4 is the solution to the system. Check the solution by substituting the pair of values found in step 4 into both of the original equations.

Example 3 illustrates two special situations you may encounter while applying the addition method.

Example 3 Using the Addition Method to Solve a System

Use the addition method to solve each system.

(a)
$$\begin{aligned} 4x + 5y &= 20 \\ 8x + 10y &= 19 \end{aligned}$$

Multiply the first equation by -2. Then

$$\begin{aligned} -8x - 10y &= -40 \\ 8x + 10y &= 19 \\ \hline 0 &= -21 \end{aligned}$$

We add the two left sides to get 0 and the two right sides to get -21.

NOTE

The graph of this system is a pair of parallel lines.

The result $0 = -21$ is a *false* statement, which means that there is no point of intersection. Therefore, the system is inconsistent, and there is no solution.

(b)
$$\begin{aligned} 5x - 7y &= 9 \\ 15x - 21y &= 27 \end{aligned}$$

Multiply the first equation by -3. We then have

$$\begin{aligned} -15x + 21y &= -27 \\ 15x - 21y &= 27 \\ \hline 0 &= 0 \end{aligned}$$

We add the two equations.

NOTE

The solution set can be written $\{(x, y) \mid 5x - 7y = 9\}$. This means the set of all ordered pairs (x, y) that make $5x - 7y = 9$ a true statement.

Both variables have been eliminated, and the result is a *true* statement. If the two original equations were graphed, we would see that the two lines coincide. Thus, there are infinitely many solutions, one for each point on that line. Recall that this is a *dependent system.*

Check Yourself 3

Use the addition method to solve each system.

(a) $3x + 2y = 8$
$9x + 6y = 11$

(b) $x - 2y = 8$
$3x - 6y = 24$

We summarize the results from Example 3.

Property

Inconsistent and Dependent Systems

When a system of two linear equations is solved:

1. If a false statement such as $3 = 4$ is obtained, then the system is inconsistent and has no solution.
2. If a true statement such as $8 = 8$ is obtained, then the system is dependent and has an infinite number of solutions.

A third method for finding the solutions of linear systems in two variables is called the **substitution method.** You may very well find the substitution method more difficult to apply in solving certain systems than the addition method, particularly when the equations involved in the substitution lead to fractions. However, the substitution method does have important extensions to systems involving higher-degree equations, as you will see in later mathematics classes.

To outline the technique, we solve one of the equations from the original system for one of the variables. That expression is then substituted into the *other* equation of the system to provide an equation in a single variable. That equation is solved, and the corresponding value for the other variable is found as before, as Example 4 illustrates.

Example 4 Using the Substitution Method to Solve a System

< Objective 2 >

(a) Use the substitution method to solve the system.

$$2x - 3y = -3$$
$$y = 2x - 1$$

Since the second equation is already solved for y, we substitute $2x - 1$ for y into the first equation.

> **NOTE**
> We now have an equation in the single variable x.

$$2x - 3(2x - 1) = -3$$

Solving for x gives

$$2x - 6x + 3 = -3$$
$$-4x = -6$$
$$x = \frac{3}{2}$$

We now substitute $x = \frac{3}{2}$ into the equation that was solved for y.

$$y = 2\left(\frac{3}{2}\right) - 1$$
$$= 3 - 1 = 2$$

The solution set for our system is $\left\{\left(\frac{3}{2}, 2\right)\right\}$.

Again, we check our proposed solution by showing that it is a solution to each equation in the system.

Check

$$2x - 3y = -3 \qquad\qquad y = 2x - 1$$

$$2\left(\frac{3}{2}\right) - 3(2) \overset{?}{=} -3 \qquad\qquad (2) \overset{?}{=} 2\left(\frac{3}{2}\right) - 1$$

$$3 - 6 \overset{?}{=} -3 \qquad\qquad 2 \overset{?}{=} 3 - 1$$

$$-3 = -3 \quad \text{True} \qquad\qquad 2 = 2 \quad \text{True}$$

NOTE

Why did we choose to solve the second equation for y? We could have solved for x, so that

$$x = \frac{y + 2}{3}$$

We simply chose the easier case to avoid fractions.

(b) Use the substitution method to solve the system.

$$2x + 3y = 16$$
$$3x - y = 2$$

We start by solving the second equation for y.

$$3x - y = 2$$
$$-y = -3x + 2$$
$$y = 3x - 2$$

Substituting into the other equation yields

$$2x + 3(3x - 2) = 16$$
$$2x + 9x - 6 = 16$$
$$11x = 22$$
$$x = 2$$

NOTE

The solution should be checked in *both* equations of the original system.

We now substitute $x = 2$ into the equation that we solved for y.

$$y = 3(2) - 2$$
$$= 6 - 2 = 4$$

The solution set for the system is $\{(2, 4)\}$. We leave the check of this result to you.

Check Yourself 4

Use the substitution method to solve each system.

(a) $2x + 3y = 6$
$x = 3y + 6$

(b) $3x + 4y = -3$
$x + 4y = 1$

We summarize the substitution method for solving a system of linear equations with an algorithm.

Step by Step

Solving by the Substitution Method

Step 1 Solve one of the equations of the original system for one of the variables.

Step 2 Substitute the expression obtained in step 1 into the *other* equation of the system to write an equation in a single variable.

Step 3 Solve the equation found in step 2.

Step 4 Substitute the value found in step 3 into the equation found in step 1 to find the corresponding value of the remaining variable.

Step 5 The ordered pair found in step 4 is the solution to the system of equations. Check the solution by substituting the pair of values found in step 4 into *both* equations of the original system.

As with the addition method, if both variables are eliminated after simplifying in step 3 and a true statement is formed, then we have a dependent system. If a false statement occurs, then the system is inconsistent.

A natural question at this point is, "How do you decide which solution method to use?" First, the graphical method can generally provide only approximate solutions. When exact solutions are necessary, one of the algebraic methods should be applied. Which method to use depends on the system.

If you can easily solve for a variable in one of the equations, the substitution method should work well. However, if solving for a variable in either equation of the system leads to fractions, you may find the addition approach more efficient.

We now apply our skills to solving applications and word problems. Being able to extend these skills to problem solving is an important goal, and the procedures developed here are used throughout the rest of the book.

Although we consider applications from a variety of areas in this section, all are approached with the same five-step strategy.

Step by Step

Solving Applications

Step 1 Read the problem carefully to determine the unknown quantities.

Step 2 Choose variables to represent the unknown quantities.

Step 3 Translate the problem to the language of algebra to form a system of equations.

Step 4 Solve the system of equations.

Step 5 Answer the question from the original problem and verify your solution by returning to the original problem.

Example 5 Solving a Mixture Problem

< Objective 3 >

A coffee merchant has two types of coffee beans, one selling for \$8 per pound and the other for \$10 per pound. The beans are to be mixed to provide 100 lb of a mixture selling for \$9.50 per pound. How much of each type of coffee bean should be used to form 100 lb of the mixture?

> **NOTES**
>
> Because we use *two* variables, we must form *two* equations.
>
> The total value of the mixture comes from 100(9.50) = 950.

Step 1 The unknowns are the amounts of the two types of beans.

Step 2 We use two variables to represent the two unknowns. Let x be the amount of the \$8 beans and y the amount of the \$10 beans.

Step 3 We want to establish a system of two equations. One equation will be based on the *total amount* of the mixture, the other on the mixture's *value*.

$$x + y = 100$$ The mixture must weigh 100 lb.

$$8x + 10y = 950$$

($8x$: Value of \$8 beans; $10y$: Value of \$10 beans; 950: Total value)

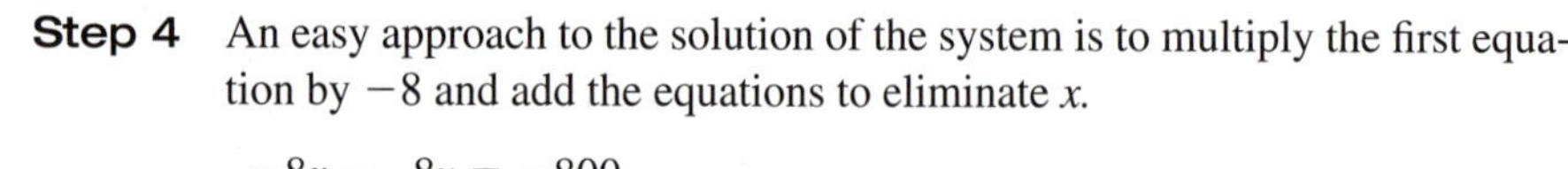

Step 4 An easy approach to the solution of the system is to multiply the first equation by -8 and add the equations to eliminate x.

$$\begin{aligned} -8x - 8y &= -800 \\ 8x + 10y &= 950 \\ \hline 2y &= 150 \\ y &= 75 \end{aligned}$$

NOTE

$$8(25) + 10(75) = 200 + 750$$
$$= 950$$
$$9.50(100) = 950$$

Substituting $y = 75$ into the first equation, $x + y = 100$, gives

$x = 25$

Step 5 We should use 25 lb of \$8 beans and 75 lb of \$10 beans.

To check the result, show that the value of the \$8 beans added to the value of the \$10 beans equals the desired value of the mixture.

Check Yourself 5

Peanuts, which sell for \$2.40 per pound, and cashews, which sell for \$6 per pound, are to be mixed to form a 60-lb mixture selling for \$3 per pound. How much of each type of nut should be used?

A related problem is illustrated in Example 6.

Example 6 Solving a Mixture Problem

A chemist has a 25% and a 50% acid solution. How much of each solution should be used to form 200 milliliters (mL) of a 35% acid solution?

Step 1 The unknowns in this case are the amounts of the 25% and 50% solutions to be used in forming the mixture.

Step 2 Again we use two variables to represent the two unknowns. Let x be the amount of the 25% solution and y the amount of the 50% solution. Let's draw a picture before proceeding to form a system of equations.

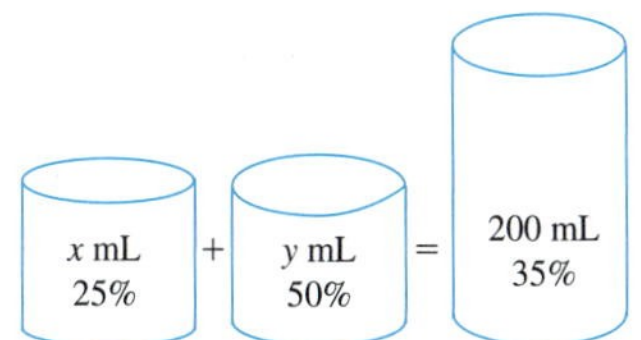

Drawing a sketch of a problem is often a valuable part of the problem-solving strategy.

Step 3 Now, to form our two equations, we want to consider two relationships: the *total amounts* combined and the *amounts of acid* combined.

$$x + y = 200 \qquad \text{Total amounts combined}$$
$$0.25x + 0.50y = 0.35(200) \qquad \text{Amounts of acid combined}$$

Step 4 Clear the second equation of decimals by multiplying it by 100. The solution then proceeds as before, with the result

$$x = 120 \qquad (25\% \text{ solution})$$
$$y = 80 \qquad (50\% \text{ solution})$$

Step 5 We need 120 mL of the 25% solution and 80 mL of the 50% solution.

To check, show that the amount of acid in the 25% solution, (0.25)(120), added to the amount in the 50% solution, (0.50)(80), equals the correct amount in the mixture, (0.35)(200). We leave that to you.

Check Yourself 6

A pharmacist wants to prepare 300 mL of a 20% alcohol solution. How much of a 30% solution and a 15% solution should be used to form the desired mixture?

Applications that involve a constant rate of travel, or speed, require the distance formula

$$d = rt$$

where

d = distance traveled

r = rate or speed

t = time

Example 7 Solving a Distance-Rate-Time Problem

A boat can travel 36 mi downstream in 2 hr. Coming back upstream, the boat takes 3 hr. What is the rate of the boat in still water? What is the rate of the current?

> **NOTE**
>
> Downstream the rate is then $x + y$
>
> Upstream, the rate is $x - y$

Step 1 We want to find the two rates.

Step 2 Let x be the rate of the boat in still water and y the rate of the current.

Step 3 To form a system, think about the following. Downstream, the rate of the boat is *increased* by the effect of the current. Upstream, the rate is *decreased.*

In many applications, it helps to lay out the information in a table. We try that strategy here.

	d	r	t
Downstream	36	$x + y$	2
Upstream	36	$x - y$	3

Since $d = rt$, from the table we can easily form two equations:

$$36 = (x + y)(2)$$
$$36 = (x - y)(3)$$

Step 4 We clear the equations of parentheses and simplify, to write the equivalent system.

$$x + y = 18$$
$$x - y = 12$$

Solving, we have

$$x = 15$$
$$y = 3$$

Step 5 The rate of the current is 3 mi/hr, and the rate of the boat in still water is 15 mi/hr.

To check, verify the $d = rt$ equation in *both* the upstream and the downstream cases. We leave that to you.

Check Yourself 7

A plane flies 480 mi in an easterly direction, with the wind, in 4 hr. Returning westerly along the same route, against the wind, the plane takes 6 hr. What is the rate of the plane in still air? What is the rate of the wind?

Systems of equations have many applications in business settings. Example 8 illustrates one such application.

Example 8 A Business and Finance Application

A manufacturer produces LCD and plasma television sets. The LCD models require 12 hr of labor to produce, while the plasma models require 18 hr. The company has 360 hr of labor available per week. The plant's capacity is a total of 25 sets per week.

If all the available time and capacity are to be used, how many of each type of set should be produced?

Step 1 The unknowns in this case are the number of LCD and plasma models that can be produced.

NOTE

The choices for x and y could be reversed.

Step 2 Let x be the number of LCD models and y the number of plasma models.

Step 3 Our system comes from the two given conditions that fix the total number of sets that can be produced and the total labor hours available.

$$x + y = 25 \leftarrow \text{Total number of sets}$$

$$12x + 18y = 360 \leftarrow \text{Total labor hours available}$$

Labor hours, LCD sets ($12x$); Labor hours, plasma sets ($18y$)

Step 4 Solving the system in step 3, we have

$x = 15$ and $y = 10$

Step 5 This tells us that to use all the available capacity, the plant should produce 15 LCD sets and 10 plasma sets per week.

We leave the check of this result to the reader.

Check Yourself 8

A manufacturer produces recorders and compact disk players. Each recorder requires 2 hr of electronic assembly, and each CD player requires 3 hr. The recorders require 4 hr of case assembly and the CD players 2 hr. The company has 120 hr of electronic assembly time available per week and 160 hr of case assembly time. How many of each type of unit can be produced each week if all available assembly time is to be used?

Here is another application that leads to a system of two equations.

Example 9 A Business and Finance Application

Two car rental agencies have different rate structures for a subcompact car.

Company A charges \$68 per day plus 15¢ per mile. Company B charges \$66 per day plus 16¢ per mile. If you rent a car for 1 day, for what number of miles will the two companies have the same total charge?

Letting c represent the total a company will charge and m the number of miles driven, we calculate the rates.

For company A:

$c = 68 + 0.15m$

For company B:

$c = 66 + 0.16m$

The system can be solved most easily by substitution. Substituting $66 + 0.16m$ for c in the first equation gives

$$66 + 0.16m = 68 + 0.15m$$
$$0.01m = 2$$
$$m = 200 \text{ mi}$$

The graph of the system is shown below.

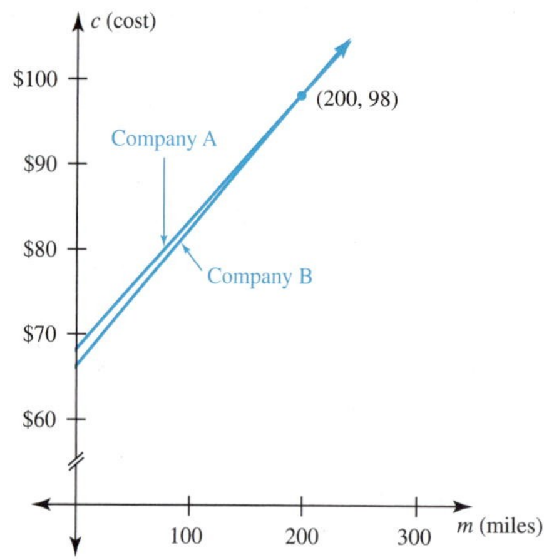

From the graph, how would you make a decision about which agency to use?

Check Yourself 9

For a compact car, the same two companies charge \$54 per day plus 20¢ per mile and \$51 per day plus 22¢ per mile. For a 2-day rental, for what mileage will the charges be the same? What is the total charge?

Check Yourself ANSWERS

1. $\left\{\left(\frac{5}{2}, -3\right)\right\}$ **2.** $\{(-3, -4)\}$

3. **(a)** Inconsistent system, no solution; **(b)** dependent system, an infinite number of solutions

4. **(a)** $\left\{\left(4, -\frac{2}{3}\right)\right\}$; **(b)** $\left\{\left(-2, \frac{3}{4}\right)\right\}$ **5.** 50 lb of peanuts and 10 lb of cashews

6. 100 mL of the 30% and 200 mL of the 15% **7.** Plane: 100 mi/hr, wind: 20 mi/hr

8. 30 recorders and 20 CD players **9.** At 300 mi, \$168 charge

Reading Your Text

These fill-in-the-blank exercises will help you understand some of the key vocabulary used in this section. The answers to these exercises are in the Answers Appendix in the back of the text.

(a) Multiplying both sides of an equation by the same nonzero number results in an ________ equation.

(b) The solution to a system can be approximated by graphing the lines and locating the ________ point.

(c) When solving a system results in a false statement such as $3 = 4$, the system has ________ solutions.

(d) The distance formula states that distance is equal to rate times ________.

Skills | Calculator/Computer | Career Applications | Above and Beyond

4.3 exercises

< Objective 1 >

Use the addition method to solve each system. If a unique solution does not exist, state whether the system is inconsistent or dependent.

1. $2x - y = 1$
$-2x + 3y = 5$

2. $x + 3y = 12$
$2x - 3y = 6$

3. $4x + 3y = -5$
$5x + 3y = -8$

4. $2x + 3y = 1$
$5x + 3y = 16$

5. $x + y = 3$
$3x - 2y = 4$

6. $x - y = -2$
$2x + 3y = 21$

7. $2x + y = 8$
$-4x - 2y = -16$

8. $2x + 3y = 11$
$x - y = 3$

9. $5x - 2y = 31$
$4x + 3y = 11$

10. $2x - y = 4$
$6x - 3y = 10$

11. $3x - 2y = 7$
$-6x + 4y = -15$

12. $3x + 4y = 0$
$5x - 3y = -29$

13. $-2x + 7y = 2$
$3x - 5y = -14$

14. $5x - 2y = 3$
$10x - 4y = 6$

< Objective 2 >

Use the substitution method to solve each system. If a unique solution does not exist, state whether the system is inconsistent or dependent.

15. $x - y = 7$
$y = 2x - 12$

16. $x - y = 4$
$x = 2y - 2$

17. $3x + 2y = -18$
$x = 3y + 5$

18. $3x - 18y = 4$
$x = 6y + 2$

19. $10x - 2y = 4$
$y = 5x - 2$

20. $5x - 2y = 8$
$y = -x + 3$

21. $3x + 4y = 9$
$y = 3x + 1$

22. $6x - 5y = 27$
$x = 5y + 2$

23. $x - 7y = 3$
$2x - 5y = 15$

24. $4x + 3y = -11$
$5x + y = -11$

25. $3x - 9y = 7$
$-x + 3y = 2$

26. $5x - 6y = 21$
$x - 2y = 5$

Use any method to solve each system.
Hint: You can use multiplication to clear equations of fractions.

27. $2x - 3y = 4$
$x = 3y + 6$

28. $5x + y = 2$
$5x - 3y = 6$

29. $4x - 3y = 0$
$5x + 2y = 23$

30. $7x - 2y = -17$
$x + 4y = 4$

31. $3x - y = 17$
$5x + 3y = 5$

32. $7x + 3y = -51$
$y = 2x + 9$

33. $\frac{1}{4}x - \frac{1}{6}y = 2$
$\frac{3}{4}x + y = 3$

34. $\frac{1}{5}x - \frac{1}{2}y = 0$
$x - \frac{3}{2}y = 4$

35. $\frac{2}{3}x + \frac{3}{5}y = -3$
$\frac{1}{3}x + \frac{2}{5}y = -3$

36. $\frac{3}{8}x - \frac{1}{2}y = -5$
$\frac{1}{4}x + \frac{3}{2}y = 4$

< Objective 3 >

Each application can be solved with a system of linear equations. Match the application with the appropriate system below.

(a) $12x + 5y = 116$
$8x + 12y = 112$

(b) $x + y = 4{,}000$
$0.03x + 0.05y = 180$

(c) $x + y = 200$
$0.20x + 0.60y = 90$

(d) $x + y = 36$
$y = 3x - 4$

(e) $2(x + y) = 36$
$3(x - y) = 36$

(f) $x + y = 200$
$6.50x + 4.50y = 980$

(g) $L = 2W + 3$
$2L + 2W = 36$

(h) $x + y = 120$
$2.20x + 5.40y = 360$

37. **Number Problem** One number is 4 less than 3 times another. If the sum of the numbers is 36, what are the two numbers?

38. **Business and Finance** Suppose a second-run movie theater sold 200 adult and student tickets for a showing with a revenue of $980. If the adult tickets cost $6.50 and the student tickets cost $4.50, how many of each type of ticket were sold?

39. **Geometry** The length of a rectangle is 3 cm more than twice its width. If the perimeter of the rectangle is 36 cm, find the dimensions of the rectangle.

40. **Business and Finance** An order of 12 dozen roller-ball pens and 5 dozen ballpoint pens costs $116. A later order for 8 dozen roller-ball pens and 12 dozen ballpoint pens costs $112. What was the cost of 1 dozen of each type of pen?

41. **Business and Finance** A candy merchant wants to mix peanuts selling at $2.20 per pound with cashews selling at $5.40 per pound to form 120 lb of a mixed-nut blend that will sell for $3 per pound. What amount of each type of nut should be used?

42. **Business and Finance** Donald has investments totaling $4,000 in two accounts—one a savings account paying 3% interest and the other a bond paying 5%. If the annual interest from the two investments was $180, how much did he have invested at each rate?

43. **Science and Medicine** A chemist wants to combine a 20% alcohol solution with a 60% solution to form 200 mL of a 45% solution. How much of each solution should be used to form the mixture?

44. **Science and Medicine** Xian was able to make a downstream trip of 36 mi in 2 hr. Returning upstream, he took 3 hr to make the trip. How fast can his boat travel in still water? What was the rate of the river's current?

Solve each application with a system of equations.

45. **Business and Finance** Suppose 750 tickets were sold for a high school concert with a total revenue of $5,300. If adult tickets were $8 and students tickets were $4.50, how many of each type of ticket were sold?

46. **Business and Finance** Theater tickets sold for $7.50 on the main floor and $5 in the balcony. The total revenue was $3,250, and there were 100 more main-floor tickets sold than balcony tickets. Find the number of each type of ticket sold.

47. **Geometry** The length of a rectangle is 3 in. less than twice its width. If the perimeter of the rectangle is 84 in., find the dimensions of the rectangle.

48. **Geometry** The length of a rectangle is 5 cm more than 3 times its width. If the perimeter of the rectangle is 74 cm, find the dimensions of the rectangle.

49. **Business and Finance** A garden store sold 8 bags of mulch and 3 bags of fertilizer for $96. The next purchase was for 5 bags of mulch and 5 bags of fertilizer. The cost of that purchase was $100. Find the cost of a single bag of mulch and a single bag of fertilizer.

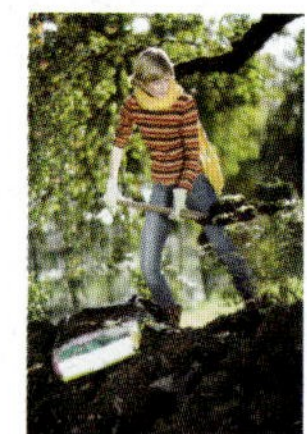

50. **Business and Finance** The cost of an order for 10 computer disks and 3 packages of paper was $22.50. The next order was for 30 disks and 5 packages of paper, and its cost was $53.50. Find the price of a single disk and a single package of paper. VIDEO

51. **Business and Finance** A coffee retailer has two grades of decaffeinated beans, one wholesales for $4 per pound and the other for $6.50 per pound. She wishes to blend the beans to form a 150-lb mixture that wholesales for $4.75 per pound. How many pounds of each grade of bean should she use in the mixture?

52. **Business and Finance** A candy merchant sells jelly beans at $3.50 per pound and gumdrops at $4.70 per pound. To form a 200-lb mixture that will sell for $4.40 per pound, how many pounds of each type of candy should be used?

53. **Business and Finance** Cheryl decided to divide $12,000 into two investments—a time deposit that pays 4% annual interest and a bond that pays 6%. If her annual interest was $640, how much did she invest at each rate?

54. **Business and Finance** Miguel has $1,500 more invested in a mutual fund paying 5% interest than in a savings account paying 3%. If he received $155 in interest for 1 year, how much did he have invested in the two accounts?

55. **Science and Medicine** A chemist mixes a 10% acid solution with a 50% acid solution to form 400 mL of a 40% solution. How much of each solution should be used in the mixture?

56. **Science and Medicine** A laboratory technician wishes to mix a 70% saline solution and a 20% saline solution to prepare 500 mL of a 40% solution. What amount of each solution should be used?

57. **Science and Medicine** A boat traveled 36 mi up a river in 3 hr. Returning downstream, the boat took 2 hr. What is the boat's rate in still water, and what is the rate of the river's current?

58. **Science and Medicine** A jet flew east a distance of 1,800 mi with the jetstream in 3 hr. Returning west, against the jetstream, the jet took 4 hr. Find the jet's speed in still air and the rate of the jetstream.

59. **Number Problem** The sum of the digits of a two-digit number is 8. If the digits are reversed, the new number is 36 more than the original number. Find the original number.

Hint: If u represents the units digit of the number and t the tens digit, the original number can be represented by $10t + u$.

60. **Number Problem** The sum of the digits of a two-digit number is 10. If the digits are reversed, the new number is 54 less than the original number. What was the original number?

61. **Business and Finance** A manufacturer produces a battery-powered calculator and a solar model. The battery-powered model requires 10 min of electronic assembly and the solar model 15 min. There are 450 min of assembly time available per day. Both models require 8 min for packaging, and 280 min of packaging time is available per day. If the manufacturer wants to use all the available time, how many of each unit should be produced per day? chapter 4 > Make the Connection

62. **Business and Finance** A small tool manufacturer produces a standard model and a cordless model power drill. The standard model takes 2 hr of labor to assemble and the cordless model 3 hr. There are 72 hr of labor available per week for the drills. Material costs for the standard drill are $10, and for the cordless drill they are $20. The company wishes to limit material costs to $420 per week. How many of each model drill should be produced in order to use all the available resources? chapter 4 > Make the Connection

63. **Business and Finance** In economics, a demand equation gives the quantity D that will be demanded by consumers at a given price p, in dollars. Suppose that $D = 210 - 4p$ for a particular product.

A supply equation gives the supply S that will be available from producers at price p. Suppose also that for the same product $S = 10p$.

The equilibrium point is that point where the supply equals the demand (where $S = D$). Use the given equations to find the equilibrium point. chapter 4 > Make the Connection

64. **Business and Finance** Suppose the demand equation for a product is $D = 150 - 3p$ and the supply equation is $S = 12p$. Find the equilibrium point for the product (where $S = D$). chapter 4 > Make the Connection

65. **Business and Finance** Two car rental agencies have different rate structures for compact cars.

Company A: $48/day and 22¢/mi

Company B: $46/day and 26¢/mi

For a 2-day rental, at what number of miles will the charges be the same?

66. **Construction** Two construction companies submit these bids.

Company A: $5,000 plus $15 per square foot of building

Company B: $7,000 plus $12.50 per square foot of building

For what number of square feet of building will the bids of the two companies be the same?

Complete each statement with **always, sometimes,** *or* **never.**

67. Both variables are __________ eliminated when the equations of a linear system are added.

68. A system is __________ both inconsistent and dependent.

69. It is __________ possible to use the addition method to solve a linear system.

70. The substitution method is __________ easier to use than the addition method.

Skills | **Calculator/Computer** | Career Applications | Above and Beyond

Adjust the viewing window on a graphing calculator so that you can see the point of intersection of the lines in each system. Then approximate the solution, expressing each coordinate to the nearest tenth.

71. $y = 2x - 3$
$2x + 3y = 1$

72. $3x - 4y = -7$
$2x + 3y = -1$

73. $5x - 12y = 8$
$7x + 2y = 44$

74. $9x - 3y = 10$
$x + 5y = 58$

Skills | Calculator/Computer | **Career Applications** | Above and Beyond

75. **Construction Technology** The beam shown in the figure is 24 feet long and has the load indicated on each end. VIDEO

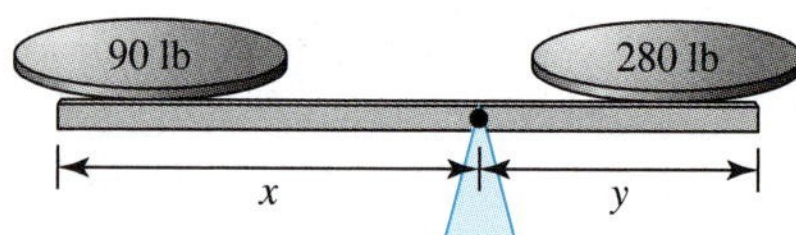

(a) Construct a system of equations in order to determine the point at which the beam balances.

Hint: The product of the length of a side and the load on that side must equal the product of the length of the other side and the load on that other side.

(b) Find the necessary lengths, x and y, in order to balance the beam. Report your results to the nearest hundredth foot.

76. **Manufacturing Technology** Production for one week is up 2,600 units over the previous week. A total of 27,200 units were produced during the 2-week span.

(a) Construct a system of equations to determine the production for each of the 2 weeks.

(b) Determine the number of units produced each week.

77. **Allied Health** A medical lab technician needs to determine how much 9% sulfuric acid (H_2SO_4) solution to mix with 2% H_2SO_4 in order to produce 75 mL of 4% solution.

(a) Construct a system of equations to solve this application using x for the quantity of 9% solution needed.

(b) Determine the amount of each solution needed. Report your results to the nearest tenth mL.

78. **Allied Health** A medical lab technician needs to determine how much 20% saline solution to mix with 5% saline in order to produce 100 mL of 12% solution.

(a) Construct a system of equations to solve this application using x for the quantity of 20% solution needed.

(b) Determine the amount of each solution needed. Report your results to the nearest tenth mL.

Certain systems that are not linear can be solved with the methods of this section if we first substitute to change variables. For instance, the system

$$\frac{1}{x} + \frac{1}{y} = 4$$
$$\frac{1}{x} - \frac{3}{y} = -6$$

can be solved by the substitutions $u = \frac{1}{x}$ *and* $v = \frac{1}{y}$*. That gives the system* $u + v = 4$ *and* $u - 3v = -6$*. The system is then solved for u and v, and the corresponding values for x and y are found. Use this method to solve each system.*

79. $\frac{1}{x} + \frac{1}{y} = 4$, $\frac{1}{x} - \frac{3}{y} = -6$

80. $\frac{1}{x} + \frac{3}{y} = 1$, $\frac{4}{x} + \frac{3}{y} = 3$

81. $\frac{2}{x} + \frac{3}{y} = 4$, $\frac{2}{x} - \frac{6}{y} = 10$

82. $\frac{4}{x} - \frac{3}{y} = -1$, $\frac{12}{x} - \frac{1}{y} = 1$

Here is another method for writing the equation of a line through two points. Given the coordinates of two points, substitute each pair of values into the equation $y = mx + b$*. This gives a system of two equations in variables m and b, which can be solved as before.*

Use this method to write an equation of the line through each pair of points.

83. (2, 1) and (4, 4)

84. (−3, 7) and (6, 1)

85. We discussed three different methods of solving a system of two linear equations in two unknowns: the graphical method, the addition method, and the substitution method. Discuss the strengths and weaknesses of each method.

86. Determine a system of two linear equations for which the solution is (3, 4). Are there other systems that have the same solution? If so, determine at least one more and explain why this can be true.

87. Suppose we have the linear system:

$$Ax + By = C$$
$$Dx + Ey = F$$

(a) Multiply the first equation by $-D$, multiply the second equation by A, and add; this allows you to eliminate x. Solve for y and indicate what must be true about the coefficients in order for a unique value for y to exist.

(b) Now return to the original system and eliminate y instead of x.

Hint: Try multiplying the first equation by E and the second equation by $-B$.

Solve for x and again indicate what must be true about the coefficients for a unique value for x to exist.

Answers

1. {(2, 3)} **3.** $\left\{\left(-3, \frac{7}{3}\right)\right\}$ **5.** {(2, 1)} **7.** Infinite number of solutions, dependent system **9.** {(5, −3)} **11.** No solutions, inconsistent system **13.** {(−8, −2)} **15.** {(5, −2)} **17.** {(−4, −3)} **19.** Infinite number of solutions, dependent system **21.** $\left\{\left(\frac{1}{3}, 2\right)\right\}$ **23.** {(10, 1)} **25.** No solutions, inconsistent system **27.** $\left\{\left(-2, -\frac{8}{3}\right)\right\}$ **29.** {(3, 4)} **31.** {(4, −5)} **33.** $\left\{\left(\frac{20}{3}, -2\right)\right\}$ **35.** {(9, −15)} **37.** (d) **39.** (g) **41.** (h) **43.** (c) **45.** 550 adult tickets; 200 student tickets **47.** 27 in. by 15 in. **49.** Mulch \$7.20; fertilizer \$12.80 **51.** 105 lb of \$4 beans; 45 lb of \$6.50 beans **53.** \$8,000 bond; \$4,000 time deposit **55.** 100 mL of 10%; 300 mL of 50% **57.** 15 mi/hr boat; 3 mi/hr current **59.** 26 **61.** 15 battery-powered calculators; 20 solar models **63.** (15, 150) **65.** 100 mi **67.** sometimes **69.** always **71.** {(1.3, −0.5)} **73.** {(5.8, 1.7)} **75.** **(a)** $x + y = 24$; $90x = 280y$; **(b)** $x = 18.16$ ft, $y = 5.84$ ft **77.** **(a)** $x + y = 75$; $0.09x + 0.02y = 3$; **(b)** 9%: 21.4 mL; 2%: 53.6 mL **79.** $\left\{\left(\frac{2}{3}, \frac{2}{5}\right)\right\}$ **81.** $\left\{\left(\frac{1}{3}, -\frac{3}{2}\right)\right\}$ **83.** $y = \frac{3}{2}x - 2$ **85.** Above and Beyond **87.** **(a)** $y = \frac{AF - CD}{AE - BD}$; $AE - BD \neq 0$; **(b)** $x = \frac{CE - BF}{AE - BD}$, $AE - BD \neq 0$

4.4 Systems of Equations in Three Variables

< 4.4 Objectives >

1 > Solve a system of linear equations in three variables

2 > Use a system with three variables to solve an application

Suppose an application involves three quantities that we want to label x, y, and z. A typical equation used for the solution might be

$$2x + 4y - z = 8$$

This is called a **linear equation in three variables.** A solution to such an equation is an **ordered triple** (x, y, z) of real numbers that satisfies the equation. For example, the ordered triple (2, 1, 0) is a solution to the equation above since substituting 2 for x, 1 for y, and 0 for z results in a true statement.

$$\begin{aligned} 2x + 4y - z &= 8 \\ 2(2) + 4(1) - (0) &\overset{?}{=} 8 \\ 4 + 4 &\overset{?}{=} 8 \\ 8 &= 8 \quad \text{True} \end{aligned}$$

NOTE

For a unique solution to exist, when *three variables* are involved, we must have at least *three equations*.

Of course, other solutions, in fact infinitely many, exist. You might want to verify that (1, 1, −2) and (3, 1, 2) are also solutions. To extend the concepts of Section 4.3, we want to consider systems of three linear equations in three variables, such as

$$\begin{aligned} x + y + z &= 5 \\ 2x - y + z &= 9 \\ x - 2y + 3z &= 16 \end{aligned}$$

The solution to such a system is the set of all ordered triples that satisfy each equation of the system. In this case, you should verify that (2, −1, 4) is a solution to the system since that ordered triple makes each equation a true statement.

To solve such a system, we choose *two pairs* of equations from the system and use the addition method to eliminate the *same variable* from each of those pairs. The method is best illustrated by example. We now proceed to see how we found the solution to the previous system.

Example 1 Solving a Linear System in Three Variables

< Objective 1 >

NOTE

The choice of which variable to eliminate is yours.

Generally, you should pick the variable that allows the easiest computation.

Solve the system.

$$\begin{aligned} x + y + z &= 5 \\ 2x - y + z &= 9 \\ x - 2y + 3z &= 16 \end{aligned}$$

First, we choose a variable to eliminate. Eliminating the y-variable seems reasonably convenient. Then we choose a pair of the equations and use the addition method to eliminate our chosen variable.

We choose to eliminate the y-variable by adding the first two equations together.

$$\begin{aligned} x + y + z &= 5 \\ 2x - y + z &= 9 \\ \hline 3x \qquad + 2z &= 14 \end{aligned}$$

Next we choose a different pair of equations and eliminate the *same* variable. This time, we take the first and third equations. We multiply the first equation by 2 so that we eliminate the y-variable by adding the equations together.

$$x + y + z = 5 \xrightarrow{\times 2} 2x + 2y + 2z = 10$$

$$\begin{array}{r} 2x + 2y + 2z = 10 \\ x - 2y + 3z = 16 \\ \hline 3x \qquad + 5z = 26 \end{array}$$

We now have a pair of equations in x and z.

$$\begin{aligned} 3x + 2z &= 14 \\ 3x + 5z &= 26 \end{aligned}$$

We can solve this system using any of the methods we studied earlier.

We choose to multiply the first equation by -1 and add the result to the second equation.

$$3x + 2z = 14 \xrightarrow{\times(-1)} -3x - 2z = -14$$

$$\begin{array}{r} -3x - 2z = -14 \\ 3x + 5z = \quad 26 \\ \hline 3z = \quad 12 \end{array}$$

Divide by 3 to produce $z = 4$.

Substitute this result into one of the equations in the two-variable system to find the x-value of this solution.

$3x + 2z = 14$ Use one of the two-variable equations.

$3x + 2(4) = 14$ Substitute $z = 4$ and solve for x.

$3x + 8 = 14$

$3x = 6$ Subtract 8 from both sides.

$x = 2$ Divide both sides by 3.

Finally, substituting $x = 2$ and $z = 4$ into any of the equations in the *original* three-variable system enables us to find the appropriate y-value.

$$\begin{aligned} x + y + z &= 5 \\ (2) + y + (4) &= 5 \\ y + 6 &= 5 \\ y &= -1 \end{aligned}$$

Therefore, $(2, -1, 4)$ is the solution to the given system of equations.

NOTES

We can use any of the original three-variable equations to find y.

You can check this result by substituting $(2, -1, 4)$ into each of the original equations and see that it is a solution in every case.

Check Yourself 1

Solve the system.

$$\begin{aligned} x - 2y + z &= 0 \\ 2x + 3y - z &= 16 \\ 3x - y - 3z &= 23 \end{aligned}$$

One or more of the equations of a system may already have a missing variable. This only makes the elimination process easier, as Example 2 illustrates.

Example 2 Solving a Linear System in Three Variables

Solve the system.

$$\begin{aligned} 2x + y - z &= -3 \\ y + z &= 2 \\ 4x - y + z &= 12 \end{aligned}$$

Since the middle equation involves only y and z, we can simply eliminate x from the other two equations. Multiply the first equation by -2 and add the result to the third equation to eliminate x.

NOTE

We now have a *second* equation in y and z.

$$\begin{aligned} -4x - 2y + 2z &= 6 \\ 4x - y + z &= 12 \\ \hline -3y + 3z &= 18 \\ y - z &= -6 \end{aligned}$$

We now form a system consisting of the pair of two-variable equations and solve as before.

$$\begin{aligned} y + z &= 2 \\ y - z &= -6 \\ \hline 2y \quad &= -4 \qquad \text{Adding eliminates } z. \\ y \quad &= -2 \end{aligned}$$

Using the equation $y + z = 2$ with $y = -2$ gives

$$\begin{aligned} (-2) + z &= 2 \\ z &= 4 \end{aligned}$$

and from the first equation, if $y = -2$ and $z = 4$,

$$\begin{aligned} 2x + y - z &= -3 \\ 2x + (-2) - (4) &= -3 \\ 2x &= 3 \\ x &= \frac{3}{2} \end{aligned}$$

The solution set for the system is

$$\left\{\left(\frac{3}{2}, -2, 4\right)\right\}$$

We check our proposed solution by substituting $x = \frac{3}{2}$, $y = -2$, and $z = 4$ into each equation in the original system and see if true statements are formed.

Check

$$\begin{aligned} 2x + y - z &= -3 \\ 2\left(\frac{3}{2}\right) + (-2) - (4) &\stackrel{?}{=} -3 \\ 3 - 2 - 4 &\stackrel{?}{=} -3 \\ -3 &= -3 \end{aligned}$$

$$\begin{aligned} y + z &= 2 \\ (-2) + (4) &\stackrel{?}{=} 2 \\ 2 &= 2 \end{aligned}$$

$$\begin{aligned} 4x - y + z &= 12 \\ 4\left(\frac{3}{2}\right) - (-2) + (4) &\stackrel{?}{=} 12 \\ 6 + 2 + 4 &\stackrel{?}{=} 12 \\ 12 &= 12 \end{aligned}$$

Check Yourself 2

Solve the system.

$$\begin{aligned} x + 2y - z &= -3 \\ x - y + z &= 2 \\ x \quad - z &= 3 \end{aligned}$$

This algorithm summarizes the addition method for finding the solutions to a linear system of three equations in three variables.

Step by Step

Solving a System of Three Equations in Three Unknowns

Step 1 Choose a pair of equations from the system and use the addition method to eliminate one of the variables.

Step 2 Choose a *different* pair of equations and eliminate the *same* variable.

Step 3 Solve the system of two equations in two variables determined in steps 1 and 2.

Step 4 Substitute the values found above into one of the original equations and solve for the remaining variable.

Step 5 The solution is the ordered triple of values found in steps 3 and 4. It can be checked by substituting into each of the equations of the original system.

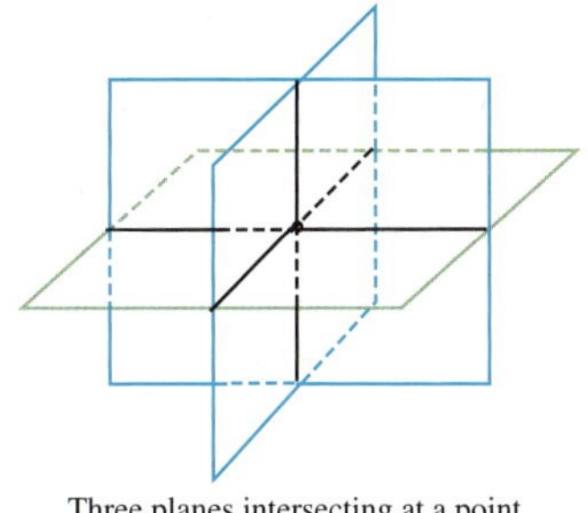

Three planes intersecting at a point

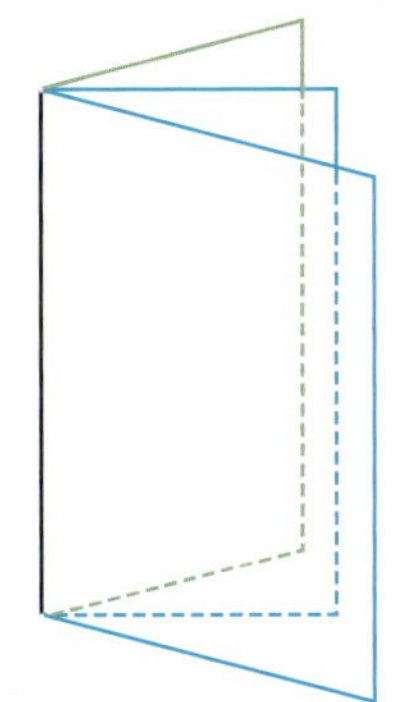

Three planes intersecting in a line

Systems of three equations in three variables may have (1) exactly one solution, (2) infinitely many solutions, or (3) no solution. Before we look at an algebraic approach in the second and third cases, you should understand the geometry involved.

The graph of a linear equation in three variables is a plane (a flat surface) in three dimensions. Two distinct planes are either parallel or they intersect in a line.

If three distinct planes intersect, that intersection will be either a single point (as in our first example) or a line (think of three pages in an open book—they intersect along the binding of the book).

Here are examples involving dependent systems with infinitely many solutions and inconsistent systems with no solutions.

Example 3 Solving a Dependent Linear System in Three Variables

Solve the system.

$$\begin{aligned} x + 2y - z &= 5 \\ x - y + z &= -2 \\ -5x - 4y + z &= -11 \end{aligned}$$

We begin as before by choosing two pairs of equations from the system and eliminating the same variable from each of the pairs. Adding the first two equations gives

$$2x + y = 3$$

Adding the first and third equations gives

$$-4x - 2y = -6$$

Now consider the system formed by these last two equations. We multiply the first equation in this new system by 2 and add again:

$$\begin{aligned} 4x + 2y &= 6 \\ -4x - 2y &= -6 \\ \hline 0 &= 0 \end{aligned}$$

NOTE

There are ways of representing the solutions to such a system, as you will see in later courses.

This true statement tells us that the system has an infinite number of solutions (lying along a straight line). Again, such a system is dependent.

Check Yourself 3

Solve the system.

$$\begin{aligned} 2x - y + 3z &= 3 \\ -x + y - 2z &= 1 \\ y - z &= 5 \end{aligned}$$

We see that a linear system in three variables can have exactly one solution (a consistent system) or, as in Example 3, infinitely many solutions (a dependent system). Consider now a third possibility: no solutions (an inconsistent system). It is illustrated in Example 4.

Example 4 Solving an Inconsistent Linear System in Three Variables

Solve the system.

$$\begin{aligned} 3x + y - 3z &= 1 \\ -2x - y + 2z &= 1 \\ -x - y + z &= 2 \end{aligned}$$

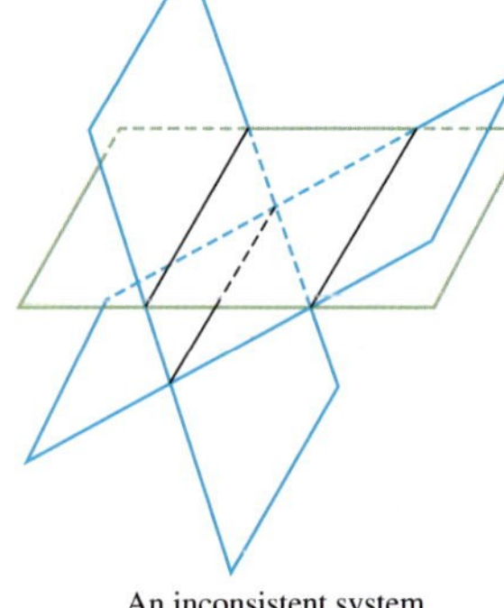

An inconsistent system

This time we eliminate variable y. Adding the first two equations gives

$x - z = 2$

Adding the first and third equations gives

$2x - 2z = 3$

Now, multiply the first equation in this new system by -2 and add the result to the second two-variable equation.

$$\begin{aligned} -2x + 2z &= -4 \\ 2x - 2z &= 3 \\ \hline 0 &= -1 \end{aligned}$$

We eliminated all of the variables and are left with the contradiction $0 = -1$. This means that the system is *inconsistent* and has no solutions. There is *no* point common to all three planes. The solution set is the empty set $\varnothing$.

Check Yourself 4

Solve the system.

$$\begin{aligned} x - y - z &= 0 \\ -3x + 2y + z &= 1 \\ 3x - y + z &= -1 \end{aligned}$$

We have not illustrated all possible types of inconsistent and dependent systems. Other possibilities involve either distinct parallel planes or planes that coincide. The solution techniques in these additional cases are similar to those illustrated in Examples 3 and 4.

In many instances, if an application involves three unknown quantities, you will find it useful to assign three variables to those quantities and then build a system of three equations from the given relationships in the problem. The extension of our problem-solving strategy is natural, as Example 5 illustrates.

Example 5 Solving a Number Problem

< Objective 2 >

The sum of the digits of a three-digit number is 12. The tens digit is 2 less than the hundreds digit, and the units digit is 4 less than the sum of the other two digits. What is the number?

NOTES

Sometimes it helps to choose variable letters that relate to the words, as is done here.

Take a moment now to go back to the original problem and pick out those conditions. That skill is a crucial part of the problem-solving strategy.

Step 1 The three unknowns are, of course, the three digits of the number.

Step 2 We now want to assign variables to each of three digits. Let u be the units digit, t the tens digit, and h the hundreds digit.

Step 3 There are three conditions given in the problem that allow us to write the necessary three equations. From those conditions

$$h + t + u = 12$$
$$t = h - 2$$
$$u = h + t - 4$$

Step 4 There are various ways to approach the solution. To use addition, write the system in the equivalent form

$$\begin{aligned} h + t + u &= 12 \\ -h + t &= -2 \\ -h - t + u &= -4 \end{aligned}$$

and solve by our earlier methods.

Step 5 The solution, which you can verify, is $h = 5$, $t = 3$, and $u = 4$. The number is 534.

To check, you should show that the digits of 534 meet each of the conditions of the original problem.

Check Yourself 5

The sum of the measures of the angles of a triangle is 180°. In a given triangle, the measure of the second angle is twice the measure of the first. The measure of the third angle is 30° less than the sum of the measures of the first two. Find the measure of each angle.

We continue by using systems to solve an investment application.

Example 6 Solving an Investment Application

Monica decided to divide a total of \$42,000 into three investments: a savings account paying 5% interest, a time deposit paying 7%, and a bond paying 9%. Her total annual interest from the three investments was \$2,600, and the interest from the savings account was \$200 less than the total interest from the other two investments. How much did she invest at each rate?

NOTE

For 1 year, the interest formula is

$I = Pr$

(interest equals principal times rate).

Step 1 The three amounts are the unknowns.

Step 2 We let s be the amount invested at 5%, t the amount at 7%, and b the amount at 9%. Note that the interest from the savings account is then $0.05s$, and so on.

A table helps with the next step.

	5%	7%	9%
Principal	s	t	b
Interest	$0.05s$	$0.07t$	$0.09b$

Step 3 Again there are three conditions in the given problem. By using the table above, they lead to these equations.

$$s + t + b = 42{,}000$$ Total invested

$$0.05s + 0.07t + 0.09b = 2{,}600$$ Total interest

$$0.05s = 0.07t + 0.09b - 200$$ The savings interest was \$200 *less than* that from the other two investments.

NOTE

Find the interest earned from each investment, and verify that the conditions of the problem are satisfied.

Step 4 We clear the equation of decimals and solve as before, with the result

Step 5 $s = \$24{,}000 \qquad t = \$11{,}000 \qquad b = \$7{,}000$

We leave the check of this solution to you.

Check Yourself 6

Glenn has a total of \$11,600 invested in three accounts: a savings account paying 6% interest, a stock paying 8%, and a mutual fund paying 10%. The annual interest from the stock and mutual fund is twice that from the savings account, and the mutual fund returned \$120 more than the stock. How much did Glenn invest in each account?

Check Yourself ANSWERS

1. $\{(5, 1, -3)\}$ **2.** $\{(1, -3, -2)\}$
3. The system is dependent (there are infinitely many solutions).
4. The system is inconsistent (there are no solutions).
5. The three angles are 35°, 70°, and 75°.
6. \$5,000 in savings, \$3,000 in stocks, and \$3,600 in the mutual fund

Reading Your Text

These fill-in-the-blank exercises will help you understand some of the key vocabulary used in this section. The answers to these exercises are in the Answers Appendix in the back of the text.

(a) The solution to a linear equation in ________ variables is an ordered triple.

(b) For a unique solution to exist to a system of equations when three variables are involved, we must have at least ________ equations.

(c) If solving a system of equations results in a true statement such as $0 = 0$, the system has an ________ number of solutions.

(d) If a system of equations has only one solution we call it a ________ system.

Skills | Calculator/Computer | Career Applications | Above and Beyond

4.4 exercises

< Objective 1 >

Solve each system of equations. If a unique solution does not exist, state whether the system is inconsistent or dependent.

1. $x - y + z = 3$
$2x + y + z = 8$
$3x + y - z = 1$

2. $x - y - z = 2$
$2x + y + z = 8$
$x + y + z = 6$

3. $x + y + z = 1$
$2x - y + 2z = -1$
$-x - 3y + z = 1$

4. $x - y - z = 6$
$-x + 3y + 2z = -11$
$3x + 2y + z = 1$

5. $x + y + z = 1$
$-2x + 2y + 3z = 20$
$2x - 2y - z = -16$

6. $x + y + z = -3$
$3x + y - z = 13$
$3x + y - 2z = 18$

7. $2x + y - z = 2$
$-x - 3y + z = -1$
$-4x + 3y + z = -4$

8. $x + 4y - 6z = 8$
$2x - y + 3z = -10$
$3x - 2y + 3z = -18$

9. $3x - y + z = 5$
$x + 3y + 3z = -6$
$x + 4y - 2z = 12$

10. $2x - y + 3z = 2$
$x - 2y + 3z = 1$
$4x - y + 5z = 5$

11. $x + 2y + z = 2$
$2x + 3y + 3z = -3$
$2x + 3y + 2z = 2$

12. $x - 4y - z = -3$
$x + 2y + z = 5$
$3x - 7y - 2z = -6$

13. $x + 3y - 2z = 8$
$3x + 2y - 3z = 15$
$4x + 2y + 3z = -1$

14. $x + y - z = 2$
$3x + 5y - 2z = -5$
$5x + 4y - 7z = -7$

15. $x + y - z = 2$
$x - 2z = 1$
$2x - 3y - z = 8$

16. $x + y + z = 6$
$x - 2y = -7$
$4x + 3y + z = 7$

17. $x - 3y + 2z = 1$
$16y - 9z = 5$
$4x + 4y - z = 8$

18. $x - 4y + 4z = -1$
$y - 3z = 5$
$3x - 4y + 6z = 1$

19. $x + 2y - 4z = 13$
$3x + 4y - 2z = 19$
$3x + 2z = 3$

20. $x + 2y - z = 6$
$-3x - 2y + 5z = -12$
$x - 2z = 3$

< Objective 2 >

Choose a variable to represent each unknown quantity, write a system of equations, and then solve the system.

21. Number Problem The sum of three numbers is 16. The largest number is equal to the sum of the other two, and 3 times the smallest number is 1 more than the largest. Find the three numbers.

22. Number Problem The sum of three numbers is 24. Twice the smallest number is 2 less than the largest number, and the largest number is equal to the sum of the other two. What are the three numbers?

23. Problem Solving A cashier has 25 coins consisting of nickels, dimes, and quarters with a value of \$4.90. If the number of dimes is 1 less than twice the number of nickels, how many of each type of coin does she have?

24. Business and Finance A high school theater has tickets at \$6 for adults, \$3.50 for students, and \$2.50 for children under 12 years old. 278 tickets were sold for one showing with a total revenue of \$1,300. If the number of adult tickets sold was 10 less than twice the number of student tickets, how many of each type of ticket were sold for the showing?

25. **Geometry** The perimeter of a triangle is 19 cm. If the length of the longest side is twice that of the shortest side and 3 cm less than the sum of the lengths of the other two sides, find the lengths of the three sides.

26. **Geometry** The measure of the largest angle of a triangle is 10° more than the sum of the measures of the other two angles and 10° less than 3 times the measure of the smallest angle. Find the measures of the three angles of the triangle.

27. **Business and Finance** Jovita divides $12,000 into three investments: a savings account paying 4% annual interest, a bond paying 6%, and a money market fund paying 9%. The annual interest from the three accounts is $860, and she has 2 times as much invested in the bond as in the savings account. What amount does she have invested in each account?

28. **Business and Finance** Adrienne has $10,000 invested in a savings account paying 5%, a time deposit paying 7%, and a bond paying 10%. She has $1,000 less invested in the bond than in her savings account, and she earned $700 in annual interest. What has she invested in each account?

29. **Number Problem** The sum of the digits of a three-digit number is 9, and the tens digit of the number is twice the hundreds digit. If the digits are reversed in order, the new number is 99 more than the original number. What is the original number?

30. **Number Problem** The sum of the digits of a three-digit number is 9. The tens digit is 3 times the hundreds digit. If the digits are reversed in order, the new number is 99 less than the original number. Find the original three-digit number.

31. **Science and Medicine** Roy, Sally, and Jeff drive a total of 50 mi to work each day. Sally drives twice as far as Roy, and Jeff drives 10 mi farther than Sally. Use a system of three equations in three unknowns to find how far each person drives each day. VIDEO

32. **Statistics** A parking lot has spaces reserved for motorcycles, cars, and vans. There are 5 more spaces reserved for vans than for motorcycles. There are 3 times as many car spaces as van and motorcycle spaces combined. If the parking lot has 180 total reserved spaces, how many of each type are there?

Determine whether each statement is **true** *or* **false.**

33. A system of three linear equations in three variables can have one unique solution, an infinite number of solutions, or no solution.

34. When solving a system, obtaining a statement such as $0 = 0$ means that the system is inconsistent.

Skills | Calculator/Computer | Career Applications | **Above and Beyond**

The solution process illustrated in this section can be extended to solving systems of more than three variables in a natural fashion. For instance, if four variables are involved, eliminate one variable in the system and then solve the resulting system in three variables as before. Substituting those three values into one of the original equations provides the value for the remaining variable and the solution to the system.

Use this procedure to solve each system.

35. $$\begin{aligned} x + 2y + 3z + w &= 0 \\ -x - y - 3z + w &= -2 \\ x - 3y + 2z + 2w &= -11 \\ -x + y - 2z + w &= 1 \end{aligned}$$

36. $$\begin{aligned} x + y - 2z - w &= 4 \\ x - y + z + 2w &= 3 \\ 2x + y - z - w &= 7 \\ x - y + 2z + w &= 2 \end{aligned}$$

In some systems of equations there are more equations than variables. We can illustrate this situation with a system of three equations in two variables. To solve this type of system, pick any two of the equations and solve this system. Then substitute the solution obtained into the third equation. If a true statement results, the solution used is the solution to the entire system. If a false statement occurs, the system has no solution.

Use this procedure to solve each system.

37. $$\begin{aligned} x - y &= 5 \\ 2x + 3y &= 20 \\ 4x + 5y &= 38 \end{aligned}$$

38. $$\begin{aligned} 3x + 2y &= 6 \\ 5x + 7y &= 35 \\ 7x + 9y &= 8 \end{aligned}$$

39. Experiments show that cars, trucks, and buses emit different amounts of air pollutants. In one such experiment, a truck emitted 1.5 lb of carbon dioxide (CO_2) per passenger-mile and 2 g of nitrogen oxide (NO) per passenger-mile. A car emitted 1.1 lb of CO_2 per passenger-mile and 1.5 g of NO per passenger-mile. A bus emitted 0.4 lb of CO_2 per passenger-mile and 1.8 g of NO per passenger-mile. A total of 85 passenger-miles was driven by the three vehicles, and 73.5 lb of CO_2 and 149.5 g of NO were collected. In the system of equations, T, C, and B represent the number of passenger-miles driven in the truck, car, and bus, respectively. Determine the passenger-miles driven by each vehicle.

$$\begin{aligned} T + C + B &= 85.0 \\ 1.5T + 1.1C + 0.4B &= 73.5 \\ 2T + 1.5C + 1.8B &= 149.5 \end{aligned}$$

40. Experiments show that cars, trucks, and trains emit different amounts of air pollutants. In one such experiment, a truck emitted 0.8 lb of carbon dioxide per passenger-mile and 1 g of nitrogen oxide per passenger-mile. A car emitted 0.7 lb of CO_2 per passenger-mile and 0.9 g of NO per passenger-mile. A train emitted 0.5 lb of CO_2 per passenger-mile and 4 g of NO per passenger-mile. A total of 141 passenger-miles was driven by the three vehicles, and 82.7 lb of CO_2 and 424.4 g of NO were collected. In the system of equations, T, C, and R represent the number of passenger-miles driven in the truck, car, and train, respectively. Determine the passenger-miles driven by each vehicle.

$$\begin{aligned} T + C + R &= 141.0 \\ 0.8T + 0.7C + 0.5R &= 82.7 \\ T + 0.9C + 4R &= 424.4 \end{aligned}$$

41. In Chapter 8 you will learn about quadratic functions and their graphs. A quadratic function has the form $y = ax^2 + bx + c$, where a, b, and c are specific numbers and $a \neq 0$. Three distinct points on the graph are enough to determine the equation.

(a) Suppose that (1, 5), (2, 10), and (3, 19) are on the graph of $y = ax^2 + bx + c$. Substituting the pair (1, 5) into this equation (that is, let $x = 1$ and $y = 5$) yields $5 = a + b + c$. Substituting each of the other ordered pairs yields $10 = 4a + 2b + c$ and $19 = 9a + 3b + c$. Solve the resulting system of equations to determine the values of a, b, and c. Then write the equation of the function.

(b) Repeat part (a), using the three points (1, 2), (2, 9), and (3, 22).

Answers

1. $\{(1, 2, 4)\}$ **3.** $\{(-2, 1, 2)\}$ **5.** $\{(-4, 3, 2)\}$ **7.** Infinite number of solutions, dependent system **9.** $\left\{\left(3, \frac{1}{2}, -\frac{7}{2}\right)\right\}$

11. $\{(3, 2, -5)\}$ **13.** $\{(2, 0, -3)\}$ **15.** $\left\{\left(4, -\frac{1}{2}, \frac{3}{2}\right)\right\}$ **17.** No solutions, inconsistent system **19.** $\left\{\left(2, \frac{5}{2}, -\frac{3}{2}\right)\right\}$

21. 3, 5, 8 **23.** 3 nickels; 5 dimes; 17 quarters **25.** 4 cm, 7 cm, 8 cm **27.** \$2,000 savings; \$4,000 bond; \$6,000 money market

29. 243 **31.** Roy 8 mi; Sally 16 mi; Jeff 26 mi **33.** True **35.** $\{(1, 2, -1, -2)\}$ **37.** $\{(7, 2)\}$

39. Truck: 20 passenger-miles; car: 25 passenger-miles; bus: 40 passenger-miles **41.** **(a)** $y = 2x^2 - x + 4$; **(b)** $y = 3x^2 - 2x + 1$

4.5 Systems of Linear Inequalities

< 4.5 Objectives >

1 > Graph systems of linear inequalities

2 > Use a system of linear inequalities to solve an application

Now that we can solve systems of equations, it is time to look at systems of inequalities and the applications that stem from this topic. With a system of equations, the solutions are any points of intersection when we graph the equations in the system. We extend this idea to solve systems of inequalities.

In this case, the solution set is the set of all ordered pairs that satisfy both inequalities. **The graph of the solution set of a system of linear inequalities** is then the intersection of the graphs of the individual inequalities. Here is an example.

Example 1 Solving a System of Linear Inequalities

< Objective 1 >

Solve the system of linear inequalities by graphing.

$x + y > 4$

$x - y < 2$

We start by graphing each inequality separately. We draw the boundary line and, using (0, 0) as a test point, we see that we should shade the half-plane above the line in both graphs.

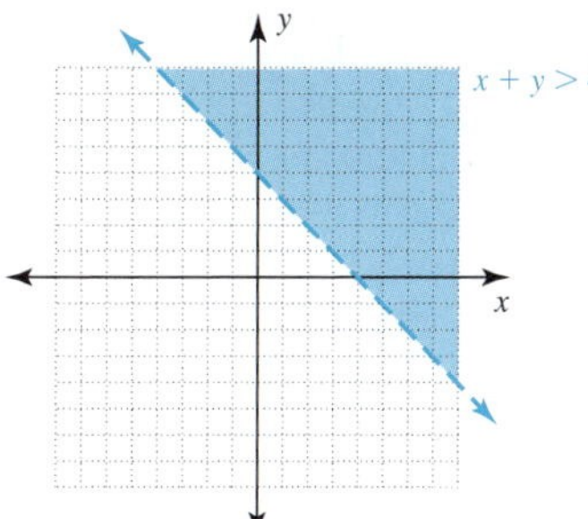

RECALL

The boundary lines are dashed to indicate that points on the lines are *not* solutions.

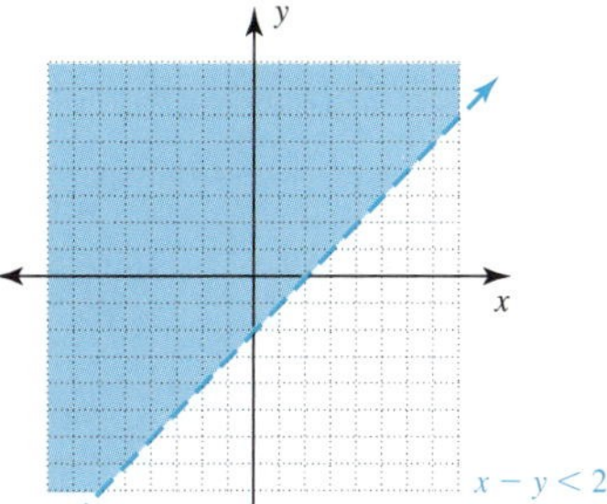

NOTES

Points on the lines are not included in the solution.

We want to show all ordered pairs that satisfy *both* statements.

We combine the two graphs on the same set of axes. The graph of the solution set is given by the intersection of the two graphs.

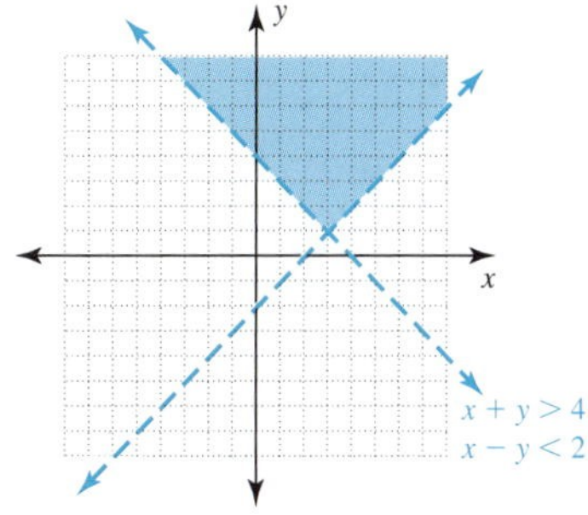

Check Yourself 1

Solve the system of linear inequalities by graphing.

$$2x - y < 4$$
$$x + y < 3$$

Most applications of systems of linear inequalities lead to **bounded regions.** This requires a system of three or more inequalities, as shown in Example 2.

Example 2 Solving a System of Linear Inequalities

Solve the system of linear inequalities by graphing.

$$x + 2y \le 6$$
$$x + y \le 5$$
$$x \ge 2$$
$$y \ge 0$$

On the same set of axes, we graph the boundary line of each of the inequalities. We then choose the appropriate half-planes (indicated by the arrow that is perpendicular to the line) in each case, and we locate the intersection of those regions for our graph.

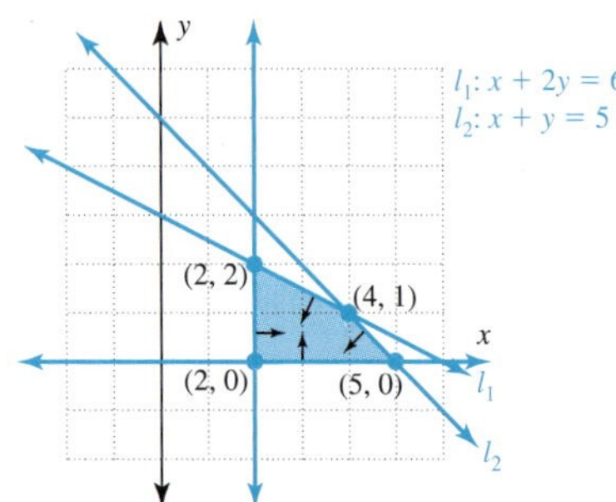

NOTE

The vertices of the shaded region are given because they have particular significance in later applications of this concept. Can you see how we found the vertices?

Check Yourself 2

Solve the system of linear inequalities by graphing.

$$2x - y \le 8$$
$$x + y \le 7$$
$$x \ge 0$$
$$y \ge 0$$

Let's expand on an application from Section 4.3 to see an application of our work with systems of linear inequalities.

Example 3 A Business and Finance Application

< Objective 2 >

A manufacturer produces LCD and plasma television sets. LCD sets require 12 hr of labor to produce and plasma sets require 18 hr. The labor available is limited to 360 hr per week. Also, the plant capacity is limited to producing a total of 25 sets per week. Draw a graph of the region representing the number of sets that can be produced, given these conditions.

As suggested earlier, we let x represent the number of LCD sets produced and y the number of plasma sets. Since the labor is limited to 360 hr, we have

$$12x + 18y \le 360$$

(12x: 12 hr per LCD set; 18y: 18 hr per plasma set)

NOTE

The total labor is limited to (or less than or equal to) 360 hr.

The total production, here $x + y$ sets, is limited to 25, so we write

$x + y \le 25$

NOTE

We have $x \ge 0$ and $y \ge 0$ because the number of sets produced cannot be negative.

For convenience in graphing, we divide both sides of the first inequality by 6, to write the equivalent system

$$\begin{aligned} 2x + 3y &\le 60 \\ x + y &\le 25 \\ x &\ge 0 \\ y &\ge 0 \end{aligned}$$

NOTE

The shaded area is called the **feasible region.** All points in the region meet the given conditions of the problem and represent possible production options.

We now graph the system of inequalities as before. The shaded area represents all possibilities in terms of the number of sets that can be produced. Only points with integer coordinates represent realistic solutions in the context of this application.

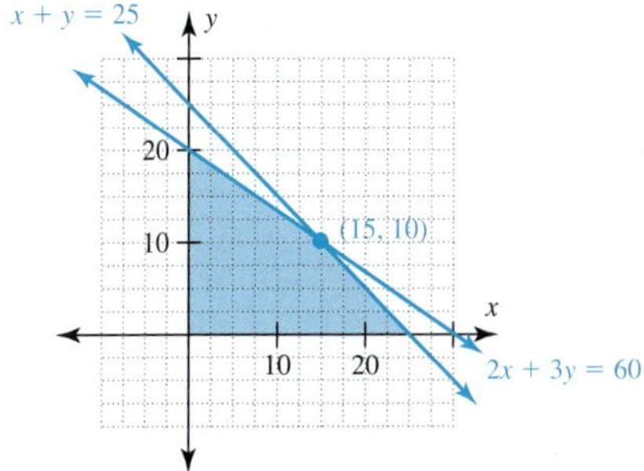

Check Yourself 3

A manufacturer produces standard TVs and HDTVs. The standard TVs require 10 hr of labor to produce and the HDTVs require 20 hr. The labor hours available are limited to 300 hr per week. Existing orders require that at least 10 standard TVs and at least 5 HDTVs be produced per week. Draw a graph of the region representing the possible production options.

Check Yourself ANSWERS

1.

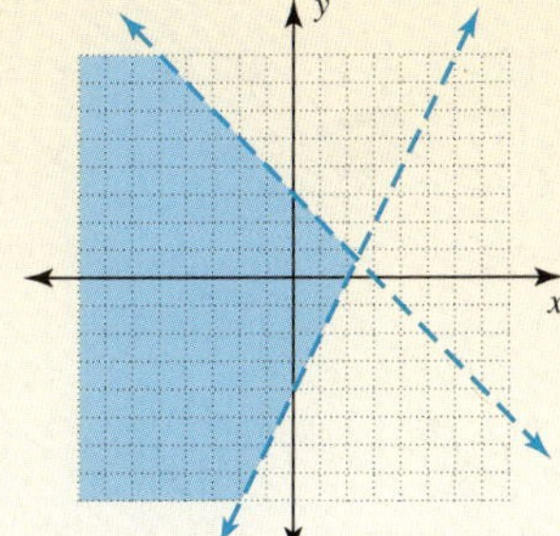

2.

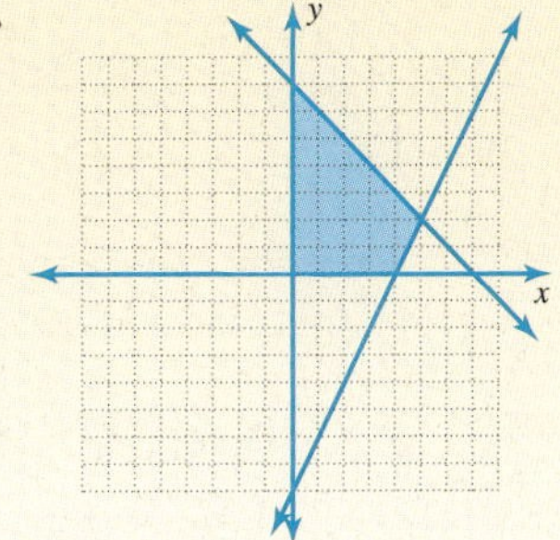

3. Let x be the number of standard TVs and y be the number of HDTVs. The system is

$$\begin{aligned} 10x + 20y &\le 300 \\ x &\ge 10 \\ y &\ge 5 \end{aligned}$$

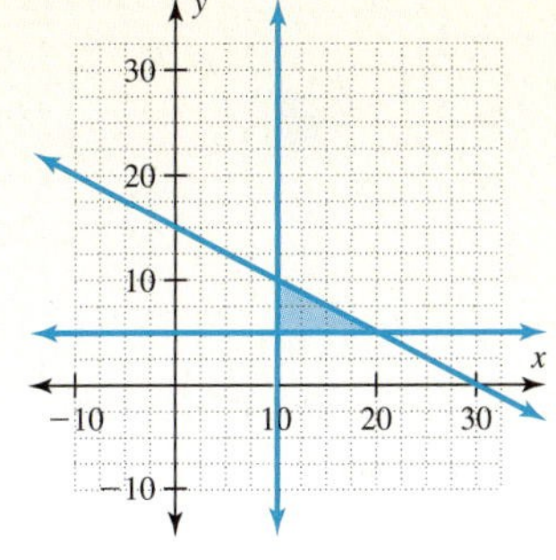

Reading Your Text

These fill-in-the-blank exercises will help you understand some of the key vocabulary used in this section. The answers to these exercises are in the Answers Appendix in the back of the text.

(a) The solution set of a system of inequalities is the set of all ________ ________ that satisfy every inequality in the system.

(b) When a boundary line of the graph of an inequality is ________, the points on the boundary line are not solutions to the inequality.

(c) When a boundary line of the graph of an inequality is ________, the points on the boundary line are solutions to the inequality.

(d) The solution set of a system of inequalities is often a ________ region.

4.5 exercises

Skills | Calculator/Computer | Career Applications | Above and Beyond

< Objective 1 >

Solve each system of linear inequalities graphically.

1. $x + 2y \leq 4$
$x - y \geq 1$

2. $3x - y > 6$
$x + y < 6$

3. $3x + y < 6$
$x + y > 4$

4. $2x + y \geq 8$
$x + y \geq 4$

5. $x + 3y \leq 12$
$2x - 3y \leq 6$

6. $x - 2y > 8$
$3x - 2y > 12$

7. $3x + 2y \leq 12$
$x \geq 2$

8. $2x + y \leq 6$
$y \geq 1$

9. $2x + y \leq 8$
$x > 1$
$y > 2$

10. $3x - y \leq 6$
$x \geq 1$
$y \leq 3$

11. $x + 2y \leq 8$
$2 \leq x \leq 6$
$y \geq 0$

12. $x + y < 6$
$0 \leq y \leq 3$
$x \geq 1$

13. $3x + y \le 6$
$x + y \le 4$
$x \ge 0$
$y \ge 0$

14. $x - 2y \ge -2$
$x + 2y \le 6$
$x \ge 0$
$y \ge 0$

15. $4x + 3y \le 12$
$x + 4y \le 8$
$x \ge 0$
$y \ge 0$

16. $2x + y \le 8$
$x + y \ge 3$
$x \ge 0$
$y \ge 0$

17. $x - 4y \le -4$
$x + 2y \le 8$
$x \ge 2$ VIDEO

18. $x - 3y \ge -6$
$x + 2y \ge 4$
$x \le 4$

< Objective 2 >

Sketch the graph representing each system of inequalities.

19. **Business and Finance** A manufacturer produces both two-slice and four-slice toasters. Each two-slice toaster takes 6 hr of labor to produce while four-slice toasters take 10 hr. The labor available is limited to 300 hr per week, and the total production capacity is 40 toasters per week. Draw a graph of the feasible region, given these conditions, where x is the number of two-slice toasters and y is the number of four-slice toasters. chapter 4 > Make the Connection VIDEO

20. **Business and Finance** A small firm produces both standard and HD radios. The standard radios take 15 hr to produce, and the HD radios take 20 hr. The number of production hours is limited to 300 hr per week. The plant's capacity is limited to a total of 18 radios per week, and existing orders require that at least 4 standard radios and at least 3 HD radios be produced per week. Draw a graph of the feasible region, given these conditions, where x is the number of standard radios and y the number of HD radios. chapter 4 > Make the Connection

Determine whether each statement is **true** *or* **false.**

21. The feasible region in an application shows all the points that meet all the conditions of the problem.

22. The graph of the solution set of a system of three linear inequalities can be unbounded.

Complete each statement with **always, sometimes,** *or* **never.**

23. The graph of the solution set of a system of two linear inequalities _________ includes the origin.

24. The graph of a nonempty solution set of a system of two linear inequalities is _________ bounded.

25. When you solve a system of linear inequalities, it is often easier to shade the region that is not part of the solution, rather than the region that is. Try this method, then describe its benefits.

26. Describe a system of linear inequalities for which there is no solution.

27. Write the system of inequalities whose graph is the shaded region.

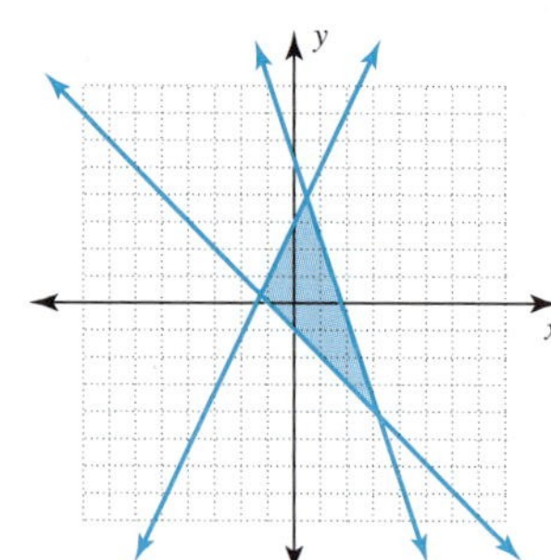

28. Write the system of inequalities whose graph is the shaded region.

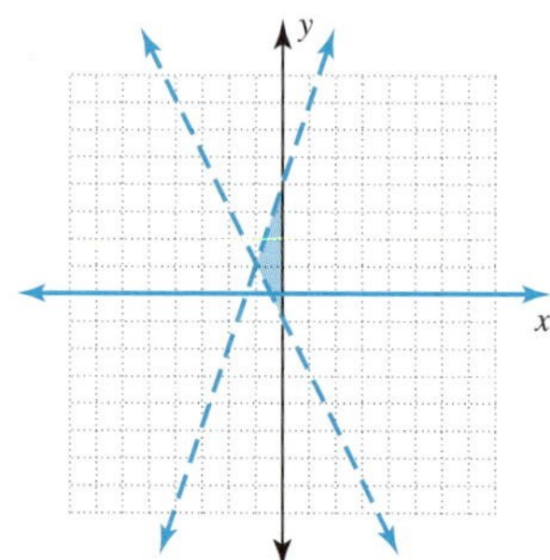

Answers

1.

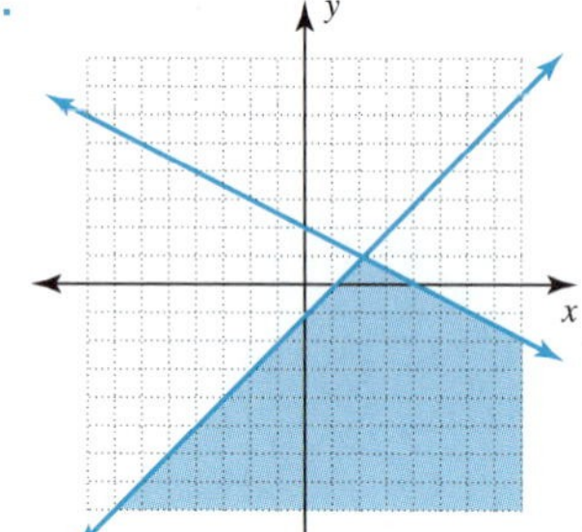

3.

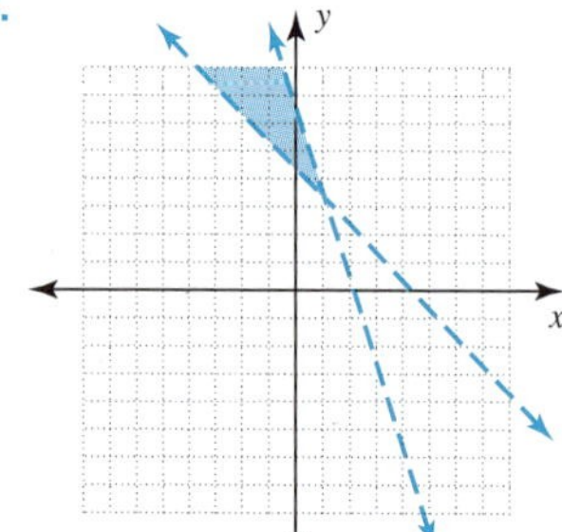

5.

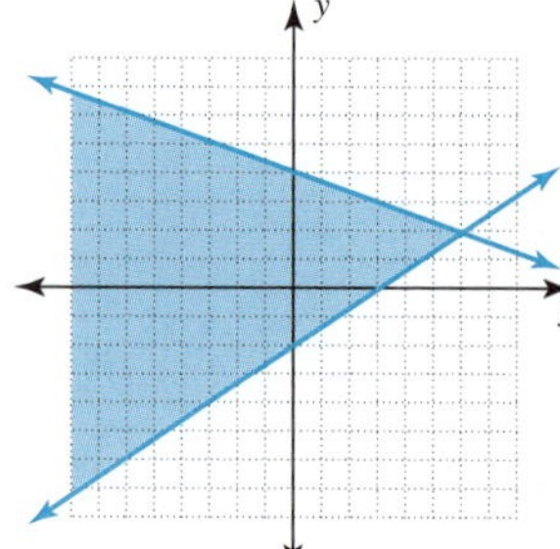

7.

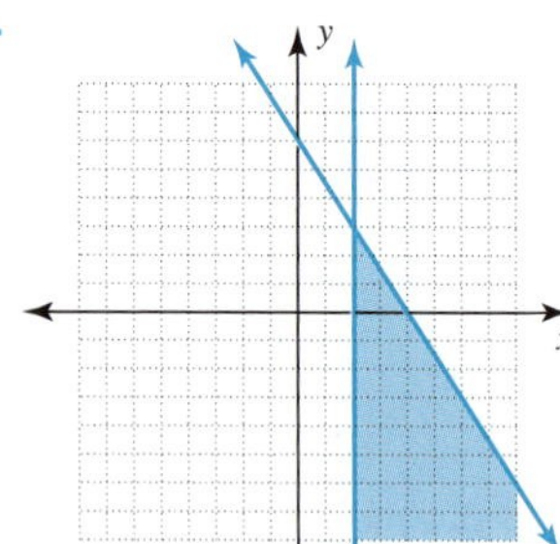

9.

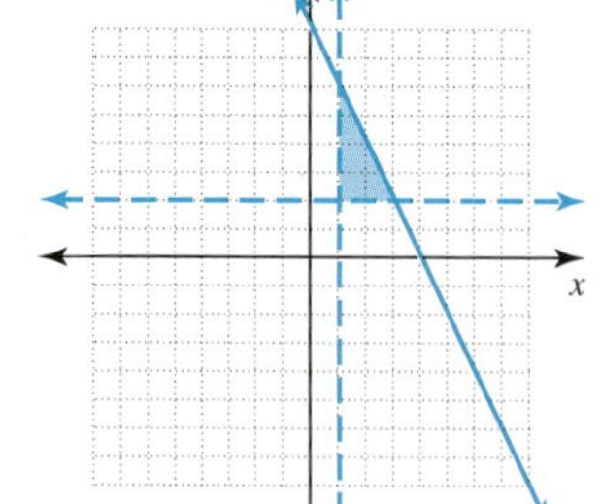

11.

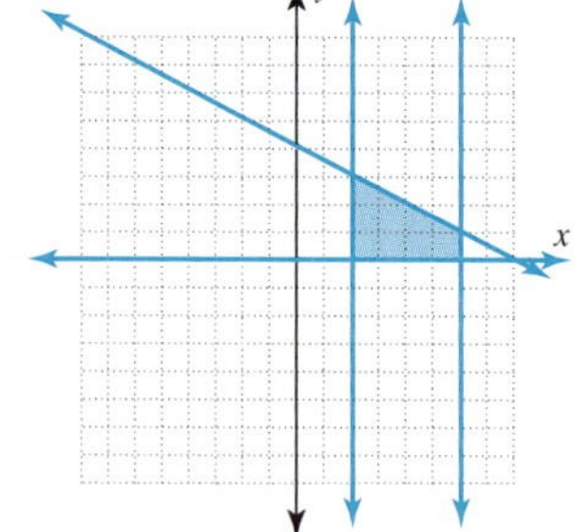

13. 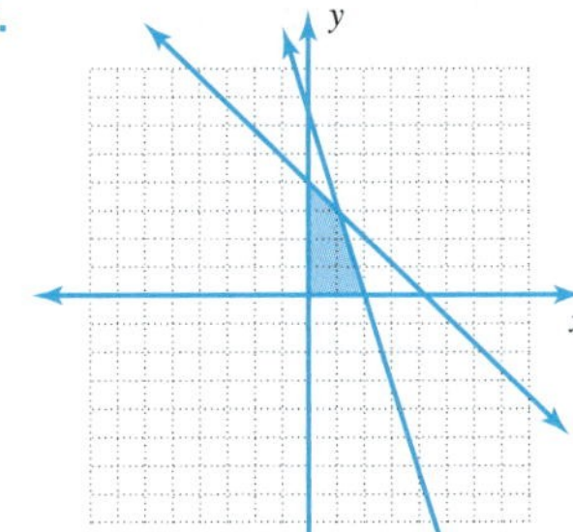

15. 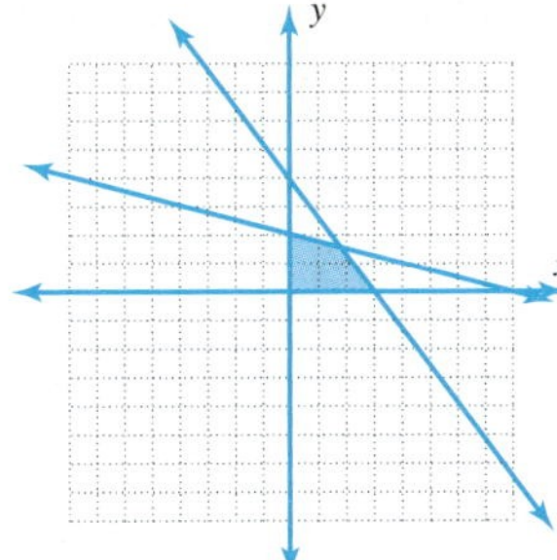

17. 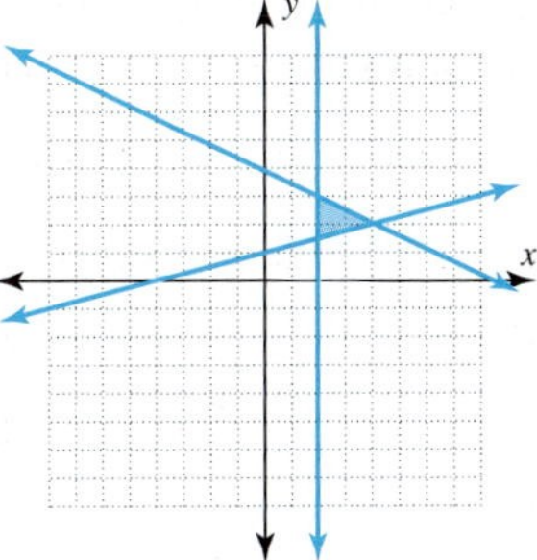

19. 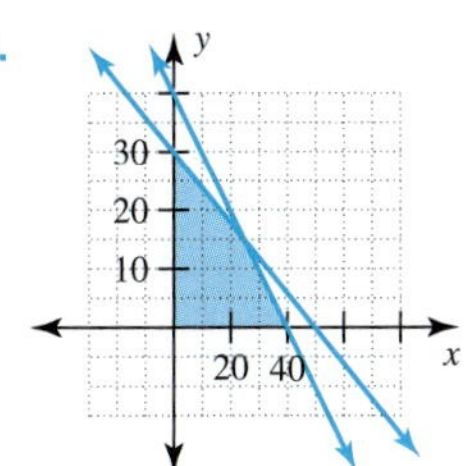

21. True

23. sometimes

25. Above and Beyond

27. $y \leq 2x + 3$
$y \leq -3x + 5$
$y \geq -x - 1$

Definition/Procedure	Example	Reference
Graphing Systems of Linear Equations		**Section 4.1**
A **system of linear equations** is two or more linear equations considered together. A solution to a system in two variables is an ordered pair of real numbers (x, y) that satisfies both equations in the system. There are three solution techniques: the graphing method, the addition method, and the substitution method.	The solution to the system $2x - y = 7$, $x + y = 2$ is $(3, -1)$. It is the only ordered pair that satisfies both equations.	*p. 338*
Solving a System Graphically Graph each equation of the system on the same set of coordinate axes. If a solution exists, it will correspond to the point of intersection of the two lines. Such a system is called a **consistent system.** If a solution does not exist, there is no point at which the two lines intersect. Such lines are parallel, and the system is called an **inconsistent system.** If there are infinitely many solutions, the lines coincide. Such a system is called a **dependent system.** You may or may not be able to determine exact solutions for the system of equations with this method. 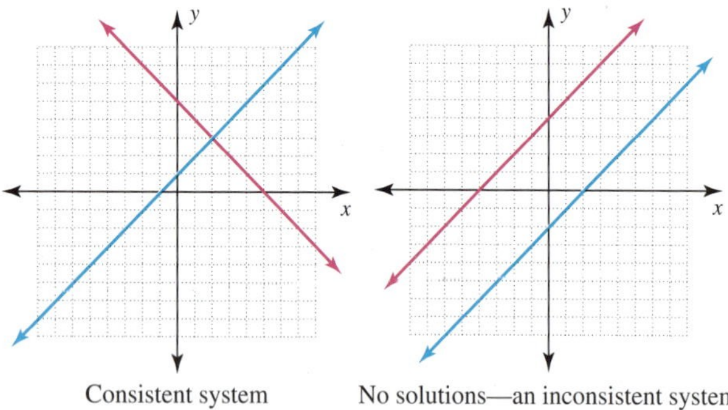 Consistent system No solutions—an inconsistent system 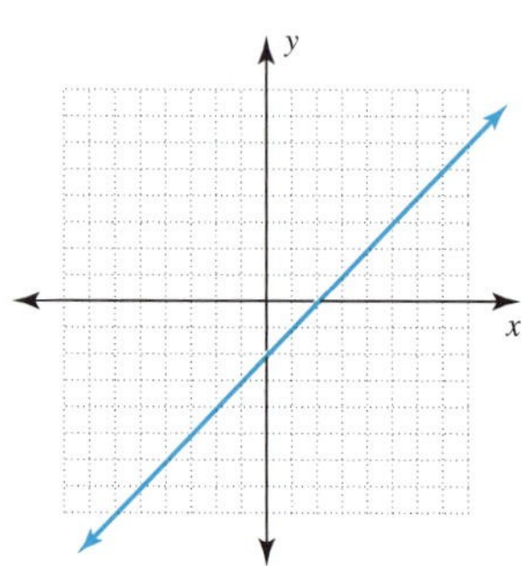 An infinite number of solutions—a dependent system	To solve the system $2x - y = 7$, $x + y = 2$ graphically 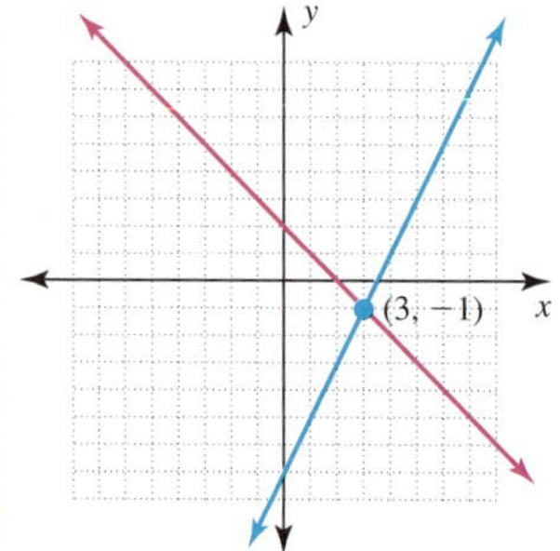	*p. 339*

Continued

Definition/Procedure	Example	Reference
Solving Equations in One Variable Graphically		Section 4.2
Finding a Solution Graphically **Step 1** Let each side of the equation represent a function of x. **Step 2** Graph the two functions on the same set of axes. **Step 3** Find the point of intersection of the two graphs. The x-value at this point represents the solution to the original equation.	To solve $2x - 6 = 8x$, let $f(x) = 2x - 6$ $g(x) = 8x$ then graph both lines. 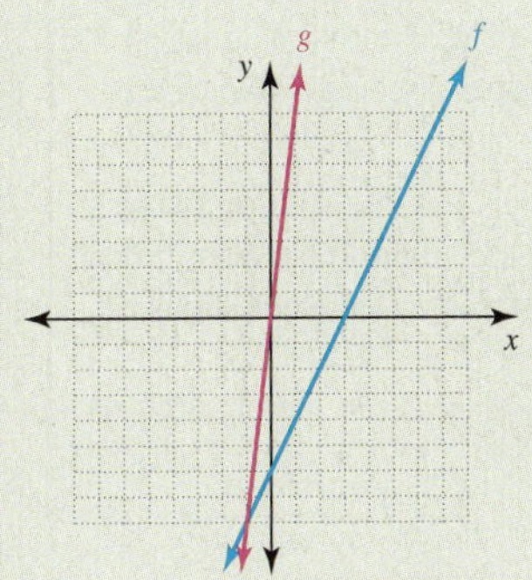 The intersection occurs when $x = -1$. The solution set is $\{-1\}$.	*p.* 353
Systems of Equations in Two Variables		Section 4.3
Solving by the Addition Method **Step 1** If necessary, multiply one or both of the equations by a constant so that one of the variables can be eliminated by addition. **Step 2** Add the equations of the equivalent system formed in step 1. **Step 3** Solve the equation found in step 2. **Step 4** Substitute the value found in step 3 into either of the equations of the original system to find the corresponding value of the remaining variable. **Step 5** The ordered pair found in step 4 is the solution to the system. Check the solution by substituting the pair of values found in step 4 into the equations of the original system.	To solve $5x - 2y = 11$ $2x + 3y = 12$ multiply the first equation by 3 and the second equation by 2. Then add to eliminate y. $19x = 57$ $x = 3$ Substituting 3 for x in the first equation gives $15 - 2y = 11$ $y = 2$ So $\{(3, 2)\}$ is the solution set.	*p.* 365

Continued

Definition/Procedure	Example	Reference
Solving by the Substitution Method **Step 1** Solve one of the equations of the original system for one of the variables. **Step 2** Substitute the expression obtained in step 1 into the other equation of the system to write an equation in a single variable. **Step 3** Solve the equation found in step 2. **Step 4** Substitute the value found in step 3 into the equation found in step 1 to find the corresponding value of the remaining variable. **Step 5** The ordered pair found in step 4 is the solution to the system of equations. Check the solution by substituting the pair of values found in step 4 into the equations of the original system.	To solve $3x - 2y = 6$ $6x + \quad y = 2$ by substitution, solve the second equation for y. $y = -6x + 2$ Substituting into the first equation gives $3x - 2(-6x + 2) = 6$ and $x = \frac{2}{3}$ Substituting $\frac{2}{3}$ for x in the equation that we solved for y gives $y = (-6)\left(\frac{2}{3}\right) + 2$ $= -2$ The solution set is $\left\{\left(\frac{2}{3}, -2\right)\right\}$	*p.* 367
Applications of Systems of Linear Equations **Step 1** Read the problem carefully to determine the unknown quantities. **Step 2** Choose variables to represent any unknowns. **Step 3** Translate the problem to the language of algebra to form a system of equations. **Step 4** Solve the system of equations by any of the methods discussed. **Step 5** Answer the question in the original problem and verify your solution by returning to the original problem.	Determine the condition that relates the unknown quantities. Use a different letter for each variable. A table or a sketch often helps in writing the equations of the system.	*p.* 368

Continued

Definition/Procedure	Example	Reference
Systems of Equations in Three Variables		Section 4.4
A solution to a system of three equations in three variables is an ordered triple of numbers (x, y, z) that satisfies each equation in the system. **Solving a System of Three Equations in Three Unknowns** **Step 1** Choose a pair of equations from the system and use the addition method to eliminate one of the variables. **Step 2** Choose a different pair of equations and eliminate the same variable. **Step 3** Solve the system of two equations in two variables determined in steps 1 and 2. **Step 4** Substitute the values found above into one of the original equations and solve for the remaining variable. **Step 5** The solution is the ordered triple of values found in steps 3 and 4. It can be checked by substituting into the other equations of the original system.	Solve. $x + y - z = 6$ $2x - 3y + z = -9$ $3x + y + 2z = 2$ Adding the first two equations gives $3x - 2y = -3$ Multiplying the first equation by 2 and adding the result to the third equation gives $5x + 3y = 14$ We solve the system consisting of the pair of two-variable equations as before to obtain $x = 1 \quad y = 3$ Substituting these values into one of the original equations gives $z = -2$ The solution set is $\{(1, 3, -2)\}$.	*p.* 381
Systems of Linear Inequalities		Section 4.5
A **system of linear inequalities** is two or more linear inequalities considered together. The **graph of the solution set** of a system of linear inequalities is the intersection of the graphs of the individual inequalities. **Solving Systems of Linear Inequalities by Graphing** **Step 1** Graph each inequality, shading the appropriate half-plane on the same set of coordinate axes. **Step 2** The graph of the system is the intersection of the regions shaded in step 1.	To solve the system $x + 2y \le 8$ $x + y \le 6$ $x \ge 0$ $y \ge 0$ graphically	*p.* 389

summary exercises :: chapter 4

This summary exercise set will help ensure that you have mastered each of the objectives of this chapter. The exercises are grouped by section. You should reread the material associated with any exercises that you find difficult. The answers to the odd-numbered exercises are in the Answers Appendix in the back of the text.

4.1 *Graph each system of equations and then solve the system.*

1. $x + y = 8$
$x - y = 4$

2. $x + 2y = 8$
$x - y = 5$

3. $2x + 3y = 12$
$2x + y = 8$

4. $x + 4y = 8$
$y = 1$

Use a graphing calculator to solve each system. Estimate your answer to the nearest hundredth. You may need to adjust the viewing window to see the point of intersection.

5. $44x + 35y = 1{,}115$
$11x - 27y = 850$

6. $15x - 48y = 935$
$25x + 51y = 1{,}051$

4.2 *Solve each equation graphically. Do not use a calculator.*

7. $3x - 6 = 0$

8. $4x + 3 = 7$

9. $3x + 5 = x + 7$

10. $4x - 3 = x - 6$

11. $\frac{6x - 1}{2} = 2(x - 1)$

12. $3x + 2 = 2x - 1$

13. $3(x - 2) = 2(x - 1)$

14. $3(x + 1) + 3 = -7(x + 2)$

4.3 *Use the addition method to solve each system. If a unique solution does not exist, state whether the given system is inconsistent or dependent.*

15. $x + 2y = 7$
$x - y = 1$

16. $x + 3y = 14$
$4x + 3y = 29$

17. $3x - 5y = 5$
$-x + y = -1$

18. $x - 4y = 12$
$2x - 8y = 24$

19. $6x + 5y = -9$
$-5x + 4y = 32$

20. $3x + y = 8$
$-6x - 2y = -10$

21. $5x - y = -17$
$4x + 3y = -6$

22. $4x - 3y = 1$
$6x + 5y = 30$

23. $x - \frac{1}{2}y = 8$
$\frac{2}{3}x + \frac{3}{2}y = -2$

24. $\frac{1}{5}x - 2y = 4$
$\frac{3}{5}x + \frac{2}{3}y = -8$

Use the substitution method to solve each system. If a unique solution does not exist, state whether the given system is inconsistent or dependent.

25. $2x + y = 23$
$x = y + 4$

26. $x - 5y = 26$
$y = x - 10$

27. $3x + y = 7$
$y = -3x + 5$

28. $2x - 3y = 13$
$x = 3y + 9$

29. $5x - 3y = 13$
$x - y = 3$

30. $4x - 3y = 6$
$x + y = 12$

31. $3x - 2y = -12$
$6x + y = 1$

32. $x - 4y = 8$
$-2x + 8y = -16$

Solve each problem by choosing a variable to represent each unknown quantity. Then write a system of equations that allows you to solve for each variable.

33. Number Problem One number is 2 more than 3 times another. If the sum of the two numbers is 30, find the two numbers.

34. Problem Solving Suppose that a cashier has 78 \$5 and \$10 bills with a value of \$640. How many of each type of bill does she have?

35. Business and Finance Tickets for a high school basketball game sold at \$7 for an adult ticket and \$4.50 for a student ticket. If the revenue from 1,200 tickets was \$7,400, how many of each type of ticket were sold?

36. Business and Finance A purchase of 8 blank DVDs and 4 blank Blu-Ray Discs (BDs) costs \$36. A second purchase of 4 DVDs and 5 BDs costs \$30. What is the price of a single DVD and of a single BD?

37. Geometry The length of a rectangle is 4 cm less than twice its width. If the perimeter of the rectangle is 64 cm, find the dimensions of the rectangle.

38. Business and Finance A grocer in charge of bulk foods wishes to combine peanuts selling for \$2.25 per pound and cashews selling for \$6 per pound. What amount of each nut should be used to form a 120-lb mixture selling for \$3 per pound?

39. Business and Finance Reggie has two investments totaling \$17,000—one a savings account paying 6%, the other a time deposit paying 8%. If his annual interest is \$1,200, what does he have invested in each account?

40. Science and Medicine A pharmacist mixes a 20% alcohol solution and a 50% alcohol solution to form 600 mL of a 40% solution. How much of each solution did he use in forming the mixture?

41. Science and Medicine A jet flying east, with the wind, makes a trip of 2,200 mi in 4 hr. Returning against the wind, the jet can travel only 1,800 mi in 4 hr. What is the plane's rate in still air? What is the rate of the wind?

42. **Business and Finance** A manufacturer produces small and large flash drives. The small drives require 20 min of component assembly time; the large drives, 25 min. The manufacturer has 500 min of component assembly time available per day. Each drive requires 30 min for packaging and testing, and 690 min of that time is available per day. How many of each of the drives should be produced daily to use all the available time?

43. **Busines and Finance** If the demand equation for a product is $D = 270 - 5p$ and the supply equation is $S = 13p$, find the equilibrium point.

44. **Business and Finance** Two car rental agencies have different rates for the rental of a compact automobile.

Company A: $46 per day plus 18¢ per mile

Company B: $49 per day plus 16¢ per mile

For a 3-day rental, at what number of miles will the charges from the two companies be the same?

4.4 *Solve each system. If a unique solution does not exist, state whether the given system is inconsistent or dependent.*

45. $\begin{aligned} x - y + z &= 0 \\ x + 4y - z &= 14 \\ x + y - z &= 6 \end{aligned}$

46. $\begin{aligned} x - y + z &= 3 \\ 3x + y + 2z &= 15 \\ 2x - y + 2z &= 7 \end{aligned}$

47. $\begin{aligned} x - y - z &= 2 \\ -2x + 2y + z &= -5 \\ -3x + 3y + z &= -10 \end{aligned}$

48. $\begin{aligned} x - y &= 3 \\ 2y + z &= 5 \\ x + 2z &= 7 \end{aligned}$

49. $\begin{aligned} x + y - z &= -1 \\ x - y - 2z &= 2 \\ -5x - y - z &= -1 \end{aligned}$

50. $\begin{aligned} 2x + 3y + z &= 7 \\ -2x - 9y + 2z &= 1 \\ 4x - 6y + 3z &= 10 \end{aligned}$

Solve each problem by choosing a variable to represent each unknown quantity.

51. **Number Problem** The sum of three numbers is 15. The largest number is 4 times the smallest number, and it is also 1 more than the sum of the other two numbers. Find the three numbers.

52. **Number Problem** The sum of the digits of a three-digit number is 16. The tens digit is 3 times the hundreds digit, and the units digit is 1 more than the hundreds digit. What is the number?

53. **Business and Finance** An amateur theater has orchestra tickets at $10, box seat tickets at $7, and balcony tickets at $5. For one performance, a total of 360 tickets was sold, and the total revenue was $3,040. If the number of orchestra tickets sold was 40 more than that of the other two types combined, how many of each type of ticket were sold for the performance?

54. **Business and Finance** Rachel divided $12,000 into three investments: a savings account paying 5%, a stock paying 7%, and a mutual fund paying 9%. Her annual interest from the investments was $800, and the amount that she had invested at 5% was equal to the sum of the amounts invested in the other accounts. How much did she have invested in each type of account?

4.5 *Solve each system of linear inequalities.*

55. $x - y < 7$
$x + y > 3$

56. $x - 2y \le -2$
$x + 2y \le 6$

57. $x - 6y < 6$
$-x + y < 4$

58. $2x + y \le 8$
$x \ge 1$
$y \ge 0$

59. $2x + y \le 6$
$x \ge 1$
$y \ge 0$

60. $4x + y \le 8$
$x \ge 0$
$y \ge 2$

61. $4x + 2y \le 8$
$x + y \le 3$
$x \ge 0$
$y \ge 0$

62. $3x + y \le 6$
$x + y \le 4$
$x \ge 0$
$y \ge 0$

CHAPTER 4

chapter test 4

Use this chapter test to assess your progress and to review for your next exam. Allow yourself about an hour to take this test. The answers to these exercises are in the Answers Appendix in the back of the text.

Solve each equation graphically.

1. $4x - 7 = 5$

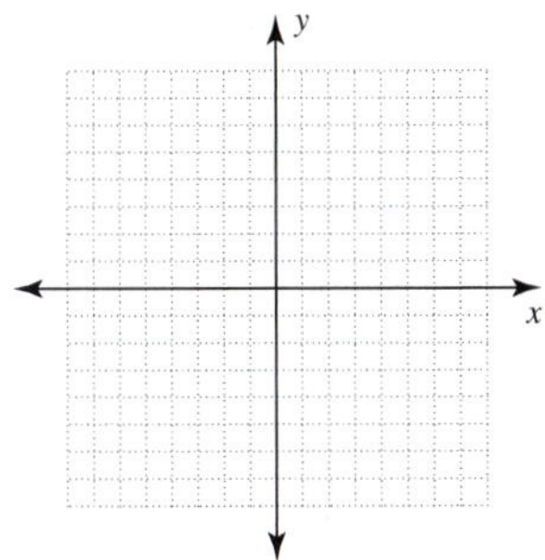

2. $6 - x = 4(x - 1)$

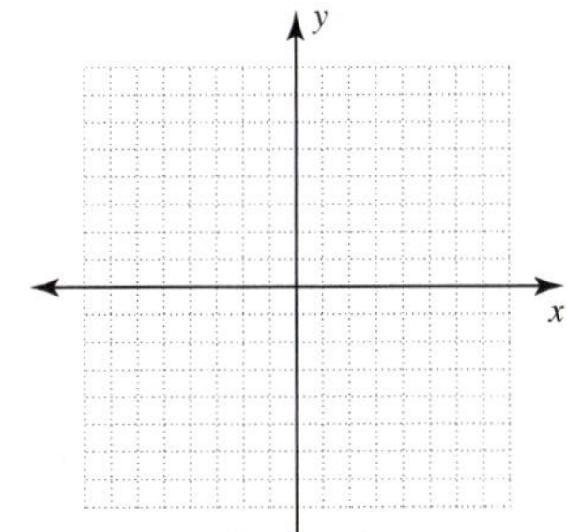

3. $-8x + 11 = 2x - 9$

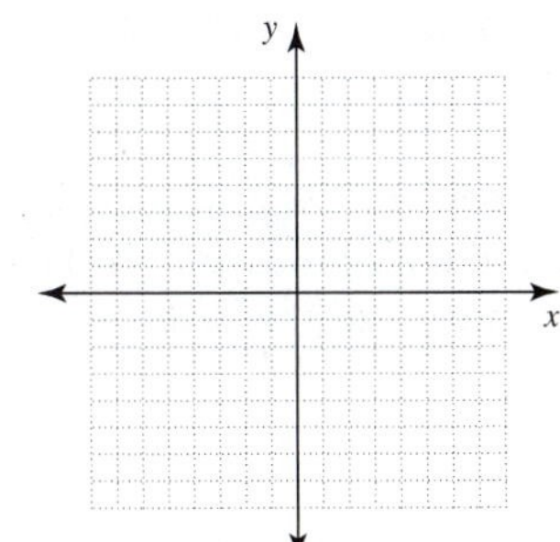

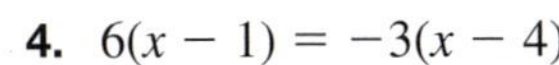

4. $6(x - 1) = -3(x - 4)$

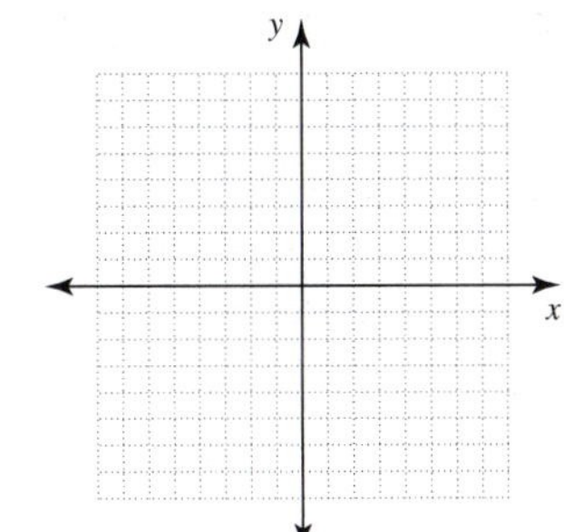

Solve each system. If a unique solution does not exist, state whether the given system is inconsistent or dependent.

5. $\begin{aligned} 3x + y &= -5 \\ 5x - 2y &= -23 \end{aligned}$

6. $\begin{aligned} 4x - 2y &= -10 \\ y &= 2x + 5 \end{aligned}$

7. $\begin{aligned} 9x - 3y &= 4 \\ -3x + y &= -1 \end{aligned}$

8. $\begin{aligned} 5x - 3y &= 5 \\ 3x + 2y &= -16 \end{aligned}$

9. $\begin{aligned} x - 2y &= 5 \\ 2x + 5y &= 10 \end{aligned}$

10. $\begin{aligned} 5x - 3y &= 20 \\ 4x + 9y &= -3 \end{aligned}$

Solve each system.

11. $\begin{aligned} x - y + z &= 1 \\ -2x + y + z &= 8 \\ x + 5z &= 19 \end{aligned}$

12. $\begin{aligned} x + 3y - 2z &= -6 \\ 3x - y + 2z &= 8 \\ -2x + 3y - 4z &= -11 \end{aligned}$

Solve each system of linear inequalities.

13. $\begin{aligned} x - 2y &< 6 \\ x + y &< 3 \end{aligned}$

14. $\begin{aligned} 3x + 4y &\ge 12 \\ x &\ge 1 \end{aligned}$

15. $\begin{aligned} x + 2y &\le 8 \\ x + y &\le 6 \\ x &\ge 0 \\ y &\ge 0 \end{aligned}$

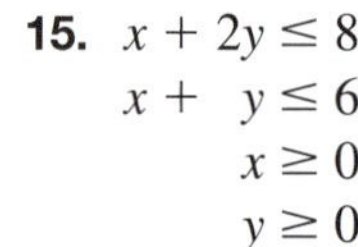

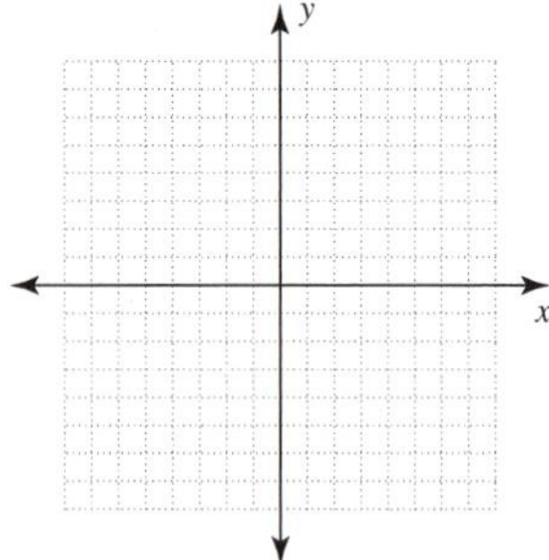

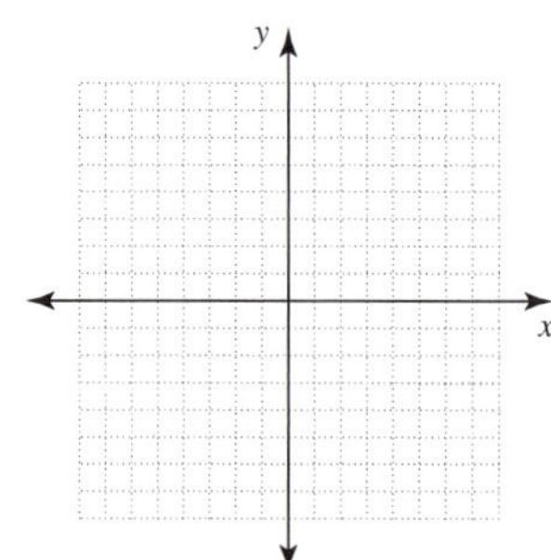

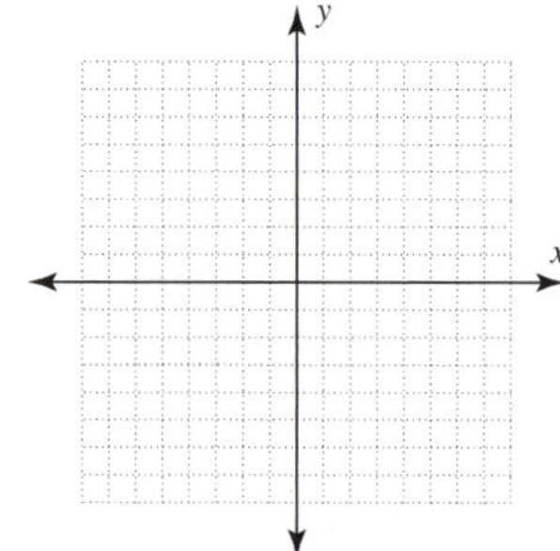

Solve each application by choosing a variable to represent each unknown quantity. Then write a system of equations that will allow you to solve for each variable.

16. **Business and Finance** An order for 30 USB cables and 12 mouse pads totaled \$147. A second order for 12 more cables and 6 additional pads cost \$66. What was the cost per USB cable and mouse pad?

17. **Business and Finance** A bulk candy wholesaler wants to combine jawbreakers selling for \$2.40 per pound and licorice selling for \$3.90 per pound to form a 100-lb mixture that will sell for \$3 per pound. How many pounds of each type of candy should the wholesaler use?

18. **Manufacturing Technology** An electronics firm assembles LCD monitors and standard monitors. LCD monitors require 9 hr of assembly time while standard models need 6 hr each. Both types also require 5 hr for packaging and testing. There are a total of 72 hr available for assembly and 50 hr for packaging and testing each week. How many of each type of set should be finished if the firm wants to use all of the available labor?

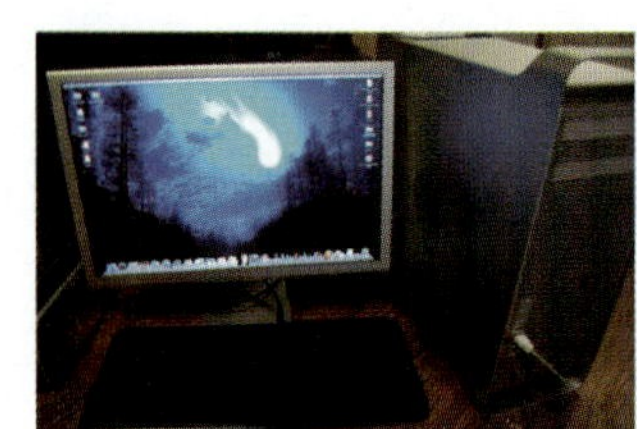

19. **Business and Finance** Hans decided to divide \$16,000 into three investments: a savings account paying 3% annual interest, a bond paying 5%, and a mutual fund paying 7%. His annual interest from the three investments was \$900, and he has as much in the mutual fund as in the savings account and bond combined. What amount did he invest in each type?

20. **Construction Technology** The fence around a rectangular yard requires 260 ft of fencing. The length is 20 ft less than twice the width. Find the dimensions of the yard.

cumulative review chapters 0–4

Use this exercise set to review concepts from earlier chapters. While it is not a comprehensive exam, it will help you identify any material that you need to review before moving on to the next chapter. The answers to these exercises are in the Answers Appendix in the back of the text.

Solve.

1. $3x - 2(x + 5) = 12 - 3x$

2. $2x - 7 < 3x - 5$

3. $x + 8 \geq 4x - 3$

4. $2x + 3(x - 2) = -4(x + 1) + 16$

Graph.

5. $5x + 7y = 35$

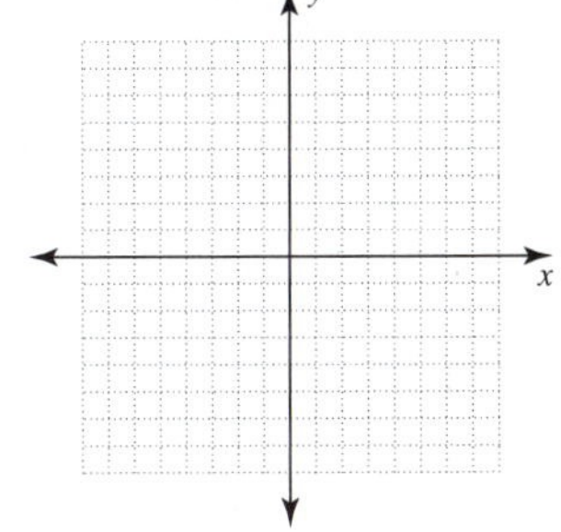

6. $2x + 3y < 6$

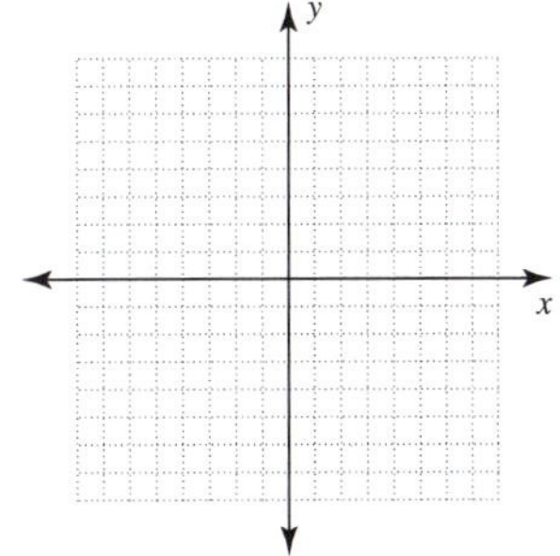

7. Solve the equation $P = P_0 + IRT$ for R.

8. Find the slope of the line connecting (4, 6) and (3, −1).

9. Write an equation of the line that passes through the points (−1, 4) and (5, −2).

10. Write an equation of the line passing through the point (3, 2) and parallel to the line $4x - 5y = 20$.

11. Find $f(-5)$ if $f(x) = 3x^2 - 4x - 5$.

Solve each system of equations.

12. $2x + 3y = 6$
$5x + 3y = -24$

13. $x + y + z = 3$
$2x - y + 2z = 0$
$-x - 3y + z = -9$

Solve each application.

14. GEOMETRY The length of a rectangle is 3 cm more than twice its width. If the perimeter of the rectangle is 54 cm, find the dimensions of the rectangle.

15. NUMBER PROBLEM The sum of the digits of a two-digit number is 10. If the digits are reversed, the number is 36 less than the original number. What was the original number?

CHAPTER

5

Exponents and Polynomials

INTRODUCTION

People in business, finance, industry, and many academic disciplines use polynomial and exponential equations to solve problems and make predictions. We emphasize investment and business applications in this chapter's exercises and activity.

We evaluate investment strategies by estimating the future value of different choices. Although the future value of any investment is subject to many variables, mathematical models allow investors to compare different investment options.

You will explore such powerful ideas as compound interest and savings by gaining experience working with polynomials and exponents.

CHAPTER 5 OUTLINE

5.1

Positive Integer Exponents

< 5.1 Objectives >

1 > Use exponential notation

2 > Simplify exponential expressions

Take a sheet of paper, $8\frac{1}{2}$ by 11 in., and cut it in half. Stack the pieces together and then cut this stack again in half. You should now have four pieces. If you continue this process, stacking and cutting, you double the number of pieces with each cut.

Cut Number	Number of Pieces
1	2
2	4
3	8
4	16
5	32
6	64
7	128
8	256

We can compute the height of the stack, for a given number of cuts, if we know the thickness of the original paper. Assuming the thickness to be $\frac{1}{500}$ in. = 0.002 in., the height after 8 cuts is

$$(256)(0.002) = 0.512$$

or a bit more than one-half inch.

If you actually try this, you know that it becomes *very* difficult to make even 8 cuts, and that the pieces become *very* small. If we continue to make cuts, we get a surprising result. How high do you think the stack would be if we could make 16 cuts? The number of pieces would be

$$\underbrace{2 \cdot 2 \cdot 2 \cdots 2}_{\text{16 times}}$$

which you can verify to be 65,536 pieces. Then the height would be

$$(65{,}536)(0.002) \approx 131 \text{ in.}$$

which is almost 11 ft high!

The calculations done in the paper-cutting exercise are best represented with **exponents.** The way in which the number of pieces (and the height of the stack) grows is often called **exponential growth.**

We first reviewed exponential notation in Section 0.5, but we give a brief review here. Exponents are a shorthand form for writing repeated multiplication. Instead of writing

$$2 \cdot 2 \cdot 2 \cdot 2 \cdot 2 \cdot 2 \cdot 2$$

we write

$$2^7$$

RECALL

We call *a* the **base** of the expression and 5 the **exponent** or **power.**

Instead of writing

$a \cdot a \cdot a \cdot a \cdot a$

we write

a^5

which we read as "*a* to the fifth power."

Definition

Exponential Expressions

For any real number *a* and any natural number *n*,

$$a^n = \underbrace{a \cdot a \cdots a}_{n \text{ factors}}$$

An expression of this type is written in **exponential notation.**

Example 1 Using Exponential Notation

< Objective 1 >

Write each expression using exponential notation.

(a) $w \cdot w \cdot w \cdot w = w^4$

(b) $5y \cdot 5y \cdot 5y = (5y)^3$

Check Yourself 1

Write each expression using exponential notation.

(a) $3z \cdot 3z \cdot 3z \cdot 3z$ **(b)** $x \cdot x \cdot x \cdot x \cdot x \cdot x$

Now consider what happens when we multiply two exponential expressions with the same base.

NOTE

We expand the expressions and then remove the parentheses.

$$a^4 \cdot a^5 = \underbrace{(a \cdot a \cdot a \cdot a)}_{4 \text{ factors}}\underbrace{(a \cdot a \cdot a \cdot a \cdot a)}_{5 \text{ factors}}$$

$$= \underbrace{a \cdot a \cdot a \cdot a \cdot a \cdot a \cdot a \cdot a \cdot a}_{9 \text{ factors}}$$

$$= a^9$$

The product is simply the original base taken to the power that is the sum of the two original exponents. This leads to our first property of exponents.

Property

Product Rule for Exponents

For any real number *a* and positive integers *m* and *n*,

$$a^m \cdot a^n = \underbrace{(a \cdot a \cdots a)}_{m \text{ factors}}\underbrace{(a \cdot a \cdots a)}_{n \text{ factors}}$$

$$= \underbrace{a \cdot a \cdots a}_{m + n \text{ factors}}$$

$$= a^{m+n}$$

The product of two exponential expressions with the same base is that base raised to the sum of the powers.

NOTE

Our first property of exponents:

$a^m \cdot a^n = a^{m+n}$

Example 2 illustrates the product rule for exponents.

Example 2 Using the Product Rule

< Objective 2 >

NOTE

In every case, the base stays the same.

Simplify each expression.

(a) $b^4 \cdot b^6 = b^{4+6} = b^{10}$

(b) $(2a)^3 \cdot (2a)^4 = (2a)^{3+4} = (2a)^7$

(c) $(-2)^5(-2)^4 = (-2)^{5+4} = (-2)^9$

(d) $(10^7)(10^{11}) = 10^{7+11} = 10^{18}$

Check Yourself 2

Simplify each expression.

(a) $(5b)^6(5b)^5$ **(b)** $(-3)^4(-3)^3$ **(c)** $10^8 \cdot 10^{12}$ **(d)** $(xy)^2(xy)^3$

Applying the commutative and associative properties of multiplication, we know that a product such as

$2x^3 \cdot 3x^2$

can be rewritten as

$(2 \cdot 3)(x^3 \cdot x^2)$

or as

$6x^5$

We expand on these ideas in Example 3.

Example 3 Using Properties of Exponents

RECALL

Multiply the coefficients and add the exponents by the product rule. With practice, you will not need to write the regrouping step.

Using the product rule for exponents together with the commutative and associative properties, simplify each expression.

(a) $(x^4)(x^2)(x^3)(x) = x^{10}$ $x = x^1$; adding exponents gives $4 + 2 + 3 + 1 = 10$.

(b) $(3x^4)(5x^2) = (3 \cdot 5)(x^4 \cdot x^2) = 15x^6$

(c) $(2x^5y)(9x^3y^4) = (2 \cdot 9)(x^5 \cdot x^3)(y \cdot y^4) = 18x^8y^5$

(d) $(-3x^2y^2)(-2x^4y^3) = (-3)(-2)(x^2 \cdot x^4)(y^2 \cdot y^3) = 6x^6y^5$

Check Yourself 3

Simplify each expression.

(a) $(x)(x^5)(x^3)$ **(b)** $(7x^5)(2x^2)$

(c) $(-2x^3y)(x^2y^2)$ **(d)** $(-5x^3y^2)(-x^2y^3)$

Now consider the quotient

$$\frac{a^6}{a^4}$$

If we write this in expanded form, we have

$$\frac{\overbrace{a \cdot a \cdot a \cdot a \cdot a \cdot a}^{\text{6 factors}}}{\underbrace{a \cdot a \cdot a \cdot a}_{\text{4 factors}}}$$

This can be simplified to

$$\frac{\overset{1}{\cancel{a}} \cdot \overset{1}{\cancel{a}} \cdot \overset{1}{\cancel{a}} \cdot \overset{1}{\cancel{a}} \cdot a \cdot a}{\underset{1}{\cancel{a}} \cdot \underset{1}{\cancel{a}} \cdot \underset{1}{\cancel{a}} \cdot \underset{1}{\cancel{a}}} \qquad \text{or} \qquad a^2$$

NOTE

Divide the numerator and denominator by the four common factors of *a*.

This means that

$$\frac{a^6}{a^4} = a^2$$

This leads to our second property of exponents.

Property

Quotient Rule for Exponents

For any nonzero real number *a*,

$$\frac{a^m}{a^n} = a^{m-n}$$

The quotient of two exponential expressions with the same base is that base raised to the difference of the powers.

Example 4 illustrates this rule.

Example 4 Using Properties of Exponents

Simplify each expression.

RECALL

$a^1 = a$; there is no need to write the exponent 1 because it is understood.

(a) $\dfrac{x^{10}}{x^4} = x^{10-4} = x^6$ Subtract the exponents, applying the quotient rule.

(b) $\dfrac{a^8}{a^7} = a^{8-7} = a$

(c) $\dfrac{63w^8}{7w^5} = 9w^{8-5} = 9w^3$ *Divide* the coefficients and subtract the exponents.

(d) $\dfrac{-32a^4b^5}{8a^2b} = -4a^{4-2}b^{5-1} = -4a^2b^4$ Divide the coefficients and subtract the exponents for *each* variable.

(e) $\dfrac{10^{16}}{10^6} = 10^{16-6} = 10^{10}$

Check Yourself 4

Simplify each expression.

(a) $\dfrac{y^{12}}{y^5}$ **(b)** $\dfrac{x^9}{x^8}$ **(c)** $\dfrac{45r^8}{-9r^6}$ **(d)** $\dfrac{49a^6b^7}{7ab^3}$ **(e)** $\dfrac{10^{13}}{10^5}$

What happens when a product such as xy is raised to a power? Consider, for example, $(xy)^4$:

$$(xy)^4 = (xy)(xy)(xy)(xy)$$
$$= (x \cdot x \cdot x \cdot x)(y \cdot y \cdot y \cdot y)$$ We use the commutative and associative properties.
$$= x^4y^4$$

This is expressed in the product-power rule.

Property

Product-Power Rule for Exponents

For any real numbers a and b,

$(ab)^n = a^nb^n$

To raise a product to a power, raise each factor to that power.

Example 5 Using the Product-Power Rule

Simplify each expression.

(a) $(2x)^3 = 2^3x^3 = 8x^3$

(b) $(-3x)^4 = (-3)^4x^4 = 81x^4$

Check Yourself 5

Simplify each expression.

(a) $(3x)^3$ **(b)** $(-2x)^4$

Now, consider the expression

$(3^2)^3$

This can be expanded to

$(3^2)(3^2)(3^2)$

and then expanded again to

$$(3 \cdot 3) \cdot (3 \cdot 3) \cdot (3 \cdot 3) = 3 \cdot 3 \cdot 3 \cdot 3 \cdot 3 \cdot 3$$
$$= 3^6$$

This leads to the power rule for exponents.

Property

Power Rule for Exponents

For any real number a,

$(a^m)^n = a^{mn}$

To raise an exponential expression to a power, raise the base to the product of the powers.

Example 6 Using the Power Rule for Exponents

NOTE

Part (c) also uses the product-power rule.

Simplify each expression.

(a) $(2^5)^3 = 2^{15}$

(b) $(x^2)^4 = x^8$

(c) $(2x^3)^3 = 2^3(x^3)^3 = 2^3x^9 = 8x^9$

Check Yourself 6

Simplify each expression.

(a) $(3^4)^3$ **(b)** $(x^2)^6$ **(c)** $(3x^3)^4$

We have one final exponent property to develop. Suppose we have a quotient raised to a power.

$$\left(\frac{x}{2}\right)^3 = \frac{x}{2} \cdot \frac{x}{2} \cdot \frac{x}{2} = \frac{x \cdot x \cdot x}{2 \cdot 2 \cdot 2} = \frac{x^3}{2^3}$$

The power 3 applies to both the numerator x and to the denominator 2. This gives our fifth property of exponents.

Property

Quotient-Power Rule for Exponents

For any real numbers a and b, where b is not equal to 0,

$$\left(\frac{a}{b}\right)^m = \frac{a^m}{b^m}$$

To raise a quotient to a power, raise the numerator and denominator to that same power.

Example 7 illustrates this property. Again, we may also have to apply the other properties when simplifying an expression.

Example 7 Using the Quotient-Power Rule for Exponents

Simplify each expression.

(a) $\left(\frac{3}{4}\right)^3 = \frac{3^3}{4^3} = \frac{27}{64}$

(b) $\left(\frac{x^3}{y^2}\right)^4 = \frac{(x^3)^4}{(y^2)^4} = \frac{x^{12}}{y^8}$

(c) $\left(\frac{r^2s^3}{t^4}\right)^2 = \frac{(r^2s^3)^2}{(t^4)^2} = \frac{(r^2)^2(s^3)^2}{(t^4)^2} = \frac{r^4s^6}{t^8}$

Check Yourself 7

Simplify each expression.

(a) $\left(\frac{2}{3}\right)^4$ **(b)** $\left(\frac{m^3}{n^4}\right)^5$ **(c)** $\left(\frac{a^2b^3}{c^5}\right)^2$

Here is a summary of the five properties of exponents discussed in this section.

Property	General Form	Example
Product Rule	$a^m a^n = a^{m+n}$	$x^2 \cdot x^3 = x^5$
Quotient Rule	$\frac{a^m}{a^n} = a^{m-n}$	$\frac{5^7}{5^3} = 5^4$
Power Rule	$(a^m)^n = a^{mn}$	$(z^5)^4 = z^{20}$
Product-Power Rule	$(ab)^m = a^m b^m$	$(4x)^3 = 4^3x^3 = 64x^3$
Quotient-Power Rule	$\left(\frac{a}{b}\right)^m = \frac{a^m}{b^m}$	$\left(\frac{2}{3}\right)^6 = \frac{2^6}{3^6} = \frac{64}{729}$

Check Yourself ANSWERS

1. (a) $(3z)^4$; **(b)** x^6 **2. (a)** $(5b)^{11}$; **(b)** $(-3)^7$; **(c)** 10^{20}; **(d)** $(xy)^5$ **3. (a)** x^9; **(b)** $14x^7$; **(c)** $-2x^5y^3$; **(d)** $5x^5y^5$ **4. (a)** y^7; **(b)** x; **(c)** $-5r^2$; **(d)** $7a^5b^4$; **(e)** 10^8 **5. (a)** $27x^3$; **(b)** $16x^4$ **6. (a)** 3^{12}; **(b)** x^{12}; **(c)** $81x^{12}$ **7. (a)** $\frac{16}{81}$; **(b)** $\frac{m^{15}}{n^{20}}$; **(c)** $\frac{a^4b^6}{c^{10}}$

Reading Your Text

These fill-in-the-blank exercises will help you understand some of the key vocabulary used in this section. The answers to these exercises are in the Answers Appendix in the back of the text.

(a) Exponents are a shorthand form for writing repeated ______________.

(b) An expression of the type a^n is said to be in ______________ form.

(c) When using the product rule for exponents, we multiply the coefficients and ______________ the exponents.

(d) To raise a quotient to a power, raise the numerator and ______________ to that same power.

5.1 exercises

Skills Calculator/Computer Career Applications Above and Beyond

< Objectives 1 and 2 >

Write each expression in simplest exponential form.

1. $x^4 \cdot x^5$

2. $x^7 \cdot x^9$

3. $x^5 \cdot x^3 \cdot x^2$

4. $x^8 \cdot x^4 \cdot x^7$

5. $3^5 \cdot 3^2$

6. $(-3)^4(-3)^6$

7. $(-2)^3(-2)^5$

8. $4^3 \cdot 4^4$

9. $4 \cdot x^2 \cdot x^4 \cdot x^7$

10. $3 \cdot x^3 \cdot x^5 \cdot x^8$

11. $\left(\frac{1}{2}\right)^2\left(\frac{1}{2}\right)^3\left(\frac{1}{2}\right)$

12. $\left(-\frac{1}{3}\right)^4\left(-\frac{1}{3}\right)\left(-\frac{1}{3}\right)^5$

13. $(-2)^2(-2)^3(x^4)(x^5)$

14. $(-3)^4(-3)^2(x)^2(x)^6$

15. $(2x)^2(2x)^3(2x)^4$

16. $(-3x)^3(-3x)^5(-3x)^7$

17. $(x^2y^3)(x^4y^2)$

18. $(x^4y)(x^2y^3)$

19. $(x^3y^2)(x^4y^2)(x^2y^3)$

20. $(x^2y^3)(x^3y)(x^4y^2)$

21. $(2x^4)(3x^3)(-4x^3)$
22. $(2x^3)(-3x)(-4x^4)$
23. $(5x^2)(3x^3)(x)(-2x^3)$
24. $(4x^2)(2x)(x^2)(2x^3)$
25. $(5xy^3)(2x^2y)(3xy)$
26. $(-3xy)(5x^2y)(-2x^3y^2)$ VIDEO
27. $(x^2yz)(x^3y^5z)(x^4yz)$
28. $(xyz)(x^8y^3z^6)(x^2yz)(xyz^4)$
29. $\frac{x^{10}}{x^7}$
30. $\frac{b^{23}}{b^{18}}$
31. $\frac{x^7y^{11}}{x^4y^3}$
32. $\frac{x^5y^9}{xy^4}$
33. $\frac{x^5y^4z^2}{xy^2z}$
34. $\frac{x^8y^6z^4}{x^3yz^3}$
35. $\frac{21x^4y^5}{7xy^2}$
36. $\frac{48x^6y^6}{12x^3y}$ VIDEO
37. $(-3x)(5x^5)$
38. $(-5x^2)(-2x^2)$
39. $(2x)^3$
40. $(-3x)^3$
41. $(x^3)^7$
42. $(-x^3)^5$
43. $(3x)(-2x)^3$
44. $(2x)(-3x)^3$
45. $(2x^3)^5$
46. $(-3x^2)^3$
47. $(-2x^2)^3(3x^2)^3$ VIDEO
48. $(-3x^2)^2(5x^2)^2$
49. $(3x^3)^2(x^2)^4$
50. $(2x^3)^4(3x^4)^2$
51. $\left(\frac{3}{4}\right)^2$
52. $\left(\frac{2}{3}\right)^2$
53. $\left(\frac{x}{5}\right)^3$
54. $\left(\frac{a}{2}\right)^4$
55. $\left(\frac{m^3}{n^2}\right)^3$
56. $\left(\frac{a^4}{b^3}\right)^4$
57. $\left(\frac{a^3b^2}{c^4}\right)^2$
58. $\left(\frac{x^5y^2}{z^4}\right)^3$
59. $\left(\frac{2x^5}{y^3}\right)^2$
60. $\left(\frac{2x^5}{3x^3}\right)^3$
61. $(-8x^2y)(-3x^4y^5)^4$
62. $(5x^5y)^2(-3x^3y^4)^3$
63. $\left(\frac{3x^4y^9}{2x^2y^7}\right)\left(\frac{x^6y^3}{x^3y^2}\right)^2$
64. $\left(\frac{6x^5y^4}{5xy}\right)\left(\frac{x^3y^5}{xy^3}\right)^3$

Determine whether each statement is **true** *or* **false.**

65. Exponents are a shorthand form for writing repeated addition.
66. If we multiply x^m by x^n, we can write x raised to the $m + n$ power.
67. When we divide x^m by x^n, we can simply divide the exponents.
68. If we raise x^m to the t power, we can write x raised to the mt power.

Skills | **Calculator/Computer** | Career Applications | Above and Beyond

As you might expect, you can use a calculator to evaluate exponential expressions.

Some calculators have a special *squared* key [x^2], which squares (exponent equals 2) the quantity that precedes it.

Calculators also have a general exponent key, which can be used for any exponent. On graphing calculators, you can find a *caret* key [^]. On scientific calculators, there is an x to the y key [x^y] or a y to the x key [y^x]. In any case, enter the base, enter the general exponent key, and then enter the exponent. If you do mathematics on a computer, you can find the caret above the 6 key (Shift+6).

To evaluate 5^4,

Graphing Calculator

5 [^] 4 [ENTER]

Scientific Calculator

5 [x^y] 4 [=]

We must be careful taking a negative number to a power. Just as we saw in Chapter 1, you must enclose the negative number in parentheses. Remember that with a scientific calculator, we enter the negative sign after the number, whereas with a graphing calculator, we enter the negative sign before the number.

To evaluate $(-3)^5$,

Graphing Calculator

((−) 3) ^ 5 ENTER

Scientific Calculator

(3 +/−) x^y 5 =

If you are using a graphing calculator, your screen may look like one of those shown (depending on the model).

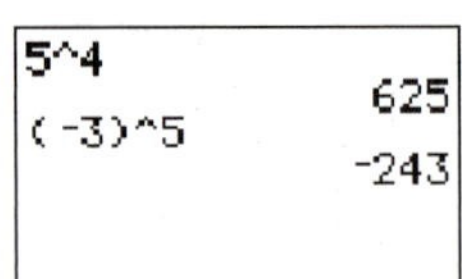

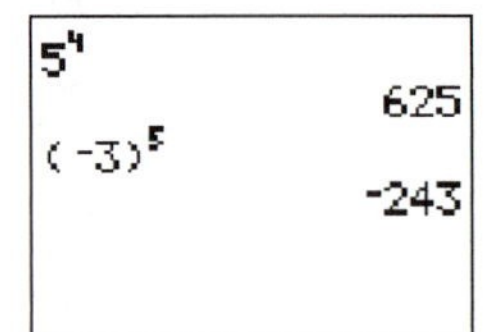

Use a calculator to evaluate each expression.

69. 4^3

70. 5^7

71. $(-3)^4$

72. $(-4)^5$

73. $2^3 \cdot 2^5$

74. $3^4 \cdot 3^6$

75. $(3x^2)(2x^4)$, if $x = 2$

76. $(4x^3)(5x^4)$, if $x = 3$

77. $(2x^4)(4x^2)$, if $x = -2$

78. $(3x^5)(2x^3)$, if $x = -3$

79. $(-2x^3)(-3x^5)$, if $x = 2$

80. $(-3x^2)(-4x^4)$, if $x = 4$

Skills | Calculator/Computer | **Career Applications** | Above and Beyond

81. **Business and Finance** The value A of a savings account that compounds interest annually is given by the formula

$$A = P(1 + r)^t$$

where

P = Original amount (principal)

r = Interest rate, in decimal form

t = Time, in years

Find the amount of money in the account after 8 years if \$2,000 was invested initially at 5% compounded annually.

82. **Business and Finance** Using the formula for compound interest in exercise 81, determine the amount of money in the account if the original investment is doubled. chapter 5 > Make the Connection

83. **Mechanical Engineering** The kinetic energy (in joules) of a falling object is given by

$$\text{KE} = \frac{1}{2}mv^2$$

in which m is the mass of the object and v is its velocity.

The velocity (in meters per second) of a falling object is given by $v = 4.9t^2$, in which t represents the time since the object was dropped.

(a) Write an equation for the kinetic energy of a 12-kg object in terms of the time since it was dropped.

(b) What is the kinetic energy of a 12-kg object 4 s after it is dropped?

84. **Construction Technology** The load on a post is given by $P = 3L^2$, in which L is the length of the post, in inches. The change in the length (in inches) of the post when loaded is given by

$$\text{Contraction} = \frac{PL}{28{,}000{,}000}$$

(a) Express the contraction of a post in terms of its length.

(b) Report, to the nearest thousandth, the contraction of a post if its length is 120 in.

85. **Mechanical Engineering** The required depth of a beam is equal to four times the cube of its length. The moment of inertia of a 4-in.-wide beam is equal to $\frac{1}{9}$ of the cube of the depth of the beam. Express the moment of inertia of the beam in terms of its length.

86. **Mechanical Engineering** In a cantilevered beam, the specifications call for the span L of the beam to equal the square of the length of the cantilever c. The bending moment in the beam can be expressed as

$$M = \frac{wL^2}{8}$$

Express the bending moment M in terms of w and c. 

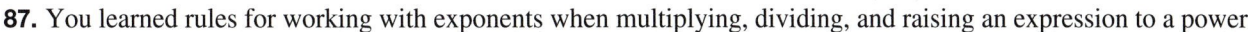

Skills	Calculator/Computer	Career Applications	**Above and Beyond**

87. You learned rules for working with exponents when multiplying, dividing, and raising an expression to a power.

(a) Explain each rule in your own words. Give numerical examples.

(b) Is there a rule for raising a *sum* to a power? That is, does $(a + b)^n = a^n + b^n$? Use numerical examples to explain why this is true in general or why it is not. Is it always true or always false?

88. Work with another student to investigate the rate of inflation. The annual rate of inflation was about 2.3% from 2001 to 2010. This means that the value of the goods that you could buy for \$100 in 2001 would cost 2.3% more in 2002, 2.3% more than that in 2003, and so forth. If a movie ticket cost \$10 in 2001, what would it cost today if movie tickets just kept up with inflation? Construct a table to solve the problem. chapter 5 > Make the Connection

Solve each problem.

89. Write x^{12} as a power of x^2.

90. Write y^{15} as a power of y^3. 

91. Write a^{16} as a power of a^2.

92. Write m^{20} as a power of m^5.

93. Write each expression as a power of 8 (remember that $8 = 2^3$): 2^{12}, 2^{18}, $(2^5)^3$, $(2^7)^6$.

94. Write each expression as a power of 9: 3^8, 3^{14}, $(3^5)^8$, $(3^4)^7$.

95. What expression, raised to the third power, is $-8x^6y^9z^{15}$?

96. What expression, raised to the fourth power, is $81x^{12}y^8z^{16}$?

The formula $(1 + R)^y = G$ gives us useful information about the growth of a population. Here R is the rate of growth expressed as a decimal, y is the time in years, and G is the growth factor. It takes 35 years for a country's population to double if its growth rate is 2%.

$(1.02)^{35} \approx 2$

97. **Social Science**

(a) With a 2% growth rate, how many doublings occur in 105 years? How much larger will the country's population be to the nearest whole number?

(b) The less-developed countries of the world had an average growth rate of 2% in 2010. If their total population was 5.9 billion, what will their population be in 105 years if this rate remains unchanged?

98. **Social Science** The United States has a growth rate of 0.7%. What will be its growth factor after 35 years (to the nearest percent)?

99. Write an explanation of why $(x^3)(x^4)$ is *not* x^{12}.

100. Your algebra study partners are confused. "Why isn't $x^2 \cdot x^3 = 2x^5$?" they ask you. Write an explanation that will convince them.

Answers

1. x^9 **3.** x^{10} **5.** 3^7 **7.** $(-2)^8$ **9.** $4x^{13}$ **11.** $\left(\frac{1}{2}\right)^6$ **13.** $(-2)^5x^9$ **15.** $(2x)^9$ **17.** x^6y^5 **19.** x^9y^7 **21.** $-24x^{10}$ **23.** $-30x^9$ **25.** $30x^4y^5$ **27.** $x^9y^7z^3$ **29.** x^3 **31.** x^3y^8 **33.** x^4y^2z **35.** $3x^3y^3$ **37.** $-15x^6$ **39.** $8x^3$ **41.** x^{21} **43.** $-24x^4$ **45.** $32x^{15}$ **47.** $-216x^{12}$ **49.** $9x^{14}$ **51.** $\frac{9}{16}$ **53.** $\frac{x^3}{125}$ **55.** $\frac{m^9}{n^6}$ **57.** $\frac{a^6b^4}{c^8}$ **59.** $\frac{4x^{10}}{y^6}$ **61.** $-648x^{18}y^{21}$ **63.** $\frac{3x^8y^4}{2}$ **65.** False **67.** False **69.** 64 **71.** 81 **73.** 256 **75.** 384 **77.** 512 **79.** 1,536 **81.** $2,954.91 **83.** **(a)** KE = $144.06t^4$; **(b)** 36,879.36 joules **85.** $M = \frac{64}{9}L^9$ **87.** **(a)** Above and Beyond; **(b)** Above and Beyond **89.** $(x^2)^6$ **91.** $(a^2)^8$ **93.** 8^4; 8^6; 8^5; 8^{14} **95.** $-2x^2y^3z^5$ **97.** **(a)** Three doublings, 8 times as large; **(b)** 47.2 billion **99.** Above and Beyond

Activity 5 ::

Wealth and Compound Interest

Suppose that when you were born, an uncle put \$500 in the bank for you. He never deposited money again, but the bank paid 5% interest on the money every year on your birthday. How much money was in the bank after 1 year? After 2 years? After 1 year, the amount is \$500 + 500(0.05), which can be written as \$500(1 + 0.05) because of the distributive property. Because 1 + 0.05 = 1.05, the amount in the bank after 1 year was 500(1.05). After 2 years, this amount was again multiplied by 1.05. How much is in the bank today? Complete the chart.

Birthday	Computation	Amount
0 (Day of birth)		\$500
1	\$500(1.05)	
2	\$500(1.05)(1.05)	
3	\$500(1.05)(1.05)(1.05)	
4	$\$500(1.05)^4$	
5	$\$500(1.05)^5$	
6		
7		
8		

(a) Write a formula for the amount in the bank on your *n*th birthday. About how many years does it take for the money to double? How many years does it take for it to double again? Can you see any connection between this and the rules for exponents? Explain why you think there may or may not be a connection.

(b) If the account earned 6% each year, how much more would it accumulate by the end of year 8? Year 21?

(c) Imagine that you start an Individual Retirement Account (IRA) at age 20, contributing \$2,500 each year for 5 years (total \$12,500) to an account that produces a return of 8% every year. You stop contributing and let the account grow. Using the information from the previous example, calculate the value of the account at age 65.

(d) Imagine that you don't start the IRA until you are 30. In an attempt to catch up, you invest \$2,500 into the same account, 8% annual return, each year for 10 years. You then stop contributing and let the account grow. What will its value be at age 65?

(e) What have you discovered as a result of these computations?

5.2 Integer Exponents and Scientific Notation

< 5.2 Objectives >

1 > Use zero as an exponent

2 > Simplify expressions with negative exponents

3 > Use scientific notation to write numbers

4 > Solve applications involving scientific notation

In Section 5.1, we reviewed some properties of exponents, but all of the exponents were positive integers. In this section, we look at zero and negative exponents. First, we extend the quotient rule so that we can define a zero exponent.

Recall the quotient rule: to divide two expressions that have the same base, we keep the base and subtract the exponents.

NOTE

We must have $a \neq 0$. The form 0^0 is called **indeterminate** and is considered in later mathematics classes.

$$\frac{a^m}{a^n} = a^{m-n} \qquad \text{if } a \neq 0$$

Now, suppose that we allow *m to equal n*. We then have

$$\frac{a^m}{a^m} = a^{m-m} = a^0 \qquad \text{if } a \neq 0$$

But we know that any nonzero number divided by itself is one, so

$$\frac{a^m}{a^m} = 1 \qquad \text{if } a \neq 0$$

Comparing these two equations, we see that the next definition is reasonable.

Definition

Zero Exponent	For any nonzero real number a, $a^0 = 1$

Throughout this section we assume all variables represent nonzero real numbers.

Example 1 The Zero Exponent

< Objective 1 >

Use the definition of the zero exponent to simplify each expression.

NOTE

In $6x^0$, the exponent 0 applies *only* to x.

(a) $17^0 = 1$

(b) $(a^3b^2)^0 = 1$

(c) $6x^0 = 6 \cdot 1 = 6$

(d) $-3y^0 = -3$

Check Yourself 1

Simplify each expression.

(a) 25^0 **(b)** $(m^4n^2)^0$ **(c)** $8s^0$ **(d)** $-7t^0$

NOTE

John Wallis (1616–1702), an English mathematician, was the first to fully discuss the meaning of 0, negative, and rational exponents. We study rational exponents in Chapter 7.

When multiplying exponential expressions with the same base, the product rule says to keep the base and add the exponents.

$a^m \cdot a^n = a^{m+n}$

Now, what if we allow one of the exponents to be negative and apply the product rule? Suppose, for instance, that $m = 3$ and $n = -3$. Then

$$a^m \cdot a^n = a^3 \cdot a^{-3} = a^{3+(-3)}$$
$$= a^0 = 1$$

so $\quad a^3 \cdot a^{-3} = 1$

RECALL

If the product of two numbers is 1, they are reciprocals of each other.

If we divide both sides by a^3, we have

$a^{-3} = \dfrac{1}{a^3}$

This is the basis for a definition of negative exponents.

Definition

Negative Integer Exponents

For any nonzero real number a,

$a^{-n} = \dfrac{1}{a^n}$

and a^{-n} is the **multiplicative inverse** of a^n.

We can also say that a^{-n} is the reciprocal of a^n. Example 2 illustrates this definition.

Example 2 **Using Properties of Exponents**

< Objective 2 >

NOTE

From this point on, to *simplify* means to write the expression with *positive exponents only*.

We restrict all variables so that they represent nonzero real numbers.

Simplify each expression.

(a) $y^{-5} = \dfrac{1}{y^5}$

(b) $4^{-2} = \dfrac{1}{4^2} = \dfrac{1}{16}$

(c) $(-3)^{-3} = \dfrac{1}{(-3)^3} = \dfrac{1}{-27} = -\dfrac{1}{27}$

Check Yourself 2

Simplify each expression.

(a) a^{-10} **(b)** 2^{-4} **(c)** $(-4)^{-2}$

Example 3 illustrates the case where coefficients besides 1 appear in an expression with negative exponents. As will be clear, we must be cautious and determine exactly what is included in the base of the exponent.

Example 3 Using Properties of Exponents

Simplify each expression.

(a) $2x^{-3} = 2 \cdot \frac{1}{x^3} = \frac{2}{x^3}$

The exponent -3 applies only to the variable x, and *not* to the coefficient 2.

(b) $4w^{-2} = 4 \cdot \frac{1}{w^2} = \frac{4}{w^2}$

(c) $(4w)^{-2} = \frac{1}{(4w)^2} = \frac{1}{16w^2}$

> CAUTION

The expressions $4w^{-2}$ and $(4w)^{-2}$ are *not* the same. Do you see why?

Check Yourself 3

Simplify each expression.

(a) $3w^{-4}$ **(b)** $10x^{-5}$ **(c)** $(2y)^{-4}$ **(d)** $-5t^{-2}$

Suppose that a variable with a negative exponent appears in the denominator of an expression. For instance, if we wish to simplify

$$\frac{1}{a^{-2}}$$

we can multiply numerator and denominator by a^2.

$$\frac{1}{a^{-2}} = \frac{1 \cdot a^2}{a^{-2} \cdot a^2} = \frac{a^2}{a^0} = \frac{a^2}{1} = a^2$$

Negative exponent in denominator — Positive exponent in numerator

So

$$\frac{1}{a^{-2}} = a^2$$

This leads to a property for negative exponents.

NOTE

$a^{-2} = \frac{1}{a^2}$

so

$\frac{1}{a^{-2}} = \frac{1}{\frac{1}{a^2}}$

We invert and multiply:

$\frac{1}{\frac{1}{a^2}} = 1 \div \frac{1}{a^2} = 1 \cdot \frac{a^2}{1} = a^2$

Property

Negative Exponents

For any nonzero real number a and integer n,

$$\frac{1}{a^{-n}} = a^n$$

Example 4 Using Properties of Exponents

Simplify each expression.

(a) $\frac{1}{y^{-3}} = y^3$

(b) $\frac{1}{2^{-5}} = 2^5 = 32$

(c) $\frac{3}{4x^{-2}} = \frac{3x^2}{4}$ The exponent -2 applies only to x, not to 4.

(d) $\frac{a^{-3}}{b^{-4}} = \frac{b^4}{a^3}$

Check Yourself 4

Simplify each expression.

(a) $\frac{1}{x^{-4}}$ **(b)** $\frac{1}{3^{-3}}$ **(c)** $\frac{2}{3a^{-2}}$ **(d)** $\frac{c^{-5}}{d^{-7}}$

RECALL

You can review these properties in Section 5.1.

The product and quotient rules for exponents apply to expressions that involve any integer exponent—positive, negative, or 0.

Example 5 Using Properties of Exponents

Simplify each expression. Use only positive exponents to express the result.

(a) $x^3 \cdot x^{-7} = x^{3+(-7)}$ Add the exponents by the product rule.

$= x^{-4} = \frac{1}{x^4}$

(b) $\frac{m^{-5}}{m^{-3}} = m^{-5-(-3)} = m^{-5+3}$ Subtract the exponents by the quotient rule.

$= m^{-2} = \frac{1}{m^2}$

(c) $\frac{x^5x^{-3}}{x^{-7}} = \frac{x^{5+(-3)}}{x^{-7}} = \frac{x^2}{x^{-7}} = x^{2-(-7)} = x^9$ We apply first the product rule and then the quotient rule.

Check Yourself 5

Simplify each expression.

(a) $x^9 \cdot x^{-5}$ **(b)** $\frac{y^{-7}}{y^{-3}}$ **(c)** $\frac{a^{-3}a^2}{a^{-5}}$

How do we simplify a rational expression raised to a negative power? As we have seen, the properties of exponents extend to negative exponents.

Suppose we wish to simplify $\left(\frac{x}{y}\right)^{-2}$.

$\left(\frac{x}{y}\right)^{-2} = \frac{x^{-2}}{y^{-2}}$ Use the quotient-power rule.

$= \frac{y^2}{x^2} = \left(\frac{y}{x}\right)^2$ Use the quotient-power rule again.

Property

Quotient Raised to a Negative Power

For any nonzero real number *a* and *b*,

$\left(\frac{a}{b}\right)^{-n} = \left(\frac{b}{a}\right)^n$

Example 6 Extending the Properties of Exponents

Simplify each expression.

(a) $\left(\frac{s^3}{t^2}\right)^{-2} = \left(\frac{t^2}{s^3}\right)^2$ Use the negative power property to begin.

$= \frac{(t^2)^2}{(s^3)^2}$ Use the quotient-power rule.

$= \frac{t^4}{s^6}$ Multiply the exponents.

NOTE

You can complete this problem in several ways. For instance, you could move n^{-2} to the denominator in the second step.

$$\left(\frac{n^{-2}}{m^2}\right)^3 = \left(\frac{1}{m^2n^2}\right)^3 = \frac{1^3}{(m^2n^2)^3} = \frac{1}{m^6n^6}$$

(b) $\left(\dfrac{m^2}{n^{-2}}\right)^{-3} = \left(\dfrac{n^{-2}}{m^2}\right)^3$ Use the negative power property to begin.

$= \dfrac{(n^{-2})^3}{(m^2)^3}$ Use the quotient-power rule.

$= \dfrac{n^{-6}}{m^6}$ Multiply the exponents.

$= \dfrac{1}{m^6n^6}$ Your final answer should have positive exponents only.

(c) $\left(\dfrac{3}{q^5}\right)^{-2} = \left(\dfrac{q^5}{3}\right)^2$

$= \dfrac{(q^5)^2}{(3)^2}$

$= \dfrac{q^{10}}{9}$

Check Yourself 6

Simplify each expression.

(a) $\left(\dfrac{a^4}{b^3}\right)^{-3}$ **(b)** $\left(\dfrac{x^5}{y^{-2}}\right)^{-4}$ **(c)** $\left(\dfrac{r^4}{5}\right)^{-2}$ **(d)** $\left(\dfrac{3t^2}{s^3}\right)^{-3}$

As you might expect, simplifying more complicated expressions often requires more than one property. Example 7 illustrates such cases.

Example 7 Using Properties of Exponents

Simplify each expression.

(a) $\dfrac{(a^2)^{-3}(a^3)^4}{(a^{-3})^3} = \dfrac{a^{-6} \cdot a^{12}}{a^{-9}}$ Apply the power rule to each factor.

$= \dfrac{a^{-6+12}}{a^{-9}} = \dfrac{a^6}{a^{-9}}$ Apply the product rule.

$= a^{6-(-9)} = a^{6+9} = a^{15}$ Apply the quotient rule.

(b) $\dfrac{8x^{-2}y^{-5}}{12x^{-4}y^3} = \dfrac{8}{12} \cdot \dfrac{x^{-2}}{x^{-4}} \cdot \dfrac{y^{-5}}{y^3}$ It helps to separate this expression into three fractions.

$= \dfrac{2}{3} \cdot x^{-2-(-4)} \cdot y^{-5-3}$

$= \dfrac{2}{3} \cdot x^2 \cdot y^{-8} = \dfrac{2x^2}{3y^8}$

>CAUTION

A *common error* is to write

$$\frac{8x^{-2}y^{-5}}{12x^{-4}y^3} = \frac{12x^4}{8x^2y^3y^5}$$

This is *not* correct.

Alternatively, we could *correctly* begin

$$\frac{8x^{-2}y^{-5}}{12x^{-4}y^3} = \frac{8x^4}{12x^2y^3y^5}$$

The coefficients should not be moved along with the variables. Keep in mind that the exponents apply *only* to the variables in this expression. The coefficients remain *where they were* in the original expression when the expression is rewritten using this approach.

(c) $\left(\dfrac{pr^3s^{-5}}{p^3r^{-3}s^{-2}}\right)^{-2} = \left(\dfrac{p^3r^{-3}s^{-2}}{pr^3s^{-5}}\right)^2$ Apply the negative part of the outside exponent first.

$= \left(\dfrac{p^3}{p} \cdot \dfrac{r^{-3}}{r^3} \cdot \dfrac{s^{-2}}{s^{-5}}\right)^2$ Write the inside as the product of three fractions.

$$= \left(p^2 \cdot \frac{1}{r^6} \cdot s^3\right)^2$$ Simplify each fraction separately.

$$= \left(\frac{p^2s^3}{r^6}\right)^2$$ Combine the fractions by multiplying.

$$= \frac{p^4s^6}{r^{12}}$$ Complete the problem by applying the outside exponent.

Check Yourself 7

Simplify each expression.

(a) $\dfrac{(x^5)^{-2}(x^2)^3}{(x^{-4})^3}$ **(b)** $\dfrac{12a^{-3}b^{-2}}{16a^{-2}b^3}$ **(c)** $\left(\dfrac{xy^{-3}z^{-5}}{x^{-4}y^{-2}z^3}\right)^{-3}$

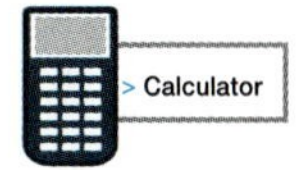


```
2300*1000
       2300000
Ans*1000
    2300000000
Ans*1000
        2.3E12
```

We now look at an important use of exponents, scientific notation.

We begin the discussion with a calculator exercise. On most calculators, if you multiply 2.3 times 1,000, the display reads

2300

Multiply by 1,000 a second time. Now you should see

2300000

Multiplying by 1,000 a third time results in the display

2.3 + 09 or 2.3E09 or 2300000000

And multiplying by 1,000 again yields

2.3 + 12 or 2.3E12

NOTE

Consider the table:

$2.3 = 2.3 \times 10^0$
$23 = 2.3 \times 10^1$
$230 = 2.3 \times 10^2$
$2{,}300 = 2.3 \times 10^3$
$23{,}000 = 2.3 \times 10^4$
$230{,}000 = 2.3 \times 10^5$

Can you see what is happening? This is the way calculators display very large numbers. The number on the left is always between 1 and 10, and the number on the right indicates the number of places the decimal point must be moved to the right to put the answer in standard (or decimal) form.

This notation is used frequently in science. It is not uncommon in scientific applications of algebra to find yourself working with very large or very small numbers. Even in the time of Archimedes (287–212 B.C.), the study of such numbers was not unusual. Archimedes estimated that the universe was 23,000,000,000,000,000 m in diameter, which is the approximate distance light travels in $2\frac{1}{2}$ years. By comparison, Polaris (the North Star) is 680 light-years from Earth. We discuss light years in Example 9.

NOTE

Scientific notation is one of the few places where we still use the multiplication symbol ×.

In scientific notation, Archimedes' estimate for the diameter of the universe is

2.3×10^{16} m

We define scientific notation as follows.

Definition

Scientific Notation

Any positive number written in the form

$a \times 10^n$

in which $1 \le a < 10$ and n is an integer, is written in **scientific notation.**

RECALL

$10^0 = 1$, so
$2.3 \times 10^0 = 2.3 \times 1$
$= 2.3$

When a number is written in scientific notation, we look at the sign of the exponent to determine if the number is large or small. If the exponent is nonnegative, then the number is greater than or equal to one. If the exponent is negative, then the number is less than one.

$2.3 \times 10^2 = 230$	$2.3 \times 10^{-2} = 0.023$
$2.3 \times 10^1 = 23$	$2.3 \times 10^{-1} = 0.23$
$2.3 \times 10^0 = 2.3$	

Example 8 Using Scientific Notation

< Objective 3 >

NOTES

Study the pattern for writing a number in scientific notation.

The exponent shows the *number of places* we move the decimal point so that the multiplier is a number between 1 and 10.

To convert back to standard or decimal form, we simply reverse the process.

Write each number in scientific notation.

(a) $120{,}000. = 1.2 \times 10^5$
5 places — The power is 5.

(b) $88{,}000{,}000. = 8.8 \times 10^7$
7 places — The power is 7.

(c) $520{,}000{,}000. = 5.2 \times 10^8$
8 places

(d) $4{,}000{,}000{,}000. = 4 \times 10^9$
9 places

(e) $0.0005 = 5 \times 10^{-4}$
4 places

(f) $0.0000000081 = 8.1 \times 10^{-9}$
9 places

Check Yourself 8

Write in scientific notation.

(a) 212,000,000,000,000,000 **(b)** 0.00079
(c) 5,600,000 **(d)** 0.0000007

Example 9 An Application of Scientific Notation

< Objective 4 >

(a) Light travels at approximately 3×10^8 meters per second (m/s). There are about 3.15×10^7 s in a year. How far does light travel in a year?

We multiply the distance traveled in 1 s by the number of seconds in a year. This yields

$$(3 \times 10^8)(3.15 \times 10^7) = (3 \cdot 3.15)(10^8 \cdot 10^7)$$
$$= 9.45 \times 10^{15}$$

Multiply the coefficients and add the exponents.

NOTE

$9.45 \times 10^{15} \approx 10 \times 10^{15}$
$= 10^{16}$

For our purposes we round the distance light travels in 1 year to 10^{16} m. This unit is called a **light-year,** and it is used to measure astronomical distances.

(b) The distance from Earth to the star Spica (in Virgo) is 2.2×10^{18} m. How many light-years is Spica from Earth?

NOTE

We divide the distance (in meters) by the number of meters in 1 light-year.

$$\frac{2.2 \times 10^{18}}{10^{16}} = 2.2 \times 10^{18-16}$$
$$= 2.2 \times 10^2 = 220 \text{ light-years}$$

Check Yourself 9

The farthest object that can be seen with the unaided eye is the Andromeda galaxy. This galaxy is 2.3×10^{22} m from Earth. What is this distance in light-years?

Check Yourself ANSWERS

1. **(a)** 1; **(b)** 1; **(c)** 8; **(d)** -7 **2.** **(a)** $\frac{1}{a^{10}}$; **(b)** $\frac{1}{16}$; **(c)** $\frac{1}{16}$ **3.** **(a)** $\frac{3}{w^4}$; **(b)** $\frac{10}{x^5}$; **(c)** $\frac{1}{16y^4}$; **(d)** $-\frac{5}{t^2}$

4. **(a)** x^4; **(b)** 27; **(c)** $\frac{2a^2}{3}$; **(d)** $\frac{d^7}{c^5}$ **5.** **(a)** x^4; **(b)** $\frac{1}{y^4}$; **(c)** a^4 **6.** **(a)** $\frac{b^9}{a^{12}}$; **(b)** $\frac{1}{x^{20}y^8}$; **(c)** $\frac{25}{r^8}$; **(d)** $\frac{s^9}{27t^6}$

7. **(a)** x^8; **(b)** $\frac{3}{4ab^5}$; **(c)** $\frac{y^3z^{24}}{x^{15}}$ **8.** **(a)** 2.12×10^{17}; **(b)** 7.9×10^{-4}; **(c)** 5.6×10^6; **(d)** 7×10^{-7}

9. 2,300,000 or 2.3×10^6 light-years

Reading Your Text

These fill-in-the-blank exercises will help you understand some of the key vocabulary used in this section. The answers to these exercises are in the Answers Appendix in the back of the text.

(a) When multiplying exponential expressions with the same base, the product rule says to keep the base and __________ the exponents.

(b) a^{-n} is the multiplicative inverse, or __________, of a^n.

(c) When a number is written in scientific notation and the exponent is nonnegative, then the number is greater than or equal to __________.

(d) Light travels at an approximate speed of 3×10^8 __________ per second.

Skills | Calculator/Computer | Career Applications | Above and Beyond

5.2 exercises

< Objectives 1 and 2 >

Simplify each expression.

1. x^{-5} **2.** 3^{-3} **3.** 5^{-2} **4.** x^{-8}

5. $(-5)^{-2}$ **6.** $(-3)^{-3}$ **7.** $(-2)^{-3}$ **8.** $(-2)^{-4}$

9. $\left(\frac{2}{3}\right)^{-3}$ **10.** $\left(\frac{3}{4}\right)^{-2}$ **11.** $3x^{-2}$ **12.** $4x^{-3}$

13. $-5x^{-4}$ **14.** $(-2x)^{-4}$ **15.** $(-3x)^{-2}$ **16.** $-5x^{-2}$

17. $\frac{1}{x^{-3}}$ **18.** $\frac{1}{x^{-5}}$ **19.** $\frac{2}{5x^{-3}}$ **20.** $\frac{3}{4x^{-4}}$

21. $\frac{x^{-3}}{y^{-4}}$ **22.** $\frac{x^{-5}}{y^{-3}}$ **23.** $x^5 \cdot x^{-3}$ **24.** $y^{-4} \cdot y^5$

25. $a^{-9} \cdot a^6$ **26.** $w^{-5} \cdot w^3$ **27.** $z^{-2} \cdot z^{-8}$ **28.** $b^{-7} \cdot b^{-1}$

29. $a^{-5} \cdot a^5$ **30.** $x^{-4} \cdot x^4$ **31.** $\frac{x^{-5}}{x^{-2}}$ VIDEO **32.** $\frac{x^{-3}}{x^{-6}}$

33. $(x^5)^3$ **34.** $(w^4)^6$ **35.** $(2x^{-3})(x^2)^4$ **36.** $(p^4)(3p^3)^2$

37. $(3a^{-4})(a^3)(a^2)$ **38.** $(5y^{-2})(2y)(y^5)$ **39.** $(x^4y)(x^2)^3(y^3)^0$ **40.** $(r^4)^2(r^2s)(s^3)^2$

41. $(ab^2c)(a^4)^4(b^2)^3(c^3)^4$
42. $(p^2qr^2)(p^2)(q^3)^2(r^2)^0$
43. $(x^5)^{-3}$
44. $(x^{-2})^{-3}$
45. $(b^{-4})^{-2}$
46. $(a^0b^{-4})^3$
47. $(x^5y^{-3})^2$
48. $(p^{-3}q^2)^{-2}$
49. $(x^{-4}y^{-2})^{-3}$
50. $(3x^{-2}y^{-2})^3$
51. $(2x^{-3}y^0)^{-5}$
52. $\dfrac{a^{-6}}{b^{-4}}$
53. $\dfrac{x^{-2}}{y^{-4}}$
54. $\left(\dfrac{x^{-3}}{y^2}\right)^{-3}$ VIDEO
55. $\dfrac{x^{-4}}{y^{-2}}$
56. $\dfrac{(3x^{-4})^2(2x^2)}{x^6}$
57. $(4x^{-2})^2(3x^{-4})$
58. $(5x^{-4})^{-4}(2x^3)^{-5}$
59. $(2x^5)^4(x^3)^2$
60. $(3x^2)^3(x^2)^4(x^2)$
61. $(2x^{-3})^3(3x^3)^2$
62. $(x^2y^3)^4(xy^3)^0$
63. $(xy^5z)^4(xyz^2)^8(x^6yz)^5$
64. $(x^2y^2z^2)^0(xy^2z)^2(x^3yz^2)$
65. $(3x^{-2})(5x^2)^2$
66. $(2a^3)^2(a^0)^5$
67. $(2w^3)^4(3w^{-5})^2$
68. $(3x^3)^2(2x^4)^5$
69. $\dfrac{3x^6}{2y^9} \cdot \dfrac{y^5}{x^3}$
70. $\dfrac{x^8}{y^6} \cdot \dfrac{2y^9}{x^3}$
71. $(-7x^2y)(-3x^5y^6)^4$
72. $\left(\dfrac{2w^5z^3}{3x^3y^9}\right)\left(\dfrac{x^5y^4}{w^4z^0}\right)^2$
73. $(2x^2y^{-3})(3x^{-4}y^{-2})$
74. $(-5a^{-2}b^{-4})(2a^5b^0)$
75. $\dfrac{(x^{-3})(y^2)}{y^{-3}}$
76. $\dfrac{6x^3y^{-4}}{24x^{-2}y^{-2}}$
77. $\dfrac{15x^{-3}y^2z^{-4}}{20x^{-4}y^{-3}z^2}$
78. $\dfrac{24x^{-5}y^{-3}z^2}{36x^{-2}y^3z^{-2}}$
79. $\dfrac{x^{-5}y^{-7}}{x^0y^{-4}}$
80. $\left(\dfrac{xy^3z^{-4}}{x^{-3}y^{-2}z^2}\right)^{-2}$
81. $\dfrac{x^{-2}y^2}{x^3y^{-2}} \cdot \dfrac{x^{-4}y^2}{x^{-2}y^{-2}}$ VIDEO
82. $\left(\dfrac{x^{-3}y^3}{x^{-4}y^2}\right)^3 \cdot \left(\dfrac{x^{-2}y^{-2}}{xy^4}\right)^{-1}$

< **Objective 3** >

Write each expression in standard notation.

83. 9×10^4
84. 8.23×10^3
85. 1.21×10^1
86. 3.815×10^2
87. 8×10^{-3}
88. 7.5×10^{-6}
89. 2.8×10^{-5}
90. 5.21×10^{-4}

Write each number in scientific notation.

91. 3,420,000
92. 4,000,000,000
93. 8
94. 40
95. 0.0005
96. 0.000003
97. 0.00037
98. 0.000051

Express each number in scientific notation.

99. The average distance from Earth to the sun: 93,000,000 mi. 
100. The diameter of a grain of sand: 0.000021 m.
101. The diameter of the sun: 130,000,000,000 cm.
102. The number of oxygen atoms in 16 grams of oxygen gas: 602,000,000,000,000,000,000,000 (Avogadro's number).
103. The mass of the sun is approximately 1.98×10^{30} kg. If this number is written in standard or decimal form, how many 0's follow the digit 8?

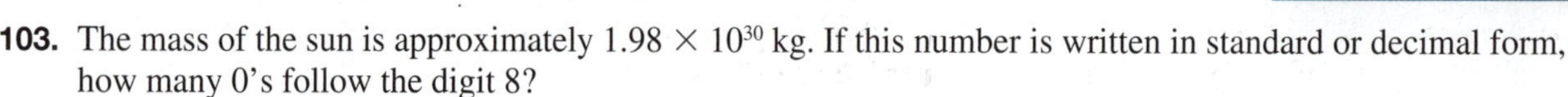

104. Archimedes estimated the universe to be 2.3×10^{19} mm in diameter. If this number is written in standard form, how many 0's follow the digit 3?

Evaluate each expression and write your result in scientific notation.

105. $(2 \times 10^5)(4 \times 10^4)$

106. $(2.5 \times 10^7)(3 \times 10^5)$

107. $\dfrac{6 \times 10^9}{3 \times 10^7}$

108. $\dfrac{4.5 \times 10^{12}}{1.5 \times 10^7}$

109. $\dfrac{(3.3 \times 10^{15})(6 \times 10^{15})}{(1.1 \times 10^8)(3 \times 10^6)}$

VIDEO

110. $\dfrac{(6 \times 10^{12})(3.2 \times 10^8)}{(1.6 \times 10^7)(3 \times 10^2)}$

111. $(4 \times 10^{-3})(2 \times 10^{-5})$

112. $(1.5 \times 10^{-6})(4 \times 10^2)$

113. $\dfrac{9 \times 10^3}{3 \times 10^{-2}}$

114. $\dfrac{7.5 \times 10^{-4}}{1.5 \times 10^2}$

< Objective 4 >

115. **Science and Medicine** Megrez, the nearest of the Big Dipper stars, is 6.6×10^{17} m from Earth. Approximately how long does it take light, traveling at 10^{16} m/yr, to travel from Megrez to Earth?

116. **Science and Medicine** Alkaid, the most distant star in the Big Dipper, is 2.1×10^{18} m from Earth. Approximately how long does it take light to travel from Alkaid to Earth?

117. **Science and Medicine** The approximate amount of water on Earth is 15,500 followed by 19 zeros, in liters. Write this number in scientific notation.

118. **Science and Medicine** Use your result from exercise 117 to find the amount of water per person on Earth if there are approximately 7.1 billion living people.

119. **Science and Medicine** If there are 7.1×10^9 people on Earth and there is enough freshwater to provide each person with 6.57×10^5 L, how much freshwater is on Earth?

120. **Science and Medicine** The United States uses an average of 2.6×10^6 L of water per person each year. If there are 3.1×10^8 people in the United States, how many liters of water does the United States use each year?

Determine whether each statement is **true** *or* **false.**

121. Any number raised to the zero power is zero.

122. 5^{-3} represents the multiplicative inverse of 5^3.

123. A fraction raised to the power $-n$ can be rewritten as the reciprocal fraction raised to the power n.

124. If we rewrite $3x^{-4}$ as a fraction, both the 3 and the x appear in the denominator.

Skills | **Calculator/Computer** | Career Applications | Above and Beyond

Social Science In 2010, the population of Earth was approximately 7 billion and doubling every 45 years. The formula for the population P (in billions) in year y for this doubling rate is

$P = 7 \times 2^{(y-2010)/45}$

125. According to the model, what will Earth's population be in 2025? Round your result to two decimal places.

126. According to the model, what was Earth's population in 1950? Round your result to two decimal places.

Social Science In 2010, the population of the United States was approximately 310 million and doubling every 60 years. The formula for the population P (in millions) in year y for this doubling rate is

$$P = 310 \times 2^{(y-2010)/60}$$

127. According to the model, what will the U.S. population be in 2025? Round your result to the nearest million.

128. According to the model, what was the U.S. population in 1980? Round your result to the nearest million.

Skills | Calculator/Computer | **Career Applications** | Above and Beyond

129. **Electronics** The resistance in a circuit is measured to be $12 \times 10^4\ \Omega$.

(a) Express the resistance in standard notation.

(b) Express the resistance in scientific notation.

130. **Information Technology** The distance from the ground to a satellite in geostationary orbit is 42,245 km. How long will it take a light beam to reach the satellite (use scientific notation to express your result)?

Recall: Light travels at approximately 3×10^8 m/s.

Allied Health Medical lab technicians can determine the concentration c of a solution, in moles per liter (mol/L), based on the absorbance A, which is a measure of the amount of light that passes through the solution, the molar absorption coefficient ε, and the length of the light path d, in centimeters, in the colorimeter using the formula

$$c = \frac{A}{\varepsilon d}$$

Use this information to complete exercises 131 and 132.

131. Determine the concentration of a solution with an absorbance of 0.254 mol/L, a molar absorption coefficient of 3.6×10^3, and a 1.4-cm light path. Use scientific notation to express your result.

132. Determine the concentration of a solution with an absorbance of 0.315 mol/L, a molar absorption coefficient of 8.2×10^6, and an 18-mm light path. Use scientific notation to express your result.

Simplify each expression.

133. $x^{2n} \cdot x^{3n}$

134. $x^{n+1} \cdot x^{3n}$

135. $\frac{x^{n+3}}{x^{n+1}}$

136. $\frac{x^{n-4}}{x^{n-1}}$

137. $(y^n)^{3n}$

138. $(x^{n+1})^n$

139. $\frac{x^{2n} \cdot x^{n+2}}{x^{3n}}$

140. $\frac{x^n \cdot x^{3n+5}}{x^{4n}}$

141. Recall the paper "cut and stack" experiment described at the beginning of this chapter. If you had a piece of paper large enough to complete the task (and assuming that you *could* complete the task!), determine the height of the stack, given that the thickness of the paper is 0.002 in.

(a) After 30 cuts

(b) After 50 cuts

142. Can $(a + b)^{-1}$ be written as $\frac{1}{a} + \frac{1}{b}$ by using the properties of exponents? If not, why not? Explain.

143. Write a short description of the difference between $(-4)^{-3}$, -4^{-3}, $(-4)^3$, and -4^3. Are any of these equal?

144. If $n > 0$, which expressions are negative?

$$-n^{-3} \quad n^{-3} \quad (-n)^{-3} \quad (-n)^3 \quad -n^3$$

If $n < 0$, which of these expressions are negative? Explain what effect a negative in the exponent has on the sign of the result when an exponential expression is simplified.

145. You are offered a 28-day job in which you have a choice of two different pay arrangements. Plan 1 offers a flat \$4,000,000 at the end of the 28th day on the job. Plan 2 offers 1¢ the first day, 2¢ the second day, 4¢ the third day, and so on, with the amount doubling each day. Make a table to decide which offer is better. Write a formula for the amount you make on the nth day and a formula for the total after n days. Which pay arrangement should you take? Why?

Answers

1. $\frac{1}{x^5}$ **3.** $\frac{1}{25}$ **5.** $\frac{1}{25}$ **7.** $-\frac{1}{8}$ **9.** $\frac{27}{8}$ **11.** $\frac{3}{x^2}$ **13.** $-\frac{5}{x^4}$ **15.** $\frac{1}{9x^2}$ **17.** x^3 **19.** $\frac{2x^3}{5}$ **21.** $\frac{y^4}{x^3}$ 23. x^2 25. $\frac{1}{a^3}$ **27.** $\frac{1}{z^{10}}$ **29.** 1 **31.** $\frac{1}{x^3}$ **33.** x^{15} **35.** $2x^5$ **37.** $3a$ **39.** $x^{10}y$ **41.** $a^{17}b^8c^{13}$ **43.** $\frac{1}{x^{15}}$ **45.** b^8 **47.** $\frac{x^{10}}{y^6}$ 49. $x^{12}y^6$ **51.** $\frac{x^{15}}{32}$ **53.** $\frac{y^4}{x^2}$ **55.** $\frac{y^2}{x^4}$ **57.** $\frac{48}{x^8}$ **59.** $16x^{26}$ **61.** $\frac{72}{x^3}$ **63.** $x^{42}y^{33}z^{25}$ **65.** $75x^2$ **67.** $144w^2$ **69.** $\frac{3x^3}{2y^4}$ **71.** $-567x^{22}y^{25}$ **73.** $\frac{6}{x^2y^5}$ **75.** $\frac{y^5}{x^3}$ **77.** $\frac{3xy^5}{4z^6}$ **79.** $\frac{1}{x^5y^3}$ **81.** $\frac{y^8}{x^7}$ **83.** 90,000 **85.** 12.1 **87.** 0.008 **89.** 0.000028 **91.** 3.42×10^6 **93.** 8×10^0 **95.** 5×10^{-4} **97.** 3.7×10^{-4} **99.** 9.3×10^7 **101.** 1.3×10^{11} **103.** 28 **105.** 8×10^9 **107** 2×10^2 **109.** 6×10^{16} **111.** 8×10^{-8} **113.** 3×10^5 **115.** 66 yr **117.** 1.55×10^{23} **119.** 4.66×10^{15} L **121.** False **123.** True **125.** 8.82 billion **127.** 369 million **129.** **(a)** 120,000 Ω; **(b)** 1.2×10^5 Ω **131.** $c \approx 5.04 \times 10^{-5}$ **133.** x^{5n} **135.** x^2 **137.** y^{3n^2} **139.** x^2 **141.** **(a)** About 33.9 mi; **(b)** about 35.5 million mi **143.** Above and Beyond **145.** $p = \frac{2^{n-1}}{100}$; $t = \frac{2^n}{100} - 0.01$

5.3 An Introduction to Polynomials

< 5.3 Objectives >

1 > Classify polynomials

2 > Find the degree of a polynomial

3 > Write polynomials in descending order

Our work in this section deals with the most common kind of algebraic expression, a *polynomial.* To define a polynomial, first recall the definition of the word *term.*

Definition

Term A **term** can be written as a number or the product of a number and one or more variables and their exponents.

NOTE

In a polynomial, terms are separated by + and − signs.

For example, x^5, $3x$, $-4xy^2$, 8, $\frac{5}{x}$, and $-14\sqrt{x}$ are terms. A **polynomial** consists of one or more terms in which the only allowable exponents for the variables are the whole numbers, 0, 1, 2, 3, and so on. These terms are connected by addition or subtraction signs. The variable is never used as a divisor and never appears under a radical sign in a term of a polynomial.

Definition

Numerical Coefficient In each term of a polynomial, the number factor is called the **numerical coefficient,** or more simply the **coefficient,** of that term.

Example 1 Identifying Polynomials

NOTE

Each sign (+ or −) is attached to the term that *follows* that sign.

(a) $x + 3$ is a polynomial. The terms are x and 3. The coefficients are 1 and 3.

(b) $3x^2 - 2x + 5$ is also a polynomial. Its terms are $3x^2$, $-2x$, and 5. The coefficients are 3, −2, and 5.

(c) $5x^3 + 2 - \frac{3}{x}$ is *not* a polynomial because of the division by x in the third term.

Check Yourself 1

Which are polynomials?

(a) $5x^2$ **(b)** $3y^3 - 2y + \frac{5}{y}$ **(c)** $4x^2 - 2x + 3$

NOTE

The prefix *mono* means 1. The prefix *bi* means 2. The prefix *tri* means 3. There are no special names for polynomials with more than three terms.

Certain polynomials are given special names because of the number of terms that they have.

Definition

Monomial, Binomial, and Trinomial

A polynomial with exactly one term is called a **monomial.**

A polynomial with exactly two terms is called a **binomial.**

A polynomial with exactly three terms is called a **trinomial.**

Example 2 Classifying Polynomials

< Objective 1 >

(a) $3x^2y$ is a monomial. It has exactly one term.

(b) $2x^3 + 5x$ is a binomial. It has exactly two terms, $2x^3$ and $5x$.

(c) $5x^2 - 4x + 3$ is a trinomial. Its three terms are $5x^2$, $-4x$, and 3.

Check Yourself 2

Classify each as a monomial, binomial, or trinomial.

(a) $5x^2 - 2x^3$ **(b)** $4x^7$ **(c)** $2x^2 + 5x - 3$

RECALL

In a polynomial, the allowable exponents of the variable are the whole numbers 0, 1, 2, 3, and so on. The degree is a whole number.

We also classify polynomials by their *degree.* The **degree** of a polynomial that has only one variable is the highest power of that variable appearing in any one term.

Example 3 Finding the Degree of a Polynomial

< Objective 2 >

(a) $5x^3 - 3x^2 + 4x$ has degree 3. (The highest power)

(b) $4x - 5x^4 + 3x^3 + 2$ has degree 4. (The highest power)

(c) $8x$ has degree 1 because $8x = 8x^1$.

(d) 7 has degree 0 because $7 = 7 \cdot 1 = 7x^0$.

RECALL

In Section 5.2, you learned that $x^0 = 1$.

Check Yourself 3

Find the degree of each polynomial.

(a) $6x^5 - 3x^3 - 2$ **(b)** $5x$ **(c)** $3x^3 + 2x^6 - 1$ **(d)** 9

Polynomials can have more than one variable, such as $4x^2y^3 + 5xy^2$. The degree is then the largest sum of the powers in any single term (here 2 + 3, or 5). In general, we work with polynomials in a single variable, such as x.

Polynomials are easier to work with if you write them in **descending order** (sometimes called *descending-exponent form*). When a polynomial has only one variable, this means that the term with the highest exponent is written first, then the term with the next-highest exponent is written, and so on.

We call the first term of a polynomial written in descending order the *leading term*. As stated above, this is the term that has the largest exponent of the variable. The power of the variable of the leading term is the same as the degree of the polynomial.

The numerical coefficient of the leading term is called the *leading coefficient*.

Example 4 Writing Polynomials in Descending Order

< Objective 3 >

The exponents get smaller from left to right.

(a) $5x^7 - 3x^4 + 2x^2$ is in descending order. The leading coefficient is 5.

(b) $4x^4 + 5x^6 - 3x^5$ is *not* in descending order. In descending order, the polynomial is written as

$$5x^6 - 3x^5 + 4x^4$$

The leading coefficient is 5.

You should see that the degree of the polynomial is 6, which is given by the exponent of the variable in the leading term.

Check Yourself 4

Write each polynomial in descending order and identify the leading coefficient.

(a) $5x^4 - 4x^5 + 7$ **(b)** $4x^3 + 9x^4 + 6x^8$

A polynomial can represent any number. Its value depends on the value given to the variable.

Example 5 Evaluating Polynomials

Given the polynomial

$$3x^3 - 2x^2 - 4x + 1$$

(a) Find the value of the polynomial when $x = 2$.

Substituting 2 for x, we have

$$3(2)^3 - 2(2)^2 - 4(2) + 1$$
$$= 3(8) - 2(4) - 4(2) + 1$$
$$= 24 - 8 - 8 + 1$$
$$= 9$$

RECALL

We apply the order of operations rules. See Section 0.5 for a review.

(b) Find the value of the polynomial at $x = -2$.

Now we substitute -2 for x.

$$3(-2)^3 - 2(-2)^2 - 4(-2) + 1$$
$$= 3(-8) - 2(4) - 4(-2) + 1$$
$$= -24 - 8 + 8 + 1$$
$$= -23$$

> CAUTION

Be particularly careful when dealing with powers of negative numbers!

Check Yourself 5

Find the value of the polynomial

$$4x^3 - 3x^2 + 2x - 1$$

when

(a) $x = 3$ **(b)** $x = -3$

Polynomials are used in almost every professional field. Many applications are related to predictions and forecasts. In allied health, polynomials are sometimes used to calculate the concentration of a medication in the bloodstream after a given amount of time. Example 6 demonstrates just such an application.

Example 6 An Allied Health Application

The concentration of digoxin, a medication prescribed for congestive heart failure, in a patient's bloodstream t hours after injection is given by the polynomial $-0.0015t^2 + 0.0845t + 0.7170$, where concentration is measured in nanograms per milliliter (ng/mL). Determine the concentration of digoxin in a patient's bloodstream 19 hr after injection.

We are asked to evaluate the polynomial at $t = 19$.

$-0.0015t^2 + 0.0845t + 0.7170$

We substitute 19 for t in the polynomial.

$-0.0015(19)^2 + 0.0845(19) + 0.7170$

$= -0.0015(361) + 1.6055 + 0.7170$

$= -0.5415 + 1.6055 + 0.7170$

$= 1.781$

The concentration is 1.781 nanograms per milliliter.

Check Yourself 6

The concentration of a sedative, in micrograms per milliliter (μg/mL), in a patient's bloodstream t hours after injection is given by the polynomial $-1.35t^2 + 10.81t + 7.38$. Determine the concentration of the sedative in the patient's bloodstream 3.5 hr after injection. Round your result to the nearest tenth μg/mL.

Check Yourself ANSWERS

1. **(a)** and **(c)** are polynomials. **2.** **(a)** binomial; **(b)** monomial; **(c)** trinomial **3.** **(a)** 5; **(b)** 1; **(c)** 6; **(d)** 0 **4.** **(a)** $-4x^5 + 5x^4 + 7$, -4; **(b)** $6x^8 + 9x^4 + 4x^3$, 6 **5.** **(a)** 86; **(b)** -142 **6.** 28.7 μg/mL

Reading Your Text

These fill-in-the-blank exercises will help you understand some of the key vocabulary used in this section. The answers to these exercises are in the Answers Appendix in the back of the text.

(a) A ________ can be written as a number or the product of a number and one or more variables and their exponents.

(b) In each term of a polynomial, the number factor is called the numerical ________.

(c) A polynomial with exactly two terms is called a ________.

(d) The prefix ________ means 3.

5.3 exercises

Skills | Calculator/Computer | Career Applications | Above and Beyond

< Objective 1 >

Which expressions are polynomials?

1. $7x^3$

2. $4x^3 - \frac{2}{x}$

3. $2x^5y^3 - 4x^2y^4$

4. 7

5. -7

6. $4x^3 + x$

7. $\frac{3 + x}{x^2}$

8. $5a^2 - 2a + 7$

For each polynomial, list the terms and the coefficients.

9. $2x^2 - 3x$

10. $5x^3 + x$

11. $4x^3 - 3x + 2$

12. $7x^2$

Classify each polynomial as a monomial, binomial, or trinomial where possible.

13. $4x^3 - 2x^2$

14. $4x^7$

15. $7y^2 + 4y + 5$

16. $3x^3$

17. $2x^4 - 3x^2 + 5x - 2$

18. $x^4 + \frac{5}{x} + 7$

19. $7x^{10}$

20. $4x^4 - 2x^2 + 5x - 7$

21. $x^5 - \frac{3}{x^2}$

22. $25x^2 - 16$

< Objectives 2 and 3 >

Write each polynomial in descending order, give its degree, and identify the leading coefficient.

23. $4x^5 - 3x^2$

24. $5x^2 + 3x^3 + 4$

25. $7x^7 - 5x^9 + 4x^3$

26. $4 - x + x^2$

27. $-9x$

28. $x^{17} - 3x^4$

29. $5x^2 - 3x^5 + x^6 - 7$

30. 15

Evaluate each polynomial for the given values of the variable.

31. $9x - 2$; $x = 1$ and $x = -1$

32. $5x - 5$; $x = 2$ and $x = -2$

33. $x^3 - 2x$; $x = 2$ and $x = -2$

34. $3x^2 + 7$; $x = 3$ and $x = -3$

35. $3x^2 + 4x - 2$; $x = 4$ and $x = -4$

36. $2x^2 - 5x + 1$; $x = 2$ and $x = -2$

37. $-x^2 - x + 12$; $x = 3$ and $x = -4$

38. $-x^2 - 5x - 6$; $x = -3$ and $x = -2$

39. **Business and Finance** The cost of typing a term paper is given as \$5 for each page plus \$35. Use x as the number of pages to be typed and write a polynomial to describe this cost. Find the cost of typing a 50-page paper.

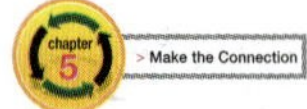

40. **Business and Finance** The cost of making suits is described as \$20 for each suit plus \$150. Use x as the number of suits and write a polynomial to describe this cost. Find the cost of making seven suits.

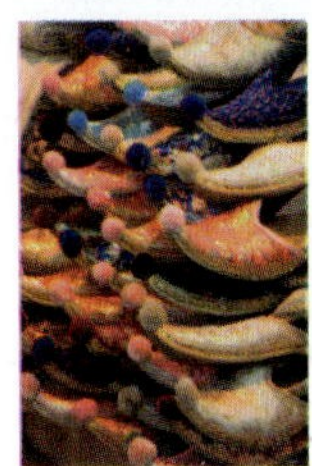

41. **Business and Finance** The revenue, in dollars, when x pairs of slippers are sold is given by $4x^2 - 95$. Find the revenue when 12 pairs are sold.

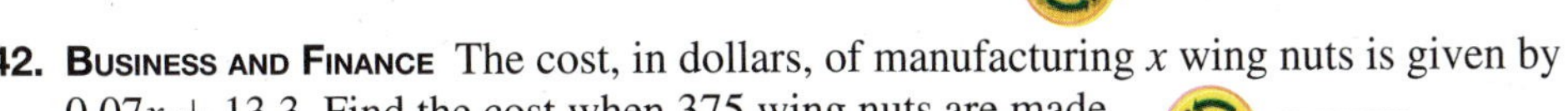

42. **Business and Finance** The cost, in dollars, of manufacturing x wing nuts is given by $0.07x + 13.3$. Find the cost when 375 wing nuts are made.

Complete each statement with **always, sometimes,** *or* **never.**

43. A monomial is __________ a polynomial.

44. A binomial is __________ a trinomial.

45. The degree of a trinomial is __________ 3.

46. A trinomial __________ has three terms.

47. A polynomial __________ has four or more terms.

48. A binomial __________ has two coefficients.

49. If x equals 0, the value of a polynomial in x __________ equals 0.

50. The coefficient of the leading term in a polynomial is __________ the largest coefficient of the polynomial.

Skills | Calculator/Computer | **Career Applications** | Above and Beyond

Electronics Many devices in our homes consume energy, even when the device is not considered "on." For example, a remote-controlled TV has to power circuitry inside the device to recognize when the user turns the TV "on" via the remote control. Similarly, many microwave ovens have integrated clocks that require power to keep and display the time.

Assume that the total energy, in watt-hours (Wh), used by a certain television per day can be described by the expression $58t + 144$, in which t is the number of hours the television is actually "on" in a day. Use this information to complete exercises 51 and 52.

51. If the TV is not turned on in an entire day, how many watt-hours are consumed in the 24-hr period?

52. If the TV is on for a total of 4.2 hr in one day, how many watt-hours are consumed that day?

53. **Allied Health** One diabetic patient's morning blood glucose level in terms of the number of days t since the patient was diagnosed can be approximated by the polynomial

$$0.472t^3 - 5.298t^2 + 11.802t + 93.143$$

Estimate the patient's morning blood glucose level on the fifth day after being diagnosed.

54. **Mechanical Engineering** The deflection of a beam is given by

$$D = \frac{x^3}{3.8 \times 10^6}$$

Find the deflection for an 18-ft-long beam (that is, $x = 18$). Use scientific notation to express your result.

Capital italic letters such as P and Q are often used in function notation to name polynomials. For example, we might write $P(x) = 3x^3 - 5x^2 + 2$ in which $P(x)$ is read "P of x." The notation permits a convenient shorthand. We write $P(2)$, read "P of 2," to indicate the value of the polynomial when $x = 2$. Here

$$\begin{aligned} P(2) &= 3(2)^3 - 5(2)^2 + 2 \\ &= 3 \cdot 8 - 5 \cdot 4 + 2 \\ &= 6 \end{aligned}$$

If $P(x) = x^3 - 2x^2 + 5$ *and* $Q(x) = 2x^2 + 3$, *evaluate each function.*

55. $P(1)$

56. $P(-1)$

57. $Q(2)$

58. $Q(-2)$

59. $P(3)$

60. $Q(-3)$

61. $P(0)$

62. $Q(0)$

63. $P(2) + Q(-1)$

64. $P(-2) + Q(3)$

65. $P(3) - Q(-3) \div Q(0)$

66. $Q(-2) \div Q(2) \cdot P(0)$

67. $|Q(4)| - |P(4)|$

68. $\dfrac{P(-1) + Q(0)}{P(0)}$

Answers

1. Polynomial **3.** Polynomial **5.** Polynomial **7.** Not a polynomial **9.** $2x^2, -3x; 2, -3$ **11.** $4x^3, -3x, 2; 4, -3, 2$ **13.** Binomial **15.** Trinomial **17.** Not classified **19.** Monomial **21.** Not a polynomial **23.** $4x^5 - 3x^2, 5, 4$ **25.** $-5x^9 + 7x^7 + 4x^3, 9, -5$ **27.** $-9x, 1, -9$ **29.** $x^6 - 3x^5 + 5x^2 - 7, 6, 1$ **31.** $7, -11$ **33.** $4, -4$ **35.** 62, 30 **37.** 0, 0 **39.** $5x + 35$; \$285 **41.** \$481 **43.** always **45.** sometimes **47.** sometimes **49.** sometimes **51.** 144 Wh **53.** 78.703 **55.** 4 **57.** 11 **59.** 14 **61.** 5 **63.** 10 **65.** 7 **67.** -2

5.4 Adding and Subtracting Polynomials

< 5.4 Objectives >

1 > Add polynomials

2 > Subtract polynomials

Addition is always a matter of combining like quantities (two apples plus three apples, four books plus five books, and so on). If you keep that basic idea in mind, adding polynomials is just a matter of combining like terms. Suppose that you want to add

$5x^2 + 3x + 4$ and $4x^2 + 5x - 6$

We can remove the parentheses when adding two polynomials.

$(5x^2 + 3x + 4) + (4x^2 + 5x - 6) = 5x^2 + 3x + 4 + 4x^2 + 5x - 6$

RECALL

Like terms have exactly the same variables to the same powers.

Because addition is *commutative,* we can add in any order we wish. In this case, we place *like terms* next to each other.

$$= \underbrace{5x^2 + 4x^2}_{\text{Like terms } (x^2)} + \underbrace{3x + 5x}_{\text{Like terms } (x)} + \underbrace{4 - 6}_{\text{Like terms (constants)}}$$

Addition is associative, so we can *group* the like terms together.

$= (5x^2 + 4x^2) + (3x + 5x) + (4 - 6)$

By combining like terms, we reach a final sum.

$= 9x^2 + 8x - 2$

Often, you can write the sum directly by locating and combining like terms mentally.

Example 1 Combining Like Terms

< Objective 1 >

Add $3x - 5$ and $2x + 3$.

Write the sum.

$(3x - 5) + (2x + 3)$

$= 3x - 5 + 2x + 3 = 5x - 2$

Like terms Like terms

NOTE

We mentally combine like terms.

$3x + 2x = 5x$ and

$-5 + 3 = -2$

Check Yourself 1

Add $6x^2 + 2x$ and $4x^2 - 7x$.

We use the same technique to find the sum of two trinomials.

Example 2 Adding Polynomials

Add $4a^2 - 7a + 5$ and $3a^2 + 3a - 4$.

Write the sum.

$(4a^2 - 7a + 5) + (3a^2 + 3a - 4)$

$= 4a^2 - 7a + 5 + 3a^2 + 3a - 4$

Like terms ($-7a$ and $3a$); Like terms ($4a^2$ and $3a^2$); Like terms (5 and -4)

$= 7a^2 - 4a + 1$

Check Yourself 2

Add $5y^2 - 3y + 7$ and $3y^2 - 5y - 7$.

When adding polynomials, there may be terms in one polynomial that do not combine with any term in the other polynomial. In this case, that term remains the same in the final result, as we see in Example 3.

Example 3 Adding Polynomials

Add $2x^2 + 7x$ and $4x - 6$.

Write the sum.

$(2x^2 + 7x) + (4x - 6)$

$= 2x^2 + 7x + 4x - 6$

These are the only like terms; $2x^2$ and -6 cannot be combined.

$= 2x^2 + 11x - 6$

Check Yourself 3

Add $5m^2 + 8$ and $8m^2 - 3m$.

As we mentioned in Section 5.3, writing polynomials in descending order usually makes the work easier.

Example 4 Adding Polynomials

Add $3x - 2x^2 + 7$ and $5 + 4x^2 - 3x$.

Write the polynomials in descending order; then add.

$(-2x^2 + 3x + 7) + (4x^2 - 3x + 5)$

$= 2x^2 + 12$

Check Yourself 4

Add $8 - 5x^2 + 4x$ and $7x - 8 + 8x^2$.

RECALL

This process is the same as simplifying an expression. You learned this in Section 1.3.

Subtracting polynomials requires an extra step since the associative and commutative properties do not apply to subtraction. Recall that we view subtraction as adding the opposite, so

$a - b = a + (-b)$

The opposite of a quantity with more than one term requires that we take the opposite of each term. For example,

$-(a + b) = -a - b$ and $-(a - b) = -a + b$

Alternatively, the negative in front of a quantity can be understood as -1. Applying the distributive property, we get the same results.

$-(a + b) = -1(a + b) = -a - b$ and $-(a - b) = -1(a - b) = -a + b$

We can now go on to subtracting polynomials.

Example 5 Removing Parentheses

NOTE

We use the distributive property.

$-(2x + 3y) = (-1)(2x + 3y)$
$= -2x - 3y$

Remove the parentheses.

(a) $-(2x + 3y) = -2x - 3y$ Change each sign to remove the parentheses.

(b) $m - (5n - 3p) = m - 5n + 3p$

Sign changes

(c) $2x - (-3y + z) = 2x + 3y - z$

Sign changes

Check Yourself 5

Remove the parentheses.

(a) $-(3m + 5n)$ **(b)** $-(5w - 7z)$
(c) $3r - (2s - 5t)$ **(d)** $5a - (-3b - 2c)$

Subtracting polynomials is just a matter of using distribution to remove the parentheses and then combining like terms. Consider Example 6.

Example 6 Subtracting Polynomials

< Objective 2 >

(a) Subtract $5x - 3$ from $8x + 2$.

Write

$(8x + 2) - (5x - 3)$

$= 8x + 2 - 5x + 3$

Sign changes

$= 3x + 5$

RECALL

The expression following *from* is written first in the problem.

(b) Subtract $4x^2 - 8x + 3$ from $8x^2 + 5x - 3$.

Write

$(8x^2 + 5x - 3) - (4x^2 - 8x + 3)$

$= 8x^2 + 5x - 3 - 4x^2 + 8x - 3$

Sign changes

$= 4x^2 + 13x - 6$

Check Yourself 6

(a) Subtract $7x + 3$ from $10x - 7$.
(b) Subtract $5x^2 - 3x + 2$ from $8x^2 - 3x - 6$.

Writing all polynomials in descending order makes locating and combining like terms much easier.

Example 7 Subtracting Polynomials

(a) Subtract $4x^2 - 3x^3 + 5x$ from $8x^3 - 7x + 2x^2$.

Write

$$(8x^3 + 2x^2 - 7x) - (-3x^3 + 4x^2 + 5x)$$
$$= 8x^3 + 2x^2 - 7x + 3x^3 - 4x^2 - 5x$$

Sign changes

$$= 11x^3 - 2x^2 - 12x$$

(b) Subtract $8x - 5$ from $-5x + 3x^2$.

Write

$$(3x^2 - 5x) - (8x - 5)$$
$$= 3x^2 \underbrace{- 5x - 8x} + 5$$

Only the like terms can be combined.

$$= 3x^2 - 13x + 5$$

Check Yourself 7

(a) Subtract $7x - 3x^2 + 5$ from $5 - 3x + 4x^2$.
(b) Subtract $3a - 2$ from $5a + 4a^2$.

If you think back to adding and subtracting in arithmetic, you may remember that we arranged the work vertically. That is, we placed the numbers being added or subtracted under one another so that each column represented the same place value. This meant that in adding or subtracting columns, you always dealt with "like quantities."

It is also possible to use a vertical method for adding or subtracting polynomials. First write the polynomials in descending order, then arrange them one under another, so that each column contains like terms. Finally, add or subtract in each column.

Example 8 Adding Using the Vertical Method

(a) Add $3x - 5$ and $x^2 + 2x + 4$.

$$\begin{array}{r} 3x - 5 \\ x^2 + 2x + 4 \\ \hline x^2 + 5x - 1 \end{array}$$

Each column has like terms.

(b) Add $2x^2 - 5x$, $3x^2 + 2$, and $6x - 3$.

$$\begin{array}{rrr} 2x^2 & - 5x & \\ 3x^2 & & + 2 \\ & 6x & - 3 \\ \hline 5x^2 & + \ x & - 1 \end{array}$$

Each column has like terms.

Check Yourself 8

Add $3x^2 + 5$, $x^2 - 4x$, and $6x + 7$.

Example 9 illustrates subtraction with the vertical method.

Example 9 Subtracting Using the Vertical Method

NOTE

Since we are subtracting the entire quantity $5x - 3$, we place parentheses around $5x - 3$. Then we remove them by distributing the negative sign.

(a) Subtract $5x - 3$ from $8x - 7$.

Write

$$\begin{array}{r} 8x - 7 \\ -(5x - 3) \\ \hline \end{array}$$

$$\begin{array}{r} 8x - 7 \\ -5x + 3 \\ \hline 3x - 4 \end{array}$$

To subtract, change each sign of $5x - 3$ to get $-5x + 3$, then add.

(b) Subtract $5x^2 - 3x + 4$ from $8x^2 + 5x - 3$.

Write

$$\begin{array}{r} 8x^2 + 5x - 3 \\ -(5x^2 - 3x + 4) \\ \hline \end{array}$$

$$\begin{array}{r} 8x^2 + 5x - 3 \\ -5x^2 + 3x - 4 \\ \hline 3x^2 + 8x - 7 \end{array}$$

To subtract, change each sign of $5x^2 - 3x + 4$ to get $-5x^2 + 3x - 4$, then add.

Check Yourself 9

Use the vertical method to subtract.

(a) $4x^2 - 3x$ from $8x^2 + 2x$ **(b)** $8x^2 + 4x - 3$ from $9x^2 - 5x + 7$

Subtraction using the vertical method takes some practice. Take time to study the method carefully. We use it in long division in Section 5.6.

Check Yourself ANSWERS

1. $10x^2 - 5x$ **2.** $8y^2 - 8y$ **3.** $13m^2 - 3m + 8$ **4.** $3x^2 + 11x$ **5.** **(a)** $-3m - 5n$; **(b)** $-5w + 7z$; **(c)** $3r - 2s + 5t$; **(d)** $5a + 3b + 2c$ **6.** **(a)** $3x - 10$; **(b)** $3x^2 - 8$ **7.** **(a)** $7x^2 - 10x$; **(b)** $4a^2 + 2a + 2$ **8.** $4x^2 + 2x + 12$ **9.** **(a)** $4x^2 + 5x$; **(b)** $x^2 - 9x + 10$

Reading Your Text

These fill-in-the-blank exercises will help you understand some of the key vocabulary used in this section. The answers to these exercises are in the Answers Appendix in the back of the text.

(a) If a ______________ sign appears in front of parentheses, simply remove the parentheses.

(b) If a minus sign appears in front of parentheses, the subtraction can be rewritten as addition by changing the ______________ of each term inside the parentheses.

(c) When we change each sign inside parentheses to subtract, we are using the ______________ property.

(d) When subtracting polynomials, the expression following the word *from* is written ______________ when writing the problem.

5.4 exercises

Skills | Calculator/Computer | Career Applications | Above and Beyond

< Objective 1 >

Add.

1. $5a - 7$ and $4a + 11$
2. $9x + 3$ and $3x - 4$
3. $8b^2 - 11b$ and $5b^2 - 7b$
4. $2m^2 + 3m$ and $6m^2 - 8m$
5. $3x^2 - 2x$ and $-5x^2 + 2x$
6. $9p^2 + 3p$ and $-13p^2 - 3p$
7. $2x^2 + 5x - 3$ and $3x^2 - 7x + 4$
8. $4d^2 - 8d + 7$ and $5d^2 - 6d - 9$
9. $3b^2 - 7$ and $2b - 7$
10. $4x - 3$ and $3x^2 - 9x$
11. $(8y^3 - 5y^2) + (5y^2 - 2y)$
12. $(9x^4 - 2x^2) + (2x^2 + 3)$
13. $(2a^2 - 4a^3) + (3a^3 + 2a^2)$
14. $(9m^3 - 2m) + (-6m - 4m^3)$
15. $(7x^2 - 5 + 4x) + (8 - 5x - 9x^2)$
16. $(5b^3 - 8b + 2b^2) + (3b^2 - 7b^3 + 5b)$

Remove the parentheses in each expression and simplify if possible.

17. $-(4a + 5b)$
18. $-(7x - 4y)$
19. $5a - (2b - 3c)$
20. $8x - (2y - 5z)$
21. $9r - (3r + 5s)$
22. $10m - (3m - 2n)$
23. $5p - (-3p + 2q)$
24. $8d - (-7c - 2d)$

< Objective 2 >

Subtract.

25. $x + 4$ from $2x - 3$
26. $5x + 1$ from $7x + 8$
27. $3m^2 - 2m$ from $4m^2 - 5m$
28. $9a^2 - 5a$ from $11a^2 - 10a$
29. $6y^2 + 5y$ from $4y^2 + 5y$
30. $9n^2 - 4n$ from $7n^2 - 4n$
31. $x^2 - 4x - 3$ from $3x^2 - 5x - 2$
32. $3x^2 - 2x + 4$ from $5x^2 - 8x - 3$
33. $3a + 7$ from $8a^2 - 9a$
34. $3x^3 + x^2$ from $4x^3 - 5x$
35. $(5b - 2b^2) - (4b^2 - 3b)$
36. $(3y^2 - 2y) - (7y - 3y^2)$
37. $(7x^2 - 11x + 31) - (5x^2 + 19 - 11x)$
38. $(4x^3 + x - 3x^2) - (4x - 2x^2 + 4x^3)$
39. $(2x^3 - 3x + 1) - (4x^2 + 8x + 6)$
40. $(4x - 8x^3 + x^2) - (6 - 5x^3 + 4x^4 + 4x)$

Perform the indicated operations.

41. Subtract $2b + 5$ from the sum of $5b - 4$ and $3b + 8$.
42. Subtract $5m - 7$ from the sum of $2m - 8$ and $9m - 2$.
43. Subtract $3x^2 + 2x - 1$ from the sum of $x^2 + 5x - 2$ and $2x^2 + 7x - 8$.
44. Subtract $4x^2 - 5x - 3$ from the sum of $x^2 - 3x - 7$ and $2x^2 - 2x + 9$.
45. Subtract $2x^2 - 3x$ from the sum of $4x^2 - 5$ and $2x - 7$.
46. Subtract $7a^2 + 8a - 15$ from the sum of $14a - 5$ and $7a^2 - 8$.
47. Subtract the sum of $3y^2 - 3y$ and $5y^2 + 3y$ from $2y^2 - 8y$.
48. Subtract the sum of $7r^3 - 4r^2$ and $-3r^3 + 4r^2$ from $2r^3 + 3r^2$.

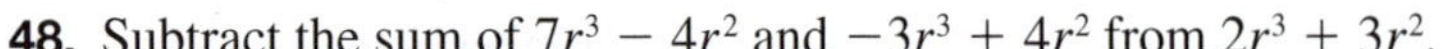

49. $[(9x^2 - 3x + 5) - (3x^2 + 2x - 1)] - (x^2 - 2x - 3)$
50. $[(5x^2 + 2x - 3) - (-2x^2 + x - 2)] - (2x^2 + 3x - 5)$

Use the vertical method to find each sum.

51. $4w^2 + 11$, $7w - 9$, and $2w^2 - 9w$

52. $3x^2 - 4x - 2$, $6x - 3$, and $2x^2 + 8$

53. $3x^2 + 3x - 4$, $4x^2 - 3x - 3$, and $2x^2 - x + 7$

54. $5x^2 + 2x - 4$, $x^2 - 2x - 3$, and $2x^2 - 4x - 3$

Use the vertical method to subtract.

55. $7a^2 - 9a$ from $9a^2 - 4a$

56. $6r^3 + 4r^2$ from $4r^3 - 2r^2$

57. $5x^2 - 6x + 7$ from $8x^2 - 5x + 7$

58. $8x^2 - 4x + 2$ from $9x^2 - 8x + 6$

59. $5x^2 - 3x$ from $8x^2 - 9$

60. $-13x^2 + 6x$ from $-11x^2 - 3$

61. **Geometry** A rectangle has sides of $8x + 9$ and $6x - 7$. Find the polynomial that represents its perimeter.

62. **Geometry** A triangle has sides $3x + 7$, $4x - 9$, and $5x + 6$. Find the polynomial that represents its perimeter.

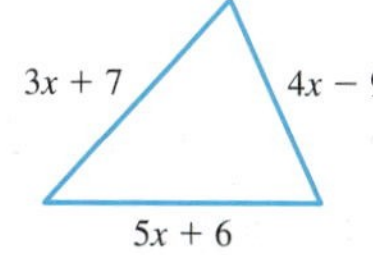

63. **Business and Finance** The cost of producing x wall clocks is $150 + 25x$. The revenue for selling x clocks is $90x - x^2$. The profit is given by the revenue minus the cost. Find the polynomial that represents the profit.

64. **Business and Finance** The revenue for selling y aloe plants is $3y^2 - 2y + 5$, and the cost of carrying y plants is $y^2 + y - 3$. Find the polynomial that represents the profit.

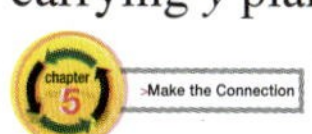

Complete each statement with **always, sometimes,** *or* **never.**

65. The sum of two trinomials is __________ another trinomial.

66. When subtracting one polynomial from a second polynomial, we __________ write the second polynomial first.

Skills | Calculator/Computer | **Career Applications** | Above and Beyond

67. **Manufacturing Technology** The shear polynomial for a polymer is

$0.4x^2 - 144x + 308$

After vulcanization, the shear polynomial is increased by $0.2x^2 - 14x + 144$. Find the shear polynomial for the polymer after vulcanization.

68. **Allied Health** A diabetic patient's morning m and evening n blood glucose levels are given in terms of the number of days t since the patient was diagnosed and can be approximated by

$m = 0.472t^3 - 5.298t^2 + 11.802t + 94.143$

$n = -1.083t^3 + 11.464t^2 - 29.524t + 117.429$

Give the difference d between the morning and evening blood glucose levels in terms of the number of days since diagnosis.

Electrical Engineering The resistance (in Ω) of conductors, such as metals, varies in a relatively linear fashion, based on temperature. The common equation for the relationship between conductors at temperatures between 0°C and 100°C is

$$R_t = R_0(1 + \alpha t)$$

in which

R_t = Resistance of the conductor at temperature t (in °C)
R_0 = Resistance of the conductor at 0°C
α = Temperature coefficient of the conductor (a constant)
t = Temperature, in °C

Use this information in exercises 69 and 70.

69. Assuming $\alpha = 3.9 \times 10^{-3}$, express R_t in terms of t, if a certain piece of a copper conductor has a total resistance of $1.72 \times 10^{-8}\,\Omega$ at 0°C.

70. What is the resistance of the conductor described in exercise 69 if the temperature is 20°C?

Skills | Calculator/Computer | Career Applications | **Above and Beyond**

Find values for a, b, c, and d so that each equation is true.

71. $3ax^4 - 5x^3 + x^2 - cx + 2 = 9x^4 - bx^3 + x^2 - 2d$

72. $(4ax^3 - 3bx^2 - 10) - 3(x^3 + 4x^2 - cx - d) = x^3 - 6x + 8$

Answers

1. $9a + 4$ **3.** $13b^2 - 18b$ **5.** $-2x^2$ **7.** $5x^2 - 2x + 1$ **9.** $3b^2 + 2b - 14$ **11.** $8y^3 - 2y$ **13.** $-a^3 + 4a^2$ **15.** $-2x^2 - x + 3$ **17.** $-4a - 5b$ **19.** $5a - 2b + 3c$ **21.** $6r - 5s$ **23.** $8p - 2q$ **25.** $x - 7$ **27.** $m^2 - 3m$ **29.** $-2y^2$ **31.** $2x^2 - x + 1$ **33.** $8a^2 - 12a - 7$ **35.** $-6b^2 + 8b$ **37.** $2x^2 + 12$ **39.** $2x^3 - 4x^2 - 11x - 5$ **41.** $6b - 1$ **43.** $10x - 9$ **45.** $2x^2 + 5x - 12$ **47.** $-6y^2 - 8y$ **49.** $5x^2 - 3x + 9$ **51.** $6w^2 - 2w + 2$ **53.** $9x^2 - x$ **55.** $2a^2 + 5a$ **57.** $3x^2 + x$ **59.** $3x^2 + 3x - 9$ **61.** $28x + 4$ **63.** $-x^2 + 65x - 150$ **65.** sometimes **67.** $0.6x^2 - 158x + 452$ **69.** $R_t = 1.72 \times 10^{-8} + 6.708 \times 10^{-11}t$ **71.** $a = 3; b = 5; c = 0; d = -1$

5.5 Multiplying Polynomials

< 5.5 Objectives >

1 > Find the product of a monomial and a polynomial

2 > Find the product of two binomials

3 > Square a binomial

4 > Find the product of two binomials that differ only in sign

You already have some experience multiplying polynomials. In Section 5.1 we used the product rule for exponents and used that rule to find the product of two monomials.

Step by Step

Multiplying Monomials

Step 1	Multiply the coefficients.
Step 2	Use the product rule for exponents to combine the variables. $ax^m \cdot bx^n = abx^{m+n}$

Here is an example in which we multiply two monomials.

Example 1 Multiplying Monomials

Multiply $3x^2y$ and $2x^3y^5$.

$(3x^2y)(2x^3y^5)$

$= (3 \cdot 2)(x^2 \cdot x^3)(y \cdot y^5)$

Multiply the coefficients. Add the exponents.

$= 6x^5y^6$

NOTE

Once again we use the commutative and associative properties to rewrite the problem.

Check Yourself 1

Multiply.

(a) $(5a^2b)(3a^2b^4)$ **(b)** $(-3xy)(4x^3y^5)$

Our next task is to find the product of a monomial and a polynomial. The distributive property gives us a rule for multiplication.

Property

Multiplying a Polynomial by a Monomial

Use the distributive property to multiply each term of the polynomial by the monomial. Simplify the result.

Example 2 Multiplying a Monomial and a Binomial

< Objective 1 >

RECALL

Distributive property:

$a(b + c) = ab + ac$

(a) Multiply $2x + 3$ by x.

$x(2x + 3)$
$= x \cdot 2x + x \cdot 3$
$= 2x^2 + 3x$

Multiply x by $2x$ and then by 3. That is, "distribute" the multiplication over the sum.

(b) Multiply $2a^3 + 4a$ by $3a^2$.

$3a^2(2a^3 + 4a)$
$= 3a^2 \cdot 2a^3 + 3a^2 \cdot 4a = 6a^5 + 12a^3$

NOTE

With practice, you will do this step mentally.

Check Yourself 2

Multiply.

(a) $2y(y^2 + 3y)$ **(b)** $3w^2(2w^3 + 5w)$

The patterns of Example 2 extend to *any* number of terms.

Example 3 Multiplying a Monomial and a Polynomial

NOTE

We show all the steps of the process. With practice, you can write the product directly and should try to do so.

Multiply.

(a) $3x(4x^3 + 5x^2 + 2)$
$= 3x \cdot 4x^3 + 3x \cdot 5x^2 + 3x \cdot 2 = 12x^4 + 15x^3 + 6x$

(b) $-5c(4c^2 - 8c)$
$= (-5c)(4c^2) - (-5c)(8c) = -20c^3 + 40c^2$

(c) $3c^2d^2(7cd^2 - 5c^2d^3)$
$= (3c^2d^2)(7cd^2) - (3c^2d^2)(5c^2d^3) = 21c^3d^4 - 15c^4d^5$

Check Yourself 3

Multiply.

(a) $3(5a^2 + 2a + 7)$ **(b)** $4x^2(8x^3 - 6)$
(c) $-5m(8m^2 - 5m)$ **(d)** $9a^2b(3a^3b - 6a^2b^4)$

We apply the distributive property twice when multiplying a polynomial by a binomial. That is, every term in the binomial must multiply each term in the polynomial. We usually write the distributive property as

$a(b + c) = ab + ac$

We use this to multiply two binomials by treating the first binomial as a and the second binomial as $(b + c)$.

$$\underbrace{(x + y)}_{a}\underbrace{(w + z)}_{b+c} = \underbrace{(x + y)}_{a}\underbrace{w}_{b} + \underbrace{(x + y)}_{a}\underbrace{z}_{c}$$

We sometimes write the distributive property as

$(a + b)c = ac + bc$

NOTE

$\underbrace{(x + y)}_{(a+b)}\underbrace{w}_{c} = \underbrace{xw}_{a\,c} + \underbrace{yw}_{b\,c}$

We use this second form of the distributive property twice in this expansion.

$$(x + y)w + (x + y)z = xw + yw + xz + yz$$

The final step is to simplify the resulting polynomial, if possible.

We use this approach to multiply binomials in Example 4, though the pattern holds true when multiplying polynomials with more than two terms, as well.

Example 4 Multiplying Binomials

< Objective 2 >

(a) Multiply $x + 2$ by $x + 3$.

We can think of $x + 2$ as a single quantity and apply the distributive property.

NOTE

This ensures that each term, x and 2, of the first binomial is multiplied by each term, x and 3, of the second binomial.

$(x + 2)(x + 3)$ Multiply $(x + 2)$ by x and then by 3.

$= (x + 2)x + (x + 2)3$

$= x \cdot x + 2 \cdot x + x \cdot 3 + 2 \cdot 3$

$= x^2 + 2x + 3x + 6$

$= x^2 + 5x + 6$

(b) Multiply $a - 3$ by $a - 4$.

$(a - 3)(a - 4)$ Think of $(a - 3)$ as a single quantity and distribute.

$= (a - 3)a - (a - 3)(4)$

$= a \cdot a - 3 \cdot a - [(a \cdot 4) - (3 \cdot 4)]$

$= a^2 - 3a - (4a - 12)$ The parentheses are needed here because a *minus sign* precedes the binomial.

$= a^2 - 3a - 4a + 12$

$= a^2 - 7a + 12$

Check Yourself 4

Multiply.

(a) $(x + 4)(x + 5)$ **(b)** $(y + 5)(y - 6)$

Fortunately, there is a pattern to this kind of multiplication that allows you to write the product of the two binomials directly without going through all these steps. We call it the **FOIL method.** The reason for this name will be clear as we look at the process in greater detail.

To multiply $(x + 2)(x + 3)$:

1. $(x + 2)(x + 3)$ → $x \cdot x$ Find the product of the *first* terms of the factors. (Remember this by F.)

2. $(x + 2)(x + 3)$ → $x \cdot 3$ Find the product of the *outer* terms. (Remember this by O.)

3. $(x + 2)(x + 3)$ → $2 \cdot x$ Find the product of the *inner* terms. (Remember this by I.)

4. $(x + 2)(x + 3)$ → $2 \cdot 3$ Find the product of the *last* terms. (Remember this by L.)

NOTE

FOIL gives you an easy way of remembering the steps: *F*irst, *O*uter, *I*nner, and *L*ast.

Combining the four steps, we have

$(x + 2)(x + 3)$
$= x^2 + 3x + 2x + 6$
$= x^2 + 5x + 6$

With practice, the FOIL method lets you write the products quickly and easily. Consider Example 5, which illustrates this approach.

Example 5 Using the FOIL Method

Use the FOIL method to find each product.

NOTE

When possible, you should combine the outer and inner products mentally and just write the final product.

O: $5 \cdot x$
F: $x \cdot x$

(a) $(x + 4)(x + 5)$

I: $4 \cdot x$
L: $4 \cdot 5$

$= x^2 + 5x + 4x + 20$
F O I L

$= x^2 + 9x + 20$

O: $3 \cdot x$
F: $x \cdot x$

(b) $(x - 7)(x + 3)$ Combine the outer and inner products as $-4x$.

I: $-7 \cdot x$
L: $(-7)(3)$

$= x^2 - 4x - 21$

Check Yourself 5

Multiply.

(a) $(x + 6)(x + 7)$ **(b)** $(x + 3)(x - 5)$ **(c)** $(x - 2)(x - 8)$

You can also find the product of binomials with leading coefficients other than 1 or with more than one variable with the FOIL method.

Example 6 Using the FOIL Method

Use the FOIL method to find each product.

O: $(4x)(2)$
F: $(4x)(3x)$

(a) $(4x - 3)(3x + 2)$

I: $(-3)(3x)$
L: $(-3)(2)$

Combine $8x + (-9x) = -x$

$= 12x^2 - x - 6$

O: $(3x)(-7y)$

F: $(3x)(2x)$

(b) $(3x - 5y)(2x - 7y)$

I: $(-5y)(2x)$

L: $(-5y)(-7y)$

Combine $-21xy + (-10xy) = -31xy$

$= 6x^2 - 31xy + 35y^2$

Check Yourself 6

Multiply.

(a) $(5x + 2)(3x - 7)$ **(b)** $(4a - 3b)(5a - 4b)$
(c) $(3m + 5n)(2m + 3n)$

Here is a summary of our work in multiplying binomials.

Step by Step

Multiplying Two Binomials

Step 1 Multiply the first terms of the binomials (F).

Step 2 Then multiply the first term of the first binomial by the second term of the second binomial (O).

Step 3 Next multiply the second term of the first binomial by the first term of the second binomial (I).

Step 4 Finally, multiply the second terms of the binomials (L).

Step 5 Form the sum of the four terms found above, combining any like terms.

The FOIL method works well for multiplying any two binomials. But what if one of the factors has three or more terms? The vertical format, shown in Example 7, works for factors with any number of terms.

Example 7 Using the Vertical Method

Multiply $x^2 - 5x + 8$ by $x + 3$.

Step 1

$$\begin{array}{r} x^2 - 5x + 8 \\ x + 3 \\ \hline 3x^2 - 15x + 24 \end{array}$$

Multiply each term of $x^2 - 5x + 8$ by 3.

Step 2

$$\begin{array}{r} x^2 - 5x + 8 \\ x + 3 \\ \hline 3x^2 - 15x + 24 \\ x^3 - 5x^2 + 8x \quad\quad \\ \hline \end{array}$$

Multiply each term by x.

We shift this line over so that like terms are in the same columns.

Step 3

$$\begin{array}{r} x^2 - 5x + 8 \\ x + 3 \\ \hline 3x^2 - 15x + 24 \\ x^3 - 5x^2 + 8x \quad\quad \\ \hline x^3 - 2x^2 - 7x + 24 \end{array}$$

Add to combine like terms to write the product.

NOTE

The vertical method ensures that each term of one factor multiplies each term of the other. That's why it works!

Check Yourself 7

Multiply $2x^2 - 5x + 3$ by $3x + 4$.

The vertical method works equally well when multiplying two trinomials, or even polynomials with more than three terms. The vertical method helps to ensure that we do not miss any of the necessary products.

Example 8 Using the Vertical Method to Multiply Trinomials

Multiply.

$(2x^2 + 3x - 5)(x^2 - 4x + 1)$

Step 1

$$\begin{array}{r}
2x^2 + 3x - 5 \\
x^2 - 4x + 1 \\
\hline
2x^2 + 3x - 5
\end{array}$$

Multiply each term of $2x^2 + 3x - 5$ by 1.

Step 2

$$\begin{array}{rrrr}
 & 2x^2 & + 3x & - 5 \\
 & x^2 & - 4x & + 1 \\
\hline
 & 2x^2 & + 3x & - 5 \\
-8x^3 & - 12x^2 & + 20x &
\end{array}$$

Now multiply each term by $-4x$.

As before, we align the resulting polynomials so that terms with the same degree line up.

Step 3

$$\begin{array}{rrrrr}
 & & 2x^2 & + 3x & - 5 \\
 & & x^2 & - 4x & + 1 \\
\hline
 & & 2x^2 & + 3x & - 5 \\
 & -8x^3 & - 12x^2 & + 20x & \\
2x^4 & + 3x^3 & - 5x^2 & &
\end{array}$$

Finally, multiply each term by x^2.

Step 4 Add the three polynomials by combining like terms.

$$\begin{array}{rrrrr}
 & & 2x^2 & + 3x & - 5 \\
 & & x^2 & - 4x & + 1 \\
\hline
 & & 2x^2 & + 3x & - 5 \\
 & -8x^3 & - 12x^2 & + 20x & \\
2x^4 & + 3x^3 & - 5x^2 & & \\
\hline
2x^4 & - 5x^3 & - 15x^2 & + 23x & - 5
\end{array}$$

Check Yourself 8

Multiply $(-2x^2 + 3x - 7)(3x^2 - 5x + 8)$.

NOTE

Squaring a binomial always results in three terms.

Certain products occur frequently enough in algebra that it is worth learning special formulas for them. First, we look at the **square of a binomial,** which is the product of two equal binomial factors.

$$(x + y)^2 = (x + y)(x + y) = x^2 + xy + xy + y^2$$
$$= x^2 + 2xy + y^2$$

$$(x - y)^2 = (x - y)(x - y) = x^2 - xy - xy + y^2$$
$$= x^2 - 2xy + y^2$$

These patterns lead to another rule.

Step by Step

Squaring a Binomial

Step 1 Find the first term of the square by squaring the first term of the binomial.

Step 2 Find the middle term of the square as twice the product of the two terms of the binomial.

Step 3 Find the last term of the square by squaring the last term of the binomial.

Example 9 Squaring a Binomial

< Objective 3 >

Multiply.

(a) $(x + 3)^2 = x^2 + 2 \cdot x \cdot 3 + 3^2$

Square of first term — Twice the product of the two terms — Square of the last term

$= x^2 + 6x + 9$

> CAUTION

A very common mistake in squaring binomials is to forget the middle term.

(b) $(3a + 4b)^2 = (3a)^2 + 2(3a)(4b) + (4b)^2$
$= 9a^2 + 24ab + 16b^2$

(c) $(y - 5)^2 = y^2 + 2 \cdot y \cdot (-5) + (-5)^2$
$= y^2 - 10y + 25$

(d) $(5c - 3d)^2 = (5c)^2 + 2(5c)(-3d) + (-3d)^2$
$= 25c^2 - 30cd + 9d^2$

Check Yourself 9

Multiply.

(a) $(2x + 1)^2$ **(b)** $(4x - 3y)^2$

Example 10 Squaring a Binomial

NOTE

You should see that $(2 + 3)^2 \neq 2^2 + 3^2$ because $5^2 \neq 4 + 9$.

Find $(y + 4)^2$.

$(y + 4)^2$ is *not* equal to $y^2 + 4^2$ or $y^2 + 16$

The correct square is

$(y + 4)^2 = y^2 + 8y + 16$

The middle term is twice the product of y and 4.

Check Yourself 10

Multiply.

(a) $(x + 5)^2$ **(b)** $(3a + 2)^2$
(c) $(y - 7)^2$ **(d)** $(5x - 2y)^2$

A second special product will be very important in Chapter 6. Suppose the form of a product is

$(x + y)(x - y)$

The two factors differ only in sign.

Look at what happens when we multiply.

$(x + y)(x - y)$

$= x^2 \underbrace{- xy + xy}_{0} - y^2$

$= x^2 - y^2$

Property

Special Product

The product of two binomials that differ only in the sign between the terms is the square of the first term minus the square of the second term.

$(a + b)(a - b) = a^2 - b^2$

Example 11 Multiplying Polynomials

< Objective 4 >

Multiply each pair of factors.

(a) $(x + 5)(x - 5) = x^2 - 5^2$

Square of the first term — Square of the second term

$= x^2 - 25$

NOTE

$(2y)^2 = (2y)(2y)$
$= 4y^2$

(b) $(x + 2y)(x - 2y) = x^2 - (2y)^2$

Square of the first term — Square of the second term

$= x^2 - 4y^2$

(c) $(3m + n)(3m - n) = 9m^2 - n^2$

(d) $(4a - 3b)(4a + 3b) = 16a^2 - 9b^2$

Check Yourself 11

Find the products.

(a) $(a - 6)(a + 6)$ **(b)** $(x - 3y)(x + 3y)$
(c) $(5n + 2p)(5n - 2p)$ **(d)** $(7b - 3c)(7b + 3c)$

When you are finding the product of three or more factors, it is useful to first look for the pattern in which two binomials differ only in their sign. If you see it, finding this product first makes it easier to find the product of all the factors.

Example 12 Multiplying Polynomials

(a) $x(x - 3)(x + 3)$ — These binomials differ only in sign.

$= x(x^2 - 9)$

$= x^3 - 9x$

(b) $(x + 1)(x - 5)(x + 5)$ — These binomials differ only in sign.

$= (x + 1)(x^2 - 25)$ — With two binomials, use the FOIL method.

$= x^3 + x^2 - 25x - 25$

(c) $(2x - 1)(x + 3)(2x + 1)$

$(2x - 1)(x + 3)(2x + 1)$
$= (x + 3)(2x - 1)(2x + 1)$
$= (x + 3)(4x^2 - 1)$
$= 4x^3 + 12x^2 - x - 3$

These two binomials differ only in the sign of the second term. We can use the commutative property to rearrange the terms.

Check Yourself 12

Multiply.

(a) $3x(x - 5)(x + 5)$ **(b)** $(x - 4)(2x + 3)(2x - 3)$
(c) $(x - 7)(3x - 1)(x + 7)$

Check Yourself ANSWERS

1. (a) $15a^4b^5$; **(b)** $-12x^4y^6$ **2. (a)** $2y^3 + 6y^2$; **(b)** $6w^5 + 15w^3$ **3. (a)** $15a^2 + 6a + 21$; **(b)** $32x^5 - 24x^2$; **(c)** $-40m^3 + 25m^2$; **(d)** $27a^5b^2 - 54a^4b^5$ **4. (a)** $x^2 + 9x + 20$; **(b)** $y^2 - y - 30$ **5. (a)** $x^2 + 13x + 42$; **(b)** $x^2 - 2x - 15$; **(c)** $x^2 - 10x + 16$ **6. (a)** $15x^2 - 29x - 14$; **(b)** $20a^2 - 31ab + 12b^2$; **(c)** $6m^2 + 19mn + 15n^2$ **7.** $6x^3 - 7x^2 - 11x + 12$ **8.** $-6x^4 + 19x^3 - 52x^2 + 59x - 56$ **9. (a)** $4x^2 + 4x + 1$; **(b)** $16x^2 - 24xy + 9y^2$ **10. (a)** $x^2 + 10x + 25$; **(b)** $9a^2 + 12a + 4$; **(c)** $y^2 - 14y + 49$; **(d)** $25x^2 - 20xy + 4y^2$ **11. (a)** $a^2 - 36$; **(b)** $x^2 - 9y^2$; **(c)** $25n^2 - 4p^2$; **(d)** $49b^2 - 9c^2$ **12. (a)** $3x^3 - 75x$; **(b)** $4x^3 - 16x^2 - 9x + 36$; **(c)** $3x^3 - x^2 - 147x + 49$

Reading Your Text

These fill-in-the-blank exercises will help you understand some of the key vocabulary used in this section. The answers to these exercises are in the Answers Appendix in the back of the text.

(a) When multiplying two monomials, we first multiply the ________.

(b) When multiplying a polynomial by a monomial, use the ________ property to multiply each term of the polynomial by the monomial.

(c) The FOIL method can only be used when multiplying two ________.

(d) Squaring a binomial always results in ________ terms.

Skills Calculator/Computer Career Applications Above and Beyond

5.5 exercises

< Objectives 1 and 2 >

Multiply.

1. $(5x^2)(3x^3)$ **2.** $(14a^4)(2a^7)$ **3.** $(-2b^2)(14b^8)$

4. $(14y^4)(-4y^6)$ **5.** $(-5p^7)(-8p^6)$ **6.** $(-6m^8)(9m^7)$

7. $(4m^5)(-3m)$ **8.** $(-5r^7)(-3r)$ **9.** $(4x^3y^2)(8x^2y)$

10. $(-3r^4s^2)(-7r^2s^2)$

11. $(-3m^5n^2)(2m^4n)$

12. $(3a^4b^2)(-14a^3b^4)$

13. $10(x + 3)$

14. $4(7b - 5)$

15. $3a(4a + 5)$

16. $5x(2x - 7)$

17. $3s^2(4s^2 - 7s)$

18. $3a^3(9a^2 + 15)$

19. $3x(5x^2 - 3x - 1)$

20. $5m(4m^3 - 3m^2 + 2)$

21. $3xy(2x^2y + xy^2 + 5xy)$

22. $5ab^2(ab - 3a + 5b)$

23. $6m^2n(3m^2n - 2mn + mn^2)$ VIDEO

24. $8pq^2(2pq - 3p + 5q)$

25. $(x + 3)(x + 2)$

26. $(a - 3)(a - 7)$

27. $(m - 5)(m - 9)$

28. $(b + 7)(b + 5)$

29. $(p - 8)(p + 7)$

30. $(x - 10)(x + 9)$

31. $(w - 7)(w + 6)$

32. $(s - 12)(s - 8)$

33. $(3x - 5)(x - 8)$

34. $(w + 5)(4w - 7)$

35. $(2x - 3)(3x + 4)$

36. $(7a - 2)(2a - 3)$

37. $(3a - b)(4a - 9b)$ VIDEO

38. $(7s - 3t)(3s + 8t)$

39. $(3p - 4q)(7p + 5q)$

40. $(5x - 4y)(2x - y)$

41. $(2x^2 - x + 8)(x^2 + 4x + 6)$

42. $(x^3 - 3x^2 + 5x + 1)(2x^2 + 4x - 1)$

< Objective 3 >

Find each square.

43. $(x + 3)^2$

44. $(y + 9)^2$

45. $(w - 6)^2$

46. $(a - 8)^2$

47. $(z + 12)^2$

48. $(p - 11)^2$

49. $(2a - 1)^2$

50. $(3x - 2)^2$

51. $(6m + 1)^2$

52. $(7b - 2)^2$

53. $(3x - y)^2$

54. $(5m + n)^2$

55. $(2r + 5s)^2$

56. $(3a - 4b)^2$

57. $(6a - 5b)^2$

58. $(7p + 6q)^2$

59. $\left(x + \frac{1}{2}\right)^2$ VIDEO

60. $\left(w - \frac{1}{4}\right)^2$

< Objective 4 >

Find each product.

61. $(x - 6)(x + 6)$

62. $(y + 8)(y - 8)$

63. $(m + 7)(m - 7)$

64. $(w - 10)(w + 10)$

65. $\left(x - \frac{1}{2}\right)\left(x + \frac{1}{2}\right)$

66. $\left(x + \frac{2}{3}\right)\left(x - \frac{2}{3}\right)$

67. $(p - 0.4)(p + 0.4)$

68. $(m - 1.1)(m + 1.1)$

69. $(a - 3b)(a + 3b)$

70. $(p + 4q)(p - 4q)$

71. $(3x + 2y)(3x - 2y)$

72. $(7x - y)(7x + y)$

73. $(8w + 5z)(8w - 5z)$

74. $(8c + 3d)(8c - 3d)$

75. $(5x - 9y)(5x + 9y)$ VIDEO

76. $(6s - 5t)(6s + 5t)$

Multiply.

77. $2x(3x - 2)(4x + 1)$

78. $3x(2x + 1)(2x - 1)$

79. $5a(4a - 3)(4a + 3)$

80. $6m(3m - 2)(3m - 7)$

81. $3s(5s - 2)(4s - 1)$

82. $7w(2w - 3)(2w + 3)$

83. $(x - 2)(x + 1)(x - 3)$

84. $(y + 3)(y - 2)(y - 4)$

85. $(a - 1)^3$

86. $(x + 1)^3$

87. $\left(\frac{x}{2} + \frac{2}{3}\right)\left(\frac{2x}{3} - \frac{2}{5}\right)$

88. $\left(\frac{x}{3} + \frac{3}{4}\right)\left(\frac{3x}{4} - \frac{3}{5}\right)$

89. $[x + (y - 2)][x - (y - 2)]$

90. $[x + (3 - y)][x - (3 - y)]$

Complete each application.

91. Geometry The length of a rectangle is given by $(3x + 5)$ centimeters (cm), and the width is given by $(2x - 7)$ cm. Express the area of the rectangle in terms of x.

92. Geometry The base of a triangle measures $(3y + 7)$ in., and the height is $(2y - 3)$ in. Express the area of the triangle in terms of y.

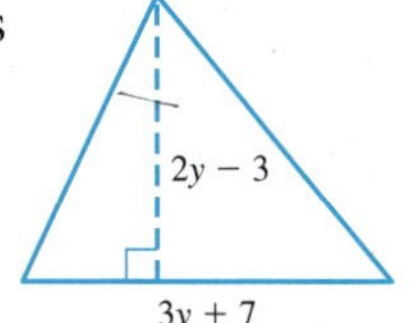

93. Business and Finance The price of a vase is given by $p = 12 - 0.05x$, where x is the number of vases sold. If the revenue generated is found by multiplying the number of vases sold by the price of each vase, find the polynomial that represents the revenue.

94. Business and Finance The price of a sofa is given by $p = 1{,}000 - 0.02x^2$, where x is the number of sofas sold. Find the polynomial that represents the revenue generated from the sale of x sofas.

chapter 5 > Make the Connection

95. Business and Finance Suppose an orchard is planted with trees in straight rows. If there are $(5x - 4)$ rows with $(5x - 4)$ trees in each row, find a polynomial that represents the number of trees in the orchard.

96. Geometry A square has sides of length $(3x - 2)$ cm. Express the area of the square as a polynomial.

97. Geometry The length and width of a rectangle are given by two consecutive odd integers. Write an expression for the area of the rectangle.

98. Geometry The length of a rectangle is 6 less than 3 times the width. Write an expression for the area of the rectangle.

Let x represent the unknown number and write an expression for each product.

99. The product of 6 more than a number and 6 less than that number

100. The square of 5 more than a number

101. The square of 4 less than a number

102. The product of 5 less than a number and 5 more than that number

Determine whether each statement is **true** *or* **false.**

103. $(x + y)^2 = x^2 + y^2$

104. $(x - y)^2 = x^2 - y^2$

105. $(x + y)^2 = x^2 + 2xy + y^2$

106. $(x - y)^2 = x^2 - 2xy + y^2$

Skills | Calculator/Computer | **Career Applications** | Above and Beyond

107. Mechanical Engineering The bending moment of a dual-span beam with cantilever is given by

$$M = \frac{wc^2}{8}$$

To account for a load on the cantilever, this expression needs to be multiplied by $\frac{c^2}{2}$. Create a single expression for the new bending moment.

108. Mechanical Engineering The maximum stress for a given allowable strain (deformation) for a certain material is given by the polynomial

$$\text{Stress} = 82.6x - 0.4x^2 + 322$$

in which x is the weight percent of nickel.

After heat-treating, the stress polynomial is multiplied by $0.08x + 0.9$. Find the stress (strength) after heat-treating.

109. Mechanical Engineering The maximum stress for a given allowable strain (deformation) for a certain material is given by the polynomial

$$\text{Stress} = 86.2x - 0.6x^2 + 258$$

in which x is the allowable strain, in micrometers.

When an alloying substance is added to the material, the strength is increased by multiplying the stress polynomial by $\frac{p}{8.6}$, in which p is the percent of the alloying material.

Find a polynomial (in two variables) that describes the allowable stress (strength) of the material after alloying (rounded to two decimal places).

110. Construction Technology The shrinkage of a post of length L under a load is given by the product of $\frac{L}{28{,}000{,}000}$ and the load. If the load is equal to 3 times the square of the length, find an expression for the shrinkage of the post.

Skills	Calculator/Computer	Career Applications	**Above and Beyond**

111. Complete the statement: $(a + b)^2$ is not equal to $a^2 + b^2$ because. . . But wait! Isn't $(a + b)^2$ *sometimes* equal to $a^2 + b^2$? What do you think?

112. Is $(a + b)^3$ ever equal to $a^3 + b^3$? Explain.

113. Identify the length, width, and area of the full square.

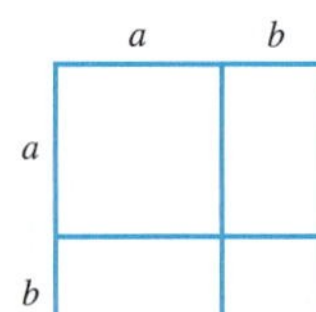

Length = ____________

Width = ____________

Area = ____________

114. Identify the length, width, and area of the full square.

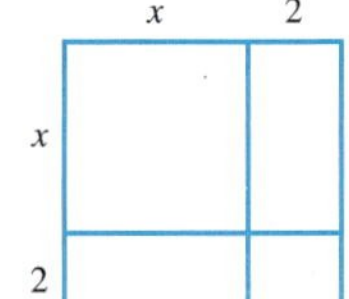

Length = ____________

Width = ____________

Area = ____________

115. The square shown is x units on a side. The area is ____________.

Draw a picture of what happens when the sides are doubled. The area is ____________.

Continue the picture to show what happens when the sides are tripled. The area is ____________.

If the sides are quadrupled, the area is ____________.

In general, if the sides are multiplied by n, the area is ____________.

If each side is increased by 3, the area is increased by ____________.

If each side is decreased by 2, the area is decreased by ____________.

In general, if each side is increased by n, the area is increased by ____________; and if each side is decreased by n, the area is decreased by ____________.

116. Work with another student to complete this table and write the polynomial that represents the volume of the box. A paper box is to be made from a piece of cardboard 20 in. wide and 30 in. long. The box will be formed by cutting squares out of each of the four corners and folding up the sides to make a box.

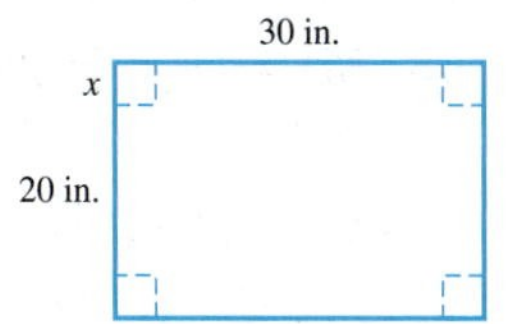

If x is the dimension of the side of the square cut out of the corner, when the sides are folded up the box will be x in. tall. You should use a piece of paper to try this to see how the box will be made. Complete the chart.

Length of Side of Corner Square	Length of Box	Width of Box	Depth of Box	Volume of Box
1 in.				
2 in.				
3 in.				
n in.				

Write general formulas for the width, length, and height of the box and a general formula for the *volume* of the box, and simplify it by multiplying. The variable will be the height, the side of the square cut out of the corners. What is the highest power of the variable in the polynomial you have written for the volume? Extend the table to decide what the dimensions are for a box with maximum volume. Draw a sketch of this box and write in the dimensions.

Note that $(28)(32) = (30 - 2)(30 + 2) = 900 - 4 = 896$. *Use this pattern to find each product.*

117. $(49)(51)$ **118.** $(27)(33)$ **119.** $(34)(26)$

120. $(98)(102)$ **121.** $(55)(65)$ **122.** $(64)(56)$ 

Answers

1. $15x^5$ **3.** $-28b^{10}$ **5.** $40p^{13}$ **7.** $-12m^6$ **9.** $32x^5y^3$ **11.** $-6m^9n^3$ **13.** $10x + 30$ **15.** $12a^2 + 15a$ **17.** $12s^4 - 21s^3$ **19.** $15x^3 - 9x^2 - 3x$ **21.** $6x^3y^2 + 3x^2y^3 + 15x^2y^2$ **23.** $18m^4n^2 - 12m^3n^2 + 6m^3n^3$ **25.** $x^2 + 5x + 6$ **27.** $m^2 - 14m + 45$ **29.** $p^2 - p - 56$ **31.** $w^2 - w - 42$ **33.** $3x^2 - 29x + 40$ **35.** $6x^2 - x - 12$ **37.** $12a^2 - 31ab + 9b^2$ **39.** $21p^2 - 13pq - 20q^2$ **41.** $2x^4 + 7x^3 + 16x^2 + 26x + 48$ **43.** $x^2 + 6x + 9$ **45.** $w^2 - 12w + 36$ **47.** $z^2 + 24z + 144$ **49.** $4a^2 - 4a + 1$ **51.** $36m^2 + 12m + 1$ **53.** $9x^2 - 6xy + y^2$ **55.** $4r^2 + 20rs + 25s^2$ **57.** $36a^2 - 60ab + 25b^2$ **59.** $x^2 + x + \frac{1}{4}$ **61.** $x^2 - 36$ **63.** $m^2 - 49$ **65.** $x^2 - \frac{1}{4}$ **67.** $p^2 - 0.16$ **69.** $a^2 - 9b^2$ **71.** $9x^2 - 4y^2$ **73.** $64w^2 - 25z^2$ **75.** $25x^2 - 81y^2$ **77.** $24x^3 - 10x^2 - 4x$ **79.** $80a^3 - 45a$ **81.** $60s^3 - 39s^2 + 6s$ **83.** $x^3 - 4x^2 + x + 6$ **85.** $a^3 - 3a^2 + 3a - 1$ **87.** $\frac{x^2}{3} + \frac{11x}{45} - \frac{4}{15}$ **89.** $x^2 - y^2 + 4y - 4$ **91.** $(6x^2 - 11x - 35)\text{ cm}^2$ **93.** $12x - 0.05x^2$ **95.** $(25x^2 - 40x + 16)$ trees **97.** $x(x + 2)$ or $x^2 + 2x$ **99.** $x^2 - 36$ **101.** $x^2 - 8x + 16$ **103.** False **105.** True **107.** $\frac{wc^4}{16}$ **109.** Stress $= -0.07x^2p + 10.02xp + 30p$ **111.** Above and Beyond **113.** Above and Beyond **115.** Above and Beyond **117.** 2,499 **119.** 884 **121.** 3,575

5.6 Dividing Polynomials

< 5.6 Objectives >

1 > Divide a polynomial by a monomial

2 > Find the quotient of two polynomials

In Section 5.1, we introduced the quotient rule for exponents to divide one monomial by another monomial.

Step by Step

Dividing a Monomial by a Monomial

Step 1 Divide the coefficients.

Step 2 Use the quotient rule for exponents to combine the variables.

Example 1 Dividing Monomials

RECALL

The quotient rule says: If x is not zero,

$\frac{x^m}{x^n} = x^{m-n}$

(a) $\frac{8x^4}{2x^2} = 4x^{4-2}$

Divide: $\frac{8}{2} = 4$

Subtract the exponents.

$= 4x^2$

(b) $\frac{45a^5b^3}{9a^2b} = 5a^3b^2$

Check Yourself 1

Divide.

(a) $\frac{16a^5}{8a^3}$

(b) $\frac{28m^4n^3}{7m^3n}$

Now look at how this can be extended to divide any polynomial by a monomial.

NOTE

This step depends on the distributive property.

$$\frac{12a^3 + 8a^2}{4a} = \frac{12a^3}{4a} + \frac{8a^2}{4a}$$

Divide each term in the numerator by the denominator $4a$.

Now do each division.

$$= 3a^2 + 2a$$

Property

Dividing a Polynomial by a Monomial

Divide each term of the polynomial by the monomial. Simplify the result.

Example 2 Dividing by Monomials

< Objective 1 >

Divide each term by 2.

(a) $\dfrac{4a^2 + 8}{2} = \dfrac{4a^2}{2} + \dfrac{8}{2}$

$= 2a^2 + 4$

Divide each term by 6y.

(b) $\dfrac{24y^3 - 18y^2}{6y} = \dfrac{24y^3}{6y} - \dfrac{18y^2}{6y}$

$= 4y^2 - 3y$

(c) $\dfrac{15x^2 + 10x}{-5x} = \dfrac{15x^2}{-5x} + \dfrac{10x}{-5x}$ Remember the rules for signs in division.

$= -3x - 2$

NOTE

With practice, you can just write the quotient.

(d) $\dfrac{14x^4 + 28x^3 - 21x^2}{7x^2} = \dfrac{14x^4}{7x^2} + \dfrac{28x^3}{7x^2} - \dfrac{21x^2}{7x^2}$

$= 2x^2 + 4x - 3$

(e) $\dfrac{9a^3b^4 - 6a^2b^3 + 12ab^4}{3ab} = \dfrac{9a^3b^4}{3ab} - \dfrac{6a^2b^3}{3ab} + \dfrac{12ab^4}{3ab}$

$= 3a^2b^3 - 2ab^2 + 4b^3$

(f) $\dfrac{4x^5 - 8x^3 + 2x^2}{2x^2} = \dfrac{4x^5}{2x^2} - \dfrac{8x^3}{2x^2} + \dfrac{2x^2}{2x^2}$

$= 2x^3 - 4x + 1$ $\quad \dfrac{2x^2}{2x^2} = 1$

Check Yourself 2

Divide.

(a) $\dfrac{20y^3 - 15y^2}{5y}$

(b) $\dfrac{8a^3 - 12a^2 + 4a}{-4a}$

(c) $\dfrac{16m^4n^3 - 12m^3n^2 + 8mn}{4mn}$

(d) $\dfrac{15x^2y - 3xy + 12xy^2}{3xy}$

We are now ready to look at dividing one polynomial by another polynomial (with more than one term). The process is very much like long division in arithmetic, as Example 3 illustrates.

Example 3 Dividing by Binomials

< Objective 2 >

Divide $x^2 + 7x + 10$ by $x + 2$.

NOTE

The first term in the dividend, x^2, is divided by the first term in the divisor, x.

Step 1 $x + 2\overline{)x^2 + 7x + 10}$ with quotient x — Divide x^2 by x to get x.

Step 2 $x + 2\overline{)x^2 + 7x + 10}$ with quotient x

$x^2 + 2x$ — Multiply the divisor $x + 2$ by x.

RECALL

To subtract $x^2 + 2x$, mentally change each sign to $-x^2 - 2x$ and then add. Take your time and be careful here. This is where most errors are made.

Step 3

$$\begin{array}{r} x \phantom{{}+10} \\ x+2\overline{)x^2+7x+10} \\ \underline{x^2+2x\phantom{{}+10}} \\ 5x+10 \end{array}$$

Subtract and bring down 10.

Step 4

$$\begin{array}{r} x+5 \\ x+2\overline{)x^2+7x+10} \\ \underline{x^2+2x\phantom{{}+10}} \\ 5x+10 \end{array}$$

Divide $5x$ by x to get 5.

Step 5

$$\begin{array}{r} x+5 \\ x+2\overline{)x^2+7x+10} \\ \underline{x^2+2x\phantom{{}+10}} \\ 5x+10 \\ \underline{5x+10} \\ 0 \end{array}$$

$-(5x + 10) = -5x - 10$

Multiply $x + 2$ by 5 and then subtract.

NOTE

We repeat the process until the degree of the remainder is less than that of the divisor or until there is no remainder.

The quotient is $x + 5$.

Check Yourself 3

Divide $x^2 + 9x + 20$ by $x + 4$.

In Example 3, we showed all the steps separately to help you see the process. In practice, the work can be shortened.

Example 4 Dividing by Binomials

Divide $x^2 + x - 12$ by $x - 3$.

$$\begin{array}{r} x+4 \\ x-3\overline{)x^2+x-12} \\ \underline{x^2-3x\phantom{{}-12}} \\ 4x-12 \\ \underline{4x-12} \\ 0 \end{array}$$

The Steps

1. Divide x^2 by x to get x, the first term of the quotient.
2. Multiply $x - 3$ by x.
3. Subtract and bring down -12. Remember to mentally change the signs to $-x^2 + 3x$ and add.
4. Divide $4x$ by x to get 4, the second term of the quotient.
5. Multiply $x - 3$ by 4 and subtract.

NOTE

You might want to write out a problem like 408 ÷ 17, to compare the steps.

The quotient is $x + 4$.

Check Yourself 4

Divide.

$(x^2 + 2x - 24) \div (x - 4)$

You may have a remainder in algebraic long division just as in arithmetic. Consider Example 5.

Example 5 Dividing by Binomials

Divide $4x^2 - 8x + 11$ by $2x - 3$.

$$\begin{array}{r} 2x - 1 \\ 2x - 3 \overline{)4x^2 - 8x + 11} \\ \underline{4x^2 - 6x } \\ -2x + 11 \\ \underline{-2x + 3} \\ 8 \end{array}$$

Quotient: $2x - 1$; Divisor: $2x - 3$; Remainder: 8

This result can be written as

$$\frac{4x^2 - 8x + 11}{2x - 3} = \underbrace{2x - 1}_{\text{Quotient}} + \frac{8 \leftarrow \text{Remainder}}{\underbrace{2x - 3}_{\text{Divisor}}}$$

Check Yourself 5

Divide.

$(6x^2 - 7x + 15) \div (3x - 5)$

The division process extends to higher-degree dividends. The steps involved in the division process are exactly the same, as Example 6 illustrates.

Example 6 Dividing by Binomials

Divide $6x^3 + x^2 - 4x - 5$ by $3x - 1$.

$$\begin{array}{r} 2x^2 + x - 1 \\ 3x - 1 \overline{)6x^3 + x^2 - 4x - 5} \\ \underline{6x^3 - 2x^2 } \\ 3x^2 - 4x \\ \underline{3x^2 - x } \\ -3x - 5 \\ \underline{-3x + 1} \\ -6 \end{array}$$

We write this result as

$$\frac{6x^3 + x^2 - 4x - 5}{3x - 1} = 2x^2 + x - 1 + \frac{-6}{3x - 1}$$

Check Yourself 6

Divide $4x^3 - 2x^2 + 2x + 15$ by $2x + 3$.

Suppose that the dividend is "missing" a term in some power of the variable. In such a case, the coefficient of the missing term is 0. Consider Example 7.

Example 7 Dividing by Binomials

NOTE

Think of $0x$ as a placeholder. Writing it helps to align like terms.

Divide $x^3 - 2x^2 + 5$ by $x + 3$.

$$
\begin{array}{r|l}
 & x^2 - 5x + 15 \\
x + 3 & x^3 - 2x^2 + 0x + 5 \\
 & \underline{x^3 + 3x^2} \\
 & \quad -5x^2 + 0x \\
 & \quad \underline{-5x^2 - 15x} \\
 & \qquad\quad 15x + 5 \\
 & \qquad\quad \underline{15x + 45} \\
 & \qquad\qquad -40
\end{array}
$$

Write $0x$ for the "missing" term in x.

This result can be written as

$$\frac{x^3 - 2x^2 + 5}{x + 3} = x^2 - 5x + 15 + \frac{-40}{x + 3}$$

Check Yourself 7

Divide.

$(4x^3 + x + 10) \div (2x - 1)$

You should always arrange the terms of the divisor and the dividend in descending order before starting the long division process, as illustrated in Example 8.

Example 8 Dividing by Binomials

Divide $5x^2 - x + x^3 - 5$ by $-1 + x^2$.

Write the divisor as $x^2 - 1$ and the dividend as $x^3 + 5x^2 - x - 5$.

$$
\begin{array}{r|l}
 & x + 5 \\
x^2 - 1 & x^3 + 5x^2 - x - 5 \\
 & \underline{x^3 \qquad\quad - x} \\
 & \qquad 5x^2 \qquad - 5 \\
 & \qquad \underline{5x^2 \qquad - 5} \\
 & \qquad\qquad\qquad 0
\end{array}
$$

Write $x^3 - x$, the product of x and $x^2 - 1$, so that like terms fall in the same columns.

Check Yourself 8

Divide.

$(5x^2 + 10 + 2x^3 + 4x) \div (2 + x^2)$

Check Yourself ANSWERS

1. **(a)** $2a^2$; **(b)** $4mn^2$ **2.** **(a)** $4y^2 - 3y$; **(b)** $-2a^2 + 3a - 1$; **(c)** $4m^3n^2 - 3m^2n + 2$; **(d)** $5x - 1 + 4y$

3. $x + 5$ **4.** $x + 6$ **5.** $2x + 1 + \frac{20}{3x - 5}$ **6.** $2x^2 - 4x + 7 + \frac{-6}{2x + 3}$

7. $2x^2 + x + 1 + \frac{11}{2x - 1}$ **8.** $2x + 5$

Reading Your Text

These fill-in-the-blank exercises will help you understand some of the key vocabulary used in this section. The answers to these exercises are in the Answers Appendix in the back of the text.

(a) When dividing two monomials, we first divide the __________ .

(b) When dividing a polynomial by a monomial, divide each __________ of the polynomial by the monomial.

(c) When doing long division of polynomials, we continue until the __________ of the remainder is less than that of the divisor.

(d) When dividing polynomials, if the dividend is missing a term in some power of the variable we use __________ as the coefficient for that term.

Skills | Calculator/Computer | Career Applications | Above and Beyond

5.6 exercises

< Objective 1 >

Divide.

1. $\frac{22x^9}{11x^5}$

2. $\frac{20a^7}{5a^5}$

3. $\frac{35m^3n^2}{7mn^2}$

4. $\frac{42x^5y^2}{6x^3y}$

5. $\frac{3a + 6}{3}$

6. $\frac{3x - 6}{3}$

7. $\frac{9b^2 - 12}{3}$

8. $\frac{10m^2 + 5m}{5}$

9. $\frac{28a^3 - 42a^2}{7a}$

10. $\frac{9x^3 + 12x^2}{3x}$

11. $\frac{12m^2 + 6m}{-3m}$

12. $\frac{20b^3 - 25b^2}{-5b}$

13. $\frac{18a^4 - 45a^3 + 63a^2}{9a^2}$

14. $\frac{21x^5 - 28x^4 + 14x^3}{7x}$

15. $\frac{20x^5 - 15x^4 + 5x^3 + 10x^2 - 5x}{5x}$

16. $\frac{2t^5 - 12t^3 + 2t^2}{2t^2}$

17. $\frac{3x^3y^2 + 6x^2y^2 - 9xy^2 - 3xy + 12x^2y^3}{3xy}$

18. $\frac{16a^3b^2 - 4a^2b^2 + 8a^2b^3}{4a^2b^2}$

19. $\frac{20x^4y^2 - 15x^2y^3 + 10x^3y}{5x^2y}$

20. $\frac{16m^3n^3 + 24m^2n^2 - 40mn^3}{8mn^2}$ VIDEO

< Objective 2 >

21. $\frac{x^2 - x - 12}{x - 4}$

22. $\frac{x^2 + 8x + 15}{x + 3}$

23. $\frac{x^2 - x - 20}{x + 4}$ 

24. $\frac{x^2 - 2x - 35}{x + 5}$

25. $\frac{x^2 + x - 30}{x - 5}$

26. $\frac{3x^2 + 20x - 32}{3x - 4}$

27. $\frac{2x^2 - 3x - 5}{x - 3}$

28. $\frac{3x^2 + 17x - 12}{x + 6}$

29. $\frac{4x^2 - 18x - 15}{x - 5}$

30. $\frac{4x^2 - 11x - 24}{x - 5}$

31. $\frac{6x^2 - x - 10}{3x - 5}$

32. $\frac{4x^2 + 6x - 25}{2x + 7}$

33. $\frac{x^3 + x^2 - 4x - 4}{x + 2}$

34. $\frac{x^3 - 2x^2 + 4x - 21}{x - 3}$

35. $\frac{4x^3 + 7x^2 + 10x + 5}{4x - 1}$

36. $\frac{2x^3 - 3x^2 + 4x + 4}{2x + 1}$

37. $\frac{x^3 - x^2 + 5}{x - 2}$

38. $\frac{x^3 + 4x - 3}{x + 3}$

39. $\frac{49x^3 - 2x}{7x - 3}$

40. $\frac{8x^3 - 6x^2 + 2x}{4x + 1}$

41. $\frac{2x^2 - 8 - 3x + x^3}{x - 2}$

42. $\frac{x^2 - 18x + 2x^3 + 32}{x + 4}$

43. $\frac{x^4 - 1}{x - 1}$

44. $\frac{x^4 + x^2 - 16}{x + 2}$

45. $\frac{x^3 - 3x^2 - x + 3}{x^2 - 1}$

46. $\frac{x^3 + 2x^2 + 3x + 6}{x^2 + 3}$

47. $\frac{x^4 + 2x^2 - 2}{x^2 + 3}$ 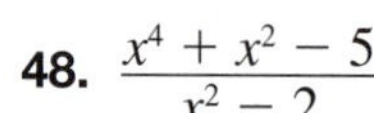

48. $\frac{x^4 + x^2 - 5}{x^2 - 2}$

49. $\frac{y^3 - 1}{y - 1}$

50. $\frac{y^3 - 8}{y - 2}$

51. $\frac{x^4 - 1}{x^2 - 1}$

52. $\frac{x^6 - 1}{x^3 - 1}$

Complete each statement with **always, sometimes,** *or* **never.**

53. When a trinomial is divided by a binomial, there is __________ a remainder.

54. A binomial divided by a binomial is __________ a binomial.

55. If a monomial exactly divides a trinomial, the result is __________ a trinomial.

56. For any positive integer n, if $x^n - 1$ is divided by $x - 1$, there is __________ a remainder.

Skills | Calculator/Computer | Career Applications | **Above and Beyond**

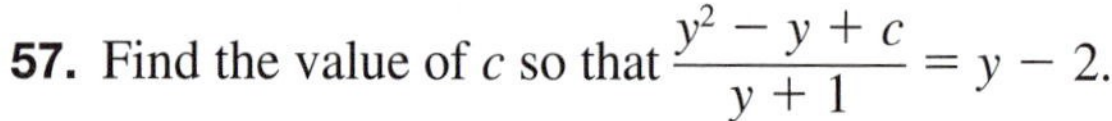

57. Find the value of c so that $\frac{y^2 - y + c}{y + 1} = y - 2$.

58. Find the value of c so that $\frac{x^3 + x^2 + x + c}{x^2 + 1} = x + 1$.

59. Write a summary of your work with polynomials. Explain how a polynomial is recognized, and explain the rules for the arithmetic of polynomials—how to add, subtract, multiply, and divide. What parts of this chapter do you feel you understand very well and what part(s) do you still have questions about? Exchange papers with another student and compare your answers.

60. An interesting (and useful) thing about division of polynomials: To find out about this interesting thing, do this division. Compare your answer with that of another student.

$(x - 2)\overline{)2x^2 + 3x - 5}$ Is there a remainder?

Now, evaluate the polynomial $2x^2 + 3x - 5$ when $x = 2$. Is this value the same as the remainder?

Try $(x + 3)\overline{)5x^2 - 2x + 1}$. Is there a remainder?

Evaluate the polynomial $5x^2 - 2x + 1$ when $x = -3$. Is this value the same as the remainder?

What happens when there is no remainder?

Try $(x - 6)\overline{)3x^3 - 14x^2 - 23x - 6}$. Is the remainder zero?

Evaluate the polynomial $3x^3 - 14x^2 - 23x - 6$ when $x = 6$. Is this value zero? Write a description of the patterns you see. Make up several more examples and test your conjecture.

61. **(a)** Divide $\frac{x^2 - 1}{x - 1}$. **(b)** Divide $\frac{x^3 - 1}{x - 1}$. **(c)** Divide $\frac{x^4 - 1}{x - 1}$.

(d) Based on your results on parts (a), (b), and (c), predict $\frac{x^{50} - 1}{x - 1}$.

62. **(a)** Divide $\frac{x^2 + x + 1}{x - 1}$. **(b)** Divide $\frac{x^3 + x^2 + x + 1}{x - 1}$. **(c)** Divide $\frac{x^4 + x^3 + x^2 + x + 1}{x - 1}$.

(d) Based on your results to (a), (b), and (c), predict $\frac{x^{10} + x^9 + x^8 + \cdots + x + 1}{x - 1}$.

Answers

1. $2x^4$ **3.** $5m^2$ **5.** $a + 2$ **7.** $3b^2 - 4$ **9.** $4a^2 - 6a$ **11.** $-4m - 2$ **13.** $2a^2 - 5a + 7$ **15.** $4x^4 - 3x^3 + x^2 + 2x - 1$ **17.** $x^2y + 2xy - 3y - 1 + 4xy^2$ **19.** $4x^2y - 3y^2 + 2x$ **21.** $x + 3$ **23.** $x - 5$ **25.** $x + 6$ **27.** $2x + 3 + \frac{4}{x - 3}$ **29.** $4x + 2 + \frac{-5}{x - 5}$ **31.** $2x + 3 + \frac{5}{3x - 5}$ **33.** $x^2 - x - 2$ **35.** $x^2 + 2x + 3 + \frac{8}{4x - 1}$ **37.** $x^2 + x + 2 + \frac{9}{x - 2}$ **39.** $7x^2 + 3x + 1 + \frac{3}{7x - 3}$ **41.** $x^2 + 4x + 5 + \frac{2}{x - 2}$ **43.** $x^3 + x^2 + x + 1$ **45.** $x - 3$ **47.** $x^2 - 1 + \frac{1}{x^2 + 3}$ **49.** $y^2 + y + 1$ **51.** $x^2 + 1$ **53.** sometimes **55.** always **57.** $c = -2$ **59.** Above and Beyond **61.** **(a)** $x + 1$; **(b)** $x^2 + x + 1$; **(c)** $x^3 + x^2 + x + 1$; **(d)** $x^{49} + x^{48} + \cdots + x + 1$

summary :: chapter 5

Definition/Procedure	Example	Reference
Positive Integer Exponents		**Section 5.1**
Properties of Exponents For any nonzero real numbers a and b **Product Rule** $a^m \cdot a^n = a^{m+n}$	$x^5 \cdot x^7 = x^{5+7} = x^{12}$	p. 409
Quotient Rule $\frac{a^m}{a^n} = a^{m-n}$	$\frac{x^7}{x^5} = x^{7-5} = x^2$	p. 411
Product-Power Rule $(ab)^n = a^n b^n$	$(2y)^3 = 2^3y^3 = 8y^3$	p. 412
Power Rule $(a^m)^n = a^{mn}$	$(2^3)^4 = 2^{12}$	p. 412
Quotient-Power Rule $\left(\frac{a}{b}\right)^m = \frac{a^m}{b^m}$	$\left(\frac{2}{3}\right)^2 = \frac{2^2}{3^2} = \frac{4}{9}$	p. 413
Integer Exponents and Scientific Notation		**Section 5.2**
Zero Exponent For any real number a where $a \neq 0$, $a^0 = 1$	$5x^0 = 5 \cdot 1 = 5$	p. 420
Negative Integer Exponents For any nonzero real number a and whole number n, $a^{-n} = \frac{1}{a^n}$ and a^{-n} is the multiplicative inverse of a^n.	$x^{-3} = \frac{1}{x^3}$ $2y^{-5} = \frac{2}{y^5}$	p. 421
Quotient Raised to a Negative Power For nonzero numbers a and b and whole number n, $\left(\frac{a}{b}\right)^{-n} = \left(\frac{b}{a}\right)^n$	$\left(\frac{x}{2}\right)^{-4} = \left(\frac{2}{x}\right)^4$ $= \frac{16}{x^4}$	p. 423
Scientific Notation Scientific notation is a useful way of expressing very large or very small numbers through the use of powers of 10. Any number written in the form $a \times 10^n$ in which $1 \le a < 10$ and n is an integer, is said to be written in scientific notation.	$38,000,000 = 3.8 \times 10^7$ 7 places	p. 425
An Introduction to Polynomials		**Section 5.3**
Term A number or the product of a number and variables and their exponents.	The terms of $4x^3 - 3x^2 + 5x$ are $4x^3$, $-3x^2$, and $5x$.	p. 432
Polynomial An algebraic expression made up of terms in which the exponents are whole numbers. These terms are connected by plus or minus signs. Each sign (+ or −) is attached to the term following that sign.	$4x^3 - 3x^2 + 5x$ is a polynomial.	p. 432

Continued

Definition/Procedure	Example	Reference
Coefficient In each term of a polynomial, the number that is multiplied by the variable(s) is called the *numerical coefficient* or, more simply, the *coefficient* of that term.	The coefficients of $4x^3 - 3x^2 + 5x$ are 4, -3, and 5.	p. 432
Types of Polynomials A polynomial can be classified according to the number of terms it has.		p. 433
A *mono*mial has exactly one term.	$2x^3$ is a monomial.	
A *bi*nomial has exactly two terms.	$3x^2 - 7x$ is a binomial.	
A *tri*nomial has exactly three terms.	$5x^5 - 5x^3 + 2$ is a trinomial.	
Degree of a Polynomial with Only One Variable The highest power of the variable appearing in any one term.	The degree of $4x^5 - 5x^3 + 3x$ is 5.	p. 433
Descending Order The form of a polynomial when it is written with the highest-degree term first, the next-highest-degree term second, and so on.	$4x^5 - 5x^3 + 3x$ is written in descending order.	p. 433
Leading Term The first term of a polynomial written in descending order. This is the term in which the variable has the largest exponent. The power of the variable of the leading term is the same as the degree of the polynomial.	The leading term is $4x^5$, so the leading coefficient is 4.	
Leading Coefficient The numerical coefficient of the leading term.		
Adding and Subtracting Polynomials		**Section 5.4**
Adding Polynomials Remove the grouping symbols and combine like terms.	$(2x + 3) + (3x - 5)$ $= 2x + 3 + 3x - 5$ $= 5x - 2$	p. 439
Subtracting Polynomials Remove the grouping symbols by changing the sign of each term in the polynomial being subtracted. Then combine any like terms.	$(3x^2 + 2x) - (2x^2 + 3x - 1)$ $= 3x^2 + 2x - 2x^2 - 3x + 1$ Sign changes $= 3x^2 - 2x^2 + 2x - 3x + 1$ $= x^2 - x + 1$	p. 441
Multiplying Polynomials		**Section 5.5**
Multiplying a Monomial by a Monomial Multiply the coefficients, and use the product rule for exponents to combine the variables. $ax^m \cdot bx^n = abx^{m+n}$	$(-2x^2y)(3x^3y)$ $= (-2)(3)(x^2x^3)(yy)$ $= -6x^5y^2$	p. 447
Multiplying a Polynomial by a Monomial Multiply each term of the polynomial by the monomial and simplify the result.	$2x(x^2 + 4)$ $= 2x^3 + 8x$	p. 447

Continued

Definition/Procedure	Example	Reference
Multiplying a Binomial by a Binomial Use the FOIL method. F O I L $(a+b)(c+d) = a\cdot c + a\cdot d + b\cdot c + b\cdot d$	$(2x-3)(3x+5)$ $= 6x^2 + 10x - 9x - 15$ F O I L $= 6x^2 + x - 15$	p. 449
Multiplying a Polynomial by a Polynomial Arrange the polynomials vertically. Multiply each term of the upper polynomial by each term of the lower polynomial, and combine like terms.	$\begin{array}{r} x^2 - 3x + 5 \\ 2x - 3 \\ \hline -3x^2 + 9x - 15 \\ 2x^3 - 6x^2 + 10x \quad \\ \hline 2x^3 - 9x^2 + 19x - 15 \end{array}$	p. 451
Squaring of a Binomial $(a+b)^2 = a^2 + 2ab + b^2$ 1. The first term of the square is the square of the first term of the binomial. 2. The middle term is twice the product of the two terms of the binomial. 3. The last term is the square of the last term of the binomial.	$(2x-5)^2$ $= (2x)^2 + 2\cdot 2x\cdot(-5) + 5^2$ $= 4x^2 - 20x + 25$	p. 453
The Product of Binomials That Differ Only in Sign Subtract the square of the second term from the square of the first term. $(a+b)(a-b) = a^2 - b^2$	$(2x-5y)(2x+5y)$ $= (2x)^2 - (5y)^2$ $= 4x^2 - 25y^2$	p. 454
Dividing Polynomials		**Section 5.6**
Dividing a Monomial by a Monomial Divide the coefficients, and use the quotient rule for exponents to combine the variables.	$\frac{28m^4n^3}{7m^3n} = \left(\frac{28}{7}\right)m^{4-3}n^{3-1}$ $= 4mn^2$	p. 460
Dividing a Polynomial by a Monomial Divide each term of the polynomial by the monomial. Simplify the result.	$\frac{9x^4 + 6x^3 - 15x^2}{3x}$ $= 3x^3 + 2x^2 - 5x$	p. 460
Dividing a Polynomial by a Polynomial Use the long division method.	$\begin{array}{r} x + 5 \\ x-3\overline{)x^2 + 2x - 7} \\ \underline{x^2 - 3x} \quad \\ 5x - 7 \\ \underline{5x - 15} \\ 8 \end{array}$ The result is $x + 5 + \frac{8}{x-3}$	p. 461

summary exercises :: chapter 5

This summary exercise set will help ensure that you have mastered each of the objectives of this chapter. The exercises are grouped by section. You should reread the material associated with any exercises that you find difficult. The answers to the odd-numbered exercises are in the Answers Appendix in the back of the text.

5.1 and 5.2 *Simplify each expression, using the properties of exponents.*

1. r^4r^9
2. $4x^{-5}$
3. $(2w)^{-3}$
4. $\dfrac{3}{m^{-4}}$
5. $y^{-5}y^2$
6. $\dfrac{w^{-7}}{w^{-3}}$
7. $\dfrac{x^{12}}{x^{15}}$
8. $(6c^0d^4)(-3c^2d^2)$
9. $(5a^2b^3)(2a^{-2}b^{-6})$
10. $\left(\dfrac{3m^2n^3}{p^4}\right)^3$
11. $\left(\dfrac{m^{-3}n^{-3}}{m^{-4}n^4}\right)^3$
12. $\left(\dfrac{r^{-5}}{s^4}\right)^{-2}$
13. $\left(\dfrac{x^3y^4}{x^6y^2}\right)^3$
14. $\left(\dfrac{a^8}{b^4}\right)\left(\dfrac{b^2}{2a^2}\right)^3$
15. $(2a^3)^0(-3a^4)^2$
16. $\left(\dfrac{2x^{-3}y^{-2}}{4x^{-5}y^{-4}}\right)^2\left(\dfrac{8x^5y^6}{4x^7y^4}\right)^3$
17. Write 0.0000425 in scientific notation.
18. Write 3.1×10^{-4} in standard notation.

Business and Finance The value of an investment that compounds interest annually is given by the formula

$$A = P(1 + r)^t$$

where

P = Original amount (principal)

r = Interest rate, in decimal form

t = time, in years

19. Determine the value of an investment after 5 years if \$2,000 was invested initially at 2.7% compounded annually.
20. Determine the value of an investment after 3 years if \$25,000 was invested initially at 6.85% compounded annually.
21. **Statistics** The United States covers approximately 3,794,101 mi^2. Write this in scientific notation, rounding to one decimal place.
22. **Statistics** Approximately 6.76% of the U.S. surface area is water. How many square miles of the United States are covered by water (see exercise 21)? Use scientific notation to report your result.

23. **Statistics** Approximately 6.76% of the U.S. surface area is water. Find the number of square miles of land mass (not covered by water) in the United States (see exercises 21 and 22). Use scientific notation to report your result.
24. **Social Science** According to the U.S. Census Bureau, the U.S. population was approximately 3.1×10^8 in 2010. Use your result from exercise 23 to find the country's population density. *Hint:* The units in your answer should be people per square mile of land mass.

5.3 *Classify each polynomial as a monomial, binomial, or trinomial, if possible.*

25. $6x^4 - 3x$

26. $7x^5$

27. $4x^5 - 8x^3 + 5$

28. $x^3 + 2x^2 - 5x + 3$

29. $-7a^4 - 9a^3$

30. 12

Arrange each polynomial in descending order and state its degree.

31. $5x^5 + 3x^2$

32. $13x^2$

33. $6x^2 + 4x^4 + 6$

34. $5 + x$

35. -8

36. $9x^4 - 3x + 7x^6$

5.4 *Add or subtract as indicated.*

37. Add $7a^2 + 3a$ and $14a^2 - 5a$

38. Add $5x^2 + 3x - 5$ and $4x^2 - 6x - 2$

39. Add $5y^3 - 3y^2$ and $4y + 3y^2$

40. Subtract $7x^2 - 23x$ from $11x^2 - 15x$

41. Subtract $2x^2 - 5x - 7$ from $7x^2 - 2x + 3$

42. Subtract $5x^2 + 3$ from $9x^2 - 4x$

Perform the indicated operations.

43. Subtract $5x - 3$ from the sum of $9x + 2$ and $-3x - 7$.

44. Subtract $5a^2 - 3a$ from the sum of $5a^2 + 2$ and $7a - 7$.

45. Subtract the sum of $16w^2 - 3w$ and $8w + 2$ from $7w^2 - 5w + 2$.

Use the vertical method to find each sum.

46. $x^2 + 5x - 3$ and $2x^2 + 4x - 3$

47. $9b^2 - 7$ and $8b + 5$

48. $x^2 + 7$, $3x - 2$, and $4x^2 - 8x$

Use the vertical method to subtract.

49. $5x^2 - 3x + 2$ from $7x^2 - 5x - 7$

50. $8m - 7$ from $9m^2 - 7$

Business and Finance Castor Books pays \$18.45 for each copy of the new *Illustrated Cars through the Ages* coffee table book (wholesale cost). They estimate that the weekly cost of selling the book is \$525. The bookstore sells each copy for \$39.95.

51. Construct a weekly cost function for Castor Books to sell this book.

52. Construct a simplified profit model for the sale of this book.

53. Find the profit Castor Books earns from *Illustrated Cars through the Ages* if they only sell 20 copies one week.

54. Find the profit Castor Books earns from *Illustrated Cars through the Ages* if they sell 250 copies one week.

5.5 *Multiply.*

55. $(3a^4)(2a^3)$

56. $(2x^2)(3x^5)$

57. $(-9p^3)(-6p^2)$

58. $(3a^2b^3)(-7a^3b^4)$

59. $5(3x - 8)$

60. $2a(5a - 3)$

61. $(-5rs)(2r^2s - 5rs)$

62. $7mn(3m^2n - 2mn^2 + 5mn)$

63. $(x + 5)(x + 4)$

64. $(w - 9)(w - 10)$

65. $(a - 9b)(a + 9b)$

66. $(p - 3q)^2$

67. $(a + 4b)(a + 3b)$

68. $(b - 8)(2b + 3)$

69. $(3x - 5y)(2x - 3y)$

70. $(5r + 7s)(3r - 9s)$

71. $(y + 2)(y^2 - 2y + 3)$

72. $(b + 3)(b^2 - 5b - 7)$

73. $(x - 2)(x^2 + 2x + 4)$

74. $(m^2 - 3)(m^2 + 7)$

75. $2x(x + 5)(x - 6)$

76. $a(2a - 5b)(2a - 7b)$

77. $(x + 7)^2$

78. $(a - 7)^2$

79. $(2w - 5)^2$

80. $(3p + 4)^2$

81. $(a + 7b)^2$

82. $(8x - 3y)^2$

83. $(x - 5)(x + 5)$

84. $(y + 9)(y - 9)$

85. $(2m + 3)(2m - 3)$

86. $(4r - 5)(4r + 5)$

87. $(5r - 2s)(5r + 2s)$

88. $(7a + 3b)(7a - 3b)$

89. $3x(x - 4)^2$

90. $3c(c + 5d)(c - 5d)$

91. $(y - 4)(y + 5)(y + 4)$

92. $(2x + 3)(3x - 5)(2x - 3)$

93. **Geometry** Report the area of the rectangle as a polynomial in descending order.

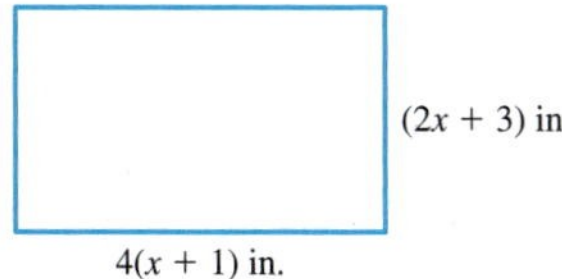

94. **Geometry** A 15- by 21-in. cardboard rectangle has four squares cut from the corners. Each square measures x in. on a side. The cardboard is folded up at the corners to create a box.

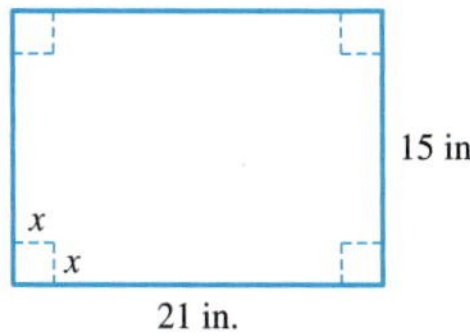

(a) Construct a polynomial function to find the volume of the box.

Hint: $V = LWH$

(b) Find the volume of the box if the corner cuts are 2-in. squares.

5.6 *Divide.*

95. $\frac{9a^5}{3a^2}$

96. $\frac{24m^4n^2}{6m^2n}$

97. $\frac{15a - 10}{5}$

98. $\frac{32a^3 + 24a}{8a}$

99. $\frac{9r^2s^3 - 18r^3s^2}{-3rs^2}$

100. $\frac{35x^3y^2 - 21x^2y^3 + 14x^3y}{7x^2y}$

Perform the indicated long division.

101. $\frac{x^2 - 2x - 15}{x + 3}$

102. $\frac{2x^2 + 9x - 35}{2x - 5}$

103. $\frac{x^2 - 8x + 17}{x - 5}$

104. $\frac{6x^2 - x - 10}{3x + 4}$

105. $\frac{3x^3 + 7x^2 - x - 3}{3x + 1}$

106. $\frac{4x^3 + x + 3}{2x - 1}$

107. $\frac{3x^2 + x^3 + 5 + 4x}{x + 2}$

108. $\frac{2x^4 - 2x^2 - 10}{x^2 - 3}$

chapter test 5

CHAPTER 5

Use this chapter test to assess your progress and to review for your next exam. Allow yourself about an hour to take this test. The answers to these exercises are in the Answers Appendix in the back of the text.

Use the properties of exponents to simplify each expression.

1. $(3x^2y)(-2xy^3)$

2. $\left(\frac{8m^2n^5}{2p^3}\right)^2$

3. $(x^4y^5)^2$

4. $\frac{9c^{-5}d^3}{18c^{-7}d^4}$

5. $(3x^2y)^3(-2xy^2)^2$

6. $(-3x^{-3}y)^2(4x^{-5}y^{-2})^{-1}$

7. $\frac{3x^0}{(2y)^0}$

Classify each polynomial as a monomial, binomial, or trinomial.

8. $6x^2 + 7x$

9. $5x^2 + 8x - 8$

Arrange the polynomial in descending order. Give the coefficient and degree of each term. Then, give the degree of the polynomial.

10. $-3x^2 + 8x^4 - 7$

Add.

11. $3x^2 - 7x + 2$ and $7x^2 - 5x - 9$

12. $7a^2 - 3a$ and $7a^3 + 4a^2$

Subtract.

13. $5x^2 - 2x + 5$ from $8x^2 + 9x - 7$

14. $5a^2 + a$ from the sum of $3a^2 - 5a$ and $9a^2 - 4a$

Use the vertical method to add or subtract.

15. Add $x^2 + 3$, $5x - 9$, and $3x^2$.

16. Subtract $3x^2 - 5$ from $5x^2 - 7x$.

Multiply.

17. $5ab(3a^2b - 2ab + 4ab^2)$

18. $(x + 3y)(4x - 5y)$

19. $(3m + 2n)^2$

Divide.

20. $\dfrac{4x^3 - 5x^2 + 7x - 9}{x - 2}$

21. STATISTICS Canada covers approximately 9,984,670 km^2. Approximately 8.92% of its surface area is water. How much of Canada's surface area is water (in square kilometers)? Use scientific notation to report your result, rounding to one decimal place.

22. GEOMETRY Report the area of the rectangle as a polynomial, in descending order.

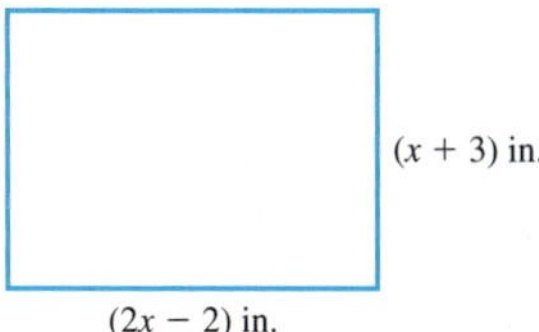

BUSINESS AND FINANCE Funghi Books pays $6.27 for each copy of *The Forager's Mushroom Cookbook* (wholesale cost). They estimate that the weekly cost of selling the book is $285. The bookstore sells each copy for $14.95.

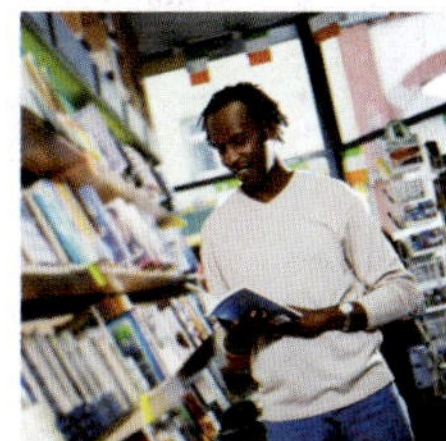

23. Construct a weekly cost function for Funghi Books to sell this book.

24. Construct a simplified profit model for the sale of this book.

25. Find the profit Funghi Books earns from *The Forager's Mushroom Cookbook* if they sell 75 copies one week.

cumulative review chapters 0–5

Use this exercise set to review concepts from earlier chapters. While it is not a comprehensive exam, it will help you identify any material that you need to review before moving on to the next chapter. The answers to these exercises are in the Answers Appendix in the back of the text.

Evaluate each expression if $x = 3$, $y = -2$, and $z = 4$.

1. $-2x^2 - 3y^2 + 5z$

2. $\dfrac{-4y + 3x^2}{5z + x - y}$

Solve each equation.

3. $11x - 7 = 10x$

4. $-\frac{2}{3}x = 24$

5. $7x - 5 = 3x + 11$

6. $7 - 3x = 5 - 6x$

7. $\frac{3}{5}x - 8 = 15 - \frac{2}{5}x$

8. $2(x - 3) + 5 = 2x - 1$

9. $4x - (2 - x) = 5x + 3$

10. $\dfrac{x + 1}{5} - \dfrac{2x - 3}{2} = 3$

Solve each inequality.

11. $7x - 5 > 8x + 10$

12. $6x - 9 < 3x + 6$

13. If $f(x) = 2x^3 - x^2 + 7$, evaluate $f(-2)$.

14. If $g(x) = 3x + 11$, solve $g(x) = 3$.

15. **Business and Finance** A hardware store sells a desk lamp for $39.95. Write a function modeling the revenue if it sells x lamps.

16. **Business and Finance** How much revenue does the hardware store earn by selling 17 of these lamps in one week?

17. Find the slope and y-intercept of the line represented by the equation $4x + y = 9$.

18. Find the slope of a line perpendicular to the line represented by the equation $3x + 9y = 10$.

Write an equation for the line that satisfies each set of conditions.

19. L has slope -2 and y-intercept of $(0, 4)$.

20. L passes through the point $(3, 2)$ and is parallel to the line $4x - 5y = 20$.

21. L passes through the points $(1, -3)$ and $(3, 5)$.

22. L is perpendicular to the line $x - 2y = 3$ and passes through the point $(1, -2)$.

Perform the indicated operations.

23. $(x^2 - 3x + 5) + (2x^2 + 5x - 9)$

24. $(3x^2 - 8x - 7) - (2x^2 - 5x + 11)$

25. $4x(3x - 5)$

26. $(2x - 5)(3x + 8)$

27. $(x + 2)(x^2 - 3x + 5)$

28. $(2x + 7)(2x - 7)$

29. $(3x - 5)^2$

30. $5x(2x - 5)^2$

31. $\dfrac{32x^2y^3 - 16x^4y^2 + 8xy^2}{8xy^2}$

32. $\dfrac{2x^3 - 15x - 7}{x - 3}$

Simplify each expression.

33. $(3x^2)^2(2x^3)$

34. $(x^4y^{-3})^4$

35. $(2x^0)^3(-3x^2y)^2$

36. $\dfrac{6x^3y^{-5}}{3x^{-2}y^6}$

37. Calculate. Write your answer in scientific notation.

$$\frac{(4.2 \times 10^7)(6.0 \times 10^{-3})}{1.2 \times 10^{-5}}$$

Solve each problem.

38. **Number Problem** If 7 times a number decreased by 9 is 47, find the number.

39. **Number Problem** The sum of two consecutive odd integers is 132. What are the two integers?

40. **Geometry** The length of a rectangle is 4 centimeters (cm) more than 5 times the width. If the perimeter is 56 cm, what are the dimensions of the rectangle?

Factoring

INTRODUCTION

Developing security codes and software is big business. Corporations all over the world sell encryption systems designed to keep data secure and safe.

In 1977, three professors from the Massachusetts Institute of Technology developed the RSA encryption system. They offered $100 to anyone who could break their security code, which was based on a number that has *129 digits*. They called the code RSA-129. For the code to be broken, the 129-digit number had to be factored into two prime numbers; that is, two prime numbers had to be found that when multiplied together give the 129-digit number. The three professors predicted that it would take *40 quadrillion* years to find the two numbers.

In April 1994, a research scientist, three computer hobbyists, and more than 600 volunteers from the Internet, using 1,600 computers, found the two numbers after 8 months of work and won the $100.

Software companies are waging a legal battle against the U.S. government because the government does not permit codes for which it does not have the key. The software firms claim that this prohibition costs them about $60 billion in lost sales because companies will not buy an encryption system knowing it can be monitored by the U.S. government.

CHAPTER 6 OUTLINE

6.1 An Introduction to Factoring

< 6.1 Objectives >

1 > Find the greatest common factor (GCF) of a set of terms

2 > Factor a monomial from a polynomial

3 > Factor a polynomial by grouping

In Chapter 5 you were given factors and asked to find a product. We are now going to reverse this process. You will be given a polynomial and asked to find its factors. This is called **factoring.**

We start with an example from arithmetic. To *multiply* $5 \cdot 7$, you write

$$5 \cdot 7 = 35$$

To *factor* 35, you write

$$35 = 5 \cdot 7$$

Factoring is the *reverse* of multiplication.

Now we look at factoring in algebra. When multiplying, we use the distributive property as

$$a(b + c) = ab + ac$$

For instance,

$$3(x + 5) = 3x + 15$$

NOTE

3 and $x + 5$ are the factors of $3x + 15$.

To use the distributive property when factoring, we apply the property in reverse.

$$ab + ac = a(b + c)$$

The distributive property lets us remove the common monomial factor a from the terms of $ab + ac$. To use this, the first step is to see whether the terms of the polynomial have a common monomial factor. In our example,

$$3x + 15 = 3 \cdot x + 3 \cdot 5$$

Common factor

So, by the distributive property,

$$3x + 15 = 3(x + 5)$$

The original terms are each divided by the greatest common factor to determine the expression in parentheses.

To check this, multiply $3(x + 5)$.

Multiplying →

$$3(x + 5) = 3x + 15$$

← Factoring

NOTES

Factoring is the reverse of multiplication.

This diagram relates multiplication and factorization.

The first step in factoring a polynomial is to identify the *greatest common factor* (GCF) of a set of terms.

When learning to simplify fractions, you probably learned to find the GCF of a set of integers. To remind you, the GCF of a set of integers is the largest positive integer that is a factor of every number in the set.

Example 1 Finding the GCF of Integers

< Objective 1 >

RECALL

One number divides a second number if there is no remainder when we divide the second number by the first.

For instance, 3 divides 9, but 3 does not divide 10.

Find the GCF of each set of integers.

(a) 9, 12

1 and 3 are the only positive integers that divide both 9 and 12. Therefore, 3 is the GCF of 9 and 12.

(b) 10, 25, 150

1 and 5 are the only positive integers that divide 10, 25, and 150. The GCF is 5.

(c) 8, 40, 56

The common factors of 8, 40, and 56 are 1, 2, 4, and 8, so 8 is the GCF.

Check Yourself 1

Find the GCF of each set.

(a) 6, 21 **(b)** 12, 30, 72 **(c)** 7, 56, 91 **(d)** 8, 14, 31

When working with variables, the basic idea is the same. We begin by looking at a single variable raised to different powers. Consider the set

x^3, x^5, x^6

If we look at these three terms in factored form, we have

$x^3 = x \cdot x \cdot x$ Three times

$x^5 = x \cdot x \cdot x \cdot x \cdot x$ Five times

$x^6 = x \cdot x \cdot x \cdot x \cdot x \cdot x$ Six times

You should see that x^3 divides all three terms, but x to any higher power does not.

In general, the GCF of a single variable to different powers is that variable to the *lowest* power in the set.

We look at each variable separately when working with products of variables to powers. For instance, consider

$x^2yz^2, x^4y^3, x^4y^2z^3$

To find the GCF, we look at each variable separately.

The GCF of x^2, x^4, and x^4 is x^2.

The GCF of y, y^3, and y^2 is y ($y = y^1$).

Since the second term does not have a z-term, there is no z-term in the GCF.

Therefore, the GCF of x^2yz^2, x^4y^3, and $x^4y^2z^3$ is x^2y. This is the product of the GCF of each variable factor with its powers.

Example 2 Finding the GCF of Variables

Find the GCF of

$x^2y^5z^2, x^3y^5z, x^3y^4z^3$

We look at each variable separately.

The GCF of x^2, x^3, and x^3 is x^2.

The GCF of y^5, y^5, y^4 is y^4.

The GCF of z^2, z, and z^3 is z.

The GCF of $x^2y^5z^2$, x^3y^5z, and $x^3y^4z^3$ is the product of the GCFs of each variable factor. In this case, we have x^2, y^4, and z, so the GCF is x^2y^4z.

Check Yourself 2

Find the GCF of each set of terms.

(a) $x^8y^2z^9$, $x^{10}y^7z^5$, $x^6y^5z^9$ **(b)** $a^4b^4c^3$, a^3c^5, $a^3b^4c^2$
(c) $x^5y^9z^{11}$, x^3yz^8, $x^4y^8z^5$

To find the GCF of a set of terms, we find the GCF of the numerical coefficients, as we did in Example 1, and the GCF of the variable factors, as we did in Example 2. The GCF of the set of terms is the product of these two GCFs.

Definition

The Greatest Common Factor (GCF)

The **greatest common factor (GCF)** of a set of terms is the product of the GCFs of the numerical factors and the variable factors.

Example 3 Finding the GCF

Find the GCF of each set of terms.

(a) $12a^3$, $18a^2$

We look at the coefficients and the variables separately.

The GCF of 12 and 18 is 6.

The GCF of a^3 and a^2 is a^2.

We take the product of these GCFs to find the GCF of the set. The GCF of $12a^3$ and $18a^2$ is $6a^2$.

(b) $9x^2y^3$, $15x^4y$, $6x^2y^5$

RECALL

x^2 is the GCF of x^2, x^4, and x^2. The GCF of y^3, y, and y^5 is y.

The GCF of 9, 15, and 6 is 3.

The GCF of x^2y^3, x^4y, and x^2y^5 is x^2y.

Therefore, $3x^2y$ is the GCF of $9x^2y^3$, $15x^4y$, and $6x^2y^5$.

(c) $5x^4$, $12x^5$, $10x^3$

The GCF of 5, 12, and 10 is 1. These numbers have no other common factors.

The GCF of x^4, x^5, and x^3 is x^3.

The GCF of the set of terms is x^3 (1 is understood to be the coefficient).

Check Yourself 3

Find the GCF of each set of terms.

(a) $10x^2$, $14x^3$, $20x^5$ **(b)** $3ab^4c^2$, $6a^3b^4c^3$, $12a^5b^3c^3$
(c) $8x^2y^3$, x^3y^2, $8y^5$ **(d)** $2x$, $3y$, $4z$

The first step when factoring a polynomial is to factor any GCF from the polynomial. To do that, we find the GCF of the terms and then use the distributive property to factor the GCF out of the polynomial.

Because factoring is the reverse of multiplying, we can check that we have factored correctly. If we multiply the GCF by the remaining polynomial, the product should be the original polynomial.

Step by Step

Factoring a Monomial from a Polynomial

Step 1 Find the *greatest common factor* of all the terms.

Step 2 Factor the GCF from each term; then apply the distributive property.

Step 3 Mentally check your factoring by multiplication.

Example 4 Factoring the GCF from a Binomial

< Objective 2 >

NOTE

You should always check your result by multiplying to make sure that you get the original polynomial. Try it here. Multiply $4x$ by $2x + 3$.

(a) Factor $8x^2 + 12x$.

The largest common numerical factor of 8 and 12 is 4, and x is the variable factor with the largest common power. So $4x$ is the GCF. Write

$$8x^2 + 12x = 4x \cdot 2x + 4x \cdot 3$$

(GCF: $4x$)

We use the distributive property to write

$$8x^2 + 12x = 4x(2x + 3)$$

NOTE

It is also true that

$6a^4 - 18a^2 = 3a(2a^3 - 6a)$

However, this is *not completely factored*. Do you see why? You want to find the common monomial factor with the *largest possible* coefficient and the *largest* exponent, in this case $6a^2$.

(b) Factor $6a^4 - 18a^2$.

The GCF in this case is $6a^2$. Write

$$6a^4 - 18a^2 = 6a^2 \cdot a^2 - 6a^2 \cdot 3$$

(GCF: $6a^2$)

Again, using the distributive property yields

$$6a^4 - 18a^2 = 6a^2(a^2 - 3)$$

You should check this by multiplying.

(c) Factor $3x^2 + 5y^2$.

The GCF of 3 and 5 is 1. The variable factors, x^2 and y^2, have no variables in common, so the GCF of this polynomial is 1. In this case, we cannot factor the polynomial any further.

NOTE

A polynomial that cannot be factored is called a **prime polynomial.**

Check Yourself 4

Factor each polynomial.

(a) $5x + 20$ **(b)** $6x^2 - 24x$ **(c)** $2x^2 + 3$

(d) $10a^3 - 15a^2$

The process is exactly the same for polynomials with more than two terms. Consider Example 5.

Example 5 Factoring the GCF from a Polynomial

NOTE

The GCF is 5.

(a) Factor $5x^2 - 10x + 15$.

$$5x^2 - 10x + 15 = 5 \cdot x^2 - 5 \cdot 2x + 5 \cdot 3$$

(GCF: 5)

$$= 5(x^2 - 2x + 3)$$

NOTE

The GCF is $3a$.

(b) Factor $6ab + 9ab^2 - 15a^2$.

$$6ab + 9ab^2 - 15a^2 = 3a \cdot 2b + 3a \cdot 3b^2 - 3a \cdot 5a$$

GCF

$$= 3a(2b + 3b^2 - 5a)$$

NOTE

The GCF is $4a^2$.

(c) Factor $4a^4 + 12a^3 - 20a^2$.

$$4a^4 + 12a^3 - 20a^2 = 4a^2 \cdot a^2 + 4a^2 \cdot 3a - 4a^2 \cdot 5$$

GCF

$$= 4a^2(a^2 + 3a - 5)$$

RECALL

In each of these examples, you should check the result by multiplying the factors.

(d) Factor $6a^2b + 9ab^2 + 3ab$.

Mentally note that 3, a, and b are factors of each term, so

$$6a^2b + 9ab^2 + 3ab = 3ab(2a + 3b + 1)$$

Check Yourself 5

Factor each polynomial.

(a) $8b^2 + 16b - 32$
(b) $4xy - 8x^2y + 12x^3$
(c) $7x^4 - 14x^3 + 21x^2$
(d) $5x^2y^2 - 10xy^2 + 15x^2y$

In Sections 5.4 and 5.5, you learned that when you multiply a polynomial by a negative number, you need to distribute the negative sign by changing the sign of every term in the polynomial. For instance,

$$-(5x^2 - 3y^2 + 2xy) = -5x^2 + 3y^2 - 2xy$$

RECALL

We write the leading negative but omit 1. The coefficient 1 is understood.

When the leading term of a polynomial has a negative coefficient, we factor out -1, along with the GCF. In this instance, we see that the GCF of the three terms is 1, but the leading term has a negative coefficient, so we factor out -1, remembering to change all of the signs when we do.

$$-5x^2 + 3y^2 - 2xy = -(5x^2 - 3y^2 + 2xy)$$

When the GCF of a polynomial is not 1, but the leading term is negative, we factor out both, the GCF and -1. For instance, the GCF of $-3x^2$, $6x$, and -9 is 3. To factor

$$-3x^2 + 6x - 9$$

we factor out the GCF 3 and the leading negative.

$$-3x^2 + 6x - 9 = -3(x^2 - 2x + 3)$$

Example 6 Factoring out the GCF With a Negative Coefficient

RECALL

Change the sign of every term in the polynomial when factoring out a negative.

In each case, factor out the GCF with a negative coefficient.

(a) $-x^2 - 5x + 7$

Here, we factor out -1.

$$-x^2 - 5x + 7 = (-1)(x^2) + (-1)(5x) - (-1)(7) \qquad -(-1)(7) = +7$$

$$= -1(x^2 + 5x - 7)$$

$$= -(x^2 + 5x - 7) \qquad \text{We do not write the 1. It is understood.}$$

It is an especially good idea to check your work when factoring out a negative coefficient because it is very easy to forget to change one of the signs in the remaining polynomial.

Check

$-(x^2 + 5x - 7) = -(x^2) - (5x) - (-7)$

$= -x^2 - 5x + 7$ This is the polynomial we started with; it checks.

(b) $-10x^2y + 5xy^2 - 20xy$

$5xy$ is the GCF of $10x^2y$, $5xy^2$, and $20xy$. Because the leading coefficient is negative, we factor out $-5xy$.

$-10x^2y + 5xy^2 - 20xy = (-5xy)(2x) - (-5xy)(y) + (-5xy)(4)$

$= -5xy(2x - y + 4)$

Check

$-5xy(2x - y + 4) = (-5xy)(2x) - (-5xy)(y) + (-5xy)(4)$

$= -10x^2y + 5xy^2 - 20xy$ The original polynomial

Check Yourself 6

In each case, factor out the GCF with a negative coefficient. Multiply to check your result.

(a) $-x^2 + 3x - 9$ **(b)** $-6a^3b^2 - 3a^2b + 12ab$

In some cases, we can factor a common binomial from an expression.

Example 7 **Finding a Common Binomial Factor**

(a) Factor $3x(x + y) + 2(x + y)$.

We see that the *binomial* $x + y$ is a common factor and can therefore be factored out.

NOTE

Because of the commutative property, we can write the factors in any order.

$3x(x + y) + 2(x + y)$

$= (x + y)(3x + 2)$

(b) Factor $3x^2(x - y) + 6x(x - y) + 9(x - y)$.

We note that here the GCF is $3(x - y)$. Factoring as before, we have

$3(x - y)(x^2 + 2x + 3)$

Check Yourself 7

Completely factor each polynomial.

(a) $7a(a - 2b) + 3(a - 2b)$ **(b)** $4x^2(x + y) - 8x(x + y) - 16(x + y)$

Some polynomials can be factored by grouping the terms and finding common factors within each group. We call this process **factoring by grouping.**

In Example 7, we looked at the expression

$3x(x + y) + 2(x + y)$

and found that we could factor out the common binomial $(x + y)$. This gave us

$(x + y)(3x + 2)$

We extend this technique in Example 8.

Example 8 Factoring by Grouping

< Objective 3 >

> **NOTE**
>
> Our example has *four* terms. That is the clue to try to factor by grouping.

Suppose we want to factor the polynomial

$ax - ay + bx - by$

As you can see, the polynomial has no common factors. However, look at what happens if we separate the polynomial into *two groups* of *two terms.*

$ax - ay + bx - by$

$= \underbrace{ax - ay} + \underbrace{bx - by}$

Now each group has a common factor, and we can write the polynomial as

$a(x - y) + b(x - y)$

In this form, we can see that $x - y$ is the GCF. Factoring out $x - y$, we get

$a(x - y) + b(x - y) = (x - y)(a + b)$

Check Yourself 8

Factor by grouping.

$x^2 - 2xy + 3x - 6y$

Be particularly careful of your treatment of addition and subtraction operations when you factor by grouping. Consider Example 9.

Example 9 Factoring by Grouping

Factor $2x^3 - 3x^2 - 6x + 9$.

> **RECALL**
>
> $9 = (-3)(-3)$

We group the terms of the polynomial as follows:

$\underbrace{2x^3 - 3x^2} \underbrace{- 6x + 9}$ Factor out the common factor of -3 from the second pair of terms.

$= x^2(2x - 3) - 3(2x - 3)$

$= (2x - 3)(x^2 - 3)$

Check Yourself 9

Factor by grouping.

$3y^3 + 2y^2 - 6y - 4$

It may also be necessary to change the order of the terms as they are grouped. Look at Example 10.

Example 10 Factoring by Grouping

Factor $x^2 - 6yz + 2xy - 3xz$.

Grouping the terms as before, we have

$\underbrace{x^2 - 6yz} + \underbrace{2xy - 3xz}$

Do you see that we have accomplished nothing because there are no common factors in the first group?

We can, however, rearrange the terms to write the original polynomial as

$$\underbrace{x^2 + 2xy}\ \underbrace{- 3xz - 6yz}$$

$$= x(x + 2y) - 3z(x + 2y)$$ We can now factor out the common factor of $x + 2y$ in both groups.

$$= (x + 2y)(x - 3z)$$

NOTES

With practice, you will learn to see how to group the terms.

Often, you can group the factors in more than one way. In any case, the factored form is the same.

Remember that you can always check your factoring by multiplying. Here we check by multiplying $(x + 2y)(x - 3z)$.

$$(x + 2y)(x - 3z) = (x)(x) + (x)(-3z) + (2y)(x) + (2y)(-3z)$$
$$= x^2 - 3xz + 2xy - 6yz$$

By rearranging terms, we see that this is equal to the original expression.

Check Yourself 10

We can write the polynomial of Example 10 as

$$x^2 - 3xz + 2xy - 6yz$$

Factor, and verify that the factored form is the same in either case.

It might happen that a polynomial *cannot* be factored. We call such a polynomial a **prime polynomial.**

Check Yourself ANSWERS

1. **(a)** 3; **(b)** 6; **(c)** 7; **(d)** 1 **2.** **(a)** $x^6y^2z^5$; **(b)** a^3c^2; **(c)** x^3yz^5 **3.** **(a)** $2x^2$; **(b)** $3ab^3c^2$; **(c)** y^2; **(d)** 1 **4.** **(a)** $5(x + 4)$; **(b)** $6x(x - 4)$; **(c)** prime polynomial; **(d)** $5a^2(2a - 3)$ **5.** **(a)** $8(b^2 + 2b - 4)$; **(b)** $4x(y - 2xy + 3x^2)$; **(c)** $7x^2(x^2 - 2x + 3)$; **(d)** $5xy(xy - 2y + 3x)$ **6.** **(a)** $-(x^2 - 3x + 9)$; **(b)** $-3ab(2a^2b + a - 4)$ **7.** **(a)** $(a - 2b)(7a + 3)$; **(b)** $4(x + y)(x^2 - 2x - 4)$ **8.** $(x - 2y)(x + 3)$ **9.** $(3y + 2)(y^2 - 2)$ **10.** $(x - 3z)(x + 2y)$

Reading Your Text

These fill-in-the-blank exercises will help you understand some of the key vocabulary used in this section. The answers to these exercises are in the Answers Appendix in the back of the text.

(a) The ________ property lets us remove the common monomial factor a from the expression $ab + ac$.

(b) The first step in factoring is to identify the greatest ________ factor.

(c) After factoring, it is always a good idea to check your result by ________.

(d) If a polynomial has four terms, try the factoring by ________ method.

6.1 exercises

Skills | Calculator/Computer | Career Applications | Above and Beyond

< Objective 1 >

Find the greatest common factor of each set of terms.

1. 20, 22 **2.** 15, 35 **3.** 16, 32, 88 **4.** 44, 66, 143

5. x^2, x^5 **6.** y^7, y^9 **7.** a^3, a^6, a^9 **8.** b^4, b^6, b^8

9. $5x^4, 10x^5$ **10.** $8y^9, 24y^3$ **11.** $4a^4, 10a^7, 12a^{14}$ **12.** $9b^3, 6b^5, 12b^4$

13. $9x^2y, 12xy^2, 15x^2y^2$ **14.** $12a^3b^2, 18a^2b^3, 6a^4b^4$ **15.** $15ab^3, 10a^2bc, 25b^2c^3$ **16.** $12x^3, 15x^2y^2, 21y^5$

17. $15a^2bc^2, 9ab^2c^2, 6a^2b^2c^2$ **18.** $18x^3y^2z^3, 27x^4y^2z^3, 81xy^2z$ **19.** $5, 12xy, 4ab$ **20.** $3a^2b, 5a^3, 9bc$

VIDEO

21. $(x + y)^2, (x + y)^5$ **22.** $12(a + b)^4, 4(a + b)^3$

< Objective 2 >

Factor each polynomial.

23. $10x + 5$ **24.** $5x - 15$ **25.** $24m - 32n$ **26.** $7p - 21q$

27. $4x - 7$ **28.** $x + 2$ **29.** $12m^2 + 8m$ **30.** $30n^2 - 35n$

31. $10s^2 + 5s$ **32.** $12y^2 - 6y$ **33.** $24x^2 - 60x$ **34.** $14b^2 - 28b$

35. $m^2 + 1$ **36.** $8x^2 + 5$ **37.** $15a^3 - 25a^2$ **38.** $36b^4 + 24b^2$

39. $6pq + 18p^2q$ **40.** $9xy - 18xy^2$ **41.** $7m^3n - 21mn^3$ **42.** $36p^2q^2 - 9pq$

43. $6x^2 - 18x + 30$ **44.** $7a^2 + 21a - 42$ **45.** $5a^3 - 15a^2 + 25a$ **46.** $5x^3 - 15x^2 + 25x$

47. $12x + 8xy - 28xy^2$ **48.** $4s + 6st - 14st^2$ **49.** $10x^2y + 15xy - 5xy^2$ **50.** $3ab^2 + 6ab - 15a^2b$

51. $10r^3s^2 + 25r^2s^2 - 15r^2s^3$ **52.** $28x^2y^3 - 35x^2y^2 + 42x^3y$ **53.** $9a^5 - 15a^4 + 21a^3 - 27a$

VIDEO

54. $8p^6 - 40p^4 + 24p^3 + 16p^2$ **55.** $15m^3n^2 - 20m^2n + 35mn^3 - 10mn$

56. $14ab^4 + 21a^2b^3 - 35a^3b^2 + 28ab^2$ **57.** $x(x - 9) + 5(x - 9)$ **58.** $y(y + 5) - 3(y + 5)$

VIDEO

59. $p(p - 2q) - q(p - 2q)$ **60.** $x(3x - 4y) - y(3x - 4y)$ **61.** $x(y - z) + 3(y - z)$ **62.** $2a(c - d) - b(c - d)$

Factor out any GCF and negative coefficient.

63. $-t^2 + 6t + 10$ **64.** $-u^2 - 4u + 9$ **65.** $-3x^2 + 4x - 5$ **66.** $-5m^2n - m + 4n^2$

67. $-4x^2 - 16x + 4$ **68.** $-6y^5 + 9y^3 + 18y + 12$ **69.** $-4m^2n^3 - 6mn^3 - 10n^2$ **70.** $-8a^4b^2 + 4a^2b^3 + 12ab^3$

< Objective 3 >

Factor each polynomial by grouping.

71. $ab - ac + b^2 - bc$ **72.** $ax + 2a + bx + 2b$ **73.** $6r^2 + 12rs - r - 2s$ **74.** $2mn - 4m^2 + 3n - 6m$

VIDEO

75. $ab^2 - 2b^2 + 3a - 6$ **76.** $r^2s^2 - 3s^2 - 2r^2 + 6$ **77.** $x^2 + 3x - 4xy - 12y$ **78.** $a^2 - 12b + 3ab - 4a$

79. $m^2 - 6n^3 + 2mn^2 - 3mn$ **80.** $r^2 - 3rs^2 - 12s^3 + 4rs$

Factor each polynomial completely.
Hint: Factor out the GCF and then factor the remaining polynomial by grouping.

81. $20x + 60 - 5xy - 15y$ **82.** $12x^2y - 18xy + 4x^2 - 6x$ **83.** $-18ab - 6b^2 + 9ac + 3bc$

84. $-12x^3 + 24x^2 - 4x^3y + 8x^2y$

Determine whether each factoring is correct.

85. $x^2 - x - 6 = (x - 3)(x + 2)$

86. $x^2 - x - 12 = (x - 4)(x + 3)$

87. $x^2 + x - 12 = (x + 6)(x - 2)$

88. $x^2 + 2x - 8 = (x + 8)(x - 1)$

89. $2x^2 - 5x - 3 = (2x + 1)(x - 3)$

90. $6x^2 - 13x + 6 = (3x - 2)(2x - 3)$

Determine whether each statement is **true** *or* **false.**

91. Factoring is the reverse of addition.

92. The key property used in factoring out the GCF is the distributive property.

Complete each statement with **always, sometimes,** *or* **never.**

93. If the GCF is factored out of a trinomial, the result is __________ the GCF times a trinomial.

94. If a four-term polynomial has no common factor (other than 1), factoring by grouping is __________ successful.

Skills | Calculator/Computer | **Career Applications** | Above and Beyond

95. **Allied Health** A patient's protein secretion amount, in milligrams per day, is recorded over several days. Based on these observations, lab technicians determine that the polynomial $-t^3 - 6t^2 + 11t + 66$ provides a good approximation of the patient's protein secretion amount t days after testing begins. Factor this polynomial.

96. **Allied Health** The concentration, in micrograms per milliliter (μg/mL), of the antibiotic chloramphenicol is given by $8t^2 - 2t^3$, in which t is the number of hours after the drug is taken. Factor this polynomial. VIDEO

97. **Manufacturing Technology** Polymer pellets need to be as perfectly round as possible. In order to avoid the formation of flat spots during the hardening process, the pellets are kept off a surface by blasts of air. The height of a pellet above the surface t seconds after a blast is given by $v_0t - 4.9t^2$. Factor this expression.

98. **Information Technology** The total time to transmit a packet is given by the expression $T = d + 2p$, in which d is the quotient of the distance and the propagation velocity, and p is the quotient of the size of the packet and the information transfer rate. How long will it take to transmit a 1,500-byte packet 10 m on an Ethernet if the information transfer rate is 100 MB per second and the propagation velocity is 2×10^8 m/s?
Hint: Use either 1,500 bytes = 0.0015 MB or 1 MB = 10^6 bytes.

Skills | Calculator/Computer | Career Applications | **Above and Beyond**

99. The GCF of $2x - 6$ is 2. The GCF of $5x + 10$ is 5. Find the greatest common factor of the product $(2x - 6)(5x + 10)$.

100. The GCF of $3z + 12$ is 3. The GCF of $4z + 8$ is 4. Find the GCF of the product $(3z + 12)(4z + 8)$.

101. The GCF of $2x^3 - 4x$ is $2x$. The GCF of $3x + 6$ is 3. Find the GCF of the product $(2x^3 - 4x)(3x + 6)$.

102. State, in a sentence, the rule that exercises 99 to 101 illustrate.

103. For the monomials x^4y^2, x^8y^6, and x^9y^4, explain how you can determine the GCF by inspecting exponents.

104. It is not possible to use the grouping method to factor $2x^3 + 6x^2 + 8x + 4$. Is it correct to conclude that the polynomial is prime? Justify your answer.

105. **Geometry** The area of a rectangle with width t is given by $33t - t^2$. Factor the expression and determine the length of the rectangle in terms of t.

106. GEOMETRY The area of a rectangle of length x is given by $3x^2 + 5x$. Find the width of the rectangle.

107. NUMBER PROBLEM For centuries, mathematicians have found factoring numbers into prime factors a fascinating subject. A prime number is a whole number greater than 1 that has no factors other than 1 and itself. The list of primes begins with 2 and then goes on: 3, 5, 7, 11, ... What are the first 10 primes? What are the primes less than 100? If you list the numbers from 1 to 100 and then cross out all numbers that are multiples of 2, 3, 5, and 7, what is left? Are all the numbers not crossed out prime? Write a paragraph to explain why this might be so. You might want to investigate the Sieve of Eratosthenes, a system from 230 B.C. for finding prime numbers.

108. NUMBER PROBLEM If we could make a list of all the prime numbers, what number would be at the end of the list? Because there are infinitely many prime numbers, there is no "largest prime number." But is there some formula that gives us all the primes? Here are some formulas proposed over the centuries:

$n^2 + n + 17 \qquad 2n^2 + 29 \qquad n^2 - n + 11$

In each expression, $n = 1, 2, 3, 4, \ldots$, that is, a positive integer beginning with 1. Investigate these expressions with a partner. Do the expressions give prime numbers when they are evaluated for these values of n? Do the expressions give *every* prime in the range of resulting numbers? Can you put in *any* positive number for n?

109. NUMBER PROBLEM How are primes used in coding messages and for security? Work together to decode the messages. The messages are coded using this code: After the numbers are factored into prime factors, the power of 2 gives the number of the letter in the alphabet. This code would be easy for a code breaker to figure out, but you might make up a code that would be more difficult to break. chapter 6 >Make the Connection

(a) 1310720, 229376, 1572864, 1760, 460, 2097152, 336

(b) 786432, 142, 4608, 278528, 1344, 98304, 1835008, 352, 4718592, 5242880

(c) Code a message, using this rule. Exchange your message with a partner to decode it.

110. One concept used in computer encryption involves factoring a large number that is the product of two prime numbers. If the original number is very large, it is extremely difficult and time-consuming to find the prime factors. Try to factor each number. (These are not considered large!) chapter 6 >Make the Connection

(a) 1,739 **(b)** 5,429 **(c)** 19,177 **(d)** 163,747

Answers

1. 2 **3.** 8 **5.** x^2 **7.** a^3 **9.** $5x^4$ **11.** $2a^4$ **13.** $3xy$ **15.** $5b$ **17.** $3abc^2$ **19.** 1 **21.** $(x + y)^2$ **23.** $5(2x + 1)$ **25.** $8(3m - 4n)$ **27.** Prime polynomial **29.** $4m(3m + 2)$ **31.** $5s(2s + 1)$ **33.** $12x(2x - 5)$ **35.** Prime polynomial **37.** $5a^2(3a - 5)$ **39.** $6pq(1 + 3p)$ **41.** $7mn(m^2 - 3n^2)$ **43.** $6(x^2 - 3x + 5)$ **45.** $5a(a^2 - 3a + 5)$ **47.** $4x(3 + 2y - 7y^2)$ **49.** $5xy(2x + 3 - y)$ **51.** $5r^2s^2(2r + 5 - 3s)$ **53.** $3a(3a^4 - 5a^3 + 7a^2 - 9)$ **55.** $5mn(3m^2n - 4m + 7n^2 - 2)$ **57.** $(x - 9)(x + 5)$ **59.** $(p - 2q)(p - q)$ **61.** $(y - z)(x + 3)$ **63.** $-(t^2 - 6t - 10)$ **65.** $-(3x^2 - 4x + 5)$ **67.** $-4(x^2 + 4x - 1)$ **69.** $-2n^2(2m^2n + 3mn + 5)$ **71.** $(b - c)(a + b)$ **73.** $(r + 2s)(6r - 1)$ **75.** $(a - 2)(b^2 + 3)$ **77.** $(x + 3)(x - 4y)$ **79.** $(m - 3n)(m + 2n^2)$ **81.** $5(x + 3)(4 - y)$ **83.** $-3(3a + b)(2b - c)$ **85.** Correct **87.** Incorrect **89.** Correct **91.** False **93.** always **95.** $-(t + 6)(t^2 - 11)$ **97.** $t(v_0 - 4.9t)$ **99.** 10 **101.** $6x$ **103.** Above and Beyond **105.** $33 - t$ **107.** Above and Beyond **109.** Above and Beyond

ISBNs and the Check Digit

If you look at the back of your textbook, you should see a long number and a bar code. The number is called the International Standard Book Number or ISBN.

The ISBN system was first developed in 1966 by Gordon Foster at Trinity College in Dublin, Ireland. When first developed, ISBNs were 9 digits long, but by 1970, an international agreement extended them to 10 digits.

In 2007, 13-digits became the standard for ISBN numbers. This is the number on the back of your text. Each ISBN has five blocks of numbers. A common form is XXX-X-XX-XXXXXX-X, though it can vary.

- The first block or set of digits is either 978 or 979. This set was added in 2007 to increase the number of ISBNs available for new books.
- The second set of digits represents the language of the book. Zero represents English.
- The third set represents the publisher. This block is usually two or three digits long.
- The fourth set is the book code and is assigned by the publisher. This block is usually five or six digits long.
- The fifth and final block is a one-digit *check digit.*

Consider the ISBN assigned to this text: 978-0-07-338446-7 (*Elementary and Intermediate Algebra,* Fifth Edition, by Baratto and Bergman). The check digit in this ISBN is the final digit, 7. It ensures that the book has a valid ISBN.

To use the check digit, we use the algorithm that follows.

Step by Step: Validating an ISBN

Step 1 Identify the first 12 digits of the ISBN (omit the check digit).

Step 2 Multiply the first digit by one, the second by 3, the third by 1, the fourth by 3, and continue alternating until each of the first twelve digits has been multiplied.

Step 3 Add all 12 of these products together.

Step 4 Take only the units digit of this sum and subtract it from 10.

Step 5 If the difference found in step 4 is the same as the check digit, then the ISBN is valid.

We can use the ISBN from this text, 978-0-07-338446, to see how this works.

To do so, we multiply the first digit by one, the second by three, the third by one, the fourth by 3, again, and so on. Then we add these products together. We call this a *weighted sum.*

$$9 \cdot 1 + 7 \cdot 3 + 8 \cdot 1 + 0 \cdot 3 + 0 \cdot 1 + 7 \cdot 3 + 3 \cdot 1 + 3 \cdot 3 + 8 \cdot 1 + 4 \cdot 3 + 4 \cdot 1 + 6 \cdot 3$$
$$= 9 + 21 + 8 + 0 + 0 + 21 + 3 + 9 + 8 + 12 + 4 + 18$$
$$= 113$$

The units digit is 3; we subtract this from 10.

$$10 - 3 = 7$$

The last digit in the ISBN 978-0-07-338446-7 is 7. This matches the difference above and so this text has a valid ISBN number.

Determine whether each set of numbers represents a valid ISBN.

1. 978-0-07-038023-6
2. 978-0-07-327374-7
3. 978-0-553-34948-1
4. 978-0-07-000317-3
5. 978-0-14-200066-3

For each valid ISBN, go online and find the book associated with that ISBN.

6.2 Factoring Special Products

< 6.2 Objectives >

1 > Factor a difference of squares

2 > Factor sums and differences of cubes

3 > Factor a perfect square trinomial

In Section 6.1, you learned that factoring out any GCF is the first step to take when factoring polynomials. We also began looking at the second step. In that section, we saw that if a polynomial had four terms, we should try to factor it by grouping.

More generally, the second step when factoring polynomials is to try to recognize some pattern that allows us to factor using some special formula or method. While four terms should make us think "grouping," what other patterns indicate specific factoring approaches?

In this section, we look at some other patterns that arise. We learned the first one when multiplying polynomials in Section 5.5. Consider the product when two binomials differ in sign.

$(a + b)(a - b) = a^2 - b^2$

Since factoring is the reverse of multiplication, we have

$a^2 - b^2 = (a + b)(a - b)$

We call a polynomial of the form $a^2 - b^2$ a **difference of squares.** It always factors into a pair of binomials that differ only in sign.

NOTES

Always see if you can factor out a GCF before looking for any special pattern.

A sum of squares

$a^2 + b^2$

never factors as a product of binomials.

Property

Factoring a Difference of Squares $a^2 - b^2 = (a + b)(a - b)$

Applying this formula requires us to recognize **perfect square** terms. In order for a term to be a perfect square, the coefficient and variable factors must be perfect squares.

The perfect-square coefficients are 1, 4, 9, 16, 25, 36, and so on, because

$1 = 1^2; 4 = 2^2; 9 = 3^2; 16 = 4^2; 25 = 5^2; 36 = 6^2; \ldots$

Variable factors are perfect squares if their exponent is a multiple of 2 because

$x^2 = (x)^2; x^4 = (x^2)^2; x^6 = (x^3)^2; x^8 = (x^4)^2; \ldots$

NOTE

The exponent must be an even number.

Example 1 Recognizing Perfect Squares

Determine whether each term is a perfect square. Rewrite the perfect squares as squared terms.

(a) $36x$

This is not a perfect square. $36 = 6^2$, but x is not a square (the exponent is 1).

(b) $24x^6$

While the variable factor is a perfect square, the coefficient is not. This is not a perfect square.

(c) $9x^4$

The coefficient is a perfect square, as is the variable factor. This is a perfect square.

$9x^4 = (3x^2)^2$

(d) $64x^6$

This is a perfect square.

$64x^6 = (8x^3)^2$

(e) $16x^9$

The exponent is odd, so this is not a perfect square.

Check Yourself 1

Determine whether each term is a perfect square. Rewrite the perfect squares as squared terms.

(a) $36x^{12}$ **(b)** $4x^6$ **(c)** $9x^7$
(d) $25x^8$ **(e)** $16x^{25}$ **(f)** $6x^2$

Example 2 Factoring a Difference of Squares

< Objective 1 >

Factor $x^2 - 16$.

Think $x^2 - 4^2$.

Because $x^2 - 16$ is a difference of squares, we have

$x^2 - 16 = (x + 4)(x - 4)$

NOTE

You could also write $(x - 4)(x + 4)$. The order does not matter because multiplication is commutative.

Check Yourself 2

Factor $m^2 - 49$.

Anytime an expression is a difference of squares, it can be factored.

Example 3 Factoring a Difference of Squares

Factor $4a^2 - 9$.

Think $(2a)^2 - 3^2$.

So $4a^2 - 9 = (2a)^2 - (3)^2$
$= (2a + 3)(2a - 3)$

Check Yourself 3

Factor $9b^2 - 25$.

The difference of squares formula works the same way, even when there is more than one variable.

Example 4 Factoring a Difference of Squares

NOTE

Think $(5a)^2 - (4b^2)^2$.

Factor $25a^2 - 16b^4$.

$$25a^2 - 16b^4 = (5a)^2 - (4b^2)^2 = (5a + 4b^2)(5a - 4b^2)$$

Check Yourself 4

Factor $49c^4 - 9d^2$.

Factoring out a GCF should always be your first factoring step, even with special products.

Example 5 Factoring a Special Product

NOTE

Step 1
Factor out the GCF.

Step 2
Factor the remaining polynomial.

Factor $32x^2y - 18y^3$.

Note that $2y$ is a common factor, so

$$32x^2y - 18y^3 = 2y(\underbrace{16x^2 - 9y^2}_{\text{Difference of squares}})$$

$$= 2y(4x + 3y)(4x - 3y)$$

Check Yourself 5

Factor $50a^3 - 8ab^2$.

On occasion, we may need to factor a difference of squares more than once to completely factor it. Consider Example 6.

Example 6 Factoring a Difference of Squares

Completely factor $m^4 - 81n^4$.

We begin by observing that this is a difference of squares,

$$m^4 = (m^2)^2 \qquad \text{and} \qquad 81n^4 = (9n^2)^2$$

Therefore, we can proceed by factoring the polynomial as a difference of squares.

NOTE

The other binomial factor

$m^2 + 9n^2$

is a sum of squares, which does not factor.

$$m^4 - 81n^4 = (m^2 + 9n^2)(m^2 - 9n^2)$$

We are not finished yet. The second binomial factor is also a difference of squares, so we continue.

$$m^4 - 81n^4 = (m^2 + 9n^2)(m^2 - 9n^2)$$
$$= (m^2 + 9n^2)(m + 3n)(m - 3n)$$

Check Yourself 6

Factor $x^4 - 16y^4$.

We do find other patterns that we can use to factor polynomials. Consider the product

$$(x + 2)(x^2 - 2x + 4)$$

Multiplying gives

$$(x + 2)(x^2 - 2x + 4) = x^3 - 2x^2 + 4x + 2x^2 - 4x + 8$$
$$= x^3 + 8 \quad \text{Combine like terms.}$$

Similarly,

$$(x - 2)(x^2 + 2x + 4) = x^3 - 8$$

RECALL

Make sure to multiply every term in the first factor by every term in the second factor.

These products are examples of **sums** and **differences of cubes.** Unlike a sum of squares, we *can* factor a sum of cubes.

Property

The Sum and Difference of Cubes

The Sum of Cubes

$a^3 + b^3 = (a + b)(a^2 - ab + b^2)$ Factoring

because

$(a + b)(a^2 - ab + b^2) = a^3 + b^3$ Multiplying

The Difference of Cubes

$a^3 - b^3 = (a - b)(a^2 + ab + b^2)$ Factoring

because

$(a - b)(a^2 + ab + b^2) = a^3 - b^3$ Multiplying

NOTE

Neither

$a^2 - ab + b^2$ nor
$a^2 + ab + b^2$

can be factored further. They are *prime polynomials*.

We are now looking for **perfect cubes.**

The perfect-cube coefficients are 1, 8, 27, 64, 125, and so on, because

$$1 = 1^3;\ 8 = 2^3;\ 27 = 3^3;\ 64 = 4^3;\ 125 = 5^3;\ \ldots$$

Variable factors are perfect cubes if their exponent is a multiple of 3 because

$$x^3 = (x)^3;\ x^6 = (x^2)^3;\ x^9 = (x^3)^3;\ x^{12} = (x^4)^3;\ \ldots$$

Example 7 Factoring the Sum or Difference of Cubes

< Objective 2 >

(a) Factor $x^3 + 27$.

The first term is the cube of x, and the second is the cube of 3, so we can apply the $a^3 + b^3$ equation. Letting $a = x$ and $b = 3$, we have

$$x^3 + 27 = (x)^3 + (3)^3 = (x + 3)(x^2 - 3x + 9)$$

(b) Factor $8w^3 - 27z^3$.

This is a difference of cubes, so use the $a^3 - b^3$ equation.

$$8w^3 - 27z^3 = (2w)^3 - (3z)^3 = (2w - 3z)[(2w)^2 + (2w)(3z) + (3z)^2]$$
$$= (2w - 3z)(4w^2 + 6wz + 9z^2)$$

(c) Factor $5a^3b - 40b^4$.

First note the common factor of $5b$.

$$5a^3b - 40b^4 = 5b(a^3 - 8b^3)$$

The binomial is a difference of cubes, so

$$= 5b[(a)^3 - (2b)^3]$$
$$= 5b(a - 2b)(a^2 + 2ab + 4b^2)$$

NOTE

We are now looking for perfect cubes—the exponents must be multiples of 3 and the coefficients perfect cubes—1, 8, 27, 64, and so on.

RECALL

Looking for a *common* factor should be your first step. Remember to write the GCF as a part of the final factored form.

Check Yourself 7

Factor completely.

(a) $27x^3 + 8y^3$ (b) $3a^4 - 24ab^3$

The final set of special products we look at in this section is the **perfect square trinomial.**

If you recall the patterns from Section 5.5, we have

$$(a + b)^2 = a^2 + 2ab + b^2 \quad \text{and} \quad (a - b)^2 = a^2 - 2ab + b^2$$

This gives us two more formulas for factoring special products.

If you can recognize when you have an example of a perfect square trinomial, you can factor it straightaway.

Property

Factoring Perfect Square Trinomials

$$a^2 + 2ab + b^2 = (a + b)^2$$

$$a^2 - 2ab + b^2 = (a - b)^2$$

Example 8 Factoring Perfect Square Trinomials

< Objective 3 >

(a) Factor $x^2 + 10x + 25$.

To determine that this is a perfect square trinomial, we observe

$x^2 = (x)^2$ The first term is a perfect square.

$25 = (5)^2$ The third term is a perfect square.

$10x = 2 \cdot x \cdot 5$ The middle term is 2 times the product of x and 5.

Since these three conditions are met, we can factor the expression.

$$x^2 + 10x + 25 = (x + 5)^2$$

We check this result.

$$(x + 5)^2 = x^2 + 2 \cdot x \cdot 5 + 5^2$$
$$= x^2 + 10x + 25 \quad \text{The original polynomial}$$

> CAUTION

In a perfect square trinomial, any constant term must follow a "plus" sign.

(b) Factor $x^2 - 6x + 9$.

The first term is the square of x, and the third term is the square of 3.

Since 2 times the product of x and 3 is $6x$, this is a perfect square trinomial.

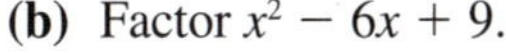

$$x^2 - 6x + 9 = (x - 3)^2$$

Check Yourself 8

Factor each polynomial.

(a) $t^2 + 14t + 49$ (b) $x^2 - 16x + 64$

As before, the process does not change when more than one variable is involved.

Example 9 Factoring a Perfect Square Trinomial

Factor $9x^2 + 30xy + 25y^2$.

We note $9x^2 = (3x)^2$

$25y^2 = (5y)^2$

And, noting that 2 times the product of $3x$ and $5y$ is

$2(3x)(5y) = 30xy$

We can factor this as

$9x^2 + 30xy + 25y^2 = (3x + 5y)^2$

RECALL

Check this result by expanding $(3x + 5y)^2$.

Check Yourself 9

Factor $4v^2 - 28vw + 49w^2$.

>CAUTION

A *sum of squares*

$a^2 + b^2$

does not factor beyond, possibly, factoring out any GCF.

We conclude with a table of the formulas for factoring special products that we have seen. Remember that factoring out any GCF is always the first step when factoring a polynomial.

Factoring Special Polynomials

Difference of squares	$a^2 - b^2 = (a + b)(a - b)$
Sum of cubes	$a^3 + b^3 = (a + b)(a^2 - ab + b^2)$
Difference of cubes	$a^3 - b^3 = (a - b)(a^2 + ab + b^2)$
Perfect square trinomial	$a^2 + 2ab + b^2 = (a + b)^2$
	$a^2 - 2ab + b^2 = (a - b)^2$

Check Yourself ANSWERS

1. **(a)** $(6x^6)^2$; **(b)** $(2x^3)^2$; **(d)** $(5x^4)^2$ **2.** $(m + 7)(m - 7)$ **3.** $(3b + 5)(3b - 5)$
4. $(7c^2 + 3d)(7c^2 - 3d)$ **5.** $2a(5a + 2b)(5a - 2b)$ **6.** $(x^2 + 4y^2)(x + 2y)(x - 2y)$
7. **(a)** $(3x + 2y)(9x^2 - 6xy + 4y^2)$; **(b)** $3a(a - 2b)(a^2 + 2ab + 4b^2)$ **8.** **(a)** $(t + 7)^2$; **(b)** $(x - 8)^2$
9. $(2v - 7w)^2$

Reading Your Text

These fill-in-the-blank exercises will help you understand some of the key vocabulary used in this section. The answers to these exercises are in the Answers Appendix in the back of the text.

(a) A ________ of squares can always be factored.

(b) When factoring, the first step is to factor out any ________.

(c) A ________ of squares never factors as the product of binomials.

(d) Sums and differences of ________ are always factorable.

Skills Calculator/Computer Career Applications Above and Beyond

6.2 exercises

< Objective 1 >

Completely factor each polynomial or write "not factorable."

1. $m^2 - n^2$ **2.** $r^2 - 9$ **3.** $x^2 - 169$ **4.** $c^2 - d^2$

5. $49 - y^2$ **6.** $196 - y^2$ **7.** $9b^2 - 16$ **8.** $36 - x^2$

9. $16w^2 - 49$ **10.** $4x^2 - 25$ **11.** $x^2 - 10$ **12.** $3x^2 - 8$

13. $x^2 + 16$ **14.** $9x^4 + 4y^2$ **15.** $4s^2 - 9r^2$ **16.** $64y^2 - x^2$

17. $9w^2 - 49z^2$ VIDEO **18.** $25x^2 - 81y^2$ **19.** $4x^2y^2 - 169y^2$ **20.** $196x^4y^2 - 225x^2$

21. $121x^8 + 64$ **22.** $9x^4 - 5y$ **23.** $49a^2 - 9b^2$ **24.** $64m^2 - 9n^2$

25. $x^4 - 36$ **26.** $y^6 - 49$ **27.** $x^2y^2 - 16$ **28.** $m^2n^2 - 64$

29. $25 - a^2b^2$ **30.** $49 - w^2z^2$ **31.** $49 - xy$ **32.** $25 + a^2b^2$

33. $r^4 - 4s^2$ **34.** $p^2 - 9q^4$ **35.** $81a^2 - 100b^6$ **36.** $64x^4 - 25y^4$

37. $18x^3 - 2xy^2$ VIDEO **38.** $50a^2b - 2b^3$ **39.** $12m^3n - 75mn^3$ **40.** $63p^4 - 7p^2q^2$

41. $16a^4 - 81b^4$ **42.** $81x^4 - y^4$

< Objective 2 >

43. $y^3 + 125$ **44.** $y^3 - 8$ **45.** $m^3 - 125$ **46.** $b^3 + 27$

47. $a^3b^3 - 27$ **48.** $p^3q^3 - 64$ **49.** $8w^3 + z^3$ **50.** $c^3 - 27d^3$

51. $r^3 - 64s^3$ VIDEO **52.** $125x^3 + y^3$ **53.** $8x^3 - 27y^3$ **54.** $64m^3 + 27n^3$

55. $x^3 + 5$ **56.** $3x^3 - 20$ **57.** $3a^3 + 81b^3$ **58.** $4x^3 - 32y^3$

< Objective 3 >

59. $x^2 - 4x + 4$ **60.** $u^2 + 18u + 81$ **61.** $y^2 + 4y + 8$ **62.** $t^2 - 6t + 36$

63. $x^2 + 2x + 1$ **64.** $x^2 - 20x + 100$ **65.** $16a^2 + 24a + 9$ **66.** $25a^2 - 20ab + 4b^2$

67. $25x^2 + 120x + 144$ **68.** $4x^2 + 100x + 625$ **69.** $9x^3 + 12x^2y + 4xy^2$ **70.** $25x^4 - 10x^3 + x^2$

71. $x^2(x + y) - y^2(x + y)$ VIDEO **72.** $a^2(b - c) - 16b^2(b - c)$ **73.** $2m^2(m - 2n) - 18n^2(m - 2n)$

74. $3a^3(2a + b) - 27ab^2(2a + b)$

Determine whether each statement is **true** *or* **false.**

75. A perfect square term has a coefficient that is a square and any variables have exponents that are factors of 2.

76. Any time an expression is the difference of squares, it can be factored.

77. Although the difference of squares can be factored, the sum of squares cannot.

78. When factoring, the middle factor is always factored out as the first step.

Complete each statement with **always, sometimes,** *or* **never.**

79. A "difference-of-squares" binomial __________ factors.

80. A "sum-of-squares" binomial of degree 2 __________ factors.

81. A "sum-of-cubes" binomial __________ factors.

82. In attempting to factor a binomial, we should __________ factor out a GCF, first, if possible.

Skills | Calculator/Computer | **Career Applications** | Above and Beyond

83. **Manufacturing Technology** The difference d in the calculated maximum deflection between two similar cantilevered beams is given by the formula

$$d = \left(\frac{w}{8EI}\right)\left(l_1^2 - l_2^2\right)\left(l_1^2 + l_2^2\right)$$

Rewrite the formula in its completely factored form.

84. **Manufacturing Technology** The work W done by a steam turbine is given by the formula

$$W = \frac{1}{2}m\left(v_1^2 - v_2^2\right)$$

Factor the right-hand side of this equation.

85. **Allied Health** A toxic chemical is introduced into a protozoan culture. The number of deaths per hour is given by the polynomial $338 - 2t^2$, in which t is the number of hours after the chemical is introduced. Factor this expression.

86. **Allied Health** Radiation therapy is one technique used to control cancer. After treatment, the total number of cancerous cells, in thousands, can be estimated by $144 - 4t^2$, in which t is the number of days of treatment. Factor this expression.

Skills | Calculator/Computer | Career Applications | **Above and Beyond**

87. Find the value for k so that $kx^2 - 25$ has the factors $2x + 5$ and $2x - 5$.

88. Find the value for k so that $9m^2 - kn^2$ has the factors $3m + 7n$ and $3m - 7n$.

89. Find the value for k so that $2x^3 - kxy^2$ has the factors $2x$, $x - 3y$, and $x + 3y$.

90. Find the value for k so that $20a^3b - kab^3$ has the factors $5ab$, $2a - 3b$, and $2a + 3b$.

91. Complete this statement: "To factor a number, you. . . ."

92. Complete this statement: "To factor an algebraic expression into prime factors means. . . ."

93. What binomial multiplied by $25x^2 - 15xy + 9y^2$ gives the sum of two cubes? What is the result of the multiplication?

94. What binomial when multiplied by $9x^2 + 6xy + 4y^2$ gives the difference of two cubes? What is the result of the multiplication?

95. What are the characteristics of a perfect cube monomial?

96. Suppose you factored the polynomial $4x^2 - 16$ as

$4x^2 - 16 = (2x + 4)(2x - 4)$

Is this completely factored? If not, what is the final form?

Answers

1. $(m + n)(m - n)$ **3.** $(x + 13)(x - 13)$ **5.** $(7 + y)(7 - y)$ **7.** $(3b + 4)(3b - 4)$ **9.** $(4w + 7)(4w - 7)$ **11.** Not factorable **13.** Not factorable **15.** $(2s + 3r)(2s - 3r)$ **17.** $(3w + 7z)(3w - 7z)$ **19.** $y^2(2x + 13)(2x - 13)$ **21.** Not factorable **23.** $(7a + 3b)(7a - 3b)$ **25.** $(x^2 + 6)(x^2 - 6)$ **27.** $(xy + 4)(xy - 4)$ **29.** $(5 + ab)(5 - ab)$ **31.** Not factorable **33.** $(r^2 + 2s)(r^2 - 2s)$ **35.** $(9a + 10b^3)(9a - 10b^3)$ **37.** $2x(3x + y)(3x - y)$ **39.** $3mn(2m + 5n)(2m - 5n)$ **41.** $(4a^2 + 9b^2)(2a + 3b)(2a - 3b)$ **43.** $(y + 5)(y^2 - 5y + 25)$ **45.** $(m - 5)(m^2 + 5m + 25)$ **47.** $(ab - 3)(a^2b^2 + 3ab + 9)$ **49.** $(2w + z)(4w^2 - 2wz + z^2)$ **51.** $(r - 4s)(r^2 + 4rs + 16s^2)$ **53.** $(2x - 3y)(4x^2 + 6xy + 9y^2)$ **55.** Not factorable **57.** $3(a + 3b)(a^2 - 3ab + 9b^2)$ **59.** $(x - 2)^2$ **61.** Not factorable **63.** $(x + 1)^2$ **65.** $(4a + 3)^2$ **67.** $(5x + 12)^2$ **69.** $x(3x + 2y)^2$ **71.** $(x + y)^2(x - y)$ **73.** $2(m - 2n)(m + 3n)(m - 3n)$ **75.** False **77.** True **79.** always **81.** always **83.** $d = \left(\frac{w}{8EI}\right)(l_1 + l_2)(l_1 - l_2)(l_1^2 + l_2^2)$ **85.** $2(13 - t)(13 + t)$ **87.** 4 **89.** 18 **91.** Above and Beyond **93.** $5x + 3y$; $125x^3 + 27y^3$ **95.** Above and Beyond

6.3 Factoring: Trial and Error

< 6.3 Objectives >

1 > Factor $x^2 + bx + c$

2 > Completely factor a trinomial

3 > Factor a trinomial that is quadratic in form

NOTE

Quadratics are second-degree polynomials in one variable.

Polynomials that can be written as

$ax^2 + bx + c$ where $a \neq 0$

are called **quadratic polynomials.**

Quadratic polynomials appear in many models. We use them to approximate the effects of gravity as well as many other physical systems.

Factoring quadratic polynomials is the main thrust of the next three sections. In this section, we use the *trial-and-error* method. This method is particularly appropriate for quadratics in which $a = 1$. That is, quadratic polynomials of the form

$x^2 + bx + c$

When factoring such a polynomial, our goal is to find a pair of binomial factors $(x + m)$ and $(x + n)$ so that

$(x + m)(x + n) = x^2 + bx + c$

You learned how to find the product of binomials in Section 5.5. Because factoring is the reverse of multiplication, we can use the FOIL pattern to find the factors of certain trinomials.

Recall that when we multiply the binomials $x + 2$ and $x + 3$, our result is

$(x + 2)(x + 3) = x^2 + 5x + 6$

The product of the first terms $(x \cdot x)$.

The sum of the products of the outer and inner terms ($3x$ and $2x$).

The product of the last terms $(2 \cdot 3)$.

> CAUTION
>
> Not every trinomial can be written as the product of two binomials.

Suppose now that you are given $x^2 + 5x + 6$ and want to find its factors. First, we know that the factors of a trinomial may be two binomials. So write

$x^2 + 5x + 6 = (\qquad)(\qquad)$

Because the first term of the trinomial is x^2, the first terms of the binomial factors must be x and x. We now have

$x^2 + 5x + 6 = (x \qquad)(x \qquad)$

NOTE

We are only interested in factoring polynomials *over the integers* (that is, with integer coefficients).

The product of the last terms must be 6. Because 6 is positive, the factors must have *like* signs. Here are the possibilities:

$$
\begin{aligned}
6 &= 1 \cdot 6 \\
&= 2 \cdot 3 \\
&= (-1)(-6) \\
&= (-2)(-3)
\end{aligned}
$$

This means, if we can factor the polynomial, the possible factors of the trinomial are

$(x + 1)(x + 6)$
$(x + 2)(x + 3)$
$(x - 1)(x - 6)$
$(x - 2)(x - 3)$

How do we tell which is the correct pair? From the FOIL pattern we know that the sum of the outer and inner products must equal the middle term of the trinomial, in this case $5x$. This is the crucial step!

Possible Factorizations	Middle Terms
$(x + 1)(x + 6)$	$7x$
$(x + 2)(x + 3)$	$5x$
$(x - 1)(x - 6)$	$-7x$
$(x - 2)(x - 3)$	$-5x$

The correct middle term!

So we know that the correct factorization is

$x^2 + 5x + 6 = (x + 2)(x + 3)$

Are there any clues so far that make this process quicker? Yes, there is an important one that you may have spotted. We started with a trinomial that had a positive middle term and a positive last term. The pairs of negative factors for 6 gave us negative middle terms. So we do not need to bother with the negative factors if the middle term and the last term of the trinomial are both positive.

Example 1 Factoring Trinomials

< Objective 1 >

(a) Factor $x^2 + 9x + 8$.

Because the middle term and the last term of the trinomial are both positive, consider only the positive factors of 8, that is, $8 = 1 \cdot 8$ or $8 = 2 \cdot 4$.

NOTE

If you are wondering why we do not list $(x + 8)(x + 1)$ as a possibility, remember that multiplication is commutative. The order doesn't matter!

Possible Factorizations	Middle Terms
$(x + 1)(x + 8)$	$9x$
$(x + 2)(x + 4)$	$6x$

Because the first pair gives the correct middle term,

$x^2 + 9x + 8 = (x + 1)(x + 8)$

We check our result by multiplying the factors. If the product is the same as the original polynomial, then our answer is correct.

Check

$(x + 1)(x + 8) = x^2 + 8x + x + 8$ FOIL

$= x^2 + 9x + 8$ The original polynomial; our answer checks.

NOTE

The factorizations of 20 are

$20 = 1 \cdot 20$
$= 2 \cdot 10$
$= 4 \cdot 5$

(b) Factor $x^2 + 12x + 20$.

Possible Factorizations	Middle Terms
$(x + 1)(x + 20)$	$21x$
$(x + 2)(x + 10)$	$12x$
$(x + 4)(x + 5)$	$9x$

So

$x^2 + 12x + 20 = (x + 2)(x + 10)$

Remember to check your result.

Check Yourself 1

Factor.

(a) $x^2 + 6x + 5$ **(b)** $x^2 + 10x + 16$

What if the middle term of the trinomial is negative but the first and last terms are still positive? Consider

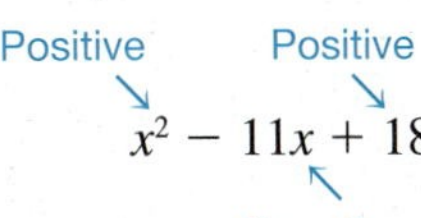

Because we want a negative middle term $(-11x)$ and a positive last term, we use *two negative factors* for 18. Recall that the product of two negative numbers is positive, and the sum of two negative numbers is negative.

Example 2 Factoring Trinomials

(a) Factor $x^2 - 11x + 18$.

Possible Factorizations	Middle Terms
$(x - 1)(x - 18)$	$-19x$
$(x - 2)(x - 9)$	$-11x$
$(x - 3)(x - 6)$	$-9x$

NOTE

The negative factorizations of 18 are

$18 = (-1)(-18)$
$= (-2)(-9)$
$= (-3)(-6)$

So

$x^2 - 11x + 18 = (x - 2)(x - 9)$

Check

$(x - 2)(x - 9) = x^2 - 9x - 2x + 18$
$= x^2 - 11x + 18$ The original polynomial; our answer checks.

(b) Factor $x^2 - 13x + 12$.

Possible Factorizations	Middle Terms
$(x - 1)(x - 12)$	$-13x$
$(x - 2)(x - 6)$	$-8x$
$(x - 3)(x - 4)$	$-7x$

NOTES

The negative factorizations of 12 are

$12 = (-1)(-12)$
$= (-2)(-6)$
$= (-3)(-4)$

We have been listing all possible factorizations. In fact, we can stop when we find the correct pair. With practice, you will learn to do much of this mentally.

So

$x^2 - 13x + 12 = (x - 1)(x - 12)$

Remember to check your result.

Check Yourself 2

Factor.

(a) $x^2 - 10x + 9$ **(b)** $x^2 - 10x + 21$

Now we look at the process of factoring a trinomial whose last term is negative. For instance, to factor $x^2 + 2x - 15$, we can start as before:

$$x^2 + 2x - 15 = (x \quad ?)(x \quad ?)$$

Note that the product of the last terms must be negative (−15 here). So we must choose factors that have different signs.

What are our choices for the factorizations of −15?

$$\begin{aligned} -15 &= (1)(-15) \\ &= (-1)(15) \\ &= (3)(-5) \\ &= (-3)(5) \end{aligned}$$

This means that the possible factorizations and the resulting middle terms are

Possible Factorizations	Middle Terms
$(x + 1)(x - 15)$	$-14x$
$(x - 1)(x + 15)$	$14x$
$(x + 3)(x - 5)$	$-2x$
$(x - 3)(x + 5)$	$2x$

So $x^2 + 2x - 15 = (x - 3)(x + 5)$.

Many students find it easier to list the number part of the factors rather than the whole algebraic factors. In Example 2, we would create a table of just the possible factors of −15 and then look at the sum of each pair. Since the middle term of the polynomial is $2x$, we want the factor pair whose sum is 2.

Factors of −15	Sum
1, −15	−14
−1, 15	14
3, −5	−2
−3, 5	2

The sum of the factor pair −3 and 5 is 2, which is the coefficient of the middle term of the polynomial $x^2 + 2x - 15$, so this is the correct pair. Our factors are $x - 3$ and $x + 5$.

$$x^2 + 2x - 15 = (x - 3)(x + 5)$$

Next, we try to factor when the third term is negative.

Example 3 Factoring Trinomials

(a) Factor $x^2 - 5x - 6$.

The constant term is −6, so we list the factor pairs whose product is −6. The middle term is −5, so we are looking for the pair whose sum is −5.

NOTE

We really only need to list the factor pairs in which the number with the larger absolute value is negative because we need their sum to be negative.

Factors of −6	Sum
1, −6	−5
−1, 6	5
2, −3	−1
−2, 3	1

Since the product of 1 and -6 is the constant term, -6, and their sum is equal to the coefficient of the middle term, -5, we have our factors.

$$x^2 - 5x - 6 = (x + 1)(x - 6)$$

Check

$$(x + 1)(x - 6) = x^2 - 6x + x - 6$$
$$= x^2 - 5x - 6$$ The original polynomial; our answer checks.

(b) Factor $x^2 + 8xy - 9y^2$.

The process is similar if two variables are involved in the trinomial. Start with

$$x^2 + 8xy - 9y^2 = (x \quad ?)(x \quad ?).$$

The product of the last terms must be $-9y^2$.

The coefficient of the last term is -9, so we need pairs of numbers whose product is -9. The coefficient of the middle term is 8, so we need the pair whose sum is 8. This lets us look only at the pairs in which the number with the larger absolute value is positive.

Since the only factor pairs of 9 are 1 and 9 or 3 and 3, we only need to consider -1 and 9. Their sum is 8, so this is our factor pair.

$$x^2 + 8xy - 9y^2 = (x - y)(x + 9y)$$

Check

$$(x - y)(x + 9y) = x^2 + 9xy - xy - 9y^2$$
$$= x^2 + 8xy - 9y^2$$ The original polynomial; our answer checks.

Check Yourself 3

Factor.

(a) $x^2 + 7x - 30$ **(b)** $x^2 - 3xy - 10y^2$

By now, you should see how the sign patterns in a quadratic affect the signs of the binomial factors.

Property Factoring Quadratic Polynomials

		Sign Pattern
$x^2 + bx + c$	Both signs are positive	$(x + m)(x + n)$
$x^2 - bx + c$	Positive constant term Negative x-term coefficient	$(x - m)(x - n)$
$x^2 + bx - c$	Negative constant term Positive x-term coefficient	$(x - m)(x + n)$ $m < n$
$x^2 - bx - c$	Negative constant term Negative x-term coefficient	$(x - m)(x + n)$ $m > n$

As we pointed out in Section 6.1, any time that there is a common factor, it should be factored out *before* we try any other factoring technique. Consider Example 4.

Example 4 Completely Factoring Trinomials

< Objective 2 >

(a) Factor $3x^2 - 21x + 18$.

$3x^2 - 21x + 18 = 3(x^2 - 7x + 6)$ Factor out the common factor of 3.

We now factor the remaining trinomial. For $x^2 - 7x + 6$:

Factors of −6	Sum
−1, −6	−7
−2, −3	−5

The correct middle term

> CAUTION
>
> A common mistake is to forget to write the 3 that was factored out as the first step.

So $3x^2 - 21x + 18 = 3(x - 1)(x - 6)$.

Check

$$3(x - 1)(x - 6) = 3(x^2 - 6x - x + 6)$$
$$= 3(x^2 - 7x + 6)$$
$$= 3x^2 - 21x + 18$$

The original polynomial; our answer checks.

(b) Factor $2x^3 + 16x^2 - 40x$.

$2x^3 + 16x^2 - 40x = 2x(x^2 + 8x - 20)$ Factor out the common factor of $2x$.

The constant term is −20, so we list the factor pairs whose product is −20. Since the constant term is negative, we know the factors must have opposite signs. The middle term is 8, so we only need to look at those pairs in which the number with the larger absolute value is positive.

NOTE

Once we have the desired middle term, we stop.

Factors of −20	Sum
−1, 20	19
−2, 10	8

We can stop when we find the correct middle term.

$2x^3 + 16x^2 - 40x = 2x(x - 2)(x + 10)$

Remember to check this result by multiplying the factors.

Check Yourself 4

Factor.

(a) $3x^2 - 3x - 36$ **(b)** $4x^3 + 24x^2 + 32x$

As in Section 6.1, when the leading coefficient is negative, we factor out −1 along with any GCF.

Example 5 Completely Factoring Trinomials

(a) Factor $-x^2 - 11x - 30$.

The GCF is 1, but the leading term is negative, so we factor out −1.

$-x^2 - 11x - 30 = -(x^2 + 11x + 30)$

To factor the remaining quadratic $x^2 + 11x + 30$, we need to find factors of 30 whose sum is 11.

Factors of 30	Sum
1, 30	31
2, 15	17
3, 10	13
5, 6	11

Since $5 + 6 = 11$, we have $x^2 + 11x + 30 = (x + 5)(x + 6)$. Therefore,

$$-x^2 - 11x - 30 = -(x + 5)(x + 6)$$

Check

$$\begin{aligned} -(x + 5)(x + 6) &= -(x^2 + 6x + 5x + 30) \\ &= -(x^2 + 11x + 30) \\ &= -x^2 - 11x - 30 \end{aligned}$$

The original polynomial; our answer checks.

(b) Factor $-2x^3 + 14x^2 + 36x$.

We begin by looking for any GCF. In this case, we have the common factor $2x$. Since the leading coefficient is negative, we factor $-2x$ from the polynomial.

$$-2x^3 + 14x^2 + 36x = -2x(x^2 - 7x - 18)$$

To factor the remaining quadratic $x^2 - 7x - 18$, we need to find factors of -18 whose sum is -7. Because the sum is negative, we only need to consider factors of -18 in which the number with the larger magnitude is negative.

RECALL

We stop when we find the correct pair.

Factors of −18	Sum
1, −18	−17
2, −9	−7

Since $2 + (-9) = -7$, we have $x^2 - 7x - 18 = (x + 2)(x - 9)$. Therefore,

$$-2x^3 + 14x^2 + 36x = -2x(x + 2)(x - 9)$$

We leave it to you to check this result.

Check Yourself 5

Completely factor each polynomial.

(a) $-x^2 - 2x + 3$ **(b)** $-3x^3y + 21x^2y - 30xy$

Occasionally, we need to factor an expression that is not quadratic, but **is quadratic in form.** Consider the expression $x^4 + 5x^2 + 6$.

Example 6 Factoring an Expression That Is Quadratic in Form

< Objective 3 >

Factor $x^4 + 5x^2 + 6$.

Observing that x^4 is the square of x^2, we factor

$$(x^2 + 2)(x^2 + 3)$$

Multiplying these binomials shows that we factored correctly.

$$(x^2 + 2)(x^2 + 3) = x^4 + 3x^2 + 2x^2 + 6 = x^4 + 5x^2 + 6$$

Since the binomials $x^2 + 2$ and $x^2 + 3$ cannot be further factored, we are done.

$x^4 + 5x^2 + 6 = (x^2 + 2)(x^2 + 3)$

Check Yourself 6

Factor $y^4 + 9y^2 + 8$.

When factoring such an expression, it is important to check that it is in fact completely factored.

Example 7 Factoring an Expression That Is Quadratic in Form

Factor $x^4 - 13x^2 + 36$.

Since x^4 is the square of x^2, we begin with

$(x^2 - \quad)(x^2 - \quad)$

Looking for two integers whose product is 36 and whose sum is 13 produces 4 and 9.

$(x^2 - 4)(x^2 - 9)$

Now we note that each binomial is a difference of squares.

$x^4 - 13x^2 + 36 = (x + 2)(x - 2)(x + 3)(x - 3)$

Check Yourself 7

Factor $t^4 - 17t^2 + 16$.

One final note: When factoring, we require that all coefficients be integers. Given this restriction, not all polynomials are factorable over the integers.

To factor $x^2 - 9x + 12$, we know that the only possible binomial factors (using integers as coefficients) are

RECALL

We refer to such polynomials as prime.

$(x - 1)(x - 12)$

$(x - 2)(x - 6)$

$(x - 3)(x - 4)$

You can verify that *none* of these pairs gives the correct middle term of $-9x$. We then say that the original trinomial is not factorable using integers as coefficients.

In Section 6.4, you will learn a test to determine if a trinomial is factorable.

Check Yourself ANSWERS

1. **(a)** $(x + 1)(x + 5)$; **(b)** $(x + 2)(x + 8)$ **2.** **(a)** $(x - 1)(x - 9)$; **(b)** $(x - 3)(x - 7)$
3. **(a)** $(x - 3)(x + 10)$; **(b)** $(x + 2y)(x - 5y)$ **4.** **(a)** $3(x + 3)(x - 4)$; **(b)** $4x(x + 2)(x + 4)$
5. **(a)** $-(x - 1)(x + 3)$; **(b)** $-3xy(x - 2)(x - 5)$ **6.** $(y^2 + 8)(y^2 + 1)$
7. $(t + 1)(t - 1)(t + 4)(t - 4)$

Reading Your Text

These fill-in-the-blank exercises will help you understand some of the key vocabulary used in this section. The answers to these exercises are in the Answers Appendix in the back of the text.

(a) ________ polynomials are second-degree polynomials in one variable.

(b) If the constant term of a quadratic polynomial is negative, the signs of any binomial factors must be ________.

(c) The first step when factoring is to factor out any ________.

(d) A trinomial is quadratic in form if the variable factor in the leading term is the ________ of the variable factor in the middle term.

Skills Calculator/Computer Career Applications Above and Beyond

6.3 exercises

Determine whether each statement is **true** *or* **false.**

1. $x^2 + 2x - 3 = (x + 3)(x - 1)$
2. $y^2 + 3y + 18 = (y - 6)(y + 3)$
3. $x^2 - 10x - 24 = (x - 6)(x + 4)$
4. $a^2 + 9a - 36 = (a - 12)(a + 3)$
5. $x^2 - 16x + 64 = (x - 8)(x - 8)$
6. $w^2 - 12w - 45 = (w + 9)(w - 5)$

< Objective 1 >

Complete each statement.

7. $x^2 - 8x + 15 = (x - 3)(\quad)$
8. $y^2 - 3y - 18 = (y - 6)(\quad)$
9. $m^2 + 8m + 12 = (m + 2)(\quad)$
10. $x^2 - 10x + 24 = (x - 6)(\quad)$
11. $p^2 - 8p - 20 = (p + 2)(\quad)$
12. $a^2 + 9a - 36 = (a + 12)(\quad)$
13. $x^2 - 16x + 64 = (x - 8)(\quad)$
14. $w^2 - 12w - 45 = (w + 3)(\quad)$
15. $x^2 - 7xy + 10y^2 = (x - 2y)(\quad)$
16. $a^2 + 18ab + 81b^2 = (a + 9b)(\quad)$

Completely factor each trinomial.

17. $x^2 + 8x + 15$
18. $x^2 - 11x + 24$
19. $x^2 - 11x + 28$
20. $y^2 - y - 20$
21. $s^2 + 13s + 30$
22. $b^2 + 14b + 33$
23. $a^2 - 2a - 48$
24. $x^2 - 17x + 60$
25. $x^2 - 8x + 7$
26. $x^2 + 7x - 18$
27. $m^2 + 3m - 28$ 
28. $a^2 + 10a + 25$
29. $x^2 - 6x - 40$
30. $x^2 - 11x + 10$
31. $x^2 - 14x + 49$
32. $s^2 - 4s - 32$
33. $p^2 - 10p - 24$
34. $x^2 - 11x - 60$
35. $x^2 + 5x - 66$
36. $a^2 + 2a - 80$
37. $c^2 + 19c + 60$
38. $t^2 - 4t - 60$
39. $x^2 + 7xy + 10y^2$
40. $x^2 - 8xy + 12y^2$
41. $a^2 - ab - 42b^2$
42. $m^2 - 8mn + 16n^2$
43. $x^2 + x + 7$
44. $x^2 - 3x + 9$
45. $x^2 - 13xy + 40y^2$
46. $r^2 - 9rs - 36s^2$
47. $x^2 - 2xy - 8y^2$
48. $u^2 + 6uv - 55v^2$
49. $s^2 - 2st - 2t^2$
50. $x^2 + 5xy + y^2$
51. $25m^2 + 10mn + n^2$
52. $64m^2 - 16mn + n^2$

< Objective 2 >

53. $3a^2 - 3a - 126$

54. $2c^2 + 2c - 60$

55. $r^3 + 7r^2 - 18r$

56. $m^3 + 5m^2 - 14m$

57. $2x^3 - 20x^2 - 48x$

58. $3p^3 + 48p^2 - 108p$

59. $x^2y - 9xy^2 - 36y^3$

60. $4s^4 - 20s^3t - 96s^2t^2$

61. $m^3 - 29m^2n + 120mn^2$

62. $2a^3 - 52a^2b + 96ab^2$

63. $-s^2 - 13s - 30$

64. $-b^2 - 11b - 28$

65. $-x^2 + 14x - 49$

66. $-s^2 + 4s + 32$

< Objective 3 >

67. $u^4 - 5u^2 + 4$

68. $y^4 - 29y^2 + 100$

69. $w^4 - 5w^2 - 36$

70. $t^4 - 15t^2 - 16$

71. $2y^4 - 12y^2 - 54$

72. $3x^4 - 24x^2 + 48$

Determine whether each statement is **true** *or* **false.**

73. Factoring is the reverse of division.

74. From the FOIL pattern, we know that the sum of the inner and outer products must equal the middle term of the trinomial.

75. The sum of two negative factors is always negative.

76. Every trinomial has integer coefficients.

Complete each statement with **always, sometimes,** *or* **never.**

77. In factoring $x^2 + bx + c$, if c is a prime number then the trinomial is __________ factorable.

78. If a GCF has already been factored out of a trinomial, we will __________ find a common factor in one (or both) of the binomial factors.

79. In factoring $x^2 + bx + c$, if c is negative then the signs in the binomial factors are __________ opposites.

80. In factoring $x^2 + bx + c$, if c is positive then the signs in the binomial factors are __________ both negative.

Skills	Calculator/Computer	**Career Applications**	Above and Beyond

81. **Mechanical Engineering** The bending stress on an overhanging beam is given by the expression $310(x^2 - 36x + 128)$. Factor this expression.

82. **Automotive Technology** The acceleration curve for low gear in a car is described by the equation $a = \frac{1}{20}(x^2 - 16x - 80)$. Rewrite this equation by factoring the right-hand side.

83. **Construction Technology** The stress-strain curve of a weld is given by the formula $s = 325 + 60l - l^2$. Rewrite this formula by factoring the right-hand side.

84. **Manufacturing Technology** The maximum stress for a given allowable strain (deformation) of a certain material is given by the polynomial equation

$$\text{Stress} = 85.8x - 0.6x^2 - 1{,}537.2$$

in which x is the allowable strain, in micrometers. Factor the right-hand side of this equation.

Hint: Factor out -0.6, and write the polynomial in descending order.

85. **Manufacturing Technology** The shape of a beam loaded with a single concentrated load is described by the expression $\frac{x^2 - 64}{200}$. Factor the numerator $x^2 - 64$.

86. **Allied Health** The concentration, in micrograms per milliliter (μg/mL), of Vancocin, an antibiotic used to treat peritonitis, is given by the negative of the polynomial $t^2 - 8t - 20$, where t is the number of hours since the drug was administered via intravenous injection. Write the given polynomial in factored form.

Skills | Calculator/Computer | Career Applications | **Above and Beyond**

Find all positive integer values for k so that each polynomial can be factored.

87. $x^2 + kx + 8$ **88.** $x^2 + kx + 9$ **89.** $x^2 - kx + 16$ **90.** $x^2 - kx + 17$

91. $x^2 - kx - 5$ **92.** $x^2 - kx - 7$ **93.** $x^2 + 3x + k$ **94.** $x^2 + 5x + k$

95. $x^2 + 2x - k$ **96.** $x^2 + x - k$

Answers

1. True **3.** False **5.** True **7.** $x - 5$ **9.** $m + 6$ **11.** $p - 10$ **13.** $x - 8$ **15.** $x - 5y$ **17.** $(x + 3)(x + 5)$ **19.** $(x - 4)(x - 7)$ **21.** $(s + 3)(s + 10)$ **23.** $(a - 8)(a + 6)$ **25.** $(x - 1)(x - 7)$ **27.** $(m + 7)(m - 4)$ **29.** $(x + 4)(x - 10)$ **31.** $(x - 7)^2$ **33.** $(p - 12)(p + 2)$ **35.** $(x + 11)(x - 6)$ **37.** $(c + 4)(c + 15)$ **39.** $(x + 2y)(x + 5y)$ **41.** $(a + 6b)(a - 7b)$ **43.** Not factorable **45.** $(x - 5y)(x - 8y)$ **47.** $(x + 2y)(x - 4y)$ **49.** Not factorable **51.** $(5m + n)^2$ **53.** $3(a + 6)(a - 7)$ **55.** $r(r - 2)(r + 9)$ **57.** $2x(x - 12)(x + 2)$ **59.** $y(x + 3y)(x - 12y)$ **61.** $m(m - 5n)(m - 24n)$ **63.** $-(s + 10)(s + 3)$ **65.** $-(x - 7)^2$ **67.** $(u + 2)(u - 2)(u + 1)(u - 1)$ **69.** $(w + 3)(w - 3)(w^2 + 4)$ **71.** $2(y + 3)(y - 3)(y^2 + 3)$ **73.** False **75.** True **77.** sometimes **79.** always **81.** $310(x - 4)(x - 32)$ **83.** $s = (5 + l)(65 - l)$ or $s = -(l + 5)(l - 65)$ **85.** $(x + 8)(x - 8)$ **87.** 6 or 9 **89.** 8 or 10 or 17 **91.** 4 **93.** 2 **95.** 3, 8, 15, 24, . . .

6.4 Factoring: The *ac* Method

< 6.4 Objectives >

1 > Use the *ac* test to determine factorability

2 > Use the *ac* method to factor a trinomial

3 > Completely factor a trinomial

RECALL

Such quadratics have the form

$x^2 + bx + c$

In Section 6.3, we learned to factor quadratic polynomials when the leading coefficient was 1. The *trial-and-error method* is helpful in such situations because it is quick and intuitive.

When the leading coefficient $a \neq 1$, it is more complicated. In this case, there are better methods than trial and error.

As before, we look to a multiplication example to provide clues on factoring.

$$(5x + 2)(2x + 3) = \underbrace{(5x)(2x)}_{\text{First}} + \underbrace{(5x)(3)}_{\text{Outer}} + \underbrace{(2)(2x)}_{\text{Inner}} + \underbrace{(2)(3)}_{\text{Last}}$$

$$= 10x^2 + 15x + 4x + 6$$

$$= 10x^2 + \underbrace{19x}_{5x \cdot 3 + 2 \cdot 2x} + 6$$

Can you see the problem? We need to look at all possible factorizations of the first term and the constant term. Then we need to look at all possible combinations of these factorizations to determine how they affect the middle term. This complication makes it much more difficult to "undo" the middle term through trial and error.

NOTE

Though not as quick as trial and error, we can use the *ac* method when $a = 1$.

In this section, we learn a systematic way of factoring such polynomials. The ***ac* method** provides us with a step-by-step approach to factoring quadratic polynomials. It has the further benefit of providing us with a way of determining if a polynomial is even factorable.

The first step toward factoring in a systematic way is to identify the coefficients in a quadratic polynomial. We always write such polynomials in descending order when identifying the coefficients. We use the standard format $ax^2 + bx + c$ and look to identify a, b, and c.

Example 1 Identifying Coefficients

Write each polynomial in the form $ax^2 + bx + c$, if necessary, and identify the coefficients a, b, and c.

(a) $2x^2 + 5x + 12$

The polynomial is already in descending order. We have

$a = 2, b = 5, c = 12$

(b) $5x^2 - 2x + 8$

The polynomial is already in descending order. We have

$a = 5, b = -2, c = 8$

RECALL

The sign travels with the term so that $b = -2$ in part (b).

(c) $-16 + 3x^2 - 6x$

We begin by writing the polynomial in descending order.

$3x^2 - 6x - 16$

Next, we identify the coefficients.

$a = 3, b = -6, c = -16$

(d) $5x + x^2 - 9$

RECALL

We understand the coefficient is 1 when no number appears before the variable.
$x^2 = 1x^2$

We write the polynomial in descending order and identify the coefficients.

$5x + x^2 - 9 = x^2 + 5x - 9$

$a = 1, b = 5, c = -9$

Check Yourself 1

Write each polynomial in the form $ax^2 + bx + c$, if necessary, and identify the coefficients *a*, *b*, and *c*.

(a) $x^2 + 6x + 11$ **(b)** $8 - x + 4x^2$
(c) $3 - 2x^2 + 8x$ **(d)** $-4x^2 + x - 1$

RECALL

Always factor out any GCF and leading negative before applying any test or other factoring methods.

Because the *ac* method provides us with a systematic approach to factoring, we use the *ac* test determine whether a polynomial is even factorable before applying the method. There are two reasons to begin with the *ac* test. If a polynomial does not factor, then we should not waste our time trying. If it does factor, then the *ac* test provides the first steps for using the *ac* method.

Property

The *ac* Test

A trinomial of the form $ax^2 + bx + c$ is factorable if, and only if, there are two integers *m* and *n* such that

$ac = mn$ and $b = m + n$

In other words, we are looking for two integers whose product is the same as $a \cdot c$ and whose sum is *b*.

In Example 2 we look for *m* and *n* to determine whether each trinomial is factorable.

Example 2 Using the *ac* Test

< Objective 1 >

Use the *ac* test to determine which trinomials can be factored. Find the values of *m* and *n* for each trinomial that can be factored.

(a) $x^2 - 3x - 18$

First, we find the values of *a*, *b*, and *c*, so that we can find *ac*.

$a = 1 \qquad b = -3 \qquad c = -18$

$ac = 1(-18) = -18$ and $b = -3$

Then we look for two integers *m* and *n* such that $mn = ac$ and $m + n = b$. In this case, that means

$mn = -18$ and $m + n = -3$

We now look at all pairs of integers with a product of -18. We then look at the sum of each pair of integers.

mn	$m + n$
$1(-18) = -18$	$1 + (-18) = -17$
$2(-9) = -18$	$2 + (-9) = -7$
$3(-6) = -18$	$3 + (-6) = -3$
$6(-3) = -18$	
$9(-2) = -18$	
$18(-1) = -18$	

We need look no further than 3 and −6.

NOTE

We could have chosen $m = -6$ and $n = 3$ as well.

The two integers with a product of ac and a sum of b are 3 and −6. We can say that

$m = 3$ and $n = -6$

Because we found values for m and n, we know that $x^2 - 3x - 18$ is factorable.

(b) $x^2 - 24x + 23$

We find that

$a = 1 \quad b = -24 \quad c = 23$

$ac = 1(23) = 23$ and $b = -24$

So $mn = 23$ and $m + n = -24$

We now calculate integer pairs, looking for two numbers with a product of 23 and a sum of −24.

mn	$m + n$
$1(23) = 23$	$1 + 23 = 24$
$-1(-23) = 23$	$-1 + (-23) = -24$

$m = -1$ and $n = -23$

So $x^2 - 24x + 23$ is factorable.

(c) $x^2 - 11x + 8$

We find that $a = 1$, $b = -11$, and $c = 8$. Therefore, $ac = 8$ and $b = -11$. Thus, $mn = 8$ and $m + n = -11$. We calculate integer pairs.

mn	$m + n$
$1(8) = 8$	$1 + 8 = 9$
$2(4) = 8$	$2 + 4 = 6$
$-1(-8) = 8$	$-1 + (-8) = -9$
$-2(-4) = 8$	$-2 + (-4) = -6$

There are no other pairs of integers with a product of 8, and none of these pairs has a sum of −11. The trinomial $x^2 - 11x + 8$ is not factorable.

(d) $2x^2 + 7x - 15$

We find that $a = 2$, $b = 7$, and $c = -15$. Therefore, $ac = 2(-15) = -30$ and $b = 7$. Thus, $mn = -30$ and $m + n = 7$. We calculate integer pairs.

mn	$m + n$
$1(-30) = -30$	$1 + (-30) = -29$
$2(-15) = -30$	$2 + (-15) = -13$
$3(-10) = -30$	$3 + (-10) = -7$
$5(-6) = -30$	$5 + (-6) = -1$
$6(-5) = -30$	$6 + (-5) = 1$
$10(-3) = -30$	$10 + (-3) = 7$

There is no need to go any further. We see that 10 and -3 have a product of -30 and a sum of 7, so

$m = 10$ and $n = -3$

Therefore, $2x^2 + 7x - 15$ is factorable.

Check Yourself 2

Use the *ac* test to determine which trinomials can be factored. Find the values of m and n for each trinomial that can be factored.

(a) $x^2 - 7x + 12$ **(b)** $x^2 + 5x - 14$
(c) $3x^2 - 6x + 7$ **(d)** $2x^2 + x - 6$

How does the *ac* test help us factor? What are m and n in this context? You might be asking these questions, so we take another look at multiplication. We look at the first polynomial we worked with in this section.

$$\begin{aligned}(5x + 2)(2x + 3) &= 10x^2 + 15x + 4x + 6 \\ &= 10x^2 + 19x + 6\end{aligned}$$

In its final form, we have

$$10x^2 + 19x + 6$$

We formed the middle term $19x$ by combining the *outer* and *inner* products when multiplying with the FOIL pattern. Now imagine that we had a way of writing the middle term as the sum of its outer and inner products.

$$10x^2 + 19x + 6 = 10x^2 + 15x + 4x + 6$$

We can factor the right side by grouping. Recall, when we factor by grouping, we partition the polynomial into two pairs of terms and factor out any GCF from each pair.

$$\underbrace{10x^2 + 15x}_{\text{GCF} = 5x} + \underbrace{4x + 6}_{\text{GCF} = 2} = 5x(2x + 3) + 2(2x + 3)$$

NOTE

We get the same result if we write

$10x^2 + 4x + 15x + 6$

Next, we factor out the common binomial factor $2x + 3$.

$$\begin{aligned}10x^2 + 15x + 4x + 6 &= 5x(2x + 3) + 2(2x + 3) \\ &= (5x + 2)(2x + 3)\end{aligned}$$ The correct factorization

The question is how do we know to write $19x = 15x + 4x$? Why didn't we write it as $10x + 9x$ or $18x + x$? We look at the results from the *ac* test on this polynomial. In $10x^2 + 19x + 6$, we have

$$a = 10, b = 19, c = 6$$

To use the *ac* test, we find the product

$$ac = (10)(6) = 60$$

NOTE

$15 \cdot 4 = 60$

$15 + 4 = 19$

The *ac* test tells us that if we can find two integers m and n whose product is $ac = 60$ and whose sum is $b = 19$ then we can factor the polynomial. In this case, those integers are 15 and 4.

This is the key to the *ac* method of factoring. We use the same m and n found in the *ac* test to rewrite the middle term as a sum.

$$ax^2 + bx + c = ax^2 + mx + nx + c$$

Example 3 Using the Results of the *ac* Test to Factor

< Objective 2 >

Rewrite the middle term as the sum of two terms, then factor by grouping.

(a) $x^2 - 3x - 18$

We have $a = 1$, $b = -3$, and $c = -18$, so $ac = -18$ and $b = -3$. We are looking for two numbers m and n where $mn = -18$ and $m + n = -3$.

In Example 2(a), we looked at every pair of integers whose product (mn) was -18, to find a pair that had a sum $(m + n)$ of -3. We found the two integers to be 3 and -6, because $3(-6) = -18$ and $3 + (-6) = -3$, so $m = 3$ and $n = -6$.

We use that result to rewrite the middle term as the sum of $3x$ and $-6x$.

$$x^2 + 3x - 6x - 18$$

We factor this by grouping.

$$\begin{aligned} x^2 + 3x - 6x - 18 &= x(x + 3) - 6(x + 3) \\ &= (x - 6)(x + 3) \end{aligned}$$

We check this result by multiplying the factors to ensure their product is the original polynomial.

Check

$$\begin{aligned} (x - 6)(x + 3) &= x^2 + 3x - 6x - 18 \\ &= x^2 - 3x - 18 \end{aligned}$$ The original polynomial

(b) $x^2 - 24x + 23$

We use the results from Example 2(b), in which we found $m = -1$ and $n = -23$, to rewrite the middle term of the equation.

$$x^2 - 24x + 23 = x^2 - x - 23x + 23$$

Then we factor by grouping.

RECALL

Check by multiplying.

$$\begin{aligned} x^2 - x - 23x + 23 &= (x^2 - x) - (23x - 23) \\ &= x(x - 1) - 23(x - 1) \\ &= (x - 23)(x - 1) \end{aligned}$$

(c) $2x^2 + 7x - 15$

From Example 2(d), we know that this trinomial is factorable, and $m = 10$ and $n = -3$. We use that result to rewrite the middle term of the trinomial.

$$\begin{aligned} 2x^2 + 7x - 15 &= 2x^2 + 10x - 3x - 15 \\ &= (2x^2 + 10x) - (3x + 15) \\ &= 2x(x + 5) - 3(x + 5) \\ &= (2x - 3)(x + 5) \end{aligned}$$

Check

$$\begin{aligned} (2x - 3)(x + 5) &= 2x^2 + 10x - 3x - 15 \\ &= 2x^2 + 7x - 15 \end{aligned}$$ The original polynomial

Careful readers will note that we did not ask you to factor Example 2(c), $x^2 - 11x + 8$. Recall that, by the *ac* method, we determined that this trinomial was not factorable.

Check Yourself 3

Use the results of Check Yourself 2 to rewrite the middle term as the sum of two terms, then factor by grouping.

(a) $x^2 - 7x + 12$ **(b)** $x^2 + 5x - 14$ **(c)** $2x^2 + x - 6$

Now look at some examples that require us to first find m and n and then factor the trinomial.

Example 4 Rewriting Middle Terms to Factor

Rewrite the middle term as the sum of two terms, and then factor by grouping.

(a) $2x^2 - 13x - 7$

We find that $a = 2$, $b = -13$, and $c = -7$, so $mn = ac = -14$ and $m + n = b = -13$.

mn	$m + n$
$1(-14) = -14$	$1 + (-14) = -13$

So $m = 1$ and $n = -14$. We rewrite the middle term of the trinomial.

$$\begin{aligned} 2x^2 - 13x - 7 &= 2x^2 + x - 14x - 7 \\ &= (2x^2 + x) - (14x + 7) \\ &= x(2x + 1) - 7(2x + 1) \\ &= (x - 7)(2x + 1) \end{aligned}$$

Check

$$\begin{aligned} (x - 7)(2x + 1) &= 2x^2 + x - 14x - 7 \\ &= 2x^2 - 13x - 7 \end{aligned}$$

The original polynomial

(b) $6x^2 - 5x - 6$

We find that $a = 6$, $b = -5$, and $c = -6$, so $mn = ac = -36$ and $m + n = b = -5$.

mn	$m + n$
$1(-36) = -36$	$1 + (-36) = -35$
$2(-18) = -36$	$2 + (-18) = -16$
$3(-12) = -36$	$3 + (-12) = -9$
$4(-9) = -36$	$4 + (-9) = -5$

So $m = 4$ and $n = -9$. We rewrite the middle term of the trinomial.

RECALL

Multiply to check.

$$\begin{aligned} 6x^2 - 5x - 6 &= 6x^2 + 4x - 9x - 6 \\ &= (6x^2 + 4x) - (9x + 6) \\ &= 2x(3x + 2) - 3(3x + 2) \\ &= (2x - 3)(3x + 2) \end{aligned}$$

Check Yourself 4

Rewrite the middle term as the sum of two terms and then factor by grouping.

(a) $2x^2 - 7x - 15$ **(b)** $6x^2 - 5x - 4$

Be certain to check trinomials and binomial factors for any common monomial factor. Example 5 shows the factoring out of monomial factors.

Example 5 Factoring Out Common Factors

< Objective 3 >

Completely factor the trinomial

$3x^2 + 12x - 15$

We first factor out the common factor of 3.

$3x^2 + 12x - 15 = 3(x^2 + 4x - 5)$

Finding m and n for the trinomial $x^2 + 4x - 5$ yields $mn = -5$ and $m + n = 4$.

NOTE

If we had not removed the GCF in the first step, we would have gotten either $(x - 1)(3x + 15)$ or $(3x - 3)(x + 5)$ after factoring. Neither of these is factored completely.

mn	$m + n$
$1(-5) = -5$	$1 + (-5) = -4$
$5(-1) = -5$	$5 + (-1) = 4$

So $m = 5$ and $n = -1$. This gives us

$$\begin{aligned} 3x^2 + 12x - 15 &= 3(x^2 + 4x - 5) \\ &= 3(x^2 + 5x - x - 5) \\ &= 3[(x^2 + 5x) - (x + 5)] \\ &= 3[x(x + 5) - (x + 5)] \\ &= 3[(x - 1)(x + 5)] \\ &= 3(x - 1)(x + 5) \end{aligned}$$

RECALL

Again, multiply this result to check.

Check Yourself 5

Completely factor the trinomial.

$6x^3 + 3x^2 - 18x$

Not all possible product pairs need to be tried to find m and n. A look at the sign pattern of the trinomial eliminates many of the possibilities. Assuming the leading coefficient is positive, there are four possible sign patterns. If the leading coefficient is negative, factor out -1 and then consider the remaining polynomial, whose leading coefficient is now positive.

Pattern	Example	Conclusion
1. b and c are both positive.	$2x^2 + 13x + 15$	m and n must both be positive.
2. b is negative and c is positive.	$x^2 - 7x + 12$	m and n must both be negative.
3. b is positive and c is negative.	$x^2 + 3x - 10$	m and n are of opposite signs. (The value with the larger absolute value is positive.)
4. b and c are both negative.	$x^2 - 3x - 10$	m and n are of opposite signs. (The value with the larger absolute value is negative.)

Check Yourself ANSWERS

1. **(a)** $a = 1, b = 6, c = 11$; **(b)** $4x^2 - x + 8; a = 4, b = -1, c = 8$; **(c)** $-2x^2 + 8x + 3; a = -2, b = 8, c = 3$; **(d)** $a = -4, b = 1, c = -1$
2. **(a)** Factorable, $m = -3, n = -4$; **(b)** factorable, $m = 7, n = -2$; **(c)** not factorable; **(d)** factorable, $m = 4, n = -3$
3. **(a)** $x^2 - 3x - 4x + 12 = (x - 4)(x - 3)$; **(b)** $x^2 + 7x - 2x - 14 = (x - 2)(x + 7)$; **(c)** $2x^2 + 4x - 3x - 6 = (2x - 3)(x + 2)$
4. **(a)** $2x^2 - 10x + 3x - 15 = (2x + 3)(x - 5)$; **(b)** $6x^2 - 8x + 3x - 4 = (3x - 4)(2x + 1)$
5. $3x(x + 2)(2x - 3)$

Reading Your Text

These fill-in-the-blank exercises will help you understand some of the key vocabulary used in this section. The answers to these exercises are in the Answers Appendix in the back of the text.

(a) The first step in learning to factor a trinomial by the *ac* method is to identify its ________.

(b) If the leading coefficient of a trinomial is positive, there are ________ possible sign patterns.

(c) To discover whether a trinomial is ________, we try the *ac* test.

(d) When factoring a polynomial, we always begin by factoring out any ________.

Skills | Calculator/Computer | Career Applications | Above and Beyond

6.4 exercises

For each trinomial, identify a, b, and c.

1. $x^2 + 7x - 5$ **2.** $x^2 + 5x + 11$ **3.** $x^2 - 3x + 8$ **4.** $x^2 + 7x - 15$

5. $5x + 3x^2 - 8$ **6.** $3x^2 + 7 - 5x$ **7.** $11 + 8x + 4x^2$ **8.** $7x - 9 + 5x^2$

9. $-7x^2 - 5x + 2$ **10.** $-7x^2 + 9x - 18$

< Objective 1 >

Use the ac test to determine which of the trinomials can be factored. Find the values of m and n for each trinomial that can be factored.

11. $x^2 + x - 6$ **12.** $x^2 - x - 6$ **13.** $x^2 + 3x - 1$ **14.** $x^2 - 3x + 7$

15. $x^2 - 5x + 6$ **16.** $x^2 - x + 2$ **17.** $2x^2 + 5x - 3$ **18.** $3x^2 - 14x - 5$

19. $6x^2 - 19x + 10$ VIDEO **20.** $4x^2 + 5x + 6$

< Objectives 2 and 3 >

Complete each statement.

21. $4x^2 - 4x - 3 = (2x + 1)(\quad)$ **22.** $3w^2 + 11w - 4 = (w + 4)(\quad)$

23. $6a^2 + 13a + 6 = (2a + 3)(\quad)$ **24.** $25y^2 - 10y + 1 = (5y - 1)(\quad)$

25. $15x^2 - 16x + 4 = (3x - 2)(\quad)$ **26.** $6m^2 + 5m - 4 = (3m + 4)(\quad)$

27. $16a^2 + 8ab + b^2 = (4a + b)(\quad)$ **28.** $6x^2 + 5xy - 4y^2 = (3x + 4y)(\quad)$ VIDEO

29. $4m^2 + 5mn - 6n^2 = (m + 2n)(\quad)$ **30.** $10p^2 - pq - 3q^2 = (5p - 3q)(\quad)$

Completely factor each polynomial.

31. $x^2 + 8x + 15$ **32.** $x^2 - 11x + 24$ **33.** $s^2 + 13s + 30$ **34.** $b^2 + 14b + 33$

35. $x^2 + 3x + 11$ **36.** $x^2 - 8x + 8$ **37.** $x^2 - 6x - 40$ **38.** $x^2 - 11x + 10$

39. $p^2 - 10p - 24$ **40.** $x^2 - 11x - 60$ **41.** $x^2 + 5x - 66$ **42.** $a^2 + 2a - 80$

43. $c^2 + 19c + 60$ **44.** $t^2 - 4t - 60$ **45.** $n^2 + 5n - 50$ **46.** $x^2 - 16x + 63$

47. $m^2 - 6m + 1$
48. $w^2 + w - 5$
49. $x^2 + 7xy + 10y^2$
50. $x^2 - 8xy + 12y^2$
51. $a^2 - ab - 42b^2$
52. $m^2 - 8mn + 16n^2$
53. $x^2 - 13xy + 40y^2$
54. $r^2 - 9rs - 36s^2$
55. $6x^2 + 19x + 10$
56. $6x^2 - 7x - 3$
57. $15x^2 + x - 6$
58. $12w^2 + 19w + 4$
59. $6m^2 + 25m - 25$
60. $8x^2 - 6x - 9$
61. $9x^2 - 12x + 4$
62. $20x^2 - 23x + 6$
63. $12x^2 - 8x - 15$
64. $16a^2 + 40a + 25$
65. $3y^2 + 7y - 6$
66. $12x^2 + 11x - 15$
67. $8x^2 - 27x - 20$

VIDEO

68. $24v^2 + 5v - 36$
69. $4x^2 + 3x + 11$
70. $6x^2 - x + 1$
71. $2x^2 + 3xy + y^2$
72. $3x^2 - 5xy + 2y^2$
73. $5a^2 - 8ab - 4b^2$
74. $5x^2 + 7xy - 6y^2$
75. $9x^2 + 4xy - 5y^2$
76. $16x^2 + 32xy + 15y^2$
77. $6m^2 - 17mn + 12n^2$
78. $15x^2 - xy - 6y^2$
79. $36a^2 - 3ab - 5b^2$
80. $3q^2 - 17qr - 6r^2$
81. $x^2 + 4xy + 4y^2$
82. $25b^2 - 80bc + 64c^2$
83. $2x^2 + 18x - 1$
84. $5x^2 - 12x - 6$
85. $20x^2 - 20x - 15$
86. $24x^2 - 18x - 6$
87. $8m^2 + 12m + 4$
88. $14x^2 - 20x + 6$
89. $15r^2 - 21rs + 6s^2$
90. $10x^2 + 5xy - 30y^2$
91. $-p^2 + 10p + 24$
92. $-x^2 + 11x + 60$
93. $-n^2 - 5n + 50$
94. $-x^2 + 16x - 63$
95. $x^2 - 2x + 12$
96. $-u^2 - 5u + 7$
97. $-3x^2 - x + 9$
98. $5x^2 - 3x + 4$
99. $-6x^2 - 19x - 10$
100. $-6x^2 + 7x + 3$
101. $2x^3 - 2x^2 - 4x$
102. $2y^3 + y^2 - 3y$
103. $2y^4 + 5y^3 + 3y^2$

VIDEO

104. $4z^3 - 18z^2 - 10z$
105. $10(x + y)^2 - 11(x + y) - 6$
106. $8(a - b)^2 + 14(a - b) - 15$
107. $5(x - 1)^2 - 15(x - 1) - 350$
108. $3(x + 1)^2 - 6(x + 1) - 45$
109. $15 + 29x - 48x^2$
110. $12 + 4a - 21a^2$
111. $-6x^2 + 19x - 15$
112. $-3s^2 - 10s + 8$

Determine whether each statement is **true** *or* **false.**

113. A trinomial can always be factored into the product of two binomials.

114. The *ac* method requires factoring by grouping.

Complete each statement with **always, sometimes,** *or* **never.**

115. A trinomial with integer coefficients is __________ factorable.

116. If a trinomial with all positive terms is factored, the signs between the terms in the binomial factors are __________ positive.

117. The product of two binomials __________ results in a trinomial.

118. If the GCF for the terms in a polynomial is not 1, it should __________ be factored out first.

Skills | Calculator/Computer | **Career Applications** | Above and Beyond

119. **Agricultural Technology** The yield Y of a crop is given by the equation

$$Y = -0.05x^2 + 1.5x + 140$$

Rewrite this equation by factoring the right-hand side.
Hint: Begin by factoring out -0.05.

120. **Construction Technology** The profit curve P for a welding shop is given by the equation

$$P = 2x^2 - 143x - 1{,}360$$

Rewrite this equation by factoring the right-hand side.

121. **Allied Health** The number N of people who are sick t days after the outbreak of a flu epidemic is given by the equation

$$N = 50 + 25t - 3t^2$$

Rewrite this equation by factoring the right-hand side.

122. **Mechanical Engineering** The flow rate through a hydraulic hose can be found using the equation

$$2Q^2 + Q - 21 = 0$$

Rewrite this equation by factoring the left-hand side.

Answers

1. $a = 1; b = 7; c = -5$ **3.** $a = 1; b = -3; c = 8$ **5.** $a = 3; b = 5; c = -8$ **7.** $a = 4; b = 8; c = 11$ **9.** $a = -7; b = -5; c = 2$ **11.** Factorable; 3, −2 **13.** Not factorable **15.** Factorable; −3, −2 **17.** Factorable; 6, −1 **19.** Factorable; −15, −4 **21.** $2x - 3$ **23.** $3a + 2$ **25.** $5x - 2$ **27.** $4a + b$ **29.** $4m - 3n$ **31.** $(x + 3)(x + 5)$ **33.** $(s + 10)(s + 3)$ **35.** Not factorable **37.** $(x - 10)(x + 4)$ **39.** $(p - 12)(p + 2)$ **41.** $(x + 11)(x - 6)$ **43.** $(c + 4)(c + 15)$ **45.** $(n + 10)(n - 5)$ **47.** Not factorable **49.** $(x + 2y)(x + 5y)$ **51.** $(a - 7b)(a + 6b)$ **53.** $(x - 5y)(x - 8y)$ **55.** $(3x + 2)(2x + 5)$ **57.** $(5x - 3)(3x + 2)$ **59.** $(6m - 5)(m + 5)$ **61.** $(3x - 2)^2$ **63.** $(6x + 5)(2x - 3)$ **65.** $(3y - 2)(y + 3)$ **67.** $(8x + 5)(x - 4)$ **69.** Not factorable **71.** $(2x + y)(x + y)$ **73.** $(5a + 2b)(a - 2b)$ **75.** $(9x - 5y)(x + y)$ **77.** $(3m - 4n)(2m - 3n)$ **79.** $(12a - 5b)(3a + b)$ **81.** $(x + 2y)^2$ **83.** Not factorable **85.** $5(2x - 3)(2x + 1)$ **87.** $4(2m + 1)(m + 1)$ **89.** $3(5r - 2s)(r - s)$ **91.** $-(p - 12)(p + 2)$ **93.** $-(n + 10)(n - 5)$ **95.** Not factorable **97.** $-(3x^2 + x - 9)$ **99.** $-(3x + 2)(2x + 5)$ **101.** $2x(x - 2)(x + 1)$ **103.** $y^2(2y + 3)(y + 1)$ **105.** $(5x + 5y + 2)(2x + 2y - 3)$ **107.** $5(x - 11)(x + 6)$ **109.** $(1 + 3x)(15 - 16x)$ or $-(3x + 1)(16x - 15)$ **111.** $-(2x - 3)(3x - 5)$ **113.** False **115.** sometimes **117.** sometimes **119.** $Y = -0.05(x + 40)(x - 70)$ **121.** $N = -(3t + 5)(t - 10)$

6.5 Factoring Strategies

< 6.5 Objectives >

1 > Recognize factoring patterns

2 > Apply appropriate factoring strategies

3 > Determine if a polynomial is prime

You have learned a variety of techniques to factor polynomials in this chapter. This section reviews those techniques and presents some guidelines for choosing an appropriate strategy or combination of strategies.

RECALL

Change the signs of every term in the remaining polynomial when factoring out a negative.

1. Always look for a greatest common factor. If the GCF is not 1, your first step is to factor out the GCF. If the leading coefficient is negative, factor out -1 along with any GCF.
   (a) To factor $5x^2y - 10xy + 25xy^2$, see that the GCF is $5xy$.

   $$5x^2y - 10xy + 25xy^2 = 5xy(x - 2 + 5y)$$

   (b) To factor $-3x^2 + 9x - 6$, see that the GCF is 3. Because the leading coefficient is negative, we factor out -1 along with 3, so we factor out -3.

   $$-3x^2 + 9x - 6 = -3(x^2 - 3x + 2)$$

RECALL

$a^2 - b^2 = (a + b)(a - b)$

The sum of squares $a^2 + b^2$ cannot be factored.

$a^3 - b^3 = (a - b)(a^2 + ab + b^2)$

$a^3 + b^3 = (a + b)(a^2 - ab + b^2)$

2. Next, consider the number of terms in the polynomial.
   (a) If the polynomial is a *binomial,* consider the special binomial formulas.

   (i) To factor $x^2 - 49y^2$, recognize a difference of squares.

   $$x^2 - 49y^2 = (x + 7y)(x - 7y)$$

   (ii) The binomial

   $$x^2 + 121$$

   is the sum of squares and cannot be further factored.

   (iii) To factor $t^3 - 64$, recognize a difference of cubes.

   $$t^3 - 64 = (t - 4)(t^2 + 4t + 16)$$

   (iv) The binomial $z^3 + 1$ is a sum of cubes.

   $$z^3 + 1 = (z + 1)(z^2 - z + 1)$$

   (b) If the polynomial is a *trinomial,* try to factor it as a product of two binomials. You can use either the trial-and-error method or the *ac* method.

   To factor $2x^2 - x - 6$, either method leads to

   $$2x^2 - x - 6 = (2x + 3)(x - 2)$$

   (c) If the polynomial has *more than three terms,* try to factor by grouping.

   To factor $2x^2 - 3xy + 10x - 15y$, group the first two terms, the last two terms, and factor out common factors.

   $$2x^2 - 3xy + 10x - 15y = x(2x - 3y) + 5(2x - 3y)$$

   Now factor out the common binomial factor $(2x - 3y)$.

   $$2x^2 - 3xy + 10x - 15y = (2x - 3y)(x + 5)$$

3. Always factor the polynomial completely. After you apply one technique, another one may be necessary.

   (a) To factor $6x^3 + 22x^2 - 40x$, first factor out the common factor $2x$.

   $$6x^3 + 22x^2 - 40x = 2x(3x^2 + 11x - 20)$$

   Now continue to factor the trinomial.

   $$6x^3 + 22x^2 - 40x = 2x(3x - 4)(x + 5)$$

   (b) To factor $x^3 - x^2y - 4x + 4y$, first we proceed by grouping.

   $$\begin{aligned} x^3 - x^2y - 4x + 4y &= x^2(x - y) - 4(x - y) \\ &= (x - y)(x^2 - 4) \end{aligned}$$

   Now because $x^2 - 4$ is a difference of squares, we continue to factor.

   $$x^3 - x^2y - 4x + 4y = (x - y)(x + 2)(x - 2)$$

4. Always check your answer by multiplying.

RECALL

Polynomials that cannot be factored are called **prime polynomials.**

5. Some polynomials are simply not factorable. If you try all of the preceding steps and still cannot break the polynomial, you should answer "not factorable."

   You can use the *ac* test to prove that a quadratic polynomial does not factor. To use the *ac* test, compute the product *ac*. Then look for factorizations of *ac* whose sum is *b*. If no such pair of numbers exists, then the quadratic is prime.

   $$2x^2 - 4x + 3$$
   $$ac = (2)(3) = 6;\ b = -4$$

   The factorizations of 6 are

Factorization	Sum
1, 6	7
2, 3	5
−1, −6	−7
−2, −3	−5

   Since none of the possible factorizations of 6 sum to −4, the quadratic polynomial is prime.

Example 1 Recognizing Factoring Patterns

< Objective 1 >

State the appropriate first step to factor each polynomial.

(a) $9x^2 - 18x - 72$

Factor out the GCF.

(b) $x^2 - 3x + 2xy - 6y$

Group the terms.

(c) $x^4 - 81y^4$

Factor the difference of squares.

(d) $3x^2 + 7x + 2$

Use the *ac* method (or trial and error).

Check Yourself 1

State the appropriate first step to factor each polynomial.

(a) $5x^2 + 2x - 3$ **(b)** $a^4b^4 - 16$
(c) $3x^2 + 3x - 60$ **(d)** $2a^2 - 5a + 4ab - 10b$

Example 2 Factoring Polynomials

< Objective 2 >

Factor $2xy + 10x + 6y + 30$.

$2xy + 10x + 6y + 30$ Factor out the GCF of 2.
$= 2(xy + 5x + 3y + 15)$

Because the polynomial has four terms, we move to step 3 and try grouping.

RECALL

Check by multiplying the factors together. You should get the original polynomial.

$= 2[(xy + 5x) + (3y + 15)]$
$= 2[x(y + 5) + 3(y + 5)]$
$= 2[(y + 5)(x + 3)]$
$= 2(y + 5)(x + 3)$

The binomial factors are first degree with 1 as their leading coefficient, so they cannot be further factored. We finish by checking our work.

$2(y + 5)(x + 3) = 2(y \cdot x + y \cdot 3 + 5 \cdot x + 5 \cdot 3)$
$= 2(xy + 3y + 5x + 15)$
$= 2xy + 6y + 10x + 30$
$= 2xy + 10x + 6y + 30$ The original polynomial

Check Yourself 2

Factor $4mn - 12m - 20n + 60$.

Be sure to keep your eyes open for factors that can be further factored. This is illustrated in Example 3.

Example 3 Factoring Polynomials

Factor $3mn^4 - 48m$.

$3mn^4 - 48m$ Factor out the GCF.
$= 3m(n^4 - 16)$ Note the difference-of-squares binomial.
$= 3m(n^2 - 4)(n^2 + 4)$ Note another difference of squares.
$= 3m(n - 2)(n + 2)(n^2 + 4)$

The only binomial that could possibly factor is $n^2 + 4$, but since this is a sum of squares, it does not. We are done.

Check Yourself 3

Factor $3x^2y - 75y$.

Remember to always start with step 1: Factor out the GCF.

Example 4 Factoring Polynomials

Factor $-6x^2y + 18xy + 60y$.

$-6x^2y + 18xy + 60y$ — We have a GCF of $6y$ and a leading negative to factor out.

$= -6y(x^2 - 3x - 10)$ — We factor the trinomial using trial and error or the ac method.

$= -6y(x - 5)(x + 2)$

Check Yourself 4

Factor $-5xy^2 - 15xy + 90x$.

Example 5 Factoring Polynomials

Factor $x^2y^2 - 9x^2 + y^2 - 9$.

There is no GCF greater than 1. Since we have four terms, we try grouping.

$x^2y^2 - 9x^2 + y^2 - 9$

$= x^2(y^2 - 9) + (y^2 - 9)$

$= (x^2 + 1)(y^2 - 9)$ — We note the difference of squares.

$= (x^2 + 1)(y - 3)(y + 3)$

Since $x^2 + 1$ is a sum of squares, we are done.

Check Yourself 5

Factor $x^2y^2 - 3x^2 - 4y^2 + 12$.

We claimed that several polynomials were not factorable. For instance, we said that a sum of squares does not factor. We also said that the trinomial factor in a sum or difference of cubes does not factor. We can use the ac test to demonstrate these claims.

Example 6 Prime Polynomials and Factoring

< Objective 3 >

Use the ac test to show that $x^2 + 1$ does not factor.

This polynomial is a quadratic with $b = 0$.

$x^2 + 1 = x^2 + 0x + 1$

Using the notation in Section 6.4, we have

$a = 1, b = 0, c = 1$

We want two numbers m and n, so that $mn = ac$ and $m + n = b$.

$ac = (1)(1) = 1$

Therefore, the product of m and n must be 1. The factor pairs of 1 are

$m = 1$ and $n = 1$ or $m = -1$ and $n = -1$

We find the sum of each factor pair. If the sum of one of the factor pairs is zero, then the polynomial factors.

$1 + 1 = 2 \neq 0 \qquad (-1) + (-1) = -2 \neq 0$

Because no factor pair of $ac = 1$ adds to $b = 0$, we conclude that $x^2 + 1$ does not factor.

Check Yourself 6

$x^3 - 8 = (x - 2)(x^2 + 2x + 4)$

Use the *ac* test to show that $x^2 + 2x + 4$ does not factor.

Check Yourself ANSWERS

1. **(a)** *ac* method (or trial and error); **(b)** factor the difference of squares; **(c)** factor out the GCF; **(d)** group the terms **2.** $4(m - 5)(n - 3)$ **3.** $3y(x - 5)(x + 5)$ **4.** $-5x(y + 6)(y - 3)$ **5.** $(x - 2)(x + 2)(y^2 - 3)$ **6.** Not factorable

Reading Your Text

These fill-in-the-blank exercises will help you understand some of the key vocabulary used in this section. The answers to these exercises are in the Answers Appendix in the back of the text.

(a) The ________ of squares is not factorable.

(b) If a polynomial consists of four terms, try to factor by ________.

(c) If we factor properly, then the product of the factors is the original ________.

(d) We can factor a trinomial using trial and error or the ________ method.

6.5 exercises

Skills | Calculator/Computer | Career Applications | Above and Beyond

< Objectives 1–3 >

State which method(s) should be applied, given the guidelines of this section. Then factor each polynomial completely.

1. $x^2 - 3x$ **2.** $4y^2 - 9$ **3.** $x^2 - 5x - 24$ **4.** $8x^3 + 10x$

5. $x(x - y) + 2(x - y)$ **6.** $5a^2 - 10a + 25$ **7.** $2x^2y - 6xy + 8y^2$ VIDEO **8.** $2p - 6q + pq - 3q^2$

9. $y^2 - 13y + 40$ **10.** $m^3 + 27m^2n$ **11.** $3b^2 + 17b - 28$ VIDEO **12.** $3x^2 + 6x - 5xy - 10y$ VIDEO

13. $3x^2 - 14xy - 24y^2$ **14.** $16c^2 - 49d^2$ **15.** $2a^2 + 11a + 12$ **16.** $m^3n^3 - mn$

17. $125r^3 + r^2$ **18.** $(x - y)^2 - 16$ **19.** $3x^2 - 30x + 63$ **20.** $3a^2 - 108$

21. $40a^2 + 5$ **22.** $4p^2 - 8p - 60$ **23.** $2w^2 - 14w - 36$ **24.** $xy^3 - 9xy$

25. $3a^2b - 48b^3$ VIDEO **26.** $12b^3 - 86b^2 + 14b$ **27.** $x^4 - 3x^2 - 10$ **28.** $m^4 - 9n^4$

29. $8p^3 - q^3r^3$ **30.** $27x^3 + 125y^3$

Completely factor each polynomial.

31. $x^2 - 5x - 14$
32. $y^2 - 3y - 40$
33. $a^2 + 10a + 24$
34. $n^2 + 11n + 18$
35. $y^2 - 10y + 21$
36. $z^2 - 12z + 20$
37. $w^3 - 125$
38. $1 - z^3$
39. $2t^2 + 9t - 5$
40. $3t^2 - 11t - 4$
41. $4y^2 - 81$
42. $9m^2 - 25n^2$
43. $3a^2 + 6a - 21$
44. $5b^2 - 15b + 30$
45. $8a^3 + 27$
46. $y^3 - 64$
47. $4t^3 - 4$
48. $4m^4 - 4$
49. $3x^2 + 30x + 75$
50. $4x^2 - 24x + 36$
51. $xy - 2x + 6y - 12$
52. $yz + 5y - 3z - 15$
53. $2x^2 + 5x + 4$
54. $4x^2 + 12x - 72$
55. $a^2 - 10a + 25$
56. $x^2 + 6xy + 9y^2$
57. $2abx - 14ax - 8bx + 56x$
58. $3ax + 12x + 3a + 12$
59. $3x^2 + x - 10$
60. $3x^2 - 2x + 4$
61. $5n^2 + 20n + 30$
62. $3ay^2 - 48a$
63. $3y^3 - 81$
64. $6t^2 - 12t + 24$
65. $3x^2y + 9xy + 9y$
66. $18mn^2 - 15mn - 12m$
67. $9x^3 - 4x$
68. $5uv^2 - 20uv + 30u$
69. $2n^4 - 16n$
70. $8x^3 - 2x$
71. $-3x^2 + 12x + 15$
72. $-5y^2 - 30y - 40$
73. $x^2y^2 - 4x^2 - 9y^2 + 36$
74. $40x^2 + 12x - 72$
75. $8x^2 + 8x + 2$
76. $m^2n^2 - 4n^2 - 25m^2 + 100$
77. $x^3 + 5x^2 - 4x - 20$
78. $x^3 - 2x^2 - 9x + 18$
79. $x^4 + 3x^2 - 10$
80. $y^4 - 2y^2 - 24$
81. $t^4 + t^2 - 20$
82. $w^4 - 10w^2 + 9$
83. $(x - 5)^2 - 169$ VIDEO
84. $(x - 7)^2 - 81$
85. $x^2 + 4xy + 4y^2 - 16$
86. $9x^2 + 12xy + 4y^2 - 25$
87. $6(x - 2)^2 + 7(x - 2) - 5$
88. $12(x + 1)^2 - 17(x + 1) + 6$

Determine whether each statement is **true** *or* **false.**

89. Factoring a polynomial may require more than one technique.

90. Every polynomial can be factored.

91. If we are trying to factor a four-term polynomial, we should try to factor by grouping.

92. No matter what type of polynomial we are trying to factor, we should always check for a GCF first.

Answers

1. GCF, $x(x - 3)$ **3.** Trial and error, $(x - 8)(x + 3)$ **5.** GCF, $(x + 2)(x - y)$ **7.** GCF, $2y(x^2 - 3x + 4y)$ **9.** Trial and error, $(y - 5)(y - 8)$ **11.** *ac* method, $(b + 7)(3b - 4)$ **13.** *ac* method, $(3x + 4y)(x - 6y)$ **15.** *ac* method, $(2a + 3)(a + 4)$ **17.** GCF, $r^2(125r + 1)$ **19.** GCF, then trial and error, $3(x - 3)(x - 7)$ **21.** GCF, $5(8a^2 + 1)$ **23.** GCF, then trial and error, $2(w - 9)(w + 2)$ **25.** GCF, then difference of squares, $3b(a + 4b)(a - 4b)$ **27.** Trial and error, $(x^2 - 5)(x^2 + 2)$ **29.** Difference of cubes, $(2p - qr)(4p^2 + 2pqr + q^2r^2)$ **31.** $(x - 7)(x + 2)$ **33.** $(a + 6)(a + 4)$ **35.** $(y - 7)(y - 3)$ **37.** $(w - 5)(w^2 + 5w + 25)$ **39.** $(2t - 1)(t + 5)$ **41.** $(2y + 9)(2y - 9)$ **43.** $3(a^2 + 2a - 7)$ **45.** $(2a + 3)(4a^2 - 6a + 9)$ **47.** $4(t - 1)(t^2 + t + 1)$ **49.** $3(x + 5)^2$ **51.** $(x + 6)(y - 2)$ **53.** Not factorable **55.** $(a - 5)^2$ **57.** $2x(a - 4)(b - 7)$ **59.** $(3x - 5)(x + 2)$ **61.** $5(n^2 + 4n + 6)$ **63.** $3(y - 3)(y^2 + 3y + 9)$ **65.** $3y(x^2 + 3x + 3)$ **67.** $x(3x - 2)(3x + 2)$ **69.** $2n(n - 2)(n^2 + 2n + 4)$ **71.** $-3(x + 1)(x - 5)$ **73.** $(x - 3)(x + 3)(y - 2)(y + 2)$ **75.** $2(2x + 1)^2$ **77.** $(x + 5)(x - 2)(x + 2)$ **79.** $(x^2 - 2)(x^2 + 5)$ **81.** $(t + 2)(t - 2)(t^2 + 5)$ **83.** $(x + 8)(x - 18)$ **85.** $(x + 2y + 4)(x + 2y - 4)$ **87.** $(2x - 5)(3x - 1)$ **89.** True **91.** True

6.6 Factoring and Problem Solving

< 6.6 Objectives >

1 > Solve quadratic equations by factoring

2 > Find the zeros of a quadratic function

3 > Use factoring to solve applications

The factoring techniques you learned provide you with the tools to solve equations that can be written in the form

$$ax^2 + bx + c = 0 \qquad a \neq 0$$

This is a quadratic equation in one variable, here x. You can recognize a quadratic equation because the highest power of the variable is two.

where a, b, and c are constants.

An equation written in the form $ax^2 + bx + c = 0$ is called a **quadratic equation in standard form.** Because of the *zero-product principle,* we can use factoring to solve quadratic equations.

Property

Zero-Product Principle

$ab = 0$ if, and only if, $a = 0$ or $b = 0$ or both.

This property says two things.

1. The product of any number and zero is *always* zero.
2. The product of two nonzero numbers is *never* zero.

We apply this principle to solve a quadratic equation in Example 1.

Example 1 Solving Equations by Factoring

< Objective 1 >

NOTE

To use the zero-product principle, 0 must be on one side of the equation.

Solve

$$x^2 - 3x - 18 = 0$$

Factoring on the left, we have

$$(x - 6)(x + 3) = 0$$

By the zero-product principle, we know that one or both of the factors must be zero. This means

$$x - 6 = 0 \qquad \text{or} \qquad x + 3 = 0$$

Solving each equation gives

$$x = 6 \qquad \text{or} \qquad x = -3$$

> CAUTION

You should see the sign changes. The factor $(x - 6)$ gives the solution $x = 6$. The factor $(x + 3)$ gives the solution $x = -3$.

The two solutions are 6 and -3.

The solutions are sometimes called the **roots** of the equation. These roots have an important connection to the graph of the function $f(x) = x^2 - 3x - 18$, also written $y = x^2 - 3x - 18$. The graph of this function forms a curve, which we will study

NOTE

Graph the equation
$y = x^2 - 3x - 18$
on your graphing calculator.
Use the **ZERO** utility to show

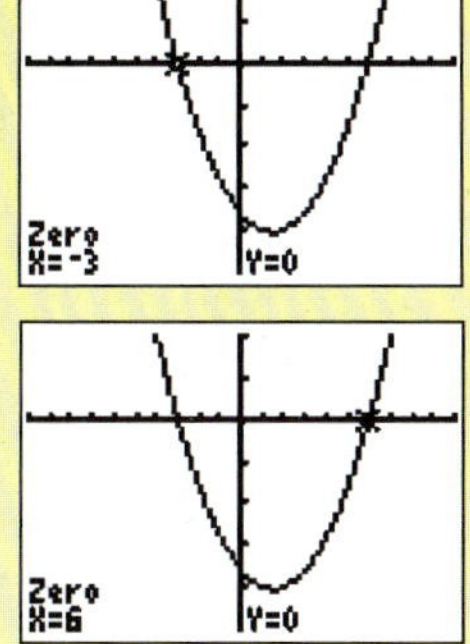

in Chapter 8. This particular curve crosses the x-axis at two points, $(-3, 0)$ and $(6, 0)$. -3 and 6 are called the **zeros** of this function, since $f(-3) = 0$ and $f(6) = 0$. So, the solutions of the equation $x^2 - 3x - 18 = 0$, -3 and 6, are zeros of the function $f(x) = x^2 - 3x - 18$, and they tell us the points where the graph of f crosses the x-axis, $(-3, 0)$ and $(6, 0)$.

We use substitution to check a solution to a quadratic equation in the same way we check any solution.

Check

$x^2 - 3x - 18 = 0$ The original equation

$(6)^2 - 3(6) - 18 \stackrel{?}{=} 0$ The solution $x = 6$.

$36 - 18 - 18 \stackrel{?}{=} 0$

$0 = 0$ True

$x^2 - 3x - 18 = 0$

$(-3)^2 - 3(-3) - 18 \stackrel{?}{=} 0$ The solution $x = -3$.

$9 + 9 - 18 \stackrel{?}{=} 0$

$0 = 0$ True

Check Yourself 1

Solve and check $x^2 - 9x + 20 = 0$.

You may need to use any of the factoring techniques to solve quadratic equations.

Example 2 Solving Equations by Factoring

>CAUTION

A *common mistake* is to forget the statement $x = 0$ when solving this type of equation. Be sure to include both answers.

(a) Solve $x^2 - 5x = 0$.

Factor the left side of the equation and apply the zero-product principle.

$x(x - 5) = 0$ Factor out the GCF.

Now

$x = 0$ or $x - 5 = 0$

$x = 5$

The two solutions are 0 and 5. The solution set is $\{0, 5\}$.

(b) Solve $x^2 - 9 = 0$.

Factoring yields

$(x + 3)(x - 3) = 0$ Use the difference of squares formula.

$x + 3 = 0$ or $x - 3 = 0$

$x = -3$ $x = 3$

The solution set is $\{-3, 3\}$, which may be written as $\{\pm 3\}$.

NOTE

The symbol $\pm$ is read "plus or minus."

Check Yourself 2

Solve by factoring.

(a) $x^2 + 8x = 0$ **(b)** $x^2 - 16 = 0$

Example 3 illustrates a crucial point. Our method depends on the zero-product principle, which means that the product of factors *must equal 0.*

Example 3 Solving Equations by Factoring

> CAUTION

Consider the equation

$x(2x - 1) = 3$

Students are sometimes tempted to write

$x = 3$ or $2x - 1 = 3$

This is *not correct.*

If $a \cdot b = 0$, then a or b **must** be 0. But if $a \cdot b = 3$, we do not know that a or b is 3, for example, it could be that $a = 6$ and $b = 1/2$ (or many other possibilities).

Solve $2x^2 - x = 3$.

The first step in solving is to write the equation in standard form (that is, with 0 on one side of the equation). Start by subtracting 3 from both sides of the equation.

$2x^2 - x - 3 = 0$ Make sure all nonzero terms are on one side of the equation. The other side must be 0.

We can now factor and use the zero-product principle.

$$(2x - 3)(x + 1) = 0$$

$$2x - 3 = 0 \quad \text{or} \quad x + 1 = 0$$

$$2x = 3 \qquad\qquad x = -1$$

$$x = \frac{3}{2}$$

The solution set is $\left\{\frac{3}{2}, -1\right\}$.

Check Yourself 3

Solve $3x^2 = 5x + 2$.

So far, each of the quadratic equations had two distinct real-number solutions. That may not always be the case.

Example 4 Solving Equations by Factoring

Solve $x^2 - 6x + 9 = 0$.

Factoring gives

$$(x - 3)(x - 3) = 0$$

and

$$x - 3 = 0 \quad \text{or} \quad x - 3 = 0$$

$$x = 3 \qquad\qquad x = 3$$

The solution set is $\{3\}$.

When an equation such as this one has two solutions that are the same number, we call 3 the **repeated** (or **double**) **solution**.

Even though a quadratic equation always has two solutions, they may not always be real numbers. You will learn more about this in Chapters 7 and 8.

Check Yourself 4

Solve $x^2 + 6x + 9 = 0$.

Remember to factor out any GCF first. This makes your work easier, as Example 5 illustrates.

Example 5 Solving Equations by Factoring

Solve $3x^2 - 3x - 60 = 0$.

First, note the common factor 3 in the quadratic expression of the equation. Factoring out the 3 gives

$3(x^2 - x - 20) = 0$

Divide both sides of the equation by 3.

$$\frac{3(x^2 - x - 20)}{3} = \frac{0}{3}$$

or $x^2 - x - 20 = 0$

We now factor and solve as before.

$(x - 5)(x + 4) = 0$

$x - 5 = 0$ or $x + 4 = 0$

$x = 5$ $\quad x = -4$ or $\{-4, 5\}$

NOTE

The advantage of dividing both sides by 3 is that the coefficients in the quadratic expression become smaller making it easier to factor.

Check Yourself 5

Solve $2x^2 - 10x - 48 = 0$.

Fractions may seem to complicate matters, but there is a nice way to eliminate them. Recall from Section 1.5 that we can choose to multiply both sides of an equation by a nonzero constant without affecting the solution set.

Example 6 Clearing Fractions from a Quadratic Equation

Solve $\frac{x^2}{5} + \frac{x}{10} - 1 = 0$.

Noting denominators of 5, 10, and 1, we choose the least common multiple of these, namely 10, and multiply.

$$10\left(\frac{x^2}{5}\right) + 10\left(\frac{x}{10}\right) - 10(1) = 10(0)$$

$2x^2 + x - 10 = 0$ Fractions have been "cleared."

$(2x + 5)(x - 2) = 0$ We factor as usual, and solve.

$2x + 5 = 0$ or $x - 2 = 0$

$x = -\frac{5}{2}$ $\quad x = 2$ or $\left\{-\frac{5}{2}, 2\right\}$

NOTE

The is often called "clearing fractions."

Check Yourself 6

Solve $x^2 - \frac{x}{6} = \frac{1}{3}$.

Here is a summary of the steps to follow when solving a quadratic equation by factoring.

Step by Step

Solving Quadratic Equations by Factoring

Step 1 Add or subtract the necessary terms on both sides of the equation so that the equation is in standard form (set equal to 0).

Step 2 Factor the quadratic expression.

Step 3 Set each factor equal to 0.

Step 4 Solve the resulting equations to find the solutions.

Step 5 Check each solution by substituting in the original equation.

We will have many occasions to work with quadratic functions. The standard form of such a function is $f(x) = ax^2 + bx + c$, where a is not 0. When working with this type of function, we often need to find the zeros. These are input values that result in an output value of 0. The zeros are also the solutions to the equation $ax^2 + bx + c = 0$. Graphically, these values are the x-coordinates of the x-intercepts of the graph of f.

Definition

Zeros of a Quadratic Function

The zeros of the function $f(x) = ax^2 + bx + c$, where $a \neq 0$, are the solutions to the equation $ax^2 + bx + c = 0$. These zeros give the x-intercepts of the graph of the function.

Example 7 Finding the Zeros of a Quadratic Function

< Objective 2 >

NOTE

The x-intercepts of the graph of the function $f(x) = x^2 - x - 2$ are $(-1, 0)$ and $(2, 0)$, so -1 and 2 are the zeros of the function.

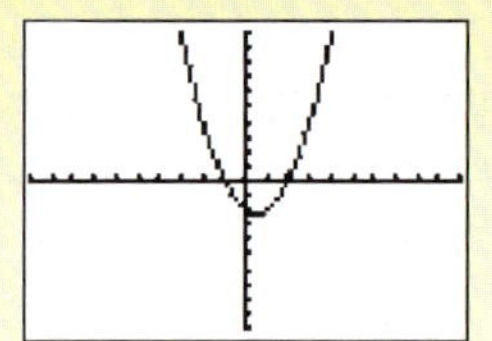

Find the zeros of the function

$f(x) = x^2 - x - 2$

To find the zeros of the function, set $f(x) = 0$, and solve.

$$x^2 - x - 2 = 0$$

$$(x - 2)(x + 1) = 0$$

$$x - 2 = 0 \quad \text{or} \quad x + 1 = 0$$

$$x = 2 \qquad\qquad x = -1$$

The zeros are -1 and 2.

Check Yourself 7

Find the zeros of $f(x) = 2x^2 - x - 3$.

Being able to solve quadratic equations allows us to solve a whole new set of applications. We use the five-step approach introduced in Section 1.4 to solve the applications that follow.

Quadratic equations can have two, one, or no distinct real-number solutions. In many cases, you will not only have to verify your solutions, but determine if they are reasonable in the context of a given application.

We begin by looking at a number problem.

Example 8 Solving a Number Application

< Objective 3 >

One integer is 3 less than twice another. If their product is 35, find the two integers.

Step 1 The unknowns are the two integers.

Step 2 Let x represent the first integer. Then

$$2x - 3$$

Twice — 3 less than

represents the second.

Step 3 Form an equation.

$$\underbrace{x(2x - 3)}_{\text{Product of the two integers}} = 35$$

Step 4 Remove the parentheses and solve.

$2x^2 - 3x = 35$ Distribute x to remove the parentheses.

$2x^2 - 3x - 35 = 0$ Subtract 35 so that the right side is 0.

Factor on the left.

$(2x + 7)(x - 5) = 0$

$2x + 7 = 0$ or $x - 5 = 0$

$2x = -7$ $x = 5$

$x = -\frac{7}{2}$

NOTE

These represent solutions to the equation, but not the answer to the original problem.

Step 5 From step 4, we have the two solutions: $-\frac{7}{2}$ and 5. Since the original problem asks for *integers,* 5 is the only solution.

Using 5 for x, we see that the other integer is $2x - 3 = 2(5) - 3 = 7$. So the desired integers are 5 and 7.

To verify, 7 is 3 less than 2 times 5, and the product of 5 and 7 is 35.

Check Yourself 8

One integer is 2 more than 3 times another. If their product is 56, what are the two integers?

Problems involving consecutive integers may also lead to quadratic equations. Recall that consecutive integers can be represented by x, $x + 1$, $x + 2$, and so on. Consecutive even (or odd) integers are represented by x, $x + 2$, $x + 4$, and so on.

Example 9 Solving a Number Application

The sum of the squares of two consecutive integers is 85. What are the two integers?

Step 1 The unknowns are the two consecutive integers.

Step 2 Let x be the first integer. Then $x + 1$ is the next consecutive integer.

Step 3 Form the equation.

$$\underbrace{x^2 + (x + 1)^2}_{\text{Sum of squares}} = 85$$

Step 4 Solve.

$x^2 + (x + 1)^2 = 85$

$x^2 + x^2 + 2x + 1 = 85$ Remove parentheses.

$$2x^2 + 2x - 84 = 0$$

$$x^2 + x - 42 = 0$$

Note the common factor 2, and divide both sides of the equation by 2.

$$(x + 7)(x - 6) = 0$$

$$x + 7 = 0 \quad \text{or} \quad x - 6 = 0$$

$$x = -7 \qquad\qquad x = 6$$

Step 5 The work in step 4 leads to two possibilities, -7 or 6. Since *both numbers are integers,* both meet the conditions of the original problem. There are then two pairs of consecutive integers that work.

-7 and -6 (where $x = -7$ and $x + 1 = -6$)

or 6 and 7 (where $x = 6$ and $x + 1 = 7$)

Check

$(-7)^2 + (-6)^2 = 49 + 36 = 85$ True

$(6)^2 + (7)^2 = 36 + 49 = 85$ True

Check Yourself 9

The sum of the squares of two consecutive even integers is 100. Find the two integers.

We proceed to geometry applications.

Example 10 Solving a Geometric Application

The length of a rectangle is 3 cm greater than its width. If the area of the rectangle is 108 cm^2, what are the dimensions of the rectangle?

Step 1 We are asked to find the dimensions (the length and the width) of the rectangle.

Step 2 Whenever geometric figures are involved in an application, start by drawing and *labeling* a sketch of the problem. Let x represent the width and $x + 3$ the length.

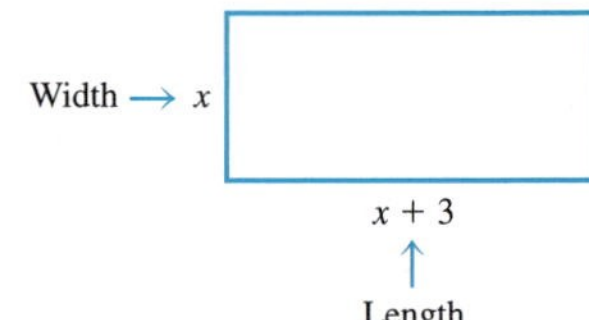

Step 3 Once the sketch is labeled, the next step should be easy. The area of a rectangle is the product of its length and width, so

$$x(x + 3) = 108$$

Step 4 Solve the equation.

$$x(x + 3) = 108$$

$$x^2 + 3x - 108 = 0$$

Multiply and write in standard form.

$$(x + 12)(x - 9) = 0$$

Factor and solve as before.

$$x + 12 = 0 \quad \text{or} \quad x - 9 = 0$$

$$x = -12 \qquad\qquad x = 9$$

Step 5 We reject -12 cm as a solution. A length cannot be negative, and so we only consider 9 cm in finding the required dimensions.

The width x is 9 cm, and the length $x + 3$ is 12 cm. Since this gives a rectangle of area 108 cm^2, the solution is verified.

Check Yourself 10

In a triangle, the base is 4 in. less than its height. If its area is 30 in.², find the length of the base and the height of the triangle.
Note: The formula for the area of a triangle is $A = \frac{1}{2}bh$.

We look at another geometric application.

Example 11 Solving a Rectangular Box Application

An open box is formed from a rectangular piece of cardboard, whose length is 2 in. more than its width, by cutting 2-in. squares from each corner and folding up the sides. If the volume of the box is 96 in.³, what are the dimensions of the original piece of cardboard?

Step 1 We are asked for the dimensions of the sheet of cardboard.

Step 2 Sketch the problem.

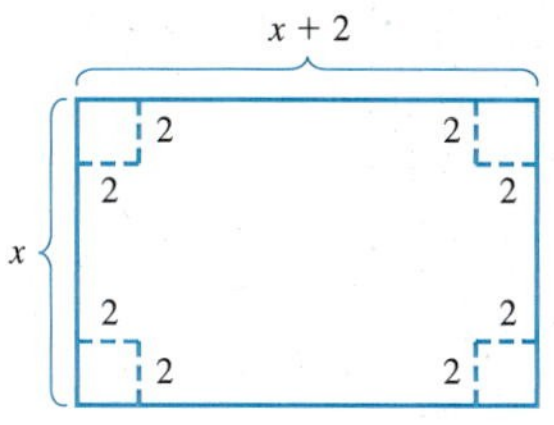

> **NOTE**
>
> The width of the original cardboard is x. Removing two 2-in. squares leaves $x - 4$ for the width of the box. Similarly, the length of the box is $x - 2$. Do you see why?

Step 3 To form an equation for volume, we sketch the completed box.

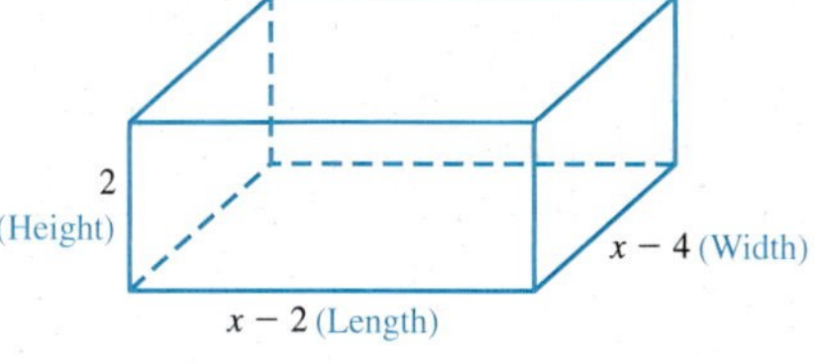

Since volume is the product of height, length, and width,

$$2(x - 2)(x - 4) = 96$$

Step 4

$$2(x - 2)(x - 4) = 96$$

$$(x - 2)(x - 4) = 48 \quad \text{Divide both sides by 2.}$$

$$x^2 - 6x + 8 = 48 \quad \text{Multiply on the left.}$$

$$x^2 - 6x - 40 = 0 \quad \text{Write in standard form.}$$

$$(x - 10)(x + 4) = 0 \quad \text{Solve as before.}$$

$$x = 10 \text{ in.} \quad \text{or} \quad x = -4 \text{ in.}$$

Step 5 Again, we only consider the positive solution. The width x of the original piece of cardboard is 10 in., and its length $x + 2$ is 12 in. The dimensions of the completed box are 6 in. by 8 in. by 2 in., which gives the required volume of 96 in.³.

Check Yourself 11

A similar box is to be made by cutting 3-in. squares from a piece of cardboard that is 4 in. longer than it is wide. If the volume is 180 in.³, what are the dimensions of the original piece of cardboard?

We now turn to another field for an application. We use quadratic equations in many physical models of motion.

Example 12 Solving a Thrown Ball Application

Suppose that a person throws a ball directly upward, releasing the ball from 5 ft above the ground. If the ball is thrown with an initial velocity of 80 ft/s, the height of the ball, in feet, above the ground after t seconds is given by

$$h = -16t^2 + 80t + 5$$

When is the ball's height 69 ft?

Step 1 The height h of the ball and the time t since the ball was released are the unknowns.

Step 2 We want to know the value(s) of t for which $h = 69$.

Step 3 Write $69 = -16t^2 + 80t + 5$.

Step 4 Solve for t.

$$69 = -16t^2 + 80t + 5$$

$$0 = -16t^2 + 80t - 64 \quad \text{We need 0 on one side.}$$

$$0 = t^2 - 5t + 4 \quad \text{Divide both sides by } -16.$$

$$0 = (t - 1)(t - 4) \quad \text{Factor.}$$

$$t - 1 = 0 \quad \text{or} \quad t - 4 = 0$$

$$t = 1 \qquad\qquad t = 4$$

Step 5 By substituting 1 for t, we confirm that $h = 69$.

$$h = -16(1)^2 + 80(1) + 5 = -16 + 80 + 5 = 69$$

You should also check that $h = 69$ when $t = 4$.

The ball's height is 69 ft at $t = 1$ s (on the way up) and at $t = 4$s (on the way down).

Check Yourself 12

A ball is thrown vertically upward from the top of a 100-m-tall building with an initial velocity of 25 m/s. After t seconds, the height h, in meters, is given by

$$h = -5t^2 + 25t + 100$$

When is the ball's height 130 m?

In Example 13, we must pay particular attention to the solutions that satisfy the given physical conditions.

Example 13 Solving a Thrown Ball Application

A ball is thrown vertically upward from the top of a 60-m-tall building with an initial velocity of 20 m/s. After t seconds, the height h is given by

$$h = -5t^2 + 20t + 60$$

(a) When is the ball's height 35 m?

Steps 1 and 2 We want to know the value(s) of t for which $h = 35$.

Step 3 Write $35 = -5t^2 + 20t + 60$.

Step 4 Solve for t.

$$35 = -5t^2 + 20t + 60$$

$$0 = -5t^2 + 20t + 25 \quad \text{We need a 0 on one side.}$$

$$0 = t^2 - 4t - 5 \quad \text{Divide through by } -5.$$

$$0 = (t - 5)(t + 1) \quad \text{Factor.}$$

$$t - 5 = 0 \quad \text{or} \quad t + 1 = 0$$

$$t = 5 \qquad t = -1$$

NOTE

When $t = 5$,
$h = -5(5)^2 + 20(5) + 60$
$= -125 + 100 + 60$
$= 35$

Step 5 $t = -1$ does not make sense in the context of this problem. (It represents 1 s *before* the ball is released!) We can easily check that when $t = 5$, $h = 35$. The ball's height is 35 m at 5 s after release.

(b) When does the ball hit the ground?

Steps 1 and 2 When the ball hits the ground, its height is 0 m.

Step 3 Write $0 = -5t^2 + 20t + 60$.

Step 4 Solve for t.

$$0 = -5t^2 + 20t + 60$$

$$0 = t^2 - 4t - 12 \quad \text{Divide through by } -5.$$

$$0 = (t - 6)(t + 2) \quad \text{Factor.}$$

$$t - 6 = 0 \quad \text{or} \quad t + 2 = 0$$

$$t = 6 \qquad t = -2$$

Step 5 As before, we reject the negative solution $t = -2$. The ball hits the ground after 6 s. You should verify that when $t = 6$, $h = 0$.

Check Yourself 13

A ball is thrown vertically upward from the top of a 150-ft-tall building with an initial velocity of 64 ft/s. After t seconds, the height h is given by

$$h = -16t^2 + 64t + 150$$

When is the ball's height 70 ft?

Stopping distance is another physical application modeled by quadratic equations. Stopping distance refers to the distance a car travels between a driver braking and the car coming to a full stop.

Example 14 Solving a Stopping Distance Application

The stopping distance d, in feet, of a car that is traveling at x mi/hr on a particular surface is approximated by the equation

$$d = \frac{x^2}{20} + x$$

If the stopping distance of this car is 240 ft, what is its speed?

Step 1 The stopping distance d, in feet, and the car's speed x, in miles per hour, are the unknowns.

Step 2 We want to know the value of x such that $d = 240$.

Step 3 Write $240 = \frac{x^2}{20} + x$.

Step 4 Solve for x.

$$240 = \frac{x^2}{20} + x$$

$4{,}800 = x^2 + 20x$ — Multiply through by 20 to clear the fraction.

$0 = x^2 + 20x - 4{,}800$ — We need a 0 on one side.

$0 = (x + 80)(x - 60)$ — Factor.

$$x + 80 = 0 \quad \text{or} \quad x - 60 = 0$$

$$x = -80 \qquad\qquad x = 60$$

Step 5 Because x represents the speed of the car, we reject the value $x = -80$. We verify that if $x = 60$, $d = 240$.

$$d = \frac{60^2}{20} + 60 = \frac{3{,}600}{20} + 60$$

$$= 180 + 60 = 240$$

The speed of the car is 60 mi/hr.

Check Yourself 14

If the stopping distance of this car is 175 ft, what is its speed? Use the equation

$$d = \frac{x^2}{20} + x$$

Check Yourself ANSWERS

1. $\{4, 5\}$ **2. (a)** $\{-8, 0\}$; **(b)** $\{-4, 4\}$ **3.** $\left\{-\frac{1}{3}, 2\right\}$ **4.** $\{-3\}$ **5.** $\{-3, 8\}$ **6.** $\left\{-\frac{1}{2}, \frac{2}{3}\right\}$ **7.** -1 and $\frac{3}{2}$ **8.** 4, 14 **9.** $-8, -6$ or 6, 8 **10.** Base 6 in.; height 10 in. **11.** 12 in. by 16 in. **12.** 2 s and 3 s **13.** 5 s **14.** 50 mi/hr

Reading Your Text

These fill-in-the-blank exercises will help you understand some of the key vocabulary used in this section. The answers to these exercises are in the Answers Appendix in the back of the text.

(a) An equation written in the form $ax^2 + bx + c = 0$ is called a ________ equation in standard form.

(b) The product of two nonzero numbers is ________ zero.

(c) To use the zero-product principle, the product of factors must be set equal to ________.

(d) The ________ of a quadratic function give the x-intercepts of the graph of the function.

Skills | Calculator/Computer | Career Applications | Above and Beyond

6.6 exercises

< Objective 1 >

Solve each quadratic equation.

1. $(x - 3)(x - 4) = 0$
2. $(x - 7)(x + 1) = 0$
3. $(3x + 1)(x - 6) = 0$
4. $(5x - 4)(x - 6) = 0$
5. $x^2 - 3x - 10 = 0$
6. $x^2 - 5x + 4 = 0$
7. $x^2 - 2x - 15 = 0$
8. $x^2 + 4x - 32 = 0$
9. $x^2 - 11x + 30 = 0$
10. $x^2 + 13x + 36 = 0$
11. $x^2 - 4x - 21 = 0$
12. $x^2 + 5x - 36 = 0$
13. $x^2 - 5x = 50$
14. $x^2 + 14x = -33$ VIDEO
15. $x^2 = 5x + 84$
16. $x^2 = 6x + 27$
17. $x^2 - 8x = 0$
18. $x^2 + 7x = 0$
19. $x^2 + 10x = 0$
20. $x^2 - 9x = 0$
21. $x^2 = 5x$
22. $x^2 = 11x$
23. $x^2 - 25 = 0$
24. $x^2 - 49 = 0$
25. $9x^2 = 25$
26. $x^2 = 169$
27. $4x^2 + 12x + 9 = 0$
28. $9x^2 - 30x + 25 = 0$
29. $2x^2 - 17x + 36 = 0$
30. $5x^2 + 17x - 12 = 0$
31. $5x^2 + 9x = 18$
32. $12x^2 = 25x - 12$
33. $6x^2 = 7x - 2$
34. $4x^2 - 3 = x$
35. $2m^2 = 12m + 54$
36. $5x^2 - 55x = 60$
37. $7x^2 - 63x = 0$
38. $6x^2 - 9x = 0$
39. $5x^2 = 15x$
40. $7x^2 = -49x$ VIDEO
41. $\frac{x^2}{8} - \frac{x}{4} - 1 = 0$
42. $\frac{x^2}{15} + \frac{x}{5} - \frac{2}{3} = 0$
43. $\frac{x^2}{6} + \frac{3x}{2} + 3 = 0$
44. $\frac{x^2}{10} - \frac{x}{2} = \frac{3}{5}$
45. $\frac{2x^2}{3} + \frac{x}{15} = \frac{1}{5}$
46. $\frac{4x^2}{5} - x - \frac{3}{10} = 0$
47. $x(x + 2) = 15$
48. $x(x + 3) = 28$
49. $x(2x - 3) = 9$
50. $x(3x + 1) = 52$
51. $2x(3x + 1) = 28$ VIDEO
52. $3x(2x - 1) = 30$
53. $(x - 3)(x - 1) = 15$
54. $(x + 3)(x - 2) = 14$
55. $(x - 5)(x + 2) = 18$
56. $(3x - 5)(x + 2) = 14$

< Objective 2 >

Find the zeros of each function.

57. $f(x) = x^2 - 6x + 5$
58. $f(x) = x^2 + 2x - 8$
59. $f(x) = x^2 - 9x$
60. $f(x) = 6x^2 - x - 2$

< Objective 3 >

Solve each application.

61. **Number Problem** One integer is 3 more than twice another. If the product of those integers is 65, find the two integers.
62. **Number Problem** One positive integer is 5 less than 3 times another, and their product is 78. What are the two integers?
63. **Number Problem** The sum of two integers is 10, and their product is 24. Find the two integers.
64. **Number Problem** The sum of two integers is 12. If the product of the two integers is 27, what are the two integers?
65. **Number Problem** The product of two consecutive integers is 72. What are the two integers?
66. **Number Problem** If the product of two consecutive odd integers is 63, find the two integers.
67. **Number Problem** The sum of the squares of two consecutive whole numbers is 61. Find the two whole numbers.

68. **Number Problem** If the sum of the squares of two consecutive even integers is 100, what are the two integers? VIDEO

69. **Number Problem** The sum of the squares of three consecutive integers is 50. Find the three integers.

70. **Number Problem** If the sum of the squares of three consecutive odd positive integers is 83, what are the three integers?

71. **Geometry** The width of a rectangle is 3 ft less than its length. If the area of the rectangle is 70 ft², what are the dimensions of the rectangle? VIDEO

72. **Geometry** The length of a rectangle is 5 cm more than its width. If the area of the rectangle is 84 cm², find the dimensions of the rectangle.

73. **Geometry** The length of a rectangle is 2 cm more than 3 times its width. If the area of the rectangle is 85 cm², find the dimensions of the rectangle.

74. **Geometry** If the length of a rectangle is 3 ft less than twice its width and the area of the rectangle is 54 ft², what are the dimensions of the rectangle?

75. **Geometry** The length of a rectangle is 1 cm more than its width. If the length of the rectangle is doubled, the area of the rectangle is increased by 30 cm². What were the dimensions of the original rectangle?

76. **Geometry** The height of a triangle is 2 in. more than the length of the base. If the base is tripled in length, the area of the new triangle is 48 in.² more than the original. Find the height and base of the original triangle.

77. **Geometry** A box is to be made from a rectangular piece of tin that is twice as long as it is wide. To accomplish this, a 10-cm square is cut from each corner, and the sides are folded up. The volume of the finished box is to be 4,000 cm³. Find the dimensions of the original piece of tin.
Hint: To solve this equation, use the given sketch of the piece of tin. The original dimensions are represented by x and $2x$. Do you see why? The volume of the resulting box is the product of the length, width, and height.

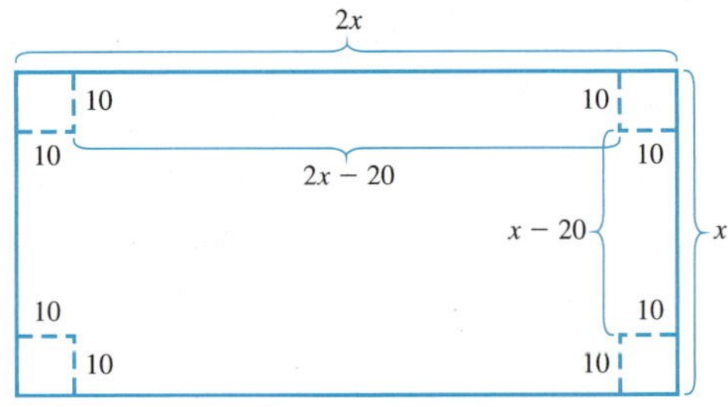

78. **Geometry** An open box is formed from a square piece of material by cutting 2-in. squares from each corner of the material and folding up the sides. If the volume of the box that is formed is to be 72 in.³, what was the size of the original piece of material? VIDEO

79. **Geometry** An open carton is formed from a rectangular piece of cardboard that is 4 ft longer than it is wide, by removing 1-ft squares from each corner and folding up the sides. If the volume of the carton is then 32 ft³, what were the dimensions of the original piece of cardboard?

80. **Geometry** A box that has a volume of 1,936 in.³ was made from a square piece of tin. The square piece cut from each corner had sides of length 4 in. What were the original dimensions of the square?

81. **Geometry** A square piece of cardboard is to be formed into a box. After 5-cm squares are cut from each corner and the sides are folded up, the resulting box will have a volume of 4,500 cm³. Find the length of a side of the original piece of cardboard.

82. **Geometry** A rectangular piece of cardboard has a length that is 2 cm longer than twice its width. If 2-cm squares are cut from each of its corners, it can be folded into a box that has a volume of 280 cm³. What were the original dimensions of the piece of cardboard?

83. **Science and Medicine** If a ball is thrown vertically upward from the ground, with an initial velocity of 64 ft/s, its height h after t seconds is given by

$$h = -16t^2 + 64t$$

How long does it take the ball to return to the ground?

84. **Science and Medicine** If a ball is thrown vertically upward from the ground, with an initial velocity of 64 ft/s, its height h after t seconds is given by

$$h = -16t^2 + 64t$$

How long does it take the ball to reach a height of 48 ft on the way up?

85. **Science and Medicine** If a ball is thrown vertically upward from the ground, with an initial velocity of 96 ft/s, its height h after t seconds is given by

$$h = -16t^2 + 96t$$

How long does it take the ball to return to the ground?

86. **Science and Medicine** If a ball is thrown vertically upward from the ground, with an initial velocity of 96 ft/s, its height h after t seconds is given by

$$h = -16t^2 + 96t$$

How long does it take the ball to pass through a height of 128 ft on the way back down to the ground?

87. **Science and Medicine** If a ball is thrown vertically upward from the roof of a building 192 ft high with an initial velocity of 64 ft/s, its approximate height h after t seconds is given by

$$h = -16t^2 + 64t + 192$$

How long does it take the ball to fall back to the ground? 

88. **Science and Medicine** If a ball is thrown vertically upward from the roof of a building 192 ft high with an initial velocity of 64 ft/s, its approximate height h after t seconds is given by

$$h = -16t^2 + 64t + 192$$

When will the ball reach a height of 240 ft?

89. **Science and Medicine** If a ball is thrown vertically upward from the roof of a building 192 ft high with an initial velocity of 96 ft/s, its approximate height h after t seconds is given by

$$h = -16t^2 + 96t + 192$$

How long does it take the ball to return to the thrower?

90. **Science and Medicine** If a ball is thrown vertically upward from the roof of a building 192 ft high with an initial velocity of 96 ft/s, its approximate height h after t seconds is given by

$$h = -16t^2 + 96t + 192$$

When will the ball reach a height of 272 ft?

91. **Science and Medicine** If a ball is thrown upward from the roof of a 105-m building with an initial velocity of 20 m/s, its approximate height h after t seconds is given by

$$h = -5t^2 + 20t + 105$$

How long will it take the ball to fall to the ground?

92. **Science and Medicine** If a ball is thrown upward from the roof of a 105-m building with an initial velocity of 20 m/s, its approximate height h after t seconds is given by

$$h = -5t^2 + 20t + 105$$

When will the ball reach a height of 80 m?

Science and Medicine *Use the equation* $d = \frac{x^2}{20} + x$, *which relates the speed of a car* x, *in miles per hour, to the stopping distance* d, *in feet, on a particular road surface.*

93. If the stopping distance of a car is 120 ft, how fast is the car traveling?

94. How fast is a car going if it requires 75 ft to stop?

95. Marcus is driving at high speed on the freeway when he spots a vehicle stopped ahead of him. If Marcus's car is 315 ft from the vehicle, what is the maximum speed at which Marcus can be traveling and still stop in time?

96. Juliana is driving at high speed on a country highway when she sees a deer in the road ahead. If Juliana's car is 400 ft from the deer, what is the maximum speed at which Juliana can be traveling and still stop in time?

Determine whether each statement is **true** *or* **false.**

97. To find the area of a rectangle, we add the length and the width.

98. If a ball is thrown upward, the ball might reach a specified height two times.

Complete each statement with **always, sometimes,** *or* **never.**

99. The product of two consecutive integers is __________ even.

100. The sum of two consecutive integers is __________ even.

101. To solve a quadratic equation by factoring, we __________ work to place zero on one side of the equation.

102. A quadratic equation in standard form __________ has three solutions.

Skills | Calculator/Computer | **Career Applications** | Above and Beyond

103. **Agricultural Technology** The height h (in feet) of a drop of water above an irrigation nozzle, in terms of the time t (in seconds) since the drop left the nozzle, is given by the formula

$$h = v_0t - 16t^2$$

If the initial velocity is $v_0 = 80$ ft/s, how many seconds need to pass for a drop to be 75 ft high?

104. **Manufacturing Technology** A piece of stainless steel warps due to the heat created during welding. The shape of the warping is approximated by

$$w = \frac{a^2 - 16}{64}$$

At what value of a is $w = 0$?

105. **Allied Health** The number N of people who are sick t days after the outbreak of a flu epidemic is given by the equation

$$N = 50 + 25t - 3t^2$$

How many days will it take until no one is infected? 

106. **Mechanical Engineering** The flow rate through a hydraulic hose can be found using the equation

$$2Q^2 + Q - 21 = 0$$

Find the flow rate by solving the equation (you need to consider only positive solutions).

Agricultural Technology The net productivity of a forested wetland is related to the amount of water moving through the wetland and can be modeled by a quadratic equation. In exercises 107 to 110, y represents the amount of wood produced and x represents the amount of water present, in cubic centimeters (cm^3). Determine where the productivity is zero in each wetland represented by the equations.

107. $y = -3x^2 + 300x$

108. $y = -4x^2 + 500x$

109. $y = -6x^2 + 792x$

110. $y = -7x^2 + 1{,}022x$

111. **Allied Health** A patient's body temperature T (°F) can be approximated by the formula

$$T = 0.4t^2 - 2.6t + 103$$

in which t is the number of hours since the patient took the analgesic acetaminophen.

Determine the amount of time before the patient's temperature rises back up to 100°F. 

112. **Allied Health** A healthy person's blood glucose level g (in mg per 100 mL) can be approximated by the formula

$$g = -480t^2 + 400t + 80$$

in which t is the number of hours since the person ate a meal. How long after eating will the person's blood glucose level return to 80 mg per 100 mL?

113. **Business and Finance** The manager of a bicycle shop knows that the cost of selling x bicycles is $C = 20x + 60$ and the revenue from selling x bicycles is $R = x^2 - 8x$. Find the break-even value of x. *Hint:* The break-even occurs when cost equals revenue.

114. **Business and Finance** A company that produces computer games has found that its daily operating cost in dollars is $C = 40x + 150$ and its daily revenue in dollars is $R = 65x - x^2$. For what value(s) of x will the company break even?

Skills | Calculator/Computer | Career Applications | **Above and Beyond**

Write an equation that has the given solution.
Hint: Write the binomial factors and then find their product.

115. $\{-4, 5\}$

116. $\{0, 5\}$

117. $\{2, 6\}$ 

118. $\{-4, 4\}$

The zero-product principle can be extended to three or more factors. If $a \cdot b \cdot c = 0$, then at least one of these factors is 0. Use this information to solve each equation.

119. $x^3 - 3x^2 - 10x = 0$

120. $x^3 + 8x^2 + 15x = 0$

121. $x^3 - 9x = 0$

122. $x^3 = 16x$

Extend the ideas in the previous exercises to find solutions for each equation.
Hint: Factor by grouping in exercises 123 and 124.

123. $x^3 + x^2 - 4x - 4 = 0$

124. $x^3 - 5x^2 - x + 5 = 0$

125. $x^4 - 10x^2 + 9 = 0$

126. $x^4 - 5x^2 + 4 = 0$

127. Write a short comparison that explains the difference between $ax^2 + bx + c$ and $ax^2 + bx + c = 0$.

128. When solving quadratic equations, some people try to solve an equation in the manner shown below, but this does not work! Write a paragraph to explain what is wrong with this approach.

$$2x^2 + 7x + 3 = 52$$
$$(2x + 1)(x + 3) = 52$$
$$2x + 1 = 52 \quad \text{or} \quad x + 3 = 52$$
$$x = \frac{51}{2} \quad \text{or} \quad x = 49$$

Answers

1. $\{3, 4\}$ **3.** $\left\{-\frac{1}{3}, 6\right\}$ **5.** $\{-2, 5\}$ **7.** $\{-3, 5\}$ **9.** $\{5, 6\}$ **11.** $\{-3, 7\}$ **13.** $\{-5, 10\}$ **15.** $\{-7, 12\}$ **17.** $\{0, 8\}$ **19.** $\{-10, 0\}$ **21.** $\{0, 5\}$ **23.** $\{-5, 5\}$ **25.** $\left\{-\frac{5}{3}, \frac{5}{3}\right\}$ **27.** $\left\{-\frac{3}{2}\right\}$ **29.** $\left\{4, \frac{9}{2}\right\}$ **31.** $\left\{-3, \frac{6}{5}\right\}$ **33.** $\left\{\frac{1}{2}, \frac{2}{3}\right\}$ **35.** $\{-3, 9\}$ **37.** $\{0, 9\}$ **39.** $\{0, 3\}$ **41.** $\{-2, 4\}$ **43.** $\{-3, -6\}$ **45.** $\left\{-\frac{3}{5}, \frac{1}{2}\right\}$ **47.** $\{-5, 3\}$ **49.** $\left\{-\frac{3}{2}, 3\right\}$ **51.** $\left\{-\frac{7}{3}, 2\right\}$ **53.** $\{-2, 6\}$ **55.** $\{-4, 7\}$ **57.** 1 and 5 **59.** 0 and 9 **61.** 5, 13 **63.** 4, 6 **65.** −9, −8 or 8, 9 **67.** 5, 6 **69.** −5, −4, −3 or 3, 4, 5 **71.** 7 ft by 10 ft **73.** 5 cm by 17 cm **75.** 5 cm by 6 cm **77.** 30 cm by 60 cm **79.** 6 ft by 10 ft **81.** 40 cm **83.** 4 s **85.** 6 s **87.** 6 s **89.** 6 s **91.** 7 s **93.** 40 mi/hr **95.** 70 mi/hr **97.** False **99.** always **101.** always **103.** 1.25 s and 3.75 s **105.** 10 days **107.** 0 cm³, 100 cm³ **109.** 0 cm³, 132 cm³ **111.** 5 hr **113.** 30 bicycles **115.** $x^2 - x - 20 = 0$ **117.** $x^2 - 8x + 12 = 0$ **119.** $\{-2, 0, 5\}$ **121.** $\{-3, 0, 3\}$ **123.** $\{-2, -1, 2\}$ **125.** $\{-3, -1, 1, 3\}$ **127.** Above and Beyond

Definition/Procedure	Example	Reference
An Introduction to Factoring		**Section 6.1**
Common Monomial Factor A single term that is a factor of every term of the polynomial. The greatest common factor (GCF) is the common factor that has the largest possible numerical coefficient and the largest possible exponents.	$4x^2$ is the greatest common monomial factor of $8x^4 - 12x^3 + 16x^2$.	p. 480
Factoring a Monomial from a Polynomial 1. Determine the greatest common factor. 2. Apply the distributive property in the form $ab + ac = a(b + c)$ ↑ The greatest common factor	$8x^4 - 12x^3 + 16x^2$ $= 4x^2(2x^2 - 3x + 4)$	p. 481
Factoring by Grouping When a polynomial has four terms, factor the first pair and factor the last pair. If these two pairs have a common binomial factor, factor that out. The result will be the product of two binomials.	$4x^2 - 6x + 10x - 15$ $= 2x(2x - 3) + 5(2x - 3)$ $= (2x - 3)(2x + 5)$	p. 484
Factoring Special Products		**Section 6.2**
Factoring a Difference of Squares Use the formula $a^2 - b^2 = (a + b)(a - b)$	To factor $16x^2 - 25y^2$, ↓ ↓ think $(4x)^2 - (5y)^2$ so, $16x^2 - 25y^2$ $= (4x + 5y)(4x - 5y)$	p. 491
Factoring a Difference of Cubes Use the formula $a^3 - b^3 = (a - b)(a^2 + ab + b^2)$	Factor $x^3 - 64$. $x^3 - 64 = x^3 - 4^3$ $= (x - 4)(x^2 + 4x + 16)$	p. 494
Factoring a Sum of Cubes Use the formula $a^3 + b^3 = (a + b)(a^2 - ab + b^2)$	Factor $x^3 + 8y^3$. $x^3 + 8y^3 = x^3 + (2y)^3$ $= (x + 2y)(x^2 - 2xy + 4y^2)$	p. 494
Factoring a Perfect Square Trinomial Use one of the formulas $a^2 + 2ab + b^2 = (a + b)^2$ $a^2 - 2ab + b^2 = (a - b)^2$	Factor $25x^2 + 40xy + 16y^2$. $25x^2 + 40xy + 16y^2$ $= (5x)^2 + 2(5x \cdot 4y) + (4y)^2$ $= (5x + 4y)^2$	p. 495
Factoring: Trial and Error		**Section 6.3**
Factoring Sign Pattern Both signs are positive $(x +\ )(x +\ )$ The constant is positive and the x-coefficient is negative $(x -\ )(x -\ )$ The constant is negative $(x +\ )(x -\ )$	 $x^2 + 5x + 6$ $x^2 - x + 6$ $x^2 - x - 12$ or $x^2 + 4x - 12$	p. 503

Continued

Definition/Procedure	Example	Reference
Factoring: The *ac* Method		**Section 6.4**
The *ac* Test A trinomial of the form $ax^2 + bx + c$ is factorable if, and only if, there are two integers m and n such that $ac = mn \qquad b = m + n$	Given $2x^2 - 5x - 3$ $a = 2 \quad b = -5 \quad c = -3$ $ac = -6 \quad b = -5$ $m = -6 \quad n = 1$ $mn = -6 \quad m + n = -5$	p. 511
To Factor a Trinomial In general, you can apply these steps. **Step 1** Write the trinomial in standard $ax^2 + bx + c$ form. **Step 2** Identify the three coefficients a, b, and c. **Step 3** Find two integers m and n such that $ac = mn$ and $b = m + n$ **Step 4** Rewrite the trinomial as $ax^2 + mx + nx + c$ **Step 5** Factor by grouping. Not all trinomials are factorable. To discover whether a trinomial is factorable, try the ***ac* test.**	$2x^2 - 6x + x - 3$ $= 2x(x - 3) + (x - 3)$ $= (2x + 1)(x - 3)$	p. 514
Factoring Strategies		**Section 6.5**
1. Always look for a greatest common factor. If you find a GCF (other than 1), factor out the GCF as your first step. If the leading coefficient is negative, factor out the GCF with a negative coefficient. **2.** Next, consider the **number of terms** in the polynomial. **(a)** If the polynomial is a *binomial,* consider the special binomial formulas. **(b)** If the polynomial is a *trinomial,* try to factor it as a product of two binomials. You can use either the trial-and-error method or the *ac* method. **(c)** If the polynomial has *more than three terms,* try factoring by grouping. **3.** You should always **factor the polynomial completely.** So after you apply one of the techniques given in part 2, another one may be necessary. **4.** You can always **check your answer by multiplying.**	Factor completely. $-2x^4 + 32$ $= -2(x^4 - 16)$ $= -2(x^2 + 4)(x^2 - 4)$ $= -2(x^2 + 4)(x + 2)(x - 2)$ To check, we multiply. $-2(x^2 + 4)(x + 2)(x - 2)$ $= -2(x^2 + 4)(x^2 - 2x + 2x - 4)$ $= -2(x^2 + 4)(x^2 - 4)$ $= -2(x^4 - 4x^2 + 4x^2 - 16)$ $= -2(x^4 - 16)$ $= -2x^4 + 32$	p. 520
Factoring and Problem Solving		**Section 6.6**
The Zero-Product Principle $ab = 0$ if, and only if, $a = 0$ or $b = 0$ or both. This property says two things. **1.** The product of any number and zero is *always* zero. **2.** The product of two nonzero numbers is *never* zero.		p. 526

Continued

Definition/Procedure	Example	Reference
Solving Quadratic Equations by Factoring		
1. Add or subtract the necessary terms on both sides of the equation so that the equation is in standard form (set equal to 0). **2.** Factor the quadratic expression. **3.** Set each factor equal to 0. **4.** Solve the resulting equations to find the solutions. **5.** Check each solution by substituting in the original equation.	To solve $x^2 + 7x = 30$ $x^2 + 7x - 30 = 0$ $(x + 10)(x - 3) = 0$ $x + 10 = 0$ or $x - 3 = 0$ $x = -10$ or $x = 3$ *Check* $(-10)^2 + 7(-10) \stackrel{?}{=} 30$ $100 + (-70) = 30$ True $(3)^2 + 7(3) \stackrel{?}{=} 30$ $9 + 21 = 30$ True Solution set: $\{-10, 3\}$	*p.* 544

summary exercises :: chapter 6

This summary exercise set will help ensure that you have mastered each of the objectives of this chapter. The exercises are grouped by section. You should reread the material associated with any exercises that you find difficult. The answers to the odd-numbered exercises are in the Answers Appendix in the back of the text.

6.1 *Completely factor each polynomial.*

1. $14a - 35$
2. $9m^2 - 21m$
3. $24s^2t - 16s^2$
4. $18a^2b + 36ab^2$
5. $27s^4 + 18s^3$
6. $3x^3 - 6x^2 + 15x$
7. $18m^2n^2 - 27m^2n + 36m^2n^3$
8. $81x^7y^6 + 63x^4y^6$
9. $8a^2b + 24ab - 16ab^2$
10. $3x^2y - 6xy^3 + 9x^3y - 12xy^2$
11. $2x(3x + 4y) - y(3x + 4y)$
12. $5(w - 3z) - w(w - 3z)$
13. $x^2 - 4x + 5x - 20$
14. $x^2 + 7x - 2x - 14$
15. $8x^2 + 6x - 20x - 15$
16. $12x^2 - 9x - 28x + 21$
17. $6x^3 + 9x^2 - 4x^2 - 6x$
18. $3x^4 + 6x^3 + 5x^3 + 10x^2$

6.2

19. $p^2 - 49$
20. $25a^2 - 16$
21. $9n^2 - 25m^2$
22. $16r^2 - 49s^2$
23. $25 - z^2$
24. $a^4 - 16b^2$
25. $25a^2 - 36b^2$
26. $x^{10} - 9y^2$
27. $3w^3 - 12wz^2$
28. $16a^4 - 49b^2$
29. $2m^2 - 72n^4$
30. $3w^3z - 12wz^3$
31. $x^2 - 18x + 81$
32. $x^2 + 12x + 36$
33. $x^2 + 14x + 49$
34. $x^2 - 10x + 25$
35. $4x^2 + 12x + 9$
36. $16x^2 - 8x + 1$
37. $x^2 - 20$
38. $4x^2 + 49$
39. $a^4 - 16b^4$
40. $m^3 - 64$
41. $8x^3 + 1$
42. $8c^3 - 27d^3$
43. $125m^3 + 64n^3$
44. $2x^4 + 54x$

6.3–6.5

45. $x^2 + 12x + 20$
46. $a^2 + a - 2$
47. $w^2 - 15w + 54$
48. $r^2 - 9r - 36$
49. $x^2 - 8xy - 48y^2$
50. $a^2 + 17ab + 30b^2$
51. $14x^2 + 43x - 21$
52. $2a^2 + 3a - 35$
53. $-x^2 - 9x - 20$
54. $x^2 - 10x + 24$
55. $a^2 - 7a + 12$
56. $w^2 + 13w + 40$
57. $x^2 + 16x + 64$
58. $-r^2 + 15r - 36$
59. $b^2 - 4bc - 21c^2$
60. $m^2n + 4mn - 32n$
61. $m^3 + 2m^2 - 35m$
62. $2x^2 - 2x - 40$
63. $3y^3 - 48y^2 + 189y$
64. $3b^3 - 15b^2 - 42b$
65. $3x^2 + 8x + 5$
66. $5w^2 + 13w - 6$
67. $2b^2 - 9b + 9$
68. $8x^2 + 2x - 3$
69. $10x^2 - 11x + 3$
70. $4a^2 + 7a - 15$
71. $16y^2 - 8xy - 15x^2$
72. $8x^2 + 14xy - 15y^2$
73. $-8x^3 + 36x^2 + 20x$
74. $9x^2 - 15x - 6$
75. $6x^3 - 3x^2 - 9x$
76. $-3x^2 + 3xy + 18y^2$

6.4 *Use the ac test to determine which trinomials can be factored. Find the values of m and n for each trinomial that can be factored.*

77. $x^2 - x - 30$
78. $x^2 + 3x + 2$
79. $2x^2 - 11x + 12$
80. $4x^2 - 23x + 15$

6.5 *Completely factor each polynomial.*

81. $5x^4 - 5x^3 - 210x^2$
82. $81n^4 - 3n$
83. $72y - 2y^3$
84. $xy - 8x - 5y + 40$
85. $10x^5 + 80x^4 + 160x^3$
86. $3x^2 + 8x - 15$
87. $x^3 + 2x^2 - 25x - 50$
88. $54m + 16m^4$

6.6 *Solve each equation.*

89. $x^2 + 5x - 6 = 0$
90. $x^2 + 2x - 8 = 0$
91. $x^2 + 7x = 30$
92. $x^2 - 6x = 40$
93. $x^2 + x = 20$
94. $x^2 = 28 - 3x$
95. $x^2 - 10x = 0$
96. $x^2 = 12x$
97. $x^2 - 25 = 0$
98. $x^2 = 225$
99. $2x^2 - x - 3 = 0$
100. $3x^2 - 4x = 15$
101. $3x^2 + 9x - 30 = 0$
102. $4x^2 + 24x = -32$
103. $x(x - 5) = 36$
104. $(x - 2)(2x + 1) = 33$
105. $x^3 - 2x^2 - 15x = 0$
106. $x^3 + x^2 - 4x - 4 = 0$

Solve each application.

107. **Number Problem** One integer is 15 more than another. If the product of the integers is -54, find the two integers.

108. **Number Problem** The sum of the squares of two consecutive odd integers is 130. What are the two integers?

109. **Geometry** The length of a rectangle is 6 ft more than its width. If the area of the rectangle is 216 ft^2, find the dimensions of the rectangle.

110. **Geometry** An open carton is formed from a rectangular piece of cardboard that is 5 in. longer than it is wide, by removing 4-in. squares from each corner and folding up the sides. If the volume of the carton is 200 in.^{3}, what are the dimensions of the original piece of cardboard?

111. **Science and Medicine** If a ball is thrown vertically upward from the roof of a building 20 ft high with an initial velocity of 80 ft/s, its approximate height h after t seconds is given by

$$h = -16t^2 + 80t + 20$$

How long does it take the ball to pass through a height of 84 ft on the way back down to the ground?

112. **Construction** A rectangular garden has dimensions 15 ft by 20 ft. The garden is to be enlarged with a strip of equal width surrounding it. If the resulting garden is to be 294 ft^2 larger than before, how wide should the strip be?

113. **Business and Finance** Suppose that the cost, in dollars, of producing x stereo systems is given by

$C(x) = 3{,}000 - 60x + 3x^2$

How many systems can be produced for $7,500?

114. **Business and Finance** The demand equation for a certain type of computer paper when sold at price p (in dollars) is predicted to be

$D = -3p + 69$

The supply equation is predicted to be

$S = -p^2 + 24p - 3$

Find the equilibrium price.

CHAPTER 6

chapter test 6

Hint: The equilibrium price occurs when demand is equal to supply.

Use this chapter test to assess your progress and to review for your next exam. Allow yourself about an hour to take this test. The answers to these exercises are in the Answers Appendix in the back of the text.

Completely factor each polynomial.

1. $32a^2b - 50b^3$
2. $x^2 + 2x - 5x - 10$
3. $7b + 42$
4. $x^4 - 81$
5. $9x^2 - 12xy + 4y^2$
6. $5x^2 - 10x + 20$
7. $16y^2 - 49x^2$
8. $8x^2 - 2xy - 3y^2$
9. $27y^3 - 8x^3$
10. $3w^2 + 10w + 7$
11. $6x^2 + 4x - 15x - 10$
12. $a^2 - 5a - 14$
13. $-6x^3 - 3x^2 + 30x$
14. $y^2 + 12yz + 20z^2$

Solve each equation.

15. $x^2 - 11x = -30$
16. $2x^2 + 16x + 30 = 0$
17. $x^2 - 2x - 3 = 0$
18. $6x^2 - 7x = 3$

Solve each application.

19. **Geometry** The length of a rectangle is 4 cm less than twice the width. If the area is 240 cm^2, what is the length of the rectangle?

20. **Science and Medicine** If a ball is thrown upward from the roof of an 18-m building with an initial velocity of 20 m/s, its approximate height h after t seconds is given by

$h = -5t^2 + 20t + 18$

When is the ball's height 38 m?

cumulative review chapters 0–6

Use this exercise set to review concepts from earlier chapters. While it is not a comprehensive exam, it will help you identify any material that you need to review before moving on to the next chapter. The answers to these exercises are in the Answers Appendix in the back of the text.

1. If $x = -3$, $y = 2$, and $z = -4$, find the value of the expression $\dfrac{3x^2 - 2z + 1}{5y + 2x}$.

Solve each equation.

2. $7a - 3 = 6a + 8$
3. $\frac{2}{3}x = -22$

Solve for the indicated variable.

4. $A = P + Prt$ for r

5. $P = 2L + 2W$ for W

Solve each inequality.

6. $2x - 7 < 5 + 4x$

7. $2x - 9 \leq 7x - 3(x - 1)$

Solve each system.

8. $3x - 5y = 5$
$-x + y = -1$

9. $2x - 3y = 13$
$x = 3y + 9$

10. Find the slope and y-intercept of the line defined by $2x + 5y = 10$.

Write an equation for the line that satisfies each set of conditions.

11. L passes through the points $(-2, 1)$ and $(1, 7)$.

12. L has y-intercept $(0, -2)$ and is parallel to the line with equation $3x + y = 5$.

Perform the indicated operations. Simplify your results.

13. $7x^2y + 3xy - 5x^2y + 2xy$

14. $(-3x^2 + 5x - 6) + (2x^2 - 3x + 5)$

15. $(5x^2 + 4x - 3) - (4x^2 - 5x - 1)$

16. $3x(x^2 - 2x + 2)$

17. $(2x - 5)(x + 1)$

18. $(x + 3)(x^2 - 3x + 9)$

19. $(4x - 3)(2x + 5)$

20. $(a - 3b)(a + 3b)$

21. $(x - 2y)^2$

22. $2x(5x + 3y)(5x - 3y)$

Simplify each expression.

23. $(-2x^{-2}y^3)^3$

24. $\frac{27a^5b^7}{9ab}$

25. $(2x^3y^{-2})^{-2}(x^{-3}y^{-1})$

26. $\frac{16x^2y^{-3}z^{-4}}{8x^{-1}y^2z^{-5}}$

Factor each expression.

27. $12x + 20$

28. $25x^2 - 49y^2$

29. $12x^2 - 15x + 8x - 10$

30. $2x^2 - 13x + 15$

Solve each quadratic equation.

31. $x^2 + 2x = 15$

32. $x^2 - 9 = 0$

Divide.

33. $\frac{24x^2y^5 - 15x^4y^2 + 9xy}{3xy}$

34. $\frac{x^2 + 3x + 2}{x - 1}$

Solve each application.

35. **Number Problem** Three times a number decreased by 5 is 46. Find the number.

36. **Business and Finance** Juan's biology text cost \$5 more than his mathematics text. Together they cost \$201. Find the cost of the biology text.

37. **Science and Medicine** The equation $d = \frac{x^2}{20} + x$ relates the speed of a car x, in miles per hour, to the stopping distance d, in feet. How fast is a car going if it requires 240 ft to stop?

38. **Business and Finance** Suppose that a manufacturer's weekly profit in dollars is given by

$P = -5x^2 + 300x$

How many items x must be produced to realize a profit of \$2,500?

CHAPTER

7

Radicals and Exponents

INTRODUCTION

We encounter applications and formulas any time we take a measurement. Time, temperature, and distance are just some of the quantities that we measure. Evaluating the results of an experiment almost always requires us to take measurements. Experiments, of course, are one of the ways we make new discoveries. Experiments in science and math provide us with some of the most interesting applications.

In the seventeenth century, the Italian scientist Galileo Galilei (1564–1642) conducted some important experiments with pendulums (he called them *pulsilogia*). While Galileo may be best known for his work in astronomy, he played major roles in the *scientific revolution* as a physicist, mathematician, astronomer, and philosopher and has even been called "the Father of Modern Science."

In Activity 7, we offer you the opportunity to conduct experiments similar to the ones Galileo performed with pendulums while a student at the University of Pisa.

CHAPTER 7 OUTLINE

7.1 Roots and Radicals

< 7.1 Objectives >

1 > Evaluate expressions containing radicals

2 > Estimate radical expressions

3 > Simplify expressions containing radicals

4 > Apply the Pythagorean theorem

5 > Use the distance formula

6 > Write the equation of a circle and sketch its graph

RECALL

Square and cube roots such as $\sqrt{9}$ and $\sqrt[3]{27}$ are examples of radical notation.

In Chapter 5, we reviewed the properties of integer exponents. In this chapter, we look to extend those properties. To achieve this, we use *radicals* to reverse the process of raising an expression to a power.

We read the statement

$$x^2 = 9$$

as "x squared equals 9."

NOTE

Negative numbers have no real square roots.

We look at nonreal roots and *imaginary numbers* in Section 7.6.

In this section, we look at the relationship between the base x and the number 9. We call that relationship the **square root** and say, equivalently, that "x is the square root of 9."

We know from experience that x must be 3 or -3 because

$$3^2 = 9 \qquad \text{and} \qquad (-3)^2 = 9$$

We see that 9 has two square roots. In fact, every positive number has two square roots, one positive and one negative. We can generalize this.

If $x^2 = a$, we say x is a *square root* of a.

We also know that

$$3^3 = 27$$

and similarly we call 3 a *cube root* of 27. Here 3 is the *only* real number with that property. Every real number (positive and negative) has exactly *one* real cube root.

Definition

Roots

In general, we state that if

$$x^n = a$$

then x is an *nth root* of a.

NOTE

The symbol $\sqrt{\ }$ first appeared in print in 1525. In Latin, "radix" means root, and this was contracted to a small *r*. The present symbol may have been used because it resembled that small *r*.

We are now ready for new notation. The symbol $\sqrt{\ }$ is called a *radical sign*. We saw earlier that 3 is the positive square root of 9.

We call 3 the *principal square root* of 9, and we write

$$\sqrt{9} = 3$$

While every positive number has two square roots, one positive and one negative, the principal square root is always the positive one.

> **NOTE**
>
> The index 2 for square roots is generally not written. We understand that
>
> $\sqrt{a}$
>
> is the principal square root of *a*.

In some cases we want to indicate the negative square root; to do so, we write

$-\sqrt{9} = -3$

If we want both square roots, we write

$\pm\sqrt{9} = \pm3$

Every radical expression contains three parts. The principal *n*th root of *a* is written as

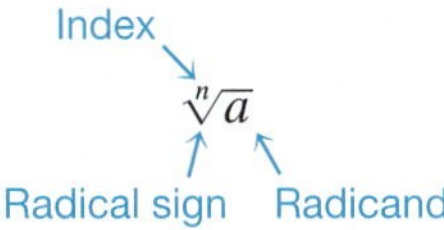

Example 1 Evaluating Radical Expressions

< Objective 1 >

Evaluate, if possible.

(a) $\sqrt{49} = 7$

(b) $-\sqrt{49} = -7$

(c) $\pm\sqrt{49} = \pm7$

(d) $\sqrt{-49}$ is not a real number.

Let's examine this more carefully. Suppose that for some real number *x*,

$x = \sqrt{-49}$

By our earlier definition, this means that

$x^2 = -49$

which is impossible. There is no real square root of -49. We call $\sqrt{-49}$ an *imaginary number.*

> **NOTE**
>
> Neither 7^2 nor $(-7)^2$ equals -49.

Check Yourself 1

Evaluate, if possible.

(a) $\sqrt{64}$ **(b)** $-\sqrt{64}$ **(c)** $\pm\sqrt{64}$ **(d)** $\sqrt{-64}$

> **NOTE**
>
> *Indices* is the plural of index

Example 2 considers cube roots and radicals with higher indices.

Example 2 Evaluating Radical Expressions

> **NOTES**
>
> The cube root of a *negative* number is *negative.*
>
> In general, an *even* root of a *negative* number is *not real;* it is *imaginary.*

Evaluate, if possible.

(a) $\sqrt[3]{64} = 4$ because $4^3 = 64$

(b) $-\sqrt[3]{64} = -4$

(c) $\sqrt[3]{-64} = -4$ because $(-4)^3 = -64$

(d) $\sqrt[4]{81} = 3$ because $3^4 = 81$

(e) $\sqrt[4]{-81}$ is not a real number.

(f) $\sqrt[5]{32} = 2$ because $2^5 = 32$

(g) $\sqrt[5]{-32} = -2$ because $(-2)^5 = -32$

Check Yourself 2

Evaluate.

(a) $\sqrt[3]{125}$ (b) $-\sqrt[3]{125}$ (c) $\sqrt[3]{-125}$ (d) $\sqrt[4]{16}$
(e) $\sqrt[5]{243}$ (f) $\sqrt[4]{-16}$ (g) $\sqrt[5]{-243}$

We usually use a calculator or computer to evaluate roots. This is because a whole number must be a perfect square in order for its square root to be a rational number. Similarly, a whole number must be a perfect cube for its cube root to be rational.

The perfect square whole numbers are

1, 4, 9, 16, 25, ...

The perfect cube whole numbers are

1, 8, 27, 64, 125, ...

Considering perfect squares, we ask about the numbers in between the perfect squares. For instance, what is $\sqrt{20}$? While we cannot write $\sqrt{20}$ as a whole or rational number, we can approximate its value.

We know that

$\sqrt{16} = 4$ because $4^2 = 16$ and

$\sqrt{25} = 5$ because $5^2 = 25$

Since $16 < 20 < 25$, we have

$\sqrt{16} < \sqrt{20} < \sqrt{25}$ or $4 < \sqrt{20} < 5$

So, $\sqrt{20}$ is between 4 and 5.

When we approximate a root without a calculator, we look for two consecutive integers that "trap" the root. In the preceding instance, 4 is the largest integer less than $\sqrt{20}$ and 5 is the smallest integer greater than $\sqrt{20}$.

Example 3 Approximating Roots

< Objective 2 >

NOTES

$3^2 = 9$ and $4^2 = 16$.

$\sqrt{50}$ is much closer to 7 than to 8.

Approximate each square root by finding consecutive integers that "trap" the root.

(a) $\sqrt{12}$

$9 < 12 < 16$ so $3 < \sqrt{12} < 4$

(b) $\sqrt{50}$

$7^2 = 49$ and $8^2 = 64$ so $7 < \sqrt{50} < 8$

(c) $\sqrt{130}$

$11^2 = 121$ and $12^2 = 144$ so $11 < \sqrt{130} < 12$

(d) $\sqrt[3]{50}$

$3^3 = 27$ and $4^3 = 64$ so $3 < \sqrt[3]{50} < 4$

Check Yourself 3

Approximate each square root by finding consecutive integers that "trap" the root.

(a) $\sqrt{6}$ (b) $\sqrt{75}$ (c) $\sqrt{200}$ (d) $\sqrt[3]{200}$

When we need a more precise approximation for the root of a number, we use a calculator or computer. We use the square-root key [$\sqrt{\ }$] to approximate the square root of a number that is not a perfect square. We then round this approximation appropriately. Consider Example 4.

Example 4 Estimating Radical Expressions

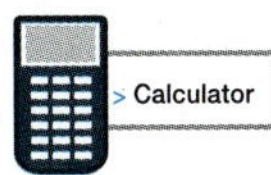

Use a calculator to find decimal approximations for each number. Round all answers to three decimal places.

(a) $\sqrt{17}$

```
√(17)
     4.123105626
√(28)
     5.291502622
```

On many calculators, the square root is shown as the "2nd function" or "inverse" of x^2. Press [2nd] [$\sqrt{\ }$] and type 17. Then type a closing parenthesis [ )] and press [ENTER]. The display should read 4.123105626. Rounded to three decimal places, the result is 4.123.

(b) $\sqrt{28}$

The display should read 5.291502622. Rounded to three decimal places, the result is 5.292.

```
ERR:NONREAL ANS
1:Quit
2:Goto
```

(c) $\sqrt{-11}$

Press [2nd] [$\sqrt{\ }$], and then type [(−)] 11 [ )], and press [ENTER]. The display will say NONREAL ANS (or something similar). This indicates that −11 does not have a real square root.

Check Yourself 4

Use a calculator to approximate each number. Round your answers to three decimal places.

(a) $\sqrt{13}$ **(b)** $\sqrt{38}$ **(c)** $\sqrt{-21}$

To evaluate roots other than square roots using a graphing calculator, we utilize the [MATH] menu, and select the symbol $\sqrt[x]{\ }$.

Example 5 Estimating Radical Expressions

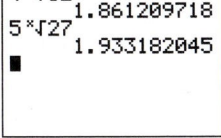

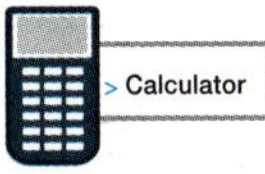

Use a calculator to approximate each number. Round your answers to three decimal places.

(a) $\sqrt[4]{12}$

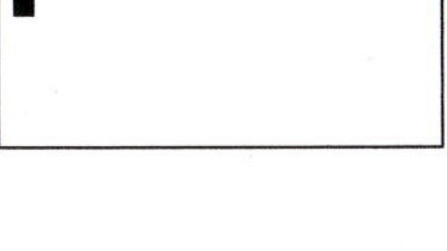

Begin by typing the index 4. Then press the [MATH] key, and choose the symbol $\sqrt[x]{\ }$. Now type 12, and press [ENTER]. The display should read 1.861209718. Rounded to three decimal places, the result is 1.861.

(b) $\sqrt[5]{27}$

Type the index 5, get the symbol $\sqrt[x]{\ }$, then type 27. Pressing [ENTER] should show 1.933182045. Rounded to three decimal places, the result is 1.933.

Check Yourself 5

Use a calculator to approximate each number. Round your answers to three decimal places.

(a) $\sqrt[4]{35}$ **(b)** $\sqrt[5]{29}$

NOTE

Radicals are grouping symbols, like parentheses and fraction bars. We perform the operations inside the radical before moving on. In the expression $\sqrt{(-2)^2}$ we apply the exponent before taking the root.

You need to be cautious when working with even roots. For example, consider the statement

$$\sqrt{x^2} = x$$

First, let $x = 2$: $\sqrt{2^2} = \sqrt{4} = 2$

Now, let $x = -2$: $\sqrt{(-2)^2} = \sqrt{4} = 2$

So the statement $\sqrt{x^2} = x$ is true when x is positive, but $\sqrt{x^2} = x$ is not true when x is negative.

In fact, if x is negative, we have

$$\sqrt{x^2} = -x$$

Putting these ideas together gives

$$\sqrt{x^2} = \begin{cases} x & \text{when } x \geq 0 \\ -x & \text{when } x < 0 \end{cases}$$

Earlier, you studied absolute value, and we make a connection here.

$|2| = 2$ and $|-2| = 2$

So, $|x| = x$ when x is positive, and $|x| = -x$ when x is negative.

Putting these ideas together, gives

$$|x| = \begin{cases} x & \text{when } x \geq 0 \\ -x & \text{when } x < 0 \end{cases}$$

NOTE

We can extend this last statement to

$\sqrt[n]{x^n} = |x|$

when n is *even*.

We summarize the discussion by writing

$$\sqrt{x^2} = |x|$$

Example 6 Evaluating Radical Expressions

Evaluate.

(a) $\sqrt{5^2} = 5$

(b) $\sqrt{(-4)^2} = |-4| = 4$

(c) $\sqrt[4]{2^4} = 2$

(d) $\sqrt[4]{(-3)^4} = |-3| = 3$

NOTE

Alternatively, we could write $\sqrt{(-4)^2} = \sqrt{16} = 4$

Check Yourself 6

Evaluate.

(a) $\sqrt{6^2}$ **(b)** $\sqrt{(-6)^2}$ **(c)** $\sqrt[4]{3^4}$ **(d)** $\sqrt[4]{(-3)^4}$

Roots with indices that are odd do *not* require absolute values. For instance,

$$\sqrt[3]{3^3} = \sqrt[3]{27} = 3$$

$$\sqrt[3]{(-3)^3} = \sqrt[3]{-27} = -3$$

and we see that

$$\sqrt[n]{x^n} = x \qquad \text{when } n \text{ is odd}$$

To summarize, we can write

$$\sqrt[n]{x^n} = \begin{cases} |x| & \text{when } n \text{ is even} \\ x & \text{when } n \text{ is odd} \end{cases}$$

We turn to an example with variables in the radicand.

Example 7 Simplifying Radical Expressions

< Objective 3 >

NOTE

We can determine the power of the variable in the root by dividing the power in the radicand by the index. In Example 7(d), $8 \div 4 = 2$.

Simplify each expression.

(a) $\sqrt[3]{a^3} = a$

(b) $\sqrt{16m^2} = 4|m|$

(c) $\sqrt[5]{32x^5} = 2x$

(d) $\sqrt[4]{x^8} = x^2$ because $(x^2)^4 = x^8$

(e) $\sqrt[3]{27y^6} = 3y^2$ Do you see why?

Check Yourself 7

Simplify.

(a) $\sqrt[4]{x^4}$ **(b)** $\sqrt{49w^2}$ **(c)** $\sqrt[5]{a^{10}}$ **(d)** $\sqrt[3]{8y^9}$

NOTE

"Distance" or "length" is always a nonnegative number.

One of the most important results in the history of mathematics uses radicals. You should recall from earlier math classes that the **Pythagorean theorem** describes the relationship between the lengths of the sides of a right triangle (a triangle with a 90° angle).

Property

The Pythagorean Theorem

NOTE

This is only true in the case of right triangles.

In any right triangle, the square of the length of the longest side (the hypotenuse) is equal to the sum of the squares of the length of the two shorter sides (the legs).

$c^2 = a^2 + b^2$

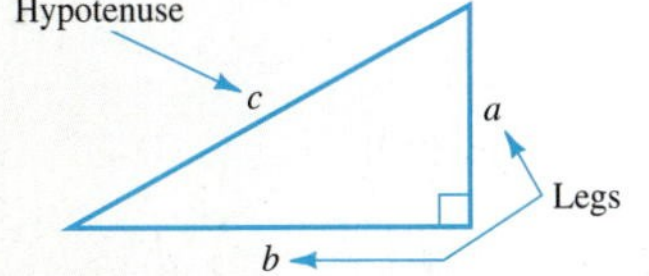

The conclusion of this theorem says that $a^2 + b^2 = c^2$. This means that $\sqrt{a^2 + b^2} = \sqrt{c^2}$. Since the lengths a, b, and c are all positive numbers, $\sqrt{c^2} = c$. So we may write $c = \sqrt{a^2 + b^2}$ in a right triangle.

Example 8 Using the Pythagorean Theorem

< Objective 4 >

NOTE

$13 < \sqrt{193} < 14$ because $13^2 = 169$ and $14^2 = 196$.

Suppose the two legs of a right triangle are 7 and 12. How long is the hypotenuse?

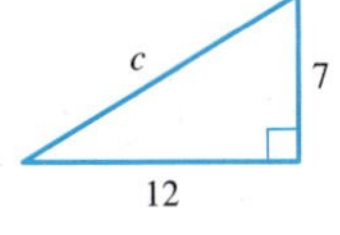

We should always draw a sketch when solving geometry problems.

The Pythagorean theorem tells us that $c = \sqrt{(12)^2 + (7)^2} = \sqrt{144 + 49} = \sqrt{193}$.

Check Yourself 8

Suppose the two legs of a right triangle are 11 and 14. How long is the hypotenuse?

As long as we know the lengths of two of the sides in a right triangle, we can determine the length of the third side. In Example 9, we know the lengths of the hypotenuse and one leg.

Example 9 Using the Pythagorean Theorem

The hypotenuse of a right triangle is 23 and one leg is 16. How long is the other leg?

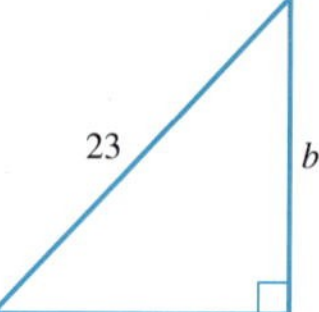

As in Example 8, drawing a sketch helps.

Let $c = 23$ and $a = 16$. Since $a^2 + b^2 = c^2$, we write

$$b^2 = c^2 - a^2$$
$$b = \sqrt{c^2 - a^2}$$

NOTE

$16 < \sqrt{273} < 17$

So, $b = \sqrt{(23)^2 - (16)^2} = \sqrt{529 - 256} = \sqrt{273}$

The length of the other leg is $\sqrt{273}$.

Check Yourself 9

The hypotenuse of a right triangle is 19 and one leg is 12. How long is the other leg?

Now suppose that we want to find the distance d between two points in the coordinate plane, say (3, 1) and (8, 7).

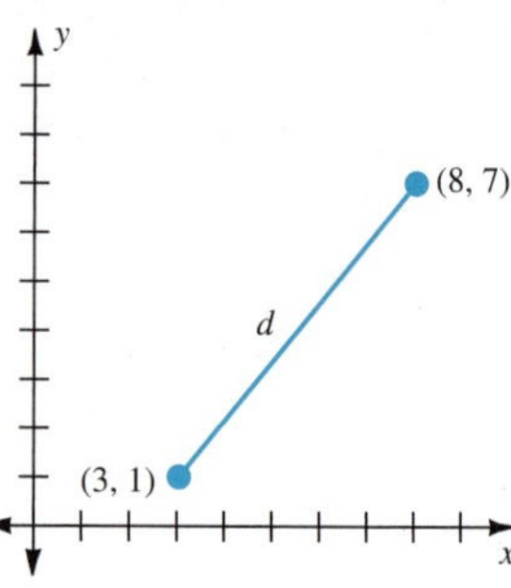

We can make a right triangle here.

The length of the horizontal leg of the triangle is $8 - 3 = 5$ units, while the length of the vertical leg of this triangle is $7 - 1 = 6$ units. Using the Pythagorean theorem, we have

$$d = \sqrt{(8 - 3)^2 + (7 - 1)^2} = \sqrt{5^2 + 6^2} = \sqrt{25 + 36} = \sqrt{61}$$

We generalize this to construct the distance formula.

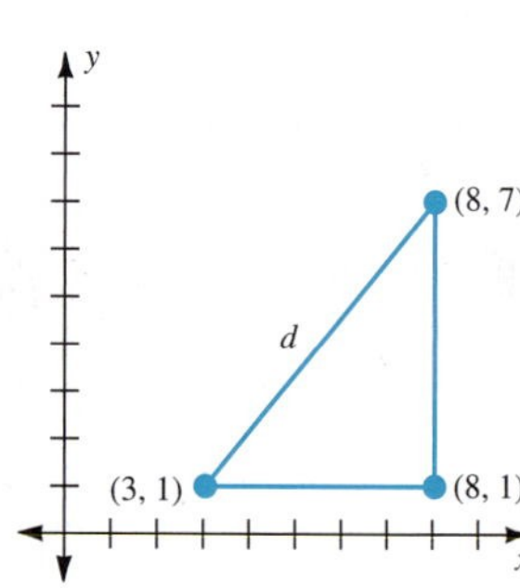

Property

The Distance Formula

The distance d between two points (x_1, y_1) and (x_2, y_2) can be found using the formula

$$d = \sqrt{(x_2 - x_1)^2 + (y_2 - y_1)^2}$$

Example 10 Finding the Distance Between Points

< Objective 5 >

Find the distance between each pair of points.

(a) $(3, -5)$ and $(-5, -5)$

Let $(x_1, y_1) = (3, -5)$ and $(x_2, y_2) = (-5, -5)$. Plugging those values into the distance formula gives

$$d = \sqrt{(-5-3)^2 + [-5-(-5)]^2} = \sqrt{(-8)^2 + 0^2} = \sqrt{64} = 8$$

The distance between the two points is 8 units.

(b) $(-4, 7)$ and $(1, 5)$

Let $(x_1, y_1) = (-4, 7)$ and $(x_2, y_2) = (1, 5)$. Plugging those values into the distance formula gives

$$d = \sqrt{[1-(-4)]^2 + (5-7)^2} = \sqrt{5^2 + (-2)^2} = \sqrt{29}$$

The distance between the two points is $\sqrt{29}$ units.

Check Yourself 10

Find the distance between each pair of points.

(a) $(-2, 7)$ and $(-5, 7)$ **(b)** $(3, -5)$ and $(7, -4)$

The distance formula gives us a method for describing the equation of a circle in the coordinate plane. Consider the definition of a circle.

Definition

Circle

A **circle** is the set of all points in the plane equidistant from a fixed point, called the **center** of the circle. The distance between the center of a circle and any point on the circle is called the **radius** of the circle.

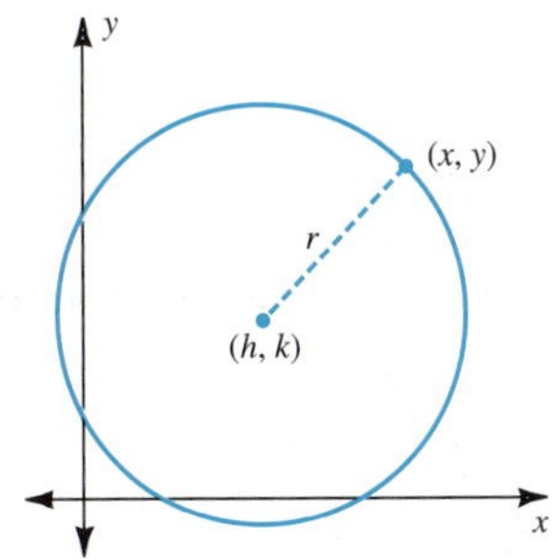

Suppose a circle has its center at a point with coordinates (h, k) and radius r. If (x, y) represents any point on the circle, then the distance from (h, k) to (x, y) is r. Applying the distance formula, we have

$$r = \sqrt{(x-h)^2 + (y-k)^2}$$

Squaring both sides gives an equation of a circle.

$$r^2 = (x-h)^2 + (y-k)^2$$

Property

Equation of a Circle

An equation of a circle with center (h, k) and radius r is

$$(x-h)^2 + (y-k)^2 = r^2$$

A special case is the circle centered at the origin with radius r. Then $(h, k) = (0, 0)$, and its equation is

$$x^2 + y^2 = r^2$$

The circle equation can be used in two ways. Given the center and radius of the circle, we can write its equation; or given its equation, we can find the center and radius of a circle.

Example 11 Finding the Equation of a Circle

< Objective 6 >

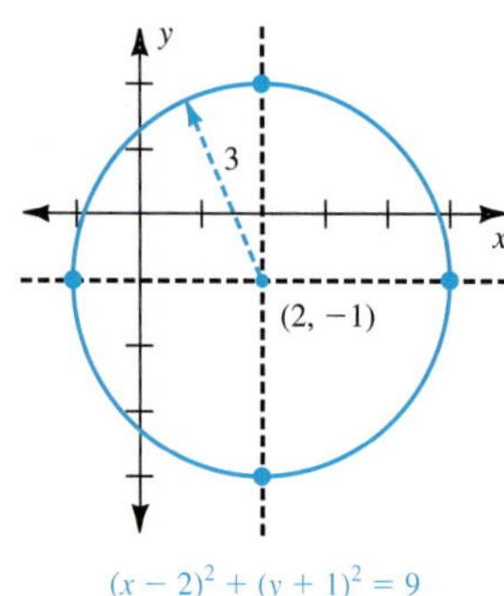

$(x - 2)^2 + (y + 1)^2 = 9$

Find an equation for the circle with center at $(2, -1)$ and radius 3. Sketch the circle.

To sketch the circle, we first locate its center. Then we determine four points 3 units to the right and left and up and down from the center of the circle. Drawing a smooth curve through those four points completes the graph.

Let $(h, k) = (2, -1)$ and $r = 3$. Applying the circle equation yields

$$(x - 2)^2 + [y - (-1)]^2 = 3^2$$

$$(x - 2)^2 + (y + 1)^2 = 9$$

Check Yourself 11

Find an equation for the circle with center at $(-2, 1)$ and radius 5. Sketch the circle.

Now, given an equation for a circle, we can also find the radius and center and then sketch the circle.

Example 12 Finding the Center and Radius of a Circle

Find the center and radius of the circle with equation

$$(x - 5)^2 + (y + 2)^2 = 16$$

Remember, the general form is

$$(x - h)^2 + (y - k)^2 = r^2$$

Our equation "fits" this form when it is written as

$y + 2 = y - (-2)$

$$(x - 5)^2 + [y - (-2)]^2 = 4^2$$

So the center is at $(5, -2)$, and the radius is 4. The graph is shown.

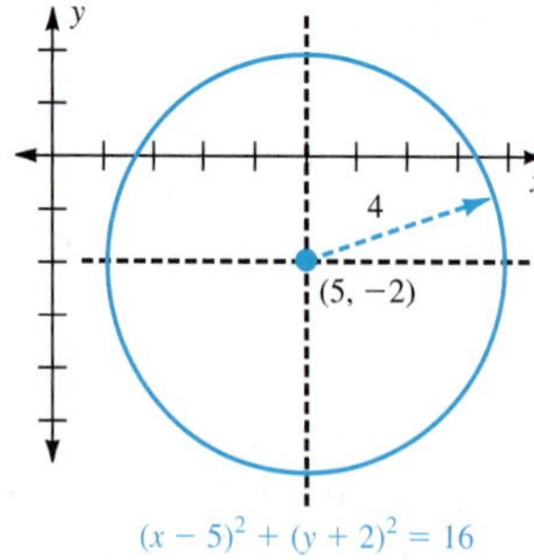

$(x - 5)^2 + (y + 2)^2 = 16$

Check Yourself 12

Find the center and radius of the circle with equation

$$(x + 3)^2 + (y - 2)^2 = 25$$

Sketch the circle.

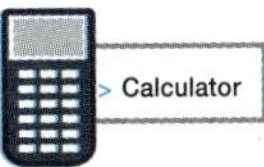

NOTE

We can graph a circle on a calculator by solving for y, then graphing both the upper half and lower half of the circle. For example, consider the circle with equation

$$(x - 1)^2 + (y + 2)^2 = 9$$

$$\begin{aligned}(x - 1)^2 + (y + 2)^2 &= 9\\ (y + 2)^2 &= 9 - (x - 1)^2\\ y + 2 &= \pm\sqrt{9 - (x - 1)^2}\\ y &= -2 \pm \sqrt{9 - (x - 1)^2}\end{aligned}$$

Now graph the two functions

$$y = -2 + \sqrt{9 - (x - 1)^2}$$

and

$$y = -2 - \sqrt{9 - (x - 1)^2}$$

on a calculator. (The viewing window may need to be squared to obtain the shape of a circle.)

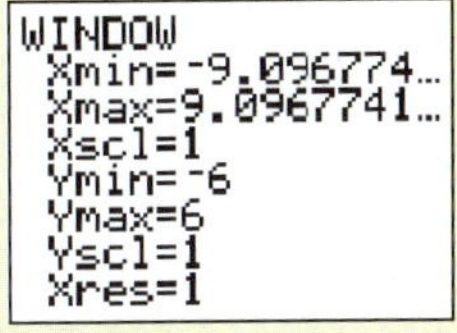

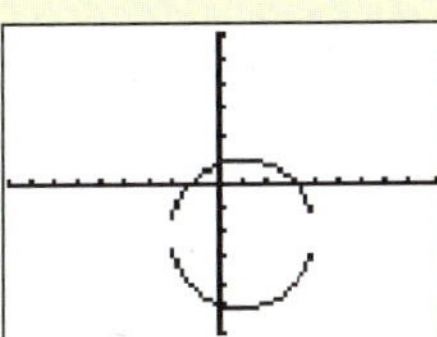

The "gaps" in the graph are a limitation of graphing calculators technology. When sketching these graphs, be sure to fill in the gaps to form a full circle.

Check Yourself ANSWERS

1. **(a)** 8; **(b)** -8; **(c)** ± 8; **(d)** not a real number **2.** **(a)** 5; **(b)** -5; **(c)** -5; **(d)** 2; **(e)** 3; **(f)** not a real number; **(g)** -3 **3.** **(a)** $2 < \sqrt{6} < 3$; **(b)** $8 < \sqrt{75} < 9$; **(c)** $14 < \sqrt{200} < 15$; **(d)** $5 < \sqrt[3]{200} < 6$ **4.** **(a)** 3.606; **(b)** 6.164; **(c)** not a real number **5.** **(a)** 2.432; **(b)** 1.961 **6.** **(a)** 6; **(b)** 6; **(c)** 3; **(d)** 3 **7.** **(a)** $|x|$; **(b)** $7|w|$; **(c)** a^2; **(d)** $2y^3$ **8.** $\sqrt{317}$ **9.** $\sqrt{217}$ **10.** **(a)** 3 units; **(b)** $\sqrt{17}$ units **11.** $(x + 2)^2 + (y - 1)^2 = 25$ **12.** Center $(-3, 2)$; radius 5

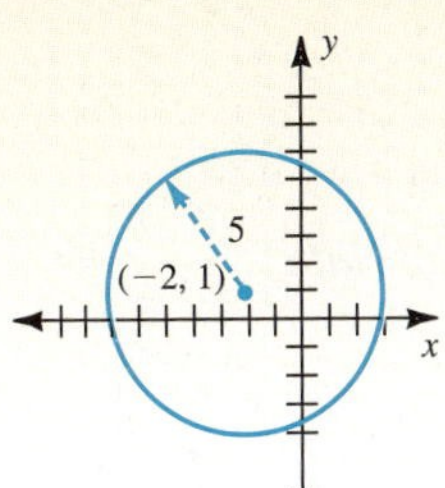

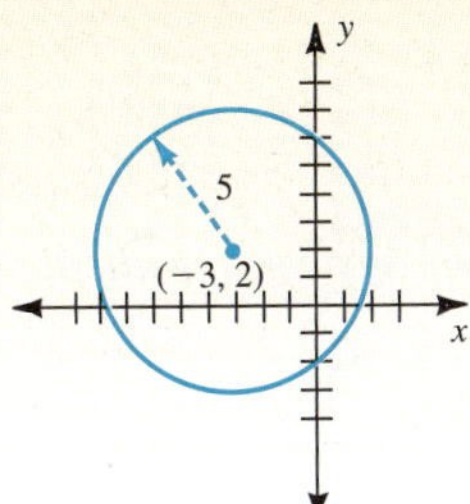

Reading Your Text

These fill-in-the-blank exercises will help you understand some of the key vocabulary used in this section. The answers to these exercises are in the Answers Appendix in the back of the text.

(a) If $x^2 = a$, we say x is a ________ of a.

(b) Every positive number has ________ square roots.

(c) The cube root of a negative number is ________.

(d) The distance between the center of a circle and any point on the circle is called the ________ of the circle.

7.1 exercises

Skills | Calculator/Computer | Career Applications | Above and Beyond

< Objective 1 >

Evaluate each root, if possible.

1. $\sqrt{49}$ **2.** $\sqrt{36}$ **3.** $-\sqrt{36}$ **4.** $-\sqrt{81}$

5. $\pm\sqrt{81}$ VIDEO **6.** $\pm\sqrt{49}$ **7.** $\sqrt{-49}$ **8.** $\sqrt{-25}$

9. $\sqrt[3]{27}$ **10.** $\sqrt[3]{64}$ **11.** $\sqrt[3]{-64}$ **12.** $-\sqrt[3]{125}$ VIDEO

13. $-\sqrt[3]{216}$ **14.** $\sqrt[3]{-27}$ **15.** $\sqrt[4]{81}$ **16.** $\sqrt[5]{32}$

17. $\sqrt[5]{-32}$ **18.** $\sqrt[4]{-81}$ **19.** $-\sqrt[4]{16}$ **20.** $\sqrt[5]{-243}$

21. $\sqrt[4]{-16}$ **22.** $-\sqrt[5]{32}$ **23.** $-\sqrt[5]{243}$ **24.** $-\sqrt[4]{625}$

25. $\sqrt{\frac{4}{9}}$ **26.** $\sqrt{\frac{9}{25}}$ VIDEO **27.** $\sqrt[3]{\frac{8}{27}}$ **28.** $\sqrt[3]{-\frac{27}{64}}$

< Objective 2 >

29. $\sqrt{6^2}$ **30.** $\sqrt{9^2}$ **31.** $\sqrt{(-3)^2}$ **32.** $\sqrt{(-5)^2}$

33. $\sqrt[3]{4^3}$ **34.** $\sqrt[3]{(-5)^3}$ **35.** $\sqrt[4]{3^4}$ **36.** $\sqrt[4]{(-2)^4}$

Approximate each square root by finding consecutive integers that "trap" the root.

37. $\sqrt{3}$ **38.** $\sqrt{18}$ **39.** $\sqrt{30}$ **40.** $\sqrt{91}$

41. $\sqrt{150}$ **42.** $\sqrt{300}$ **43.** $\sqrt{500}$ **44.** $\sqrt{1{,}000}$

45. $\sqrt[3]{100}$ **46.** $\sqrt[3]{500}$ **47.** $\sqrt[3]{-20}$ **48.** $\sqrt[3]{-1{,}500}$

< Objective 3 >

Simplify each root.

49. $\sqrt{x^2}$ **50.** $\sqrt[3]{w^3}$ **51.** $\sqrt[5]{y^5}$ **52.** $\sqrt[7]{z^7}$

53. $\sqrt{9x^2}$ **54.** $\sqrt{81y^2}$ **55.** $\sqrt{a^4b^6}$ **56.** $\sqrt{w^6z^{10}}$

57. $\sqrt{16x^4}$ VIDEO **58.** $\sqrt{49y^6}$

< Objective 4 >

Find the missing length in each triangle. Express your answer in radical form where appropriate.

59.

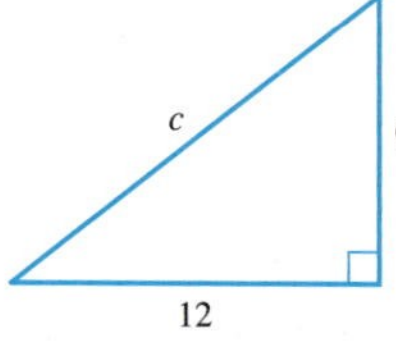

60.

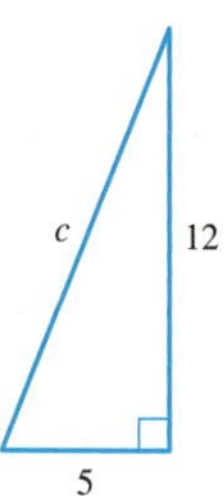

61.

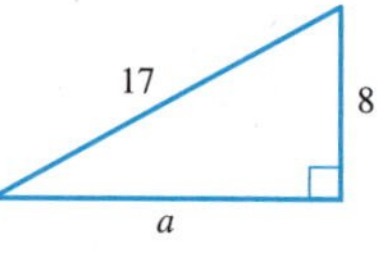

62.

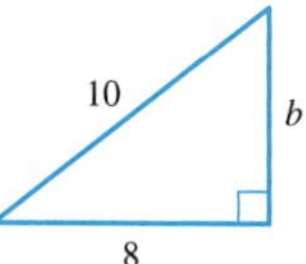

63.

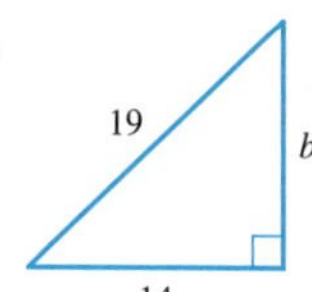

64.

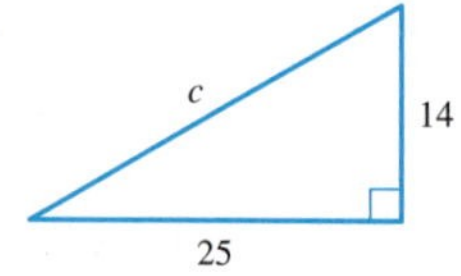

< Objective 5 >

Find the distance between each pair of points.

65. $(2, -6)$ and $(2, -9)$

66. $(-3, 7)$ and $(4, 7)$

67. $(-7, 1)$ and $(-4, 0)$

68. $(-17, -5)$ and $(-12, -3)$

< Objective 6 >

Find the center and radius of each circle. Then graph it.

69. $x^2 + y^2 = 4$

70. $x^2 + y^2 = 25$

71. $(x - 1)^2 + y^2 = 9$

72. $x^2 + (y + 2)^2 = 16$

73. $(x - 4)^2 + (y + 1)^2 = 16$

74. $(x + 3)^2 + (y + 2)^2 = 25$

Write the equation of each circle.

75.

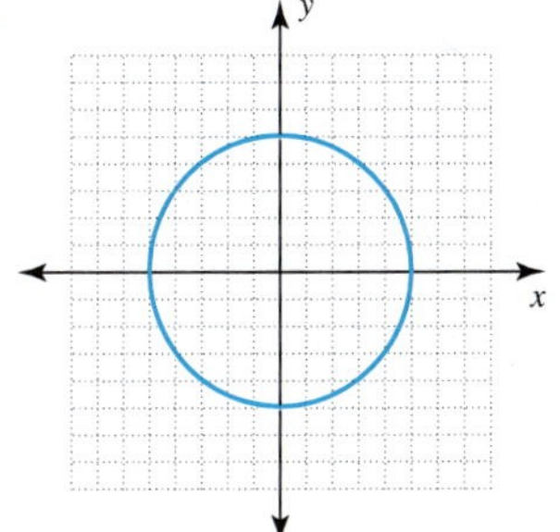

76.

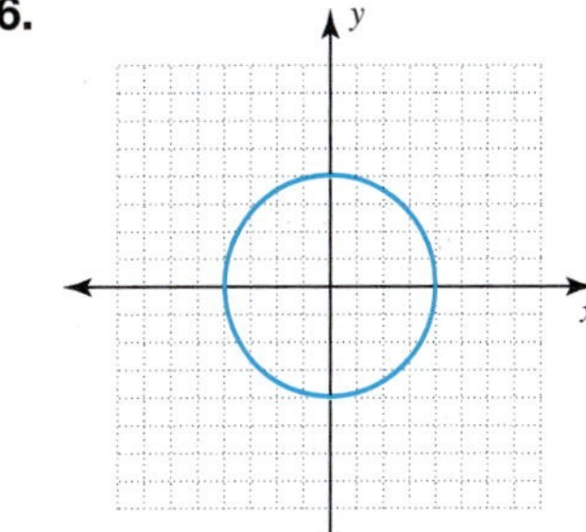

77.

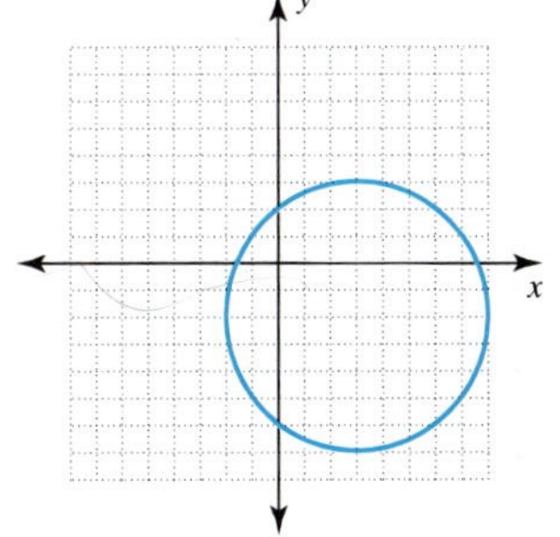

78.

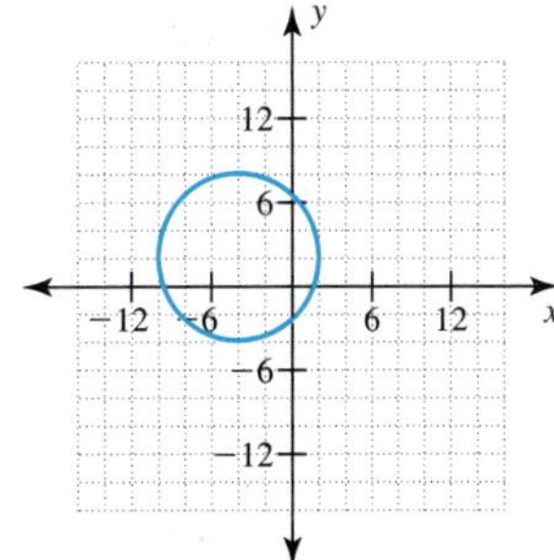

Find the two equivalent expressions in each set.

79. $\sqrt{-16}, -\sqrt{16}, -4$

80. $-\sqrt{25}, -5, \sqrt{-25}$

81. $\sqrt[3]{-125}, -\sqrt[3]{125}, |-5|$

82. $\sqrt[5]{-32}, -\sqrt[5]{32}, |-2|$

83. $\sqrt[4]{10{,}000}, 100, \sqrt[3]{1{,}000}$

84. $10^2, \sqrt{10{,}000}, \sqrt[3]{100{,}000}$

Determine whether each statement is **true** *or* **false.**

85. $\sqrt{16x^{16}} = 4x^4$

86. $\sqrt{(x-4)^2} = x - 4$

87. $\sqrt{16x^{-4}y^{-4}}$ is a real number

88. $\sqrt{x^2 + y^2} = x + y$

89. $\dfrac{\sqrt{x^2 - 25}}{x - 5} = \sqrt{x + 5}$

90. $\sqrt{2} + \sqrt{6} = \sqrt{8}$

91. Given the equation $x^2 + y^2 = 9$, the graph is a circle with center at (0, 0).

92. Given the equation $x^2 + y^2 = 9$, the graph is a circle with radius 9.

Complete each statement with **always, sometimes,** *or* **never.**

93. The principal square root of a number is __________ negative.

94. The cube root of a number is __________ a real number.

Skills | **Calculator/Computer** | Career Applications | Above and Beyond

Use a calculator to evaluate each root. Round your answers to three decimal places.

95. $\sqrt{15}$

96. $\sqrt{29}$

97. $\sqrt{156}$

98. $\sqrt{213}$

99. $\sqrt{-15}$

100. $\sqrt{-79}$

101. $\sqrt[3]{83}$

102. $\sqrt[3]{97}$

103. $\sqrt[5]{123}$

104. $\sqrt[5]{283}$

105. $\sqrt[3]{-15}$

106. $\sqrt[5]{-29}$

Find the length of a side of each square given the area shown. Round your answer to the nearest hundredth of a foot.

107.

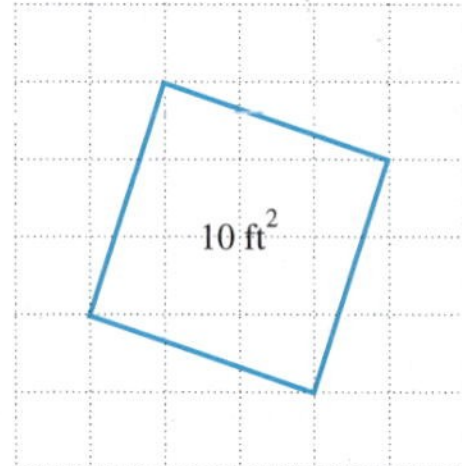

108.

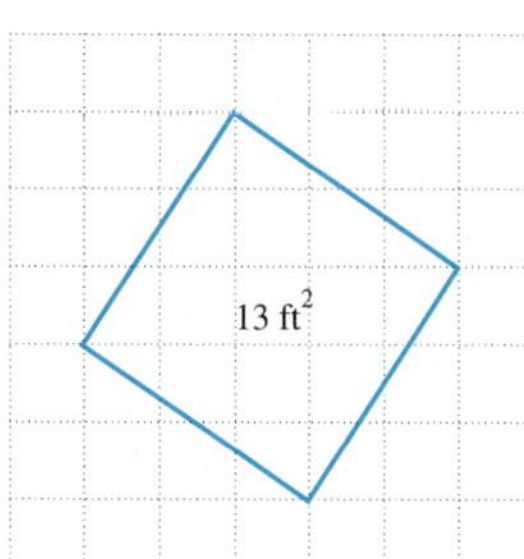

109.

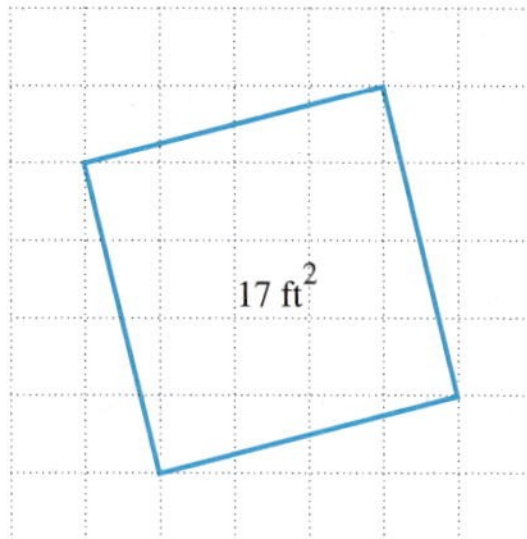

110.

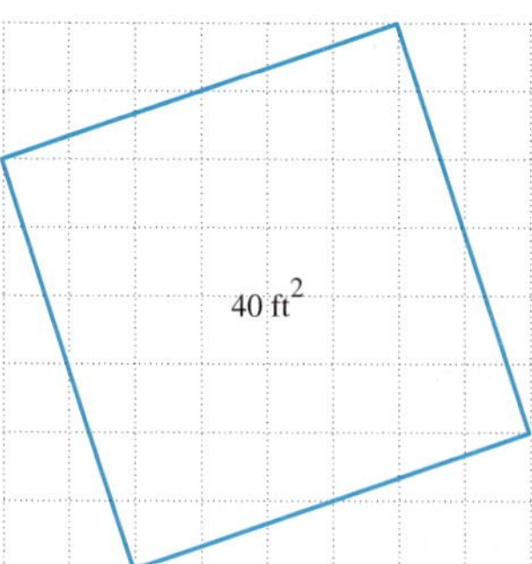

111. **Geometry** The area of a square is 32 ft². Find the length of a side to the nearest hundredth.

112. **Geometry** The area of a square is 83 ft². Find the length of a side to the nearest hundredth.

113. **Geometry** The area of a circle is 147 ft². Find the radius to the nearest hundredth.

114. **Geometry** If the area of a circle is 72 cm², find the radius to the nearest hundredth.

115. **Mechanical Engineering** The time in seconds that it takes for an object to fall from rest is given by $t = \frac{1}{4}\sqrt{d}$, in which d is the distance fallen (in feet).

Find the time required for an object to fall to the ground from an 800-ft tall building. Report your result to the nearest hundredth second.

116. **Mechanical Engineering** Use the information in exercise 115 to find the time required for an object to fall to the ground from a 1,400-ft tall building. Report your result to the nearest hundredth second.

Skills | Calculator/Computer | Career Applications | **Above and Beyond**

Each equation defines a relation. Give the domain and the range of each relation.

117. $(x + 3)^2 + (y - 2)^2 = 16$

118. $(x - 1)^2 + (y - 5)^2 = 9$

119. $x^2 + (y - 3)^2 = 25$ 

120. $(x + 2)^2 + y^2 = 36$

121. Is there any prime number whose square root is an integer? Explain your answer.

122. Find two consecutive integers whose square roots are also consecutive integers.

123. Use a calculator to complete each exercise.

(a) Choose a number greater than 1 and find its square root. Then find the square root of the result and continue in this manner, observing the successive square roots. Do these numbers seem to be approaching a certain value? If so, what?

(b) Choose a number greater than 0 but less than 1 and find its square root. Then find the square root of the result, and continue in this manner, observing successive square roots. Do these numbers seem to be approaching a certain value? If so, what?

124. **(a)** Can a number be equal to its own square root?

(b) Other than the number(s) found in part (a), is a number always greater than its square root? Investigate.

125. Let a and b be positive numbers. If a is greater than b, is it always true that the square root of a is greater than the square root of b? Investigate.

126. Suppose that a weight is attached to a string of length L, and the other end of the string is held fixed. If we pull the weight and then release it, allowing the weight to swing back and forth, we can observe the behavior of a simple pendulum. The period T is the time required for the weight to complete a full cycle, swinging forward and then back. The formula below describes the relationship between T and L.

$$T = 2\pi\sqrt{\frac{L}{g}}$$

If L is expressed in centimeters, then $g = 980 \text{ cm/s}^2$. For each string length, calculate the corresponding period. Round to the nearest tenth of a second. chapter 7 > Make the Connection

(a) 30 cm **(b)** 50 cm **(c)** 70 cm **(d)** 90 cm **(e)** 110 cm

127. In parts (a) through (f), evaluate when possible.

(a) $\sqrt{4 \cdot 9}$ **(b)** $\sqrt{4} \cdot \sqrt{9}$ **(c)** $\sqrt{9 \cdot 16}$ **(d)** $\sqrt{9} \cdot \sqrt{16}$ **(e)** $\sqrt{(-4)(-25)}$ **(f)** $\sqrt{-4} \cdot \sqrt{-25}$

(g) Based on parts (a) through (f), make a general conjecture concerning $\sqrt{ab}$. Be careful to specify any restrictions on possible values for a and b.

128. In parts (a) through (d), evaluate when possible.

(a) $\sqrt{9+16}$ **(b)** $\sqrt{9}+\sqrt{16}$ **(c)** $\sqrt{36+64}$ **(d)** $\sqrt{36}+\sqrt{64}$

(e) Based on parts (a) through (d), what can you say about $\sqrt{a+b}$ and $\sqrt{a}+\sqrt{b}$?

Answers

1. 7 **3.** -6 **5.** ± 9 **7.** Not a real number **9.** 3 **11.** -4 **13.** -6 **15.** 3 **17.** -2 **19.** -2 **21.** Not a real number **23.** -3 **25.** $\frac{2}{3}$ **27.** $\frac{2}{3}$ **29.** 6 **31.** 3 **33.** 4 **35.** 3 **37.** 1, 2 **39.** 5, 6 **41.** 12, 13 **43.** 22, 23 **45.** 4, 5 **47.** $-3, -2$ **49.** $|x|$ **51.** y **53.** $3|x|$ **55.** $|a^2b^3|$ **57.** $4x^2$ **59.** 15 **61.** 15 **63.** $\sqrt{165}$ **65.** 3 units **67.** $\sqrt{10}$ units

69. Center: (0, 0); radius: 2

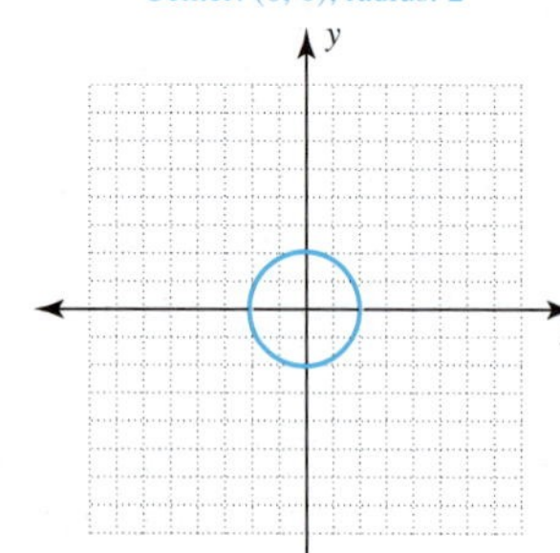

71. Center: (1, 0); radius: 3

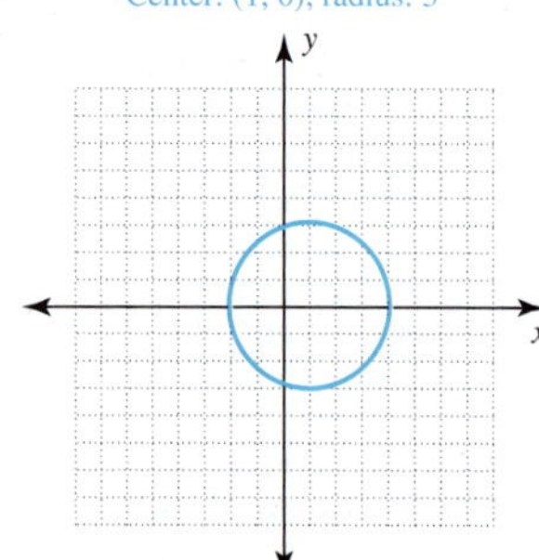

73. Center: (4, -1); radius: 4

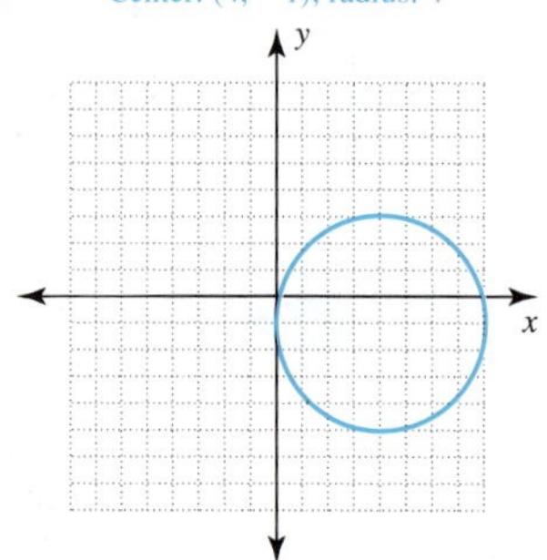

75. $x^2+y^2=25$ **77.** $(x-3)^2+(y+2)^2=25$ **79.** $-\sqrt{16}, -4$ **81.** $\sqrt[3]{-125}, -\sqrt[3]{125}$ **83.** $\sqrt[4]{10{,}000}, \sqrt[3]{1{,}000}$ **85.** False **87.** True **89.** False **91.** True **93.** never **95.** 3.873 **97.** 12.49 **99.** Not a real number **101.** 4.362 **103.** 2.618 **105.** -2.466 **107.** 3.16 ft **109.** 4.12 ft **111.** 5.66 ft **113.** 6.84 ft **115.** 7.07 s **117.** Domain: $\{x \mid -7 \le x \le 1\}$; range: $\{y \mid -2 \le y \le 6\}$ **119.** Domain: $\{x \mid -5 \le x \le 5\}$; range: $\{y \mid -2 \le y \le 8\}$ **121.** No **123.** Above and Beyond **125.** Above and Beyond **127.** **(a)** 6; **(b)** 6; **(c)** 12; **(d)** 12; **(e)** 10; **(f)** not possible; **(g)** Above and Beyond

Activity 7 ::

The Swing of a Pendulum

The action of a pendulum seems simple. Scientists have studied the characteristics of a swinging pendulum and found them to be quite useful. In 1851, in Paris, Jean Foucault (pronounced "Foo-koh") used a pendulum to clearly demonstrate the rotation of Earth about its own axis.

A pendulum can be as simple as a string or cord with a weight fastened to one end. The other end is fixed, and the weight is allowed to swing. We define the **period** of a pendulum to be the amount of time required for the pendulum to make one complete swing (back and forth). The question we pose is: How does the *period* of a pendulum relate to the *length* of the pendulum?

For this activity, you need a piece of string that is approximately 1 m long. Fasten a weight (such as a small hexagonal nut) to one end, and then place clear marks on the string every 10 cm up to 70 cm, measured from the center of the weight. You also need a stopwatch.

1. Working with one or two partners, hold the string at the mark that is 10 cm from the weight. Pull the weight to the side with your other hand and let it swing freely. To estimate the period, let the weight swing through 30 periods, record the time in the table, and then divide by 30. Round your result to the nearest hundredth of a second and record it.

   Note: If you are unable to perform the experiment and collect your own data, you can use the sample data collected in this manner and presented at the end of this activity.

   Repeat the described procedure for each length indicated in the table.

Length of string, cm	10	20	30	40	50	60	70
Time for 30 periods, s							
Time for 1 period, s							

2. Let L represent the length of the pendulum and T represent the time period that results from swinging that pendulum. Fill out the table.

L	10	20	30	40	50	60	70
T							

3. Which variable, L or T, is viewed here as the independent variable?
4. On graph paper, draw horizontal and vertical axes, but plan to graph the data points in the first quadrant only. Explain why this is reasonable.
5. With the independent variable marked on the horizontal axis, scale the axes appropriately, keeping an eye on your data.
6. Plot your data points. Should you connect them with a smooth curve?
7. What period T would correspond to a string length of 0? Include this point on your graph.
8. Use your graph to predict the period for a string length of 80 cm.

9. Verify your prediction by measuring the period when the string is held at 80 cm (as described in step 1). How close did your experimental estimate come to the prediction made in step 8?

You created a graph showing T as a function of L. The shape of the graph may not be familiar to you yet. In fact, the shape of your pendulum graph fits that of a square-root function.

Sample Data

Length of string, cm	10	20	30	40	50	60	70
Time for 30 periods, s	19	27	33	38	42	46	49

7.2 Simplifying Radical Expressions

< 7.2 Objectives >

1 > Use the product property to simplify radical expressions

2 > Use the quotient property to simplify radical expressions

3 > Rationalize a denominator

In most cases, and especially in later math classes, you need to write radical expressions in *simplest form*. There are three conditions that must be satisfied for a radical expression to be simplified.

Property

Simplifying Radical Expressions

A radical expression is *simplified* if

1. The radicand has no factor raised to a power greater than or equal to the index.
2. There are no fractions under the radical.
3. There are no radicals in the denominator of a fraction.

For instance,

$\sqrt{17}$ is simplified because 17 has no perfect-square factors and there are no fractions in the expression.
$\sqrt{12}$ is not simplified because 4 is a perfect-square factor of 12.

To simplify radical expressions, we need to develop two properties.
To develop our first property, consider an expression such as

$$\sqrt{25 \cdot 4}$$

One approach to simplify the expression would be

$$\sqrt{25 \cdot 4} = \sqrt{100} = 10$$

Now what happens if we separate the original radical?

$$\begin{aligned}\sqrt{25 \cdot 4} &= \sqrt{25} \cdot \sqrt{4} \\ &= 5 \cdot 2 = 10\end{aligned}$$

The result is the same, and this suggests our first property for radicals.

Property

Product Property for Radicals

The root of a product is equal to the product of the roots.

$$\sqrt[n]{ab} = \sqrt[n]{a} \cdot \sqrt[n]{b}$$

Note: If n is an even integer, a and b must be positive real numbers.

The second property we need is similar.

Property

Quotient Property for Radicals

The root of a quotient is equal to the quotient of the roots.

$$\sqrt[n]{\frac{a}{b}} = \frac{\sqrt[n]{a}}{\sqrt[n]{b}}$$

Note: If n is an even integer, a and b must be positive real numbers.

An example of the preceding property is

$$\sqrt{\frac{100}{4}} = \frac{\sqrt{100}}{\sqrt{4}}$$

Check to see that each side equals 5.

Our initial example deals with satisfying the first condition. We want to find the largest perfect-square factor (in the case of a square root) in the radicand and then apply the product property to simplify the expression.

Example 1 Simplifying Radical Expressions

< Objective 1 >

NOTES

The largest perfect-square factor of 18 is 9.

The largest perfect-square factor of 75 is 25.

The largest perfect-square factor of $27x^3$ is $9x^2$. The exponent must be *even* to be a perfect square.

Simplify each expression.

(a) $\sqrt{18} = \sqrt{9 \cdot 2}$

$= \sqrt{9} \cdot \sqrt{2}$ Apply the product property.

$= 3\sqrt{2}$

(b) $\sqrt{75} = \sqrt{25 \cdot 3}$

$= \sqrt{25} \cdot \sqrt{3}$

$= 5\sqrt{3}$

(c) $\sqrt{27x^3} = \sqrt{9x^2 \cdot 3x}$

$= \sqrt{9x^2} \cdot \sqrt{3x} = 3x\sqrt{3x}$

(d) $\sqrt{72a^3b^4} = \sqrt{36a^2b^4 \cdot 2a}$

$= \sqrt{36a^2b^4} \cdot \sqrt{2a}$

$= 6ab^2\sqrt{2a}$

RECALL

We assume that all variables represent positive real numbers when the index of a radical is even.

Check Yourself 1

Simplify each expression.

(a) $\sqrt{45}$ **(b)** $\sqrt{200}$ **(c)** $\sqrt{75p^5}$ **(d)** $\sqrt{98m^3n^4}$

> **CAUTION**

The root of a product is equal to the product of the roots. The root of a sum **is not** equal to the sum of the roots.

Recall that, in our study of exponents, $(a + b)^n \neq a^n + b^n$ in general.

Be careful! Even though

$\sqrt{a \cdot b} = \sqrt{a} \cdot \sqrt{b}$

the expression $\sqrt{a + b}$ is *not the same* as $\sqrt{a} + \sqrt{b}$

Let $a = 4$ and $b = 9$, and substitute.

$\sqrt{a + b} = \sqrt{4 + 9} = \sqrt{13}$

$\sqrt{a} + \sqrt{b} = \sqrt{4} + \sqrt{9} = 2 + 3 = 5$

Because $\sqrt{13} \neq 5$, we see that the expressions $\sqrt{a + b}$ and $\sqrt{a} + \sqrt{b}$ are not, in general, the same.

Simplifying a cube root requires that we find perfect-cube factors in the radicand. We can extend this idea to simplify a radical expression with any index.

Example 2 Simplifying Radical Expressions

Simplify each expression.

(a) $\sqrt[3]{48} = \sqrt[3]{8 \cdot 6}$ We factor 48 as $8 \cdot 6$ because 8 is a perfect cube.

$= \sqrt[3]{8} \cdot \sqrt[3]{6}$ Apply the product property.

$= 2\sqrt[3]{6}$ $\sqrt[3]{8} = 2$ because $2^3 = 8$.

NOTE

To be a perfect cube, the exponent must be a *multiple of* 3.

(b) $\sqrt[3]{24x^4} = \sqrt[3]{8x^3 \cdot 3x}$ $\quad 8x^3$ is a perfect cube.

$= \sqrt[3]{8x^3} \cdot \sqrt[3]{3x}$

$= 2x\sqrt[3]{3x}$ $\quad (2x)^3 = 8x^3$

(c) $\sqrt[3]{54a^7b^4} = \sqrt[3]{27a^6b^3 \cdot 2ab}$

$= \sqrt[3]{27a^6b^3} \cdot \sqrt[3]{2ab}$

$= 3a^2b\sqrt[3]{2ab}$

Check Yourself 2

Simplify each expression.

(a) $\sqrt[3]{128w^4}$ **(b)** $\sqrt[3]{40x^5y^7}$ **(c)** $\sqrt[4]{48a^8b^5}$

Satisfying our second condition for a radical to be in simplified form (no fractions should appear inside the radical) requires the second property for radicals. Consider Example 3.

Example 3 Simplifying Radical Expressions

< Objective 2 >

NOTE

Apply the quotient property.

Simplify each expression.

(a) $\sqrt{\dfrac{5}{9}} = \dfrac{\sqrt{5}}{\sqrt{9}}$

$= \dfrac{\sqrt{5}}{3}$

(b) $\sqrt{\dfrac{a^4}{25}} = \dfrac{\sqrt{a^4}}{\sqrt{25}} = \dfrac{a^2}{5}$

(c) $\sqrt[3]{\dfrac{5x^2}{8}} = \dfrac{\sqrt[3]{5x^2}}{\sqrt[3]{8}} = \dfrac{\sqrt[3]{5x^2}}{2}$

Check Yourself 3

Simplify each expression.

(a) $\sqrt{\dfrac{7}{16}}$ **(b)** $\sqrt{\dfrac{3}{25a^2}}$ **(c)** $\sqrt[3]{\dfrac{5x}{27}}$

In Example 3, each denominator is a perfect square or cube. When that is not the case, we need another technique so we do not violate the third condition, that no radical appears in the denominator of a fraction. Consider the expression

$$\frac{1}{\sqrt{2}}$$

There is a radical in the denominator. However, we know that

$$\sqrt{2} \cdot \sqrt{2} = 2$$

Therefore, if we multiply the denominator by $\sqrt{2}$, we no longer have a radical in the denominator. Of course, we must multiply the whole fraction by 1, otherwise we change its value. In this case, we use

$$1 = \frac{\sqrt{2}}{\sqrt{2}}$$

and multiply.

$$\frac{1}{\sqrt{2}} = \frac{1}{\sqrt{2}} \cdot 1 = \frac{1}{\sqrt{2}} \cdot \frac{\sqrt{2}}{\sqrt{2}} = \frac{\sqrt{2}}{2}$$

We call this technique **rationalizing the denominator.**

Example 4 Simplifying Radical Expressions

< Objective 3 >

Simplify each expression.

NOTE

We begin by applying the quotient property.

(a) $\sqrt{\frac{1}{3}} = \frac{\sqrt{1}}{\sqrt{3}} = \frac{1}{\sqrt{3}}$

$\frac{1}{\sqrt{3}}$ is not simplified because of the radical in the denominator. To solve this problem, we multiply the numerator and denominator by $\sqrt{3}$. The denominator becomes

$\sqrt{3} \cdot \sqrt{3} = \sqrt{9} = 3$

We then have

RECALL

We can do this because we are multiplying the fraction by $\frac{\sqrt{3}}{\sqrt{3}}$ or 1, which does not change its value.

$$\frac{1}{\sqrt{3}} = \frac{1 \cdot \sqrt{3}}{\sqrt{3} \cdot \sqrt{3}} = \frac{\sqrt{3}}{3}$$

The expression $\frac{\sqrt{3}}{3}$ is simplified because all three of our conditions are satisfied.

NOTE

$\sqrt{2} \cdot \sqrt{5} = \sqrt{2 \cdot 5} = \sqrt{10}$
$\sqrt{5} \cdot \sqrt{5} = 5$

(b) $\sqrt{\frac{2}{5}} = \frac{\sqrt{2}}{\sqrt{5}}$

$= \frac{\sqrt{2} \cdot \sqrt{5}}{\sqrt{5} \cdot \sqrt{5}}$

$= \frac{\sqrt{10}}{5}$

and the expression is in simplest form because our three conditions are satisfied.

(c) $\sqrt{\frac{3x}{20}} = \frac{\sqrt{3x}}{\sqrt{20}}$ Apply the quotient property.

$= \frac{\sqrt{3x}}{\sqrt{4 \cdot 5}}$ 4 is a perfect square.

$= \frac{\sqrt{3x}}{\sqrt{4} \cdot \sqrt{5}}$ Apply the product property to the denominator.

$= \frac{\sqrt{3x}}{2\sqrt{5}}$ Simplify the denominator.

$= \frac{\sqrt{3x}}{2\sqrt{5}} \cdot \frac{\sqrt{5}}{\sqrt{5}}$ We need to multiply the denominator by $\sqrt{5}$ to rationalize it.

$= \frac{\sqrt{3x \cdot 5}}{2 \cdot 5}$ $\sqrt{3x} \cdot \sqrt{5} = \sqrt{3x \cdot 5}$; $\sqrt{5} \cdot \sqrt{5} = 5$.

$= \frac{\sqrt{15x}}{10}$ Simplified.

Check Yourself 4

Simplify each expression.

(a) $\sqrt{\frac{3}{7}}$ **(b)** $\sqrt{\frac{4y}{5}}$ **(c)** $\sqrt{\frac{18}{125}}$

We continue to rationalize denominators along the way to simplifying radical expressions.

Example 5 Simplifying Radicals

Simplify each expression.

(a) $\dfrac{3}{\sqrt{8}} = \dfrac{3 \cdot \sqrt{2}}{\sqrt{8} \cdot \sqrt{2}}$ Multiply numerator and denominator by $\sqrt{2}$.

$$= \frac{3\sqrt{2}}{\sqrt{16}} = \frac{3\sqrt{2}}{4}$$

(b) $\sqrt[3]{\dfrac{5}{4}} = \dfrac{\sqrt[3]{5}}{\sqrt[3]{4}}$

Multiplying the numerator and denominator by $\sqrt[3]{2}$ produces a perfect cube inside the radical in the denominator.

$$\frac{\sqrt[3]{5}}{\sqrt[3]{4}} = \frac{\sqrt[3]{5} \cdot \sqrt[3]{2}}{\sqrt[3]{4} \cdot \sqrt[3]{2}}$$

$$= \frac{\sqrt[3]{10}}{\sqrt[3]{8}} = \frac{\sqrt[3]{10}}{2}$$

NOTE

Why did we use $\sqrt[3]{2}$?

$\sqrt[3]{4} \cdot \sqrt[3]{2} = \sqrt[3]{2^2} \cdot \sqrt[3]{2}$

$= \sqrt[3]{2^3}$

and the exponent is a multiple of 3.

Check Yourself 5

Simplify each expression.

(a) $\dfrac{5}{\sqrt{12}}$ **(b)** $\sqrt[3]{\dfrac{2}{9}}$

In Example 6, we rationalize denominators that contain variables.

Example 6 Rationalizing Variable Denominators

Simplify each expression.

(a) $\sqrt{\dfrac{8x^3}{3y}}$

By the quotient property, we have

$$\sqrt{\frac{8x^3}{3y}} = \frac{\sqrt{8x^3}}{\sqrt{3y}}$$

Because the numerator can be simplified, we start there.

$$\frac{\sqrt{8x^3}}{\sqrt{3y}} = \frac{\sqrt{4x^2} \cdot \sqrt{2x}}{\sqrt{3y}} = \frac{2x\sqrt{2x}}{\sqrt{3y}}$$

Multiplying the numerator and denominator by $\sqrt{3y}$ rationalizes the denominator.

$$\frac{2x\sqrt{2x} \cdot \sqrt{3y}}{\sqrt{3y} \cdot \sqrt{3y}} = \frac{2x\sqrt{6xy}}{\sqrt{9y^2}} = \frac{2x\sqrt{6xy}}{3y}$$

(b) $\dfrac{2}{\sqrt[3]{3x}}$

To satisfy the third condition, we must remove the radical from the denominator. For this we need a perfect cube inside the radical in the denominator. Multiplying the numerator and denominator by $\sqrt[3]{9x^2}$ provides the perfect cube.

$$\frac{2\sqrt[3]{9x^2}}{\sqrt[3]{3x} \cdot \sqrt[3]{9x^2}} = \frac{2\sqrt[3]{9x^2}}{\sqrt[3]{27x^3}}$$

$$= \frac{2\sqrt[3]{9x^2}}{3x}$$

NOTE

$\sqrt[3]{9x^2} = \sqrt[3]{3^2x^2}$

so

$\sqrt[3]{3x} \cdot \sqrt[3]{9x^2} = \sqrt[3]{3^3x^3}$

and each exponent is a multiple of 3.

Check Yourself 6

Simplify each expression.

(a) $\sqrt{\dfrac{12a^3}{5b}}$ **(b)** $\dfrac{3}{\sqrt[3]{2w^2}}$

We summarize our work to this point in simplifying radical expressions.

Step by Step

Simplifying Radical Expressions

NOTE

In the case of a cube root, steps 1 and 2 refer to perfect cubes.

Step 1 To satisfy the first condition, determine the largest perfect-square factor of the radicand. Apply the product property to "remove" that factor from inside the radical.

Step 2 To satisfy the second condition, use the quotient property to write the expression in the form

$$\frac{\sqrt{a}}{\sqrt{b}}$$

If b is a perfect square, simplify the radical in the denominator. If not, proceed to step 3.

Step 3 Multiply the numerator and denominator of the radical expression by an appropriate radical to simplify and remove the radical in the denominator. Simplify the resulting expression when necessary.

Check Yourself ANSWERS

1. **(a)** $3\sqrt{5}$; **(b)** $10\sqrt{2}$; **(c)** $5p^2\sqrt{3p}$; **(d)** $7mn^2\sqrt{2m}$ **2.** **(a)** $4w\sqrt[3]{2w}$; **(b)** $2xy^2\sqrt[3]{5x^2y}$; **(c)** $2a^2b\sqrt[4]{3b}$ **3.** **(a)** $\frac{\sqrt{7}}{4}$; **(b)** $\frac{\sqrt{3}}{5a}$; **(c)** $\frac{\sqrt[3]{5x}}{3}$ **4.** **(a)** $\frac{\sqrt{21}}{7}$; **(b)** $\frac{2\sqrt{5y}}{5}$; **(c)** $\frac{3\sqrt{10}}{25}$ **5.** **(a)** $\frac{5\sqrt{3}}{6}$; **(b)** $\frac{\sqrt[3]{6}}{3}$ **6.** **(a)** $\frac{2a\sqrt{15ab}}{5b}$; **(b)** $\frac{3\sqrt[3]{4w}}{2w}$

Reading Your Text

These fill-in-the-blank exercises will help you understand some of the key vocabulary used in this section. The answers to these exercises are in the Answers Appendix in the back of the text.

(a) The root of a product is equal to the _________ of the roots.

(b) We use the _________ property to rewrite expressions that contain fractions under a radical.

(c) To be simplified, radicals cannot remain in the _________ of a fraction.

(d) When a radical remains in the denominator of a fraction, we must _________ the denominator.

Skills | Calculator/Computer | Career Applications | Above and Beyond

7.2 exercises

< Objective 1 >

Simplify each expression. Assume all variables represent positive real numbers.

1. $\sqrt{12}$
2. $\sqrt{24}$
3. $\sqrt{50}$
4. $\sqrt{28}$
5. $-\sqrt{108}$ VIDEO
6. $\sqrt{32}$
7. $\sqrt{52}$
8. $-\sqrt{96}$
9. $\sqrt{60}$
10. $\sqrt{150}$
11. $-\sqrt{125}$
12. $\sqrt{128}$
13. $\sqrt[3]{16}$
14. $\sqrt[3]{-54}$
15. $\sqrt[3]{-48}$
16. $\sqrt[3]{250}$
17. $\sqrt[3]{135}$
18. $\sqrt[3]{-160}$
19. $\sqrt[4]{32}$
20. $\sqrt[4]{96}$
21. $\sqrt{18z^2}$
22. $\sqrt{45a^2}$
23. $\sqrt{63x^4}$
24. $\sqrt{54w^4}$
25. $\sqrt{98m^3}$
26. $\sqrt{75a^5}$ VIDEO
27. $\sqrt{80x^2y^3}$
28. $\sqrt{108p^5q^2}$
29. $\sqrt[3]{40b^3}$
30. $\sqrt[3]{16x^3}$
31. $\sqrt[3]{48p^9}$
32. $\sqrt[3]{-80a^6}$
33. $\sqrt[3]{54m^7}$
34. $\sqrt[3]{250x^{13}}$
35. $\sqrt[3]{56x^6y^5z^4}$
36. $-\sqrt[3]{250a^4b^{15}c^9}$
37. $\sqrt[4]{32x^8}$
38. $\sqrt[4]{96w^5z^{13}}$
39. $\sqrt[4]{128a^{12}b^{17}}$
40. $\sqrt[5]{64w^{10}}$

< Objective 2 >

41. $\sqrt{\frac{5}{16}}$
42. $\sqrt{\frac{a^6}{49}}$
43. $\sqrt{\frac{5}{9y^4}}$
44. $\sqrt{\frac{7}{25x^2}}$
45. $\sqrt[3]{\frac{5}{8}}$
46. $\sqrt[3]{\frac{4x^2}{27}}$ VIDEO

< Objective 3 >

47. $\frac{5}{\sqrt{7}}$
48. $\sqrt{\frac{5}{8}}$
49. $\frac{7}{\sqrt{12}}$
50. $\frac{\sqrt{5}}{\sqrt{11}}$
51. $\frac{2\sqrt{3}}{\sqrt{10}}$
52. $\frac{3\sqrt{5}}{\sqrt{3}}$
53. $\sqrt[3]{\frac{7}{4}}$
54. $\sqrt[3]{\frac{5}{9}}$
55. $\frac{5}{\sqrt[3]{16}}$
56. $\sqrt{\frac{3}{x}}$ VIDEO
57. $\sqrt{\frac{12}{w}}$
58. $\frac{\sqrt{18}}{\sqrt{a}}$
59. $\frac{\sqrt{8m^3}}{\sqrt{5n}}$
60. $\sqrt{\frac{24x^5}{7y}}$
61. $\sqrt[3]{\frac{5}{y}}$
62. $\sqrt[3]{\frac{7}{x^2}}$
63. $\frac{3}{\sqrt[3]{2x}}$
64. $\frac{5}{\sqrt[3]{3a}}$
65. $\sqrt[3]{\frac{2}{5x^2}}$
66. $\sqrt[3]{\frac{5}{7w^2}}$
67. $\frac{\sqrt[3]{5}}{\sqrt[3]{4a^2}}$
68. $\frac{\sqrt[3]{2}}{\sqrt[3]{9m^2}}$
69. $\sqrt[3]{\frac{a^5}{b^7}}$
70. $\sqrt[3]{\frac{w^7}{z^{10}}}$

Use the distance formula (Section 7.1) to find the distance between each pair of points.

71. $(-7, 1)$ and $(0, 0)$
72. $(-18, -5)$ and $(-12, -3)$
73. $(22, -13)$ and $(18, -9)$ VIDEO
74. $(-12, -17)$ and $(-9, -11)$

Determine whether each statement is **true** *or* **false**.

75. $\sqrt{16x^{16}} = 4x^8$
76. $\sqrt{x^2 + y^2} = x + y$
77. $\frac{\sqrt{x^2 - 25}}{\sqrt{x - 5}} = \sqrt{x + 5}$
78. $\sqrt[3]{x^6} \cdot \sqrt[3]{x^3 - 1} = x^2\sqrt[3]{x - 1}$
79. $\sqrt[3]{(8b^6)^2} = (\sqrt[3]{8b^6})^2$
80. $\frac{\sqrt[3]{8x^3}}{\sqrt[3]{2x}} = \sqrt[3]{4x^2}$
81. For nonnegative numbers a and b, $\sqrt{ab} = \sqrt{a}\sqrt{b}$.
82. For nonnegative numbers a and b, $\sqrt{a + b} = \sqrt{a} + \sqrt{b}$.

83. For positive numbers a and b, $\sqrt{\frac{a}{b}} = \frac{\sqrt{a}}{\sqrt{b}}$.

84. For nonnegative numbers a and b, $\sqrt{a - b} = \sqrt{a} - \sqrt{b}$.

Skills | Calculator/Computer | **Career Applications** | Above and Beyond

Mechanical Engineering *In the Chapter 7 Activity, "The Swing of a Pendulum," you worked with the relationship between the period and length of a pendulum. The general model for this relationship is given by*

$$T = k\sqrt{L}$$

in which k is a gravitational constant.

Use this information to complete exercises 85 to 87.

85. Compute the value of k for each given T-value. Report your results to the nearest thousandth.

L	10	20	30	40	50	60	70
T	0.633	0.9	1.1	1.267	1.4	1.533	1.633
k							

86. Find the average (mean) of the values you found for k in exercise 85.

87. Fill in the row for time T in the table below using your own results from the Chapter 7 Activity. Then use your results for T to complete the row for k and find the mean k-value.

L	10	20	30	40	50	60	70
T							
k							

88. Physicists have found that the time of a pendulum's period as a function of its length can be modeled with

$$T = 2\pi\sqrt{\frac{L}{g}}$$

in which g is the gravitational constant $g = 980\ \frac{\text{cm}}{\text{s}^2}$, L is measured in centimeters, and T is in seconds.

(a) Use the properties of radicals to simplify the radical and rewrite it in the form

$$T = k\sqrt{L}$$

Round k to the nearest thousandth.

(b) How does the theoretical k found in part (a) compare with the experimental period found in exercise 86? Exercise 87?

Skills | Calculator/Computer | Career Applications | **Above and Beyond**

Simplify.

89. $\dfrac{7\sqrt{x^2y^4} \cdot \sqrt{36xy}}{6\sqrt{x^{-6}y^{-2}} \cdot \sqrt{49x^{-1}y^{-3}}}$

90. $\dfrac{3\sqrt[3]{32c^{12}d^2} \cdot \sqrt[3]{2c^5d^4}}{4\sqrt[3]{9c^8d^{-2}} \cdot \sqrt[3]{3c^{-3}d^{-4}}}$

Decide whether each expression is simplified. If it is not, explain what needs to be done.

91. $\sqrt{10mn}$

92. $\sqrt{18ab}$

93. $\sqrt{\dfrac{98x^2y}{7x}}$

94. $\dfrac{\sqrt{6xy}}{3x}$

95. Find the area and perimeter of this square:

One of these measures, the area, is a rational number, and the other, the perimeter, is an irrational number. Explain how this happened. Will the area always be a rational number? Explain.

96. **(a)** Evaluate the three expressions $\frac{n^2-1}{2}$, n, $\frac{n^2+1}{2}$ using odd values of n: 1, 3, 5, 7, and so forth. Complete the table shown.

n	$a = \frac{n^2-1}{2}$	$b = n$	$c = \frac{n^2+1}{2}$	a^2	b^2	c^2
1						
3						
5						
7						
9						
11						
13						
15						

(b) Check for each of these sets of three numbers to see whether this statement is true: $\sqrt{a^2+b^2} = \sqrt{c^2}$. For how many of your sets of three did this work? Sets of three numbers for which this statement is true are called *Pythagorean triples* because $a^2 + b^2 = c^2$. Can the radical equation be written as $\sqrt{a^2+b^2} = a + b$? Explain your answer.

97. Explain the difference between a pair of binomials in which the middle sign is changed and the opposite of a binomial. To illustrate, use $4 - \sqrt{7}$.

98. Find the missing binomial.

$(\sqrt{3} - 2)(\qquad) = -1$

99. Use a calculator to evaluate the expression in parts (a) through (d). Round your answers to the nearest hundredth.

(a) $3\sqrt{5} + 4\sqrt{5}$ **(b)** $7\sqrt{5}$ **(c)** $2\sqrt{6} + 3\sqrt{6}$ **(d)** $5\sqrt{6}$

(e) Based on parts (a) through (d), make a conjecture concerning $a\sqrt{m} + b\sqrt{m}$. Check your conjecture on an example of your own that is similar to parts (a) through (d).

Answers

1. $2\sqrt{3}$ **3.** $5\sqrt{2}$ **5.** $-6\sqrt{3}$ **7.** $2\sqrt{13}$ **9.** $2\sqrt{15}$ **11.** $-5\sqrt{5}$ **13.** $2\sqrt[3]{2}$ **15.** $-2\sqrt[3]{6}$ **17.** $3\sqrt[3]{5}$ **19.** $2\sqrt[4]{2}$ **21.** $3z\sqrt{2}$ **23.** $3x^2\sqrt{7}$ **25.** $7m\sqrt{2m}$ **27.** $4xy\sqrt{5y}$ **29.** $2b\sqrt[3]{5}$ **31.** $2p^3\sqrt[3]{6}$ **33.** $3m^2\sqrt[3]{2m}$ **35.** $2x^2yz\sqrt[3]{7y^2z}$ **37.** $2x^2\sqrt[4]{2}$ **39.** $2a^3b^4\sqrt[4]{8b}$ **41.** $\frac{\sqrt{5}}{4}$ **43.** $\frac{\sqrt{5}}{3y^2}$ **45.** $\frac{\sqrt[3]{5}}{2}$ **47.** $\frac{5\sqrt{7}}{7}$ **49.** $\frac{7\sqrt{3}}{6}$ **51.** $\frac{\sqrt{30}}{5}$ **53.** $\frac{\sqrt[3]{14}}{2}$ **55.** $\frac{5\sqrt[3]{4}}{4}$ **57.** $\frac{2\sqrt{3w}}{w}$ **59.** $\frac{2m\sqrt{10mn}}{5n}$ **61.** $\frac{\sqrt[3]{5y^2}}{y}$ **63.** $\frac{3\sqrt[3]{4x^2}}{2x}$ **65.** $\frac{\sqrt[3]{50x}}{5x}$ **67.** $\frac{\sqrt[3]{10a}}{2a}$ **69.** $\frac{a\sqrt[3]{a^2b^2}}{b^3}$ **71.** $5\sqrt{2}$ units **73.** $4\sqrt{2}$ units **75.** True **77.** True **79.** True **81.** True **83.** True **85.**

L	10	20	30	40	50	60	70
T	0.633	0.9	1.1	1.267	1.4	1.533	1.633
k	0.2	0.201	0.201	0.2	0.198	0.198	0.195

87. Answers will vary. **89.** x^5y^5 **91.** Simplified **93.** Remove perfect-square factors from the radical and simplify. **95.** Above and Beyond **97.** Above and Beyond **99.** **(a)** 15.65; **(b)** 15.65; **(c)** 12.25; **(d)** 12.25 **(e)** Above and Beyond

7.3 Operations on Radicals

< 7.3 Objectives >

1 > Add and subtract radical expressions

2 > Multiply radical expressions

3 > Divide radical expressions

Adding and subtracting radical expressions exactly parallels our earlier work with polynomials containing like terms.

To add $3x^2 + 4x^2$, we have

$$3x^2 + 4x^2 = (3 + 4)x^2 = 7x^2$$

RECALL

This uses the distributive property.

Keep in mind that we are able to simplify or combine the above expressions because of like terms in x^2. (Recall that like terms have the same variable factor raised to the same power.)

We *cannot* combine terms such as

$4a^3 + 3a^2$ or $3x - 5y$

By extending these ideas, we conclude that radical expressions can be combined *only* if they are *similar,* that is, if the expressions contain the same radicand with the same index.

Two simplified radicals that have the same index and the same radicand (the expression inside the radical) are called **like radicals.** For example,

$2\sqrt{3}$ and $5\sqrt{3}$ are like radicals.

$\sqrt{2}$ and $\sqrt{5}$ are not like radicals because they have different radicands.

$\sqrt{5}$ and $\sqrt[3]{5}$ are not like radicals because they have different indices (2 and 3, representing a square root and a cube root).

Like radicals can be added (or subtracted) in the same way as like terms. We apply the distributive property and then combine the coefficients.

$$2\sqrt{5} + 3\sqrt{5} = (2 + 3)\sqrt{5} = 5\sqrt{5}$$

Example 1 **Adding and Subtracting Radical Expressions**

< Objective 1 >

Add or subtract as indicated.

(a) $3\sqrt{7} + 2\sqrt{7} = (3 + 2)\sqrt{7} = 5\sqrt{7}$

(b) $7\sqrt{3} - 4\sqrt{3} = (7 - 4)\sqrt{3} = 3\sqrt{3}$

(c) $5\sqrt{10} - 3\sqrt{10} + 2\sqrt{10} = (5 - 3 + 2)\sqrt{10} = 4\sqrt{10}$

(d) $2\sqrt{5} + 3\sqrt{3}$ cannot be combined or further simplified.

(e) $\sqrt{7} + \sqrt[3]{7}$ cannot be simplified.

NOTES

Apply the distributive property.

In (d), the expressions have different radicands, 5 and 3.

In (e), the expressions have different indices, 2 and 3.

(f) $5\sqrt{x} + 2\sqrt{x} = (5 + 2)\sqrt{x}$
$= 7\sqrt{x}$

(g) $5\sqrt{3ab} - 2\sqrt{3ab} + 3\sqrt{3ab} = (5 - 2 + 3)\sqrt{3ab} = 6\sqrt{3ab}$

(h) $\sqrt[3]{3x^2} + \sqrt[3]{3x}$ cannot be simplified. The radicands are *not* the same.

Check Yourself 1

Add or subtract as indicated.

(a) $5\sqrt{3} + 2\sqrt{3}$ **(b)** $7\sqrt{5} - 2\sqrt{5} + 3\sqrt{5}$
(c) $2\sqrt{3} + 3\sqrt{2}$ **(d)** $\sqrt{2y} + 5\sqrt{2y} - 3\sqrt{2y}$
(e) $2\sqrt[3]{3m} - 5\sqrt[3]{3m}$ **(f)** $\sqrt{5x} - \sqrt[3]{5x}$

We may need to simplify radical expressions before combining them. Example 2 illustrates this idea.

Example 2 Adding and Subtracting Radicals

Add or subtract as indicated.

(a) $\sqrt{48} + 2\sqrt{3}$

In this form, the radicals cannot be combined. However, we can simplify the first radical because 16 is a perfect-square factor of 48.

$\sqrt{48} = \sqrt{16 \cdot 3} = 4\sqrt{3}$

With this result we can proceed.

$$\sqrt{48} + 2\sqrt{3} = 4\sqrt{3} + 2\sqrt{3}$$
$$= (4 + 2)\sqrt{3} = 6\sqrt{3}$$

NOTE

$\sqrt{50} = \sqrt{25 \cdot 2}$

$\sqrt{32} = \sqrt{16 \cdot 2}$

$\sqrt{98} = \sqrt{49 \cdot 2}$

(b) $\sqrt{50} - \sqrt{32} + \sqrt{98} = 5\sqrt{2} - 4\sqrt{2} + 7\sqrt{2}$
$= (5 - 4 + 7)\sqrt{2} = 8\sqrt{2}$

(c) $x\sqrt{2x} + 3\sqrt{8x^3}$

We know

$$3\sqrt{8x^3} = 3\sqrt{4x^2 \cdot 2x}$$
$$= 3\sqrt{4x^2} \cdot \sqrt{2x}$$
$$= 3 \cdot 2x\sqrt{2x} = 6x\sqrt{2x}$$

So

$$x\sqrt{2x} + 3\sqrt{8x^3} = x\sqrt{2x} + 6x\sqrt{2x}$$
$$= (x + 6x)\sqrt{2x} = 7x\sqrt{2x}$$

NOTE

$\sqrt[3]{16a} = \sqrt[3]{8 \cdot 2a}$

$\sqrt[3]{54a} = \sqrt[3]{27 \cdot 2a}$

(d) $\sqrt[3]{2a} - \sqrt[3]{16a} + \sqrt[3]{54a} = \sqrt[3]{2a} - 2\sqrt[3]{2a} + 3\sqrt[3]{2a}$
$= 2\sqrt[3]{2a}$

Check Yourself 2

Add or subtract as indicated.

(a) $\sqrt{125} + 3\sqrt{5}$ **(b)** $\sqrt{75} - \sqrt{27} + \sqrt{48}$
(c) $5\sqrt{24y^3} - y\sqrt{6y}$ **(d)** $\sqrt[3]{81x} - \sqrt[3]{3x} + \sqrt[3]{24x}$

We may also need to apply the quotient property before combining rational expressions as shown in Example 3.

Example 3 Adding and Subtracting Radicals

Add or subtract as indicated.

(a) $2\sqrt{6} + \sqrt{\frac{2}{3}}$

NOTE

Multiply by $\frac{\sqrt{3}}{\sqrt{3}}$.

We apply the quotient property to the *second term* and rationalize the denominator.

$$\sqrt{\frac{2}{3}} = \frac{\sqrt{2}}{\sqrt{3}} = \frac{\sqrt{2}\cdot\sqrt{3}}{\sqrt{3}\cdot\sqrt{3}} = \frac{\sqrt{6}}{3} \qquad \frac{\sqrt{3}}{\sqrt{3}} = 1$$

So

$$2\sqrt{6} + \sqrt{\frac{2}{3}}$$

NOTE

Note that $\frac{\sqrt{6}}{3}$ and $\frac{1}{3}\sqrt{6}$ are equivalent.

$$= 2\sqrt{6} + \frac{\sqrt{6}}{3} \qquad \text{Factor out } \sqrt{6} \text{ from these two terms.}$$

$$= \left(2 + \frac{1}{3}\right)\sqrt{6} = \frac{7}{3}\sqrt{6}$$

(b) $\sqrt{20x} - \sqrt{\frac{x}{5}}$

NOTE

$\sqrt{20x} = \sqrt{4\cdot 5x}$
$= \sqrt{4}\sqrt{5x} = 2\sqrt{5x}$

Again, we simplify the two expressions first.

$$\sqrt{20x} - \sqrt{\frac{x}{5}} = 2\sqrt{5x} - \frac{\sqrt{x}\cdot\sqrt{5}}{\sqrt{5}\cdot\sqrt{5}}$$

$$= 2\sqrt{5x} - \frac{\sqrt{5x}}{5}$$

$$= 2\sqrt{5x} - \frac{1}{5}\sqrt{5x}$$

$$= \left(2 - \frac{1}{5}\right)\sqrt{5x} = \frac{9}{5}\sqrt{5x}$$

Check Yourself 3

Add or subtract as indicated.

(a) $3\sqrt{7} + \sqrt{\frac{1}{7}}$ **(b)** $\sqrt{40x} - \sqrt{\frac{2x}{5}}$

As Example 4 illustrates, we may need to do more work when there are fractions.

Example 4 Adding Radicals

Add $\frac{\sqrt{5}}{3} + \frac{2}{\sqrt{5}}$.

First, we rationalize the denominator in the second term

$$\frac{\sqrt{5}}{3} + \frac{2}{\sqrt{5}} = \frac{\sqrt{5}}{3} + \frac{2\sqrt{5}}{\sqrt{5}\cdot\sqrt{5}} \qquad \text{Multiply the second term by } \frac{\sqrt{5}}{\sqrt{5}}.$$

$$= \frac{\sqrt{5}}{3} + \frac{2\sqrt{5}}{5} \qquad \text{Simplify.}$$

The LCD of the fractions is 15, so we rewrite each fraction with that denominator.

$$\frac{\sqrt{5}\cdot 5}{3\cdot 5}+\frac{2\sqrt{5}\cdot 3}{5\cdot 3}=\frac{5\sqrt{5}+6\sqrt{5}}{15}$$
$$=\frac{11\sqrt{5}}{15}$$

Check Yourself 4

Subtract $\frac{3}{\sqrt{10}}-\frac{\sqrt{10}}{5}$.

The product and quotient properties give us ways to multiply and divide radical expressions.

Property

Multiplying Radicals

The product of roots is equal to the root of the product.

$$\sqrt[n]{a}\cdot\sqrt[n]{b}=\sqrt[n]{ab}$$

Note: If n is an even integer, a and b must be positive real numbers.

NOTE

This is simply the product property (Section 7.2)

$\sqrt[n]{ab}=\sqrt[n]{a}\cdot\sqrt[n]{b}$

written in reverse.

We illustrate the multiplication property in Example 5. We assume that all variables represent positive real numbers.

Example 5 Multiplying Radicals

< Objective 2 >

Multiply.

(a) $\sqrt{7}\cdot\sqrt{5}=\sqrt{7\cdot 5}=\sqrt{35}$

(b) $\sqrt{3x}\cdot\sqrt{10y}=\sqrt{3x\cdot 10y}$
$=\sqrt{30xy}$

(c) $\sqrt[3]{4x}\cdot\sqrt[3]{7x}=\sqrt[3]{4x\cdot 7x}$
$=\sqrt[3]{28x^2}$

NOTE

Just multiply the radicands.

Check Yourself 5

Multiply.

(a) $\sqrt{6}\cdot\sqrt{7}$ (b) $\sqrt{5a}\cdot\sqrt{11b}$ (c) $\sqrt[3]{3y}\cdot\sqrt[3]{5y}$

Keep in mind that all radical expressions should be simplified. Often we have to simplify a product once it has been formed.

Example 6 Multiplying Radicals

> **NOTE**
>
> $\sqrt{18}$ is *not* simplified. 9 is a perfect-square factor of 18.

Multiply and simplify.

(a) $\sqrt{3} \cdot \sqrt{6} = \sqrt{18}$

$= \sqrt{9 \cdot 2} = \sqrt{9}\sqrt{2}$

$= 3\sqrt{2}$

(b) $\sqrt{5x}\sqrt{15x} = \sqrt{5x \cdot 15x}$ — Multiply the radicals.

$= \sqrt{75x^2}$

$= \sqrt{25x^2 \cdot 3}$ — Factor: $25x^2$ is a perfect square.

$= \sqrt{25x^2} \cdot \sqrt{3}$ — Use the product property to simplify.

$= 5x\sqrt{3}$ — $(5x)^2 = 25x^2$

> **NOTE**
>
> In (c), we want a perfect-cube factor.

(c) $\sqrt[3]{4a^2b} \cdot \sqrt[3]{10a^2b^2} = \sqrt[3]{40a^4b^3}$ — Multiply the radicands.

$= \sqrt[3]{8a^3b^3 \cdot 5a}$ — Factor: $8a^3b^3$ is a perfect cube.

$= \sqrt[3]{8a^3b^3} \cdot \sqrt[3]{5a}$

$= 2ab\sqrt[3]{5a}$ — $(2ab)^3 = 8a^3b^3$

Check Yourself 6

Multiply and simplify.

(a) $\sqrt{10} \cdot \sqrt{20}$ **(b)** $\sqrt{6x} \cdot \sqrt{15x}$ **(c)** $\sqrt[3]{9p^2q^2} \cdot \sqrt[3]{6pq^2}$

We are now ready to combine multiplication with the techniques for adding and subtracting radicals. This allows us to multiply radical expressions with more than one term. Consider Examples 7 and 8.

Example 7 Using the Distributive Property

> **NOTES**
>
> We distribute $\sqrt{2}$ over the sum $\sqrt{5} + \sqrt{7}$ to multiply.
>
> Distribute $\sqrt{3}$.
> $\sqrt{18} = \sqrt{9 \cdot 2} = 3\sqrt{2}$
> $\sqrt{45} = \sqrt{9 \cdot 5} = 3\sqrt{5}$
>
> Alternatively, we could simplify $\sqrt{8x}$ in the original expression as our first step. We leave it to the reader to verify that the result is the same.

Multiply and simplify.

(a) $\sqrt{2}(\sqrt{5} + \sqrt{7})$

Distributing $\sqrt{2}$, we have

$\sqrt{2} \cdot \sqrt{5} + \sqrt{2} \cdot \sqrt{7} = \sqrt{10} + \sqrt{14}$

The expression cannot be simplified further.

(b) $\sqrt{3}(\sqrt{6} + 2\sqrt{15}) = \sqrt{3} \cdot \sqrt{6} + \sqrt{3} \cdot 2\sqrt{15}$

$= \sqrt{18} + 2\sqrt{45}$

$= 3\sqrt{2} + 6\sqrt{5}$

(c) $\sqrt{x}(\sqrt{2x} + \sqrt{8x}) = \sqrt{x} \cdot \sqrt{2x} + \sqrt{x} \cdot \sqrt{8x}$

$= \sqrt{2x^2} + \sqrt{8x^2}$

$= x\sqrt{2} + 2x\sqrt{2} = 3x\sqrt{2}$

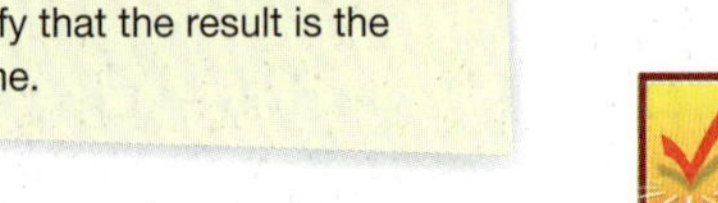

Check Yourself 7

Multiply and simplify.

(a) $\sqrt{3}(\sqrt{10} + \sqrt{2})$ **(b)** $\sqrt{2}(3 + 2\sqrt{6})$ **(c)** $\sqrt{a}(\sqrt{3a} + \sqrt{12a})$

If both of the radical expressions involved in a multiplication have two terms, we must apply the patterns for multiplying polynomials. Here is an example to illustrate.

Example 8 Multiplying Radical Binomials

Multiply and simplify.

(a) $(\sqrt{3} + 1)(\sqrt{3} + 5)$

To write the desired product, we use the FOIL pattern for multiplying binomials.

$$(\sqrt{3} + 1)(\sqrt{3} + 5)$$

$$= \underbrace{\sqrt{3} \cdot \sqrt{3}}_{\text{First}} + \underbrace{5 \cdot \sqrt{3}}_{\text{Outer}} + \underbrace{1 \cdot \sqrt{3}}_{\text{Inner}} + \underbrace{1 \cdot 5}_{\text{Last}}$$

$$= 3 + 6\sqrt{3} + 5$$

$$= 8 + 6\sqrt{3}$$

NOTE

Combine the outer and inner products.

(b) $(\sqrt{6} + \sqrt{2})(\sqrt{6} - \sqrt{2})$

Multiplying as before, we have

$$\sqrt{6} \cdot \sqrt{6} - \sqrt{6} \cdot \sqrt{2} + \sqrt{6} \cdot \sqrt{2} - \sqrt{2} \cdot \sqrt{2} = 6 + 0 - 2 = 4$$

NOTE

The product $(a + b)(a - b)$ is $a^2 - b^2$

When a and b are square roots, the product is rational.

Two binomial radical expressions that differ *only* in the sign of the second term are called *conjugates* of each other. So

$\sqrt{6} + \sqrt{2}$ and $\sqrt{6} - \sqrt{2}$

are conjugates, and their product does *not* contain a radical—the product is a rational number. That is always the case with two conjugates and has particular significance later in this section.

(c) $(\sqrt{2} + \sqrt{5})^2 = (\sqrt{2} + \sqrt{5})(\sqrt{2} + \sqrt{5})$

Multiplying as before, we have

$$(\sqrt{2} + \sqrt{5})^2 = \sqrt{2} \cdot \sqrt{2} + \sqrt{2} \cdot \sqrt{5} + \sqrt{2} \cdot \sqrt{5} + \sqrt{5} \cdot \sqrt{5}$$
$$= 7 + 2\sqrt{10}$$

NOTE

We apply the multiplication pattern for binomials.

$(\sqrt{2} + \sqrt{5})^2$ can also be handled using our formula for the square of a binomial

$$(a + b)^2 = a^2 + 2ab + b^2$$

in which $a = \sqrt{2}$ and $b = \sqrt{5}$.

Check Yourself 8

Multiply and simplify.

(a) $(\sqrt{2} + 3)(\sqrt{2} + 5)$ **(b)** $(\sqrt{5} - \sqrt{3})(\sqrt{5} + \sqrt{3})$ **(c)** $(\sqrt{7} - \sqrt{3})^2$

We are now ready to state the basic property for dividing radical expressions. Again, it is simply a restatement of our earlier quotient property.

Property

Dividing Radicals

The quotient of roots is equal to the root of the quotient.

$$\frac{\sqrt[n]{a}}{\sqrt[n]{b}} = \sqrt[n]{\frac{a}{b}}$$

Note: If n is an even integer, a and b must be positive real numbers.

NOTE

This is simply the quotient property (Section 7.2)

$$\sqrt[n]{\frac{a}{b}} = \frac{\sqrt[n]{a}}{\sqrt[n]{b}}$$

written in reverse.

Dividing rational expressions is most often carried out by rationalizing the denominator. This process can be separated into two types of problems: those with a monomial divisor and those with binomial divisors.

Example 9 Dividing Radical Expressions

< Objective 3 >

Simplify each expression. Assume that all variables represent positive real numbers.

(a) $\frac{3}{\sqrt{5}} = \frac{3 \cdot \sqrt{5}}{\sqrt{5} \cdot \sqrt{5}} = \frac{3\sqrt{5}}{5}$

We multiply the numerator and denominator by $\sqrt{5}$ to rationalize the denominator.

(b) $\frac{\sqrt{7x}}{\sqrt{10y}} = \frac{\sqrt{7x} \cdot \sqrt{10y}}{\sqrt{10y} \cdot \sqrt{10y}}$

$= \frac{\sqrt{70xy}}{10y}$

NOTE

$\sqrt[3]{2} \cdot \sqrt[3]{4} = \sqrt[3]{2} \cdot \sqrt[3]{2^2}$

$= \sqrt[3]{2^3}$

$= 2$

(c) $\frac{3}{\sqrt[3]{2}} = \frac{3\sqrt[3]{4}}{\sqrt[3]{2} \cdot \sqrt[3]{4}}$

$= \frac{3\sqrt[3]{4}}{2}$

In this case we want a perfect cube in the denominator, so we multiply the numerator and denominator by $\sqrt[3]{4}$.

These division problems are similar to those we saw in Section 7.2 when we simplified radical expressions. They are shown here to illustrate this case of division with radicals.

Check Yourself 9

Simplify each expression.

(a) $\frac{5}{\sqrt{7}}$ **(b)** $\frac{\sqrt{3a}}{\sqrt{5b}}$ **(c)** $\frac{5}{\sqrt[3]{9}}$

The division property is particularly useful when the radicands in the numerator and denominator have common factors. Consider Example 10.

Example 10 Dividing Radicals

NOTE

5 is a common factor of the radicands in the numerator and denominator.

Simplify

$$\frac{\sqrt{10}}{\sqrt{15a}}$$

We apply the division property so that the radicand can be reduced as a fraction.

$$\frac{\sqrt{10}}{\sqrt{15a}} = \sqrt{\frac{10}{15a}} = \sqrt{\frac{2}{3a}}$$ In the radicand, divide numerator and denominator by 5.

Now we use the quotient property and rationalize the denominator.

$$\sqrt{\frac{2}{3a}} = \frac{\sqrt{2}}{\sqrt{3a}} = \frac{\sqrt{2}\cdot\sqrt{3a}}{\sqrt{3a}\cdot\sqrt{3a}}$$ Multiply numerator and denominator by $\sqrt{3a}$.

$$= \frac{\sqrt{6a}}{3a}$$ Use the multiplication property in the numerator and in the denominator.

Check Yourself 10

Simplify $\dfrac{\sqrt{15}}{\sqrt{18x}}$.

We now turn our attention to a second type of division problem involving radical expressions. Here the divisors (the denominators) are binomials. This uses the idea of conjugates that we saw in Example 8.

Example 11 Rationalizing Denominators

Rationalize each denominator.

(a) $\dfrac{6}{\sqrt{6}+\sqrt{2}}$

The product of the denominator, $\sqrt{6}+\sqrt{2}$ and its *conjugate*, $\sqrt{6}-\sqrt{2}$ is always rational.

NOTE

If there is a sum or difference containing a radical in the denominator, multiply the numerator and denominator by the conjugate of the denominator. This rationalizes the denominator.

$$\frac{6}{\sqrt{6}+\sqrt{2}}\cdot\frac{\sqrt{6}-\sqrt{2}}{\sqrt{6}-\sqrt{2}} = \frac{6(\sqrt{6}-\sqrt{2})}{(\sqrt{6}+\sqrt{2})(\sqrt{6}-\sqrt{2})}$$ Multiply by the conjugate of the denominator.

$$= \frac{6(\sqrt{6}-\sqrt{2})}{(\sqrt{6})^2-(\sqrt{2})^2}$$ $(a+b)(a-b) = a^2 - b^2$

$$= \frac{6(\sqrt{6}-\sqrt{2})}{6-2}$$ $(\sqrt{6})^2 = 6$; $(\sqrt{2})^2 = 2$

$$= \frac{6(\sqrt{6}-\sqrt{2})}{4}$$

$$= \frac{3(\sqrt{6}-\sqrt{2})}{2}$$ Simplify; we could also write $\frac{3}{2}(\sqrt{6}-\sqrt{2})$

(b) $\dfrac{\sqrt{5}+\sqrt{3}}{\sqrt{5}-\sqrt{3}}$

We need to multiply the numerator and denominator by the conjugate of the denominator. The denominator is $\sqrt{5}-\sqrt{3}$, so its conjugate is $\sqrt{5}+\sqrt{3}$.

$$\frac{\sqrt{5}+\sqrt{3}}{\sqrt{5}-\sqrt{3}}\cdot\frac{\sqrt{5}+\sqrt{3}}{\sqrt{5}+\sqrt{3}} = \frac{(\sqrt{5}+\sqrt{3})(\sqrt{5}+\sqrt{3})}{(\sqrt{5}-\sqrt{3})(\sqrt{5}+\sqrt{3})}$$ Multiply by the conjugate of the denominator.

$$= \frac{5+\sqrt{15}+\sqrt{15}+3}{5-3}$$ FOIL in the numerator; $(a+b)(a-b) = a^2 - b^2$

$$= \frac{8+2\sqrt{15}}{2}$$ Combine like terms.

$$= \frac{2(4+\sqrt{15})}{2}$$ Factor: $8 + 2\sqrt{15} = 2(4+\sqrt{15})$

$$= 4+\sqrt{15}$$ Simplify.

Check Yourself 11

Rationalize each denominator.

(a) $\dfrac{4}{\sqrt{3} - \sqrt{2}}$ (b) $\dfrac{\sqrt{6} + \sqrt{3}}{\sqrt{6} - \sqrt{3}}$

Check Yourself ANSWERS

1. (a) $7\sqrt{3}$; (b) $8\sqrt{5}$; (c) cannot be simplified; (d) $3\sqrt{2y}$; (e) $-3\sqrt[3]{3m}$; (f) cannot be simplified
2. (a) $8\sqrt{5}$; (b) $6\sqrt{3}$; (c) $9y\sqrt{6y}$; (d) $4\sqrt[3]{3x}$ 3. (a) $\frac{22}{7}\sqrt{7}$; (b) $\frac{9}{5}\sqrt{10x}$ 4. $\frac{\sqrt{10}}{10}$
5. (a) $\sqrt{42}$; (b) $\sqrt{55ab}$; (c) $\sqrt[3]{15y^2}$ 6. (a) $10\sqrt{2}$; (b) $3x\sqrt{10}$; (c) $3pq\sqrt[3]{2q}$
7. (a) $\sqrt{30} + \sqrt{6}$; (b) $3\sqrt{2} + 4\sqrt{3}$; (c) $3a\sqrt{3}$ 8. (a) $17 + 8\sqrt{2}$; (b) 2; (c) $10 - 2\sqrt{21}$
9. (a) $\frac{5\sqrt{7}}{7}$; (b) $\frac{\sqrt{15ab}}{5b}$; (c) $\frac{5\sqrt[3]{3}}{3}$ 10. $\frac{\sqrt{30x}}{6x}$ 11. (a) $4(\sqrt{3} + \sqrt{2})$; (b) $3 + 2\sqrt{2}$

Reading Your Text

These fill-in-the-blank exercises will help you understand some of the key vocabulary used in this section. The answers to these exercises are in the Answers Appendix in the back of the text.

(a) Radical expressions with the same indices and the same radicands are called ________ radicals.

(b) We use the ________ property to combine like radicals.

(c) The product of roots is equal to the ________ of the product.

(d) Binomial radical expressions that differ only in the sign of the second term are called ________.

7.3 exercises

Skills | Calculator/Computer | Career Applications | Above and Beyond

< Objective 1 >

Add or subtract as indicated. Assume that all variables represent positive real numbers.

1. $3\sqrt{5} + 4\sqrt{5}$
2. $5\sqrt{6} + 3\sqrt{6}$
3. $11\sqrt{3a} - 8\sqrt{3a}$
4. $2\sqrt{5w} + 3\sqrt{5w}$
5. $7\sqrt{m} + 6\sqrt{n}$
6. $8\sqrt{a} - 6\sqrt{b}$
7. $2\sqrt[3]{2} + 7\sqrt[3]{2}$
8. $5\sqrt[4]{3} - 2\sqrt[4]{3}$
9. $8\sqrt{6} - 2\sqrt{6} + 3\sqrt{6}$
10. $8\sqrt{3} + 2\sqrt{3} - 7\sqrt{3}$

Simplify the radical expressions when necessary. Then add or subtract as indicated. Assume that all variables represent positive real numbers.

11. $\sqrt{20} + \sqrt{5}$
12. $\sqrt{27} + \sqrt{3}$
13. $4\sqrt{28} - \sqrt{63}$
14. $2\sqrt{40} + \sqrt{90}$
15. $\sqrt{98} - \sqrt{18} + \sqrt{8}$
16. $\sqrt{108} - \sqrt{27} + \sqrt{75}$
17. $\sqrt[3]{81} + \sqrt[3]{3}$
18. $\sqrt[3]{16} - \sqrt[3]{2}$
19. $\sqrt{54w} - \sqrt{24w}$
20. $\sqrt{27p} + \sqrt{75p}$
21. $\sqrt{18x^3} + \sqrt{8x^3}$
22. $\sqrt{125y^3} - \sqrt{20y^3}$

23. $\sqrt[3]{16w^5} + 2w\sqrt[3]{2w^2} - \sqrt[3]{2w^5}$

24. $\sqrt[4]{2z^7} - z\sqrt[4]{32z^3} + \sqrt[4]{162z^7}$

25. $\sqrt{3} + \sqrt{\frac{1}{3}}$

26. $\sqrt{6} - \sqrt{\frac{1}{6}}$

27. $\sqrt[3]{48} - \sqrt[3]{\frac{3}{4}}$

28. $\sqrt[3]{96} + \sqrt[3]{\frac{4}{9}}$

29. $\frac{\sqrt{6}}{2} + \frac{1}{\sqrt{6}}$

30. $\frac{\sqrt{10}}{2} - \frac{1}{\sqrt{10}}$

31. $\frac{\sqrt{12}}{3} - \frac{1}{\sqrt{3}}$ VIDEO

32. $\frac{\sqrt{20}}{5} + \frac{2}{\sqrt{5}}$

< Objective 2 >

Multiply each expression and simplify.

33. $\sqrt{a} \cdot \sqrt{11}$

34. $\sqrt{10} \cdot \sqrt{w}$

35. $\sqrt{3} \cdot \sqrt{7} \cdot \sqrt{2}$

36. $\sqrt{5} \cdot \sqrt{7} \cdot \sqrt{3}$

37. $\sqrt[3]{4} \cdot \sqrt[3]{9}$

38. $\sqrt[3]{5} \cdot \sqrt[3]{7}$

39. $\sqrt{3} \cdot \sqrt{12}$

40. $\sqrt{5} \cdot \sqrt{20}$

41. $\sqrt[3]{9p^2} \cdot \sqrt[3]{6p}$

42. $\sqrt[3]{25x^2} \cdot \sqrt[3]{10x^2}$

43. $\sqrt[3]{4x^2y} \cdot \sqrt[3]{10xy^3}$ VIDEO

44. $\sqrt[3]{18r^2s^2} \cdot \sqrt[3]{9r^2s}$

45. $\sqrt{2}(\sqrt{3} + 5)$

46. $\sqrt{3}(\sqrt{5} - 7)$

47. $\sqrt{3}(5\sqrt{2} - \sqrt{18})$

48. $\sqrt{2}(2\sqrt{10} + \sqrt{40})$

49. $\sqrt{x}(\sqrt{3x} + \sqrt{27x})$

50. $\sqrt{y}(\sqrt{8y} - \sqrt{2y})$

51. $\sqrt[3]{4}(\sqrt[3]{4} + \sqrt[3]{32})$

52. $\sqrt[3]{6}(\sqrt[3]{32} - \sqrt[3]{4})$

53. $(\sqrt{2} + 3)(\sqrt{2} - 4)$

54. $(\sqrt{3} - 1)(\sqrt{3} + 5)$

55. $(\sqrt{2} + 3\sqrt{5})(\sqrt{2} - 2\sqrt{5})$

56. $(\sqrt{6} - 2\sqrt{3})(\sqrt{6} - 3\sqrt{3})$

57. $(\sqrt{5} + 3)(\sqrt{5} - 3)$

58. $(\sqrt{10} + 2)(\sqrt{10} - 2)$

59. $(\sqrt{a} + \sqrt{3})(\sqrt{a} - \sqrt{3})$

60. $(\sqrt{m} + \sqrt{7})(\sqrt{m} - \sqrt{7})$

61. $(\sqrt{3} - 5)^2$

62. $(\sqrt{5} + \sqrt{2})^2$

63. $(\sqrt{a} + 3)^2$

64. $(\sqrt{x} - 4)^2$

65. $(\sqrt{x} + \sqrt{y})^2$

66. $(\sqrt{r} - \sqrt{s})^2$

< Objective 3 >

Rationalize the denominator in each expression. Simplify when necessary.

67. $\frac{\sqrt{3}}{\sqrt{7}}$

68. $\frac{\sqrt{5}}{\sqrt{3}}$

69. $\frac{\sqrt{2a}}{\sqrt{3b}}$

70. $\frac{\sqrt{5x}}{\sqrt{6y}}$

71. $\frac{3}{\sqrt[3]{4}}$

72. $\frac{2}{\sqrt[3]{9}}$

73. $\frac{1}{2 + \sqrt{3}}$

74. $\frac{2}{3 - \sqrt{2}}$

75. $\frac{8}{3 - \sqrt{5}}$

76. $\frac{20}{4 + \sqrt{6}}$

77. $\frac{\sqrt{6} + \sqrt{3}}{\sqrt{6} - \sqrt{3}}$ VIDEO

78. $\frac{\sqrt{7} - \sqrt{5}}{\sqrt{7} + \sqrt{5}}$

79. $\frac{\sqrt{w} + 3}{\sqrt{w} - 3}$

80. $\frac{\sqrt{x} - 5}{\sqrt{x} + 5}$

81. $\frac{\sqrt{x} - \sqrt{y}}{\sqrt{x} + \sqrt{y}}$

82. $\frac{\sqrt{m} + \sqrt{n}}{\sqrt{m} - \sqrt{n}}$

Simplify each expression.

83. $x\sqrt[3]{8x^4} + 4\sqrt[3]{27x^7}$

84. $\sqrt[3]{8x^2} - \sqrt[3]{27x^2}$

85. $\frac{\sqrt{2x^2 + 3x}}{\sqrt{x}}$

86. $\frac{\sqrt{x^2 - 9}}{\sqrt{x + 3}}$

87. **Geometry** Find the perimeter of the triangle shown in the figure. Write your answer in radical form.

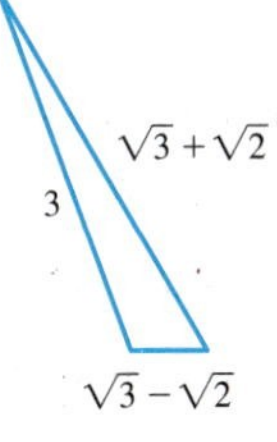

88. GEOMETRY Find the perimeter of the triangle shown in the figure. Write your answer in radical form.

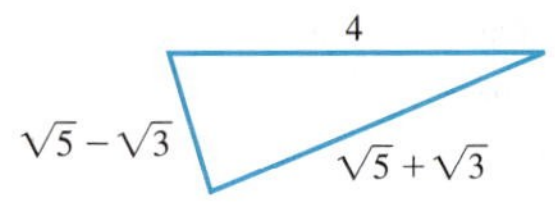

89. GEOMETRY Consider the rectangle with dimensions given in feet.

(a) Find the perimeter of the rectangle.

(b) Find the area of the rectangle.

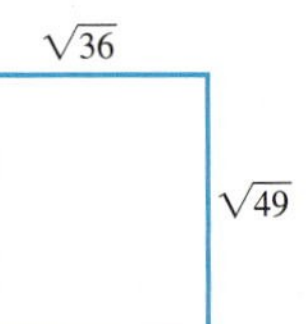

90. GEOMETRY Consider the rectangle with dimensions given in feet.

(a) Find the perimeter of the rectangle. Write your result in radical form.

(b) Find the area of the rectangle.

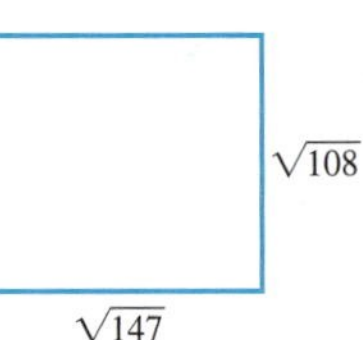

91. GEOMETRY Consider the rectangle with dimensions given in feet.

(a) Find the perimeter of the rectangle. Write your result in radical form.

(b) Find the area of the rectangle. Write your result in radical form.

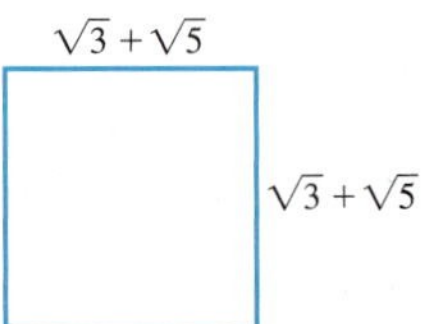

92. GEOMETRY Consider the rectangle with dimensions given in feet.

(a) Find the perimeter of the rectangle. Write your result in radical form.

(b) Find the area of the rectangle. Write your result in radical form.

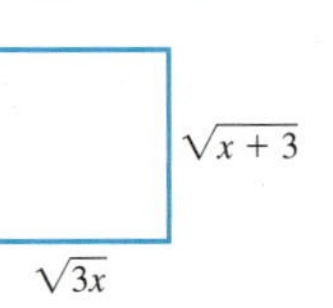

Determine whether each statement is **true** *or* **false.**

93. Two radical expressions can be added only if they have the same radicand and the same index.

94. Two radical expressions can be multiplied only if they have the same radicand and the same index.

95. Unlike radicals can sometimes be added or subtracted.

96. To add or subtract like radicals, we use the distributive property.

97. The square root of a quotient is equal to the quotient of the square roots.

98. The square root of a sum is equal to the sum of the square roots.

Complete each statement with **always, sometimes,** *or* **never.**

99. Conjugate square-root expressions __________ have a product that contains no radical sign.

100. When we multiply two radicals, the product is __________ rational.

Skills | **Calculator/Computer** | Career Applications | Above and Beyond

Use a calculator to approximate each expression. Round your answer to the nearest hundredth.

101. $\sqrt{3}-\sqrt{2}$ **102.** $\sqrt{7}+\sqrt{11}$ **103.** $\sqrt{5}+\sqrt{3}$ **104.** $\sqrt{17}-\sqrt{13}$

105. $4\sqrt{3}-7\sqrt{5}$ **106.** $8\sqrt{2}+3\sqrt{7}$ **107.** $5\sqrt{7}+8\sqrt{13}$ **108.** $7\sqrt{2}-4\sqrt{11}$

Skills | Calculator/Computer | **Career Applications** | Above and Beyond

Mechanical Engineering *In the Chapter 7 Activity, "The Swing of a Pendulum," you worked with the relationship between the period and length of a pendulum. The general model for this relationship is given by*

$$T = 2\pi\sqrt{\frac{L}{g}}$$

in which g is a gravitational constant.

Use this information to complete exercises 109 and 110.

109. If we measure T in seconds and L in feet, then $g = 32\,\frac{\text{ft}}{s^2}$. Use this value for g and simplify the pendulum period function (rationalize the denominator).

110. If L is measured in inches and T in seconds, then g must be expressed differently. Use this new value for g and simplify the pendulum period function.

Skills | Calculator/Computer | Career Applications | **Above and Beyond**

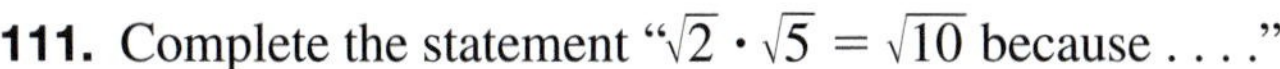

111. Complete the statement "$\sqrt{2} \cdot \sqrt{5} = \sqrt{10}$ because . . . ."

112. Explain why $2\sqrt{3} + 5\sqrt{3} = 7\sqrt{3}$ but $7\sqrt{3} + 3\sqrt{5} \neq 10\sqrt{8}$.

113. When you look out over an unobstructed landscape or seascape, the distance to the visible horizon depends on your height above the ground. The equation

$$d = \sqrt{\frac{3}{2}h}$$

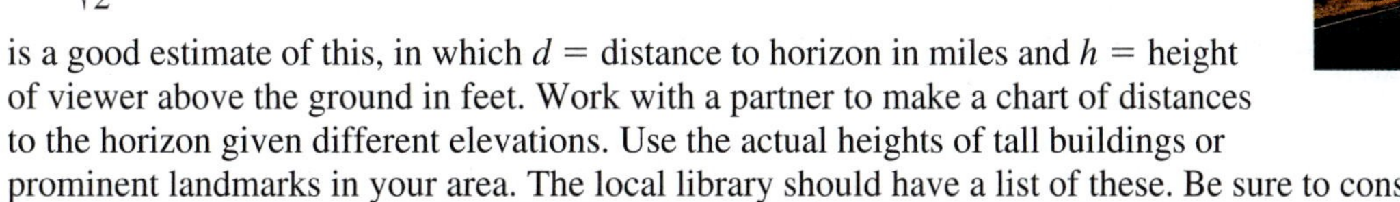

is a good estimate of this, in which d = distance to horizon in miles and h = height of viewer above the ground in feet. Work with a partner to make a chart of distances to the horizon given different elevations. Use the actual heights of tall buildings or prominent landmarks in your area. The local library should have a list of these. Be sure to consider the view to the horizon you get when flying in a plane. What would your elevation have to be to see from one side of your city or town to the other? From one side of your state or county to the other?

Answers

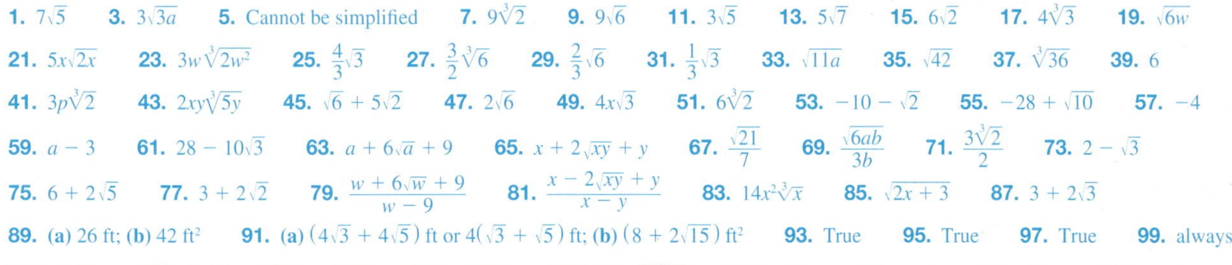

1. $7\sqrt{5}$ **3.** $3\sqrt{3a}$ **5.** Cannot be simplified **7.** $9\sqrt[3]{2}$ **9.** $9\sqrt{6}$ **11.** $3\sqrt{5}$ **13.** $5\sqrt{7}$ **15.** $6\sqrt{2}$ **17.** $4\sqrt[3]{3}$ **19.** $\sqrt{6w}$ **21.** $5x\sqrt{2x}$ **23.** $3w\sqrt[3]{2w^2}$ **25.** $\frac{4}{3}\sqrt{3}$ **27.** $\frac{3}{2}\sqrt[3]{6}$ **29.** $\frac{2}{3}\sqrt{6}$ **31.** $\frac{1}{3}\sqrt{3}$ **33.** $\sqrt{11a}$ **35.** $\sqrt{42}$ **37.** $\sqrt[3]{36}$ **39.** 6 **41.** $3p\sqrt[3]{2}$ **43.** $2xy\sqrt[3]{5y}$ **45.** $\sqrt{6} + 5\sqrt{2}$ **47.** $2\sqrt{6}$ **49.** $4x\sqrt{3}$ **51.** $6\sqrt[3]{2}$ **53.** $-10 - \sqrt{2}$ **55.** $-28 + \sqrt{10}$ **57.** -4 **59.** $a - 3$ **61.** $28 - 10\sqrt{3}$ **63.** $a + 6\sqrt{a} + 9$ **65.** $x + 2\sqrt{xy} + y$ **67.** $\frac{\sqrt{21}}{7}$ **69.** $\frac{\sqrt{6ab}}{3b}$ **71.** $\frac{3\sqrt[3]{2}}{2}$ **73.** $2 - \sqrt{3}$ **75.** $6 + 2\sqrt{5}$ **77.** $3 + 2\sqrt{2}$ **79.** $\frac{w + 6\sqrt{w} + 9}{w - 9}$ **81.** $\frac{x - 2\sqrt{xy} + y}{x - y}$ **83.** $14x^2\sqrt[3]{x}$ **85.** $\sqrt{2x + 3}$ **87.** $3 + 2\sqrt{3}$ **89.** (a) 26 ft; (b) 42 ft^2 **91.** (a) $(4\sqrt{3} + 4\sqrt{5})$ ft or $4(\sqrt{3} + \sqrt{5})$ ft; (b) $(8 + 2\sqrt{15})$ ft^2 **93.** True **95.** True **97.** True **99.** always **101.** 0.32 **103.** 3.97 **105.** -8.72 **107.** 42.07 **109.** $T = \frac{\pi}{4}\sqrt{2L}$ **111.** Above and Beyond **113.** Above and Beyond

7.4 Solving Radical Equations

< 7.4 Objectives >

1 > Solve an equation containing a radical expression

2 > Solve an equation containing two radicals

3 > Solve an equation containing a cube root

We are now ready to look at equations containing radical expressions or **radical equations**. The basic idea is to raise both sides of the equation to a power in order to remove the radical, leaving us with an equation that we already know how to solve.

When raising the sides of an equation to some power, we need to be cautious. While this technique allows us to find potential solutions, it also tends to introduce *extraneous solutions*. It is imperative that we check our results when solving radical equations.

For example, suppose we begin with the equation $x = 1$. Squaring both sides gives us

$$x^2 = 1$$
$$x^2 - 1 = 0$$
$$(x + 1)(x - 1) = 0$$

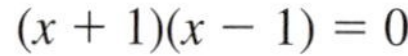

so the solutions appear to be 1 and -1.

Clearly -1 is not a solution to the original equation, $x = 1$; -1 is an extraneous solution.

We must be aware of the possibility of extraneous solutions anytime we raise both sides of an equation to any *even power*. Having said that, we are now prepared to introduce the power property of equality.

> CAUTION

Always check your answers when applying the power property to solve an equation that contains radical expressions.

Property

The Power Property of Equality

Given any two expressions a and b and any positive integer n,

if $a = b$ then $a^n = b^n$

While you never lose a solution by applying the power property, you often gain an extraneous one. Because of this, it is very important that you *check all solutions* when you use the power property to solve an equation.

Example 1 Solving a Radical Equation

< Objective 1 >

Solve $\sqrt{x + 2} = 3$.

Squaring each side, we have

$$(\sqrt{x + 2})^2 = 3^2$$
$$x + 2 = 9$$
$$x = 7$$

NOTE

$(\sqrt{x + 2})^2 = x + 2$

Squaring both sides of the equation removes the radical.

Check
Substituting 7 into the original equation, we find

$$\sqrt{7+2} \stackrel{?}{=} 3$$
$$\sqrt{9} \stackrel{?}{=} 3$$
$$3 = 3$$

Because this is a true statement, 7 is the solution to the equation.

Check Yourself 1

Solve the equation $\sqrt{x-5} = 4$.

Example 2 Solving a Radical Equation

NOTE

Applying the power property only removes the radical if that radical is isolated on one side of the equation.

Solve $\sqrt{4x+5} + 1 = 0$.

We must *first isolate the radical* on one side.

$$\sqrt{4x+5} = -1$$

At this point, we should recognize that there are no solutions to this equation. The radical sign means "the positive square root of," so its value cannot be -1. If we fail to notice this, we would continue as follows.

Squaring both sides, we have

$$(\sqrt{4x+5})^2 = (-1)^2$$
$$4x + 5 = 1$$

and solving for x, we find that

$$x = -1$$

Now we check the solution by substituting -1 for x in the original equation.

NOTE

This is clearly a false statement, so -1 is *not* a solution to the original equation.

Check

$$\sqrt{4(-1)+5} + 1 \stackrel{?}{=} 0$$
$$\sqrt{1} + 1 \stackrel{?}{=} 0$$
$$2 \neq 0$$

Because -1 is an extraneous solution, there are *no solutions* to the original equation.

Check Yourself 2

Solve $\sqrt{3x-2} + 2 = 0$.

Now consider an example where we have to square a binomial.

Example 3 Solving a Radical Equation

NOTE

These problems can also be solved graphically. With a graphing utility, plot the two graphs $Y_1 = \sqrt{x+3}$ and $Y_2 = x + 1$. The graphs have one point of intersection, where $x = 1$.

Solve $\sqrt{x+3} = x + 1$.

We square each side, as before.

$$(\sqrt{x+3})^2 = (x+1)^2$$
$$x + 3 = x^2 + 2x + 1$$

We use the FOIL pattern to square the binomial on the right.

Simplifying this gives us the quadratic equation

$$x^2 + x - 2 = 0$$

NOTE

We solved similar equations in Section 6.6.

Factoring, we have

$(x - 1)(x + 2) = 0$

which gives us two possible solutions

$x = 1 \quad \text{or} \quad x = -2$

These are the only two possible solutions to the original equation, but we do not know if either of them are actually solutions or if we introduced extraneous solutions through our algebraic techniques.

Check

$\sqrt{x + 3} = x + 1$ — Always return to the original equation to check.

$\sqrt{(1) + 3} \stackrel{?}{=} (1) + 1$ — Check the solution $x = 1$.

$\sqrt{4} \stackrel{?}{=} 2$

$2 = 2$ — True: 1 is a solution.

Now check $x = -2$.

$\sqrt{(-2) + 3} \stackrel{?}{=} (-2) + 1$ — Substitute $x = -2$.

$\sqrt{1} \stackrel{?}{=} -1$

$1 \neq -1$ — -2 is not a solution.

So, 1 is a solution, but -2 is not. The solution set is $\{1\}$.

> CAUTION

Be careful! Sometimes (as in Example 3), one side of the equation contains a binomial. In that case, we must remember the middle term when we square the binomial. The square of a binomial is *always a trinomial.*

Check Yourself 3

Solve $\sqrt{x - 5} = x - 7$.

It is not always the case that one of the solutions is extraneous. We may have zero, one, or two valid solutions when we generate a quadratic from a radical equation.

In Example 4 we see a case in which both of the derived solutions satisfy the equation.

Example 4 Solving a Radical Equation

NOTE

With a graphing utility plot $Y_1 = \sqrt{7x + 1} - 1$ and $Y_2 = 2x$. Where do they intersect?

Solve $\sqrt{7x + 1} - 1 = 2x$.

First, *we isolate the radical term.*

$\sqrt{7x + 1} = 2x + 1$

Next, we square both sides of the equation.

$7x + 1 = 4x^2 + 4x + 1$

Write the quadratic equation in standard form.

$4x^2 - 3x = 0$

Factoring gives

$x(4x - 3) = 0$

which yields two possible solutions

$x = 0 \quad \text{or} \quad x = \frac{3}{4}$

Checking the solutions by substitution, we find that both values for x give true statements.

Letting $x = 0$, we have

$$\sqrt{7(0)+1} - 1 \stackrel{?}{=} 2(0)$$
$$\sqrt{1} - 1 \stackrel{?}{=} 0$$
$$0 = 0 \quad \text{True!}$$

Letting $x = \frac{3}{4}$, we have

$$\sqrt{7\left(\frac{3}{4}\right)+1} - 1 \stackrel{?}{=} 2\left(\frac{3}{4}\right)$$
$$\sqrt{\frac{25}{4}} - 1 \stackrel{?}{=} \frac{3}{2} \quad 7\left(\frac{3}{4}\right) + 1 = \frac{21}{4} + \frac{4}{4} = \frac{25}{4}$$
$$\frac{5}{2} - 1 \stackrel{?}{=} \frac{3}{2}$$
$$\frac{3}{2} = \frac{3}{2} \quad \text{True!}$$

Since both of our proposed solutions check, there are two solutions to the original equation. The solution set is $\left\{0, \frac{3}{4}\right\}$.

Check Yourself 4

Solve $\sqrt{5x+1} - 1 = 3x$.

When an equation has more than one radical term, we usually have to apply the power property more than once. In such a case, it is generally best to avoid having to work with two radicals on the same side of the equation. Example 5 illustrates one approach to solving such equations.

Example 5 Solving an Equation Containing Two Radicals

< Objective 2 >

Solve $\sqrt{x-2} - \sqrt{2x-6} = 1$.

First we isolate $\sqrt{x-2}$ by adding $\sqrt{2x-6}$ to both sides of the equation. This gives

$$\sqrt{x-2} = 1 + \sqrt{2x-6}$$

Squaring both sides removes the radical on the left. We use the FOIL pattern to square the binomial on the right.

NOTE

$1 + \sqrt{2x-6}$ is a binomial of the form $a + b$, in which $a = 1$ and $b = \sqrt{2x-6}$. The square on the right then has the form $a^2 + 2ab + b^2$.

$$(\sqrt{x-2})^2 = (1 + \sqrt{2x-6})^2$$
$$x - 2 = 1 + \sqrt{2x-6} + \sqrt{2x-6} + 2x - 6 \quad \text{FOIL on the right.}$$
$$x - 2 = 2x - 5 + 2\sqrt{2x-6} \quad \text{Simplify by combining like terms.}$$

Our equation now has a single radical. We isolate the radical and square both sides again.

$$x - 2 = 2x - 5 + 2\sqrt{2x-6}$$
$$-x + 3 = 2\sqrt{2x-6} \quad \text{Subtract } 2x \text{ and add 5 to isolate the radical.}$$
$$(-x+3)^2 = (2\sqrt{2x-6})^2 \quad \text{Square both sides.}$$
$$x^2 - 6x + 9 = 4(2x-6) \quad \text{We square the 2 outside the radical, as well.}$$
$$x^2 - 6x + 9 = 8x - 24 \quad \text{Distribute to remove parentheses.}$$
$$x^2 - 14x + 33 = 0 \quad \text{Write the quadratic equation in standard form.}$$
$$(x-3)(x-11) = 0 \quad \text{Factor.}$$

Therefore, 3 and 11 are the only possible solutions. Check these to see that 3 is a solution but not 11.

Check Yourself 5

Solve $\sqrt{x + 3} - \sqrt{2x + 4} + 1 = 0$.

Earlier in this section, we noted that extraneous roots are possible whenever we raise both sides of the equation to an *even power*. In Example 6, we raise both sides of the equation to an odd power. We will still check the solutions, but in this case it is simply a check of our work and not a search for extraneous solutions.

Example 6 Solving a Radical Equation

< Objective 3 >

NOTE

Because a *cube root* is involved, we *cube* both sides to remove the radical.

NOTE

Raising both sides to an odd-numbered power does not introduce extraneous solutions.

Solve $\sqrt[3]{x^2 + 23} = 3$.

Cubing each side gives

$x^2 + 23 = 27$

which results in the quadratic equation

$x^2 - 4 = 0$

This has two solutions

$x = 2 \qquad \text{or} \qquad x = -2$

Checking the solutions, we find that both result in true statements. Again you should verify this result.

Check Yourself 6

Solve $\sqrt[3]{x^2 - 8} - 2 = 0$.

We summarize our work in this section with an algorithm for solving radical equations.

Step by Step

Solving Radical Equations

Step 1 Isolate a radical term on one side of the equation.

Step 2 Raise each side of the equation to the smallest power that eliminates the isolated radical.

Step 3 If any radicals remain in the equation, repeat step 1.

Step 4 Solve the resulting equation to determine any possible solutions.

Step 5 Check all solutions and discard any extraneous ones.

Check Yourself ANSWERS

1. {21} **2.** No solutions **3.** {9} **4.** $\left\{0, -\frac{1}{9}\right\}$ **5.** {6} **6.** {4, −4}

Reading Your Text

These fill-in-the-blank exercises will help you understand some of the key vocabulary used in this section. The answers to these exercises are in the Answers Appendix in the back of the text.

(a) We must look for ____________ solutions anytime we raise both sides of an equation to an even power.

(b) Given any two expressions a and b and any positive ____________ n, if $a = b$ then $a^n = b^n$.

(c) Applying the power property only removes a radical if that radical term is ____________ on one side of the equation.

(d) We may have zero, one, or two valid ____________ when we generate a quadratic equation from a radical equation.

Skills Calculator/Computer Career Applications Above and Beyond

7.4 exercises

< Objectives 1–3 >

Solve and check each equations.

1. $\sqrt{x} = 2$

2. $\sqrt{x} - 3 = 0$

3. $2\sqrt{y} - 1 = 0$

4. $3\sqrt{2z} = 9$

5. $\sqrt{m + 5} = 3$

6. $\sqrt{y + 7} = 5$

7. $\sqrt{2x + 4} - 4 = 0$

8. $\sqrt{3x + 3} - 6 = 0$

9. $\sqrt{3x - 2} + 2 = 0$

10. $\sqrt{4x + 1} + 3 = 0$

11. $\sqrt{x - 1} = \sqrt{1 - x}$

12. $\sqrt{x + 1} = \sqrt{1 + x}$
Hint: Both radicands must be nonnegative.

13. $\sqrt{w + 3} = \sqrt{3 + w}$
Hint: Both radicands must be nonnegative.

14. $\sqrt{w - 3} = \sqrt{3 - w}$

15. $\sqrt{2x - 3} + 1 = 3$

16. $\sqrt{3x + 1} - 2 = -1$

17. $2\sqrt{3z + 2} - 1 = 5$

18. $3\sqrt{4q - 1} - 2 = 7$

19. $\sqrt{15 - 2x} = x$

20. $\sqrt{48 - 2y} = y$

21. $\sqrt{x + 5} = x - 1$

22. $\sqrt{2x - 1} = x - 8$

23. $\sqrt{3m - 2} + m = 10$

24. $\sqrt{2x + 1} + x = 7$

25. $\sqrt{t + 9} + 3 = t$

26. $\sqrt{2y + 7} + 4 = y$

27. $\sqrt{6x + 1} - 1 = 2x$

28. $\sqrt{7x + 1} - 1 = 3x$

29. $\sqrt[3]{x - 5} = 3$

30. $\sqrt[3]{x + 6} = 2$

31. $\sqrt[3]{x^2 - 1} = 2$

32. $\sqrt[3]{x^2 + 11} = 3$

33. $\sqrt{2x} = \sqrt{x + 1}$

34. $\sqrt{3x} = \sqrt{5x - 1}$

35. $2\sqrt{3r} = \sqrt{r + 11}$

36. $5\sqrt{2q - 7} = \sqrt{15q}$

37. $\sqrt{x + 2} + 1 = \sqrt{x + 4}$ 

38. $\sqrt{x + 5} - 1 = \sqrt{x + 3}$

39. $\sqrt{4m - 3} - 2 = \sqrt{2m - 5}$

40. $\sqrt{2c - 1} = \sqrt{3c + 1} - 1$

41. $\sqrt{x + 1} + \sqrt{x} = 1$

42. $\sqrt{z - 1} - \sqrt{6 - z} = 1$

43. $\sqrt{5x + 6} - \sqrt{x + 3} = 3$

44. $\sqrt{5y + 6} - \sqrt{3y + 4} = 2$

45. $\sqrt{y^2 + 12y} - 3\sqrt{5} = 0$

46. $\sqrt{x^2 + 2x} - 2\sqrt{6} = 0$

47. $\sqrt{\dfrac{x - 3}{x + 2}} = \dfrac{2}{3}$

48. $\dfrac{\sqrt{x - 2}}{x - 2} = \dfrac{x - 5}{\sqrt{x - 2}}$

49. $\sqrt{\sqrt{t} + 5} = 3$

50. $\sqrt{\sqrt{s} - 1} = \sqrt{s - 7}$

Solve each application.

51. Number Problem The sum of an integer and its square root is 12. Find the integer.

52. Number Problem The difference between an integer and its square root is 12. What is the integer?

53. Number Problem The sum of an integer and twice its square root is 24. What is the integer? 

54. Number Problem The sum of an integer and 3 times its square root is 40. Find the integer.

Science and Medicine *The function $d = \sqrt{2h}$ can be used to estimate the distance d to the horizon (in miles) from a given height (in feet).*

55. If a plane flies at 30,000 ft, how far away is the horizon?

56. Janine was looking out across the ocean from her hotel room on the beach. Her eyes were 250 ft above the ground. She saw a ship on the horizon. Approximately how far was the ship from her?

57. Given a distance d to the horizon, give the height in terms of d that would allow you to see that far.

58. Use the result of exercise 57 to estimate the height required to see 100 miles to the horizon.

Science and Medicine *When a car comes to a sudden stop, you can determine the skidding distance (in feet) for a given speed (in miles per hour) by using the formula $s = 2\sqrt{5x}$, in which s is skidding distance and x is speed. Calculate the skidding distance to the nearest foot for each speed.*

59. 55 mi/hr

60. 65 mi/hr

61. 75 mi/hr

62. 40 mi/hr

63. Given the skidding distance s, what formula would allow you to calculate the speed in miles per hour?

64. Use the formula obtained in exercise 63 to determine the speed of a car in miles per hour if the skid marks were 35 ft long.

Complete each statement with **always, sometimes,** *or* **never.**

65. When we raise both sides of an equation to an even power, we ____________ obtain extraneous solutions.

66. Before applying the power property, we ____________ isolate a radical term on one side of the equation.

67. If we raise both sides of an equation to an odd power, we ____________ obtain extraneous solutions.

68. When applying the power property of equality, you ____________ lose a solution.

69. The square of a binomial is ____________ a trinomial.

70. If we generate a quadratic equation from a radical, we ____________ find more than one solution.

Skills	**Calculator/Computer**	Career Applications	Above and Beyond

7.4 exercises

Use a graphing calculator to solve each equation. Express solutions to the nearest hundredth.

Hint: Define Y_1 by the expression on the left side of the equation, and define Y_2 by the expression on the right side. Graph these functions and locate any intersection points. For each such point, the x-value represents a solution.

71. $\sqrt{x+4} = x - 3$

72. $\sqrt{2-x} = x + 4$

73. $3 - 2\sqrt{x+4} = 2x - 5$

74. $5 - 3\sqrt{2-x} = 3 - 4x$

Skills	Calculator/Computer	**Career Applications**	Above and Beyond

MECHANICAL ENGINEERING *In the Chapter 7 Activity, "The Swing of a Pendulum," you worked with the relationship between the period and length of a pendulum. The general model for this relationship is given by*

$$T = 2\pi\sqrt{\frac{L}{g}}$$

in which g is a gravitational constant.

Use this information to complete exercises 75 and 76. 

75. Express the length of a pendulum as a function of the time of its period (that is, solve the model for L).

76. The Foucault (pronounced "foo-koh") pendulum in the Smithsonian Institution (Washington, D.C.) had an 8-s period. Use $g = 980\ \frac{\text{cm}}{s^2}$ to determine the pendulum's length, to the nearest cm.

77. CONSTRUCTION TECHNOLOGY Plans for the dimensions of a rectangular room are shown here (in feet). Find the length of the room's diagonal (give a simplified exact-value result).

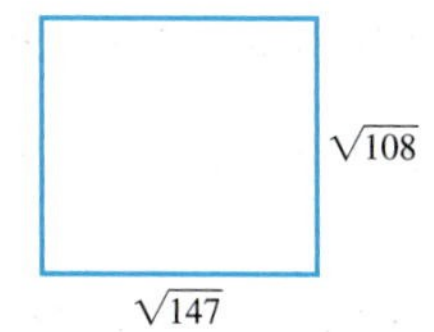

78. CONSTRUCTION TECHNOLOGY Plans call for the dimensions of a rectangular room to be given in terms of an unknown x, as shown here. Find the length of the room's diagonal.

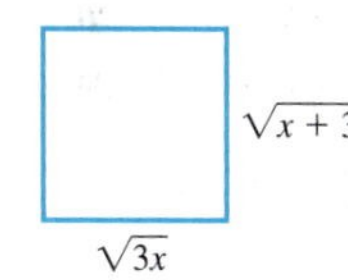

Skills	Calculator/Computer	Career Applications	**Above and Beyond**

79. For what values of x is $\sqrt{(x-1)^2} = x - 1$ a true statement?

80. For what values of x is $\sqrt[3]{(x-1)^3} = x - 1$ a true statement?

Solve for the indicated variable.

81. $h = \sqrt{pq}$ for q

82. $c = \sqrt{a^2 + b^2}$ for a

83. $v = \sqrt{2gR}$ for R

84. $v = \sqrt{2gR}$ for g

85. $r = \sqrt{\frac{S}{2\pi}}$ for S VIDEO

86. $r = \sqrt{\frac{3V}{4\pi}}$ for V

87. $r = \sqrt{\frac{2V}{\pi h}}$ for V

88. $r = \sqrt{\frac{2V}{\pi h}}$ for h

89. $d = \sqrt{(x-1)^2 + (y-2)^2}$ for x

90. $d = \sqrt{(x-1)^2 + (y-2)^2}$ for y

Science and Medicine *A weight suspended on the end of a string is a* pendulum. *The most common example of a pendulum (this side of Edgar Allan Poe) is the kind found in many clocks.*

The regular back-and-forth motion of the pendulum is periodic, *and one such cycle of motion is called a* period. *The time, in seconds, that it takes for one period is given by the radical equation*

$$T = 2\pi\sqrt{\frac{L}{g}}$$

in which g is the force of gravity ($10\ m/s^2$) and L is the length of the pendulum, in meters.

91. Find the period (to the nearest hundredth of a second) if the pendulum is 0.9 m long.

92. Find the period if the pendulum is 0.049 m long.

93. Solve the equation for length L.

94. How long is a pendulum if its period is 1 s?

Answers

1. $\{4\}$ **3.** $\left\{\frac{1}{4}\right\}$ **5.** $\{4\}$ **7.** $\{6\}$ **9.** No solutions **11.** $\{1\}$ **13.** $\{w \mid w \geq -3\}$ **15.** $\left\{\frac{7}{2}\right\}$ **17.** $\left\{\frac{7}{3}\right\}$ **19.** $\{3\}$ **21.** $\{4\}$ **23.** $\{6\}$ **25.** $\{7\}$ **27.** $\left\{0, \frac{1}{2}\right\}$ **29.** $\{32\}$ **31.** $\{3, -3\}$ **33.** $\{1\}$ **35.** $\{1\}$ **37.** $\left\{-\frac{7}{4}\right\}$ **39.** $\{3, 7\}$ **41.** $\{0\}$ **43.** $\{6\}$ **45.** $\{-15, 3\}$ **47.** $\{7\}$ **49.** $\{16\}$ **51.** 9 **53.** 16 **55.** $\approx$245 mi **57.** $\frac{d^2}{2}$ **59.** 33 ft **61.** 39 ft **63.** $x = \frac{s^2}{20}$ **65.** sometimes **67.** never **69.** always **71.** $\{6.19\}$ **73.** $\{1.63\}$ **75.** $L = \frac{g}{4\pi^2}T^2$ **77.** $\sqrt{255}$ ft **79.** $\{x \mid x \geq 1\}$ **81.** $q = \frac{h^2}{p}$ **83.** $R = \frac{v^2}{2g}$ **85.** $S = 2\pi r^2$ **87.** $V = \frac{\pi h r^2}{2}$ **89.** $x = 1 \pm \sqrt{d^2 - (y-2)^2}$ **91.** 1.88 s **93.** $L = \frac{T^2 g}{4\pi^2}$

7.5

Rational Exponents

< 7.5 Objectives >

1 > Simplify expressions with rational exponents

2 > Use a calculator to estimate the value of an expression with rational exponents

3 > Convert between radical and exponential notation

In Section 7.1, we discussed roots and radical notation. In this section, we develop notation using exponents to provide an alternate way of writing roots.

This notation involves **rational numbers as exponents.** To start the development, we extend all the previous properties of exponents to include rational exponents.

Given that extension, suppose that

$$a = 4^{1/2}$$

Squaring both sides of the equation yields

$$a^2 = (4^{1/2})^2$$

or

$$a^2 = 4^1$$

$$a^2 = 4$$

NOTE

We will see later in this section that the property $(x^m)^n = x^{mn}$ holds for rational numbers m and n.

From this last equation we see that a is the number whose square is 4; that is, a is the principal square root of 4. Using radical notation, we can write

$$a = \sqrt{4}$$

But we began with

$$a = 4^{1/2}$$

To be consistent, we must have

$$4^{1/2} = \sqrt{4}$$

NOTE

$4^{1/2}$ is the principal square root of 4

This argument can be repeated for any exponent of the form $\frac{1}{n}$, so it seems reasonable to construct a definition.

Definition

Rational Exponents

If a is any real number and n is a positive integer ($n > 1$), then

$$a^{1/n} = \sqrt[n]{a}$$

We restrict a so that a is nonnegative when n is even.

$a^{1/n}$ is the principal nth root of a.

In Example 1, we use rational exponents to represent roots.

Example 1 Writing Expressions in Radical Form

< Objective 1 >

Write each expression in radical form and simplify.

(a) $25^{1/2} = \sqrt{25} = 5$

(b) $27^{1/3} = \sqrt[3]{27} = 3$

(c) $-36^{1/2} = -\sqrt{36} = -6$

(d) $(-36)^{1/2} = \sqrt{-36}$ is not a real number.

(e) $32^{1/5} = \sqrt[5]{32} = 2$

NOTES

$27^{1/3}$ is the *cube root* of 27.

$32^{1/5}$ is the *fifth root* of 32.

Check Yourself 1

Write each expression in radical form and simplify.

(a) $8^{1/3}$ (b) $-64^{1/2}$ (c) $81^{1/4}$

NOTE

The two radical forms for $a^{m/n}$ are equivalent, and the choice of which form to use generally depends on whether we are evaluating numerical expressions or rewriting expressions containing variables in radical form.

We extend exponent notation to allow *any* rational exponent, assuming that the exponent properties are valid.

$$a^{m/n} = (a^{1/n})^m = (a^m)^{1/n} \qquad \text{because} \qquad \frac{m}{n} = (m)\left(\frac{1}{n}\right) = \left(\frac{1}{n}\right)(m)$$

We know that $a^{1/n} = \sqrt[n]{a}$. Combining this with the above, we offer a definition of $a^{m/n}$.

Definition

Rational Exponents

For any real number a and positive integers m and n with $n > 1$,

$$a^{m/n} = (\sqrt[n]{a})^m = \sqrt[n]{a^m}$$

Note: If $n = 2$, indicating a square root, we include it in exponential form, but omit the index when using radicals to write the expression.

We apply this extension of rational exponents in Example 2.

Example 2 Simplifying Expressions with Rational Exponents

Simplify each expression.

(a) $9^{3/2} = (9^{1/2})^3 = (\sqrt{9})^3$
$= 3^3 = 27$

(b) $\left(\frac{16}{81}\right)^{3/4} = \left[\left(\frac{16}{81}\right)^{1/4}\right]^3 = \left(\sqrt[4]{\frac{16}{81}}\right)^3$
$= \left(\frac{2}{3}\right)^3 = \frac{8}{27}$

(c) $(-8)^{2/3} = [(-8)^{1/3}]^2 = (\sqrt[3]{-8})^2$
$= (-2)^2 = 4$

NOTE

This illustrates why we use $(\sqrt[n]{a})^m$ for $a^{m/n}$ when evaluating numerical expressions. The numbers are smaller and easier to work with.

In (a) we could also have evaluated the expression as

$9^{3/2} = \sqrt{9^3} = \sqrt{729}$
$= 27$

Check Yourself 2

Simplify each expression.

(a) $16^{3/4}$ **(b)** $\left(\frac{8}{27}\right)^{2/3}$ **(c)** $(-32)^{3/5}$

Now we want to extend rational-exponent notation. Using the definition of negative exponents, we can write

$$a^{-m/n} = \frac{1}{a^{m/n}}$$

Example 3 illustrates negative rational exponents.

Example 3 Simplifying Expressions with Rational Exponents

Simplify each expression.

(a) $16^{-1/2} = \frac{1}{16^{1/2}} = \frac{1}{4}$ $16^{1/2} = \sqrt{16} = 4$

(b) $27^{-2/3} = \frac{1}{27^{2/3}} = \frac{1}{(\sqrt[3]{27})^2} = \frac{1}{3^2} = \frac{1}{9}$

Check Yourself 3

Simplify each expression.

(a) $16^{-1/4}$ **(b)** $81^{-3/4}$

You can use a graphing calculator to evaluate expressions containing rational exponents by using the [∧] and parentheses keys.

Example 4 Using a Calculator to Estimate Powers

< Objective 2 >

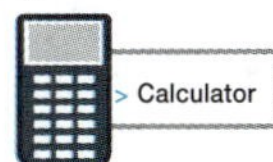

RECALL

Fractions indicate division.

Use a graphing calculator to evaluate each expression. Round all answers to three decimal places.

(a) $45^{2/5}$

Enter 45 and press the [∧] key. Then use the keystrokes

[(] 2 [÷] 5 [)] You *must* use parentheses for the entire exponent.

Press [ENTER], and the display reads 4.584426407. Rounded to three decimal places, the result is 4.584.

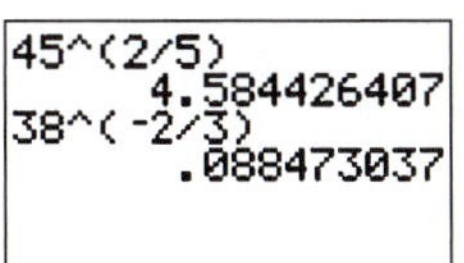

(b) $38^{-2/3}$

Enter 38 and press the [∧] key. Then use the keystrokes

[(] [(−)] 2 [÷] 3 [)]

Press [ENTER], and the display reads 0.088473037. Rounded to three decimal places, the result is 0.088.

Check Yourself 4

Use a calculator to evaluate each expression. Round each answer to three decimal places.

(a) $23^{3/5}$ **(b)** $18^{-4/7}$

As we mentioned earlier in this section, we assume that all our previous exponent properties continue to hold for rational exponents. Those properties are restated here.

Property

Properties of Exponents

For any nonzero real numbers a and b and rational numbers m and n,

1. Product rule $a^m \cdot a^n = a^{m+n}$
2. Quotient rule $\frac{a^m}{a^n} = a^{m-n}$
3. Power rule $(a^m)^n = a^{mn}$
4. Product-power rule $(ab)^m = a^m b^m$
5. Quotient-power rule $\left(\frac{a}{b}\right)^m = \frac{a^m}{b^m}$

We restrict a and b to being nonnegative real numbers when m or n indicates an even root.

Example 5 illustrates these properties when simplifying expressions with rational exponents. We assume that all variables represent positive real numbers.

Example 5 Simplifying Expressions

NOTES

Product rule—add the exponents.

Quotient rule—subtract the exponents.

Power rule—multiply the exponents.

Simplify each expression.

(a) $x^{2/3} \cdot x^{1/2} = x^{2/3+1/2}$

$= x^{4/6+3/6} = x^{7/6}$

(b) $\frac{w^{3/4}}{w^{1/2}} = w^{3/4-1/2}$

$= w^{3/4-2/4} = w^{1/4}$

(c) $(a^{2/3})^{3/4} = a^{(2/3)(3/4)}$ $\quad \frac{2}{3} \cdot \frac{3}{4} = \frac{2}{4} = \frac{1}{2}$

$= a^{1/2}$

Check Yourself 5

Simplify each expression.

(a) $z^{3/4} \cdot z^{1/2}$ **(b)** $\frac{x^{5/6}}{x^{1/3}}$ **(c)** $(b^{5/6})^{2/5}$

As you might expect, we often need to use more than one property of exponents to simplify an expression.

Example 6 Simplifying Expressions

Simplify each expression.

(a) $(x^{2/3} \cdot y^{5/6})^{3/2}$

$= (x^{2/3})^{3/2} \cdot (y^{5/6})^{3/2}$ Product-power rule

$= x^{(2/3)(3/2)} \cdot y^{(5/6)(3/2)} = xy^{5/4}$ Power rule

(b) $\left(\dfrac{r^{-1/2}}{s^{1/3}}\right)^6 = \dfrac{(r^{-1/2})^6}{(s^{1/3})^6}$ Quotient-power rule

$= \dfrac{r^{-3}}{s^2} = \dfrac{1}{r^3s^2}$ Power rule

(c) $\left(\dfrac{4a^{-2/3} \cdot b^2}{a^{1/3} \cdot b^{-4}}\right)^{1/2} = \left(\dfrac{4b^2 \cdot b^4}{a^{1/3} \cdot a^{2/3}}\right)^{1/2} = \left(\dfrac{4b^6}{a}\right)^{1/2}$ Simplify inside the parentheses first.

$= \dfrac{(4b^6)^{1/2}}{a^{1/2}} = \dfrac{4^{1/2}(b^6)^{1/2}}{a^{1/2}}$

$= \dfrac{2b^3}{a^{1/2}}$

Check Yourself 6

Simplify each expression.

(a) $(a^{3/4} \cdot b^{1/2})^{2/3}$ **(b)** $\left(\dfrac{w^{1/2}}{z^{-1/4}}\right)^4$ **(c)** $\left(\dfrac{8x^{-3/4}y}{x^{1/4} \cdot y^{-5}}\right)^{1/3}$

We can also use the relationships between rational exponents and radicals to write expressions containing rational exponents as radicals and vice versa.

Example 7 Writing Expressions in Radical Form

< Objective 3 >

Write each expression in radical form.

(a) $a^{3/5} = \sqrt[5]{a^3}$ $a^{m/n} = \sqrt[n]{a^m}$

(b) $(mn)^{3/4} = \sqrt[4]{(mn)^3}$

$= \sqrt[4]{m^3n^3}$

(c) $2y^{5/6} = 2\sqrt[6]{y^5}$ The exponent applies *only* to the variable *y*.

(d) $(2y)^{5/6} = \sqrt[6]{(2y)^5}$ Now the exponent applies to $2y$ because of the parentheses.

$= \sqrt[6]{32y^5}$

Check Yourself 7

Write each expression in radical form.

(a) $(ab)^{2/3}$ **(b)** $3x^{3/4}$ **(c)** $(3x)^{3/4}$

Example 8 Writing Expressions in Exponential Form

Use rational exponents to write each expression and simplify.

(a) $\sqrt[3]{5x} = (5x)^{1/3}$

(b) $\sqrt{9a^2b^4} = (9a^2b^4)^{1/2}$

$= 9^{1/2}(a^2)^{1/2}(b^4)^{1/2} = 3ab^2$

(c) $\sqrt[4]{16w^{12}z^8} = (16w^{12}z^8)^{1/4}$

$= 16^{1/4}(w^{12})^{1/4}(z^8)^{1/4} = 2w^3z^2$

Check Yourself 8

Use rational exponents to write each expression and simplify.

(a) $\sqrt{7a}$ (b) $\sqrt[3]{27p^6q^9}$ (c) $\sqrt[4]{81x^8y^{16}}$

Our final example applies various multiplication patterns to terms containing rational exponents.

Example 9 Multiplying Terms with Rational Exponents

NOTES

The expression in (a) cannot be simplified further.

The pattern in (b) results in the difference of two squares.

Use the appropriate property to find each product.

(a) $a^{1/3}(a^{2/3} + a^{1/2}) = a^{3/3} + a^{5/6} = a + a^{5/6}$ Use the distributive property.

(b) $(a^{2/3} - a^{1/2})(a^{2/3} + a^{1/2}) = a^{4/3} - a^{2/2} = a^{4/3} - a$ This is a difference of squares.

Check Yourself 9

Use the appropriate properties to find each product.

(a) $b^{1/3}(b^{2/3} + b^{1/3})$ (b) $(b^{2/3} - b^{1/2})^2$

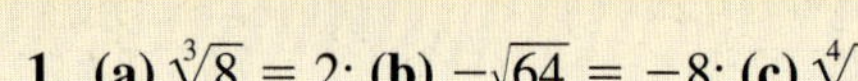

Check Yourself ANSWERS

1. (a) $\sqrt[3]{8} = 2$; (b) $-\sqrt{64} = -8$; (c) $\sqrt[4]{81} = 3$ 2. (a) 8; (b) $\frac{4}{9}$; (c) -8 3. (a) $\frac{1}{2}$; (b) $\frac{1}{27}$
4. (a) 6.562; (b) 0.192 5. (a) $z^{5/4}$; (b) $x^{1/2}$; (c) $b^{1/3}$ 6. (a) $a^{1/2}b^{1/3}$; (b) w^2z; (c) $\frac{2y^2}{x^{1/3}}$
7. (a) $\sqrt[3]{a^2b^2}$; (b) $3\sqrt[4]{x^3}$; (c) $\sqrt[4]{27x^3}$ 8. (a) $(7a)^{1/2}$; (b) $(27p^6q^9)^{1/3} = 3p^2q^3$; (c) $(81x^8y^{16})^{1/4} = 3x^2y^4$
9. (a) $b + b^{2/3}$; (b) $b^{4/3} - 2b^{7/6} + b$

Reading Your Text

These fill-in-the-blank exercises will help you understand some of the key vocabulary used in this section. The answers to these exercises are in the Answers Appendix in the back of the text.

(a) If a is any ________ number and n is a positive integer greater than one, then $a^{1/n} = \sqrt[n]{a}$, where a is nonnegative when n is even.

(b) $a^{1/n}$ indicates the ________ nth root of a.

(c) We can write expressions containing rational exponents as ________.

(d) In the expression $2y^{1/3}$, the ________ applies only to the variable y.

Skills | Calculator/Computer | Career Applications | Above and Beyond

7.5 exercises

< Objective 1 >

Evaluate each expression.

1. $49^{1/2}$
2. $100^{1/2}$
3. $-25^{1/2}$
4. $(-81)^{1/2}$
5. $(-49)^{1/2}$
6. $-49^{1/2}$
7. $27^{1/3}$
8. $(-64)^{1/3}$ VIDEO
9. $81^{1/4}$
10. $-81^{1/4}$
11. $\left(\frac{4}{9}\right)^{1/2}$
12. $\left(\frac{27}{8}\right)^{1/3}$
13. $27^{2/3}$
14. $16^{3/2}$
15. $(-8)^{4/3}$
16. $64^{2/3}$
17. $32^{2/5}$
18. $-81^{3/4}$
19. $81^{3/2}$
20. $(-243)^{3/5}$
21. $\left(\frac{8}{27}\right)^{2/3}$
22. $\left(\frac{16}{9}\right)^{3/2}$

< Objective 2 >

Evaluate each expression. Use a calculator to check your results.

23. $49^{-1/2}$
24. $27^{-1/3}$
25. $81^{-1/4}$
26. $121^{-1/2}$
27. $9^{-3/2}$ VIDEO
28. $16^{-3/4}$
29. $64^{-5/6}$
30. $16^{-3/2}$
31. $\left(\frac{4}{25}\right)^{-1/2}$
32. $\left(\frac{27}{8}\right)^{-2/3}$

Simplify each expression. Assume all variables represent positive real numbers.

33. $x^{1/2}x^{1/2}$
34. $a^{2/3}a^{1/3}$
35. $y^{5/7}y^{1/7}$
36. $m^{1/4}m^{5/4}$
37. $b^{2/3}b^{3/2}$
38. $p^{5/6}p^{2/3}$
39. $\frac{x^{2/3}}{x^{1/3}}$
40. $\frac{a^{5/6}}{a^{1/6}}$
41. $\frac{s^{7/5}}{s^{2/5}}$
42. $\frac{z^{9/2}}{z^{3/2}}$
43. $\frac{w^{7/6}}{w^{1/2}}$
44. $\frac{b^{7/6}}{b^{2/3}}$
45. $(x^{3/4})^{4/3}$
46. $(y^{4/3})^{3/4}$
47. $(a^{2/5})^{3/2}$
48. $(p^{3/4})^{2/3}$
49. $(y^{-2/3})^{6}$
50. $(w^{-2/3})^{6}$
51. $(a^{2/3}b^{3/2})^{6}$ VIDEO
52. $(p^{3/4}q^{5/2})^{4}$
53. $(x^{2/7}y^{3/7})^{7}$
54. $(3m^{3/4}n^{5/4})^{4}$
55. $(s^{3/4}t^{1/4})^{4/3}$
56. $(x^{5/2}y^{5/7})^{2/5}$
57. $(8p^{3/2}q^{5/2})^{2/3}$
58. $(16a^{1/3}b^{2/3})^{3/4}$
59. $(x^{3/5}y^{3/4}z^{3/2})^{2/3}$
60. $(p^{5/6}q^{2/3}r^{5/3})^{3/5}$
61. $\frac{a^{5/6}b^{3/4}}{a^{1/3}b^{1/2}}$
62. $\frac{x^{2/3}y^{3/4}}{x^{1/2}y^{1/2}}$
63. $\frac{(r^{-1}s^{1/2})^{3}}{rs^{-1/2}}$
64. $\frac{(w^{-2}z^{-1/4})^{6}}{w^{-8}z^{1/2}}$
65. $\left(\frac{x^{12}}{y^{8}}\right)^{1/4}$
66. $\left(\frac{q^{12}}{p^{28}}\right)^{1/4}$
67. $\left(\frac{m^{-1/4}}{n^{1/2}}\right)^{4}$
68. $\left(\frac{r^{1/5}}{s^{-1/2}}\right)^{10}$
69. $\left(\frac{r^{-1/2}s^{3/4}}{t^{1/4}}\right)^{4}$
70. $\left(\frac{a^{1/3}b^{-1/6}}{c^{-1/6}}\right)^{6}$
71. $\left(\frac{8x^{3}y^{-6}}{z^{-9}}\right)^{1/3}$
72. $\left(\frac{16p^{-4}q^{6}}{r^{2}}\right)^{-1/2}$
73. $\left(\frac{16m^{-3/5}n^{2}}{m^{1/5}n^{-2}}\right)^{1/4}$ VIDEO
74. $\left(\frac{27x^{5/6}y^{-4/3}}{x^{-7/6}y^{5/3}}\right)^{1/3}$
75. $\left(\frac{x^{3/2}y^{1/2}}{z^{2}}\right)^{1/2}\left(\frac{x^{3/4}y^{3/2}}{z^{-3}}\right)^{1/3}$
76. $\left(\frac{p^{1/2}q^{4/3}}{r^{-4}}\right)^{3/4}\left(\frac{p^{15/8}q^{-3}}{r^{6}}\right)^{1/3}$

< Objective 3 >

Write each expression in radical form. Do not simplify.

77. $a^{3/4}$
78. $m^{5/6}$
79. $2x^{2/3}$
80. $3m^{-2/5}$
81. $3x^{2/5}$
82. $2y^{-3/4}$
83. $(3x)^{2/5}$
84. $(2y)^{-3/4}$

Write each expression using rational exponents, and simplify where necessary.

85. $\sqrt{7a}$ **86.** $\sqrt{25w^4}$ **87.** $\sqrt[3]{8m^6n^9}$ **88.** $\sqrt[5]{32r^{10}s^{15}}$

Simplify each expression and write your answer in scientific notation.

89. $(4 \times 10^8)^{1/2}$ **90.** $(8 \times 10^6)^{1/3}$ **91.** $(16 \times 10^{-12})^{1/4}$ **92.** $(9 \times 10^{-4})^{1/2}$

93. $(16 \times 10^{-8})^{1/2}$ **94.** $(16 \times 10^{-8})^{3/4}$ **95.** $(27 \times 10^6)^{1/3}$ **96.** $(64 \times 10^6)^{-1/3}$

Apply the appropriate multiplication patterns and simplify your result.

97. $a^{1/2}(a^{3/2} + a^{3/4})$ **98.** $2x^{1/4}(3x^{3/4} - 5x^{-1/4})$ **99.** $(a^{1/2} + 2)(a^{1/2} - 2)$ **100.** $(w^{1/3} - 3)(w^{1/3} + 3)$

101. $(m^{1/2} + n^{1/2})(m^{1/2} - n^{1/2})$ **102.** $(x^{1/3} + y^{1/3})(x^{1/3} - y^{1/3})$ **103.** $(x^{1/2} + 2)^2$ **104.** $(a^{1/3} - 3)^2$

VIDEO

105. $(r^{1/2} + s^{1/2})^2$ **106.** $(p^{1/2} - q^{1/2})^2$

Determine whether each statement is **true** *or* **false.**

107. $a^{m/n}$ means the same thing as $a^{n/m}$. **108.** If $a > 0$, then $(a^m)^n$ is equal to $(a^n)^m$.

Complete each statement with **always, sometimes,** *or* **never.**

109. A negative number raised to a rational power is __________ a real number.

110. An expression with a rational exponent can __________ be rewritten as a radical.

Skills | **Calculator/Computer** | Career Applications | Above and Beyond

Use a calculator to evaluate each expression. Round each answer to three decimal places.

111. $46^{3/5}$ **112.** $23^{2/7}$ **113.** $12^{-2/5}$ **114.** $36^{-3/4}$

Skills | Calculator/Computer | **Career Applications** | Above and Beyond

AGRICULTURAL TECHNOLOGY *One formula that researchers encounter when investigating rainfall runoff in regions of semi-arid farmland is*

$$t = C\left(\frac{L}{xy^2}\right)^{1/3}$$

Use this information to complete exercises 115 and 116.

115. Evaluate t when $C = 20$, $L = 600$, $x = 3$, and $y = 5$. **116.** Solve the given formula for L.

AGRICULTURAL TECHNOLOGY *The average velocity of water in an open irrigation ditch is given by the formula*

$$V = \frac{1.5x^{2/3}y^{1/2}}{z}$$

Use this information to complete exercises 117 and 118.

117. Find the average velocity when $x = 27$, $y = 16$, and $z = 12$.

118. Rewrite the average velocity formula using radicals in place of rational exponents and fractions in place of decimals.

Perform the indicated operations. Assume that n represents a positive integer and that the denominators are not zero.

119. $x^{3n}x^{2n}$ **120.** $p^{1-n}p^{n+3}$ **121.** $(y^2)^{2n}$ **122.** $(a^{3n})^3$

123. $\dfrac{r^{n+2}}{r^n}$ **124.** $\dfrac{w^n}{w^{n-3}}$ **125.** $(a^3b^2)^{2n}$ **126.** $(c^4d^2)^{3m}$

127. $\left(\dfrac{x^{n+2}}{x^n}\right)^{1/2}$ **128.** $\left(\dfrac{b^n}{b^{n-3}}\right)^{1/3}$

Write each expression in exponential form, simplify, and give the result as a single radical.

129. $\sqrt{\sqrt{x}}$ **130.** $\sqrt[3]{\sqrt{a}}$ **131.** $\sqrt[4]{\sqrt{y}}$ **132.** $\sqrt{\sqrt[3]{w}}$

Certain expressions containing rational exponents are factorable. For instance, to factor $x^{2/3} - x^{1/3} - 6$, let $u = x^{1/3}$. Note that $x^{2/3} = (x^{1/3})^2 = u^2$.

Substituting, we have $u^2 - u - 6$, and factoring yields $(u - 3)(u + 2)$ or $(x^{1/3} - 3)(x^{1/3} + 2)$.

Use this technique to factor each expression.

133. $x^{2/3} + 4x^{1/3} + 3$ **134.** $y^{2/5} - 2y^{1/5} - 8$ **135.** $a^{4/5} - 7a^{2/5} + 12$

136. $w^{4/3} + 3w^{2/3} - 10$ **137.** $x^{4/3} - 4$ **138.** $x^{2/5} - 16$

139. **Business and Finance** The geometric mean is used to measure average inflation rates or interest rates. If prices increased by 15% over 5 years, then the average *annual* rate of inflation is obtained by taking the fifth root of 1.15.

$(1.15)^{1/5} \approx 1.0283$ or $\approx 2.8\%$

1 is added to 0.15 because we are taking the original price and adding 15% of that price. We could write that as $P + 0.15P$

Factoring gives

$$P + 0.15P = P(1 + 0.15)$$
$$= P(1.15)$$

Between December 2000 and March 2012, the Bureau of Labor Statistics computed an inflation rate of 34.8%, which is equivalent to an annual growth rate of 2.47%. From December 2000 through March 2012 is 147 months. To what exponent was 1.348 raised to obtain this average annual growth rate?

140. On a calculator, try evaluating $(-9)^{4/2}$ two ways:

(a) $[(-9)^4]^{1/2}$ **(b)** $[(-9)^{1/2}]^4$

Discuss the results.

141. Describe the difference between x^{-2} and $x^{1/2}$.

142. Some rational exponents, such as $\frac{1}{2}$, can easily be rewritten as terminating decimals (0.5). Others, such as $\frac{1}{3}$, cannot. What is it that determines which rational numbers can be rewritten as terminating decimals?

143. Use the properties of exponents to decide what x should be to make each statement true. Explain your choices regarding which properties of exponents you decide to use.

(a) $(a^{2/3})^x = a$ **(b)** $(a^{5/6})^x = \dfrac{1}{a}$ **(c)** $a^{2x} \cdot a^{3/2} = 1$ **(d)** $\left(\sqrt{a^{2/3}}\right)^x = a$

Answers

1. 7 **3.** -5 **5.** Not a real number **7.** 3 **9.** 3 **11.** $\frac{2}{3}$ **13.** 9 **15.** 16 **17.** 4 **19.** 729 **21.** $\frac{4}{9}$ **23.** $\frac{1}{7}$ **25.** $\frac{1}{3}$ **27.** $\frac{1}{27}$ **29.** $\frac{1}{32}$ **31.** $\frac{5}{2}$ **33.** x **35.** $y^{6/7}$ **37.** $b^{13/6}$ **39.** $x^{1/3}$ **41.** s **43.** $w^{2/3}$ **45.** x **47.** $a^{3/5}$ **49.** $\frac{1}{y^4}$ **51.** a^4b^9 **53.** x^2y^3 **55.** $st^{1/3}$ **57.** $4pq^{5/3}$ **59.** $x^{2/5}y^{1/2}z$ **61.** $a^{1/2}b^{1/4}$ **63.** $\frac{s^2}{r^4}$ **65.** $\frac{x^3}{y^2}$ **67.** $\frac{1}{mn^2}$ **69.** $\frac{s^3}{r^2t}$ **71.** $\frac{2xz^3}{y^2}$ **73.** $\frac{2n}{m^{1/5}}$ **75.** $xy^{3/4}$ **77.** $\sqrt[4]{a^3}$ **79.** $2\sqrt[3]{x^2}$ **81.** $3\sqrt[5]{x^2}$ **83.** $\sqrt[5]{9x^2}$ **85.** $(7a)^{1/2}$ **87.** $2m^2n^3$ **89.** 2×10^4 **91.** 2×10^{-3} **93.** 4×10^{-4} **95.** 3×10^2 **97.** $a^2 + a^{5/4}$ **99.** $a - 4$ **101.** $m - n$ **103.** $x + 4x^{1/2} + 4$ **105.** $r + 2r^{1/2}s^{1/2} + s$ **107.** False **109.** sometimes **111.** 9.946 **113.** 0.37 **115.** 40 **117.** 4.5 **119.** x^{5n} **121.** y^{4n} **123.** r^2 **125.** $a^{6n}b^{4n}$ **127.** x **129.** $\sqrt[4]{x}$ **131.** $\sqrt[8]{y}$ **133.** $(x^{1/3} + 1)(x^{1/3} + 3)$ **135.** $(a^{2/5} - 3)(a^{2/5} - 4)$ **137.** $(x^{2/3} - 2)(x^{2/3} + 2)$ **139.** $\frac{4}{49}$ **141.** Above and Beyond **143.** (a) $\frac{3}{2}$; (b) $-\frac{6}{5}$; (c) $-\frac{3}{4}$; (d) 3

7.6 Complex Numbers

< 7.6 Objectives >

1 > Use the imaginary number i

2 > Add and subtract complex numbers

3 > Multiply and divide complex numbers

Radicals such as

$$\sqrt{-4} \qquad \text{and} \qquad \sqrt{-49}$$

are *not* real numbers because no real number squared produces a negative number. Our work in this section extends our number system to include these **imaginary numbers** which allows us to consider radicals such as $\sqrt{-4}$.

First we offer a definition.

Definition

The Imaginary Number i

The number i is defined as

$$i = \sqrt{-1}$$

so that

$$i^2 = -1$$

This definition of the number i gives us an alternate means of indicating the square root of a negative number.

Property

Writing an Imaginary Number

When a is a positive real number,

$$\sqrt{-a} = \sqrt{a}i \quad \text{or} \quad i\sqrt{a}$$

Example 1 Using the Number i

< Objective 1 >

Write each expression as a multiple of i.

(a) $\sqrt{-4} = \sqrt{4}i = 2i$

(b) $-\sqrt{-9} = -\sqrt{9}i = -3i$

(c) $\sqrt{-8} = \sqrt{8}i = 2\sqrt{2}i$ or $2i\sqrt{2}$

(d) $\sqrt{-7} = \sqrt{7}i \quad \text{or} \quad i\sqrt{7}$

NOTE

We simplify $\sqrt{8}$ as $2\sqrt{2}$. We write the i *in front of* the radical to make it clear that i is *not part of* the radicand.

NOTE

Some courses do not cover complex numbers at this level. The use of complex numbers in ensuing sections will be indicated with the symbol $\mathbb{C}$ so that they may be skipped if desired.

Check Yourself 1

Write each radical as a multiple of i.

(a) $\sqrt{-25}$ **(b)** $\sqrt{-24}$

We define complex numbers in terms of the number i.

Definition

Complex Number

NOTE

The term *imaginary number* was introduced by René Descartes in 1637. Euler used i to indicate $\sqrt{-1}$ in 1748, but it was not until 1832 that Gauss used the term *complex number*.

A **complex number** is any number that can be written in the form

$a + bi$

in which a and b are real numbers and

$i = \sqrt{-1}$ so that $i^2 = -1$

NOTES

Charles Steinmetz (1865–1923) used complex numbers to explain the behavior of electric circuits.

$5i$ is also called a **pure imaginary** number.

The real numbers are a subset of the set of complex numbers.

The form $a + bi$ is called the **standard form** of a complex number. We call a the **real part** of the complex number and b the **imaginary part.** Some examples follow.

$3 + 7i$ is an example of a complex number with real part 3 and imaginary part 7.

$5i$ is also a complex number because it can be written as $0 + 5i$.

-3 is a complex number because it can be written as $-3 + 0i$.

Adding and subtracting complex numbers is simply a matter of combining like terms. The real parts are like terms and the imaginary terms are each like terms.

Property

Adding and Subtracting Complex Numbers

For the complex numbers $a + bi$ and $c + di$,

$$(a + bi) + (c + di) = (a + c) + (b + d)i$$

$$(a + bi) - (c + di) = (a - c) + (b - d)i$$

We add or subtract the real parts and the imaginary parts of the complex numbers separately.

Example 2 illustrates these properties.

Example 2 Adding and Subtracting Complex Numbers

< Objective 2 >

Perform the indicated operations.

(a) $(5 + 3i) + (6 - 7i) = (5 + 6) + (3 - 7)i$
$= 11 - 4i$

Combine the real and imaginary parts separately.

(b) $5 + (7 - 5i) = (5 + 7) + (-5i)$
$= 12 - 5i$

NOTE

Regrouping is a matter of combining like terms.

(c) $(8 - 2i) - (3 - 4i) = (8 - 3) + [-2 - (-4)]i$
$= 5 + 2i$

Distribute the − sign.

Check Yourself 2

Perform the indicated operations.

(a) $(4 - 7i) + (3 - 2i)$ **(b)** $-7 + (-2 + 3i)$
(c) $(-4 + 3i) - (-2 - i)$

Because complex numbers are binomials, the product of two complex numbers is found by applying the multiplication pattern for binomials.

Example 3 Multiplying Complex Numbers

< Objective 3 >

Multiply.

(a) $(2 + 3i)(3 - 4i)$

$= 2 \cdot 3 + 2(-4i) + (3i)3 + (3i)(-4i)$ FOIL.

$= 6 + (-8i) + 9i + (-12i^2)$

$= 6 - 8i + 9i + (-12)(-1)$

$= 6 + i + 12$ Combine the imaginary terms.

$= 18 + i$ Combine the real terms.

NOTE

We use the definition of i to replace i^2 with −1 and then simplify.

(b) $(1 - 2i)(3 - 4i)$

$= 1 \cdot 3 + 1(-4i) + (-2i)3 + (-2i)(-4i)$

$= 3 + (-4i) + (-6i) + 8i^2$

$= 3 - 10i + 8(-1)$

$= 3 - 10i - 8$

$= -5 - 10i$

Check Yourself 3

Multiply $(2 - 5i)(3 - 2i)$.

Example 3 suggests a pattern when multiplying complex numbers.

Property

Multiplying Complex Numbers

For the complex numbers $a + bi$ and $c + di$,

$$(a + bi)(c + di) = ac + adi + bci + bdi^2$$
$$= ac + adi + bci - bd$$
$$= (ac - bd) + (ad + bc)i$$

You can memorize the formula for the product of complex numbers, however, it is much easier to use the multiplication pattern for binomials and simplify the result.

One particular product should seem very familiar. We call $a + bi$ and $a - bi$ **complex conjugates.** For instance,

$3 + 2i$ and $3 - 2i$

are complex conjugates.

Consider the product

$$(3 + 2i)(3 - 2i) = 3^2 - (2i)^2$$
$$= 9 - 4i^2 = 9 - 4(-1)$$
$$= 9 + 4 = 13$$

The product of $3 + 2i$ and $3 - 2i$ is a real number. In general, we can write the product of complex conjugates as

$$(a + bi)(a - bi) = a^2 + b^2$$

The fact that this product is always a real number is very useful when dividing complex numbers.

Example 4 Multiplying Complex Numbers

Multiply.

$$(7 + 4i)(7 - 4i) = (7)^2 - (4i)^2 \qquad (a + b)(a - b) = a^2 - b^2$$
$$= 49 - 4^2 i^2$$
$$= 49 - 16(-1) \qquad i^2 = -1$$
$$= 49 + 16$$
$$= 65$$

NOTE

We get the same result when we apply the formula with $a = 7$ and $b = 4$.

Check Yourself 4

Multiply $(5 + 3i)(5 - 3i)$.

Next, we look to divide complex numbers. The problem we run into is how to divide by i. For instance, to compute $\frac{6}{i}$, we would have to know how many times i goes into 6.

Instead, we multiply the numerator and denominator by the conjugate of the denominator. This makes the denominator a real number and we can distribute, if necessary, to complete the division. Consider Example 5.

Example 5 Dividing Complex Numbers

Divide.

(a) $\dfrac{6 + 9i}{3i}$

$$\frac{6 + 9i}{3i} = \frac{(6 + 9i)(-3i)}{(3i)(-3i)}$$

The conjugate of $3i$ is $-3i$, so we multiply the numerator and denominator by $-3i$.

$$= \frac{-18i - 27i^2}{-9i^2}$$

$$= \frac{-18i - 27(-1)}{(-9)(-1)}$$

$$= \frac{27 - 18i}{9} = \frac{27}{9} - \frac{18i}{9}$$

$$= 3 - 2i$$

(b) $\dfrac{3 - i}{3 + 2i} = \dfrac{(3 - i)(3 - 2i)}{(3 + 2i)(3 - 2i)}$

$$= \frac{9 - 6i - 3i + 2i^2}{9 - 4i^2}$$

$$= \frac{9 - 9i - 2}{9 + 4}$$

$$= \frac{7 - 9i}{13} = \frac{7}{13} - \frac{9}{13}i$$

(c) $\dfrac{2 + i}{4 - 5i} = \dfrac{(2 + i)(4 + 5i)}{(4 - 5i)(4 + 5i)}$

$$= \frac{8 + 10i + 4i + 5i^2}{16 - 25i^2}$$

$$= \frac{8 + 14i - 5}{16 + 25}$$

$$= \frac{3 + 14i}{41} = \frac{3}{41} + \frac{14}{41}i$$

NOTES

Think of $3i$ as $0 + 3i$ and its conjugate as $0 - 3i$, or $-3i$.

Multiplying the numerator and denominator in the original expression by i yields the same result. Try it yourself.

We multiply by $\frac{3 - 2i}{3 - 2i}$, which equals 1.

To write a complex number in standard form, we separate the real and imaginary parts.

Check Yourself 5

Divide.

(a) $\dfrac{5 + i}{5 - 3i}$ (b) $\dfrac{4 + 10i}{2i}$

We have seen that $i = \sqrt{-1}$ and $i^2 = -1$. We can use these two values to develop a table for the powers of i.

$i = i$

$i^2 = -1$

$i^3 = i^2 \cdot i = -1 \cdot i = -i$

$i^4 = i^2 \cdot i^2 = -1 \cdot (-1) = 1$

$i^5 = i^4 \cdot i = 1 \cdot i = i$

$i^6 = i^4 \cdot i^2 = 1 \cdot (-1) = -1$

$i^7 = i^4 \cdot i^3 = 1 \cdot (-i) = -i$

$i^8 = i^4 \cdot i^4 = 1 \cdot 1 = 1$

NOTE

The symbol $\mathbb{C}$ is often used to denote the set of complex numbers.

This pattern $i, -1, -i, 1$ repeats forever. You will see it again in the exercise set (and then in many subsequent math classes!).

We conclude this section with a diagram summarizing the structure of the system of complex numbers.

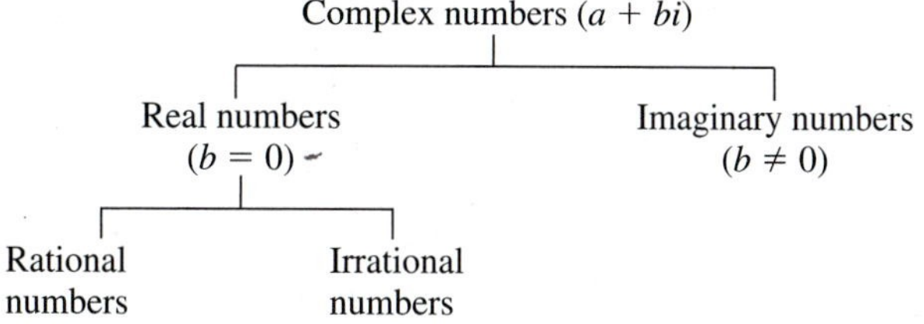

Check Yourself ANSWERS

1. **(a)** $5i$; **(b)** $2i\sqrt{6}$ **2.** **(a)** $7 - 9i$; **(b)** $-9 + 3i$; **(c)** $-2 + 4i$ **3.** $-4 - 19i$ **4.** 34

5. **(a)** $\frac{11}{17} + \frac{10}{17}i$; **(b)** $5 - 2i$

Reading Your Text

These fill-in-the-blank exercises will help you understand some of the key vocabulary used in this section. The answers to these exercises are in the Answers Appendix in the back of the text.

(a) The ________ number i is defined as $i = \sqrt{-1}$ so that $i^2 = -1$.

(b) A ________ number is any number that can be written in the form $a + bi$.

(c) To add two complex numbers, add the real parts and add the ________ parts.

(d) $a + bi$ and $a - bi$ are called complex ________.

Skills Calculator/Computer Career Applications Above and Beyond

7.6 exercises

< Objective 1 >

Write each expression as a multiple of i. Simplify your results where possible.

1. $\sqrt{-16}$

2. $\sqrt{-36}$

3. $-\sqrt{-121}$

4. $-\sqrt{-25}$

5. $\sqrt{-21}$

6. $\sqrt{-23}$

7. $\sqrt{-12}$

8. $\sqrt{-24}$

9. $-\sqrt{-108}$ VIDEO

10. $-\sqrt{-192}$

< Objective 2 >

Perform the indicated operations.

11. $(5 + 4i) - (6 + 5i)$

12. $(2 + 3i) + (4 + 5i)$

13. $(-3 - 2i) + (2 + 3i)$

14. $(-5 - 3i) + (-2 + 7i)$

15. $(5 + 4i) - (3 + 2i)$

16. $(7 + 6i) - (3 + 5i)$

17. $(8 - 5i) - (3 + 2i)$

18. $(7 - 3i) - (-2 - 5i)$

19. $(3 + i) + (4 + 5i) - 7i$

20. $(3 - 2i) + (2 + 3i) + 7i$

21. $(2 + 3i) - (3 - 5i) + (4 + 3i)$ 

22. $(5 - 7i) + (7 + 3i) - (2 - 7i)$

23. $(7 + 3i) - [(3 + i) - (2 - 5i)]$

24. $(8 - 2i) - [(4 + 3i) - (-2 + i)]$

25. $(5 + 3i) + (-5 - 3i)$

26. $(9 - 11i) + (-9 + 11i)$

< Objective 3 >

Find each product.

27. $3i(3 + 5i)$

28. $2i(7 + 3i)$

29. $6i(2 - 5i)$

30. $2i(6 + 3i)$

31. $-2i(4 - 3i)$

32. $-5i(2 - 7i)$

33. $6i\left(\frac{2}{3} + \frac{5}{6}i\right)$

34. $4i\left(\frac{1}{2} + \frac{3}{4}i\right)$

35. $(4 + 3i)(4 - 3i)$

36. $(5 - 2i)(3 - i)$

37. $(4 - 3i)(2 + 5i)$

38. $(7 + 2i)(3 - 2i)$

39. $(-2 - 3i)(-3 + 4i)$

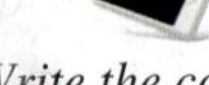

40. $(-5 - i)(-3 - 4i)$

41. $(7 - 3i)^2$

42. $(3 + 7i)^2$

Write the conjugate of each complex number. Then find the product of the given number and its conjugate.

43. $3 - 2i$

44. $5 + 2i$

45. $3 + 2i$

46. $7 - i$

47. $-3 - 2i$

48. $-5 - 7i$

49. $5i$

50. $-3i$

Find each quotient, and write your answer in standard form.

51. $\frac{3 + 2i}{i}$

52. $\frac{5 - 3i}{-i}$

53. $\frac{5 - 2i}{3i}$

54. $\frac{8 + 12i}{-4i}$

55. $\frac{3}{2 + 5i}$

56. $\frac{5}{2 - 3i}$

57. $\frac{13}{2 + 3i}$

58. $\frac{-17}{3 + 5i}$

59. $\frac{2 + 3i}{4 + 3i}$

60. $\frac{4 - 2i}{5 - 3i}$

61. $\frac{3 - 4i}{3 + 4i}$

62. $\frac{7 + 2i}{7 - 2i}$

Complete each statement with **always, sometimes,** *or* **never.**

63. The product of two complex numbers is __________ a real number.

64. If $a \neq 0$ and $b \neq 0$, the square of a complex number $a + bi$ is __________ a real number.

65. The product of a complex number and its conjugate is __________ a real number.

66. A real number can __________ be viewed as a complex number in $a + bi$ form.

Skills	Calculator/Computer	Career Applications	**Above and Beyond**

67. The first application of complex numbers was suggested by the Norwegian surveyor Caspar Wessel in 1797. He found that complex numbers could be used to represent distance and direction on a two-dimensional grid. Why would a surveyor care about such a thing?

68. To what sets of numbers does 1 belong?

We defined $\sqrt{-4} = \sqrt{4}i = 2i$ in the process of expressing the square root of a negative number as a multiple of i. Particular care must be taken with products where two negative radicands are involved. For instance,

$$\sqrt{-3} \cdot \sqrt{-12} = (i\sqrt{3})(i\sqrt{12})$$
$$= i^2\sqrt{36} = (-1)\sqrt{36} = -6$$

is correct. However, if we try to apply the product property for radicals, we have

$\sqrt{-3} \cdot \sqrt{-12} \stackrel{?}{=} \sqrt{(-3)(-12)} = \sqrt{36} = 6$

which is not *correct. The property* $\sqrt{a} \cdot \sqrt{b} = \sqrt{ab}$ *is not applicable in the case where a and b are both negative. Radicals such as* $\sqrt{-a}$ *must be written in the standard form* $i\sqrt{a}$ before *multiplying to use the rules for real-valued radicals.*

Find each product.

69. $\sqrt{-5} \cdot \sqrt{-7}$

70. $\sqrt{-3} \cdot \sqrt{-10}$

71. $\sqrt{-2} \cdot \sqrt{-18}$

72. $\sqrt{-4} \cdot \sqrt{-25}$

73. $\sqrt{-6} \cdot \sqrt{-15}$

74. $\sqrt{-5} \cdot \sqrt{-30}$

75. $\sqrt{-10} \cdot \sqrt{-10}$

76. $\sqrt{-11} \cdot \sqrt{-11}$

Simplify each power of i.

77. i^{10}

78. i^{9}

79. i^{20}

80. i^{15}

81. i^{38}

82. i^{40}

83. i^{51}

84. i^{61}

85. Show that a square root of i is $\frac{\sqrt{2}}{2} + \frac{\sqrt{2}}{2}i$. That is, $\left(\frac{\sqrt{2}}{2} + \frac{\sqrt{2}}{2}i\right)^2 = i$.

86. You know that 2 is a cube root of 8, but did you know that there are two more cube roots of 8? Now that you have studied multiplication of complex numbers, show that $-1 + i\sqrt{3}$ and $-1 - i\sqrt{3}$ are also cube roots of 8.

Answers

1. $4i$ **3.** $-11i$ **5.** $i\sqrt{21}$ **7.** $2i\sqrt{3}$ **9.** $-6i\sqrt{3}$ **11.** $-1 - i$ **13.** $-1 + i$ **15.** $2 + 2i$ **17.** $5 - 7i$ **19.** $7 - i$ **21.** $3 + 11i$ **23.** $6 - 3i$ **25.** $0 + 0i$ or 0 **27.** $-15 + 9i$ **29.** $30 + 12i$ **31.** $-6 - 8i$ **33.** $-5 + 4i$ **35.** 25 **37.** $23 + 14i$ **39.** $18 + i$ **41.** $40 - 42i$ **43.** $3 + 2i$; 13 **45.** $3 - 2i$; 13 **47.** $-3 + 2i$; 13 **49.** $-5i$; 25 **51.** $2 - 3i$ **53.** $-\frac{2}{3} - \frac{5}{3}i$ **55.** $\frac{6}{29} - \frac{15}{29}i$ **57.** $2 - 3i$ **59.** $\frac{17}{25} + \frac{6}{25}i$ **61.** $-\frac{7}{25} - \frac{24}{25}i$ **63.** sometimes **65.** always **67.** Above and Beyond **69.** $-\sqrt{35}$ **71.** -6 **73.** $-3\sqrt{10}$ **75.** -10 **77.** -1 **79.** 1 **81.** -1 **83.** $-i$ **85.** Above and Beyond

summary :: chapter 7

Definition/Procedure	Example	Reference
Roots and Radicals		**Section 7.1**
Square Roots Every positive number has two square roots. The positive or principal square root of a number a denoted $\sqrt{a}$ The negative square root is written as $-\sqrt{a}$	$\sqrt{25} = 5$ 5 is the principal square root of 25 because $5^2 = 25$. $-\sqrt{49} = -7$	p. 550
Higher Roots Cube roots, fourth roots, and so on are denoted by using an index and a radical. The principal nth root of a is written as Index → $\sqrt[n]{a}$ Radical sign, Radicand	$\sqrt[3]{27} = 3$ $\sqrt[3]{-64} = -4$ $\sqrt[4]{81} = 3$ $\sqrt{(-5)^2} = 5$ $\sqrt[3]{(-3)^3} = -3$	p. 551
Radicals Containing Variables $\sqrt[n]{x^n} = \begin{cases} \lvert x \rvert & \text{if } n \text{ is even} \\ x & \text{if } n \text{ is odd} \end{cases}$	$\sqrt{m^2} = \lvert m \rvert$ $\sqrt[3]{27x^3} = 3x$	p. 554
Pythagorean Theorem In a right triangle, $c^2 = a^2 + b^2$. (c, a, b)	If $c = 17$ and $b = 14$, $c^2 = a^2 + b^2$ $17^2 = a^2 + 14^2$ $17^2 - 14^2 = a^2$ $93 = a^2$ $a - \sqrt{93}$	p. 555
Distance Formula The distance d between two points (x_1, y_1) and (x_2, y_2) is $d = \sqrt{(x_2 - x_1)^2 + (y_2 - y_1)^2}$	Given (3, 9) and (−5, −2), $d = \sqrt{(-5 - 3)^2 + (-2 - 9)^2}$ $= \sqrt{(-8)^2 + (-11)^2}$ $= \sqrt{64 + 121} = \sqrt{185}$	p. 556
Circles The standard form for the equation of a circle with center (h, k) and radius r is $(x - h)^2 + (y - k)^2 = r^2$ Determining the center and radius of the circle from its equation allows us to easily graph the circle.	Given the equation $(x - 2)^2 + (y + 3)^2 = 4$ we see that the center is at $(2, -3)$ and the radius is 2. (y, x, $r = 2$, $(2, -3)$)	p. 557

Continued

Definition/Procedure	Example	Reference
Simplifying Radical Expressions		Section 7.2
Simplifying radical expressions entails applying two properties for radicals.		
Product Property $\sqrt[n]{ab} = \sqrt[n]{a} \cdot \sqrt[n]{b}$ (*a* and *b* nonnegative)	$\sqrt{35} = \sqrt{5 \cdot 7}$ $= \sqrt{5} \cdot \sqrt{7}$	p. 567
Quotient Property $\sqrt[n]{\frac{a}{b}} = \frac{\sqrt[n]{a}}{\sqrt[n]{b}}$ $b \neq 0$ (*a* and *b* nonnegative)	$\sqrt{\frac{2}{5}} = \frac{\sqrt{2}}{\sqrt{5}}$	p. 567
Simplifying Radical Expressions A radical is *simplified* if three conditions are satisfied. **1.** The radicand has no factor raised to a power greater than or equal to the index. **2.** No fraction appears in the radical. **3.** No radical appears in a denominator. Satisfying the third condition may require *rationalizing the denominator.*	$\sqrt{18x^3} = \sqrt{9x^2 \cdot 2x}$ $= \sqrt{9x^2} \cdot \sqrt{2x}$ $= 3x\sqrt{2x}$ $\sqrt{\frac{5}{9}} = \frac{\sqrt{5}}{\sqrt{9}} = \frac{\sqrt{5}}{3}$ $\sqrt{\frac{3}{7x}} = \frac{\sqrt{3}}{\sqrt{7x}} = \frac{\sqrt{3} \cdot \sqrt{7x}}{\sqrt{7x} \cdot \sqrt{7x}}$ $= \frac{\sqrt{21x}}{\sqrt{49x^2}} = \frac{\sqrt{21x}}{7x}$	p. 567
Operations on Radicals		Section 7.3
Radical expressions may be combined by using addition or subtraction only if they are *similar*, that is, if they have the same radicand with the same index. Similar radicals are combined by application of the distributive property.	$8\sqrt{5} + 3\sqrt{5} = (8 + 3)\sqrt{5}$ $= 11\sqrt{5}$ $2\sqrt{18} - 4\sqrt{2}$ $= 2\sqrt{9 \cdot 2} - 4\sqrt{2}$ $= 2\sqrt{9} \cdot \sqrt{2} - 4\sqrt{2}$ $= 2 \cdot 3\sqrt{2} - 4\sqrt{2}$ $= 6\sqrt{2} - 4\sqrt{2} = (6 - 4)\sqrt{2}$ $= 2\sqrt{2}$	p. 576
Multiplication To multiply two radical expressions, we use $\sqrt[n]{a} \cdot \sqrt[n]{b} = \sqrt[n]{ab}$ and simplify the product. If binomial expressions are involved, we use the distributive property or the FOIL method.	$\sqrt{3x} \cdot \sqrt{6x^2} = \sqrt{18x^3}$ $= \sqrt{9x^2 \cdot 2x}$ $= \sqrt{9x^2} \cdot \sqrt{2x}$ $= 3x\sqrt{2x}$ $\sqrt{2}(5 + \sqrt{8}) = \sqrt{2} \cdot 5 + \sqrt{2} \cdot \sqrt{8}$ $= 5\sqrt{2} + 4$ $(3 + \sqrt{2})(5 - \sqrt{2})$ $= 15 - 3\sqrt{2} + 5\sqrt{2} - 2$ $= 13 + 2\sqrt{2}$	p. 579

Continued

Definition/Procedure	Example	Reference
Division To divide two radical expressions, rationalize the denominator by multiplying the numerator and denominator by the appropriate radical. If the divisor (the denominator) is a binomial, multiply the numerator and denominator by the conjugate of the denominator.	$\frac{5}{\sqrt{8}} = \frac{5 \cdot \sqrt{2}}{\sqrt{8} \cdot \sqrt{2}} = \frac{5\sqrt{2}}{\sqrt{16}}$ $= \frac{5\sqrt{2}}{4}$ $3 + \sqrt{5}$ is the conjugate of $3 - \sqrt{5}$. $\frac{2}{3 - \sqrt{5}} = \frac{2(3 + \sqrt{5})}{(3 - \sqrt{5})(3 + \sqrt{5})}$ $= \frac{2(3 + \sqrt{5})}{4}$ $= \frac{3 + \sqrt{5}}{2}$	p. 582
Solving Radical Equations		**Section 7.4**
Power Property of Equality If $a = b$ then $a^n = b^n$	If $\sqrt{x + 1} = 5$ then $(\sqrt{x + 1})^2 = 5^2$ $x + 1 = 25$ $x = 24$	p. 588
Solving Radical Equations **Step 1** Isolate a radical term on one side of the equation. **Step 2** Raise each side of the equation to the smallest power that eliminates the isolated radical. **Step 3** If any radicals remain in the equation, repeat step 1. **Step 4** Solve the resulting equation to determine any possible solutions. **Step 5** Check all solutions and discard any extraneous ones.	Solve $\sqrt{x} + \sqrt{x + 7} = 7$. $\sqrt{x} = 7 - \sqrt{x + 7}$ $x = 49 - 14\sqrt{x + 7} + (x + 7)$ $x = 56 + x - 14\sqrt{x + 7}$ $-56 = -14\sqrt{x + 7}$ $4 = \sqrt{x + 7}$ $16 = x + 7$ $x = 9$ Check: $\sqrt{9} + \sqrt{9 + 7} \stackrel{?}{=} 7$ $3 + 4 = 7$ ✓ Solution set: $\{9\}$	p. 592
Rational Exponents		**Section 7.5**
Rational exponents are an alternate way of indicating roots. If a is any real number and n is a positive integer ($n > 1$), $a^{1/n} = \sqrt[n]{a}$ We restrict a so that a is nonnegative when n is even.	$36^{1/2} = \sqrt{36} = 6$ $-27^{1/3} = -\sqrt[3]{27} = -3$ $243^{1/5} = \sqrt[5]{243} = 3$ $25^{-1/2} = \frac{1}{\sqrt{25}} = \frac{1}{5}$	p. 597
For any real number a and positive integers m and n, with $n > 1$, then $a^{m/n} = (\sqrt[n]{a})^m = \sqrt[n]{a^m}$	$27^{2/3} = (\sqrt[3]{27})^2$ $= 3^2 = 9$ $(a^4b^8)^{3/4} = \sqrt[4]{(a^4b^8)^3}$ $= \sqrt[4]{a^{12}b^{24}}$ $= a^3b^6$	p. 598
Properties of Exponents The five properties for exponents continue to hold for rational exponents. **Product Rule** $a^m \cdot a^n = a^{m+n}$	$x^{1/2} \cdot x^{1/3} = x^{1/2 + 1/3} = x^{5/6}$	p. 600

Continued

Definition/Procedure	Example	Reference
Quotient Rule $\frac{a^m}{a^n} = a^{m-n}$	$\frac{x^{3/2}}{x^{1/2}} = x^{3/2-1/2} = x^{2/2} = x$	p. 600
Power Rule $(a^m)^n = a^{m \cdot n}$	$(x^{1/3})^5 = x^{1/3 \cdot 5} = x^{5/3}$	p. 600
Product-Power Rule $(ab)^m = a^m b^m$	$(2xy)^{1/2} = 2^{1/2}x^{1/2}y^{1/2}$	p. 600
Quotient-Power Rule $\left(\frac{a}{b}\right)^m = \frac{a^m}{b^m}$	$\left(\frac{x^{1/3}}{3}\right)^2 = \frac{(x^{1/3})^2}{3^2}$ $= \frac{x^{2/3}}{9}$	p. 600
Complex Numbers		**Section 7.6**
The number i is defined as $i = \sqrt{-1}$ so that $i^2 = -1$	$\sqrt{-16} = 4i$ $\sqrt{-8} = 2i\sqrt{2}$	p. 606
A **complex number** is any number that can be written in the form $a + bi$ in which a and b are real numbers.		p. 607
Addition and Subtraction For the complex numbers $a + bi$ and $c + di$, $(a + bi) + (c + di) = (a + c) + (b + d)i$ and $(a + bi) - (c + di) = (a - c) + (b - d)i$	$(2 + 3i) + (-3 - 5i)$ $= (2 - 3) + (3 - 5)i$ $= -1 - 2i$ $(5 - 2i) - (3 - 4i)$ $= (5 - 3) + [-2 - (-4)]i$ $= 2 + 2i$	p. 607
Multiplication For the complex numbers $a + bi$ and $c + di$, $(a + bi)(c + di) = (ac - bd) + (ad + bc)i$ It is generally easier to use the FOIL multiplication pattern and the definition of i than to apply the formula.	$(2 + 5i)(3 - 4i)$ $= 6 - 8i + 15i - 20i^2$ $= 6 + 7i - 20(-1)$ $= 26 + 7i$	p. 609
Division To divide complex numbers, we multiply the numerator and denominator by the complex conjugate of the denominator and write the result in standard form.	$\frac{3 + 2i}{3 - 2i} = \frac{(3 + 2i)(3 + 2i)}{(3 - 2i)(3 + 2i)}$ $= \frac{9 + 6i + 6i + 4i^2}{9 - 4i^2}$ $= \frac{9 + 12i + 4(-1)}{9 - 4(-1)}$ $= \frac{5 + 12i}{13}$ $= \frac{5}{13} + \frac{12}{13}i$	p. 610

summary exercises :: chapter 7

This summary exercise set will help ensure that you have mastered each of the objectives of this chapter. The exercises are grouped by section. You should reread the material associated with any exercises that you find difficult. The answers to the odd-numbered exercises are in the Answers Appendix in the back of the text.

7.1 *Evaluate each root over the set of real numbers.*

1. $\sqrt{121}$ **2.** $-\sqrt{64}$ **3.** $\sqrt{-81}$

4. $\sqrt[3]{64}$ **5.** $\sqrt[3]{-64}$ **6.** $\sqrt[4]{81}$

7. $\sqrt{\dfrac{9}{16}}$ **8.** $\sqrt[3]{-\dfrac{8}{27}}$ **9.** $\sqrt{8^2}$

Simplify each expression. Assume that all variables represent positive real numbers.

10. $\sqrt{4x^2}$ **11.** $\sqrt{a^4}$ **12.** $\sqrt{36y^2}$

13. $\sqrt{49w^4z^6}$ **14.** $\sqrt[3]{x^9}$ **15.** $\sqrt[3]{-27b^6}$

16. $\sqrt[3]{8r^3s^9}$ **17.** $\sqrt[4]{16x^4y^8}$ **18.** $\sqrt[5]{32p^5q^{15}}$

Find the distance between each pair of points.

19. $(4,8)$ and $(6, 2)$ **20.** $(-4, 3)$ and $(-1, 1)$ **21.** $(-1, -2)$ and $(1, 3)$ **22.** $(-2, -5)$ and $(3, -1)$

Find the center and radius of the circle given by each equation.

23. $x^2 + y^2 = 81$ **24.** $(x - 3)^2 + y^2 = 36$ **25.** $(x + 2)^2 + (y - 1)^2 = 25$

Graph each equation.

26. $x^2 + y^2 = 9$ **27.** $(x - 2)^2 + y^2 = 9$ **28.** $(x + 3)^2 + (y + 3)^2 = 25$

7.2 *Simplify each expression.*

29. $\sqrt{45}$ **30.** $-\sqrt{75}$ **31.** $\sqrt{60x^2}$ **32.** $\sqrt{108a^3}$

33. $\sqrt[3]{32}$ **34.** $\sqrt[3]{-80w^4z^3}$ **35.** $\sqrt{\dfrac{9}{16}}$ **36.** $\sqrt{\dfrac{7}{36}}$

37. $\sqrt{\dfrac{y^4}{49}}$ **38.** $\sqrt{\dfrac{2x}{9}}$ **39.** $\sqrt{\dfrac{5}{16x^2}}$ **40.** $\sqrt[3]{\dfrac{5a^2}{27}}$

7.3

41. $7\sqrt{10} + 4\sqrt{10}$ **42.** $5\sqrt{3x} - 2\sqrt{3x}$ **43.** $7\sqrt[3]{2x} + 3\sqrt[3]{2x}$ **44.** $8\sqrt{10} - 3\sqrt{10} + 2\sqrt{10}$

45. $\sqrt{72} + \sqrt{50}$ **46.** $9\sqrt{7} - 2\sqrt{63}$ **47.** $\sqrt{20} - \sqrt{45} + 2\sqrt{125}$ **48.** $2\sqrt[3]{16} + 3\sqrt[3]{54}$

49. $\sqrt{27w^3} - w\sqrt{12w}$ **50.** $\sqrt[3]{128a^5} + 6a\sqrt[3]{2a^2}$ **51.** $\sqrt{20} + \frac{3}{\sqrt{5}}$ **52.** $\sqrt{72x} - \sqrt{\frac{x}{2}}$

53. $\sqrt[3]{81a^4} - a\sqrt[3]{\frac{a}{9}}$ **54.** $\frac{\sqrt{15}}{3} - \frac{1}{\sqrt{15}}$ **55.** $\sqrt{3x} \cdot \sqrt{7y}$ **56.** $\sqrt{6x^2} \cdot \sqrt{18}$

57. $\sqrt[3]{4a^2b} \cdot \sqrt[3]{ab^2}$ **58.** $\sqrt{5}(\sqrt{3} + 2)$ **59.** $\sqrt{6}(\sqrt{8} - \sqrt{2})$ **60.** $\sqrt{a}(\sqrt{5a} + \sqrt{125a})$

61. $(\sqrt{3} + 5)(\sqrt{3} - 7)$ **62.** $(\sqrt{7} - \sqrt{2})(\sqrt{7} + \sqrt{3})$ **63.** $(\sqrt{5} - 2)(\sqrt{5} + 2)$ **64.** $(\sqrt{7} - \sqrt{3})(\sqrt{7} + \sqrt{3})$

65. $(2 + \sqrt{3})^2$ **66.** $(\sqrt{5} - \sqrt{2})^2$ **67.** $\sqrt{\frac{3}{7}}$ **68.** $\frac{\sqrt{12}}{\sqrt{x}}$

69. $\frac{\sqrt{10a}}{\sqrt{5b}}$ **70.** $\sqrt[3]{\frac{3}{a^2}}$ **71.** $\frac{2}{\sqrt[3]{3x}}$ **72.** $\frac{\sqrt[3]{x^2}}{\sqrt[3]{y^5}}$

73. $\frac{1}{3 + \sqrt{2}}$ **74.** $\frac{11}{5 - \sqrt{3}}$ **75.** $\frac{\sqrt{5} - 2}{\sqrt{5} + 2}$ **76.** $\frac{\sqrt{x} - 3}{\sqrt{x} + 3}$

7.4 *Solve and check each equation.*

77. $\sqrt{x - 5} = 4$ **78.** $\sqrt{3x - 2} + 2 = 5$ **79.** $\sqrt{y + 7} = y - 5$ **80.** $\sqrt{2x - 1} + x = 8$

81. $\sqrt[3]{5x + 2} = 3$ **82.** $\sqrt[3]{x^2 + 2} - 3 = 0$ **83.** $\sqrt{z + 7} = 1 + \sqrt{z}$ **84.** $\sqrt{4x + 5} - \sqrt{x - 1} = 3$

7.5 *Evaluate each expression.*

85. $49^{1/2}$ **86.** $-100^{1/2}$ **87.** $(-27)^{1/3}$ **88.** $16^{1/4}$

89. $64^{2/3}$ **90.** $25^{3/2}$ **91.** $\left(\frac{4}{9}\right)^{3/2}$ **92.** $\left(\frac{27}{125}\right)^{2/3}$

93. $49^{-1/2}$ **94.** $81^{-3/4}$

Simplify each expression.

95. $x^{3/2} \cdot x^{5/2}$ **96.** $b^{2/3} \cdot b^{3/2}$ **97.** $\frac{r^{8/5}}{r^{3/5}}$ **98.** $\frac{a^{5/4}}{a^{1/2}}$

99. $(x^{3/5})^{2/3}$ **100.** $(y^{-4/3})^6$ **101.** $(x^{4/5}y^{3/2})^{10}$ **102.** $(16x^{1/3}y^{2/3})^{3/4}$

103. $\left(\frac{x^{-2}y^{-1/6}}{x^{-4}y}\right)^3$ **104.** $\left(\frac{27y^3z^{-6}}{x^{-3}}\right)^{1/3}$

Write each expression in radical form.

105. $x^{3/4}$ **106.** $(w^2z)^{2/5}$ **107.** $3a^{2/3}$ **108.** $(3a)^{2/3}$

Write each expression using rational exponents and simplify when necessary.

109. $\sqrt[5]{7x}$ **110.** $\sqrt{16w^4}$ **111.** $\sqrt[3]{27p^3q^9}$ **112.** $\sqrt[4]{16a^8b^{16}}$

7.6 *Write each root as a multiple of i. Simplify your result.*

113. $\sqrt{-49}$ **114.** $\sqrt{-13}$ **115.** $\sqrt{-48}$ **116.** $-\sqrt{-60}$

Perform the indicated operations.

117. $(2 + 3i) + (3 - 5i)$ **118.** $(7 - 3i) + (-3 - 2i)$ **119.** $(5 - 3i) - (2 + 5i)$ **120.** $(-4 + 2i) - (-1 - 3i)$

Find each product.

121. $4i(7 - 2i)$ **122.** $(5 - 2i)(3 + 4i)$ **123.** $(3 - 4i)^2$ **124.** $(2 - 3i)(2 + 3i)$

Find each quotient. Write your answers in standard form.

125. $\frac{5 - 15i}{5i}$ **126.** $\frac{10}{3 - 4i}$ **127.** $\frac{3 - 2i}{3 + 2i}$ **128.** $\frac{5 + 10i}{2 + i}$

chapter test 7

CHAPTER 7

Use this chapter test to assess your progress and to review for your next exam. Allow yourself about an hour to take this test. The answers to these exercises are in the Answers Appendix in the back of the text.

Assume that all variables represent positive real numbers in all problems.

Simplify each expression.

1. $\sqrt[3]{9p^7q^5}$ **2.** $\sqrt{\frac{5x}{8y}}$ **3.** $\sqrt{6x}(\sqrt{18x} - \sqrt{2x})$ **4.** $\sqrt{49a^4}$

5. $\frac{7x}{\sqrt{64y^2}}$ **6.** $\frac{\sqrt{6} - \sqrt{3}}{\sqrt{6} + \sqrt{3}}$ **7.** $(16x^4)^{3/2}$ **8.** $\left(\frac{4x^{1/5}y^{1/5}}{x^{-7/5}y^{3/5}}\right)^{5/2}$

9. $\sqrt[3]{-27w^6z^9}$ **10.** $\sqrt[3]{4x^5}\,\sqrt[3]{8x^6}$

Use a calculator to evaluate each root. Round your answers to the nearest tenth.

11. $\sqrt{43}$ **12.** $\sqrt[3]{\frac{73}{27}}$

Solve and check each equation.

13. $\sqrt{x - 7} - 2 = 0$ **14.** $\sqrt{3w + 4} + w = 8$

Simplify each expression.

15. $\frac{3}{\sqrt[3]{9x}}$ **16.** $\sqrt{7x^3}\,\sqrt{2x^4}$

17. $\sqrt[3]{54m^4} + m\sqrt[3]{16m}$ **18.** $\sqrt{3x^3} + x\sqrt{75x} - \sqrt{27x^3}$

Write the expression in radical form and simplify.

19. $(a^7b^3)^{2/5}$

Write the expression using rational exponents and simplify.

20. $\sqrt[3]{125p^9q^6}$

Simplify each expression.

21. $5\sqrt{20} + 4\sqrt{45}$

22. $\sqrt{32x^5y^7z^6}$

23. $(27m^{3/2}n^{-6})^{2/3}$

24. $\left(\dfrac{16r^{-1/3}s^{5/3}}{rs^{-7/3}}\right)^{3/4}$

25. $(\sqrt{11} - \sqrt{5})^2$

Find the distance between each pair of points. Express your answer in radical form and as a decimal, rounded to the nearest thousandth.

26. $(-2, -3)$ and $(5, 2)$

27. $(-5, 8)$ and $(6, 0)$

Give the center and radius of the circle whose equation is given.

28. $(x - 4)^2 + (y + 1)^2 = 49$

Simplify each expression and write your answer in standard form.

29. $(-5 + 3i)(7 - 6i)$

30. $\dfrac{4 + 6i}{2 - 3i}$

cumulative review chapters 0–7

Use this exercise set to review concepts from earlier chapters. While it is not a comprehensive exam, it will help you identify any material that you need to review before moving on to the next chapter. The answers to these exercises are in the Answers Appendix in the back of the text.

1. Solve the equation $7x - 6(x - 1) = 2(5 + x) + 11$.

2. If $f(x) = 3x^6 - 4x^3 + 9x^2 - 11$, find $f(-1)$.

3. Find the slope-intercept equation of the line that has a y-intercept of $(0, -6)$ and is parallel to the line given by the equation $6x - 4y = 18$.

4. The sum of three consecutive integers is 54. Find the three integers.

Simplify each expression.

5. $5x^2 - 8x + 11 - (-3x^2 - 2x + 8) - (-2x^2 - 4x + 3)$

6. $(5x + 3)(2x - 9)$

Factor each expression completely.

7. $2x^3 + x^2 - 3x$

8. $9x^4 - 36y^4$

9. $4x^2 + 8xy - 5x - 10y$

10. $x^4 - 13x^2 - 48$

11. Find the slope of the line whose equation is $4x + 3y = 7$.

12. Write the slope-intercept equation of the line that passes through the point $(-4, 2)$ and is perpendicular to the line with equation $y = 4x + 5$.

13. A company that produces computer games has found that its daily operating cost in dollars is $C = 30x + 500$ and its daily revenue in dollars is $R = 75x - x^2$. For what value(s) of x will the company break even?

Simplify each expression.

14. $\sqrt{3x^3y}\,\sqrt{4x^5y^6}$

15. $(\sqrt{3} - 5)(\sqrt{2} + 3)$

Graph each equation.

16. $y = 3x - 5$

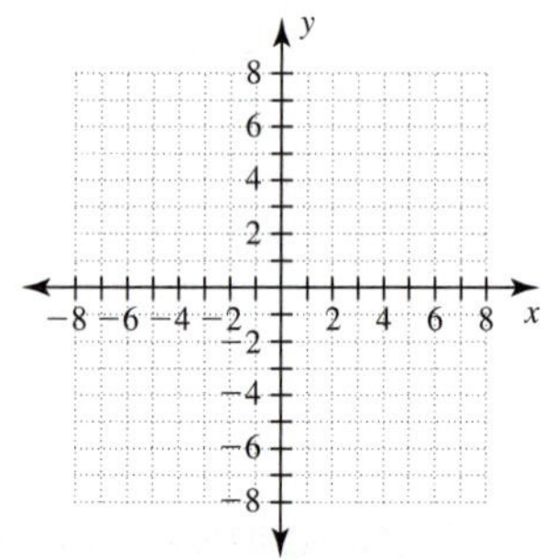

17. $x = -5$

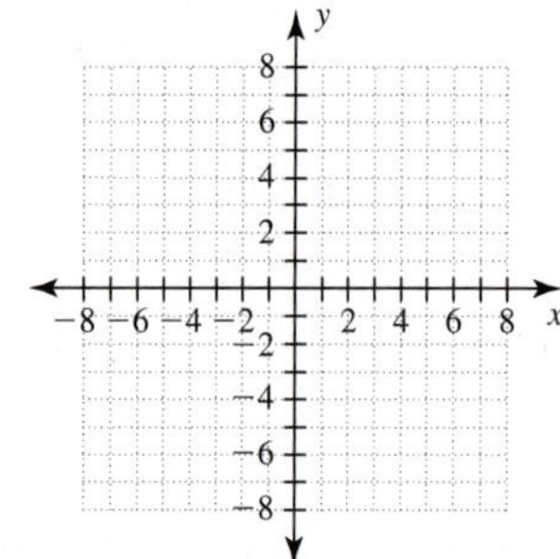

18. $2x - 3y = 12$

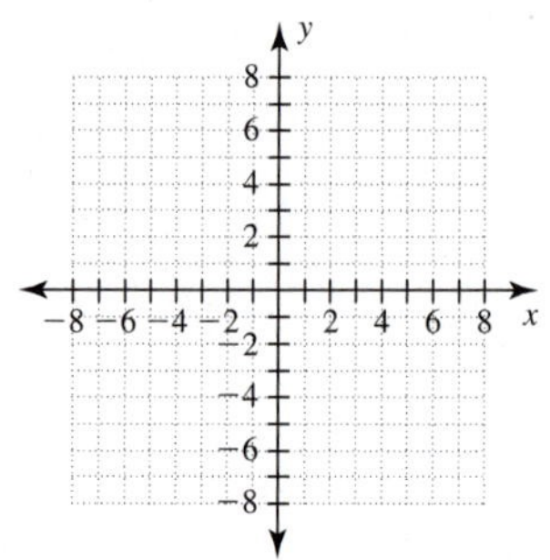

Solve each system of equations.

19. $4x - 3y = 15$
$x + y = 2$

20. $6x - 5y = 27$
$x = 5y + 2$

Simplify.

21. $\left(\dfrac{x^2y^{-3}}{x^5y^4}\right)^{-2}$

22. $(-12x^3y^2)(-18xy^3)$

23. $\left(\dfrac{64x^3y^9}{27}\right)^{1/3}$

Solve.

24. $\sqrt{x + 3} = 2$

25. Solve the inequality.

$5x - (2 - 3x) \geq 6 + 10x$

26. Find the zeros of the function $f(x) = 2x^2 + 9x - 5$.

27. The length of a rectangle is 3 inches less than twice its width. If the perimeter of the rectangle is 96 inches, find the dimensions of the rectangle.

CHAPTER

8

Quadratic Functions

INTRODUCTION

Large cities often commission fireworks artists to choreograph elaborate displays on holidays. Such displays look like beautiful paintings in the sky, in which the fireworks seem to dance to well-known popular and classical music. The displays are feats of engineering and very accurate timing. Suppose the designer wants a second set of rockets of a certain color and shape to be released after the first set reaches a specific height and explodes. He or she must know the strength of the initial liftoff and use a quadratic equation to determine the proper time for setting off the second round.

The equation $h = -16t^2 + 100t$ gives the height in feet t seconds after the rockets are shot into the air if the initial velocity is 100 ft/s. Using this equation, the designer knows how high the rocket will ascend and when it will begin to fall. The designer can time the next round to achieve the desired effect. Displays that involve large banks of fireworks in shows that last up to an hour are programmed using computers, but quadratic equations are at the heart of the mechanism that creates the beautiful effects.

CHAPTER 8 OUTLINE

8.1 Solving Quadratic Equations

< 8.1 Objectives >

1 > Use the square-root method to solve quadratic equations

2 > Solve quadratic equations by completing the square

3 > Use quadratic equations to solve applications

RECALL

A quadratic equation is in *standard form* when it is written as

$ax^2 + bx + c = 0$

Our work with radicals in Chapter 7 provides us with powerful tools to solve quadratic equations.

In Chapter 6, we defined quadratic equations as those that can be written as

$$ax^2 + bx + c = 0 \quad \text{where } a \neq 0$$

You learned to solve quadratic equations by factoring, which we begin by reviewing. We then extend our problem-solving techniques by taking advantage of our knowledge of square roots.

Example 1 Factoring Quadratic Equations

Solve each quadratic equation.

(a) $x^2 + 3x - 10 = 0$

We begin by factoring the left side.

RECALL

We want the pair of numbers whose product is -10 and whose sum is 3.

$$x^2 + 3x - 10 = 0$$

$$(x + 5)(x - 2) = 0 \qquad x^2 + 3x - 10 = (x + 5)(x - 2)$$

Next, we use the zero-product principle to construct two equations.

$$x + 5 = 0 \quad \text{or} \quad x - 2 = 0$$

$$x = -5 \quad \text{or} \quad x = 2$$

Finally, we check our results.

Check

$x = -5$

$$x^2 + 3x - 10 = 0$$

$$(-5)^2 + 3(-5) - 10 \stackrel{?}{=} 0$$

$$25 - 15 - 10 \stackrel{?}{=} 0$$

$$0 = 0 \quad \text{True}$$

$x = 2$

$$x^2 + 3x - 10 = 0$$

$$(2)^2 + 3(2) - 10 \stackrel{?}{=} 0$$

$$4 + 6 - 10 \stackrel{?}{=} 0$$

$$0 = 0 \quad \text{True}$$

Both check, so the solution set is $\{-5, 2\}$.

(b) $x^2 = -7x - 12$

First, we write the equation in standard form.

NOTE

Add $7x$ and 12 to both sides of the equation. The quadratic expression must be *set equal* to 0.

$$x^2 + 7x + 12 = 0$$

Once the equation is in standard form, we can factor the quadratic expression.

$$(x + 3)(x + 4) = 0$$

Finally, using the zero-product principle, we solve the equations $x + 3 = 0$ and $x + 4 = 0$ to obtain

$$x = -3 \quad \text{or} \quad x = -4$$

We leave it to you to check the solutions. The solution set is $\{-3, -4\}$.

(c) $x^2 = 16$

Again, we write the equation in standard form.

$x^2 - 16 = 0$

Factoring, we have

$(x + 4)(x - 4) = 0$

The solutions are

$x = -4$ or $x = 4$

The solution set is $\{-4, 4\}$ or $\{\pm 4\}$.

NOTE

We factor the quadratic expression as a difference of squares.

Check Yourself 1

Solve each quadratic equation.

(a) $2x^2 - x - 3 = 0$ **(b)** $x^2 - 4x = 45$ **(c)** $w^2 = 25$

Consider the equation in Example 1(c).

$x^2 = 16$

If we take the square root of both sides, we have

$\sqrt{x^2} = \sqrt{16}$

We know this last equation is equivalent to

$|x| = 16$

So, $x = 4$ or $x = -4$. We usually write this solution set as $\{\pm 4\}$.

RECALL

$\sqrt{x^2} = |x|$

Solving a quadratic equation of the form $x^2 = k$ by taking the square root of both sides is called the **square-root method**. Be careful to include both the positive and negative roots of k.

Property

Square-Root Property

If $x^2 = k$, then

$x = \sqrt{k}$ or $x = -\sqrt{k}$

Example 2 Using the Square-Root Method

< Objective 1 >

Use the square-root method to solve each equation.

(a) $x^2 = 9$

By the square-root property,

$x = \sqrt{9}$ or $x = -\sqrt{9}$

$= 3$ $\quad = -3$ or $\{\pm 3\}$

(b) $x^2 - 17 = 0$

Add 17 to both sides of the equation.

$x^2 = 17$

$x = \sqrt{17}$ or $x = -\sqrt{17}$ or $\{-\sqrt{17}, \sqrt{17}\}$ or $\{\pm\sqrt{17}\}$

NOTE

With a calculator, we can approximate $\sqrt{17} \approx 4.123$ (rounded to three decimal places).

(c) $4x^2 - 3 = 0$

$$4x^2 = 3$$

$$x^2 = \frac{3}{4}$$

$$x = \pm\sqrt{\frac{3}{4}} = \pm\frac{\sqrt{3}}{\sqrt{4}}$$

$$x = \pm\frac{\sqrt{3}}{2} \quad \text{or} \quad \left\{\pm\frac{\sqrt{3}}{2}\right\}$$

NOTE

In Example 2(d), we see that complex-number solutions may result.

(d) $x^2 + 1 = 0$ $(\mathbb{C})$

$$x^2 = -1$$

$$x = \pm\sqrt{-1}$$

$$x = \pm i \quad \text{or} \quad \{\pm i\}$$ Nonreal

Check Yourself 2

Solve each equation.

(a) $x^2 = 5$ **(b)** $x^2 - 2 = 0$ **(c)** $9x^2 - 8 = 0$ **(d)** $x^2 + 9 = 0$ $(\mathbb{C})$

We can even use this approach to solve an equation like

$$(x + 3)^2 = 16$$

We say that the quantity inside the parentheses, $x + 3$, must be equal to 4 or -4.

$$x + 3 = \sqrt{16} \quad \text{or} \quad x + 3 = -\sqrt{16}$$

$$x + 3 = 4 \quad \text{or} \quad x + 3 = -4$$

Solving for x yields

$$x = -3 + 4 \quad \text{or} \quad x = -3 - 4$$

$$= 1 \qquad\qquad = -7$$

The solution set is $\{-7, 1\}$.

To check, substitute into the original equation:

$$[(-7) + 3]^2 \stackrel{?}{=} 16$$

$$(-4)^2 \stackrel{?}{=} 16$$

$$16 = 16$$ True

and

$$[(1) + 3]^2 \stackrel{?}{=} 16$$

$$(4)^2 \stackrel{?}{=} 16$$

$$16 = 16$$ True

Example 3 Using the Square-Root Method

NOTE

The two solutions $5 + \sqrt{5}$ and $5 - \sqrt{5}$ are abbreviated as $5 \pm \sqrt{5}$.

Use the square-root method to solve each equation.

(a) $(x - 5)^2 - 5 = 0$

$$(x - 5)^2 = 5$$ Add 5 to both sides.

$$x - 5 = \pm\sqrt{5}$$ Use the square-root property.

$$x = 5 \pm\sqrt{5} \quad \text{or} \quad \{5 \pm \sqrt{5}\}$$ Add 5 to both sides.

(b) $9(y + 1)^2 - 2 = 0$

$9(y + 1)^2 = 2$ Add 2 to both sides.

$(y + 1)^2 = \frac{2}{9}$ Divide both sides by 9.

$y + 1 = \pm\sqrt{\frac{2}{9}}$ Use the square-root property.

$y + 1 = \pm\frac{\sqrt{2}}{3}$ Use the quotient property for radicals.

$y = -1 \pm \frac{\sqrt{2}}{3}$ Add -1 to both sides.

$= \frac{-3}{3} \pm \frac{\sqrt{2}}{3}$ Use a common denominator of 3.

$= \frac{-3 \pm \sqrt{2}}{3}$ Combine the fractions.

The solution set is $\left\{\frac{-3 \pm \sqrt{2}}{3}\right\}$.

Check Yourself 3

Use the square-root method to solve each equation.

(a) $(x - 2)^2 - 3 = 0$ **(b)** $4(x - 1)^2 = 3$

We summarize our work so far.

Step by Step

Using the Square-Root Method

Step 1 Use the addition property to isolate the squared term on one side of the equation.

Step 2 Use the multiplication property to isolate the squared factor. If the squared factor equals a negative number, the equation has no real solutions. Otherwise, continue to step 3.

Step 3 Take the square root of both sides of the equation. Remember that every positive number has two square roots.

Step 4 Use the addition and multiplication properties to solve the remaining equation for x, if necessary.

Graphing Calculator Option

Using the Memory Feature to Check Solutions

In Section 1.2, you learned how to use the memory features of a graphing calculator to evaluate expressions. We can use that same approach to check complicated solutions to equations.

In Example 3(b), we solved the equation

$9(y + 1)^2 - 2 = 0$

and obtained the solution set $\left\{\frac{-3 \pm \sqrt{2}}{3}\right\}$.

To check, store the value $\frac{-3 + \sqrt{2}}{3}$ in memory location **Y**.

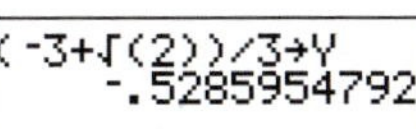

Then evaluate the expression on the left side of the original equation, entering memory cell **Y** where you see y in the equation. We expect the result to be 0.

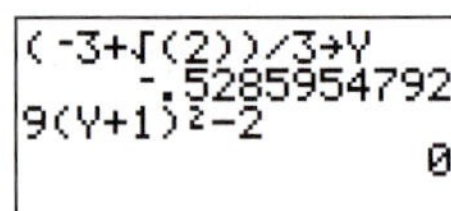

The solution $\frac{-3+\sqrt{2}}{3}$ checks.

Graphing Calculator Check

Check the other solution, $\frac{-3-\sqrt{2}}{3}$, for this same equation.

ANSWERS

It checks. When $\frac{-3-\sqrt{2}}{3}$ is substituted into $9(y+1)^2 - 2$, the output is 0.

NOTE

If $(x-h)^2 = k$, then

$x - h = \pm\sqrt{k}$

and

$x = h \pm \sqrt{k}$

Not all quadratic equations can be solved directly by factoring or using the square-root method. We must extend our techniques.

The square-root method is useful in this process because any quadratic equation can be written in the form

$$(x-h)^2 = k$$

which yields the solutions

$$x = h \pm \sqrt{k}$$

Changing an equation in standard form

$$ax^2 + bx + c = 0$$

to the form

$$(x-h)^2 = k$$

is called **completing the square,** and is based on the relationship between the middle term and the last term of any perfect-square trinomial.

We look at three perfect-square trinomials to see whether we can detect a pattern.

$$x^2 + 4x + 4 = (x+2)^2$$
$$x^2 - 6x + 9 = (x-3)^2$$
$$x^2 + 8x + 16 = (x+4)^2$$

NOTE

This relationship is true *only* if the leading, or x^2-, coefficient is 1.

In each case the last (or constant) term is the square of one-half of the coefficient of x in the middle (or linear) term. For example, in the second equation,

$$x^2 - 6x + 9 = (x-3)^2$$

$\frac{1}{2}$ of this coefficient is -3, and $(-3)^2 = 9$, the constant.

Use the third equation, $x^2 + 8x + 16 = (x+4)^2$, to verify this relationship for yourself. To summarize, in perfect-square trinomials, the constant is always the square of one-half the coefficient of x.

In the next example, we complete the square to produce perfect-square trinomials.

Example 4 Completing the Square

Determine the constant that must be added to the given expression to produce a perfect-square trinomial.

(a) $x^2 + 10x$

The coefficient of the linear or x-term is 10.

Half of 10 is 5.

$5^2 = 25$

Therefore, if we add 25 to $x^2 + 10x$, we have a perfect square.

$x^2 + 10x + 25 = (x + 5)^2$ True

(b) $x^2 - 12x$

The coefficient of the linear or x-term is -12.

Half of -12 is -6.

$(-6)^2 = 36$

Therefore, if we add 36 to $x^2 - 12x$, we have a perfect square.

$x^2 - 12x + 36 = (x - 6)^2$ True

(c) $x^2 + 7x$

We have to work a little harder in this case.

The coefficient of the linear or x-term is 7.

Half of 7 is $\frac{7}{2}$.

$\left(\frac{7}{2}\right)^2 = \frac{49}{4}$.

Therefore, if we add $\frac{49}{4}$ to $x^2 + 7x$, we have a perfect square.

$x^2 + 7x + \frac{49}{4} = \left(x + \frac{7}{2}\right)^2$. True

Check Yourself 4

In each case, determine the constant needed to produce a perfect-square trinomial. Then write the trinomial as a binomial squared.

(a) $x^2 - 16x$ **(b)** $x^2 + 3x$

In Example 5, we solve a quadratic equation by **completing the square**.

Example 5 Solving an Equation by Completing the Square

< Objective 2 >

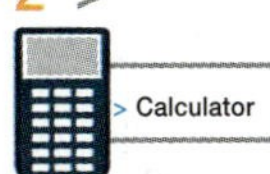

NOTE

If we graph the related function $y = x^2 + 8x - 7$ in the standard viewing window, we see that the x-intercepts are just to the right of -9 and just to the left of 1.

Solve $x^2 + 8x - 7 = 0$ by completing the square.

First, we rewrite the equation with the constant on the *right-hand side*.

$x^2 + 8x = 7$

Our objective is to have a perfect-square trinomial on the left-hand side. We know that we must add the square of one-half of the x-coefficient to complete the square. In this case, that value is 16, so now we add 16 to each side of the equation.

$x^2 + 8x + 16 = 7 + 16$ $\frac{1}{2} \cdot 8 = 4$ and $4^2 = 16$

Factor the perfect-square trinomial on the left, and add on the right.

$(x + 4)^2 = 23$

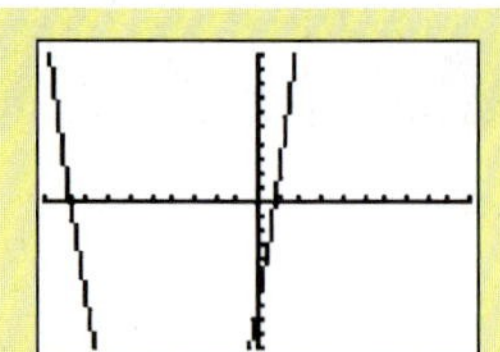

Be certain that you see how these points relate to the exact solutions $-4 + \sqrt{23}$ and $-4 - \sqrt{23}$

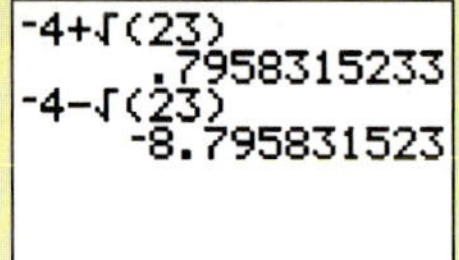

Now use the square-root property.

$$x + 4 = \pm\sqrt{23}$$

Subtracting 4 from both sides of the equation gives

$$x = -4 \pm \sqrt{23} \quad \text{or} \quad \{-4 \pm \sqrt{23}\}$$

Check Yourself 5

Solve $x^2 - 6x - 2 = 0$ by completing the square.

Step by Step

Completing the Square

Step 1 Isolate the constant on the right side of the equation.

Step 2 Divide both sides of the equation by the coefficient of the x^2-term if that coefficient is not equal to 1.

Step 3 Add the square of one-half of the coefficient of the linear or x-term to both sides of the equation. This gives a perfect-square trinomial on the left side of the equation.

Step 4 Write the left side of the equation as the square of a binomial and simplify the right side.

Step 5 Use the square-root property and solve the resulting linear equations.

Example 6 Solving an Equation by Completing the Square

NOTES

Add the square of one-half of the x-coefficient to both sides of the equation.

$$\frac{1}{2} \cdot 5 = \frac{5}{2}$$

$$3 + \left(\frac{5}{2}\right)^2 = 3 + \frac{25}{4}$$

$$= \frac{12}{4} + \frac{25}{4} = \frac{37}{4}$$

Solve $x^2 + 5x - 3 = 0$ by completing the square.

$$x^2 + 5x - 3 = 0$$

$$x^2 + 5x = 3 \qquad \text{Add 3 to both sides.}$$

$$x^2 + 5x + \left(\frac{5}{2}\right)^2 = 3 + \left(\frac{5}{2}\right)^2 \qquad \text{Make the left side a perfect square.}$$

$$\left(x + \frac{5}{2}\right)^2 = \frac{37}{4}$$

$$x + \frac{5}{2} = \pm\frac{\sqrt{37}}{2} \qquad \text{Use the square-root property.}$$

$$x = \frac{-5 \pm \sqrt{37}}{2} \quad \text{or} \quad \left\{\frac{-5 \pm \sqrt{37}}{2}\right\}$$

Check Yourself 6

Solve $x^2 + 3x - 7 = 0$ by completing the square.

Some equations have nonreal or complex solutions, as Example 7 illustrates.

Example 7 Solving an Equation by Completing the Square

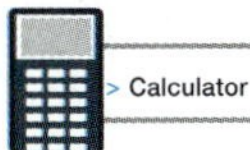

Solve $x^2 + 4x + 13 = 0$ by completing the square. ($\mathbb{C}$)

$$
\begin{aligned}
x^2 + 4x + 13 &= 0 \\
x^2 + 4x &= -13 && \text{Subtract 13 from both sides.} \\
x^2 + 4x + 4 &= -13 + 4 && \text{Add } \left[\tfrac{1}{2}(4)\right]^2 \text{ to both sides.} \\
(x + 2)^2 &= -9 && \text{Factor the left-hand side.} \\
x + 2 &= \pm\sqrt{-9} && \text{Use the square-root property.} \\
x + 2 &= \pm i\sqrt{9} && \text{Simplify the radical.} \\
x + 2 &= \pm 3i \\
x &= -2 \pm 3i \quad \text{or} \quad \{-2 \pm 3i\} && \text{Nonreal}
\end{aligned}
$$

NOTE

The graph of $y = x^2 + 4x + 13$ does not intersect the x-axis.

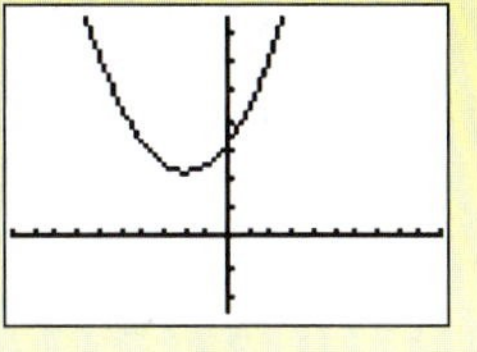

Check Yourself 7

Solve $x^2 + 10x + 41 = 0$. ($\mathbb{C}$)

Example 8 illustrates a situation in which the leading coefficient of the quadratic expression is not 1. An extra step is required in such cases.

Example 8 Solving an Equation by Completing the Square

> CAUTION

Before you can complete the square on the left, the coefficient of x^2 must be 1. If it is not, we *divide* both sides of the equation by that coefficient.

Solve $4x^2 + 8x - 7 = 0$ by completing the square.

$$
\begin{aligned}
4x^2 + 8x - 7 &= 0 \\
4x^2 + 8x &= 7 && \text{Add 7 to both sides.} \\
x^2 + 2x &= \frac{7}{4} && \text{Divide both sides by 4.} \\
x^2 + 2x + 1 &= \frac{7}{4} + 1 && \text{Complete the square on the left.} \\
(x + 1)^2 &= \frac{11}{4} \\
x + 1 &= \pm\sqrt{\frac{11}{4}} && \text{Use the square-root property.} \\
x &= -1 \pm \sqrt{\frac{11}{4}} \\
&= -1 \pm \frac{\sqrt{11}}{2} \\
&= \frac{-2 \pm \sqrt{11}}{2} \quad \text{or} \quad \left\{\frac{-2 \pm \sqrt{11}}{2}\right\}
\end{aligned}
$$

Check Yourself 8

Solve $4x^2 - 8x + 3 = 0$ by completing the square.

RECALL

You learned to use the Pythagorean theorem in Section 7.1.

As we saw in Chapter 6, we use quadratic equations to model many applications. Problems in geometry, construction, physics, and economics all lend themselves to the techniques of this section.

We begin by applying the Pythagorean theorem to a problem.

Example 9 A Construction Application

< Objective 3 >

How long must a guy wire be to reach from the top of a 30-ft pole to a point on the ground 20 ft from the base of the pole? Round to the nearest tenth of a foot.

We start by drawing a sketch of the problem.

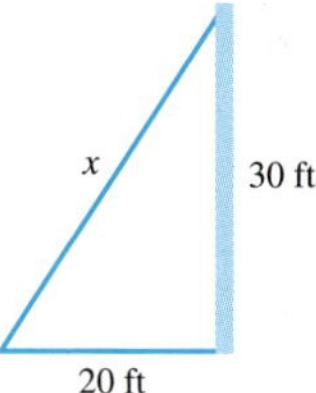

Using the Pythagorean theorem, we write

$$x^2 = 20^2 + 30^2$$

$$x^2 = 400 + 900$$

$$x^2 = 1{,}300$$

We use the square-root method to solve this equation.

$$x = \pm\sqrt{1{,}300}$$

NOTE

Always check to see if your final answer is reasonable.

Since x must be positive, we reject $-\sqrt{1{,}300}$ and keep $\sqrt{1{,}300}$. To the nearest tenth of a foot, we have $x = 36.1$ ft.

Check Yourself 9

A 16-ft ladder leans against a wall with its base 4 ft from the wall. How far above the floor, to the nearest tenth of a foot, is the top of the ladder?

We close this section with an application from the field of economics. Related to the production of a certain item are two important equations: a **supply** equation and a **demand** equation. Each is a function of the price of the item. The **equilibrium** price is the price for which supply equals demand.

Example 10 A Business Application

The demand equation for a certain computer chip is given by

$$D = -4p + 50$$

The supply equation is predicted to be

$$S = -p^2 + 20p - 6$$

Find the equilibrium price.

We wish to know what price p results in equal supply and demand, so we write

$$-p^2 + 20p - 6 = -4p + 50$$

$$-56 = p^2 - 24p$$

or $\quad p^2 - 24p = -56$

Then, completing the square,

$$p^2 - 24p + 144 = -56 + 144$$

$$(p - 12)^2 = 88$$

RECALL

We can simplify

$12 \pm \sqrt{88} = 12 \pm 2\sqrt{22}$

but if we are finding the price using a calculator, there is no need to simplify the radical.

$$p - 12 = \pm\sqrt{88}$$
$$p = 12 \pm \sqrt{88}$$
$$p = 21.38 \text{ or } 2.62$$

You should confirm that when $p = 2.62$, supply and demand are positive.

Here we must be careful: Supply and demand must be positive. When $p = 21.38$, they are negative, so the equilibrium price is \$2.62.

Check Yourself 10

The demand equation for a certain item is given by

$D = -2p + 32$

The supply equation is predicted to be

$S = -p^2 + 18p - 5$

Find the equilibrium price.

Check Yourself ANSWERS

1. **(a)** $\left\{-1, \frac{3}{2}\right\}$; **(b)** $\{-5, 9\}$; **(c)** $\{\pm 5\}$ **2.** **(a)** $\{\pm\sqrt{5}\}$; **(b)** $\{\pm\sqrt{2}\}$; **(c)** $\left\{\pm\frac{2\sqrt{2}}{3}\right\}$; **(d)** $\{\pm 3i\}$ nonreal **3.** **(a)** $\{2 \pm \sqrt{3}\}$; **(b)** $\left\{\frac{2 \pm \sqrt{3}}{2}\right\}$ **4.** **(a)** constant: 64; $x^2 - 16x + 64 = (x - 8)^2$; **(b)** constant: $\frac{9}{4}$; $x^2 + 3x + \frac{9}{4} = \left(x + \frac{3}{2}\right)^2$ **5.** $\{3 \pm \sqrt{11}\}$ **6.** $\left\{\frac{-3 \pm \sqrt{37}}{2}\right\}$ **7.** $\{-5 \pm 4i\}$ nonreal **8.** $\left\{\frac{1}{2}, \frac{3}{2}\right\}$ **9.** 15.5 ft **10.** \$2.06

Reading Your Text

These fill-in-the-blank exercises will help you understand some of the key vocabulary used in this section. The answers to these exercises are in the Answers Appendix in the back of the text.

(a) A ________ equation is an equation that can be written as $ax^2 + bx + c = 0$, in which a is not equal to zero.

(b) The ________ property states that, if $x^2 = k$, then $x = \sqrt{k}$ or $x = -\sqrt{k}$.

(c) The process of changing an equation in standard form to the form $(x + h)^2 = k$ is called ________ the square.

(d) When completing the square, we first isolate the ________ on the right side of the equation.

Skills | Calculator/Computer | Career Applications | Above and Beyond

8.1 exercises

Solve each equation by factoring.

1. $x^2 + 9x + 14 = 0$

2. $x^2 + 5x + 6 = 0$

3. $z^2 - 2z - 35 = 0$

4. $q^2 - 5q - 24 = 0$

5. $2x^2 - 5x - 3 = 0$

6. $3x^2 + 10x - 8 = 0$

7. $6y^2 - y - 2 = 0$

8. $21z^2 + z - 2 = 0$

< Objective 1 >

Use the square-root method to solve each equation.

9. $x^2 = 121$

10. $x^2 = 144$

11. $y^2 = 7$

12. $p^2 = 18$

13. $2x^2 - 12 = 0$

14. $5x^2 = 65$

15. $2t^2 + 12 = 4$ (ℂ)

16. $3u^2 - 5 = -32$ (ℂ)

17. $(x + 1)^2 = 12$ VIDEO

18. $(2x - 3)^2 = 5$

19. $(3z + 1)^2 - 5 = 0$

20. $(3p - 4)^2 + 9 = 0$ (ℂ)

Find the constant that must be added to each binomial expression to form a perfect-square trinomial.

21. $x^2 + 18x$

22. $r^2 - 14r$

23. $y^2 - 8y$

24. $w^2 + 16w$

25. $x^2 - 3x$ VIDEO

26. $z^2 + 7z$

27. $n^2 + n$

28. $x^2 - x$

29. $x^2 + \frac{1}{5}x$

30. $x^2 - \frac{1}{3}x$

31. $x^2 - \frac{1}{6}x$

32. $y^2 - \frac{1}{4}y$

< Objective 2 >

Solve each equation by completing the square.

33. $x^2 + 10x = 4$

34. $x^2 - 14x - 7 = 0$

35. $y^2 - 2y = 8$

36. $z^2 + 4z - 72 = 0$

37. $x^2 - 2x - 5 = 0$ VIDEO

38. $x^2 - 3x = 10$

39. $x^2 + 10x + 13 = 0$

40. $x^2 + 3x - 17 = 0$

41. $z^2 - 5z - 7 = 0$

42. $q^2 - 8q + 20 = 0$ (ℂ)

43. $m^2 - 3m - 5 = 0$

44. $y^2 + y - 5 = 0$

Solve each equation.

45. $x^2 + \frac{1}{2}x = 1$

46. $x^2 - \frac{1}{3}x = 2$

47. $2x^2 + 2x - 1 = 0$

48. $5x^2 - 6x = 3$

49. $3x^2 - 8x = 2$

50. $4x^2 + 8x - 1 = 0$

51. $3x^2 - 2x + 12 = 0$ (ℂ)

52. $7y^2 - 2y + 3 = 0$ (ℂ)

53. $x^2 + 10x + 28 = 0$ (ℂ)

54. $x^2 - 2x + 10 = 0$ (ℂ)

< Objective 3 >

Solve each application. Round your results to the nearest thousandth when necessary.

55. **Construction** How long must a guy wire be to run from the top of a 20-ft pole to a point on the ground 8 ft from the base of the pole?

56. **Construction** How long must a guy wire be to run from the top of a 16-ft pole to a point on the ground 6 ft from the base of the pole?

57. **Construction** The base of a 15-ft ladder is 5 ft away from a wall. How far above the floor is the top of the ladder?

58. **Construction** The base of an 18-ft ladder is 4 ft away from a wall. How far above the floor is the top of the ladder?

59. **Geometry** One leg of a right triangle is twice the length of the other. The hypotenuse is 8 m long. Find the length of each leg. Round your results to the nearest hundredth.

60. **Number Problem** The square of a number decreased by 2 is equal to the opposite of the number. Find the numbers.

61. **Number Problem** The square of 2 more than a number is 64. Find the numbers.

62. **Number Problem** The square of the sum of a number and 5 is 36. Find the numbers.

63. **Geometry** The length of one leg of a right triangle is 4 in. more than the other. If the length of the hypotenuse is 8 in., what are the lengths of the two legs?

64. **Geometry** The length of a rectangle is 2 cm longer than its width. If the diagonal of the rectangle is 4 cm, what are the dimensions of the rectangle?

65. **Geometry** The width of a rectangle is 6 ft less than its length. If the area of the rectangle is 75 ft^2, what are the dimensions of the rectangle?

66. **Geometry** The length of a rectangle is 8 cm more than its width. If the area of the rectangle is 90 cm^2, find the dimensions.

67. **Science and Medicine** The equation

$$h = -16t^2 - 32t + 320$$

gives the height of a ball thrown downward from the top of a 320-ft building with an initial velocity of 32 ft/s. Find the time it takes for the ball to reach a height of 160 ft. chapter 8 > Make the Connection

68. **Science and Medicine** The equation

$$h = -16t^2 - 32t + 320$$

gives the height of a ball thrown downward from the top of a 320-ft building with an initial velocity of 32 ft/s. Find the time it takes for the ball to reach a height of 64 ft. chapter 8 > Make the Connection

69. **Business and Finance** The demand equation for a certain type of printer is given by

$$D = -200p + 35{,}000$$

The supply equation is predicted to be

$$S = -p^2 + 400p - 20{,}000$$

Find the equilibrium price.

70. **Business and Finance** The demand equation for a certain type of printer is given by

$$D = -80p + 7{,}000$$

The supply equation is predicted to be

$$S = -p^2 + 220p - 8{,}000$$

Find the equilibrium price.

71. **Number Problem** If the square of 3 more than an integer is 9, find the number(s).

72. **Number Problem** If the square of 2 less than an integer is 16, find the number(s).

73. **Business and Finance** The revenue for selling x units of a product is given by

$$R = x\left(25 - \frac{1}{2}x\right)$$

Find the number of units sold if the revenue is \$294.50.

74. **Number Problem** Find two consecutive positive integers such that the sum of their squares is 85.

Determine whether each statement is **true** *or* **false.**

75. Not all quadratic expressions are factorable.

76. Some quadratic equations have no real-number solutions.

77. To complete the square when the coefficient of x^2 is not 1, one of the first steps is to divide both sides of the equation by that coefficient.

78. If the coefficient of x is odd, the completing-the-square method will not work.

Complete each statement with **always, sometimes,** *or* **never.**

79. An equation of the form $(x - h)^2 = k$, where k is positive, __________ has two distinct solutions.

80. An equation of the form $x^2 = k$ __________ has real-number solutions.

81. When completing the square for $x^2 + bx$, the number added is __________ negative.

82. A quadratic equation can __________ be solved by completing the square.

Skills | **Calculator/Computer** | Career Applications | Above and Beyond

Use a graphing calculator to graph each equation. Approximate the x-intercepts of each graph. (You may have to adjust the viewing window to see both intercepts.) Round your answers to the nearest tenth.

83. $y = x^2 + 12x - 2$

84. $y = x^2 - 14x - 7$

85. $y = x^2 - 2x - 8$

86. $y = x^2 + 4x - 72$

87. On a graphing calculator, view the graph of $f(x) = x^2 + 1$.

(a) What can you say about the x-intercepts of the graph?

(b) Determine the zeros of the function, using the square-root method. (ℂ)

(c) How does your answer to part (a) relate to your answer to part (b)?

88. On a graphing calculator, view the graph of $f(x) = x^2 + 4$.

(a) What can you say about the x-intercepts of the graph?

(b) Determine the zeros of the function, using the square-root method. (ℂ)

(c) How does your answer to part (a) relate to your answer to part (b)?

89. **Mechanical Engineering** The rotational moment in a shaft is given by the formula

$$M = 2x^2 - 30x$$

Find the positive value of x when the moment is 152. 

90. **MECHANICAL ENGINEERING** The deflection d of a loaded beam is described by the equation

$$d = \frac{x^2 - 64}{200}$$

Find the location x (in feet from center) if $d = 0.085$ in. 

91. **AGRICULTURAL TECHNOLOGY** The volume V of structural lumber (in cubic feet) that can be harvested from a conifer tree is given by the formula

$$V = \frac{D^2H}{350} - \frac{1}{5}$$

where H is the height of the tree (in feet) and D is the diameter (in inches) at its base. Find the necessary base diameter of a 48-ft-tall tree if we need the volume to be 70 ft^3.

92. **MANUFACTURING TECHNOLOGY** Suppose that the cost C, in dollars, of producing x chairs is given by

$$C = 2x^2 - 40x + 2{,}400$$

How many chairs can be produced for \$4,650?

93. **MANUFACTURING TECHNOLOGY** Suppose that the profit P, in dollars, of producing and selling x appliances is given by

$$P = -3x^2 + 240x - 1{,}800$$

How many appliances must be produced and sold to achieve a profit of \$2,325?

94. **MANUFACTURING TECHNOLOGY** A small manufacturer's weekly profit P, in dollars, is given by

$$P = -3x^2 + 270x$$

Find the number of items x that must be produced to realize a profit of \$5,775.

95. **MANUFACTURING TECHNOLOGY** The demand equation for a certain computer chip is given by

$$D = -5p + 62$$

The supply equation is predicted to be

$$S = -p^2 + 23p - 11$$

Find the equilibrium price.

96. **ALLIED HEALTH** A toxic chemical is introduced into a protozoan culture. The number of deaths per hour N is given by the equation

$$N = 363 - 3t^2$$

in which t is the number of hours after the chemical's introduction. How long will it take before the protozoa stop dying?

97. **ALLIED HEALTH** One technique of controlling cancer is to use radiation therapy. After such a treatment, the total number of cancerous cells N, in thousands, can be estimated by the equation

$$N = 121 - 4t^2$$

where t is the number of days of treatment. How many days of treatment are required to kill all the cancer cells?

98. **ALLIED HEALTH** An experimental drug is being tested on a bacteria colony. It is found that t days after the colony is treated, the number N of bacteria per cubic centimeter is given by the equation

$$N = -20t^2 - 120t + 1{,}000$$

In how many days will the colony be reduced to 200 bacteria?

99. GEOMETRY Consider this representation of "completing the square": Suppose we wish to complete the square for $x^2 + 10x$. A square with dimensions x by x has area equal to x^2.

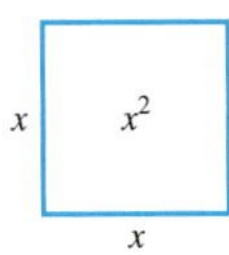

We divide the quantity $10x$ by 2 and get $5x$. If we extend the base x by 5 units and draw the rectangle attached to the square, the rectangle's dimensions are 5 by x with an area of $5x$.

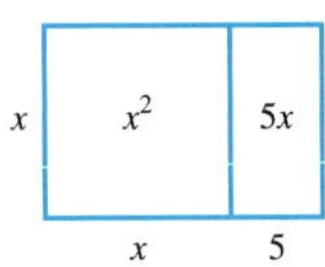

Now we extend the height by 5 units, and we draw another rectangle whose area is $5x$.

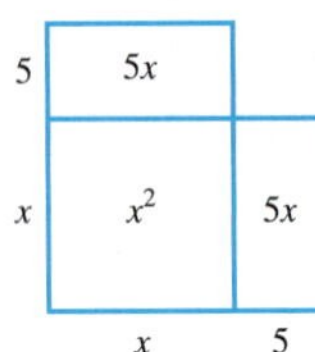

(a) What is the total area represented in the figure so far?

(b) How much area must be added to the figure to "complete the square"?

(c) Write the area of the completed square as a binomial squared.

100. GEOMETRY Repeat the process described in exercise 99 with $x^2 + 16x$.

GEOMETRY *In this section, you solved quadratic equations by "extracting roots," taking the square root of both sides after writing one side as the square of a binomial. But what if the algebraic expression cannot be written this way? Work with another student to decide what needs to be added to each expression to make it a "perfect-square trinomial." Label the dimensions of the squares and the area of each section.*

101. $x^2 + 6x +$ ______ $= (x + ?)^2$

102. $a^2 + a +$ ______ $= (a + ?)^2$

103. $x^2 + 12x +$ ______ $= (x + ?)^2$

104. $x^2 + 20x +$ ______ $= (x + ?)^2$

105. Why must the leading coefficient of the quadratic expression be set equal to 1 before you can use the technique of completing the square?

106. What relationship exists between the solution(s) of a quadratic equation and the graph of a quadratic function?

Find the constant that must be added to each binomial to form a perfect-square trinomial. Let x be the variable; other letters represent constants.

107. $x^2 + 2ax$ **108.** $x^2 + 2abx$ **109.** $x^2 + 3ax$

110. $x^2 + abx$ **111.** $a^2x^2 + 2ax$ **112.** $a^2x^2 + 4abx$

Solve each equation by completing the square.

113. $x^2 + 2ax = 4$ **114.** $x^2 + 2ax - 8 = 0$

Answers

1. $\{-7, -2\}$ **3.** $\{-5, 7\}$ **5.** $\left\{-\frac{1}{2}, 3\right\}$ **7.** $\left\{-\frac{1}{2}, \frac{2}{3}\right\}$ **9.** $\{\pm 11\}$ **11.** $\{\pm\sqrt{7}\}$ **13.** $\{\pm\sqrt{6}\}$ **15.** $\{\pm 2i\}$ nonreal **17.** $\{-1 \pm 2\sqrt{3}\}$ **19.** $\left\{\frac{-1 \pm \sqrt{5}}{3}\right\}$ **21.** 81 **23.** 16 **25.** $\frac{9}{4}$ **27.** $\frac{1}{4}$ **29.** $\frac{1}{100}$ **31.** $\frac{1}{144}$ **33.** $\{-5 \pm \sqrt{29}\}$ **35.** $\{-2, 4\}$ **37.** $\{1 \pm \sqrt{6}\}$ **39.** $\{-5 \pm 2\sqrt{3}\}$ **41.** $\left\{\frac{5 \pm \sqrt{53}}{2}\right\}$ **43.** $\left\{\frac{3 \pm \sqrt{29}}{2}\right\}$ **45.** $\left\{\frac{-1 \pm \sqrt{17}}{4}\right\}$ **47.** $\left\{\frac{-1 \pm \sqrt{3}}{2}\right\}$ **49.** $\left\{\frac{4 \pm \sqrt{22}}{3}\right\}$ **51.** $\left\{\frac{1 \pm i\sqrt{35}}{3}\right\}$ or $\left\{\frac{1}{3} \pm \frac{\sqrt{35}}{3}i\right\}$ nonreal **53.** $\{-5 \pm i\sqrt{3}\}$ nonreal **55.** 21.541 ft **57.** 14.142 ft **59.** 3.58 m, 7.16 m **61.** $-10, 6$ **63.** 3.292 in., 7.292 in. **65.** 6.165 ft by 12.165 ft **67.** 2.317 s **69.** \$112.92 **71.** $-6, 0$ **73.** 19, 31 **75.** True **77.** True **79.** always **81.** never **83.**

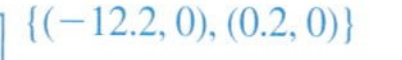

$\{(-12.2, 0), (0.2, 0)\}$ **85.**

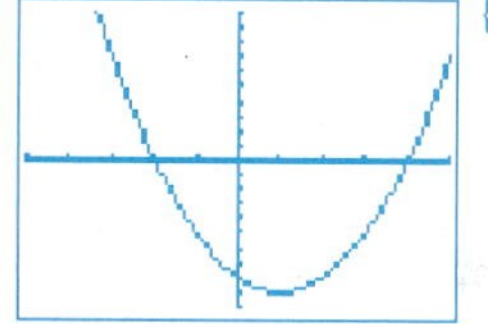

$\{(-2.0, 0), (4, 0)\}$

87. **(a)** There are none; **(b)** $x = \pm i$ nonreal; **(c)** If the graph of $f(x)$ has no x-intercepts, the zeros of the function are not real. **89.** 19 **91.** 22.62 in. **93.** 22 or 55 appliances **95.** \$2.91 **97.** 5.5 days **99.** **(a)** $x^2 + 5x + 5x$; **(b)** 25; **(c)** $x^2 + 10x + 25 = (x + 5)^2$ **101.** 9; 3 **103.** 36, 6 **105.** Above and Beyond **107.** a^2 **109.** $\frac{9}{4}a^2$ **111.** 1 **113.** $\{-a \pm \sqrt{4 + a^2}\}$

Activity 8 ::

The Gravity Model

This activity requires two to three people and an open area outside. You will also need a ball (preferably a baseball or some other small, heavy ball), a stopwatch, and a tape measure.

1. Designate one person as the "ball thrower."
2. The ball thrower should throw the ball straight up into the air, as high as possible. While this is happening, another group member should take note of exactly where the ball thrower releases the ball. Be careful that no one gets hit by the ball as it comes back down.
3. Use the tape measure to determine the height above the ground that the thrower released the ball. Convert your measurement to feet, using decimals to indicate parts of a foot. For example, 3 in. is 0.25 ft. Record this as the "initial height."
4. Repeat steps 2 and 3 to ensure that you have the correct initial height.
5. The thrower should now throw the ball straight up, as high as possible. The person with the stopwatch should time the ball until it lands.
6. Repeat step 5 twice more, recording the time.
7. Take the average (mean) of your three recorded times. We will use this number as the "hang time" of the ball for the remainder of this activity.

According to Sir Isaac Newton (1642–1727), the height of an object with initial velocity v_0 and initial height s_0 is given by

$$s = -16t^2 + v_0t + s_0$$

in which the height s is measured in feet and t represents the time in seconds.

The initial velocity v_0 is positive when the object is thrown upward, and it is negative when the object is thrown downward.

For example, the height of a ball thrown upward from a height of 4 ft with an initial velocity of 50 ft/s is given by

$$s = -16t^2 + 50t + 4$$

To determine the height of the ball 2 s after release, we evaluate the polynomial for $t = 2$.

$$\begin{aligned} s(2) &= -16(2)^2 + 50(2) + 4 \\ &= 40 \end{aligned}$$

The ball is 40 ft high after 2 s.

Take note that the height of the ball when it lands is zero.

8. Substitute the time found in step 7 for t, your initial height for s_0, and 0 for s in the height equation. Solve the resulting equation for the initial velocity, v_0.
9. Now write your height equation using the initial velocity found in step 8 along with the initial height. Your equation should have two variables, s and t.
10. What is the height of the ball after 1 s?
11. The maximum height of the ball will be attained when $t = \frac{v_0}{32}$. Determine this time and find the maximum height of the thrown ball.
12. If a ball is dropped from a height of 256 ft, how long will it take to hit the ground? *Hint:* If a ball is dropped, its initial velocity is 0 ft/s.

8.2

The Quadratic Formula

< 8.2 Objectives >

1 > Use the quadratic formula to solve equations

2 > Use the discriminant of a quadratic equation to classify its solutions

3 > Use quadratic equations to solve applications

We are now ready for the **quadratic formula.** We can use this formula to solve every quadratic equation, even those with nonreal solutions. We use our work completing the square in Section 8.1 to derive the quadratic formula.

To use the quadratic formula to solve an equation, we write the quadratic equation in *standard form.*

$ax^2 + bx + c = 0 \quad \text{where} \quad a \neq 0$

We begin by reviewing the standard form of a quadratic equation and identifying the coefficients.

Example 1 Writing Quadratic Equations in Standard Form

Write each quadratic equation in standard form. Identify the coefficients a, b, and c.

(a) $2x^2 - 5x + 3 = 0$

The equation is already in standard form.

$a = 2 \qquad b = -5 \qquad \text{and} \qquad c = 3$

(b) $5x^2 + 3x = 5$

The equation is *not* in standard form. Rewrite it by subtracting 5 from both sides.

$5x^2 + 3x - 5 = 0$ Standard form

$a = 5 \qquad b = 3 \qquad \text{and} \qquad c = -5$

Check Yourself 1

Write each quadratic equation in standard form. Identify the coefficients a, b, and c.

(a) $x^2 - 3x = 5$ **(b)** $3x^2 = 7 - 2x$

Once a quadratic equation is written in standard form, we can find its solutions. Remember that a solution is a value for x that makes the equation true.

We now derive the quadratic formula, which we will use to solve quadratic equations.

Step by Step

Deriving the Quadratic Formula

Step 1 Isolate the constant on the right side of the equation. $ax^2 + bx = -c$

Step 2 Divide both sides by the coefficient of the x^2-term. $x^2 + \frac{b}{a}x = -\frac{c}{a}$

Step 3 Add the square of one-half the x-coefficient to both sides. $x^2 + \frac{b}{a}x + \frac{b^2}{4a^2} = -\frac{c}{a} + \frac{b^2}{4a^2}$

Step 4 Factor the left side to write it as the square of a binomial. $\left(x + \frac{b}{2a}\right)^2 = \frac{-4ac + b^2}{4a^2}$

Then apply the square-root property. $x + \frac{b}{2a} = \pm\sqrt{\frac{b^2 - 4ac}{4a^2}}$

Step 5 Solve the resulting linear equations. $x = -\frac{b}{2a} \pm \frac{\sqrt{b^2 - 4ac}}{2a}$

Step 6 Simplify. $= \frac{-b \pm \sqrt{b^2 - 4ac}}{2a}$

We use this result to state the quadratic formula.

Property

The Quadratic Formula

Given any quadratic equation in the form

$ax^2 + bx + c = 0$ in which $a \neq 0$

the two solutions to the equation are found by using the formula

$$x = \frac{-b \pm \sqrt{b^2 - 4ac}}{2a}$$

Now we can use the quadratic formula to solve some equations.

 **Example 2** Using the Quadratic Formula to Solve an Equation

< Objective 1 >

Use the quadratic formula to solve $x^2 - 5x + 4 = 0$.

The equation is in standard form, so first identify a, b, and c.

$$x^2 - 5x + 4 = 0$$

$a = 1$ $b = -5$ $c = 4$

NOTE

The leading coefficient is 1, so $a = 1$.

We now substitute the values for a, b, and c into the formula.

$$x = \frac{-b \pm \sqrt{b^2 - 4ac}}{2a}$$

$$x = \frac{-(-5) \pm \sqrt{(-5)^2 - 4(1)(4)}}{2(1)}$$

$$= \frac{5 \pm \sqrt{25 - 16}}{2}$$

$$= \frac{5 \pm \sqrt{9}}{2}$$

$$= \frac{5 \pm 3}{2}$$

NOTE

We could have solved this particular equation by factoring.

Therefore,

$x = \frac{5 + 3}{2}$ or $x = \frac{5 - 3}{2}$

$= 4$ $\qquad$ $= 1$

We check our result by substituting the solutions into the original quadratic equation and seeing if that results in true statements.

Check

$x = 4$

$$x^2 - 5x + 4 = 0$$
$$(4)^2 - 5(4) + 4 \stackrel{?}{=} 0$$
$$16 - 20 + 4 \stackrel{?}{=} 0$$
$$0 = 0 \quad \text{True}$$

$x = 1$

$$x^2 - 5x + 4 = 0$$
$$(1)^2 - 5(1) + 4 \stackrel{?}{=} 0$$
$$1 - 5 + 4 \stackrel{?}{=} 0$$
$$0 = 0 \quad \text{True}$$

Both check so the solution set is $\{1, 4\}$.

Check Yourself 2

Use the quadratic formula to solve $x^2 - 2x - 8 = 0$. Check your result.

We solve a quadratic equation in which $a \neq 1$ in Example 3.

Example 3 Using the Quadratic Formula

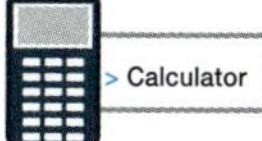

Use the quadratic formula to solve.

$6x^2 - 7x - 3 = 0$

First, we find a, b, and c.

$a = 6 \qquad b = -7 \qquad c = -3$

Substituting those values into the quadratic formula gives

$$x = \frac{-(-7) \pm \sqrt{(-7)^2 - 4(6)(-3)}}{2(6)}$$

$$= \frac{7 \pm \sqrt{49 + 72}}{12} \quad \text{Simplify.}$$

$$= \frac{7 \pm \sqrt{121}}{12}$$

$$= \frac{7 \pm 11}{12} \quad \sqrt{121} = 11$$

Therefore,

$x = \frac{7 + 11}{12}$ or $x = \frac{7 - 11}{12}$

$= \frac{3}{2}$ $\qquad$ $= -\frac{1}{3}$

We leave it to you to check both solutions. The solution set is $\left\{-\frac{1}{3}, \frac{3}{2}\right\}$.

NOTES

Because $b^2 - 4ac = 121$ is a perfect square, the two solutions are rational numbers.

Compare these solutions to the x-intercepts of

$y = 6x^2 - 7x - 3$

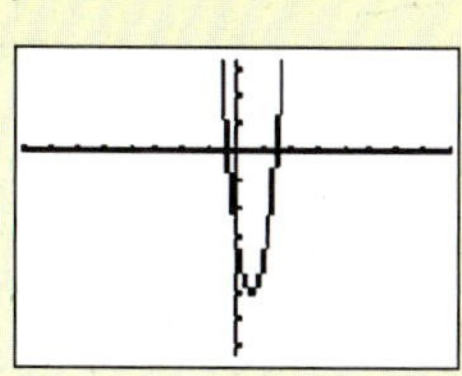

Since the solutions are rational, we could have factored the original equation. We use the solutions to see

$6x^2 - 7x - 3 = (3x + 1)(2x - 3)$

Check Yourself 3

Use the quadratic formula to solve.

$3x^2 + 2x - 8 = 0$

To use the quadratic formula, we must write the equation in standard form.

Example 4 Using the Quadratic Formula

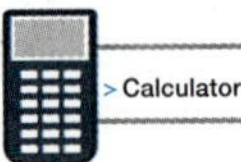

NOTE

The equation *must be in standard form* to determine *a*, *b*, and *c*.

Solve

$9x^2 = 12x - 4$

We write the equation in standard form.

$9x^2 - 12x + 4 = 0$

So,

$a = 9 \qquad b = -12 \qquad c = 4$

Substitute these values into the quadratic formula.

$$x = \frac{-(-12) \pm \sqrt{(-12)^2 - 4(9)(4)}}{2(9)}$$

$$= \frac{12 \pm \sqrt{0}}{18}$$

Simplifying yields

$$x = \frac{2}{3} \qquad \text{or} \qquad \left\{\frac{2}{3}\right\}$$

NOTE

The graph of

$y = 9x^2 - 12x + 4$

intersects the x-axis only at the point $\left(\frac{2}{3}, 0\right)$.

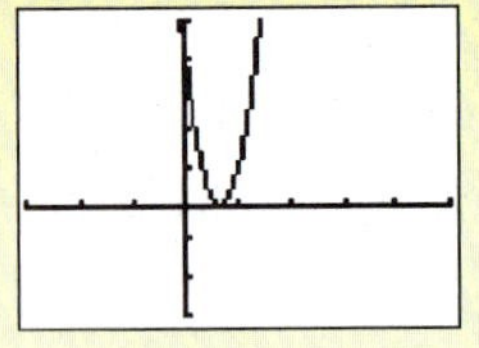

Look again at the original equation, and now try factoring.

$$9x^2 = 12x - 4$$

$$9x^2 - 12x + 4 = 0 \quad \text{The trinomial on the left is a perfect-square trinomial.}$$

$$(3x - 2)(3x - 2) = 0$$

At this point, it is clear that there is exactly one solution, $x = \frac{2}{3}$.

Since the factor $3x - 2$ is repeated and solutions are also called *roots,* we say that $\frac{2}{3}$ is a **repeated root** of the equation.

We always find a *repeated root* if the quadratic equation, placed in standard form, has a perfect-square trinomial on one side. If we use the quadratic formula to solve such an equation, the value of the radicand $b^2 - 4ac$ is 0.

Check Yourself 4

Solve

$4x^2 - 4x = -1$

Most of the time, we use the quadratic formula when the solutions are irrational.

Example 5 Using the Quadratic Formula

Solve

$x^2 - 3x = 5$

We write the equation in standard form.

$x^2 - 3x - 5 = 0$

We find a, b, and c and substitute.

$$x = \frac{-(-3) \pm \sqrt{(-3)^2 - 4(1)(-5)}}{2(1)}$$

$a = 1, b = -3, c = -5$

Simplifying gives

$$x = \frac{3 \pm \sqrt{29}}{2} \quad \text{or} \quad \left\{\frac{3 \pm \sqrt{29}}{2}\right\}$$

In Section 8.1, you learned to check irrational solutions with a calculator.

NOTE

We can use a calculator to approximate the solutions. To the nearest hundredth, we have −1.19 and 4.19.

```
(3-√(29))/2
        -1.192582404
(3+√(29))/2
         4.192582404
```

Check Yourself 5

Solve $2x^2 = x + 7$.

Simplifying your result may require extra steps, as we see in Example 6.

Example 6 Using the Quadratic Formula

Solve

$3x^2 - 6x + 2 = 0$

We have $a = 3$, $b = -6$, and $c = 2$. Substituting gives

$$x = \frac{-(-6) \pm \sqrt{(-6)^2 - 4(3)(2)}}{2(3)}$$

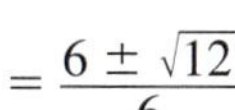

$$= \frac{6 \pm \sqrt{12}}{6}$$

We now look for the largest perfect-square factor of 12, the radicand.

Simplifying, we know $\sqrt{12}$ is equal to $\sqrt{4 \cdot 3}$, or $2\sqrt{3}$. We write the solutions as

$$x = \frac{6 \pm 2\sqrt{3}}{6} = \frac{2(3 \pm \sqrt{3})}{6} = \frac{3 \pm \sqrt{3}}{3}$$

> CAUTION

Students are sometimes tempted to simplify this result to

$$\frac{6 \pm 2\sqrt{3}}{6} \stackrel{?}{=} 1 \pm 2\sqrt{3}$$

This is *not a valid step*. We must divide *each term* in the numerator by 2 when simplifying the expression.

Check Yourself 6

Solve

$x^2 - 4x = 6$

Next, we examine a case in which the solutions are nonreal or complex numbers.

Example 7 Using the Quadratic Formula (ℂ)

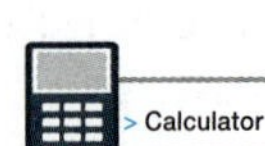

Use the quadratic formula to solve

$x^2 - 2x + 2 = 0$

Labeling the coefficients, we find that

$a = 1 \qquad b = -2 \qquad c = 2$

Applying the quadratic formula, we have

$$x = \frac{2 \pm \sqrt{-4}}{2}$$

and noting that $\sqrt{-4}$ is $2i$, we can simplify to

$x = 1 \pm i$ or $\{1 \pm i\}$ Nonreal

NOTES

The solutions are nonreal anytime $b^2 - 4ac$ is negative.

The graph of $y = x^2 - 2x + 2$ does not intersect the x-axis, so there are no real solutions.

Check Yourself 7

Solve by using the quadratic formula.

$x^2 - 4x + 6 = 0$ (ℂ)

You have now learned four different methods for solving quadratic equations: Factoring; the square-root method; completing the square; and the quadratic formula.

When trying to solve a quadratic equation (in standard form), we believe that you should see if you can factor the quadratic expression first.

If you cannot easily factor, we suggest the quadratic formula, unless the square-root method is easy.

We generally do not solve a quadratic equation by graphing. Usually, we can only derive approximations when graphing.

We complete the square to generate the quadratic formula, but it is not an efficient method to use to solve quadratic equations. In a future math class, you will manipulate equations of *conic sections* by completing the square.

Step by Step

Using the Quadratic Formula

Step 1 Write the equation in standard form (one side is equal to 0).

$ax^2 + bx + c = 0$

Step 2 Find a, b, and c.

Step 3 Substitute those values into the quadratic formula.

$$x = \frac{-b \pm \sqrt{b^2 - 4ac}}{2a}$$

Step 4 Simplify.

Quadratic equations can have zero, one, or two real solutions. Looking at the quadratic formula, it should be clear that the radicand determines the nature of the solution set.

For a quadratic equation in standard form, we call the radicand $b^2 - 4ac$ the **discriminant.**

> **NOTE**
>
> The graph of the related quadratic function has the same number of x-intercepts as the quadratic equation has real solutions.

If the discriminant is negative, then it does not have any real square roots. In such a case, the quadratic equation has two distinct complex roots.

If the discriminant is zero, then its square root is zero and there is no difference between adding and subtracting zero. In this case, there is one (repeated) solution.

If the discriminant is positive, then we have two distinct real solutions. If the discriminant is a perfect square, then the solutions are rational. We can factor the quadratic expression when the discriminant is a perfect square.

We summarize this discussion in our definition of the discriminant.

Definition

The Discriminant

> **NOTE**
>
> Although the solutions are not necessarily distinct or real, every second-degree equation has two solutions.

Given the equation $ax^2 + bx + c = 0$, the quantity $b^2 - 4ac$ is called the **discriminant.**

$$\text{If } b^2 - 4ac \begin{cases} < 0 & \text{there are } \textit{no real solutions,} \text{ but two nonreal solutions} \\ = 0 & \text{there is } \textit{one real solution} \text{ (a repeated root)} \\ > 0 & \text{there are } \textit{two distinct real solutions} \end{cases}$$

When the discriminant is a nonzero perfect square, the solutions are rational and the quadratic expression is factorable.

Example 8 Analyzing the Discriminant

< Objective 2 >

> **RECALL**
>
> We can use the quadratic formula to find any solutions.

How many real solutions are there to each quadratic equation?

(a) $x^2 + 7x - 15 = 0$

The discriminant $(7)^2 - 4(1)(-15)$ is 109. This indicates that there are two real solutions.

(b) $3x^2 - 5x + 7 = 0$

The discriminant $b^2 - 4ac = -59$ is negative. There are no real solutions.

(c) $9x^2 - 12x + 4 = 0$

The discriminant is 0. There is exactly one real solution (a repeated root).

Check Yourself 8

How many real solutions are there to each quadratic equation?

(a) $2x^2 - 3x + 2 = 0$ **(b)** $3x^2 + x - 11 = 0$
(c) $4x^2 - 4x + 1 = 0$ **(d)** $x^2 = -5x - 7$

Often, we solve a quadratic equation and the solutions contain radicals. When working in algebra, we usually leave the solution in radical form. We call the radical form of a solution an **exact value.**

When we solve a quadratic equation in order to solve an application, we usually use a calculator to approximate the solutions. We may even discard one or both solutions as not feasible in the context of an application.

We consider such cases in Examples 9 and 10.

Example 9 A Physics Application

< Objective 3 >

If a ball is launched from the ground with an initial velocity of 80 ft/s, the height h of the ball after t seconds is given by

$h = 80t - 16t^2$

Find the time it takes the ball to reach a height of 48 ft.

> **NOTE**
>
> Here, h measures the height above the ground, in feet, t seconds after the ball is thrown upward.

We substitute 48 for h, and then we rewrite the equation in standard form.

$$(48) = 80t - 16t^2$$
$$16t^2 - 80t + 48 = 0$$

To simplify the computation, we divide both sides of the equation by the common factor 16.

$t^2 - 5t + 3 = 0$

We use the quadratic formula to solve for t.

$$t = \frac{5 \pm \sqrt{13}}{2}$$

> **NOTE**
>
> There are two solutions because the ball reaches the height *twice*, once on the way up and once on the way down.

This gives us two solutions, $\frac{5 + \sqrt{13}}{2}$ and $\frac{5 - \sqrt{13}}{2}$. But, because we have specified units of time, we generally estimate the answer to the nearest tenth or hundredth of a second.

In this case, estimating to the nearest tenth of a second gives solutions of 0.7 s and 4.3 s.

Check Yourself 9

The equation to find the height h of a ball thrown with an initial velocity of 64 ft/s is

$h(t) = 64t - 16t^2$

Find the time it takes the ball to reach a height of 32 ft (to the nearest tenth of a second).

Example 10 Solving a Thrown-Ball Application

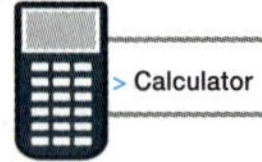

The height h of a ball thrown downward from the top of a 240-ft building with an initial velocity of 64 ft/s is given by

$h(t) = 240 - 64t - 16t^2$

When will the ball's height be 176 ft?

Let $h(t) = 176$, and write the equation in standard form.

$$(176) = 240 - 64t - 16t^2$$
$$0 = 64 - 64t - 16t^2$$
$$16t^2 + 64t - 64 = 0$$

Divide both sides of the equation by 16 to simplify the computation.

$t^2 + 4t - 4 = 0$

NOTE

The graph of $h(t) = 240 - 64t - 16t^2$ shows the height h at any time t.

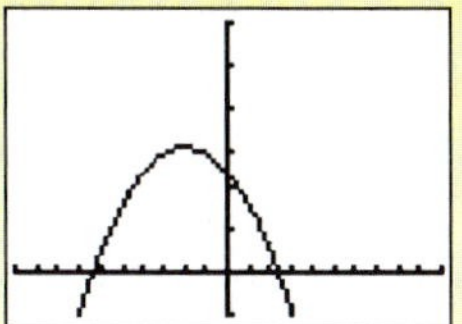

The ball's height is 176 ft after approximately 0.8 s.

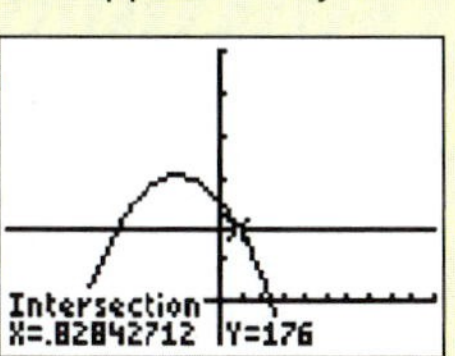

Applying the quadratic formula with $a = 1$, $b = 4$, and $c = -4$ yields

$t = -2 \pm 2\sqrt{2}$

Estimating these solutions, we have $t = -4.8$ s and $t = 0.8$ s, but of these two values only the *positive value* makes any sense. (To accept the negative solution would be to say that the ball reached the specified height before it was thrown.)

Check Yourself 10

The height h of a ball thrown upward from the top of a 96-ft building with an initial velocity of 16 ft/s is given by

$h(t) = 96 + 16t - 16t^2$

When will the ball's height be 32 ft? (Estimate your answer to the nearest tenth of a second.)

Recall from Section 7.1 that the **Pythagorean theorem** gives an important relationship between the lengths of the sides of a right triangle (a triangle with a 90° angle).

We restate this important theorem here.

Definition

The Pythagorean Theorem

In any right triangle, the square of the lengths of the longest side (the hypotenuse) is equal to the sum of the squares of the lengths of the two shorter sides (the legs).

$c^2 = a^2 + b^2$

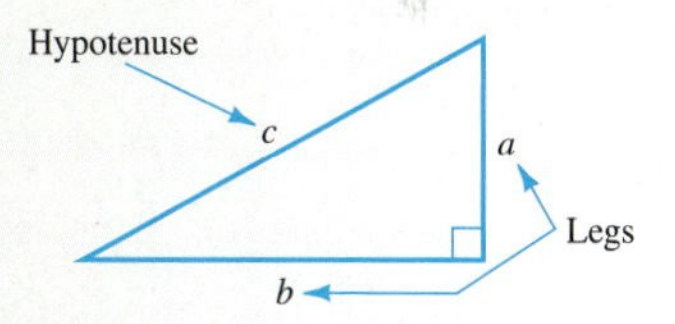

We use the Pythagorean theorem to solve a geometry problem in Example 11. Since we are solving an application, we approximate the solution. Checking our answer lets us determine if it is *reasonable*, but we cannot use the decimal approximation to check it exactly.

Example 11 A Geometry Application

One leg of a right triangle is 4 cm longer than the other leg. The length of the hypotenuse of the triangle is 12 cm. Find the length of the two legs, rounded to the nearest hundredth of a centimeter.

As in any geometric problem, a sketch of the information helps us visualize the problem.

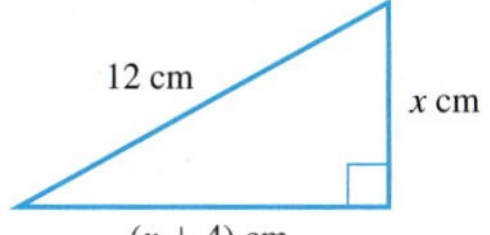

We assign the variable x to the shorter leg and $x + 4$ to the other leg.

Now we apply the Pythagorean theorem to write an equation for the solution.

$$x^2 + (x + 4)^2 = (12)^2$$

$$x^2 + x^2 + 8x + 16 = 144$$

or $$2x^2 + 8x - 128 = 0$$

NOTES

Dividing both sides of a quadratic equation by a common factor is always a prudent step. It simplifies your work with the quadratic formula.

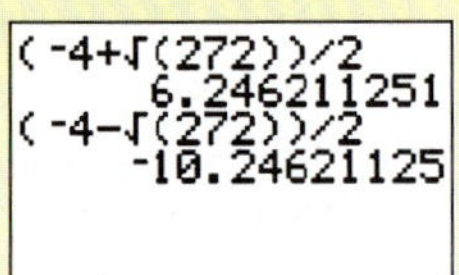

Do you see why we should not expect an "exact" check here?

Dividing both sides by 2 gives

$$x^2 + 4x - 64 = 0$$

Using the quadratic formula, we get

$$x = \frac{-4 \pm \sqrt{272}}{2}$$

We are interested in lengths rounded to the nearest hundredth, so there is no need to simplify the radical expression. Using a calculator, we can reject one solution (do you see why?) and find $x \approx 6.25$ cm for the other.

This means that the two legs have lengths of 6.25 and 10.25 cm. How do we check? The sides must "reasonably" satisfy the Pythagorean theorem:

$$6.25^2 + 10.25^2 \approx 12^2$$
$$144.125 \approx 144$$

This indicates that our answer is reasonable.

Check Yourself 11

One leg of a right triangle is 2 cm longer than the other. The hypotenuse is 1 cm less than twice the length of the shorter leg. Find the length of each side of the triangle, rounded to the nearest tenth.

We conclude with one more application. Sketching the problem in Example 12 makes it much easier to create an equation to model the application.

Example 12 Solving a Construction Application

A rectangular garden is to be surrounded by a walkway of constant width. The garden's dimensions are 5 m by 8 m. The total area, garden plus walkway, is to be 100 m^2. How wide does the walkway need to be? Round to the nearest hundredth of a meter.

First we sketch the situation.

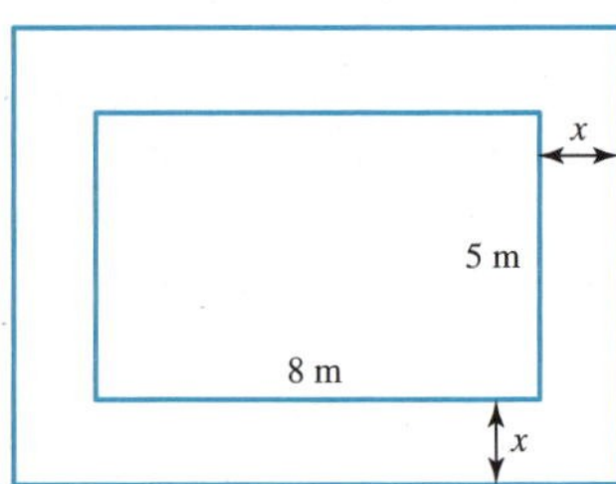

The area of the entire figure may be represented by

$$\underbrace{(8 + 2x)}_{\text{Length}}\underbrace{(5 + 2x)}_{\text{Width}} \qquad \text{(Do you see why?)}$$

Since the area is to be 100 m^2, we write

$$(8 + 2x)(5 + 2x) = 100$$
$$40 + 16x + 10x + 4x^2 = 100$$
$$4x^2 + 26x - 60 = 0$$
$$2x^2 + 13x - 30 = 0$$

$$x = \frac{-(13) \pm \sqrt{(13)^2 - 4(2)(-30)}}{2(2)}$$

$$= \frac{-13 \pm \sqrt{409}}{4}$$

$$x \approx 1.8059 \qquad \text{or} \qquad \approx -8.3059$$

Since x must be positive, we have (rounded) $x = 1.81$ m.

Check Yourself 12

A rectangular swimming pool is 16 ft by 40 ft. A tarp to go over the pool also covers a strip of equal width surrounding the pool. If the area of the tarp is 1,100 ft^2, how wide is the covered strip around the pool? Round to the nearest tenth of a foot.

Check Yourself ANSWERS

1. **(a)** $x^2 - 3x - 5 = 0$; $a = 1, b = -3, c = -5$; **(b)** $3x^2 + 2x - 7 = 0$; $a = 3, b = 2, c = -7$
2. $\{-2, 4\}$ **3.** $\left\{-2, \frac{4}{3}\right\}$ **4.** $\left\{\frac{1}{2}\right\}$ **5.** $\left\{\frac{1 \pm \sqrt{57}}{4}\right\}$ **6.** $\{2 \pm \sqrt{10}\}$ **7.** $\{2 \pm i\sqrt{2}\}$ nonreal
8. **(a)** None; **(b)** two; **(c)** one; **(d)** none **9.** 0.6 s and 3.4 s **10.** 2.6 s
11. Approximately 4.3 cm, 6.3 cm, and 7.7 cm **12.** 3.6 ft

Reading Your Text

These fill-in-the-blank exercises will help you understand some of the key vocabulary used in this section. The answers to these exercises are in the Answers Appendix in the back of the text.

(a) Every quadratic equation can be solved by using the quadratic ____________.

(b) If the solutions to a quadratic equation are rational, the original equation can be solved by ____________.

(c) Given a quadratic equation in standard form, $b^2 - 4ac$ is called the ____________.

(d) Given a quadratic equation in standard form, if $b^2 - 4ac < 0$ there are ____________ real solutions.

Skills Calculator/Computer Career Applications Above and Beyond

8.2 exercises

< Objective 1 >

Solve each quadratic equation first by factoring and then with the quadratic formula.

1. $x^2 - 5x - 14 = 0$ **2.** $x^2 - 2x - 35 = 0$ **3.** $t^2 + 8t - 65 = 0$

4. $q^2 + 3q - 130 = 0$ **5.** $3x^2 + x - 10 = 0$ **6.** $3x^2 + 2x - 1 = 0$

7. $16t^2 - 24t + 9 = 0$ **8.** $6m^2 - 23m + 10 = 0$

Solve each quadratic equation **(a)** *by completing the square and* **(b)** *with the quadratic formula.*

9. $x^2 - 4x - 7 = 0$ **10.** $x^2 + 6x - 1 = 0$ **11.** $x^2 + 3x - 27 = 0$

12. $t^2 + 4t - 7 = 0$ **13.** $3x^2 - 5x + 1 = 0$ **14.** $2x^2 - 6x + 1 = 0$

15. $2q^2 - 2q - 1 = 0$ **16.** $3r^2 - 2r + 4 = 0$ (ℂ) **17.** $3x^2 - x - 2 = 0$

18. $2x^2 - 5x + 3 = 0$ **19.** $2y^2 - y - 5 = 0$ **20.** $3m^2 + 2m - 1 = 0$

Use the quadratic formula to solve each equation.

21. $x^2 - 4x + 3 = 0$ VIDEO **22.** $x^2 - 7x + 3 = 0$ **23.** $p^2 - 8p + 16 = 0$

24. $u^2 + 7u - 30 = 0$ **25.** $-x^2 + 3x + 5 = 0$ VIDEO **26.** $2x^2 - 3x - 7 = 0$

27. $-3s^2 + 2s - 1 = 0$ (ℂ) **28.** $5t^2 - 2t - 2 = 0$

Hint: Clear the equations of fractions or remove grouping symbols, as needed.

29. $2x^2 - \frac{1}{2}x - 5 = 0$ **30.** $3x^2 + \frac{1}{3}x - 3 = 0$ **31.** $5t^2 - 2t - \frac{2}{3} = 0$

32. $3y^2 + 2y + \frac{3}{4} = 0$ (ℂ) **33.** $(x - 2)(x + 3) = 4$ VIDEO **34.** $(x + 1)(x - 8) = 3$

35. $(t + 1)(2t - 4) - 7 = 0$ **36.** $(2w + 1)(3w - 2) = 1$ **37.** $\frac{2x^2}{3} - \frac{7x}{3} = 1$

38. $\frac{x^2}{3} + x = \frac{1}{3}$ **39.** $t^2 - \frac{3}{2} = \frac{3t}{2}$ **40.** $p^2 - \frac{1}{4} = \frac{3p}{2}$

41. $5 + 2y = y^2$ **42.** $6 - 2x = x^2$

Solve each quadratic equation. Use any applicable method.

43. $x^2 - 8x + 16 = 0$ **44.** $4x^2 + 12x + 9 = 0$ **45.** $3t^2 - 7t + 1 = 0$

46. $2z^2 - z + 5 = 0$ (ℂ) **47.** $5y^2 - 2y = 0$ **48.** $7z^2 - 6z - 2 = 0$

49. $(x - 1)(2x + 7) = -6$ **50.** $4x^2 - 3 = 0$ **51.** $x^2 + 9 = 0$ (ℂ)

52. $(4x - 5)(x + 2) = 1$ **53.** $x^2 - 5x = 10 - 2x$ **54.** $x^2 + x = -2 - x$ (ℂ)

< Objective 2 >

For each quadratic equation, find the value of the discriminant and give the number of real solutions.

55. $2x^2 - 5x = 0$ **56.** $3x^2 + 8x = 0$ **57.** $m^2 - 18m + 81 = 0$

58. $4p^2 + 12p + 9 = 0$ **59.** $3x^2 - 7x + 1 = 0$ VIDEO **60.** $2x^2 - x + 5 = 0$

61. $2w^2 - 5w + 11 = 0$ **62.** $6q^2 - 5q + 2 = 0$

< Objective 3 >

63. SCIENCE AND MEDICINE The function

$h(t) = 112t - 16t^2$

gives the height of an arrow shot upward from the ground with an initial velocity of 112 ft/s, in which t is the time in seconds after the arrow leaves the ground. chapter 8 > Make the Connection

(a) Find the time it takes for the arrow to reach a height of 112 ft.

(b) Find the time it takes for the arrow to reach a height of 144 ft.

Express your answers to the nearest tenth of a second.

64. Science and Medicine The function

$h(t) = 320 - 32t - 16t^2$

gives the height of a ball thrown downward from the top of a 320-ft building with an initial velocity of 32 ft/s in which t is the time after the ball is thrown down from the top of the building.

(a) Find the time it takes for the ball to reach a height of 240 ft.

(b) Find the time it takes for the ball to reach a height of 96 ft.

Express your answers to the nearest tenth of a second.

65. Number Problem The product of two consecutive integers is 72. What are the two integers?

66. Number Problem The sum of the squares of two consecutive whole numbers is 61. Find the two whole numbers.

67. Geometry The width of a rectangle is 3 ft less than its length. If the area of the rectangle is 70 ft^2, what are the dimensions of the rectangle?

68. Geometry The length of a rectangle is 5 cm more than its width. If the area of the rectangle is 84 cm^2, find the dimensions.

69. Geometry One leg of a right triangle is twice the length of the other. The hypotenuse is 6 m long. Find the length of each leg. Round to the nearest tenth of a meter.

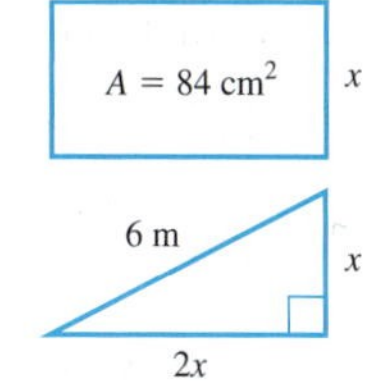

70. Geometry One leg of a right triangle is 2 ft longer than the shorter side. If the length of the hypotenuse is 14 ft, how long is each leg (nearest tenth of a foot)?

71. Science and Medicine If a ball is thrown vertically upward from the ground with an initial velocity of 64 ft/s, its height h after t seconds is given by $h(t) = 64t - 16t^2$.

(a) How long does it take the ball to return to the ground?

Hint: Let $h(t) = 0$.

(b) How long does it take the ball to reach a height of 48 ft on the way up?

72. Science and Medicine If a ball is thrown vertically upward from the ground with an initial velocity of 96 ft/s, its height h after t seconds is given by $h(t) = 96t - 16t^2$.

(a) How long does it take the ball to return to the ground?

(b) How long does it take the ball to pass through a height of 128 ft on the way back down to the ground?

73. Business and Finance Suppose that the cost $C(x)$, in dollars, of producing x chairs is given by

$C(x) = 2{,}400 - 40x + 2x^2$

How many chairs can be produced for \$5,400?

74. Business and Finance Suppose that the profit $T(x)$, in dollars, of producing and selling x microwave ovens is given by

$T(x) = -3x^2 + 240x - 1{,}800$

How many microwaves must be produced and sold to achieve a profit of \$3,000?

75. Geometry One leg of a right triangle is 1 in. shorter than the other leg. The hypotenuse is 3 in. longer than the shorter side. Find the length of each side, to the nearest tenth of an inch.

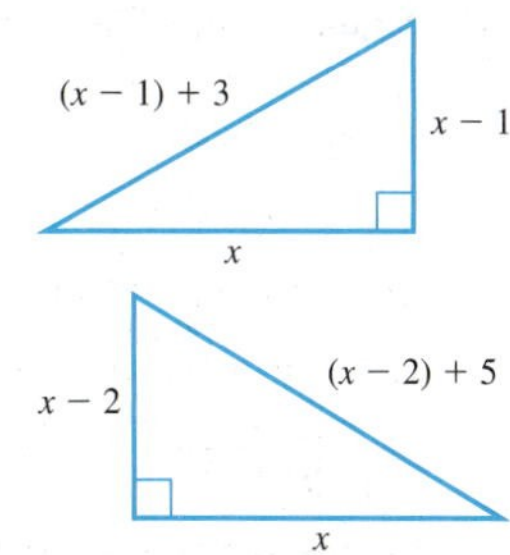

76. Geometry The hypotenuse of a right triangle is 5 cm longer than the shorter leg. The length of the shorter leg is 2 cm less than the length of the longer leg. Find the lengths of the three sides, to the nearest tenth of a centimeter.

77. **Business and Finance** A small manufacturer's weekly profit from Blu-ray players, in dollars, is given by

$P(x) = -3x^2 + 270x$

Find the number of Blu-ray players x that must be produced to realize a profit of \$5,100.

78. **Business and Finance** Suppose the profit from Blu-ray players, in dollars, is given by

$P(x) = -2x^2 + 240x$

How many Blu-ray players must be sold to realize a profit of \$5,100?

79. **Business and Finance** The demand equation for a certain computer chip is given by

$D = -2p + 14$

The supply equation is predicted to be

$S = -p^2 + 16p - 2$

Find the equilibrium price.

80. **Business and Finance** The demand equation for a certain type of printer is predicted to be

$D = -200p + 36{,}000$

The supply equation is predicted to be

$S = -p^2 + 400p - 24{,}000$

Find the equilibrium price.

81. **Science and Medicine** If a ball is thrown upward from the roof of a building 70 m tall with an initial velocity of 15 m/s, its approximate height h after t seconds is given by

$h(t) = 70 + 15t - 5t^2$ chapter 8 > Make the Connection

Note: The difference between this equation and the ones we used in Examples 9 and 10 has to do with the units used. When we use feet, the t^2-coefficient is -16 (because the acceleration due to gravity is approximately 32 ft/s^2). When we use meters as the height, the t^2-coefficient is -5 because that same acceleration is approximately 10 m/s^2. Use this information to complete each exercise. Express your answers to the nearest tenth of a second.

(a) How long does it take the ball to fall back to the ground?

(b) When will the ball reach a height of 80 m?

82. **Science and Medicine** Changing the initial velocity to 25 m/s only changes the t-coefficient.

$h(t) = 70 + 25t - 5t^2$ 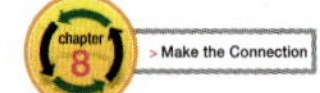

(a) How long will it take the ball to return to the thrower?

(b) When will the ball reach a height of 85 m?

Express your answers to the nearest tenth of a second.

The only part of the height equation that we have not discussed is the constant. You have probably noticed that the constant is always equal to the initial height of the ball (70 m in our previous exercises). Now, we ask you to develop an equation.

83. **Science and Medicine** A ball is thrown upward from the roof of a 100-m building with an initial velocity of 20 m/s. Use this information to complete each exercise. Express your answers to the nearest tenth of a meter. chapter 8 > Make the Connection

(a) Find the equation for the height h of the ball after t seconds.

(b) How long will it take the ball to fall back to the ground?

(c) When will the ball reach a height of 75 m?

(d) Will the ball ever reach a height of 125 m?

Hint: Check the discriminant.

84. **Science and Medicine** A ball is thrown upward from the roof of a 100-ft building with an initial velocity of 20 ft/s. Use this information to complete each exercise. Express your answers to the nearest tenth of a foot. chapter 8 > Make the Connection

(a) Find the height h of the ball after t seconds.

(b) How long will it take the ball to fall back to the ground?

(c) When will the ball reach a height of 80 ft?

(d) Will the ball ever reach a height of 120 ft? Explain.

85. **Construction** A rectangular field is 300 ft by 500 ft. A roadway of width x ft is to be built just inside the edge of the field. What is the widest roadway that can be built and still leave a 100,000-ft^2 field? Round your result to 3 decimal places.

86. **Construction** A rectangular garden is to be surrounded by a walkway of constant width. The garden's dimensions are 20 ft by 28 ft. The total area, garden plus walkway, is to be 1,100 ft^2. How wide does the walkway need to be? Round your result to three decimal places.

Complete each statement with **always, sometimes,** *or* **never.**

87. The quadratic formula can __________ be used to solve a quadratic equation.

88. If the value of $b^2 - 4ac$ is negative, the equation $ax^2 + bx + c = 0$ __________ has real solutions.

89. The solutions to a quadratic equation are __________ irrational.

90. To use the quadratic formula, a quadratic equation must __________ be written in standard form.

Skills | **Calculator/Computer** | Career Applications | Above and Beyond

91. **(a)** Use the quadratic formula to solve $x^2 - 3x - 5 = 0$. For each solution give a decimal approximation to the nearest tenth.

(b) Graph the function $f(x) = x^2 - 3x - 5$ on a graphing calculator. Estimate the x-intercepts to the nearest tenth.

(c) Describe the connection between parts (a) and (b).

92. **(a)** Solve the equation using any appropriate method.

$$x^2 - 2x = 3$$

(b) Graph the functions on a graphing calculator.

$$f(x) = x^2 - 2x \qquad \text{and} \qquad g(x) = 3$$

Estimate the x-coordinates of the points of intersection of f and g.

(c) Describe the connection between parts (a) and (b).

Skills | Calculator/Computer | **Career Applications** | Above and Beyond

93. **Agricultural Technology** The Scribner log rule is used to calculate the volume, in cubic feet, of a 16-ft log given the diameter (inside the bark) of the smaller end, in inches. The formula for the log rule is

$$V = 0.79D^2 - 2D - 4$$

Find the diameter of the small end of a log if its volume is 272 ft^3.

94. **Allied Health** Radiation therapy is one technique used to control cancer. After such a treatment, the number of cancerous cells N, in thousands, that remain in a particular patient can be estimated by the formula

$$N = -3t^2 - 6t + 140$$

in which t is the number of days of treatment. According to the model, how many days of treatment are required to kill all of a patient's cancer cells? VIDEO

95. **Allied Health** The concentration C, in nanograms per milliliter (ng/mL), of digoxin, a medication prescribed for congestive heart failure, is given by the equation

$$C = -0.0015t^2 + 0.0845t + 0.7170$$

where t is the number of hours since the drug was taken orally. For a drug to have a beneficial effect, its concentration in the bloodstream must exceed a certain value, the minimum therapeutic level, which for digoxin is 0.8 ng/mL.

(a) How long will it take for the drug to start having an effect?

(b) How long does it take for the drug to stop having an effect?

96. **Allied Health** The concentration C, in micrograms per milliliter (μg/mL), of phenobarbital, an anticonvulsant medication, is given by the equation

$$C = -1.35t^2 + 10.81t + 7.38$$

where t is the number of hours since the drug was taken orally. For a drug to have a beneficial effect, its concentration in the bloodstream must exceed a certain value, the minimum therapeutic level, which for phenobarbital is 10 μg/mL.

(a) How long will it take for the drug to start having an effect?

(b) How long does it take for the drug to stop having an effect?

Skills | Calculator/Computer | Career Applications | **Above and Beyond**

97. Can the solution of a quadratic equation with integer coefficients include one real and one imaginary number? Justify your answer.

98. Explain how the discriminant is used to predict the nature of the solutions of a quadratic equation.

Solve each equation for x.

99. $x^2 + y^2 = z^2$

100. $2x^2y^2z^2 = 1$

101. $x^2 - 36a^2 = 0$

102. $ax^2 - 9b^2 = 0$

103. $2x^2 + 5ax - 3a^2 = 0$

104. $3x^2 - 16bx + 5b^2 = 0$

105. $2x^2 + ax - 2a^2 = 0$

106. $3x^2 - 2bx - 2b^2 = 0$

107. Given that the polynomial $x^3 - 3x^2 - 15x + 25 = 0$ has as one of its solutions $x = 5$, find the other two solutions.

Hint: If you divide the given polynomial by $x - 5$, the quotient will be a quadratic expression. The remaining solutions will be the solutions for the resulting equation.

108. Given that $2x^3 + 2x^2 - 5x - 2 = 0$ has as one of its solutions $x = -2$, find the other two solutions.

Hint: In this case, divide the original polynomial by $x + 2$.

109. Find all the zeros of the function $f(x) = x^3 + 1$. (ℂ)

110. Find the zeros of the function $f(x) = x^2 + x + 1$. $(\mathbb{C})$

111. Find all six solutions to the equation $x^6 - 1 = 0$.

Hint: Factor the left-hand side of the equation first as the difference of squares, then as the sum and difference of cubes. $(\mathbb{C})$

112. Find all six solutions to $x^6 = 64$. $(\mathbb{C})$

113. Work with a partner to decide all values of b in each equation that will give one or more real solutions.

(a) $3x^2 + bx - 3 = 0$

(b) $5x^2 + bx + 1 = 0$

(c) $-3x^2 + bx - 3 = 0$

(d) Write a rule for judging whether an equation has solutions by looking at it in standard form.

114. Which method of solving a quadratic equation seems simplest to you? Which method should you try first?

115. Complete the statement: "You can tell an equation is quadratic and not linear by . . . ."

Answers

1. $\{-2, 7\}$ **3.** $\{-13, 5\}$ **5.** $\left\{-2, \frac{5}{3}\right\}$ **7.** $\left\{\frac{3}{4}\right\}$ **9.** $\{2 \pm \sqrt{11}\}$ **11.** $\left\{\frac{-3 \pm 3\sqrt{13}}{2}\right\}$ **13.** $\left\{\frac{5 \pm \sqrt{13}}{6}\right\}$ **15.** $\left\{\frac{1 \pm \sqrt{3}}{2}\right\}$ **17.** $\left\{-\frac{2}{3}, 1\right\}$ **19.** $\left\{\frac{1 \pm \sqrt{41}}{4}\right\}$ **21.** $\{1, 3\}$ **23.** $\{4\}$ **25.** $\left\{\frac{3 \pm \sqrt{29}}{2}\right\}$ **27.** $\left\{\frac{1 \pm i\sqrt{2}}{3}\right\}$ or $\left\{\frac{1}{3} \pm \frac{\sqrt{2}}{3}i\right\}$ nonreal **29.** $\left\{\frac{1 \pm \sqrt{161}}{8}\right\}$ **31.** $\left\{\frac{3 \pm \sqrt{39}}{15}\right\}$ **33.** $\left\{\frac{-1 \pm \sqrt{41}}{2}\right\}$ **35.** $\left\{\frac{1 \pm \sqrt{23}}{2}\right\}$ **37.** $\left\{\frac{7 \pm \sqrt{73}}{4}\right\}$ **39.** $\left\{\frac{3 \pm \sqrt{33}}{4}\right\}$ **41.** $\{1 \pm \sqrt{6}\}$ **43.** $\{4\}$ **45.** $\left\{\frac{7 \pm \sqrt{37}}{6}\right\}$ **47.** $\left\{0, \frac{2}{5}\right\}$ **49.** $\left\{\frac{-5 \pm \sqrt{33}}{4}\right\}$ **51.** $\{\pm 3i\}$ nonreal **53.** $\{-2, 5\}$ **55.** 25, two **57.** 0, one **59.** 37, two **61.** −63, none **63.** **(a)** 1.2 s or 5.8 s; **(b)** 1.7 s or 5.3 s **65.** −9, −8 or 8, 9 **67.** 7 ft by 10 ft **69.** 2.7 cm and 5.4 cm **71.** **(a)** 4 s; **(b)** 1 s **73.** 50 chairs **75.** 5.5 in., 6.5 in., 8.5 in. **77.** 27 or 63 Blu-ray players **79.** \$0.94 **81.** **(a)** 5.5 s; **(b)** 1 s, 2 s **83.** **(a)** $h(t) = 100 + 20t - 5t^2$; **(b)** 6.9 s; **(c)** 5 s; **(d)** no **85.** 34.169 ft **87.** always **89.** sometimes **91.** **(a)** $\{-1.2, 4.2\}$; **(b)** $(-1.2, 0)$ and $(4.2, 0)$; **(c)** The solutions to the quadratic equation are the x-intercepts of the graph. **93.** 20 in. **95.** **(a)** 1 hr; **(b)** 55.3 hr **97.** Above and Beyond **99.** $\left\{\pm\sqrt{z^2 - y^2}\right\}$ **101.** $\{\pm 6i\}$ **103.** $\left\{-3a, \frac{a}{2}\right\}$ **105.** $\left\{\frac{-a \pm |a|\sqrt{17}}{4}\right\}$ **107.** $\{-1 \pm \sqrt{6}\}$ **109.** $\left\{-1, \frac{1 \pm i\sqrt{3}}{2}\right\}$ or $\left\{-1, \frac{1}{2} \pm \frac{\sqrt{3}}{2}i\right\}$ two nonreal zeros **111.** $\left\{-1, 1 \frac{1 \pm i\sqrt{3}}{2}, \frac{-1 \pm i\sqrt{3}}{2}\right\}$ or $\left\{-1, 1, \frac{1}{2} \pm \frac{\sqrt{3}}{2}i, -\frac{1}{2} \pm \frac{\sqrt{3}}{2}i\right\}$ four nonreal solutions **113.** Above and Beyond **115.** Above and Beyond

8.3 An Introduction to Parabolas

< 8.3 Objectives >

1 > Find the axis of symmetry and vertex of a parabola

2 > Graph a parabola

In Section 3.1, you learned to graph a linear equation. We discovered that the graph of every linear equation in two variables is a straight line. In this section, we consider the graph of a quadratic equation in two variables.

Consider a quadratic equation,

$$y = ax^2 + bx + c \qquad a \neq 0$$

This equation defines a quadratic function with x as the input variable and y as the output variable. The graph of a quadratic function is a curve called a **parabola.**

Property

Shape of a Parabola

The graph of a function of the form

$$f(x) = ax^2 + bx + c \qquad a \neq 0$$

is always a parabola.

1. If $a > 0$, the parabola opens *upward*.
2. If $a < 0$, the parabola opens *downward*.

$$f(x) = ax^2 + bx + c$$

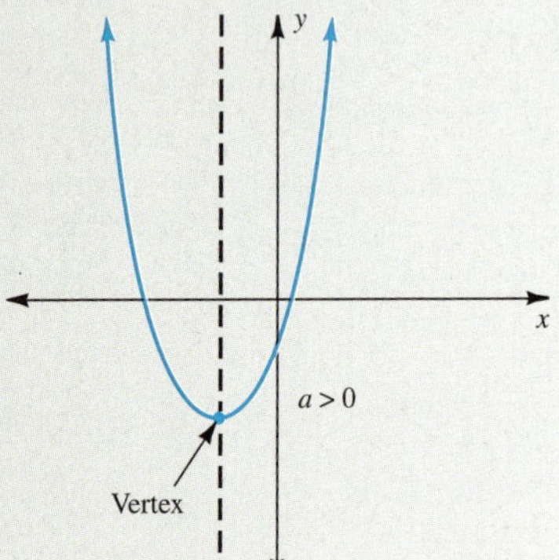

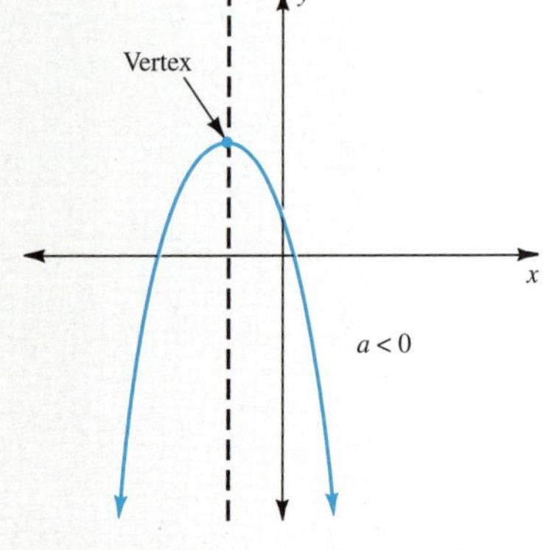

NOTE

You can think of the vertex as the **turning point** of the parabola.

We observe two properties of parabolas.

1. There is always a **minimum** (or lowest) point on a parabola that opens upward. There is always a **maximum** (or highest) point on a parabola that opens downward. In either case, that maximum or minimum value occurs at the **vertex** of the parabola.
2. Every parabola has an **axis of symmetry.** In the case of parabolas that open upward or downward, the axis of symmetry is the vertical line that splits the graph

into two pieces, each a mirror image of the other. The axis of symmetry always passes through the vertex.

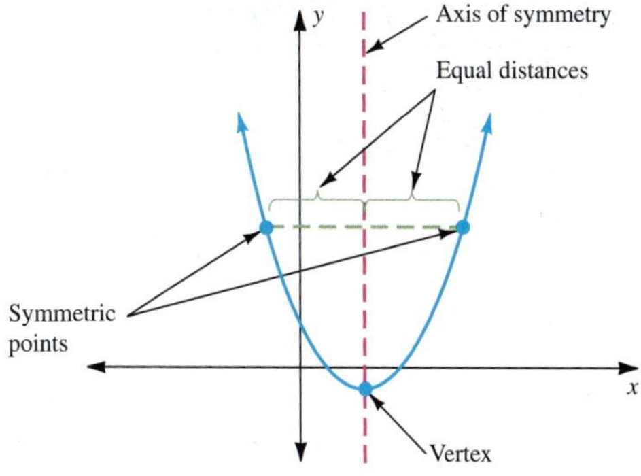

Our objective is to be able to quickly sketch a parabola. We can do this with *as few as three points* if we choose the points carefully. For this purpose we want to find the vertex and two symmetric points.

First, we see how to find the vertex from the standard equation

$$y = ax^2 + bx + c$$

If $x = 0$, then $y = c$, so $(0, c)$ gives the point where the parabola intersects the y-axis (the y-intercept).

NOTE

For this discussion, we assume that the vertex is not on the y-axis.

If the vertex *is* on the y-axis, the axis of symmetry is the y-axis itself, and its equation is $x = 0$.

Consider the sketch. To find the coordinates of the symmetric point (x_1, c), we note that it lies along the horizontal line $y = c$.

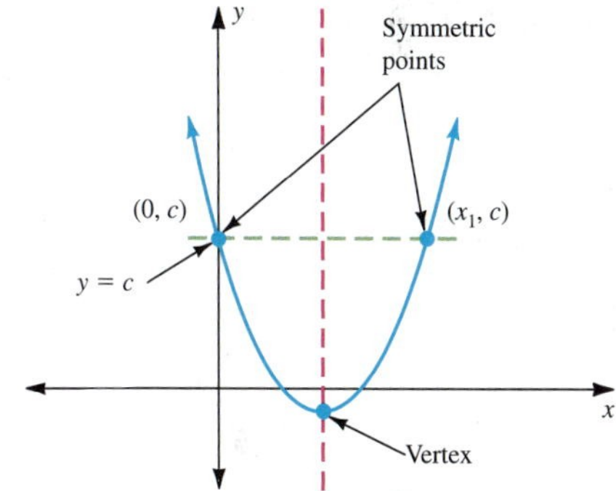

Therefore, let $y = c$ in the quadratic equation.

$c = ax^2 + bx + c$

$0 = ax^2 + bx$ — Subtract c from both sides.

$0 = x(ax + b)$ — Factor and solve.

and

$$x = 0 \qquad \text{or} \qquad x = -\frac{b}{a}$$

We now know that

$$(0, c) \qquad \text{and} \qquad \left(-\frac{b}{a}, c\right)$$

are the coordinates of the symmetric points shown. Since the axis of symmetry must be midway between these points, the x-value along that axis is given by

$$x = \frac{0 + (-b/a)}{2} = -\frac{b}{2a}$$

Since the vertex of any parabola lies on the axis of symmetry, we know that the x-coordinate of the vertex is $-\frac{b}{2a}$, and the corresponding y-coordinate can be found by evaluating the function at $x = -\frac{b}{2a}$.

Property

Axis of Symmetry and Vertex of a Parabola

If $f(x) = ax^2 + bx + c \qquad a \neq 0$

then the equation of the axis of symmetry is

$$x = -\frac{b}{2a}$$

and the coordinates of the vertex of the graph of f are

$$\left(-\frac{b}{2a}, f\left(-\frac{b}{2a}\right)\right)$$

If we can find two symmetric points, then we are well on our way to producing a graph. Perhaps the simplest case occurs when the quadratic expression is factorable. In such cases, the two x-intercepts (determined by the factors) give two symmetric points that are easily found. Example 1 illustrates such a case.

Example 1 Graphing a Parabola

< Objectives 1 and 2 >

Graph

$$f(x) = x^2 + 2x - 8$$

First, find the axis of symmetry. In this equation, $a = 1$, $b = 2$, and $c = -8$ so we have

$$x = -\frac{b}{2a} = -\frac{(2)}{2 \cdot (1)} = -\frac{2}{2} = -1$$

Thus, $x = -1$ is the axis of symmetry.

Second, find the vertex. Since the vertex of the parabola lies on the axis of symmetry, evaluate the function at $x = -1$.

$$f(-1) = (-1)^2 + 2(-1) - 8 = -9$$

So $(-1, -9)$ is the vertex of the parabola.

Third, find two symmetric points. In this case, the quadratic expression is factorable. Setting $f(x) = 0$ gives two symmetric points (the x-intercepts).

$$0 = x^2 + 2x - 8$$
$$= (x + 4)(x - 2)$$

So when $f(x) = 0$,

$$x + 4 = 0 \qquad \text{or} \qquad x - 2 = 0$$
$$x = -4 \qquad\qquad\qquad x = 2$$

and the x-intercepts are $(-4, 0)$ and $(2, 0)$.

Fourth, draw a smooth curve connecting the points, to form the parabola.

You could find additional pairs of symmetric points at this time if necessary. For instance, the symmetric points $(0, -8)$ and $(-2, -8)$ are easily located.

NOTES

Sketch the information to help solve the problem. Begin by drawing—as a dashed line—the axis of symmetry.

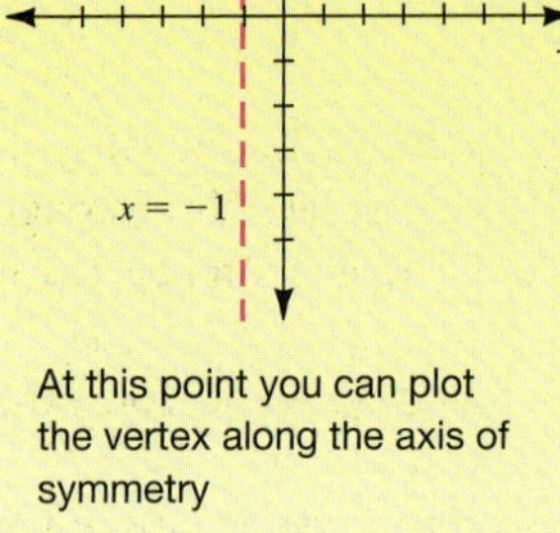

At this point you can plot the vertex along the axis of symmetry

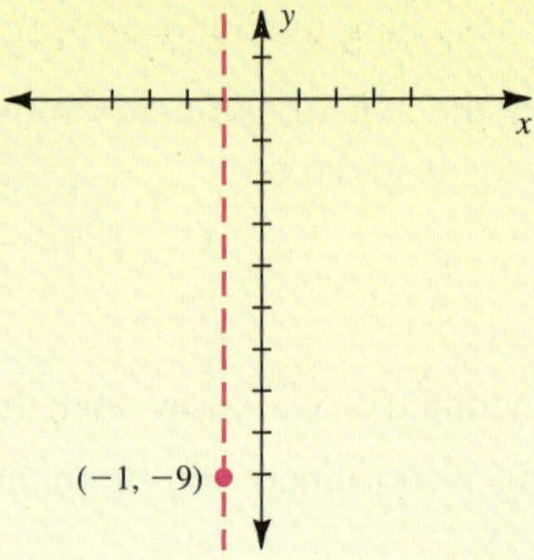

Check Yourself 1

Graph the function

$$f(x) = -x^2 - 2x + 3$$

Hint: The parabola opens downward because the coefficient of x^2 is negative.

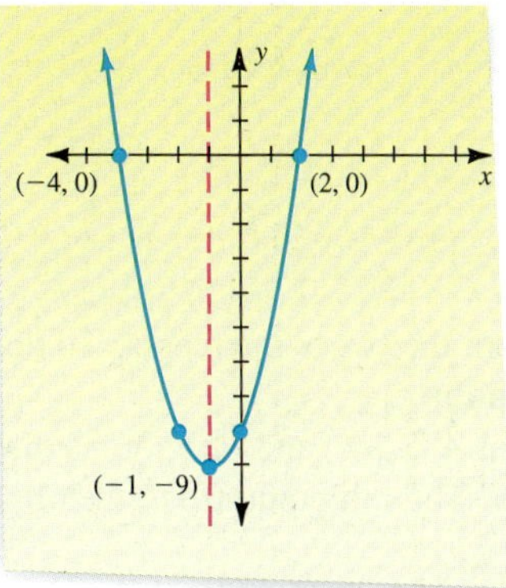

A similar process works if the quadratic expression is *not* factorable. In that case, one of two things happens:

1. The x-intercepts are irrational and therefore not particularly helpful when graphing.
2. The x-intercepts do not exist.

Consider Example 2.

Example 2 Graphing a Parabola

Graph the function

$f(x) = x^2 - 6x + 3$

First, find the axis of symmetry. Here $a = 1$, $b = -6$, and $c = 3$. So

$$x = -\frac{b}{2a} = -\frac{(-6)}{2(1)} = \frac{6}{2} = 3$$

Thus, $x = 3$ is the axis of symmetry.

Second, find the vertex. If $x = 3$,

$f(3) = (3)^2 - 6 \cdot (3) + 3 = -6$

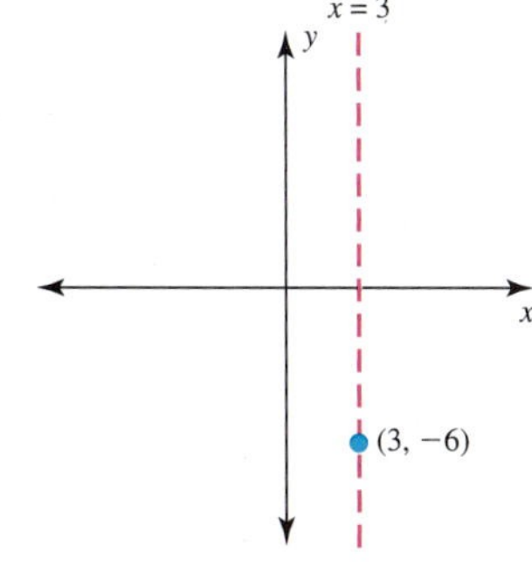

So $(3, -6)$ is the vertex of the parabola.

Third, find two symmetric points. This quadratic expression is not factorable, so we need to find another pair of symmetric points.

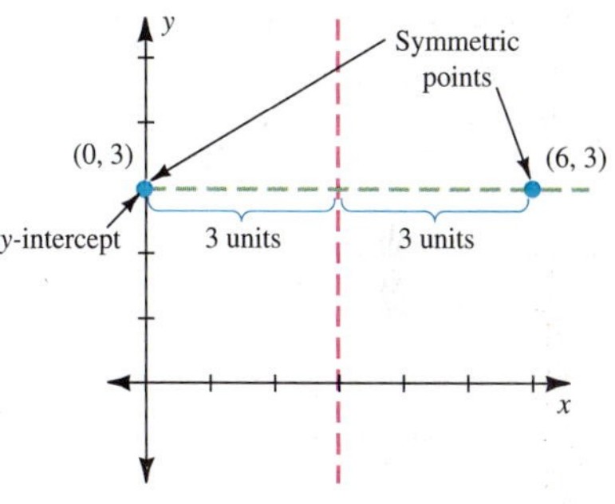

You should see that $(0, 3)$ is the y-intercept of the parabola. We found the axis of symmetry at $x = 3$ first. The point symmetric to $(0, 3)$ lies along the horizontal line through the y-intercept at the same distance (3 units) from the axis of symmetry. Hence, $(6, 3)$ is our symmetric point.

Fourth, draw a smooth curve connecting the points found above to form the parabola.

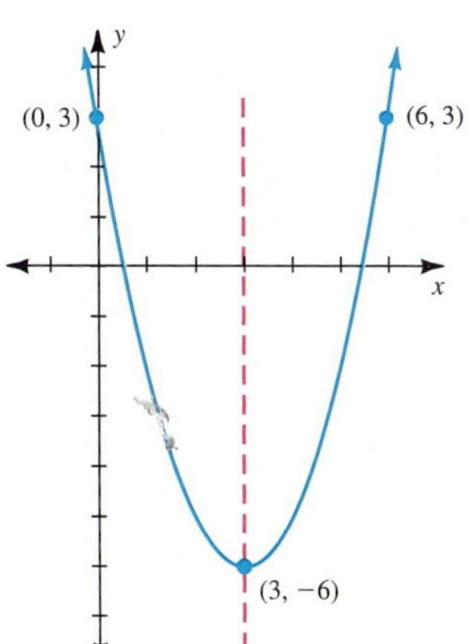

An alternate method is available when looking for symmetric points. Observing that $(0, 3)$ is the y-intercept and that the symmetric point lies along the line $y = 3$, set $f(x) = 3$ in the original equation.

$3 = x^2 - 6x + 3$

$0 = x^2 - 6x$

$0 = x(x - 6)$

so

$x = 0 \quad$ or $\quad x - 6 = 0$

$x = 6$

and $(0, 3)$ and $(6, 3)$ are the desired symmetric points.

Check Yourself 2

Graph the function

$f(x) = x^2 + 4x + 5$

We can also find the axis of symmetry directly from the two x-intercepts. To do this we must first introduce a new idea, that of the **midpoint.**

Every pair of points has a *midpoint*. The midpoint of $A\ (x_1, y_1)$ and $B\ (x_2, y_2)$ is the point on line AB that is an equal distance from A and B. The midpoint M is given by the formula

$$M = \left(\frac{x_1 + x_2}{2}, \frac{y_1 + y_2}{2}\right)$$

The x-coordinate of the midpoint is the average (mean) of the two x-values, and the y-coordinate of the midpoint is the average of the two y-values.

Example 3 Finding Midpoints

(a) Find the midpoint of (2, 0) and (10, 0).

$$M = \left(\frac{2 + 10}{2}, \frac{0 + 0}{2}\right) = \left(\frac{12}{2}, \frac{0}{2}\right) = (6, 0)$$

(b) Find the midpoint of (5, 7) and (−1, −3).

$$M = \left(\frac{5 + (-1)}{2}, \frac{7 + (-3)}{2}\right) = \left(\frac{4}{2}, \frac{4}{2}\right) = (2, 2)$$

(c) Find the midpoint of (−7, 3) and (8, 6).

$$M = \left(\frac{-7 + 8}{2}, \frac{3 + 6}{2}\right) = \left(\frac{1}{2}, \frac{9}{2}\right)$$

Check Yourself 3

Find the midpoint of each pair of points.

(a) (2, 7) and (−4, −1) **(b)** (−3, −6) and (5, −2)
(c) (−2, 3) and (3, 8)

We can use the midpoint to find the axis of symmetry.

Example 4 Graphing a Parabola

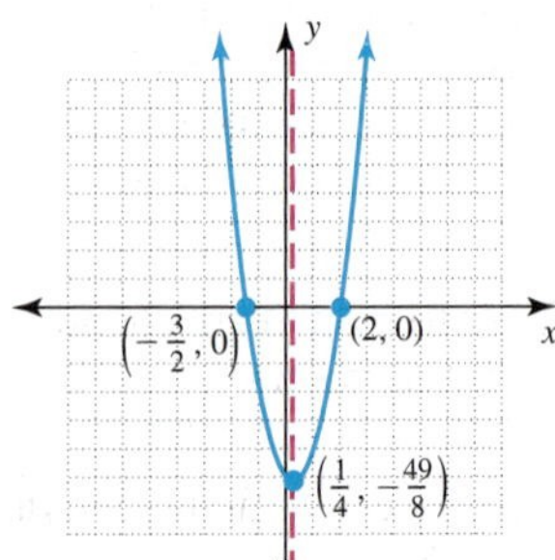

Graph the function $f(x) = 2x^2 - x - 6$.

$f(x) = 2x^2 - x - 6$

$f(x) = (2x + 3)(x - 2)$ Factor the expression.

$0 = (2x + 3)(x - 2)$ Find the x-intercepts.

$x = -\frac{3}{2}$ or $x = 2$

The x-intercepts are $\left(-\frac{3}{2}, 0\right)$ and (2, 0).

The midpoint is

$$\left(\frac{-\frac{3}{2}+2}{2}, \frac{0+0}{2}\right) = \left(\frac{-\frac{3}{2}+\frac{4}{2}}{2}, 0\right) = \left(\frac{1}{4}, 0\right)$$

The axis of symmetry is $x = \frac{1}{4}$. The vertex is $\left(\frac{1}{4}, -\frac{49}{8}\right)$.

Check Yourself 4

Graph the function

$$f(x) = 2x^2 + 5x - 3$$

It is not typical for quadratic expressions to be factorable. As a result, we generally use the formula $x = -\frac{b}{2a}$ to find the axis of symmetry.

Example 5 Graphing a Parabola

Graph the function $f(x) = 3x^2 - 6x + 5$.

First, find the axis of symmetry.

$$x = -\frac{b}{2a} = -\frac{(-6)}{2(3)} = \frac{6}{6} = 1$$

Second, find the vertex. If $x = 1$,

$$f(1) = 3(1)^2 - 6 \cdot 1 + 5 = 2$$

So (1, 2) is the vertex.

Third, find symmetric points. Again the quadratic expression is not factorable, so we use the y-intercept (0, 5) and its symmetric point (2, 5).

Fourth, connect the points with a smooth curve to form the parabola. Compare this curve to those in previous examples. This parabola is "tighter" about the axis of symmetry because the x^2-coefficient is larger.

Check Yourself 5

Graph the function

$$f(x) = \frac{1}{2}x^2 - 3x - 1$$

This algorithm summarizes our work.

Step by Step

Graphing a Parabola

Step 1 Find the axis of symmetry.

Step 2 Find the vertex.

Step 3 Determine two symmetric points.

Note: Use the x-intercepts if the quadratic expression is factorable. If the vertex is not on the y-axis, you can use the y-intercept and its symmetric point. Another option is to simply choose an x-value that does not match the axis of symmetry, compute the corresponding y-value, and then locate its symmetric point.

Step 4 Draw a smooth curve connecting the points found to form the parabola. You may choose to find additional pairs of symmetric points.

When we use a parabola to help solve an application, we often need to find the vertex of the graph.

Example 6 A Biology Application

> **RECALL**
>
> On a parabola, the vertex represents either a minimum point (if $a > 0$) or a maximum point (if $a < 0$).

An experimental drug is being tested on a bacteria colony. It is found that t days after the colony is treated, the number N of bacteria per cubic centimeter is given by the equation

$N = 15t^2 - 100t + 700$

(a) When will the bacteria colony be at a minimum?

We know that the graph of this equation is a parabola, and that the graph opens up. (Why?) So the vertex represents a minimum point for the graph.

To find the vertex, we first find the equation of the axis of symmetry.

$$t = -\frac{b}{2a} = -\frac{(-100)}{2(15)} = \frac{100}{30} = \frac{10}{3}$$

This means the bacteria colony will be at a minimum after $\frac{10}{3}$, or $3\frac{1}{3}$, days.

(b) What is that minimum number?

We now find N that corresponds to this t-value.

> **NOTE**
>
> The vertex of the parabola is $\left(3\frac{1}{3}, 533\frac{1}{3}\right)$.

$$N = 15\left(\frac{10}{3}\right)^2 - 100\left(\frac{10}{3}\right) + 700$$

$$N = 533\frac{1}{3}.$$

So the minimum number of bacteria is approximately 533.

Check Yourself 6

The number of people who are sick t days after the outbreak of a flu epidemic is given by the equation

$P = -t^2 + 120t + 20$

(a) On what day will the maximum number of people be sick?
(b) How many people will be sick on that day?

Check Yourself ANSWERS

1.

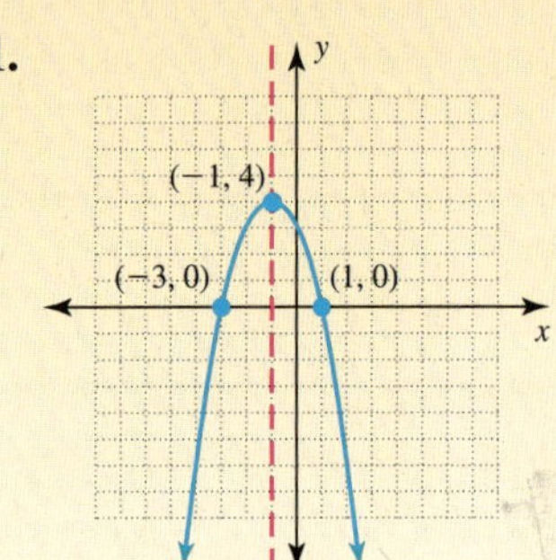

2.

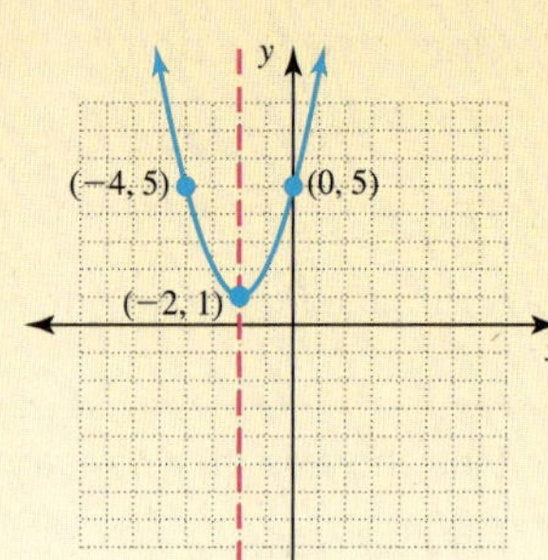

3. **(a)** $(-1, 3)$; **(b)** $(1, -4)$; **(c)** $\left(\frac{1}{2}, \frac{11}{2}\right)$

4.

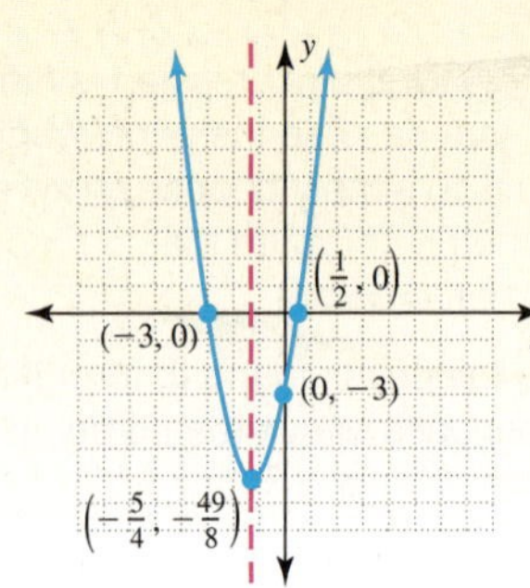

5.

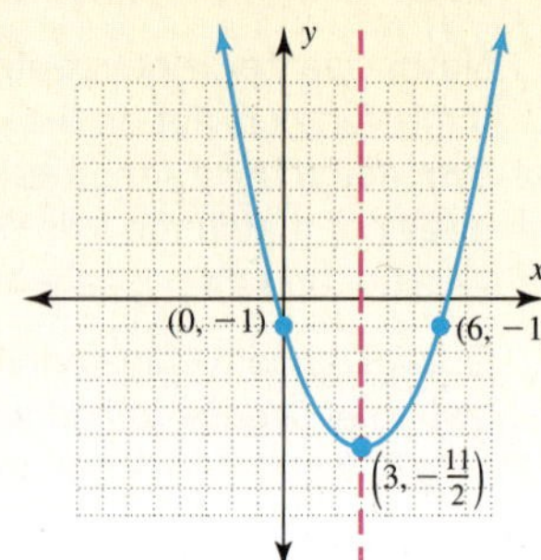

6. **(a)** Day 60; **(b)** 3,620 people

Reading Your Text

These fill-in-the-blank exercises will help you understand some of the key vocabulary used in this section. The answers to these exercises are in the Answers Appendix in the back of the text.

(a) In an equation of the form $y = ax^2 + bx + c$, the parabola opens ______________ if $a < 0$.

(b) There is always a ______________ point on a parabola if it opens upward.

(c) The vertex of a parabola lies on the ______________ of symmetry.

(d) The axis of symmetry of a parabola is the ______________ line that splits the graph into two pieces, each a mirror image of the other.

Skills Calculator/Computer Career Applications Above and Beyond

8.3 exercises

Match each graph with one of the equations.

(a) $y = x^2 + 2$ **(b)** $y = 2x^2 - 1$ **(c)** $y = 2x + 1$ **(d)** $y = x^2 - 3x$

(e) $y = -x^2 - 4x$ **(f)** $y = -2x + 1$ **(g)** $y = x^2 + 2x - 3$ **(h)** $y = -x^2 + 6x - 8$

1.

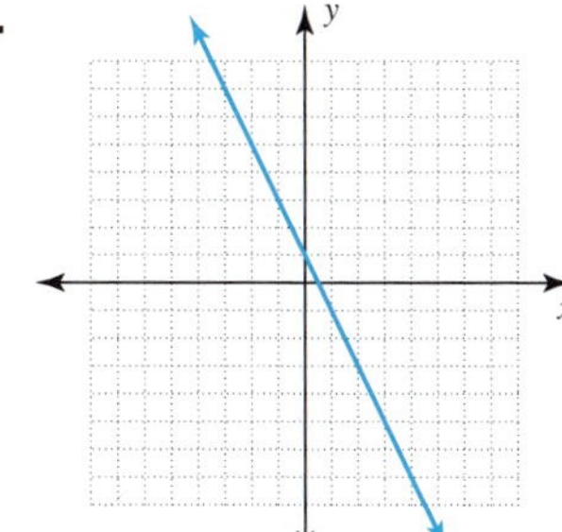

2.

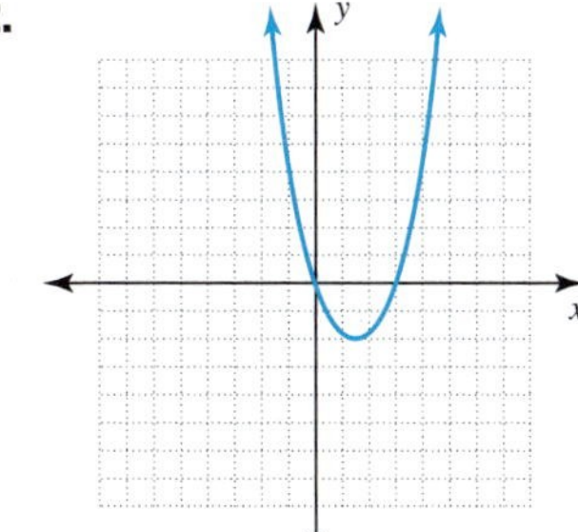

3.

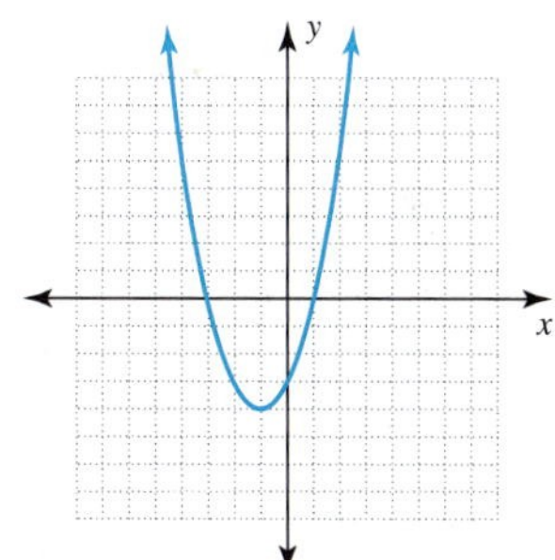

4.

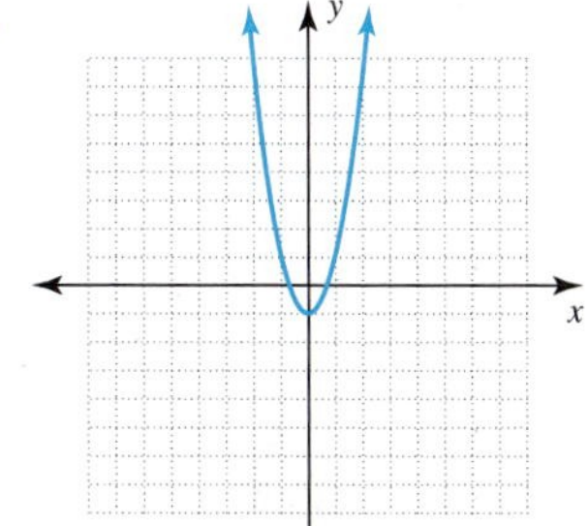

5.

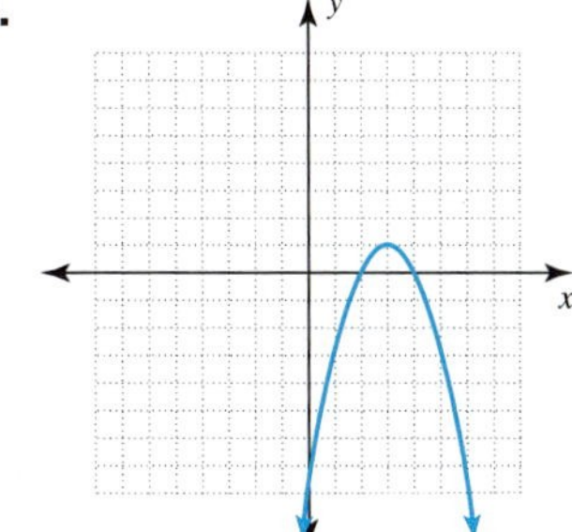

6.

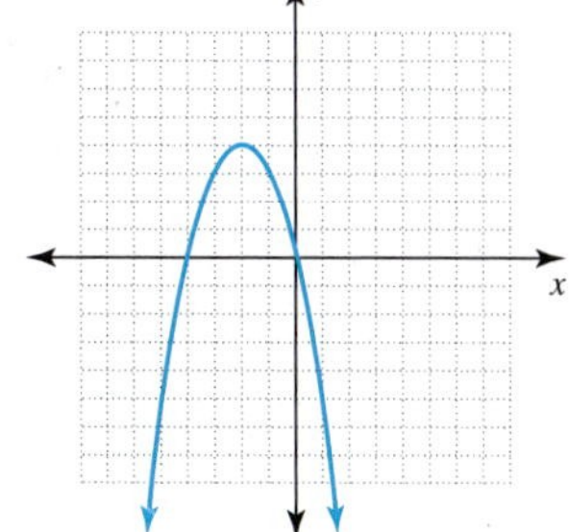

7. 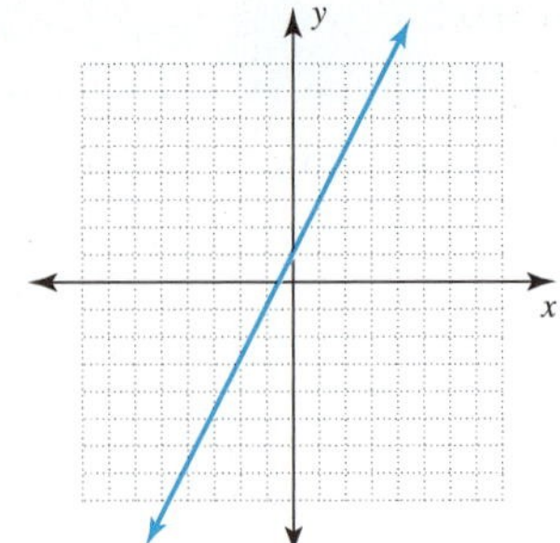

8. 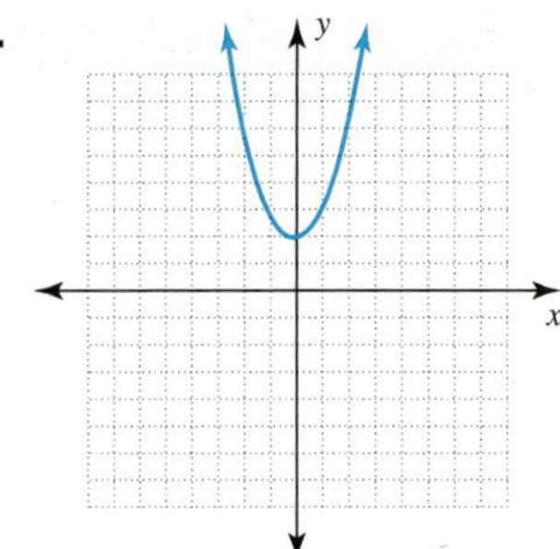

Which conditions apply to the graphs of each equation? Note that more than one condition may apply.

(a) The parabola opens upward.

(b) The parabola opens downward.

(c) The parabola has two x-intercepts.

(d) The parabola has one x-intercept.

(e) The parabola has no x-intercept.

9. $y = x^2 - 3$

10. $y = -x^2 + 4x$

11. $y = x^2 - 3x - 4$

12. $y = x^2 - 2x + 2$

13. $y = -x^2 - 3x + 10$

14. $y = x^2 - 8x + 16$

Find the midpoint of each pair of points.

15. $(-2, 6)$ and $(1, 7)$

16. $\left(\frac{1}{2}, -2\right)$ and $\left(-3, \frac{1}{2}\right)$

Given a point on a parabola and the axis of symmetry, locate the point symmetric to the given one.

17. $(6, 5)$; axis of symmetry: $x = 4$

18. $(-5, 2)$; axis of symmetry: $x = -2$

19. $(-1, -7)$; axis of symmetry: $x = 3$

20. $(4, -2)$; axis of symmetry: $x = -1$

Find the equation of the axis of symmetry, given two points on a parabola.

21. $(-1, 5)$ and $(5, 5)$

22. $(-3, 0)$ and $(6, 0)$

23. $(-6, 0)$ and $(-1, 0)$

24. $(-4, -6)$ and $(10, -6)$

< Objectives 1 and 2 >

Find the equation of the axis of symmetry, the coordinates of the vertex, and the x-intercepts. Sketch the graph of each function.

25. $f(x) = -x^2 + 4$ 

26. $f(x) = x^2 + 2x$

27. $f(x) = -x^2 - 2x$

28. $f(x) = -x^2 - 3x$

29. $f(x) = x^2 - 6x + 5$

30. $f(x) = x^2 + x - 6$

31. $f(x) = x^2 - 5x + 6$ 

32. $f(x) = x^2 + 6x + 5$

33. $f(x) = x^2 - 6x + 8$

34. $f(x) = -x^2 - 3x + 4$

35. $f(x) = -x^2 - 6x - 5$

36. $f(x) = -x^2 + 6x - 8$

Find the equation of the axis of symmetry, the coordinates of the vertex, and at least two symmetric points. Sketch the graph of each function.
Note: *A sample answer is provided for the two symmetric points.*

37. $f(x) = x^2 - 2x - 1$

38. $f(x) = x^2 + 4x + 6$

39. $f(x) = x^2 - 4x - 1$

40. $f(x) = -x^2 + 6x - 5$

41. $f(x) = -x^2 + 3x - 3$

42. $f(x) = x^2 + 5x + 3$

43. $f(x) = 2x^2 + 4x - 1$ 

44. $f(x) = \frac{1}{2}x^2 - x - 1$

45. $f(x) = -\frac{1}{3}x^2 + x - 3$

46. $f(x) = -2x^2 - 4x - 1$

47. $f(x) = 3x^2 + 12x + 5$

48. $f(x) = -3x^2 + 6x + 1$

Complete each statement with **always, sometimes,** *or* **never.**

49. The vertex of a parabola is ______________ located on its axis of symmetry.

50. The vertex of a parabola is ______________ the highest point on the graph.

51. The graph of $y = ax^2 + bx + c$ ______________ intersects the x-axis.

52. The graph of $y = ax^2 + bx + c$ ______________ has more than two x-intercepts.

53. The graph of $y = ax^2 + bx + c$ ______________ intersects the y-axis.

54. The graph of $y = ax^2 + bx + c$ ______________ intersects the y-axis more than once.

Skills | Calculator/Computer | **Career Applications** | Above and Beyond

55. **Allied Health** The concentration C, in nanograms per milliliter (ng/mL), of digoxin, a medication prescribed for congestive heart failure, is given by the equation

$$C = -0.0015t^2 + 0.0845t + 0.7170$$

where t is the number of hours since the drug was taken orally.

(a) When will the drug's concentration be at a maximum?

(b) Determine the drug's maximum concentration in the bloodstream.

56. **Allied Health** A patient's body temperature T, in degrees Fahrenheit, t hours after taking acetaminophen, a commonly prescribed analgesic, can be approximated by the equation

$$T = 0.4t^2 - 2.6t + 103$$

(a) When will the patient's temperature be at a minimum?

(b) Determine the patient's minimum temperature.

Answers

1. (f) **3.** (g) **5.** (h) **7.** (c) **9.** (a), (c) **11.** (a), (c) **13.** (b), (c) **15.** $\left(-\frac{1}{2}, \frac{13}{2}\right)$
17. (2, 5) **19.** (7, −7) **21.** $x - 2$ **23.** $x = \frac{7}{2}$

25.

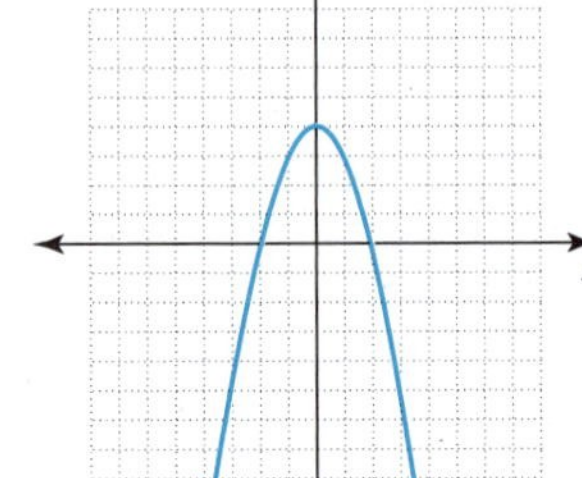

$x = 0$; vertex (0, 4); (−2, 0) and (2, 0)

27.

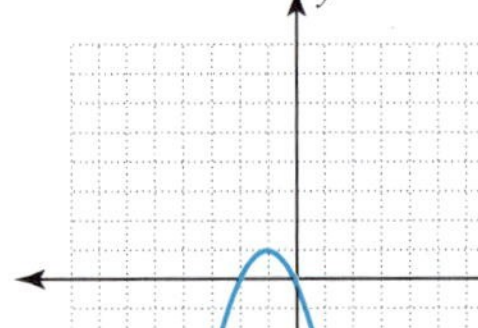

$x = -1$; vertex (−1, 1); (−2, 0) and (0, 0)

29.

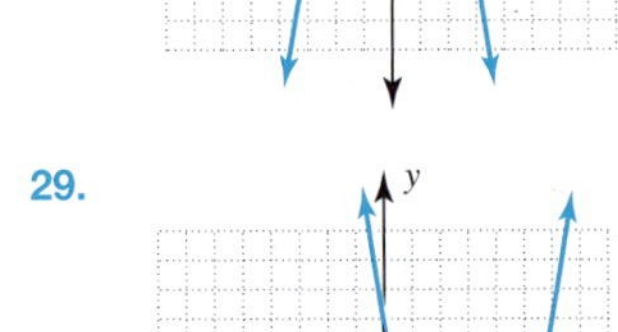

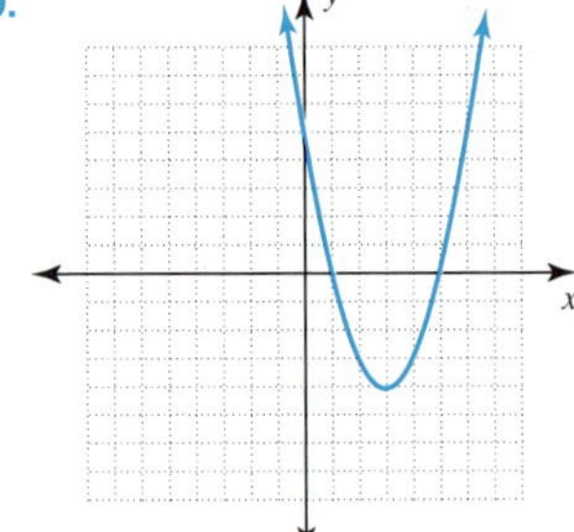

$x = 3$; vertex (3, −4); (1, 0) and (5, 0)

31.

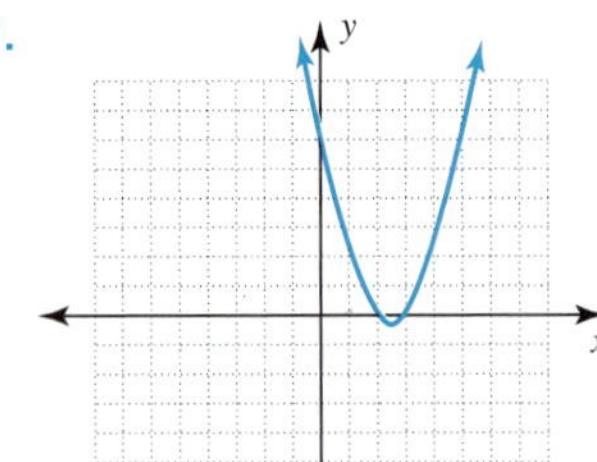

$x = \frac{5}{2}$; vertex $\left(\frac{5}{2}, -\frac{1}{4}\right)$; (2, 0) and (3, 0)

33. 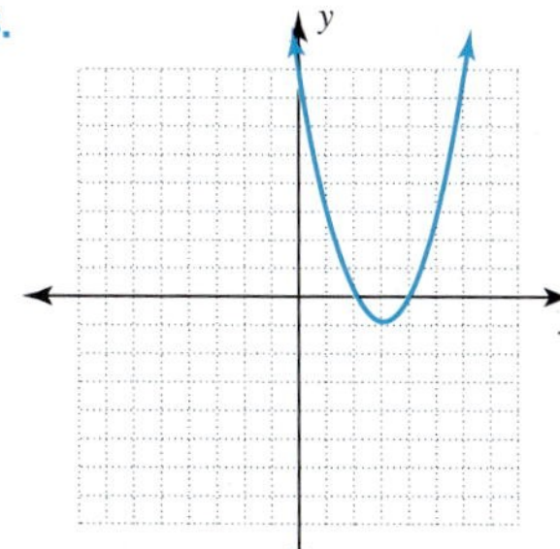
$x = 3$; vertex $(3, -1)$; $(2, 0)$ and $(4, 0)$

35. 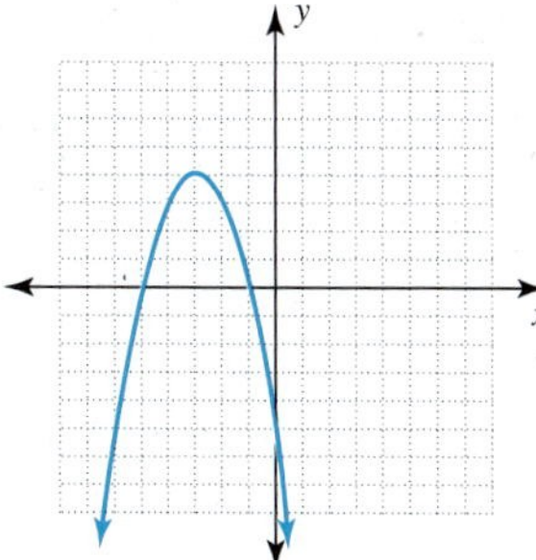
$x = -3$; vertex $(-3, 4)$; $(-5, 0)$ and $(-1, 0)$

37. 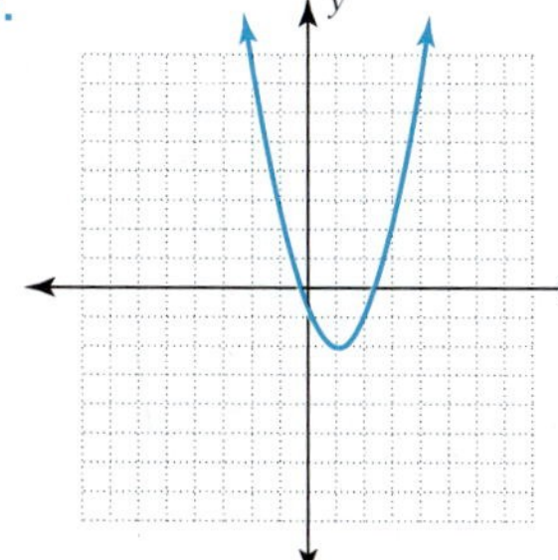
$x = 1$; vertex $(1, -2)$; $(0, -1)$ and $(2, -1)$

39. 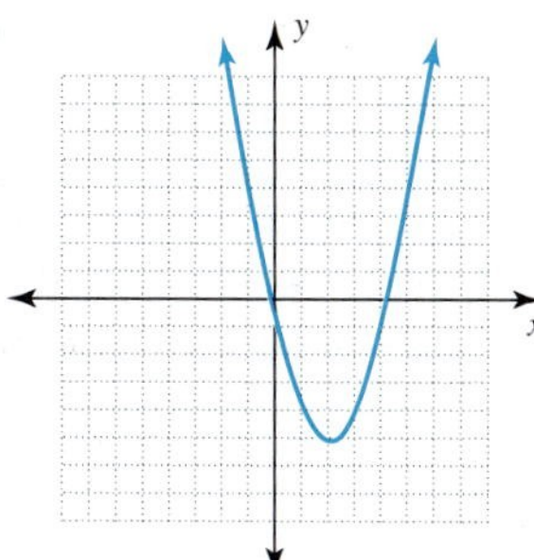
$x = 2$; vertex $(2, -5)$; $(0, -1)$ and $(4, -1)$

41. 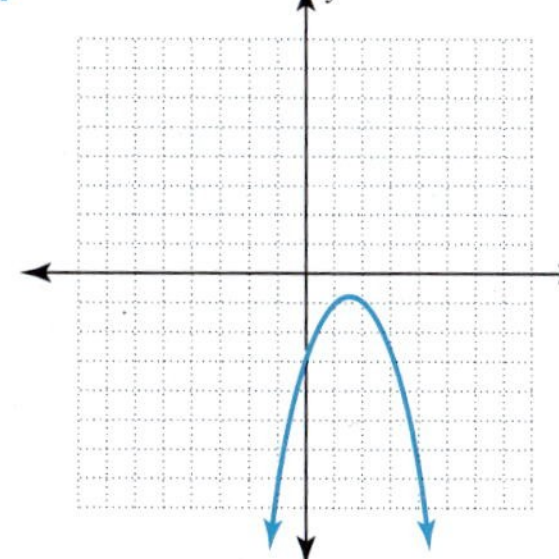
$x = \frac{3}{2}$; vertex $\left(\frac{3}{2}, -\frac{3}{4}\right)$; $(0, -3)$ and $(3, -3)$

43. 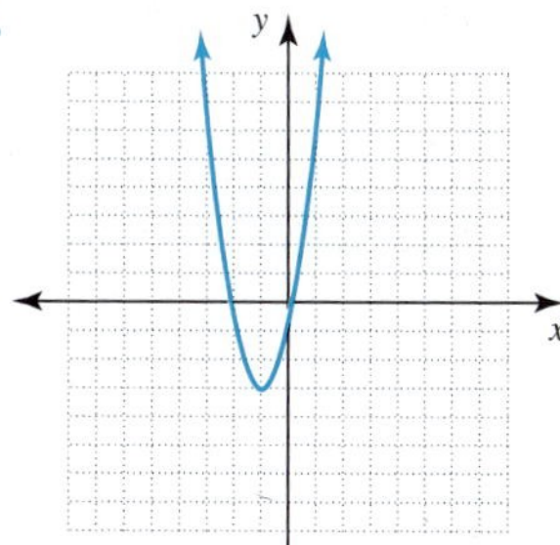
$x = -1$; vertex $(-1, -3)$; $(-2, -1)$ and $(0, -1)$

45. 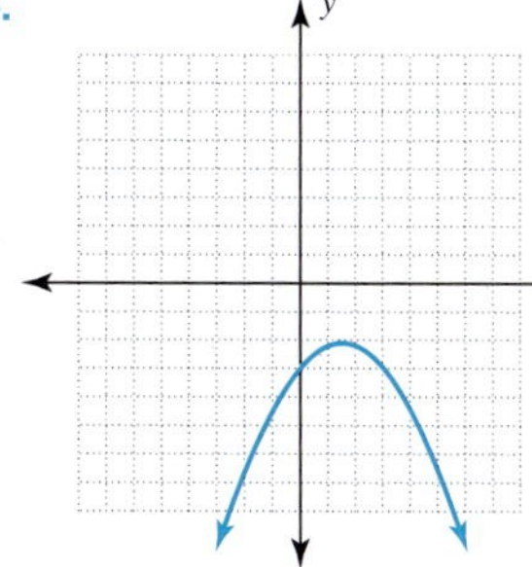
$x = \frac{3}{2}$; vertex $\left(\frac{3}{2}, -\frac{9}{4}\right)$; $(0, -3)$ and $(3, -3)$

47. 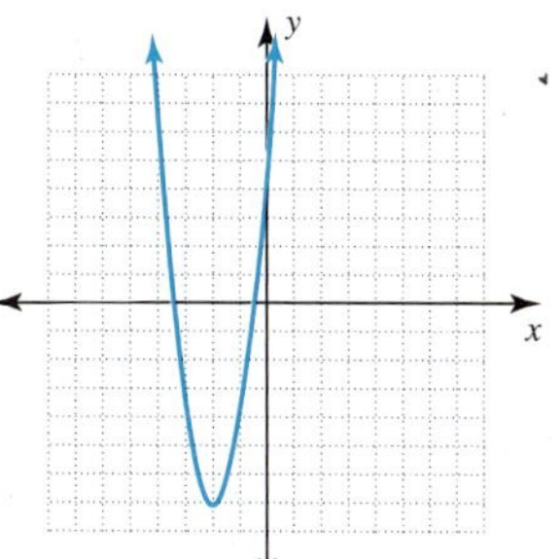
$x = -2$; vertex $(-2, -7)$; $(-4, 5)$ and $(0, 5)$

49. always **51.** sometimes **53.** always **55.** **(a)** 28.2 hr; **(b)** 1.91 ng/mL

8.4 Quadratic Equations and Problem Solving

< 8.4 Objectives >

1 > Solve quadratic equations by graphing

2 > Use quadratic equations to solve applications

3 > Solve equations that are quadratic in form

In this section, we look more closely at using graphs to solve quadratic equations. The limitation of graphs is that they only yield approximate solutions. Of course, finding solutions graphically is easy, especially with graphing calculators and computers. Further, there are many situations in which an approximation is sufficient, especially when solving applications.

Our general technique is to write the quadratic equation in standard form, define

$$f(x) = ax^2 + bx + c$$

and look for the x-intercepts of the graph. These are the points at which

$$ax^2 + bx + c = 0$$

Example 1 Solving a Quadratic Equation Graphically

< Objective 1 >

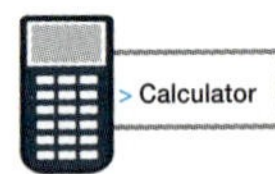

Use a graphing calculator to solve the equation. Round your answers to the nearest thousandth.

$$0.4x^2 - x - 2.5 = 0$$

In the calculator, we define $Y_1 = 0.4x^2 - x - 2.5$, and we view the graph in the standard viewing window.

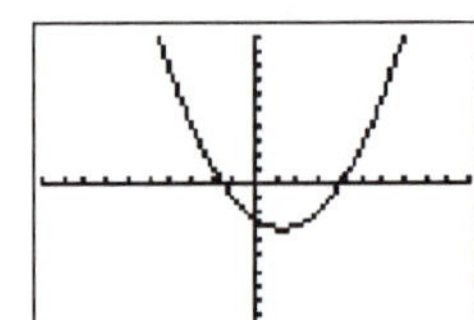

We are interested in the x-intercepts. Use the "ZERO" or "ROOT" command to approximate the x-intercepts.

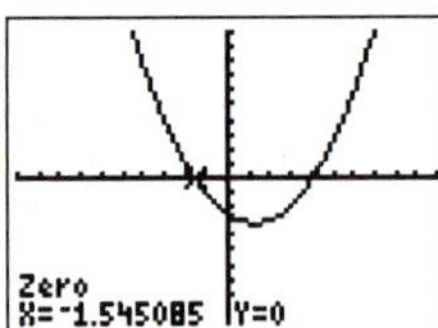

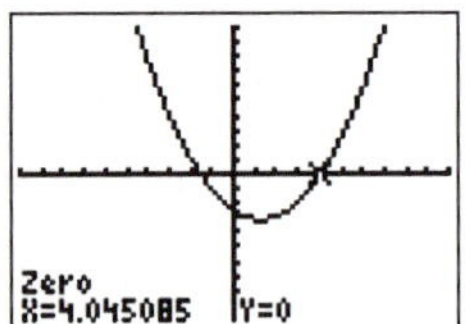

So, to the nearest thousandth, the solution set is $\{-1.545, 4.045\}$.

Check Yourself 1

Use a graphing calculator to solve the equation. Round your answers to the nearest thousandth.

$$-0.3x^2 - 0.4x + 2.75 = 0$$

Often, we apply our knowledge of parabolas when using a calculator.

Example 2 Solving a Quadratic Equation Graphically

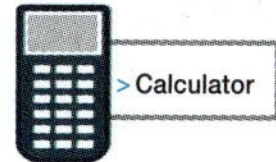

Use a graphing calculator to solve the equation. Round your answers to the nearest thousandth.

$-x^2 + 16x + 160 = 0$

When we define $Y_1 = -x^2 + 16x + 160$ and view the graph in the standard viewing window, we see

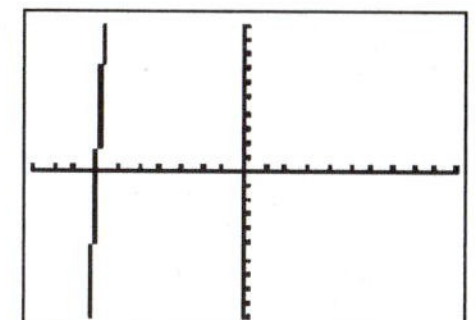

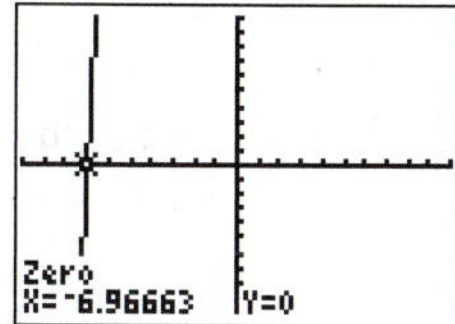

We expect to see a parabola, and we expect that it opens down. (Do you see why?) Therefore, we must be looking at the left portion of the parabola. Further, we know that the graph climbs for a while, turns, and comes back down to cross the x-axis somewhere to the right. While we have this view, we find the x-intercept (to the nearest thousandth) to be $(-6.967, 0)$.

We use the TABLE utility to try values to the right of 10 because the parabola must intersect the x-axis somewhere to the right.

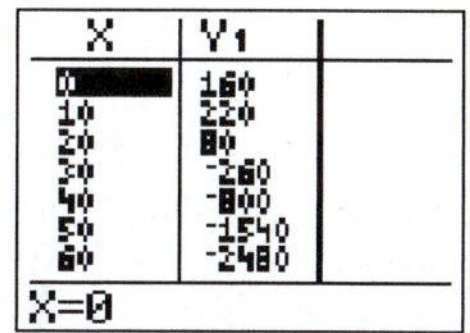

X	Y1
0	160
10	220
20	80
30	-260
40	-800
50	-1540
60	-2480

X=0

When $x = 20$, the y-value is 80; when $x = 30$, the y-value is -260.

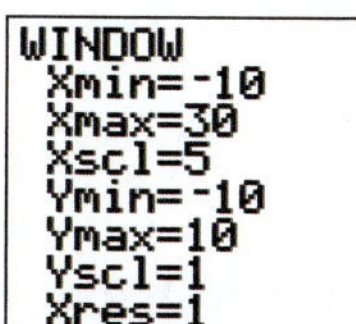

We conclude that the graph must cross the x-axis somewhere between $x = 20$ and $x = 30$. If we simply change the window to $-10 \le x \le 30$, with x-scale of 5, we see

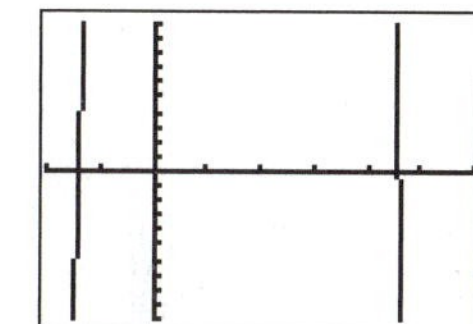

We use the ZERO command to find the right-hand x-intercept to be $(22.967, 0)$. To the nearest thousandth, the solution set is $\{-6.967, 22.967\}$.

Check Yourself 2

Use a graphing calculator to solve the equation. Round your answers to the nearest thousandth.

$x^2 + 24x - 265 = 0$

In many applications, the vertex of a parabola corresponds to a solution to the problem. If the parabola opens up ($a > 0$), then the vertex is the lowest point on the graph. If the parabola opens down ($a < 0$), then the vertex is the highest point on the graph.

We usually want to find both the x- and y-coordinates of the vertex. For instance, in thrown ball applications, the x-coordinate of the vertex tells us how long it takes for the ball to reach its maximum height, whereas the y-coordinate tells us what that height is.

We need to find vertices to solve applications in both Examples 3 and 4.

Example 3 A Business and Finance Application

< Objective 2 >

A software company sells a word-processing program. It has found that the monthly profit P, in dollars, from selling x copies of the program is approximated by

$$P(x) = -0.3x^2 + 90x - 1{,}500$$

Find the number of copies of the program that should be sold in order to maximize the profit and find that maximum profit.

Since the profit function is quadratic, the graph must be a parabola. Also, since the coefficient of x^2 is negative, the parabola must open downward, so the vertex gives the maximum value for the profit P. To find the vertex,

NOTE

View a graph of $P(x)$ using this window: $0 \le x \le 300$ and $-1{,}500 \le y \le 6{,}000$. Then use the MAXIMUM utility in your calculator to find the vertex.

$$x = -\frac{b}{2a} = -\frac{(90)}{2(-0.3)} = 150$$

The maximum profit occurs when $x = 150$, so we evaluate the function at that point.

$$P(150) = -0.3(150)^2 + (90)(150) - 1{,}500$$
$$= \$5{,}250$$

The maximum profit occurs when 150 copies are sold in a month. That profit would be $5,250.

Check Yourself 3

A company that sells portable radios finds that its weekly profit P, in dollars, and the number of radios sold x are related by

$$P(x) = -0.2x^2 + 40x - 100$$

Find the number of radios that should be sold to have the largest weekly profit and find the amount of that profit.

Example 4 A Geometry Application

A farmer has 3,600 ft of fence to enclose a rectangular lot. Find the largest possible area that can be enclosed.

As usual, when dealing with geometric figures, we start by drawing a sketch of the problem.

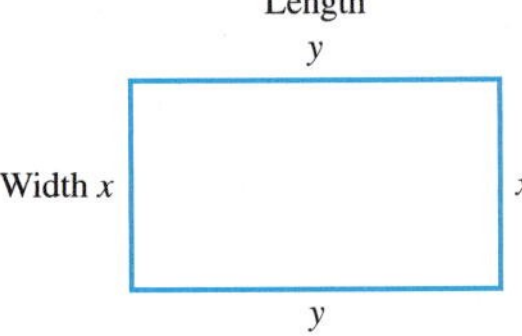

First, we write the area A as

$$A = xy$$

RECALL

Area = length × width

The perimeter of the region is $2x + 2y$

Since there are 3,600 ft of fence, we know

$$2x + 2y = 3{,}600$$
$$2y = 3{,}600 - 2x$$
$$y = 1{,}800 - x$$

Substituting for y in the area formula, we have

$$A = xy$$
$$A = x(1{,}800 - x)$$
$$A = 1{,}800x - x^2$$
$$A = -x^2 + 1{,}800x$$

NOTE

The width x is 900 ft, so we have

$y = 1{,}800 - 900$
$= 900$

Therefore, the length is also 900 ft. The desired region is a square.

We have a quadratic function where x is the input variable and A is the output variable.

$$A(x) = -x^2 + 1{,}800x$$

Again, the graph of A is a parabola opening downward, and the largest possible area occurs at the vertex. We find the vertex.

$$x = -\frac{1{,}800}{2(-1)} = 900$$

and the largest possible area is

$$A(900) = -(900)^2 + 1{,}800(900) = 810{,}000 \text{ ft}^2$$

Check Yourself 4

We want to enclose three sides of the largest possible rectangular area by using 900 ft of fence. Assume that an existing wall makes the fourth side. What are the dimensions of the rectangle?

NOTE

This is an example of a fourth-degree polynomial equation. You will study these in a later math course. Expect to find up to four solutions.

We turn our attention now to solving equations that are not quadratic, but are **quadratic in form.**

Recall that in Section 6.3 we factored expressions that are quadratic in form. Now suppose that we need to solve the equation

$$x^4 - 5x^2 + 6 = 0$$

The key to observing a "quadratic in form" situation is to note that x^4 is the square of x^2.

We use substitution. Choose another variable, say u, and let $u = x^2$. This implies that $u^2 = x^4$. Changing to an equation with u gives

$$u^2 - 5u + 6 = 0$$

This certainly looks like a quadratic equation! Consider Example 5.

Example 5 Solving an Equation That Is Quadratic in Form

< Objective 3 >

Solve

$$x^4 - 5x^2 + 6 = 0$$

We let $u = x^2$, which implies that $u^2 = x^4$. Rewriting the equation in terms of u gives

$$u^2 - 5u + 6 = 0$$

We solve for u by factoring.

$$(u - 2)(u - 3) = 0$$

So $u = 2$ or $u = 3$.

NOTE

This is often called "back substituting."

But we want to solve for x (in our original equation), so we now return to equations involving x.

$x^2 = 2$ or $x^2 = 3$ — Remember that $u = x^2$.

Solving these by the square-root method, we have

$x = \pm\sqrt{2}$ or $x = \pm\sqrt{3}$ — We need both positive and negative roots.

Each of these four values can be checked in the original equation. They all work! The solution set is $\{\pm\sqrt{2}, \pm\sqrt{3}\}$.

Check Yourself 5

Solve $x^4 - 6x^2 + 8 = 0$.

It is quite possible that such an equation has nonreal solutions as well as real solutions.

Example 6 Solving an Equation That Is Quadratic in Form (ℂ)

Solve $x^4 - 4x^2 - 12 = 0$.

This equation is quadratic in form (x^4 is the square of x^2).

Let $u = x^2$. Then $u^2 = x^4$, and

$$u^2 - 4u - 12 = 0$$
$$(u - 6)(u + 2) = 0$$
$$u = 6 \quad \text{or} \quad u = -2$$

Back-substituting,

$$x^2 = 6 \quad \text{or} \quad x^2 = -2$$

Solving these equations yields two real solutions and two nonreal solutions

$$x = \pm\sqrt{6} \quad \text{or} \quad x = \pm\sqrt{-2} = \pm i\sqrt{2}$$

The solution set is $\{\sqrt{6}, -\sqrt{6}, i\sqrt{2}, -i\sqrt{2}\}$ or $\{\pm\sqrt{6}, \pm i\sqrt{2}\}$.

NOTE

Again we find four solutions to a fourth-degree polynomial equation.

Check Yourself 6 (ℂ)

Solve $x^4 + 5x^2 - 14 = 0$.

Example 7 shows a radical equation that could be solved by the methods studied in Section 7.4. However, we can also solve it using a "quadratic in form" approach.

Example 7 Solving an Equation That Is Quadratic in Form

Solve $x - 3\sqrt{x} - 10 = 0$.
In $x - 3\sqrt{x} - 10$, x is the square of $\sqrt{x}$.
So, we let $u = \sqrt{x}$, which means $u^2 = x$.

$$u^2 - 3u - 10 = 0$$
$$(u - 5)(u + 2) = 0$$
$$u = 5 \quad \text{or} \quad u = -2$$

Back-substituting,

$$\sqrt{x} = 5 \quad \text{or} \quad \sqrt{x} = -2$$

Since $\sqrt{x}$ cannot be negative, we only need to solve the first of these.

If $\sqrt{x} = 5$, then $x = 25$.

Checking this value in the original equation, we see

$$(25) - 3\sqrt{(25)} - 10 \stackrel{?}{=} 0$$
$$25 - 3(5) - 10 \stackrel{?}{=} 0$$
$$25 - 15 - 10 = 0 \qquad \text{True}$$

The solution set is $\{25\}$.

If we had not noticed that the statement $\sqrt{x} = -2$ has no solution, we would have proceeded as shown.

$$\sqrt{x} = -2$$

$$(\sqrt{x})^2 = (-2)^2 \quad \text{We square both sides.}$$

$$x = 4$$

Checking this value in the original equation, we see

$$(4) - 3\sqrt{(4)} - 10 \stackrel{?}{=} 0$$

$$4 - 3(2) - 10 \stackrel{?}{=} 0$$

$$4 - 6 - 10 \stackrel{?}{=} 0$$

$$-12 = 0 \quad \text{False}$$

This means that $x = 4$ is an extraneous solution, and we conclude that the solution set is $\{25\}$.

Check Yourself 7

Solve $x - 11\sqrt{x} + 24 = 0$.

If an equation is quadratic in form, but is not factorable, we use the quadratic formula to find solutions. This is likely to yield some "messy" solutions in radical form, so we emphasize decimal approximations here. We use a calculator to check such messy solutions.

Example 8 Solving an Equation That Is Quadratic in Form

Solve $2x^4 - 3x^2 - 8 = 0$, finding any real-number solutions to the nearest thousandth. We let $u = x^2$, so that $u^2 = x^4$.

$$2u^2 - 3u - 8 = 0$$

The quadratic expression does not factor, so

$$u = \frac{-(-3) \pm \sqrt{(-3)^2 - 4(2)(-8)}}{2(2)} = \frac{3 \pm \sqrt{9 + 64}}{4} = \frac{3 \pm \sqrt{73}}{4}$$

$$u = \frac{3 + \sqrt{73}}{4} \quad \text{or} \quad u = \frac{3 - \sqrt{73}}{4}$$

Back-substituting,

$$x^2 = \frac{3 + \sqrt{73}}{4} \quad \text{or} \quad x^2 = \frac{3 - \sqrt{73}}{4}$$

With a calculator, we see that $\frac{3 - \sqrt{73}}{4}$ is negative, so we only obtain real solutions from the first equation, $x^2 = \frac{3 + \sqrt{73}}{4}$.

```
√((3+√(73))/4)
     1.698823398
```

Thus, $x = \pm\sqrt{\frac{3 + \sqrt{73}}{4}}$. This is what we mean by messy!

To the nearest thousandth, $x = \pm 1.699$. We use the "store" key STO➡ on a graphing calculator to check these values.

```
√((3+√(73))/4)→X
     1.698823398
2X^4-3X²-8
               0
```

First, compute the value of $\sqrt{\frac{3 + \sqrt{73}}{4}}$, and store it in a memory location of your choosing, say X. Then type the expression $2X^4 - 3X^2 - 8$ and the result should be 0 (or very nearly 0). Check the value $-\sqrt{\frac{3 + \sqrt{73}}{4}}$ in the same manner.

Check Yourself 8

Solve $3x^4 - 5x^2 - 6 = 0$, finding any real solutions, rounded to the nearest thousandth.

When using a calculator to check an approximation, the calculator's result may not appear to be exactly what you expect. For example, consider this screenshot.

```
√((5+√(97))/6)→X
                1.573152769
3X^4-5X²-6
                     -1E-12
```

The expected result was 0, but we see $-1\text{E}-12$. This is calculator notation for -1×10^{-12}, which, as a decimal, is -0.000000000001, which is extremely close to 0. Remember that an irrational number like $\sqrt{97}$ has a decimal representation that never ends and never repeats. A calculator can carry only a finite number of decimal places (such as 16). Therefore, the value for $\sqrt{97}$ in a calculator is approximate, not exact. When it is then used in further calculations, **approximation errors** occur. To use a calculator wisely, we must recognize these situations, and realize that, in the preceding example, the value entered for X does check!

We conclude this chapter by using a graphing calculator to approximate data points with a quadratic function. This is very similar to the linear regression analysis we performed in Section 3.4.

Graphing Calculator Option

Applying Quadratic Regression

Suppose we collect some data in the form of ordered pairs, and, when plotted, the scatterplot indicates that a parabola might fit the data pretty well. You can use the quadratic regression utility in a graphing calculator to find the equation for such a quadratic function.

Consider how the number of Blackberry subscribers has increased through several years.

Year	2006	2007	2008	2009	2010	2011
Subscribers (millions)	6	9.2	16.3	27.3	39.8	58.6

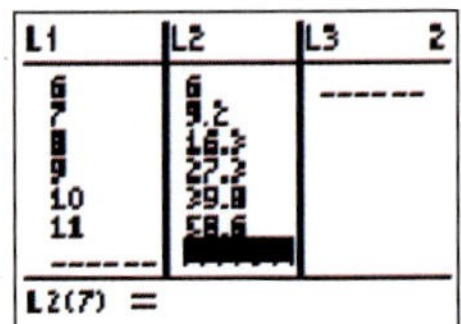

We begin by clearing the data lists.

STAT **4:ClrList** 2nd [L1] , 2nd [L2] ENTER

Then we enter the data into lists 1 and 2.

STAT **1:Edit** and type in the numbers (we enter the years as 6, 7, 8, 9, 10, and 11).

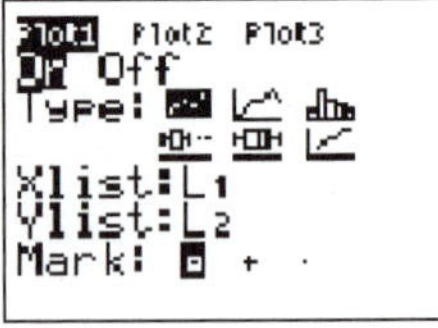

To make a scatterplot,

2nd [STAT PLOT] ENTER

Make sure "On" is selected, use the first icon for "Type" and check that [L1] is the "Xlist" and [L2] is the "Ylist."

Deselect any functions in the Y = menu and improve the viewing window with ZOOM **9:ZoomStat**

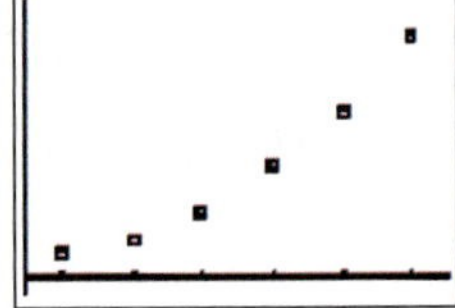

We find the "best fitting" quadratic function by
STAT **CALC 5:QuadReg** 2nd [L1] , 2nd [L2] ENTER

To four decimal places, we have

$y = 1.7786x^2 - 19.7843x + 60.6771$

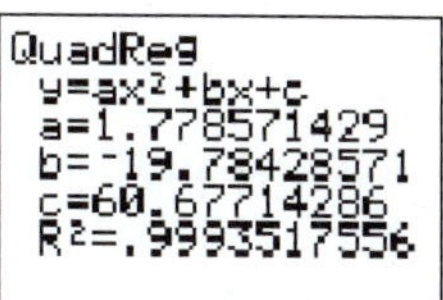

We can use this function to predict the number of subscribers in other years. However, predicting the future by **extrapolation** is always risky. We do not know that a trend, such as the growth in Blackberry subscribers, will continue or if some newer technology (iPhone, anyone) will change the curve. You may wish to review extrapolation in Chapter 3.

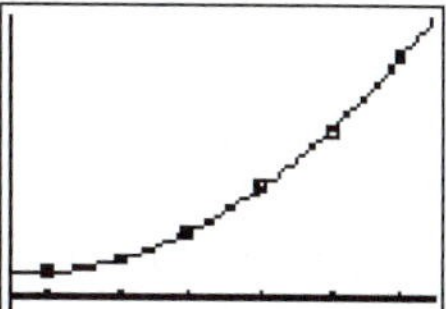

Graphing Calculator Check

The table gives the number of retail prescription drug sales (in millions) in the United States over a recent 8-year period. Use the quadratic regression utility on a graphing calculator to fit a quadratic function to this data. Round your result to four decimal places.

Year	1	2	3	4	5	6	7	8
Prescriptions (millions)	2,316	2,481	2,707	2,865	3,009	3,139	3,215	3,274

ANSWER

$y = -12.3929x^2 + 252.2024x + 2{,}056.8571$

Check Yourself ANSWERS

1. $\{-3.767, 2.434\}$ **2.** $\{-32.224, 8.224\}$ **3.** 100 radios, $1,900 **4.** Width 225 ft, length 450 ft **5.** $\{\pm\sqrt{2}, \pm 2\}$ **6.** $\{\pm\sqrt{2}, \pm i\sqrt{7}\}$ two nonreal solutions **7.** $\{9, 64\}$ **8.** $\{\pm 1.573\}$

Reading Your Text

These fill-in-the-blank exercises will help you understand some of the key vocabulary used in this section. The answers to these exercises are in the Answers Appendix in the back of the text.

(a) Graphing equations yields only __________ solutions.

(b) When graphing quadratic equations of the form $y = ax^2 + bx + c$, we know that if $a > 0$, then the vertex is the __________ point on the graph.

(c) With a fourth-degree polynomial equation, expect to find __________ solutions.

(d) If an equation is quadratic in form, but is not factorable, we can use the __________.

8.4 exercises

Skills | Calculator/Computer | Career Applications | Above and Beyond

< Objective 2 >

Solve each application.

1. **Business and Finance** A company's weekly profit P is related to the number of items sold by $P(x) = -0.3x^2 + 60x - 400$. Find the number of items that should be sold each week in order to maximize the profit. Then find the amount of that weekly profit.

2. **Business and Finance** A company's monthly profit P is related to the number of items sold by $P(x) = -0.2x^2 + 50x - 800$. How many items should be sold each month to obtain the largest possible profit? What is the amount of that profit?

3. **Construction** A builder wants to enclose the largest possible rectangular area with 2,000 ft of fencing. What should be the dimensions of the rectangle, and what is the area of that rectangle?

4. **Construction** A farmer wants to enclose a rectangular area along a river on three sides. If 1,600 ft of fencing is to be used, what dimensions give the maximum enclosed area? Find that maximum area.

5. **Science and Medicine** A ball is thrown upward into the air with an initial velocity of 96 ft/s. If h gives the height of the ball at time t, then the equation relating h and t is

$h = -16t^2 + 96t$

Find the maximum height the ball will attain.

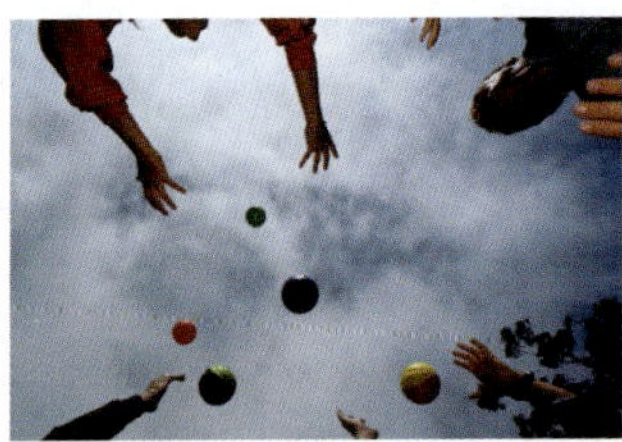

6. **Science and Medicine** A ball is thrown upward into the air with an initial velocity of 64 ft/s. If h gives the height of the ball at time t, then the equation relating h and t is

$h = -16t^2 + 64t$

Find the maximum height the ball will attain.

< Objective 3 >

Solve.

7. $x^4 - 14x^2 + 45 = 0$

8. $x^4 - 18x^2 + 32 = 0$

9. $6x^4 - 7x^2 + 2 = 0$

10. $12x^4 - 7x^2 + 1 = 0$

11. $x^4 + x^2 - 20 = 0$ (ℂ)

12. $x^4 - 6x^2 - 27 = 0$ (ℂ)

13. $x^4 + 11x^2 + 28 = 0$ (ℂ)

14. $x^4 + 11x^2 + 18 = 0$ (ℂ)

15. $x - 8\sqrt{x} + 15 = 0$

16. $x - 10\sqrt{x} + 24 = 0$

17. $x - 4\sqrt{x} - 21 = 0$

18. $x - 6\sqrt{x} - 16 = 0$

Find all real solutions. Round your results to the nearest thousandth.

19. $x^4 - 7x^2 - 4 = 0$

20. $x^4 - 5x^2 + 3 = 0$

21. $2x^4 - 7x^2 + 4 = 0$

22. $2x^4 - 9x^2 - 3 = 0$

< Objective 1 >

Use a graph to estimate the solutions to each equation. Round answers to the nearest thousandth.

23. $0 = x^2 + x - 12$ **24.** $0 = x^2 + 3x + 2$ **25.** $0 = 6x^2 - 19x$ **26.** $0 = 7x^2 - 15x$

27. $0 = 9x^2 + 12x - 7$ **28.** $0 = 3x^2 + 9x + 5$ **29.** $0 = x^2 + 2x - 7$ **30.** $0 = x^2 - 8x + 11$

For each quadratic function, use a graphing calculator to find **(a)** *the vertex of the parabola and* **(b)** *the range of the function.*

31. $f(x) = 2(x - 3)^2 + 1$ **32.** $g(x) = 3(x + 4)^2 + 2$ **33.** $f(x) = -(x - 1)^2 + 2$

34. $g(x) = -(x + 2)^2 - 1$ **35.** $f(x) = 3(x + 1)^2 - 2$ **36.** $g(x) = -2(x - 4)^2$

Science and Medicine *Each table shows a relationship between speed (miles per hour) and gasoline consumption (miles per gallon, MPG) for a vehicle. In each case, use quadratic regression to find a quadratic function that best fits the data. Round coefficients to the nearest thousandth.*

37. Oldsmobile

Speed	5	10	15	20	25	30	35
MPG	5.1	7.9	11.4	12.5	15.6	19.0	21.2

Speed	40	45	50	55	60	65	70	75
MPG	23.0	23.0	27.3	29.1	28.2	25.0	22.9	21.6

38. Chevrolet

Speed	5	10	15	20	25	30	35
MPG	7.9	18.0	16.3	19.9	22.7	26.3	24.3

Speed	40	45	50	55	60	65	70	75
MPG	26.7	27.3	26.3	25.1	22.6	21.8	20.1	18.1

39. Jeep

Speed	5	10	15	20	25	30	35
MPG	8.2	11.2	17.5	24.7	21.8	21.6	25.0

Speed	40	45	50	55	60	65	70	75
MPG	25.5	25.4	24.8	24.0	23.2	21.3	20.0	19.1

40. Honda

Speed	5	10	15	20	25	30	35
MPG	11.2	16.1	21.4	25.1	27.3	28.0	28.7

Speed	40	45	50	55	60	65	70	75
MPG	29.5	30.1	30.2	29.9	28.3	27.1	23.8	23.1

Skills	Calculator/Computer	**Career Applications**	Above and Beyond

Allied Health *The number of people infected t days after the outbreak of a flu epidemic is modeled by the equation*

$P = -t^2 + 120t + 20$

Use this model to complete exercises 41 and 42.

41. How many days after the outbreak will the maximum number of people be sick?

42. What is the maximum number of people that will be infected at one time?

Allied Health *A patient's body temperature (T°F) t hours after taking the analgesic acetaminophen can be approximated by the formula*

$T = 0.4t^2 - 2.6t + 103$

Use this model to complete exercises 43 and 44.

43. When will the patient's temperature reach its minimum?

44. What will the patient's minimum temperature be? (Round your answer to the nearest tenth.)

Skills	Calculator/Computer	Career Applications	**Above and Beyond**

Find a viewing window that includes the vertex and all intercepts for the graph of each function.

45. $f(x) = 3x^2 - 25$ **46.** $f(x) = 9x^2 - 5x - 7$ **47.** $f(x) = -2x^2 + 5x - 7$ **48.** $f(x) = -5x^2 + 2x + 7$

49. Explain how to determine the domain and range of the function $f(x) = a(x - h)^2 + k$.

Answers

1. 100 items, \$2,600 **3.** 500 ft by 500 ft; 250,000 ft² **5.** 144 ft **7.** $\{\pm 3, \pm\sqrt{5}\}$ **9.** $\left\{\pm\frac{\sqrt{2}}{2}, \pm\frac{\sqrt{6}}{3}\right\}$
11. $\{\pm 2, \pm i\sqrt{5}\}$ two nonreal solutions **13.** $\{\pm 2i, \pm i\sqrt{7}\}$ four nonreal solutions **15.** $\{9, 25\}$ **17.** $\{49\}$ **19.** $\{\pm 2.744\}$
21. $\{\pm 1.668, \pm 0.848\}$ **23.** $\{-4, 3\}$ **25.** $\{0, 3.167\}$ **27.** $\{-1.772, 0.439\}$ **29.** $\{-3.828, 1.828\}$ **31.** **(a)** (3, 1); **(b)** $y \geq 1$
33. **(a)** (1, 2); **(b)** $y \leq 2$ **35.** **(a)** (−1, −2); **(b)** $y \geq -2$ **37.** $y = -0.008x^2 + 0.946x - 1.174$ **39.** $y = -0.01x^2 + 0.899x + 5.424$
41. 60 days **43.** 3.25 hr **45.** $-3 \leq x \leq 3$; $-25 \leq y \leq 0$ (answers will vary) **47.** $-2 \leq x \leq 4$; $-10 \leq y \leq 0$ (answers will vary)
49. Above and Beyond

Definition/Procedure	Example	Reference
Solving Quadratic Equations		**Section 8.1**
Square-Root Property If $x^2 = k$, when k is any real number, then $x = \sqrt{k}$ or $x = -\sqrt{k}$.	$(x-3)^2 = 5$ $x - 3 = \pm\sqrt{5}$ $x = 3 \pm \sqrt{5}$	*p.* 625
Completing the Square **Step 1** Isolate the constant on the right side of the equation. **Step 2** Divide both sides of the equation by the coefficient of the x^2-term if that coefficient is not equal to 1. **Step 3** Add the square of one-half of the coefficient of the linear or x-term to both sides of the equation. This gives a perfect-square trinomial on the left side of the equation. **Step 4** Write the left side of the equation as the square of a binomial and simplify the right side. **Step 5** Use the square-root property and solve the resulting linear equations.	$x^2 + x = \frac{1}{2}$ $x^2 + x + \left(\frac{1}{2}\right)^2 = \frac{1}{2} + \left(\frac{1}{2}\right)^2$ $\left(x + \frac{1}{2}\right)^2 = \frac{3}{4}$ $x + \frac{1}{2} = \pm\sqrt{\frac{3}{4}}$ $x = \frac{-1 \pm \sqrt{3}}{2}$	*p.* 630
The Quadratic Formula		**Section 8.2**
Step 1 Write the equation in standard form (set it equal to 0). $ax^2 + bx + c = 0$ **Step 2** Find a, b, and c. **Step 3** Substitute those values into the quadratic formula $x = \frac{-b \pm \sqrt{b^2 - 4ac}}{2a}$ **Step 4** Simplify the solutions.	$x^2 - 2x = 4$ write the equation as $x^2 - 2x - 4 = 0$ $a = 1 \quad b = -2 \quad c = -4$ $x = \frac{-(-2) \pm \sqrt{(-2)^2 - 4(1)(-4)}}{2 \cdot (1)}$ $= \frac{2 \pm \sqrt{20}}{2}$ $= \frac{2 \pm 2\sqrt{5}}{2}$ $= 1 \pm \sqrt{5}$	*p.* 642
The Discriminant The expression $b^2 - 4ac$ is called the **discriminant** of a quadratic equation. **1.** If $b^2 - 4ac < 0$, there are no real solutions (but two nonreal solutions). **2.** If $b^2 - 4ac = 0$, there is one real solution (a repeated root). **3.** If $b^2 - 4ac > 0$, there are two distinct real solutions.	Given $2x^2 - 5x + 3 = 0$ $a = 2 \quad b = -5 \quad c = 3$ $b^2 - 4ac = 25 - 4(2)(3)$ $= 25 - 24$ $= 1$ There are two distinct real solutions.	*p.* 647
An Introduction to Parabolas		**Section 8.3**
Axis of Symmetry The axis of symmetry is the vertical line midway between any pair of symmetric points on a parabola. The axis of symmetry passes through the vertex of the parabola.	Graph the function $f(x) = x^2 - 4x - 12$	*p.* 660
Vertex of a Parabola If $f(x) = ax^2 + bx + c \qquad a \neq 0$ then the coordinates of the vertex of the graph of f are $\left(-\frac{b}{2a}, f\left(-\frac{b}{2a}\right)\right)$	**1.** Find the axis of symmetry. $x = -\frac{b}{2a} = -\frac{(-4)}{2(1)} = 2$ so $x = 2$ is the axis of symmetry.	*p.* 660

Continued

Definition/Procedure	Example	Reference
Graphing a Parabola **Step 1** Find the axis of symmetry. **Step 2** Find the vertex. **Step 3** Determine two symmetric points. **Note:** Use the x-intercepts if the quadratic expression is factorable. If the vertex is not on the y-axis, you can use the y-intercept and its symmetric point. Another option is to simply choose an x-value that does not match the axis of symmetry, compute the corresponding y-value, and then locate its symmetric point. **Step 4** Draw a smooth curve connecting the points found in step 3 to form the parabola. You may choose to find additional pairs of symmetric points.	**2.** Find the vertex. Let $x = 2$ in the original equation. $f(2) = (2)^2 - 4(2) - 12$ $= 4 - 8 - 12$ $= -16$ The vertex is $(2, -16)$. **3.** Find two symmetric points. $0 = x^2 - 4x - 12$ $0 = (x - 6)(x + 2)$ $x - 6 = 0 \quad x + 2 = 0$ $x = 6 \quad x = -2$ Two symmetric points are $(6, 0)$ and $(-2, 0)$. **4.** Draw a smooth curve connecting the points found. 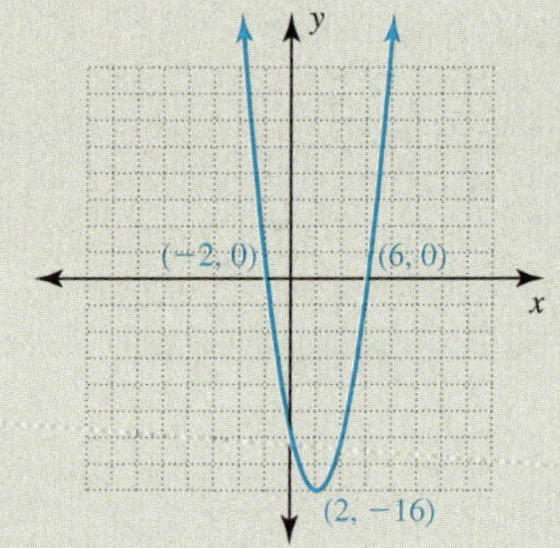 	p. 663
Quadratic Equations and Problem Solving		**Section 8.4**
Solving Quadratic Equations Graphically To solve the equation $ax^2 + bx + c = 0$ **1.** Graph the function $Y = ax^2 + bx + c$ **2.** Use the ZERO or ROOT utility to determine the x-intercepts of the graph. These values are the solutions to the original equation.	Solve the equation graphically. $0.5x^2 + 3x - 2 = 0$ 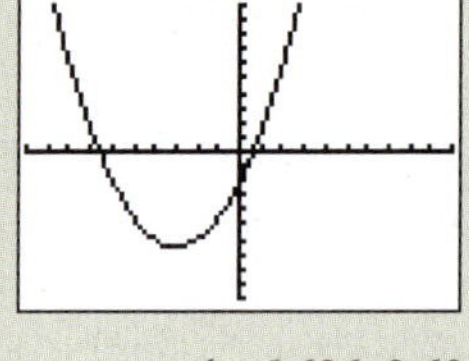 $\{-6.606, 0.606\}$	p. 670

Continued

Definition/Procedure	Example	Reference
Solving Equations Quadratic in Form Given an equation with a trinomial on one side and 0 on the other, if the variable part of the first term is the square of the variable part of the second term, then the trinomial is "quadratic in form," and may be solved by a substitution technique.	Solve: $x^4 - 14x^2 + 45 = 0$ Let $u = x^2$. Then $u^2 = x^4$. Rewrite in terms of u. $u^2 - 14u + 45 = 0$ Solve for u. $(u - 5)(u - 9) = 0$ $u = 5$ or $u = 9$ Substituting, $x^2 = 5$ or $x^2 = 9$ Solve for x. $x = \pm\sqrt{5}$ or $x = \pm 3$ Solution set: $\{\pm\sqrt{5}, \pm 3\}$.	p. 673

summary exercises :: chapter 8

This summary exercise set will help ensure that you have mastered each of the objectives of this chapter. The exercises are grouped by section. You should reread the material associated with any exercises that you find difficult. The answers to the odd-numbered exercises are in the Answers Appendix in the back of the text.

8.1 *Use the square-root method to solve each equation.*

1. $x^2 - 12 = 0$

2. $3y^2 - 15 = 0$

3. $(x - 2)^2 = 20$

4. $(3x - 2)^2 - 15 = 0$

Find the constant that must be added to each binomial to form a perfect-square trinomial.

5. $x^2 - 14x$

6. $y^2 + 3y$

Solve each equation by completing the square.

7. $x^2 - 4x - 5 = 0$

8. $x^2 - 8x - 9 = 0$

9. $w^2 - 10w - 3 = 0$

10. $y^2 + 3y - 1 = 0$

11. $2x^2 - 6x + 1 = 0$

12. $3x^2 + 4x - 1 = 0$

8.2 *Solve each equation by using the quadratic formula.*

13. $x^2 - 5x - 24 = 0$

14. $w^2 + 10w + 25 = 0$

15. $x^2 = 5x - 2$

16. $2y^2 - 5y + 2 = 0$

17. $3y^2 + 4y = 1$

18. $3y^2 + 4y + 7 = 0$ (ℂ)

19. $(x - 5)(x + 3) = 13$

20. $x^2 - 4x + 1 = 0$

21. $3x^2 + 2x + 5 = 0$ (ℂ)

22. $(x - 1)(2x + 3) = -5$ (ℂ)

For each quadratic equation, use the discriminant to determine the number of real solutions.

23. $x^2 - 3x + 3 = 0$ **24.** $x^2 + 4x = 2$ **25.** $4x^2 - 12x + 9 = 0$ **26.** $2x^2 + 3 = 3x$

27. **Number Problem** The sum of two integers is 12, and their product is 32. Find the two integers.

28. **Number Problem** The product of two consecutive positive even integers is 80. What are the two integers?

29. **Number Problem** Twice the square of a positive integer is 10 more than 8 times that integer. Find the integer.

30. **Geometry** The length of a rectangle is 2 ft more than its width. If the area of the rectangle is 80 ft^2, what are the dimensions of the rectangle?

31. **Geometry** The length of a rectangle is 3 cm less than twice its width. The area of the rectangle is 35 cm^2. Find the length and width of the rectangle.

32. **Geometry** An open box is formed by cutting 3-in. squares from each corner of a rectangular piece of cardboard that is 3 in. longer than it is wide. If the box is to have a volume of 120 in.^{3}, what must be the size of the original piece of cardboard?

Find the real zeros of each function.

33. $f(x) = x^2 - x - 2$ **34.** $f(x) = 6x^2 + 7x + 2$

35. $f(x) = -2x^2 - 7x - 6$ **36.** $f(x) = -x^2 - 1$

8.3 *Find the equation of the axis of symmetry and the coordinates of the vertex of each quadratic function.*

37. $f(x) = x^2$ **38.** $f(x) = x^2 + 2$ **39.** $f(x) = x^2 - 5$ **40.** $f(x) = x^2 - 6x + 9$

41. $f(x) = x^2 + 4x + 4$ **42.** $f(x) = -x^2 + 6x - 9$ **43.** $f(x) = x^2 + 6x + 10$ **44.** $f(x) = -x^2 - 4x - 7$

45. $f(x) = -(x - 5)^2 - 2$ **46.** $f(x) = 2(x - 2)^2 - 5$ **47.** $f(x) = -x^2 + 2x$ **48.** $f(x) = x^2 - 4x + 3$

49. $f(x) = -x^2 - x + 6$ **50.** $f(x) = x^2 + 4x + 5$

Graph each function.

51. $f(x) = x^2$ **52.** $f(x) = x^2 + 2$ **53.** $f(x) = x^2 - 5$ **54.** $f(x) = (x - 3)^2$

55. $f(x) = (x + 2)^2$ **56.** $f(x) = -x^2 + 6x - 9$ **57.** $f(x) = x^2 + 6x + 10$ **58.** $f(x) = -x^2 - 4x - 7$

59. $f(x) = x^2 - 4x$

60. $f(x) = -x^2 + 2x$

61. $f(x) = x^2 + 2x - 3$

62. $f(x) = x^2 - 4x + 3$

63. $f(x) = -x^2 - x + 6$

64. $f(x) = -x^2 + 3x + 4$

65. $f(x) = x^2 + 4x + 5$

66. $f(x) = x^2 - 6x + 4$

67. $f(x) = x^2 - 2x + 4$

68. $f(x) = -x^2 + 2x - 2$

69. $f(x) = 2x^2 - 4x + 1$

70. $f(x) = \frac{1}{2}x^2 - 4x$

8.4 *Use a graphing calculator to estimate the solutions to each equation. Round answers to the nearest thousandth.*

71. $x^2 - 2x - 5 = 0$

72. $2x^2 + 5x - 9 = 0$

73. $x^2 - 5x + 5 = 0$

74. $2x^2 - 4x - 5 = 0$

Solve.

75. $x^4 - 13x^2 + 40 = 0$

76. $x^4 - 13x^2 + 12 = 0$

77. $x^4 - 5x^2 - 36 = 0$ (ℂ)

78. $x^4 + 2x^2 - 63 = 0$ (ℂ)

79. $x - 7\sqrt{x} + 12 = 0$

80. $x - 6\sqrt{x} - 27 = 0$

Solve. Find real number solutions to the nearest thousandth.

81. $x^4 - 6x^2 - 10 = 0$

82. $x^4 - 9x^2 + 5 = 0$

83. **Business and Finance** Suppose that a manufacturer's weekly profit P is given by

$$P = -3x^2 + 240x$$

where x is the number of patio chairs manufactured and sold. Find the number of patio chairs that must be manufactured and sold if the profit is to be at least $4,500.

84. **Science and Medicine** If a ball is thrown vertically upward from the ground with an initial velocity of 64 ft/s, its approximate height is given by

$$h(t) = -16t^2 + 64t$$

When will the ball's height be at least 48 ft?

85. **Geometry** The length of a rectangle is 1 cm more than twice its width. If the length is doubled, the area of the new rectangle is 36 cm^2 more than that of the old. Find the dimensions of the original rectangle.

86. **Geometry** One leg of a right triangle is 4 in. longer than the other. The hypotenuse of the triangle is 8 in. longer than the shorter leg. What are the lengths of the three sides of the triangle?

87. **Geometry** The diagonal of a rectangle is 9 ft longer than the width of the rectangle, and the length is 7 ft more than its width. Find the dimensions of the rectangle.

88. **Science and Medicine** If a ball is thrown vertically upward from the ground with an initial velocity of 128 ft/s, the height h after t seconds is given by

$h = 128t - 16t^2$

(a) How long does it take the ball to return to the ground?

(b) How long does it take the ball to reach a height of 240 ft on the way up?

89. **Geometry** One leg of a right triangle is 2 m longer than the other. If the length of the hypotenuse is 8 m, find the length of the other two legs.

90. **Science and Medicine** Suppose that the height (in meters) of a golf ball is approximated by

$h(t) = -5t^2 + 10t + 10$

t seconds after the ball is hit. When will the ball hit the ground?

chapter test 8

CHAPTER 8

Use this chapter test to assess your progress and to review for your next exam. Allow yourself about an hour to take this test. The answers to these exercises are in the Answers Appendix in the back of the text.

Find the equation of the axis of symmetry and the coordinates of the vertex of each function.

1. $f(x) = -3x^2 - 12x - 11$

2. $f(x) = x^2 - 4x - 5$

3. $f(x) = -2x^2 + 6x - 3$

4. $f(x) = x^2 - 6x + 7$

5. $f(x) = (x - 3)^2 - 7$

Solve each equation by completing the square.

6. $m^2 + 3m - 1 = 0$

7. $2x^2 - 10x + 3 = 0$

8. Find the zeros of the function $f(x) = 3x^2 - 10x - 8$.

Use a graphing calculator to estimate the solutions to each equation. Round your answers to the nearest thousandth.

9. $x^2 + 3x - 7 = 0$

10. $4x^2 + 2x - 5$

Graph each function.

11. $x^2 - 10x + 25 = 0$

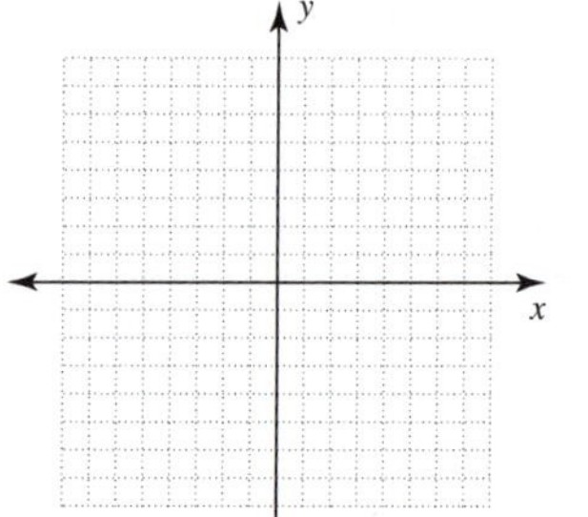

12. $f(x) = x^2 + 4x + 1$

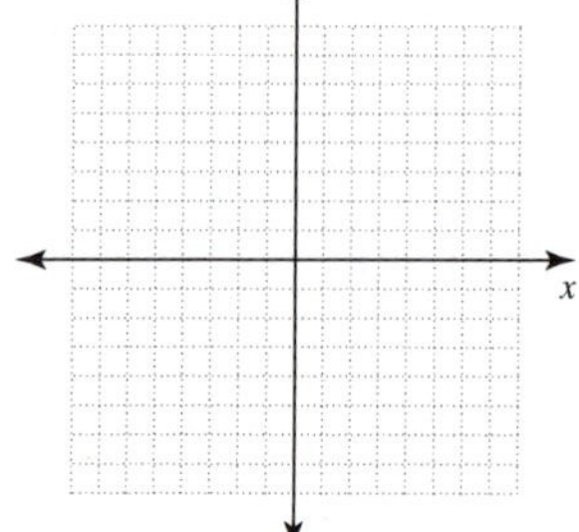

13. $f(x) = -2x^2 + 12x - 19$

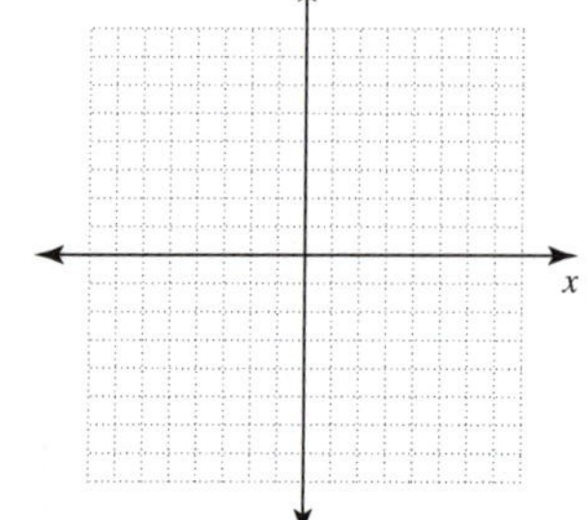

14. $f(x) = 3x^2 + 9x + 2$

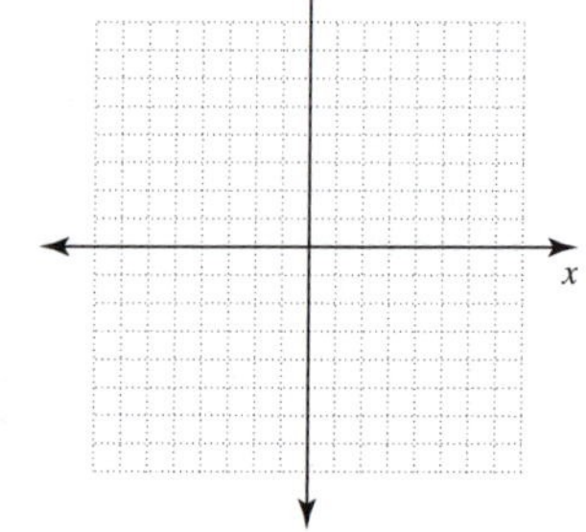

Solve each equation.

15. $2x^2 + 7x + 3 = 0$

16. $6x^2 = 10 - 11x$

17. $4x^3 - 9x = 0$

Solve.

18. The product of two consecutive positive odd integers is 63. Find the two integers.

19. Suppose that the height (in feet) of a ball thrown upward with an initial velocity of 32 ft/s from a raised platform is approximated by

$h(t) = -16t^2 + 32t + 32$

t seconds after the ball is released. How long will it take the ball to hit the ground?

Use the quadratic formula to solve each equation.

20. $x^2 - 5x - 3 = 0$

21. $x^2 + 4x = 7$

22. $15x^2 = 2x + 8$

23. $2x^2 + 2x + 5 = 0$ $(\mathbb{C})$

Solve.

24. $x^4 - 15x^2 + 36 = 0$

25. $x^4 - 4x^2 - 32 = 0$ $(\mathbb{C})$

26. $x - 11\sqrt{x} + 30 = 0$

Use the square-root method to solve each equation.

27. $4w^2 - 20 = 0$

28. $(x - 1)^2 = 10$

29. $4(x - 1)^2 = 23$

Solve. Round your answers to the nearest thousandth.

30. $2x^4 - 7x^2 - 1 = 0$

cumulative review chapters 0–8

Use this exercise set to review concepts from earlier chapters. While it is not a comprehensive exam, it will help you identify any material that you need to review before moving on to the next chapter. The answers to these exercises are in the Answers Appendix in the back of the text.

Graph each equation.

1. $2x - 3y = 6$

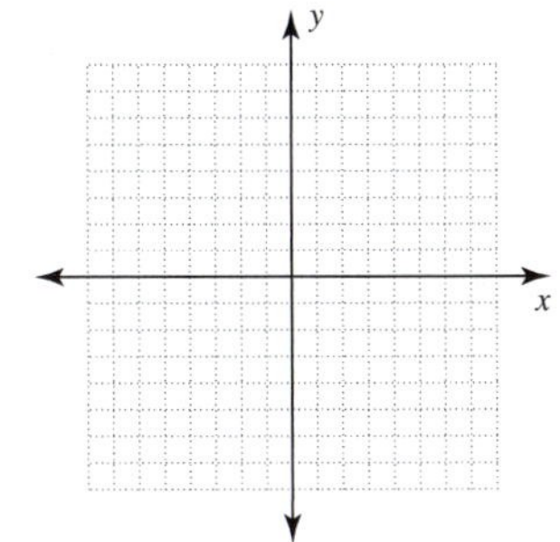

2. $y = -\frac{1}{3}x - 2$

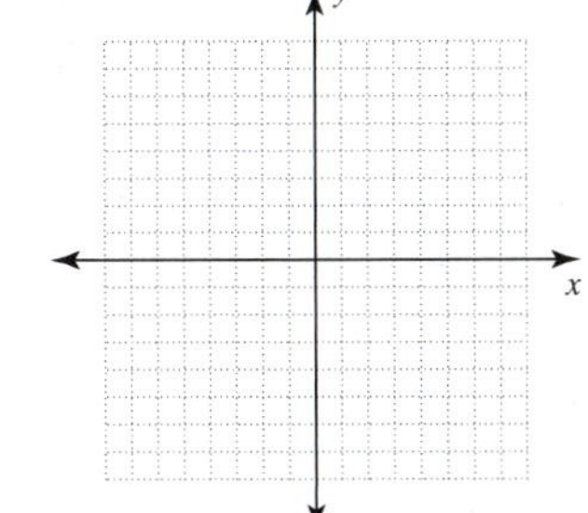

3. $y = 4$

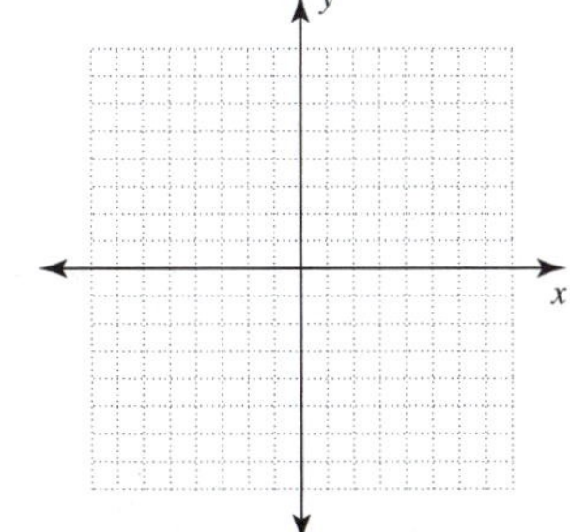

Find the slope of the line through each pair of points.

4. $(-4, 7)$ and $(-3, 4)$

5. $(-2, 3)$ and $(-5, -1)$

6. Let $f(x) = 6x^2 - 5x + 1$. Evaluate $f(-2)$.

7. Simplify the function $f(x) = (x^2 - 1)(x + 3)$.

8. Completely factor the expression $x^3 + x^2 - 6x$.

9. Simplify the expression $\sqrt{\frac{2}{3}} + 7\sqrt{6}$.

10. Simplify the expression $\sqrt{72x^3y^5}$.

Solve each equation.

11. $2x - 7 = 0$

12. $3x - 5 = 5x + 3$

13. $0 = (x - 3)(x + 5)$

14. $x^2 - 3x + 2 = 0$

15. $x^2 + 7x - 30 = 0$

16. $x^2 - 3x - 3 = 0$

17. $(x - 3)^2 = 5$

18. $x^3 - 2x^2 = 15x$

19. $\frac{x}{3} - \frac{4}{9} = \frac{5}{18}$

20. $3 - \sqrt{2x + 2} = x$

Solve the inequality.

21. $x - 2 \le 7$

22. Find the distance between $(-1, -4)$ and $(6, 1)$.

Solve each word problem. Show the equation used for the solution.

23. Five times a number decreased by 7 is -72. Find the number.

24. One leg of a right triangle is 4 ft longer than the shorter leg. If the hypotenuse is 28 ft, how long is each leg?

25. Suppose that a manufacturer's weekly profit P is given by

$$P = -4x^2 + 320x$$

where x is the number of receivers manufactured and sold. Find the number of receivers that must be manufactured and sold for a profit of \$4,956.

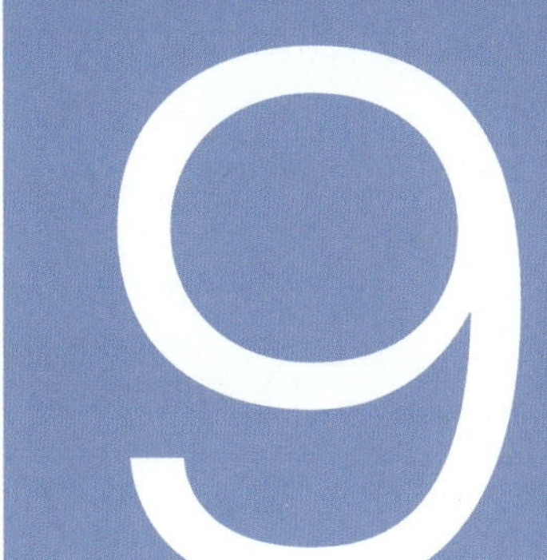

CHAPTER 9

Rational Expressions

INTRODUCTION

There are many things to learn in college and not all of them directly apply to a field of study. While specific skills such as simplifying rational expressions, using the valence number of an atom, or even grafting a plant scion to rootstock for propagation are important, there are a whole host of other, equally important skills.

Communicating technical information involves many *transferable skills.* Throughout this course, you are improving your technical communication skills while also learning math. When you work through a definition or understand a property, you are learning to read and understand technical writing. When you give the answer to an application as a full sentence or interpret a result in the context of an application, you are learning to convey technical information.

You will apply these skills when your employer upgrades a system or you assess the terms of a mortgage. Continue to improve these skills in your math class and you will be better prepared to navigate such terrain. We help you learn and practice the "art" of communicating mathematical ideas and information in Activity 9.

CHAPTER 9 OUTLINE

9.1 Simplifying Rational Expressions

< 9.1 Objectives >

1 > Evaluate rational expressions

2 > Find the excluded values of a rational expression

3 > Simplify rational expressions

4 > Identify rational functions

5 > Use rational functions to solve applications

We are now ready to learn about **rational expressions.** What is a rational expression? Roughly speaking, it is a fraction that may have variables. (We give a more precise definition later.) Your experience with fractions will help you with rational expressions.

Some examples of rational expressions are

$$\frac{8}{x} \qquad \frac{2x-3}{x+5} \qquad \frac{x+6}{x^2-2x-15} \qquad \frac{3}{4}$$

Recall that a **rational number** is a number of the form $\frac{a}{b}$, where a and b are integers and b is not 0. Just as a rational number can be thought of as $\frac{\text{integer}}{\text{integer}}$, a rational expression can be thought of as $\frac{\text{polynomial}}{\text{polynomial}}$.

Definition

Rational Expression

A **rational expression** is an expression of the form $\frac{P}{Q}$, where P and Q are polynomials and Q cannot be 0.

We often need to find the value of a rational expression for a given variable value.

Example 1 Evaluating a Rational Expression

< Objective 1 >

Evaluate each expression for the value of the variable.

(a) $\frac{3x}{2x-5}$ for $x = -3$

$$\frac{3(-3)}{2(-3)-5} = \frac{-9}{-6-5} = \frac{-9}{-11} = \frac{9}{11}$$ Substitute -3 for x.

(b) $\frac{2x+7}{x^2-x-6}$ for $x = -2$

$$\frac{2(-2)+7}{(-2)^2-(-2)-6} = \frac{-4+7}{4+2-6} = \frac{3}{0}$$ Undefined

This expression is undefined for $x = -2$.

Check Yourself 1

Evaluate each expression for the given value of the variable.

(a) $\frac{5x}{3x-2}$ for $x = 4$ **(b)** $\frac{x^2-9}{x^2-1}$ for $x = -5$

In part (b) of Example 1, we saw that the given expression is undefined when $x = -2$. This is because the denominator polynomial, $x^2 - x - 6$, is 0 when $x = -2$. We cannot divide by 0.

You should notice the emphasis we place on the idea that the denominator cannot be 0, whether we are speaking of rational numbers or rational expressions.

To review why division by 0 is undefined, think of division using a "fits into" concept. For example,

$\frac{8}{2} = 4$ How many times does 2 "fit into" 8? 4 times.

$\frac{8}{1} = 8$ How many times does 1 "fit into" 8? 8 times.

$\frac{8}{\frac{1}{2}} = 16$ How many times does $\frac{1}{2}$ "fit into" 8? 16 times.

$\frac{8}{0.1} = 80$ How many times does 0.1 "fit into" 8? 80 times.

$\frac{8}{0.01} = 800$ How many times does 0.01 "fit into" 8? 800 times.

As the denominator becomes smaller, approaching 0, the quotient gets larger. Ask yourself: How many times does 0 "fit into" 8? Your answer would have to be an *infinitely large* number! This is one reason why $\frac{8}{0}$ is undefined.

Because of this, when we work with rational expressions we must take care to avoid division by 0. We ask the question "For what values of the variable is the denominator polynomial equal to 0?" These are values that cause the value of the rational expression to be undefined.

Example 2 Finding Excluded Values

< Objective 2 >

NOTE

A fraction is undefined when its denominator is equal to 0.

When $x = 5$, $\frac{x}{x-5}$ becomes $\frac{(5)}{(5)-5}$, or $\frac{5}{0}$.

For what values of x are the expressions undefined?

(a) $\frac{x}{x-5}$

To answer this question, we must find where the denominator is 0.

$x - 5 = 0$

$x = 5$

The expression $\frac{x}{x-5}$ is undefined for $x = 5$.

(b) $\frac{3}{x+5}$

Again, set the denominator equal to 0.

$x + 5 = 0$

$x = -5$

The expression $\frac{3}{x+5}$ is undefined for $x = -5$.

Check Yourself 2

For what values of the variable are the expressions undefined?

(a) $\frac{1}{r+7}$ **(b)** $\frac{5}{2x-9}$

It may be necessary to factor the denominator to determine the values of the variable for which the expression is undefined.

Example 3 Finding Excluded Values

For each rational expression, find the values such that the expression is undefined.

(a) $\dfrac{x + 6}{x^2 + 2x - 15}$

$= \dfrac{x + 6}{(x + 5)(x - 3)}$ Factor the denominator.

Set the denominator equal to 0 to find the "problem" values, and solve.

$(x + 5)(x - 3) = 0$

$x + 5 = 0$ or $x - 3 = 0$

$x = -5$ or $x = 3$

We find that when $x = -5$ or $x = 3$, the expression is undefined.

(b) $\dfrac{x^2 + x - 12}{3x^2 + x - 2}$

$= \dfrac{x^2 + x - 12}{(3x - 2)(x + 1)}$ We factor the denominator to find the "problem" values.

$(3x - 2)(x + 1) = 0$ Set the denominator equal to 0 and solve.

$3x - 2 = 0$ or $x + 1 = 0$

$x = \dfrac{2}{3}$ or $x = -1$

The expression is undefined when $x = \dfrac{2}{3}$ or $x = -1$.

Check Yourself 3

Find the values for which the expression is undefined.

$\dfrac{x^2 - 2x - 3}{2x^2 - 3x - 20}$

Generally, we want to write rational expressions in the simplest possible form. Your past experience with fractions will help you here. Recall that

$\dfrac{3}{5} = \dfrac{3 \cdot 2}{5 \cdot 2} = \dfrac{6}{10}$

so $\dfrac{3}{5}$ and $\dfrac{6}{10}$ are equivalent fractions.

Similarly,

$\dfrac{10}{15} = \dfrac{5 \cdot 2}{5 \cdot 3} = \dfrac{2}{3}$

so $\dfrac{10}{15}$ and $\dfrac{2}{3}$ name equivalent fractions.

We can always multiply or divide the numerator and denominator of a fraction by the same nonzero number. This idea is also true in algebra.

Property

Fundamental Principle of Rational Expressions

For polynomials P, Q, and R,

$\dfrac{P}{Q} = \dfrac{PR}{QR}$ where $Q \neq 0$ and $R \neq 0$

This property can be used in two ways. We can multiply or divide the numerator and denominator of a rational expression by the same nonzero polynomial. The result is always equivalent to the original expression.

NOTE

Many methods in this chapter depend on factoring polynomials.

When simplifying fractions, we used this principle to divide the numerator and denominator by all common factors. With arithmetic fractions, those common factors are generally easy to recognize. Given rational expressions where the numerator and denominator are polynomials, we must determine those factors as our first step. The most important tools for simplifying expressions are the factoring techniques in Chapter 6.

Example 4 Simplifying Rational Expressions

< Objective 3 >

RECALL

This is the same as dividing both the numerator and denominator of $\frac{18}{30}$ by 6.

Simplify each rational expression.

(a) $\frac{18}{30}$

$$\frac{18}{30} = \frac{2 \cdot 3 \cdot 3}{2 \cdot 3 \cdot 5} = \frac{\overset{1}{\cancel{2}} \cdot \overset{1}{\cancel{3}} \cdot 3}{\underset{1}{\cancel{2}} \cdot \underset{1}{\cancel{3}} \cdot 5} = \frac{3}{5}$$

Divide by the GCF.

(b) $\frac{4x^3}{6x}$

$$\frac{4x^3}{6x} = \frac{\overset{1}{\cancel{2}} \cdot 2 \cdot \overset{1}{\cancel{x}} \cdot x \cdot x}{\underset{1}{\cancel{2}} \cdot 3 \cdot \underset{1}{\cancel{x}}} = \frac{2x^2}{3}$$

(c) $\frac{15x^3y^2}{20xy^4}$

$$\frac{15x^3y^2}{20xy^4} = \frac{3 \cdot \overset{1}{\cancel{5}} \cdot \overset{1}{\cancel{x}} \cdot x \cdot x \cdot \overset{1}{\cancel{y}} \cdot \overset{1}{\cancel{y}}}{2 \cdot 2 \cdot \underset{1}{\cancel{5}} \cdot \underset{1}{\cancel{x}} \cdot \underset{1}{\cancel{y}} \cdot \underset{1}{\cancel{y}} \cdot y \cdot y} = \frac{3x^2}{4y^2}$$

We can also simplify directly by finding the GCF. In this case, we have

$$\frac{15x^3y^2}{20xy^4} = \frac{(5xy^2)(3x^2)}{(5xy^2)(4y^2)} = \frac{3x^2}{4y^2}$$

(d) $\frac{3a^2b}{9a^3b^2}$

NOTE

With practice you will be able to simplify these terms without writing out the factorizations.

$$\frac{3a^2b}{9a^3b^2} = \frac{3a^2b}{(3a^2b)(3ab)} = \frac{1}{3ab}$$

(e) $\frac{10a^5b^4}{2a^2b^3}$

$$\frac{10a^5b^4}{2a^2b^3} = \frac{(2a^2b^3)(5a^3b)}{2a^2b^3} = \frac{5a^3b}{1} = 5a^3b$$

Check Yourself 4

Simplify each expression.

(a) $\frac{30}{66}$ **(b)** $\frac{5x^4}{15x}$ **(c)** $\frac{12xy^4}{18x^3y^2}$

(d) $\frac{5m^2n}{10m^3n^3}$ **(e)** $\frac{12a^4b^6}{2a^3b^4}$

Example 4 provides us with a brief algorithm for simplifying rational expressions. Here are these steps.

Step by Step

Simplifying Rational Expressions

Step 1 Completely factor the numerator and denominator of the expression.

Step 2 Divide the numerator and denominator by the GCF. The resulting expression will be simplified.

We use the same techniques when there are trinomials in a rational expression.

Example 5 Simplifying Rational Expressions

Simplify each rational expression.

(a) $\dfrac{5x^2 - 5}{x^2 - 4x - 5}$

$= \dfrac{5(x^2 - 1)}{x^2 - 4x - 5}$ Completely factor the expressions in both the numerator and denominator.

$= \dfrac{5(x - 1)\cancel{(x + 1)}}{(x - 5)\cancel{(x + 1)}}$ Divide by the common factor.

$= \dfrac{5(x - 1)}{x - 5}$

NOTE

Divide by the common factor $x + 1$, using the fact that

$\dfrac{x + 1}{x + 1} = 1$

if $x \neq -1$.

(b) $\dfrac{2x^2 + x - 6}{2x^2 - x - 3}$

$= \dfrac{(x + 2)\cancel{(2x - 3)}}{(x + 1)\cancel{(2x - 3)}}$

$= \dfrac{x + 2}{x + 1}$

> CAUTION

Pick any value, other than 0, for x and substitute. You will quickly see that

$\dfrac{x + 2}{x + 1} \neq \dfrac{2}{1}$

For example, if $x = 4$,

$\dfrac{4 + 2}{4 + 1} = \dfrac{6}{5}$

Be careful! The expression $\dfrac{x + 2}{x + 1}$ is already simplified. Students are often tempted to continue.

$\dfrac{\cancel{x} + 2}{\cancel{x} + 1}$ is *not equal* to $\dfrac{2}{1}$

The x's are *terms* in the numerator and denominator. They *cannot* be divided out. Only *factors* can be divided. The fraction

$\dfrac{x + 2}{x + 1}$

is simplified.

(c) $\dfrac{x^3 + 2x^2 - 3x - 6}{x^3 + 8}$

$= \dfrac{x^2(x + 2) - 3(x + 2)}{(x + 2)(x^2 - 2x + 4)}$

$= \dfrac{\cancel{(x + 2)}(x^2 - 3)}{\cancel{(x + 2)}(x^2 - 2x + 4)}$

$= \dfrac{x^2 - 3}{x^2 - 2x + 4}$

NOTE

In part (c) we factor the numerator by grouping and use the sum of cubes formula for the denominator.

Check Yourself 5

Simplify each expression.

(a) $\dfrac{5x - 15}{x^2 - 9}$ **(b)** $\dfrac{a^2 - 5a + 6}{3a^2 - 6a}$

(c) $\dfrac{3x^2 + 14x - 5}{3x^2 + 2x - 1}$ **(d)** $\dfrac{5p - 15}{p^2 - 4}$

NOTE

In this text, we usually place the negative sign outside the fraction or rational expression, as in

$-\frac{2}{3}$ and $-\frac{x+5}{x-1}$

Remember the rules for signs in division. The quotient of a positive number and a negative number is always negative. Thus, there are three equivalent ways to write such a quotient. For instance,

$$\frac{-2}{3} = \frac{2}{-3} = -\frac{2}{3}$$

The quotient of two positive numbers or two negative numbers is always positive. For example,

$$\frac{-2}{-3} = \frac{2}{3}$$

Example 6 Simplifying Rational Expressions

Simplify each expression.

(a) $\frac{6x^2}{-3xy} = \frac{2 \cdot \cancel{3} \cdot \cancel{x} \cdot x}{(-1) \cdot \cancel{3} \cdot \cancel{x} \cdot y} = \frac{2x}{-y} = -\frac{2x}{y}$

(b) $\frac{-5a^2b}{-10b^2} = \frac{\cancel{(-1)} \cdot \cancel{5} \cdot a \cdot a \cdot \cancel{b}}{\cancel{(-1)} \cdot 2 \cdot \cancel{5} \cdot \cancel{b} \cdot b} = \frac{a^2}{2b}$

Check Yourself 6

Simplify each expression.

(a) $\frac{8x^3y}{-4xy^2}$ (b) $\frac{-16a^4b^2}{-12a^2b^5}$

Simplifying certain rational expressions is easier if you recognize a particular pattern. First, verify for yourself that

$$5 - 8 = -(8 - 5)$$

More generally,

$$a - b = -(b - a)$$

If we take this equation and divide both sides by $b - a$, we get

$$\frac{a-b}{b-a} = \frac{-(b-a)}{b-a} = \frac{-1}{1} = -1 \qquad a \neq b$$

Therefore, we have the result

$$\frac{a-b}{b-a} = -1$$

Property

Polynomial Opposites

$$\frac{a-b}{b-a} = \frac{-(b-a)}{b-a} = -1 \quad \text{if} \quad a \neq b$$

We use this property to complete Example 7.

Example 7 Simplifying Rational Expressions

NOTE

$\frac{x-2}{2-x} = -1$

Simplify each expression.

(a) $\frac{2x-4}{4-x^2} = \frac{2\overset{-1}{\cancel{(x-2)}}}{(2+x)\underset{1}{\cancel{(2-x)}}}$

$= \frac{2(-1)}{2+x} = \frac{-2}{2+x}$ or $-\frac{2}{x+2}$

(b) $\dfrac{9 - x^2}{x^2 + 2x - 15} = \dfrac{(3 + x)\overset{-1}{\cancel{(3 - x)}}}{(x + 5)\underset{1}{\cancel{(x - 3)}}}$

$= \dfrac{(3 + x)(-1)}{x + 5} = \dfrac{-x - 3}{x + 5}$ or $-\dfrac{x + 3}{x + 5}$

Check Yourself 7

Simplify each rational expression.

(a) $\dfrac{5x - 20}{16 - x^2}$ **(b)** $\dfrac{x^2 - 6x - 27}{81 - x^2}$

Definition

Rational Function

A **rational function** is a function that is defined by a rational expression. It can be written as

$$f(x) = \frac{P}{Q}$$

where *P* and *Q* are polynomials. The function is *not* defined for any value of *x* for which $Q = 0$.

Example 8 Identifying Rational Functions

< Objective 4 >

Identify the rational functions.

(a) $f(x) = 3x^3 - 2x + 5$

This is a rational function; it can be written as $\dfrac{3x^3 - 2x + 5}{1}$.

(b) $f(x) = \dfrac{3x^2 - 5x + 2}{2x - 1}$

This is a rational function; it is the ratio of two polynomials.

(c) $f(x) = 3x^3 + 3\sqrt{x}$

This is not a rational function. Since $3\sqrt{x} = 3x^{1/2}$, as seen in Section 7.5, $3\sqrt{x}$ cannot be a term of a polynomial.

RECALL

In Chapter 5, you learned that the exponents in a polynomial must always be whole numbers.

Check Yourself 8

Identify the rational functions.

(a) $f(x) = x^5 - 2x^4 - 1$ **(b)** $f(x) = \dfrac{x^2 - x + 7}{\sqrt{x} - 1}$

(c) $f(x) = \dfrac{3x^3 + 3x}{2x + 1}$

If we determine the values of *x* for which a rational function is undefined, then we can describe the domain of *f* as the set of all real numbers *except* those identified "problem" values.

Example 9 Simplifying a Rational Function

(a) Determine the values of *x* for which $f(x) = \dfrac{x^2 - 2x - 24}{2x^2 + 7x - 4}$ is undefined, and write the domain of *f*.

Factoring, we have

$$f(x) = \frac{x^2 - 2x - 24}{2x^2 + 7x - 4} = \frac{(x - 6)(x + 4)}{(2x - 1)(x + 4)}$$

Focusing on the denominator, the values of *x* for which *f* is undefined are $x = \dfrac{1}{2}$ and $x = -4$.

The domain of f is $\left\{x \,\middle|\, x \neq \frac{1}{2} \text{ or } -4\right\}$.

(b) Write the function in simplified form, including the domain.

Dividing by the common factor $x + 4$, we have

$$f(x) = \frac{x-6}{2x-1}, \text{ where } x \neq \frac{1}{2} \text{ or } -4$$

Check Yourself 9

Simplify $f(x) = \dfrac{x^2 + 7x + 10}{x^2 - 5x - 14}$ and give its domain.

Earlier, we constructed total cost models. Recall that for many products, the cost to produce or sell x items consists of a *variable* or *marginal cost* and a *fixed cost.*

In these situations, we can model the average cost per unit with a rational expression. We simply divide the total cost of producing or selling some quantity of a product by the quantity produced or sold. We call this an *average cost model.*

The average cost model provides producers and retailers with very important information. First, and foremost, the average cost helps the seller determine the price that should be charged in order to recoup their fixed costs along with the per-unit cost when selling a product.

Example 10 A Business and Finance Application

< Objective 5 >

The Coastal Cabinetry Company estimates that each Revolutionary Sideboard Buffet costs them \$175 to produce. They also determine that their monthly fixed costs associated with this line of buffets come to \$1,600.

(a) Construct a monthly total cost formula for the production of this buffet.

The variable cost is \$175 and the fixed cost is \$1,600, therefore, the monthly cost to produce x buffets is given by

$$C = 175x + 1{,}600$$

(b) Find the total cost if they produce 25 buffets in a month.

We evaluate the formula when $x = 25$.

$$\begin{aligned} C &= 175x + 1{,}600 \\ &= 175(25) + 1{,}600 \\ &= 5{,}975 \end{aligned}$$

It costs \$5,975 to produce 25 buffets in a month.

(c) Construct a monthly average cost formula for the production of this buffet.

We divide the total cost by the number of units produced x.

$$\overline{C} = \frac{175x + 1{,}600}{x}$$

NOTE

It is common to use $\overline{C}$ to indicate an average cost formula.

(d) Find the average cost per unit if 25 buffets are produced in a month.

Again, we evaluate for $x = 25$.

$$\begin{aligned} \overline{C} &= \frac{175x + 1{,}600}{x} \\ &= \frac{175(25) + 1{,}600}{(25)} \\ &= \frac{5{,}975}{25} \\ &= 239 \end{aligned}$$

If they produce 25 buffets in a month, their costs average out to \$239 per buffet.

Check Yourself 10

Floors and More pays \$45 for a 25-ft^2 package of ECO Bamboo Floor hardwood floor. They figure out that their weekly fixed cost to carry and sell this line of flooring is \$675.

(a) Give a formula describing their total weekly cost to sell the ECO Bamboo Floor hardwood floor packages.
(b) Compute the total cost of selling 75 25-ft^2 packages.
(c) Give a formula for the average cost of selling these packages.
(d) Compute the average cost per package, if they sell 75 packages in a week.

Check Yourself ANSWERS

1. **(a)** 2; **(b)** $\frac{2}{3}$ **2.** **(a)** $r = -7$; **(b)** $x = \frac{9}{2}$ **3.** $x = -\frac{5}{2}$ or $x = 4$ **4.** **(a)** $\frac{5}{11}$; **(b)** $\frac{x^3}{3}$; **(c)** $\frac{2y^2}{3x^2}$; **(d)** $\frac{1}{2mn^2}$; **(e)** $6ab^2$ **5.** **(a)** $\frac{5}{x+3}$; **(b)** $\frac{a-3}{3a}$; **(c)** $\frac{x+5}{x+1}$; **(d)** $\frac{5(p-3)}{(p+2)(p-2)}$ **6.** **(a)** $-\frac{2x^2}{y}$; **(b)** $\frac{4a^2}{3b^3}$ **7.** **(a)** $-\frac{5}{x+4}$; **(b)** $\frac{-x-3}{x+9}$ or $-\frac{x+3}{x+9}$ **8.** **(a)** A rational function; **(b)** not a rational function; **(c)** a rational function **9.** $f(x) = \frac{x+5}{x-7}$, $x \neq -2$ or 7 **10.** **(a)** $C = 45x + 675$; **(b)** \$4,050; **(c)** $\overline{C} = \frac{45x + 675}{x}$; **(d)** \$54 per package

Reading Your Text

These fill-in-the-blank exercises will help you understand some of the key vocabulary used in this section. The answers to these exercises are in the Answers Appendix in the back of the text.

(a) A rational expression is an expression that can be written as the ratio of two ________.

(b) A fraction or rational expression is undefined when its ________ is equal to zero.

(c) To simplify a rational expression, divide the numerator and denominator by the ________.

(d) When the numerator and denominator have no common factors other than 1, we say the expression is ________.

9.1 exercises

Skills | Calculator/Computer | Career Applications | Above and Beyond

< Objective 1 >

Evaluate each expression for the value of the variable.

1. $\frac{3x}{2x-1}$ for $x = 5$

2. $\frac{4x}{5x-6}$ for $x = 2$

3. $\frac{3x+10}{x+2}$ for $x = -4$

4. $\frac{4x-7}{2x-1}$ for $x = -2$

5. $\frac{x^2+x}{x^2+2x}$ for $x = -2$

6. $\frac{4x-5}{2x^2-x+3}$ for $x = -1$

7. $\frac{2-3x}{x^2-4}$ for $x = -1$

8. $\frac{3x+1}{x^2-5x+6}$ for $x = 3$

< Objective 2 >

For what values of the variable is each rational expression undefined?

9. $\frac{x}{x-3}$ **10.** $\frac{y}{y+7}$ **11.** $\frac{x+5}{3}$ **12.** $\frac{x-6}{4}$

13. $\frac{2x-3}{2x-1}$ **14.** $\frac{4x-5}{5x+2}$ **15.** $\frac{2x+5}{x}$ **16.** $\frac{3x-7}{x}$

17. $\frac{x(x+1)}{x+2}$ **18.** $\frac{x+2}{3x-7}$ **19.** $\frac{5-3x}{2x}$ **20.** $\frac{2x+7}{3x+\frac{1}{3}}$

< Objective 3 >

Simplify each expression. Assume the denominators are not 0.

21. $\frac{14}{21}$ **22.** $\frac{45}{75}$ **23.** $\frac{4x^5}{6x^2}$ **24.** $\frac{30x^8}{25x^3}$

25. $\frac{10x^2y^5}{25xy^2}$ **26.** $\frac{18a^2b^3}{24a^4b^3}$ **27.** $\frac{-36x^5y^3}{21x^2y^5}$ **28.** $\frac{-15x^3y^3}{-20xy^2}$

29. $\frac{28a^5b^3c^2}{84a^2bc^4}$ **30.** $\frac{-52p^5q^3r^2}{39p^3q^5r^2}$ **31.** $\frac{6x-24}{x^2-16}$ **32.** $\frac{x^2-25}{3x-15}$

33. $\frac{x^2+2x+1}{6x+6}$ **34.** $\frac{5y^2-10y}{y^2+y-6}$ **35.** $\frac{x^2-13x+36}{x^2-81}$

36. $\frac{2m^2+11m-21}{4m^2-9}$ **37.** $\frac{3b^2-7b-6}{b-3}$ **38.** $\frac{a^2-9b^2}{a^2+8ab+15b^2}$

39. $\frac{2y^2+3yz-5z^2}{2y^2+11yz+15z^2}$ **40.** $\frac{6x^2-x-2}{3x^2-5x+2}$ **41.** $\frac{x^3-64}{x^2-16}$

42. $\frac{r^2-rs-6s^2}{r^3+8s^3}$ **43.** $\frac{a^4-81}{a^2+5a+6}$ **44.** $\frac{x^4-625}{x^2-2x-15}$

45. $\frac{xy-2x+3y-6}{x^2+8x+15}$ **46.** $\frac{cd-3c+5d-15}{d^2-7d+12}$ **47.** $\frac{x^2+3x-18}{x^3-3x^2-2x+6}$

48. $\frac{y^2+2y-35}{y^2-8y+15}$ **49.** $\frac{2m-10}{25-m^2}$ **50.** $\frac{5x-20}{16-x^2}$

51. $\frac{121-x^2}{2x^2-21x-11}$ **52.** $\frac{2x^2-7x+3}{9-x^2}$ **53.** $\frac{2(x+h)-2x}{(x+h)-x}$

54. $\frac{-3(x+h)-(-3x)}{(x+h)-x}$ **55.** $\frac{3(x+h)-3-(3x-3)}{(x+h)-x}$ **56.** $\frac{2(x+h)+5-(2x+5)}{(x+h)-x}$

57. $\frac{(x+h)^2-x^2}{(x+h)-x}$ **58.** $\frac{(x+h)^3-x^3}{(x+h)-x}$ **59.** $\frac{xy-2y+4x-8}{2y+6-xy-3x}$ 

60. $\frac{ab-3a+5b-15}{15+3a^2-5b-a^2b}$

< Objective 4 >

Identify the rational functions.

61. $f(x) = -7x^2 + 2x - 5$ **62.** $f(x) = \frac{x^3-2x^2+7}{\sqrt{x}+2}$ **63.** $f(x) = \frac{x^2-x-1}{x+2}$

64. $f(x) = \frac{\sqrt{x}-x+3}{x-2}$ **65.** $f(x) = 5x^2 - \sqrt[3]{x}$ **66.** $f(x) = \frac{x^2-x+5}{x}$

Simplify each function and give its domain.

67. $f(x) = \dfrac{x^2 - x - 2}{x + 1}$

68. $f(x) = \dfrac{x^2 + x - 12}{x + 4}$

69. $f(x) = \dfrac{3x^2 + 5x - 2}{x + 2}$

70. $f(x) = \dfrac{2x^2 - 7x + 5}{2x - 5}$

71. $f(x) = \dfrac{x^2 + 4x + 4}{5(x + 2)}$

72. $f(x) = \dfrac{x^2 - 6x + 9}{7(x - 3)}$

73. $f(x) = \dfrac{x^2 - 2x - 8}{x^2 - x - 6}$

74. $f(x) = \dfrac{x^2 + 4x - 5}{x^2 + 9x + 20}$

75. $f(x) = \dfrac{x^2 + 4x + 3}{x^2 + 7x + 6}$

76. $f(x) = \dfrac{x^2 + 7x + 10}{x^2 - 6x - 16}$

77. $f(x) = \dfrac{x^2 - 4x + 3}{x^2 - 1}$

78. $f(x) = \dfrac{x^2 - 6x + 8}{x^2 - 16}$

< Objective 5 >

79. **Business and Finance** A company has a setup cost of \$3,500 for the production of a new product. The cost to produce a single unit is \$8.75.

(a) Write a rational function that gives the average cost per unit when x units are produced.

(b) Find the average cost when 50 units are produced.

80. **Business and Finance** The total revenue from the sale of a popular video is approximated by the rational function

$$R(x) = \frac{300x^2}{x^2 + 9}$$

where x is the number of months since the video has been released and $R(x)$ gives the total revenue in hundreds of dollars.

(a) Find the total revenue generated by the end of the first month.

(b) Find the total revenue generated by the end of the second month.

(c) Find the total revenue generated by the end of the third month.

(d) Find the revenue in the second month only.

81. **Geometry** The area of the rectangle is represented by $6x^2 + 19x + 10$. What is the length?

82. **Geometry** The volume of the box is represented by $(x^2 + 5x + 6)(x + 5)$. Find the polynomial that represents the area of the bottom of the box.

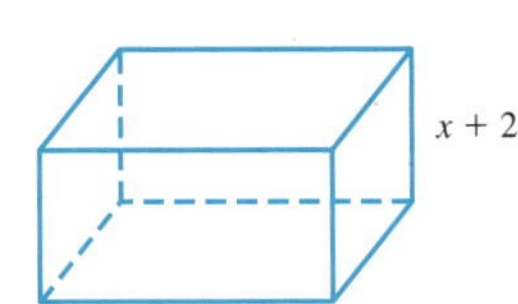

Determine whether each statement is **true** *or* **false.**

83. If we multiply both numerator and denominator by the same nonzero expression, we obtain an equivalent rational expression.

84. If we add the same nonzero expression to both numerator and denominator, we obtain an equivalent rational expression.

Complete each statement with **always, sometimes,** *or* **never.**

85. A rational expression is ______________ the ratio of two polynomials.

86. A value that causes the denominator to be zero can ______________ be used for the variable in a rational expression.

87. If we view the graph of a rational function on a graphing calculator, we often see "unusual" behavior near x-values for which the function is undefined. Consider the rational function

$$f(x) = \frac{1}{x-3}$$

chapter 9 > Make the Connection

(a) For what value(s) of x is the function undefined?

(b) Complete the table.

x	$f(x)$
4	
3.1	
3.01	
3.001	
3.0001	

(c) What do you observe concerning $f(x)$ as x is chosen close to 3, but slightly larger than 3?

(d) Complete the table.

x	$f(x)$
2	
2.9	
2.99	
2.999	
2.9999	

(e) What do you observe concerning $f(x)$ as x is chosen close to 3, but slightly smaller than 3?

(f) Graph the function on a graphing calculator. Describe the behavior of the graph of f near $x = 3$.

88. If we view the graph of a rational function on a graphing calculator, we often see "unusual" behavior near x-values for which the function is undefined. Consider the rational function

$$f(x) = \frac{1}{x+2}$$

chapter 9 > Make the Connection

(a) For what value(s) of x is the function undefined?

(b) Complete the table.

x	$f(x)$
−1	
−1.9	
−1.99	
−1.999	
−1.9999	

(c) What do you observe concerning $f(x)$ as x is chosen close to -2, but slightly larger than -2?

(d) Complete the table.

x	$f(x)$
−3	
−2.1	
−2.01	
−2.001	
−2.0001	

(e) What do you observe concerning $f(x)$ as x is chosen close to -2, but slightly smaller than -2?

(f) Graph the function on a graphing calculator. Describe the behavior of the graph of f near $x = -2$.

89. **Manufacturing Technology** The safe load of a drop-hammer-style pile driver is given from the formula

$$p = \frac{6whs + 6wh}{3s^2 + 6s + 3}$$

Simplify the rational expression.

90. **Mechanical Engineering** The shape of a beam loaded with a single concentrated load is described by the expression

$$\frac{x^2 - 64}{200}$$

Factor the numerator of this expression.

91. **Allied Health** A 4-year-old child is upset because his 9-year-old sister tells him that he will never catch up to her in age. Write an expression for the ratio of the younger child's age x to the older child's age.

92. **Allied Health** Use the expression constructed in exercise 91 to argue that the significance of the difference in their ages reduces with time. chapter 9 > Make the Connection

Skills | Calculator/Computer | Career Applications | **Above and Beyond**

93. Explain why this statement is false. 

$$\frac{6m^2 + 2m}{2m} = 6m^2 + 1$$

94. State and explain the fundamental principle of rational expressions. 

95. The rational expression $\frac{x^2 - 4}{x + 2}$ can be simplified to $x - 2$. Is this reduction true for all values of x? Explain. chapter 9 > Make the Connection

96. What is meant by a rational expression in lowest terms? 

97. To work with rational expressions correctly, it is important to understand the difference between a *factor* and a *term* of an expression. In your own words, write definitions for both, explaining the difference between the two. chapter 9 > Make the Connection

98. Give some examples of terms and factors in rational expressions and explain how both are affected when a fraction is simplified. chapter 9 > Make the Connection

99. Explain the reasoning involved in each step when simplifying the fraction $\frac{42}{56}$. chapter 9 > Make the Connection

100. Describe why $\frac{3}{5}$ and $\frac{27}{45}$ are *equivalent fractions.* 

101. Show how to simplify the rational expression

$$\frac{x^2 - 9}{4x + 12}$$

Your simplified fraction is equivalent to the given fraction. Are there other rational expressions equivalent to this one? Write another rational expression that you think is equivalent to this one. Exchange papers with another student. Do you agree that the other student's fraction is equivalent to yours? Why or why not?

Answers

1. $\frac{5}{3}$ **3.** 1 **5.** Undefined **7.** $-\frac{5}{3}$ **9.** 3 **11.** Never undefined **13.** $\frac{1}{2}$ **15.** 0 **17.** -2 **19.** 0 **21.** $\frac{2}{3}$ **23.** $\frac{2x^3}{3}$ **25.** $\frac{2xy^3}{5}$ **27.** $-\frac{12x^3}{7y^2}$ **29.** $\frac{a^3b^2}{3c^2}$ **31.** $\frac{6}{x+4}$ **33.** $\frac{x+1}{6}$ **35.** $\frac{x-4}{x+9}$ **37.** $3b + 2$ **39.** $\frac{y-z}{y+3z}$ **41.** $\frac{x^2+4x+16}{x+4}$ **43.** $\frac{(a^2+9)(a-3)}{a+2}$ **45.** $\frac{y-2}{x+5}$ **47.** $\frac{x+6}{x^2-2}$ **49.** $-\frac{2}{m+5}$ **51.** $\frac{-11-x}{2x+1}$ or $-\frac{x+11}{2x+1}$ **53.** 2 **55.** 3 **57.** $2x + h$ **59.** $-\frac{y+4}{y+3}$ **61.** Rational **63.** Rational **65.** Not rational **67.** $f(x) = x - 2; x \neq -1$ **69.** $f(x) = 3x - 1; x \neq -2$ **71.** $f(x) = \frac{x+2}{5}; x \neq -2$ **73.** $f(x) = \frac{x-4}{x-3}; x \neq -2, 3$ **75.** $f(x) = \frac{x+3}{x+6}; x \neq -6, -1$ **77.** $f(x) = \frac{x-3}{x+1}; x \neq -1, 1$ **79.** (a) $R(x) = \frac{3{,}500 + 8.75x}{x}$; (b) \$78.75 per unit **81.** $2x + 5$ **83.** True **85.** always

87. (a) 3; (b)

x	$f(x)$
4	1
3.1	10
3.01	100
3.001	1,000
3.0001	10,000

(c) Answers will vary; (d)

x	$f(x)$
2	-1
2.9	-10
2.99	-100
2.999	$-1{,}000$
2.9999	$-10{,}000$

(e) Answers will vary;

(f) 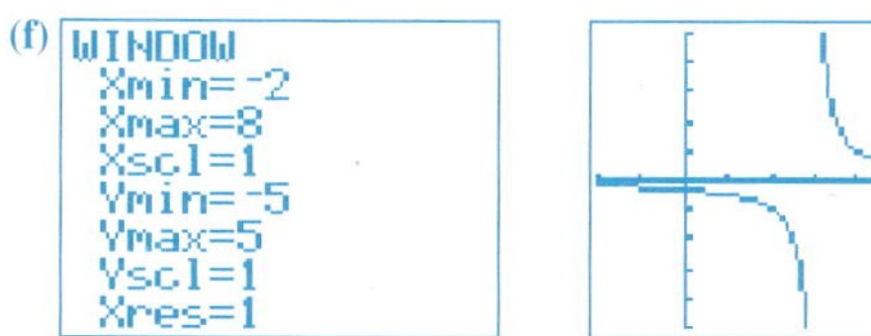

89. $\frac{2wh}{s+1}$ **91.** $\frac{x}{x+5}$ **93.** Above and Beyond **95.** Above and Beyond **97.** Above and Beyond **99.** Above and Beyond **101.** Above and Beyond

Activity 9 ::

Communicating Mathematical Ideas

Organizations concerned with mathematics education, such as NCTM (National Council of Teachers of Mathematics) and AMATYC (American Mathematical Association of Two-Year Colleges), have long recognized the importance of communication. Your drive to explore and investigate, your ability to solve problems, and your skill in presenting your findings are all key factors for success in today's world.

This activity is designed to help you practice and develop these skills. As you work through the problems below, focus on effectively communicating your work to others.

1. A rectangle with an area of 30 cm^2 has length l and width w.

(a) Use the area constraint to write the perimeter of the rectangle as a function of its width w.

(b) Based on the physical constraints of this application, what is the domain of the function?

(c) Find the minimum perimeter of the rectangle.

(d) Find the dimensions that yield this minimum perimeter.

2. In general, for a cylinder of radius r and height h, the volume and surface area are given by

$$V_{\text{Cyl}} = \pi r^2 h$$

$$S_{\text{Cyl}} = 2\pi rh + 2\pi r^2$$

We wish to manufacture a metal tank (cylinder) that holds 80 cu ft of fluid.

(a) Use the volume formula to express the height of the tank as a function of its radius.

(b) Use (a) to express the surface area as a function of the radius (substitute for the height).

(c) Provide a graph of the surface area function found in (b).

(d) Use your calculator to approximate (one decimal place) the radius that produces the minimum surface area (that is, uses the minimum amount of metal).

(e) What is the minimum surface area of a cylinder that holds 80 cu ft of fluid (use (b) or use the graph)?

(f) Find the height of this minimum surface area cylinder (use the formula from (a)).

3. Consider the rational function $f(x) = \dfrac{2x + 3}{x - 3}$.

(a) Give the domain of the function.

(b) Give any y-intercepts of the function (write any answers as ordered pairs).

(c) Give any x-intercepts of the function (write any answers as ordered pairs).

9.2 Multiplying and Dividing Rational Expressions

< 9.2 Objectives >

1 > Multiply and simplify rational expressions

2 > Divide and simplify rational expressions

3 > Multiply and divide rational functions

We turn to an example from arithmetic to begin our discussion of multiplying rational expressions. Recall that to multiply two fractions, we multiply the numerators and multiply the denominators. For instance,

$$\frac{2}{5} \cdot \frac{3}{7} = \frac{2 \cdot 3}{5 \cdot 7} = \frac{6}{35}$$

In algebra, the pattern is exactly the same.

Property

Multiplying Rational Expressions

For polynomials P, Q, R, and S,

$$\frac{P}{Q} \cdot \frac{R}{S} = \frac{PR}{QS} \quad \text{where } Q \neq 0 \text{ and } S \neq 0$$

Example 1 Multiplying Rational Expressions

< Objective 1 >

NOTE

We assume the denominator is not zero throughout this section.

Multiply.

$$\frac{2x^3}{5y^2} \cdot \frac{10y}{3x^2} = \frac{20x^3y}{15x^2y^2}$$

$$= \frac{5x^2y \cdot 4x}{5x^2y \cdot 3y}$$ Divide by the common factor $5x^2y$ to simplify.

$$= \frac{4x}{3y}$$

Check Yourself 1

Multiply.

$$\frac{9a^2b^3}{5ab^4} \cdot \frac{20ab^2}{27ab^3}$$

NOTE

Divide by the common factors of 3 and 4. The alternative is to multiply *first*.

$$\frac{3}{8} \cdot \frac{4}{9} = \frac{12}{72}$$

and then use the GCF to simplify.

$$\frac{12}{72} = \frac{1}{6}$$

It is easier to divide the numerator and denominator by any common factors *before* multiplying. Consider

$$\frac{3}{8} \cdot \frac{4}{9} = \frac{\overset{1}{\cancel{3}} \cdot \overset{1}{\cancel{4}}}{\underset{2}{\cancel{8}} \cdot \underset{3}{\cancel{9}}} = \frac{1}{6}$$

We multiply rational expressions in exactly the same way.

Example 2 Multiplying Rational Expressions

NOTE

We use the factoring methods in Chapter 6 to simplify rational expressions.

Multiply and simplify.

(a) $\frac{x}{x^2 - 3x} \cdot \frac{6x - 18}{9x}$ Factor.

$= \frac{\overset{1}{\cancel{x}}}{\underset{1}{\cancel{x}}\underset{1}{\cancel{(x-3)}}} \cdot \frac{\overset{2}{\cancel{6}}\overset{1}{\cancel{(x-3)}}}{\underset{3}{\cancel{9}}x}$ Divide by the common factors of 3, x, and $x - 3$.

$= \frac{2}{3x}$

(b) $\frac{x^2 - y^2}{5x^2 - 5xy} \cdot \frac{10xy}{x^2 + 2xy + y^2}$ Factor and divide by the common factors of 5, x, $x - y$, and $x + y$.

$= \frac{\overset{1}{\cancel{(x+y)}}\overset{1}{\cancel{(x-y)}}}{\underset{1}{\cancel{5}}\underset{1}{\cancel{x}}\underset{1}{\cancel{(x-y)}}} \cdot \frac{\overset{2}{\cancel{10}}\overset{1}{\cancel{x}}y}{\underset{1}{\cancel{(x+y)}}(x+y)}$

$= \frac{2y}{x + y}$

RECALL

$\frac{2 - x}{x - 2} = -1$

(c) $\frac{4}{x^2 - 2x} \cdot \frac{10x - 5x^2}{8x + 24}$

$= \frac{\overset{1}{\cancel{4}}}{\underset{1}{\cancel{x}}\underset{1}{\cancel{(x-2)}}} \cdot \frac{5\overset{1}{\cancel{x}}\overset{-1}{\cancel{(2-x)}}}{\underset{2}{\cancel{8}}(x+3)}$

$= \frac{-5}{2(x+3)} = -\frac{5}{2(x+3)}$

Check Yourself 2

Multiply and simplify.

(a) $\frac{x^2 - 5x - 14}{4x^2} \cdot \frac{8x + 56}{x^2 - 49}$ (b) $\frac{x}{2x - 6} \cdot \frac{3x - x^2}{2}$

This algorithm summarizes our work in multiplying rational expressions.

Step by Step

Multiplying Rational Expressions

Step 1 Completely factor the numerators and denominators.

Step 2 Divide by (or "cancel") any common factors.

Step 3 Multiply as needed to form the product.

RECALL

To divide fractions, we multiply by the reciprocal of the divisor. That is, invert the *divisor* (the second fraction) and multiply.

We use arithmetic to see how to divide rational expressions, as well. Recall that

$$\frac{3}{5} \div \frac{2}{3} = \frac{3}{5} \cdot \frac{3}{2} = \frac{9}{10}$$

Once more, the pattern in algebra is identical.

Property

Dividing Rational Expressions

For polynomials P, Q, R, and S,

$$\frac{P}{Q} \div \frac{R}{S} = \frac{P}{Q} \cdot \frac{S}{R} = \frac{PS}{QR}$$

where $Q \neq 0$, $R \neq 0$, and $S \neq 0$.

Example 3 Dividing Rational Expressions

< Objective 2 >

Divide and simplify.

(a) $\frac{3x^2}{8x^3y} \div \frac{9x^2y^2}{4y^4} = \frac{3x^2}{8x^3y} \cdot \frac{4y^4}{9x^2y^2} = \frac{y}{6x^3}$ Do not simplify until after inverting the divisor.

NOTE

Invert the divisor (or second expression) and multiply.

(b) $\frac{2x^2 + 4xy}{9x - 18y} \div \frac{4x + 8y}{3x - 6y} = \frac{2x^2 + 4xy}{9x - 18y} \cdot \frac{3x - 6y}{4x + 8y}$

$$= \frac{2x(x + 2y)}{9(x - 2y)} \cdot \frac{3(x - 2y)}{4(x + 2y)} = \frac{x}{6}$$

> CAUTION

Invert (or "flip") the divisor, then factor.

(c) $\frac{2x^2 - x - 6}{4x^2 + 6x} \div \frac{x^2 - 4}{4x} = \frac{2x^2 - x - 6}{4x^2 + 6x} \cdot \frac{4x}{x^2 - 4}$

$$= \frac{(2x + 3)(x - 2)}{2x(2x + 3)} \cdot \frac{4x}{(x + 2)(x - 2)}$$

$$= \frac{2}{x + 2}$$

Check Yourself 3

Divide and simplify.

(a) $\frac{5xy}{7x^3} \div \frac{10y^2}{14x^3}$ (b) $\frac{3x - 9y}{2x + 10y} \div \frac{x^2 - 3xy}{4x^2 + 20xy}$

(c) $\frac{x^2 - 9}{x^3 - 27} \div \frac{x^2 - 2x - 15}{2x^2 - 10x}$

Here is an algorithm summarizing our work in dividing rational expressions.

Step by Step

Dividing Rational Expressions

Step 1 Invert the divisor (the *second* rational expression) to make a multiplication problem.

Step 2 Multiply and simplify the rational expressions.

The product of two rational functions is always a rational function. Given two rational functions $f(x)$ and $g(x)$, we can rename the product, so

$$h(x) = f(x) \cdot g(x)$$

This is always true for values of x for which both f and g are defined. So, for example, $h(1) = f(1) \cdot g(1)$ as long as both $f(1)$ and $g(1)$ exist. Example 4 illustrates this concept.

Example 4 Multiplying Rational Functions

< Objective 3 >

Consider the rational functions

$$f(x) = \frac{x^2 - 3x - 10}{x + 1} \quad \text{and} \quad g(x) = \frac{x^2 - 4x - 5}{x - 5}$$

(a) $f(0) \cdot g(0)$

Because $f(0) = -10$ and $g(0) = 1$, we have $f(0) \cdot g(0) = (-10)(1) = -10$.

NOTES

$f(0) = \frac{(0)^2 - 3(0) - 10}{(0) + 1}$

$= \frac{-10}{1} = -10$

$g(0) = \frac{(0)^2 - 4(0) - 5}{(0) - 5}$

$= \frac{-5}{-5} = 1$

$f(x)$ is undefined for $x = -1$, and $g(x)$ is undefined for $x = 5$. Therefore, $h(x)$ is undefined for both of these values.

(b) $f(5) \cdot g(5)$

Although we can find $f(5)$, $g(5)$ is undefined. The number 5 is excluded from the domain of the function. Therefore, $f(5) \cdot g(5)$ is undefined.

(c) $h(x) = f(x) \cdot g(x)$

$$= \frac{x^2 - 3x - 10}{x + 1} \cdot \frac{x^2 - 4x - 5}{x - 5}$$

$$= \frac{(x - 5)(x + 2)}{(x + 1)} \cdot \frac{(x + 1)(x - 5)}{(x - 5)}$$

Factor the numerators, and divide by the common factors.

$$= (x - 5)(x + 2) \qquad x \neq -1, x \neq 5$$

(d) $h(0)$

$h(0) = (0 - 5)(0 + 2) = -10$

(e) $h(5)$

Although the temptation is to substitute 5 for x in part (c), the function is undefined when x is -1 or 5. As was true in part (b), the function is undefined at that point.

Check Yourself 4

Consider

$$f(x) = \frac{x^2 - 2x - 8}{x + 2} \quad \text{and} \quad g(x) = \frac{x^2 - 3x - 10}{x - 4}$$

(a) $f(0) \cdot g(0)$ **(b)** $f(4) \cdot g(4)$ **(c)** $h(x) = f(x) \cdot g(x)$

(d) $h(0)$ **(e)** $h(4)$

When we divide two rational functions to create a third rational function, we must be certain to exclude values for which the polynomial in the denominator is equal to zero, as Example 5 illustrates.

Example 5 Dividing Rational Functions

Consider the rational functions

$$f(x) = \frac{x^3 - 2x^2}{x + 2} \quad \text{and} \quad g(x) = \frac{x^2 - 3x + 2}{x - 4}$$

(a) Find $\frac{f(0)}{g(0)}$.

Because $f(0) = 0$ and $g(0) = -\frac{1}{2}$, we have

$$\frac{f(0)}{g(0)} = \frac{0}{-\frac{1}{2}} = 0$$

(b) Find $\frac{f(1)}{g(1)}$.

Although we can find both $f(1)$ and $g(1)$, $g(1) = 0$, so division is undefined. The value 1 is excluded from the domain of the quotient.

(c) Find $h(x) = \dfrac{f(x)}{g(x)}$.

$$h(x) = \frac{f(x)}{g(x)}$$

−2 is excluded from the domain of f and 4 is excluded from the domain of g.

$$= \frac{\dfrac{x^3 - 2x^2}{x + 2}}{\dfrac{x^2 - 3x + 2}{x - 4}}$$

Invert and multiply.

$$= \frac{x^3 - 2x^2}{x + 2} \cdot \frac{x - 4}{x^2 - 3x + 2}$$

$$= \frac{x^2\overset{1}{\cancel{(x - 2)}}}{x + 2} \cdot \frac{x - 4}{(x - 1)\underset{1}{\cancel{(x - 2)}}}$$

Because $(x - 1)(x - 2)$ is part of the denominator, 1 and 2 are excluded from the domain of h.

$$= \frac{x^2(x - 4)}{(x + 2)(x - 1)} \qquad x \neq -2, 1, 2, 4$$

(d) For which values of x is $h(x)$ undefined?

The function $h(x)$ will be undefined for any value of x that would cause division by zero. So $h(x)$ is undefined for the values −2, 1, 2, and 4.

Check Yourself 5

Given the rational functions

$$f(x) = \frac{x^2 - 2x + 1}{x + 3} \quad \text{and} \quad g(x) = \frac{x^2 - 5x + 4}{x - 2}$$

(a) Find $\dfrac{f(0)}{g(0)}$. **(b)** Find $\dfrac{f(1)}{g(1)}$. **(c)** Find $h(x) = \dfrac{f(x)}{g(x)}$.

(d) For which values of x is $h(x)$ undefined?

Check Yourself ANSWERS

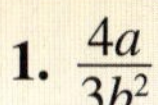

1. $\dfrac{4a}{3b^2}$ **2. (a)** $\dfrac{2(x + 2)}{x^2}$; **(b)** $-\dfrac{x^2}{4}$ **3. (a)** $\dfrac{x}{y}$; **(b)** 6; **(c)** $\dfrac{2x}{x^2 + 3x + 9}$

4. (a) −10; **(b)** undefined; **(c)** $h(x) = (x - 5)(x + 2)$, $x \neq -2$, $x \neq 4$; **(d)** −10; **(e)** undefined

5. (a) $-\dfrac{1}{6}$; **(b)** undefined; **(c)** $h(x) = \dfrac{(x - 1)(x - 2)}{(x + 3)(x - 4)}$; **(d)** $x \neq -3, 1, 2, 4$

Reading Your Text

These fill-in-the-blank exercises will help you understand some of the key vocabulary used in this section. The answers to these exercises are in the Answers Appendix in the back of the text.

(a) To find the product of rational expressions, __________ the numerators and the denominators.

(b) When multiplying rational expressions, you should completely __________ each polynomial.

(c) To divide fractions, we multiply by the __________ of the divisor.

(d) The product of two rational expressions is __________ a rational expression.

9.2 exercises

Skills | Calculator/Computer | Career Applications | Above and Beyond

< Objectives 1 and 2 >

Compute, as indicated. Simplify your results.

1. $\frac{x^2}{3} \cdot \frac{6x}{x^4}$

2. $\frac{-y^3}{10} \cdot \frac{15y}{y^6}$

3. $\frac{x}{8x^4} \div \frac{x^5}{24}$

4. $\frac{p^5}{8} \div \left(-\frac{p^2}{12p}\right)$

5. $\frac{4xy^2}{15x^3} \cdot \frac{25xy}{16y^3}$

6. $\frac{3x^3y}{10xy^3} \cdot \frac{5xy^2}{-9xy^3}$

7. $\frac{8b^3}{15ab} \div \frac{2ab^2}{20ab^3}$

8. $\frac{5x^3y^3}{8x^5} \div \frac{15x^3y}{32x^3y^2}$

9. $\frac{m^3n}{2mn} \cdot \frac{6mn^2}{m^3n} \div \frac{3mn}{5m^2n}$

10. $\frac{4cd^2}{5cd} \cdot \frac{3c^3d}{2c^2d} \div \frac{9cd}{20cd^3}$

11. $\frac{6x + 18}{4x} \cdot \frac{16x^3}{3x + 9}$

12. $\frac{a^2 - 3a}{5a} \cdot \frac{20a^2}{3a - 9}$

13. $\frac{3b - 15}{6b} \div \frac{4b - 20}{9b^2}$

14. $\frac{7m^2 + 28m}{4m} \div \frac{5m + 20}{12m^2}$

15. $\frac{x^2 - 3x - 10}{5x} \cdot \frac{15x^2}{3x - 15}$

16. $\frac{y^2 - 8y}{4y} \cdot \frac{12y^2}{y^2 - 64}$

17. $\frac{c^2 + 2c - 8}{6c} \div \frac{5c + 20}{18c}$

18. $\frac{m^2 - 64}{6m^2} \div \frac{2m - 16}{24m^5}$

19. $\frac{x^2 - x - 12}{3x - 12} \cdot \frac{15x^3}{x^2 - 9}$

20. $\frac{y^2 + 7y + 10}{y^2 + 5y} \cdot \frac{2y}{y^2 - 4}$

21. $\frac{d^2 - 3d - 18}{16d - 96} \div \frac{d^2 - 9}{20d}$

22. $\frac{b^2 + 2b - 8}{b^2 - 2b} \div \frac{b^2 - 16}{4b}$

23. $\frac{2x^2 - x - 3}{3x^2 + 7x + 4} \cdot \frac{3x^2 - 11x - 20}{4x^2 - 9}$

24. $\frac{4p^2 - 1}{2p^2 - 9p - 5} \cdot \frac{3p^2 - 13p - 10}{9p^2 - 4}$

25. $\frac{a^2 - 9}{2a^2 - 6a} \div \frac{2a^2 + 5a - 3}{4a^2 - 1}$

26. $\frac{2x^2 - 5x - 7}{4x^2 - 9} \div \frac{5x^2 + 5x}{2x^2 + 3x}$

27. $\frac{2w - 6}{w^2 + 2w} \cdot \frac{3w}{3 - w}$

28. $\frac{3y - 15}{y^2 + 3y} \cdot \frac{4y}{5 - y}$

29. $\frac{a - 7}{2a + 6} \div \frac{21 - 3a}{a^2 + 3a}$

30. $\frac{x - 5}{x^2 + 3x} \div \frac{25 - 5x}{2x + 6}$

31. $\frac{x^2 - 9y^2}{2x^2 - xy - 15y^2} \cdot \frac{4x + 10y}{x^2 + 3xy}$

32. $\frac{2a^2 - 7ab - 15b^2}{2ab - 10b^2} \cdot \frac{2a^2 - 3ab}{4a^2 - 9b^2}$

33. $\frac{3m^2 - 5mn + 2n^2}{9m^2 - 4n^2} \div \frac{m^3 - m^2n}{9m^2 + 6mn}$

34. $\frac{2x^2y - 5xy^2}{4x^2 - 25y^2} \div \frac{4x^2 + 20xy}{2x^2 + 15xy + 25y^2}$

35. $\frac{x^3 + 8}{x^2 - 4} \cdot \frac{5x - 10}{x^3 - 2x^2 + 4x}$

36. $\frac{a^3 - 27}{a^2 - 9} \div \frac{a^3 + 3a^2 + 9a}{3a^3 + 9a^2}$

< Objective 3 >

37. Let $f(x) = \frac{x^2 - 3x - 4}{x + 2}$ and $g(x) = \frac{x^2 - 2x - 8}{x - 4}$.
Find **(a)** $f(0) \cdot g(0)$; **(b)** $f(4) \cdot g(4)$; **(c)** $h(x) = f(x) \cdot g(x)$; **(d)** $h(0)$; and **(e)** $h(4)$.

38. Let $f(x) = \frac{x^2 - 4x + 3}{x + 5}$ and $g(x) = \frac{x^2 + 7x + 10}{x - 3}$.
Find **(a)** $f(1) \cdot g(1)$; **(b)** $f(3) \cdot g(3)$; **(c)** $h(x) = f(x) \cdot g(x)$; **(d)** $h(1)$; and **(e)** $h(3)$.

39. Let $f(x) = \frac{2x^2 - 3x - 5}{x + 2}$ and $g(x) = \frac{3x^2 + 5x - 2}{x + 1}$.
Find **(a)** $f(1) \cdot g(1)$; **(b)** $f(-2) \cdot g(-2)$; **(c)** $h(x) = f(x) \cdot g(x)$; **(d)** $h(1)$; and **(e)** $h(-2)$.

40. Let $f(x) = \frac{x^2 - 1}{x - 3}$ and $g(x) = \frac{x^2 - 9}{x - 1}$.
Find **(a)** $f(2) \cdot g(2)$; **(b)** $f(3) \cdot g(3)$; **(c)** $h(x) = f(x) \cdot g(x)$; **(d)** $h(2)$; and **(e)** $h(3)$.

41. Let $f(x) = \frac{3x^2 + x - 2}{x - 2}$ and $g(x) = \frac{x^2 - 4x - 5}{x + 4}$.
Find **(a)** $\frac{f(0)}{g(0)}$; **(b)** $\frac{f(1)}{g(1)}$; **(c)** $h(x) = \frac{f(x)}{g(x)}$; and **(d)** the values of x for which $h(x)$ is undefined.

42. Let $f(x) = \frac{x^2 + x}{x - 5}$ and $g(x) = \frac{x^2 - x - 6}{x - 5}$.
Find **(a)** $\frac{f(0)}{g(0)}$; **(b)** $\frac{f(2)}{g(2)}$; **(c)** $h(x) = \frac{f(x)}{g(x)}$; and **(d)** the values of x for which $h(x)$ is undefined.

43. Construction Plans call for the dimensions of a rectangular room to be given in terms of an unknown x. Find the area of the room shown, in terms of x. VIDEO

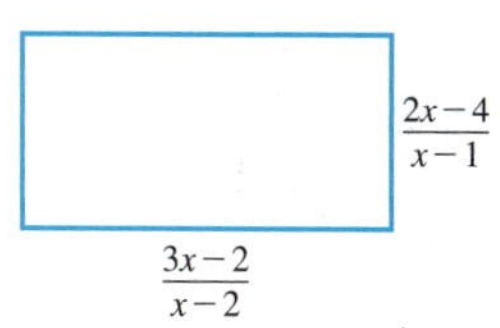

44. Construction Plans call for the dimensions of a rectangular room to be given in terms of an unknown x. Find the area of the room shown, in terms of x.

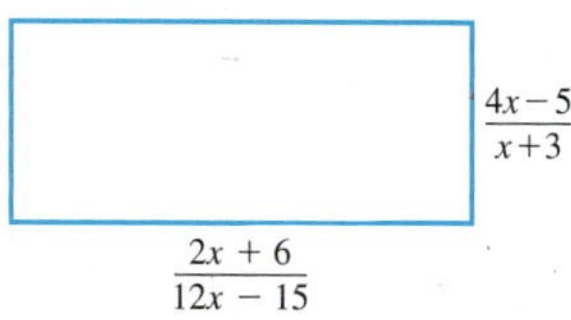

Determine whether each statement is **true** *or* **false.**

45. When we multiply two rational expressions, we multiply the numerators together and we multiply the denominators together.

46. When we divide two rational expressions, we invert the second rational expression, and then we divide.

47. The product of three negative values is negative.

48. Order of operations states that we multiply and divide before applying powers.

49. Division by zero results in a quotient of zero.

50. A fraction can always be simplified if a variable in the numerator also appears in the denominator.

Skills | **Calculator/Computer** | Career Applications | Above and Beyond

You can use a graphing calculator to check your work when multiplying and dividing rational expressions. In the [Y =] menu, enter the first expression as Y_1 and the second as Y_2. Use the [VARS] menu, move to the [Y-Vars] submenu, and enter the [1: Function] submenu to define Y_3 as the product $Y_1 * Y_2$ or the quotient Y_1/Y_2. Then enter the result of your work as Y_4. Next, deselect Y_1 and Y_2 by moving the cursor over the equal sign of each function and pressing [ENTER].

The graph and table utilities will only show Y_3 and Y_4. If you simplified correctly, the two functions will give the same graph and table entries.

Consider

$$\frac{x^2-1}{x+2} \cdot \frac{x^2+2x}{x-1}$$

We multiply and simplify.

$$\frac{x^2-1}{x+2} \cdot \frac{x^2+2x}{x-1} = \frac{(x+1)(x-1)}{x+2} \cdot \frac{x(x+2)}{x-1} \quad \text{Factor.}$$

$$= \frac{(x+1)\cancel{(x-1)}}{\cancel{x+2}} \cdot \frac{x\cancel{(x+2)}}{\cancel{x-1}} \quad \text{Simplify.}$$

$$= x(x+1)$$

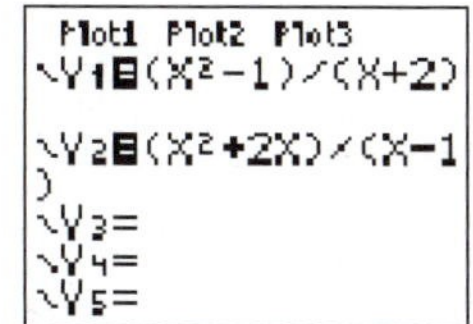

Enter the expressions. Use parentheses to contain each numerator and denominator.

With the cursor in Y_3, press [VARS].

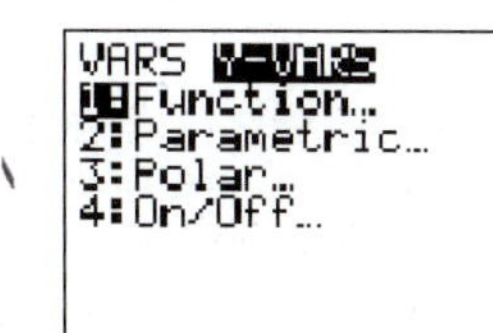

Use the [▶] to access the [Y-VARS] submenu.

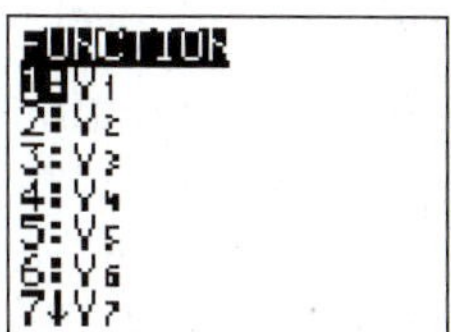

Select the [1: Function] submenu and select Y_1.

Plot1 Plot2 Plot3
\Y1=(X²-1)/(X+2)
\Y2=(X²+2X)/(X-1)
\Y3=Y1
\Y4=
\Y5=

Y_1 appears in the Y_3 field.

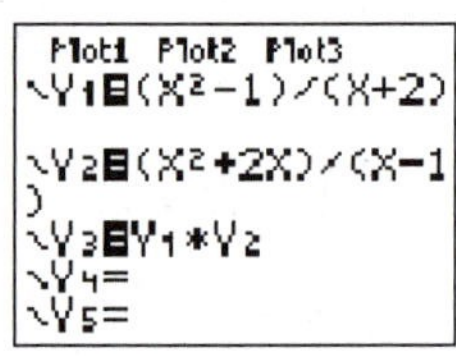

Press [×] and repeat to enter Y_2.

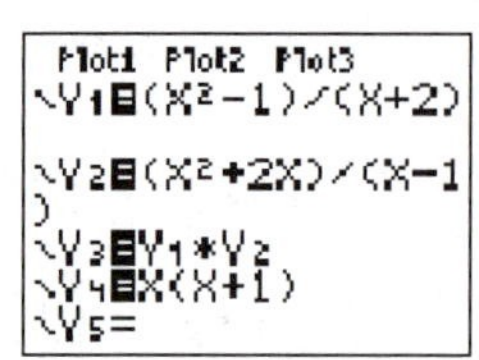

Enter our result, $x(x + 1)$ as Y_4.

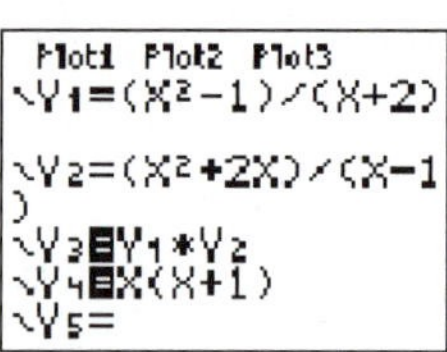

Deselect Y_1 and Y_2.

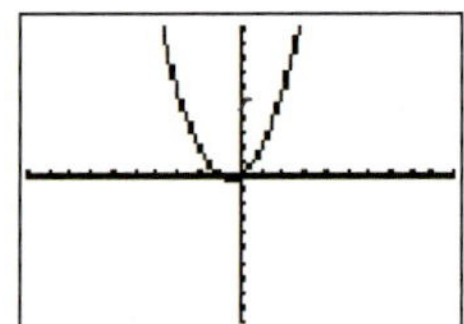

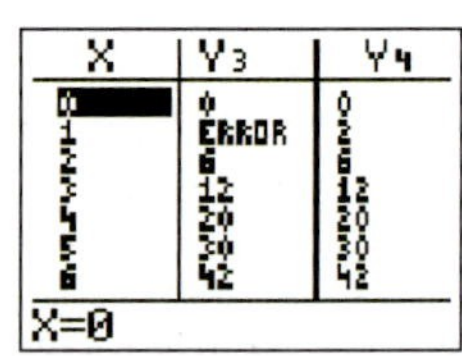

X	Y3	Y4
0	0	0
1	ERROR	2
2	6	6
3	12	12
4	20	20
5	30	30
6	42	42

X=0

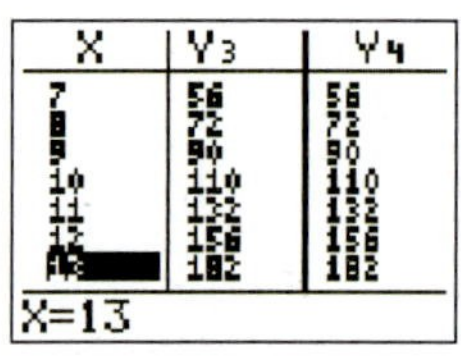

X	Y3	Y4
7	56	56
8	72	72
9	90	90
10	110	110
11	132	132
12	156	156
13	182	182

X=13

Only one graph shows because Y_3 and Y_4 produce the same graph.

Two looks at the TABLE utility show that the two functions are nearly identical. The only discrepancy is at $x = 1$. The function in Y_2 is not defined at $x = 1$.

Note: If you looked at $x = -2$, you would also see an error in the Y_3 column.

Complete exercises 51 to 54 in the usual manner and then use a graphing calculator to check your work.

51. $\dfrac{x^3 - 3x^2 + 2x - 6}{x^2 - 9} \cdot \dfrac{5x^2 + 15x}{20x}$

52. $\dfrac{3a^3 + a^2 - 9a - 3}{15a^2 + 5a} \cdot \dfrac{3a^2 + 9}{a^4 - 9}$

53. $\dfrac{x^4 - 16}{x^2 + x - 6} \div (x^3 + 4x)$

54. $\dfrac{w^3 + 27}{w^2 + 2w - 3} \div (w^3 - 3w^2 + 9w)$

Skills | Calculator/Computer | **Career Applications** | Above and Beyond

55. Agricultural Technology Herbicides constitute approximately $\frac{2}{3}$ of all pesticides used in the United States. Insecticides account for another $\frac{1}{4}$ of the pesticides used in the United States. Write a simplified expression for the ratio of herbicides to insecticides used in the United States.

56. Agricultural Technology Fungicides account for approximately $\frac{1}{10}$ of the pesticides used in the United States. Insecticides account for another $\frac{1}{4}$ of the pesticides used. Write a simplified expression for the ratio of fungicides to insecticides used in the United States.

Simplify each expression.

57. $\dfrac{x^2 + 5x}{3x - 6} \cdot \dfrac{x^2 - 4}{3x^2 + 15x} \cdot \dfrac{6x}{x^2 + 6x + 8}$

58. $\dfrac{m^2 - n^2}{m^2 - mn} \cdot \dfrac{6m}{2m^2 + mn - n^2} \cdot \dfrac{8m - 4n}{12m^2 + 12mn}$

59. $\dfrac{x^2 - 2x - 8}{2x - 8} \cdot \dfrac{x^2 + 5x}{x^2 + 5x + 6} \div \dfrac{x^2 + 2x - 15}{x^2 - 9}$

60. $\dfrac{14x - 7}{x^2 + 3x - 4} \cdot \dfrac{x^2 + 6x + 8}{2x^2 + 5x - 3} \div \dfrac{x^2 + 2x}{x^2 + 2x - 3}$

Answers

1. $\frac{2}{x}$ **3.** $\frac{3}{x^8}$ **5.** $\frac{5}{12x}$ **7.** $\frac{16b^3}{3a}$ **9.** $5mn$ **11.** $8x^2$ **13.** $\frac{9b}{8}$ **15.** $x(x + 2)$ **17.** $\frac{3(c - 2)}{5}$ **19.** $\frac{5x^3}{x - 3}$ **21.** $\frac{5d}{4(d - 3)}$ **23.** $\frac{x - 5}{2x + 3}$ **25.** $\frac{2a + 1}{2a}$ **27.** $-\frac{6}{w + 2}$ **29.** $-\frac{a}{6}$ **31.** $\frac{2}{x}$ **33.** $\frac{3}{m}$ **35.** $\frac{5}{x}$

37. **(a)** -4; **(b)** undefined; **(c)** $h(x) = (x + 1)(x - 4), x \neq -2, 4$; **(d)** -4; **(e)** undefined

39. **(a)** -6; **(b)** undefined; **(c)** $h(x) = (2x - 5)(3x - 1), x \neq -2, -1$; **(d)** -6; **(e)** undefined

41. **(a)** $-\frac{4}{5}$; **(b)** $\frac{5}{4}$; **(c)** $h(x) = \frac{(3x - 2)(x + 4)}{(x - 2)(x - 5)}$; **(d)** $-4, -1, 2, 5$ **43.** $\frac{2(3x - 2)}{x - 1}$ **45.** True **47.** True **49.** False **51.** $\frac{x^2 + 2}{4}$

53. $\frac{x + 2}{x(x + 3)}$ **55.** $\frac{8}{3}$ **57.** $\frac{2x}{3(x + 4)}$ **59.** $\frac{x}{2}$

9.3 Adding and Subtracting Rational Expressions

< 9.3 Objectives >

1 > Add and subtract like rational expressions

2 > Find the LCD of unlike rational expressions

3 > Add and subtract unlike rational expressions

4 > Add and subtract rational functions

NOTE

Similar to fractions, rational expressions with the same denominator are called **like rational expressions**.

Recall that adding or subtracting fractions with the same denominator is straightforward. The same is true in algebra. To add or subtract rational expressions with the same denominator, we add or subtract their numerators and then write that sum or difference over the common denominator.

Property

Adding and Subtracting Rational Expressions

$$\frac{P}{R} + \frac{Q}{R} = \frac{P + Q}{R}$$

and $\frac{P}{R} - \frac{Q}{R} = \frac{P - Q}{R}$

where $R \neq 0$.

Example 1 Adding and Subtracting Rational Expressions

< Objective 1 >

Perform the indicated operations.

$$\frac{3}{2a^2} - \frac{1}{2a^2} + \frac{5}{2a^2} = \frac{3 - 1 + 5}{2a^2}$$

$$= \frac{7}{2a^2}$$

NOTE

Since we have a common denominator, we simply perform the indicated operations on the numerators.

Check Yourself 1

Perform the indicated operations.

$$\frac{5}{3y^2} + \frac{4}{3y^2} - \frac{7}{3y^2}$$

We always express the sum or difference of rational expressions in simplest form as shown in Example 2.

Example 2 Adding and Subtracting Rational Expressions

Add or subtract as indicated.

(a) $\frac{5x}{x^2 - 9} + \frac{15}{x^2 - 9}$

$= \frac{5x + 15}{x^2 - 9}$ Add the numerators.

$= \frac{5(x + 3)}{(x - 3)(x + 3)} = \frac{5}{x - 3}$ Factor and simplify.

RECALL

Always distribute the negative sign when subtracting polynomials with more than one term.

(b) $\frac{3x + y}{2x} - \frac{x - 3y}{2x} = \frac{(3x + y) - (x - 3y)}{2x}$ Be sure to *enclose the second numerator* in parentheses.

$= \frac{3x + y - x + 3y}{2x}$ Remove the parentheses by *changing each sign.*

$= \frac{2x + 4y}{2x} = \frac{2(x + 2y)}{2x}$ Factor and simplify.

$= \frac{x + 2y}{x}$

Check Yourself 2

Perform the indicated operations.

(a) $\frac{6a}{a^2 - 2a - 8} + \frac{12}{a^2 - 2a - 8}$ **(b)** $\frac{5x - y}{3y} - \frac{2x - 4y}{3y}$

NOTE

By **inspection,** we mean you look at the denominators and find the LCD.

Now, what if our rational expressions *do not* have a common denominator? In that case, we find the least common denominator (LCD). The **least common denominator** is the simplest polynomial that is divisible by each of the individual denominators. Each expression is "built up" to an equivalent expression having that LCD as a denominator. We can then add or subtract as before.

Although in many cases we can find the LCD by inspection, we can state an algorithm for finding the LCD that is similar to the one used in arithmetic.

Step by Step

Finding the Least Common Denominator

Step 1 Write each of the denominators in completely factored form.

Step 2 Write the LCD as the product of each prime factor to the highest power to which it appears in the factored form of any of the individual denominators.

Example 3 illustrates the procedure.

Example 3 Finding the LCD of Two Rational Expressions

< Objective 2 >

Find the LCD of each pair of rational expressions.

(a) $\frac{3}{4x^2}$ and $\frac{5}{6xy}$

NOTE

You may be able to find this LCD by inspecting the numerical coefficients and the variable factors.

Factor the denominators.

$4x^2 = 2^2 \cdot x^2$

$6xy = 2 \cdot 3 \cdot x \cdot y$

The LCD must have the factors

$2^2 \cdot 3 \cdot x^2 \cdot y$

so $12x^2y$ is the LCD.

(b) $\frac{7}{x - 3}$ and $\frac{2}{x + 5}$

NOTE

It is generally best to leave the LCD in factored form.

Here, neither denominator can be factored. The LCD must have the factors $x - 3$ and $x + 5$. The LCD is

$(x - 3)(x + 5)$

Check Yourself 3

Find the LCD of each pair of rational expressions.

(a) $\frac{3}{8a^3}$ and $\frac{5}{6a^2}$ **(b)** $\frac{4}{x+7}$ and $\frac{3}{x-5}$

In Example 4, we see how to apply factoring techniques.

Example 4 Finding the LCD of Two Rational Expressions

Find the LCD of each pair of rational expressions.

(a) $\frac{2}{x^2 - x - 6}$ and $\frac{1}{x^2 - 9}$

> **NOTE**
>
> The LCD must contain *each* of the factors appearing in the original denominators.

Factoring the denominators gives

$x^2 - x - 6 = (x + 2)(x - 3)$

and

$x^2 - 9 = (x + 3)(x - 3)$

The LCD of the denominators is

$(x + 2)(x - 3)(x + 3)$

(b) $\frac{5}{x^2 - 4x + 4}$ and $\frac{3}{x^2 + 2x - 8}$

> **NOTE**
>
> The LCD must contain $(x - 2)^2$ as a factor since $x - 2$ appears *twice* as a factor in the first denominator.

Again, we factor.

$x^2 - 4x + 4 = (x - 2)^2$

$x^2 + 2x - 8 = (x - 2)(x + 4)$

The LCD is

$(x - 2)^2(x + 4)$

Check Yourself 4

Find the LCD of each pair of rational expressions.

(a) $\frac{3}{x^2 - 2x - 15}$ and $\frac{5}{x^2 - 25}$

(b) $\frac{5}{y^2 + 6y + 9}$ and $\frac{3}{y^2 - y - 12}$

In Example 5, we add and subtract unlike rational expressions.

Example 5 Adding and Subtracting Rational Expressions

< Objective 3 >

Add or subtract as indicated.

(a) $\frac{5}{4xy} + \frac{3}{2x^2}$

For the denominators $4xy$ and $2x^2$, the LCD is $4x^2y$. We rewrite each rational expression with the LCD as the denominator.

NOTE

In each case, we are multiplying by 1 $\left(\frac{x}{x} \text{ in the first fraction and } \frac{2y}{2y} \text{ in the second fraction}\right)$, which is why the resulting fractions are equivalent to the original ones.

$$\frac{5}{4xy} + \frac{3}{2x^2} = \frac{5 \cdot x}{4xy \cdot x} + \frac{3 \cdot 2y}{2x^2 \cdot 2y}$$
$$= \frac{5x}{4x^2y} + \frac{6y}{4x^2y} = \frac{5x + 6y}{4x^2y}$$

Multiply the first rational expression by $\frac{x}{x}$ and the second by $\frac{2y}{2y}$ to form the LCD of $4x^2y$.

(b) $\frac{3}{a-3} - \frac{2}{a}$

For the denominators $a - 3$ and a, the LCD is $a(a - 3)$. We rewrite each rational expression with that LCD as the denominator.

$$\frac{3}{a-3} - \frac{2}{a}$$

$$= \frac{3a}{a(a-3)} - \frac{2(a-3)}{a(a-3)}$$

$$= \frac{3a - 2(a-3)}{a(a-3)}$$ Subtract the numerators.

$$= \frac{3a - 2a + 6}{a(a-3)} = \frac{a+6}{a(a-3)}$$ Distribute in the numerator and combine like terms.

Check Yourself 5

Perform the indicated operations.

(a) $\frac{3}{2ab} + \frac{4}{5b^2}$ **(b)** $\frac{5}{y+2} - \frac{3}{y}$

We now proceed to Example 6, where we factor to find the LCD.

Example 6 Adding and Subtracting Rational Expressions

Add or subtract as indicated.

(a) $\frac{-5}{x^2 - 3x - 4} + \frac{8}{x^2 - 16}$

We first factor the two denominators.

$$x^2 - 3x - 4 = (x + 1)(x - 4)$$
$$x^2 - 16 = (x + 4)(x - 4)$$

We see that the LCD must be

$$(x + 1)(x + 4)(x - 4)$$

Rewriting the original expressions with factored denominators gives

NOTE

$$\frac{x+4}{x+4} = 1$$

and $\frac{x+1}{x+1} = 1$

$$\frac{-5}{(x+1)(x-4)} + \frac{8}{(x-4)(x+4)}$$

$$= \frac{-5(x+4)}{(x+1)(x-4)(x+4)} + \frac{8(x+1)}{(x-4)(x+4)(x+1)}$$

$$= \frac{-5(x+4) + 8(x+1)}{(x+1)(x-4)(x+4)}$$ Add the numerators.

$$= \frac{-5x - 20 + 8x + 8}{(x+1)(x-4)(x+4)}$$

$= \dfrac{3x - 12}{(x + 1)(x - 4)(x + 4)}$ Combine like terms in the numerator.

$= \dfrac{3(x - 4)}{(x + 1)(x - 4)(x + 4)}$ Factor.

$= \dfrac{3}{(x + 1)(x + 4)}$ Divide by the common factor $x - 4$.

(b) $\dfrac{5}{x^2 - 5x + 6} - \dfrac{3}{4x - 12}$

Factor the denominators to find the LCD.

$$x^2 - 5x + 6 = (x - 2)(x - 3)$$
$$4x - 12 = 4(x - 3)$$

The LCD is $4(x - 2)(x - 3)$. Proceeding as before, we have

$$\frac{5}{(x - 2)(x - 3)} - \frac{3}{4(x - 3)}$$

$$= \frac{5 \cdot 4}{4(x - 2)(x - 3)} - \frac{3(x - 2)}{4(x - 2)(x - 3)}$$

$= \dfrac{20 - 3(x - 2)}{4(x - 2)(x - 3)}$ Subtract the numerators.

$= \dfrac{20 - 3x + 6}{4(x - 2)(x - 3)}$ Simplify the numerator.

$$= \frac{-3x + 26}{4(x - 2)(x - 3)} = -\frac{3x - 26}{4(x - 2)(x - 3)}$$

Check Yourself 6

Add or subtract as indicated.

(a) $-\dfrac{4}{x^2 - 4} + \dfrac{7}{x^2 - 3x - 10}$ **(b)** $\dfrac{5}{3x - 9} - \dfrac{2}{x^2 - 9}$

Example 7 looks slightly different from those you have seen so far, but the reasoning is the same.

Example 7 Subtracting Rational Expressions

Subtract.

$$3 - \frac{5}{2x - 1}$$

To perform the subtraction, remember that 3 is equivalent to the fraction $\frac{3}{1}$, so

$$3 - \frac{5}{2x - 1} = \frac{3}{1} - \frac{5}{2x - 1}$$

For the denominators 1 and $2x - 1$, the LCD is just $2x - 1$. We now rewrite the first expression with that denominator.

$$3 - \frac{5}{2x - 1} = \frac{3(2x - 1)}{2x - 1} - \frac{5}{2x - 1}$$

$= \dfrac{3(2x - 1) - 5}{2x - 1}$ Subtract the numerators.

$= \dfrac{6x - 8}{2x - 1}$ Simplify the numerator.

Check Yourself 7

Subtract.

$$\frac{4}{3x+1} - 3$$

Example 8 uses an observation from Section 9.1. Recall that

$$a - b = -(b - a)$$

Example 8 Adding and Subtracting Rational Expressions

Add.

$$\frac{x^2}{x-5} + \frac{3x+10}{5-x}$$

Your first thought might be to use a denominator of $(x - 5)(5 - x)$. However, we can simplify our work considerably if we multiply the numerator and denominator of the second fraction by -1 to find a common denominator.

NOTES

Use

$$\frac{-1}{-1} = 1$$

Because

$(-1)(5 - x) = x - 5$

The fractions now have a common denominator, so we can add as before.

$$\frac{x^2}{x-5} + \frac{3x+10}{5-x}$$

$$= \frac{x^2}{x-5} + \frac{(-1)(3x+10)}{(-1)(5-x)}$$

$$= \frac{x^2}{x-5} + \frac{-3x-10}{x-5}$$

$$= \frac{x^2-3x-10}{x-5}$$ Add the numerators.

$$= \frac{(x+2)(x-5)}{x-5}$$ Factor the numerator.

$$= x + 2$$ Simplify.

Check Yourself 8

Add.

$$\frac{x^2}{x-7} + \frac{10x-21}{7-x}$$

We construct an algorithm to follow when adding and subtracting rational expressions.

Step by Step

Adding and Subtracting Rational Expressions

Case 1 Adding and Subtracting Like Rational Expressions

The rational expressions have the same denominator.

Step 1 Add or subtract the numerators.

Step 2 Write the sum or difference over the common denominator.

Step 3 Simplify the result.

Case 2 Adding and Subtracting Unlike Rational Expressions

The rational expressions have different denominators.

Step 1 Find the LCD of the expressions.

Step 2 Write each expression as an equivalent expression with the LCD as the denominator.

Step 3 Add or subtract the numerators.

Step 4 Write the sum or difference over the LCD.

Step 5 Simplify the result.

The sum of two rational functions is always a rational function. Given two rational functions $f(x)$ and $g(x)$, we can rename the sum, so $h(x) = f(x) + g(x)$. This is always true for values of x for which both f and g are defined. So, for example,

$$h(-2) = f(-2) + g(-2)$$

as long as both $f(-2)$ and $g(-2)$ exist.

Example 9 Adding Rational Functions

< Objective 4 >

Consider.

$$f(x) = \frac{3x}{x+5} \quad \text{and} \quad g(x) = \frac{x}{x-4}$$

NOTE

$$f(1) = \frac{3(1)}{(1)+5} = \frac{3}{6} = \frac{1}{2}$$

$$g(1) = \frac{(1)}{(1)-4} = \frac{1}{-3} = -\frac{1}{3}$$

(a) Find $f(1) + g(1)$.

Because $f(1) = \frac{1}{2}$ and $g(1) = -\frac{1}{3}$, we have

$$f(1) + g(1) = \frac{1}{2} + \left(-\frac{1}{3}\right)$$

$$= \frac{3}{6} + \left(-\frac{2}{6}\right) = \frac{1}{6}$$

(b) Find $h(x) = f(x) + g(x)$.

$$h(x) = f(x) + g(x)$$

$$= \frac{3x}{x+5} + \frac{x}{x-4}$$

$$= \frac{3x(x-4) + x(x+5)}{(x+5)(x-4)} = \frac{3x^2 - 12x + x^2 + 5x}{(x+5)(x-4)}$$

$$= \frac{4x^2 - 7x}{(x+5)(x-4)} \qquad x \neq -5, 4$$

(c) Find the ordered pair $(1, h(1))$.

$$h(1) = \frac{-3}{-18} = \frac{1}{6}$$

The ordered pair is $\left(1, \frac{1}{6}\right)$.

Check Yourself 9

Given

$$f(x) = \frac{x}{2x - 5} \quad \text{and} \quad g(x) = \frac{2x}{3x - 1}$$

(a) Find $f(1) + g(1)$. **(b)** Find $h(x) = f(x) + g(x)$.

(c) Find the ordered pair $(1, h(1))$.

When subtracting rational functions, we must be careful with the signs in the numerator.

Example 10 Subtracting Rational Functions

Consider

$$f(x) = \frac{3x}{x + 5} \quad \text{and} \quad g(x) = \frac{x - 2}{x - 4}$$

(a) Find $f(1) - g(1)$.

Because $f(1) = \frac{1}{2}$ and $g(1) = \frac{1}{3}$,

$$\begin{aligned} f(1) - g(1) &= \frac{1}{2} - \frac{1}{3} \\ &= \frac{3}{6} - \frac{2}{6} \\ &= \frac{3 - 2}{6} \\ &= \frac{1}{6} \end{aligned}$$

(b) Find $h(x) = f(x) - g(x)$.

$$\begin{aligned} h(x) &= \frac{3x}{x + 5} - \frac{x - 2}{x - 4} \\ &= \frac{3x(x - 4)}{(x + 5)(x - 4)} - \frac{(x - 2)(x + 5)}{(x - 4)(x + 5)} \\ &= \frac{3x(x - 4) - (x - 2)(x + 5)}{(x + 5)(x - 4)} \\ &= \frac{(3x^2 - 12x) - (x^2 + 3x - 10)}{(x + 5)(x - 4)} \\ &= \frac{2x^2 - 15x + 10}{(x + 5)(x - 4)} \quad x \neq -5, 4 \end{aligned}$$

Subtract numerators.

Combine like terms.

(c) Find the ordered pair $(1, h(1))$.

$$h(1) = \frac{-3}{-18} = \frac{1}{6}$$

The ordered pair is $\left(1, \frac{1}{6}\right)$.

Check Yourself 10

Given

$$f(x) = \frac{x}{2x - 5} \quad \text{and} \quad g(x) = \frac{2x - 1}{3x - 1}$$

(a) Find $f(1) - g(1)$. **(b)** Find $h(x) = f(x) - g(x)$.

(c) Find the ordered pair $(1, h(1))$.

Check Yourself ANSWERS

1. $\frac{2}{3y^2}$ **2. (a)** $\frac{6}{a-4}$; **(b)** $\frac{x+y}{y}$ **3. (a)** $24a^3$; **(b)** $(x+7)(x-5)$

4. (a) $(x-5)(x+5)(x+3)$; **(b)** $(y+3)^2(y-4)$ **5. (a)** $\frac{8a+15b}{10ab^2}$; **(b)** $\frac{2y-6}{y(y+2)}$

6. (a) $\frac{3}{(x-2)(x-5)}$; **(b)** $\frac{5x+9}{3(x+3)(x-3)}$ **7.** $\frac{-9x+1}{3x+1}$ **8.** $x-3$

9. (a) $\frac{2}{3}$; **(b)** $h(x) = \frac{7x^2-11x}{(2x-5)(3x-1)}, x \neq \frac{5}{2}, \frac{1}{3}$; **(c)** $\left(1, \frac{2}{3}\right)$ **10. (a)** $-\frac{5}{6}$;

(b) $h(x) = \frac{-x^2+11x-5}{(2x-5)(3x-1)}, x \neq \frac{5}{2}, \frac{1}{3}$; **(c)** $\left(1, -\frac{5}{6}\right)$

Reading Your Text

These fill-in-the-blank exercises will help you understand some of the key vocabulary used in this section. The answers to these exercises are in the Answers Appendix in the back of the text.

(a) Rational expressions with the same denominator are called __________ rational expressions.

(b) Rational expressions can be simplified if the numerator and denominator have a common __________.

(c) To find the LCD of a pair of rational expressions, first completely __________ the denominators.

(d) When subtracting polynomials, we always __________ the negative sign to every term of the second polynomial.

Skills Calculator/Computer Career Applications Above and Beyond

9.3 exercises

< Objective 1 >

Perform the indicated operations and simplify your results.

1. $\frac{9}{4x^3} + \frac{3}{4x^3}$

2. $\frac{11}{3b^3} - \frac{2}{3b^3}$

3. $\frac{5}{3a+7} + \frac{2}{3a+7}$

4. $\frac{6}{5x+3} - \frac{3}{5x+3}$

5. $\frac{2x}{x-3} - \frac{6}{x-3}$ VIDEO

6. $\frac{6w}{w+4} + \frac{24}{w+4}$

7. $\frac{y^2}{2y+8} + \frac{3y-4}{2y+8}$

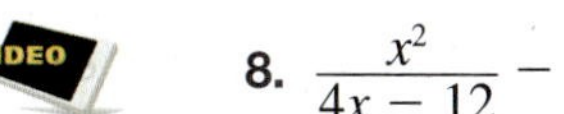

8. $\frac{x^2}{4x-12} - \frac{9}{4x-12}$

9. $\frac{5m-2}{m-6} - \frac{3m+10}{m-6}$

10. $\frac{3b-8}{b-6} + \frac{b-16}{b-6}$

11. $\frac{x-7}{x^2-x-6} + \frac{2x-2}{x^2-x-6}$

12. $\frac{5x-12}{x^2-8x+15} - \frac{3x-2}{x^2-8x+15}$

< Objectives 2 and 3 >

13. $\frac{3}{2x} + \frac{4}{5x}$

14. $\frac{4}{5w} - \frac{3}{4w}$

15. $\frac{6}{a} + \frac{3}{a^2}$

16. $\frac{3}{p} - \frac{7}{p^2}$

17. $\frac{2}{m} - \frac{2}{n}$

18. $\frac{5}{x} + \frac{10}{y}$

19. $\frac{3}{4b^2} - \frac{5}{3b^3}$

20. $\frac{4}{5x^3} - \frac{3}{2x^2}$

21. $\frac{3}{b} - \frac{1}{b-3}$

22. $\frac{4}{c} + \frac{3}{c+1}$

23. $\frac{2}{x+1} + \frac{3}{x+2}$ VIDEO

24. $\frac{4}{y-1} + \frac{2}{y+3}$

25. $\frac{5}{y-3} - \frac{1}{y+1}$

26. $\frac{4}{x+5} - \frac{3}{x-1}$

27. $\frac{3w}{w-6} + \frac{4w}{w-2}$

28. $\frac{3n}{n+5} + \frac{n}{n-4}$

29. $\frac{3x}{3x-2} - \frac{2x}{2x+1}$

30. $\frac{5c}{5c-1} + \frac{2c}{2c-3}$

31. $\frac{5}{x-9} + \frac{4}{9-x}$

32. $\frac{5}{a-5} - \frac{3}{5-a}$

33. $\frac{3}{x^2-16} + \frac{2}{x-4}$

34. $\frac{5}{y^2+5y+6} + \frac{2}{y+2}$

35. $\frac{4m}{m^2-3m+2} - \frac{1}{m-2}$ VIDEO

36. $\frac{x}{x^2-1} - \frac{2}{x-1}$

Evaluate each expression for the variable value(s).

37. $\frac{5x+5}{x^2+3x+2} - \frac{x-3}{x^2+5x-6}, \quad x = -4$

38. $\frac{y-3}{y^2-6y+8} + \frac{2y-6}{y^2-4}, \quad y = 3$

39. $\frac{2m+2n}{m^2-n^2} + \frac{m-2n}{m^2+2mn+n^2}, \quad m = 3, n = 2$

40. $\frac{w-3z}{w^2-2wz+z^2} - \frac{w+2z}{w^2-z^2}, \quad w = 2, z = 1$

41. $\frac{1}{a-3} - \frac{1}{a+3} + \frac{2a}{a^2-9}, \quad a = 4$

42. $\frac{1}{m+1} + \frac{1}{m-3} - \frac{4}{m^2-2m-3}, \quad m = -2$

43. $\frac{3w^2+16w-8}{w^2+2w-8} + \frac{w}{w+4} - \frac{w-1}{w-2}, \quad w = 3$

44. $\frac{4x^2-7x-45}{x^2-6x+5} - \frac{x+2}{x-1} - \frac{x}{x-5}, \quad x = -3$

45. $\frac{a^2-9}{2a^2-5a-3} \cdot \left(\frac{1}{a-2} + \frac{1}{a+3}\right), \quad a = -3$

46. $\frac{m^2+2mn+n^2}{m^2+2mn-3n^2} \cdot \left(\frac{2}{m-n} - \frac{1}{m+n}\right), \quad m = 4, n = -3$

< Objective 4 >

Find **(a)** $f(1) + g(1)$; **(b)** $h(x) = f(x) + g(x)$; *and* **(c)** *the ordered pair* $(1, h(1))$.

47. $f(x) = \frac{3x}{x+1}$ and $g(x) = \frac{2x}{x-3}$

48. $f(x) = \frac{4x}{x-4}$ and $g(x) = \frac{x+4}{x+1}$

49. $f(x) = \frac{x}{x+1}$ and $g(x) = \frac{1}{x^2+2x+1}$

50. $f(x) = \frac{x+2}{x-4}$ and $g(x) = \frac{x+3}{x+4}$

Find **(a)** $f(1) - g(1)$; **(b)** $h(x) = f(x) - g(x)$; *and* **(c)** *the ordered pair* $(1, h(1))$.

51. $f(x) = \dfrac{x + 5}{x - 5}$ and $g(x) = \dfrac{x - 5}{x + 5}$

52. $f(x) = \dfrac{2x}{x - 4}$ and $g(x) = \dfrac{3x}{x + 7}$

53. $f(x) = \dfrac{x + 9}{4x - 36}$ and $g(x) = \dfrac{x - 9}{x^2 - 18x + 81}$

54. $f(x) = \dfrac{4x + 1}{x + 5}$ and $g(x) = -\dfrac{2}{x}$,

55. GEOMETRY Find the perimeter of the figure.

56. GEOMETRY Find the perimeter of the figure.

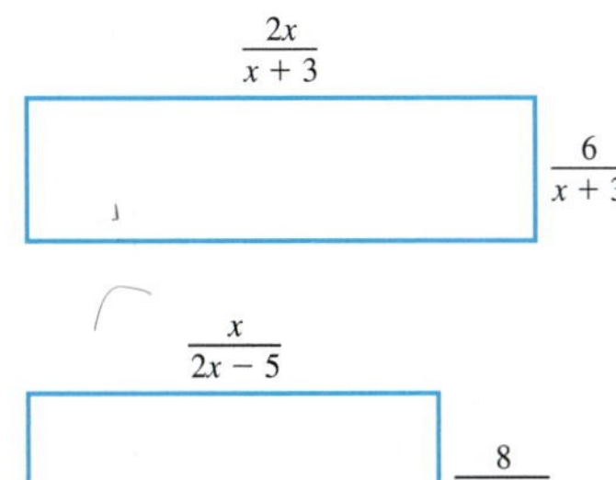

57. NUMBER PROBLEM Represent the sum of the reciprocals of two consecutive even integers.

58. NUMBER PROBLEM One number is two less than another. Represent the sum of the reciprocals of the two numbers.

59. GEOMETRY Find the perimeter of the figure.

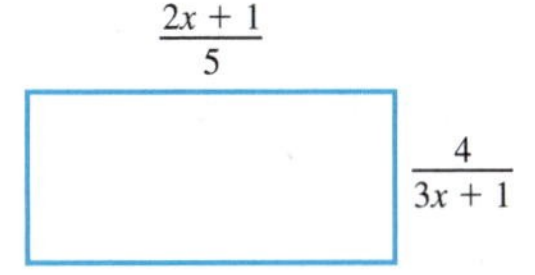

60. GEOMETRY Find the perimeter of the figure.

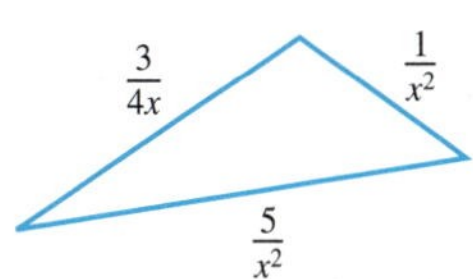

Determine whether each statement is **true** *or* **false.**

61. The expression $a - b$ is the opposite of $b - a$.

62. The expression $a - b$ is the opposite of $a + b$.

63. We find the GCF of the denominators in order to add unlike rational expressions.

64. When adding like rational expressions, add the denominators and place the sum under the common numerator.

Complete each statement with **always, sometimes,** *or* **never.**

65. When adding rational expressions, we __________ make sure the denominators are the same.

66. When multiplying rational expressions, we __________ have to ensure that the denominators are the same.

67. We __________ find the LCD of two rational expressions by taking the product of their denominators.

68. To add rational expressions, we find the sum of the numerators and __________ find the sum of the denominators.

You learned to use a graphing calculator to check your results when multiplying and dividing rational expressions in the exercise set of Section 9.2. As you might expect, you can also use a graphing calculator to check your work when adding and subtracting rational expressions.

Simply enter the sum $Y_1 + Y_2$ or the difference $Y_1 - Y_2$ into Y_3 rather than the product or quotient. Otherwise, everything remains the same.

Complete exercises 69 to 74 in the usual manner and then use a graphing calculator to check your work.

69. $\dfrac{6y}{y^2 - 8y + 15} + \dfrac{9}{y - 3}$

70. $\dfrac{8a}{a^2 - 8a + 12} + \dfrac{4}{a - 2}$

71. $\dfrac{6x}{x^2 - 10x + 24} - \dfrac{18}{x - 6}$

72. $\dfrac{21p}{p^2 - 3p - 10} - \dfrac{15}{p - 5}$

73. $\dfrac{2}{z^2 - 4} + \dfrac{3}{z^2 + 2z - 8}$

74. $\dfrac{5}{x^2 - 3x - 10} + \dfrac{2}{x^2 - 25}$

Answers

1. $\frac{3}{x^3}$ **3.** $\frac{7}{3a+7}$ **5.** 2 **7.** $\frac{y-1}{2}$ **9.** 2 **11.** $\frac{3}{x+2}$ **13.** $\frac{23}{10x}$ **15.** $\frac{3(2a+1)}{a^2}$ **17.** $\frac{2(n-m)}{mn}$ **19.** $\frac{9b-20}{12b^3}$ **21.** $\frac{2b-9}{b(b-3)}$ **23.** $\frac{5x+7}{(x+1)(x+2)}$ **25.** $\frac{4(y+2)}{(y-3)(y+1)}$ **27.** $\frac{w(7w-30)}{(w-6)(w-2)}$ **29.** $\frac{7x}{(3x-2)(2x+1)}$ **31.** $\frac{1}{x-9}$ **33.** $\frac{2x+11}{(x+4)(x-4)}$ **35.** $\frac{3m+1}{(m-1)(m-2)}$ **37.** $-\frac{16}{5}$ **39.** $\frac{49}{25}$ **41.** 2 **43.** 8 **45.** Undefined **47.** (a) $\frac{1}{2}$; (b) $h(x) = \frac{5x^2-7x}{(x+1)(x-3)}$, $x \neq -1, 3$; (c) $\left(1, \frac{1}{2}\right)$ **49.** (a) $\frac{3}{4}$; (b) $h(x) = \frac{x^2+x+1}{(x+1)^2}$, $x \neq -1$; (c) $\left(1, \frac{3}{4}\right)$ **51.** (a) $-\frac{5}{6}$; (b) $h(x) = \frac{20x}{(x-5)(x+5)}$, $x \neq 5, -5$; (c) $\left(1, -\frac{5}{6}\right)$ **53.** (a) $-\frac{3}{16}$; (b) $h(x) = \frac{x+5}{4(x-9)}$, $x \neq 9$; (c) $\left(1, -\frac{3}{16}\right)$ **55.** 4 **57.** $\frac{2(x+1)}{x(x+2)}$ **59.** $\frac{2(6x^2+5x+21)}{5(3x+1)}$ **61.** True **63.** False **65.** always **67.** sometimes **69.** $\frac{15}{y-5}$ **71.** $-\frac{12}{x-4}$ **73.** $\frac{5z+14}{(z+2)(z-2)(z+4)}$

9.4 Complex Rational Expressions

< 9.4 Objectives >

1 > Use the fundamental priniciple to simplify complex rational expressions

2 > Use division to simplify complex rational expressions

3 > Use complex rational expressions to solve applications

We have two different methods to simplify complex fractions. A **complex fraction** or **complex rational expression** is a rational expression that also contains nontrivial rational expressions in the numerator or denominator. In other words, there is at least one fraction within the fraction.

Some examples of complex fractions are

$$\frac{\frac{5}{6}}{\frac{3}{4}} \qquad \frac{\frac{4}{x}}{\frac{3}{x+1}} \qquad \frac{1+\frac{1}{x}}{1-\frac{1}{x}}$$

RECALL

Fundamental principle:

$\frac{P}{Q} = \frac{PR}{QR}$

if $Q \neq 0$ and $R \neq 0$.

Two methods can be used to simplify complex fractions. Method 1 involves the fundamental principle, and Method 2 uses division techniques.

Recall that by the *fundamental principle* we can always multiply the numerator and denominator of a fraction by the same nonzero quantity. In simplifying a complex fraction, we multiply the numerator and denominator by the LCD of all fractions that appear within the complex fraction.

Here is an example with denominators 5 and 10, so we write

NOTE

We are multiplying by $\frac{10}{10}$ or 1.

$$\frac{\frac{3}{5}}{\frac{7}{10}} = \frac{\frac{3}{5}\cdot 10}{\frac{7}{10}\cdot 10} = \frac{6}{7}$$

Our second approach interprets the complex fraction as division and applies our earlier work in dividing fractions in which we *invert and multiply*.

$$\frac{\frac{3}{5}}{\frac{7}{10}} = \frac{3}{5} \div \frac{7}{10} = \frac{3}{5}\cdot\frac{10}{7} = \frac{6}{7}$$ Invert and multiply.

Which method is better? The answer depends on the expression you are trying to simplify. Both approaches are effective and you should be familiar with both. With practice, you will be able to tell which method may be easier to use in a particular situation.

Consider the two methods as we apply them to a complex fraction.

Example 1 Simplifying Complex Fractions

< Objectives 1 and 2 >

Simplify $\frac{\frac{3}{4}}{\frac{5}{8}}$.

Method 1 The fundamental principle

The LCD of $\frac{3}{4}$ and $\frac{5}{8}$ is 8. So multiply the numerator and denominator by 8.

$$\frac{\frac{3}{4}}{\frac{5}{8}} = \frac{\frac{3}{4} \cdot 8}{\frac{5}{8} \cdot 8} = \frac{3 \cdot 2}{5 \cdot 1} = \frac{6}{5}$$

Method 2 Division

We treat the fraction bar as division.

$$\frac{\frac{3}{4}}{\frac{5}{8}} = \frac{3}{4} \div \frac{5}{8} = \frac{3}{4} \times \frac{8}{5} = \frac{6}{5}$$

Check Yourself 1

(a) Use the fundamental principle to simplify $\frac{\frac{4}{7}}{\frac{3}{7}}$.

(b) Use division to simplify $\frac{\frac{3}{8}}{\frac{5}{6}}$.

We use the same methods when there are variables in the expression. Consider Example 2.

Example 2 Simplifying Complex Rational Expressions

Simplify $\frac{\frac{5}{x}}{\frac{10}{x^2}}$.

Method 1 The fundamental principle

The LCD of $\frac{5}{x}$ and $\frac{10}{x^2}$ is x^2, so multiply the numerator and denominator by x^2.

NOTE

Be sure to write the result in simplest form.

$$\frac{\frac{5}{x}}{\frac{10}{x^2}} = \frac{\left(\frac{5}{x}\right)x^2}{\left(\frac{10}{x^2}\right)x^2} = \frac{5x}{10} = \frac{x}{2}$$

Method 2 Division

$$\frac{\frac{5}{x}}{\frac{10}{x^2}} = \frac{5}{x} \div \frac{10}{x^2} = \frac{5}{x} \cdot \frac{x^2}{10} = \frac{x}{2}$$

Check Yourself 2

(a) Use the fundamental principle to simplify $\frac{\frac{6}{x^3}}{\frac{9}{x^2}}$.

(b) Use division to simplify $\frac{\frac{m^4}{15}}{\frac{m^3}{20}}$.

We may also have a sum or a difference in the numerator or denominator of a complex fraction. The simplification steps are exactly the same. Consider Example 3.

Example 3 Simplifying Complex Rational Expressions

Simplify.

$$\frac{1 + \frac{2x}{y}}{2 - \frac{x}{y}}$$

Method 1 The LCD of 1, $\frac{2x}{y}$, 2, and $\frac{x}{y}$ is y. So we multiply the numerator and denominator by y.

$$\frac{1 + \frac{2x}{y}}{2 - \frac{x}{y}} = \frac{\left(1 + \frac{2x}{y}\right) \cdot y}{\left(2 - \frac{x}{y}\right) \cdot y}$$

$$= \frac{1 \cdot y + \frac{2x}{y} \cdot y}{2 \cdot y - \frac{x}{y} \cdot y}$$ Distribute y over the numerator and denominator.

$$= \frac{y + 2x}{2y - x}$$ Simplify.

Method 2 In this approach, we must *first work separately* in the numerator and denominator to form single fractions.

NOTE

Make sure you understand the steps in forming single fractions in the numerator and denominator.

$$\frac{1 + \frac{2x}{y}}{2 - \frac{x}{y}} = \frac{\frac{y}{y} + \frac{2x}{y}}{\frac{2y}{y} - \frac{x}{y}} = \frac{\frac{y + 2x}{y}}{\frac{2y - x}{y}}$$

$$= \frac{y + 2x}{y} \cdot \frac{y}{2y - x}$$ Invert the divisor and multiply.

$$= \frac{y + 2x}{2y - x}$$

Check Yourself 3

Simplify.

$$\frac{\frac{x}{y} - 1}{\frac{2x}{y} + 2}$$

This algorithm summarizes our work with complex fractions.

Step by Step

Simplifying Complex Rational Expressions

Method 1

Step 1 Multiply the numerator and denominator of the expression by the LCD of all the fractions that appear within the numerator and denominator.

Step 2 Simplify the resulting rational expression.

Method 2

Step 1 Write the numerator and denominator of the complex fraction as single fractions, if necessary.

Step 2 Invert the denominator and multiply. Simplify your result.

Again, simplifying a complex rational expression means writing an equivalent simple fraction in lowest terms, as Example 4 illustrates.

Example 4 Simplifying Complex Rational Expressions

Simplify.

$$\frac{1 - \frac{2y}{x} + \frac{y^2}{x^2}}{1 - \frac{y^2}{x^2}}$$

We choose the first method of simplification in this case. The LCD of all the fractions that appear is x^2. So we multiply the numerator and denominator by x^2.

$$\frac{1 - \frac{2y}{x} + \frac{y^2}{x^2}}{1 - \frac{y^2}{x^2}} = \frac{\left(1 - \frac{2y}{x} + \frac{y^2}{x^2}\right) \cdot x^2}{\left(1 - \frac{y^2}{x^2}\right) \cdot x^2}$$

$$= \frac{x^2 - 2xy + y^2}{x^2 - y^2}$$ Distribute x^2 over the numerator and denominator.

$$= \frac{(x - y)(x - y)}{(x + y)(x - y)} = \frac{x - y}{x + y}$$ Factor and simplify.

Check Yourself 4

Simplify.

$$\frac{1 + \frac{5}{x} + \frac{6}{x^2}}{1 - \frac{9}{x^2}}$$

In Example 5, we illustrate the second method of simplification for purposes of comparison.

Example 5 Simplifying Complex Rational Expressions

NOTES

Take time to make sure you understand how the numerator and denominator are rewritten as single fractions.

Method 2 is probably the more efficient in this case. The LCD of the denominators would be $(x + 2)(x - 1)$, leading to a more complicated process if we use Method 1.

Simplify.

$$\frac{1 - \frac{1}{x + 2}}{x - \frac{2}{x - 1}}$$

$$\frac{1 - \frac{1}{x + 2}}{x - \frac{2}{x - 1}} = \frac{\frac{x + 2}{x + 2} - \frac{1}{x + 2}}{\frac{x(x - 1)}{x - 1} - \frac{2}{x - 1}} = \frac{\frac{x + 2 - 1}{x + 2}}{\frac{x(x - 1) - 2}{x - 1}} = \frac{\frac{x + 1}{x + 2}}{\frac{x^2 - x - 2}{x - 1}}$$

$$= \frac{x + 1}{x + 2} \cdot \frac{x - 1}{x^2 - x - 2} = \frac{\cancel{x + 1}^{1}}{x + 2} \cdot \frac{x - 1}{(x - 2)\cancel{(x + 1)}_{1}} = \frac{x - 1}{(x + 2)(x - 2)}$$

Check Yourself 5

Simplify.

$$\frac{2 + \frac{5}{x - 3}}{x - \frac{1}{2x + 1}}$$

Complex fractions can show up when we solve certain problems known as "work" problems.

Example 6 Solving a Work Problem

< Objective 3 >

> **NOTE**
>
> We study work problems more in Section 9.6.

If one person can install a storm door in u hours, and a second person can do the same installation in v hours, then the time required to install a door working together is given by the expression

$$\frac{1}{\frac{1}{u} + \frac{1}{v}}$$

Simplify this expression.

Using Method 1, we multiply the numerator and denominator by uv:

$$\frac{1}{\frac{1}{u} + \frac{1}{v}} = \frac{1 \cdot uv}{\left(\frac{1}{u} + \frac{1}{v}\right) \cdot uv}$$

$$= \frac{uv}{\left(\frac{1}{u}\right)uv + \left(\frac{1}{v}\right)uv}$$ Distribute the multiplication of uv in the denominator.

$$= \frac{uv}{v + u}$$

Check Yourself 6

Simplify.

$$\frac{5}{\frac{1}{t} - \frac{1}{u}}$$

You may recall the expression

$$\frac{f(x + h) - f(x)}{h}$$

from earlier exercises. This expression is called the *difference quotient of a function*. It is very important in later math classes. When $f(x)$ is a rational expression, the difference quotient is a complex rational expression. Consider Example 7.

Example 7 Simplifying a Difference Quotient

Let $f(x) = \frac{2}{x}$.

Simplify the expression $\frac{f(x + h) - f(x)}{h}$.

First we evaluate

$$f(x+h) = \frac{2}{x+h}$$

So,

$$\frac{f(x+h) - f(x)}{h} = \frac{\frac{2}{x+h} - \frac{2}{x}}{h}$$

Using Method 1, we multiply the numerator and denominator by $x(x + h)$.

$$\frac{\left(\frac{2}{x+h} - \frac{2}{x}\right) \cdot [x(x+h)]}{h \cdot [x(x+h)]}$$

$$= \frac{\left(\frac{2}{x+h}\right) \cdot [x(x+h)] - \left(\frac{2}{x}\right) \cdot [x(x+h)]}{h \cdot [x(x+h)]}$$ Distribute $x(x + h)$ in the numerator.

$$= \frac{\frac{(2)(x)(x+h)}{x+h} - \frac{(2)(x)(x+h)}{x}}{hx(x+h)}$$ Multiply in the numerator.

> **NOTE**
>
> Try using Method 2. You should get the same result.

$$= \frac{2x - 2(x+h)}{hx(x+h)}$$ Cancel common factors in the numerator.

$$= \frac{2x - 2x - 2h}{hx(x+h)} = \frac{-2h}{hx(x+h)} = -\frac{2}{x(x+h)}$$ Simplify the numerator, and divide by the common factor h.

Check Yourself 7

Let $f(x) = \frac{3}{x}$.

Simplify the expression $\frac{f(x+h) - f(x)}{h}$.

Check Yourself ANSWERS

1. **(a)** $\frac{4}{3}$; **(b)** $\frac{9}{20}$ **2.** **(a)** $\frac{2}{3x}$; **(b)** $\frac{4m}{3}$ **3.** $\frac{x - y}{2(x + y)}$ **4.** $\frac{x + 2}{x - 3}$ **5.** $\frac{2x + 1}{(x - 3)(x + 1)}$ **6.** $\frac{5tu}{u - t}$ **7.** $-\frac{3}{x(x + h)}$

Reading Your Text

These fill-in-the-blank exercises will help you understand some of the key vocabulary used in this section. The answers to these exercises are in the Answers Appendix in the back of the text.

(a) A ________ fraction is a fraction that has a fraction in its numerator or denominator (or both).

(b) By the ________ principle we can always multiply the numerator and denominator of a fraction by the same nonzero quantity.

(c) When dividing fractions, we ________ the second fraction and multiply.

(d) No matter which method we use, the final step requires that we write the result in ________ terms.

< Objectives 1 and 2 >

Simplify each complex rational expression.

1. $\dfrac{\frac{2}{3}}{\frac{3}{4}}$

2. $\dfrac{\frac{5}{6}}{\frac{2}{3}}$

3. $\dfrac{\frac{3}{4}+\frac{1}{3}}{\frac{1}{6}+\frac{3}{4}}$

4. $\dfrac{\frac{3}{4}+\frac{1}{2}}{\frac{7}{8}-\frac{1}{4}}$ VIDEO

5. $\dfrac{2+\frac{1}{3}}{3-\frac{1}{5}}$

6. $\dfrac{1+\frac{3}{4}}{2-\frac{1}{8}}$

7. $\dfrac{\frac{x}{8}}{\frac{x^2}{4}}$

8. $\dfrac{\frac{x^2}{12}}{\frac{x^5}{18}}$

9. $\dfrac{\frac{3}{m}}{\frac{6}{m^2}}$

10. $\dfrac{\frac{15}{x^2}}{\frac{20}{x^3}}$

11. $\dfrac{\frac{y+1}{y}}{\frac{y-1}{2y}}$

12. $\dfrac{\frac{x+3}{4x}}{\frac{x-3}{2x}}$

13. $\dfrac{\frac{a+2b}{3a}}{\frac{a^2+2ab}{9b}}$

14. $\dfrac{\frac{m-3n}{4m}}{\frac{m^2-3mn}{8n}}$

15. $\dfrac{\frac{x-3}{x^2-25}}{\frac{x^2+x-12}{x^2+5x}}$

16. $\dfrac{\frac{x+5}{x^2-6x}}{\frac{x^2-25}{x^2-36}}$ VIDEO

17. $\dfrac{2-\frac{1}{x}}{2+\frac{1}{x}}$

18. $\dfrac{3+\frac{1}{b}}{3-\frac{1}{b}}$

19. $\dfrac{\frac{1}{x}-\frac{1}{y}}{\frac{1}{xy}}$

20. $\dfrac{\frac{4}{xy}}{\frac{1}{y}-\frac{1}{x}}$

21. $\dfrac{\frac{x^2}{y^2}-1}{\frac{x}{y}+1}$

22. $\dfrac{\frac{m}{n}+2}{\frac{m^2}{n^2}-4}$

23. $\dfrac{1+\frac{3}{a}-\frac{4}{a^2}}{1+\frac{2}{a}-\frac{3}{a^2}}$ VIDEO

24. $\dfrac{1-\frac{2}{x}-\frac{8}{x^2}}{1-\frac{1}{x}-\frac{6}{x^2}}$

25. $\dfrac{\frac{x^2}{y}+2x+y}{\frac{1}{y^2}-\frac{1}{x^2}}$

26. $\dfrac{\frac{a}{b}+1-\frac{2b}{a}}{\frac{1}{b^2}-\frac{4}{a^2}}$

27. $\dfrac{2-\frac{2}{x+1}}{2+\frac{2}{x+1}}$

28. $\dfrac{3-\frac{4}{m+2}}{3+\frac{4}{m+2}}$

29. $\dfrac{1-\frac{1}{y-1}}{y-\frac{8}{y+2}}$

30. $\dfrac{1+\frac{1}{x+2}}{x-\frac{18}{x-3}}$

31. $\dfrac{\frac{1}{x-3}+\frac{1}{x+3}}{\frac{1}{x-3}-\frac{1}{x+3}}$

32. $\dfrac{\frac{2}{m-2}+\frac{1}{m-3}}{\frac{2}{m-2}-\frac{1}{m-3}}$

33. $\dfrac{\frac{x}{x+1}+\frac{1}{x-1}}{\frac{x}{x-1}-\frac{1}{x+1}}$ VIDEO

34. $\dfrac{\frac{y}{y-4}+\frac{1}{y+2}}{\frac{4}{y-4}-\frac{1}{y+2}}$

35. $\dfrac{\frac{a+1}{a-1}-\frac{a-1}{a+1}}{\frac{a+1}{a-1}+\frac{a-1}{a+1}}$

36. $\dfrac{\frac{x+2}{x-2}-\frac{x-2}{x+2}}{\frac{x+2}{x-2}+\frac{x-2}{x+2}}$

37. $1+\dfrac{1}{1+\frac{1}{x}}$

38. $1+\dfrac{1}{1-\frac{1}{y}}$

39. $1+\dfrac{1}{1+\dfrac{1}{1+\frac{1}{x}}}$

40. **(a)** Extend the "continued fraction" patterns in exercises 37 and 39 to write the next complex fraction.

(b) Simplify the complex fraction in part (a).

For each function, find and simplify the expression $\dfrac{f(x+h)-f(x)}{h}$.

41. $f(x)=\frac{5}{x}$

42. $f(x)=\frac{1}{2x}$

43. $f(x)=-\frac{1}{x}$

44. $f(x)=-\frac{3}{x}$

< Objective 3 >

Solve each application.

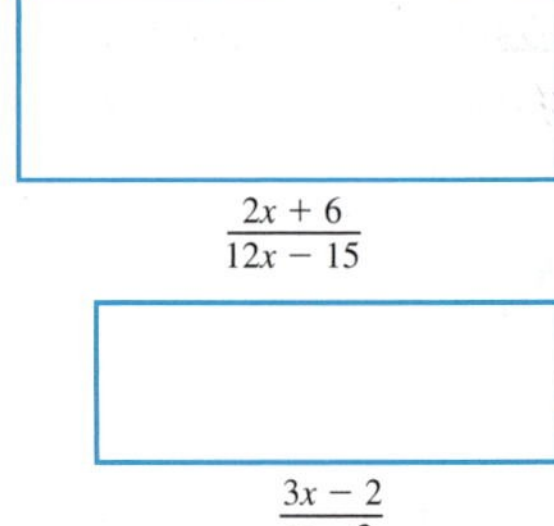

45. **Geometry** The area of the rectangle shown is $\frac{2}{3}$. Find the width.

46. **Geometry** The area of the rectangle shown is $\frac{2(3x-2)}{x-1}$. Find the width.

Determine whether each statement is **true** *or* **false.**

47. We can always rewrite a complex fraction as a simple fraction.

48. The complex fraction $\frac{2}{\frac{3}{4}}$ is the same as the complex fraction $\frac{\frac{2}{3}}{4}$.

Skills | **Calculator/Computer** | Career Applications | Above and Beyond

Use the table utility on a graphing calculator to complete each table. Comment on the equivalence of the two expressions. chapter 9 > Make the Connection

49.

x	−3	−2	−1	0	1	2	3
$\dfrac{1-\frac{2}{x}}{1-\frac{4}{x^2}}$							
$\dfrac{x}{x+2}$							

50.

x	−3	−2	−1	0	1	2	3
$\dfrac{-8+\frac{20}{x}}{4-\frac{25}{x^2}}$							
$-\dfrac{4x}{2x+5}$							

Skills | Calculator/Computer | **Career Applications** | Above and Beyond

Allied Health *Total compliance (C_T) for a patient is based on lung compliance (C_L) and chest-wall compliance (C_{CW}). It is measured in centimeters of water (cm H_2O) and computed using the formula*

$$C_T = \frac{1}{\frac{1}{C_L} + \frac{1}{C_{CW}}}$$

Use this formula to complete exercises 51 and 52.

51. Simplify the total compliance formula. 

52. Determine the total compliance for a patient whose lung compliance is 0.15 cm H_2O and whose chest-wall compliance is 0.20 cm H_2O. Round your result to three decimal places.

Electrical Engineering *Generally, the wiring in buildings is arranged so all electric devices are in parallel. This way, if one device is disconnected, the current to the other devices is not interrupted. The equivalent single resistance (measured in ohms, Ω) for two devices connected in parallel (see figure) is given by the formula*

$$R_{eq} = (R_1^{-1} + R_2^{-1})^{-1}$$

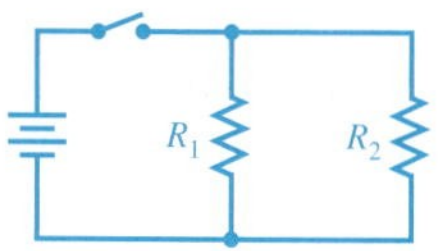

Use the formula to complete exercises 53 and 54.

53. Write the resistance formula without exponents. Simplify the resistance formula.

54. Use the resistance formula to determine the equivalent total resistance of the parallel circuit shown. Report your result to the nearest ohm.

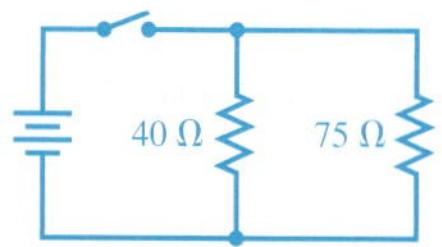

Skills | Calculator/Computer | Career Applications | **Above and Beyond**

Suppose you drive at 40 *mi/hr from city A to city B. You then return along the same route from city B to city A at* 50 *mi/hr. What is your average rate for the round trip? Your obvious guess would be* 45 *mi/hr, but you are in for a surprise.*

Suppose that the cities are 200 *mi apart. Your time from city A to city B is the distance divided by the rate, or*

$$\frac{200 \text{ mi}}{40 \text{ mi/hr}} = 5 \text{ hr}$$

Similarly, your time from city B to city A is

$$\frac{200 \text{ mi}}{50 \text{ mi/hr}} = 4 \text{ hr}$$

The total time is then 9 hr, and now using rate equals distance divided by time, we have

$$\frac{400 \text{ mi}}{9 \text{ hr}} = \frac{400}{9} \text{ mi/hr} = 44\tfrac{4}{9} \text{ mi/hr}$$

Note that the rate for the round trip is independent of the distance involved. For instance, try the previous computations if cities A and B are 400 *mi apart.*

The answer to the problem is the complex fraction

$$R = \frac{2}{\frac{1}{R_1} + \frac{1}{R_2}}$$

where R_1 = rate going
R_2 = rate returning
R = rate for round trip

Use this information to solve exercises 55 to 58.

55. Verify that if $R_1 = 40$ mi/hr and $R_2 = 50$ mi/hr, then $R = 44\frac{4}{9}$ mi/hr, by simplifying the complex fraction *after* substituting those values.

56. Simplify the given complex fraction first. *Then* substitute 40 for R_1 and 50 for R_2 to calculate R.

57. Repeat exercise 55, where $R_1 = 50$ mi/hr and $R_2 = 60$ mi/hr.

58. Use the procedure in exercise 56 with the above values for R_1 and R_2.

59. Outline the two different methods used to simplify a complex rational expression. What are the advantages of each method? chapter 9 > Make the Connection

60. Does the expression $\frac{x^2 + y^2}{x + y}$ simplify to $\frac{x^2}{x} + \frac{y^2}{y}$? If not, can it be simplified? 

61. Write and simplify the complex expression that is the reciprocal of $x + \frac{6}{x - 1}$.

62. Let $f(x) = \frac{3}{x}$. Write and simplify the complex expression whose numerator is $f(3 + h) - f(3)$ and whose denominator is h.

63. Write and simplify a complex expression that is the mean (or average) of $\frac{1}{x}$ and $\frac{1}{x - 1}$.

64. Use the Internet to find the first six Fibonnacci numbers. Compare this set of numbers to your results in exercises 37, 39, and 40.

65. Mathematicians have shown that there are situations in which the method used to determine the number of U.S. representatives each state gets may not be fair, and a state may not get its basic quota of representatives. They give the table below of a hypothetical seven states and their populations as an example.

State	Population	Exact Quota	Rounded Quota	Actual Number of Reps.
A	325	1.625	2	2
B	788	3.940	4	4
C	548	2.740	3	3
D	562	2.810	3	3
E	4,263	21.315	21	21
F	3,219	16.095	16	15
G	295	1.475	1	2
Total	10,000	50		50

In this case, the total population of all states is 10,000, and there are 50 representatives in all, so there should be no more than 10,000/50, or 200, people per representative. The quotas are found by dividing the population by 200. Whether a state A should get an additional representative before another state E should get one is decided in this method by using the simplified inequality below. For each state, we find the ratio of the state's population to the square root of the product of the actual number of representatives (a and e, in this case) and one more than this integer. We then compare these ratios for each two states. If

$$\frac{A}{\sqrt{a(a + 1)}} > \frac{E}{\sqrt{e(e + 1)}}$$

is true, then A gets an extra representative before E does.

(a) If you go through the process of comparing the inequality for each pair of states, state F loses a representative to state G. Do you see how this happens? Will state F complain?

(b) Alexander Hamilton, one of the signers of the Constitution, proposed that the extra representative positions be given one at a time to states with the largest remainder until all the "extra" positions were filled. How would this affect the table? Do you agree or disagree?

66. In Italy in the 1500s, Pietro Antonio Cataldi expressed square roots as infinite continued fractions. It is not a difficult process to follow. For instance, if you want the square root of 5, then let

$$x + 1 = \sqrt{5}$$

Squaring both sides gives

$$(x + 1)^2 = 5 \qquad \text{or} \qquad x^2 + 2x + 1 = 5$$

which can be written

$$x(x + 2) = 4$$

$$x = \frac{4}{x + 2}$$

One can continue replacing the x on the right with $\frac{4}{2+x}$.

$$x = \cfrac{4}{2 + \cfrac{4}{2 + \cfrac{4}{2 + \cfrac{4}{2 + \cdots}}}}$$

to obtain

$\sqrt{5} - 1$

(a) Evaluate the complex fraction above (ignore the three dots) and then add 1, and see how close it is to the square root of 5. What should you put where the ellipsis (. . .) is? Try a number you think is close to $\sqrt{5} - 1$. How far would you have to go to get the square root correct to the nearest hundredth?

(b) Develop an infinite complex fraction for $\sqrt{10} - 1$.

67. There are many other interesting patterns associated with complex fractions. Work with a partner to evaluate each complex fraction in the sequence below. This is an interesting sequence of fractions because the numerators and denominators are a famous sequence of whole numbers, and the fractions get closer and closer to a number called "the golden mean."

$$1, \quad 1 + \frac{1}{1}, \quad 1 + \cfrac{1}{1 + \cfrac{1}{1}}, \quad 1 + \cfrac{1}{1 + \cfrac{1}{1 + \cfrac{1}{1}}}, \quad 1 + \cfrac{1}{1 + \cfrac{1}{1 + \cfrac{1}{1 + \cfrac{1}{1}}}}, \ldots$$

After you have evaluated these first five, you no doubt will see a pattern in the resulting fractions that allows you to go on indefinitely without having to evaluate more complex fractions. Write each of these fractions as decimals. Write your observations about the sequence of fractions and about the sequence of decimal fractions.

68. Here is yet another method for simplifying a complex fraction. Suppose we want to simplify

$$\cfrac{\frac{3}{5}}{\frac{7}{10}}$$

Multiply the numerator and denominator of the complex fraction by $\frac{10}{7}$.

(a) What principle allows you to do this?

(b) Why was $\frac{10}{7}$ chosen?

(c) When learning to divide fractions, you may have heard the saying "Yours is not to reason why . . . just invert and multiply." How does this method serve to explain the "reason why" we invert and multiply?

chapter 9 >Make the Connection

Answers

1. $\frac{8}{9}$ **3.** $\frac{13}{11}$ **5.** $\frac{5}{6}$ **7.** $\frac{1}{2x}$ **9.** $\frac{m}{2}$ **11.** $\frac{2(y+1)}{y-1}$ **13.** $\frac{3b}{a^2}$ **15.** $\frac{x}{(x-5)(x+4)}$ **17.** $\frac{2x-1}{2x+1}$ **19.** $y - x$ **21.** $\frac{x-y}{y}$ **23.** $\frac{a+4}{a+3}$ **25.** $\frac{x^2y(x+y)}{x-y}$ **27.** $\frac{x}{x+2}$ **29.** $\frac{y+2}{(y-1)(y+4)}$ **31.** $\frac{x}{3}$ **33.** 1 **35.** $\frac{2a}{a^2+1}$ **37.** $\frac{2x+1}{x+1}$ **39.** $\frac{3x+2}{2x+1}$ **41.** $-\frac{5}{x(x+h)}$ **43.** $\frac{1}{x(x+h)}$ **45.** $\frac{4x-5}{x+3}$ **47.** True

49. The expressions are almost equivalent.

x	-3	-2	-1	0	1	2	3
$\dfrac{1-\frac{2}{x}}{1-\frac{4}{x^2}}$	3	Error	-1	Error	0.3333	Error	0.6
$\dfrac{x}{x+2}$	3	Error	-1	0	0.3333	0.5	0.6

51. $C_T = \frac{C_L C_{CW}}{C_L + C_{CW}}$ **53.** $R_{eq} = \frac{R_1 R_2}{R_1 + R_2}$ **55.** Above and Beyond **57.** $54\frac{6}{11}$ mi/hr **59.** Above and Beyond **61.** $\frac{x-1}{x^2-x+6}$ **63.** $\frac{2x-1}{2x(x-1)}$ **65.** Above and Beyond **67.** Above and Beyond

9.5 Graphing Rational Functions

< 9.5 Objectives >

1 > Find the domain and intercepts of a rational function

2 > Graph a rational function

3 > Find the asymptotes of a rational function

RECALL

The graph of a linear function is always a straight line.

The graph of a quadratic function is always a parabola.

You already know how to graph linear and quadratic functions. One feature of such functions is that their graphs are predictable. The graph of a rational function is not so straightforward. Instead, they vary widely.

By confining our study to a simpler subgroup of rational functions, we can identify some consistent patterns and characteristics. In this introduction to graphing rational functions, we do just that.

We begin by focusing on the domain of a rational function and consider any intercepts of its graph.

To remind you, the domain of a rational function consists of all real numbers except those that make the denominator 0. To find those numbers, we set the denominator equal to 0 and solve. We then exclude any solutions from the domain.

RECALL

A fraction equals 0 if, and only if, its numerator equals 0.

To find any y-intercept, evaluate $f(0)$. If there is a y-intercept, it has the coordinates $(0, f(0))$. If 0 is not in the domain, then there is no y-intercept.

Any x-intercepts are found by solving $f(x) = 0$. In the case of a rational function, this only happens when the numerator is 0, so we set the numerator equal to 0 and solve.

Example 1 — Finding the Domain and Intercepts of a Rational Function

< Objective 1 >

Find the domain and intercept(s) of each function.

(a) $f(x) = \dfrac{8}{x}$

Domain
Set the denominator equal to 0 and solve.

$x = 0$

The only solution is 0, so the domain is all real numbers except 0.

$\{x \mid x \neq 0\}$

RECALL

A function can have at most one y-intercept.

y-intercept
To find any y-intercept, we evaluate $f(0)$. Since 0 is not in the domain, this function has no y-intercept.

x-intercepts
To find any x-intercepts, we set the numerator equal to 0 and solve.

$8 = 0$

This equation has no solutions, so there are no x-intercepts.

(b) $f(x) = \dfrac{6}{x - 2}$

Domain
Set the denominator equal to 0 and solve.

$$x - 2 = 0$$
$$x = 2$$

The only solution is 2, so the domain is all real numbers except 2.

$\{x \mid x \neq 2\}$

y-intercept
To find any y-intercept, we evaluate $f(0)$.

$$f(0) = \frac{6}{(0) - 2} = \frac{6}{-2}$$
$$= -3$$

Therefore the y-intercept is $(0, -3)$.

x-intercepts
To find any x-intercepts, we set the numerator equal to 0 and solve.

$$6 = 0$$

This equation has no solutions, so there are no x-intercepts.

Check Yourself 1

Find the domain and intercepts of each function.

(a) $f(x) = \dfrac{5}{x}$ **(b)** $f(x) = -\dfrac{3}{x + 2}$

As we said, a fraction equals 0 if, and only if, the numerator equals 0. In Example 1, neither numerator could equal 0. When searching for x-intercepts, we need only determine when the numerator has the value 0.

Let us now consider a slightly more complicated function.

Example 2 Finding the Domain and Intercepts of a Rational Function

Find the domain and intercept(s) of $f(x) = \dfrac{x - 5}{x - 3}$.

Domain
Set the denominator equal to 0 and solve.

$$x - 3 = 0$$
$$x = 3$$

The only solution is 3, so the domain is all real numbers except 3.

$\{x \mid x \neq 3\}$

y-intercept

To find any y-intercept, we evaluate $f(0)$.

$$f(0) = \frac{(0) - 5}{(0) - 3} = \frac{-5}{-3}$$
$$= \frac{5}{3}$$

Therefore, the y-intercept is $\left(0, \frac{5}{3}\right)$.

x-intercepts

To find any x-intercepts, we set the numerator equal to 0 and solve.

$$x - 5 = 0$$
$$x = 5$$

The only solution is 5, so the only x-intercept is (5, 0).

Check Yourself 2

Find the domain and intercept(s) of $f(x) = \frac{x + 4}{x - 2}$.

We turn our attention now to the graph of a rational function.

Example 3 Graphing a Rational Function

< Objective 2 >

Graph $f(x) = \frac{8}{x}$.

From Example 1, we know the function is undefined at $x = 0$ and that there are no intercepts.

Making a table of some easy-to-compute points, we have

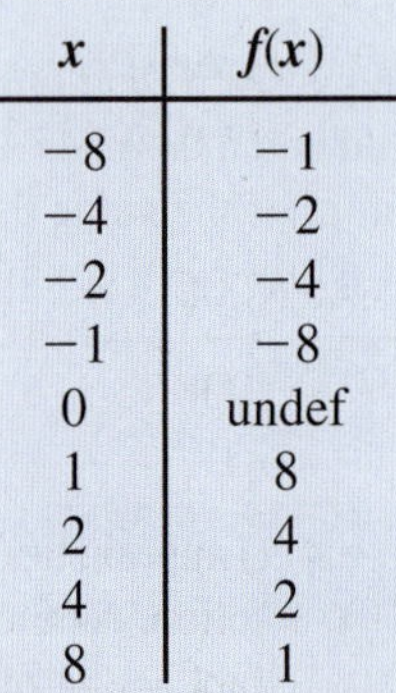

x	$f(x)$
−8	−1
−4	−2
−2	−4
−1	−8
0	undef
1	8
2	4
4	2
8	1

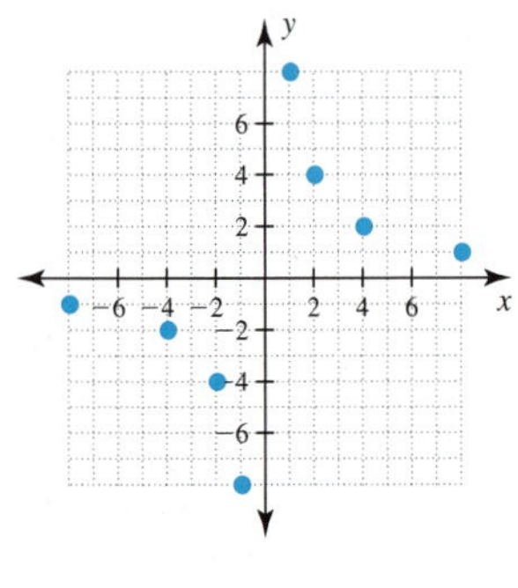

RECALL

We select convenient values for x (left column) and evaluate the function at each of those values to complete the second column.

Do you see why we chose only even numbers for x?

Some of our choices give

$x = -8$

$$f(-8) = \frac{8}{(-8)} = -1;$$

$x = -2$

$$f(-2) = \frac{8}{(-2)} = -4;$$

$x = 4$

$$f(4) = \frac{8}{(4)} = 2$$

We connect the points on each side of 0 with smooth curves. Since the graph does not pass through $x = 0$, the result is two separate and distinct curves.

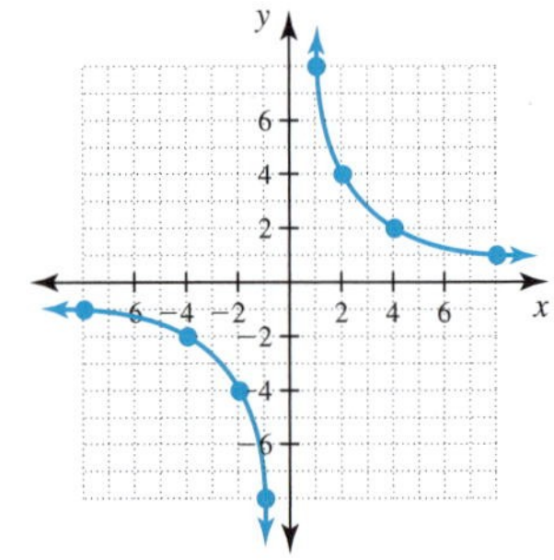

Check Yourself 3

Graph $f(x) = -\frac{6}{x}$.

Example 3 highlights two questions.

1. How does the graph behave near the undefined location ($x = 0$, in this case)?
2. How does the graph behave far to the left or right (that is, for x-values with a large magnitude)?

To answer the first question, we look at what happens when we choose x-values close to 0.

x	1	0.1	0.01	0.001	0.0001
$f(x)$	8	80	800	8,000	80,000

We see that as x gets closer to 0, $f(x)$ gets *large*. This is true of the function $f(x) = \frac{8}{x}$ because x is the denominator. As the denominator of a fraction gets smaller, the fraction gets larger.

When we repeat this exercise by taking x-values close to 0 from the left or negative side, we get

x	-1	-0.1	-0.01	-0.001	-0.0001
$f(x)$	-8	-80	-800	$-8,000$	$-80,000$

Again, we see that as x gets closer to 0, the magnitude of $f(x)$ gets *large*, this time in the negative direction.

Overall, we see that from both sides of 0, the magnitude of $f(x)$ gets infinitely large and the graph approaches the y-axis (in different directions), but never reaches it. The y-axis is given by the vertical line $x = 0$ and the graph of f approaches this line. We call this line a **vertical asymptote.**

In order to determine what happens to the graph as we move farther out to the left or right, we build tables of values in which the magnitude of x gets large.

x	10	100	1,000	10,000	1,000,000
$f(x)$	0.8	0.08	0.008	0.0008	0.000008

We see that as x gets large, $f(x)$ approaches 0. This makes sense in light of the fact that x is the denominator of the fraction. As the denominator of a fraction gets larger, the fraction gets closer to zero. Graphically, the function approaches the x-axis from above.

Next, we see what happens as x moves in the negative direction.

x	-10	-100	$-1,000$	$-10,000$	$-1,000,000$
$f(x)$	-0.8	-0.08	-0.008	-0.0008	-0.000008

The table shows that as the magnitude of x grows large in the negative direction, $f(x)$ approaches 0 from the negative side. This is equivalent to the graph approaching the x-axis from below.

As the magnitude of x grows large in either direction, the graph approaches the x-axis, which is given by the line $y = 0$. When a function approaches a horizontal line this way, we call the line a **horizontal asymptote.** In this case, $y = 0$ is the horizontal asymptote of the function.

Example 4 Finding the Asymptotes of a Rational Function

< Objective 3 >

Identify the asymptotes of each rational function.

(a) $f(x) = \dfrac{2}{x + 3}$

Vertical Asymptote(s)

Any vertical asymptotes would occur where the denominator is 0, so we set the denominator equal to 0 and solve.

$$x + 3 = 0$$
$$x = -3$$

NOTE

The graph may have a "hole" instead of a vertical asymptote at $x = -3$. You will learn more about them in a future math class.

The only solution is $x = -3$. We already know that this means the domain of the function is given by $\{x \mid x \neq -3\}$. Whether we have a vertical asymptote or some other phenomenon remains to be seen.

We can use a graphing calculator to quickly build a table of values or graph and investigate the behavior of the function near $x = -3$.

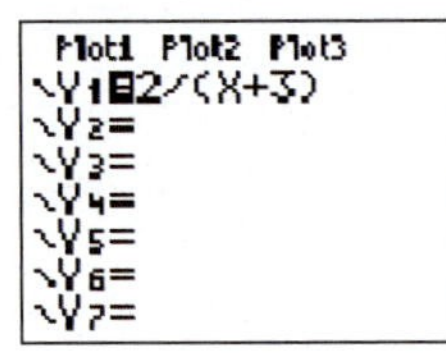

Use parentheses to contain the denominator.

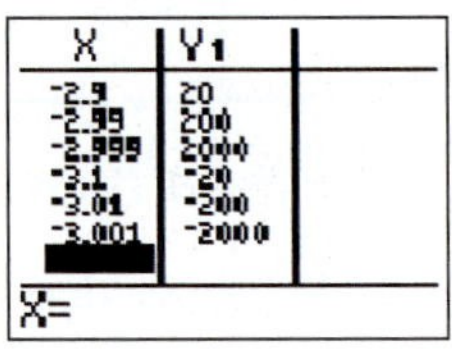

The function gets large as x gets close to -3.

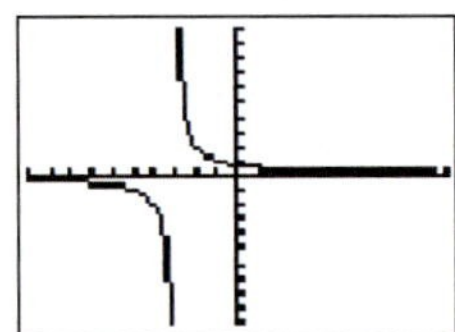

The graph indicates a vertical asymptote at $x = -3$.

Based on this evidence, we conclude that $x = -3$ is a vertical asymptote.

Horizontal Asymptote(s)

To find any horizontal asymptotes, we look at what happens when the magnitude of x increases. From the graph, it looks like there might be a horizontal asymptote at $y = 0$ (the x-axis).

RECALL

Calculators use scientific notation to display numbers too small or large to fit in their display.

$1\text{E}6 = 1 \times 10^6 = 1{,}000{,}000$

$2\text{E}-6 = 2 \times 10^{-6} = 0.000002$

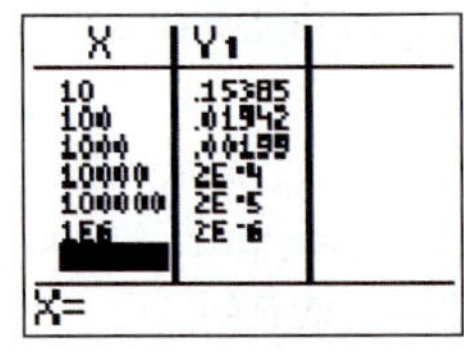

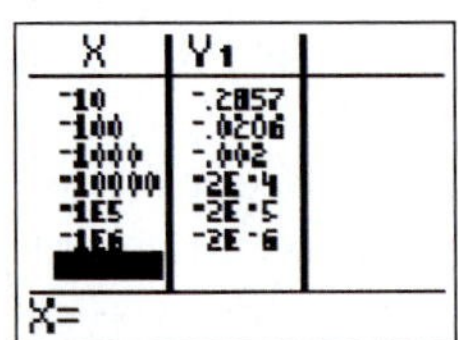

Since the function approaches 0 as the magnitude of x gets large, we conclude that $y = 0$ is a horizontal asymptote.

(b) $f(x) = \dfrac{x + 4}{x - 1}$

Vertical Asymptote(s)

Set the denominator equal to 0 and solve.

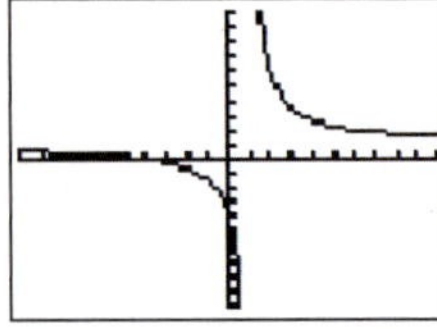

$$x - 1 = 0$$
$$x = 1$$

Any vertical asymptote would have to occur at $x = 1$. Looking at the graph, we see that this is exactly what occurs.

Horizontal Asymptote(s)

From the graph, we see that the left side crosses the x-axis, so $y = 0$ is probably not a horizontal asymptote. However, the graph does show *asymptotic* behavior just above the x-axis.

When we look at what happens as the magnitude of x increases, we see that the function approaches 1 and conclude that $y = 1$ is a horizontal asymptote.

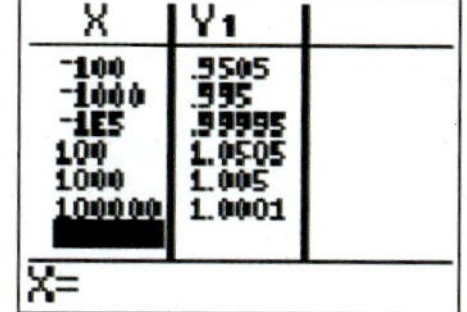

X	Y1
-100	.9505
-1000	.995
-1E5	.99995
100	1.0505
1000	1.005
100000	1.0001

X=

Check Yourself 4

Identify the asymptotes of each rational function.

(a) $f(x) = \dfrac{4}{x+2}$ **(b)** $f(x) = \dfrac{x+5}{x+1}$

Property

Vertical and Horizontal Asymptotes

A **vertical asymptote** is a vertical line that the graph of a function approaches, but does not touch. Its equation is of the form

$x = c$

As x-values are chosen close to c, output values grow infinitely large (positive or negative).

A **horizontal asymptote** is a horizontal line that the graph of a function approaches, for large positive or negative values of x. Its equation is of the form

$y = b$

As large positive or negative x-values are chosen, output values get closer to b.

We use these steps to put together a sketch of a rational function f.

Step by Step

Graphing Rational Functions

Step 1 Determine the domain of f.

Step 2 Find the intercepts of the graph of f. Plot these.

Step 3 Locate vertical and horizontal asymptotes. Draw these as dashed lines.

Step 4 Choose a few easy-to-compute points to plot.

Step 5 Connect plotted points with a smooth curve, allowing the curve to approach the asymptotes.

Example 5 Graphing Rational Functions

Graph each function.

(a) $f(x) = \dfrac{4}{x+2}$

The domain of f is $\{x \mid x \neq -2\}$.

Since $f(0) = 2$, the y-intercept is $(0, 2)$. There are no x-intercepts.

There is a vertical asymptote with equation $x = -2$.

There is a horizontal asymptote with equation $y = 0$.

So far, we have

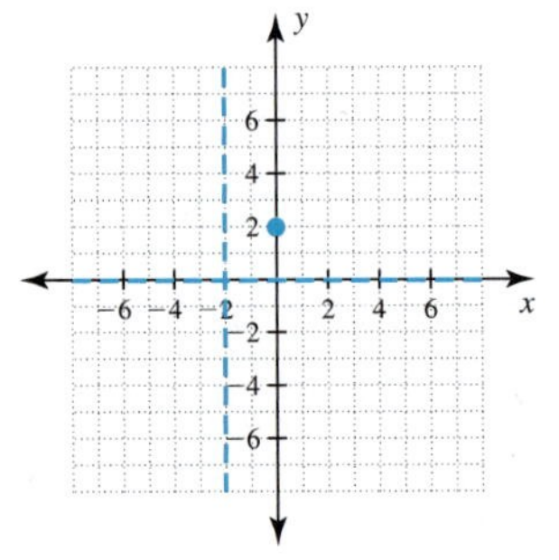

We choose convenient x-values and find some points.

x	-6	-4	-3	-1	2	6
$f(x)$	-1	-2	-4	4	1	0.5

Plotting these, and connecting with a smooth curve gives

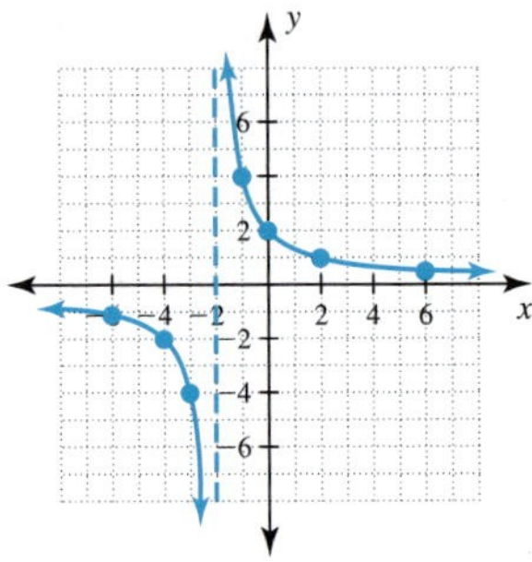

(b) $f(x) = \dfrac{x+5}{x+1}$

The domain of f is $\{x \mid x \neq -1\}$. Since $f(0) = 5$, the y-intercept is $(0, 5)$. The x-intercept is $(-5, 0)$. We have a vertical asymptote at $x = -1$. We have a horizontal asymptote at $y = 1$.

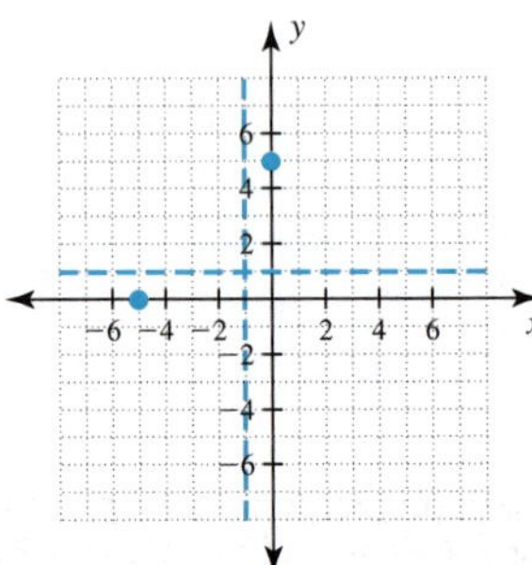

Next choose convenient x-values.

x	-9	-3	-2	1	3	7
$f(x)$	0.5	-1	-3	3	2	1.5

RECALL

You can use a graphing calculator's TABLE utility to find "nice" points.

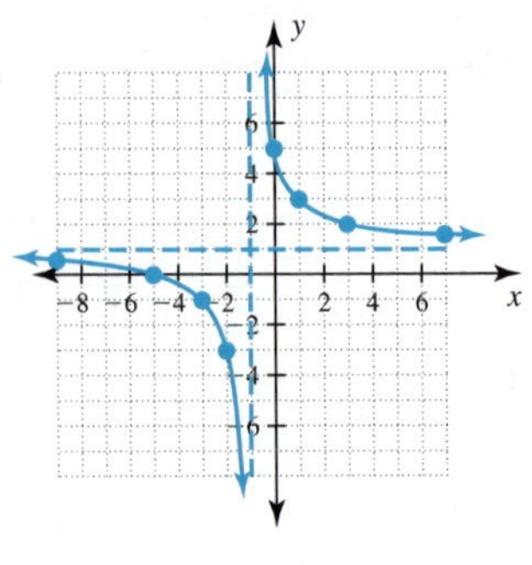

(c) $f(x) = \dfrac{2x - 6}{x + 2}$

The domain of f is $\{x \mid x \neq -2\}$. Since $f(0) = -3$, the y-intercept is $(0, -3)$. To find an x-intercept, set $2x - 6$ equal to 0 and solve.

$$2x - 6 = 0$$
$$x = 3$$

So $(3, 0)$ is an x-intercept.

There is a vertical asymptote with equation $x = -2$.

Letting x get "huge," we find that output values approach 2. So there is a horizontal asymptote with equation $y = 2$.

We choose convenient values for x to plot some additional points.

x	−7	−6	−4	−1	2	8
$f(x)$	4	4.5	7	−8	−0.5	1

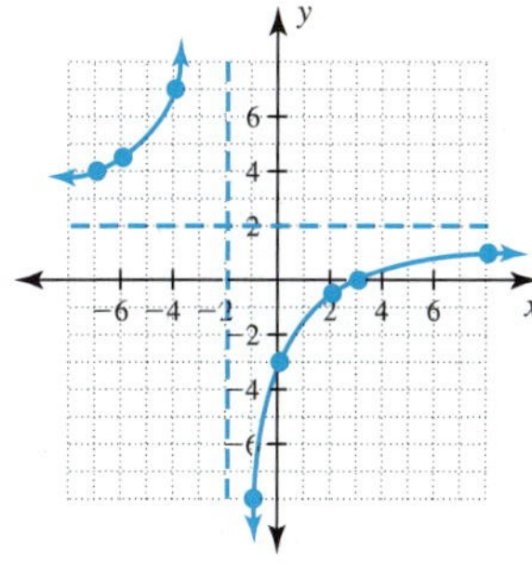

Check Yourself 5

Graph $f(x) = \dfrac{3 - x}{x - 2}$.

If you look back over the examples of this section, you see that each function has been one of two forms:

1. $f(x) = \dfrac{a}{x - c}$ (assuming $a \neq 0$)

2. $f(x) = \dfrac{a(x - b)}{x - c}$ (assuming $a \neq 0$, $b \neq c$)

Before studying the box below, try to establish on your own the intercepts and asymptotes for each of the two forms.

Property

Intercepts and Asymptotes of Rational Functions

For the rational function $f(x) = \dfrac{a}{x - c}$, where $a \neq 0$, there are no x-intercepts, and the y-intercept is $\left(0, -\dfrac{a}{c}\right)$ if $c \neq 0$.

There is a vertical asymptote with equation $x = c$.

There is a horizontal asymptote with equation $y = 0$.

For the rational function $f(x) = \dfrac{a(x - b)}{x - c}$, where $a \neq 0$ and $b \neq c$:

The y-intercept is $\left(0, \dfrac{ab}{c}\right)$ if $c \neq 0$.

The x-intercept is $(b, 0)$.

There is a vertical asymptote with equation $x = c$.

There is a horizontal asymptote with equation $y = a$.

In future courses, you will have the opportunity to study the graphs of rational functions to a greater extent. Those covered in this section should provide you with a good introduction.

Check Yourself ANSWERS

1. **(a)** Domain: $\{x \mid x \neq 0\}$; no intercepts; **(b)** domain: $\{x \mid x \neq -2\}$; y-intercept: $\left(0, -\frac{3}{2}\right)$; no x-intercept
2. Domain: $\{x \mid x \neq 2\}$; y-intercept: $(0, -2)$; x-intercept: $(-4, 0)$
3. 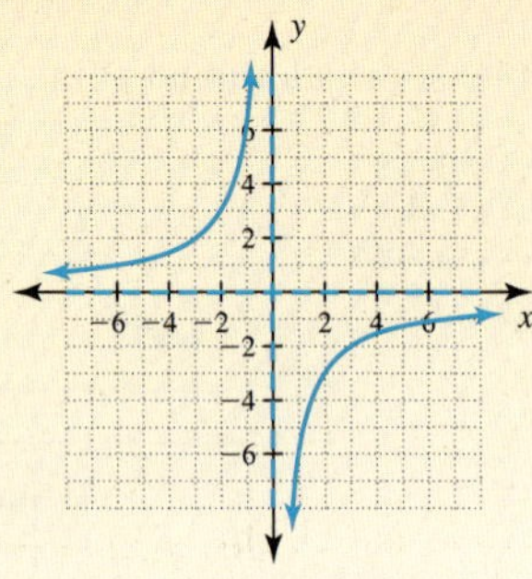
4. **(a)** Vertical asymptote: $x = -2$; horizontal asymptote: $y = 0$; **(b)** vertical asymptote: $x = -1$; horizontal asymptote: $y = 1$
5. 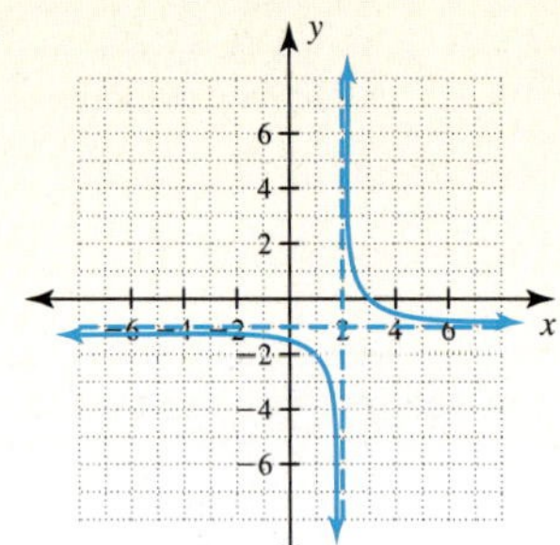

Reading Your Text

These fill-in-the-blank exercises will help you understand some of the key vocabulary used in this section. The answers to these exercises are in the Answers Appendix in the back of the text.

(a) To find the ________, we input 0 for x and find the resulting output value.

(b) To find the domain, focus on the ________.

(c) To search for ________ asymptotes, we focus on the denominator.

(d) To find ________ asymptotes, we choose large input values of x.

< Objective 1 >

Find the domain and any intercepts of each function.

1. $f(x) = \frac{7}{x}$

2. $f(x) = -\frac{5}{x}$

3. $f(x) = \frac{3}{x - 9}$

4. $f(x) = \frac{12}{x - 3}$

5. $f(x) = \frac{8}{2 - x}$

6. $f(x) = \frac{6}{3 - x}$

7. $f(x) = \frac{x - 6}{x - 4}$

8. $f(x) = \frac{x + 9}{x + 3}$

9. $f(x) = \frac{2 - x}{x - 4}$

10. $f(x) = \frac{10 - x}{x + 2}$

11. $f(x) = \frac{2x - 10}{x + 4}$

12. $f(x) = \frac{3x + 9}{x - 1}$

13. $f(x) = \frac{2 - 4x}{x - 3}$

14. $f(x) = \frac{3 - 5x}{x + 1}$

< Objective 3 >

Identify the asymptotes of each function.

15. $f(x) = \frac{7}{x}$

16. $f(x) = -\frac{5}{x}$

17. $f(x) = \frac{3}{x - 9}$

18. $f(x) = \frac{12}{x - 3}$

19. $f(x) = \frac{8}{2 - x}$

20. $f(x) = \frac{6}{3 - x}$

21. $f(x) = \frac{x - 6}{x - 4}$

22. $f(x) = \frac{x + 9}{x + 3}$

23. $f(x) = \frac{2 - x}{x - 4}$

24. $f(x) = \frac{10 - x}{x + 2}$

25. $f(x) = \frac{2x - 10}{x + 4}$

26. $f(x) = \frac{3x + 9}{x - 1}$

27. $f(x) = \frac{2 - 4x}{x - 3}$

28. $f(x) = \frac{3 - 5x}{x + 1}$

< Objective 2 >

Graph each function.

29. $f(x) = \frac{6}{x}$

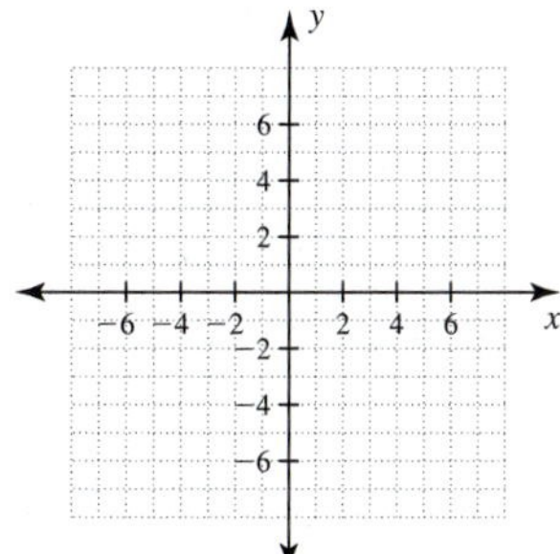

30. $f(x) = \frac{12}{x}$

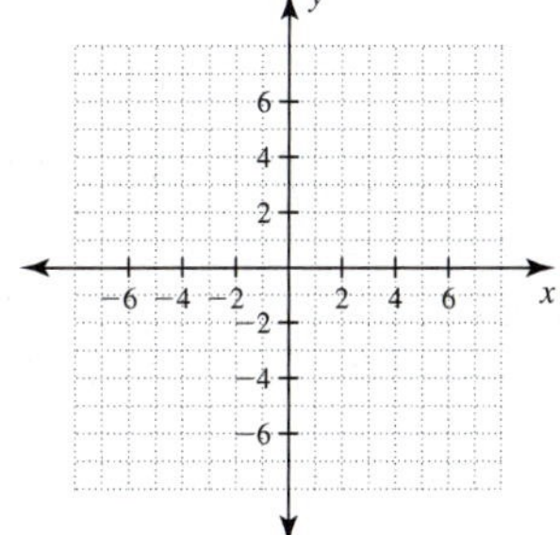

31. $f(x) = -\frac{8}{x}$

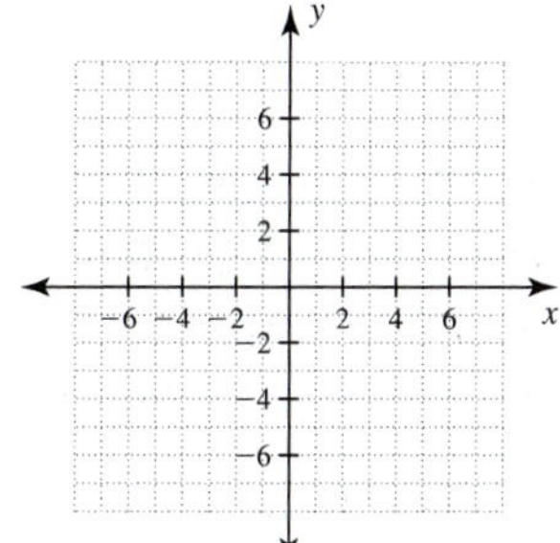

32. $f(x) = -\frac{10}{x}$

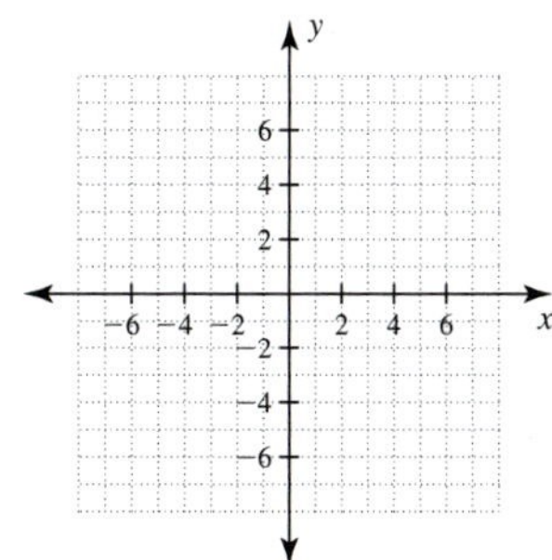

33. $f(x) = \frac{10}{x - 2}$

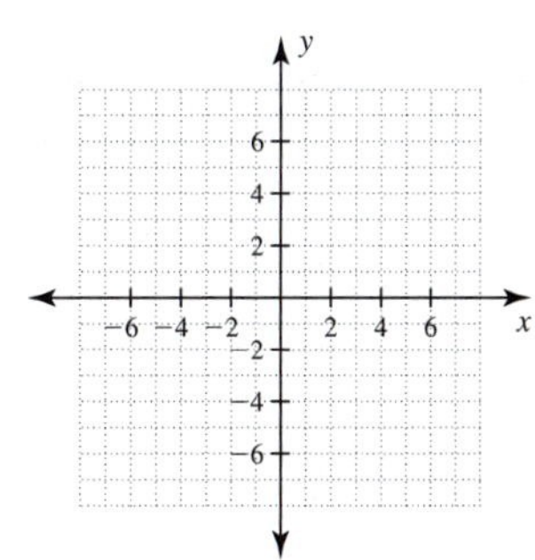

34. $f(x) = \frac{12}{x - 3}$

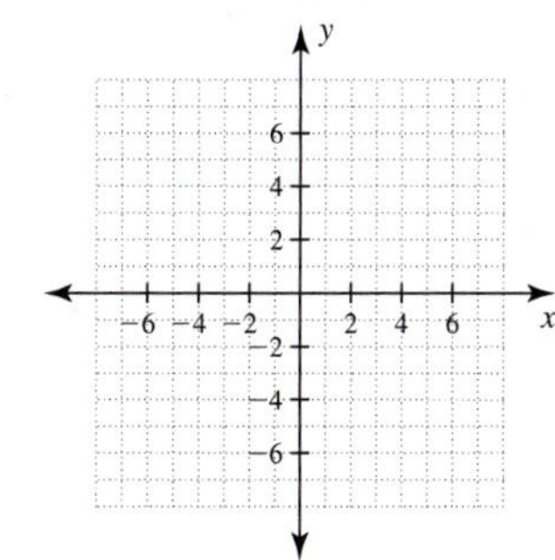

35. $f(x) = \dfrac{6}{3 - x}$

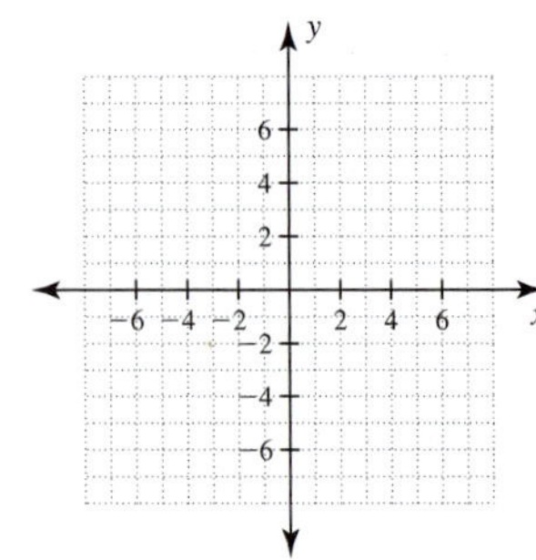

36. $f(x) = \dfrac{8}{2 - x}$

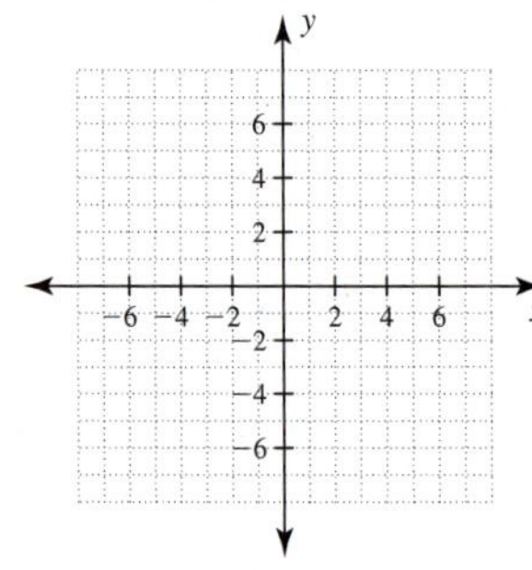

37. $f(x) = \dfrac{x + 5}{x - 2}$

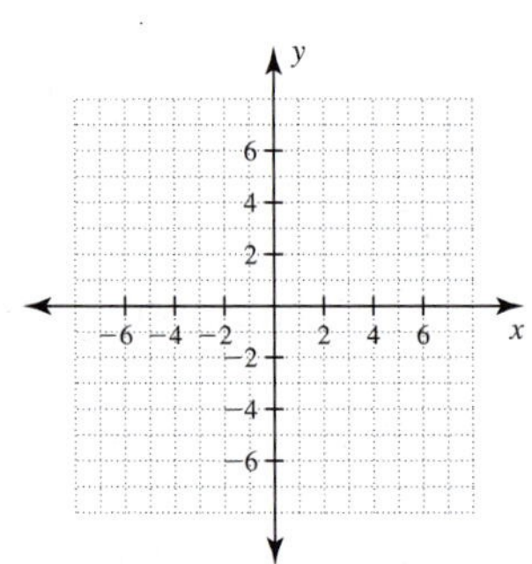

38. $f(x) = \dfrac{x - 6}{x - 1}$

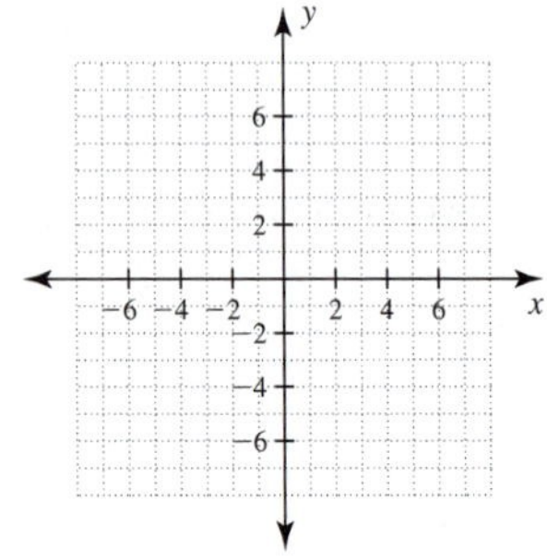

39. $f(x) = \dfrac{4 - x}{x + 2}$

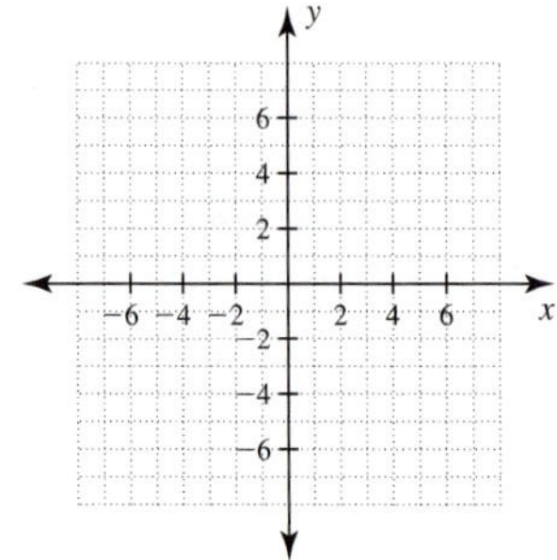

40. $f(x) = \dfrac{5 - x}{x - 1}$

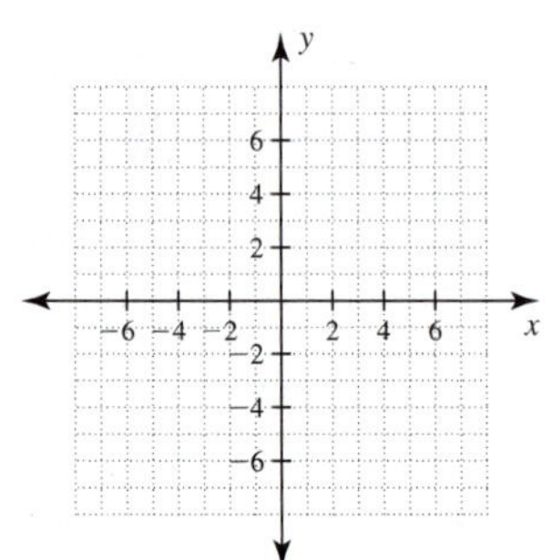

41. $f(x) = \dfrac{2x - 6}{x + 4}$

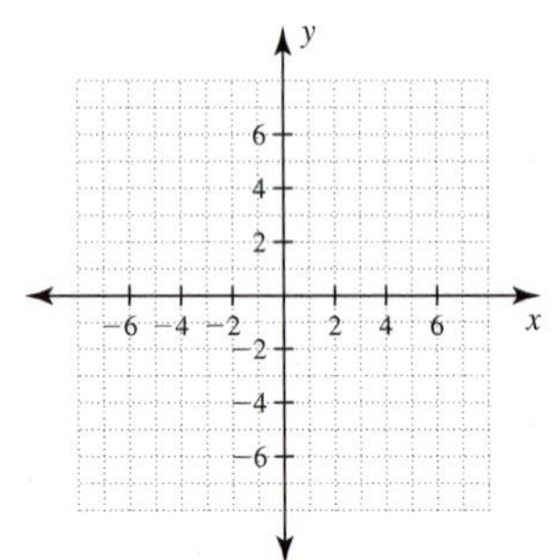

42. $f(x) = \dfrac{3x + 9}{x - 1}$

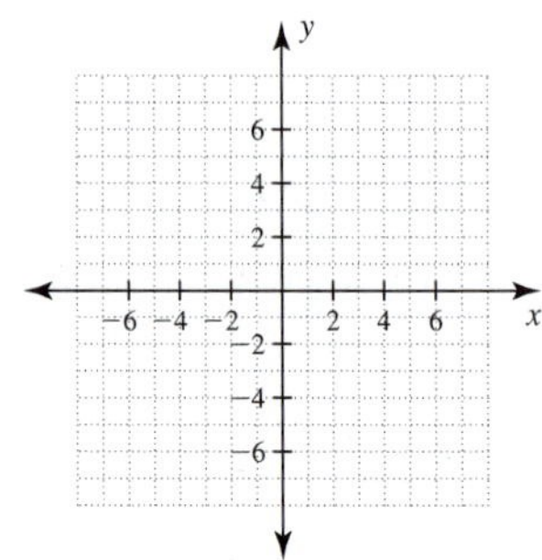

43. $f(x) = \dfrac{2 - 4x}{x - 3}$

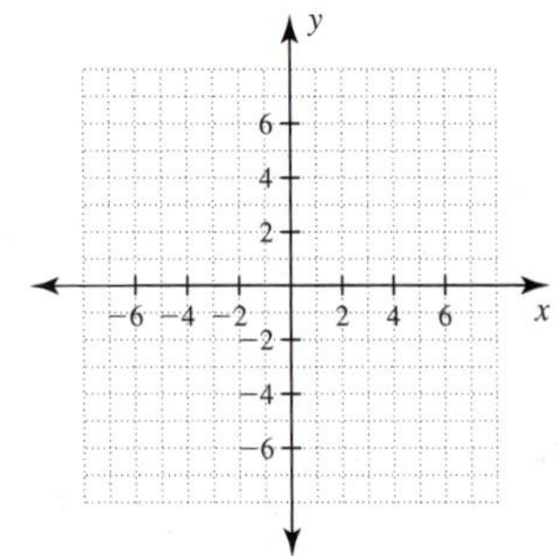

44. $f(x) = \dfrac{3 - 5x}{x + 1}$

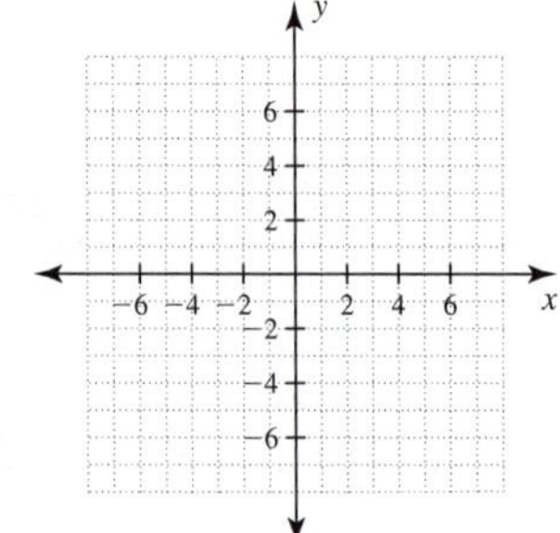

Complete each statement with **always, sometimes,** *or* **never.**

45. The graph of $f(x) = \frac{a}{x}$, $a \neq 0$, ________ has an x-intercept.

46. The graph of $f(x) = \frac{a}{x}$, $a \neq 0$, ________ has a y-intercept.

47. The graph of $f(x) = \frac{a(x - b)}{x - c}$, $a \neq 0$, $c \neq 0$, and $b \neq c$, ________ has an x-intercept.

48. The graph of $f(x) = \frac{a(x - b)}{x - c}$, $a \neq 0$, $c \neq 0$, and $b \neq c$, ________ has a y-intercept.

49. The graph of $f(x) = \frac{a}{x - c}$, $a \neq 0$, ________ has a vertical asymptote that lies exactly on the y-axis.

50. If $a > 0$, the graph of $f(x) = \frac{a}{x}$ ________ has points in Quadrants II and IV.

Skills | **Calculator/Computer** | Career Applications | Above and Beyond

Each function has a horizontal asymptote. Use a calculator to complete the table, rounding answers to four decimal places. Based on these values, give the equation of the horizontal asymptote.

51. $f(x) = \frac{7x - 5}{2x + 3}$

x	100	1,000	10,000	−100	−1,000	−10,000
$f(x)$						

52. $f(x) = \frac{8x + 6}{3x - 5}$

x	100	1,000	10,000	−100	−1,000	−10,000
$f(x)$						

53. $f(x) = \frac{3x + 7}{8 - 9x}$

x	100	1,000	10,000	−100	−1,000	−10,000
$f(x)$						

54. $f(x) = \frac{2x - 9}{5 - 4x}$

x	100	1,000	10,000	−100	−1,000	−10,000
$f(x)$						

55. Based on your answers to the previous exercises, give the equation of the horizontal asymptote for $f(x) = \frac{ax + b}{cx + d}$ (assuming there is no common factor in the numerator and denominator).

We have not yet considered what happens to the graph of a rational function if there is a common factor in the numerator and denominator. In each exercise, be sure to begin by factoring the numerator and denominator. Discuss the behavior of each graph, starting with the domain.

56. $f(x) = \dfrac{x^2 - 2x - 8}{x + 2}$

Use a calculator to check output values close to $x = -2$. What happens to the graph at $x = -2$? View the graph on the calculator.

57. $f(x) = \dfrac{3x - 15}{x^2 - 9x + 20}$

Use a calculator to check output values close to $x = 5$. What happens to the graph at $x = 5$? View the graph on the calculator.

58. $f(x) = \dfrac{x^2 - 4x + 3}{x - 1}$

Use a calculator to check output values close to $x = 1$. What happens to the graph at $x = 1$? View the graph on the calculator.

59. $f(x) = \dfrac{5x + 10}{x^2 + 7x + 10}$

Use a calculator to check output values close to $x = -2$. What happens to the graph at $x = -2$? View the graph on the calculator.

Answers

1. $\{x \mid x \neq 0\}$; no intercepts **3.** $\{x \mid x \neq 9\}$; y-int: $\left(0, -\frac{1}{3}\right)$; no x-int **5.** $\{x \mid x \neq 2\}$; y-int: (0, 4); no x-int **7.** $\{x \mid x \neq 4\}$; y-int: $\left(0, \frac{3}{2}\right)$; x-int: (6, 0) **9.** $\{x \mid x \neq 4\}$; y-int: $\left(0, -\frac{1}{2}\right)$; x-int: (2, 0) **11.** $\{x \mid x \neq -4\}$; y int: $\left(0, -\frac{5}{2}\right)$; x-int: (5, 0) **13.** $\{x \mid x \neq 3\}$; y-int: $\left(0, -\frac{2}{3}\right)$; x-int: $\left(\frac{1}{2}, 0\right)$ **15.** Vert: $x = 0$; horiz: $y = 0$ **17.** Vert: $x = 9$; horiz: $y = 0$ **19.** Vert: $x = 2$; horiz: $y = 0$ **21.** Vert: $x = 4$; horiz: $y = 1$ **23.** Vert: $x = 4$; horiz: $y = -1$ **25.** Vert: $x = -4$; horiz: $y = 2$ **27.** Vert: $x = 3$; horiz: $y = -4$

29.

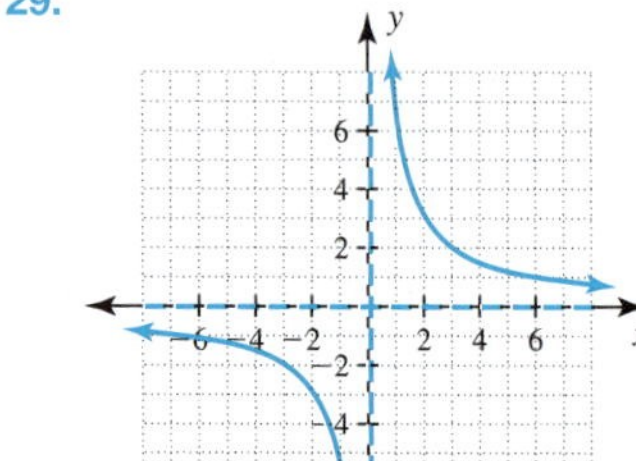

31.

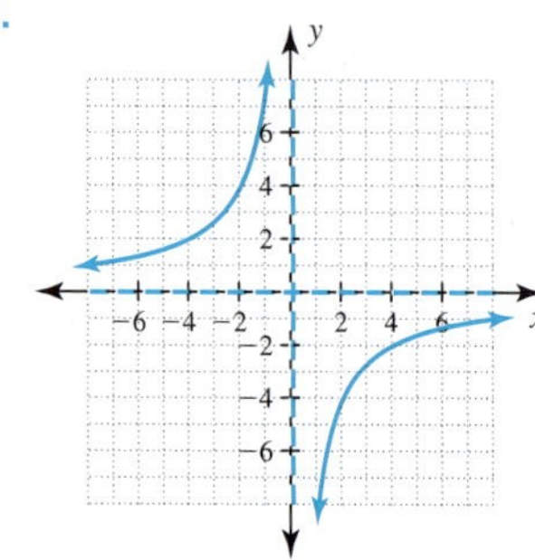

33.

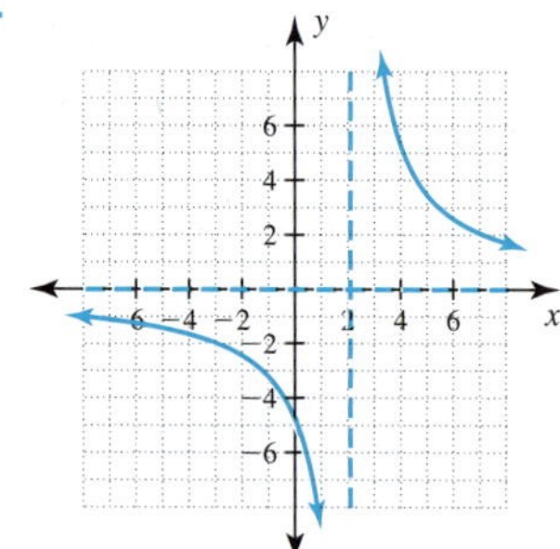

35.

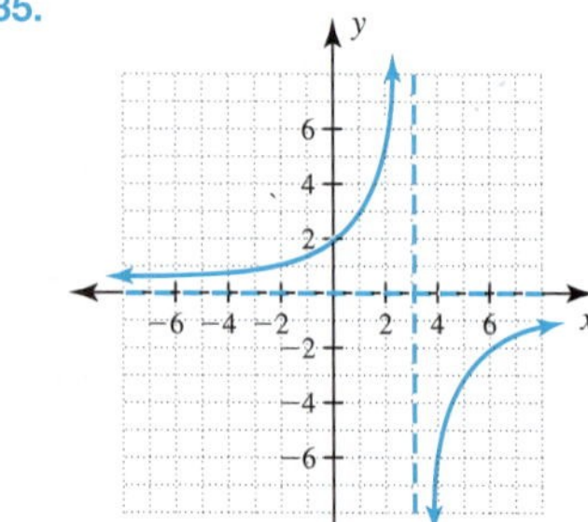

37.

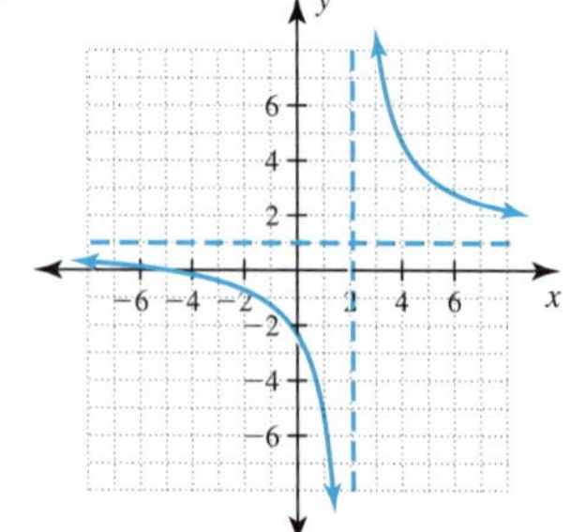

39.

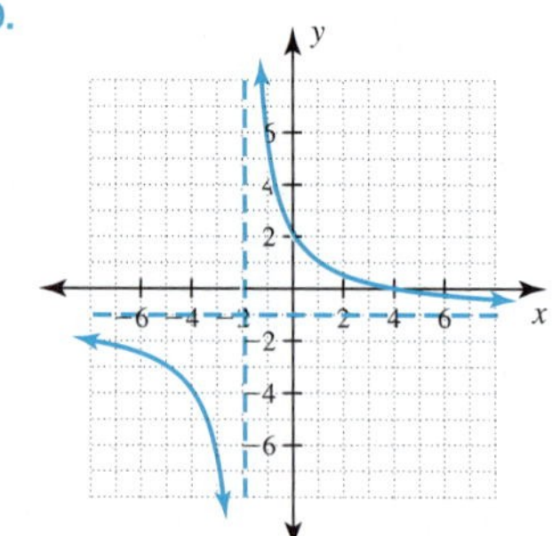

41. 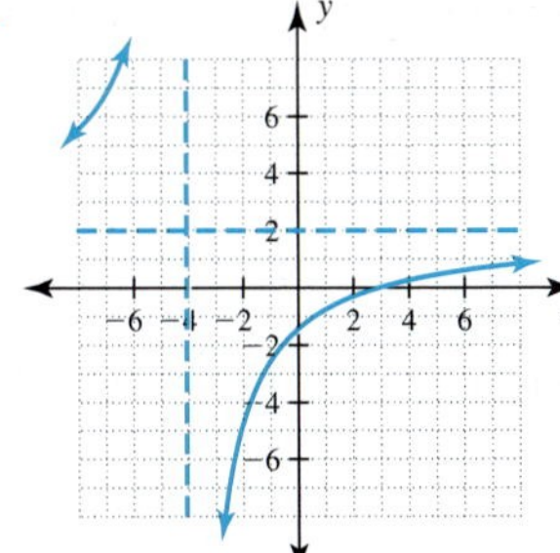

43. 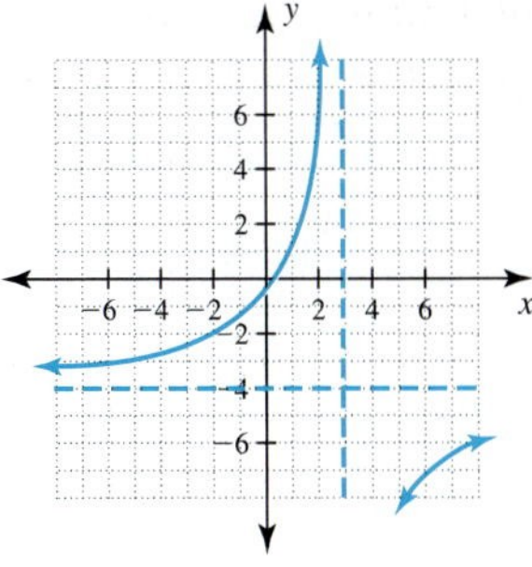

45. never **47.** always **49.** sometimes

51.

x	100	1,000	10,000	−100	−1,000	−10,000
$f(x)$	3.4236	3.4923	3.4992	3.5787	3.5078	3.5008

$y = \frac{7}{2}$

53.

x	100	1,000	10,000	−100	−1,000	−10,000
$f(x)$	−0.3442	−0.3344	−0.3334	−0.3227	−0.3323	−0.3332

$y = -\frac{1}{3}$

55. $y = \frac{a}{c}$ **57.** Above and Beyond **59.** Above and Beyond

9.6 Rational Equations and Problem Solving

< 9.6 Objectives >

1 > Solve a rational equation

2 > Solve a proportion

3 > Solve a formula for a variable

4 > Use rational equations to solve applications

NOTE

A **rational equation** is an equation that contains one or more rational expressions.

We use rational expressions to model many applications and situations. In this section, we look to solve *rational equations.*

Most of the time, we solve rational equations by clearing the denominators. We do this by multiplying both sides of the equation by the LCD of all the rational expressions in the equation. This leaves us with an equation without fractions.

Finally, we use whichever techniques are appropriate to solve the equation. We illustrate this process in Example 1.

Example 1 Clearing Equations of Fractions

< Objective 1 >

Solve.

$$\frac{2x}{3} + \frac{x}{5} = 13$$

For the denominators 3 and 5, the LCD is 15. Multiplying both sides of the equation by 15 gives

$$15\left(\frac{2x}{3} + \frac{x}{5}\right) = 15 \cdot 13$$

$$15 \cdot \frac{2x}{3} + 15 \cdot \frac{x}{5} = 15 \cdot 13 \quad \text{Distribute 15 on the left.}$$

$$\frac{\overset{5}{\cancel{15}} \cdot 2x}{\underset{1}{\cancel{3}}} + \frac{\overset{3}{\cancel{15}} \cdot x}{\underset{1}{\cancel{5}}} = 195$$

$$10x + 3x = 195 \quad \text{Simplify. The equation is now cleared of fractions.}$$

$$13x = 195$$

$$x = 15$$

The solution set is $\{15\}$.

To check, substitute 15 in the original equation.

$$\frac{2x}{3} + \frac{x}{5} = 13$$

$$\frac{2(15)}{3} + \frac{(15)}{5} \stackrel{?}{=} 13$$

$$10 + 3 \stackrel{?}{=} 13$$

$$13 = 13 \quad \text{A true statement}$$

So 15 is the solution to the equation.

Check Yourself 1

Solve.

$$\frac{3x}{2} - \frac{x}{3} = 7$$

> CAUTION

A common mistake is to confuse an *equation* such as

$$\frac{2x}{3} + \frac{x}{5} = 13$$

and an *expression* such as

$$\frac{2x}{3} + \frac{x}{5}$$

You should keep in mind the difference between an expression and an equation. Consider the equation in Example 1.

Equation: $\frac{2x}{3} + \frac{x}{5} = 13$

Here we want to *solve the equation for x*, as in Example 1. We multiply both sides by the LCD to clear fractions and continue to solve the equation.

Expression: $\frac{2x}{3} + \frac{x}{5}$

Here we want to find *a third fraction* that is equivalent to the given expression. We write each fraction as an equivalent fraction with the LCD as a common denominator.

$$\frac{2x}{3} + \frac{x}{5} = \frac{2x \cdot 5}{3 \cdot 5} + \frac{x \cdot 3}{5 \cdot 3}$$

$$= \frac{10x}{15} + \frac{3x}{15} = \frac{10x + 3x}{15}$$

$$= \frac{13x}{15}$$

We can multiply both sides of an equation to clear it of fractions. With an expression, we can only simplify it. As such, we are usually left with a fraction.

The process is similar when variables are in the denominators. Consider Example 2.

Example 2 Solving Rational Equations

Solve.

$$\frac{7}{4x} - \frac{3}{x^2} = \frac{1}{2x^2}$$

NOTE

We assume that x cannot be 0. Do you see why?

For $4x$, x^2, and $2x^2$, the LCD is $4x^2$. So, multiplying both sides by $4x^2$, we have

$$4x^2\left(\frac{7}{4x} - \frac{3}{x^2}\right) = 4x^2 \cdot \frac{1}{2x^2}$$

$$4x^2 \cdot \frac{7}{4x} - 4x^2 \cdot \frac{3}{x^2} = 4x^2 \cdot \frac{1}{2x^2}$$ Distribute $4x^2$ on the left side.

$$\frac{\overset{x}{\cancel{4x^2}} \cdot 7}{\underset{1}{\cancel{4x}}} - \frac{\overset{4}{\cancel{4x^2}} \cdot 3}{\underset{1}{\cancel{x^2}}} = \frac{\overset{2}{\cancel{4x^2}} \cdot 1}{\underset{1}{\cancel{2x^2}}}$$ Simplify.

$$7x - 12 = 2$$

$$7x = 14$$

$$x = 2$$

The solution set is $\{2\}$.

We leave it to you to check the solution 2. Be sure to return to the original equation and substitute 2 for x.

Check Yourself 2

Solve.

$$\frac{5}{2x} - \frac{4}{x^2} = \frac{7}{2x^2}$$

This algorithm summarizes our work in solving equations containing rational expressions.

Step by Step

Solving Rational Equations

Step 1 Clear the equation of fractions by multiplying both sides of the equation by the LCD of all the fractions that appear.

Step 2 Solve the equation resulting from step 1.

Step 3 Check all solutions by substitution in the original equation. Discard any extraneous solutions.

Example 3 illustrates the same process when there are binomials in the denominators.

Example 3 Solving Rational Equations

NOTE

We assume that x is not -2 or 3.

Solve.

$$\frac{4}{x+2} + 3 = \frac{3x}{x-3}$$

The LCD is $(x + 2)(x - 3)$. Multiplying by that LCD gives

$$(x+2)(x-3)\left(\frac{4}{x+2}\right) + (x+2)(x-3)(3) = (x+2)(x-3)\left(\frac{3x}{x-3}\right)$$

Simplifying each term gives

$$4(x-3) + 3(x+2)(x-3) = 3x(x+2)$$

We now clear the parentheses and proceed as before.

$$\begin{aligned} 4x - 12 + 3x^2 - 3x - 18 &= 3x^2 + 6x \\ 3x^2 + x - 30 &= 3x^2 + 6x \\ x - 30 &= 6x \\ -5x &= 30 \\ x &= -6 \end{aligned}$$

The solution set is $\{-6\}$.

Check

$$\frac{4}{(-6)+2} + 3 \stackrel{?}{=} \frac{3(-6)}{(-6)-3}$$

$$\frac{-4}{-4} + 3 \stackrel{?}{=} \frac{-18}{-9}$$

$$-1 + 3 = 2$$

$$2 = 2 \quad \text{True}$$

Check Yourself 3

Solve.

$$\frac{5}{x-4} + 2 = \frac{2x}{x-3}$$

Factoring plays an important role when solving rational equations.

Example 4 Solving Rational Equations

Solve.

$$\frac{3}{x-3} - \frac{7}{x+3} = \frac{2}{x^2-9}$$

The denominator on the right side factors as $(x - 3)(x + 3)$, which forms the LCD. We multiply each term by that LCD.

$$(x-3)(x+3)\left(\frac{3}{x-3}\right) - (x-3)(x+3)\left(\frac{7}{x+3}\right) = (x-3)(x+3)\left[\frac{2}{(x-3)(x+3)}\right]$$

Simplifying yields

$$3(x+3) - 7(x-3) = 2$$
$$3x + 9 - 7x + 21 = 2$$
$$-4x = -28$$
$$x = 7$$

The solution set is $\{7\}$.

Be sure to check this result by substitution in the original equation.

Check Yourself 4

Solve $\dfrac{4}{x-4} - \dfrac{3}{x+1} = \dfrac{5}{x^2-3x-4}$.

RECALL

Equivalent equations have exactly the same solution set.

Extraneous solutions are values that do not check. They are not solutions to the original problem.

We must be careful when mulitplying both sides of an equation by an expression containing a variable. If a solution makes that expression 0, then we do not produce an *equivalent equation*. In such a case, we may produce *extraneous solutions*.

To deal with this problem, it is important to check your work after completing the exercise. If a proposed solution does not check, then the value is not actually a solution to the original problem. Discard all such values.

Example 5 Solving Rational Equations

NOTES

We must assume that $x \neq 2$.

Multiply all three terms by $x - 2$.

Solve.

$$\frac{x}{x-2} - 7 = \frac{2}{x-2}$$

The LCD is $x - 2$, and multiplying, we have

$$\left(\frac{x}{x-2}\right)(x-2) - 7(x-2) = \left(\frac{2}{x-2}\right)(x-2)$$

Simplifying yields

$$x - 7(x - 2) = 2$$
$$x - 7x + 14 = 2$$
$$-6x = -12$$
$$x = 2$$

To check this result, substitute 2 for x,

$$\frac{(2)}{(2) - 2} - 7 \stackrel{?}{=} \frac{2}{(2) - 2}$$
$$\frac{2}{0} - 7 \stackrel{?}{=} \frac{2}{0}$$

> CAUTION

Because division by 0 is undefined, we conclude that 2 is *not a solution* to the original equation. It is an extraneous solution. The original equation has no solutions.

There are no solutions to this equation. The solution set is empty. The set is written { } or ∅.

Check Yourself 5

Solve $\frac{x - 3}{x - 4} = 4 + \frac{1}{x - 4}$.

Rational equations may lead to quadratic equations, as illustrated in Example 6.

Example 6 Solving Rational Equations

NOTE

Assume $x \neq 3$ and $x \neq 4$.

Solve.

$$\frac{x}{x - 4} = \frac{15}{x - 3} - \frac{2x}{x^2 - 7x + 12}$$

After factoring the denominator on the right, we see the LCD of the denominators $x - 3$, $x - 4$, and $x^2 - 7x + 12$ is $(x - 3)(x - 4)$. Multiplying by that LCD, we have

$$(x - 3)(x - 4)\left(\frac{x}{x - 4}\right) = (x - 3)(x - 4)\left(\frac{15}{x - 3}\right) - (x - 3)(x - 4)\left[\frac{2x}{(x - 3)(x - 4)}\right]$$

Simplifying yields

$$x(x - 3) = 15(x - 4) - 2x$$
$$x^2 - 3x = 15x - 60 - 2x \qquad \text{Remove the parentheses.}$$
$$x^2 - 16x + 60 = 0 \qquad \text{Write in standard form and factor.}$$
$$(x - 6)(x - 10) = 0$$

So $x = 6$ or $x = 10$

Verify that 6 and 10 are both solutions to the original equation. The solution set is {6, 10}.

Check Yourself 6

Solve $\frac{3x}{x + 2} - \frac{2}{x + 3} = \frac{36}{x^2 + 5x + 6}$.

Now consider a special kind of rational equation such as

$$\frac{135}{t} = \frac{180}{t + 1}$$

An equation of the form $\frac{a}{b} = \frac{c}{d}$ is called a **proportion.** This type of equation occurs often enough in algebra that it is worth developing some special methods to solve it. First, we need some definitions.

A **ratio** is a means of comparing two quantities. A ratio can be written as a fraction. For instance, the ratio of 2 to 3 can be written as $\frac{2}{3}$. A statement that two ratios are equal is called a *proportion*. A proportion has the form

$$\frac{a}{b} = \frac{c}{d}$$

In this proportion, *a* and *d* are called the *extremes* of the proportion, and *b* and *c* are called the *means*.

A useful property of proportions is easily developed. If

NOTE

bd is the LCD of the denominators.

$$\frac{a}{b} = \frac{c}{d}$$ We multiply both sides by $b \cdot d$.

$$\left(\frac{a}{b}\right)bd = \left(\frac{c}{d}\right)bd \qquad \text{or} \qquad ad = bc$$

Property

Proportions

If $\frac{a}{b} = \frac{c}{d}$, then $ad = bc$.

In any proportion, the product of the extremes (ad) is equal to the product of the means (bc).

This rule shows us an approach to solving proportions.

Example 7 Solving a Proportion

< Objective 2 >

Solve each equation.

(a) $\frac{x}{5} = \frac{12}{15}$

NOTE

The extremes are *x* and 15. The means are 5 and 12.

Set the product of the extremes equal to the product of the means.

$$15x = 5 \cdot 12$$
$$15x = 60$$
$$x = 4$$

Our solution is 4. You can check as before, by substituting in the original proportion.

(b) $\frac{x+3}{10} = \frac{x}{7}$

Set the product of the extremes equal to the product of the means. Be certain to use parentheses when the numerator has more than one term.

$$7(x + 3) = 10x$$
$$7x + 21 = 10x$$
$$21 = 3x$$
$$7 = x$$

We leave it to you to check this result.

Check Yourself 7

Solve each equation.

(a) $\frac{x}{8} = \frac{3}{4}$ **(b)** $\frac{x-1}{9} = \frac{x+1}{12}$

Many applications can be modeled with proportions. Typically these applications involve a common ratio, such as miles to gallons or miles to hours, and they can usually be solved with three basic steps.

Step by Step

Using a Proportion to Solve an Application

Step 1	Assign a variable to represent the unknown quantity.
Step 2	Write a proportion using the known and unknown quantities. Be sure each ratio uses the same units.
Step 3	Solve the proportion written in step 2 for the unknown quantity.

Example 8 illustrates this approach.

Example 8 Using Proportions

A car uses 3 gal of gas to travel 105 mi. At that rate, how many gallons will be used on a 385-mi trip?

Step 1 Assign a variable to represent the unknown quantity. Let x be the number of gallons of gas that will be used on the 385-mi trip.

Step 2 Write a proportion. Note that the ratio of miles to gallons must stay the same.

Miles → Miles

$$\frac{105}{3} = \frac{385}{x}$$

Gallons → Gallons

Step 3 Solve the proportion. The product of the extremes is equal to the product of the means.

> **NOTE**
>
> To verify your solution, return to the original problem and check that the two ratios are equivalent.

$$105x = 3 \cdot 385$$

$$105x = 1{,}155$$

$$\frac{105x}{105} = \frac{1{,}155}{105}$$

$$x = 11$$

So, 11 gal of gas will be used on a 385-mi trip.

Check Yourself 8

A car uses 8 L of gasoline when traveling 100 km. At that rate, how many liters of gas will be used on a 250-km trip?

Proportions can also be used to solve problems in which a quantity is divided using a specific ratio.

Example 9 Using Proportions

A 60-in.-long piece of wire is to be cut into two pieces whose lengths have the ratio 5 to 7. Find the length of each piece.

Step 1 Let x represent the length of the shorter piece. Then $60 - x$ is the length of the longer piece.

> **RECALL**
>
> A picture of the problem always helps.

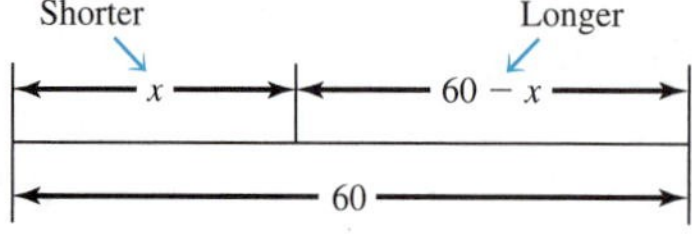

Step 2 The two pieces have the ratio $\frac{5}{7}$, so

$$\frac{x}{60 - x} = \frac{5}{7}$$

Step 3 Solving as before, we get

$$7x = (60 - x)5$$
$$7x = 300 - 5x$$
$$12x = 300$$
$$x = 25 \quad \text{Shorter piece}$$
$$60 - x = 35 \quad \text{Longer piece}$$

The pieces have lengths of 25 in. and 35 in.

Check Yourself 9

A 21-ft-long board is to be cut into two pieces so that the ratio of their lengths is 3 to 4. Find the lengths of the two pieces.

NOTE

Two triangles are similar if corresponding angles have the same measure.

Two triangles are **similar** if they have the same shape. With such triangles, the ratios of the corresponding sides are equal. In the triangles shown, side a corresponds to side d and sides b and e correspond.

Because the ratios of corresponding sides are equal, we have a natural proportion when trying to find the missing sides of similar triangles.

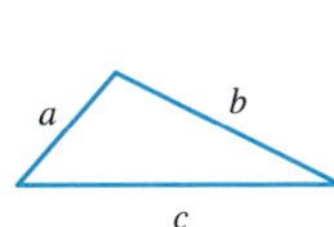

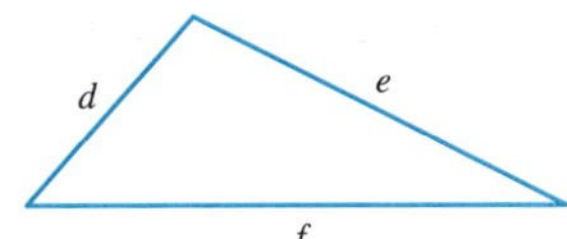

Because these are similar triangles, we know that $\frac{a}{b} = \frac{d}{e}$.

Example 10 Solving Triangles

Given that these two triangles are similar, find the lengths of the indicated sides.

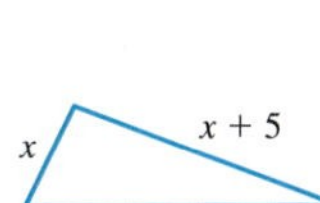

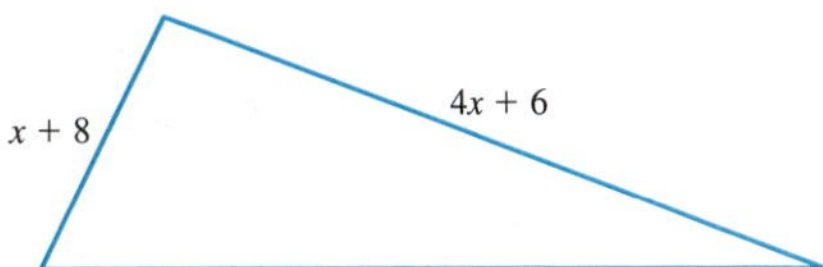

Since $\frac{a}{b} = \frac{d}{e}$, we have

$$\frac{x}{x + 5} = \frac{x + 8}{4x + 6}$$

We use the proportion property to rewrite the equation.

$$x(4x + 6) = (x + 8)(x + 5)$$
$$4x^2 + 6x = x^2 + 13x + 40$$
$$3x^2 - 7x - 40 = 0$$
$$(3x + 8)(x - 5) = 0$$
$$x = -\frac{8}{3} \quad \text{or} \quad x = 5$$

We know that x represents the length of one side of a triangle, so it must be a positive number. We can now find the four indicated lengths.

$$x = 5$$
$$x + 5 = 10$$
$$x + 8 = 13$$
$$4x + 6 = 26$$

Check Yourself 10

Given that the two triangles below are similar, find the lengths of the indicated sides.

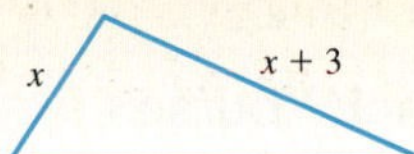

We can use what we have learned to solve some formulas (or literal equations) for a specified variable. Consider Example 11.

Example 11 Solving a Formula

< Objective 3 >

NOTE

This is a parallel electric circuit. The symbol for a resistor is

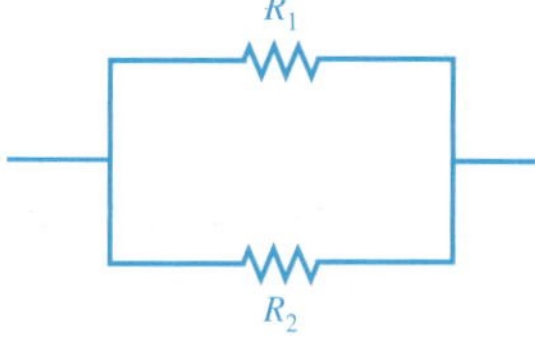

If two resistors with resistances R_1 and R_2 are connected in parallel, the combined resistance R can be found from

$$\frac{1}{R} = \frac{1}{R_1} + \frac{1}{R_2}$$

Solve the formula for R.

First, the LCD is RR_1R_2, so we multiply.

RECALL

The numbers 1 and 2 are *subscripts*. We read R_1 as "R sub 1" and R_2 as "R sub 2."

$$RR_1R_2 \cdot \frac{1}{R} = RR_1R_2 \cdot \frac{1}{R_1} + RR_1R_2 \cdot \frac{1}{R_2}$$

Simplifying yields

$$R_1R_2 = RR_2 + RR_1$$

$$R_1R_2 = R(R_2 + R_1)$$ Factor out R on the right.

$$\frac{R_1R_2}{R_2 + R_1} = R \quad \text{or} \quad R = \frac{R_1R_2}{R_1 + R_2}$$ Divide by $R_2 + R_1$ to isolate R.

NOTE

This formula gives the focal length of a convex lens.

Check Yourself 11

Solve for D_1.

$$\frac{1}{F} = \frac{1}{D_1} + \frac{1}{D_2}$$

Motion problems often involve rational expressions. Recall that the key equation for solving motion problems relates the distance traveled, the speed or rate, and the time.

Property

The Distance Formula

$d = r \cdot t$

Often we use this equation in different forms by solving for r or for t.

$r = \frac{d}{t}$ and $t = \frac{d}{r}$

Example 12 Solving a Motion Problem

< Objective 4 >

NOTE

Compare this to the motion problems you solved in Section 2.1.

RECALL

Because $d = rt$ we know that $t = \frac{d}{r}$.

A boat can travel 16 mi/hr in still water. If the boat can travel 5 mi downstream in the same time it takes to travel 3 mi upstream, what is the rate of the river's current?

Step 1 We want to find the rate of the current.

Step 2 Let c be the rate of the current. Then $16 + c$ is the rate of the boat going downstream and $16 - c$ is the rate going upstream.

Step 3 We use a chart to help us set up an appropriate equation.

	d	r	t
Downstream	5 mi	$16 + c$	$\frac{5}{16 + c}$
Upstream	3 mi	$16 - c$	$\frac{3}{16 - c}$

The key to finding our equation is noting that the time is the same upstream and downstream, so

$$\frac{5}{16 + c} = \frac{3}{16 - c}$$

Step 4 We use the proportion rule to rewrite the equation.

$$5(16 - c) = 3(16 + c)$$
$$80 - 5c = 48 + 3c$$
$$32 = 8c$$
$$c = 4$$

Step 5 The current is moving at 4 mi/hr.

To check, verify that $\frac{5}{16 + 4} = \frac{3}{16 - 4}$.

Check Yourself 12

A plane flew 540 mi into a steady 30 mi/hr wind. The pilot then returned along the same route with a tailwind. If the entire trip took $7\frac{1}{2}$ hr, what would his speed have been in still air?

The final application we look at in this section is the **work problem.**

Work, as an application, relates to every field. The basic question asked is how long will it take to complete a job.

For example, if two people are painting a house and we know how fast they each work, we can determine how long it will take them to paint the house wo-

> **NOTES**
>
> We chose hours in this description but the principles are true for any time unit. You must be consistent with your units of time.
>
> We assume the two entities do not "slow each other down."

together. Similarly, we can determine how long it will take two printers working at different speeds to complete an order.

We begin tackling this problem by establishing some basic principles.

If a job takes t hours to complete, then $\frac{1}{t}$ represents the portion of the job that can be completed in 1 hour.

This agrees with our intuition. If a job takes 4 hours to complete, then $\frac{1}{4}$ of the job is complete after 1 hour of work.

If two entities are working on the same job, and the first takes a hours to complete the job alone, and the second takes b hours, then

$$\frac{1}{a} + \frac{1}{b}$$

represents the portion of the job completed each hour, working together.

So, if the first entity completes the job in 3 hours, then after 1 hour, that entity completes $\frac{1}{3}$ of the job. If a second entity takes 4 hours to complete the job, and it completes $\frac{1}{4}$ of the job in an hour, then working together, the two entities complete $\frac{1}{3} + \frac{1}{4}$ of the job in an hour.

Property

The Work Relationship

If two entities are working on the same job, and the first takes a hours to complete the job alone, and the second takes b hours, and if the job takes the two entities t hours to complete together, then

$$\frac{1}{a} + \frac{1}{b} = \frac{1}{t}$$

> **NOTE**
>
> We combine the principles into one formula to describe the work relationship.

We use the work relationship in Example 13.

Example 13 Solving a Work Problem

Jason and Hilger are required to paint over the graffiti on a wall. If Jason worked alone, it would take him 20 hr to complete the job; Hilger could do the job in 15 hr. How long will it take them to complete the job working together?

Apply the formula with $a = 20$ and $b = 15$, and solve for t.

$$\frac{1}{a} + \frac{1}{b} = \frac{1}{t}$$

$$\frac{1}{(20)} + \frac{1}{(15)} = \frac{1}{t}$$

The LCD of 20, 15, and t is $60t$, so we multiply both sides by $60t$.

> **NOTE**
>
> We could also add the fractions on the left side and use the proportion rule to solve this equation.

$$60t\left(\frac{1}{20} + \frac{1}{15}\right) = \frac{1}{t} \cdot 60t$$

$$\overset{3}{\cancel{60}}t\left(\frac{1}{\cancel{20}}\right) + \overset{4}{\cancel{60}}t\left(\frac{1}{\cancel{15}}\right) = \frac{60\cancel{t}}{\cancel{t}}$$ Distribute on the left.

$$3t + 4t = 60$$ Simplify the fractions.

$$7t = 60$$ Combine like terms.

$$\frac{7t}{7} = \frac{60}{7}$$ Divide both sides by 7 to isolate the variable.

$$t = \frac{60}{7} = 8\frac{4}{7} \text{ hr} \approx 8 \text{ hr } 34 \text{ min}$$

Working together, they can complete the job in 8 hr 34 min.

Check Yourself 13

(a) One computer prints a company's paychecks in 40 min, while an older printer takes 80 min to complete the same job. If both printers are working, how long will it take to print the paychecks? *Hint:* Measure work times in minutes rather than hours.

(b) Filling a hot tub with a hose takes 4 hr. Draining the same tub takes 7 hr. If the drain is open, how long will it take to fill the tub? *Hint:* Treat the draining time as a negative number.

In Example 14, we are given the total time it takes two people to work an application and how fast they are, relative each other.

Example 14 Solving a Work Application

For every three electric meters that Carmen reads and records, Dwight reads and records four. If the two, working together, can canvass a community in 8 hr, how long would it take Carmen, working alone, to accomplish the task?

NOTE

Think about this!

Dwight reads 4 meters for every 3 that Carmen reads, so he can complete the job in $\frac{3}{4}$ the time it would take Carmen to finish.

We apply the work formula again, but this time we know t. We can reduce a and b to a single variable.

Let $t = 8$ and a be the time it takes Carmen to canvass the community alone. Then the time it takes Dwight to canvass the community alone is given by

$$b = \frac{3}{4}a$$

$$\frac{1}{a} + \frac{1}{b} = \frac{1}{t}$$

$$\frac{1}{a} + \frac{1}{\left(\frac{3}{4}a\right)} = \frac{1}{(8)}$$

We simplify the second term on the left side.

$$\frac{1}{\frac{3}{4}a} = 1 \div \frac{3a}{4} = 1 \times \frac{4}{3a} = \frac{4}{3a}$$

We can now complete the problem.

$$\frac{1}{a} + \frac{1}{\left(\frac{3}{4}a\right)} = \frac{1}{(8)}$$

$$\frac{1}{a} + \frac{4}{3a} = \frac{1}{8} \qquad \text{We simplified this second term above.}$$

$$24a\left(\frac{1}{a} + \frac{4}{3a}\right) = \overset{3}{\cancel{24}}a\left(\frac{1}{\cancel{8}}\right) \qquad \text{The LCD of } a\text{, } 3a\text{, and 8 is 24.}$$

$$24\cancel{a}\left(\frac{1}{\cancel{a}}\right) + \overset{8}{\cancel{24}}\cancel{a}\left(\frac{4}{\cancel{3a}}\right) = 3a \qquad \text{Distribute on the left side.}$$

$$24 + 32 = 3a$$

$$56 = 3a$$

$$\frac{56}{3} = t_1 \qquad \text{Divide both sides by 3.}$$

Therefore, it would take Carmen $\frac{56}{3}$ hr or $18\frac{2}{3}$ hr (18 hr 40 min) to finish the job alone.

Check Yourself 14

Using new street-cleaning trucks, a route takes two-thirds as long to complete than it did with the old trucks. Working together, the route is cleaned in 3 hr. How long would it take either truck, working alone?

Check Yourself ANSWERS

1. {6} **2.** {3} **3.** {9} **4.** {−11} **5.** { } or ∅ **6.** $\left\{-5, \frac{8}{3}\right\}$ **7. (a)** 6; **(b)** 7
8. 20 L **9.** 9 ft and 12 ft **10.** The lengths are 6 and 9 on the first triangle and 8 and 12 on the second.
11. $D_1 = \frac{FD_2}{D_2 - F}$ **12.** 150 mi/hr **13. (a)** $26\frac{2}{3}$ min or 26 min 40 s; **(b)** $9\frac{1}{3}$ hr or 9 hr 20 min
14. New: 5 hr; old: 7 hr 30 min

Reading Your Text

These fill-in-the-blank exercises will help you understand some of the key vocabulary used in this section. The answers to these exercises are in the Answers Appendix in the back of the text.

(a) To solve a rational equation, multiply both sides by the ________ of any fractions in the equation.

(b) Always check the solution to a rational equation by substituting the answer into the ________ equation.

(c) We use the ________ formula to solve applications relating distance, rate, and time.

(d) Two triangles are ________ if corresponding angles have the same measure.

9.6 exercises

Skills Calculator/Computer Career Applications Above and Beyond

Decide whether each is an expression or an equation. If it is an equation, find a solution. If it is an expression, write it as a single fraction.

1. $\frac{x}{4} - \frac{x}{5} = 2$ **2.** $\frac{x}{4} - \frac{x}{7} = 3$ **3.** $\frac{x}{2} - \frac{x}{5}$ **4.** $\frac{x}{7} - \frac{x}{14}$

5. $\frac{3x+1}{4} = x - 1$ **6.** $\frac{3x-1}{2} - \frac{x}{5} - \frac{x+3}{4}$ **7.** $\frac{x}{4} = \frac{x}{12} + \frac{1}{2}$ **8.** $\frac{2x-1}{3} + \frac{x}{2}$

< Objective 1 >

Solve each equation.

9. $\frac{x}{3} + \frac{3}{2} = \frac{x}{6} + \frac{7}{3}$ **10.** $\frac{x}{12} - \frac{2}{3} = \frac{x}{6} + \frac{3}{4}$ **11.** $\frac{4}{x} + \frac{3}{4} = \frac{10}{x}$ VIDEO

12. $\frac{3}{x} = \frac{5}{3} - \frac{7}{x}$

13. $\frac{5}{4x} - \frac{1}{2} = \frac{1}{2x}$

14. $\frac{7}{6x} - \frac{1}{3} = \frac{1}{2x}$

15. $\frac{4}{x+5} = \frac{3}{x+3}$

16. $\frac{5}{x-2} = \frac{4}{x+1}$

17. $\frac{9}{x} + 2 = \frac{2x}{x+3}$

18. $\frac{6}{x} + 3 = \frac{3x}{x+1}$

19. $\frac{3}{x+2} - \frac{5}{x} = \frac{13}{x+2}$

20. $\frac{7}{x} - \frac{2}{x-3} = \frac{6}{x}$

21. $\frac{3}{2} + \frac{2}{2x-4} = \frac{1}{x-2}$

22. $\frac{10}{2x+6} + \frac{2}{x+3} = \frac{1}{2}$

23. $\frac{x}{3x+12} + \frac{x-1}{x+4} = \frac{5}{3}$

24. $\frac{x}{4x-12} - \frac{x-4}{x-3} = \frac{1}{8}$

25. $\frac{16-3x}{x+6} + \frac{x+3}{5} = \frac{3x-2}{15}$

26. $\frac{x+1}{x-2} - \frac{x+3}{x} = \frac{6}{x^2-2x}$ 

27. $\frac{1}{x-2} - \frac{2}{x+2} = \frac{2}{x^2-4}$

28. $\frac{1}{x+4} + \frac{1}{x-4} = \frac{12}{x^2-16}$

29. $\frac{7}{x+5} - \frac{1}{x-5} = \frac{x}{x^2-25}$

30. $\frac{2}{x-2} = \frac{3}{x+2} + \frac{x}{x^2-4}$

31. $\frac{11}{x+2} - \frac{5}{x^2-x-6} = \frac{1}{x-3}$

32. $\frac{3}{x-4} - \frac{4}{x^2-3x-4} = \frac{1}{x+1}$

33. $\frac{5}{x-2} - \frac{3}{x+3} = \frac{24}{x^2+x-6}$

34. $\frac{3}{x+1} - \frac{5}{x+6} = \frac{2}{x^2+7x+6}$

35. $\frac{x}{x-3} - 2 = \frac{3}{x-3}$

36. $\frac{x}{x-5} + 2 = \frac{5}{x-5}$

37. $\frac{2}{x^2-3x} - \frac{1}{x^2+2x} = \frac{2}{x^2-x-6}$

38. $\frac{2}{x^2-x} - \frac{4}{x^2+5x-6} = \frac{3}{x^2+6x}$

39. $\frac{2}{x^2-4x+3} - \frac{3}{x^2-9} = \frac{2}{x^2+2x-3}$

40. $\frac{2}{x^2-4} - \frac{1}{x^2+x-2} = \frac{3}{x^2-3x+2}$

41. $\frac{7}{x-5} - \frac{3}{x+5} = \frac{40}{x^2-25}$

42. $\frac{3}{x-3} - \frac{18}{x^2-9} = \frac{5}{x+3}$

43. $\frac{2x}{x-3} + \frac{2}{x-5} = \frac{3x}{x^2-8x+15}$

44. $\frac{x}{x-4} = \frac{5x}{x^2-x-12} - \frac{3}{x+3}$

45. $\frac{2x}{x+2} = \frac{5}{x^2-x-6} - \frac{1}{x-3}$

46. $\frac{3x}{x-1} = \frac{2}{x-2} - \frac{2}{x^2-3x+2}$

47. $\frac{7}{x-2} + \frac{16}{x+3} = 3$

48. $\frac{5}{x-2} + \frac{6}{x+2} = 2$

49. $\frac{11}{x-3} - 1 = \frac{10}{x+3}$

50. $\frac{17}{x-4} - 2 = \frac{10}{x+2}$

< Objective 2 >

51. $\frac{x}{11} = \frac{12}{33}$

52. $\frac{4}{x} = \frac{16}{20}$

53. $\frac{5}{8} = \frac{20}{x}$

54. $\frac{x}{10} = \frac{9}{30}$

55. $\frac{x+1}{5} = \frac{20}{25}$

56. $\frac{2}{5} = \frac{x-2}{20}$

57. $\frac{3}{5} = \frac{x-1}{20}$

58. $\frac{5}{x-3} = \frac{15}{21}$

59. $\frac{x}{6} = \frac{x+5}{16}$

60. $\frac{x-2}{x+2} = \frac{12}{20}$

61. $\frac{x}{x+7} = \frac{10}{17}$

62. $\frac{x}{10} = \frac{x+6}{30}$

63. $\frac{2}{x-1} = \frac{6}{x+9}$

64. $\frac{3}{x-3} = \frac{4}{x-5}$

65. $\frac{1}{x+3} = \frac{7}{x^2-9}$

66. $\frac{1}{x+5} = \frac{4}{x^2+3x-10}$

Find the lengths of the indicated sides in each pair of similar triangles.

67. 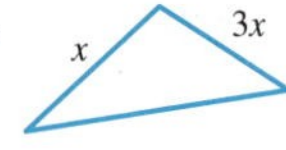

x + 5
5x + 9

68. 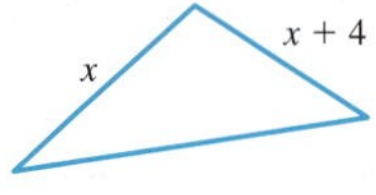

2x − 2
x + 8

< Objective 3 >

Solve each equation for the indicated variable.

69. $\frac{1}{x} = \frac{1}{a} - \frac{1}{b}$ for x 

70. $\frac{1}{x} = \frac{1}{a} + \frac{1}{b}$ for a

71. $\frac{1}{R} = \frac{1}{R_1} + \frac{1}{R_2}$ for R_1

72. $\frac{1}{F} = \frac{1}{D_1} + \frac{1}{D_2}$ for D_2

73. $y = \frac{x + 1}{x - 1}$ for x

74. $y = \frac{x - 3}{x - 2}$ for x

< Objective 4 >

Solve each application.

75. **Number Problem** One number is 4 times another number. The sum of the reciprocals of the numbers is $\frac{5}{24}$. Find the two numbers.

76. **Number Problem** The sum of the reciprocals of two consecutive even integers is equal to 10 times the reciprocal of the product of those integers. Find the two integers.

77. **Number Problem** If the same number is subtracted from the numerator and denominator of $\frac{11}{15}$, the result is $\frac{1}{3}$. Find that number.

78. **Number Problem** If the numerator of $\frac{8}{9}$ is multiplied by a number and that same number is subtracted from the denominator, the result is 10. What is that number?

79. **Science and Medicine** A motorboat can travel 20 mi/hr in still water. If the boat can travel 3 mi downstream on a river in the same time it takes to travel 2 mi upstream, what is the rate of the river's current?

80. **Science and Medicine** Janet and Michael took a canoe trip, traveling 6 mi upstream along a river, against a 2 mi/hr current. They then returned downstream to the starting point of their trip. If their entire trip took 4 hr, what was their rate in still water? VIDEO

81. **Science and Medicine** A plane flew 720 mi with a steady 30 mi/hr tailwind. The pilot then returned to the starting point, flying against the same wind. If the round-trip flight took 10 hr, what was the plane's airspeed?

82. **Science and Medicine** A small jet has an airspeed (the rate in still air) of 300 mi/hr. During one day's flights, the pilot noted that the plane could fly 85 mi with a tailwind in the same time it took to fly 65 mi against the same wind. What was the rate of the wind?

83. **Business and Finance** One computer printer can print a company's weekly payroll checks in 60 min. A second printer would take 90 min to complete the job. How long would it take the two printers operating together to print the checks?

84. **Business and Finance** An electrician can wire a house in 20 hr. If she works with an apprentice, the same job can be completed in 12 hr. How long would it take the apprentice working alone to wire the house?

85. **Science and Medicine** Po Ling can bicycle 75 mi in the same time it takes her to drive 165 mi. If her driving rate is 30 mi/hr faster than her rate on the bicycle, find each rate.

86. **Science and Medicine** A passenger train can travel 275 mi in the same time a freight train takes to travel 225 mi. If the speed of the passenger train is 10 mi/hr more than that of the freight train, find the speed of each train.

87. **Science and Medicine** A hydroplane took 1 hr longer to fly 540 mi on the first portion of a trip than to fly 360 mi on the second. If the rate was the same for each portion, what was the flying time for each leg of the trip?

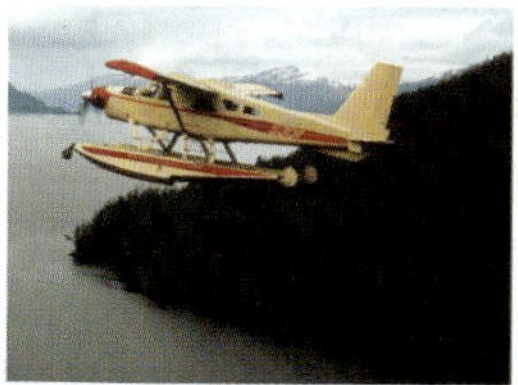

88. **Science and Medicine** Gilbert took 2 hr longer to drive 240 mi on the first day of a business trip than to drive 144 mi on the second day. If his rate was the same both days, what was his driving time for each day?

89. **Science and Medicine** One road crew can pave a section of highway in 15 hr. A second crew, working with newer equipment, can do the same job in 10 hr. How long would it take to pave that same section of highway if both crews worked together?

90. **Science and Medicine** A landscaper can prepare and seed a new lawn in 12 hr. If he works with an assistant, the job takes 8 hr. How long would it take the assistant working alone to complete the job?

91. **Science and Medicine** An experienced roofer can work twice as fast as her helper. Working together, they can shingle a new section of roof in 4 hr. How long would it take the experienced roofer working alone to complete the same job?

92. **Science and Medicine** Virginia can complete her company's monthly report in 5 hr less time than Carl. If they work together, the report will take them 6 hr to finish. How long would it take Virginia working alone?

93. **Science and Medicine** A speed of 60 mi/hr corresponds to 88 ft/s. If a light plane's speed is 150 mi/hr, what is its speed in feet per second?

94. **Business and Finance** If 342 cups of coffee can be made from 9 lb of coffee, how many cups can be made from 6 lb of coffee?

95. **Social Science** A car uses 5 gal of gasoline on a trip of 160 mi. At the same mileage rate, how much gasoline will a 384-mi trip require?

96. **Social Science** A car uses 12 L of gasoline in traveling 150 km. At that rate, how many liters of gasoline will be used in a trip of 400 km?

97. **Business and Finance** Sveta earns $13,500 commission in 20 weeks in her new sales position. At that rate, how much will she earn in 1 year (52 weeks)?

98. **Business and Finance** Kevin earned $165 interest for 1 year on an investment of $1,500. At the same rate, what amount of interest would be earned by an investment of $2,500?

99. **Social Science** A company is selling a natural insect control that mixes ladybug beetles and praying mantises in the ratio of 7 to 4. If there are a total of 110 insects per package, how many of each type of insect is in a package?

100. **Social Science** A woman casts a 4-ft shadow. At the same time, a 72-ft building casts a 48-ft shadow. How tall is the woman?

101. **Business and Finance** A brother and sister are to divide an inheritance of \$12,000 in the ratio of 2 to 3. What amount will each receive?

102. **Business and Finance** In Bucks County, the property tax rate is \$25.32 per \$1,000 of assessed value. If a house and property have a value of \$128,000, find the tax the owner will have to pay.

Determine whether each statement is **true** *or* **false.**

103. When solving a rational equation, we need to multiply both numerator and denominator of a rational expression by the same nonzero quantity.

104. When solving a rational equation, we usually multiply each side of the equation by the same nonzero quantity, to clear it of fractions.

Complete each statement with **always, sometimes,** *or* **never.**

105. We must ________ check a possible solution to a rational equation, in case we have an extraneous solution.

106. After clearing fractions in a rational equation, we ________ obtain a quadratic equation to solve.

107. The value of the term $\frac{5}{x}$, $x \neq 0$, is ____________ equal to 1.

108. The value of the term $\frac{0}{x}$, $x \neq 0$, is ____________ equal to 1.

109. The value of the term $\frac{5}{x}$, $x \neq 0$, is ____________ equal to 0.

110. The value of the term $\frac{0}{x}$, $x \neq 0$, is ____________ equal to 0.

Skills | Calculator/Computer | **Career Applications** | Above and Beyond

Electrical Engineering *One formula for determining the equivalent resistance in a parallel circuit is*

$$\frac{1}{R_{eq}} = \frac{1}{R_1} + \frac{1}{R_2}$$

Use this formula to complete exercises 111 and 112.

111. Find the unknown resistance in the figure shown, to the nearest hundredth of an ohm, if the total (equivalent) resistance is 0.41 Ω.

112. Find the unknown resistance in the figure shown, to the nearest ohm, if the total (equivalent) resistance is 26 Ω.

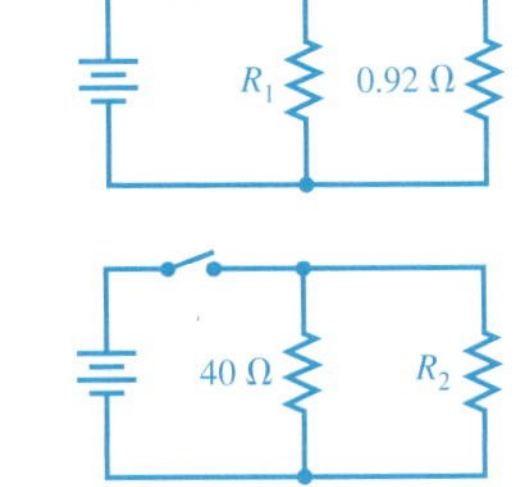

Skills | Calculator/Computer | Career Applications | **Above and Beyond**

113. What special considerations must be made when an equation contains rational expressions with variables in a denominator?

chapter 9 > Make the Connection

Answers

1. Equation, $\{40\}$ **3.** Expression, $\frac{3x}{10}$ **5.** Equation, $\{5\}$ **7.** Equation, $\{3\}$ **9.** $\{5\}$ **11.** $\{8\}$ **13.** $\left\{\frac{3}{2}\right\}$ **15.** $\{3\}$ **17.** $\left\{-\frac{9}{5}\right\}$ **19.** $\left\{-\frac{2}{3}\right\}$ **21.** No solution or $\varnothing$ **23.** $\{-23\}$ **25.** $\{9\}$ **27.** $\{4\}$ **29.** $\{8\}$ **31.** $\{4\}$ **33.** $\left\{\frac{3}{2}\right\}$ **35.** No solution or $\varnothing$ **37.** $\{7\}$ **39.** $\{5\}$ **41.** $\left\{-\frac{5}{2}\right\}$ **43.** $\left\{-\frac{1}{2}, 6\right\}$ **45.** $\left\{-\frac{1}{2}\right\}$ **47.** $\left\{-\frac{1}{3}, 7\right\}$ **49.** $\{-8, 9\}$ **51.** 4 **53.** 32 **55.** 3 **57.** 13 **59.** 3 **61.** 10 **63.** 6 **65.** 10 **67.** Sides are 3 and 9 on the first triangle and 8 and 24 on the second. **69.** $x = \frac{ab}{b - a}$ **71.** $R_1 = \frac{RR_2}{R_2 - R}$ **73.** $x = \frac{y + 1}{y - 1}$ **75.** 6, 24 **77.** 9 **79.** 4 mi/hr **81.** 150 mi/hr **83.** 36 min **85.** Bicycling: 25 mi/hr; driving: 55 mi/hr **87.** First leg: 3 hr; second leg: 2 hr **89.** 6 hr **91.** 6 hr **93.** 220 ft/s **95.** 12 gal **97.** \$35,100 **99.** 70 ladybugs, 40 praying mantises **101.** Brother \$4,800, sister \$7,200 **103.** False **105.** always **107.** sometimes **109.** never **111.** 0.74 Ω **113.** Above and Beyond

summary :: chapter 9

Definition/Procedure	Example	Reference
Simplifying Rational Expressions		**Section 9.1**
Rational expressions have the form $\frac{P}{Q}$ in which P and Q are polynomials and $Q \neq 0$.	$\frac{x^2 - 5x}{x - 3}$ is a rational expression. The variable x cannot have the value 3.	*p.* 690
Fundamental Principle of Rational Expressions For polynomials P, Q, and R, $\frac{P}{Q} = \frac{PR}{QR}$ where $Q \neq 0$ and $R \neq 0$. This principle can be used in two ways. We can multiply or divide the numerator and denominator of a rational expression by the same nonzero polynomial.	$\frac{R}{R} = 1$ when $R \neq 0$.	*p.* 692
Simplifying Rational Expressions **Step 1** Completely factor both the numerator and the denominator of the expression. **Step 2** Divide the numerator and denominator by the GCF. The resulting expression is in simplest form (or in lowest terms).	$\frac{x^2 - 4}{x^2 - 2x - 8}$ $= \frac{(x - 2)(x + 2)}{(x - 4)(x + 2)}$ $= \frac{x - 2}{x - 4}$	*p.* 694
Identifying Rational Functions A rational function is a function that is defined by a rational expression. It can be written as $f(x) = \frac{P(x)}{Q(x)}$ where $P(x)$ and $Q(x)$ are polynomial functions, $Q(x) \neq 0$.	$f(x) = \frac{2x^2 - 3x}{x + 1}$ and $g(x) = \frac{3}{x^2 - 3}$ are both rational functions	*p.* 696
Multiplying and Dividing Rational Expressions		**Section 9.2**
Multiplying Rational Expressions For polynomials P, Q, R, and S, $\frac{P}{Q} \cdot \frac{R}{S} = \frac{PR}{QS}$ where $Q \neq 0, S \neq 0$		*p.* 705
In practice, we apply this algorithm to multiply two rational expressions. **Step 1** Completely factor the numerators and denominators. **Step 2** Divide (or "cancel") by any common factors. **Step 3** Multiply as needed to form the product.	$\frac{2x - 6}{x^2 - 9} \cdot \frac{x^2 + 3x}{6x + 24}$ $= \frac{2(x - 3)}{(x - 3)(x + 3)} \cdot \frac{x(x + 3)}{6(x + 4)}$ $= \frac{x}{3(x + 4)}$	*p.* 706
Dividing Rational Expressions For polynomials P, Q, R, and S, $\frac{P}{Q} \div \frac{R}{S} = \frac{P}{Q} \cdot \frac{S}{R} = \frac{PS}{QR}$ where $Q \neq 0, R \neq 0, S \neq 0$. To divide two rational expressions, **Step 1** Invert the divisor (the *second* rational expression) to make a multiplication problem. **Step 2** Multiply and simplify the rational expressions.	$\frac{5y}{2y - 8} \div \frac{10y^2}{y^2 - y - 12}$ $= \frac{5y}{2y - 8} \cdot \frac{y^2 - y - 12}{10y^2}$ $= \frac{5y}{2(y - 4)} \cdot \frac{(y - 4)(y + 3)}{10y^2}$ $= \frac{y + 3}{4y}$	*p.* 706

Continued

Definition/Procedure	Example	Reference
Adding and Subtracting Rational Expressions		**Section 9.3**
Adding and Subtracting Rational Expressions To add or subtract rational expressions with the same denominator, add or subtract their numerators and then write that sum over the common denominator. The result should be written in lowest terms. In symbols, $\frac{P}{R}+\frac{Q}{R}=\frac{P+Q}{R}$ and $\frac{P}{R}-\frac{Q}{R}=\frac{P-Q}{R}$ where $R \neq 0$.	$\frac{5w}{w^2-16}-\frac{20}{w^2-16}$ $=\frac{5w-20}{w^2-16}$ $=\frac{5(w-4)}{(w+4)(w-4)}$ $=\frac{5}{w+4}$	p. 713
Least Common Denominator The least common denominator (LCD) of a group of rational expressions is the simplest polynomial that is divisible by each of the individual denominators of the rational expressions. To find the LCD, **Step 1** Write each of the denominators in completely factored form. **Step 2** Write the LCD as the product of each prime factor, to the highest power to which it appears in the factored form of any of the individual denominators.	To find the LCD for $\frac{2}{x^2+2x+1}$ and $\frac{3}{x^2+x}$ factor the denominators. $x^2+2x+1=(x+1)(x+1)$ $x^2+x=x(x+1)$ The LCD is $x(x+1)(x+1)$	p. 714
To add or subtract rational expressions with different denominators, we first find the LCD by the procedure outlined above. We then write each of the rational expressions with that LCD as a common denominator. Then we can add or subtract as before.	$\frac{2}{(x+1)^2}-\frac{3}{x(x+1)}$ $=\frac{2\cdot x}{(x+1)^2x}-\frac{3(x+1)}{x(x+1)(x+1)}$ $=\frac{2x-3(x+1)}{x(x+1)(x+1)}$ $=-\frac{x+3}{x(x+1)(x+1)}$	p. 716
Complex Rational Expressions		**Section 9.4**
A complex fraction or **complex rational expression** is a rational expression that also contains nontrivial rational expressions in the numerator or denominator. In other words, there is at least one fraction within the fraction. There are two commonly used methods for simplifying complex fractions. **Method 1** 1. Multiply the numerator and denominator of the expression by the LCD of all the fractions that appear within the numerator and denominator. 2. Simplify the resulting rational expression. **Method 2** 1. Write the numerator and denominator of the complex fraction as single fractions, if necessary. 2. Invert the denominator and multiply. Simplify your result.	Simplify $\frac{1-\frac{2}{x}}{1-\frac{4}{x^2}}$. Method 1: $\frac{\left(1-\frac{2}{x}\right)x^2}{\left(1-\frac{4}{x^2}\right)x^2}$ $=\frac{x^2-2x}{x^2-4}=\frac{x(x-2)}{(x+2)(x-2)}$ $=\frac{x}{x+2}$ Method 2: $\frac{\frac{x-2}{x}}{\frac{x^2-4}{x^2}}$ $=\frac{x-2}{x}\cdot\frac{x^2}{x^2-4}$ $=\frac{x-2}{x}\cdot\frac{x^2}{(x+2)(x-2)}$ $=\frac{x}{x+2}$	p. 725

Continued

Definition/Procedure	Example	Reference
Graphing Rational Functions		Section 9.5
To draw the graph of a rational function: **Step 1** Determine the domain of *f*. **Step 2** Find the intercepts of the graph of *f*. Plot these. **Step 3** Locate vertical and horizontal asymptotes. Draw these as dashed lines. **Step 4** Choose a few easy-to-compute points to plot. **Step 5** Connect plotted points with a smooth curve, allowing the curve to approach the asymptotes.	Graph $f(x) = \frac{6 - 2x}{x + 3}$. Domain: $\{x \mid x \neq -3\}$ y-intercept: (0, 2) x-intercept: (3, 0) Vertical asymptote: $x = -3$ Horizontal asymptote: $y = -2$	p. 738
To find a vertical asymptote: **(1)** locate any x-value for which f is undefined by setting the denominator equal to 0; **(2)** investigate values of $f(x)$ close to an undefined x-value. To find a horizontal asymptote, investigate $f(x)$ values for large positive and negative values of x.	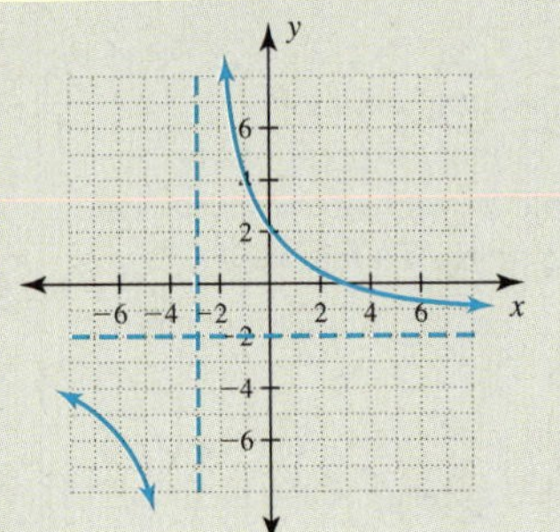	
Rational Equations and Problem Solving		Section 9.6
To solve an equation involving rational expressions, **Step 1** Clear the equation of fractions by multiplying both sides of the equation by the LCD of all the fractions that appear. **Step 2** Solve the equation resulting from step 1. **Step 3** Check all solutions by substitution in the original equation. Discard any extraneous solutions.	Solve. $\frac{3}{x-3} - \frac{2}{x+2} = \frac{19}{x^2 - x - 6}$ Multiply by the LCD $(x - 3)(x + 2)$. $3(x + 2) - 2(x - 3) = 19$ $3x + 6 - 2x + 6 = 19$ $x = 7$ *Check* $\frac{3}{4} - \frac{2}{9} \stackrel{?}{=} \frac{19}{36}$ $\frac{19}{36} = \frac{19}{36}$ True	p. 752

summary exercises :: chapter 9

This summary exercise set will help ensure that you have mastered each of the objectives of this chapter. The exercises are grouped by section. You should reread the material associated with any exercises that you find difficult. The answers to the odd-numbered exercises are in the Answers Appendix in the back of the text.

9.1 *For what value(s) of the variable is each rational expression defined?*

1. $\frac{x}{2}$ **2.** $\frac{3}{y}$ **3.** $\frac{2}{x - 5}$ **4.** $\frac{4x}{3x - 2}$

Simplify each rational expression.

5. $\frac{18x^5}{24x^3}$ **6.** $\frac{15m^3n}{-5mn^2}$ **7.** $\frac{7y - 49}{y - 7}$ **8.** $\frac{5x - 20}{x^2 - 16}$

9. $\frac{9 - x^2}{x^2 + 2x - 15}$

10. $\frac{3w^2 + 8w - 35}{2w^2 + 13w + 15}$

11. $\frac{6a^2 - ab - b^2}{9a^2 - b^2}$

12. $\frac{6w - 3z}{8w^3 - z^3}$

Simplify each function. Indicate any value for x for which the function is undefined.

13. $f(x) = \frac{x^2 - 3x - 4}{x + 1}$

14. $f(x) = \frac{x^2 + x - 6}{x - 2}$

9.2 *Multiply or divide as indicated. Simplify your results.*

15. $\frac{x^7}{36} \cdot \frac{24}{x^4}$

16. $\frac{a^3b}{4ab^2} \div \frac{ab}{12ab^2}$

17. $\frac{6y - 18}{9y} \cdot \frac{10}{5y - 15}$

18. $\frac{m^2 - 3m}{m^2 - 5m + 6} \cdot \frac{m^2 - 4}{m^2 + 7m + 10}$

19. $\frac{a^2 - 2a}{a^2 - 4} \div \frac{2a^2}{3a + 6}$

20. $\frac{r^2 + 2rs}{r^3 - r^2s} \div \frac{5r + 10s}{r^2 - 2rs + s^2}$

21. $\frac{x^2 - 2xy - 3y^2}{x^2 - xy - 2y^2} \cdot \frac{x^2 - 4y^2}{x^2 - 8xy + 15y^2}$

22. $\frac{w^3 + 3w^2 + 2w + 6}{w^4 - 4} \div (w^3 + 27)$

23. Let $f(x) = \frac{x^2 - 16}{x - 5}$ and $g(x) = \frac{x^2 - 25}{x + 4}$. Find **(a)** $f(3) \cdot g(3)$; **(b)** $h(x) = f(x) \cdot g(x)$; and **(c)** $h(3)$.

24. Let $f(x) = \frac{2x^2 - 5x - 3}{x - 4}$ and $g(x) = \frac{x^2 - 3x - 4}{2x^2 + 5x + 2}$. Find **(a)** $f(3) \cdot g(3)$; **(b)** $h(x) = f(x) \cdot g(x)$; and **(c)** $h(3)$.

9.3 *Perform the indicated operations. Simplify your results.*

25. $\frac{5x + 7}{x + 4} - \frac{2x - 5}{x - 4}$

26. $\frac{5}{6x^2} + \frac{3}{4x}$

27. $\frac{2}{x - 5} - \frac{1}{x}$

28. $\frac{2}{y + 5} + \frac{3}{y + 4}$

29. $\frac{2}{3m - 3} - \frac{5}{2m - 2}$

30. $\frac{7}{x - 3} - \frac{5}{3 - x}$

31. $\frac{6}{3x + 3} - \frac{6}{5x - 5}$

32. $\frac{2a}{a^2 - 9a + 20} + \frac{8}{a - 4}$

33. $\frac{2}{s - 1} - \frac{6s}{s^2 + s - 2}$

34. $\frac{4}{x^2 - 9} - \frac{3}{x^2 - 4x + 3}$

35. $\frac{x^2 - 14x - 8}{x^2 - 2x - 8} + \frac{2x}{x - 4} - \frac{3}{x + 2}$

36. $\frac{w^2 + 2wz + z^2}{w^2 - wz - 2z^2} \cdot \left(\frac{3}{w + z} - \frac{1}{w - z}\right)$

37. Let $f(x) = \frac{2x}{x - 2}$ and $g(x) = \frac{x}{x - 3}$. Find **(a)** $f(4) + g(4)$; **(b)** $h(x) = f(x) + g(x)$; and **(c)** the ordered pair $(4, h(4))$.

38. Let $f(x) = \frac{x + 2}{x - 2}$ and $g(x) = \frac{x + 1}{x - 7}$. Find **(a)** $f(3) + g(3)$; **(b)** $h(x) = f(x) + g(x)$; and **(c)** the ordered pair $(3, h(3))$.

9.4 *Simplify each complex rational expression.*

39. $\dfrac{\dfrac{x^3}{15}}{\dfrac{x^5}{10}}$

40. $\dfrac{\dfrac{y-1}{y^2-4}}{\dfrac{y^2-1}{y^2-y-2}}$

41. $\dfrac{1+\dfrac{a}{b}}{1-\dfrac{a}{b}}$

42. $\dfrac{2-\dfrac{x}{y}}{4-\dfrac{x^2}{y^2}}$

43. $\dfrac{\dfrac{1}{r^2}-\dfrac{1}{s^2}}{\dfrac{1}{r}-\dfrac{1}{s}}$

44. $\dfrac{1-\dfrac{1}{x+2}}{1+\dfrac{1}{x+2}}$

45. $\dfrac{1-\dfrac{2}{x-1}}{x+\dfrac{3}{x-4}}$

46. $\dfrac{\dfrac{w}{w+1}-\dfrac{1}{w-1}}{\dfrac{w}{w-1}+\dfrac{1}{w+1}}$

47. $\dfrac{1}{1-\dfrac{1}{1-\dfrac{1}{y-1}}}$

48. $1-\dfrac{1}{1+\dfrac{1}{1-\dfrac{1}{x}}}$

49. $\dfrac{1-\dfrac{1}{x-1}}{x-\dfrac{8}{x+2}}$

50. $\dfrac{1}{1-\dfrac{1}{1+\dfrac{1}{y+1}}}$

9.5 *Find the domain of each function.*

51. $f(x)=\dfrac{x-3}{x+4}$

52. $f(x)=\dfrac{7}{5-x}$

53. $f(x)=\dfrac{8x}{x-2}$

54. $f(x)=\dfrac{x-6}{x}$

55. $f(x)=\dfrac{2x-6}{x+1}$

56. $f(x)=\dfrac{4-2x}{x-4}$

Find all intercepts of each function.

57. $f(x)=\dfrac{x-3}{x+4}$

58. $f(x)=\dfrac{7}{5-x}$

59. $f(x)=\dfrac{8x}{x-2}$

60. $f(x)=\dfrac{x-6}{x}$

61. $f(x)=\dfrac{2x-6}{x+1}$

62. $f(x)=\dfrac{4-2x}{x-4}$

Identify all asymptotes of each function.

63. $f(x)=\dfrac{x-3}{x+4}$

64. $f(x)=\dfrac{7}{5-x}$

65. $f(x)=\dfrac{8x}{x-2}$

66. $f(x)=\dfrac{x-6}{x}$

67. $f(x)=\dfrac{2x-6}{x+1}$

68. $f(x)=\dfrac{4-2x}{x-4}$

Graph each function.

69. $f(x)=\dfrac{x-3}{x+4}$

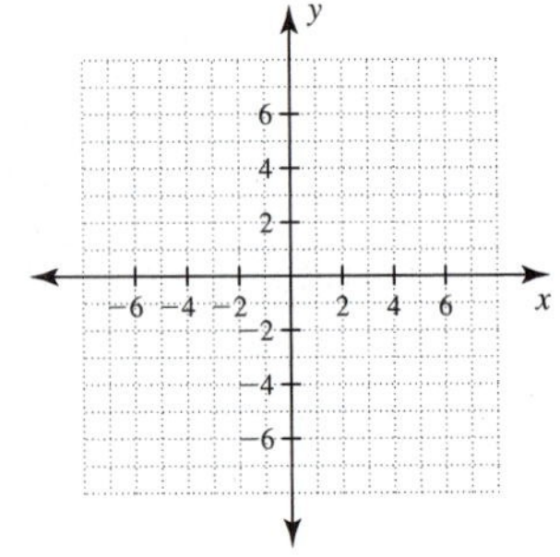

70. $f(x)=\dfrac{x-6}{x}$

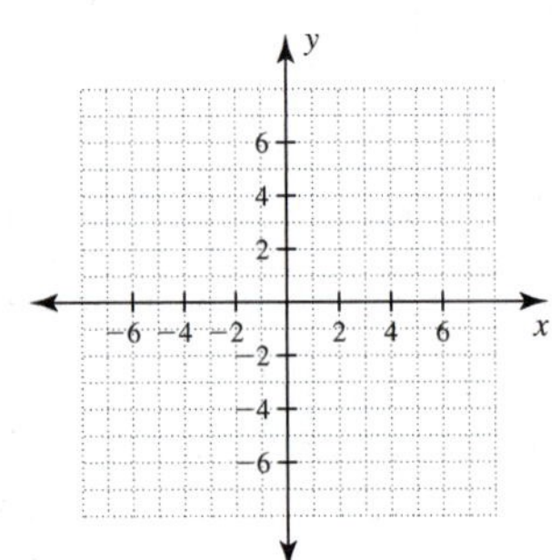

71. $f(x) = \frac{2x - 6}{x + 1}$

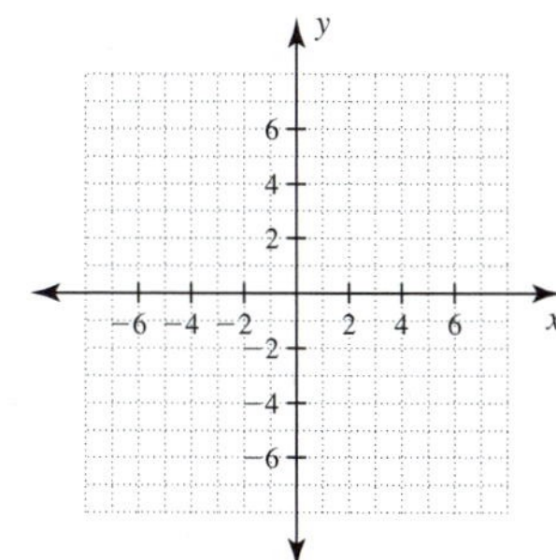

72. $f(x) = \frac{4 - 2x}{x - 4}$

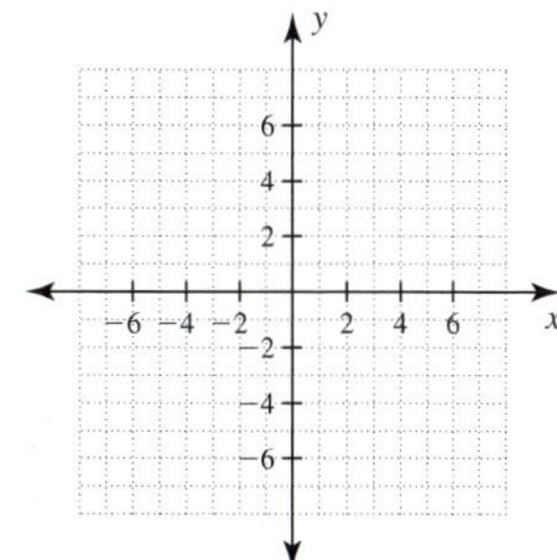

9.6 *Solve each equation.*

73. $\frac{1}{2x} + \frac{1}{3x} = \frac{1}{6}$

74. $\frac{5}{2x^2} - \frac{1}{4x} = \frac{1}{x}$

75. $\frac{x}{x - 2} + 1 = \frac{x + 4}{x - 2}$

76. $\frac{2x - 1}{x - 3} - \frac{5}{x - 3} = 1$

77. $\frac{2}{3x + 1} = \frac{1}{x + 2}$

78. $\frac{5}{x + 1} + \frac{1}{x - 2} = \frac{7}{x + 1}$

79. $\frac{4}{x - 1} - \frac{5}{3x - 7} = \frac{3}{x - 1}$

80. $\frac{7}{x} - \frac{1}{x - 3} = \frac{9}{x^2 - 3x}$

81. $\frac{2}{x - 3} - \frac{11}{x^2 - 9} = \frac{3}{x + 3}$

82. $\frac{5}{x + 3} + \frac{1}{x - 5} = \frac{1}{x + 3}$

83. $\frac{2}{x - 4} = \frac{x}{x - 2} - \frac{x + 4}{x^2 - 6x + 8}$

84. $\frac{x}{x - 5} = \frac{3x}{x^2 - 7x + 10} + \frac{8}{x - 2}$

In exercises 85 and 86, two similar triangles are given. Find the lengths of the indicated sides.

85.

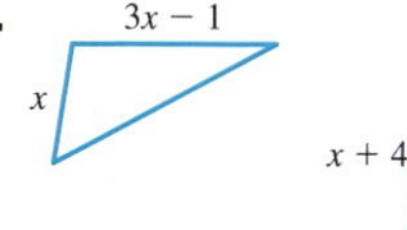

86.

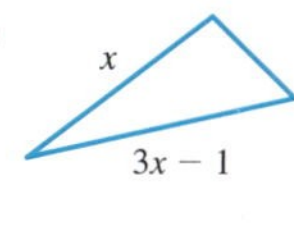

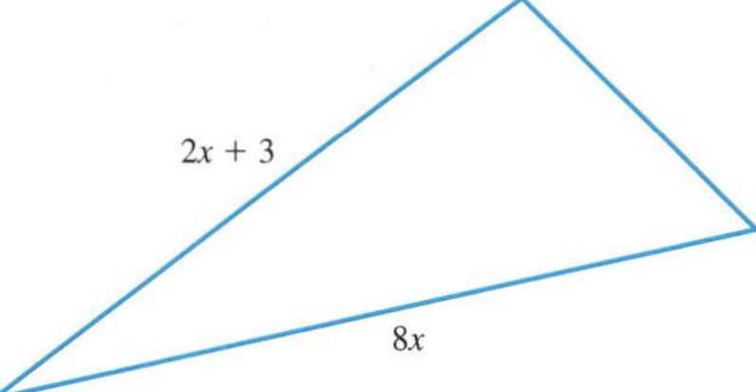

Solve.

87. Number Problem The sum of the reciprocals of two consecutive integers is equal to 11 times the reciprocal of the product of those two integers. What are the two integers?

88. Science and Medicine Karl drove 224 mi on the expressway for a business meeting. On his return, he decided to use a shorter route of 200 mi, but road construction slowed his speed by 6 mi/hr. If the trip took the same time each way, what was his average speed in each direction?

89. Construction An electrician can wire a certain model home in 20 hr while it would take her apprentice 30 hr to wire the same model. How long would it take the two of them working together to wire the house?

90. Science and Medicine A light plane took 1 hr longer to fly 540 mi on the first portion of a trip than to fly 360 mi on the second. If the rate was the same for each portion, what was the flying time for each leg of the trip?

chapter test 9

Use this chapter test to assess your progress and to review for your next exam. Allow yourself about an hour to take this test. The answers to these exercises are in the Answers Appendix in the back of the text.

Perform the indicated operations. Simplify all results.

1. $\dfrac{m^2 - 3m}{m^2 - 9} \div \dfrac{4m}{m^2 - m - 12}$

2. $\dfrac{2}{x + 3} + \dfrac{12}{x^2 - 9}$

3. $\dfrac{3 - \frac{x}{y}}{9 - \frac{x^2}{y^2}}$

4. $\dfrac{3ab^2}{5ab^3} \cdot \dfrac{20a^2b}{21b}$

5. $\dfrac{3}{x^2 - 3x - 4} + \dfrac{5}{x^2 - 16}$

6. $\dfrac{x^2 - 3x}{5x^2} \cdot \dfrac{10x}{x^2 - 4x + 3}$

7. $\dfrac{9x^2 - 9x - 4}{6x^2 - 11x + 3} \cdot \dfrac{15 - 10x}{3x - 4}$

8. $\dfrac{1 - \dfrac{10}{z + 3}}{2 - \dfrac{12}{z - 1}}$

9. $\dfrac{x^2 + 3xy}{2x^3 - x^2y} \div \dfrac{x^2 + 6xy + 9y^2}{4x^2 - y^2}$

10. $\dfrac{6x}{x^2 - x - 2} - \dfrac{2}{x + 1}$

11. $\dfrac{5}{x - 2} - \dfrac{1}{x}$

Solve.

12. $\dfrac{5}{x} - \dfrac{x - 3}{x + 2} = \dfrac{22}{x^2 + 2x}$

Simplify.

13. $\dfrac{3w^2 + w - 2}{3w^2 - 8w + 4}$

14. $\dfrac{x^3 + 2x^2 - 3x}{x^3 - 3x^2 + 2x}$

15. $-\dfrac{21x^5y^3}{28xy^5}$

16. If the numerator of $\frac{4}{7}$ is multiplied by a number and that same number is added to the denominator, the result is $\frac{6}{5}$. What was that number?

Graph each function.

17. $f(x) = \dfrac{3x}{x + 1}$

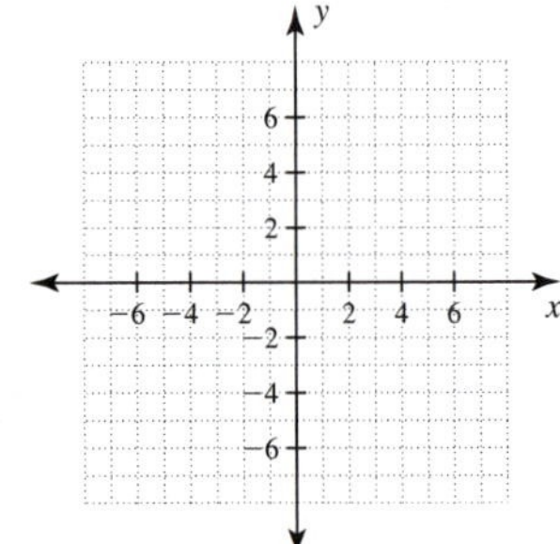

18. $f(x) = -\dfrac{10}{x}$

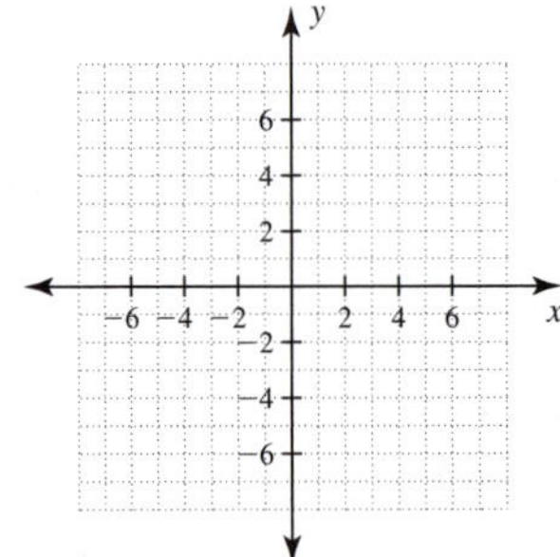

Simplify. Indicate any value of x for which the function is undefined.

19. $f(x) = \dfrac{x^2 - 5x + 4}{x - 4}$

20. **(a)** Find the intercepts of the function $f(x) = \dfrac{4x - 1}{3x + 2}$.

(b) Identify the asymptotes of f.

cumulative review chapters 0–9

Use this exercise set to review concepts from earlier chapters. While it is not a comprehensive exam, it will help you identify any material that you need to review before moving on to the next chapter. The answers to these exercises are in the Answers Appendix in the back of the text.

1. Solve the equation $5x - 3(2x + 6) = 4 - (3x - 2)$.

2. If $f(x) = 5x^4 - 3x^2 + 7x - 9$, find $f(-1)$.

3. Find an equation for the line that is parallel to the line $6x + 7y = 42$ and has a y-intercept of $(0, -3)$.

4. Find the x- and y-intercepts of the equation $7x - 6y = -42$.

Simplify each polynomial function.

5. $f(x) = 3x - 2[x - (3x - 1)] + 6x(x - 2)$

6. $f(x) = x(2x - 1)(x + 3)$

7. Find the domain of the function $f(x) = \dfrac{x^2 + x}{x}$.

8. Evaluate the expression $6^2 - (16 \div 8 \cdot 2) - 4^2$.

Factor each polynomial completely.

9. $6x^3 + 7x^2 - 3x$

10. $16x^{16} - 9y^8$

Simplify each rational expression.

11. $\dfrac{5}{x - 1} - \dfrac{2x + 6}{x^2 + 2x - 3}$

12. $\dfrac{x + 1}{x^2 - 5x - 6} \div \dfrac{x^2 - 1}{x - 6}$

13. $\dfrac{1 - \dfrac{3}{x + 3}}{\dfrac{1}{x^2 - 9}}$

14. The height of a ball thrown into the air from a platform can be determined by the function

$h(t) = -16t^2 + 58t + 15$

To the nearest hundredth of a second, when will the ball be at a height of 50 ft?

Solve each equation.

15. $7x + (x - 10) = -12(x - 5)$

16. $x^4 - 18x^2 + 32 = 0$

17. $-4(7x + 6) = 8(5x + 12)$

18. $6 - 2\sqrt{x - 3} = x$

19. $\dfrac{5}{x} = \dfrac{2}{x + 3}$

Solve the inequality.

20. $-4(-2x - 7) > -6x$

21. If the hypotenuse of a right triangle has length 22 cm, and one leg has length 15 cm, find the length of the other leg to the nearest tenth of a centimeter.

22. Identify the asymptotes of $f(x) = \frac{5 - 2x}{x - 5}$.

23. Simplify the expression $\left(\frac{a^{-2}b}{a^3b^{-2}}\right)^2$.

Solve each application.

24. When each works alone, Barry can mow a lawn in 3 hr less time than Don. When they work together, it takes 2 hr. How long does it take each to do the job by himself?

25. The length of a rectangle is 2 cm less than twice the width. The area of the rectangle is 180 cm^2. Find the length and width of the rectangle.

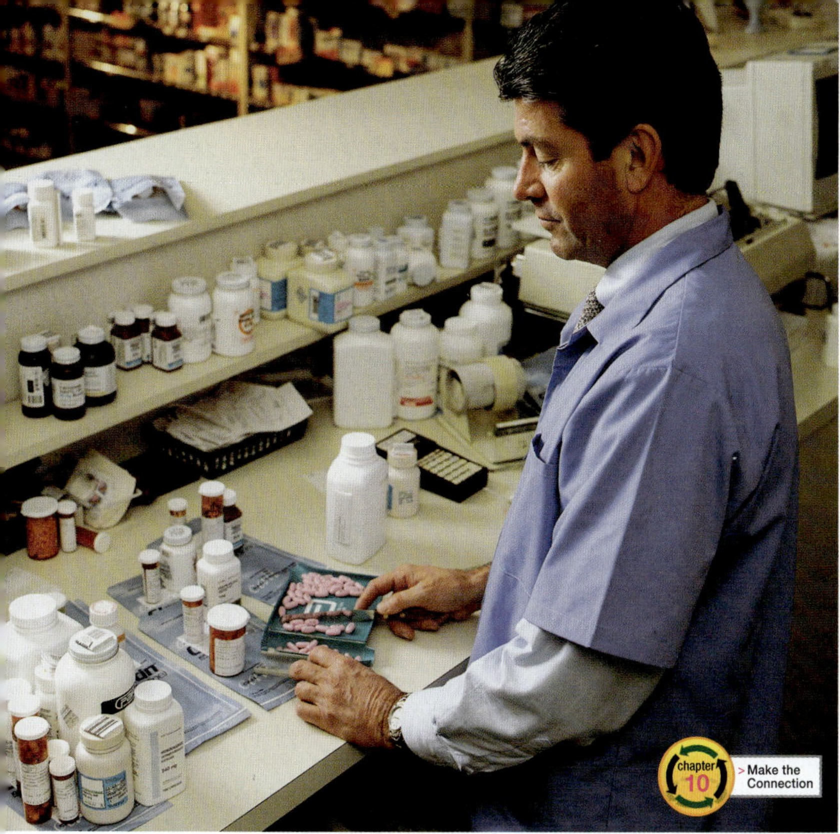

CHAPTER

10

Exponential and Logarithmic Functions

INTRODUCTION

You undoubtedly realize that the sciences apply mathematics in many ways. Chemistry, physics, and pharmacology are all scientific fields that use exponential and logarithmic functions to model phenomena.

For instance, pharmacologists use exponential functions to model drug absorption and elimination. When a patient takes a medication, it is usually distributed throughout the body by the circulatory or respiratory system. In order to be effective, there must be enough medication in a patient's body to achieve the desired effect, but not enough to cause harm.

Therapeutic levels are maintained by taking the right dosage at timed intervals. The correct dosage and amount of time between dosages are determined by the rate at which the body absorbs and eliminates the medication.

You will have the opportunity to investigate this elimination by experimenting with decay in the Chapter 10 Activity.

CHAPTER 10 OUTLINE

10.1 Algebra of Functions

< 10.1 Objectives >

1 > Find the sum and difference of functions

2 > Find the domain of a combination function

3 > Find the product of functions

4 > Find the quotient of functions

From the growth of an investment or population to the rate at which medicine leaves your body, there are a whole slew of applications that we tackle in this chapter. In order to model such applications properly, we need a new family of functions.

You will learn about *exponential* and *logarithmic functions* and use them to model applications and solve problems. Before doing that, we look to gain a deeper and more general understanding of functions.

We begin by seeing the ways that we combine functions algebraically. You have already worked with some examples. For instance, we construct profit models by finding the difference between revenue and cost functions.

$P(x) = R(x) - C(x)$

We model many applications by combining two or more functions. In this section, we look at the properties we use to combine functions.

Definition

Adding Functions

The **sum of the functions** f and g is written as $f + g$ and is defined as

$(f + g)(x) = f(x) + g(x)$

for every value in the domain of both f and g.

Definition

Subtracting Functions

The **difference of the functions** f and g is written as $f - g$ and is defined as

$(f - g)(x) = f(x) - g(x)$

for every value in the domain of both f and g.

Example 1 Adding and Subtracting Functions

< Objective 1 >

Suppose the functions f and g are defined by the tables.

x	$f(x)$
-4	-8
0	6
2	5
1	-2

x	$g(x)$
-4	3
0	-5
2	7
3	1

RECALL

From the first table, we know that $f(-4) = -8$. From the second table we see that $g(-4) = 3$

(a) Evaluate $(f + g)(-4)$.

$$\begin{aligned}(f + g)(-4) &= f(-4) + g(-4)\\ &= -8 + 3\\ &= -5\end{aligned}$$

(b) Evaluate $(f - g)(0)$.

$$\begin{aligned}(f - g)(0) &= f(0) - g(0)\\ &= 6 - (-5) \qquad f(0) = 6;\ g(0) = -5\\ &= 11\end{aligned}$$

(c) Evaluate $(f + g)(3)$.

$$(f + g)(3) = f(3) + g(3)$$

From the second table, we see that $g(3) = 1$. However, 3 is not an x-value in the first table, so 3 is not in the domain of f. Therefore, 3 is not part of the domain of $f + g$ and $(f + g)(3)$ does not exist.

(d) Find the domain of $f + g$.

We need to find all values of x that are in the domains of *both f and g*.

$D = \{-4, 0, 2\}$

Check Yourself 1

Suppose the functions f and g are defined by the tables.

x	$f(x)$
−3	7
−1	0
5	−3
6	2

x	$g(x)$
−3	−4
2	8
5	−6
7	0

(a) Evaluate $(f + g)(5)$.
(b) Evaluate $(f - g)(-3)$.
(c) Evaluate $(f - g)(-1)$.
(d) Find the domain of $f - g$.

When we combine functions, we are creating a new function. As you saw in part (d) of Example 1, this new function has its own domain. Based on our work in the example, we can even construct a table for this function.

Example 2 Building a Combination Function

Use the functions f and g in Example 1 to build a table of values for the function

$$h(x) = (f + g)(x)$$

In Example 1, we found the domain of $f + g$ to be $\{-4, 0, 2\}$ and computed the output for $x = -4$.

$$\begin{aligned}h(2) &= (f + g)(2) &&h \text{ is defined as } f + g.\\ &= f(2) + g(2) &&h(2) = f(2) + g(2)\\ &= 5 + 7 &&\text{From the tables, we have } f(2) = 5 \text{ and } g(2) = 7.\\ &= 12\end{aligned}$$

Similarly,

$$h(0) = f(0) + g(0)$$
$$= 6 + (-5)$$
$$= 1$$

We can now construct a table for $h = f + g$.

x	$h(x)$
−4	−5
0	1
2	12

Check Yourself 2

Use the functions f and g from Check Yourself 1 to build a table of values for the function

$$h(x) = (f + g)(x)$$

In Example 3, we look at functions that are defined by equations rather than tables.

Example 3 Combining Functions

Let $f(x) = 2x - 1$ and $g(x) = -3x + 4$.

(a) Find $(f + g)(x)$.

$$(f + g)(x) = f(x) + g(x)$$
$$= (2x - 1) + (-3x + 4) = -x + 3$$

(b) Find $(f - g)(x)$.

$$(f - g)(x) = f(x) - g(x)$$
$$= (2x - 1) - (-3x + 4) = 5x - 5$$

(c) Evaluate $(f + g)(2)$.

If we use the definition of the sum of the functions, we find that

$$(f + g)(2) = f(2) + g(2) \qquad f(2) = 2(2) - 1 = 3$$
$$= 3 + (-2) = 1 \qquad g(2) = -3(2) + 4 = -2$$

Alternatively, use part (a).

$$(f + g)(x) = -x + 3$$

therefore,

$$(f + g)(2) = -(2) + 3 = 1$$

Check Yourself 3

Let $f(x) = -2x - 3$ and $g(x) = 5x - 1$.

(a) Find $(f + g)(x)$. **(b)** Find $(f - g)(x)$.
(c) Evaluate $(f + g)(2)$.

We saw two ways of completing part (c). We can use the definition of the sum of functions to write

$$(f + g)(2) = f(2) + g(2)$$

or we can construct the function

$$h(x) = (f + g)(x)$$
$$= -x + 3 \qquad \text{From part (a).}$$

and use this new construct to evaluate the sum when $x = 2$.

Both methods always work when we need to evaluate a combination function. If we only need to evaluate the combination at one point, we usually use the first method. If we need to use the combination function to evaluate several points, then we opt for the second method.

In any case, it is helpful to determine where a combination function exists (its domain). We illustrate this in Example 4.

Example 4 Finding the Domain of a Sum

< Objective 2 >

Let $f(x) = 2x - 4$ and $g(x) = \frac{1}{x}$.

(a) Find $(f + g)(x)$.

$(f + g)(x) = (2x - 4) + \left(\frac{1}{x}\right) = 2x - 4 + \frac{1}{x}$

RECALL

We sometimes use $\mathbb{R}$ to denote the set of real numbers.

(b) Find the domain of $f + g$.

The domain of $f + g$ is the set of all numbers in the domain of f and also in the domain of g. The domain of f consists of all real numbers. The domain of g consists of all real numbers except 0 because we cannot divide by 0. The domain of $f + g$ is the set of all real numbers except 0, $D = \{x \mid x \neq 0\}$.

Check Yourself 4

Let $f(x) = -3x + 1$ and $g(x) = \frac{1}{x - 2}$.

(a) Find $(f + g)(x)$. **(b)** Find the domain of $f + g$.

NOTE

As with addition and subtraction, you create a new function when multiplying or dividing functions.

So far, we have added and subtracted functions. As you might expect, we can also multiply and divide functions. We define these combinations in a manner similar to the definitions of addition and subtraction.

Definition

Multiplying Functions

The **product of the functions** f and g is written as $f \cdot g$ and is defined as

$(f \cdot g)(x) = f(x) \cdot g(x)$

for every value in the domain of both f and g.

Definition

Dividing Functions

The **quotient of the functions** f and g is written as $f \div g$ or $\frac{f}{g}$ and is defined as

$(f \div g)(x) = f(x) \div g(x)$ or $\left(\frac{f}{g}\right)(x) = \frac{f(x)}{g(x)}$

for every value in the domain of both f and g, such that $g(x) \neq 0$.

Example 5 Multiplying and Dividing Functions

< Objectives 3 and 4 >

Let the functions f and g be defined by the tables

x	$f(x)$
-3	7
-1	0
5	-3
7	2

x	$g(x)$
-3	-4
2	8
5	-6
7	0

(a) Evaluate $(f \cdot g)(-3)$.

$$\begin{aligned}(f \cdot g)(-3) &= f(-3) \cdot g(-3)\\ &= (7)(-4)\\ &= -28\end{aligned}$$

NOTE

Usually, we write

$$\left(\frac{f}{g}\right)(5) = \frac{f(5)}{g(5)} = \frac{(-3)}{(-6)} = \frac{1}{2}$$

(b) Evaluate $(f \div g)(5)$.

$$\begin{aligned}(f \div g)(5) &= f(5) \div g(5)\\ &= (-3) \div (-6)\\ &= \frac{1}{2}\end{aligned}$$

(c) Evaluate $\left(\frac{f}{g}\right)(7)$.

$$\begin{aligned}\left(\frac{f}{g}\right)(7) &= \frac{f(7)}{g(7)}\\ &= \frac{(2)}{(0)}\end{aligned}$$

Division by 0 is undefined, so 7 is not in the domain of $f \div g$.

$\left(\frac{f}{g}\right)(7)$ does not exist.

(d) Find the domain of $f \cdot g$.

We want all values of x that are in the domains *both* of f and of g.

$D = \{-3, 5, 7\}$

(e) Find the domain of $f \div g$.

Again we want all values of x that are in the domains of *both* f and g, but we must exclude any x-value such that $g(x) = 0$. Since $g(7) = 0$, 7 cannot be in the domain of $f \div g$. Thus,

$D = \{-3, 5\}$

Check Yourself 5

Let the functions f and g be defined by the tables

x	$f(x)$
-5	-1
-2	3
0	4
3	0

x	$g(x)$
-5	0
0	-6
1	7
3	2

(a) Find $(f \cdot g)(0)$.
(b) Find $(f \div g)(3)$.
(c) Find the domain of $f \cdot g$.
(d) Find the domain of $f \div g$.

As with addition and subtraction, we create new functions when we multiply or divide functions. We can construct tables of values for the product and quotient functions just like we did in Example 2.

Example 6 Building Combination Functions

(a) Use the functions f and g in Example 5 to build a table of values for the function $h(x) = (f \cdot g)(x)$

In Example 5(d), we determined that the domain of $f \cdot g$ is $\{-3, 5, 7\}$. We found $(f \cdot g)(-3)$ in part (a), so we still need to find $(f \cdot g)(5)$ and $(f \cdot g)(7)$.

$$\begin{aligned} h(5) &= (f \cdot g)(5) && h \text{ is defined as } f \cdot g. \\ &= f(5) \cdot g(5) && h(5) = f(5) \cdot g(5) \\ &= (-3)(-6) && \text{From the tables, we have } f(5) = -3 \text{ and } g(5) = -6. \\ &= 18 \end{aligned}$$

$$\begin{aligned} h(7) &= (f \cdot g)(7) \\ &= f(7) \cdot g(7) \\ &= (2)(0) && f(7) = 2;\ g(7) = 0 \\ &= 0 \end{aligned}$$

We construct a table of values for $h = f \cdot g$.

x	$h(x)$
-3	-28
5	18
7	0

(b) Use the functions f and g in Example 5 to build a table of values for the function $k(x) = \left(\frac{f}{g}\right)(x)$

In Example 5(e), we determined that the domain of $\frac{f}{g}$ is $\{-3, 5\}$. We found $\left(\frac{f}{g}\right)(5)$ in part (b), but we still need to find $\left(\frac{f}{g}\right)(-3)$.

$$\begin{aligned} k(-3) &= \left(\frac{f}{g}\right)(-3) && k \text{ is defined as } \frac{f}{g} \\ &= \frac{f(-3)}{g(-3)} && k(-3) = \frac{f(-3)}{g(-3)}. \\ &= \frac{(7)}{(-4)} && f(-3) = 7;\ g(-3) = -4 \\ &= -\frac{7}{4} \end{aligned}$$

We construct a table of values for $k = \frac{f}{g}$.

x	$k(x)$
-3	$-\frac{7}{4}$
5	$\frac{1}{2}$

Check Yourself 6

Use the functions f and g from Check Yourself 5 to build tables of values for the functions

(a) $h(x) = (f \cdot g)(x)$ **(b)** $k(x) = \left(\frac{f}{g}\right)(x)$

The same ideas apply when our functions are defined by formulas or equations instead of tables.

Example 7 Multiplying Functions

RECALL

You learned to multiply binomials in Section 5.5.

Let $f(x) = x - 1$ and $g(x) = x + 5$. Find $(f \cdot g)(x)$.

$(f \cdot g)(x) = f(x) \cdot g(x)$ By definition.

$= (x - 1)(x + 5)$ Substitute the formulas for each function.

$= x^2 + 5x - x - 5$ FOIL to find the product.

$= x^2 + 4x - 5$ Simplify.

Check Yourself 7

Given $f(x) = x - 3$ and $g(x) = x + 2$, find $(f \cdot g)(x)$.

Example 8 Dividing Functions

Let $f(x) = x - 1$ and $g(x) = x + 5$.

(a) Find $\left(\frac{f}{g}\right)(x)$.

$$\left(\frac{f}{g}\right)(x) = \frac{f(x)}{g(x)}$$

$$= \frac{x - 1}{x + 5}$$

(b) Find the domain of $\frac{f}{g}$.

The domain of $f(x) = x - 1$ is all real numbers because we can subtract 1 from any number. The domain of $g(x) = x + 5$ is also all real numbers because we can add 5 to any number.

However, when $x = -5$ we have $g(-5) = (-5) + 5 = 0$. Since the denominator cannot be 0, -5 is not in the domain of $\frac{f}{g}$.

$D = \{x \mid x \neq -5\}$

Check Yourself 8

Let $f(x) = x - 3$ and $g(x) = x + 2$.

(a) Find $\left(\frac{f}{g}\right)(x)$. **(b)** Find the domain of $\left(\frac{f}{g}\right)(x)$.

We construct combination functions in many applications.

Example 9 A Business and Finance Application

One store sells a line of teddy bears for \$21.95. They purchase the stuffed animals for \$14.70 each. They approximate their weekly fixed cost to be \$75 from carrying and selling teddy bears.

(a) Construct this store's weekly revenue $R(x)$ and cost $C(x)$ functions for teddy bears.

They sell the bears for \$21.95 each, so their revenue is

$R(x) = 21.95x$

The bears cost them \$14.70 apiece and their weekly fixed costs are \$75.

$C(x) = 14.7x + 75$

(b) If the store sells 15 of these teddy bears in a week, how much revenue do they earn? What is their cost?

We evaluate the revenue and cost functions for $x = 15$.

$$\begin{aligned} R(15) &= 21.95(15) \\ &= 329.25 \end{aligned}$$

$$\begin{aligned} C(15) &= 14.7(15) + 75 \\ &= 295.5 \end{aligned}$$

If they sell 15 bears, they earn \$329.25 while incurring \$295.50 in costs.

(c) Use part (b) to determine the store's profit if they sell 15 teddy bears.

$\$329.25 - \$295.50 = \$33.75$

Their profit is \$33.75 if they sell 15 teddy bears in a week.

(d) Construct the store's profit function for teddy bears by finding the difference between the revenue and cost functions.

$$\begin{aligned} P(x) &= R(x) - C(x) \\ &= 21.95x - (14.7x + 75) && \text{Subtract the entire cost function.} \\ &= 21.95x - 14.7x - 75 && \text{Distribute to remove the parentheses.} \\ &= 7.25x - 75 && \text{Simplify.} \end{aligned}$$

(e) Use the profit function to find the profit from selling 15 teddy bears.

$$\begin{aligned} P(15) &= 7.25(15) - 75 \\ &= 33.75 \end{aligned}$$

As in part (c), the store's profit from 15 bears is \$33.75.

Check Yourself 9

A sporting goods store sells a two-person river raft for \$349.95. They purchase the rafts for \$233.30, wholesale. They approximate their weekly fixed costs to be \$450 from these rafts.

(a) Construct this store's weekly revenue $R(x)$ and cost $C(x)$ functions for these rafts.
(b) If the store sells six rafts in a week, how much revenue do they earn? What is their cost?
(c) Use part (b) to determine their profit if they sell six rafts.
(d) Construct the store's profit function for the rafts by finding the difference between the revenue and cost functions.
(e) Use the profit function to find the profit from selling six rafts.

Check Yourself ANSWERS

1. **(a)** -9; **(b)** 11; **(c)** does not exist; **(d)** $D = \{-3, 5\}$

2.

x	$h(x)$
-3	3
5	-9

3. **(a)** $3x - 4$; **(b)** $-7x - 2$; **(c)** 2 **4.** **(a)** $-3x + 1 + \frac{1}{x-2}$; **(b)** $D = \{x \mid x \neq 2\}$

5. **(a)** -24; **(b)** 0; **(c)** $D = \{-5, 0, 3\}$; **(d)** $D = \{0, 3\}$

6. **(a)**

x	$h(x)$
-5	0
0	-24
3	0

(b)

x	$k(x)$
0	$-\frac{2}{3}$
3	0

7. $x^2 - x - 6$

8. **(a)** $\frac{x-3}{x+2}$; **(b)** $D = \{x \mid x \neq -2\}$ **9.** **(a)** $R(x) = 349.95x$, $C(x) = 233.3x + 450$;
(b) Revenue: \$2,099.70; Cost: \$1,849.80; **(c)** \$249.90; **(d)** $P(x) = 116.65x - 450$; **(e)** \$249.90

Reading Your Text

These fill-in-the-blank exercises will help you understand some of the key vocabulary used in this section. The answers to these exercises are in the Answers Appendix in the back of the text.

(a) The sum of two functions can be defined for every value x that is in the ________ of both functions.

(b) The ________ of the functions f and g is written as $f \cdot g$.

(c) If $g(x) = 0$, then x is not in the ________ of $\frac{f}{g}$.

(d) We construct profit models by finding the difference between ________ and cost functions.

< Objectives 1 and 2 >

Use the tables to find the desired values.

x	$f(x)$
−3	−5
0	7
2	3
5	−3

x	$g(x)$
0	8
1	−3
2	4
7	−1

x	$h(x)$
−4	−1
0	−7
2	0
5	6

x	$k(x)$
−5	4
0	7
3	0
7	−3

1. $(f + g)(2)$ 

2. $(f + h)(5)$

3. $(k - g)(7)$

4. $(h - f)(5)$

5. $(f + k)(2)$ 

6. $(g - f)(0)$

7. Find the domain of $f + g$.

8. Find the domain of $h + k$.

9. Find the domain of $g - h$.

10. Find the domain of $k - f$.

Find **(a)** $(f + g)(x)$; **(b)** $(f - g)(x)$; **(c)** $(f + g)(3)$; *and* **(d)** $(f - g)(2)$.

11. $f(x) = -4x + 5;\ g(x) = 7x - 4$

12. $f(x) = 9x - 3;\ g(x) = -3x + 5$

13. $f(x) = 8x - 2;\ g(x) = -5x + 6$

14. $f(x) = -7x + 9;\ g(x) = 2x - 1$

15. $f(x) = x^2 + x - 1;\ g(x) = -3x^2 - 2x + 5$

16. $f(x) = -3x^2 - 2x + 5;\ g(x) = 5x^2 + 3x - 6$

17. $f(x) = -x^3 - 5x + 8;\ g(x) = 2x^2 + 3x - 4$ 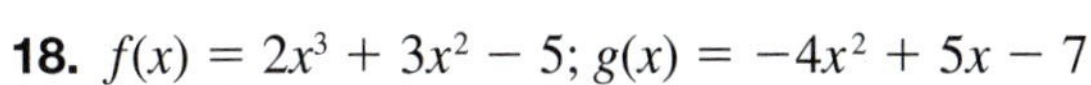

18. $f(x) = 2x^3 + 3x^2 - 5;\ g(x) = -4x^2 + 5x - 7$

Find **(a)** $(f + g)(x)$ *and* **(b)** *the domain of* $f + g$.

19. $f(x) = -9x + 11;\ g(x) = 15x - 7$

20. $f(x) = -11x + 3;\ g(x) = 8x - 5$

21. $f(x) = 3x + 2;\ g(x) = \dfrac{1}{x - 2}$

22. $f(x) = -2x + 5;\ g(x) = \dfrac{3}{x + 1}$

23. $f(x) = x^2 + x - 5;\ g(x) = \dfrac{2}{3x + 1}$

24. $f(x) = 3x^2 - 5x + 1;\ g(x) = -\dfrac{2}{2x - 3}$

< Objectives 3 and 4 >

Use the tables to find the desired values.

x	$f(x)$
−3	15
0	−4
2	5
4	9

x	$g(x)$
−4	18
0	12
2	−3
5	4

x	$h(x)$
−4	6
−3	−3
−2	−4
5	9

x	$k(x)$
−2	2
0	0
2	3
4	18

25. $(f \cdot g)(2)$

26. $(h \cdot k)(-2)$

27. $(f \cdot k)(0)$

28. $(g \cdot h)(5)$

29. $(f \div h)(-3)$

30. $(g \div h)(-4)$

31. $\left(\dfrac{g}{k}\right)(0)$

32. $\left(\dfrac{f}{k}\right)(4)$

33. Find the domain of $f \div g$.

34. Find the domain of $\dfrac{h}{k}$.

Find **(a)** $(f \cdot g)(x)$; **(b)** $\left(\frac{f}{g}\right)(x)$; *and* **(c)** *the domain of* $\left(\frac{f}{g}\right)(x)$.

35. $f(x) = 2x - 1$; $g(x) = x - 3$

36. $f(x) = -x + 3$; $g(x) = x + 4$

37. $f(x) = 3x + 2$; $g(x) = 2x - 1$ 

38. $f(x) = -3x + 5$; $g(x) = -x + 2$

39. $f(x) = 2 - x$; $g(x) = 5 + 2x$

40. $f(x) = x + 5$; $g(x) = 1 - 3x$

Business and Finance *The profit P from selling x units is equal to the difference between the revenue R and the cost C. Construct profit functions for the revenue and cost functions given in exercises 41 and 42.*

41. $R(x) = 25x$; $C(x) = x^2 + 4x + 50$

42. $R(x) = 20x$; $C(x) = x^2 + 2x + 30$

Science and Medicine *The velocity V of an object thrown in the air is given by a combination of three functions. It is the sum of the initial velocity V_0 (a constant function) the acceleration due to gravity g (another constant function), and the amount of time passed t.*

$V(t) = g \cdot t + V_0$

Construct velocity functions given the initial velocities and gravitational constants in exercises 43 and 44.

43. $V_0 = 10$ m/s; $g = -9.8$ m/s^2

44. $V_0 = 64$ ft/s; $g = -32$ ft/s^2

Business and Finance *The revenue R that a store earns from selling a particular item is given by the product of the number of units sold x, and the price per unit p.*

Construct revenue functions for the price functions given in exercises 45 and 46.

45. $p(x) = 119 - 6x$

46. $p(x) = 1{,}190 - 36x$

47. **Business and Finance** A kitchen store sells a particular ice cream maker for \$89.95. They purchase each ice cream maker for \$58.47 wholesale. They approximate the weekly fixed cost associated with the ice cream makers to be \$225.

(a) Construct this store's weekly revenue $R(x)$ and cost $C(x)$ functions for these ice cream makers.

(b) If the store sells 15 ice cream makers in a week, how much revenue do they earn? What is their cost?

(c) Use part (b) to determine their profit if they sell 15 ice cream makers.

(d) Construct the store's profit function for the ice cream makers by finding the difference between the revenue and cost functions.

(e) Use the profit function to find the profit from selling 15 ice cream makers.

48. **Business and Finance** A shoe store sells a particular pair of open-toed sandals for \$169.95. They purchase each pair of sandals for \$76.48 wholesale. They approximate the weekly fixed cost associated with the sandals to be \$765.

(a) Construct this store's weekly revenue $R(x)$ and cost $C(x)$ functions for these sandals.

(b) If the store sells 12 pairs of sandals in a week, how much revenue do they earn? What is their cost?

(c) Use part (b) to determine their profit if they sell 12 pairs of sandals.

(d) Construct the store's profit function for the sandals by finding the difference between the revenue and cost functions.

(e) Use the profit function to find the profit from selling 12 sandals.

49. Business and Finance A sporting goods store sells pairs of 28-in. rattan escrima sticks for \$20.99. They purchase each pair of sticks for \$8.40 wholesale. They approximate the weekly fixed cost associated with the sticks to be \$295.

(a) Construct this store's weekly revenue $R(x)$ and cost $C(x)$ functions for these sticks.

(b) If the store sells 30 pairs of sticks in a week, how much revenue do they earn? What is their cost?

(c) Use part (b) to determine their profit if they sell 30 pairs of sticks.

(d) Construct the store's profit function for the sticks by finding the difference between the revenue and cost functions.

(e) Use the profit function to find the profit from selling 30 sticks.

50. Business and Finance A tire store sells a particular snow tire for \$164.98 each. They purchase each tire for \$65.99 wholesale. They approximate the weekly fixed cost associated with the snow tires to be \$2,750.

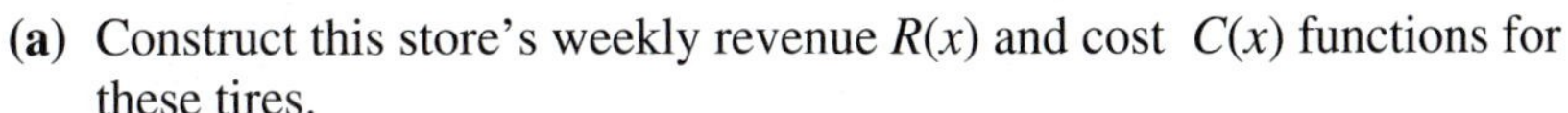

(a) Construct this store's weekly revenue $R(x)$ and cost $C(x)$ functions for these tires.

(b) If the store sells 45 snow tires in a week, how much revenue do they earn? What is their cost?

(c) Use part (b) to determine their profit if they sell 45 snow tires.

(d) Construct the store's profit function for the tires by finding the difference between the revenue and cost functions.

(e) Use the profit function to find the profit from selling 45 snow tires.

Complete each statement with **always, sometimes,** *or* **never.**

51. The domain of $f + g$ is __________ the same as the domain of f.

52. If $g(a) = 0$, then the domain of $f \div g$ __________ contains a.

53. If f and g are first-degree functions, then $f \cdot g$ is __________ a second-degree (quadratic) function.

54. If f and g are first-degree functions, then $f + g$ is __________ a second-degree (quadratic) function.

Answers

1. 7 **3.** -2 **5.** Does not exist **7.** $\{0, 2\}$ **9.** $\{0, 2\}$ **11.** (a) $3x + 1$; (b) $-11x + 9$; (c) 10; (d) -13 **13.** (a) $3x + 4$; (b) $13x - 8$; (c) 13; (d) 18 **15.** (a) $-2x^2 - x + 4$; (b) $4x^2 + 3x - 6$; (c) -17; (d) 16 **17.** (a) $-x^3 + 2x^2 - 2x + 4$; (b) $-x^3 - 2x^2 - 8x + 12$; (c) -11; (d) -20 **19.** (a) $6x + 4$; (b) $\mathbb{R}$ **21.** (a) $3x + 2 + \frac{1}{x - 2}$; (b) $\{x | x \neq 2\}$ **23.** (a) $x^2 + x - 5 + \frac{2}{3x + 1}$; (b) $\left\{x \middle| x \neq -\frac{1}{3}\right\}$ **25.** -15 **27.** 0 **29.** -5 **31.** Undefined **33.** $\{0, 2\}$ **35.** (a) $2x^2 - 7x + 3$; (b) $\frac{2x - 1}{x - 3}$; (c) $\{x | x \neq 3\}$ **37.** (a) $6x^2 + x - 2$; (b) $\frac{3x + 2}{2x - 1}$; (c) $\left\{x \middle| x \neq \frac{1}{2}\right\}$ **39.** (a) $-2x^2 - x + 10$; (b) $\frac{2 - x}{2x + 5}$ or $-\frac{x - 2}{2x + 5}$; (c) $\left\{x \middle| x \neq -\frac{5}{2}\right\}$ **41.** $P(x) = -x^2 + 21x - 50$ **43.** $V(t) = 10 - 9.8t$ **45.** $R(x) = 119x - 6x^2$ **47.** (a) $R(x) = 89.95x$, $C(x) = 58.47x + 225$; (b) Revenue: \$1,349.25; Cost: \$1,102.05; (c) \$247.20; (d) $P(x) = 31.48x - 225$; (e) \$247.20 **49.** (a) $R(x) = 20.99x$, $C(x) = 8.4x + 295$; (b) Revenue: \$629.70; Cost: \$547; (c) \$82.70; (d) $P(x) = 12.59x - 295$; (e) \$82.70 **51.** sometimes **53.** always

Activity 10 ::

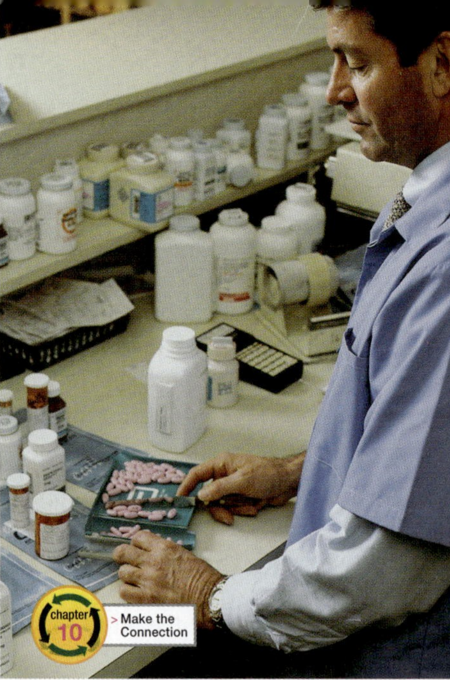

Half-Life and Decay

You should complete Section 10.4 before you and a group of students complete this Chapter 10 Activity.

You may have encountered the idea of half-life in connection with radioactive substances. Given an initial amount of some radioactive material, half of that material will remain after an amount of time known as the **half-life** of the material. As the material continues to decay, the amount that remains may be modeled by an **exponential** function.

You can simulate the decay of radioactive material. Working with two or three partners, obtain approximately 20 wooden cubes, and place a marker on just one side of each cube. (You can use dice: simply choose one number to be the marked side.)

1. Count the number of cubes you have. This is your *initial amount of radioactive material.* Record this number.
2. Roll the entire set of cubes. The cube(s) that show the marked side up have decayed. Remove these, and record the number that are still active.
3. Roll the remaining radioactive cubes, remove those that have decayed, and record the amount remaining.
4. Continue in this manner, filling out a table like that shown, until all cubes have decayed. Note that the variable x represents the number of rolls, and y represents the number of cubes still active.

x	0	1	2	3	4	5	6	7	8	9	10	11	12	13	14	15	16	17	18	19	20
y	1																				

If dice are not available, use the sample data on the next page to complete exercises 5–9.

5. Draw a scatterplot of the ordered pairs (x, y) in your table.
6. Repeat the entire procedure (with the same set of cubes), completing a new table. Plot this set of ordered pairs on the *same* coordinate system you made in step 5.
7. Do this a third time, again adding the points to your scatterplot.
8. Now, draw a smooth curve that seems (to you) to fit the points best.
9. Use your graph to determine the approximate half-life for your cubes. For example, if you began with 22 cubes, see how many rolls it took for 11 to remain. Confirm your estimate of the half-life by checking elsewhere on the graph. For example, estimate the number of rolls corresponding to 16 cubes, and see about how many rolls it took from that point for 8 to remain.

Sample Data

x	0	1	2	3	4	5	6	7	8	9	10	11	12	13	14	15	16	17	18	19	20
y	22	19	16	14	10	8	6	5	4	4	4	3	3	3	3	3	3	3	2	2	0

x	0	1	2	3	4	5	6	7	8	9	10	11	12	13	14	15	16	17	18
y	22	20	15	13	10	9	9	7	5	1	1	1	1	1	1	0			

x	0	1	2	3	4	5	6	7	8	9	10	11	12	13	14	15	16	17	18
y	22	15	12	8	7	7	5	5	5	4	4	4	3	2	2	0			

10.2 Composition of Functions

< 10.2 Objectives >

1 > Evaluate the composition of functions defined by tables

2 > Evaluate the composition of functions defined by expressions

3 > Decompose functions

4 > Use composition to solve applications

You probably learned the four basic arithmetic operations (addition, subtraction, multiplication, and division) as far back as grade school. In all the time since then, you may not have learned any new operations. Instead, you learned to apply these operations in new ways such as working with fractions, variables, and in Section 10.1, functions.

In this section, we introduce the first new operation many students learn since learning division. *Composition* is an operation we apply to functions rather than numbers. As such, you could not be introduced to this operation until learning to combine functions.

Definition

Composition of Functions

The **composition** of functions f and g is the function $f \circ g$, where

$$(f \circ g)(x) = f(g(x))$$

The domain of the composition is the set of all elements x in the domain of g for which $g(x)$ is in the domain of f.

Composition may be thought of as a chaining together of functions. To understand the meaning of $(f \circ g)(x)$, note that first the function g acts on x, producing $g(x)$, and then the function f acts on $g(x)$.

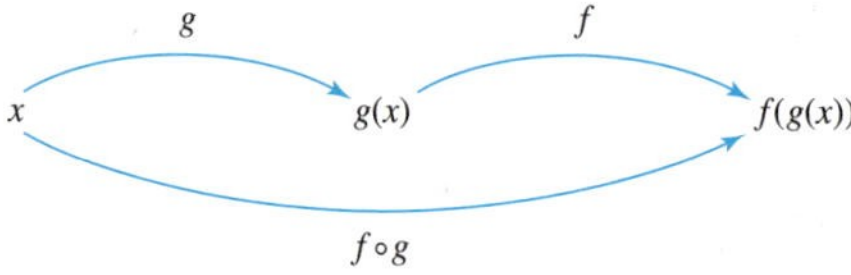

In Example 1, we look to compose functions defined by tables.

Example 1 **Composing Functions**

< Objective 1 >

Let f and g be defined by the tables

x	$f(x)$
-4	8
0	6
2	5
1	-2

x	$g(x)$
-2	5
1	0
3	-4
8	2

(a) Evaluate $(f \circ g)(1)$.

NOTE

Think, what does g do to 1? And then, what does f do to the result?

By definition, we have

$$(f \circ g)(1) = f(g(1))$$

What this notation means is that we evaluate $g(1)$ and then evaluate f at the result. From the table, we see $g(1) = 0$, so

$$f(g(1)) = f(0)$$

The table defining f gives $f(0) = 6$. Putting this together gives

$$\begin{aligned}(f \circ g)(1) &= f(g(1)) && \text{By definition}\\ &= f(0) && g(1) = 0\\ &= 6 && f(0) = 6\end{aligned}$$

> **CAUTION**

When evaluating $(f \circ g)(x)$, the first function to act is g, not f.

(b) Evaluate $(f \circ g)(8)$.

$$\begin{aligned}(f \circ g)(8) &= f(g(8)) && \text{By definition}\\ &= f(2) && g(8) = 2\text{, according to the table defining } g.\\ &= 5 && f(2) = 5\text{, according to the table defining } f.\end{aligned}$$

(c) Evaluate $(f \circ g)(-2)$.

$$\begin{aligned}(f \circ g)(-2) &= f(g(-2))\\ &= f(5) && g(-2) = 5\text{, according to the table defining } g.\end{aligned}$$

We run into a "snag" at this point. The table defining f does not give a result for $x = 5$, so 5 is not in the domain of f.

Therefore, $f(5)$ does not exist, so we cannot evaluate $(f \circ g)(-2)$. Thus, -2 is not in the domain of $f \circ g$.

NOTE

-2 is not in the domain of $f \circ g$.

(d) Evaluate $(f \circ g)(3)$.

$$\begin{aligned}(f \circ g)(3) &= f(g(3))\\ &= f(-4)\\ &= 8\end{aligned}$$

Check Yourself 1

Using the functions given in Example 1, evaluate

(a) $(g \circ f)(-4)$ **(b)** $(g \circ f)(1)$ **(c)** $(g \circ f)(2)$

Most of the time, we define a function with an expression or equation rather than a table. In Example 2, we look at composing functions defined this way. One advantage to using functions defined algebraically is that we can find a formula for the composition $f \circ g$, as we see in part (c).

Example 2 Composing Functions

< Objective 2 >

Let $f(x) = x^2 - 2$ and $g(x) = x + 3$.

(a) Evaluate $(f \circ g)(0)$.

By definition, we have

$(f \circ g)(0) = f(g(0))$

Therefore, our first step is to find $g(0)$.

$$\begin{aligned} g(x) &= x + 3 \\ g(0) &= (0) + 3 \\ &= 3 \end{aligned}$$

NOTE

The function g turns 0 into 3. The function f then turns 3 into 7.

We use this result in our next computation.

$$\begin{aligned} (f \circ g)(0) &= f(g(0)) && \text{By definition} \\ &= f(3) && g(0) = 3 \\ &= (3)^2 - 2 && f(x) = x^2 - 2 \\ &= 7 \end{aligned}$$

(b) Evaluate $(f \circ g)(4)$.

By definition, we have

$(f \circ g)(4) = f(g(4))$

Therefore, our first step is to find $g(4)$.

$$\begin{aligned} g(x) &= x + 3 \\ g(4) &= (4) + 3 \\ &= 7 \end{aligned}$$

We use this result in our next computation.

$$\begin{aligned} (f \circ g)(4) &= f(g(4)) && \text{By definition} \\ &= f(7) && g(4) = 7 \\ &= (7)^2 - 2 && f(x) = x^2 - 2 \\ &= 47 \end{aligned}$$

$(f \circ g)(4) = 47$

(c) Find $(f \circ g)(x)$.

We approach this problem in the same way. Begin with the definition.

$(f \circ g)(x) = f(g(x))$

We know that $g(x) = x + 3$, so

$$\begin{aligned} (f \circ g)(x) &= f(g(x)) \\ &= f(x + 3) \end{aligned}$$

The function f accepts an input, squares it, and subtracts 2.

$$\begin{aligned} (f \circ g)(x) &= f(g(x)) \\ &= f(x + 3) \\ &= (x + 3)^2 - 2 \\ &= x^2 + 6x + 9 - 2 \\ &= x^2 + 6x + 7 \end{aligned}$$

Check Yourself 2

Let $f(x) = x^2 + x$ and $g(x) = x - 1$. Evaluate each composition.

(a) $(f \circ g)(0)$ **(b)** $(f \circ g)(-2)$ **(c)** $(f \circ g)(x)$

If the domain of either function is limited, then the domain of their composition may be limited as well. Consider Example 3.

Example 3 Composing Functions

Let $f(x) = \sqrt{x}$ and $g(x) = 3 - x$. Evaluate each composition.

(a) $(f \circ g)(1) = f(g(1))$

$= f(2)$ $\quad g(1) = 3 - (1) = 2$

$= \sqrt{2}$

(b) $(f \circ g)(-1) = f(g(-1))$

$= f(4)$ $\quad g(-1) = 3 - (-1) = 4$

$= \sqrt{4}$

$= 2$

(c) $(f \circ g)(7) = f(g(7))$

$= f(-4)$ $\quad g(7) = 3 - (7) = -4$

$= \sqrt{-4}$

Since this is not a real number, $(f \circ g)(7)$ does not exist and 7 is not in the domain of $f \circ g$.

(d) $(f \circ g)(x) = f(g(x))$

$= f(3 - x)$

$= \sqrt{3 - x}$

This function produces real-number values only if $3 - x \geq 0$ or $x \leq 3$. This is the domain of $f \circ g$.

NOTE

Graph the function $Y = \sqrt{3 - x}$ in a graphing calculator. The graph exists only for $x \leq 3$.

Check Yourself 3

Let $f(x) = \frac{1}{x}$ and $g(x) = x^2 - 4$. Evaluate each composition.

(a) $(f \circ g)(0)$ **(b)** $(f \circ g)(-2)$ **(c)** $(f \circ g)(x)$

The order of composition is important. In general, $(f \circ g)(x) \neq (g \circ f)(x)$. Using the functions given in Example 2, we saw that $(f \circ g)(x) = x^2 + 6x + 7$, whereas

$$(g \circ f)(x) = g(f(x)) = g(x^2 - 2) = x^2 - 2 + 3 = x^2 + 1$$

which is not the same as $(f \circ g)(x)$.

Often it is convenient to write a given function as the composition of simpler functions. While the choice of simpler functions is not unique, a good choice makes many applications easier to solve.

Example 4 Writing a Function as a Composition

< Objective 3 >

Use $f(x) = x + 3$ and $g(x) = x^2$ to express the given function as a composition of f and g.

NOTE

Can you see how we are using the order of operations here?

(a) $h(x) = (x + 3)^2$

The function h takes an input and adds 3 to it before squaring the sum. The function f adds three to an input, so f needs to come first in the composition.

The function g squares an input, g needs to act after f for the composition to do everything that h does.

$$h(x) = g(f(x)) = (g \circ f)(x)$$

We can easily check to see if our composition produces the original function h.

$$\begin{aligned}(g \circ f)(x) &= g(f(x)) && \text{By definition}\\ &= g(x + 3) && f(x) = x + 3\\ &= (x + 3)^2 && g(x) = x^2\\ &= h(x)\end{aligned}$$

(b) $k(x) = x^2 + 3$

This time, the function k squares an input before adding 3. In order for a composition of f and g to equal k, we need to apply the squaring function g before applying the add 3 function f.

$$k(x) = f(g(x)) = (f \circ g)(x)$$

Check

$$\begin{aligned}(f \circ g)(x) &= f(g(x))\\ &= f(x^2)\\ &= x^2 + 3\\ &= k(x)\end{aligned}$$

Check Yourself 4

Use $f(x) = \sqrt{x}$ and $g(x) = x + 2$ to express each function as a composition of f and g.

(a) $h(x) = \sqrt{x + 2}$ **(b)** $k(x) = \sqrt{x} + 2$

There are many examples that involve the composition of functions, as illustrated in Example 5.

Example 5 A Business and Finance Application

< Objective 4 >

At Kinky's Duplication Salon, customers pay \$2 plus 4¢ per page copied. Duplication consultant Vinny makes a 5% commission for each job.

(a) Express a customer's bill B as a function of the number of pages copied p.

$B(p) = 0.04p + 2$ The bill is \$0.04 times the number of pages, plus \$2.

(b) Express Vinny's commission V as a function of each bill B.

$V(B) = 0.05B$ Vinny's commission is 5% of the bill.

(c) Use function composition to express Vinny's commission V as a function of the number of pages a customer has copied p.

Since "commission" is a function of "bill," and "bill" is a function of "pages," the composition creates "commission" as a function of "pages."

$$\begin{aligned} V(B) &= V(B(p)) && \text{Substitute } B(p) \text{ for } B\text{, creating a composition.} \\ &= V(0.04p + 2) && \text{Since } B(p) = 0.04p + 2 \\ &= 0.05(0.04p + 2) && \text{Input the quantity } 0.04p + 2 \text{ into the function } V. \\ &= 0.002p + 0.1 \end{aligned}$$

So $V(p) = 0.002p + 0.1$.

(d) Use the function in part (c) to find Vinny's commission on a 2,000-page job.

$$\begin{aligned} V(p) &= 0.002(2{,}000) + 0.1 \\ &= 4 + 0.1 = 4.1 \end{aligned}$$

Vinny's commission is \$4.10.

Check Yourself 5

On his regular route, Gonzalo averages 62 mi/hr between Charlottesville and Lawrenceville. His van averages 24 mi/gal, and his gas tank holds 12 gal of fuel. Assume that his tank is full when he starts the trip from Charlottesville to Lawrenceville.

(a) Express the fuel left in the tank as a function of n, the number of gallons used.
(b) Express the number of gallons used as a function of m, the number of miles driven.
(c) Express the fuel left in the tank as a function of m, the number of miles driven.

Check Yourself ANSWERS

1. **(a)** 2; **(b)** 5; **(c)** does not exist **2.** **(a)** 0; **(b)** 6; **(c)** $x^2 - x$ **3.** **(a)** $-\frac{1}{4}$; **(b)** does not exist; **(c)** $\frac{1}{x^2 - 4}$ **4.** **(a)** $(f \circ g)(x)$; **(b)** $(g \circ f)(x)$ **5.** **(a)** $f(n) = 12 - n$; **(b)** $g(m) = \frac{m}{24}$; **(c)** $(f \circ g)(m) = 12 - \frac{m}{24}$

Reading Your Text

These fill-in-the-blank exercises will help you understand some of the key vocabulary used in this section. The answers to these exercises are in the Answers Appendix in the back of the text.

(a) ________ functions can be thought of as a chaining together of the functions.

(b) When evaluating $(f \circ g)(x)$, the first function to act on x is ________.

(c) Often it is convenient to write a given function as the composition of ________ functions.

(d) When evaluating a function, we use the order of ________.

10.2 exercises

Skills | Calculator/Computer | Career Applications | Above and Beyond

< Objective 1 >

Use the functions defined by the tables to evaluate the compositions in exercises 1 to 12.

x	$f(x)$
−3	−1
−1	7
2	6
3	−3

x	$g(x)$
−2	4
1	−2
4	3
6	2

x	$h(x)$
−2	5
0	0
1	−2
3	6

x	$k(x)$
−2	0
0	4
2	3
3	−4

1. $(f \circ g)(4)$ **2.** $(g \circ f)(2)$ **3.** $(h \circ g)(1)$ **4.** $(g \circ h)(1)$

5. $(g \circ h)(3)$ **6.** $(k \circ h)(0)$ **7.** $(h \circ k)(0)$ **8.** $(k \circ g)(1)$

9. $(f \circ h)(3)$ **10.** $(k \circ g)(4)$ **11.** $(k \circ k)(2)$ **12.** $(f \circ f)(-3)$

< Objective 2 >

Evaluate each composition.

13. $f(x) = x - 3$ and $g(x) = 2x + 1$
(a) $(f \circ g)(0)$ **(b)** $(f \circ g)(-2)$ **(c)** $(f \circ g)(3)$ **(d)** $(f \circ g)(x)$

14. $f(x) = x - 1$ and $g(x) = 3x + 4$
(a) $(f \circ g)(0)$ **(b)** $(f \circ g)(-2)$ **(c)** $(f \circ g)(3)$ **(d)** $(f \circ g)(x)$

15. $f(x) = 3x + 1$ and $g(x) = 4x - 3$
(a) $(f \circ g)(0)$ **(b)** $(f \circ g)(-2)$ **(c)** $(g \circ f)(3)$ **(d)** $(g \circ f)(x)$

16. $f(x) = 4x - 2$ and $g(x) = -2x + 5$
(a) $(f \circ g)(0)$ **(b)** $(f \circ g)(-2)$ **(c)** $(g \circ f)(3)$ **(d)** $(g \circ f)(x)$

17. $f(x) = x^2$ and $g(x) = x + 3$
(a) $(f \circ g)(0)$ **(b)** $(f \circ g)(-2)$ **(c)** $(g \circ f)(3)$ **(d)** $(g \circ f)(x)$

18. $f(x) = x^2 + 3$ and $g(x) = 3x$
(a) $(f \circ g)(0)$ **(b)** $(f \circ g)(-2)$ **(c)** $(g \circ f)(3)$ **(d)** $(g \circ f)(x)$

19. $f(x) = 2x^2 - 1$ and $g(x) = -2x$
(a) $(g \circ f)(0)$ **(b)** $(g \circ f)(-2)$ **(c)** $(f \circ g)(3)$ **(d)** $(f \circ g)(x)$

20. $f(x) = x^2 + 3$ and $g(x) = 3x$
(a) $(g \circ f)(0)$ **(b)** $(g \circ f)(-2)$ **(c)** $(f \circ g)(3)$ **(d)** $(f \circ g)(x)$

< Objective 3 >

Write h as a composite of f and g.

21. $f(x) = 3x$ $g(x) = x + 2$ $h(x) = 3x + 2$

22. $f(x) = x - 4$ $g(x) = 7x$ $h(x) = 7x - 4$

23. $f(x) = x + 5$ $g(x) = \sqrt{x}$ $h(x) = \sqrt{x + 5}$

24. $f(x) = x + 5$ $g(x) = \sqrt{x}$ $h(x) = \sqrt{x} + 5$

25. $f(x) = x^2$ $g(x) = x - 5$ $h(x) = x^2 - 5$

26. $f(x) = x^2$ $g(x) = x - 5$ $h(x) = (x - 5)^2$

27. $f(x) = x - 3$ $g(x) = \frac{2}{x}$ $h(x) = \frac{2}{x - 3}$

28. $f(x) = x - 3$ $g(x) = \frac{2}{x}$ $h(x) = \frac{2}{x} - 3$

29. $f(x) = x - 1$ $g(x) = x^2 + 2$ $h(x) = x^2 + 1$ 

30. $f(x) = x - 1$ $g(x) = x^2 + 2$ $h(x) = x^2 - 2x + 3$

< Objective 4 >

31. Carine and Jacob are getting married at East Fork Estates. The wedding will cost $1,000 plus $40 per guest. They have read that typically 80% of the people invited actually attend a wedding.

(a) Write a function to represent the number of people N expected to attend if v are invited.

(b) Write a function to represent the cost C of the wedding for N guests.

(c) Write a function to represent the cost C of the wedding if v people are invited.

32. On his regular route, Gonzalo averages 62 mi/hr between Charlottesville and Lawrenceville. His van averages 24 mi/gal, and his gas tank holds 12 gal of fuel. Assume that his tank is full when he starts the trip from Charlottesville to Lawrenceville.

(a) Express the number of miles driven as a function of t, the time on the road.

(b) Express the number of gallons used as a function of M, the miles driven.

(c) Express the number of gallons used as a function of t, the time on the road.

33. When she arrives in London, Bichvan receives an exchange rate of 0.6221 British pound for each U.S. dollar. In Reykjavik, she receives an exchange rate of 197.3749 Icelandic kronas for each British pound. When she returns to the United States, how much (in U.S. dollars) should she expect to receive in exchange for 12,000 Icelandic kronas?

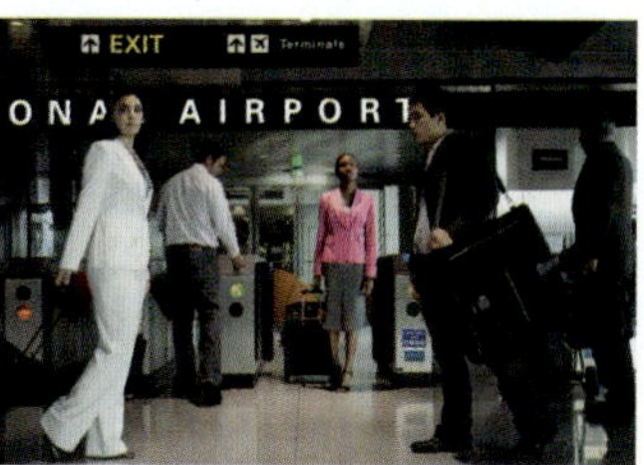

34. If the exchange rate for Japanese yen is 78.425 and the exchange rate for Indian rupees is 52.82 (both from U.S. dollars), then what is the exchange rate from Japanese yen to Indian rupees?

Determine whether each statement is **true** *or* **false.**

35. $(f \circ g)(x)$ is the same as $f(x) \cdot g(x)$.

36. $(f \circ g)(x)$ is always the same as $(g \circ f)(x)$.

Complete each statement with **always, sometimes,** *or* **never.**

37. To evaluate $(f \circ g)(5)$, we __________ find $g(5)$ first.

38. To evaluate $(f \circ g)(5)$, $g(5)$ must __________ be in the domain of f.

Answers

1. -3 **3.** 5 **5.** 2 **7.** Does not exist **9.** Does not exist **11.** -4 **13.** **(a)** -2; **(b)** -6; **(c)** 4; **(d)** $2x - 2$ **15.** **(a)** -8; **(b)** -32; **(c)** 37; **(d)** $12x + 1$ **17.** **(a)** 9; **(b)** 1; **(c)** 12; **(d)** $x^2 + 3$ **19.** **(a)** 2; **(b)** -14; **(c)** 71; **(d)** $8x^2 - 1$ **21.** $h(x) = (g \circ f)(x)$ **23.** $h(x) = (g \circ f)(x)$ **25.** $h(x) = (g \circ f)(x)$ **27.** $h(x) = (g \circ f)(x)$ **29.** $h(x) = (f \circ g)(x)$ **31.** **(a)** $N(v) = 0.8v$; **(b)** $C(N) = 40N + 1{,}000$; **(c)** $C(v) = 32v + 1{,}000$ **33.** \$97.73 **35.** False **37.** always

10.3 Inverse Functions

< 10.3 Objectives >

1 > Find the inverse of a function defined by an equation

2 > Find the inverse of a function defined by a table

3 > Determine if a function is invertible

Function composition leads us to the question: Can we chain together (i.e., compose) two functions in such a way that one function "undoes" the other?

Suppose, for example, that $f(x) = \frac{x-5}{3}$ and $g(x) = 3x + 5$. Pick a convenient x-value for f, say, $x = 8$. $f(8) = \frac{8-5}{3} = \frac{3}{3} = 1$. The function f turns 8 into 1. Now let g act on this result: $g(1) = 3(1) + 5 = 8$. The function g turns 1 back into 8.

When we view the composition of g and f, acting on 8, we see g "undoing" f's actions.

$$(g \circ f)(8) = g(f(8)) = g(1) = 8$$

In general, a function g that undoes the action of f is called the **inverse** of f.

Definition

Inverse Functions

Functions f and g are inverse functions if

$(g \circ f)(x) = x$ for all x in the domain of f

and $(f \circ g)(x) = x$ for all x in the domain of g

> CAUTION

f^{-1} is not the same as x^{-1}. f^{-1} is the inverse of f under composition. It *does not* mean $\frac{1}{f}$.

If g is the inverse of f, we denote the function g as f^{-1}.

A natural question now is, given a function f, how do we find the inverse function f^{-1}? One way to find such a function f^{-1} is to analyze the actions of f, noting the order of operations involved, and then define f^{-1} by using the *opposite* operations *in the reverse order*.

Example 1 Finding the Inverse of a Function

< Objective 1 >

Given $f(x) = \frac{x-5}{3}$, find its inverse function f^{-1}.

The function f subtracts 5 from an input and then divides the result by 3. Finding an inverse function, we need to use the *opposite* operations in *reverse order*.

So, the inverse function needs to multiply an input by 3 and then add 5 to that result. We write this as

$$f^{-1}(x) = 3x + 5$$

To verify that we do have an inverse, we need

$$(f^{-1} \circ f)(x) = x \quad \text{and} \quad (f \circ f^{-1})(x) = x$$

$$(f^{-1} \circ f)(x) = f^{-1}(f(x))$$

$$= f^{-1}\left(\frac{x-5}{3}\right) \qquad f(x) = \frac{x-5}{3}$$

$= 3\left(\frac{x-5}{3}\right) + 5$ $\quad f^{-1}(x) = 3x + 5$

$= x - 5 + 5$ $\quad$ Simplify.

$= x$ $\quad$ The result we want.

$$\begin{aligned}(f \circ f^{-1})(x) &= f(f^{-1}(x)) \\ &= f(3x + 5) \\ &= \frac{(3x + 5) - 5}{3} \\ &= \frac{3x}{3} \\ &= x\end{aligned}$$

Since both $(f^{-1} \circ f)(x) = x$ and $(f \circ f^{-1})(x) = x$, we know that f and f^{-1} are inverses.

Check Yourself 1

Given $f(x) = \frac{x+1}{4}$, find f^{-1}.

To develop another technique for finding inverses, we revisit functions defined by tables.

Example 2 Finding the Inverse of a Function

< Objective 2 >

Find the inverse of the function f.

x	$f(x)$
−4	8
0	6
2	5
1	−2

The inverse of f is easily found by reversing the order of the input values and output values.

x	$f^{-1}(x)$
8	−4
6	0
5	2
−2	1

While f turns −4 into 8, for example, f^{-1} turns 8 back into −4.

Check Yourself 2

Find the inverse of the function f.

x	$f(x)$
−2	5
1	0
4	−4
8	2

In Example 2, if we write y in place of $f(x)$, we see that we are just interchanging the roles of x and y in order to create the inverse f^{-1}. This suggests the following technique for finding the inverse of a function that is given in equation form.

Step by Step

Finding the Inverse of a Function

Step 1 Given a function $f(x)$, write y in place of $f(x)$.

Step 2 Switch the variables x and y.

Step 3 Solve for y.

Step 4 Write $f^{-1}(x)$ in place of y.

Step 5 Check that $f^{-1}(f(x)) = x$ and $f(f^{-1}(x)) = x$.

Example 3 Finding the Inverse of a Function

Find the inverse of $f(x) = 2x - 4$.

Step 1 Replace $f(x)$ with y.

$$y = 2x - 4$$

Step 2 Switch x and y.

$$x = 2y - 4$$

Step 3 Solve for y.

$$x = 2y - 4$$

$$x + 4 = 2y \quad \text{Add 4 to both sides.}$$

$$\frac{x+4}{2} = y \quad \text{Divide by 2.}$$

Step 4 Replace y with $f^{-1}(x)$.

$$f^{-1}(x) = \frac{x+4}{2} \quad \text{or} \quad f^{-1}(x) = \frac{1}{2}x + 2$$

RECALL

$$\frac{x+4}{2} = \frac{x}{2} + \frac{4}{2} = \frac{1}{2}x + 2$$

Step 5 Check.

$$\begin{aligned}(f^{-1} \circ f)(x) &= f^{-1}(f(x)) \\ &= f^{-1}(2x - 4) \\ &= \frac{1}{2}(2x - 4) + 2 \\ &= x - 2 + 2 \\ &= x\end{aligned}$$

$$\begin{aligned}(f \circ f^{-1})(x) &= f(f^{-1}(x)) \\ &= f\left(\frac{1}{2}x + 2\right) \\ &= 2\left(\frac{1}{2}x + 2\right) - 4 \\ &= x + 4 - 4 \\ &= x\end{aligned}$$

Check Yourself 3

Find the inverse of $f(x) = \dfrac{x - 7}{5}$.

The graphs of relations and their inverses are connected in an interesting way because the points (a, b) and (b, a) are always symmetric with respect to the line $y = x$.

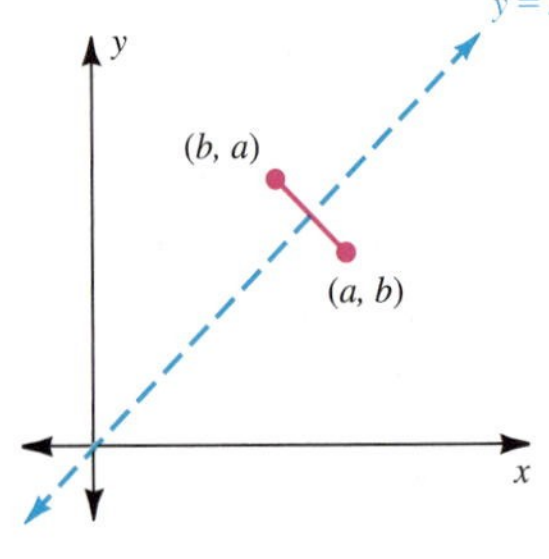

With this symmetry in mind, consider Example 4.

Example 4 Graphing a Function and Its Inverse

Graph the function f from Example 3 along with its inverse.

$f(x) = 2x - 4$

and

$f^{-1}(x) = \frac{1}{2}x + 2$

The graphs of f and f^{-1} are

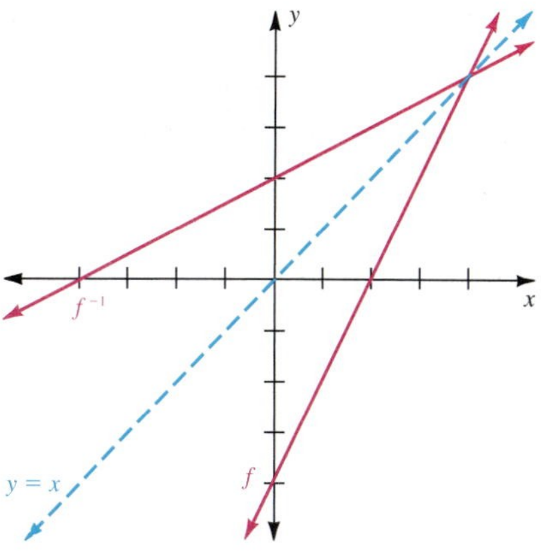

The graphs of f and f^{-1} are symmetric about the line $y = x$. That symmetry follows from our earlier observation about the pairs (a, b) and (b, a) because we simply reversed the roles of x and y in forming the inverse relation.

Check Yourself 4

Find the inverse of the function $f(x) = 3x + 6$ and graph both $f(x)$ and $f^{-1}(x)$.

In our work so far, we have seen techniques for finding the inverse of a function. However, it is quite possible that the inverse may not be a function.

Example 5 Finding the Inverse of a Function

NOTE

Interchange the elements of the ordered pairs.

Find the inverse of each function.

(a) $f = \{(1, 3), (2, 4), (3, 9)\}$

Its inverse is

$\{(3, 1), (4, 2), (9, 3)\}$

which is also a function.

NOTE

It is not a function because 6 maps to both 2 and 3.

(b) $g = \{(1, 3), (2, 6), (3, 6)\}$

Its inverse is

$\{(3, 1), (6, 2), (6, 3)\}$

which is *not* a function.

Check Yourself 5

Write the inverse of each function. Which of the inverses are also functions?

(a) $\{(-1, 2), (0, 3), (1, 4)\}$ **(b)** $\{(2, 5), (3, 7), (4, 5)\}$

Can we predict in advance whether the inverse of a function is also a function? The answer is yes.

We already know that for a relation to be a function, no element in its domain can be associated with more than one element in its range. Since, in creating an inverse, the x-values and y-values are interchanged, the inverse of a function f will also be a function only if *no element in the range of f can be associated with more than one element in its domain.* That is, no two ordered pairs of f can have the same y-coordinate. This leads us to a definition.

Definition

One-to-One Function

A function is **one-to-one** if no two distinct domain elements are paired with the same range element.

That is, if two inputs are different, then their outputs must also be different.

We then have this property.

Property

Inverse of a Function

The inverse of a function f is also a function if f is one-to-one.

In Example 5(a),

$f = \{(1, 3), (2, 4), (3, 9)\}$

is a one-to-one function and its inverse is also a function. However, the function in Example 5(b)

$g = \{(1, 3), (2, 6), (3, 6)\}$

is *not* a one-to-one function, so its inverse is *not* a function.

Our result regarding a one-to-one function and its inverse also has a convenient graphical interpretation. Here we graph the function g from Example 5.

$g = \{(1, 3), (2, 6), (3, 6)\}$

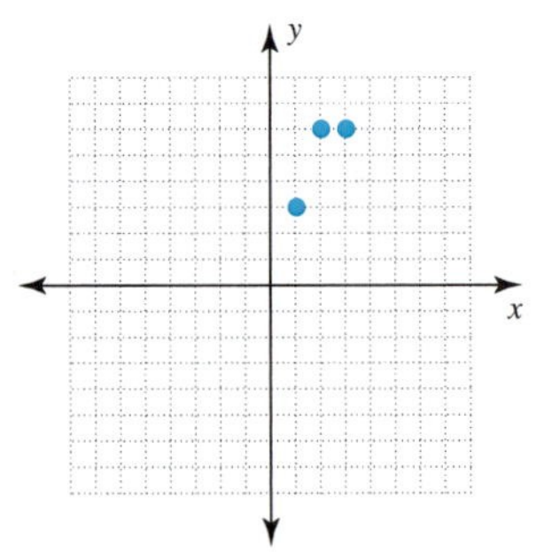

Again, g is *not* a one-to-one function, because two points, (2, 6) and (3, 6), have the same y-coordinate. As a result, a horizontal line can be drawn that passes through two points.

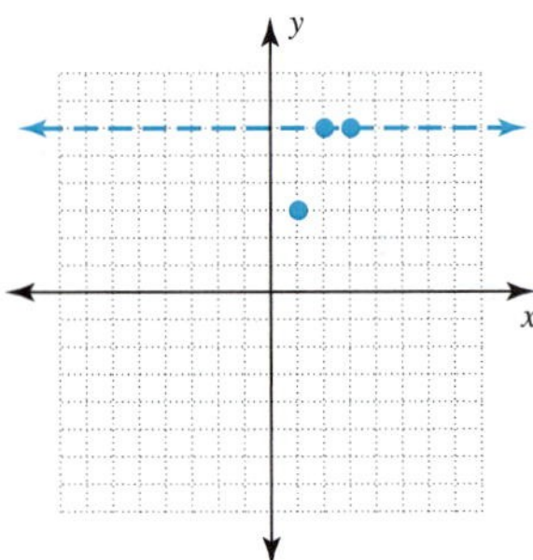

NOTE

In Section 2.6, we used the *vertical line test* to determine whether a relation was a function. The *horizontal line test* tells us whether a function is one-to-one.

This means that when we form the inverse by reversing the coordinates, the resulting relation is *not* a function. Points (6, 2) and (6, 3) are part of the inverse, and its graph fails the vertical line test.

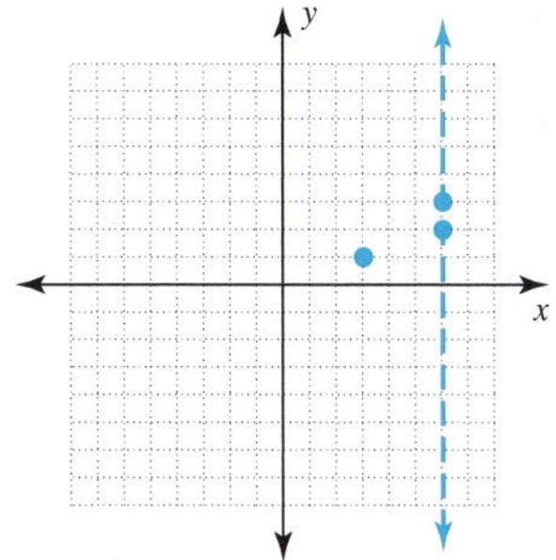

This leads to the **horizontal line test.**

Property

Horizontal Line Test

A function is one-to-one if no horizontal line passes through two or more points on its graph.

Now we have a graphical way to determine whether the inverse of a function f is a function.

Property

Inverse of a Function

The inverse of a function f is also a function if the graph of f passes the horizontal line test.

This is a very useful property, as Example 6 illustrates.

Example 6 Identifying One-to-One Functions

< Objective 3 >

For each function, determine **(i)** whether the function is one-to-one and **(ii)** whether the inverse is also a function.

(a)

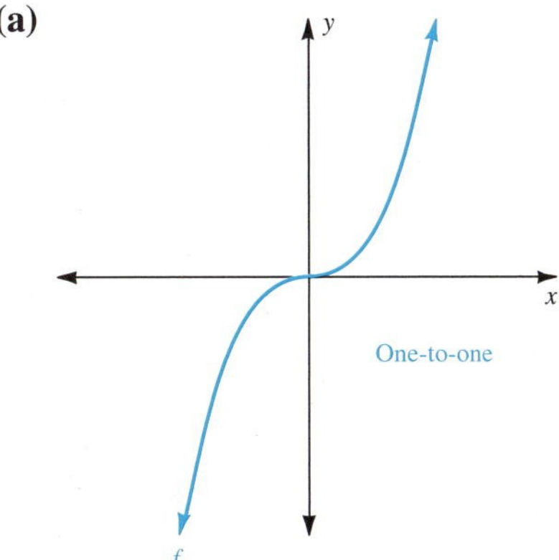

(i) Because no horizontal line passes through two or more points of the graph, f is one-to-one.

(ii) Because f is one-to-one, its inverse is also a function.

(b)

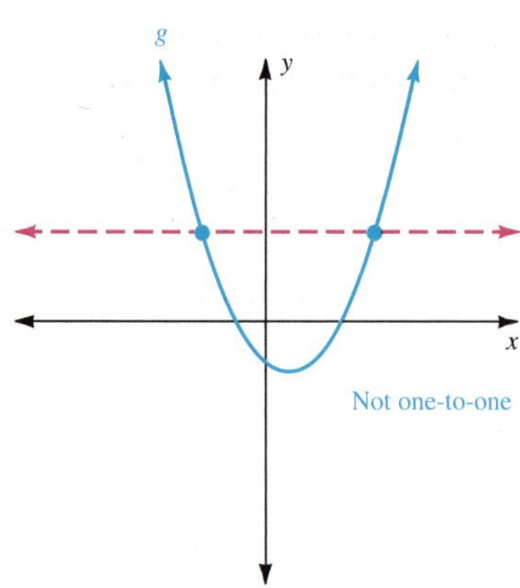

RECALL

The parabola given by g represents a function. It passes the vertical line test.

(i) Because a horizontal line can meet the graph of g at two points, g is *not* a one-to-one function.

(ii) Because g is not one-to-one, its inverse is not a function.

Check Yourself 6

For each function, determine **(i)** whether the function is one-to-one and **(ii)** whether the inverse is also a function.

(a)

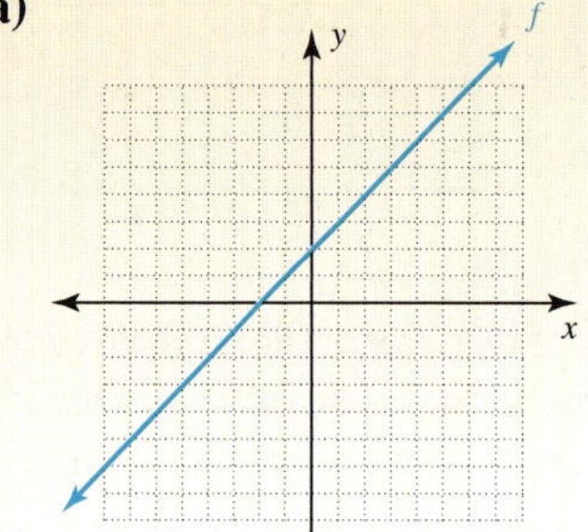

(b)

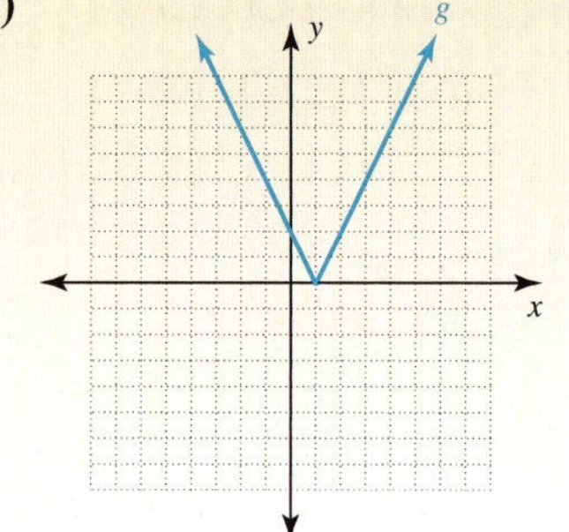

When a function is not one-to-one, we can restrict the domain of the function so that it is one-to-one (and, as a result, the inverse is a function).

Example 7 Restricting the Domain of a Function

Consider the function $f(x) = x^2 + 3$.

(a) Restrict the domain of f so that f is one-to-one.

The graph of f is a parabola.

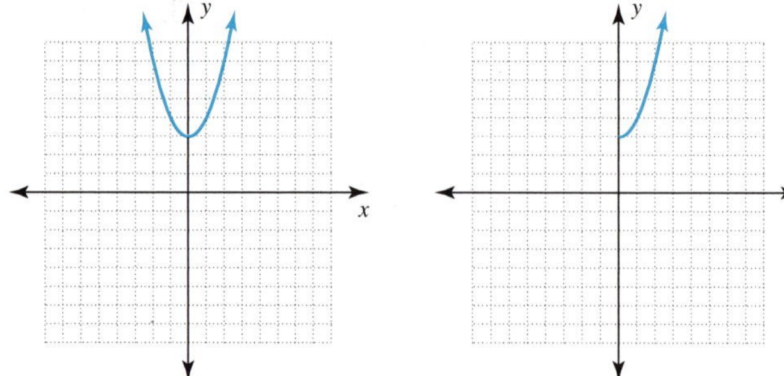

NOTE

We could also restrict the domain of f so that

$D = \{x | x \le 0\}$

in order to pass the horizontal line test.

We restrict the domain of f so that $D = \{x \mid x \ge 0\}$ and the graph passes the horizontal line test. The restricted f is stated as

$f(x) = x^2 + 3 \qquad x \ge 0$

(b) Using the restricted f, find f^{-1}.

Letting y replace $f(x)$, we write

$y = x^2 + 3$

$x = y^2 + 3$ Switch x and y.

Solve for y.

$x - 3 = y^2$

$y = \pm\sqrt{x - 3}$

We choose $y = \sqrt{x - 3}$ (do you see why?) and write

$f^{-1}(x) = \sqrt{x - 3}$

(c) Graph the restricted f and f^{-1} on the same axes.

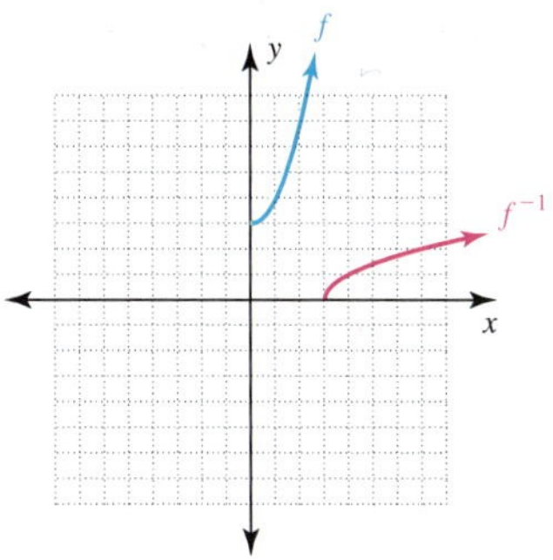

NOTE

Observe that the graphs of f (restricted) and f^{-1} are symmetric with respect to the line $y = x$.

The domain of the restricted f (all real numbers greater than or equal to 0) is the same as the range of f^{-1}. Further, the range of the restricted f (all real numbers greater than or equal to 3) is the same as the domain of f^{-1}.

Check Yourself 7

Given the function $f(x) = x^2 - 2$,

(a) Restrict the domain of f so that f is one-to-one.
(b) Using the restricted f, find f^{-1}.
(c) Graph the restricted f and f^{-1} on the same axes.

Check Yourself ANSWERS

1. $f^{-1}(x) = 4x - 1$

2.

x	$f^{-1}(x)$
5	−2
0	1
−4	4
2	8

3. $f^{-1}(x) = 5x + 7$

4. $f^{-1}(x) = \frac{x-6}{3}$ or $f^{-1}(x) = \frac{1}{3}x - 2$

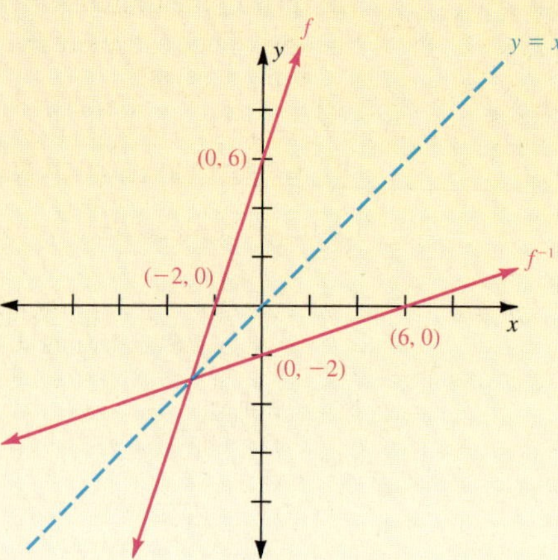

5. **(a)** {(2, −1), (3, 0), (4, 1)}; a function; **(b)** {(5, 2), (7, 3), (5, 4)}; not a function
6. **(a)** One-to-one; the inverse is a function; **(b)** Not one-to-one; the inverse is not a function
7. **(a)** $f(x) = x^2 - 2, x \geq 0$; **(b)** $f^{-1}(x) = \sqrt{x+2}$; **(c)**

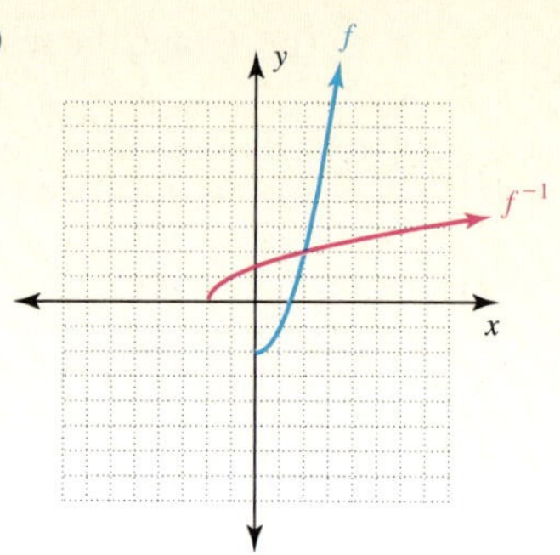

Reading Your Text

These fill-in-the-blank exercises will help you understand some of the key vocabulary used in this section. The answers to these exercises are in the Answers Appendix in the back of the text.

(a) In general, a function *g* that undoes the action of *f* is called the ________ of *f*.

(b) The graphs of f^{-1} and *f* are ________ about the line $y = x$.

(c) The inverse of the function *f* is also a function if *f* is ________.

(d) A function is one-to-one if no ________ line passes through two or more points on its graph.

10.3 exercises

Skills | Calculator/Computer | Career Applications | Above and Beyond

< Objective 1 >

Find the inverse f^{-1} of each function f.

1. $f(x) = 3x + 5$
2. $f(x) = -3x - 7$
3. $f(x) = \frac{x-1}{2}$
4. $f(x) = \frac{x+1}{3}$
5. $f(x) = 2x - 3$ 
6. $f(x) = -5x + 3$
7. $f(x) = \frac{x+4}{3}$
8. $f(x) = \frac{x-5}{7}$

9. $f(x) = \frac{x}{3} + 5$ 

10. $f(x) = \frac{2x}{5} - 7$

11. $f(x) = \frac{6 - 5x}{3}$

12. $f(x) = \frac{4x - 7}{3}$

13. $f(x) = \frac{11x}{5} + 2$

14. $f(x) = 5 - \frac{7x}{3}$

< Objective 2 >

Find the inverse of each function. In each case, determine whether the inverse is also a function.

15.

x	$f(x)$
−2	−5
2	6
3	4
4	3

16.

x	$f(x)$
−5	2
−3	3
1	3
5	2

17.

x	$f(x)$
−4	3
−2	7
3	5
7	−4

18.

x	$f(x)$
−5	2
−3	−4
0	2
6	−4

19. $f = \{(2, 3), (3, 4), (4, 5)\}$

20. $g = \{(1, 4), (2, 3), (3, 4)\}$

21. $f = \{(1, 5), (2, 5), (3, 5)\}$

22. $g = \{(4, 7), (2, 6), (6, 9)\}$

23. $f = \{(2, 3), (3, 5), (4, 7)\}$

24. $g = \{(-1, 0), (2, 0), (0, -1)\}$

For each function f, find its inverse f^{-1}. Then graph both on the same set of axes.

25. $f(x) = 3x - 6$

26. $f(x) = 4x + 8$

27. $f(x) = -2x + 6$ 

28. $f(x) = -3x - 6$

< Objective 3 >

Determine whether each function is one-to-one and state whether the inverse is a function.

29. $f = \{(-3, 5), (-2, 3), (0, 2), (1, 4), (6, 5)\}$

30. $g = \{(-3, 7), (0, 4), (2, 5), (4, 1)\}$

31.

x	$f(x)$
−3	4
0	−3
2	1
6	2
8	0

32.

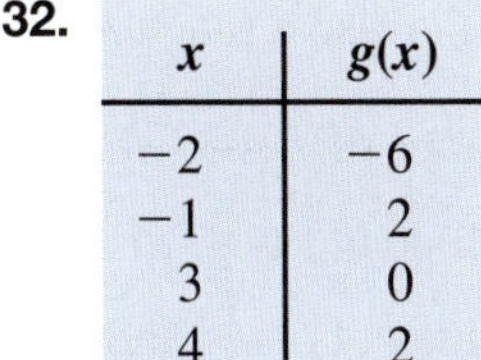

x	$g(x)$
−2	−6
−1	2
3	0
4	2

33.

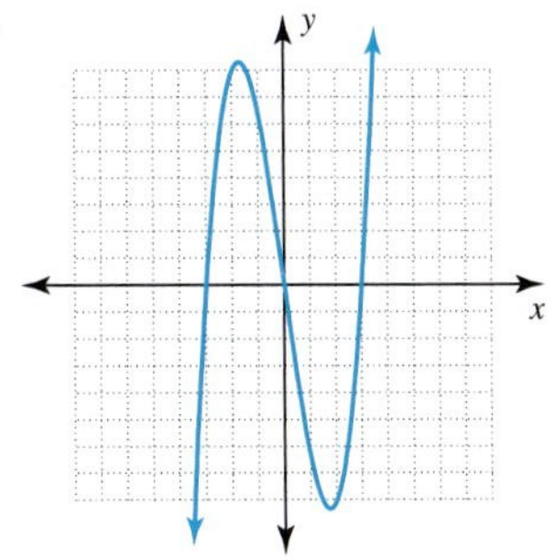

34.

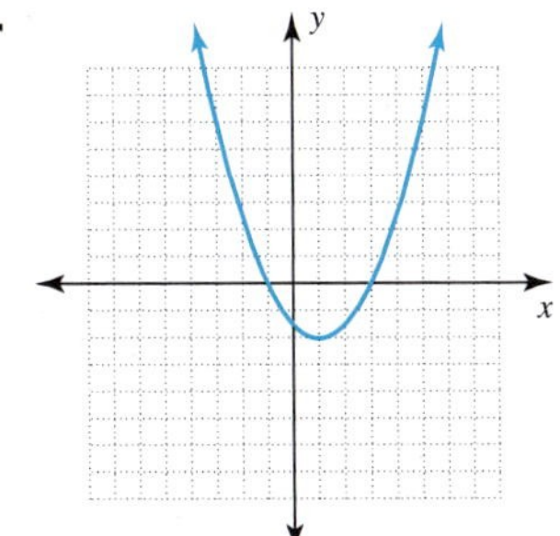

35. 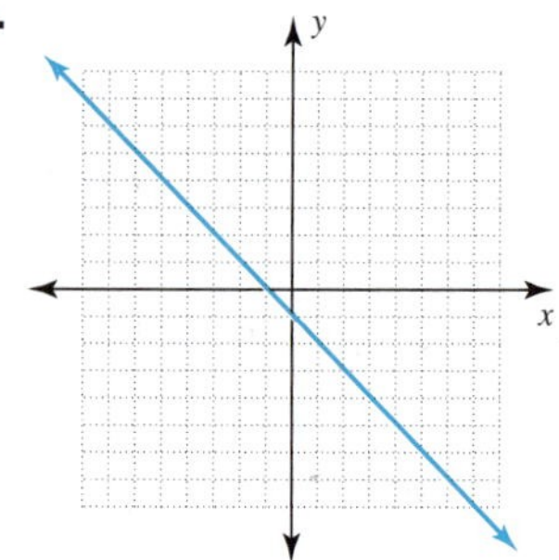

36. 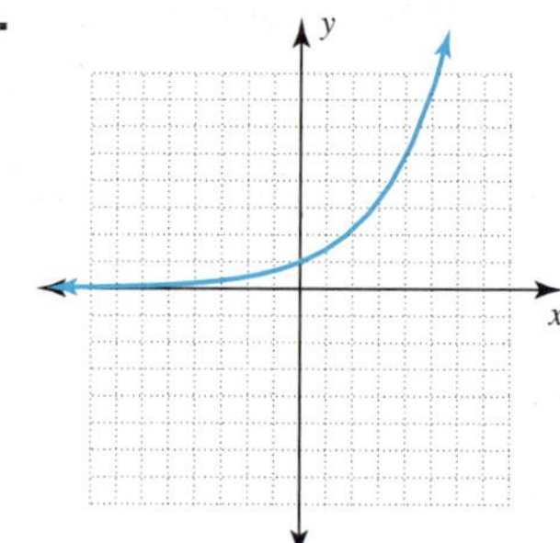

If $f(x) = 3x - 6$, then $f^{-1}(x) = \frac{1}{3}x + 2$. Evaluate as indicated.

37. $f(6)$

38. $f^{-1}(6)$

39. $f(f^{-1}(6))$

40. $f^{-1}(f(6))$

41. $f(f^{-1}(x))$

42. $f^{-1}(f(x))$

If $g(x) = \frac{x+1}{2}$, then $g^{-1}(x) = 2x - 1$. Evaluate as indicated.

43. $g(3)$

44. $g^{-1}(3)$

45. $g(g^{-1}(3))$

46. $g^{-1}(g(3))$

47. $g(g^{-1}(x))$

48. $g^{-1}(g(x))$

If $h(x) = 2x + 8$, then $h^{-1}(x) = \frac{1}{2}x - 4$. Evaluate as indicated.

49. $h(4)$

50. $h^{-1}(4)$

51. $h(h^{-1}(4))$

52. $h^{-1}(h(4))$

53. $h(h^{-1}(x))$

54. $h^{-1}(h(x))$

Suppose that f and g are one-to-one functions.

55. If $f(5) = 7$, find $f^{-1}(7)$.

56. If $g^{-1}(4) = 9$, find $g(9)$.

Skills | Calculator/Computer | Career Applications | **Above and Beyond**

Let f be a linear function; that is, let $f(x) = mx + b$.

57. Find $f^{-1}(x)$.

58. Based on exercise 57, if the slope of f is 3, what is the slope of f^{-1}?

59. Based on exercise 57, if the slope of f is $\frac{2}{5}$, what is the slope of f^{-1}?

60. Based on exercise 57, if the slope of f is m, then what is the slope of f^{-1} ?

61. Based on exercise 57, if the slope of f is -2 and its y-intercept $(0, 5)$, what is the y-intercept of f^{-1}?

62. Based on exercise 57, if the slope of f is m and its y-intercept is $(0, b)$, what is the y-intercept of f^{-1}?

Determine whether each statement is **true** *or* **false.**

63. The inverse of a linear function with nonzero slope is itself a function.

64. The inverse of a quadratic function is itself a function.

65. The graphs of a function and its inverse are symmetric with respect to the y-axis.

66. If f has an inverse function f^{-1} and $f(a) = b$, then $f^{-1}(b) = a$.

Complete each statement with **always, sometimes,** *or* **never.**

67. The inverse of a function is __________ a function.

68. If the graph of a function passes the horizontal line test, then the graph of the inverse __________ passes the vertical line test.

69. An inverse process is an operation that undoes a procedure. If the procedure is wrapping a present, describe in detail the inverse process.

70. If the procedure is the series of steps that take you from home to your classroom, describe the inverse process.

Answers

1. $f^{-1}(x) = \frac{x-5}{3}$ **3.** $f^{-1}(x) = 2x + 1$ **5.** $f^{-1}(x) = \frac{x+3}{2}$ **7.** $f^{-1}(x) = 3x - 4$ **9.** $f^{-1}(x) = 3x - 15$ **11.** $f^{-1}(x) = \frac{6-3x}{5}$
13. $f^{-1}(x) = \frac{5x-10}{11}$ **15.** $\{(-5, -2), (6, 2), (4, 3), (3, 4)\}$; function **17.** $\{(3, -4), (7, -2), (5, 3), (-4, 7)\}$; function
19. $\{(3, 2), (4, 3), (5, 4)\}$; function **21.** $\{(5, 1), (5, 2), (5, 3)\}$; not a function **23.** $\{(3, 2), (5, 3), (7, 4)\}$; function
25. $f^{-1}(x) = \frac{x+6}{3} = \frac{1}{3}x + 2$

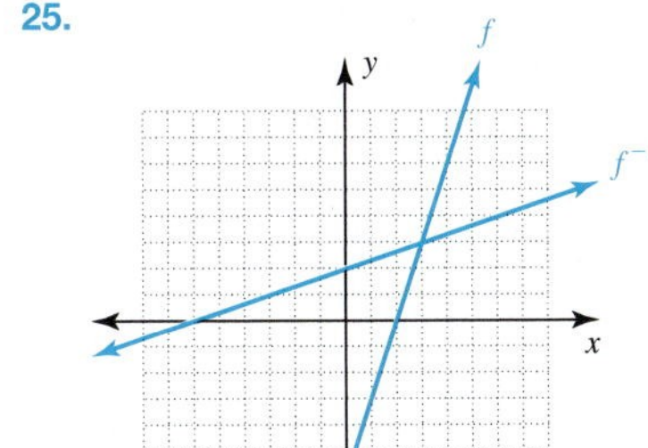

27. $f^{-1}(x) = \frac{6-x}{2} = -\frac{1}{2}x + 3$

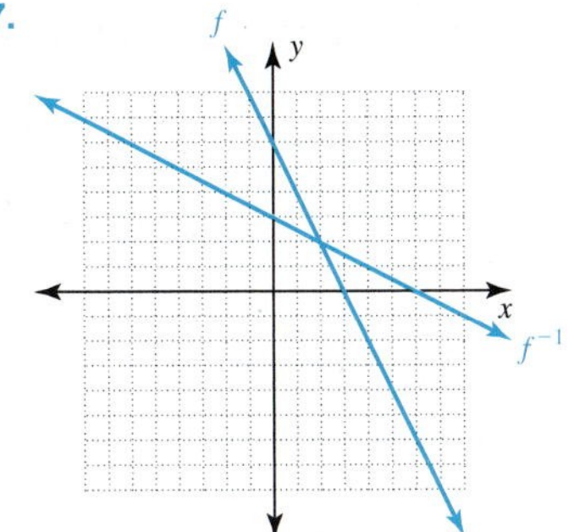

29. Not one-to-one; inverse is not a function **31.** One-to-one; inverse is a function **33.** Not one-to-one; inverse is not a function
35. One-to-one; inverse is a function **37.** 12 **39.** 6 **41.** x **43.** 2 **45.** 3 **47.** x **49.** 16 **51.** 4 **53.** x **55.** 5
57. $f^{-1}(x) = \frac{x-b}{m}$ **59.** $\frac{5}{2}$ **61.** $\left(0, \frac{5}{2}\right)$ **63.** True **65.** False **67.** sometimes **69.** Above and Beyond

10.4 Exponential Functions

< 10.4 Objectives >

1 > Graph an exponential function

2 > Use exponential functions to solve applications

3 > Solve an exponential equation

We turn to a new family of functions called *exponential functions*. In an exponential function, the variable is in the exponent rather than in the base, as in a polynomial.

Exponential functions allow us to better model many applications including population growth, radioactive decay, and compound interest.

Definition

Exponential Functions

An **exponential function** is a function that can be written as

$f(x) = a \cdot b^x$

in which $a \neq 0$, $b > 0$, and $b \neq 1$.

We call b the **base** of the exponential function.

If $a = 1$, the function $f(x) = b^x$ may be referred to as an **elementary exponential function.**

NOTE

In these examples, f, g, and h are elementary exponential functions. The function k is simply an exponential function.

Examples of exponential functions include

$f(x) = 2^x \qquad g(x) = 3^x \qquad h(x) = \left(\frac{1}{2}\right)^x \qquad k(x) = 14 \cdot 10^x$

As we have done with other new functions, we begin by finding some function values, then use that information to graph the function.

Example 1 Graphing an Elementary Exponential Function

< Objective 1 >

RECALL

$2^{-2} = \frac{1}{2^2} = \frac{1}{4}$

Graph the exponential function

$f(x) = 2^x$

First, choose convenient values for x.

$f(0) = 2^0 = 1 \qquad f(1) = 2^1 = 2 \qquad f(-1) = 2^{-1} = \frac{1}{2}$

$f(2) = 2^2 = 4 \qquad f(-2) = 2^{-2} = \frac{1}{4} \qquad f(3) = 2^3 = 8 \qquad f(-3) = 2^{-3} = \frac{1}{8}$

Next, form a table from these values. Then plot the corresponding points and connect them with a smooth curve for the desired graph.

x	$f(x)$
−3	0.125
−2	0.25
−1	0.5
0	1
1	2
2	4
3	8

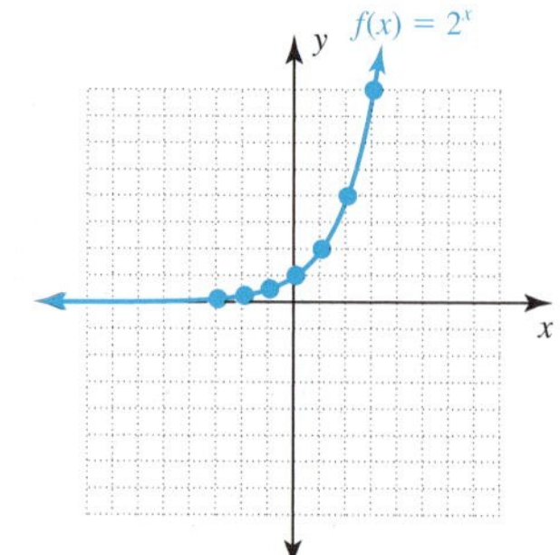

Check Yourself 1

Sketch the graph of the exponential function

$g(x) = 3^x$

NOTES

There is no value for x such that

$2^x = 0$

so the graph never touches the x-axis.

We call $y = 0$ (or the x-axis) the **horizontal asymptote.**

Let's examine some characteristics of the graph of the exponential function in Example 1. The vertical line test shows that this is indeed the graph of a function. The horizontal line test shows that the function is one-to-one.

The graph *approaches* the x-axis on the left, but it does *not intersect* the x-axis. The y-intercept is (0, 1) because $2^0 = 1$. To the right, the function values get larger. We say that the values *grow without bound.* This same language may be applied to linear or quadratic functions.

We now look at an example in which the base of the function is less than 1.

Example 2 Graphing an Exponential Function

RECALL

$\left(\frac{1}{2}\right)^x = 2^{-x}$

Graph the exponential function

$f(x) = \left(\frac{1}{2}\right)^x$

First, choose convenient values for x.

$f(0) = \left(\frac{1}{2}\right)^0 = 1$ $\quad f(1) = \left(\frac{1}{2}\right)^1 = \frac{1}{2}$ $\quad f(-1) = \left(\frac{1}{2}\right)^{-1} = 2$

$f(2) = \left(\frac{1}{2}\right)^2 = \frac{1}{4}$ $\quad f(-2) = \left(\frac{1}{2}\right)^{-2} = 4$

$f(3) = \left(\frac{1}{2}\right)^3 = \frac{1}{8}$ $\quad f(-3) = \left(\frac{1}{2}\right)^{-3} = 8$

Again, form a table of values and graph the desired function.

NOTE

By the vertical and horizontal line tests, this is the graph of a one-to-one function.

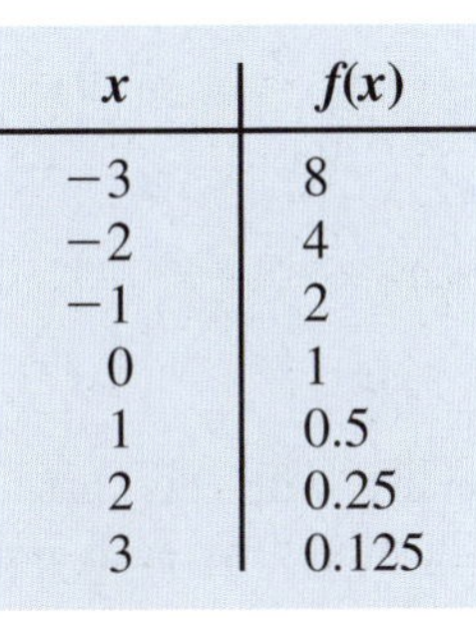

x	$f(x)$
−3	8
−2	4
−1	2
0	1
1	0.5
2	0.25
3	0.125

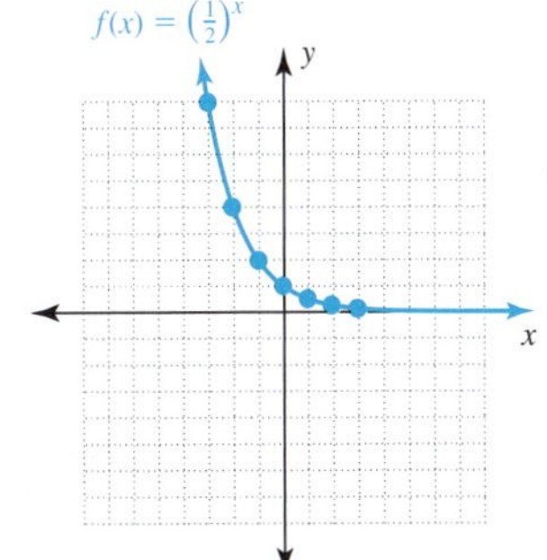

Check Yourself 2

Sketch the graph of the exponential function

$g(x) = \left(\frac{1}{3}\right)^x$

NOTES

The base of a *growth function* is *greater than* 1.

The base of a *decay function* is *less than* 1 but greater than 0.

We see from the graph that the function in Example 2 is also one-to-one. As in Example 1, the graph does not intersect the x-axis but approaches that axis, here on the right. This function also grows without bound, but this time on the left. The y-intercept for both graphs occurs at (0, 1).

The graph of Example 1 is *increasing* (going up) as we move from left to right. That function is an example of a **growth function.**

The graph of Example 2 is *decreasing* (going down) as we move from left to right. It is an example of a **decay function.**

This algorithm summarizes our work in this section so far.

Step by Step

Graphing an Elementary Exponential Function

Step 1 Create a table of values for $y = b^x$.

Step 2 Plot the points from the table and connect them with a smooth curve.

NOTE

The letter *e* as a base originated with Leonhard Euler (1707–1783), and *e* is sometimes called *Euler's number* for that reason.

The graphs of elementary exponential functions $f(x) = b^x$ have these properties.

(a) The y-intercept is (0, 1).

(b) The graphs approach, but do not touch, the x-axis.

(c) The graphs represent one-to-one functions.

(d) If $b > 1$, the graph increases from left to right. If $0 < b < 1$, the graph decreases from left to right.

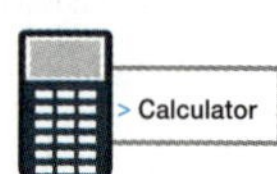

We used the bases 2 and $\frac{1}{2}$ for the exponential functions in our examples because they provided convenient computations. A far more important base for an exponential function is the irrational number e. In fact, when e is used as a base, the function

$$f(x) = e^x$$

is called *the* exponential function.

NOTE

Graph $y = e^x$ on a calculator. You may find the [e^x] key to be the second (or inverse) function to the LN key.

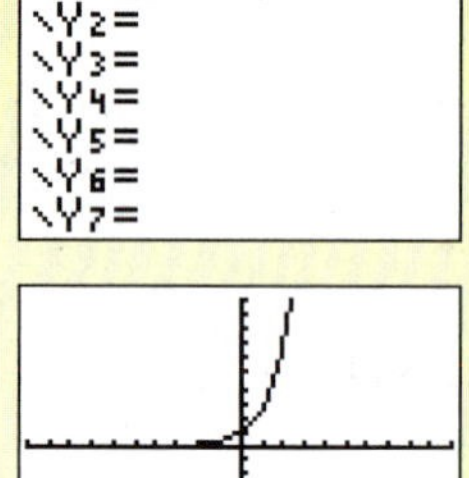

The significance of the number e will be made clear in later courses, particularly calculus. Since e is irrational, we can only approximate it with a decimal.

$$e \approx 2.71828$$

Here is the graph of $f(x) = e^x$. Of course, it is very similar to the graphs seen earlier in this section.

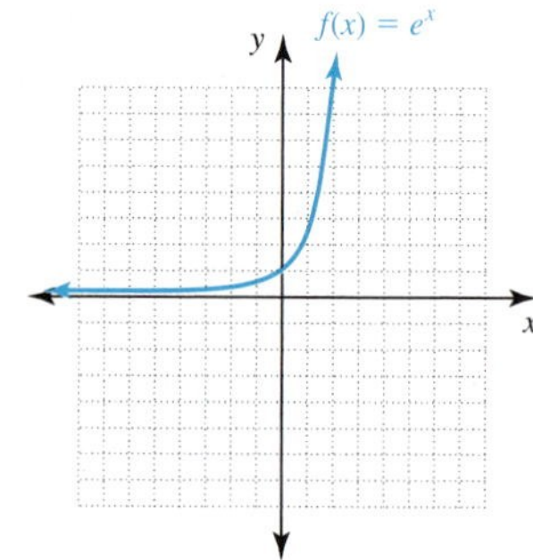

Exponential expressions with the base e occur frequently in real-world applications. Example 3 illustrates two such applications.

Example 3 An Exponential Application

< Objective 2 >

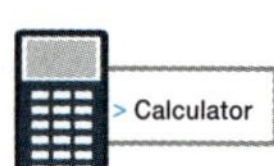

(a) If a town's population is 20,000 and it is expected to grow continuously at a rate of 5% per year, we can model the town's population after t years with the function

$$P(t) = 20{,}000e^{0.05t}$$

We can predict the population 5 yr later by evaluating the function at $t = 5$ with a calculator.

$$P(5) = 20{,}000e^{0.05(5)}$$
$$\approx 25{,}861$$

We expect 25,861 people in the town in 5 yr.

NOTES

On some calculators, you need to enclose the entire exponent in parentheses or else the calculator will misinterpret your intended order of operations.

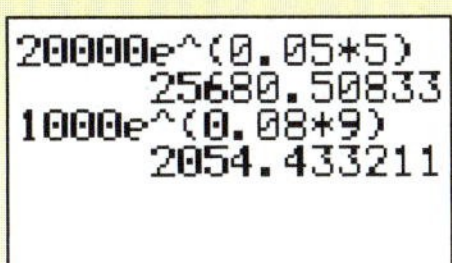

With newer models, you may not need the parentheses. The calculator keeps the cursor in the exponent until you use the right arrow key ▶ to exit the exponent.

$20000e^{0.05*5}$
25680.50833
$1000e^{0.08*9}$
2054.433211

(b) If \$1,000 is invested in an account that earns 8% annually, compounded continuously, then the function

$A(t) = 1{,}000e^{0.08t}$

can be used to model the account balance after t years.

To find the account balance after 9 yr, we use a calculator to evaluate the function when $t = 9$.

$$A(9) = 1{,}000e^{0.08(9)}$$
$$\approx 2{,}054$$

After 9 yr, the account balance is a little more than *double* the original principal.

Continuous compounding gives the highest accumulation of interest at any rate. However, daily compounding results in an amount of interest that is only slightly less.

Check Yourself 3

If \$1,000 is invested in an account that earns 6% annually, compounded continuously, then the function

$A(t) = 1{,}000e^{0.06t}$

can be used to model the account balance after t years.

Find the account balance after 12 yr.

In some applications, we see a variation of the basic exponential function $f(x) = b^x$. What happens to the graph of such a function if we add (or subtract) a constant? The graph of the new function $f(x) = b^x + k$ is a familiar graph translated vertically k units. Consider the next example.

Example 4 Graphing an Exponential Function

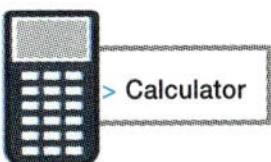

Graph the function $f(x) = 2^x - 3$.

Again, we create a table of values and connect points with a smooth curve.

x	$f(x)$
−3	−2.875
−2	−2.75
−1	−2.5
0	−2
1	−1
2	1
3	5

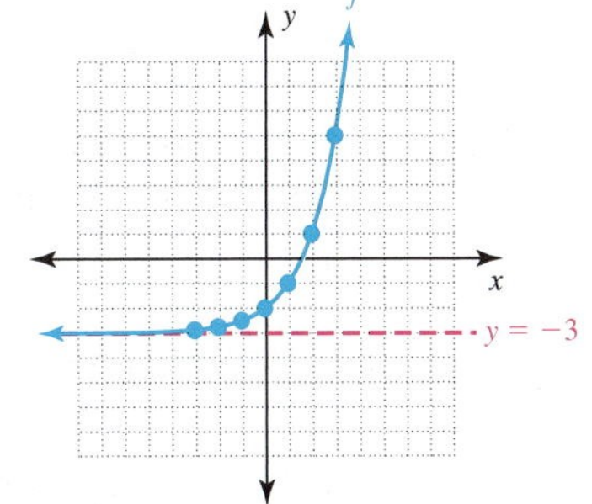

NOTE

Graph this function on a calculator. Graph the line given by $y = -3$ on the same screen.

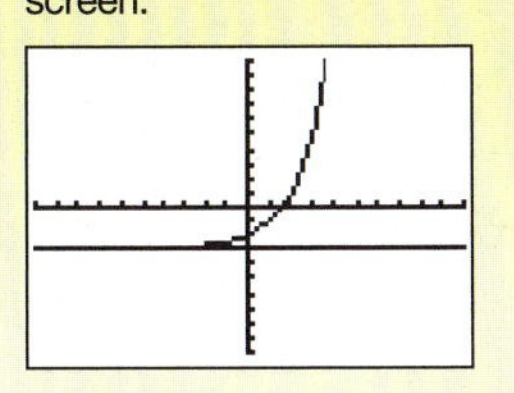

The graph of f appears to be the familiar graph of $g(x) = 2^x$ shifted down 3 units. Instead of a y-intercept of (0, 1), we see a y-intercept of (0, −2). Instead of a horizontal asymptote of $y = 0$, we see a horizontal asymptote of $y = -3$.

Check Yourself 4

Sketch the graph of the function $f(x) = \left(\frac{1}{2}\right)^x + 2$.

We generalize the previous result.

Property

Graphing $y = b^x + k$

All such graphs have these properties.

1. The y-intercept is $(0, k + 1)$.
2. There is a horizontal asymptote with equation $y = k$.

As we observed, exponential functions are always one-to-one. This yields an important property that can be used to solve certain types of equations involving exponents.

Property

Property of Exponential Equations

If $b > 0$ and $b \neq 1$, then

$b^m = b^n$ if and only if $m = n$

Example 5 Solving an Exponential Equation

< Objective 3 >

(a) Solve $2^x = 8$.

We recognize that 8 is a power of 2, and we can write the equation as

$2^x = 2^3$ Write with equal bases.

Applying the property above, we have

$x = 3$ Set exponents equal.

The solution set is $\{3\}$.

(b) Solve $3^{2x} = 81$.

Since $81 = 3^4$, we can write

$$3^{2x} = 3^4$$
$$2x = 4$$
$$x = 2$$

2 is the solution to the equation.

> **NOTE**
>
> The answer can easily be checked by substitution. Letting $x = 2$ gives
>
> $3^{2(2)} = 3^4 = 81$

(c) Solve $2^{x+1} = \frac{1}{16}$.

Again, we write $\frac{1}{16}$ as a power of 2, so that

$$2^{x+1} = 2^{-4}$$

Then $x + 1 = -4$

$$x = -5$$

The solution set is $\{-5\}$.

> **NOTE**
>
> $\frac{1}{16} = \frac{1}{2^4} = 2^{-4}$
>
> Verify the solution.
>
> $2^{-5+1} \stackrel{?}{=} \frac{1}{16}$
>
> $2^{-4} \stackrel{?}{=} \frac{1}{16}$
>
> $\frac{1}{16} = \frac{1}{16}$ True

Check Yourself 5

Solve each equation.

(a) $2^x = 16$ **(b)** $4^{x+1} = 64$ **(c)** $3^{2x} = \frac{1}{81}$

You are now ready to complete Activity 10 at the end of Section 10.1.

Graphing Calculator Option

Exponential Regression

In addition to linear and quadratic regression, graphing calculators can model exponential functions through regression analysis. Most calculators use the basic model $y = a \cdot b^x$ and approximate a and b from a data set.

Suppose we collect some data that suggests a pattern of exponential growth. Finding an exponential function that best fits the data was once a tedious and time-consuming process. It is now quite simple and quick on a graphing calculator.

We consider the population of California, in millions, over a 120-yr period.

Year	1890	1910	1930	1950	1970	1990	2010
Years Since 1890	0	20	40	60	80	100	120
Population (millions)	1.21	2.38	5.68	10.59	19.97	29.76	37.25

Source: U.S. Census Bureau

We begin by clearing the data lists.

STAT **4:ClrList** 2nd [L1] , 2nd [L2] ENTER

Then we enter the data into lists 1 and 2.

STAT **1:Edit** and type in the numbers.

To make a scatterplot,

2nd [STAT PLOT] ENTER

Make sure "On" is selected, use the first icon for "Type" and check that [L1] is the "Xlist" and [L2] is the "Ylist."

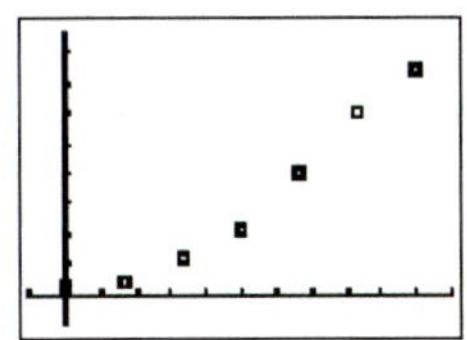

Deselect any functions in the Y= menu and improve the viewing window with

ZOOM **9:ZoomStat**

We find the "best fitting" exponential function by

STAT **CALC 0:ExpReg** ENTER

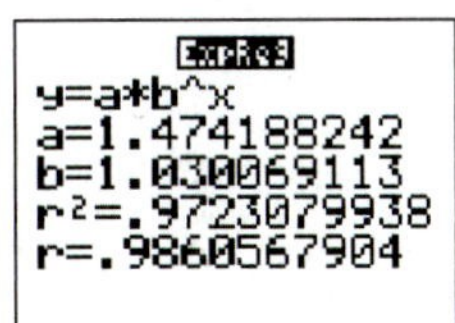

To four decimal places, we have

$y = 1.4742(1.0301)^x$

To view the graph of this function on the scatterplot, enter its equation on the Y= screen and press GRAPH.

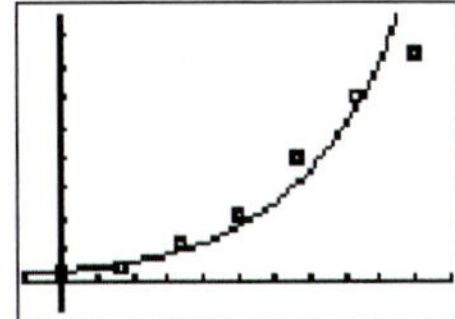

This function may be used to predict the population of California in the year 2020, for example. We must warn again that it is risky to predict too far beyond the scope of the data.

Graphing Calculator Check

The table gives the U.S. public debt (in billions of dollars) since 1950. Use a graphing calculator to find an exponential function that best fits the data. Round your results to four decimal places.
Hint: Let $x = 0$ correspond to 1950.

Year	1950	1960	1970	1980	1990	2000	2005	2010
Debt (billions)	$257.4	290.2	389.2	930.2	3,233	5,674	7,933	13,051

Source: U.S. Treasury

ANSWER

$y = 155.3622(1.0737)^x$

Check Yourself ANSWERS

1.

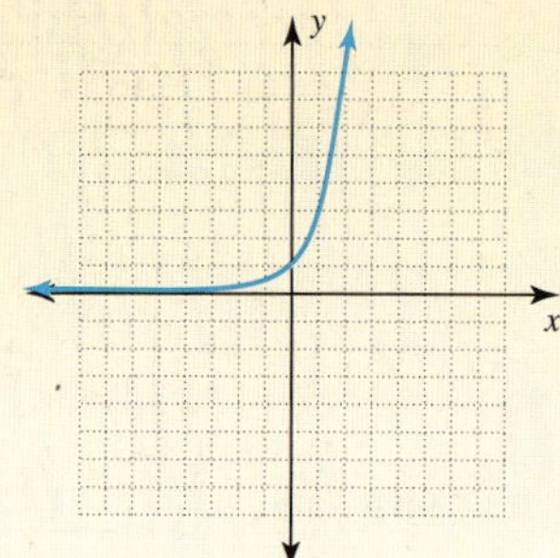

2.

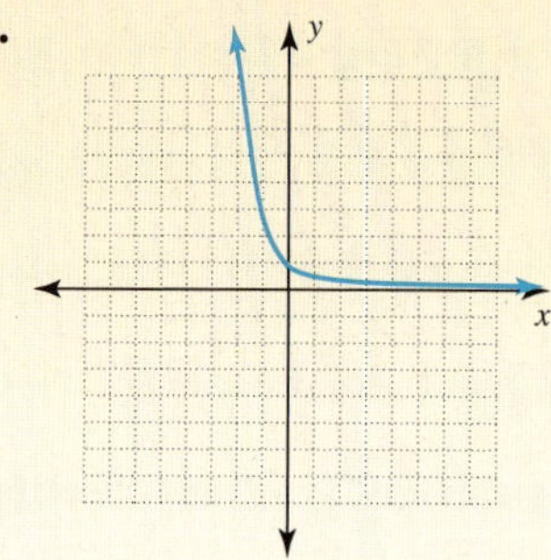

3. $2,054.43

4.

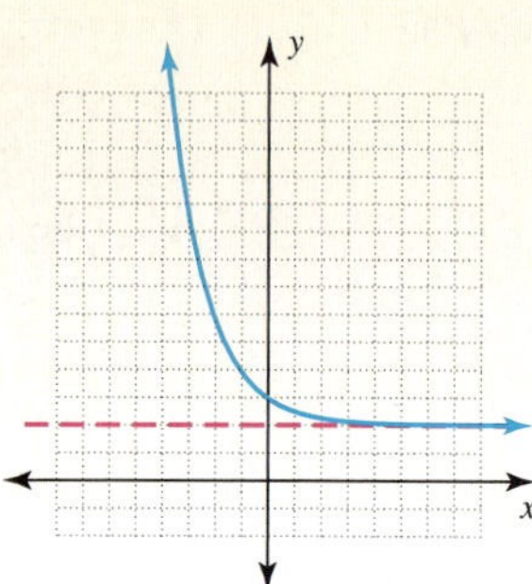

5. **(a)** $\{4\}$; **(b)** $\{2\}$; **(c)** $\{-2\}$

Reading Your Text

These fill-in-the-blank exercises will help you understand some of the key vocabulary used in this section. The answers to these exercises are in the Answers Appendix in the back of the text.

(a) An __________ function is a function that can be written as $f(x) = a \cdot b^x$.

(b) Given $f(x) = a \cdot b^x$, we call b the __________ of the function.

(c) The base of a growth function is __________ than one.

(d) When used as a mathematical constant, the letter e is sometimes called __________ number.

< Objective 1 >

Match the graphs in exercises 1 to 8 with the appropriate equation.

(a) $y = \left(\frac{1}{2}\right)^x$ **(b)** $y = 2x - 1$ **(c)** $y = 2^x$ **(d)** $y = x^2$

(e) $y = 1^x$ **(f)** $y = 5^x$ **(g)** $y = 1 - 2x$ **(h)** $y = 2^x - 1$

1.

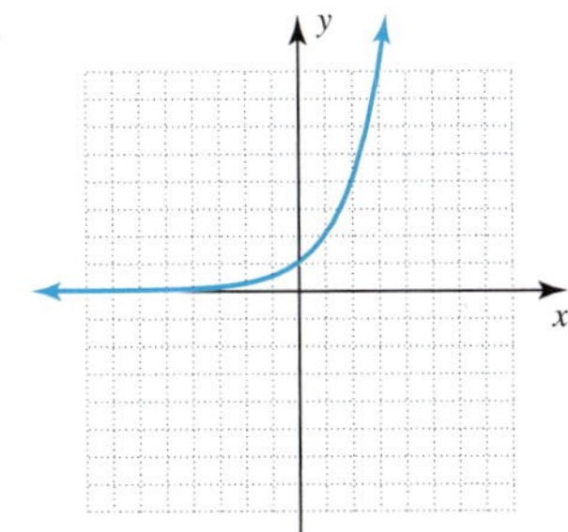

2.

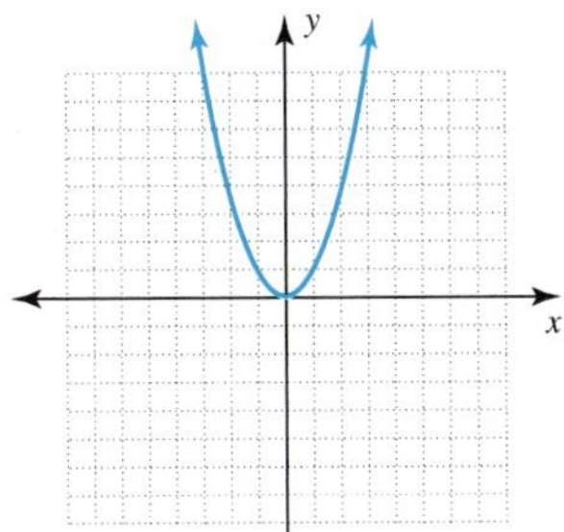

3.

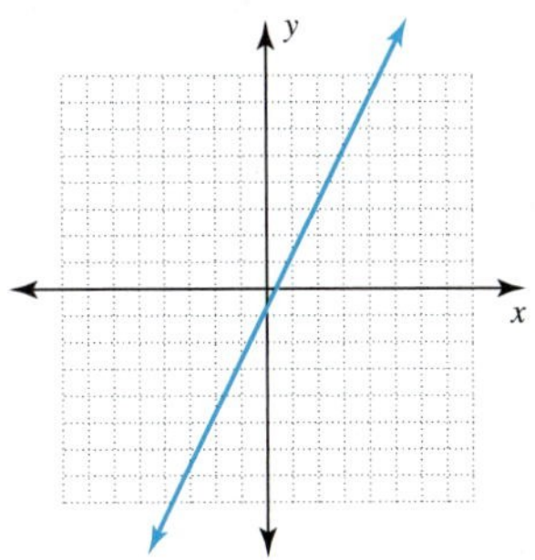

4.

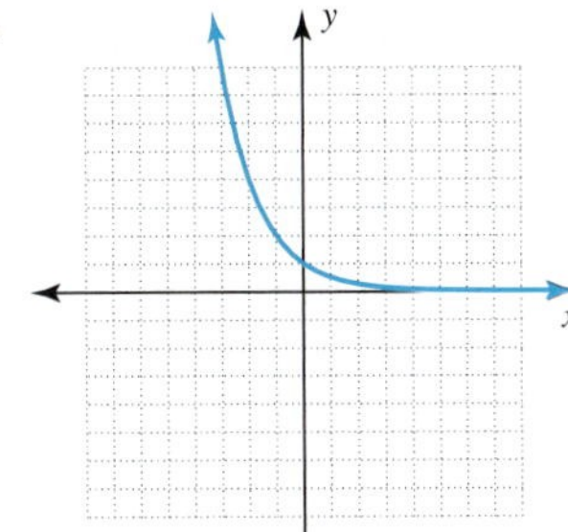

5.

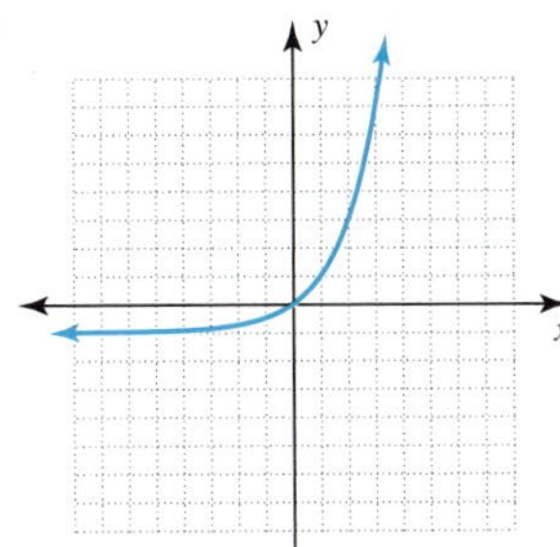

6.

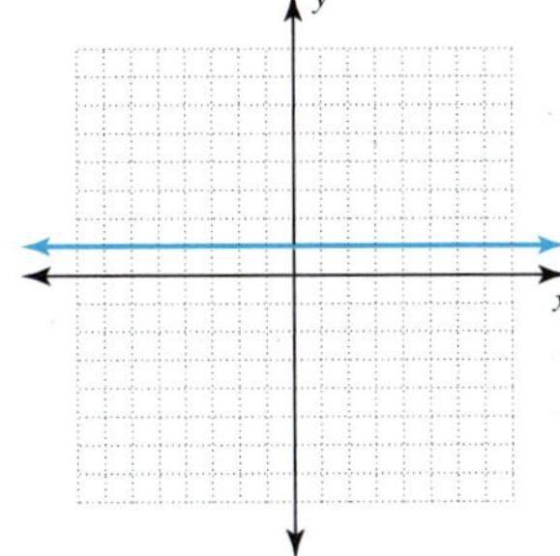

7.

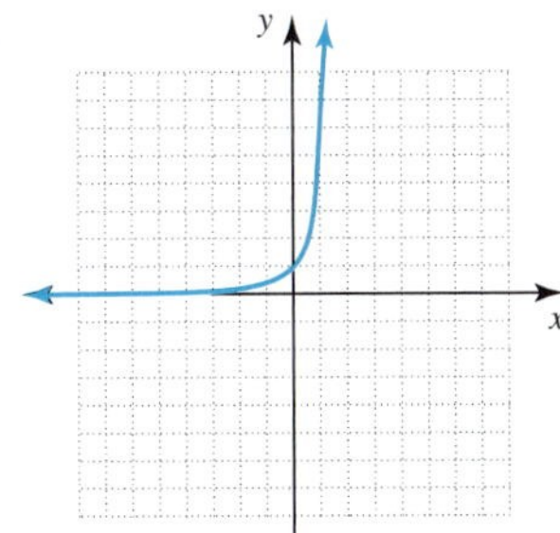

8.

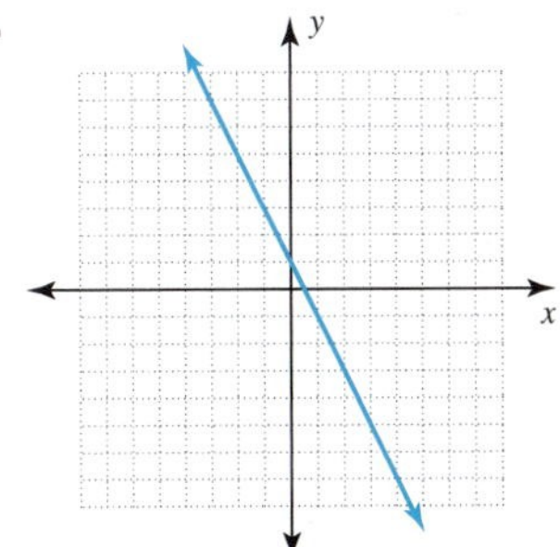

Let $f(x) = 4^x$ *and evaluate.*

9. $f(0)$ **10.** $f(4)$ **11.** $f(2)$ **12.** $f(-2)$

Let $g(x) = 4^{x+1}$ *and evaluate.*

13. $g(-1)$ **14.** $g(1)$ **15.** $g(2)$ **16.** $g(-2)$

Let $h(x) = 4^x + 1$ *and evaluate.*

17. $h(-1)$ **18.** $h(1)$ **19.** $h(2)$ VIDEO **20.** $h(-2)$

Let $f(x) = \left(\frac{1}{4}\right)^x$ *and evaluate.*

21. $f(1)$ **22.** $f(-1)$ **23.** $f(0)$ **24.** $f(2)$

Graph each exponential function.

25. $y = 4^x$

26. $y = \left(\frac{1}{4}\right)^x$

27. $y = \left(\frac{2}{3}\right)^x$

28. $y = \left(\frac{3}{2}\right)^x$

29. $y = 3 \cdot 2^x$

30. $y = 2 \cdot 3^x$

31. $y = 3^x$

32. $y = 2^{x-1}$

33. $y = 2^{2x}$

34. $y = \left(\frac{1}{2}\right)^{2x}$ VIDEO

35. $y = e^{-x}$

36. $y = e^{2x}$

Graph each function. Sketch each horizontal asymptote as a dashed line.

37. $y = 3^x + 2$

38. $y = \left(\frac{1}{3}\right)^x - 4$

< Objective 3 >

Solve each exponential equation.

39. $2^x = 128$

40. $4^x = 64$

41. $10^x = 10{,}000$

42. $5^x = 625$

43. $3^x = \frac{1}{9}$

44. $2^x = \frac{1}{16}$

45. $4^{2x} = 64$ VIDEO

46. $3^{2x} = 81$

47. $3^{x-1} = 81$

48. $4^{x-1} = 16$

49. $3^{x-1} = \frac{1}{27}$

50. $2^{x-3} = \frac{1}{16}$

< Objective 2 >

Science and Medicine *Suppose it takes 1 hr for a bacterial culture to double by dividing. If there are 100 bacteria in the culture to start, then the number of bacteria in the culture after x hours is given by $N(x) = 100 \cdot 2^x$. Use this function to complete exercises 51 to 54.*

51. Find the number of bacteria in the culture after 2 hr.

52. Find the number of bacteria in the culture after 3 hr.

53. Find the number of bacteria in the culture after 5 hr.

54. Graph the relationship between the number of bacteria in the culture and the number of hours. Be sure to choose an appropriate scale for the *N*-axis.

Science and Medicine *The half-life of radium is 1,690 yr. That is, after a 1,690-yr period, one-half of the original amount of radium decays into another substance. If the original amount of radium was 64 g, the formula relating the amount of radium left after time t is given by $R(t) = 64 \cdot 2^{-t/1,690}$. Use this formula to complete exercises 55 to 58.* chapter 10 > Make the Connection

55. Find the amount of radium left after 1,690 yr.

56. Find the amount of radium left after 3,380 yr.

57. Find the amount of radium left after 5,070 yr.

58. Graph the relationship between the amount of radium remaining and time. Be sure to use appropriate scales for the R- and t-axes.

Skills | **Calculator/Computer** | Career Applications | Above and Beyond

Business and Finance *If \$1,000 is invested in a savings account with an interest rate of 8%, compounded annually, the amount in the account after t years is given by $A(t) = 1,000(1 + 0.08)^t$. Use a calculator to complete each exercise.*

59. Find the amount in the account after 2 yr.

60. Find the amount in the account after 5 yr.

61. Find the amount in the account after 9 yr.

62. Graph the relationship between the amount in the account and time. Be sure to choose appropriate scales for the A- and t-axes.

Science and Medicine *The so-called learning curve in psychology applies to learning a skill, such as typing, in which the performance level progresses rapidly at first and then levels off with time. One can approximate N, the number of words per minute (wpm) that a person can type after t weeks of training, with the equation $N = 80(1 - e^{-0.06t})$. Use a calculator to complete exercises 63 and 64.*

63. **(a)** N after 10 weeks; **(b)** N after 20 weeks; **(c)** N after 30 weeks

64. Graph the relationship between the number of words per minute N and the number of weeks of training t.

65. **Business and Finance** Organic food sales have grown steadily. The table gives the annual sales, in billions of dollars, of organic foods in the United States. Use the exponential regression utility on a graphing calculator to fit an exponential function to the data. Round your results to four decimal places.

Hint: Let x be the number of years since 2003.

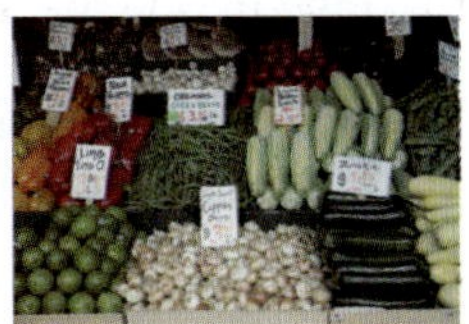

Year	2003	2004	2005	2006	2007	2008	2009	2010
Sales (in billions)	10.38	11.90	13.83	16.72	19.81	22.93	24.80	26.71

Source: Organic Trade Association

66. **Science and Medicine** The table shows the declining temperature of some coffee in degrees Celsius (°C), as 50 minutes passed. Use the exponential regression utility on a graphing calculator to fit an exponential function to the data. Round your results to four decimal places.

Minutes	0	5	8	11	15	18	24
Temp (°C)	83	76.5	70.5	65	61	57.5	52.5

Minutes	25	30	34	38	42	45	50
Temp (°C)	51	47.5	45	43	41	39.5	38

67. **Science and Medicine** The stopping distance for a car depends on (among other things) the speed that the car is traveling. The tables show the stopping distances of a certain car for various speeds. Use the exponential regression utility on a graphing calculator to fit an exponential function to the data. Round your results to four decimal places.

Speed (mi/hr)	5	15	25	35	45
Distance (ft)	7	24	48	72	90

Speed (mi/hr)	55	65	75	85	95
Distance (ft)	146	207	336	603	840

68. **Science and Medicine** After a drug is introduced into a patient's bloodstream, the concentration of the drug begins to drop as time passes. The table shows concentration levels in mg/ml of a certain drug over an 8-hr period. Use the exponential regression utility on a graphing calculator to fit an exponential function to the data. Round your results to four decimal places.

Hour (hr)	0	1	2	3	4	5	6	7	8
Concentration (mg/ml)	1.50	0.95	0.60	0.40	0.25	0.15	0.10	0.07	0.04

Assume $b > 0$, $b \neq 1$. Complete each statement with **always, sometimes,** *or* **never.**

69. The function $g(x) = b^x$ is ________ a decay function.

70. The graph of $f(x) = b^x$ ________ intersects the y-axis.

71. The graph of $f(x) = b^x$ ________ intersects the x-axis.

72. The function $g(x) = b^x$ is ________ one-to-one.

Skills | Calculator/Computer | **Career Applications** | Above and Beyond

73. **Agricultural Technology** The biomass per acre of a cornfield grows exponentially over time. The amount of biomass is given by the function

$B(t) = (1.132)^t$

where B is the biomass per acre, in pounds, and t represents the growing time, in days.

(a) How much biomass is in a 1-acre field after 80 days of growth (nearest pound)?

(b) By how much will the biomass increase between day 80 and day 90 (nearest pound)?

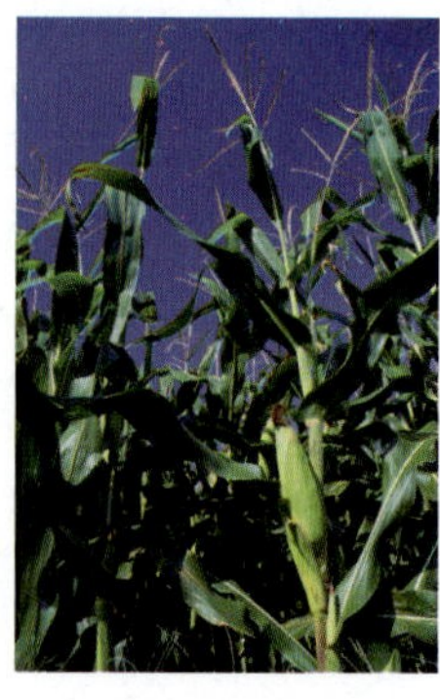

74. **Mechanical Engineering** The intensity of light transmitted through a certain material is reduced by 6% per mm of thickness. The percentage of light transmitted is found using the function

$P(T) = (0.94)^T$

where T represents the material's thickness, in mm.

(a) Find the percentage (to the nearest whole percent) of light transmitted if the material is 34 mm thick.

(b) If the material is 11.2 mm thick, how much light is transmitted?

75. **Electronics** Capacitors are used to store energy. When a capacitor reaches the point when it cannot store anymore energy, it is said to be *charged.* The charging process is not instantaneous. Instead, it can be modeled by the function

$V_c(t) = V_{ps}(1 - e^{-t/\tau})$

in which V_c is the stored voltage measured across the capacitor, V_{ps} is the voltage based on the power supply, t is the time the capacitor has been charging, in seconds, and τ is a (time) constant.

Find the stored capacitive voltage (to the nearest whole volt) after 150 s if the supply voltage in a circuit is 14 volts and has a time constant of 103 s. chapter 10 > Make the Connection

76. **Construction Technology** The temperature of a piece of metal, as it cools, is given by the formula

$T(t) = T_r + (T_m - T_r)e^{-t/\tau}$

in which $T(t)$ gives the temperature of the metal t minutes after it begins to cool. T_r represents the ambient (room) temperature, T_m is the temperature that the metal was heated to, and τ is a (time) constant specific to a particular metal.

For a metal that has a time constant of $\tau = 3.1$, what will its temperature be, to the nearest hundredth degree F, 20 min after being heated if the room temperature is 72°F, and the metal is heated to 280°F? chapter 10 > Make the Connection VIDEO

Skills	Calculator/Computer	Career Applications	**Above and Beyond**

77. Find two different calculators that have $\boxed{e^x}$ keys. Describe how to use the function on each of the calculators.

78. Are there any values of x for which e^x produces an exact answer on the calculator? Why are other answers not exact?

A possible calculator sequence for evaluating the expression

$\left(1 + \frac{1}{n}\right)^n$

where $n = 10$ *is*

[(] 1 [+] 1 [÷] 10 [)] [^] 10 [ENTER]

which yields approximately 2.5937.

Find $\left(1 + \frac{1}{n}\right)^n$ *for each value of n. Round your answers to the nearest hundred-thousandth.*

79. $n = 100$

80. $n = 1{,}000$

81. $n = 10{,}000$

82. $n = 100{,}000$

83. $n = 1{,}000{,}000$

84. What did you observe from the results of exercises 79 to 83?

85. Graph the exponential function defined by $y = 2^x$.

86. Graph the function defined by $x = 2^y$ on the same set of axes as the previous graph. What do you observe?
Hint: To graph $x = 2^y$, choose convenient values for y and then compute the corresponding values for x.

87. Suppose you have a large piece of paper whose thickness is 0.003 in. If you tear the paper in half and stack the pieces, the height of the stack is (0.003)(2) in., or 0.006 in. If you now tear the stack in half again and then stack the pieces, the stack is $(0.003)(2)(2) = (0.003)(2^2)$ in., or 0.012 in. high. chapter 10 > Make the Connection

(a) Define a function that gives the height h of the stack (in inches) after n tears.

(b) After which tear will the stack's height exceed 8 in.?

(c) Compute the height of the stack after the 15th tear. You will need to convert your answer to the appropriate units.

88. Use a graphing calculator to find all three points of intersection of the graphs of $f(x) = x^2$ and $g(x) = 2^x$. Give coordinates rounded to two decimal places.

Answers

1. (c) **3.** (b) **5.** (h) **7.** (f) **9.** 1 **11.** 16 **13.** 1 **15.** 64 **17.** $\frac{5}{4}$ **19.** 17 **21.** $\frac{1}{4}$ **23.** 1

25.

27.

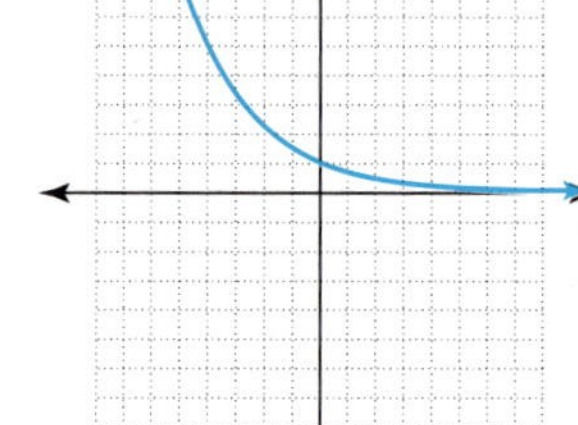

29.

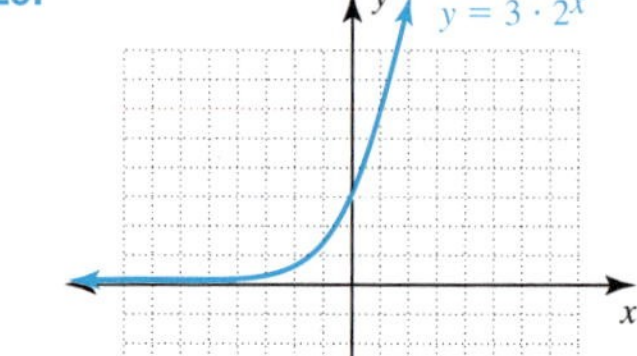
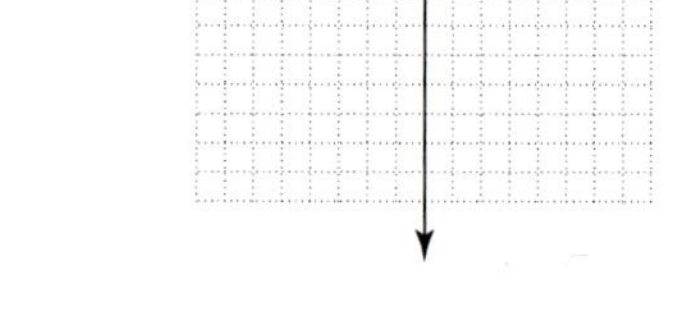

31.

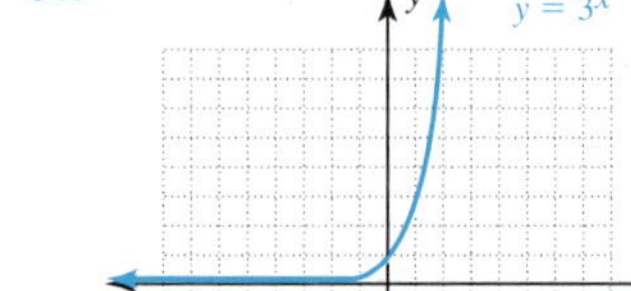

33.

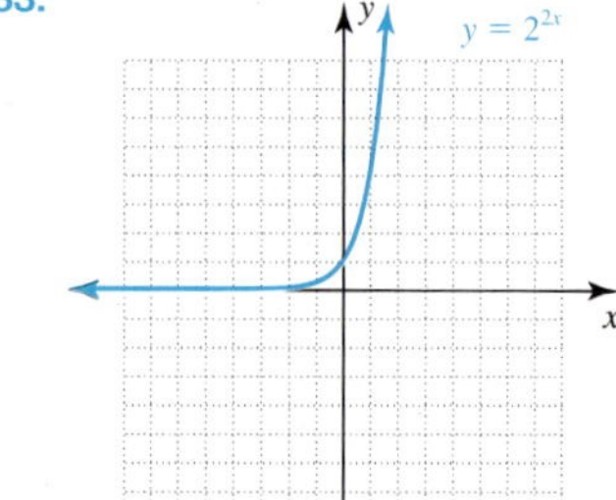

35.

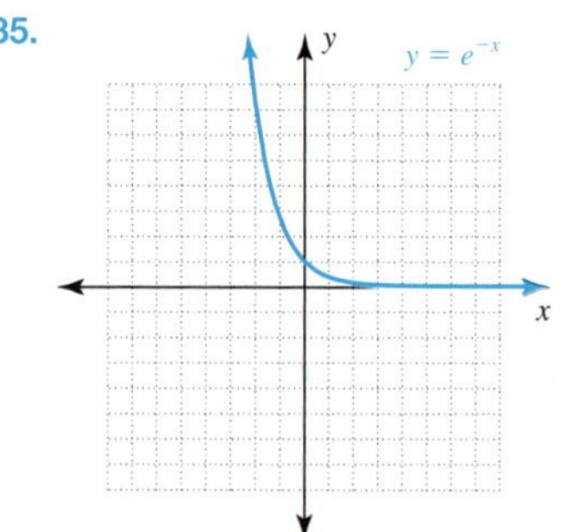

37.

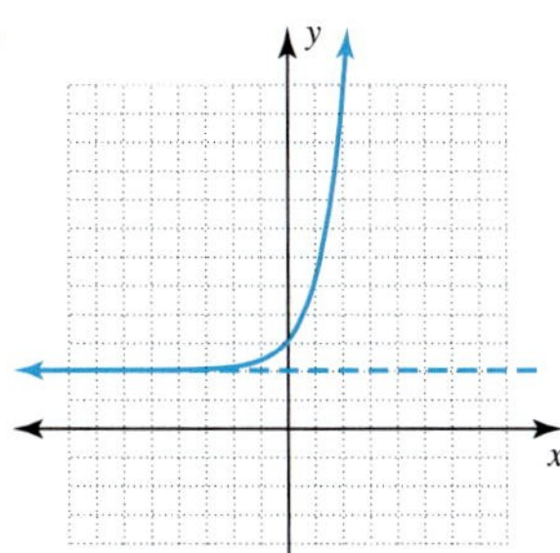

39. $\{7\}$ **41.** $\{4\}$ **43.** $\{-2\}$ **45.** $\left\{\frac{3}{2}\right\}$ **47.** $\{5\}$ **49.** $\{-2\}$ **51.** 400 bacteria

53. 3,200 bacteria **55.** 32 g **57.** 8 g **59.** \$1,166.40 **61.** \$1,999 **63.** **(a)** 36; **(b)** 56; **(c)** 67 **65.** $y = 10.595(1.1532)^x$ **67.** $y = 10.0645(1.0491)^x$ **69.** sometimes **71.** never **73.** **(a)** 20,310 lb; **(b)** 49,864 lb **75.** 11 V **77.** Above and Beyond **79.** 2.70481 **81.** 2.71815 **83.** 2.71828 **85.**

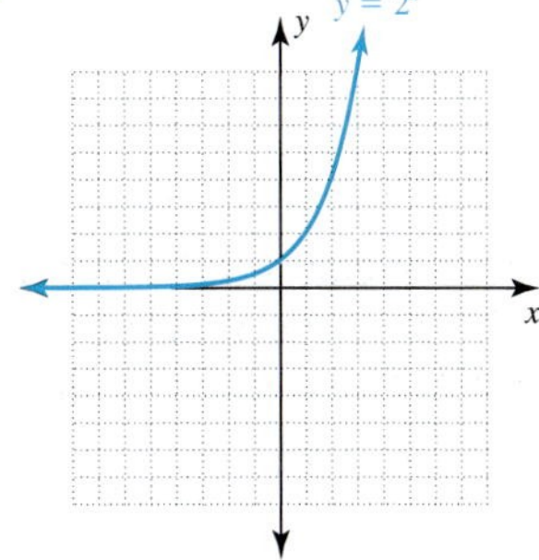

87. **(a)** $h = 0.003(2^n)$; **(b)** 12; **(c)** 8.192 ft

10.5 Logarithmic Functions

< 10.5 Objectives >

1 > Graph a logarithmic function

2 > Convert between logarithmic and exponential equations

3 > Evaluate a logarithmic expression

4 > Solve a logarithmic equation

5 > Use logarithmic functions to solve applications

Given our experience with exponential functions in Section 10.4 and inverse functions in Section 10.3, we can now introduce the logarithmic function.

The Scottish mathematician John Napier (1550–1617) is credited with inventing logarithms. The development of the logarithm grew out of a desire to ease the work involved in numerical computations, particularly in the field of astronomy. Today the availability of calculators has made logarithms unnecessary as a computational tool.

NOTE

Napier coined the word *logarithm* from the Greek words "logos" (a ratio) and "arithmos" (a number).

However, the concept of the logarithm and the properties of logarithmic functions are still very important in calculus and the applied sciences.

Again, the applications for this new function are numerous. The Richter scale for measuring the intensity of an earthquake and the decibel scale for measuring the intensity of sound both use logarithms.

To develop the idea of a logarithmic function, we return to the definition of an exponential function.

$$f(x) = b^x \qquad b > 0, b \neq 1$$

RECALL

If *f* is a one-to-one function, then its inverse is also a function.

Letting y replace $f(x)$ and interchanging the roles of x and y, we have the inverse function

$$b^y = x$$

Presently, we have no way to solve the equation $b^y = x$ for y. So, to write the inverse in a more useful form, we offer a definition.

Definition

Logarithm of *x* with base *b*

For $b > 0$ and $b \neq 1$,

$y = \log_b x$ if, and only if, $b^y = x$

We call $\log_b x$ the **logarithm of *x* with base *b*.**

NOTE

The restrictions on the base are the same as those for the exponential function.

Using function notation, for $b > 0$ and $b \neq 1$, if $f(x) = b^x$, then $f^{-1}(x) = \log_b x$. Since the exponential function is one-to-one, we know its inverse is also a function. While we could write the **logarithmic function** as $y = \log_b (x)$, we usually drop the parentheses and simply write $y = \log_b x$.

The important meaning here is that y is the power (or exponent) we place on b to produce x. For example, $3 = \log_2 8$ means 3 is the exponent we place on 2 to produce 8. So, $y = \log_b x$ is equivalent to $b^y = x$.

Property

Logarithmic and Exponential Functions

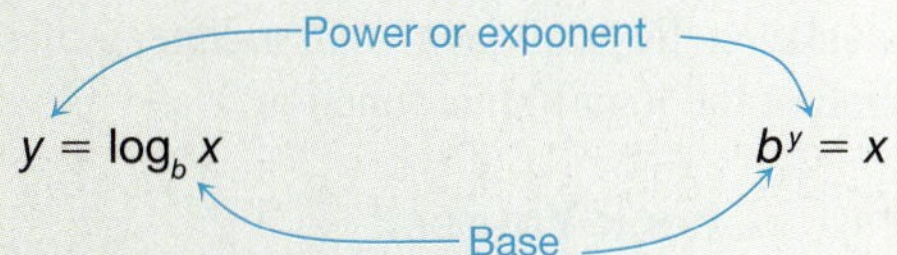

The logarithm y is the power to which we must raise b to get x. In other words, *a logarithm is simply a power or an exponent.*

We begin our work by graphing a typical logarithmic function.

Example 1 Graphing a Logarithmic Function

< Objective 1 >

Graph the logarithmic function

$y = \log_2 x$

Since $y = \log_2 x$ is equivalent to the exponential form

$2^y = x$

we can find ordered pairs satisfying this equation by choosing convenient values for y and calculating the corresponding values for x.

Letting y take on values from -3 to 3 yields the table of values shown here. As before, we plot points from the ordered pairs and connect them with a smooth curve to form the graph of the function.

NOTE

The base is 2, and the logarithm or power is y.

NOTE

What do the vertical and horizontal line tests tell you about this graph?

x	y
$\frac{1}{8}$	-3
$\frac{1}{4}$	-2
$\frac{1}{2}$	-1
1	0
2	1
4	2
8	3

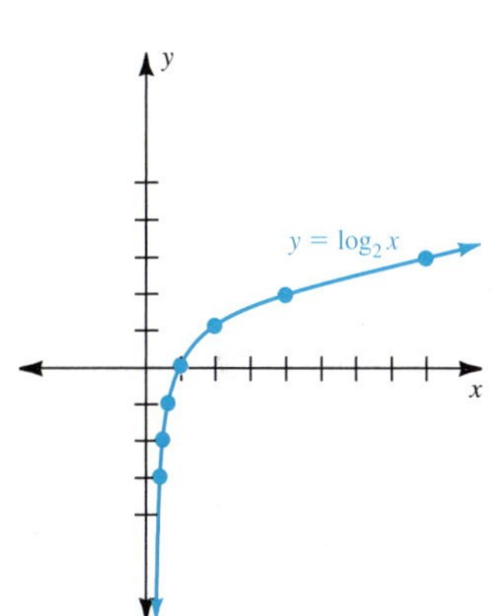

We observe that the graph represents a one-to-one function whose domain is $\{x \mid x > 0\}$ and whose range is the set of all real numbers.

For base 2 (or for any base greater than 1), the function increases over its domain.

Recall from Section 10.3 that the graphs of a function and its inverse are always reflections of each other about the line $y = x$. Since we have defined the logarithmic function as the inverse of an exponential function, we can anticipate the same relationship.

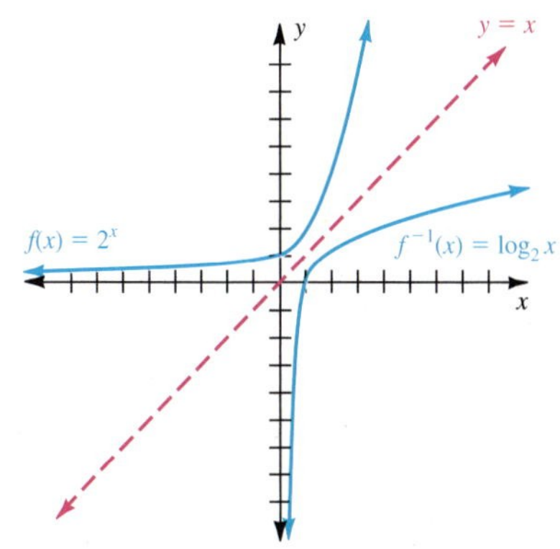

We see that the graphs of f and f^{-1} are indeed reflections of each other about the line $y = x$. In fact, this relationship provides an alternate method of sketching $y = \log_b x$. We can sketch the graph of $y = b^x$ and then reflect that graph about line $y = x$ to form the graph of the logarithmic function.

Check Yourself 1

Graph the logarithmic function defined by

$y = \log_3 x$

Hint: Consider the equivalent form $3^y = x$.

Let us summarize some facts regarding logarithmic and exponential functions.

Property

Inverse Functions

$y = b^x$ and $y = \log_b x$ are inverse functions.

1. Their graphs are symmetric with respect to the line $y = x$.
2. Because the point (0, 1) is on the graph of $y = b^x$ (it is the y-intercept), the point (1, 0) is on the graph of $y = \log_b x$ (it is the x-intercept).
3. Because the line $y = 0$ is the asymptote for $y = b^x$ (it is horizontal), the line $x = 0$ is the asymptote for $y = \log_b x$ (it is vertical).
4. The domain of $y = b^x$ is the set of all real numbers, and the range is the set of all positive real numbers.
5. The domain of $y = \log_b x$ is the set of all positive real numbers, and the range is the set of all real numbers.

It is often necessary to convert between exponential and logarithmic forms. The conversion is straightforward. You need only keep in mind the basic relationship

$y = \log_b x$ means $b^y = x$

Example 2 **Writing Equations in Logarithmic Form**

< Objective 2 >

Write each exponential equation as a logarithmic equation.

(a) $3^4 = 81$ is equivalent to $\log_3 81 = 4$.

(b) $10^3 = 1{,}000$ is equivalent to $\log_{10} 1{,}000 = 3$.

(c) $2^{-3} = \frac{1}{8}$ is equivalent to $\log_2 \frac{1}{8} = -3$.

(d) $9^{1/2} = 3$ is equivalent to $\log_9 3 = \frac{1}{2}$.

Check Yourself 2

Write each exponential equation as a logarithmic equation.

(a) $4^3 = 64$ **(b)** $10^{-2} = 0.01$

(c) $3^{-3} = \frac{1}{27}$ **(d)** $27^{1/3} = 3$

In Example 3, we reverse this process and write logarithmic equations as exponential equations.

Example 3 Writing Equations in Exponential Form

NOTE

In (a), the base is 2; the logarithm or power is 3.

Write each logarithmic equation as an exponential equation.

(a) $\log_2 8 = 3$ is equivalent to $2^3 = 8$.

(b) $\log_{10} 100 = 2$ is equivalent to $10^2 = 100$.

(c) $\log_3 \frac{1}{9} = -2$ is equivalent to $3^{-2} = \frac{1}{9}$.

(d) $\log_{25} 5 = \frac{1}{2}$ is equivalent to $25^{1/2} = 5$.

Check Yourself 3

Write each logarithmic equation as an exponential equation.

(a) $\log_2 32 = 5$ **(b)** $\log_{10} 1{,}000 = 3$

(c) $\log_4 \frac{1}{16} = -2$ **(d)** $\log_{27} 3 = \frac{1}{3}$

Certain logarithms can be calculated by changing an expression to the equivalent exponential form, as Example 4 illustrates.

Example 4 Evaluating Logarithmic Expressions

< Objective 3 >

RECALL

$b^m = b^n$ if and only if $m = n$.

(a) Evaluate $\log_3 27$.

If $x = \log_3 27$, in exponential form we have

$$3^x = 27$$

$$3^x = 3^3$$

$$x = 3$$

We have $\log_3 27 = 3$.

(b) Evaluate $\log_{10} \frac{1}{10}$.

NOTE

Rewrite each side as a power of the same base.

If $x = \log_{10} \frac{1}{10}$, we can write

$$10^x = \frac{1}{10}$$

$$= 10^{-1}$$

We have $x = -1$ and

$$\log_{10} \frac{1}{10} = -1$$

Check Yourself 4

Evaluate each logarithm.

(a) $\log_2 64$ **(b)** $\log_3 \frac{1}{27}$

The relationship between exponents and logarithms also allows us to solve logarithmic equations if two of the quantities in the equation $y = \log_b x$ are known.

Example 5 Solving Logarithmic Equations

< Objective 4 >

(a) Solve $\log_5 x = 3$ for x.

Since $\log_5 x = 3$, in exponential form we have

$$5^3 = x$$
$$x = 125$$

The solution set is $\{125\}$.

(b) Solve $y = \log_4 \frac{1}{16}$ for y.

The original equation is equivalent to

$$4^y = \frac{1}{16}$$
$$= 4^{-2}$$

$y = -2$ gives the solution.

(c) Solve $\log_b 81 = 4$ for b.

In exponential form the equation becomes

$$b^4 = 81$$
$$b = 3$$

The solution set is $\{3\}$.

NOTE

Keep in mind that the base must be *positive*, so we do not consider the possible solution $b = -3$.

Check Yourself 5

Solve each equation.

(a) $\log_4 x = 4$ **(b)** $\log_b \frac{1}{8} = -3$ **(c)** $y = \log_9 3$

NOTES

Loudness can be measured in **bels (B),** a unit named for Alexander Graham Bell. This unit is rather large, so a more practical unit is the **decibel (dB),** a unit that is one-tenth as large.

The constant I_0 is the intensity of the minimum sound level detectable by the human ear.

We use the **decibel scale** to measure the loudness of various sounds. If I represents the intensity of a sound and I_0 represents the intensity of a "threshold sound," then the decibel (dB) rating L of a noise is given by

$$L = 10 \log_{10} \frac{I}{I_0}$$

where $I_0 = 10^{-16}$ watt per square centimeter (W/cm²).

Example 6 A Decibel Application

< Objective 5 >

NOTE

To evaluate $\log_{10} 10^2$, think "To what power must we raise 10 to obtain 10^2?" Answer: 2.

(a) The intensity of a whisper is $I = 10^{-14}$. Its decibel rating is

$$L = 10 \log_{10} \frac{10^{-14}}{10^{-16}}$$
$$= 10 \log_{10} 10^2$$
$$= 10 \cdot 2$$
$$= 20$$

NOTE

Again, think "10 to what power produces 10^{12}?" Answer: 12.

(b) The intensity of a concert is $I = 10^{-4}$. Its decibel rating is

$$L = 10 \log_{10} \frac{10^{-4}}{10^{-16}}$$
$$= 10 \log_{10} 10^{12}$$
$$= 10 \cdot 12$$
$$= 120$$

Check Yourself 6

The intensity of ordinary conversation is $I = 10^{-12}$. Find its rating on the decibel scale.

NOTES

The scale is named after Charles Richter, a U.S. geologist.

A *zero-level* earthquake is the quake of least intensity that is measurable by a seismograph.

Geologists use the **Richter scale** to convert seismographic readings, which give the intensity of the shock waves of an earthquake, to a measure of the magnitude of that earthquake.

The magnitude M of an earthquake is given by

$$M = \log_{10} \frac{a}{a_0}$$

where a is the intensity of its shock waves and a_0 is the intensity of the shock wave of a zero-level earthquake.

Example 7 A Richter Scale Application

How many times stronger is an earthquake measuring 5 on the Richter scale than one measuring 4 on the Richter scale?

Suppose a_1 is the intensity of the earthquake with magnitude 5 and a_2 is the intensity of the earthquake with magnitude 4. We want to find $\frac{a_1}{a_2}$. Then

RECALL

The ratio of a_1 to a_2 is $\frac{a_1}{a_2}$

$$5 = \log_{10} \frac{a_1}{a_0} \qquad \text{and} \qquad 4 = \log_{10} \frac{a_2}{a_0}$$

We convert these logarithmic expressions to exponential form.

$$10^5 = \frac{a_1}{a_0} \qquad \text{and} \qquad 10^4 = \frac{a_2}{a_0}$$

or

$$a_1 = a_0 \cdot 10^5 \qquad \text{and} \qquad a_2 = a_0 \cdot 10^4$$

NOTE

$\frac{a_1}{a_2} = 10$ is equivalent to $a_1 = 10a_2$, which says that a_1 is 10 times the size of a_2.

We want the ratio of the intensities of the two earthquakes, so

$$\frac{a_1}{a_2} = \frac{a_0 \cdot 10^5}{a_0 \cdot 10^4} = 10^1 = 10$$

A magnitude 5 earthquake is *10 times stronger* than a magnitude 4 earthquake.

Check Yourself 7

How many times stronger is an earthquake of magnitude 6 than one of magnitude 4?

Check Yourself ANSWERS

1. $y = \log_3 x$

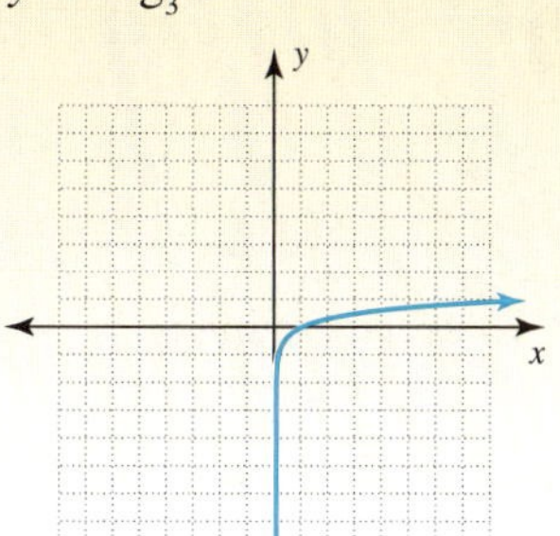

2. **(a)** $\log_4 64 = 3$; **(b)** $\log_{10} 0.01 = -2$; **(c)** $\log_3 \frac{1}{27} = -3$; **(d)** $\log_{27} 3 = \frac{1}{3}$ **3.** **(a)** $2^5 = 32$; **(b)** $10^3 = 1{,}000$; **(c)** $4^{-2} = \frac{1}{16}$; **(d)** $27^{1/3} = 3$ **4.** **(a)** $\log_2 64 = 6$; **(b)** $\log_3 \frac{1}{27} = -3$ **5.** **(a)** $\{256\}$; **(b)** $\{2\}$; **(c)** $\left\{\frac{1}{2}\right\}$ **6.** 40 dB **7.** 100 times

Reading Your Text

These fill-in-the-blank exercises will help you understand some of the key vocabulary used in this section. The answers to these exercises are in the Answers Appendix in the back of the text.

(a) The ________ of x to the base b is denoted $y = \log_b x$.

(b) Given $y = \log_b x$, the logarithm y is the ________ to which we must raise b to get x.

(c) The point (0, 1) is on the graph of $y = b^x$. It is the ________.

(d) ________ can be measured in bels.

10.5 exercises

Skills | Calculator/Computer | Career Applications | Above and Beyond

< Objective 1 >

Graph each function.

1. $y = \log_4 x$

2. $y = \log_{10} x$

3. $y = \log_2 (x - 1)$

4. $y = \log_3 (x + 1)$

5. $y = \log_8 x$

6. $y = \log_3 x + 1$

< Objective 2 >

Write each equation in logarithmic form.

7. $2^5 = 32$

8. $3^5 = 243$

9. $10^2 = 100$

10. $5^3 = 125$

11. $3^0 = 1$

12. $10^0 = 1$

13. $4^{-2} = \frac{1}{16}$ VIDEO

14. $3^{-4} = \frac{1}{81}$

15. $10^{-2} = \frac{1}{100}$

16. $3^{-3} = \frac{1}{27}$

17. $16^{1/2} = 4$

18. $125^{1/3} = 5$

19. $64^{-1/3} = \frac{1}{4}$

20. $36^{-1/2} = \frac{1}{6}$

21. $27^{2/3} = 9$

22. $9^{3/2} = 27$

23. $27^{-2/3} = \frac{1}{9}$

24. $16^{-3/2} = \frac{1}{64}$

Write each equation in exponential form.

25. $\log_2 16 = 4$

26. $\log_5 5 = 1$

27. $\log_5 1 = 0$

28. $\log_3 27 = 3$

29. $\log_{10} 10 = 1$ VIDEO

30. $\log_2 32 = 5$

31. $\log_5 125 = 3$

32. $\log_{10} 1 = 0$

33. $\log_3 \frac{1}{27} = -3$

34. $\log_5 \frac{1}{25} = -2$

35. $\log_{10} 0.001 = -3$

36. $\log_{10} \frac{1}{1{,}000} = -3$

37. $\log_{16} 4 = \frac{1}{2}$

38. $\log_{125} 5 = \frac{1}{3}$

39. $\log_8 4 = \frac{2}{3}$

40. $\log_9 27 = \frac{3}{2}$

41. $\log_{25} \frac{1}{5} = -\frac{1}{2}$

42. $\log_{64} \frac{1}{16} = -\frac{2}{3}$

< Objective 3 >

Evaluate each logarithm.

43. $\log_2 64$

44. $\log_3 81$

45. $\log_4 64$ VIDEO

46. $\log_{10} 10{,}000$

47. $\log_3 \frac{1}{81}$

48. $\log_4 \frac{1}{64}$

49. $\log_{10} \frac{1}{100}$

50. $\log_5 \frac{1}{25}$

51. $\log_{25} 5$

52. $\log_{27} 3$

< Objective 4 >

Solve each equation.

53. $y = \log_5 25$

54. $\log_2 x = 4$

55. $\log_b 256 = 4$

56. $y = \log_3 1$

57. $\log_{10} x = 2$

58. $\log_b 125 = 3$

59. $y = \log_5 5$

60. $y = \log_3 81$

61. $\log_{3/2} x = 3$

62. $\log_b \frac{4}{9} = 2$

63. $\log_b \frac{1}{25} = -2$ VIDEO

64. $\log_3 x = -3$

65. $\log_{10} x = -3$ VIDEO

66. $y = \log_2 \frac{1}{16}$

67. $y = \log_8 \frac{1}{64}$

68. $\log_b \frac{1}{100} = -2$

69. $\log_{27} x = \frac{1}{3}$

70. $y = \log_{100} 10$

71. $\log_b 5 = \frac{1}{2}$

72. $\log_{64} x = \frac{2}{3}$

73. $y = \log_{27} \frac{1}{9}$

74. $\log_b \frac{1}{8} = -\frac{3}{4}$

< Objective 5 >

Science and Medicine *Use the decibel formula*

$$L = 10 \log_{10} \frac{I}{I_0}$$

where $I_0 = 10^{-16}$ W/cm^2 *to solve each problem.*

75. A television commercial has a volume with intensity $I = 10^{-11}$ W/cm^2. Find its rating in decibels.

76. The sound of a jet plane on takeoff has an intensity $I = 10^{-2}$ W/cm^2. Find its rating in decibels.

77. The sound of a computer printer has an intensity of $I = 10^{-9}$ W/cm^2. Find its rating in decibels.

78. The sound of a busy street has an intensity of $I = 10^{-8}$ W/cm^2. Find its rating in decibels.

Science and Medicine *The formula for the decibel rating L can be solved for the intensity of the sound as* $I = 10^{-16} \cdot 10^{L/10}$. *Use this formula in exercises 79 to 83.*

79. Find the intensity of the sound in an airport waiting area if the decibel rating is 80.

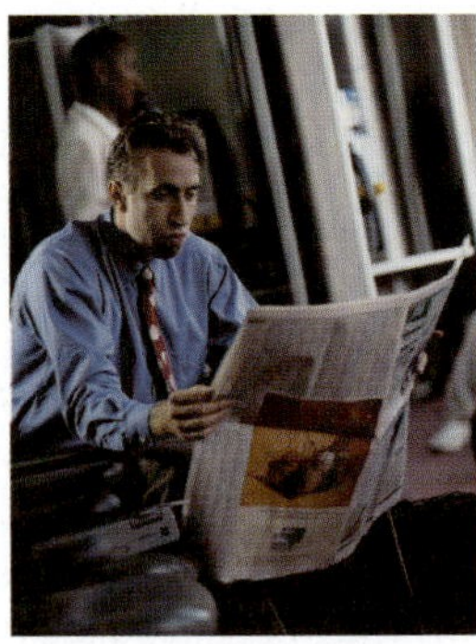

80. Find the intensity of the sound of conversation in a crowded room if the decibel rating is 70.

81. What is the ratio of intensity of a sound of 80 dB to that of 70 dB?

82. What is the ratio of intensity of a sound of 60 dB to one measuring 40 dB?

83. What is the ratio of intensity of a sound of 70 dB to one measuring 40 dB?

84. Derive the formula for intensity provided above.
Hint: First divide both sides of the decibel formula by 10. Then write the equation in exponential form.

Science and Medicine *Use the earthquake formula*

$$M = \log_{10} \frac{a}{a_0}$$

85. An earthquake has an intensity a of $10^6 \cdot a_0$, where a_0 is the intensity of the zero-level earthquake. What was its magnitude?

86. The great San Francisco earthquake of 1906 had an intensity of $10^{8.3} \cdot a_0$. What was its magnitude?

87. An earthquake can begin causing damage to buildings with a magnitude of 5 on the Richter scale. Find its intensity in terms of a_0.

88. An earthquake may cause moderate building damage with a magnitude of 6 on the Richter scale. Find its intensity in terms of a_0.

Determine whether each statement is **true** *or* **false.**

89. $m = \log_k n$ is equivalent to $k^n = m$.

90. The inverse of an exponential function is a logarithmic function.

Complete each statement with **always, sometimes,** *or* **never.**

91. The graph of $f(x) = \log_b x$ __________ has a vertical asymptote.

92. The graph of $f(x) = \log_b x$ __________ passes through the point (1, 0).

Allied Health *The molar concentration of hydrogen ions* $[H^+]$ *in an aqueous solution is equal to the product of the normality N and the percent ionization (in decimal form). The acidity level, or pH, of a solution is a function of the molar concentration and is given by the formula*

$$\text{pH} = -\log([H^+])$$

Use this information to complete exercises 93 and 94.
Hints: When "log" is written without a base, we always assume the base is 10. The [LOG] key on a graphing calculator is the log base 10 function. To find $\log_{10} 42$, for example, press [LOG] 42 [)] [ENTER]. The result should be 1.623, to the nearest thousandth.

93. Determine the pH (to the nearest thousandth) of a 0.15-N acid solution that is 65% ionized.

94. Find the pH (to the nearest tenth) of a chemical that contains 3.8×10^{-4} moles of H^+ per liter.

95. **Electrical Engineering** One formula for sound level, in decibels, is

$$\text{db} = 10 \log_{10}\left(\frac{I}{I_0}\right)$$

in which I is the sound intensity, in watts per sq m, and I_0 is the base sound level, 10^{-12} W/m^2 (the lowest sound discernible to most people).

Find the decibel level in a factory in which the sound intensity is 3.2×10^{-9} W/m^2. Report your result to the nearest decibel (dB).

96. **Construction Technology** The strength of concrete depends on its curing time. The longer it takes to cure, the stronger the concrete will be. If the desired strength of the concrete is known (to a maximum of 3,000 psi after 48 hr), the required curing time is found using the formula

$$t = 48 \log_e\left(\frac{s}{3{,}000}\right) + 48$$

in which s is the desired strength and t is measured in hours.

Determine the curing time necessary to produce concrete with a strength of 2,241 psi.
Hint: When "log" is written with base e, this may be calculated by using the [LN] key on a graphing calculator. To find $\log_e 42$, for example, press [LN] 42 [)] [ENTER]. The result should be 3.738, to the nearest thousandth.

97. The **learning curve** describes the relationship between learning and time. Its graph is a logarithmic curve in the first quadrant. Describe that curve as it relates to learning.

98. In what other scientific fields would you expect to encounter a discussion of logarithms?

The half-life *of a radioactive substance is the time it takes for one-half of the original amount of the substance to decay to a nonradioactive element. The half-life of radioactive waste is very important in figuring how long the waste must be kept isolated from the environment in some sort of storage facility. Half-lives of various radioactive waste products vary from fractions of a second to millions of years. It usually takes at least 10 half-lives for a radioactive waste product to be considered safe.*

The half-life of a radioactive substance can be determined by the formula

$$\log_e \frac{1}{2} = -\lambda x$$

where λ = radioactive decay constant
x = half-life

Approximate the half-lives of these important radioactive waste products, given the radioactive decay constant (RDC). Report your results to the nearest year. chapter 10 > Make the Connection

(To compute a logarithm base e on a calculator, see the hint in exercise 96.)

99. Plutonium-239, RDC = 0.000029

100. Strontium-90, RDC = 0.024755

101. Thorium-230, RDC = 0.000009

102. Cesium-135, RDC = 0.00000035

103. How many years will it be before each waste product is considered safe? 

104. **(a)** Evaluate $\log_2(4 \cdot 8)$. **(b)** Evaluate $\log_2 4$ and $\log_2 8$. **(c)** Write an equation that connects the result of part (a) with the results of part (b).

105. **(a)** Evaluate $\log_3(9 \cdot 81)$. **(b)** Evaluate $\log_3 9$ and $\log_3 81$. **(c)** Write an equation that connects the result of part (a) with the results of part (b).

106. Based on exercises 104 and 105, propose a statement that connects $\log_a(mn)$ with $\log_a m$ and $\log_a n$.

Answers

1.

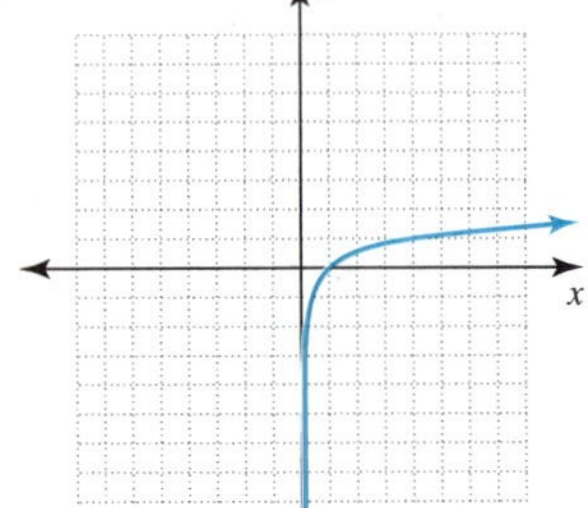

3.

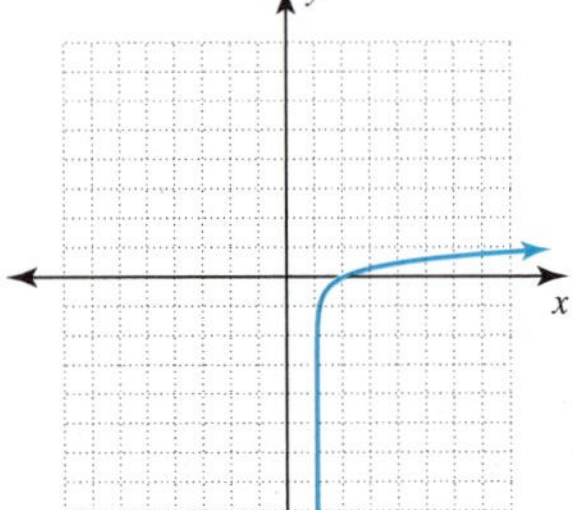

5.

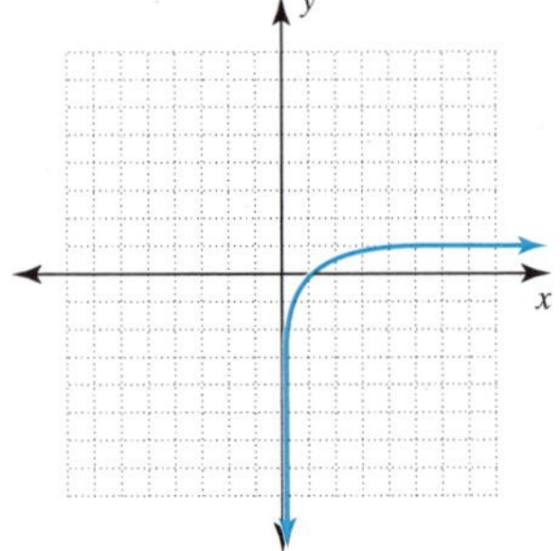

7. $\log_2 32 = 5$

9. $\log_{10} 100 = 2$ **11.** $\log_3 1 = 0$ **13.** $\log_4 \frac{1}{16} = -2$ **15.** $\log_{10} \frac{1}{100} = -2$ **17.** $\log_{16} 4 = \frac{1}{2}$ **19.** $\log_{64} \frac{1}{4} = -\frac{1}{3}$ **21.** $\log_{27} 9 = \frac{2}{3}$

23. $\log_{27} \frac{1}{9} = -\frac{2}{3}$ **25.** $2^4 = 16$ **27.** $5^0 = 1$ **29.** $10^1 = 10$ **31.** $5^3 = 125$ **33.** $3^{-3} = \frac{1}{27}$ **35.** $10^{-3} = 0.001$ **37.** $16^{1/2} = 4$

39. $8^{2/3} = 4$ **41.** $25^{-1/2} = \frac{1}{5}$ **43.** 6 **45.** 3 **47.** −4 **49.** −2 **51.** $\frac{1}{2}$ **53.** {2} **55.** {4} **57.** {100} **59.** {1}

61. $\left\{\frac{27}{8}\right\}$ **63.** {5} **65.** $\left\{\frac{1}{1{,}000}\right\}$ **67.** {−2} **69.** {3} **71.** {25} **73.** $\left\{-\frac{2}{3}\right\}$ **75.** 50 dB **77.** 70 dB **79.** 10^{-8} W/cm^2

81. 10 **83.** 1,000 **85.** 6 **87.** $10^5 \cdot a_0$ **89.** False **91.** always **93.** 1.011 **95.** 35 dB **97.** Above and Beyond

99. 23,902 yr **101.** 77,016 yr **103.** Pu-239: 239,020 yr; Sr-90: 280 yr; Th-230: 770,160 yr; Cs-135: 19,804,210 yr

105. **(a)** 6; **(b)** 2, 4; **(c)** $\log_3(9 \cdot 81) = \log_3 9 + \log_3 81$

10.6 Properties of Logarithms

< 10.6 Objectives >

1 > Apply the properties of logarithms

2 > Evaluate logarithmic expressions

3 > Use logarithms to solve applications

4 > Estimate an antilogarithm

5 > Change the base of a logarithmic expression

There are a number of properties of logarithms that are useful when working with equations and applications. We derive many of these from the properties of exponents.

We start with two basic facts that follow immediately from the definition of the logarithm.

Property

Properties of Logarithms

For $b > 0$ and $b \neq 1$,

1. $\log_b b = 1$ Since $b^1 = b$

2. $\log_b 1 = 0$ Since $b^0 = 1$

We know that the logarithmic function $y = \log_b x$ and the exponential function $y = b^x$ are inverses of each other. So, for $f(x) = b^x$, we have $f^{-1}(x) = \log_b x$.

For any one-to-one function f,

$$f^{-1}(f(x)) = x \qquad \text{for any } x \text{ in domain of } f$$

and $$f(f^{-1}(x)) = x \qquad \text{for any } x \text{ in domain of } f^{-1}$$

NOTE

The inverse "undoes" what f does to x.

Since $f(x) = b^x$ is a one-to-one function, we can apply these results to the case where

$$f(x) = b^x \qquad \text{and} \qquad f^{-1}(x) = \log_b x$$

to derive some additional properties.

Property

Properties of Logarithms

3. $\log_b b^x = x$

4. $b^{\log_b x} = x$ for $x > 0$

Since logarithms are exponents, we can again turn to the familiar exponent rules to derive some further properties of logarithms.

We know that

$$\log_b M = x \qquad \text{if and only if} \qquad b^x = M$$

and $$\log_b N = y \qquad \text{if and only if} \qquad b^y = N$$

Then $M \cdot N = b^x \cdot b^y = b^{x+y}$

From this last equation we see that $x + y$ is the power to which we must raise b to get the product MN. In logarithmic form, that becomes

$$\log_b MN = x + y$$

Now, since $x = \log_b M$ and $y = \log_b N$, we can substitute and write

$$\log_b MN = \log_b M + \log_b N$$

This is the first of the basic logarithmic properties presented here. The remaining properties may all be proved by arguments similar to those presented above.

Property

Properties of Logarithms

Product Property

$\log_b MN = \log_b M + \log_b N$

Quotient Property

$\log_b \frac{M}{N} = \log_b M - \log_b N$

Power Property

$\log_b M^p = p \log_b M$

NOTE

M, N, and b are all positive and $b \neq 1$.

Many applications of logarithms require using these properties to write a single logarithmic expression as the sum or difference of simpler expressions, as Example 1 illustrates.

Example 1 Using the Properties of Logarithms

< Objective 1 >

Use the properties of logarithms to expand each expression.

(a) $\log_b xy = \log_b x + \log_b y$ Product property

(b) $\log_b \frac{xy}{z} = \log_b xy - \log_b z$ Quotient property

$= \log_b x + \log_b y - \log_b z$ Product property

(c) $\log_{10} x^2y^3 = \log_{10} x^2 + \log_{10} y^3$ Product property

$= 2 \log_{10} x + 3 \log_{10} y$ Power property

RECALL

$\sqrt{a} = a^{1/2}$

(d) $\log_b \sqrt{\frac{x}{y}} = \log_b \left(\frac{x}{y}\right)^{1/2}$ Definition of exponent

$= \frac{1}{2} \log_b \frac{x}{y}$ Power property

$= \frac{1}{2} (\log_b x - \log_b y)$ Quotient property

Check Yourself 1

Expand each expression.

(a) $\log_b x^2y^3z$ (b) $\log_{10}\sqrt{\frac{xy}{z}}$

In some cases, we reverse the process and use the properties to write a single logarithm, given a sum or difference of logarithmic expressions.

Example 2 Using the Properties of Logarithms

Write each expression as a single logarithm with coefficient 1.

(a) $2 \log_b x + 3 \log_b y$

$= \log_b x^2 + \log_b y^3$ Power property

$= \log_b x^2y^3$ Product property

(b) $5 \log_{10} x + 2 \log_{10} y - \log_{10} z$

$= \log_{10} x^5y^2 - \log_{10} z$

$= \log_{10} \frac{x^5y^2}{z}$ Quotient property

(c) $\frac{1}{2}(\log_2 x - \log_2 y)$

$= \frac{1}{2}\left(\log_2 \frac{x}{y}\right)$

$= \log_2 \left(\frac{x}{y}\right)^{1/2}$ Power property

$= \log_2 \sqrt{\frac{x}{y}}$

Check Yourself 2

Write each expression as a single logarithm with coefficient 1.

(a) $3 \log_b x + 2 \log_b y - 2 \log_b z$ **(b)** $\frac{1}{3}(2 \log_2 x - \log_2 y)$

Example 3 illustrates the basic concept of the use of logarithms as a computational aid.

Example 3 Approximating Logarithms

< Objective 2 >

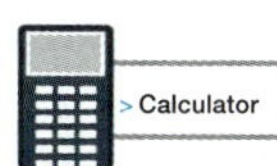

NOTES

We wrote the logarithms rounded to three decimal places and will follow this practice throughout the remainder of this chapter.

Keep in mind, however, that this is an approximation and that $10^{0.301}$ only approximates 2. Verify this with a calculator.

Suppose $\log_{10} 2 = 0.301$ and $\log_{10} 3 = 0.477$. Evaluate, as indicated.

(a) $\log_{10} 6$

Since $6 = 2 \cdot 3$,

$\log_{10} 6 = \log_{10} (2 \cdot 3)$

$= \log_{10} 2 + \log_{10} 3$

$= 0.301 + 0.477$

$= 0.778$

(b) $\log_{10} 18$

Since $18 = 2 \cdot 3 \cdot 3$,

$\log_{10} 18 = \log_{10} (2 \cdot 3 \cdot 3)$

$= \log_{10} 2 + \log_{10} 3 + \log_{10} 3$

$= 1.255$

(c) $\log_{10} \frac{1}{9}$

Since $\frac{1}{9} = \frac{1}{3^2}$,

$\log_{10} \frac{1}{9} = \log_{10} \frac{1}{3^2}$

$= \log_{10} 1 - \log_{10} 3^2$ $\log_b 1 = 0$ for any base b.

$= 0 - 2 \log_{10} 3$

$= -0.954$

(d) $\log_{10} 16$

Since $16 = 2^4$,

$\log_{10} 16 = \log_{10} 2^4 = 4 \log_{10} 2$

$= 1.204$

NOTE

Verify each answer with a calculator.

```
log(1/9)
       -.9542425094
log(16)
        1.204119983
log(√(3))
        .2385606274
```

(e) $\log_{10} \sqrt{3}$

Since $\sqrt{3} = 3^{1/2}$,

$$\log_{10} \sqrt{3} = \log_{10} 3^{1/2} = \frac{1}{2} \log_{10} 3$$
$$= 0.239$$

Check Yourself 3

Given the values for $\log_{10} 2$ and $\log_{10} 3$, evaluate as indicated.

(a) $\log_{10} 12$ **(b)** $\log_{10} 27$ **(c)** $\log_{10} \sqrt[3]{2}$

When "log" is written without a base, we always assume the base is 10. The LOG key on a calculator is the log base 10 function. To find $\log_{10}16$, for example, press LOG 16) ENTER. The result should be 1.204, to the nearest thousandth. In fact, there are two logarithm functions built into a graphing calculator, both of which are frequently used in mathematics.

Logarithms to base 10

Logarithms to base *e*

Of course, logarithms to base 10 are convenient because our number system has base 10. We call logarithms to base 10 **common logarithms,** and it is customary to omit the base in writing a common (or base-10) logarithm. So

Definition

The Common Logarithm log

$\log N$ means $\log_{10} N$

NOTE

When no base for a log is written, it is assumed to be 10.

The table shows the common logarithms for various powers of 10.

Exponential Form	Logarithmic Form
$10^3 = 1{,}000$	$\log 1{,}000 = 3$
$10^2 = 100$	$\log 100 = 2$
$10^1 = 10$	$\log 10 = 1$
$10^0 = 1$	$\log 1 = 0$
$10^{-1} = 0.1$	$\log 0.1 = -1$
$10^{-2} = 0.01$	$\log 0.01 = -2$
$10^{-3} = 0.001$	$\log 0.001 = -3$

Example 4 Approximating Logarithms with a Calculator

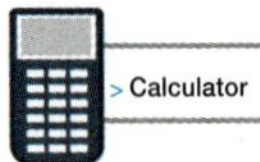

NOTE

The number 4.8 lies between 1 and 10, so log 4.8 is between 0 and 1.

Verify each with a calculator.

(a) $\log 4.8 = 0.681$

(b) $\log 48 = 1.681$

(c) $\log 480 = 2.681$

(d) $\log 4{,}800 = 3.681$

(e) $\log 0.48 = -0.319$

```
log(4.8)
          .6812412374
log(48)
          1.681241237
log(480)
          2.681241237
```

```
log(4800)
          3.681241237
log(0.48)
         -.3187587626
```

NOTES

$480 = 4.8 \times 10^2$

and

$\log(4.8 \times 10^2)$
$= \log 4.8 + \log 10^2$
$= \log 4.8 + 2$
$= 2 + \log 4.8$

The value of $\log 0.48$ is really $-1 + 0.681$. A calculator combines the signed numbers.

Check Yourself 4

Use a calculator to evaluate each logarithm, rounded to three decimal places.

(a) $\log 2.3$ **(b)** $\log 23$ **(c)** $\log 230$
(d) $\log 2{,}300$ **(e)** $\log 0.23$ **(f)** $\log 0.023$

Now we look at an application of common logarithms from chemistry. Common logarithms are used to define the pH of a solution. This is a scale that measures whether a solution is acidic or basic.

The pH of a solution is defined as

$$\text{pH} = -\log[\text{H}^+]$$

where $[\text{H}^+]$ is the hydrogen ion concentration, in moles per liter (mol/L), in the solution.

Example 5 A Chemistry Application

< Objective 3 >

NOTES

A solution with pH = 7 is neutral. It is **acidic** if the pH is less than 7 and **basic** if the pH is greater than 7.

In general, $\log_b b^x = x$, so $\log 10^{-7} = -7$.

Find the pH of each substance. Determine whether each is a base or an acid.

(a) Rainwater: $[\text{H}^+] = 1.6 \times 10^{-7}$

From the definition,

$$\begin{aligned}\text{pH} &= -\log[\text{H}^+]\\ &= -\log(1.6 \times 10^{-7})\\ &= -(\log 1.6 + \log 10^{-7}) \quad \text{Use the product rule.}\\ &\approx -[0.204 + (-7)]\\ &\approx -(-6.796) = 6.796\end{aligned}$$

Rain is slightly acidic.

(b) Household ammonia: $[\text{H}^+] = 2.3 \times 10^{-8}$

$$\begin{aligned}\text{pH} &= -\log(2.3 \times 10^{-8})\\ &= -(\log 2.3 + \log 10^{-8})\\ &\approx -[0.362 + (-8)]\\ &\approx 7.638\end{aligned}$$

Ammonia is slightly basic.

(c) Vinegar: $[\text{H}^+] = 2.9 \times 10^{-3}$

$$\begin{aligned}\text{pH} &= -\log(2.9 \times 10^{-3})\\ &= -(\log 2.9 + \log 10^{-3})\\ &\approx 2.538\end{aligned}$$

Vinegar is very acidic.

Check Yourself 5

Find the pH for each solution. Are they acidic or basic?

(a) Orange juice: $[\text{H}^+] = 6.8 \times 10^{-5}$
(b) Drain cleaner: $[\text{H}^+] = 5.2 \times 10^{-13}$

Given the logarithm of a number, we must be able to find that number. The process is straightforward.

Example 6 Solving a Logarithmic Equation

< Objective 4 >

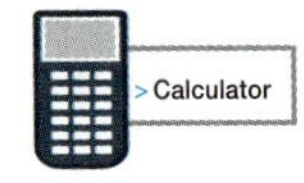


```
10^(2.1567)
       143.4498175
10^2.1567
       143.4498175
```

or

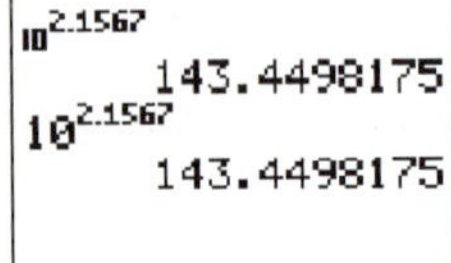

Suppose that $\log x = 2.1567$. We want to find a number x whose logarithm is 2.1567.

Rewriting in exponential form,

$\log_{10} x = 2.1567$ is equivalent to $10^{2.1567} = x$

On a graphing calculator, use the inverse function for [LOG], [10^x]. So, you can press

[2nd] [10^x] 2.1567 [)] [ENTER]

or type

10 [^] 2.1567 [ENTER]

Both give the result 143.450, rounded to the nearest thousandth, sometimes called the **antilogarithm** of 2.1567.

It is important to keep in mind that $y = \log x$ and $y = 10^x$ are inverse functions.

Check Yourself 6

In each case, find x to the nearest thousandth.

(a) $\log x = 0.828$ **(b)** $\log x = 1.828$
(c) $\log x = 2.828$ **(d)** $\log x = -0.172$

Now we return to a chemistry application that requires us to find an antilogarithm.

Example 7 A Chemistry Application

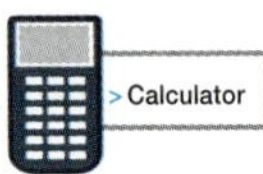


```
10^-6.2
       6.309573445E-7
```

Suppose that the pH of tomato juice is 6.2. Find the hydrogen ion concentration $[H^+]$.

Recall from our earlier formula that

$pH = -\log [H^+]$

In this case, we have

$6.2 = -\log [H^+]$

or

$\log [H^+] = -6.2$

The desired value for $[H^+]$ is the antilogarithm of -6.2. To find $[H^+]$, type [2nd] [10^x] [(-)] 6.2 [)] [ENTER].

The result is 0.00000063, and we can write

$[H^+] = 6.3 \times 10^{-7}$

NOTE

Natural logarithms are also called **Napierian logarithms** after Napier. The importance of this system of logarithms was not fully understood until later developments in the calculus.

Check Yourself 7

The pH of eggs is 7.8. Find $[H^+]$ for eggs.

As we mentioned, there are two systems of logarithms in common use. The second type of logarithm uses the number e as a base, and we call logarithms to base e **natural logarithms.** As with common logarithms, a convenient notation has developed.

Definition

The Natural Logarithm ln

The **natural logarithm** is the logarithm to base *e*, and it is denoted $\ln x$, where

$\ln x = \log_e x$ The restrictions on the domain of the natural logarithmic function are the same as before. The function is defined only if $x > 0$.

Since $y = \ln x$ means $y = \log_e x$, we can easily convert this to $e^y = x$, which leads us directly to these facts.

$\ln 1 = 0$ Because $e^0 = 1$

$\ln e = 1$ Because $e^1 = e$

$\ln e^2 = 2$

$\ln e^3 = 3$

$\ln e^{-5} = -5$

Because $\ln e^x = e^x$

We want to emphasize the inverse relationship that exists between logarithmic functions and exponential functions.

Property

Inverse Functions

For any base *b*,

$\log_b b^x = x$ (for all real *x*)

$b^{\log_b x} = x$ (for $x > 0$)

So, for common logarithms,

$\log 10^x = x$

$10^{\log x} = x$

And, for natural logarithms,

$\ln e^x = x$

$e^{\ln x} = x$

Example 8 Using the Properties of Inverses

> **NOTE**
>
> You can confirm each of these on a calculator, but should recognize the forms.

Simplify.

(a) $\log 10^8 = 8$

(b) $\ln e^{-6} = -6$

(c) $10^{\log 7} = 7$

(d) $e^{\ln 4} = 4$

Check Yourself 8

Simplify.

(a) $\ln e^{1.2}$ **(b)** $10^{\log 4.5}$ **(c)** $\log 10^{-5}$ **(d)** $e^{\ln 3.7}$

Example 9 Approximating Logarithms with a Calculator

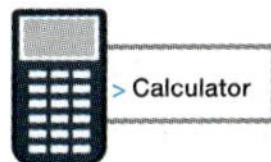

To evaluate natural logarithms, we use a calculator. To find the value of ln 2, use the sequence

[ln] 2 [)] [ENTER]

NOTE

```
ln(2)
      .6931471806
```

The result is 0.693 (to three decimal places).

Check Yourself 9

Use a calculator to evaluate each logarithm. Round to the nearest thousandth.

(a) ln 3 **(b)** ln 6 **(c)** ln 4 **(d)** $\ln \sqrt{3}$

The properties of logarithms apply in the same way, no matter what the base.

Example 10 Approximating Logarithms Using Properties

RECALL

$\log_b MN = \log_b M + \log_b N$

$\log_b M^p = p \log_b M$

If ln 2 = 0.693 and ln 3 = 1.099, evaluate each logarithm.

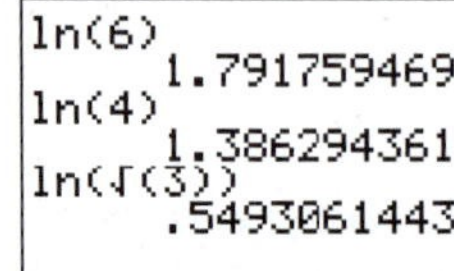

```
ln(6)
        1.791759469
ln(4)
        1.386294361
ln(√(3))
        .5493061443
```

(a) $\ln 6 = \ln (2 \cdot 3) = \ln 2 + \ln 3 = 1.792$

(b) $\ln 4 = \ln 2^2 = 2 \ln 2 = 1.386$

(c) $\ln \sqrt{3} = \ln 3^{1/2} = \frac{1}{2} \ln 3 = 0.550$

Verify these results with a calculator.

Check Yourself 10

Use ln 2 = 0.693 and ln 3 = 1.099 to evaluate each logarithm.

(a) ln 12 **(b)** ln 27

It may also be necessary to find x, given ln x. The key here is to remember that $y = \ln x$ and $y = e^x$ are inverse functions.

Example 11 Solving a Logarithmic Equation

```
e^(4.1685)
        64.6184517
e^4.1685
        64.6184517
```

or

```
e^4.1685
        64.6184517
```

Suppose that $\ln x = 4.1685$. We want to find a number x whose logarithm, base e, is 4.1685. Rewriting in exponential form,

$\ln x = 4.1685$ is equivalent to $e^{4.1685} = x$

On a graphing calculator, use the inverse function for [LN], [e^x]. Press

[2nd] [e^x] 4.1685 [)] [ENTER]

or

[2nd] [÷] [^] 4.1685 [ENTER]

Both give the result 64.618, rounded to the nearest thousandth.

Check Yourself 11

In each case, find x to the nearest thousandth.

(a) $\ln x = 2.065$ **(b)** $\ln x = -2.065$ **(c)** $\ln x = 7.293$

The natural logarithm function plays an important role in both theoretical and applied mathematics. Example 12 illustrates just one of the many applications of this function.

Example 12 A Learning Curve Application

NOTE

We read $S(t)$ as "S of t."

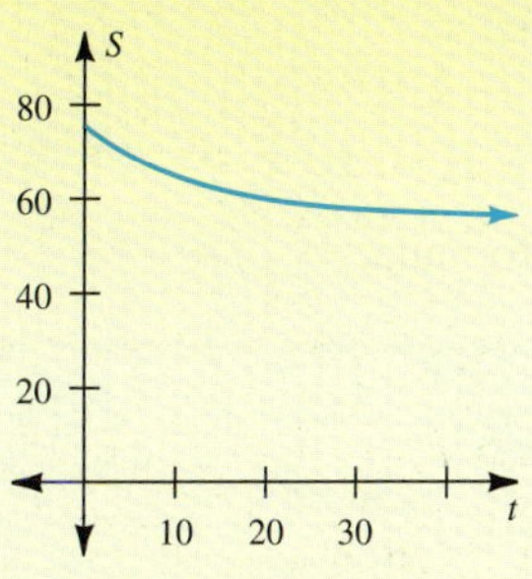

This is an example of a **forgetting curve.** Note how it drops more rapidly at first. Compare this curve to the learning curve in Section 10.4, exercise 64.

A class of students took a mathematics examination and received an average score of 76. In a psychological experiment, the students are retested at weekly intervals over the same material. If t is measured in weeks, then the new average score after t weeks is given by

$$S(t) = 76 - 5 \ln (t + 1)$$

(a) Find the average score after 10 weeks.

$$S(10) = 76 - 5 \ln (10 + 1)$$
$$= 76 - 5 \ln 11 \approx 64$$

(b) Find the average score after 20 weeks.

$$S(20) = 76 - 5 \ln (20 + 1) \approx 61$$

(c) Find the average score after 30 weeks.

$$S(30) = 76 - 5 \ln (30 + 1) \approx 59$$

Check Yourself 12

The average score for a group of biology students, retested after time t (in months), is given by

$$S(t) = 83 - 9 \ln (t + 1)$$

Find the average score rounded to the nearest tenth after

(a) 3 months **(b)** 6 months

We conclude this section with one final property of logarithms. This property allows us to quickly find the logarithm of a number to any base. Although work with logarithms with bases other than 10 or e is relatively infrequent, the relationship between logarithms of different bases is interesting in itself.

Suppose we want to find $\log_2 5$. This means we want to find the power to which 2 should be raised to produce 5. Now, if there were a log base 2 function ($\log_2 x$) on the calculator, we could obtain this directly. But since there is not, we must take another approach. If we write

$$\log_2 5 = x$$

then we have $2^x = 5$.

Taking the logarithm to base 10 of both sides of the equation yields

$$\log 2^x = \log 5$$

or $\quad x \log 2 = \log 5$ Use the power property of logarithms.

Now, dividing both sides of this equation by log 2 gives

$$x = \frac{\log 5}{\log 2}$$

> CAUTION

Do not cancel the logs.

NOTE

This says $2^{2.322} \approx 5$.

We can now find a value for x with the calculator. Dividing with the calculator log 5 by log 2, we get an approximate answer of 2.322.

Since $x = \log_2 5$ and $x = \dfrac{\log 5}{\log 2}$, then

$$\log_2 5 = \frac{\log 5}{\log 2}$$

We could also have taken the logarithm, base e, of both sides.

$$2^x = 5$$
$$\ln 2^x = \ln 5$$
$$x \ln 2 = \ln 5$$
$$x = \frac{\ln 5}{\ln 2} \approx 2.322$$

So, $\log_2 5 = \dfrac{\log_e 5}{\log_e 2} = \dfrac{\log_{10} 5}{\log_{10} 2}$.

Generalizing our result gives us the change-of-base formula.

Property

Change-of-Base Formula

For positive real numbers a and x,

$$\log_a x = \frac{\log_b x}{\log_b a}$$

The logarithm on the left side has base a while the logarithms on the right side have base b. This allows us to calculate the logarithm to base a of any positive number, using the corresponding logarithms to base b (or any other base), as Example 13 illustrates.

Example 13 Using the Change-of-Base Formula

< Objective 5 >

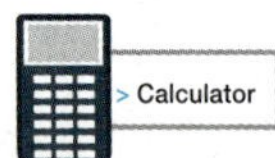

NOTES

We wrote $\log_{10} 15$ rather than log 15 to emphasize the change-of-base formula.

$\log_5 5 = 1$ and $\log_5 25 = 2$, so the result for $\log_5 15$ must be between 1 and 2.

We could choose base e so that $\log_5 15 = \dfrac{\ln 15}{\ln 5}$ instead.

```
log(15)/log(5)
       1.682606194
ln(15)/ln(5)
       1.682606194
```

Remember to close the parentheses in the numerator when entering these expressions into a calculator.

Find $\log_5 15$.

From the change-of-base formula with $a = 5$ and $b = 10$,

$$\log_5 15 = \frac{\log_{10} 15}{\log_{10} 5}$$
$$\approx 1.683$$

The graphing calculator sequence for the computation is

[log] 15 [)] [÷] [log] 5 [)] [ENTER]

The result is 1.683, rounded to the nearest thousandth.

Check Yourself 13

Use the change-of-base formula to find $\log_8 32$.

> **CAUTION**

A couple of cautions are in order.

1. We cannot "cancel" logs. There is the temptation to write

$$\frac{\log 15}{\log 5} = \frac{15}{5} = 3$$

This is quite wrong!

2. There is also the temptation to write

$$\frac{\log 15}{\log 5} = \log 15 - \log 5$$

This is also quite wrong.

It is true that $\log\left(\frac{15}{5}\right) = \log 15 - \log 5$ (the quotient property), but this is very different from $\frac{\log 15}{\log 5}$.

Graphing Calculator Option

Applying Logarithmic Regression

A general form of logarithmic functions is available as a regression model in a graphing calculator: $y = a + b \ln x$.

Suppose we have collected some data that suggest a pattern of logarithmic growth, such as

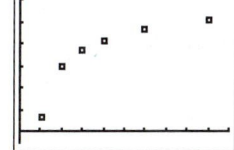

There is relatively rapid growth for small values of x, followed by growth that seems to be slowing. Look at the data, which show how the time (in seconds) for a dropped tennis ball to complete its third bounce varies according to the height (in inches) of the drop.

Height of drop (in.)	40	45	50	55	60
Time to third bounce (s)	1.75	1.87	1.99	2.07	2.12

We begin by clearing the data lists.

[STAT] **4:ClrList** [2nd] [L1] [,] [2nd] [L2] [ENTER]

Then we enter the data into lists 1 and 2.

[STAT] **1:Edit** and type in the numbers.

To make a scatterplot,

[2nd] [STAT PLOT] [ENTER]

Make sure "On" is selected, use the first icon for "Type" and check that [L1] is the "Xlist" and [L2] is the "Ylist."

Deselect any functions in the [Y=] menu and improve the viewing window with [ZOOM] **9:ZoomStat**

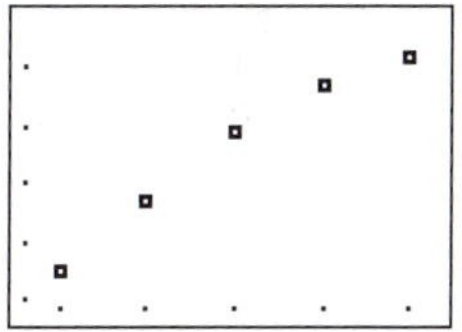

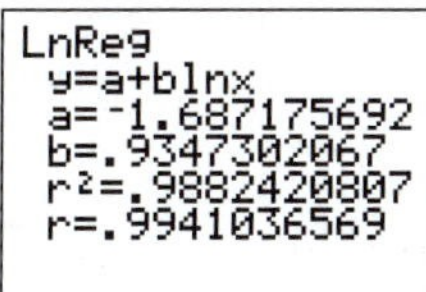

We find the "best fitting" logarithmic function by

[STAT] **CALC 9:LnReg** [ENTER]

To four decimal places, we have

$y = -1.6872 + 0.9347 \ln x$

To view the graph of this function on the scatterplot, enter its equation on the [Y=] screen and press [GRAPH].

Graphing Calculator Check

The table shows the systolic blood pressure p (in mm of Hg) for children of varying weights w (in pounds). Use the logarithmic regression utility on a graphing calculator to fit a logarithmic function to the data. Round your results to four decimal places.

Weight, w	44	61	81	113	131
Blood pressure, p	91	98	103	110	112

ANSWER

$p = 17.9243 + 19.385 \ln w$

Check Yourself ANSWERS

1. **(a)** $2 \log_b x + 3 \log_b y + \log_b z$; **(b)** $\frac{1}{2}(\log_{10} x + \log_{10} y - \log_{10} z)$ **2.** **(a)** $\log_b \frac{x^3y^2}{z^2}$; **(b)** $\log_2 \sqrt[3]{\frac{x^2}{y}}$
3. **(a)** 1.079; **(b)** 1.431; **(c)** 0.100 **4.** **(a)** 0.362; **(b)** 1.362; **(c)** 2.362; **(d)** 3.362; **(e)** −0.638; **(f)** −1.638 **5.** **(a)** 4.167, acidic; **(b)** 12.284, basic **6.** **(a)** 6.730; **(b)** 67.298; **(c)** 672.977; **(d)** 0.673 **7.** $[H^+] = 1.6 \times 10^{-8}$ **8.** **(a)** 1.2; **(b)** 4.5; **(c)** −5; **(d)** 3.7
9. **(a)** 1.099; **(b)** 1.792; **(c)** 1.386; **(d)** 0.549 **10.** **(a)** 2.485; **(b)** 3.297 **11.** **(a)** 7.885; **(b)** 0.127; **(c)** 1,469.974 **12.** **(a)** 70.5; **(b)** 65.5 **13.** $\log_8 32 = \frac{\log 32}{\log 8} \approx 1.667$

Reading Your Text

These fill-in-the-blank exercises will help you understand some of the key vocabulary used in this section. The answers to these exercises are in the Answers Appendix in the back of the text.

(a) By definition, a logarithm is an ________.

(b) The logarithmic property $\log_b M^p = p \log_b M$ is called the ________ property.

(c) We call logarithms to the base 10 ________ logarithms.

(d) A solution whose pH = 7 is ________.

10.6 exercises

Skills | Calculator/Computer | Career Applications | Above and Beyond

< Objective 1 >

Use the properties of logarithms to expand each expression.

1. $\log_b 5x$ **2.** $\log_3 7x$ **3** $\log_6 \frac{x}{7}$ **4.** $\log_b \frac{2}{y}$

5. $\log_3 a^2$ **6.** $\log_5 y^4$ **7.** $\log_5 \sqrt{x}$ **8.** $\log \sqrt[3]{z}$

9. $\log_b x^2y^4$ **10.** $\log_7 x^3z^2$ **11.** $\log_4 y^2\sqrt{x}$ **12.** $\log_b x^3\sqrt[3]{z}$

13. $\log_b \frac{x^2y}{z}$ **14.** $\log_5 \frac{3}{xy}$ **15.** $\log \frac{xy^2}{\sqrt{z}}$ VIDEO **16.** $\log_4 \frac{x^3\sqrt{y}}{z^2}$

17. $\log_5 \sqrt[3]{\frac{xy}{z^2}}$ **18.** $\log_b \sqrt[4]{\frac{x^2y}{z^3}}$

Write each expression as a single logarithm.

19. $\log_b x + \log_b y$ **20.** $\log_5 x - \log_5 y$ **21.** $3 \log_5 x - 2 \log_5 y$ **22.** $3 \log_b x + \log_b z$

23. $\log_b x + \frac{1}{2} \log_b y$ **24.** $\frac{1}{2} \log_b x - 3 \log_b z$ **25.** $\log_b x - 2 \log_b y - \log_b z$

26. $2 \log_5 x - (3 \log_5 y + \log_5 z)$ **27.** $\frac{1}{2} \log_6 y - 3 \log_6 z$ VIDEO

28. $\log_b x - \frac{1}{3}\log_b y - 4\log_b z$

29. $\frac{1}{3}(2\log_b x + \log_b y - \log_b z)$

30. $\frac{1}{5}(2\log_4 x - \log_4 y + 3\log_4 z)$

< Objective 2 >

Given that $\log 2 = 0.301$ *and* $\log 3 = 0.477$, *evaluate each logarithm.*

31. $\log 24$ **32.** $\log 36$ **33.** $\log 8$ **34.** $\log 81$

35. $\log \sqrt{2}$ VIDEO **36.** $\log \sqrt[3]{3}$ **37.** $\log \frac{1}{4}$ **38.** $\log \frac{1}{27}$

Simplify each expression.

39. $10^{\log 8.2}$ **40.** $\log 10^{-1.3}$ **41.** $\ln e^{5.8}$ **42.** $e^{\ln 2.6}$

Estimate each logarithm by "trapping" it between consecutive integers. For instance, to estimate $\log_4 52$, *we use* $4^2 = 16$ *and* $4^3 = 64$, *so* $\log_4 52$ *must lie between 2 and 3.*

43. $\log_3 25$ **44.** $\log_5 30$ **45.** $\log_2 70$

46. $\log_2 19$ **47.** $\log 680$ **48.** $\log 6{,}800$

Use the properties of logarithms to evaluate each expression.

49. $\log 5 + \log 2$ **50.** $\log 25 + \log 4$ **51.** $\log_3 45 - \log_3 5$ **52.** $10\log_4 2$

Determine whether each statement is **true** *or* **false.**

53. $\log_m x^n = n\log_m x$

54. $(\log_m x)\cdot(\log_m y) = \log_m xy$

55. $\log_m m = 0$

56. $\log_m x - \log_m y = \log_m \frac{x}{y}$

Skills | **Calculator/Computer** | Career Applications | Above and Beyond

Use a calculator to evaluate each logarithm.

57. $\log 7.3$ **58.** $\log 68$ **59.** $\log 680$ **60.** $\log 6{,}800$

61. $\log 0.72$ **62.** $\log 0.068$ **63.** $\ln 2$ **64.** $\ln 3$

65. $\ln 10$ **66.** $\ln 30$

< Objective 4 >

Solve. Round to the nearest thousandth.

67. $\log x = 0.749$ **68.** $\log x = 1.749$ **69.** $\log x = 3.749$ **70.** $\log x = -0.251$

71. $\ln x = 1.238$ **72.** $\ln x = 3.141$ **73.** $\ln x = -0.786$ **74.** $\ln x = -3.141$

< Objective 3 >

You are given the hydrogen ion concentration $[H^+]$ *for each solution. Use the formula* $pH = -\log[H^+]$ *to find each* pH. *Are the solutions acidic or basic?*

75. Blood: $[H^+] = 3.8 \times 10^{-8}$

76. Lemon juice: $[H^+] = 6.4 \times 10^{-3}$

Given the pH of the solutions, approximate the hydrogen ion concentration $[H^+]$.

77. Wine: pH = 4.7 

78. Household ammonia: pH = 7.8

The average score on a final examination for a group of psychology students, retested after time t (in weeks), is given by

$S = 85 - 8 \ln (t + 1)$

Find the average score on the retests.

79. After 3 weeks

80. After 12 weeks

< Objective 5 >

Use the change-of-base formula to approximate each logarithm.

81. $\log_3 25$

82. $\log_5 30$

83. $\log_2 10$

84. $\log_2 25$

85. $\log_{12} 8$

86. $\log_{1/2} 20$

87. The table shows measurements taken for several trees of the same species. The measurements are diameter at the base in centimeters and height in meters. Use the logarithmic regression utility on a graphing calculator to fit a logarithmic function to the data. Round your results to four decimal places.

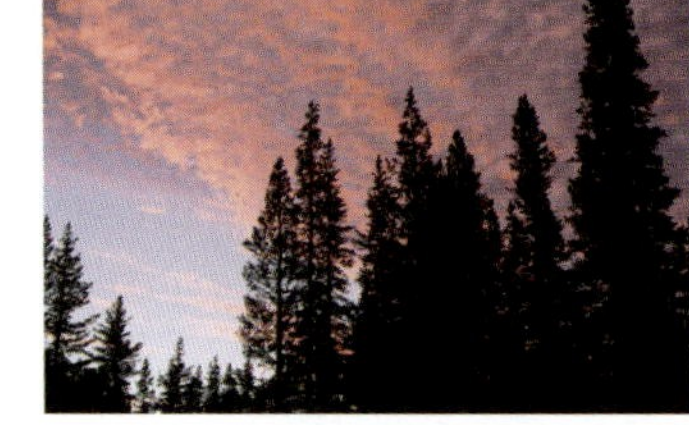

x (diameter)	2.6	4.6	9.8	14.5	15.8	27.0
y (height)	1.93	4.15	11.50	11.85	13.25	15.80

88. The table shows measurements taken for several trees of the same species. The measurements are diameter at the base, in centimeters, and crown width, in meters. Use the logarithmic regression utility on a graphing calculator to fit a logarithmic function to the data. Round your results to four decimal places.

x (diameter)	2.6	4.6	9.8	14.5	15.8	27.0
y (crown width)	0.5	1.6	3.6	3.7	4.0	6.5

Skills | Calculator/Computer | Career Applications | **Above and Beyond**

The amount of a radioactive substance remaining after time t is given by

$A = e^{\lambda t + \ln A_0}$

where A is the amount remaining after time t, A_0 *is the original amount of the substance, and* λ *is the radioactive decay constant. Assume t is measured in years.* chapter 10 > Make the Connection

89. How much plutonium-239 will remain after 50,000 yr if 24 kg was originally stored? Plutonium-239 has a radioactive decay constant of −0.000029.

90. How much plutonium-241 will remain after 100 yr if 52 kg was originally stored? Plutonium-241 has a radioactive decay constant of −0.053319.

91. How much strontium-90 will remain after 56 yr if 60 kg was originally stored? Strontium-90 has a radioactive decay constant of −0.024755.

92. How much cesium-137 will remain after 90 yr if 160 kg was originally stored? Cesium-137 has a radioactive decay constant of -0.023105.

93. Which keys on your calculator are function keys and which are operation keys? What is the difference?

94. How is the pH factor relevant to your selection of a hair-care product?

95. **(a)** Use the change-of-base formula to write $\log_3 8$ in terms of base-10 logarithms. Then use a calculator to find $\log_3 8$ rounded to three decimal places.

(b) Use the change-of-base formula to write $\log_3 8$ in terms of base-e logarithms. Then use a calculator to find $\log_3 8$ rounded to three decimal places.

(c) Compare your answers to parts (a) and (b).

Answers

1. $\log_b 5 + \log_b x$ **3.** $\log_6 x - \log_6 7$ **5.** $2 \log_3 a$ **7.** $\frac{1}{2} \log_5 x$ **9.** $2 \log_b x + 4$ lo g$_b\, y$ **11.** $2 \log_4 y + \frac{1}{2} \log_4 x$
13. $2 \log_b x + \log_b y - \log_b z$ **15.** $\log x + 2 \log y - \frac{1}{2} \log z$ **17.** $\frac{1}{3}(\log_5 x + \log_5 y - 2 \log_5 z)$ **19.** $\log_b xy$ **21.** $\log_5 \frac{x^3}{y^2}$ **23.** $\log_b x\sqrt{y}$
25. $\log_b \frac{x}{y^2 z}$ **27.** $\log_6 \frac{\sqrt{y}}{z^3}$ **29.** $\log_b \sqrt[3]{\frac{x^2 y}{z}}$ **31.** 1.38 **33.** 0.903 **35.** 0.151 **37.** -0.602 **39.** 8.2 **41.** 5.8
43. Between 2 and 3 **45.** Between 6 and 7 **47.** Between 2 and 3 **49.** 1 **51.** 2 **53.** True **55.** False **57.** 0.863
59. 2.833 **61.** -0.143 **63.** 0.693 **65.** 2.303 **67.** 5.61 **69.** 5,610.48 **71.** 3.449 **73.** 0.456 **75.** 7.42, basic
77. 2×10^{-5} **79.** 74 **81.** 2.93 **83.** 3.322 **85.** 0.837 **87.** $y = -4.2465 + 6.222 \ln x$ **89.** 5.6 kg **91.** 15 kg
93. Above and Beyond **95.** **(a)** $\frac{\log 8}{\log 3}$, 1.893; **(b)** $\frac{\ln 8}{\ln 3}$, 1.893; **(c)** They are the equal.

10.7 Logarithmic and Exponential Equations

< 10.7 Objectives >

1 > Solve logarithmic equations

2 > Solve exponential equations

3 > Use exponential equations to solve applications

We use the properties of logarithms to solve exponential and logarithmic equations. In this section, we solve both types of equations.

We solved some examples in Section 10.5. Recall that to solve $\log_3 x = 4$ for x, we simply convert the logarithmic equation to exponential form. Here,

$$3^4 = x$$

so $\quad x = 81$

and $\{81\}$ is the solution set for the given equation.

Now, what if the logarithmic equation involves more than one logarithmic term? Example 1 illustrates how to apply the properties of logarithms.

Example 1 Solving a Logarithmic Equation

< Objective 1 >

> **NOTE**
>
> We apply the product rule for logarithms.
>
> $\log_b M + \log_b N = \log_b MN$

Solve each equation.

(a) $\log_5 x + \log_5 3 = 2$

We use the product property to write the equation as

$$\log_5 3x = 2$$

We write the equation in the equivalent exponential form.

$$5^2 = 3x$$

$$3x = 25$$

$$x = \frac{25}{3}$$

To check this, we substitute $\frac{25}{3}$ into the original equation.

$$\log_5\left(\frac{25}{3}\right) + \log_5(3) \stackrel{?}{=} 2$$

To proceed, we either simplify the left side using logarithm properties, or we convert these logarithms to base 10 logs or base e logs and use a calculator.

> **RECALL**
>
> When no base is written, it is the common log (base 10).

$$\frac{\log\left(\frac{25}{3}\right)}{\log(5)} + \frac{\log(3)}{\log(5)} \stackrel{?}{=} 2$$

```
log(25/3)/log(5)
+log(3)/log(5)
                2
```

So $\left\{\frac{25}{3}\right\}$ is the solution set.

(b) $\log x + \log (x - 3) = 1$

Write the equation as

$$\log[x(x - 3)] = 1$$

or $\quad 10^1 = x(x - 3)$

We now have

$$x^2 - 3x = 10$$
$$x^2 - 3x - 10 = 0$$
$$(x - 5)(x + 2) = 0$$

Possible solutions are $x = 5$ or $x = -2$.

Substituting -2 into the original equation gives

$$\log(-2) + \log(-5) = 1$$

Since logarithms of negative numbers are *not* defined, -2 is an extraneous solution and we reject it. Substituting 5 gives

$$\log 5 + \log(5 - 3) \stackrel{?}{=} 1$$
$$\log 5 + \log 2 \stackrel{?}{=} 1$$

On a calculator, this checks.

The only solution for the original equation is 5.

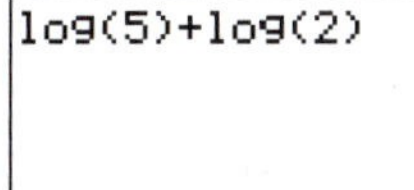

NOTES

Checking possible solutions is particularly important here.

To check $x = -2$

```
log(-2)+log(-5)
```

```
ERR:NONREAL ANS
1:Quit
2:Goto
```

Check Yourself 1

Solve $\log_2 x + \log_2 (x + 2) = 3$ for x.

We use the quotient property in a similar way when solving logarithmic equations. Consider Example 2.

Example 2 Solving a Logarithmic Equation

Solve each equation.

(a) $\log_5 x - \log_5 2 = 2$

Rewrite the original equation as

$$\log_5 \frac{x}{2} = 2$$

Now, $\quad 5^2 = \frac{x}{2}$

$$\frac{x}{2} = 25$$
$$x = 50$$

Check

$$\log_5 50 - \log_5 2 \stackrel{?}{=} 2$$

Using change-of-base,

$$\frac{\log 50}{\log 5} - \frac{\log 2}{\log 5} \stackrel{?}{=} 2$$

The solution set is $\{50\}$.

NOTE

We apply the quotient rule for logarithms.

$$\log_b M - \log_b N = \log_b \frac{M}{N}$$

```
log(50)/log(5)-l
og(2)/log(5)
               2
```

(b) $\log_3 (x + 1) - \log_3 x = 3$

$$\log_3 \left(\frac{x + 1}{x}\right) = 3$$

$$3^3 = \frac{x + 1}{x}$$

$$27x = x + 1$$

$$26x = 1$$

$$x = \frac{1}{26}$$

Check

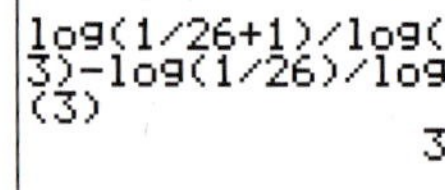

$$\log_3\left(\frac{1}{26} + 1\right) - \log_3\left(\frac{1}{26}\right) \stackrel{?}{=} 3$$

$$\frac{\log\left(\frac{1}{26} + 1\right)}{\log 3} - \frac{\log\left(\frac{1}{26}\right)}{\log 3} \stackrel{?}{=} 3$$

$\left\{\frac{1}{26}\right\}$ is the solution set.

Check Yourself 2

Solve $\log_5 (x + 3) - \log_5 x = 2$ for x.

Solving certain types of logarithmic equations calls for the one-to-one property of the logarithmic function.

Property

One-to-One Property of Logarithms

If $\log_b M = \log_b N$

then $M = N$

Example 3 Solving a Logarithmic Equation

Solve.

$\log (x + 2) - \log 2 = \log x$

Again, we rewrite the left-hand side of the equation. So

$$\log \left(\frac{x + 2}{2}\right) = \log x$$

Since the logarithmic function is one-to-one, this is equivalent to

$$\frac{x + 2}{2} = x$$

$$x + 2 = 2x$$

or $x = 2$

$\{2\}$ is the solution set.

The check is left for you.

Check Yourself 3

Solve.

$$\log (x + 3) - \log 3 = \log x$$

This algorithm summarizes our work in solving logarithmic equations.

Step by Step

Solving Logarithmic Equations

Step 1 Use the properties of logarithms to combine terms containing logarithmic expressions into a single term.

Step 2 Write the equation in exponential form.

Step 3 Solve.

Step 4 Check your solutions to make sure that possible solutions do not result in the logarithms of negative numbers.

Now we look at **exponential equations,** which are equations in which the variable appears as an exponent.

We solved some elementary exponential equations in Section 10.4. In solving an equation such as

$$3^x = 81$$

we wrote the right-hand member as a power of 3, so that

$$3^x = 3^4$$

or $\quad x = 4$

NOTE

We write both sides as a power of the same base, here 3.

This technique works only when both sides of the equation can be expressed as powers of the same base. If that is not the case, we use logarithms to solve the equation, as illustrated in Example 4.

Example 4 Solving an Exponential Equation

< Objective 2 >

Solve $3^x = 5$, rounded to the nearest thousandth.

We begin by taking the common logarithm of both sides of the original equation.

$$\log 3^x = \log 5$$

Now we apply the power property so that the variable becomes a coefficient on the left.

$$x \log 3 = \log 5$$

Dividing both sides of the equation by log 3 isolates x, and we have

$$x = \frac{\log 5}{\log 3} \qquad \text{"Logs" cannot be canceled!}$$

$$\approx 1.465 \qquad \text{(to three decimal places)}$$

RECALL

If $\quad M = N$

then $\quad \log_b M = \log_b N$

> CAUTION

This is *not* log 5 − log 3, a common error.

The solution 1.465 is not exact. You can verify the approximate solution on a calculator. Raise 3 to power 1.465. You should see a result close to 5, but not exactly 5.

NOTE

```
log(5)/log(3)
          1.464973521
3^1.465
          5.000145454
3^1.464973521
          [illegible]
```

Check Yourself 4

Solve $2^x = 10$, rounded to the nearest thousandth.

Example 5 shows how to solve an equation with a more complicated exponent.

Example 5 Solving an Exponential Equation

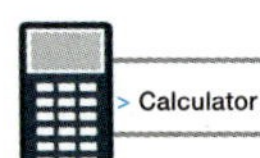
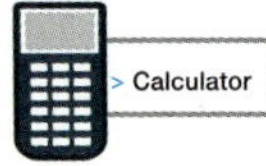

NOTES

On the left, we apply

$\log_b M^p = p \log_b M$

On a graphing calculator, the sequence is

[(] [log] 8 [)] [÷] [log] 5 [)] [−] 1 [)] [÷] 2 [ENTER]

```
(log(8)/log(5)-1
)/2
      .1460148371
```

Solve $5^{2x+1} = 8$.

Begin the same way as in Example 4.

$$\log 5^{2x+1} = \log 8$$

$$(2x + 1)\log 5 = \log 8$$

$$2x + 1 = \frac{\log 8}{\log 5}$$

$$2x = \frac{\log 8}{\log 5} - 1$$

$$x = \frac{1}{2}\left(\frac{\log 8}{\log 5} - 1\right)$$

$$x \approx 0.146$$

The solution set is $\{0.146\}$.

Check Yourself 5

Solve $3^{2x-1} = 7$.

The procedure is similar if the variable appears as an exponent in more than one term of the equation.

Example 6 Solving an Exponential Equation

Solve $3^x = 2^{x+1}$.

$$\log 3^x = \log 2^{x+1}$$

$$x \log 3 = (x + 1)\log 2 \quad \text{Apply the power property.}$$

$$x \log 3 = x \log 2 + \log 2 \quad \text{Distribute } x + 1.$$

$$x \log 3 - x \log 2 = \log 2 \quad \text{Gather terms with } x \text{ on the left side.}$$

$$x(\log 3 - \log 2) = \log 2 \quad \text{Factor out } x.$$

$$x = \frac{\log 2}{\log 3 - \log 2} \quad \text{Divide by } (\log 3 - \log 2).$$

$$\approx 1.71$$

The solution set is $\{1.71\}$.

NOTES

Use the power property to write the variables as coefficients.

Isolate x on the left.

To check the reasonableness of this result, use a calculator to verify that

$3^{1.71} \approx 2^{2.71}$

```
3^1.71
      6.544513163
2^2.71
      6.543216468
```

Check Yourself 6

Solve $5^{x+1} = 3^{x+2}$.

Here is an algorithm summarizing our work with solving exponential equations.

Step by Step

Solving Exponential Equations

Step 1 Try to write each side of the equation as a power of the same base. Then equate the exponents to form an equation.

Step 2 If the above procedure is not applicable, take the logarithm of both sides of the original equation.

Step 3 Use the power rule for logarithms to write an equivalent equation with the variables as coefficients.

Step 4 Solve the resulting equation.

Step 5 Check the solutions.

There are many applications of our work with exponential equations. We look at a financial application in Example 7.

Example 7 A Finance Application

< Objective 3 >

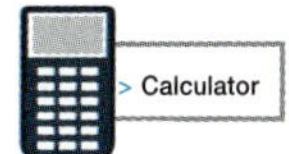

If an investment of P dollars earns interest at an annual interest rate r and the interest is compounded n times per year, then the amount in the account after t years is given by

$$A = P\left(1 + \frac{r}{n}\right)^{nt}$$

If $1,000 is placed in an account with an interest rate of 6%, find out how long it will take the money to double when interest is compounded annually and how long when compounded quarterly.

(a) Compound interest annually.

> **NOTE**
>
> Since the interest is compounded *once* per year, $n = 1$.

Using the formula with $A = 2{,}000$ (we want the original $1,000 to double), $P = 1{,}000$, $r = 0.06$, and $n = 1$, we have

$$2{,}000 = 1{,}000(1 + 0.06)^t$$

Dividing both sides by 1,000 yields

$$2 = (1.06)^t$$

We now have an exponential equation that we can solve.

> **NOTE**
>
> From accounting, we have the **rule of 72,** which states that the doubling time is approximately 72 divided by the interest rate as a percentage.
>
> $\frac{72}{6} = 12$ yr

$$\log 2 = \log (1.06)^t$$
$$= t \log 1.06$$
$$\text{or} \quad t = \frac{\log 2}{\log 1.06}$$
$$\approx 11.9 \text{ yr}$$

It takes just a little less than 12 yr for the money to double.

(b) Compound interest quarterly.

> **NOTE**
>
> Since the interest is compounded 4 times per year, $n = 4$.

Now $n = 4$ in the formula, so

$$2{,}000 = 1{,}000\left(1 + \frac{0.06}{4}\right)^{4t}$$
$$2 = (1.015)^{4t}$$
$$\log 2 = \log (1.015)^{4t}$$
$$\log 2 = 4t \log 1.015$$
$$\frac{\log 2}{4 \log 1.015} = t$$
$$t \approx 11.6 \text{ yr}$$

The doubling time is reduced by about $3\frac{1}{2}$ months by the more frequent compounding.

Check Yourself 7

Find the doubling time in Example 7 if the interest is compounded monthly.

Problems involving rates of growth or decay can also be solved by using exponential equations.

Example 8 A Population Application

A town's population is 10,000. Given a projected continuous growth rate of 7% per year, t years from now the population P is given by

$P = 10{,}000e^{0.07t}$

In how many years will the town's population double?

We want the time t when P will be 20,000 (doubled in size). So

$20{,}000 = 10{,}000e^{0.07t}$ Divide both sides by 10,000.

$2 = e^{0.07t}$

In this case, we take the *natural logarithm* of both sides of the equation because e is the base.

$\ln 2 = \ln e^{0.07t}$

$\ln 2 = 0.07t \ln e$ Apply the power property.

$\ln 2 = 0.07t$

$\frac{\ln 2}{0.07} = t$

$t \approx 9.9$ yr

RECALL

$\ln e = 1$

The population will double in approximately 9.9 yr.

Check

10000 [2nd] [e^x] 0.07 [×] 9.9 [)] [ENTER] or 10000 [2nd] [e^x] 0.07 [×] 9.9 [ENTER]

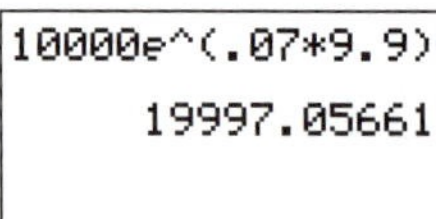

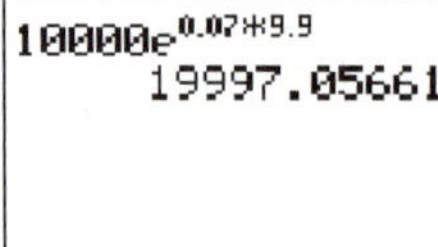

which is close to 20,000.

Check Yourself 8

If \$1,000 is invested in an account with an annual interest rate of 6%, compounded continuously, the amount A in the account after t years is given by

$A = 1{,}000e^{0.06t}$

Find the time t that it will take for the amount to double ($A = 2{,}000$). Compare this time with the result of the Check Yourself 7 Exercise. Which is shorter? Why?

Check Yourself ANSWERS

1. {2} **2.** $\left\{\frac{1}{8}\right\}$ **3.** $\left\{\frac{3}{2}\right\}$ **4.** {3.322} **5.** {1.386} **6.** {1.151}

7. 11.58 yr **8.** 11.55 yr. The doubling time is shorter, because interest is compounded more frequently.

Reading Your Text

These fill-in-the-blank exercises will help you understand some of the key vocabulary used in this section. The answers to these exercises are in the Answers Appendix in the back of the text.

(a) A logarithmic ________ is an equation that contains a logarithmic expression.

(b) If no base for a logarithm is written, it is assumed to be ________.

(c) Equations in which the ________ appears as an exponent are called exponential equations.

(d) The final step in solving logarithmic equations is to check for ________ solutions.

Skills Calculator/Computer Career Applications Above and Beyond

10.7 exercises

< Objective 1 >

Solve each equation.

1. $\log_5 x = 3$

2. $\log_3 x = -2$

3. $\log (x + 1) = 2$

4. $\log_5 (3x + 2) = 3$

5. $\log_2 x + \log_2 8 = 6$

6. $\log 5 + \log x = 2$

7. $\log_3 x - \log_3 6 = 3$

8. $\log_4 x - \log_4 8 = 3$

9. $\log_2 x + \log_2 (x + 2) = 3$

10. $\log_3 x + \log_3 (2x + 3) = 2$

11. $\log_7 (x + 1) + \log_7 (x - 5) = 1$

12. $\log_2 (x + 2) + \log_2 (x - 5) = 3$

13. $\log x - \log (x - 2) = 1$

14. $\log_5 (x + 5) - \log_5 x = 2$

15. $\log_3 (x + 1) - \log_3 (x - 2) = 2$ VIDEO

16. $\log (x + 2) - \log (2x - 1) = 1$

17. $\log (x + 5) - \log (x - 2) = \log 5$

18. $\log_3 (x + 12) - \log_3 (x - 3) = \log_3 6$

19. $\log_2 (x^2 - 1) - \log_2 (x - 2) = 3$

20. $\log (x^2 + 1) - \log (x - 2) = 1$

< Objective 2 >

Solve each equation. If your solution is an approximation, round to three decimal places.

21. $6^x = 1{,}296$

22. $4^x = 64$

23. $2^{x+1} = \frac{1}{8}$

24. $9^x = 3$

25. $8^x = 2$

26. $3^{2x-1} = 27$

27. $3^x = 7$ VIDEO

28. $5^x = 30$

29. $4^{x+1} = 12$

30. $3^{2x} = 5$

31. $7^{3x} = 50$

32. $6^{x-3} = 21$

33. $5^{3x-1} = 15$

34. $8^{2x+1} = 20$

35. $4^x = 3^{x+1}$ 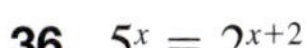

36. $5^x = 2^{x+2}$

37. $2^{x+1} = 3^{x-1}$

38. $3^{2x+1} = 5^{x+1}$

< Objective 3 >

Use the formula

$$A = P\left(1 + \frac{r}{n}\right)^{nt}$$

to complete exercises 39 to 42. Round your answers to two decimal places.

39. **Business and Finance** If \$5,000 is placed in an account with an annual interest rate of 9%, how long will it take the amount to double if the interest is compounded annually?

40. Repeat exercise 39 if the interest is compounded semiannually.

41. Repeat exercise 39 if the interest is compounded quarterly.

42. Repeat exercise 39 if the interest is compounded monthly.

Suppose the number of bacteria present in a culture after t hours is given by $N(t) = N_0 \cdot 2^{t/2}$*, where* N_0 *is the initial number of bacteria. Use the formula to complete exercises 43 to 46.*

43. How long will it take the bacteria to increase from 12,000 to 20,000?

44. How long will it take the bacteria to increase from 12,000 to 50,000?

45. How long will it take the bacteria to triple?
Hint: Let $N = 3N_0$.

46. How long will it take the culture to increase to 5 times its original size?
Hint: Let $N = 5N_0$.

Science and Medicine *The radioactive element strontium-90 has a half-life of approximately* 28 *yr. That is, in a 28-yr period, one-half of the initial amount will have decayed into another substance. If* A_0 *is the initial amount of the element, then the amount A remaining after t years is given by*

$$A = A_0\left(\frac{1}{2}\right)^{t/28}$$

Use the formula to complete exercises 47 to 50. 

47. If the initial amount of the element is 100 g, in how many years will 60 g remain?

48. If the initial amount of the element is 100 g, in how many years will 20 g remain?

49. In how many years will 75% of the original amount remain?
Hint: Let $A = 0.75A_0$.

50. In how many years will 10% of the original amount remain?
Hint: Let $A = 0.1A_0$.

Given projected growth, t years from now a city's population P can be approximated by $P(t) = 25{,}000e^{0.045t}$*. Use the formula to complete exercises 51 and 52.*

51. How long will it take the city's population to reach 35,000?

52. How long will it take the population to double?

The number of bacteria in a culture after t hours is given by $N(t) = N_0e^{0.03t}$*, where* N_0 *is the initial number of bacteria in the culture. Use the formula to complete exercises 53 and 54.*

53. In how many hours will the size of the culture double? 

54. In how many hours will the culture grow to 4 times its original population?

The atmospheric pressure P, in inches of mercury (in. Hg), at an altitude h ft above sea level is approximated by $P(t) = 30e^{-0.00004h}$. *Use the formula to complete exercises 55 and 56.*

55. Find the altitude at which the pressure is 25 in. Hg.

56. Find the altitude at which the pressure is 20 in. Hg.

Carbon-14 dating is used to determine the age of specimens and is based on the radioactive decay of the element carbon-14. This decay begins once a plant or animal dies. If A_0 *is the initial amount of carbon-14, then the amount remaining after t years is* $A(t) = A_0e^{-0.000124t}$. *Use the formula to complete exercises 57 and 58.* chapter 10 > Make the Connection

57. Estimate the age of a specimen if 70% of the original amount of carbon-14 remains.

58. Estimate the age of a specimen if 20% of the original amount of carbon-14 remains.

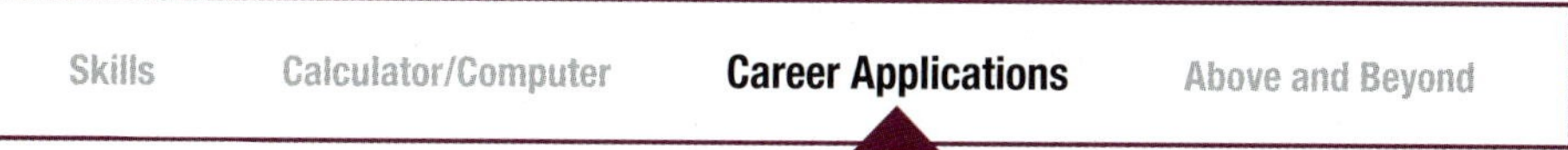

59. Allied Health Chemists assign a pH-value (a measure of a solution's acidity) as a function of the concentration of hydrogen ions (H^+, measured in moles per liter M) according to Sorenson's 1909 model,

$$pH = -\log([H^+])$$

The most acidic rainfall ever measured occurred in Scotland in 1974. What was the hydrogen ion concentration given that its pH was 2.4? (Report your results with two decimal places in scientific notation.)

60. Automotive Technology One formula for noise level N (in dB) is

$$N = 10\log\left(\frac{I}{10^{-12}\,W/m^2}\right)$$

One city ordinance requires that the maximum noise level for a car exhaust be 100 dB. What is the maximum sound intensity I (in W/m^2) allowed?

61. Information Technology One problem associated with using the Internet to perform research is the broken or "dead link." This refers to Web pages that include links to other pages that yield an error message "file not found" or direct the user to a website that is different from the original one. Although links cited in research articles generally last longer than those on the Web, they still may not last very long.

Researchers at the University of Iowa examined links cited in articles accepted by the Communications-Technology division of the Association for Education in Journalism and Mass Communication. They found that such links have a half-life of approximately 1.25 yr. This means that 1.25 yr (15 months) after the initial linkage, half the links are no longer valid.

Using this information, we can build a function to estimate the number of valid links t years after the initial citation.

$$L(t) = a(0.57435)^t$$

in which a represents the total number of links cited in an article, and $L(t)$ gives the number still valid after t years. chapter 10 > Make the Connection

(a) If an article cites 30 Internet links, how many would you expect to be "live" after 6 months? 5 yr?

(b) If you peruse an article that is 2 yr old, and 12 links are still valid, how many links would you expect to be invalid?

62. Manufacturing Technology Aceto Balsamico Tradizionale (authentic, traditional balsamic vinegar) from the Modena and Reggio regions of Italy's Emilia-Romagna province sells for anywhere from \$50 to \$600 per ounce.

The producers of this vinegar typically fill 60-liter barrels to 75% capacity. After 10 yr, only 60% of the original contents (by volume) are present in the barrel.

A bottle of this vinegar is 100 mL. Aged 12 yr (the minimum allowed), a bottle sells for roughly \$75. Aged 20 yr, a bottle sells for roughly \$110. Aged 30 yr, a bottle sells for roughly \$200. Further aging can bring the price of a bottle as high as \$600. chapter 10 > Make the Connection

(a) How much vinegar is initially placed in a 60-liter barrel?

(b) How much vinegar is left in the barrel after 10 yr? 20 yr? Report your results to the nearest liter. **Note:** In fact, barrels are changed each year with a mixing of newer and older varieties of vinegar.

(c) Construct a function to model the amount of vinegar in a barrel t years after its initial fill.

(d) How many bottles does a barrel produce after 12 yr? 20 yr? 30 yr?

(e) How much time needs to pass before the original barrel produces only one bottle of vinegar?

63. In some of the earlier exercises, we talked about bacteria cultures that double in size every few minutes. Can this go on forever? Explain.

64. The population of the United States has been doubling every 45 yr. Is it reasonable to assume that this rate will continue? What factors will start to limit that growth?

Use a calculator to describe the graph of each equation, then explain the result.

65. $y = \log 10^x$

66. $y = 10^{\log x}$

67. $y = \ln e^x$

68. $y = e^{\ln x}$

69. In this section, we solved the equation $3^x = 2^{x+1}$ by first applying the logarithmic function base 10 to each side of the equation. Try this again, but this time apply the natural logarithm function to each side of the equation. Compare the solutions that result from the two approaches.

Answers

1. {125} **3.** {99} **5.** {8} **7.** {162} **9.** {2} **11.** {6} **13.** $\left\{\frac{20}{9}\right\}$ **15.** $\left\{\frac{19}{8}\right\}$ **17.** $\left\{\frac{15}{4}\right\}$ **19.** {3, 5} **21.** {4} **23.** {−4} **25.** $\left\{\frac{1}{3}\right\}$ **27.** {1.771} **29.** {0.792} **31.** {0.67} **33.** {0.894} **35.** {3.819} **37.** {4.419} **39.** 8.04 yr **41.** 7.79 yr **43.** 1.47 hr **45.** 3.17 hr **47.** 20.6 yr **49.** 11.6 yr **51.** 7.5 yr **53.** 23.1 hr **55.** 4,558 ft **57.** 2,876 yr **59.** 3.98×10^{-3} M **61.** **(a)** 23; 2; **(b)** 24 **63.** Above and Beyond **65.** The graph is that of $y = x$. The two functions undo each other. **67.** The graph is that of $y = x$. The two functions undo each other. **69.** Above and Beyond

Definition/Procedure	Example	Reference
Algebra of Functions		**Section 10.1**
The **sum of two functions** f and g is written $f + g$. It is defined as $(f + g)(x) = f(x) + g(x)$	Let $f(x) = 2x + 1$ and $g(x) = 3x^2$. $(f + g)(x) = (2x + 1) + (3x^2)$ $= 3x^2 + 2x + 1$	p. 778
The **difference of two functions** f and g is written $f - g$. It is defined as $(f - g)(x) = f(x) - g(x)$	$(f - g)(x) = (2x + 1) - (3x^2)$ $= -3x^2 + 2x + 1$	p. 778
The **product of two functions** f and g is written $f \cdot g$. It is defined as $(f \cdot g)(x) = f(x) \cdot g(x)$	$(f \cdot g)(x) = (2x + 1))3x^2)$ $= 6x^3 + 3x^2$	p. 781
The **quotient of two functions** f and g is written as $f \div g$ or $\frac{f}{g}$ and is defined as $(f \div g)(x) = f(x) \div g(x)$ or $\left(\frac{f}{g}\right)(x) = \frac{f(x)}{g(x)}$ for $g(x) \neq 0$	$(f \div g)(x) = (2x + 1) \div (3x^2)$ $= \frac{2x + 1}{3x^2}$ or $\left(\frac{f}{g}\right)(x) = \frac{2x + 1}{3x^2}$ for $x \neq 0$.	p. 781
Composition of Functions		**Section 10.2**
The composition of two functions f and g is written $f \circ g$. It is defined as $(f \circ g)(x) = f(g(x))$	Let $f(x) = 2x + 1$ and $g(x) = 3x^2$ $(f \circ g)(x) = 2(3x^2) + 1$ $= 6x^2 + 1$	p. 792
Inverse Functions		**Section 10.3**
The **inverse** of a relation is formed by interchanging the components of each ordered pair in the given relation. If a relation (or function) is specified by an equation, switch x and y in the defining equation to form the inverse.	The inverse of the relation $\{(1, 2), (2, 3), (4, 3)\}$ is $\{(2, 1), (3, 2), (3, 4)\}$ To find the inverse of $f(x) = 4x - 8$ $y = 4x - 8$ change y to x and x to y $x = 4y - 8$ so $4y = x + 8$ $y = \frac{1}{4}(x + 8)$ $y = \frac{1}{4}x + 2$ $f^{-1}(x) = \frac{1}{4}x + 2$	p. 800

Continued

Definition/Procedure	Example	Reference
Finding Inverse Relations and Functions 1. Switch the x- and y-coordinates of the ordered pairs of the given relation or the roles of x and y in the defining equation. 2. If the relation was described in equation form, solve the defining equation of the inverse for y. 3. If desired, graph the relation and its inverse on the same set of axes. The two graphs will be symmetric about the line $y = x$.	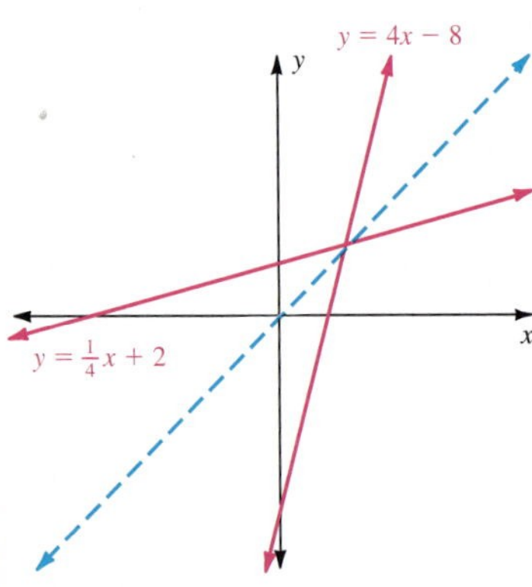	*p.* 802
The inverse of a function f may or may not be a function. If the inverse *is* also a function, we denote that inverse as f^{-1}, read "the inverse of f." A function f has an inverse f^{-1}, which is also a function, if and only if f is a **one-to-one** function. That is, no two ordered pairs in the function have the same second component. The **horizontal line test** can be used to determine whether a function is one-to-one.	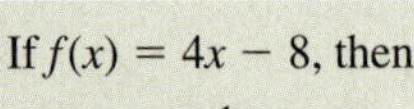 If $f(x) = 4x - 8$, then $f^{-1}(x) = \frac{1}{4}x + 2$ 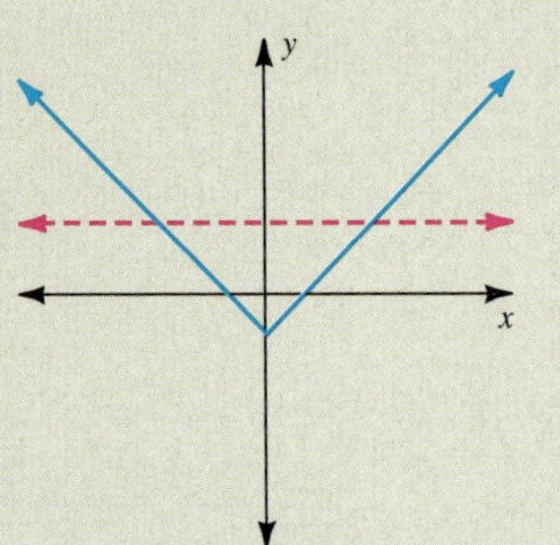 *Not* one-to-one	*p.* 804
Exponential Functions		**Section 10.4**
An **exponential function** is a function that can be written as $f(x) = a \cdot b^x$ in which $a \neq 0$, $b > 0$, and $b \neq 1$. We call b the **base** of the exponential function. If $a = 1$, the function $f(x) = b^x$ may be referred to as an **elementary exponential function.** If b is greater than 1, the function is increasing (a **growth function**). If b is less than 1, the function is decreasing (a **decay function**). In both cases, the exponential function is one-to-one. The domain is the set of all real numbers and the range is the set of positive real numbers. The function defined by $f(x) = e^x$, in which e is an irrational number (approximately 2.71828), is called *the* exponential function.	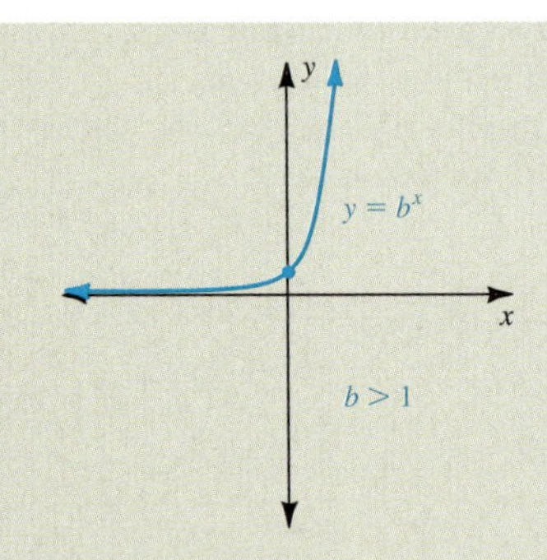	*p.* 812

Continued

Definition/Procedure	Example	Reference
Graphing an Exponential Function **Step 1** Create a table of values for $y = b^x$. **Step 2** Plot the points from the table and connect them with a smooth curve. **Properties of Elementary Exponential Graphs** 1. If $b > 1$, the graph increases from left to right. If $0 < b < 1$, the graph decreases from left to right. 2. All elementary exponential graphs have these properties in common. (a) The y-intercept is (0, 1). (b) The graph approaches, but does not touch, the x-axis. (c) The graphs represent one-to-one functions.	y $y = b^x$ x $0 < b < 1$	p. 814
Logarithmic Functions		**Section 10.5**
For $b > 0$ and $b \neq 1$, $y = \log_b x$ if, and only if, $b^y = x$ We call $\log_b x$ the **logarithm of x with base b.** A logarithm is an exponent or a power. The logarithm of x to base b is the power to which we must raise b to get x. A **logarithmic function** is any function of the form $f(x) = \log_b x \quad b > 0, b \neq 1$ The logarithm function is the inverse of the corresponding exponential function. The function is one-to-one with domain $\{x \mid x > 0\}$ and range composed of the set of all real numbers.	$\log_3 9 = 2$ is in logarithmic form. $3^2 = 9$ is the exponential form. $\log_3 9 = 2$ is equivalent to $3^2 = 9$. 2 is the power to which we must raise 3 to get 9. y $y = \log_b x$ x $b > 1$	p. 826
Properties of Logarithms		**Section 10.6**
If M, N, and b are positive real numbers with $b \neq 1$ and if p is any real number, then we can state these properties of logarithms. 1. $\log_b b = 1$ 2. $\log_b 1 = 0$ 3. $b^{\log_b x} = x$ 4. $\log_b b^x = x$	$\log 10 = 1$ $\log_2 1 = 0$ $3^{\log_3 2} = 2$ $\log_5 5^x = x$	p. 837
Product Property $\log_b MN = \log_b M + \log_b N$	$\log_3 x + \log_3 y = \log_3 xy$	p. 838
Quotient Property $\log_b \frac{M}{N} = \log_b M - \log_b N$	$\log_5 8 - \log_5 3 = \log_5 \frac{8}{3}$	p. 838
Power Property $\log_b M^p = p \log_b M$	$\log 3^2 = 2 \log 3$	p. 838

Continued

Definition/Procedure	Example	Reference
Common logarithms are logarithms to base 10. For convenience, we omit the base when writing common logarithms. $\log M = \log_{10} M$	$\log_{10} 1{,}000 = \log 1{,}000$ $= \log 10^3 = 3$	p. 840
Natural logarithms are logarithms to base e. $\ln M = \log_e M$	$\ln 3 = \log_e 3$	p. 843
Logarithmic and Exponential Equations		**Section 10.7**
A **logarithmic equation** is an equation that contains a logarithmic expression. $\log_2 x = 5$ is a logarithmic equation.	To solve $\log_2 x = 5$, write the equation as an equivalent exponential equation. $x = 2^5$ or $x = 32$	p. 854
Solving Logarithmic Equations **Step 1** Use the properties of logarithms to combine terms containing logarithmic expressions into a single term. **Step 2** Write the equation in exponential form. **Step 3** Solve. **Step 4** Check your solutions to make sure that possible solutions do not result in the logarithms of negative numbers or zero.	$\log_4 x + \log_4 (x - 6) = 2$ $\log_4 x(x - 6) = 2$ $x(x - 6) = 4^2$ $x^2 - 6x - 16 = 0$ $(x - 8)(x + 2) = 0$ $x = 8$ or $x = -2$ Because substituting -2 for x in the original equation results in the logarithm of a negative number, we reject that answer. The only solution is 8.	p. 855
An **exponential equation** is an equation in which the variable appears as an exponent.	To solve $4^x = 64$: Because $64 = 4^3$, write $4^x = 4^3$ or $x = 3$	p. 855
Solving Exponential Equations **Step 1** Try to write each side of the equation as a power of the same base. Then equate the exponents to form an equation. **Step 2** If the above procedure is not applicable, take the logarithm of both sides of the original equation. **Step 3** Use the power rule for logarithms to write an equivalent equation with the variables as coefficients. **Step 4** Solve the resulting equation. **Step 5** Check the solutions.	$2^{x+3} = 5^x$ $\log 2^{x+3} = \log 5^x$ $(x + 3) \log 2 = x \log 5$ $x \log 2 + 3 \log 2 = x \log 5$ $x \log 2 - x \log 5 = -3 \log 2$ $x (\log 2 - \log 5) = -3 \log 2$ $x = \dfrac{-3 \log 2}{\log 2 - \log 5} \approx 2.269$	p. 857

summary exercises :: chapter 10

This summary exercise set will help ensure that you have mastered each of the objectives of this chapter. The exercises are grouped by section. You should reread the material associated with any exercises that you find difficult. The answers to the odd-numbered exercises are in the Answers Appendix in the back of the text.

10.1 *In exercises 1 to 8, use the tables to find the desired values.*

x	$f(x)$
−2	3
−1	8
3	0
7	−6
8	−4

x	$g(x)$
−4	5
−2	−1
3	2
5	−3
7	0

1. $(f + g)(-2)$

2. $(g - f)(7)$

3. $(f - g)(-1)$

4. $(f \cdot g)(3)$

5. $(f \div g)(7)$

6. $\left(\frac{g}{f}\right)(-2)$

7. Find the domain of $g - f$.

8. Find the domain of $\frac{g}{f}$.

Find $(f + g)(x)$.

9. $f(x) = 4x^2 + 5x - 3$ and $g(x) = -2x^2 + x - 5$

10. $f(x) = -3x^3 + 2x^2 - 5$ and $g(x) = 4x^3 - 4x^2 + 5x + 6$

11. $f(x) = 2x^4 + 4x^2 + 5$ and $g(x) = x^3 - 5x^2 + 6x$

12. $f(x) = 3x^3 + 5x - 5$ and $g(x) = -2x^3 + 2x^2 + 5x$

Find $(f - g)(x)$.

13. $f(x) = 7x^2 - 2x + 3$ and $g(x) = 2x^2 - 5x - 7$

14. $f(x) = 9x^2 - 4x$ and $g(x) = 5x^2 + 3$

15. $f(x) = 8x^2 + 5x$ and $g(x) = 4x^2 - 3x$

16. $f(x) = -2x^2 - 3x$ and $g(x) = -3x^2 + 4x - 5$

Find each product $(f \cdot g)(x)$.

17. $f(x) = 2x$ and $g(x) = 3x - 5$

18. $f(x) = x + 1$ and $g(x) = -3x$

19. $f(x) = 3x$ and $g(x) = x^2$

20. $f(x) = 2x$ and $g(x) = x^2 - 5$

Find each quotient $\left(\frac{f}{g}\right)(x)$ *and state the domain of the resulting function.*

21. $f(x) = 2x$ and $g(x) = x - 3$

22. $f(x) = x + 1$ and $g(x) = 2x - 4$

23. $f(x) = 3x$ and $g(x) = x^2$

24. $f(x) = 2x^2$ and $g(x) = x - 5$

10.2 *In exercises 25 to 30, use the tables to find the desired values.*

x	$f(x)$
−2	3
−1	8
3	0
7	−6
8	−4

x	$g(x)$
−4	5
−2	−1
3	2
5	−3
7	0

25. $(f \circ g)(-2)$

26. $(g \circ f)(8)$

27. $(g \circ f)(-2)$

28. $(f \circ g)(-4)$

29. $(g \circ g)(-4)$

30. $(f \circ f)(-2)$

Evaluate the indicated composite functions in each part.

31. $f(x) = x - 3$ and $g(x) = 3x + 1$
(a) $(f \circ g)(0)$ **(b)** $(f \circ g)(-2)$ **(c)** $(f \circ g)(3)$ **(d)** $(f \circ g)(x)$

32. $f(x) = 5x - 1$ and $g(x) = -4x + 5$
(a) $(f \circ g)(0)$ **(b)** $(f \circ g)(-2)$ **(c)** $(g \circ f)(3)$ **(d)** $(g \circ f)(x)$

33. $f(x) = x^2$ and $g(x) = x - 5$
(a) $(f \circ g)(0)$ (b) $(f \circ g)(-2)$ **(c)** $(g \circ f)(3)$ **(d)** $(g \circ f)(x)$

34. $f(x) = x^2 + 3$ and $g(x) = -2x$
(a) $(g \circ f)(0)$ (b) $(g \circ f)(-2)$ **(c)** $(f \circ g)(3)$ **(d)** $(f \circ g)(x)$

Rewrite the function h as a composite of functions f and g.

35. $f(x) = -2x$, $g(x) = x + 2$, $h(x) = -2x - 4$

36. $f(x) = 6x$, $g(x) = x^2 + 5$, $h(x) = 36x^2 + 5$

10.3 *Find the inverse function f^{-1} of each function.*

37. $f(x) = -2x + 3$

38. $f(x) = 4x + 5$

39. $f(x) = \frac{x - 3}{4}$

40. $f(x) = \frac{x}{4} - 3$

Find the inverse of each function. In each case, determine whether the inverse is also a function.

41.

x	$f(x)$
−3	1
−1	2
2	3
3	4

42.

x	$f(x)$
−2	5
0	4
3	5
5	4

43. $\{(1, 5), (2, 7), (3, 9)\}$

44. $\{(3, 1), (5, 1), (7, 1)\}$

45. $\{(2, 4), (4, 3), (6, 4)\}$

46. $\{(-2, 6), (0, 0), (3, 8)\}$

For each function f, find its inverse f^{-1}. Then graph both on the same set of axes.

47. $f(x) = 5x + 3$

48. $f(x) = -3x + 9$

49. $f(x) = \dfrac{x - 4}{5}$

50. $f(x) = \dfrac{3 - x}{2}$

Determine whether the given function is one-to-one. In each case determine whether the inverse is a function.

51. $f = \{(-1, 2), (1, 3), (2, 5), (4, 7)\}$

52. $f = \{(-3, 2), (0, 2), (1, 2), (3, 2)\}$

53.

x	$f(x)$
−1	2
3	4
4	5
5	6

54.

x	$f(x)$
1	3
2	4
3	5
4	3

55.

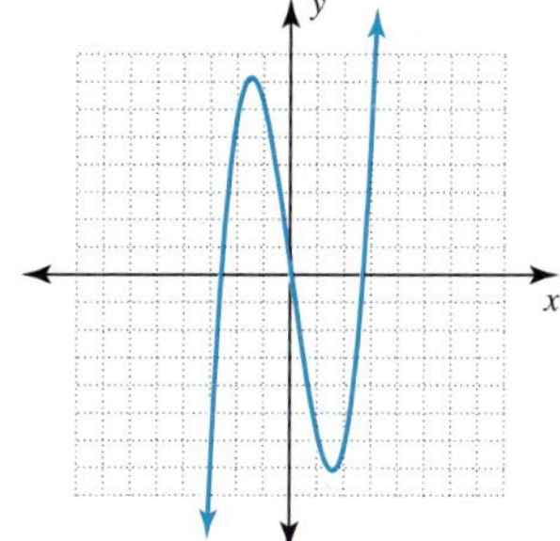

56.

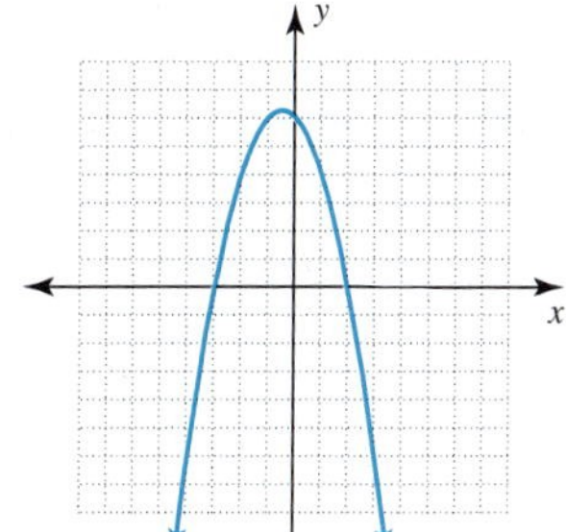

If $f(x) = 4x + 12$, then $f^{-1}(x) = \frac{x}{4} - 3$. Evaluate each expression.

57. $f(-2)$

58. $f^{-1}(4)$

59. $f(f^{-1}(8))$

60. $f^{-1}(f(-4))$

61. $f(f^{-1}(x))$

62. $f^{-1}(f(x))$

10.4 *Graph each function.*

63. $y = 3^x$

64. $y = \left(\frac{3}{4}\right)^x$

Solve each equation.

65. $5^x = 125$

66. $2^{2x+1} = 32$

67. $3^{x-1} = \frac{1}{9}$

Science and Medicine *If it takes 2 hr for the population of a certain bacterial culture to double (by dividing in half), then the number N of bacteria in the culture after t hours is given by $N = 1{,}000 \cdot 2^{t/2}$, where the initial population of the culture was 1,000. Using this formula, find the number in the culture.*

68. After 4 hr

69. After 12 hr

70. After 15 hr

10.5 *Graph each function.*

71. $y = \log_3 x$

72. $y = \log_2(x - 1)$

Convert each statement to logarithmic form.

73. $2^5 = 32$

74. $10^3 = 1{,}000$

75. $5^0 = 1$

76. $5^{-2} = \frac{1}{25}$

77. $25^{1/2} = 5$

78. $16^{3/4} = 8$

Convert each statement to exponential form.

79. $\log_4 64 = 3$

80. $\log 100 = 2$

81. $\log_{81} 9 = \frac{1}{2}$

82. $\log_5 25 = 2$

83. $\log 0.001 = -3$

84. $\log_{32} \frac{1}{2} = -\frac{1}{5}$

Solve each equation.

85. $y = \log_5 125$

86. $\log_b \frac{1}{9} = -2$

87. $\log_7 x = 2$

88. $y = \log_5 1$

89. $\log_b 3 = \frac{1}{2}$

90. $y = \log_9 3$

91. $y = \log_8 2$

92. $\log_b 32 = 2$

Science and Medicine *The decibel (dB) rating for the loudness of a sound is given by*

$$L = 10 \log \frac{I}{I_0}$$

where I is the intensity of that sound in watts per square centimeter and I_0 is the intensity of the "threshold" sound $I_0 = 10^{-16}$ W/cm². Find the decibel rating of each sound.

93. A table saw in operation with intensity $I = 10^{-6}$ W/cm²

94. The sound of a passing car horn with intensity $I = 10^{-8}$ W/cm²

Science and Medicine *The formula for the decibel rating of a sound can be solved for the intensity of the sound as*

$$I = I_0 \cdot 10^{L/10}$$

where L is the decibel rating of the given sound.

95. What is the ratio of intensity of a 60-dB sound to one of 50 dB?

96. What is the ratio of intensity of a 60-dB sound to one of 40 dB?

Science and Medicine *The magnitude of an earthquake on the Richter scale is given by*

$$M = \log \frac{a}{a_0}$$

where a is the intensity of the shock wave of the given earthquake and a_0 is the intensity of the shock wave of a zero-level earthquake.

97. The Alaskan earthquake of 1964 had an intensity of $10^{8.4}a_0$. What was its magnitude on the Richter scale?

98. Find the ratio of intensity of an earthquake of magnitude 7 to an earthquake of magnitude 6.

10.6 *Use the properties of logarithms to expand each expression.*

99. $\log_b x^2y$

100. $\log_4 \frac{y^3}{5}$

101. $\log_5 \frac{x^2y}{z^3}$

102. $\log_5 x^3yz^2$

103. $\log \frac{xy}{\sqrt{z}}$

104. $\log_b \sqrt[3]{\frac{x^2y}{z}}$

Use the properties of logarithms to write each expression as a single logarithm.

105. $\log x + 2 \log y$

106. $3 \log_b x - 2 \log_b z$

107. $\log_b x + \log_b y - \log_b z$

108. $2 \log_5 x - 3 \log_5 y - \log_5 z$

109. $\log x - \frac{1}{2} \log y$

110. $\frac{1}{3}(\log_b x - 2 \log_b y)$

Given that $\log 2 = 0.301$ *and* $\log 3 = 0.477$, *find each logarithm. Verify your results with a calculator.*

111. $\log 18$

112. $\log 16$

113. $\log \frac{1}{8}$

114. $\log \sqrt{3}$

Science and Medicine *Use a calculator to find the pH of each solution, given the hydrogen ion concentration* $[H^+]$ *for each solution, where*

$$\text{pH} = -\log [H^+]$$

Are the solutions acidic or basic?

115. Coffee: $[H^+] = 5 \times 10^{-6}$

116. Household detergent: $[H^+] = 3.2 \times 10^{-10}$

Science and Medicine *Given the pH of these solutions, find the hydrogen ion concentration* $[H^+]$.

117. Lemonade: pH = 3.5

118. Ammonia: pH = 10.2

Social Science *The average score on a final examination for a group of chemistry students, retested after time t (in weeks), is given by*

$$S(t) = 81 - 6 \ln (t + 1)$$

Find the average score on the retests after the given times.

119. After 5 weeks

120. After 10 weeks

121. After 15 weeks

122. Graph these results.

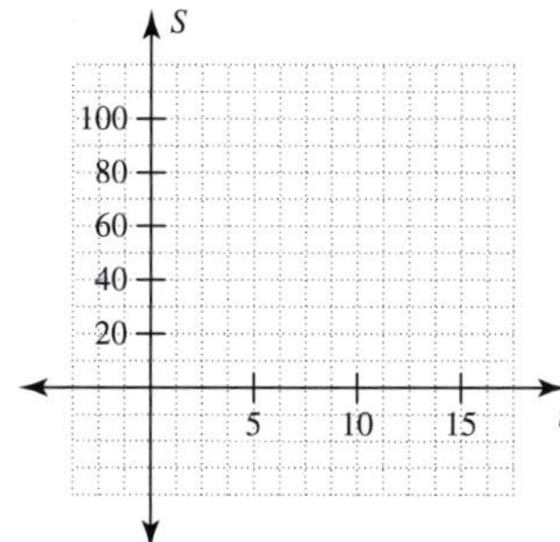

The formula for converting from a logarithm with base a to a logarithmic expression with base b is

$$\log_a x = \frac{\log_b x}{\log_b a}$$

Use this formula to find each logarithm.

123. $\log_4 20$

124. $\log_8 60$

10.7 *Solve each logarithmic equation.*

125. $\log_3 x + \log_3 5 = 3$

126. $\log_5 x - \log_5 10 = 2$

127. $\log_3 x + \log_3 (x + 6) = 3$

128. $\log_5 (x + 3) + \log_5 (x - 1) = 1$

129. $\log x - \log (x - 1) = 1$

130. $\log_2 (x + 3) - \log_2 (x - 1) = \log_2 3$

Solve each exponential equation. Give your results rounded to three decimal places.

131. $3^x = 243$

132. $5^x = \frac{1}{25}$

133. $5^x = 10$

134. $4^{x-1} = 8$

135. $6^x = 2^{2x+1}$

136. $2^{x+1} = 3^{x-1}$

Business and Finance *If an investment of P dollars earns interest at an annual rate of 12% and the interest is compounded n times per year, then the amount A in the account after t years is*

$$A(t) = P\left(1 + \frac{0.12}{n}\right)^{nt}$$

Use this formula to complete each exercise.

137. If \$1,000 is invested and the interest is compounded quarterly, how long will it take the amount in the account to double?

138. If \$3,000 is invested and the interest is compounded monthly, how long will it take the amount in the account to reach \$8,000?

Science and Medicine *A certain radioactive element has a half-life of 50 years. The amount A of the substance remaining after t years is given by*

$$A(t) = A_0 \cdot 2^{-t/50}$$

where A_0 is the initial amount of the substance. Use this formula to complete each exercise.

139. If the initial amount of the substance is 100 mg, after how long will 40 mg remain?

140. After how long will only 10% of the original amount of the substance remain?

Social Science *A city's population is presently 50,000. Given the projected growth, t years from now the population P will be given by $P(t) = 50{,}000e^{0.08t}$. Use this formula to complete each exercise.*

141. How long will it take the population to reach 70,000?

142. How long will it take the population to double?

Science and Medicine *The atmospheric pressure, in inches of mercury, at an altitude h miles above the surface of the earth, is approximated by $P(h) = 30e^{-0.021h}$. Use this formula to complete each exercise.*

143. Find the altitude at the top of Mt. McKinley in Alaska if the pressure is 27.7 in. Hg.

144. Find the altitude of an airplane if the pressure outside is 26.1 in. Hg.

chapter test 10

Use this chapter test to assess your progress and to review for your next exam. Allow yourself about an hour to take this test. The answers to these exercises are in the Answers Appendix in the back of the text.

Convert each statement to logarithmic form.

1. $10^4 = 10{,}000$

2. $27^{2/3} = 9$

If $f(x) = 5x - 7$, *then* $f^{-1}(x) = \frac{x + 7}{5}$. *Evaluate each expression.*

3. $f(f^{-1}(x))$

4. $f^{-1}(f(x))$

Use the properties of logarithms to expand the expression.

5. $\log_5\sqrt{\frac{xy^2}{z}}$

Let $f(x) = x^2 - 1$ *and* $g(x) = 3x - 2$.

6. Find $(f \cdot g)(x)$ and state the domain of the resulting function.

7. Find $(g \div f)(x)$ and state the domain of the resulting function.

8. Find $(f \circ g)(x)$.

9. Find $(g \circ f)(x)$.

Solve each equation.

10. $5^x = \frac{1}{25}$

11. $3^{2x-1} = 81$

Convert each statement to exponential form.

12. $\log_5 125 = 3$

13. $\log 0.01 = -2$

Determine whether each function is one-to-one and state whether the inverse is a function.

14. $f = \{(3, 4), (5, -1), (6, 2), (7, -1)\}$

15.

x	$f(x)$
1	3
3	8
5	11
7	5

Write the expression as a single logarithm.

16. $\frac{1}{3}(\log_b x - 2\log_b z)$

Solve each equation. Round results to three decimal places.

17. $3^{x+1} = 4$

18. $5^x = 3^{x+1}$

For the given $f(x)$ and $g(x)$, find **(a)** $(f + g)(x)$ *and* **(b)** $(f - g)(x)$.

19. $f(x) = -3x^3 + 5x^2 - 2x - 7$ and $g(x) = -2x^2 + 7x - 2$

20. Graph $y = \log_4 x$.

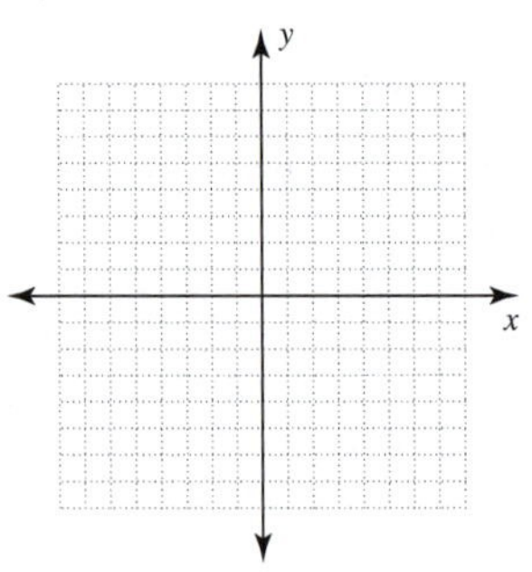

Solve each equation.

21. $\log_6(x + 1) + \log_6(x - 4) = 2$

22. $\log(2x + 1) - \log(x - 1) = 1$

Find the inverse function f^{-1} for the given function.

23. $f(x) = \dfrac{x - 3}{5}$

24.

x	$f(x)$
−3	−2
1	2
4	5
5	6

25. $\{(-3, 1), (4, 2), (5, -2)\}$

Graph each function.

26. $y = 4^x$

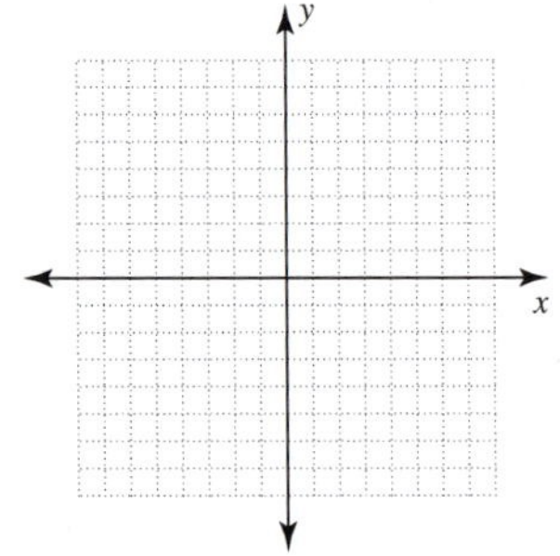

27. $y = \left(\dfrac{2}{3}\right)^x$

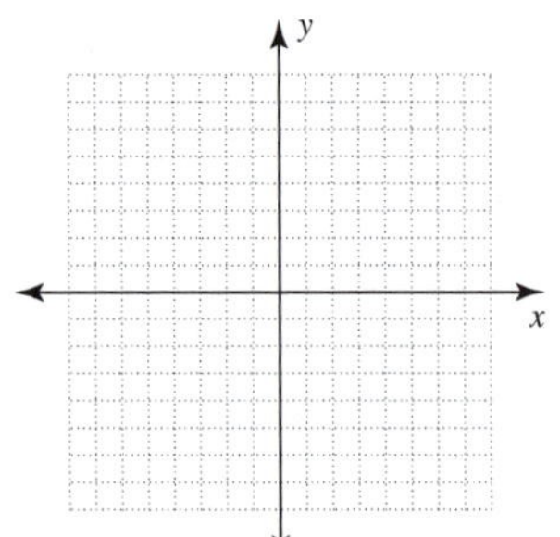

Solve each equation.

28. $y = \log_2 64$

29. $\log_b \dfrac{1}{16} = -2$

30. $\log_{25} x = \dfrac{1}{2}$

cumulative review chapters 0–10

Use this exercise set to review concepts from earlier chapters. While it is not a comprehensive exam, it will help you identify any material that you need to review before taking your final exam. The answers to these exercises are in the Answers Appendix in the back of the text.

Solve each equation.

1. $2x - 3(x + 2) = 4(5 - x) + 7$

2. $2^{3x} = 32$

3. $\log x - \log(x - 1) = 1$

Graph.

4. $5x - 3y = 15$

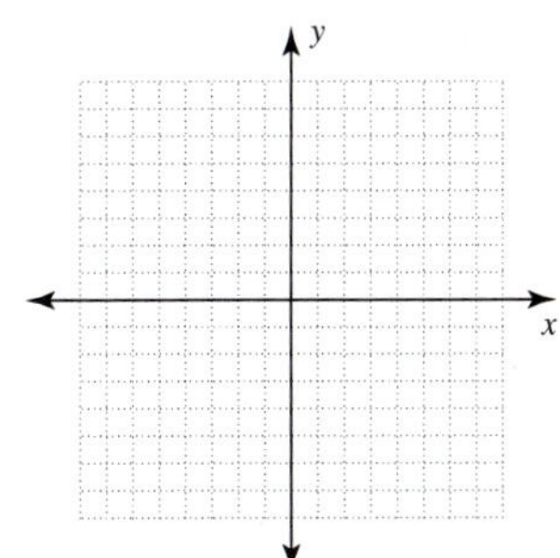

5. $-8(2 - x) \geq y$

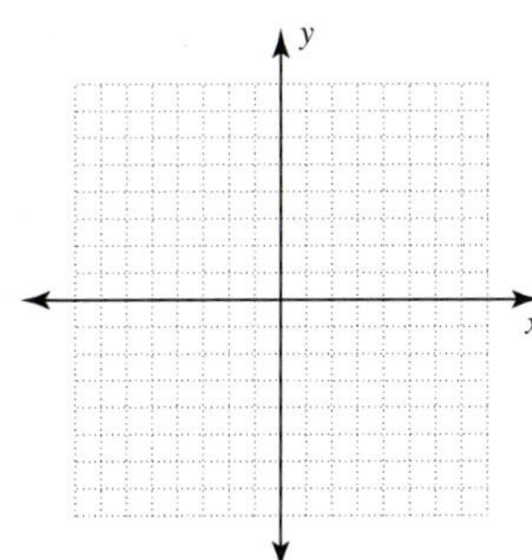

6. Find an equation of the line that passes through the points $(2, -1)$ and $(-3, 5)$.

7. Solve the linear inequality

$3x - 2(x - 5) \geq 20$

Simplify each expression.

8. $4x^2 - 3x + 8 - 2(x^2 + 5) - 3(x - 1)$

9. $(3x + 1)(2x - 5)$

Completely factor each expression.

10. $2x^2 - x - 10$

11. $25x^3 - 16xy^2$

Perform the indicated operations.

12. $\dfrac{2}{x - 4} - \dfrac{3}{x - 5}$

13. $\dfrac{x^2 - x - 6}{x^2 + 2x - 15} \div \dfrac{x - 2}{x + 5}$

Simplify each radical expression.

14. $\sqrt{18} + \sqrt{50} - 3\sqrt{32}$

15. $(3\sqrt{2} + 2)(3\sqrt{2} + 2)$

16. $\dfrac{5}{\sqrt{5} - \sqrt{2}}$

17. Find three consecutive odd integers whose sum is 237.

Solve each equation.

18. $x^2 + x - 2 = 0$

19. $2x^2 - 6x - 5 = 0$

20. Solve the equation for R.

$$\frac{1}{R} = \frac{1}{R_1} + \frac{1}{R_2}$$

If $f(x) = 3x - 1$ *and* $g(x) = x^2 - x - 6$:

21. Find $(f \cdot g)(x)$ including the domain.

22. Find $(f \div g)(x)$ including the domain.

If $f(x) = 2x - 1$ *and* $g(x) = -3x + 2$:

23. Find **(a)** $(f \circ g)(x)$ and **(b)** $(f \circ g)(5)$.

24. Find **(a)** $(g \circ f)(x)$ and **(b)** $(g \circ f)(5)$.

Factor each polynomial completely.

25. $14a^2b^2 - 21a^2b + 35ab^2$

26. $x^2 - 3xy + 5x - 15y$

27. $25c^2 - 64d^2$

28. $27x^3 - 1$

29. $16a^4 + 2ab^3$

30. $x^2 - 2x - 48$

31. $10x^2 - 39x + 14$

32. $6x^3 + 3x^2 - 45x$

Simplify.

33. $\frac{x^3y}{4xy^2} \div \frac{xy}{12xy^2}$

34. $\frac{7}{3-y} - \frac{5}{y-3}$

35. $\frac{4}{m^2-9} - \frac{3}{m^2-4m+3}$

36. $\frac{2x}{x^2-9x+20} + \frac{8}{x-4}$

37. **Geometry** The length of the longer leg of a right triangle is 4 cm more than twice the length of the shorter leg. The length of the longer leg is 2 cm shorter than the length of the hypotenuse. Find the lengths of all three sides.

Solve each equation.

38. $\log_4 x + \log_4 (x - 6) = 2$

39. $\sqrt{2x-1} + x = 8$

40. Find the center and the radius of the circle whose equation is

$(x + 5)^2 + (y - 2)^2 = 16$

A.1 Solving Inequalities in One Variable Graphically

< A.1 Objective >

1 > Solve linear inequalities in one variable graphically

In Section 4.2, we solved linear equations graphically. In this appendix, we use the graphs of linear functions to solve linear inequalities.

Linear inequalities in one variable are obtained from linear equations by replacing the symbol for equality ($=$) with one of the inequality symbols ($<$, $>$, $\le$, $\ge$).

The general form of a linear inequality in one variable is

$$x < a$$

where the symbol $<$ can be replaced with $>$, $\le$, or $\ge$. Examples of linear inequalities in one variable include

$$x \ge -3 \qquad 2x + 5 > 7 \qquad 2x - 3 \le 5x + 6$$

Recall that the solution set for an equation is the set of all values for the variable (or ordered pair) that make the equation a true statement. Similarly, the solution set for an inequality is the set of all values that make the inequality a true statement. In Example 1, we look at the graph of an inequality.

Example 1 Solving a Linear Inequality Graphically

< Objective 1 >

Use a graph to find the solution set to the inequality

$$2x + 5 > 7$$

First, rewrite the inequality as a comparison of two functions. Here $f(x) > g(x)$, in which $f(x) = 2x + 5$ and $g(x) = 7$.

Now graph the two functions on a single set of axes.

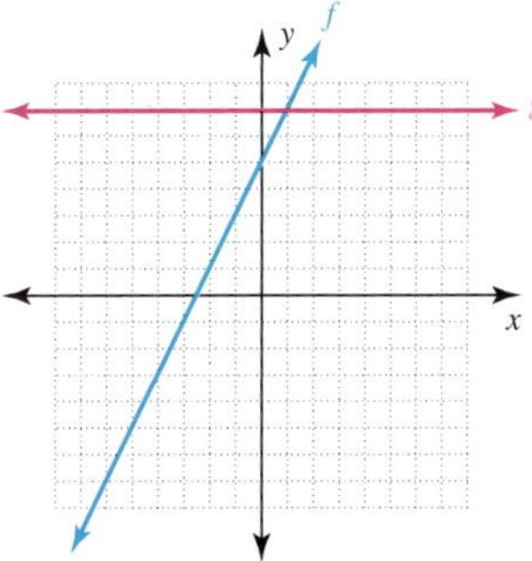

Here we ask the question, For what values of x is the graph of f above the graph of g?

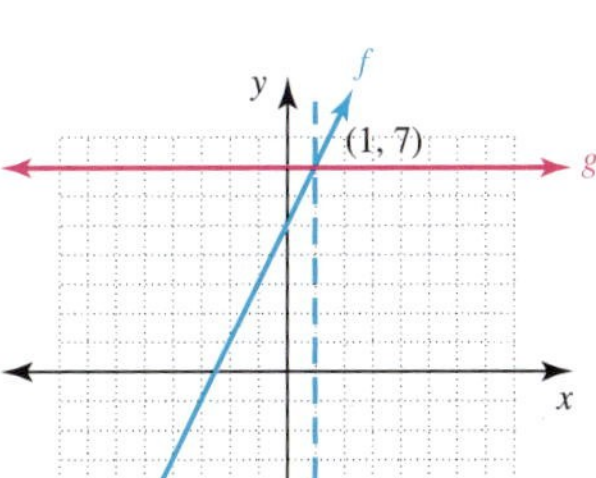

Next, draw a vertical dashed line through the point of intersection of the two functions. In this case, there is a vertical line through the point (1, 7).

NOTE

We use a dashed line to indicate that the x-value of 1 is not included.

> **NOTE**
>
> The solution set contains the x-values that make the original statement $2x + 5 > 7$ true.

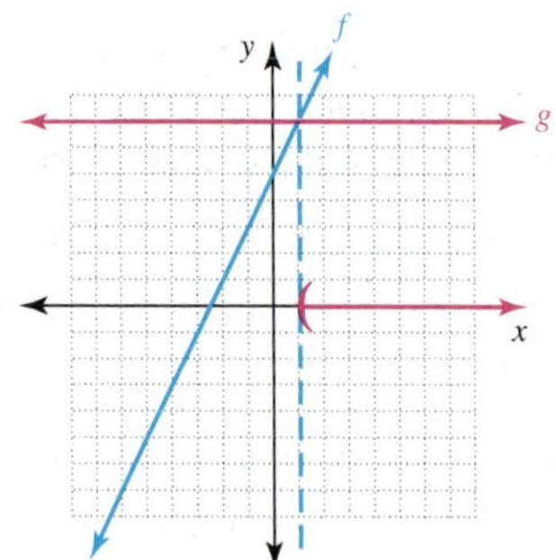

The solution set is every x-value that results in $f(x)$ being greater than $g(x)$, which is every x-value to the right of the dashed line.

We can express the solution set as $\{x \mid x > 1\}$.

In Example 1, the function $g(x) = 7$ resulted in a horizontal line. In Example 2, we see that the same method works for comparing any two functions.

Example 2 Solving an Inequality Graphically

Solve the inequality graphically.

$2x - 3 \geq 5x$

First, rewrite the inequality as a comparison of two functions. Here, $f(x) \geq g(x)$, $f(x) = 2x - 3$, and $g(x) = 5x$. Now graph the two functions on a single set of axes.

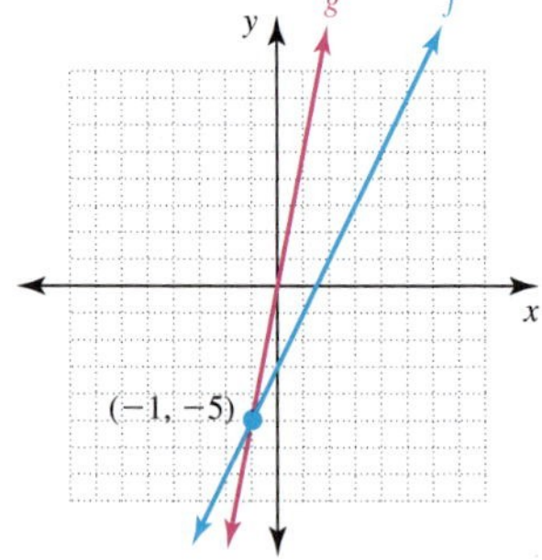

As in Example 1, draw a vertical line through the point of intersection of the two functions. The vertical line goes through $(-1, -5)$. In this case, the line is included (greater than or *equal to*), so we make the line solid, rather than dashed.

Again, we need to mark every x-value that makes the statement true. In this case, that is every x for which the line representing $f(x)$ is above or intersects the line representing $g(x)$. That is the region in which $f(x)$ is greater than or equal to $g(x)$. We mark the x-values to the left of the line, but we also want to include the x-value on the line, so we use a bracket rather than a parenthesis.

> **NOTE**
>
> We use a solid line to indicate that the x-value -1 is included.

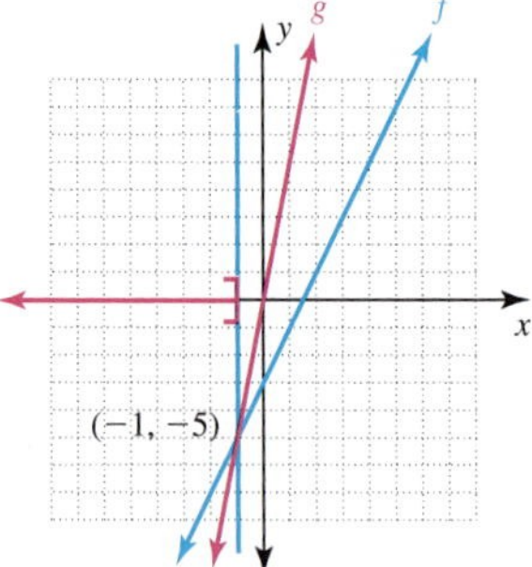

Finally, we express the solutions in set notation. We see that the solution set is every x-value less than or equal to -1, so we write

$\{x \mid x \leq -1\}$

The algorithm summarizes our work in this section.

Step by Step

Solving an Inequality in One Variable Graphically

Step 1 Rewrite the inequality as a comparison of two functions.

$f(x) < g(x)$ $\quad f(x) > g(x)$ $\quad f(x) \leq g(x)$ $\quad f(x) \geq g(x)$

Step 2 Graph the two functions on a single set of axes.

Step 3 Draw a vertical line through the point of intersection of the two graphs. Use a dashed line if equality is not included ($<$ or $>$). Use a solid line if equality is included ($\le$ or $\ge$).

Step 4 Mark the x-values that make the inequality a true statement.

Step 5 Write the solutions in set-builder notation.

It is possible for a linear inequality to have no solutions, or to be true for all real numbers. Using graphical methods makes these situations clear.

Example 3 Solving a Linear Inequality Graphically

Solve each inequality graphically.

(a) $5 + \frac{1}{2}x > \frac{x + 4}{2}$

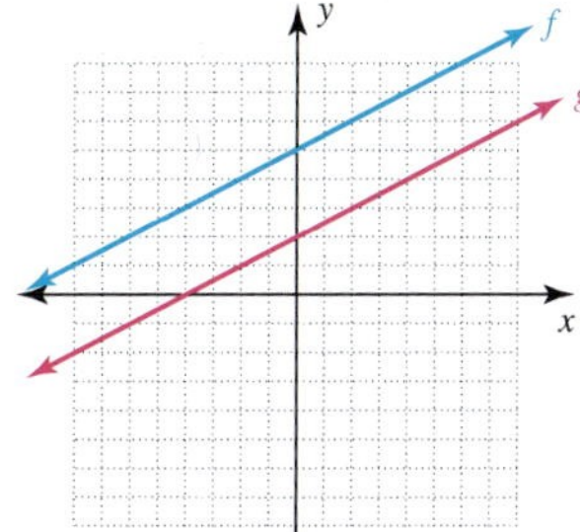

Let

$$f(x) = 5 + \frac{1}{2}x = \frac{1}{2}x + 5$$

and $$g(x) = \frac{x + 4}{2} = \frac{1}{2}x + 2$$

The graphs of f and g have the same slope of $\frac{1}{2}$ and therefore are parallel.

We ask, for what values of x is the graph of f above the graph of g? Clearly, f is *always* above g. So the original statement is true for *all* real numbers, and the solution set is $\mathbb{R}$.

(b) $\frac{9 - x}{3} \le \frac{-x - 6}{3}$

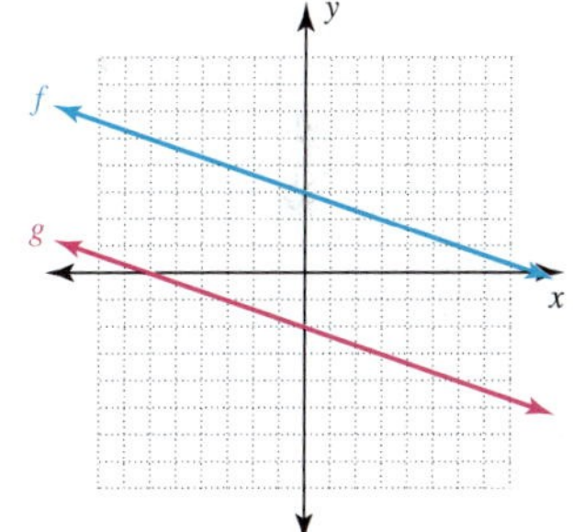

Let $$f(x) = \frac{9 - x}{3} = 3 - \frac{x}{3} = -\frac{1}{3}x + 3$$

and $$g(x) = \frac{-x - 6}{3} = -\frac{1}{3}x - 2$$

Again, the graphs have the same slope.

Now we ask, for what values of x is the graph of f below (or equal to) the graph of g? The answer here is "Never!" The original statement is never true, and the solution set is empty, or $\varnothing$.

Solving a business application graphically can be handled effectively with a graphing calculator. This is illustrated in Example 4.

Example 4 A Business and Finance Application

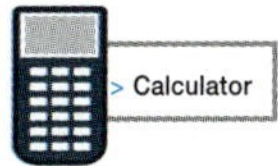

It costs a company \$14.29 to manufacture each pair sandals in their new line. They estimate their weekly fixed costs to be \$1,735. How many pairs of sandals do they need to manufacture and sell each week in order to make a profit if they receive \$25.98 per pair?

The company's cost function is

$C(x) = 14.29x + 1{,}735$

and their revenue function is

$R(x) = 25.98x$

where x is the number of pairs of sandals made and sold per week.

We want to know when revenue exceeds cost, so we need to solve the inequality

$$R(x) > C(x)$$
$$25.98x > 14.29x + 1{,}735$$

We define the functions in a calculator.

$$Y_1 = 25.98x$$
$$Y_2 = 14.29x + 1{,}735$$

We use the TABLE utility to approximate the intersection point.

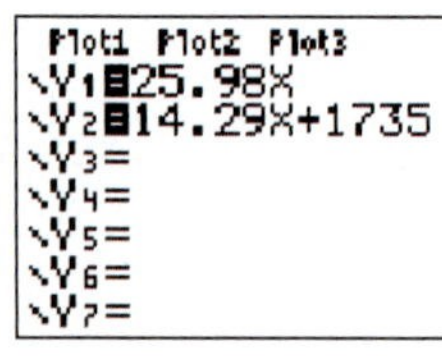

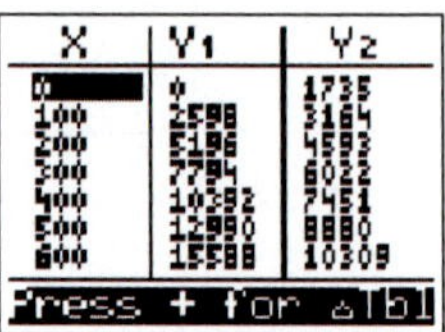

We see from the table that when $x = 100$, Y_1 (revenue) is lower than Y_2 (cost). But when $x = 200$, Y_1 is higher than Y_2. So the graphs must intersect between 100 and 200 on the x-axis, and a suitable viewing window could be $100 \le x \le 200$ and $2{,}600 \le y \le 5{,}200$. Using this, we see

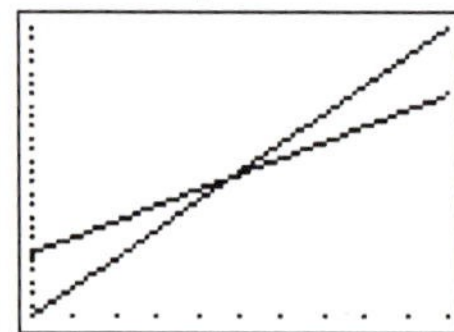

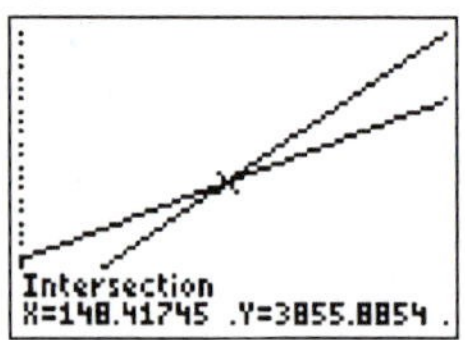

and we find the intersection at approximately (148.41745, 3,855.8854). Since x needs to be a whole number, we conclude that the company makes a profit when x is at least 149, or when $x \ge 149$.

They need to sell at least 149 pairs of sandals each week to make a profit.

A.1 exercises

< Objective 1 >

Graphically solve each inequality.

1. $2x < 8$

2. $-x < 4$

3. $\dfrac{x + 3}{2} < -1$

4. $\dfrac{-3x + 3}{4} > -3$

5. $7x - 7 < -2x + 2$

6. $7x + 2 > x - 4$

7. $6(1 + x) \ge 2(3x - 5)$

8. $2(x - 5) \ge 2x - 1$

9. $7x > \frac{9x - 5}{2}$

10. $-4x - 12 < x + 8$

Answers

1. 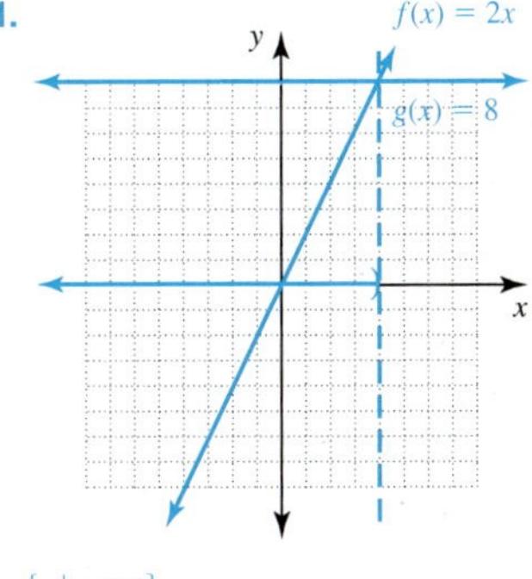

$\{x \mid x < 4\}$

3. 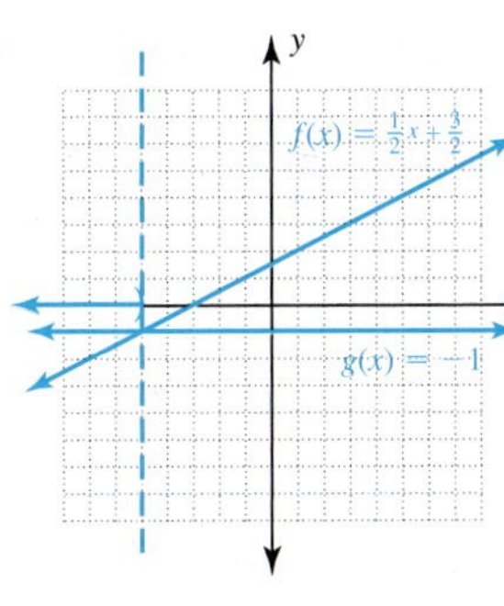

$\{x \mid x < -5\}$

5. 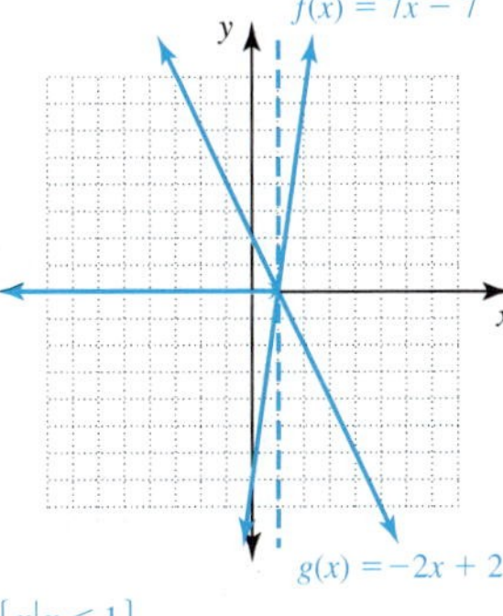

$\{x \mid x < 1\}$

7. 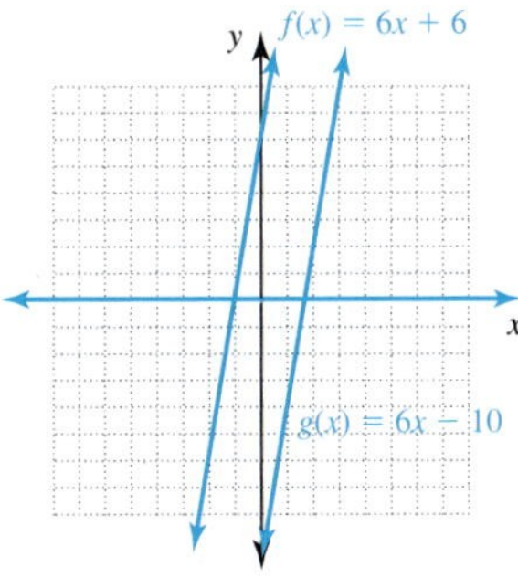

$\{x \mid x \in \mathbb{R}\}$

9. 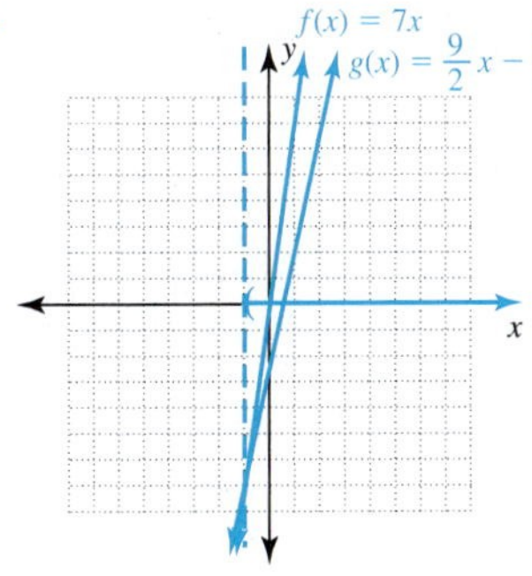

$\{x \mid x > -1\}$

A.2 Solving Absolute-Value Equations

< A.2 Objectives >

1 > Find the absolute value of an expression

2 > Solve an absolute-value equation

In this appendix, we look to solve equations containing absolute values. First, we review the concept of absolute value.

The **absolute value** of a number is its distance from 0. Because absolute value is a distance, it is always positive.

Definition

Absolute Value

The **absolute value** of a number x is given by

$$|x| = \begin{cases} -x & \text{if } x < 0 \\ x & \text{if } x \geq 0 \end{cases}$$

Example 1 Finding the Absolute Value of a Number

< Objective 1 >

Find the absolute value for each expression.

(a) $|-3|$ **(b)** $|7 - 2|$ **(c)** $|-7 - 2|$

(a) Because $-3 < 0$, $|-3| = -(-3) = 3$.

(b) $|7 - 2| = |5|$ Because $5 \geq 0$, $|5| = 5$.

(c) $|-7 - 2| = |-9|$ Because $-9 < 0$, $|-9| = -(-9) = 9$.

> CAUTION

The constant p in the property box below must be positive because an equation such as $|x| = -3$ has no solutions. The absolute value of number is always equal to a nonnegative number.

Given an equation such as

$$|x| = 5$$

there are two possible solutions. The value of x could be 5 or -5. In either case, the absolute value is 5. This can be generalized with a property of absolute-value equations.

Property

Absolute-Value Equations

For any positive number p, if

$$|x| = p$$

then

$$x = p \quad \text{or} \quad x = -p$$

We use this property in Examples 2–4.

Example 2 Solving an Absolute-Value Equation

< Objective 2 >

> CAUTION

A common mistake is to solve only the equation $x - 3 = 4$. We solve *both* equations and find **two** solutions.

Solve the equation

$|x - 3| = 4$

The property tells us that the expression inside the absolute-value bars, $x - 3$, must equal either 4 or -4. We set up two equations and solve them both.

$$(x - 3) = 4 \quad \text{or} \quad (x - 3) = -4$$
$$x - 3 = 4 \qquad x - 3 = -4$$
$$x = 7 \qquad x = -1 \qquad \text{Add 3 to both sides of the equation.}$$

We arrive at the solution set, $\{-1, 7\}$.

Example 3 Solving an Absolute-Value Equation

Solve for x.

$|3x - 2| = 4$

We know that $|3x - 2| = 4$ is equivalent to the equations

$$3x - 2 = 4 \quad \text{or} \quad 3x - 2 = -4$$
$$3x = 6 \qquad 3x = -2 \qquad \text{Add 2.}$$
$$x = 2 \qquad x = -\frac{2}{3} \qquad \text{Divide by 3.}$$

We can check the solutions by replacing x with $-\frac{2}{3}$ and 2 in the original equation.

Check $x = -\frac{2}{3}$.

$$|3x - 2| = 4 \qquad \text{The original equation}$$
$$\left|3\left(-\frac{2}{3}\right) - 2\right| \stackrel{?}{=} 4 \qquad \text{Substitute } x = -\frac{2}{3}.$$
$$|-2 - 2| \stackrel{?}{=} 4 \qquad \cancel{3}\left(-\frac{2}{\cancel{3}}\right) = -2$$
$$|-4| \stackrel{?}{=} 4 \qquad -2 - 2 = -4$$
$$4 = 4 \qquad \text{True } (|-4| = 4)$$

Check $x = 2$.

$$|3x - 2| = 4 \qquad \text{The original equation}$$
$$|3(2) - 2| \stackrel{?}{=} 4 \qquad \text{Substitute } x = 2.$$
$$|6 - 2| \stackrel{?}{=} 4$$
$$|4| \stackrel{?}{=} 4$$
$$4 = 4 \qquad \text{True}$$

Both solutions check, the solution set is $\left\{-\frac{2}{3}, 2\right\}$.

We may need to rewrite an equation so that the absolute-value expression is isolated on one side. Then we can apply the property to solve the equation.

Example 4 Solving an Absolute-Value Equation

Solve for x.

$|2 - 3x| + 5 = 10$

We begin by subtracting 5 from both sides to isolate the absolute-value expression on one side of the equation.

$|2 - 3x| = 5$

We can now proceed as before.

$$2 - 3x = 5 \quad \text{or} \quad 2 - 3x = -5$$
$$-3x = 3 \qquad\qquad -3x = -7 \qquad \text{Subtract 2.}$$
$$x = -1 \qquad\qquad x = \frac{7}{3} \qquad \text{Divide by } -3.$$

The solution set is $\left\{-1, \frac{7}{3}\right\}$.

In some applications, there is more than one absolute value in an equation. Consider an equation of the form

$$|x| = |y|$$

Since the absolute values of x and y are equal, x and y are the same distance from 0, which means they are either *equal* or *opposite in sign*. This leads to a second general property of absolute-value equations.

Property

Absolute-Value Equations

If $|x| = |y|$

then $x = y$ or $x = -y$

We apply this second property in Example 5.

Example 5 Solving Equations with Two Absolute-Value Expressions

Solve for x.

$$|3x - 4| = |x + 2|$$

We write

$$3x - 4 = x + 2 \quad \text{or} \quad 3x - 4 = -(x + 2)$$
$$3x - 4 = -x - 2$$
$$3x = x + 6 \qquad 3x = -x + 2 \qquad \text{Add 4 to both sides.}$$
$$2x = 6 \qquad 4x = 2 \qquad \text{Isolate the } x\text{-term.}$$
$$x = 3 \qquad x = \frac{1}{2} \qquad \text{Divide by the coefficient of the variable.}$$

The solution set is $\left\{\frac{1}{2}, 3\right\}$.

A.2 exercises

< Objective 1 >

Find the absolute value of each expression.

1. $|15|$
2. $|-18|$
3. $|8 - 3|$
4. $|-23 - 11|$
5. $|-12 - 19|$
6. $|-13| - |-12|$
7. $|-13| + |12|$
8. $|-13 - 12|$
9. $-|-13| - |-12|$
10. $-|(-13) - (-12)|$

< Objective 2 >

Solve each equation.

11. $|x| = 5$

12. $|x - 2| = 6$

13. $|2x - 1| = 6$

14. $2\,|3x - 5| = 12$

15. $|5x - 2| = -3$

16. $|x + 5| - 2 = 5$

17. $8 - |x - 4| = 5$

18. $|3x + 1| = |2x - 3|$

19. $|5x - 2| = |2x - 4|$

20. $\left|2 + \frac{5}{8}x\right| = \frac{3}{8}$

Answers

1. 15 **3.** 5 **5.** 31 **7.** 25 **9.** -25 **11.** $\{5, -5\}$ **13.** $\left\{-\frac{5}{2}, \frac{7}{2}\right\}$ **15.** No solutions **17.** $\{7, 1\}$ **19.** $\left\{-\frac{2}{3}, \frac{6}{7}\right\}$

A.3 Solving Absolute-Value Equations Graphically

< A.3 Objectives >

1 > Graph an absolute-value function

2 > Solve absolute-value equations in one variable graphically

Now that we know how to solve some absolute-value equations algebraically, we look at solving them graphically. We begin with the graph of the elementary absolute-value function

$f(x) = |x|$

The graph of most any absolute-value function is a transformation of this graph.

You can work with absolute values on a graphing calculator. On our calculator, we find the absolute-value function in the **NUM** submenu of the MATH menu.

MATH ▶ **1:abs(**

If you have an older model calculator, you need to close the parentheses when working with absolute values. On newer models, we use the right-arrow key ▶ to exit the absolute value bars.

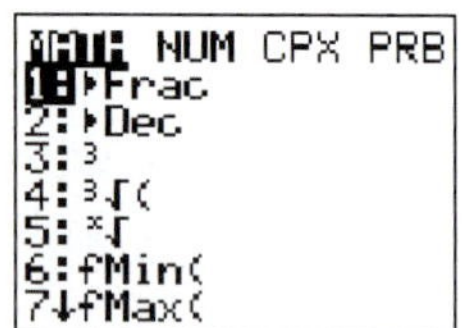

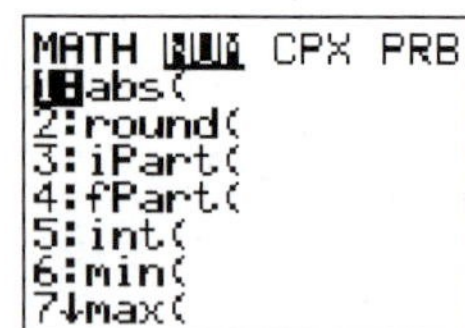

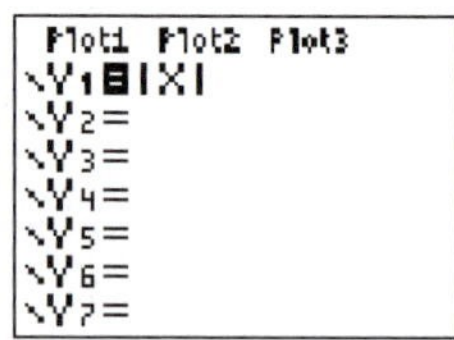

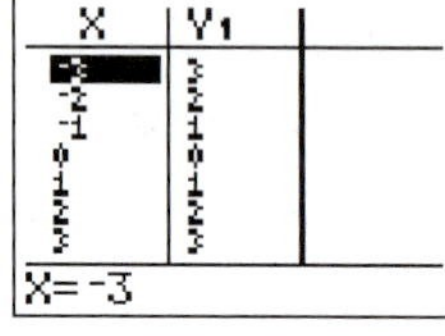

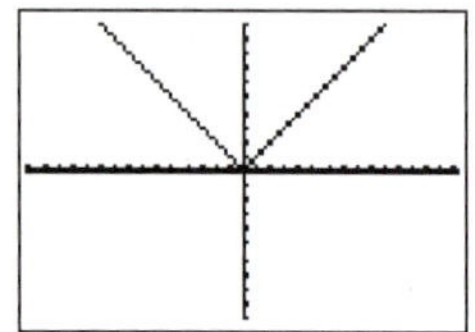

In Example 1, we look at what happens to the graph when we add or subtract some constant inside the absolute-value bars.

Example 1 Graphing an Absolute-Value Function

< Objective 1 >

Graph each function.

(a) $f(x) = |x - 3|$

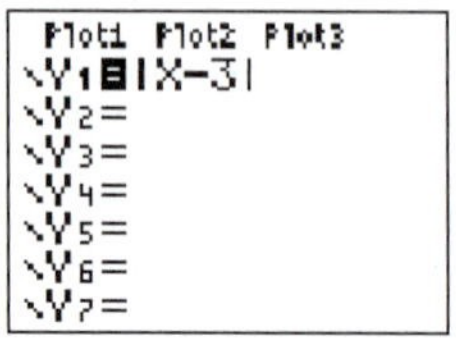

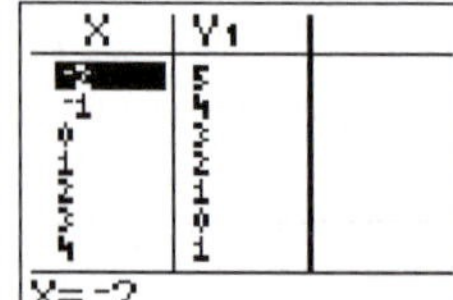

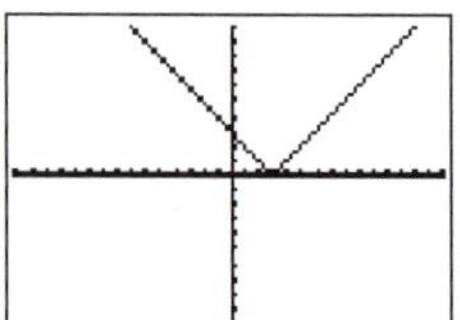

The graph of the function $f(x) = |x - 3|$ is the same shape as the graph of the function $f(x) = |x|$; it has just shifted to the right 3 units.

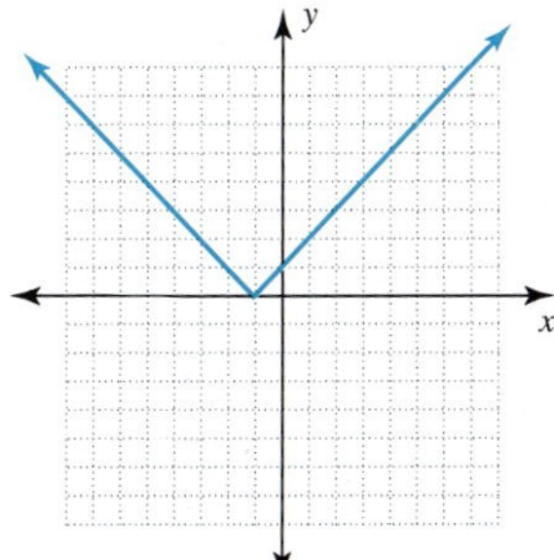

(b) $f(x) = |x + 1|$

We begin with a table of values.

x	$f(x)$
−2	1
−1	0
0	1
1	2
2	3
3	4

The graph of $f(x) = |x + 1|$ is the same shape as the graph of the function $f(x) = |x|$, except that it has shifted 1 unit to the left.

We summarize what we have discovered about the horizontal shift of the graph of an absolute-value function.

Property

Horizontal Shifts of Absolute-Value Functions

The graph of the function $f(x) = |x - a|$ is the same shape as the graph of $f(x) = |x|$ except it is shifted a units

to the right if a is positive.

to the left if a is negative.

If a is negative, $x - a$ is x plus some positive number.

We use these methods to solve absolute-value equations.

Example 2 Solving an Absolute-Value Equation Graphically

< Objective 2 >

Solve graphically

$|x - 3| = 4$

We graph the functions associated with each side of the equation.

$f(x) = |x - 3|$ and $g(x) = 4$

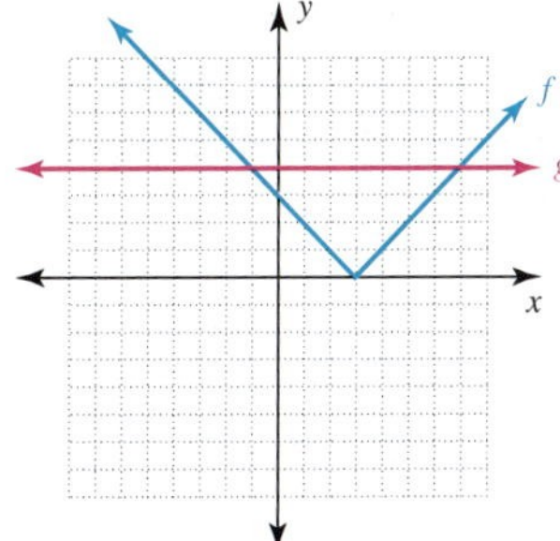

Then we draw a vertical line through each of the intersection points.

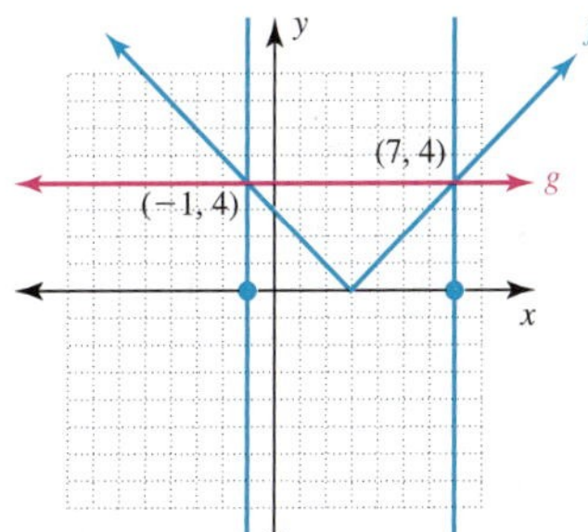

We ask the question: For what values of x do f and g coincide?

Looking at the x-values of the two vertical lines, we find the solutions to the original equation. There are two x-values that make the statement true: -1 and 7. The solution set is $\{-1, 7\}$.

Example 3 illustrates a case with an absolute-value expression and a linear (not simply constant) expression.

Example 3 Solving an Absolute-Value Equation Graphically

Solve the equation graphically.

$$|x - 2| = 4 - \frac{1}{2}x$$

Let $f(x) = |x - 2|$

and $g(x) = 4 - \frac{1}{2}x = -\frac{1}{2}x + 4$

We graph each function.

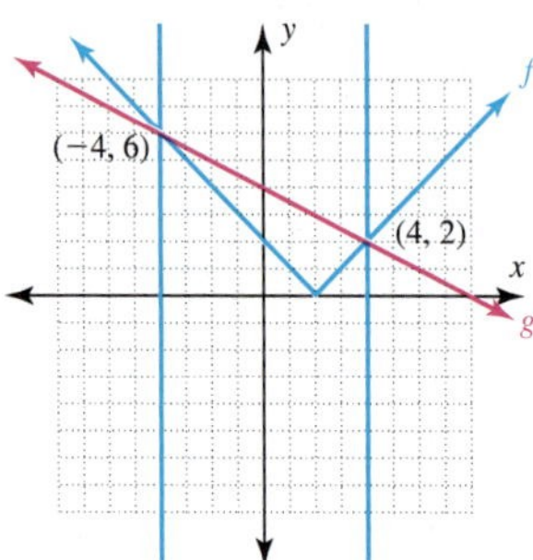

Draw a vertical line through each point of intersection. The vertical lines hit the x-axis when $x = -4$ and when $x = 4$. There are therefore two solutions, and the solution set is $\{-4, 4\}$.

In each example, we found two solutions. It is quite possible for an absolute-value equation to have one solution, no solution, or even an infinite number of solutions. Fortunately, the graph makes the situation clear.

Example 4 Solving an Absolute-Value Equation Graphically

Solve the equation graphically.

$$|x + 2| = 2x + 7$$

Let $f(x) = |x + 2|$

and $g(x) = 2x + 7$

Graph the functions.

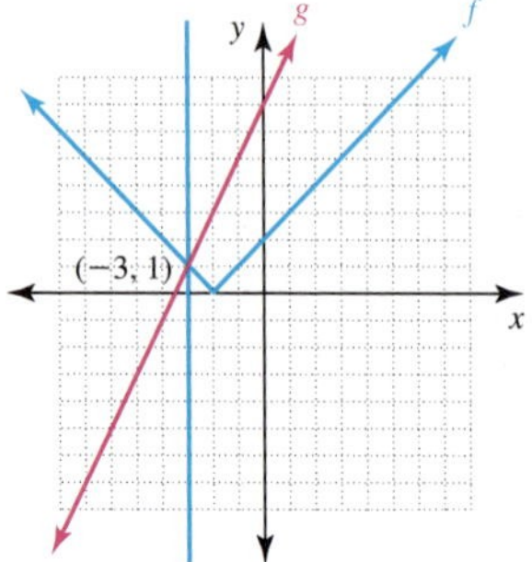

NOTES

When $x < -2$, the slope of f is -1.

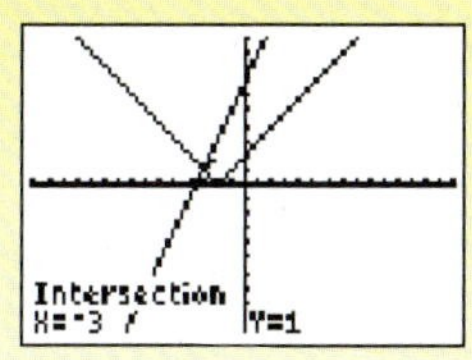

The slope of g is 2, whereas the slope of f is 1 when $x \geq -2$. Since g is increasing faster than f, we know that the graphs only intersect once. The two functions intersect at $(-3, 1)$. The x-coordinate gives the solution set $\{-3\}$.

As we learned in Chapter 4, it may be difficult to locate the intersection point of two graphs. Graphing calculators can help us approximate an intersection point, even with "messy" equations. You may wish to review the Graphing Calculator Option feature in Section 4.1.

A.3 exercises

< Objective 1 >

Graph each function.

1. $f(x) = |x - 3|$

2. $f(x) = |x + 2|$ VIDEO

3. $f(x) = |x + 3|$

4. $f(x) = |x - (-5)|$

< Objective 2 >

Solve equation graphically.

5. $|x| = 3$

6. $|x| = 5$

7. $|x - 2| = 5$ VIDEO

8. $|x + 4| = 2$

9. $|x - 3| = 5 - \frac{1}{3}x$ VIDEO

10. $|x + 1| = \frac{1}{3}x + 5$

Determine the function represented by each graph.

11.

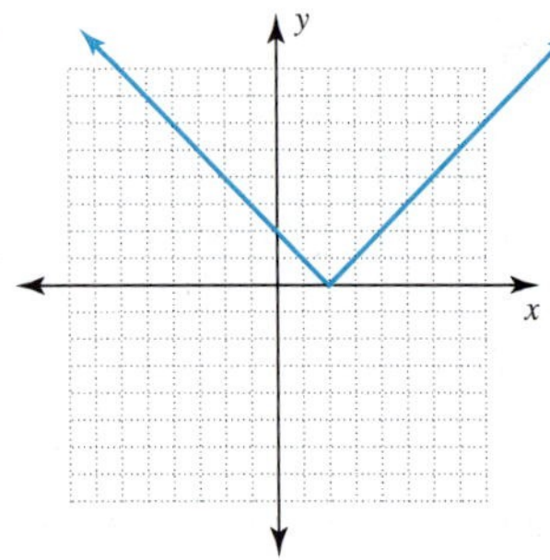

12.

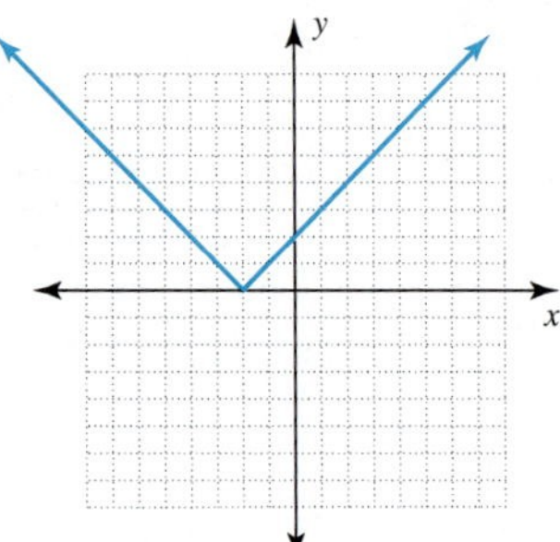

Answers

1.

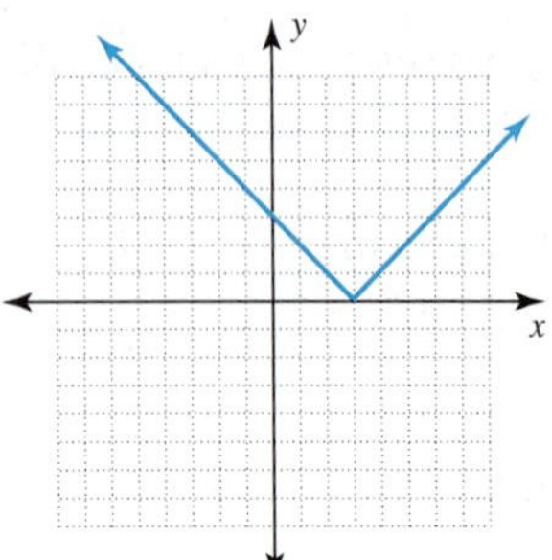

3.

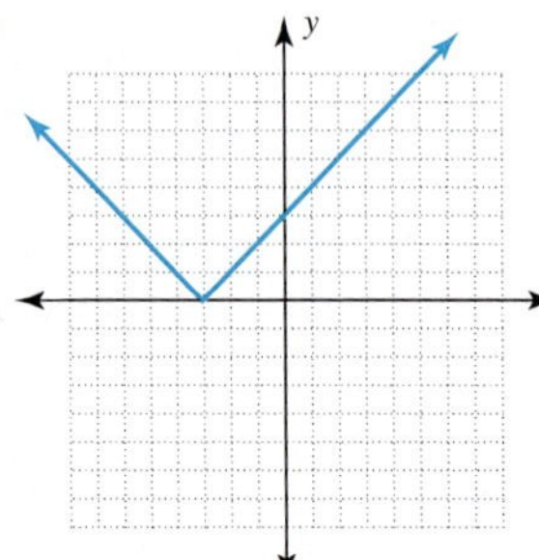

5.

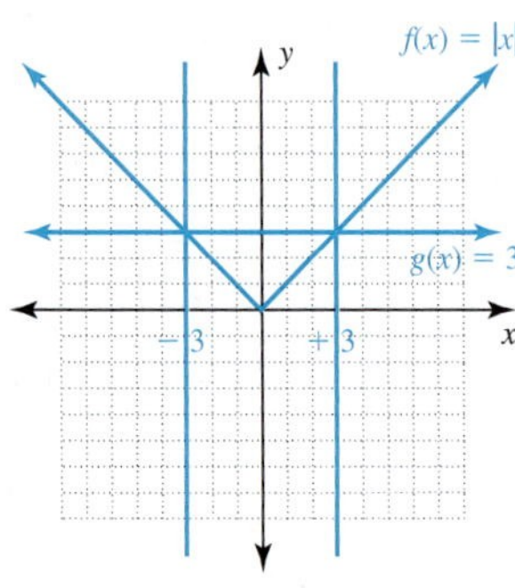

{−3, 3}

7. $f(x) = |x - 2|$

$g(x) = 5$

−3

7

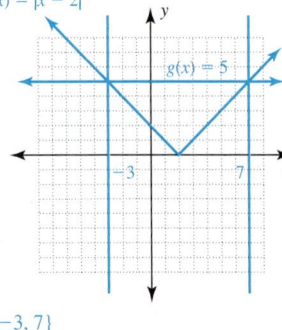

{−3, 7}

9.

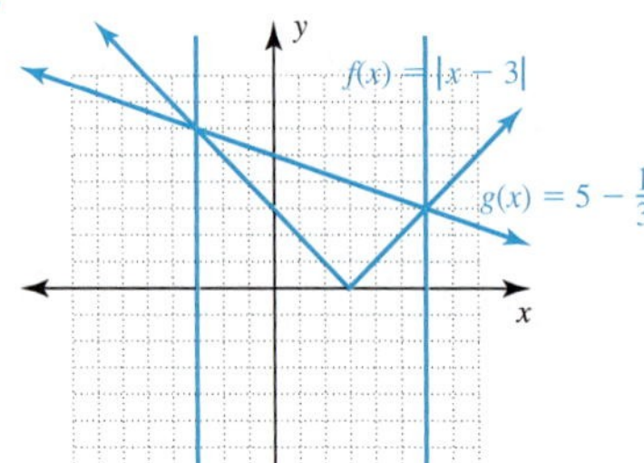

{−3, 6}

11. $f(x) = |x - 2|$

A.4 Solving Absolute-Value Inequalities

< A.4 Objectives >

1 > Solve a compound inequality

2 > Solve an absolute-value inequality

In many applications, it is not enough to simply constrain an input on one side. For instance, a statement such as $x > 4$ may provide valuable information, but there is often an upper bound as well. For instance we might actually have

$x > 4$ and $x < 20$

For convenience, we usually write such pairings as a single **compound inequality**.

$4 < x < 20$

Doing this ensures that we retain site of the pair of constraints simultaneously.

Our goal, when beginning with a compound inequality such as

$-3 \le 2x + 1 \le 7$

is to find an equivalent compound inequality in which the variable is isolated in the middle. We illustrate in Example 1.

Example 1 Solving a Compound Inequality

< Objective 1 >

Solve and graph the compound inequality.

$-3 \le 2x + 1 \le 7$

First, we subtract 1 from each of the three expressions the compound inequality.

$-3 - 1 \le 2x + 1 - 1 \le 7 - 1$

or

$-4 \le 2x \le 6$

We now divide by 2 to isolate the variable x.

$$-\frac{4}{2} \le \frac{2x}{2} \le \frac{6}{2}$$

$-2 \le x \le 3$

The solution set consists of all numbers between -2 and 3, including -2 and 3, and is written

$\{x \mid -2 \le x \le 3\}$

Its graph is

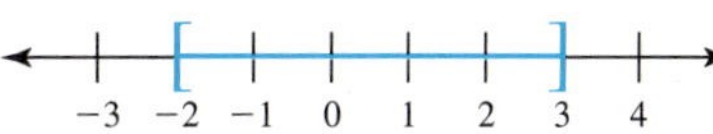

The solution set is equivalent to

$\{x \mid x \ge -2 \text{ and } x \le 3\}$

Consider the individual graphs.

$\{x \mid x \ge -2\}$

NOTES

We are really applying the addition property to each of the *two* inequalities that make up the compound inequality.

When we divide by a positive number, the direction of the inequality is preserved.

In interval notation, we write $[-2, 3]$.

> **NOTE**
>
> Using set-builder notation, we write
>
> $\{x|x \geq -2\} \cap \{x|x \leq 3\}$

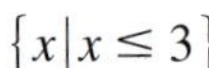

$\{x|x \leq 3\}$

$\{x|x \geq -2 \text{ and } x \leq 3\}$

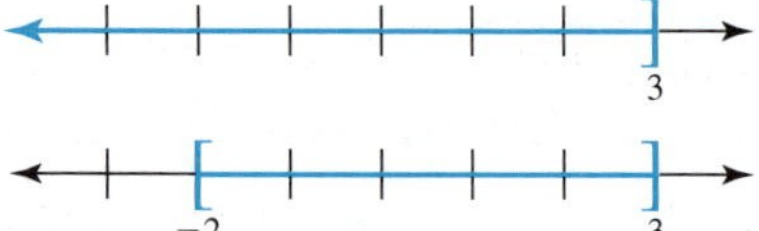

Because the connecting word is *and,* we want the *intersection* of the sets, that is, those numbers common to both sets.

A compound inequality may also consist of two inequality statements connected by the word *or*. Example 2 illustrates how to solve that type of compound inequality.

Example 2 Simplifying a Compound Inequality

Solve and graph the inequality

$2x - 3 < -5 \quad \text{or} \quad 2x - 3 > 5$

In this case, we must work with each of the inequalities *separately*.

$2x - 3 < -5 \quad \text{or} \quad 2x - 3 > 5$ Add 3.

$2x < -2 \qquad 2x > 8$ Divide by 2.

$x < -1 \qquad x > 4$

> **NOTES**
>
> In interval notation, we write $(-\infty, -1) \cup (4, \infty)$.
>
> In set-builder notation, we can write
>
> $\{x|x < -1\} \cup \{x|x > 4\}$

The graph of the solution set, $\{x|x < -1 \text{ or } x > 4\}$, is shown.

$\{x|x < -1 \text{ or } x > 4\}$

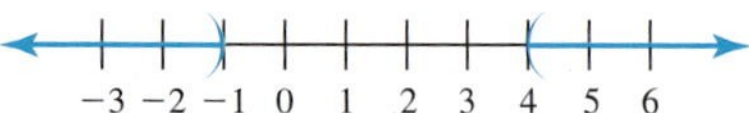

The connecting word is *or* in this case, so the solution set of the original inequality is the *union* of the two sets. That is, all of the numbers that belong to either or both of the sets.

There are two families of absolute-value inequalities. The first family consists of inequalities of the form

$|x| < p \quad \text{and} \quad |x| \leq p$

In the second family are the absolute-value inequalities

$|x| > p \quad \text{and} \quad |x| \geq p$

These families correlate to the two types of compound inequalities you just worked with. The first family corresponds to

$-p < x < p \quad (\text{as well as } -p \leq x \leq p)$

Whereas, the second family gives

$x < -p \quad \text{or} \quad x > p \quad (\text{as well as } x \leq -p \text{ or } x \geq p)$

We use the next rule to solve absolute-value inequalities from the first family.

Property

Absolute-Value Inequalities

For any positive number p, if

$|x| < p$

then

$-p < x < p$

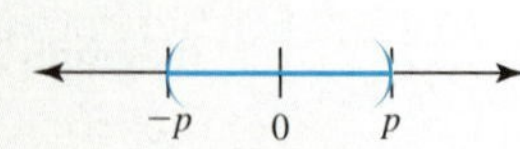

and, if

$|x| \leq p$

then

$-p \leq x \leq p$

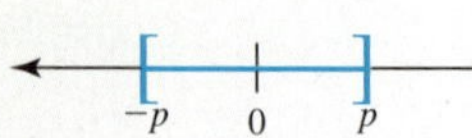

Example 3 Solving an Absolute-Value Inequality

< Objective 2 >

Solve the inequality; then graph the solution set.

$|x - 5| < 9$

According to the property, we have

$$|x - 5| < 9$$
$$-9 < x - 5 < 9 \quad \text{Add 5 to each expression.}$$
$$-9 + 5 < x < 9 + 5$$
$$-4 < x < 14$$

NOTE

In interval notation, we write (−4, 14).

Graphing the solution set, $\{x \mid -4 < x < 14\}$, we get

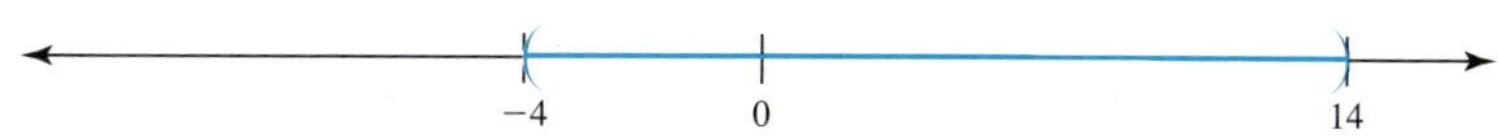

What about the second family of absolute-value inequalities?

Property

Absolute-Value Inequalities

For any positive number p, if

$|x| > p$

then

$x < -p$ or $x > p$

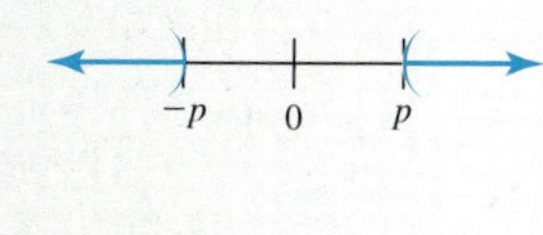

and, if

$|x| \geq p$

then

$x \leq -p$ or $x \geq p$

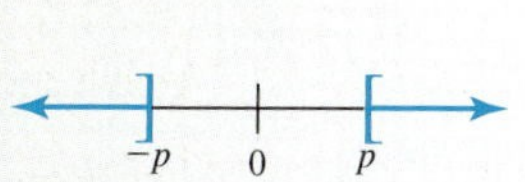

Example 4 Solving an Absolute-Value Inequality

Solve the inequality; then graph the solution set.

$|x - 7| > 19$

By the second property,

$x - 7 > 19$ or $x - 7 < -19$

Solving each inequality by adding 7 to each side, we get

$x > 26$ or $x < -12$

Now we can graph the solution set.

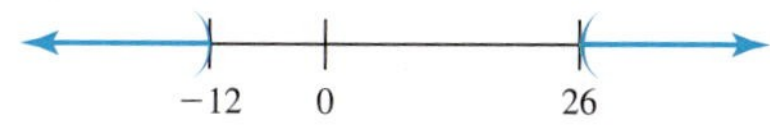

A.4 exercises

< Objective 1 >

Solve and graph each inequality.

1. $4 \le x + 1 \le 7$

2. $-6 \le 3x \le 9$

3. $1 \le 2x - 3 \le 6$ 

4. $-7 \le 3 + 2x \le 8$

5. $x - 1 < -3$ or $x - 1 > 3$

6. $2x + 5 < -3$ or $2x + 5 > 3$

< Objective 2 >

7. $|x| < 5$

8. $|x| \le 4$ VIDEO

9. $|x + 6| \le 4$

10. $|5 - x| < 3$

11. $|3x + 4| \ge 5$

12. $|2x + 3| \le 9$

Answers

1. $\{x \mid 3 \le x \le 6\}$

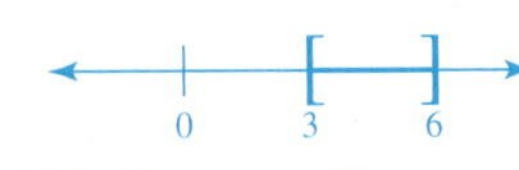

3. $\left\{x \mid 2 \le x \le \frac{9}{2}\right\}$

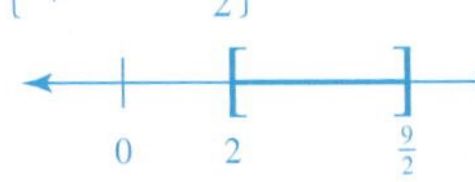

5. $\{x \mid x < -2 \text{ or } x > 4\}$

7. $\{x \mid -5 < x < 5\}$

9. $\{x \mid -10 \le x \le -2\}$

11. $\left\{x \mid x \le -3 \text{ or } x \ge \frac{1}{3}\right\}$

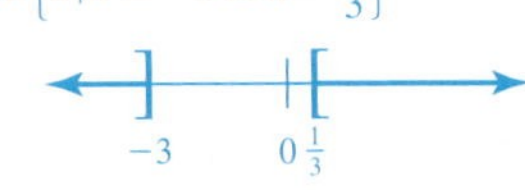

A.5 Solving Absolute-Value Inequalities Graphically

< A.5 Objectives >

1 > Solve absolute-value inequalities graphically

2 > Solve absolute-value inequalities algebraically

We already know how to solve absolute-value equations graphically, so now we look to use graphs to solve absolute-value inequalities.

Absolute-value inequalities, in one variable, are similar to equations, except we have $>$, $<$, $\leq$, or $\geq$ instead of $=$.

The general form of such an inequality is

$|ax - b| < p$ (any of the symbols $>$, $\leq$, or $\geq$ can be in the place of $<$)

Examples of absolute-value inequalities include

$|x| < 6 \qquad |x - 4| \geq 2 \qquad |3x - 5| \leq 8$

Example 1 Solving an Absolute-Value Inequality Graphically

< Objective 1 >

Graphically solve

$|x| < 6$

As we did in previous sections, we begin by letting each side of the inequality represent a function. Here

$f(x) = |x| \qquad \text{and} \qquad g(x) = 6$

Now we graph both functions on the same set of axes.

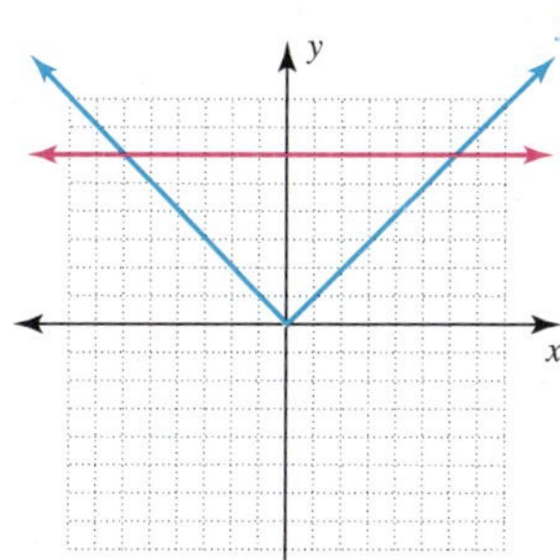

We now ask the question: For what values of x is f *below* g?

We draw a vertical dashed line (equality is not included) through the points of intersection of the two graphs.

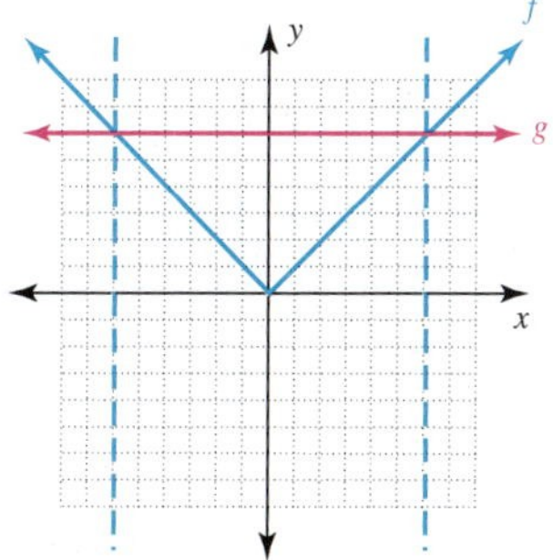

The solution set is any value of x for which the graph of $f(x)$ is below the graph of $g(x)$.

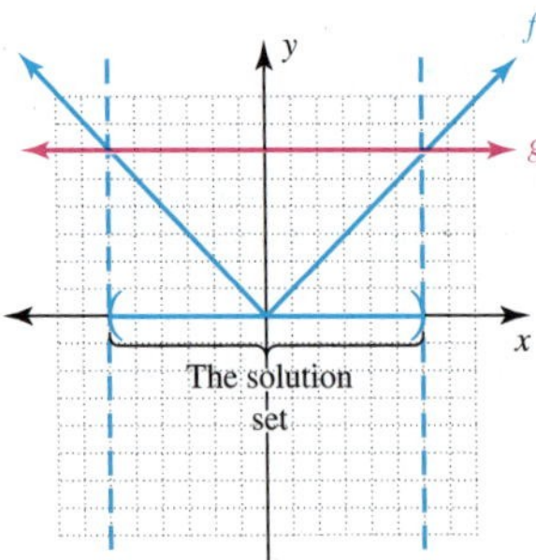

Keep in mind that we are searching for x-values that make the original statement true.

In set-builder notation, we write $\{x \mid -6 < x < 6\}$.

The graphical method of Example 1 relates to the general statement from Appendix A.4.

Property

Absolute-Value Inequalities

For any positive number p, if

$|x| < p$

then $\quad -p < x < p$

Before we continue with a graphical approach, we review this property in solving an absolute-value inequality.

Example 2 Solving an Absolute-Value Inequality Algebraically

< Objective 2 >

Solve the inequality algebraically and graph the solution set on a number line.

$|x - 3| < 5$

From this property, we know that the given absolute-value inequality is equivalent to the compound inequality

$-5 < x - 3 < 5$

Solve as before.

$-5 < x - 3 < 5$ Add 3 to all three parts.

$-2 < x < 8$

The solution set is

$\{x \mid -2 < x < 8\}$

The graph of the solution set is

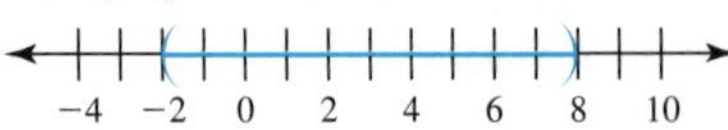

NOTE

With this property we can *translate* an absolute-value inequality to a compound inequality *not* containing an absolute value, which can be solved by our earlier methods.

NOTE

The solution set is an open interval on the number line.

In Example 3, we look at a graphical method for solving the same inequality.

Example 3 Solving an Absolute-Value Inequality Graphically

Solve the inequality graphically, and graph the solution set on a number line.

$|x - 3| < 5$

Let $f(x) = |x - 3|$ and $g(x) = 5$, and graph both functions on the same set of axes.

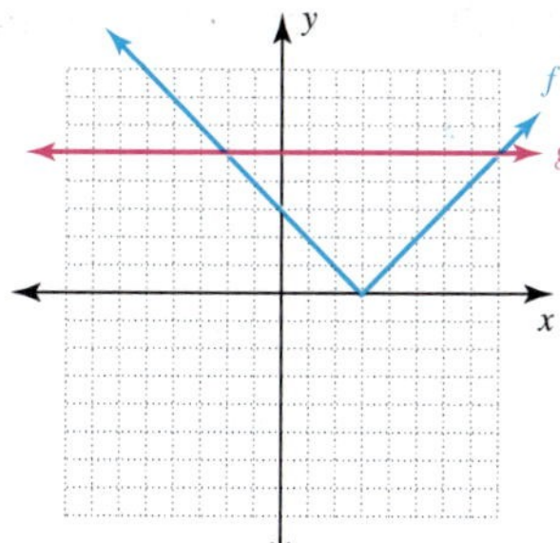

Here again, we ask: For what values of x is f below g?

Drawing a vertical dashed line through the intersection points, we find the set of x-values for which $f(x) < g(x)$.

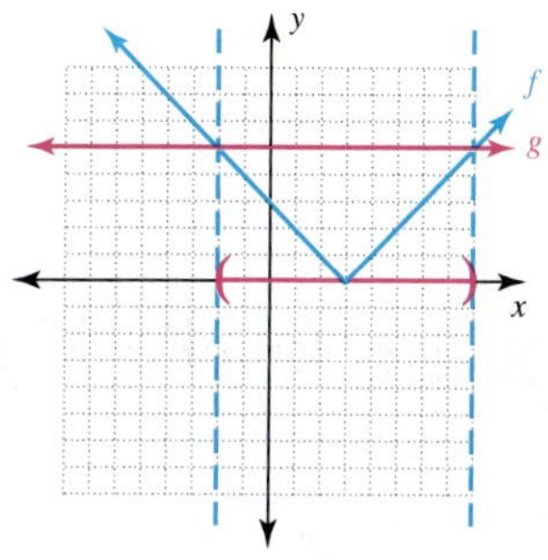

We see that the desired x-values are those that lie between -2 and 8. The solution set is $\{x \mid -2 < x < 8\}$. The graph of the solution set is

The graph of the solution set shown here is precisely the portion of the x-axis that has been marked in the previous two-dimensional graph.

We know the solution set for the statement $|x| < 6$ is the set of all numbers between -6 and 6. Now, how does the result change for the statement $|x| > 6$? Solving graphically makes this clear.

Example 4 Solving an Absolute-Value Inequality Graphically

Solve the inequality graphically, and graph the solution set on a number line.

$|x| > 6$

As before, we define

$f(x) = |x| \qquad \text{and} \qquad g(x) = 6$

We graph both functions on the same set of axes, and we draw vertical dashed lines through the points of intersection.

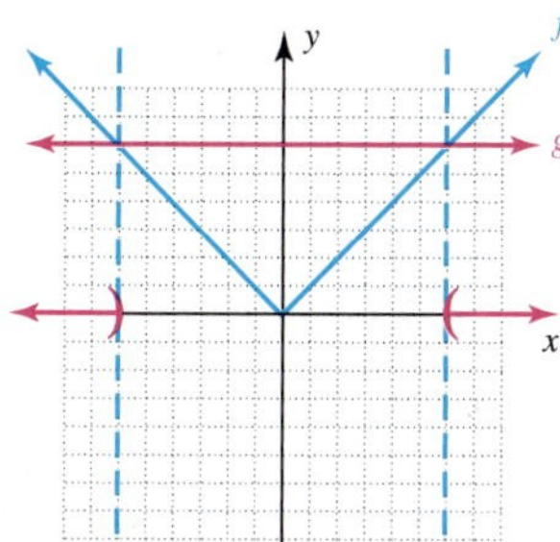

Now we ask: For what values of x is the graph of f *above* the graph of g?

We see that f is *above* g when $x < -6$ or when $x > 6$. The solution set is $\{x \mid x < -6 \text{ or } x > 6\}$. The graph of the solution set is

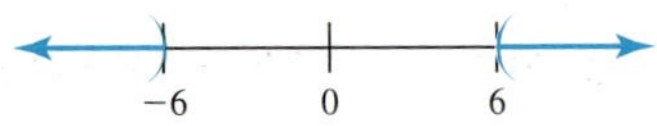

The solution set for the statement $|x| > 6$ is the set of all numbers that are greater than 6 or less than -6. We can describe these numbers with the compound inequality

$x < -6$ or $x > 6$

This relates to the second property in Appendix A.4.

Property

Absolute-Value Inequalities

For any positive number p, if

$|x| > p$

then $x < -p$ or $x > p$

We review the use of this property in solving an absolute-value inequality.

Example 5 Solving an Absolute-Value Inequality Algebraically

NOTE

Again we *translate* the absolute-value inequality to the compound inequality *not* containing an absolute value.

Solve the inequality algebraically, and graph the solution set on a number line.

$|2 - x| > 8$

From our second property, we know that the given absolute-value inequality is equivalent to the compound inequality

$2 - x < -8$ or $2 - x > 8$

Solving as before, we have

$$\begin{aligned} 2 - x &< -8 & \text{or} \quad 2 - x &> 8 \\ -x &< -10 & -x &> 6 \\ x &> 10 & x &< -6 \end{aligned}$$

When we divide by a negative number, we reverse the direction of the inequality.

The solution set is $\{x \mid x < -6 \text{ or } x > 10\}$, and the graph of the solution set is shown here.

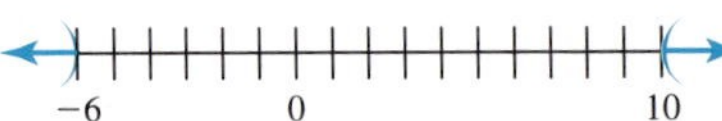

A property that can be useful in working with absolute values is given in the next box.

Property

Absolute-Value Expressions

For any real numbers a and b

$|a - b| = |b - a|$

Our final example looks at a graphical approach to solving the same inequality studied in Example 5.

Example 6 Solving an Absolute-Value Inequality Graphically

Solve the inequality graphically, and graph the solution set on a number line.

$|2 - x| > 8$

Let $f(x) = |2 - x|$ and $g(x) = 8$. Using Property 3, we know that $|2 - x| = |x - 2|$, so we define f as

$f(x) = |x - 2|$

Graphing f and g on the same set of axes, we have

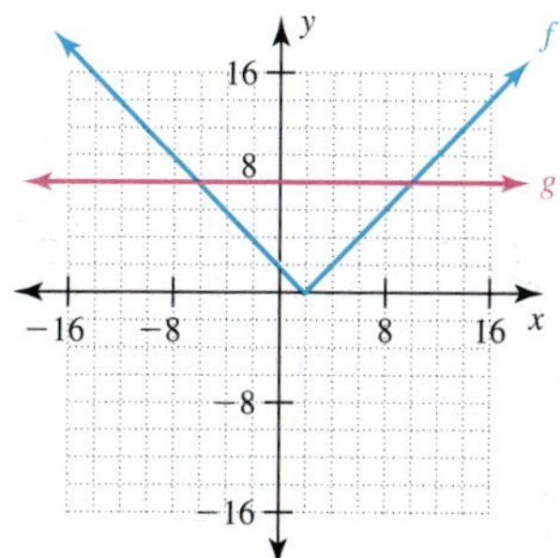

Drawing a vertical dashed line through each intersection point, we mark all x-values for which $f(x) > g(x)$.

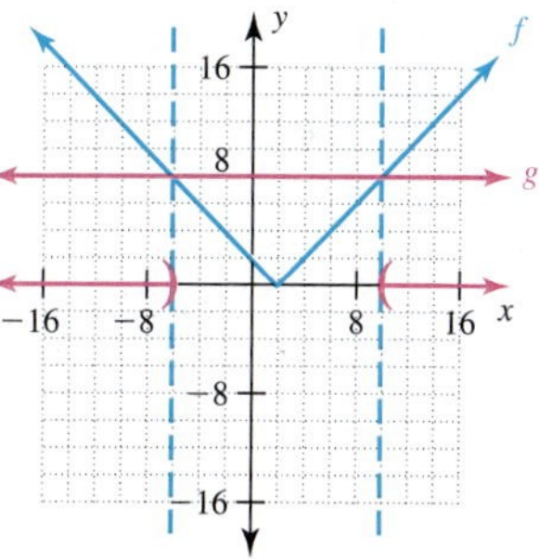

The solution set is therefore $\{x|x < -6 \text{ or } x > 10\}$, and the graph of the solution set is

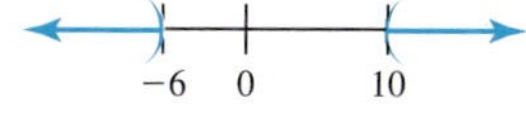

A.5 exercises

< Objective 2 >

Solve each inequality algebraically. Graph the solution set on a number line.

1. $|x| < 5$ 

2. $|x + 5| < 3$

3. $|x + 6| \le 4$ 

4. $|x + 5| \ge 0$

5. $|3x + 4| \ge 5$ VIDEO

6. $|2x + 3| \le 9$

< Objective 1 >

Solve each inequality graphically.

7. $|x| < 4$

8. $|x| \geq 2$

9. $|x - 3| < 4$

10. $|x + 2| > 4$

11. $|x + 1| \leq 5$

12. $|x - 4| > -1$

Answers

1. $\{x \mid -5 < x < 5\}$

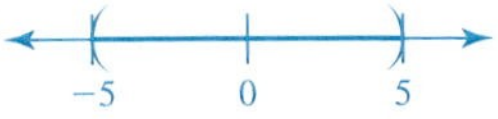

3. $\{x \mid -10 \leq x \leq -2\}$

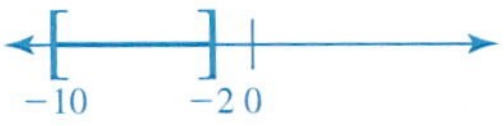

5. $\left\{x \mid x \leq -3 \text{ or } x \geq \frac{1}{3}\right\}$

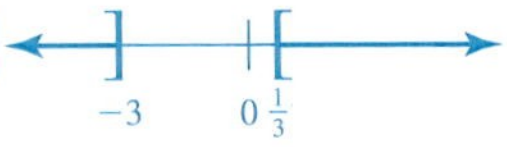

7. $f(x) = |x|$

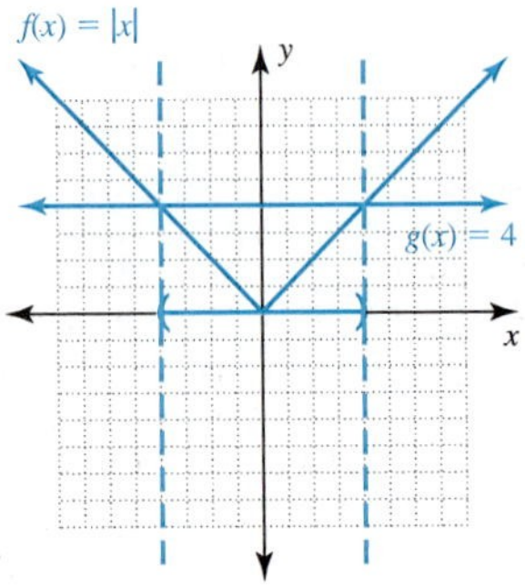

$\{x \mid -4 < x < 4\}$

9. $f(x) = |x - 3|$

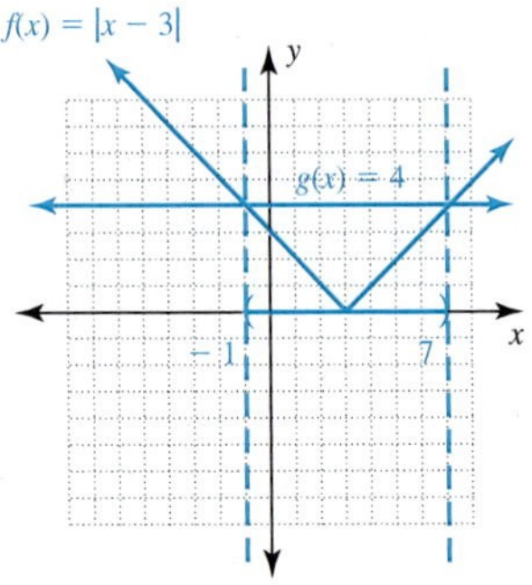

$\{x \mid -1 < x < 7\}$

11. $f(x) = |x + 1|$

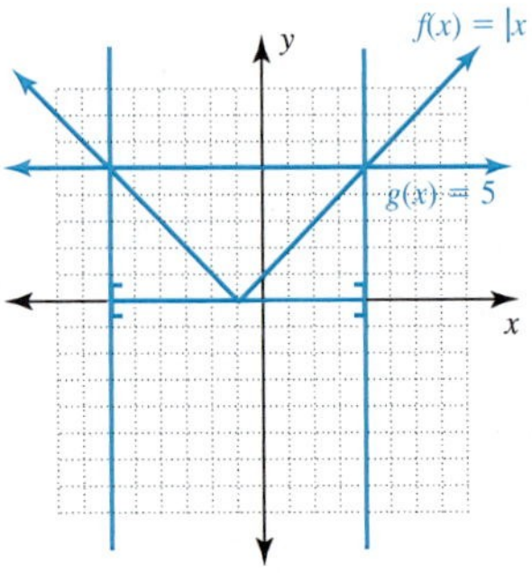

$\{x \mid -6 \leq x \leq 4\}$

answers

Answers to Reading Your Text Exercises, Summary Exercises, Chapter Tests, Cumulative Reviews, and Final Exam

Reading Your Text for Chapter 0

Section 0.1 (**a**) natural; (**b**) one; (**c**) product; (**d**) difference
Section 0.2 (**a**) whole; (**b**) negative; (**c**) integers; (**d**) magnitude
Section 0.3 (**a**) magnitudes; (**b**) order; (**c**) associative; (**d**) zero
Section 0.4 (**a**) Multiplication; (**b**) different; (**c**) positive; (**d**) area
Section 0.5 (**a**) factor; (**b**) grouping; (**c**) exponential; (**d**) exponent

Summary Exercises for Chapter 0

1. $\frac{10}{14}, \frac{20}{28}, \frac{50}{70}$ **3.** $\frac{8}{18}, \frac{16}{36}, \frac{24}{54}$ **5.** $\frac{1}{9}$ **7.** $\frac{2}{3}$ **9.** $\frac{3}{2}$
11. $\frac{7}{18}$ **13.** \$198 **15.** 48 lots **17.** 12 **19.** -3
21. -4 **23.** 16 **25.** $<$ **27.** $<$ **29.** 8 **31.** -11
33. $-\frac{31}{39}$ **35.** 4 **37.** -5 **39.** 5 **41.** $\frac{13}{2}$ **43.** 1
45. 0 **47.** \$467.66 **49.** 36 **51.** -15 **53.** 16
55. -360 **57.** -4 **59.** -24 **61.** -4 **63.** Undefined
65. $\frac{7}{6}$ **67.** -2 **69.** \$42,000 **71.** $3 \cdot 3 \cdot 3$
73. $2 \cdot 2 \cdot 2 \cdot 2 \cdot 2 \cdot 2$ **75.** 12 **77.** 48 **79.** 41 **81.** 20
83. 324 **85.** 11 **87.** 62 points

Chapter 0 Test

1. $\frac{3}{11}$ **2.** $\frac{25}{16}$ **3.** $\frac{29}{30}$ **4.** $\frac{21}{80}$ **5.** -3 **6.** -12
7. -7 **8.** 4 **9.** -35 **10.** 54 **11.** -12 **12.** 11
13. 11 **14.** 113 **15.** $\frac{9}{44}$ **16.** $\frac{14}{5}$ **17.** $-7 < -5$
18. $8 + (-3)^2 > 8 - (-3)$ **19.** 32 lots **20.** \$1,775

Reading Your Text for Chapter 1

Section 1.1 (**a**) variables; (**b**) sum; (**c**) multiplication; (**d**) expression
Section 1.2 (**a**) evaluating; (**b**) positive; (**c**) operations; (**d**) grouping
Section 1.3 (**a**) first; (**b**) term; (**c**) factors; (**d**) (numerical) coefficient
Section 1.4 (**a**) fresh; (**b**) expressions; (**c**) solution; (**d**) one
Section 1.5 (**a**) nonzero; (**b**) reciprocal; (**c**) original; (**d**) reciprocal
Section 1.6 (**a**) divide; (**b**) Multiplying; (**c**) before; (**d**) simplified
Section 1.7 (**a**) greater; (**b**) equivalent; (**c**) positive; (**d**) negative

Summary Exercises for Chapter 1

1. $y + 8$ **3.** $8a$ **5.** $x(x - 7)$ **7.** $\frac{a + 2}{a - 2}$ **9.** $\frac{9}{x}$
11. -7 **13.** 15 **15.** 25 **17.** 1 **19.** -20
21. $4a^3, -3a^2$ **23.** $5m^2, -4m^2, m^2$ **25.** $16x$ **27.** $3xy$
29. $19a + b$ **31.** $3x^3 + 9x^2$ **33.** $10a^3$ **35.** $(37 - x)$ ft
37. $(x + 4)$ m **39.** $x, 25 - x$ **41.** $(2x + 8)$ m **43.** Yes
45. Yes **47.** 2 **49.** -5 **51.** 1 **53.** -7 **55.** 7
57. -4 **59.** 27 **61.** -2 **63.** $\frac{7}{2}$ **65.** 18 **67.** $\frac{2}{5}$
69. 6 **71.** 6 **73.** 4 **75.** $\frac{1}{2}$ **77.** $\frac{5}{3}$ **79.** 12
81. 3 **83.** 3 hr **85.** 5 **87.** (number line: 0, 5)

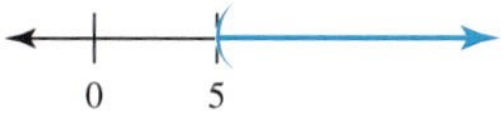

89. (number line: 0, 9)

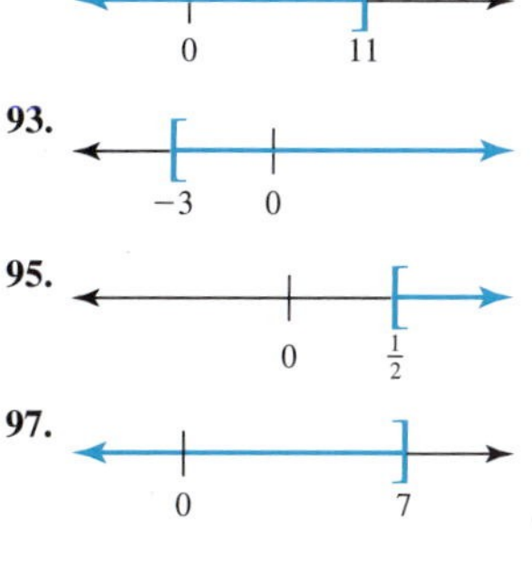

Chapter 1 Test

1. $x + y$ **2.** $m - n$ **3.** ab **4.** $\frac{p}{q - 3}$ **5.** $c - 5$
6. $3(2x - 3y)$ **7.** $3(m - n)$ **8.** $x \le 8$ **9.** 4 **10.** 36
11. -4 **12.** 5 **13.** $3a - b$ **14.** $2x^2 + 8$ **15.** No
16. Yes **17.** 12 **18.** 30 **19.** 2 **20.** $\frac{35}{4}$
21. (number line: 0, 14)
22. (number line: −4, 0)
23. 7 **24.** Juwan 6, Jan 12, Rick 17 **25.** \$2.07

Reading Your Text for Chapter 2

Section 2.1 (**a**) formula; (**b**) coefficient; (**c**) original; (**d**) distance
Section 2.2 (**a**) elements; (**b**) empty; (**c**) set-builder; (**d**) interval
Section 2.3 (**a**) solution; (**b**) infinite; (**c**) ordered-pair; (**d**) dependent
Section 2.4 (**a**) x-axis; (**b**) y-axis; (**c**) origin; (**d**) quadrants
Section 2.5 (**a**) relation; (**b**) domain; (**c**) range; (**d**) independent
Section 2.6 (**a**) function; (**b**) relation; (**c**) domain; (**d**) outputs

Summary Exercises for Chapter 2

1. $W = \frac{V}{LH}$ **3.** $y = \frac{c - ax}{b}$ **5.** $t = \frac{A - P}{Pr}$ **7.** 6
9. 42, 43 **11.** 54 mi/hr, 48 mi/hr; 216 mi **13.** 150 pairs
15. $\{1, 3\}$ **17.** $\{-1, 0, 1, 2, 3\}$ **19.** $\{1, 3\}$

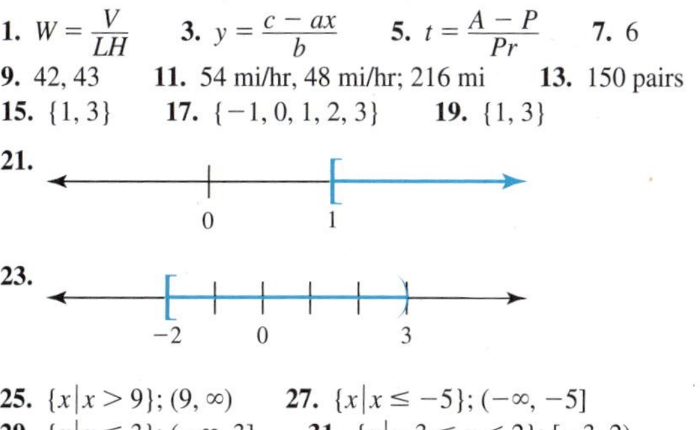

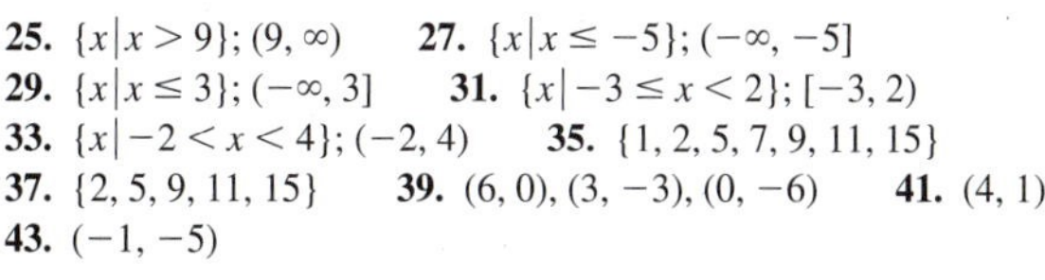

25. $\{x|x > 9\}; (9, \infty)$ **27.** $\{x|x \le -5\}; (-\infty, -5]$
29. $\{x|x \le 3\}; (-\infty, 3]$ **31.** $\{x|-3 \le x < 2\}; [-3, 2)$
33. $\{x|-2 < x < 4\}; (-2, 4)$ **35.** $\{1, 2, 5, 7, 9, 11, 15\}$
37. $\{2, 5, 9, 11, 15\}$ **39.** $(6, 0), (3, -3), (0, -6)$ **41.** $(4, 1)$
43. $(-1, -5)$

45 and 47.

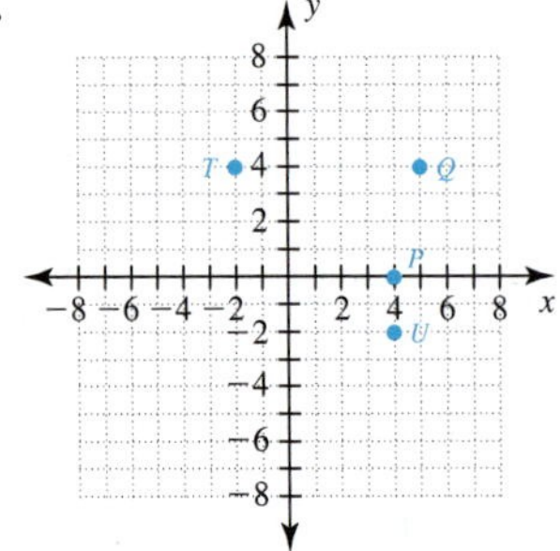

49.

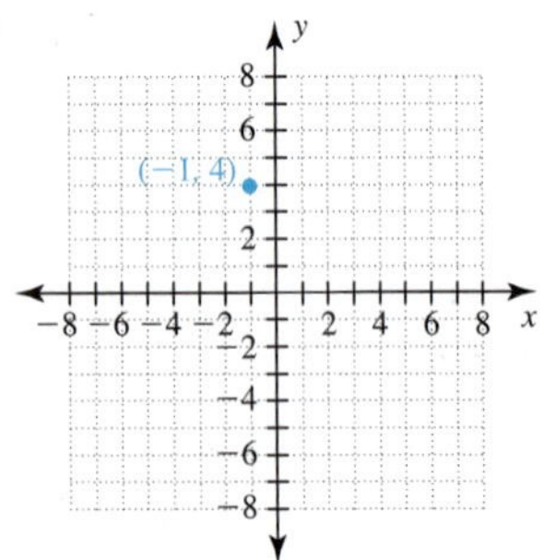

51.

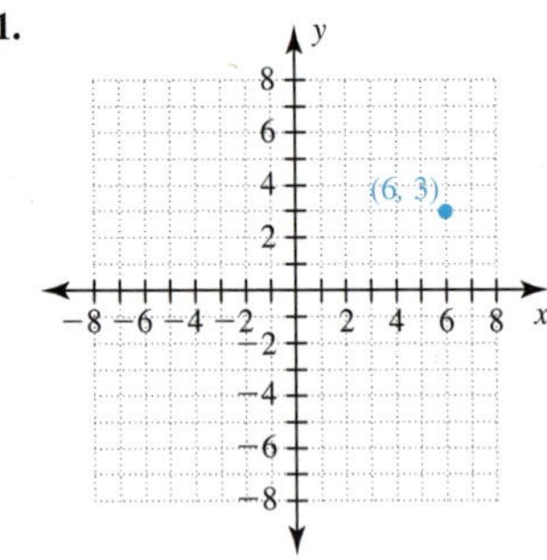

53. I **55.** III **57.** x-axis

59. $D = \{2001, 2004, 2007, 2010\}$; $R = \{$Halle Berry, Hilary Swank, Marion Cotillard, Natalie Portman$\}$

61. $D = \{$Brushed Nickel, Jeweled Bronze, Iron Amber, Walnut Silver$\}$; $R = \{\$129.99, \$199.99, \$349.99, \$399.99\}$

63. $D = \{1, 3, 4, 7, 8\}$; $R = \{1, 2, 3, 5, 6\}$

65. $D = \{1\}$; $R = \{3, 5, 7, 9, 10\}$ **67.** Function

69. Not a function **71.** Function **73.** Not a function

75. **(a)** 5; **(b)** 9; **(c)** 3 **77.** **(a)** 5; **(b)** 5; **(c)** 5

79. **(a)** 9; **(b)** 1; **(c)** 3 **81.** $f(x) = -2x + 5$

83. $f(x) = -\frac{2}{3}x + 2$ **85.** $f(x) = \frac{3}{4}x + 3$ **87.** $-\frac{3}{4}t + 2$

89. $-\frac{3}{4}x - \frac{3}{4}h + 2$ **91.** $3a - 2$ **93.** $3x + 3h - 2$

95.

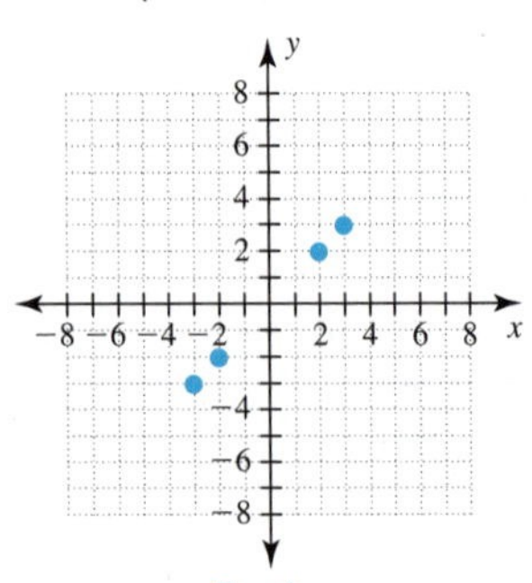

Function

97.

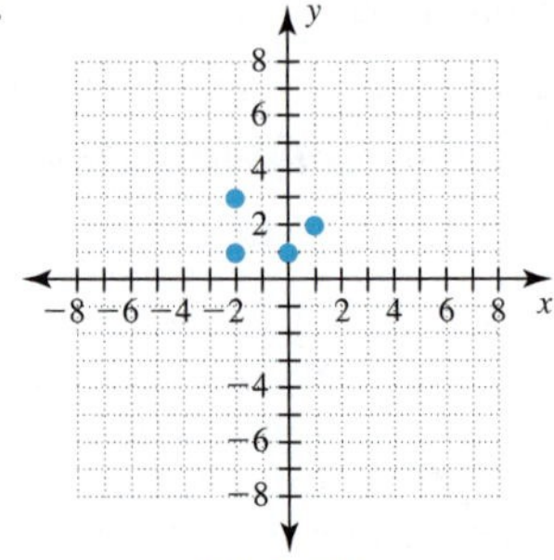

Not a function

99. Function; $D = \mathbb{R}$; $R = \{y \mid y \geq -4\} = [-4, \infty)$

101. Not a function; $D = \{x \mid -6 \leq x \leq 6\} = [-6, 6]$; $R = \{y \mid -6 \leq y \leq 6\} = [-6, 6]$

103. **(a)** 0; **(b)** -3; **(c)** -5; **(d)** 5; **(e)** $-3.3, -2, 3.4$; **(f)** $-4, 1$

105. **(a)** -2; **(b)** 12; **(c)** $-5, 4$; **(d)** -2; **(e)** 8; **(f)** 12

Chapter 2 Test

1.

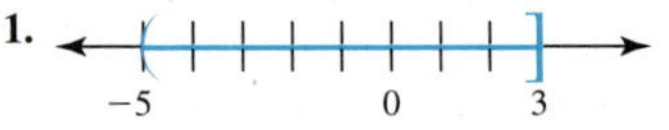

2. **(a)** $\{x \mid -3 < x \leq 2\}$; **(b)** $(-3, 2]$ **3.** $(3, 6)$ **4.** $(4, -2)$

5. $(0, -7)$

6–8.

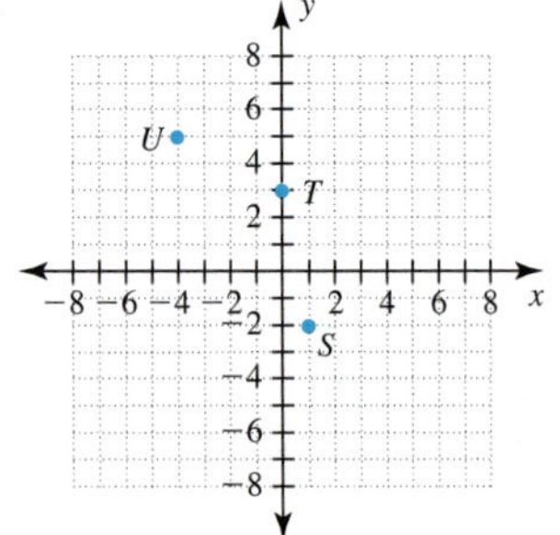

9. Function

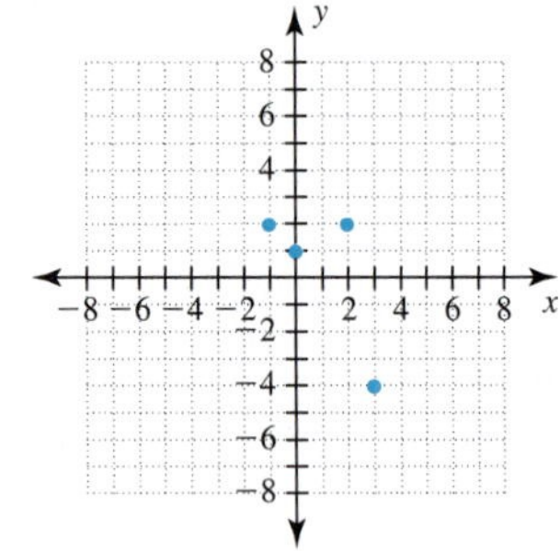

10. $(4, 0), (5, 4)$ **11.** $(3, 0), (0, 4), \left(\frac{3}{4}, 3\right)$

12. **(a)** 6; **(b)** 12; **(c)** 2 **13.** **(a)** 2; **(b)** -5

14. **(a)** 3; **(b)** 2; **(c)** 1; **(d)** $-5, 5$

15. **(a)** $D = \{-3, 1, 2, 3, 4\}$; $R = \{-2, 0, 1, 5, 6\}$; **(b)** $D = \{$United States, Germany, Russia, China$\}$; $R = \{50, 63, 65, 101\}$

16. **(a)** Function; $D = \{-4, -1, 0, 2\}$; $R = \{2, 5, 6\}$;

(b) not a function; $D = \{-3, 0, 1, 2\}$; $R = \{0, 1, 2, 4, 7\}$

17. Function **18.** Not a function

19. **(a)** $\{1, 2, 3, 5, 7\}$; **(b)** $\{5\}$ **20.** $f(x) = \frac{3}{7}x - 4$

21. $B = \frac{3V}{h}$ **22.** 7 **23.** Juwan 6, Jan 12, Rick 17

24. 3:30 P.M. **25.** 600 notepads

Cumulative Review for Chapters 0–2

1. $\frac{7}{11}$ **2.** $\frac{6}{5}$ **3.** 2 **4.** 80 **5.** 7 **6.** 7 **7.** -16
8. 4 **9.** 63 **10.** -16 **11.** 0 **12.** -5 **13.** 9
14. $\frac{1}{3}$ **15.** $\frac{19}{12}$ **16.** $\frac{7}{10}$ **17.** -9 **18.** 13 **19.** 3
20. 7 **21.** $15x - 9y$ **22.** $10x^2 - 13x + 2$ **23.** 4
24. 71 **25.** 4 **26.** $\{x \mid x \le -4\}$ **27.** $\{x \mid -3 \le x < 3\}$
28. $r = \frac{I}{Pt}$ **29.** $h = \frac{2A}{b}$ **30.** $y = \frac{c - ax}{b}$
31. $W = \frac{P}{2} - L$ or $W = \frac{P - 2L}{2}$ **32.** -5 **33.** -2
34. 5 **35.** 13; $4x - 7 = 45$ **36.** 42, 43; $x + (x + 1) = 85$
37. 7; $3x = (x + 2) + 12$ **38.** \$420; $x + (x + 120) = 720$
39. 5 cm, 17 cm; $2x + 2(3x + 2) = 44$
40. 8 in., 13 in., 16 in.; $x + (x + 5) + 2x = 37$

Reading Your Text for Chapter 3

Section 3.1 **(a)** solutions; **(b)** vertical; **(c)** constant; **(d)** zero
Section 3.2 **(a)** slope; **(b)** horizontal; **(c)** rise; **(d)** negative
Section 3.3 **(a)** parallel; **(b)** perpendicular; **(c)** undefined; **(d)** zero
Section 3.4 **(a)** one; **(b)** slope; **(c)** zero; **(d)** line
Section 3.5 **(a)** half-plane; **(b)** test; **(c)** solid; **(d)** false

Summary Exercises for Chapter 3

1.

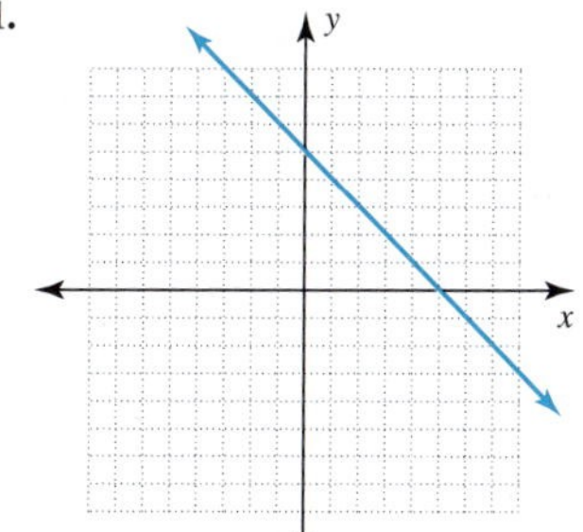

3.

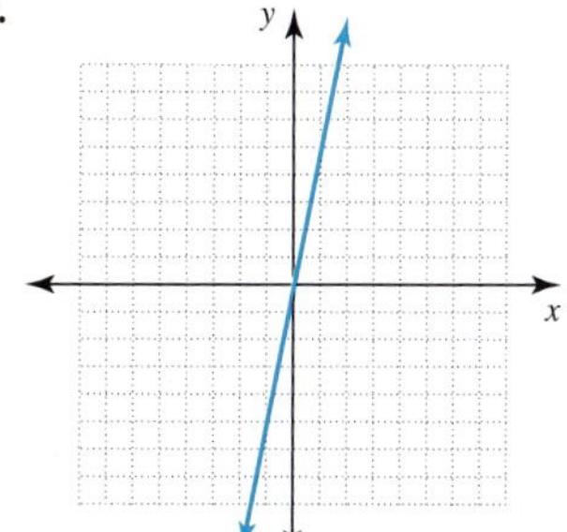

5.

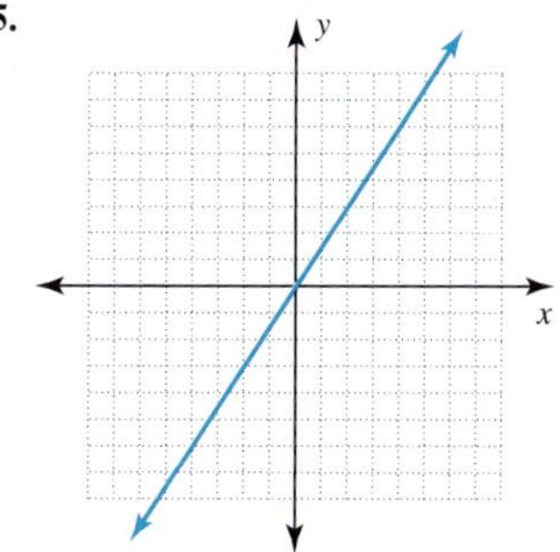

7.

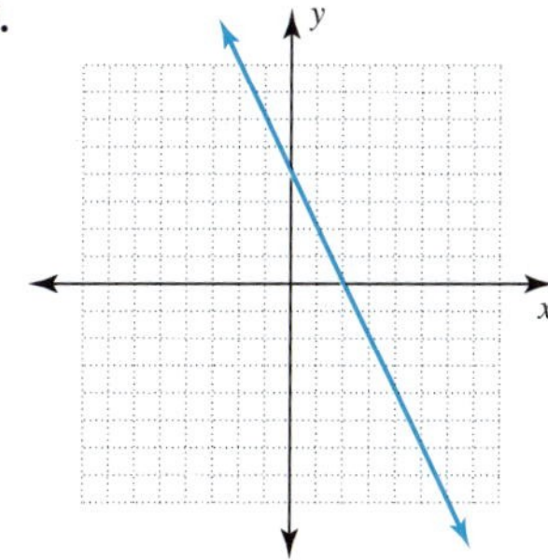

9.

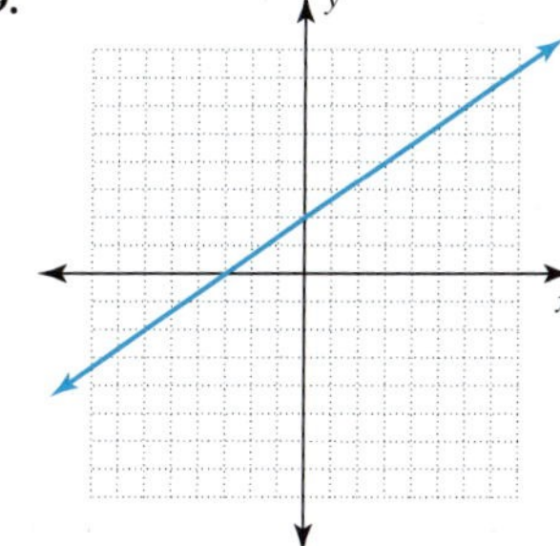

11.

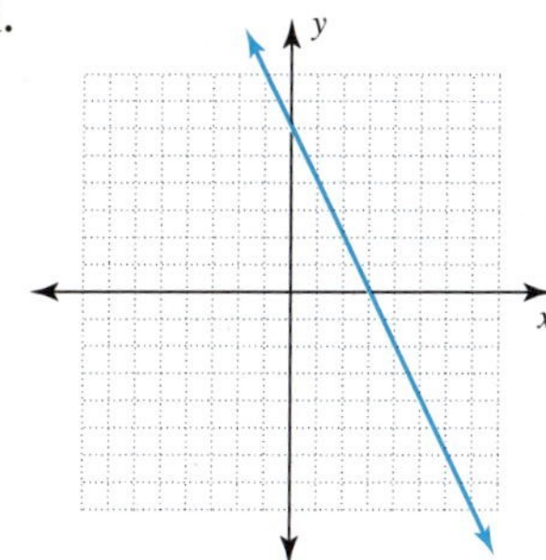

13.

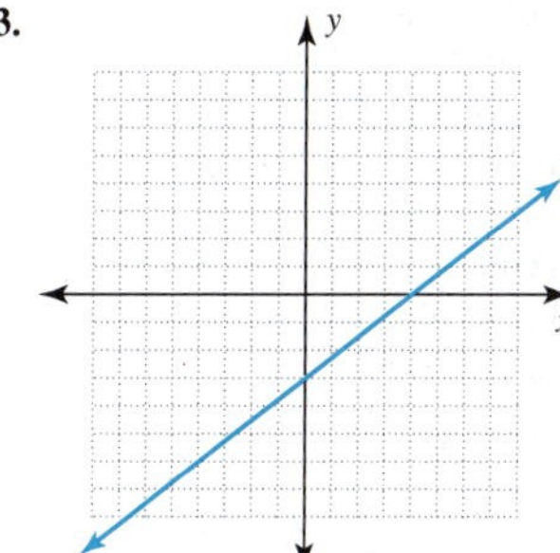

15.

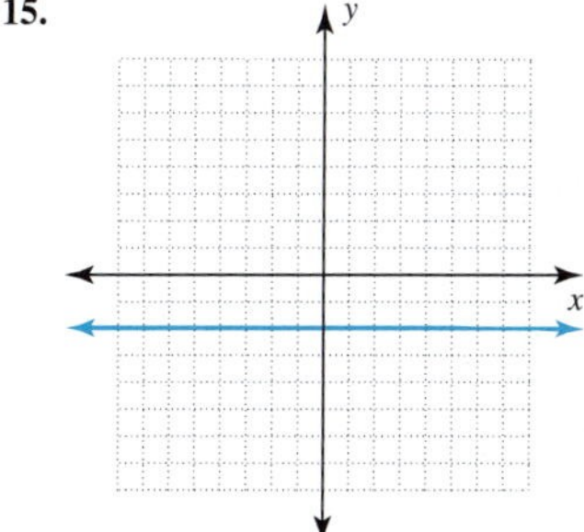

17.

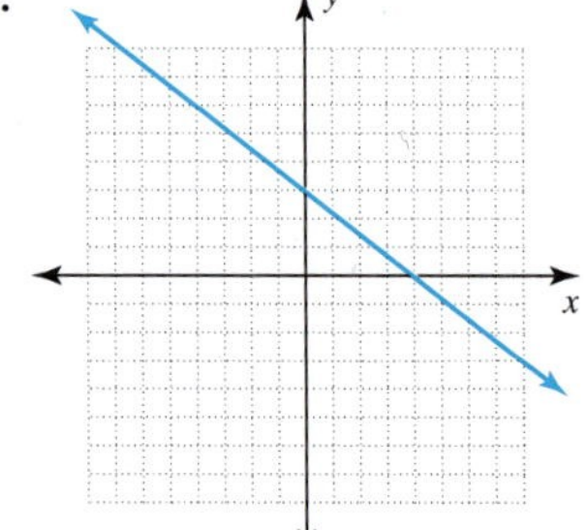

19.

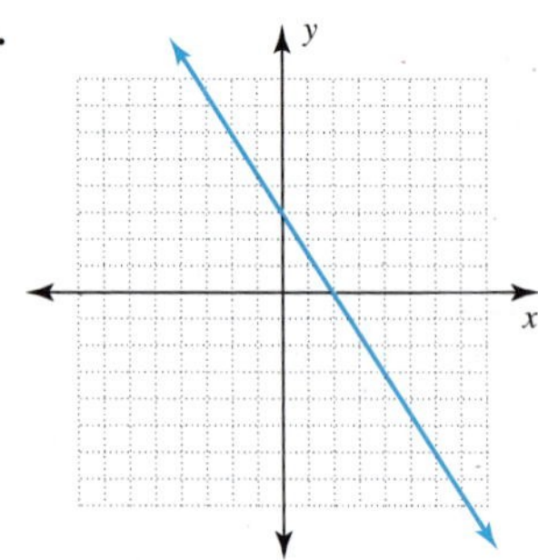

21. 2 **23.** $-\frac{1}{2}$ **25.** 0 **27.** $\frac{1}{2}$

29. Slope 2, y-intercept (0, 5) **31.** Slope $-\frac{3}{4}$, y-intercept (0, 0)

33. Slope $-\frac{2}{3}$, y-intercept (0, 2) **35.** Slope 0, y-intercept (0, −3)

37. $y = 2x + 3$ **39.** $y = -\frac{2}{3}x + 2$ **41.** Parallel

43. Neither **45.** $y = \frac{2}{3}x - 5$ **47.** $y = 3x - 3$

49. $y = \frac{5}{3}x - 7$ **51.** $y = -\frac{5}{2}x - 9$ **53.** $y = -5$

55. $y = \frac{2}{3}x + 1$ **57.** $y = \frac{3}{4}x + 3$ **59.** $y = 3x - 4$

61. $y = -\frac{2}{3}x + \frac{1}{3}$ **63.** $C(x) = 1.75x + 30$ **65.** 68 gyros

67. \$66.25 **69.** $f(x) = \frac{2}{63}x$

71. Each mile driven requires $\frac{2}{63} \approx 0.03$ gal of gas.

73. Gas is not used when not driving.

75.

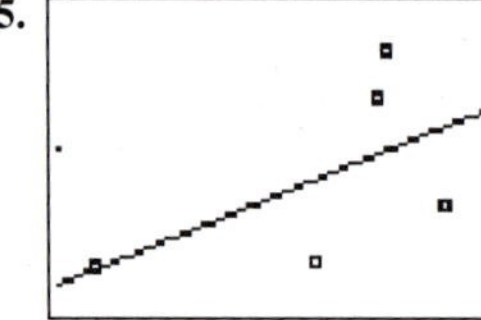

77. 5.73 **79.** 109 books

81.

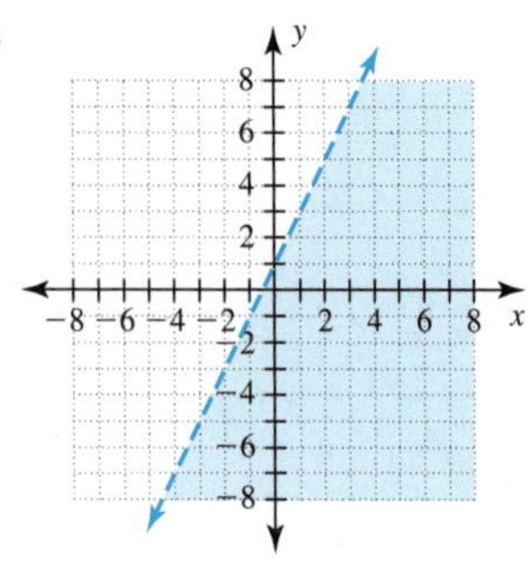

83.

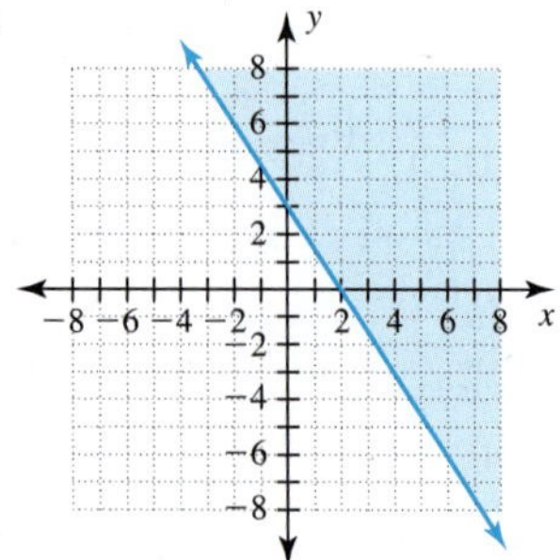

85.

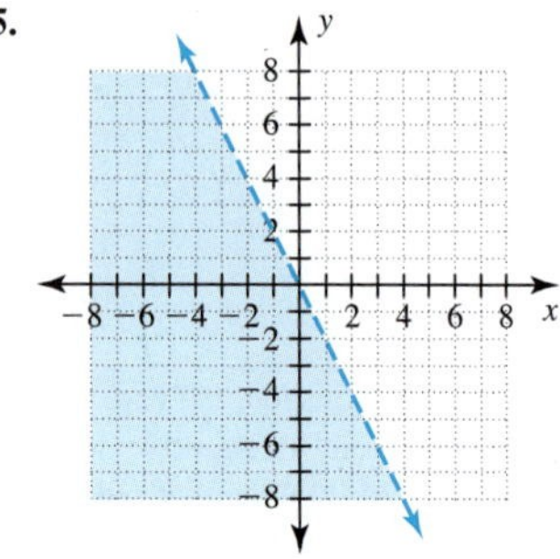

87.

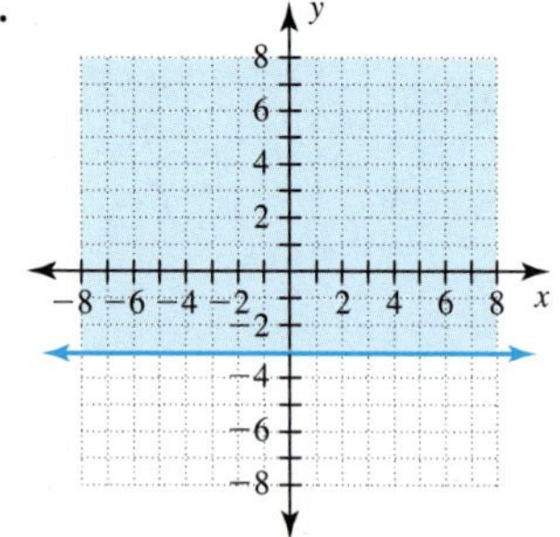

Chapter 3 Test

1. 1 **2.** $\frac{3}{7}$ **3.** Slope −5; y-intercept (0, −9)

4. Slope $-\frac{6}{5}$; y-intercept (0, 6) **5.** Slope 0; y-intercept (0, 5)

6. Undefined slope; no y-intercept

7. $y = -3x + 6$

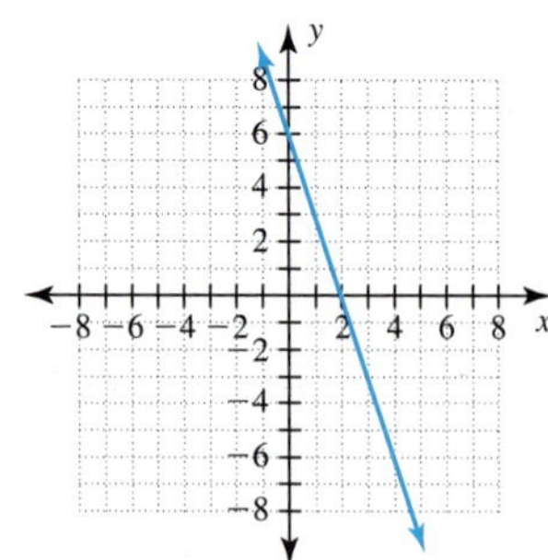

8. $y = \frac{2}{5}x - 3$

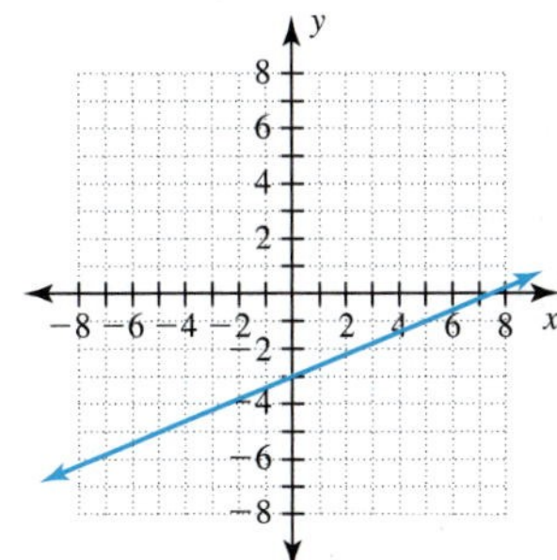

9. $y = 5x - 2$ **10.** $y = -4x - 16$

11. $y = 4x + 3$ **12.** $y = -\frac{5}{2}x - 17$

13.

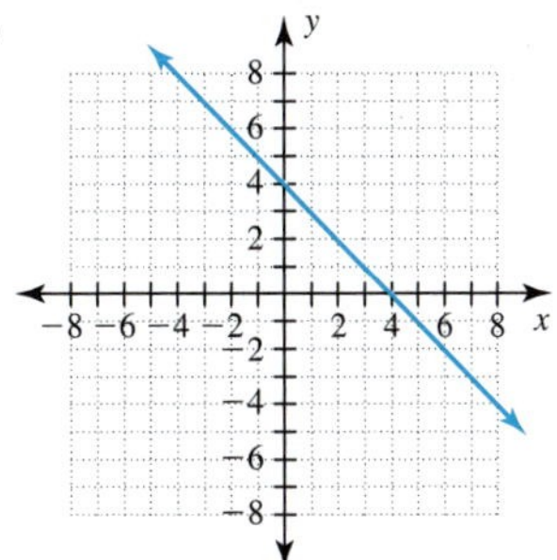

14.

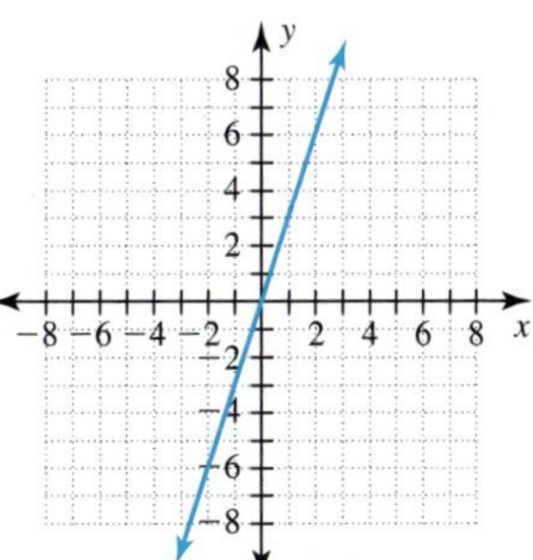

15.

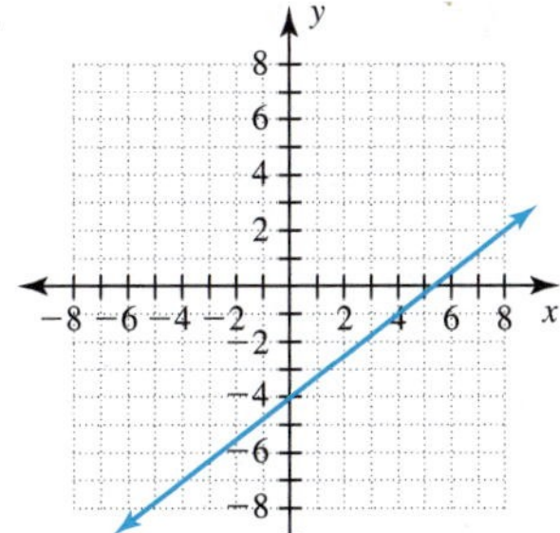

16.

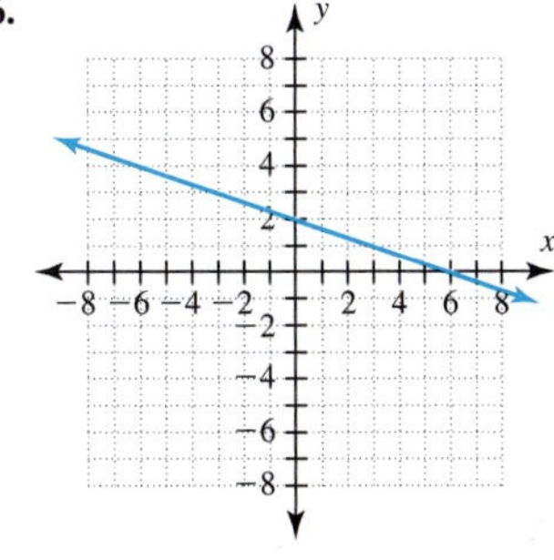

17.

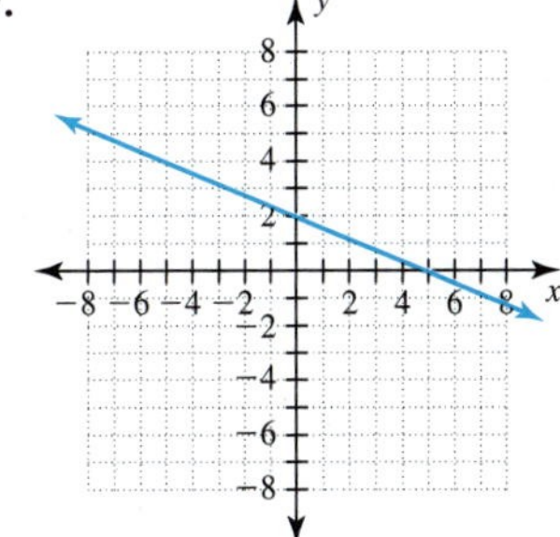

18.

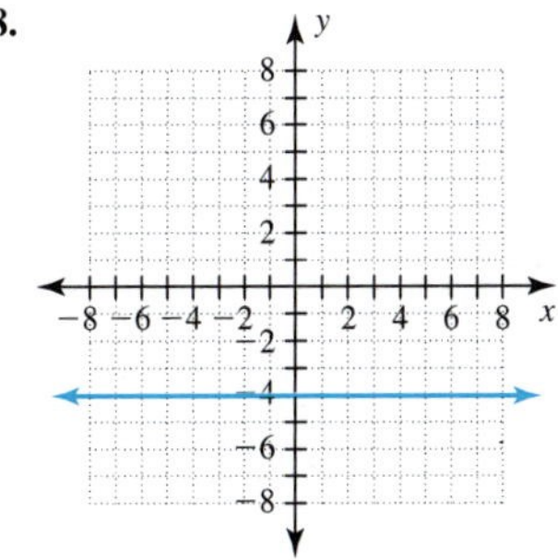

19.

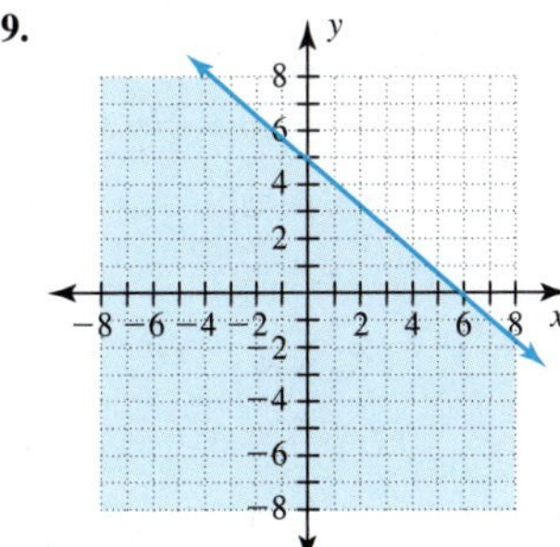

20.

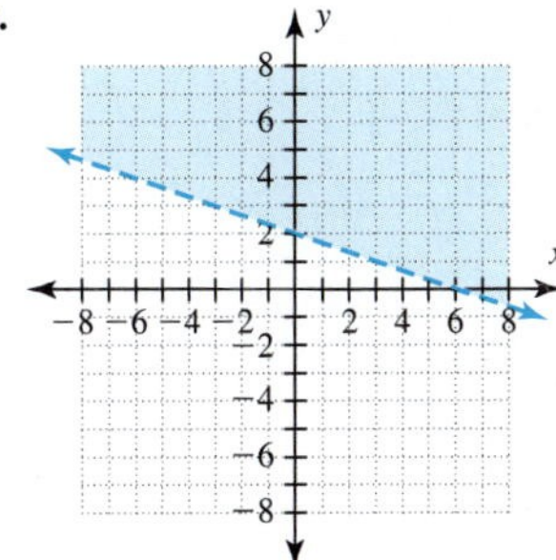

21.

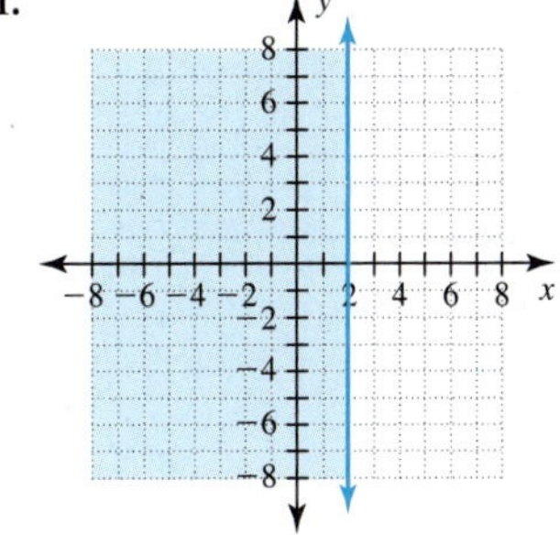

22.

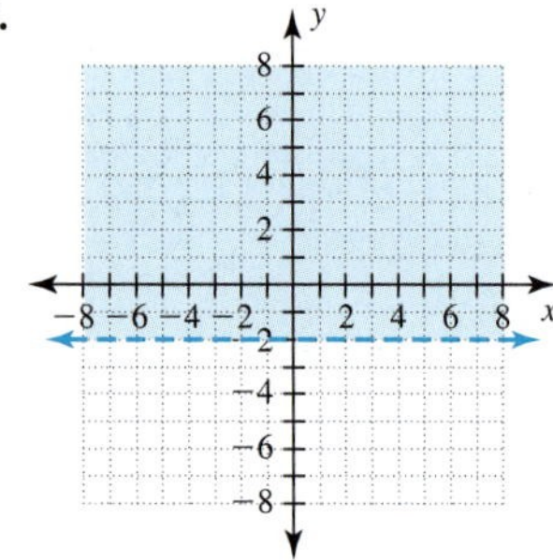

23. $y = \frac{1}{4}x + \frac{3}{2}$ **24.** $5\frac{1}{2}$ hr

25.

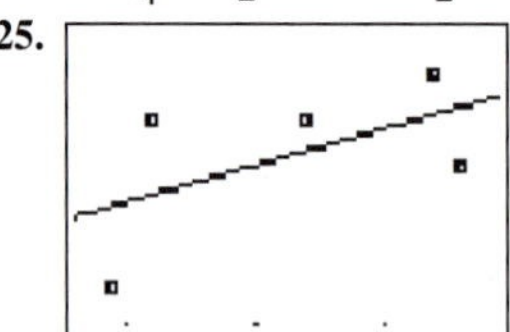

26. $y = 0.15x + 67.41$

Cumulative Review for Chapters 0–3

1. $-\frac{1}{3}$ **2.** $\frac{2}{3}$ **3.** 17 **4.** 8 **5.** 1 **6.** $-\frac{33}{5}$

7. $6x + 3y$ **8.** $-x - 4$ **9.** $-6x^2 + x + 2$

10. $-8x^2 - 5x + 25$ **11.** -11 **12.** 100

13. $-\frac{7}{2}$ **14.** $-\frac{3}{2}$

15. $\{x \mid x < 4\}$ 0 4

16. $\{x \mid x > -4\}$ −4 0

17. $\{x \mid 5 \le x \le 9\}$ 0 5 9

18. $\{x \mid x < 2 \text{ or } x > 7\}$ 0 2 7

19. $C = \frac{5}{9}(F - 32)$ **20.** $h = \frac{3V}{\pi r^2}$ **21.** -2 **22.** 0

23. Slope -4; y-intercept $(0, 9)$ **24.** Slope $\frac{2}{5}$; y-intercept $(0, -2)$

25. Slope 0; y-intercept $(0, 9)$ **26.** Slope undefined; no y-intercept

27. $y = 5x - 6$ **28.** $x + 10y = 86$ **29.** $y = -\frac{2}{3}x + 6$

30. $5x + 4y = 26$ **31.** $y = \frac{3}{2}x - 3$ **32.** $y = 3x + 10$

33. 9 **34.** 7 **35.** Coach \$450; first-class \$900

36. 12 m, 24 m, 28 m

37.

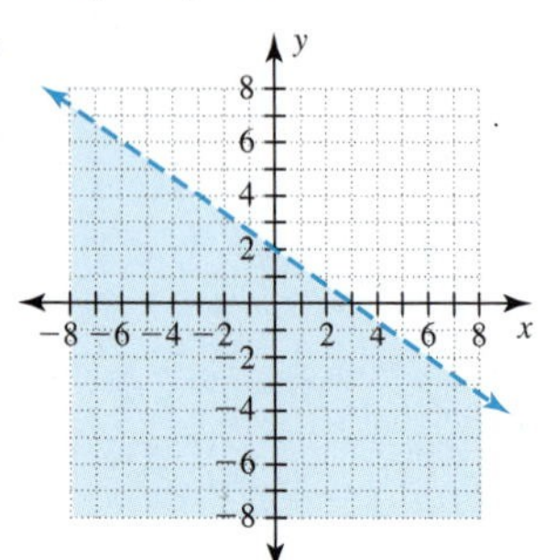

38. $f(x) = 3.95x + 30.77$ **39.** \$78.17

40. Each additional pound adds \$3.95 to the shipping costs.

Reading Your Text for Chapter 4

Section 4.1 (**a**) related; (**b**) ordered pair; (**c**) consistent; (**d**) inconsistent

Section 4.2 (**a**) solution; (**b**) check; (**c**) intersect; (**d**) original

Section 4.3 (**a**) equivalent; (**b**) intersection; (**c**) no; (**d**) time

Section 4.4 (**a**) three; (**b**) three; (**c**) infinite; (**d**) consistent

Section 4.5 (**a**) ordered pairs; (**b**) dashed; (**c**) solid; (**d**) bounded

Summary Exercises for Chapter 4

1. $\{(6, 2)\}$

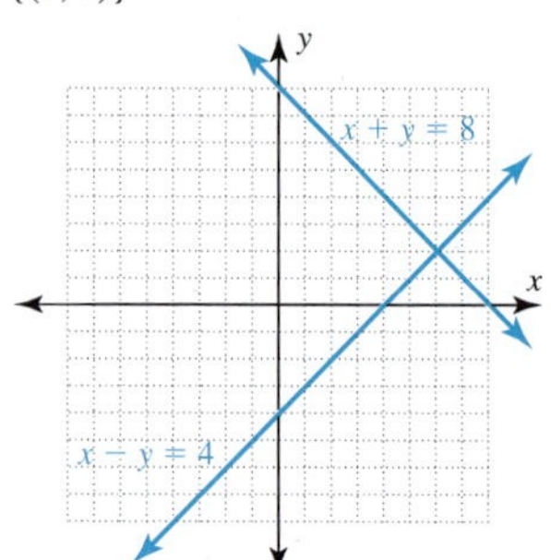

3. $\{(3, 2)\}$

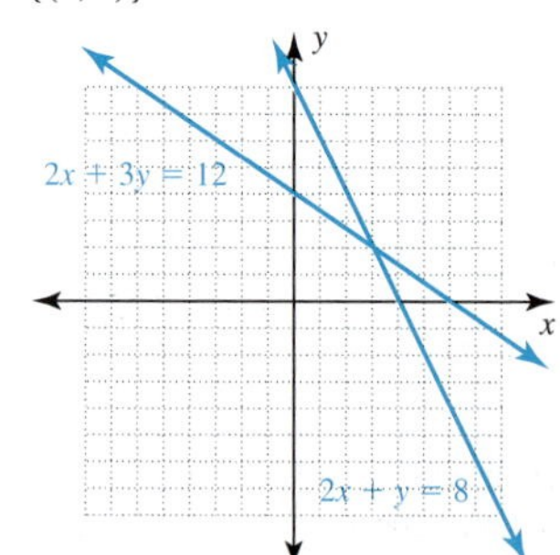

5. $\{(38.05, -15.98)\}$

7. $\{2\}$

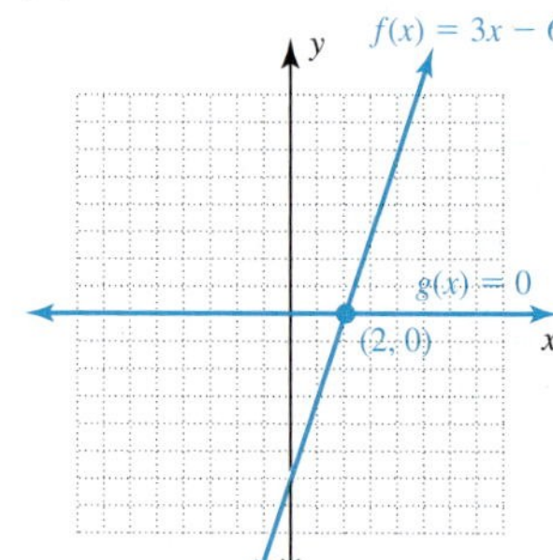

9. $\{1\}$

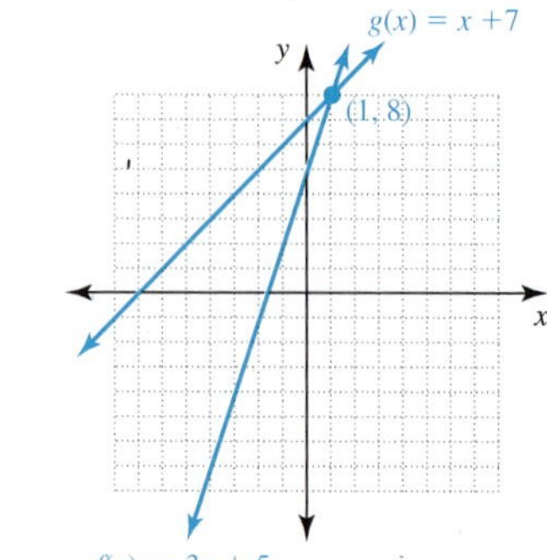

11. $\left\{-\frac{3}{2}\right\}$

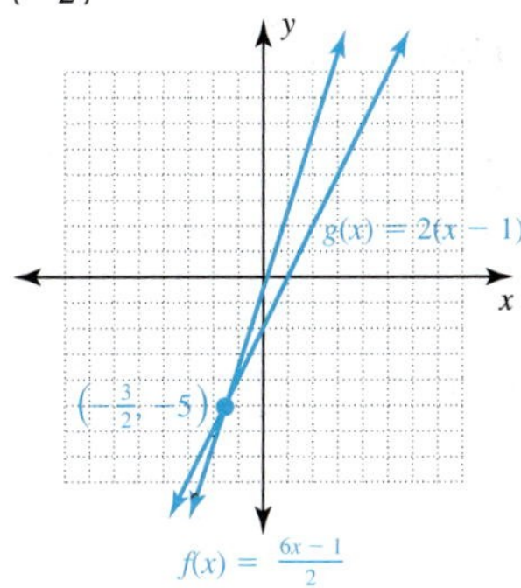

13. {4}

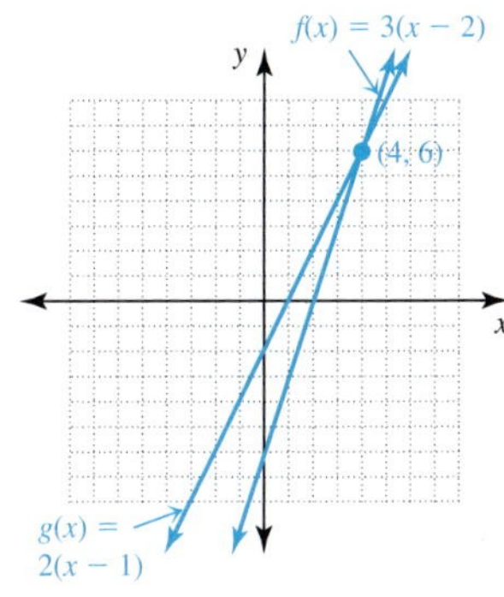

15. {(3, 2)} **17.** {(0, −1)} **19.** {(−4, 3)} **21.** {(−3, 2)}
23. {(6, −4)} **25.** {(9, 5)}
27. No solutions, inconsistent system **29.** {(2, −1)}
31. $\left\{\left(-\frac{2}{3}, 5\right)\right\}$ **33.** 7, 23
35. 800 adult tickets; 400 student tickets **37.** 20 cm, 12 cm
39. \$8,000 in savings; \$9,000 in time deposit
41. Jet: 500 mi/hr; wind: 50 mi/hr **43.** (15, 195)
45. $\left\{\left(3, \frac{8}{3}, -\frac{1}{3}\right)\right\}$ **47.** Inconsistent **49.** $\left\{\left(\frac{1}{2}, -\frac{3}{2}, 0\right)\right\}$
51. 2, 5, 8 **53.** 200 orchestra; 40 balcony; 120 box seats

55.

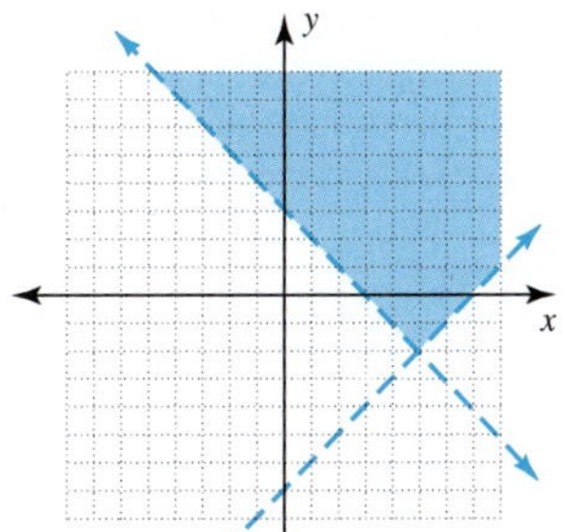

57.

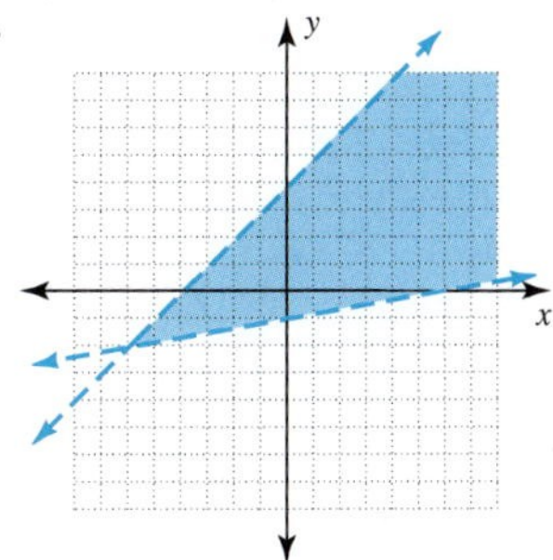

59.

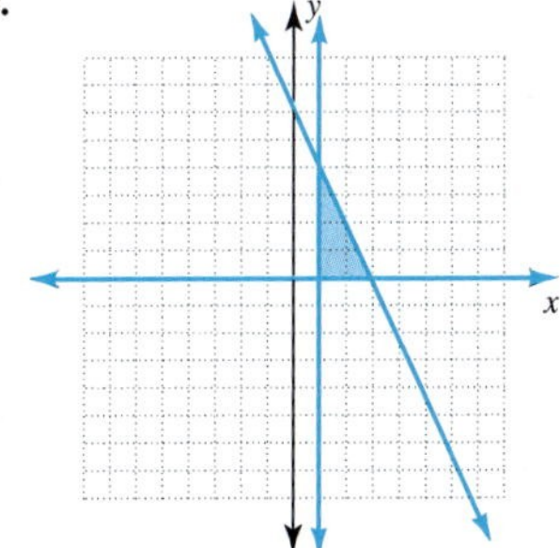

61.

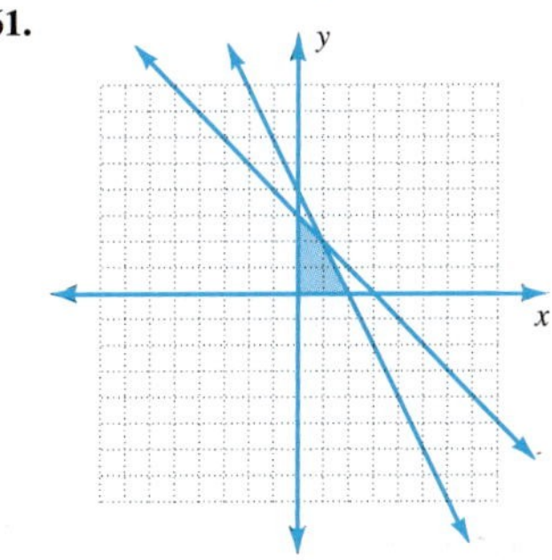

Chapter 4 Test

1. {3}

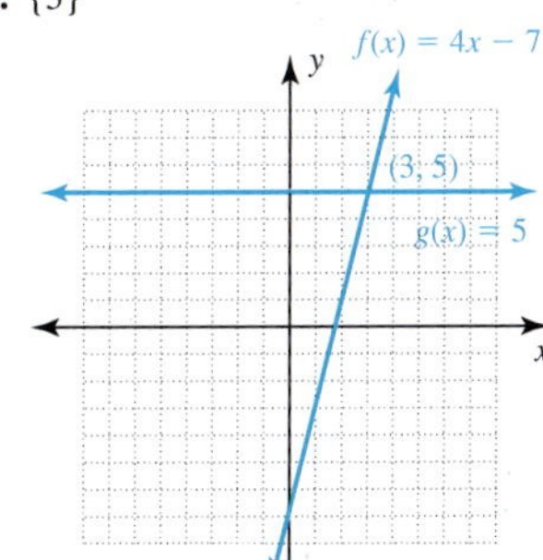

2. {2}

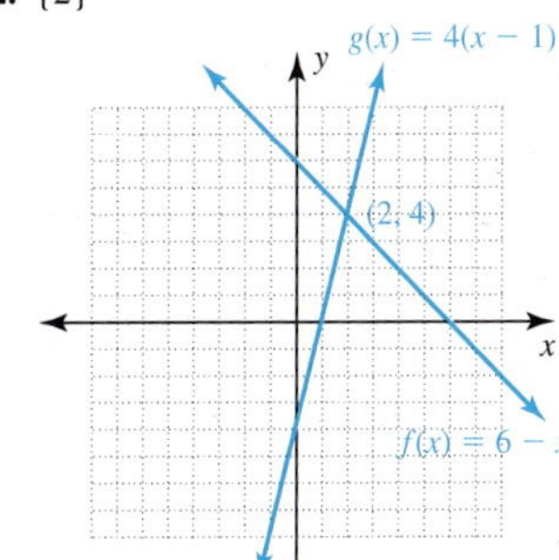

3. {2}

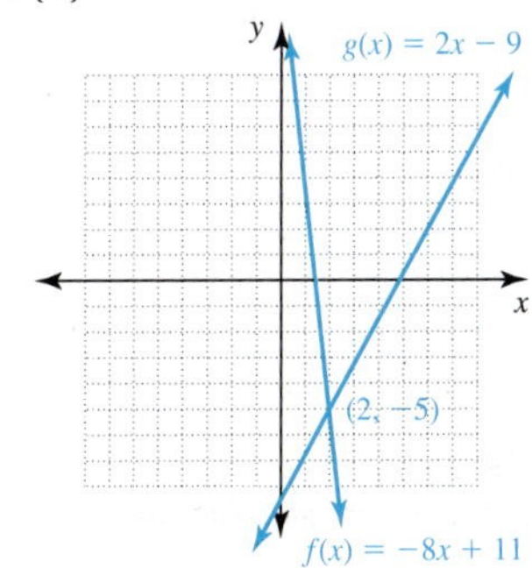

4. $\{2\}$

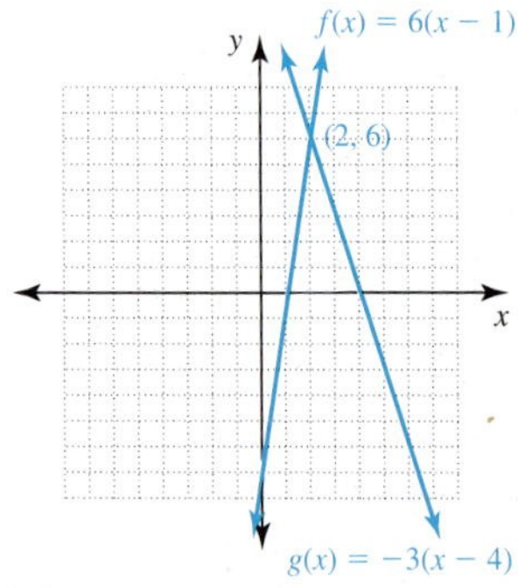

5. $\{(-3, 4)\}$ **6.** Infinite number of solutions, dependent system **7.** No solutions, inconsistent system **8.** $\{(-2, -5)\}$

9. $\{(5, 0)\}$ **10.** $\left\{\left(3, -\frac{5}{3}\right)\right\}$ **11.** $\{(-1, 2, 4)\}$

12. $\left\{\left(2, -3, -\frac{1}{2}\right)\right\}$

13.

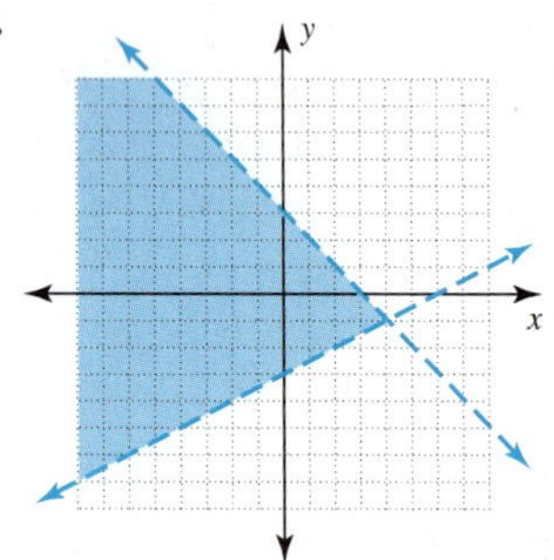

14.

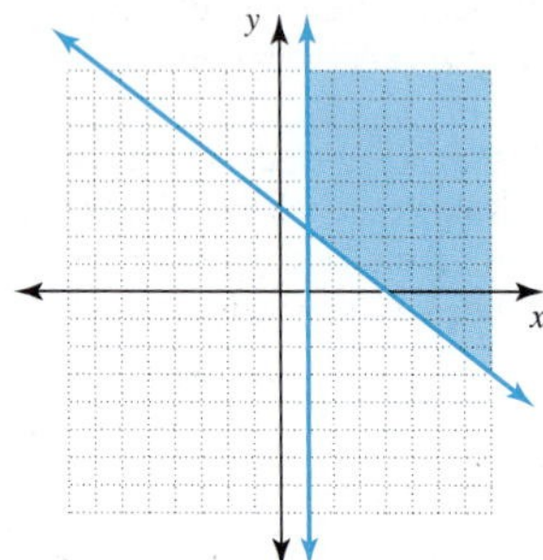

15.

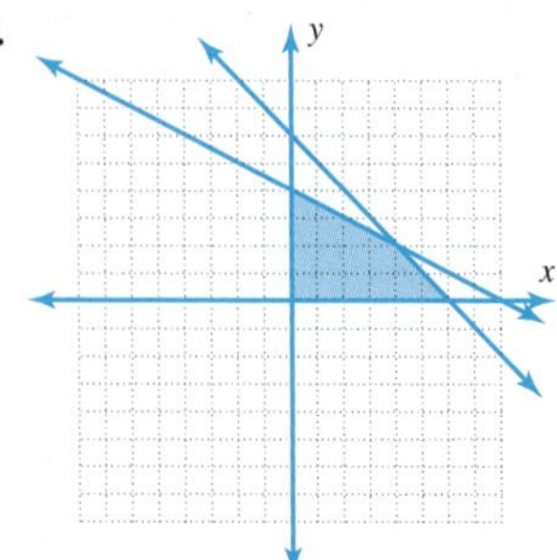

16. USB cable: \$2.50; mouse pad: \$6
17. 60 lb jawbreakers; 40 lb licorice
18. Four LCD monitors; six standard monitors
19. \$3,000 savings; \$5,000 bond; \$8,000 mutual fund
20. 50 ft by 80 ft

Cumulative Review for Chapters 0–4

1. $\left\{\frac{11}{2}\right\}$ **2.** $\{x \mid x > -2\}$ **3.** $\left\{x \mid x \le \frac{11}{3}\right\}$ **4.** $\{2\}$

5.

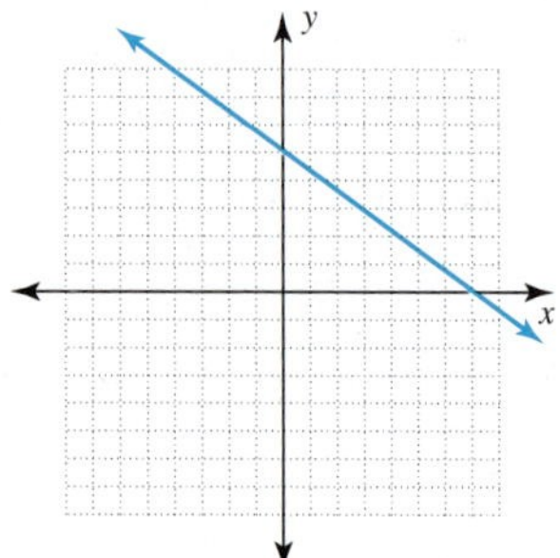

6.

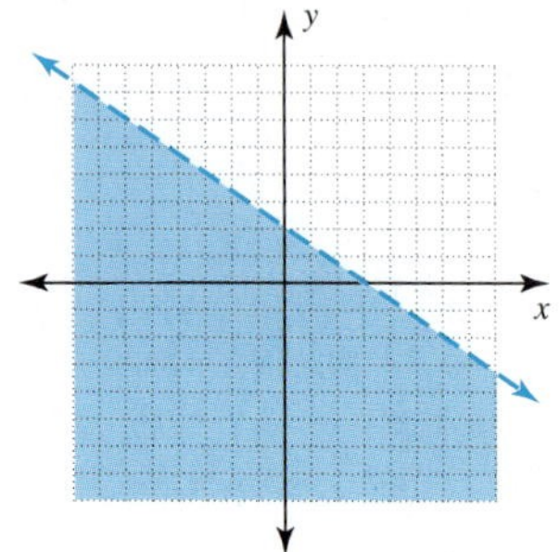

7. $R = \frac{P - P_0}{IT}$ **8.** 7 **9.** $y = -x + 3$ **10.** $y = \frac{4}{5}x - \frac{2}{5}$
11. 90 **12.** $\left\{\left(-10, \frac{26}{3}\right)\right\}$ **13.** $\{(2, 2, -1)\}$
14. 8 cm by 19 cm **15.** 73

Reading Your Text for Chapter 5

Section 5.1 **(a)** multiplication; **(b)** exponential; **(c)** add; **(d)** denominator
Section 5.2 **(a)** add; **(b)** reciprocal; **(c)** one; **(d)** meters
Section 5.3 **(a)** term; **(b)** coefficient; **(c)** binomial; **(d)** tri-
Section 5.4 **(a)** plus; **(b)** sign; **(c)** distributive; **(d)** first
Section 5.5 **(a)** coefficients; **(b)** distributive; **(c)** binomials; **(d)** three
Section 5.6 **(a)** coefficients; **(b)** term; **(c)** degree; **(d)** zero

Summary Exercises for Chapter 5

1. r^{13} **3.** $\frac{1}{8w^3}$ **5.** $\frac{1}{y^3}$ **7.** $\frac{1}{x^3}$ **9.** $\frac{10}{b^3}$ **11.** $\frac{m^3}{n^{21}}$
13. $\frac{y^6}{x^9}$ **15.** $9a^8$ **17.** 4.25×10^{-5} **19.** \$2,284.98
21. 3.8×10^6 mi^2 **23.** 3.5×10^6 mi^2 **25.** Binomial
27. Trinomial **29.** Binomial **31.** $5x^5 + 3x^2$; 5
33. $4x^4 + 6x^2 + 6$; 4 **35.** -8; 0 **37.** $21a^2 - 2a$
39. $5y^3 + 4y$ **41.** $5x^2 + 3x + 10$ **43.** $x - 2$
45. $-9w^2 - 10w$ **47.** $9b^2 + 8b - 2$ **49.** $2x^2 - 2x - 9$
51. $C(x) = 18.45x + 525$ **53.** –\$95 (loss) **55.** $6a^7$
57. $54p^5$ **59.** $15x - 40$ **61.** $-10r^3s^2 + 25r^2s^2$
63. $x^2 + 9x + 20$ **65.** $a^2 - 81b^2$ **67.** $a^2 + 7ab + 12b^2$
69. $6x^2 - 19xy + 15y^2$ **71.** $y^3 - y + 6$ **73.** $x^3 - 8$
75. $2x^3 - 2x^2 - 60x$ **77.** $x^2 + 14x + 49$ **79.** $4w^2 - 20w + 25$
81. $a^2 + 14ab + 49b^2$ **83.** $x^2 - 25$ **85.** $4m^2 - 9$
87. $25r^2 - 4s^2$ **89.** $3x^3 - 24x^2 + 48x$ **91.** $y^3 + 5y^2 - 16y - 80$
93. $(8x^2 + 20x + 12)$ in.^{2} **95.** $3a^3$ **97.** $3a - 2$
99. $-3rs + 6r^2$ **101.** $x - 5$ **103.** $x - 3 + \frac{2}{x - 5}$
105. $x^2 + 2x - 1 + \frac{-2}{3x + 1}$ **107.** $x^2 + x + 2 + \frac{1}{x + 2}$

Chapter 5 Test

1. $-6x^3y^4$ **2.** $\frac{16m^4n^{10}}{p^6}$ **3.** x^8y^{10} **4.** $\frac{c^2}{2d}$ **5.** $108x^8y^7$
6. $\frac{9y^4}{4x}$ **7.** 3 **8.** Binomial **9.** Trinomial
10. $8x^4 - 3x^2 - 7$; 8, 4; -3, 2; -7, 0; 4 **11.** $10x^2 - 12x - 7$
12. $7a^3 + 11a^2 - 3a$ **13.** $3x^2 + 11x - 12$ **14.** $7a^2 - 10a$
15. $4x^2 + 5x - 6$ **16.** $2x^2 - 7x + 5$
17. $15a^3b^2 - 10a^2b^2 + 20a^2b^3$ **18.** $4x^2 + 7xy - 15y^2$
19. $9m^2 + 12mn + 4n^2$ **20.** $4x^2 + 3x + 13 + \frac{17}{x-2}$
21. 8.9×10^5 km^2 **22.** $(2x^2 + 4x - 6)$ in.^{2}
23. $C(x) = 6.27x + 285$ **24.** $P(x) = 8.68x - 285$ **25.** \$366

Cumulative Review for Chapters 0–5

1. -10 **2.** $\frac{7}{5}$ **3.** $\{7\}$ **4.** $\{-36\}$ **5.** $\{4\}$ **6.** $\left\{-\frac{2}{3}\right\}$
7. $\{23\}$ **8.** $\mathbb{R}$ **9.** $\varnothing$ **10.** $\left\{-\frac{13}{8}\right\}$ **11.** $\{x \mid x < -15\}$
12. $\{x \mid x < 5\}$ **13.** -13 **14.** $\left\{-\frac{8}{3}\right\}$ **15.** $R(x) = 39.95x$
16. \$679.15 **17.** Slope: -4; y-intercept: (0, 9) **18.** 3
19. $y = -2x + 4$ **20.** $y = \frac{4}{5}x - \frac{2}{5}$ **21.** $y = 4x - 7$
22. $y = -2x$ **23.** $3x^2 + 2x - 4$ **24.** $x^2 - 3x - 18$
25. $12x^2 - 20x$ **26.** $6x^2 + x - 40$ **27.** $x^3 - x^2 - x + 10$
28. $4x^2 - 49$ **29.** $9x^2 - 30x + 25$ **30.** $20x^3 - 100x^2 + 125x$
31. $4xy - 2x^3 + 1$ **32.** $2x^2 + 6x + 3 + \frac{2}{x-3}$ **33.** $18x^7$
34. $\frac{x^{16}}{y^{12}}$ **35.** $72x^4y^2$ **36.** $\frac{2x^5}{y^{11}}$ **37.** 2.1×10^{10} **38.** 8
39. 65, 67 **40.** 4 cm by 24 cm

Reading Your Text for Chapter 6

Section 6.1 **(a)** distributive; **(b)** common; **(c)** multiplying; **(d)** grouping
Section 6.2 **(a)** difference; **(b)** GCF; **(c)** sum; **(d)** cubes
Section 6.3 **(a)** Quadratic; **(b)** different; **(c)** GCF; **(d)** square
Section 6.4 **(a)** coefficients; **(b)** four; **(c)** factorable; **(d)** GCF
Section 6.5 **(a)** sum; **(b)** grouping; **(c)** polynomial; **(d)** *ac*
Section 6.6 **(a)** quadratic; **(b)** never; **(c)** zero; **(d)** zeros

Summary Exercises for Chapter 6

1. $7(2a - 5)$ **3.** $8s^2(3t - 2)$ **5.** $9s^3(3s + 2)$
7. $9m^2n(2n - 3 + 4n^2)$ **9.** $8ab(a + 3 - 2b)$
11. $(3x + 4y)(2x - y)$ **13.** $(x - 4)(x + 5)$ **15.** $(4x + 3)(2x - 5)$
17. $x(2x + 3)(3x - 2)$ **19.** $(p + 7)(p - 7)$
21. $(3n + 5m)(3n - 5m)$ **23.** $(5 + z)(5 - z)$
25. $(5a + 6b)(5a - 6b)$ **27.** $3w(w + 2z)(w - 2z)$
29. $2(m + 6n^2)(m - 6n^2)$ **31.** $(x - 9)^2$ **33.** $(x + 7)^2$
35. $(2x + 3)^2$ **37.** Not factorable
39. $(a^2 + 4b^2)(a + 2b)(a - 2b)$ **41.** $(2x + 1)(4x^2 - 2x + 1)$
43. $(5m + 4n)(25m^2 - 20mn + 16n^2)$ **45.** $(x + 10)(x + 2)$
47. $(w - 6)(w - 9)$ **49.** $(x - 12y)(x + 4y)$
51. $(7x - 3)(2x + 7)$ **53.** $-(x + 4)(x + 5)$ **55.** $(a - 4)(a - 3)$
57. $(x + 8)^2$ **59.** $(b - 7c)(b + 3c)$ **61.** $m(m + 7)(m - 5)$
63. $3y(y - 7)(y - 9)$ **65.** $(3x + 5)(x + 1)$ **67.** $(2b - 3)(b - 3)$
69. $(5x - 3)(2x - 1)$ **71.** $(4y - 5x)(4y + 3x)$
73. $-4x(2x + 1)(x - 5)$ **75.** $3x(2x - 3)(x + 1)$
77. Factorable; $m = -6, n = 5$ **79.** Factorable; $m = -3, n = -8$
81. $5x^2(x - 7)(x + 6)$ **83.** $2y(6 - y)(6 + y)$ **85.** $10x^3(x + 4)^2$
87. $(x - 5)(x + 5)(x + 2)$ **89.** $\{-6, 1\}$ **91.** $\{-10, 3\}$
93. $\{-5, 4\}$ **95.** $\{0, 10\}$ **97.** $\{\pm 5\}$ **99.** $\left\{-1, \frac{3}{2}\right\}$
101. $\{-5, 2\}$ **103.** $\{-4, 9\}$ **105.** $\{0, -3, 5\}$
107. -6 and 9; -9 and 6 **109.** Width 12 ft, length 18 ft
111. 4 s **113.** 50 systems

Chapter 6 Test

1. $2b(4a + 5b)(4a - 5b)$ **2.** $(x + 2)(x - 5)$
3. $7(b + 6)$ **4.** $(x^2 + 9)(x + 3)(x - 3)$ **5.** $(3x - 2y)^2$
6. $5(x^2 - 2x + 4)$ **7.** $(4y + 7x)(4y - 7x)$
8. $(4x - 3y)(2x + y)$ **9.** $(3y - 2x)(9y^2 + 6xy + 4x^2)$
10. $(3w + 7)(w + 1)$ **11.** $(3x + 2)(2x - 5)$
12. $(a - 7)(a + 2)$ **13.** $-3x(2x + 5)(x - 2)$
14. $(y + 10z)(y + 2z)$ **15.** $\{5, 6\}$ **16.** $\{-5, -3\}$
17. $\{-1, 3\}$ **18.** $\left\{-\frac{1}{3}, \frac{3}{2}\right\}$ **19.** 20 cm **20.** 2 s

Cumulative Review for Chapters 0–6

1. 9 **2.** $\{11\}$ **3.** $\{-33\}$ **4.** $r = \frac{A - P}{Pt}$
5. $W = \frac{P - 2L}{2}$ **6.** $\{x \mid x > -6\}$ **7.** $\{x \mid x \geq -6\}$
8. $\{(0, -1)\}$ **9.** $\left\{\left(4, -\frac{5}{3}\right)\right\}$
10. Slope: $-\frac{2}{5}$; y-intercept: (0, 2) **11.** $y = 2x + 5$
12. $y = -3x - 2$ **13.** $2x^2y + 5xy$ **14.** $-x^2 + 2x - 1$
15. $x^2 + 9x - 2$ **16.** $3x^3 - 6x^2 + 6x$ **17.** $2x^2 - 3x - 5$
18. $x^3 + 27$ **19.** $8x^2 + 14x - 15$ **20.** $a^2 - 9b^2$
21. $x^2 - 4xy + 4y^2$ **22.** $50x^3 - 18xy^2$ **23.** $-\frac{8y^9}{x^6}$ **24.** $3a^4b^6$
25. $\frac{y^3}{4x^9}$ **26.** $\frac{2x^3z}{y^5}$ **27.** $4(3x + 5)$ **28.** $(5x + 7y)(5x - 7y)$
29. $(4x - 5)(3x + 2)$ **30.** $(x - 5)(2x - 3)$ **31.** $\{-5, 3\}$
32. $\{-3, 3\}$ **33.** $8xy^4 - 5x^3y + 3$ **34.** $x + 4 + \frac{6}{x-1}$
35. 17 **36.** \$103 **37.** 60 mi/hr **38.** 10 or 50 items

Reading Your Text for Chapter 7

Section 7.1 **(a)** square root; **(b)** two; **(c)** negative; **(d)** radius
Section 7.2 **(a)** product; **(b)** quotient; **(c)** denominator; **(d)** rationalize
Section 7.3 **(a)** like; **(b)** distributive; **(c)** root; **(d)** conjugates
Section 7.4 **(a)** extraneous; **(b)** integer; **(c)** isolated; **(d)** solutions
Section 7.5 **(a)** real; **(b)** principal; **(c)** radicals; **(d)** exponent
Section 7.6 **(a)** imaginary; **(b)** complex; **(c)** imaginary; **(d)** conjugates

Summary Exercises for Chapter 7

1. 11 **3.** Not a real number **5.** -4 **7.** $\frac{3}{4}$ **9.** 8
11. a^2 **13.** $7w^2z^3$ **15.** $-3b^2$ **17.** $2xy^2$ **19.** $2\sqrt{10}$
21. $\sqrt{29}$ **23.** Center: (0, 0); radius: 9
25. Center: $(-2, 1)$; radius: 5
27.

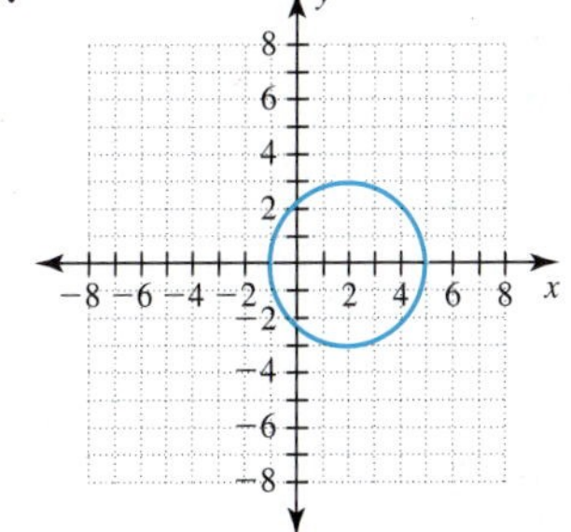

29. $3\sqrt{5}$ **31.** $2x\sqrt{15}$ **33.** $2\sqrt[3]{4}$ **35.** $\frac{3}{4}$ **37.** $\frac{y^2}{7}$
39. $\frac{\sqrt{5}}{4x}$ **41.** $11\sqrt{10}$ **43.** $10\sqrt[3]{2x}$ **45.** $11\sqrt{2}$ **47.** $9\sqrt{5}$
49. $w\sqrt{3w}$ **51.** $\frac{13}{5}\sqrt{5}$ **53.** $\frac{8a\sqrt[3]{3a}}{3}$ **55.** $\sqrt{21xy}$
57. $ab\sqrt[3]{4}$ **59.** $2\sqrt{3}$ **61.** $-32 - 2\sqrt{3}$ **63.** 1

65. $7 + 4\sqrt{3}$ **67.** $\frac{\sqrt{21}}{7}$ **69.** $\frac{\sqrt{2ab}}{b}$ **71.** $\frac{2\sqrt[3]{9x^2}}{3x}$
73. $\frac{3 - \sqrt{2}}{7}$ **75.** $9 - 4\sqrt{5}$ **77.** $\{21\}$ **79.** $\{9\}$ **81.** $\{5\}$
83. 9 **85.** 7 **87.** -3 **89.** 16 **91.** $\frac{8}{27}$ **93.** $\frac{1}{7}$
95. x^4 **97.** r **99.** $x^{2/5}$ **101.** x^8y^{15} **103.** $\frac{x^6}{y^{7/2}}$
105. $\sqrt[4]{x^3}$ **107.** $3\sqrt[3]{a^2}$ **109.** $(7x)^{1/5}$ **111.** $3pq^3$ **113.** $7i$
115. $4i\sqrt{3}$ **117.** $5 - 2i$ **119.** $3 - 8i$ **121.** $8 + 28i$
123. $-7 - 24i$ **125.** $-3 - i$ **127.** $\frac{5}{13} - \frac{12}{13}i$

Chapter 7 Test

1. $p^2q\sqrt[3]{9pq^2}$ **2.** $\frac{\sqrt{10xy}}{4y}$ **3.** $4x\sqrt{3}$ **4.** $7a^2$ **5.** $\frac{7x}{8y}$
6. $3 - 2\sqrt{2}$ **7.** $64x^6$ **8.** $\frac{32x^4}{y}$ **9.** $-3w^2z^3$
10. $2x^3\sqrt[3]{4x^2}$ **11.** 6.6 **12.** 1.4 **13.** $\{11\}$ **14.** $\{4\}$
15. $\frac{\sqrt[3]{3x^2}}{x}$ **16.** $x^3\sqrt{14x}$ **17.** $5m\sqrt[3]{2m}$ **18.** $3x\sqrt{3x}$
19. $a^2b\sqrt[5]{a^4b}$ **20.** $(125p^9q^6)^{1/3} = 5p^3q^2$ **21.** $22\sqrt{5}$
22. $4x^2y^3z^3\sqrt{2xy}$ **23.** $\frac{9m}{n^4}$ **24.** $\frac{8s^3}{r}$ **25.** $16 - 2\sqrt{55}$
26. $\sqrt{74}$; 8.602 **27.** $\sqrt{185}$; 13.601 **28.** Center $(4, -1)$; radius 7
29. $-7 + 51i$ **30.** $-\frac{10}{13} + \frac{24}{13}i$

Cumulative Review for Chapters 0–7

1. $\{-15\}$ **2.** 5 **3.** $y = \frac{3}{2}x - 6$ **4.** 17, 18, 19
5. $10x^2 - 2x$ **6.** $10x^2 - 39x - 27$ **7.** $x(x - 1)(2x + 3)$
8. $9(x^2 + 2y^2)(x^2 - 2y^2)$ **9.** $(x + 2y)(4x - 5)$
10. $(x + 4)(x - 4)(x^2 + 3)$ **11.** $-\frac{4}{3}$ **12.** $y = -\frac{1}{4}x + 1$
13. 20, 25 **14.** $2x^4y^3\sqrt{3y}$ **15.** $\sqrt{6} + 3\sqrt{3} - 5\sqrt{2} - 15$

16.

17.

18.

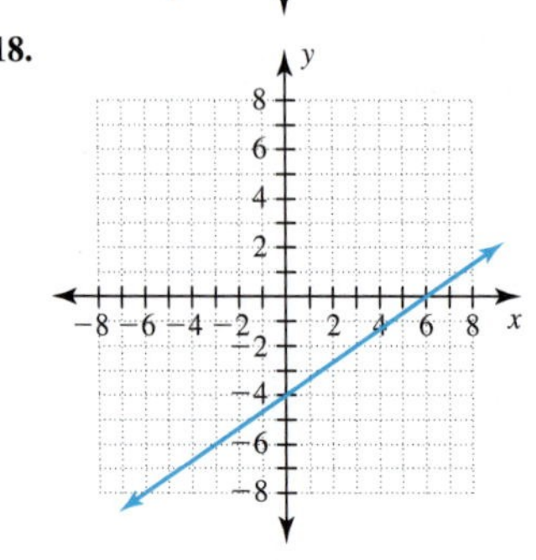

19. $\{(3, -1)\}$ **20.** $\left\{\left(5, \frac{3}{5}\right)\right\}$ **21.** x^6y^{14} **22.** $216x^4y^5$
23. $\frac{4xy^3}{3}$ **24.** $\{1\}$ **25.** $\{x \mid x \le -4\}$ **26.** $\frac{1}{2}, -5$
27. 17 in. by 31 in.

Reading Your Text for Chapter 8

Section 8.1 **(a)** quadratic; **(b)** square-root; **(c)** completing; **(d)** constant
Section 8.2 **(a)** formula; **(b)** factoring; **(c)** discriminant; **(d)** no
Section 8.3 **(a)** downward; **(b)** minimum; **(c)** axis; **(d)** vertical
Section 8.4 **(a)** approximate; **(b)** lowest; **(c)** four; **(d)** quadratic formula

Summary Exercises for Chapter 8

1. $\{\pm 2\sqrt{3}\}$ **3.** $\{2 \pm 2\sqrt{5}\}$ **5.** 49 **7.** $\{-1, 5\}$
9. $\{5 \pm 2\sqrt{7}\}$ **11.** $\left\{\frac{3 \pm \sqrt{7}}{2}\right\}$ **13.** $\{-3, 8\}$
15. $\left\{\frac{5 \pm \sqrt{17}}{2}\right\}$ **17.** $\left\{\frac{-2 \pm \sqrt{7}}{3}\right\}$ **19.** $\{1 \pm \sqrt{29}\}$
21. $\left\{\frac{-1 \pm i\sqrt{14}}{3}\right\}$ or $\left\{-\frac{1}{3} \pm \frac{\sqrt{14}}{3}i\right\}$ nonreal **23.** None
25. One **27.** 4, 8 **29.** 5 **31.** Width: 5 cm; length: 7 cm
33. $\{-1, 2\}$ **35.** $\left\{-2, -\frac{3}{2}\right\}$ **37.** $x = 0$; $(0, 0)$
39. $x = 0$; $(0, -5)$ **41.** $x = -2$; $(-2, 0)$ **43.** $x = -3$; $(-3, 1)$
45. $x = 5$; $(5, -2)$ **47.** $x = 1$; $(1, 1)$ **49.** $x = -\frac{1}{2}$; $\left(-\frac{1}{2}, \frac{25}{4}\right)$

51.

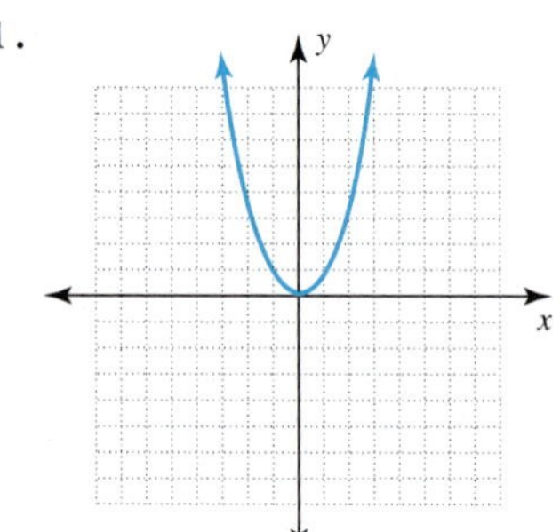

53.

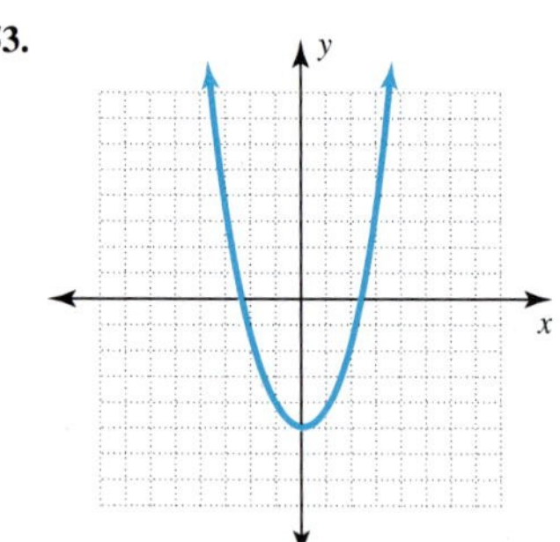

55.

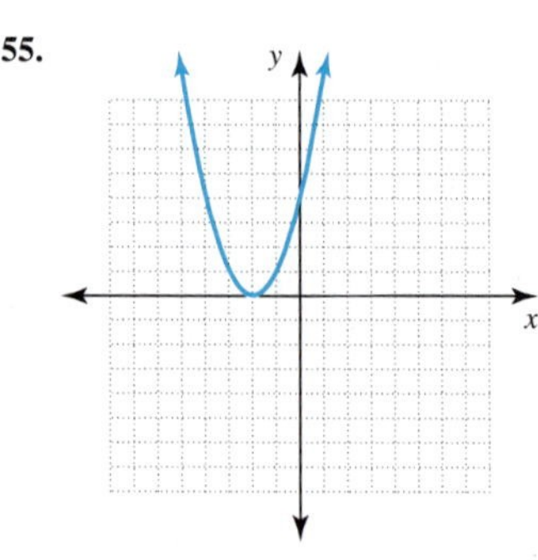

57.

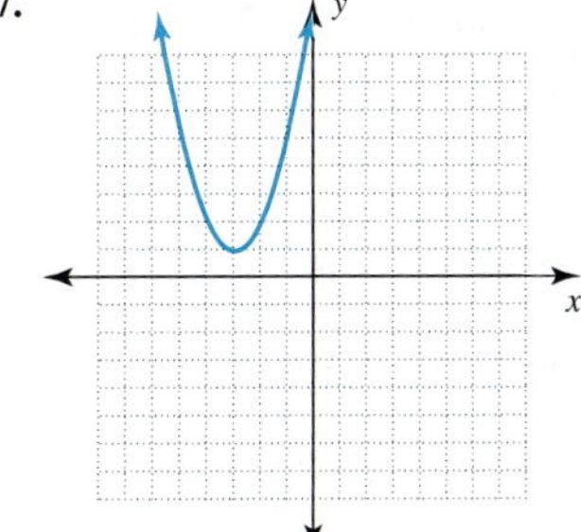

59.

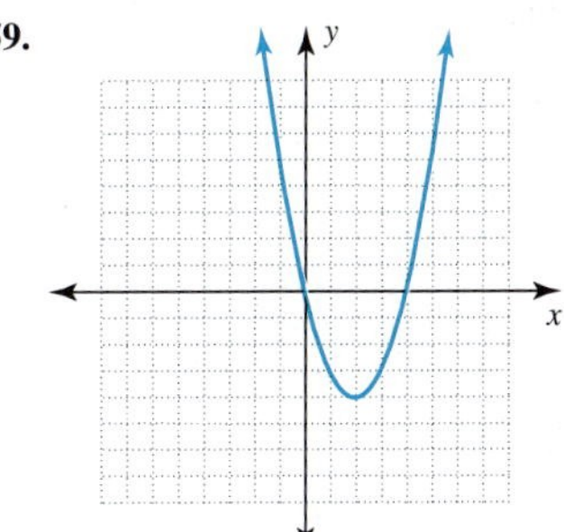

61.

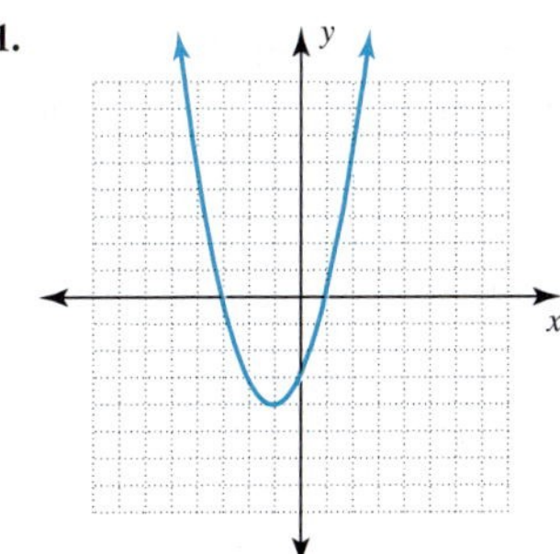

63.

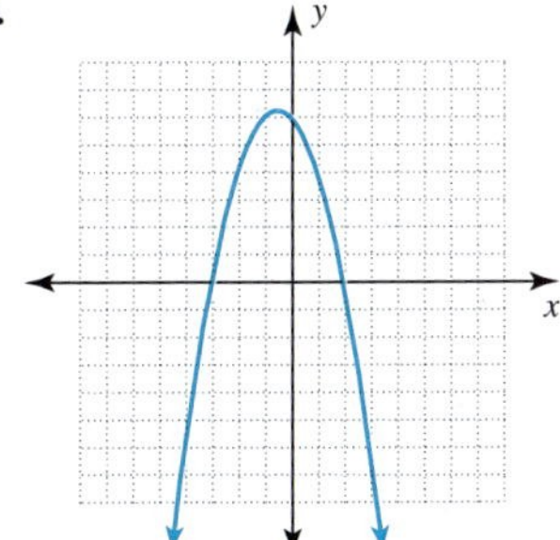

65.

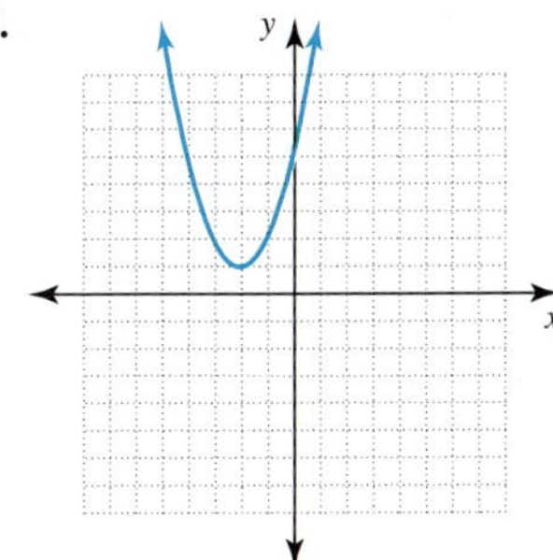

67.

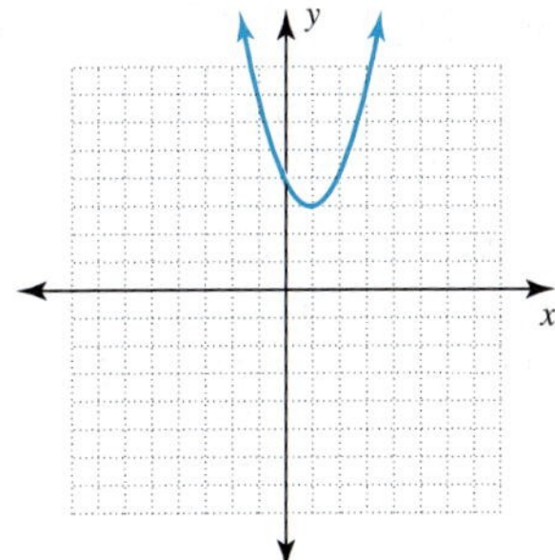

69.

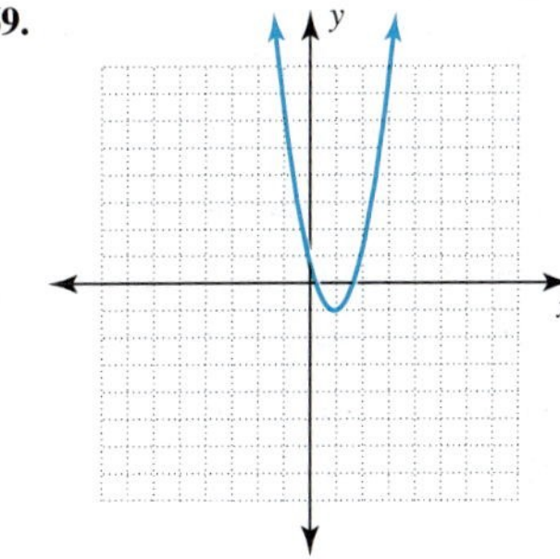

71. $\{-1.449, 3.449\}$ **73.** $\{1.382, 3.618\}$ **75.** $\{\pm 2\sqrt{2}, \pm\sqrt{5}\}$
77. $\{\pm 3, \pm 2i\}$ two nonreal solutions **79.** $\{9, 16\}$
81. $\{\pm 2.713\}$ **83.** $30 \le x \le 50$
85. Width: 4 cm; length: 9 cm **87.** Width: 8 ft; length: 15 ft
89. $(-1 + \sqrt{31})$ m, $(1 + \sqrt{31})$ m or 4.6 m, 6.6 m

Chapter 8 Test

1. $x = -2$; $(-2, 1)$ **2.** $x = 2$; $(2, -9)$ **3.** $x = \frac{3}{2}$; $\left(\frac{3}{2}, \frac{3}{2}\right)$
4. $x = 3$; $(3, -2)$ **5.** $x = 3$; $(3, -7)$ **6.** $\left\{\frac{-3 \pm \sqrt{13}}{2}\right\}$
7. $\left\{\frac{5 \pm \sqrt{19}}{2}\right\}$ **8.** $\left\{-\frac{2}{3}, 4\right\}$ **9.** $\{-4.541, 1.541\}$
10. $\{-1.396, 0.896\}$
11.

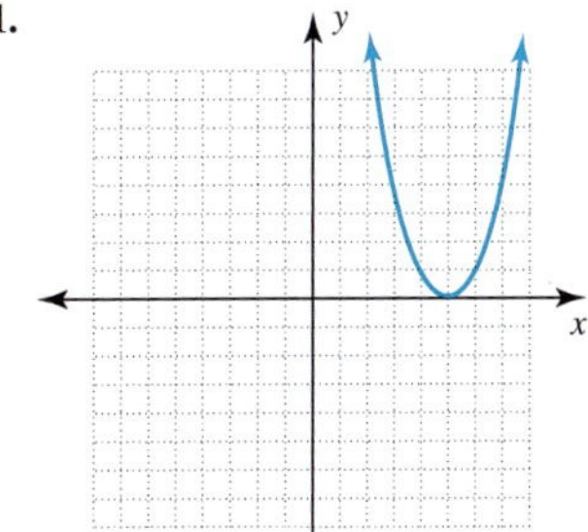

12.

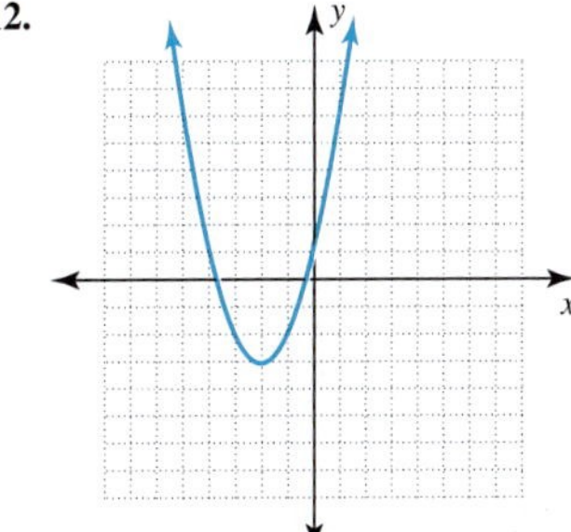

13.

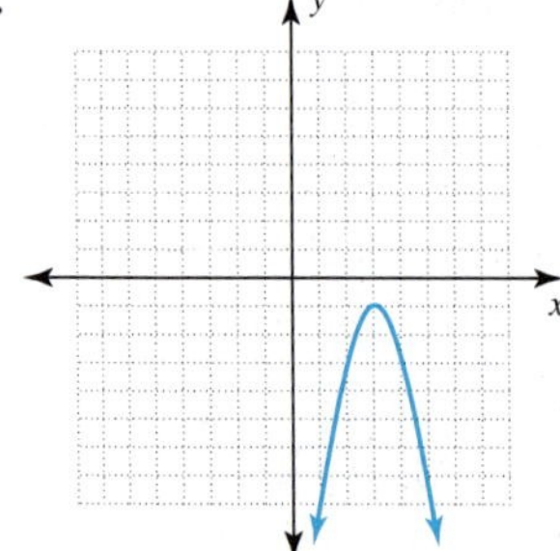

14.

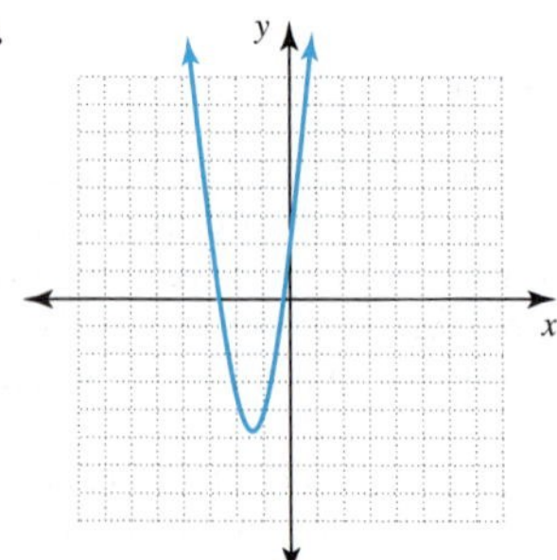

15. $\left\{-3, -\frac{1}{2}\right\}$ 16. $\left\{-\frac{5}{2}, \frac{2}{3}\right\}$ 17. $\left\{0, \pm\frac{3}{2}\right\}$ 18. 7, 9

19. 2.7 s 20. $\left\{\frac{5 \pm \sqrt{37}}{2}\right\}$ 21. $\{-2 \pm \sqrt{11}\}$

22. $\left\{-\frac{2}{3}, \frac{4}{5}\right\}$ 23. $\left\{\frac{-1 \pm 3i}{2}\right\}$ or $\left\{-\frac{1}{2} \pm \frac{3}{2}i\right\}$ nonreal

24. $\{\pm 2\sqrt{3}, \pm\sqrt{3}\}$

25. $\{\pm 2\sqrt{2}, \pm 2i\}$ two nonreal solutions 26. $\{25, 36\}$

27. $\{\pm\sqrt{5}\}$ 28. $\{1 \pm \sqrt{10}\}$ 29. $\left\{\frac{2 \pm \sqrt{23}}{2}\right\}$

30. $\{\pm 1.907\}$

Cumulative Review for Chapters 0–8

1.

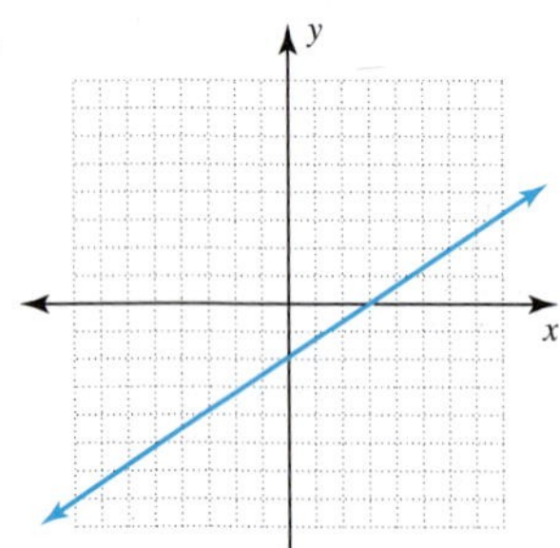

2.

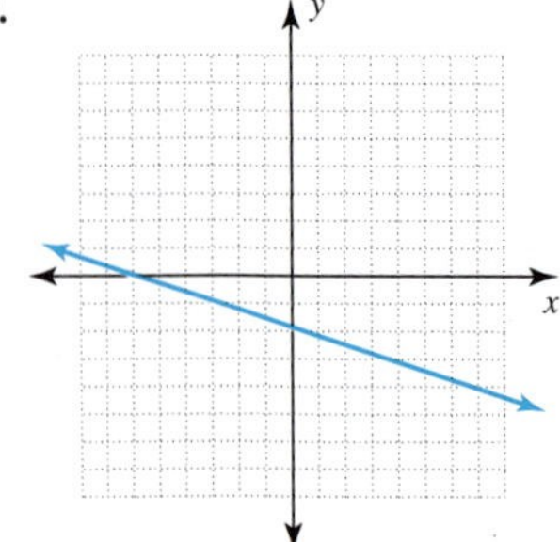

3.

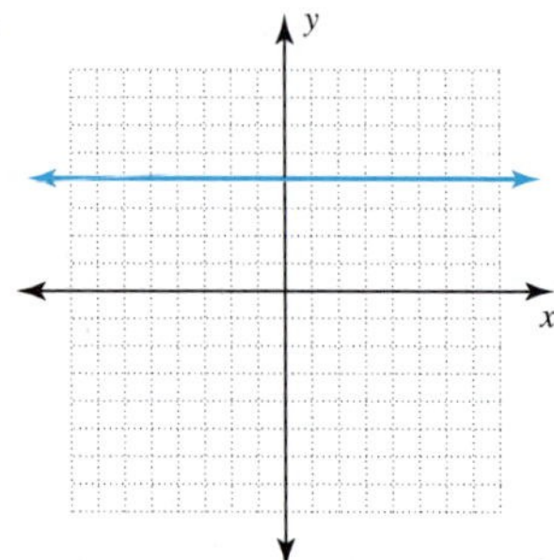

4. -3 5. $\frac{4}{3}$ 6. 35 7. $x^3 + 3x^2 - x - 3$

8. $x(x + 3)(x - 2)$ 9. $\frac{22\sqrt{6}}{3}$ 10. $6xy^2\sqrt{2xy}$ 11. $\left\{\frac{7}{2}\right\}$

12. $\{-4\}$ 13. $\{-5, 3\}$ 14. $\{1, 2\}$ 15. $\{-10, 3\}$

16. $\left\{\frac{3 \pm \sqrt{21}}{2}\right\}$ 17. $\{3 \pm \sqrt{5}\}$ 18. $\{-3, 0, 5\}$ 19. $\left\{\frac{13}{6}\right\}$

20. $\{1\}$ 21. $\{x \mid x \le 9\}$ 22. $\sqrt{74}$ 23. $5x - 7 = 72; -13$

24. $x^2 + (x + 4)^2 = 28^2$; $(-2 + 2\sqrt{97})$ ft and $(2 + 2\sqrt{97})$ ft or 17.7 ft or 21.7 ft 25. 21 or 59 receivers

Reading Your Text for Chapter 9

Section 9.1 **(a)** polynomials; **(b)** denominator; **(c)** GCF; **(d)** simplified

Section 9.2 **(a)** multiply; **(b)** factor; **(c)** reciprocal; **(d)** always

Section 9.3 **(a)** like; **(b)** factor; **(c)** factor; **(d)** distribute

Section 9.4 **(a)** complex; **(b)** fundamental; **(c)** invert; **(d)** simplest

Section 9.5 **(a)** y-intercept; **(b)** denominator; **(c)** vertical; **(d)** horizontal

Section 9.6 **(a)** LCD; **(b)** original; **(c)** motion; **(d)** similar

Summary Exercises for Chapter 9

1. Never undefined 3. $x \ne 5$ 5. $\frac{3x^2}{4}$ 7. 7

9. $-\frac{x + 3}{x + 5}$ 11. $\frac{2a - b}{3a - b}$ 13. $f(x) = x - 4, x \ne -1$

15. $\frac{2x^3}{3}$ 17. $\frac{4}{3y}$ 19. $\frac{3}{2a}$ 21. $\frac{x + 2y}{x - 5y}$

23. **(a)** -8; **(b)** $h(x) = (x + 5)(x - 4), x \ne -4, 5$; **(c)** -8

25. $\frac{3x^2 - 16x - 8}{(x + 4)(x - 4)}$ 27. $\frac{x + 5}{x(x - 5)}$ 29. $-\frac{11}{6(m - 1)}$

31. $\frac{4(x - 4)}{5(x + 1)(x - 1)}$ 33. $-\frac{4}{s + 2}$ 35. $\frac{3x - 1}{x + 2}$

37. **(a)** 8; **(b)** $h(x) = \frac{x(3x - 8)}{(x - 2)(x - 3)}$; **(c)** (4, 8) 39. $\frac{2}{3x^2}$

41. $-\frac{a + b}{a - b}$ 43. $\frac{r + s}{rs}$ 45. $\frac{x - 4}{(x - 1)^2}$ 47. $-y + 2$

49. $\frac{x + 2}{(x - 1)(x + 4)}$ 51. $\{x \mid x \ne -4\}$ 53. $\{x \mid x \ne 2\}$

55. $\{x \mid x \ne -1\}$ 57. y-intercept: $\left(0, -\frac{3}{4}\right)$; x-intercept: (3, 0)

59. y-intercept: (0, 0); x-intercept: (0, 0)

61. y-intercept: (0, −6); x-intercept: (3, 0)

63. Vertical: $x = -4$; horizontal: $y = 1$

65. Vertical: $x = 2$; horizontal: $y = 8$

67. Vertical: $x = -1$; horizontal: $y = 2$

69.

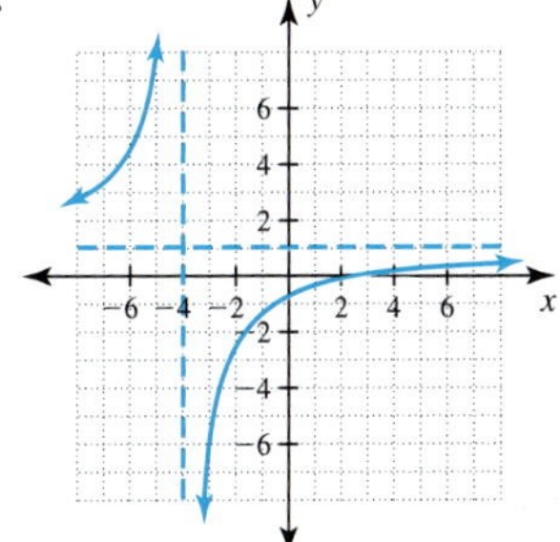

71.

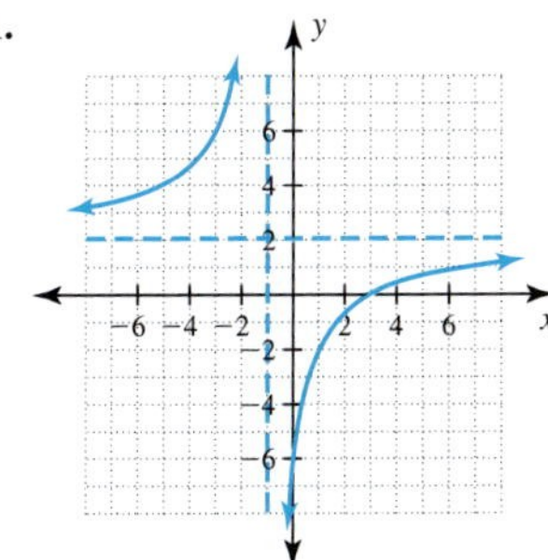

73. $\{5\}$ **75.** $\{6\}$ **77.** $\{3\}$ **79.** $\{-1\}$ **81.** $\{4\}$
83. $\{0, 7\}$ **85.** First triangle: 4 and 11; second triangle: 8 and 22
87. 5, 6 **89.** 12 hr

Chapter 9 Test

1. $\frac{m-4}{4}$ **2.** $\frac{2}{x-3}$ **3.** $\frac{y}{3y+x}$ **4.** $\frac{4a^2}{7b}$
5. $\frac{8x+17}{(x-4)(x+1)(x+4)}$ **6.** $\frac{2}{x-1}$ **7.** $-\frac{5(3x+1)}{3x-1}$
8. $\frac{z-1}{2(z+3)}$ **9.** $\frac{2x+y}{x(x+3y)}$ **10.** $\frac{4}{x-2}$ **11.** $\frac{2(2x+1)}{x(x-2)}$
12. $\{2, 6\}$ **13.** $\frac{w+1}{w-2}$ **14.** $\frac{x+3}{x-2}$ **15.** $-\frac{3x^4}{4y^2}$ **16.** 3
17.

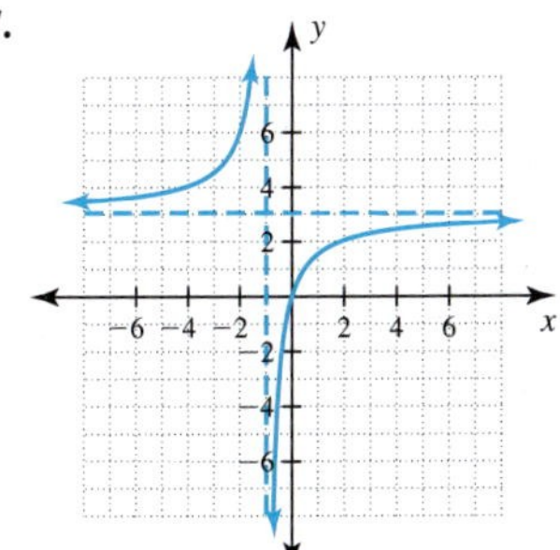

18.

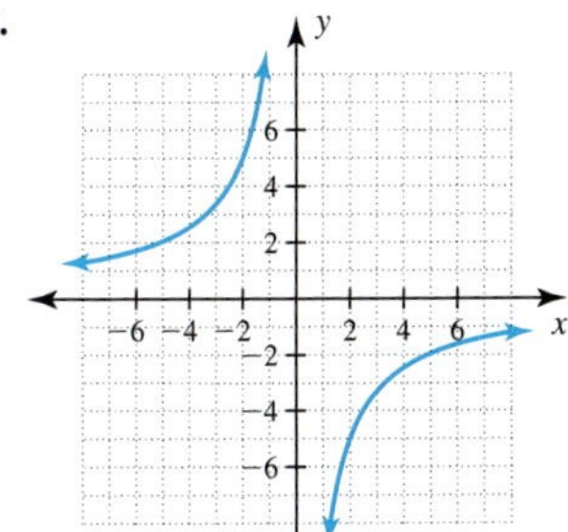

19. $f(x) = x - 1, x \neq 4$ **20.** **(a)** x-intercept: $\left(\frac{1}{4}, 0\right)$; y-intercept: $\left(0, -\frac{1}{2}\right)$; **(b)** vertical: $x = -\frac{2}{3}$, horizontal: $y = \frac{4}{3}$

Cumulative Review for Chapters 0–9

1. $\{12\}$ **2.** -14 **3.** $6x + 7y = -21$
4. x-intercept $(-6, 0)$; y-intercept $(0, 7)$ **5.** $f(x) = 6x^2 - 5x - 2$
6. $f(x) = 2x^3 + 5x^2 - 3x$ **7.** $\{x \mid x \neq 0\}$ **8.** 16
9. $x(2x + 3)(3x - 1)$ **10.** $(4x^8 + 3y^4)(4x^8 - 3y^4)$ **11.** $\frac{3}{x-1}$
12. $\frac{1}{(x+1)(x-1)}$ **13.** $x(x - 3)$ **14.** 0.76 s and 2.86 s
15. $\left\{\frac{7}{2}\right\}$ **16.** $\{\pm 4, \pm\sqrt{2}\}$ **17.** $\left\{-\frac{30}{17}\right\}$ **18.** $\{4\}$
19. $\{-5\}$ **20.** $\{x \mid x > -2\}$ **21.** 16.1 cm
22. Vertical: $x = 5$, horizontal: $y = -2$ **23.** $\frac{b^6}{a^{10}}$
24. Don 6 hr, Barry 3 hr **25.** Length 18 cm, width 10 cm

Reading Your Text for Chapter 10

Section 10.1 **(a)** domain; **(b)** product; **(c)** domain; **(d)** revenue
Section 10.2 **(a)** Composing; **(b)** g; **(c)** simpler; **(d)** operations
Section 10.3 **(a)** inverse; **(b)** symmetric; **(c)** one-to-one; **(d)** horizontal
Section 10.4 **(a)** exponential; **(b)** base; **(c)** greater; **(d)** Euler's
Section 10.5 **(a)** logarithm; **(b)** exponent or power; **(c)** y-intercept; **(d)** Loudness
Section 10.6 **(a)** exponent; **(b)** power; **(c)** common; **(d)** neutral
Section 10.7 **(a)** equation; **(b)** 10; **(c)** variable; **(d)** extraneous

Summary Exercises for Chapter 10

1. 2 **3.** Does not exist **5.** Does not exist **7.** $\{-2, 3, 7\}$
9. $2x^2 + 6x - 8$ **11.** $2x^4 + x^3 - x^2 + 6x + 5$
13. $5x^2 + 3x + 10$ **15.** $4x^2 + 8x$ **17.** $6x^2 - 10x$ **19.** $3x^3$
21. $\frac{2x}{x-3}$; $D = \{x \mid x \neq 3\}$ **23.** $\frac{3}{x}$; $D = \{x \mid x \neq 0\}$
25. 8 **27.** 2 **29.** -3 **31.** **(a)** -2; **(b)** -8; **(c)** 7; **(d)** $3x - 2$
33. **(a)** 25; **(b)** 49; **(c)** 4; **(d)** $x^2 - 5$ **35.** $(f \circ g)(x)$
37. $f^{-1}(x) = \frac{3-x}{2} = -\frac{x-3}{2}$ **39.** $f^{-1}(x) = 4x + 3$
41. $\{(1, -3), (2, -1), (3, 2), (4, 3)\}$; function
43. $\{(5, 1), (7, 2), (9, 3)\}$; function
45. $\{(4, 2), (3, 4), (4, 6)\}$; not a function
47. $f^{-1}(x) = \frac{x-3}{5}$

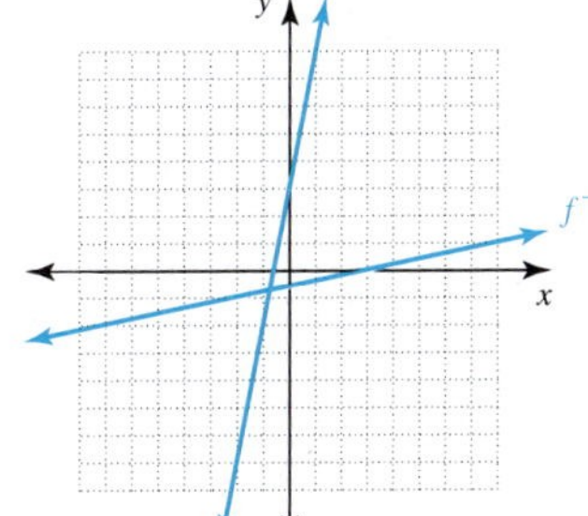

49. $f^{-1}(x) = 5x + 4$

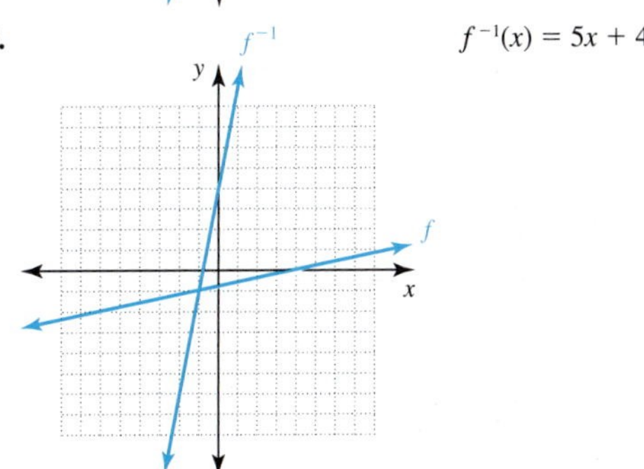

51. One-to-one; f^{-1}: $\{(2, -1), (3, 1), (5, 2), (7, 4)\}$; function
53. One-to-one; f^{-1}: $\{(2, -1), (4, 3), (5, 4), (6, 5)\}$; function
55. Not one-to-one; f^{-1} is not a function
57. 4 **59.** 8 **61.** x

63.

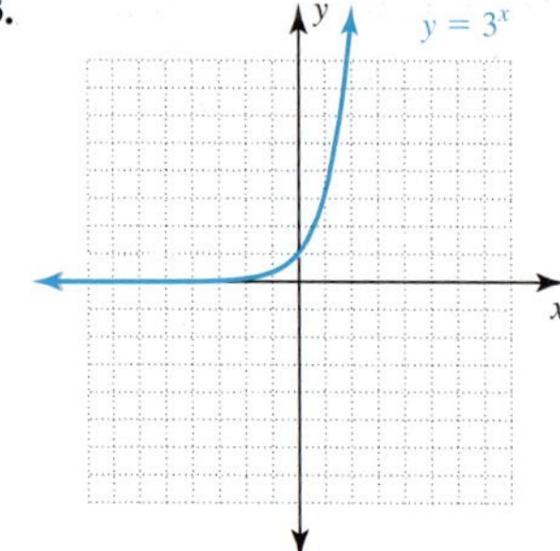

65. {3} **67.** {−1} **69.** 64,000

71.

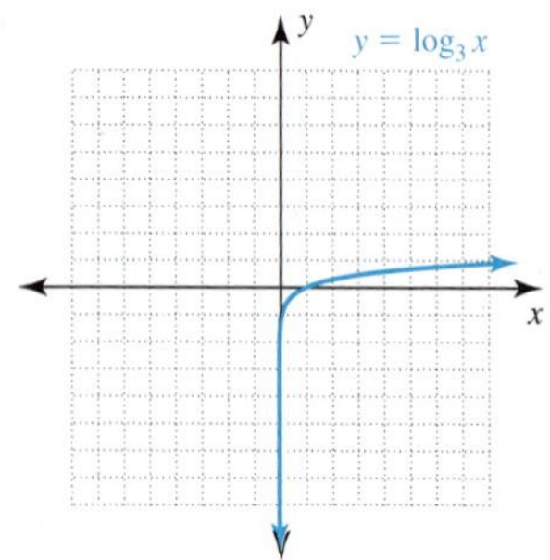

73. $\log_2 32 = 5$ **75.** $\log_5 1 = 0$ **77.** $\log_{25} 5 = \frac{1}{2}$
79. $4^3 = 64$ **81.** $81^{1/2} = 9$ **83.** $10^{-3} = 0.001$ **85.** {3}
87. {49} **89.** {9} **91.** $\left\{\frac{1}{3}\right\}$ **93.** 100 dB **95.** 10
97. 8.4 **99.** $2\log_b x + \log_b y$ **101.** $2\log_5 x + \log_5 y - 3\log_5 z$
103. $\log x + \log y - \frac{1}{2}\log z$ **105.** $\log xy^2$ **107.** $\log_b \frac{xy}{z}$
109. $\log \frac{x}{\sqrt{y}}$ **111.** 1.255 **113.** −0.903 **115.** 5.301, acidic
117. 3.2×10^{-4} **119.** 70 **121.** 64 **123.** 2.161
125. $\left\{\frac{27}{5}\right\}$ **127.** {3} **129.** $\left\{\frac{10}{9}\right\}$ **131.** {5} **133.** {1.431}
135. {1.71} **137.** 5.86 yr **139.** 66 yr **141.** 4.2 yr
143. 3.8 mi

Chapter 10 Test

1. $\log 10{,}000 = 4$ **2.** $\log_{27} 9 = \frac{2}{3}$ **3.** x
4. x **5.** $\frac{1}{2}(\log_5 x + 2\log_5 y - \log_5 z)$
6. $(f \cdot g)(x) = 3x^3 - 2x^2 - 3x + 2; D = \mathbb{R}$
7. $(g \div f)(x) = \frac{3x-2}{x^2-1}; D = \{x \mid x \neq \pm 1\}$
8. $(f \circ g)(x) = 9x^2 - 12x + 3$ **9.** $(g \circ f)(x) = 3x^2 - 5$
10. {−2} **11.** $\left\{\frac{5}{2}\right\}$ **12.** $5^3 = 125$ **13.** $10^{-2} = 0.01$
14. Not one-to-one; the inverse is not a function.
15. One-to-one; the inverse is a function. **16.** $\log_b \sqrt[3]{\frac{x}{z^2}}$
17. {0.262} **18.** {2.151}
19. **(a)** $-3x^3 + 3x^2 + 5x - 9$; **(b)** $-3x^3 + 7x^2 - 9x - 5$

20.

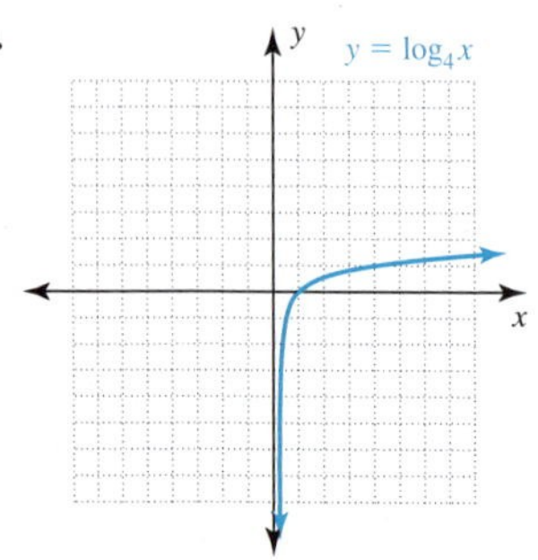

21. {8} **22.** $\left\{\frac{11}{8}\right\}$ **23.** $f^{-1}(x) = 5x + 3$
24. $f^{-1} = \{(-2, -3), (2, 1), (5, 4), (6, 5)\}$
25. $f^{-1} = \{(1, -3), (2, 4), (-2, 5)\}$

26.

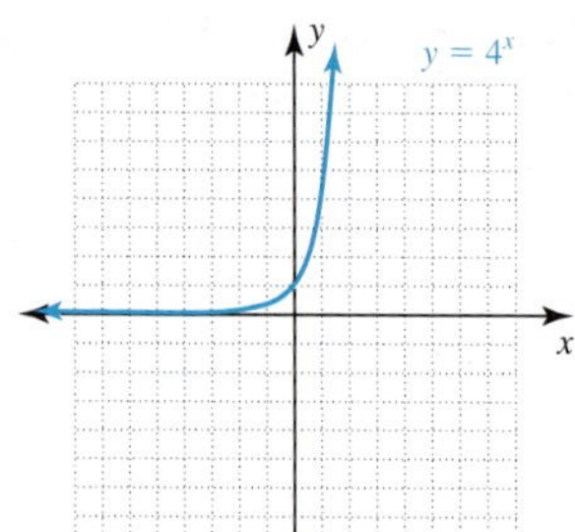

27.

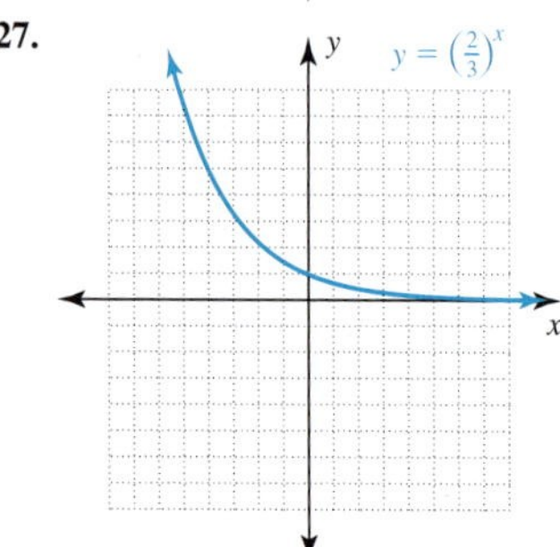

28. {6} **29.** {4} **30.** {5}

Cumulative Review for Chapters 0–10

1. {11} **2.** $\left\{\frac{5}{3}\right\}$ **3.** $\left\{\frac{10}{9}\right\}$

4.

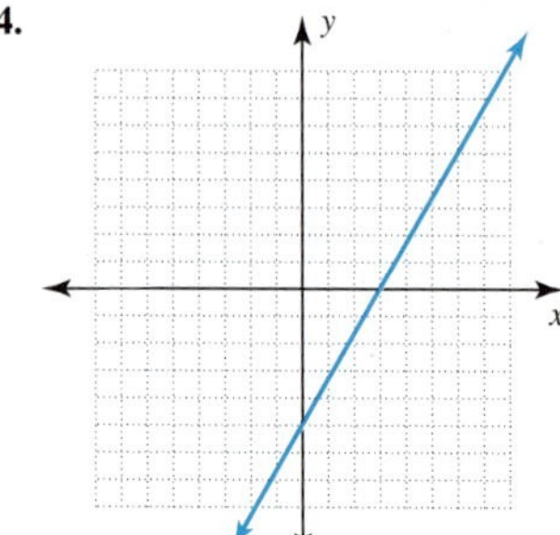

5.

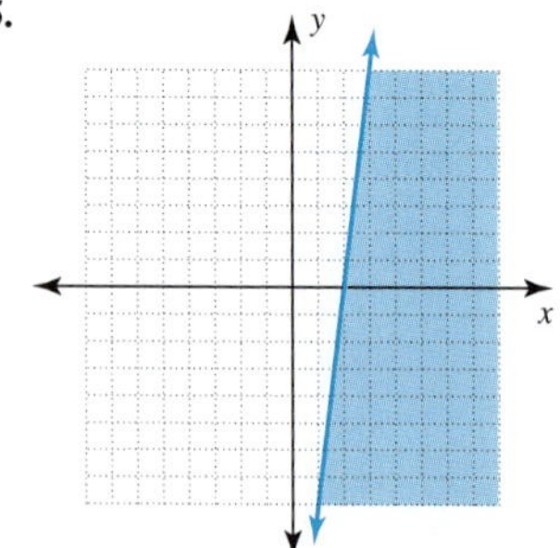

6. $6x + 5y = 7$ **7.** $\{x \mid x \geq 10\}$ **8.** $2x^2 - 6x + 1$
9. $6x^2 - 13x - 5$ **10.** $(2x - 5)(x + 2)$
11. $x(5x + 4y)(5x - 4y)$ **12.** $\frac{-x+2}{(x-4)(x-5)}$ **13.** $\frac{x+2}{x-2}$
14. $-4\sqrt{2}$ **15.** $22 + 12\sqrt{2}$ **16.** $\frac{5}{3}(\sqrt{5} + \sqrt{2})$
17. 77, 79, 81 **18.** {−2, 1} **19.** $\left\{\frac{3 \pm \sqrt{19}}{2}\right\}$
20. $R = \frac{R_1R_2}{R_1 + R_2}$ **21.** $3x^3 - 4x^2 - 17x + 6; D = \mathbb{R}$

22. $\dfrac{3x-1}{x^2-x-6}$; $D = \{x \mid x \neq -2, x \neq 3\}$
23. **(a)** $-6x + 3$; **(b)** -27 **24.** **(a)** $-6x + 5$; **(b)** -25
25. $7ab(2ab - 3a + 5b)$ **26.** $(x - 3y)(x + 5)$
27. $(5c - 8d)(5c + 8d)$ **28.** $(3x - 1)(9x^2 + 3x + 1)$
29. $2a(2a + b)(4a^2 - 2ab + b^2)$ **30.** $(x - 8)(x - 6)$
31. $(5x - 2)(2x - 7)$ **32.** $3x(x + 3)(2x - 5)$ **33.** $3x^2$
34. $\dfrac{12}{3-y}$ **35.** $\dfrac{m-13}{(m+3)(m-3)(m-1)}$ **36.** $\dfrac{10}{x-5}$
37. 10 cm, 24 cm, and 26 cm **38.** $\{8\}$ **39.** $\{5\}$
40. Center: $(-5, 2)$; radius 4

Photo Credits

Chapter 0

Opener: © 1998 Copyright IMS Communications Ltd./Capstone Design. All Rights Reserved; p. 9: © Ingram Publishing RF; p. 11: © Comstock Images/Getty RF; p. 13: © Corbis RF; p. 20: © McGraw-Hill Education. Barry Barker, photographer; p. 29: © Photodisc/Getty RF; p. 38: © Comstock/Jupiterimages RF; p. 40: © Getty RF; p. 47: © Thinkstock Getty RF; p. 49: © Comstock RF.

Chapter 1

Opener: © Flat Earth Images RF; p. 65: © Image Source/PunchStock RF; p. 67: © Digital Vision/Getty RF; p. 101: © Getty RF; p. 104: © Ingram Publishing RF; p. 128: © Comstock/Getty RF; p. 129(top): © Getty RF; p. 129(bottom): USDA; p. 143: © Digital Vision/PunchStock; p. 144: © BananaStock/Jupiterimages RF.

Chapter 2

Opener: © Vol. 80/PhotoDisc/Getty RF; p. 163: © Corbis RF; p. 165: © 2006 Texas Instruments; p. 187: © Alamy RF; p. 212: © Dynamic Graphics/JupiterImages RF; p. 214(top): © Getty RF; p. 214(bottom): © Ingram Publishing/AGE Fotostock RF; p. 236: © Photodisc/Punchstock RF.

Chapter 3

Opener: © PictureQuest RF; p. 266: © 2006 Texas Instruments; p. 281: © Ingram Publishing RF; p. 291: © Stockbyte/Getty RF; p. 314: © John Foxx/Stockbyte/Getty RF; p. 331: © Lars A. Niki RF; p. 334: © Purestock/SuperStock RF; p. 336: © Getty RF.

Chapter 4

Opener: © The McGraw-Hill Companies, Inc. Mark Dierker, photographer; p. 352: © The McGraw-Hill Companies, Inc. Mark Dierker, photographer; p. 368: © I. Rozenbaum & F. Cirou/PhotoAlto RF; P. 374(top): © Getty RF; p. 374(bottom): © Corbis RF; p. 375: © 2006 Texas Instruments; p. 386: © Ingram Publishing RF; p. 393: © Ingram Publishing/Fotosearch; p. 401: © Design Pics/Bilderbuch RF; p. 402: © Dave Moyer RF; p. 404: © Denise McCullough RF.

Chapter 5

Opener: © Getty RF; p. 419: © Getty RF; p. 426: © Corbis RF; p. 428, p. 429: © Design Pics/Don Hammond RF; p. 437(top): © Design Pics/Carson Ganci RF; p. 437(bottom): © P. Ughetto/PhotoAlto RF; p. 445: © Ingram Publishing RF; p. 457(top): © Ingram Publishing RF; p. 457(bottom): © Corbis RF; p. 471: © Design Pics/Richard Wear RF; p. 475: © Getty RF.

Chapter 6

Opener: © Getty RF; p. 489: © Getty RF; p. 536: © Design Pics/Don Hammond RF; p. 540: © PBNJ Productions/Blend Images LLC RF

Chapter 7

Opener: © Brand X RF; p. 565: © Brand X RF; p. 587: © Robert Glusic/Getty RF; p. 596: © Dinodia/age fotostock RF.

Chapter 8

Opener: © Corbis RF; p. 640: © ImageSource/Punchstock RF; p. 678: © Vol. 41/PhotoDisc/Getty RF.

Chapter 9

Opener: © Getty RF; p. 704, p. 764: © Getty RF; p. 765(top): © Medioimages/Superstock RF; p. 765(middle): © Peter Steiner/Alamy RF; p. 765(bottom): © I. Rozenbaum & F. Cirou/PhotoAlto RF.

Chapter 10

Opener: © Corbis RF; p. 785: © Getty RF; p. 788(top) © Corbis RF; p. 788(bottom): © Getty RF; p. 789: © Imagestate Media RF; p. 790: © Corbis RF; p. 799: © Glow Images RF; p. 821, p. 822: © Getty RF; p. 834: © Vol. 122/Corbis RF; p. 850: © Pixtal/AGE Fotostock RF.